Retinal and Choroidal Vascular Diseases of the Eye

Retinal and Choroidal Vascular Diseases of the Eye

Edited by

Jay Chhablani
Department of Ophthalmology, University of Pittsburgh School of Medicine,
Pittsburgh, PA, United States

ELSEVIER

ACADEMIC PRESS
An imprint of Elsevier

Academic Press is an imprint of Elsevier
125 London Wall, London EC2Y 5AS, United Kingdom
525 B Street, Suite 1650, San Diego, CA 92101, United States
50 Hampshire Street, 5th Floor, Cambridge, MA 02139, United States

Notices

Knowledge and best practice in this field are constantly changing. As new research and experience broaden our understanding, changes in research methods, professional practices, or medical treatment may become necessary.

Practitioners and researchers must always rely on their own experience and knowledge in evaluating and using any information, methods, compounds, or experiments described herein. In using such information or methods they should be mindful of their own safety and the safety of others, including parties for whom they have a professional responsibility.

To the fullest extent of the law, neither the Publisher nor the authors, contributors, or editors, assume any liability for any injury and/or damage to persons or property as a matter of products liability, negligence or otherwise, or from any use or operation of any methods, products, instructions, or ideas contained in the material herein.

ISBN: 978-0-443-15583-3

For information on all Academic Press publications visit our website at https://www.elsevier.com/books-and-journals

Publisher: Stacy Masucci
Acquisitions Editor: Michelle Fisher
Editorial Project Manager: Pat Gonzalez
Production Project Manager: Omer Mukthar
Cover Designer: Matthew Limbert

Typeset by TNQ Technologies

Contents

Section VII
Vascular diseases associated with systemic conditions

List of contributors

Reza Mirshahi, Eye Research Center, Department of Ophthalmology, Iran University of Medical Sciences, Tehran, Iran

Julia Fallon, New York Eye and Ear Infirmary of Mount Sinai, Department of Ophthalmology, New York, NY, United States

Burak Acar, Gazi University School of Medicine, Ankara, Turkey

Isha Acharya, Narayana Nethralaya, Department of Retina and Vitreous, Bangalore, Karnataka, India

Aniruddha Agarwal, Department of Ophthalmology, Cleveland Clinic Abu Dhabi, Abu Dhabi, United Arab Emirates; Cleveland Clinic Lerner College of Medicine, Case Western Reserve University, Cleveland, OH, United States; Department of Ophthalmology, Maastricht University Medical Center+, Maastricht, the Netherlands

Kushal Agrawal, Jupiter Hospital, Thane, Maharashtra, India

Sofia Ahsanuddin, New York Eye and Ear Infirmary of Mount Sinai, Department of Ophthalmology, New York, NY, United States

Socorro Alforja, Hospital Clínic de Barcelona, Barcelona, Spain; Institut d'Investigacions Biomediques August Pi i Sunyer (IDIBAPS), Barcelona, Spain

Shaikha Aljneibi, Eye Institute, Cleveland Clinic Abu Dhabi, Abu Dhabi, United Arab Emirates

J. Fernando Arévalo, The Johns Hopkins University School of Medicine, Edmund F. and Virginia B. Ball, Department of Ophthalmology, Johns Hopkins Bayview Medical Center, Retina Division, Wilmer Eye Institute, Baltimore, MD, United States; Wilmer Eye Institute, Johns Hopkins University School of Medicine, Baltimore, MD, United States

Apoorva Ayachit, Department of Vitreoretina, M. M. Joshi Eye Institute, Hubballi, Karnataka, India

Guruprasad Ayachit, Department of Vitreoretina, M. M. Joshi Eye Institute, Hubballi, Karnataka, India

Jiwon Baek, The Catholic University of Korea, College of Medicine, Ophthalmology, Seoul, South Korea; Bucheon St.Mary's Hospital, The Catholic University of Korea, Gyeonggi-do, South Korea

Francesco Bandello, School of Medicine, Vita-Salute San Raffaele University, Milan, Italy; Division of Head and Neck, Ophthalmology Unit, IRCCS San Raffaele Scientific Institute, Milan, Italy

Caroline R. Baumal, Tufts Medical Center/New England Eye Center, Boston, MA, United States

Carolina Bernal-Morales, Hospital Clínic de Barcelona, Barcelona, Spain; Institut d'Investigacions Biomediques August Pi i Sunyer (IDIBAPS), Barcelona, Spain

Pietro M. Bertelli, Wellcome-Wolfson Institute for Experimental Medicine, Faculty of Medicine, Health, and Life Sciences, Queen's University Belfast, Belfast, United Kingdom

Abhinaya Bharath, Drexel University College of Medicine, Philadelphia, PA, United States

Jyotirmay Biswas, Department of Uveitis and Ocular Pathology, Sankara Nethralaya, Chennai, Tamil Nadu, India

Søren Leer Blindbæk, Department of Ophthalmology, Odense University Hospital, Odense, Denmark; Department of Clinical Research, University of Southern Denmark, Odense, Denmark

Gabrielle Bonhomme, Division of Ophthalmology, University of Pittsburgh Medical Center, Pittsburgh, PA, United States

Alan Chew Bonilla, Retina Department, Institute of Ophhtalmology Fundacion Conde de Valenciana, Mexico City, Mexico

Sophie Bonnin, Department of Ophthalmology, Rothschild Foundation Hospital, Paris, France

Enrico Borrelli, Department of Ophthalmology, University of Turin, Italy; School of Medicine, Vita-Salute San Raffaele University, Milan, Italy

Elodie Bousquet, Retinal Disorders and Ophthalmic Genetics Division, Stein Eye Institute, University of California Los Angeles, Los Angeles, CA, United States; Paris Cité University, Hôpital Lariboisière, AP-HP, Ophthalmology, Paris, France

Jonathan T. Caranfa, Tufts Medical Center/New England Eye Center, Boston, MA, United States

Nathalie Cassoux, Université Paris Cité, Paris, France; Institut Curie, Department of Ocular Oncology, Paris, France

Irini Chatziralli, National and Kapodistrian University of Athens, 2nd Department of Ophthalmology, Athens, Greece

Lisa Checchin, School of Medicine, Vita-Salute San Raffaele University, Milan, Italy; Division of Head and Neck, Ophthalmology Unit, IRCCS San Raffaele Scientific Institute, Milan, Italy

Jay Chhablani, Department of Ophthalmology, University of Pittsburgh School of Medicine, Pittsburgh, PA, United States

Peranut Chotcomwongse, Department of Ophthalmology, Rajavithi Hospital, College of Medicine, Rangsit University, Bangkok, Thailand

Netan Choudhry, Octane Imaging Lab, Toronto, ON, Canada; Vitreous Retina Macula Specialists of Toronto, Toronto, ON, Canada; Department of Ophthalmology & Vision Sciences, University of Toronto, Toronto, ON, Canada; Cleveland Clinic Canada, Toronto, ON, Canada

Toco Chui, New York Eye and Ear Infirmary of Mount Sinai, Department of Ophthalmology, New York, NY, United States

Miguel Cruz-Pimentel, Department of Ophthalmology and Vision Sciences, University of Toronto, Toronto, ON, Canada

Chiara M. Eandi, University of Lauanne, Jules Gonin Eye Hospital, Fondation Asile des Aveugles, Lausanne, Switzerland; University of Torino, Department of Surgical Sciences, Torino, Italy

Zeinab El Moussawi, Department of Ophthalmology, American University of Beirut, Beirut, Lebanon

Marie-Helene Errera, Division of Ophthalmology, University of Pittsburgh Medical Center, Pittsburgh, PA, United States

Francisco Espejo-Arjona, Department of Ophthalmology, University Hospital Virgen Macarena, Sevilla, Spain

Jessica Eyre, Wellcome-Wolfson Institute for Experimental Medicine, Faculty of Medicine, Health, and Life Sciences, Queen's University Belfast, Belfast, United Kingdom

Khalil Ghasemi Falavarjani, Eye Research Center, Department of Ophthalmology, Iran University of Medical Sciences, Tehran, Iran

Roberto Gallego-Pinazo, Unit of Macula and Clinical Trials, Oftalvist Clinic, Valencia, Spain; Department of Macula and Clinical Trials, Oftalvist Clinic, Valencia, Spain

Alain Gaudric, Université Paris Cité, Department of Ophthalmology, Hôpital Lariboisière, APHP, Paris, France

Monica P. Gonzalez, Department of Ophthalmology, Medical College of Wisconsin, Milwaukee, WI, United States

Jakob Grauslund, Department of Ophthalmology, Odense University Hospital, Odense, Denmark; Department of Clinical Research, University of Southern Denmark, Odense, Denmark; Steno Diabetes Center Odense, Odense University Hospital, Odense, Denmark; Department of Ophthalmology, Vestfold Hospital Trust, Tønsberg, Norway

Jasenka Guduric-Fuchs, Wellcome-Wolfson Institute for Experimental Medicine, Faculty of Medicine, Health, and Life Sciences, Queen's University Belfast, Belfast, United Kingdom

Ricardo Leitão Guerra, Leitão Guerra Oftalmologia, Salvador, BA, Brazil

Rosa Gutiérrez-Bonet, Ophthalmology Department, Gregorio Marañon University Hospital, Madrid, Spain

Ashit Handa, Narayana Nethralaya, Department of Retina and Vitreous, Bangalore, Karnataka, India

Claudio Iovino, Eye Clinic, Multidisciplinary Department of Medical, Surgical and Dental Sciences, University of Campania Luigi Vanvitelli, Naples, Italy

Anthony A. Jones, Department of Ophthalmology, Medical College of Wisconsin, Milwaukee, WI, United States

Shrinivas Joshi, Department of Vitreoretina, M. M. Joshi Eye Institute, Hubballi, Karnataka, India

Judy E. Kim, Department of Ophthalmology, Medical College of Wisconsin, Milwaukee, WI, United States

Laurent Kodjikian, Service d'Ophtalmologie, Hôpital Universitaire de la Croix-Rousse, Hospices Civils de Lyon, Lyon, France; Laboratoire UMR-CNRS 5510 MATEIS, INSA, Université Lyon 1, Villeurbanne, France

Alexey N. Kulikov, S.M.Kirov Military Medical Academy, Ophthalmology Department, Saint-Petersburg, Russia

Sara Labay-Tejado, Hospital Clínic de Barcelona, Barcelona, Spain

Gerardo Ledesma-Gil, Retina Department, Institute of Ophhtalmology Fundacion Conde de Valenciana, Mexico City, Mexico

John K. Lippincott, Department of Ophthalmology, Medical College of Wisconsin, Milwaukee, WI, United States

Noemi Lois, Wellcome-Wolfson Institute for Experimental Medicine, Faculty of Medicine, Health, and Life Sciences, Queen's University Belfast, Belfast, United Kingdom

Marco Lupidi, Eye Clinic, Department of Experimental and Clinical Medicine, Polytechnic University of Marche, Ancona, Italy; Fondazione per la Macula Onlus, Di.N.O.G.Mi., University Eye Clinic, Genova, Italy

Martin A. Mainster, Department of Ophthalmology, University of Kansas School of Medicine, Kansas City, KS, United States

Denis Malaise, Institut Curie, Department of Ocular Oncology, Paris, France

Dmitrii S. Maltsev, S.M.Kirov Military Medical Academy, Ophthalmology Department, Saint-Petersburg, Russia

Danny A. Mammo, Cole Eye Institute, Cleveland Clinic, Cleveland, OH, United States

Rubble Mangla, Narayana Nethralaya, Department of Retina and Vitreous, Bangalore, Karnataka, India

Hana A. Mansour, Department of Ophthalmology, American University of Beirut, Beirut, Lebanon; Wills Eye Hospital, Mid Atlantic Retina, Thomas Jefferson University, Philadelphia, PA, United States

Ahmad M. Mansour, Department of Ophthalmology, Retina Service, American University of Beirut, Beirut, Lebanon

Cesare Mariotti, Eye Clinic, Department of Experimental and Clinical Medicine, Polytechnic University of Marche, Ancona, Italy

Alexandre Matet, Université Paris Cité, Paris, France; Institut Curie, Department of Ocular Oncology, Paris, France

Thibaud Mathis, Service d'Ophtalmologie, Hôpital Universitaire de la Croix-Rousse, Hospices Civils de Lyon, Lyon, France; Laboratoire UMR-CNRS 5510 MATEIS, INSA, Université Lyon 1, Villeurbanne, France

Reinhold J. Medina, Wellcome-Wolfson Institute for Experimental Medicine, Faculty of Medicine, Health, and Life Sciences, Queen's University Belfast, Belfast, United Kingdom; Department of Eye and Vision Science, Institute of Life Course and Medical Sciences, University of Liverpool, Liverpool, United Kingdom

Isabela Martins Melo, Department of Ophthalmology, St. Michael's Hospital Unity Health Toronto, Toronto, ON, Canada

Covadonga Menendez-Acebal, Hospital Clínic de Barcelona, Barcelona, Spain

Paul S. Micevych, University of California San Francisco, Department of Ophthalmology, San Francisco, CA, United States

Andrea Montesel, University of Lauanne, Jules Gonin Eye Hospital, Fondation Asile des Aveugles, Lausanne, Switzerland

Santiago Montolío-Marzo, Department of Macula and Clinical Trials, Oftalvist Clinic, Valencia, Spain

Rajeev H. Muni, Department of Ophthalmology, St. Michael's Hospital Unity Health Toronto, Toronto, ON, Canada; Department of Ophthalmology and Vision Sciences, University of Toronto, Toronto, ON, Canada

Javier Muñoz-Solano, Unit of Macula and Clinical Trials, Oftalvist Clinic, Valencia, Spain

Alessio Muzi, Department of Ophthalmology, Humanitas Gradenigo, Torino, Italy

Laurentino Biccas Neto, Ocular Vitória, Vitória, ES, Brazil

Joshua Ong, Michigan Medicine, University of Michigan, Ann Arbor, MI, United States

Sally S. Ong, Department of Ophthalmology, Wake Forest School of Medicine, Winston–Salem, NC, United States

Samantha Orr, Octane Imaging Lab, Toronto, ON, Canada; Vitreous Retina Macula Specialists of Toronto, Toronto, ON, Canada

Şengül Özdek, Gazi University School of Medicine, Ankara, Turkey

Michel Pacques, Paris Eye Imaging Group, Centre Hospitalier National d'Ophtalmologie des Quinze-Vingts, INSERM-DHOS Clinical Investigation Center, Paris, France

Konstantinos Pappelis, National and Kapodistrian University of Athens, 2nd Department of Ophthalmology, Athens, Greece

Varun Pathak, Wellcome-Wolfson Institute for Experimental Medicine, Faculty of Medicine, Health, and Life Sciences, Queen's University Belfast, Belfast, United Kingdom

Niveditha Pattathil, Octane Imaging Lab, Toronto, ON, Canada; Vitreous Retina Macula Specialists of Toronto, Toronto, ON, Canada

Thomas M. Pearce, Division of Neuropathology, University of Pittsburgh Medical Center, Pittsburgh, PA, United States

Fernando M. Penha, Universidade Regional de Blumenau, Blumenau, SC, Brazil

Francesco Pichi, Eye Institute, Cleveland Clinic Abu Dhabi, Abu Dhabi, United Arab Emirates; Cleveland Clinic Lerner College of Medicine, Case Western Reserve University, Cleveland, OH, United States

Matthieu Poireir, Eastern Virginia Medical School, Norfolk, VA, United States

Vishma Prabhu, Narayana Nethralaya, Department of Retina and Vitreous, Bangalore, Karnataka, India

Giuseppe Querques, School of Medicine, Vita-Salute San Raffaele University, Milan, Italy; Division of Head and Neck, Ophthalmology Unit, IRCCS San Raffaele Scientific Institute, Milan, Italy

Aleksandra V. Rachitskaya, Cole Eye Institute, Cleveland Clinic, Cleveland, OH, United States

Kimia Rahimi Ardali, Student Research Committee, Qazvin University of Medical Sciences, Qazvin, Iran

Vishal Raval, Anant Bajaj Retina Institute, LV Prasad Eye Institute, KAR Campus, Hyderabad, Telangana, India

Nilesh Raval, New York Eye and Ear Infirmary of Mount Sinai, Department of Ophthalmology, New York, NY, United States

Nikitha Gurram Reddy, Anant Bajaj Retina Institute, L.V. Prasad Eye Hospital, Kallam Anji Reddy Campus, Hyderabad, Telangana, India

Michele Reibaldi, Department of Ophthalmology, University of Turin, Italy

Emma Reid, School of Cardiovascular & Metabolic Health, University of Glasgow, Glasgow, United Kingdom

Federico Ricardi, Department of Ophthalmology, University of Turin, Italy

Richard Rosen, New York Eye and Ear Infirmary of Mount Sinai, Department of Ophthalmology, New York, NY, United States

Laurence Rosier, Centre Rétine Gallien, Bordeaux, France

Paisan Ruamviboonsuk, Department of Ophthalmology, Rajavithi Hospital, College of Medicine, Rangsit University, Bangkok, Thailand

Jorge Ruiz-Medrano, Ophthalmology Department, Puerta de Hierro-Majadahonda University Hospital, Madrid, Spain; Ocular Microsurgery Institute IMO, Madrid, Spain; Department of Ophthalmology, University Hospital Puerta de Hierro, Madrid, Spain

José M. Ruiz-Moreno, Ophthalmology Department, Puerta de Hierro-Majadahonda University Hospital, Madrid, Spain; Ocular Microsurgery Institute IMO, Madrid, Spain; VISSUM, Alicante, Spain; Ophthalmology Department, University of Castilla La Mancha (UCLM), Ciudad Real, Spain

Riccardo Sacconi, School of Medicine, Vita-Salute San Raffaele University, Milan, Italy; Division of Head and Neck, Ophthalmology Unit, IRCCS San Raffaele Scientific Institute, Milan, Italy

Ahmad Santina, Retinal Disorders and Ophthalmic Genetics Division, Stein Eye Institute, University of California Los Angeles, Los Angeles, CA, United States

David Sarraf, Retinal Disorders and Ophthalmic Genetics Division, Stein Eye Institute, University of California Los Angeles, Los Angeles, CA, United States

Carol E. Schwartz, Sunnybrook Hospital, University of Toronto, Toronto, ON, Canada

Adrienne W. Scott, Retina Division, Wilmer Eye Institute, Johns Hopkins University School of Medicine, Baltimore, MD, United States

Abdulla R. Shaheen, Department of Ophthalmology, Bascom Palmer Eye Institute, University of Miami Miller School of Medicine, Miami, FL, United States

Sukhum Silpa-archa, Department of Ophthalmology, Rajavithi Hospital, College of Medicine, Rangsit University, Bangkok, Thailand

Sumit Randhir Singh, Sri Sai Eye Hospital, Patna, Bihar, India

Meera D. Sivalingam, Wills Eye Hospital, Mid Atlantic Retina, Thomas Jefferson University, Philadelphia, PA, United States

Jayanth Sridhar, Department of Ophthalmology, Bascom Palmer Eye Institute, University of Miami Miller School of Medicine, Miami, FL, United States

Jay M. Stewart, University of California San Francisco, Department of Ophthalmology, San Francisco, CA, United States

Alan W. Stitt, Wellcome-Wolfson Institute for Experimental Medicine, Faculty of Medicine, Health, and Life Sciences, Queen's University Belfast, Belfast, United Kingdom

Mongkol Tadarati, Department of Ophthalmology, Rajavithi Hospital, College of Medicine, Rangsit University, Bangkok, Thailand

Juliette Thariat, Laboratoire de Physique Corpusculaire/IN2P3-CNRS UMR 6534 – ARCHADE, Unicaen – Université de Normandie, Caen, France

Burak Turgut, Onsekiz Mart University, Faculty of Medicine, Department of Ophthalmology, Canakkale, Turkey

Ramesh Venkatesh, Narayana Nethralaya, Department of Retina and Vitreous, Bangalore, Karnataka, India

Anna Stage Vergmann, Department of Ophthalmology, Odense University Hospital, Odense, Denmark; Department of Clinical Research, University of Southern Denmark, Odense, Denmark

Lourdes Vidal-Oliver, Department of Retina, Fisabio Clinic, Valencia, Spain

Stela Vujosevic, Department of Biomedical Surgical and Dental Sciences, University of Milano, Milano, Italy; Eye Clinic, IRCCS MultiMedica, Milano, Italy

Taku Wakabayashi, Wills Eye Hospital, Mid Atlantic Retina, Thomas Jefferson University, Philadelphia, PA, United States

Kristine Wang, Wills Eye Hospital, Mid Atlantic Retina, Thomas Jefferson University, Philadelphia, PA, United States

Lihteh Wu, Asociados de Macula, Vitreo y Retina de Costa Rica, Primer Piso Torre Mercedes Paseo Colón, San José, Costa Rica; Illinois Eye and Ear Infirmary, Department of Ophthalmology and Visual Sciences, University of Illinois Chicago College of Medicine, Chicago, IL, United States

Andres Wu, Asociados de Macula, Vitreo y Retina de Costa Rica, Primer Piso Torre Mercedes Paseo Colón, San José, Costa Rica

Max Wu, Asociados de Macula, Vitreo y Retina de Costa Rica, Primer Piso Torre Mercedes Paseo Colón, San José, Costa Rica

Yoshihiro Yonekawa, Wills Eye Hospital, Mid Atlantic Retina, Thomas Jefferson University, Philadelphia, PA, United States

Arman Zarnegar, Department of Ophthalmology, University of Pittsburgh School of Medicine, Pittsburgh, PA, United States

Javier Zarranz-Ventura, Hospital Clínic de Barcelona, Barcelona, Spain; Institut d'Investigacions Biomediques August Pi i Sunyer (IDIBAPS), Barcelona, Spain; Universitat de Barcelona, Barcelona, Spain

Ece Ozdemir Zeydanli, Ankara Retina Clinic, Ankara, Turkey

Xinyuan Zhang, Beijing Tongren Eye Center, Tongren Hospital, Capital Medical University, Beijing, China; Beijing Retinal and Choroidal Vascular Disorders Study Group, Beijing, China

Ilaria Zucchiatti, School of Medicine, Vita-Salute San Raffaele University, Milan, Italy; Division of Head and Neck, Ophthalmology Unit, IRCCS San Raffaele Scientific Institute, Milan, Italy

Foreword: Lesson learned in retinal vascular diseases

As ophthalmologists, we are in the fortunate situation to have ophthalmoscopy at our disposal, which allows us to evaluate retinal vessels and diagnose numerous vascular conditions through "old school" clinical examination. But much of what is essential is invisible to our naked eye. Recent advances in imaging technology have revolutionized our understanding of retinal and choroidal vasculature anatomy, physiology, and pathologies, and enabled us to diagnose and treat vascular conditions more accurately and efficiently than ever before.

As early as the 1950s and 1960s, Toussaint and Michaelson had already described the composition of three major laminar plexuses in different retinal depths in ex vivo human models. However, our ability to visualize retinal capillaries using dye-based fluorescein angiography was limited to superficial vessels, as deeper capillary plexuses are obstructed by scattering from more superficial retinal layers. With the introduction of optical coherence tomography (OCT) angiography, we can now acquire in vivo depth-resolved imaging and gain a detailed visualization of the different retinal capillary plexuses and the communications between them. The example of paracentral acute middle maculopathy (PAMM), as first described by Sarraf in 2013, illustrates how imaging technologies have revolutionized our ability to detect and understand retinal ischemia. Unlike ischemia of the inner retinal layers, which is easily visible on clinical exam, PAMM is diagnosed through OCT as an isolated hyperreflective band of the inner nuclear layer. The pathological basis for these ischemic lesions was not yet understood, and only with the advent of OCT angiography have we come to learn that hypoperfusion of the deep capillary plexus leads to inner nuclear layer infarction.

The advent of modern imaging technologies has yielded more groundbreaking discoveries, fundamentally altering our understanding of enigmatic conditions, such as central serous chorioretinopathy. The detection of pachyvessels, seen as thickened Haller choroidal layer with attenuation of the Sattler and choriocapillaris layers, led to the establishment of a whole new entity of pachychoroid-related conditions. While it was well accepted that choroidal vascular hyperpermeability is the main factor for subretinal fluid, recent findings have shaken prior conceptions and suggested new ones which include increased scleral thickness, vortex vein asymmetry, intervortex venous anastomotic connections, and abnormal venous emptying via vortex vein occlusion.

We are thrilled to have the opportunity to contribute to this textbook on *Retinal and Choroidal Vascular Diseases of the Eye*. Esteemed experts and key opinion leaders in their field have curated a comprehensive compendium that encompasses state-of-the-art knowledge and latest advancements. This book provides a comprehensive resource for clinicians seeking to stay up to date, from fundamentals to latest innovations in the field.

Dinah Zur and Anat Loewenstein

Foreword: Retina and choroidal vascular diseases of the eye: Exploring pathways of sight and beyond

The intricate network of blood vessels that nourishes the retina and choroid is fundamental for visual function. However, when these delicate pathways are disrupted, devastating conditions can emerge, impairing vision and impacting overall ocular health. The exploration of these conditions, their underlying mechanisms, diagnostic approaches, and treatment modalities is the focus of this comprehensive book: *Retinal and Choroidal Vascular Diseases of the Eye*.

In this book, the authors delve deep into the captivating realm of retinal and choroidal vascular diseases, providing a comprehensive overview of the lessons learned, advances made, and frontiers yet to be explored. The wide-ranging content covers the basics of retinal and choroidal vasculature, encompassing its development, physiology, pathogenic mechanisms, and essential animal models used in research. By establishing a solid foundation, the authors equip readers with the necessary knowledge to comprehend the complex mechanisms that underlie the subsequent chapters.

Imaging techniques nowadays play a pivotal role in the diagnosis and monitoring of vascular diseases. This book extensively covers various imaging modalities employed in clinical practice and research. From traditional fundus photography to cutting-edge techniques like optical coherence tomography angiography, readers will gain insights into the evolving landscape of retinal and choroidal vascular imaging. Moreover, this book explores emerging invasive and noninvasive imaging modalities that hold promise for enhanced diagnostics and treatment monitoring.

The subsequent chapters are dedicated to dissecting specific retinal and choroidal vascular diseases. Starting with well-recognized conditions such as retinal vein occlusion, retinal artery occlusion, and diabetic retinopathy, this book unravels the underlying pathophysiology, clinical presentations, and management strategies. The dissertation also sheds light on ocular ischemic syndrome and intriguing new entities, such as Paracentral Acute Middle Maculopathy (PAMM) and Acute Macular Neuroretinopathy (AMN). Moving further, this book delves into the realm of choroidal vascular diseases, including age-related macular degeneration, polypoidal choroidal vasculopathy, and the enigmatic and protean group of pachychoroid diseases.

Recognizing the intricate interplay between the eye and systemic health, this book explores vascular diseases associated with systemic inflammatory (such as systemic lupus erythematosus, Behcet's disease, multiple sclerosis, sarcoidosis, and systemic vasculitis), hematologic (such as sickle cell retinopathy), and neoplastic conditions. Finally, this book ventures into retinal and choroidal tumors, including cavernous hemangioma, choroidal hemangioma, choroidal melanoma, and vasoproliferative tumors. These rare but clinically significant entities warrant a thorough understanding for accurate diagnosis and management.

In summary, *Retinal and Choroidal Vascular Diseases of the Eye* brings together a wealth of knowledge, clinical expertise, and research insights to empower readers with a comprehensive understanding of retinal and choroidal vascular diseases. By blending scientific rigor with practical applications, this book serves as a valuable resource for ophthalmologists, researchers, and healthcare professionals seeking to navigate the intricate landscape of ocular vascular pathologies. Together, let us embark on a journey to unlock the mysteries surrounding these captivating disorders and pave the way for improved patient outcomes.

Francesco Bandello

Preface

With advancing imaging techniques, our understanding of vascular diseases of the eye has revolutionized. With newer imaging biomarkers and evolving treatment approaches, management for our patients with retinal and choroidal vascular diseases has significantly changed. The last book focused on this topic was more than a decade ago. The current book consists of the most updated knowledge on pathogenesis, imaging, therapeutic options, and future directions, written by world experts on each topic.

The book titled *Retinal and Choroidal Vascular Diseases of the Eye* starts with lessons learned in the last few decades by experienced retinal specialists. There are sections on basics, imaging, retinal vascular diseases, choroidal vascular diseases, vascular abnormalities, vascular diseases associated with systemic inflammatory conditions, common systemic conditions, and vascular tumors. The imaging section includes not only routine imaging modalities but also brings in the futuristic imaging modalities for retinal and choroidal vasculature. The section on choroidal vascular diseases comprises new knowledge in the field and discusses the changing treatment paradigms. The last section of this book includes miscellaneous conditions such as COVID-19-related vascular conditions and vascular changes after vitreoretinal surgeries. This book also includes a chapter on vascular regeneration to discuss future directions in this field. All the chapters are written in a reader-friendly manner and supplemented with many illustrations and figures. I hope this book helps readers improve patient care, teaching, and research projects.

I would like to thank all the experts who contributed their best work to make this unique blend of basic and clinical science. I hope the readers would enjoy reading this book and this helps in better patient management. I would like to thank Patricia Gonzalez, Elsevier staff, who helped throughout the process. In the end, I would like to thank our patients, colleagues, and families who supported us to create this book to enhance our knowledge.

Jay Chhablani
Pittsburgh, PA

Section I

Basics of retinal and choroidal vasculature

Chapter 1

Development and physiology

Matthieu Poireir[1] and Jay Chhablani[2]

[1]*Eastern Virginia Medical School, Norfolk, VA, United States;* [2]*Department of Ophthalmology, University of Pittsburgh School of Medicine, Pittsburgh, PA, United States*

General anatomy of the retinal and choroidal vessels

The blood flow to the eye is derived from the ophthalmic artery and its branches. The ophthalmic artery itself derives from the internal carotid artery, branching distal to the cavernous sinus and traveling through the optic canal to reach the eye and its contents.[1] After traveling from the internal carotid artery, the ophthalmic artery generally bifurcates to supply two divisions or categories: orbital branches and optical branches.[2] The orbital branches distribute vessels to the orbit and surrounding parts. They include the lacrimal, supraorbital, and ethmoidal arteries.[2] The optical branches distribute vessels to the eye itself and its muscles. They include the ciliary arteries, the central retinal artery, and the superior and inferior muscular arteries.[2]

The central retinal artery represents the first branch of the ophthalmic artery which supplies blood to the inner layer of the retina, roughly the inner two-thirds.[1] It travels within the optic nerve sheath, supplying blood therein.[2] It further bifurcates into superior and inferior arcades, which form the blood—retina barrier.[2] The ciliary arteries include the anterior, short, and long posterior ciliary arteries.[2] The long posterior ciliary arteries pierce the posterior sclera after traveling with the optic nerve to provide blood to the choroid and ciliary muscle; it then joins the major arterial circle of the iris therein distributing arterial branches to the iris and ciliary body.[3] The short posterior ciliary arteries pass forward with the optic nerve to reach the eye; they pierce the sclera around the entry of the optic nerve and supply the ciliary processes and the optic disk.[2] The number of short posterior ciliary arteries often ranges between 6 and 12 arteries; they all branch off the ophthalmic artery as it crosses the optic nerve.[2] The short posterior ciliary arteries give perpendicular terminal branches to supply the outer retina and regions between the retina and the choroid.[2] The anterior ciliary arteries, of which there are seven, branch from the muscular arteries that are derived from the ophthalmic artery and run to supply the extraocular muscles as well as the conjunctiva and sclera. They also contribute to the major arterial circle of the iris with the long posterior ciliary arteries.[2]

Retinal and choroidal vascular supply

The retinal vasculature metabolically sustains the inner part of the retina via a network of capillaries that permeates the neural tissue. The outer retina contains photoreceptors and is avascular; it is supplied by the choroidal vasculature that is separated from the neural retina by the retinal pigment epithelium (RPE). As discussed previously, the retina is supplied by the central retinal artery and the short posterior ciliary arteries. The central retinal artery travels with the optic nerve to supply the layers of the inner retina closest to the vitreous compartment.[2] The short posterior ciliary arteries (typically 6—12) also pierce the sclera around the optic nerve and then assimilate to form the outer layer of the vessels of the choroid.[2] The arterioles they give off supply the choriocapillaris that supplies the RPE and the outer segments of the photoreceptor.

The choroid is the vascular layer of the eye and is part of the uvea. It lies between the retina and the sclera.[4] The human choroid is thickest at the far extreme rear of the eye while narrowing in the outlying areas.[5] The choroid provides oxygen and nourishment to the outer layers of the retina, receiving its vascular supply from the short and long posterior ciliary arteries.[2,3] Its organization can be considered in five layers. Starting from the inner region closest to the retina, there is the Bruch's membrane, choriocapillaris, Sattler's layer, Haller's layer, and suprachoroidal (the zone between choroid and

Retinal and Choroidal Vascular Diseases of the Eye. https://doi.org/10.1016/B978-0-443-15583-3.00001-9

sclera).[5] The Bruch's membrane is a thin connective tissue sheet that separates the retina from the choroid, supplying arterial contents to the outer retina and innermost layer of the choroid.[6] The choriocapillaris provides a highly anastomosed network of capillaries to supply the RPE and photoreceptors as discussed previously.[7] Large diameter capillaries of 20−25 μm in diameter are found here, allowing passage of multiple red blood cells at any moment in time.[5] Additionally, pores of 700-nm in diameter allow rapid transport of molecules. The choriocapillaris is about 10 μm thick at the fovea (densest area of capillaries) and thins to about 7 μm in the periphery.[5] Sattler's layer contains intermediate and small-sized blood vessels; arterioles from this layer supply the choriocapillaris in a lobular fashion to form a patch-like network.[7] Hattler's layer contains larger diameter arteries in the choroid. Surrounding extravascular tissue of the choroid is formed by collagen, elastic fibers, fibroblasts, nonmuscular smooth muscle cells, melanocytes, macrophages, mast cells, and lymphocytes.[8]

The development of the choroid and retinal vasculature

During early development, the ocular vascular system emerges from the mesoderm surrounding the newly formed optic cup.[9] Undifferentiated cells derived from mesenchyme, the embryonic mesoderm cells that later form connective tissue, invade the future vitreous space through an opening near the optic cup before it closes at 5.5 weeks of gestation.[7,10] Some cells entering through are composed entirely of erythroblasts surrounded by mesenchymal cells to form erythroblast islands, which are compartments of particular environments within which erythroblasts can proliferate and differentiate.[11] Cells in these islands may express epsilon hemoglobin (Hbε), the embryonic globin made between 2 and 9 weeks gestation in erythroblasts, notably in the yolk sac from 2 to 6 weeks of gestation.[12] Other endothelial cell markers like VEGFR2 (vascular endothelial cell receptor-2), CD31 (PECAM), and CD34 are expressed in these primordial cells.[13,14] The entering cells also express endoglin, a TGFβ receptor produced during hemangioblast specification and commitment; these are the multipotent precursor cells that can differentiate into both hematopoietic and endothelial cells in vasculature formation and form blood islands, the structures around or in the embryo which lead to many different parts of the circulatory system.[15] At about 5.5 weeks of gestation, areas of blood vessels begin forming in the vitreous near the lens capsule; the cells contributing to the formation of the aforementioned erythroblasts and mesenchymal cells.[9] Over time, undifferentiated mesenchymal cells that invade vitreous differentiate into vascular precursor cells such as prevascular cells, hemangioblasts, and fibrocytes.[9] The presence of such vascular precursors suggests that the fetal vasculature of vitreous forms by hemo-vasculogenesis: the development of the blood system from precursor cells.[16] This is in contrast to other vascular structures of the eye which develop by sprouting new blood vessels from older ones, or angiogenesis.[17]

The formation and characteristics of the hyaloid artery

The hyaloid artery is a branch of the ophthalmic artery which provides blood to the embryonic eye; it passes through the embryonic fissure into the optic cup, running from the center of the optic disc through the vitreous to the posterior lens.[18] It forms in segments at 6 weeks gestation and invades the vitreous space by 7 weeks of gestation.[7] At this stage, there is no angiogenesis spurting from the hyaloid artery, yet cells express NG2, a membrane protein expressed in the developing and adult central nervous system and pericytes, the cells present along the walls of capillaries, indicating the vascular nature of this structure.[19] As the hyaloid artery matures, cells on the outer surface express higher levels of NG2.[19] Concurrently, the inner cells express endothelial cell markers like CD31; as the hyaloid vasculature matures, CD31 and other endothelial cell markers such as vWf increase while progenitor capillary markers like CXCR4 decline in expression.[7,19] After emerging from the optic nerve head, the hyaloid artery branches over the posterior lens as the tunica vasculosa lentis (TVL) and forms the pupillary membrane (PM) anteriorly over the lens.[20] The vasa hyaloidea propria emerges from the hyaloid artery as well and anastomoses with the TVL; it initially overlies the retina yet gradually lifts off to join the TVL near the equatorial region of the eye.[7,21] The TVL, PM and vasa hyaloidea propria will help supply the fetal eye before more mature vascular structures form.[7] The TVL supplies the posterior and lateral portions of the lens; the vasa hyaloidea propria will supply blood to the inner retina until a retinal vasculature forms.[22] These fetal vasculatures are characterized by cells that are major histocompatibility complex (MHC)-1 and -2 positive and also express CD45+, a transmembrane protein present on all differentiated hematopoietic cells and is involved in cell differentiation/growth.[10,23] These fetal vasculatures, when persistent past development, may cause pathology such as retinopathy of prematurity.[20] The hyaloid vasculature remains intact until 13 weeks of gestation; at that time, regression and apoptosis of vascular elements occur through the work of macrophages, leukocytes, and hyalocytes, the stellate cells of the vitreous body.[23] In mice, VEGFR2 appears critical for the regression of the hyaloid vasculature, as mice lacking VEGFR2 through knockout in retinal neurons have a denser network of hyaloid vessels.[24]

The development of the choroidal vasculature

The uveal vascular system includes the choroid and begins development as early as 4 weeks of gestation when the neural tube closes and optic vesicles become apparent.[25] There is the formation of a large plexus of primitive vessels originating from the neural tube's vascular system, which forms around the outside aspect of the optic cup.[7] This early layer expands and increases in density; it spreads in the same way as the pigmentation spreads in the RPE, starting from the anterior portions of the posterior pole to the optic cup.[9] Choroidal capillaries will continue development to encircle the optic cup; however, they remain separated from the retina by the basement membrane of the RPE.[7] Anastomoses will form at the anterior portion of the optic cup to form the annular vessel; the vessels of the choroid then join to the posterior ciliary arteries.[9] The venous aspect of the circulation forms during gestation at the same time the choriocapillaris becomes definite.[5,26] During the 15th week of gestation, the development of the choroid becomes more concrete, as both arterioles and venules can be seen localized more anteriorly outside the choriocapillaris.[5] Veins and arteries are not observable until the 22nd week of gestation, the definitive marker of arteries being a continuous layer of smooth muscle cells, often accompanied by a second layer as well.[5] The subendothelial space is characterized by patches of electron-dense material, likely an early stage of an internal elastic lamina, the outermost aspect of the tunica intima.[5] Early choroidal arteries may also be distinguished from adult human eye vasculature in this area by the occurrence of glycogen in smooth muscle cells.[5]

Choroidal endothelial cells originate from mesenchyme, just as the cells that provide for early vitreous vasculature do.[27] Other cells of the choroid, such as stromal cells, melanocytes, and pericytes, however, are derived from the neural crest.[28] These neural crest cells migrate around the optic vesicles, the pair of pouch-like structures derived from the lateral aspects of the forebrain. These undifferentiated mesenchymal cells contribute to melanocytes and stroma of the choroid.[28] Melanin, which serves as the pigmentation for the eye, appears between 6 and 7 months and is completed at birth; the pigmentation of the choroid will help to absorb light and limit reflections within the eye that could degrade vision.[29]

Although the specific molecular mechanisms that govern the formation of the choroid are largely unknown at this time, the development of the choroid's vasculature appears to depend on the presence of differentiated RPE in addition to inductive signals produced by the RPE.[30] The unmitigated expression of fibroblast growth factor (FGF)-9 has been shown to convert most of the RPE into neural retina; overexpression also inhibits the development of the choroid.[31] Accordingly in human patients, failure of RPE differentiation leads to defective development of the choroid and sclera.[27] Furthermore, in vitro, RPE cells may stimulate tube formation of choroidal cells, an effect that is inhibited by the administration of basic fibroblast growth factor (bFGF) and vascular endothelial growth factor (VEGF) neutralizing antibodies.[30] The importance of fibroblast growth factor signaling has also been shown in mice, as the overexpression of a dominant negative FGF-Receptor1 in the RPE leads to the failure of choroidal vessels to completely develop, remaining in an immature stage, implicating a necessary balance of bFGF in choroidal development.[32] The RPE and underlying mesenchyme express VEGF and its receptor at high levels during the period of choriocapillaris formation both in humans[33] and rodents.[31,34] Together these findings suggest that RPE expression of angiogenic factors such as VEGF and bFGF may play a role concerning the specific molecular mechanisms in the development of the choroid (Fig. 1.1).

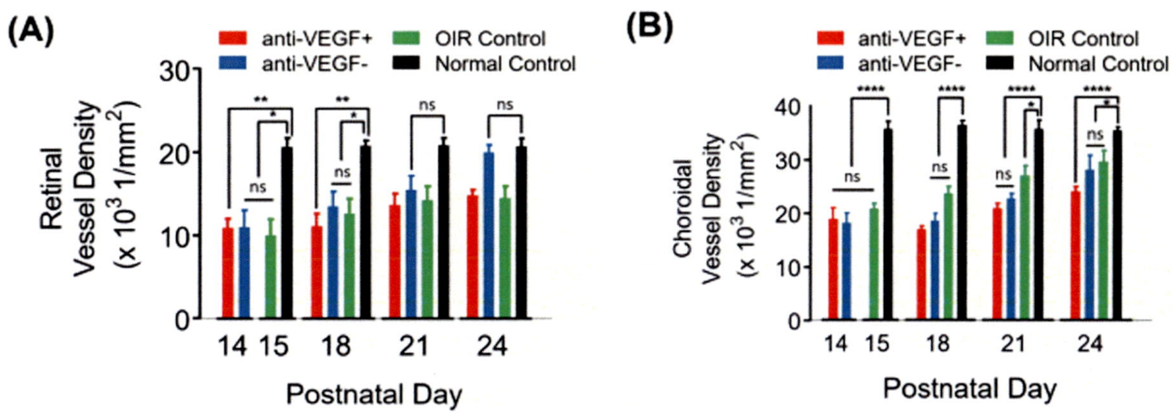

FIGURE 1.1 The effect of anti-VEGF on retinal vessel density: OIR (oxygen-induced retinopathy) mimics retinopathy of prematurity. No changes were made to this figure: the Creative Commons license link: https://creativecommons.org/licenses/by/4.0/. *From Kim Y, Park JR, Hong HK, et al. In vivo imaging of the hyaloid vascular regression and retinal and choroidal vascular development in rat eyes using optical coherence tomography angiography.* Scientific Reports. *2020;10(1). https://doi.org/10.1038/s41598-020-69765-7.*

Development of the choriocapillaris

The choriocapillaris, the innermost structure of the choroid that nourishes the RPE and photoreceptors, begins differentiation in conjunction with the development of the RPE during the fourth and fifth week of gestation, maturing until the end of the first trimester.[26] Specifically, when the optic fissure closes at 5–7 weeks of gestation, RPE cells have formed a distinct layer within which junctional complexes allow cell–cell communication.[35] At the time when mesenchymal cells invade the vitreous space (5–6 weeks of gestation), cells expressing CD34; CD31; CD39; and VEGFR-2, all markers of the endothelium, may be observed traversing to the area that will become the choriocapillaris, posterior to the RPE.[26] Their cells and their organization are similar to the blood island-like structures observed in vitreous as previously discussed; with some not surrounded and others surrounded by mesenchymal precursors. In addition, some cells express Hbε, indicating their role in early vascular development.[36] CD34/Ki67+, markers that denote angiogenesis as the primary method of vascularization, are absent at 6 weeks of gestation in the choriocapillaris, further implicating the role of hemangioblasts and the importance of precursor cells for the development of its vessels, similar to the fetal vasculature of the vitreous.[26]

Up to the seventh week of gestation, vessel lumina are still small in diameter, capable of passing one erythroblast only.[5] The early endothelial lining is characterized by a marked thickness in comparison to its later development, likely due to a higher number of intracellular vesicles that transport membrane material to the plasma membrane of the endothelial cells.[5] As the endothelium flattens, pores become observable, which most often face Bruch's membrane.[7–9] The basement membrane of the choriocapillaris develops by the beginning of the ninth week of gestation, likely involving endothelial cells or pericytes that localize to the area by the sixth week of gestation. Four out of five layers of Bruch's membrane can be distinguished during this ninth week of gestation, and only the central, elastic layer is not fully developed by the 13th week.[5] Fibroblasts, which likely help produce membrane in the Bruch's membrane, can be observed at this time as well.[32]

NG2, a molecule that helps stabilize cell-substratum interactions on endothelial basement membranes, begins to be expressed in large amounts at roughly 12 weeks of gestation, though some expression occurs as early as 7 weeks of gestation.[19,26] NG2 expression correlates with the observable maturation of the structural components of the choriocapillaris, and this maturation completes at roughly 22 weeks.[5] Crucial to this inference are characteristics of the adult choriocapillaris, including three layers of vasculature in the submacula; alpha-smooth muscle actin (αSMA), NG2, and PV-1 (present in endothelial pores).[13,26] Intermediate blood vessels form at roughly 12 weeks of gestation as well, at which time they associate with a single layer of capillaries.[26] Ki67+ cells, which are classically associated with angiogenesis, appear at roughly 11–12 weeks of gestation within the outer aspect of the immature choriocapillaris; expression lies in concord with the developing intermediate blood vessels that bud off the choriocapillaris.[9,13] These findings underline the role of angiogenesis as the main pathway through which choroidal vessels expand into the deeper stroma, as opposed to vascular precursor cells. There are no budding, larger blood vessels observable from which the choriocapillaris originates from; this implies that this vasculature around the RPE is derived from periocular mesenchyme cells, whose role has been discussed previously.[36]

The development of the retinal vasculature

The retinal vasculature metabolically sustains the inner part of the retina closest to the anterior portion of the eye.[36] There are up to four retinal vascular networks in the macula, demonstrated in humans ex vivo using confocal microscopy and in vivo using speckle variance OCT[37,38]: the superficial vascular plexus (SVP), intermediate and deep plexuses surrounding the inner nuclear layer (ICP and DCP, respectively), and the radial peripapillary capillary plexus (RPCP) (Fig. 1.2).[37] The SVP is supplied by the central retinal artery and is composed of blood vessels primarily in the ganglion cell layer (GCL). The ICP and DCP are supplied by vertical anastomoses from the SVP.[39] The RPCP is unique; its blood vessels are organized longitudinally and parallel with the nerve fiber layer axons. Moreover, the RPCP supplies the nerve fiber layer (NFL) bundles in this region.[40–42]

Direct vascularization of the inner retina begins with transient vessels stemming from the hyaloid artery but vascular remodeling later leads to regression of these transient vitreal vessels at 13 weeks of gestation (called the hyaloid vasculature) as discussed previously.[23] New vessels grow into the retina from an existing capillary ring at the optic nerve head that forms a flat, primary plexus at 15 weeks of gestation in the nerve fiber layer. Subsequently, the primary plexus sprouts into the retina in four main lobes.[43,44] Primary plexus development is characterized by the invasion of migrating astrocytes following regression of the hyaloid artery. The astrocytes emerge from the optic nerve head and spread across the inner surface of the retina forming a template from which blood vessels will form.[45,46] The importance of the astrocytes is evident from the strict correlation between their presence and blood vessels: no retinal astrocytes can be found in the avascular aspects of the retina.[44]

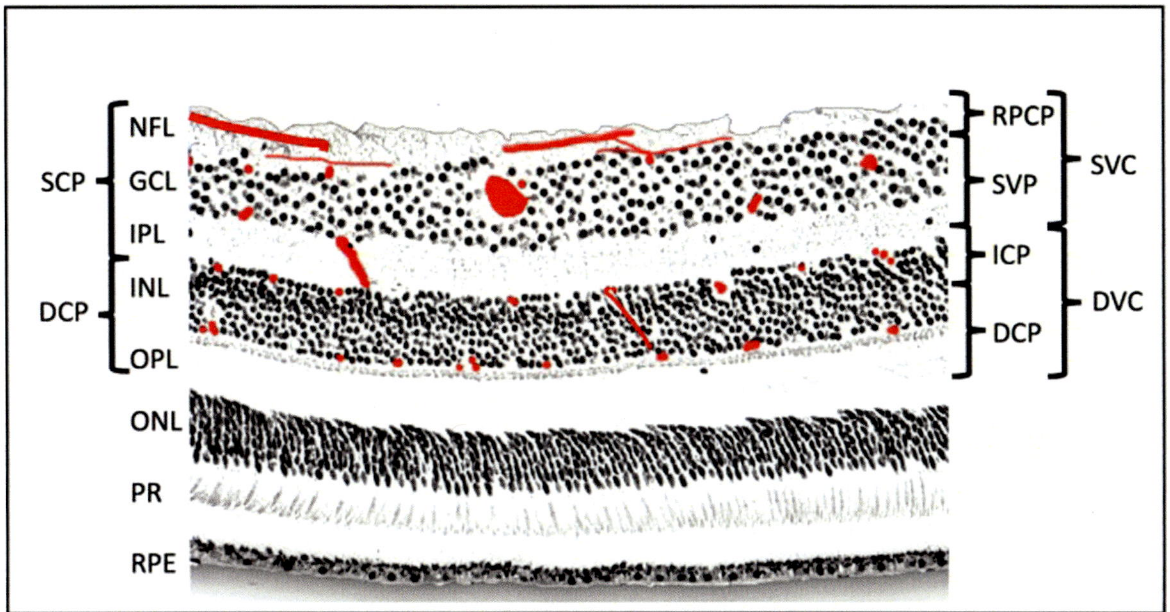

Current OCTA Nomenclature	Anatomic Layers	Proposed OCTA Nomenclature	
SCP	NFL	RPCP	SVC
	GCL	SVP	
	IPL	ICP	DVC
DCP	INL		
	OPL	DCP	

FIGURE 1.2 Depicted above are the various plexuses of vasculature that reside in the retina in relation to the underlying RPE, nerve fiber layer, ganglion cell layer (GLC), photoreceptor layer (PR), inner nuclear layer (INL), inner plexiform layer (IPL), outer nuclear layer (ONL), and outer plexiform layer (OPL). No changes were made to the figure; the Creative Commons license link: https://creativecommons.org/licenses/by/4.0/. *From Campbell JP, Zhang M, Hwang TS, et al. Detailed vascular anatomy of the human retina by projection-resolved optical coherence tomography angiography.* Scientific Reports. *2017;7. https://doi.org/10.1038/srep42201.*

These retinal astrocytes come from cells derived from the optic nerve that express Pax2, a transcription factor that acts to instruct the embryonic formation of many organs including the eyes.[47,48] Optic nerve astrocytes and retinal astrocytes, the two pedigrees that form from the common precursor, have distinct characteristics. The optic nerve astrocytes will stay to nourish the optic nerve, providing nutrients to optic nerve ganglion axons. Platelet-derived growth factor receptor alpha (PDGFRα) is the first marker that separates two groups (retinal astrocytes express the receptor in the optic nerve head several days before they start to invade the retina).[49] The ligand for PDGFRα is secreted by retinal ganglion cells and helps to match retinal astrocytes to neuronal cells.[46,50] Once the immature retina reaches the hypoxic peripheral aspects, they proliferate to establish a network; the hypoxia encourages the expression of vascular endothelial growth factor (VEGF), a key stimulus for angiogenesis, and also regulates the number of astrocytes that migrate to each area.[51–53] The orphan nuclear receptor Tlx also allows for the formation and regulation of the retinal astrocytes: immature retinal astrocytes express this protein and downregulate expression as they mature. When knocked out in mice, retinal astrocytes fail to achieve maturity and vascularization is delayed.[54,55]

As the primary plexus develops, it reaches the retinal periphery nasally at 36 WG and temporally at around 40 WG. Deeper plexus components form by angiogenic sprouting from the primary plexus veins.[39,56,57] The sprouting from the primary plexus moves to penetrate the retina and establish two vascular networks on either side of the inner nuclear layer from the ICP and IDP.[58] Like the primary plexus, the deeper plexus vessel growth occurs from the center to the periphery. Unlike the rest of the retina, the foveal zone remains completely devoid of vessels throughout development.[59] Retinal vascularization develops fully by about 38–40 weeks of gestation.[56]

There is evidence that the different plexuses of the retinal vasculature form by different mechanisms; the deeper networks of the retinal vasculature by sprouting angiogenesis and the primary inner vascular plexus by hemo-vasculogenesis, the process characterized by the formation of blood vessels from an endothelial precursor.[45,60–62] These vascular pioneers form early structures which then progress into a lumen.[17] Naturally, this theory posits that precursor cells exist ahead of the primary vascular network; in the mouse retina, no such precursors have been found which complicates our understanding of which method pervades.[57,63] However, cells positive for ADPase/CD39 and CXCR4 exist in the human retina at 12 weeks of gestation on nerve fibers after migrating from the optic nerve and have been suggested to be the elusive vascular precursor cells responsible for hemo-vasculogenesis.[18,64] Salient to this hypothesis, CXCR4 is a marker expressed on endothelial cells, hemangioblasts, and vascular progenitors while endothelial cell CD39/ecto-ADPase serves to regulate and maintain vascular homeostasis.[65–67] Furthermore, CXCR4's ligand, stromal-derived factor-1 (SDF-1), is a hypoxia-inducible molecule and is found highly expressed on the innermost retina in a gradient fashion, allowing for precursor cells to adjust and establish blood vessel formation accordingly to need based on the hypoxic nature of their environment.[68] The early vitreous and choroid vasculatures also express CD39/ecto-ADPase which, as we have seen, develop from hemo-vasculogenesis, suggesting that the vasculogenic model may be applicable in the retina.[69] Identification of endothelial precursor cells is still controversial and current research aims to identify the VEGF receptor 2, one of the earliest markers of developing endothelial cells, in early retina cells; this would provide additional evidence of endothelial precursor in the retina to aid in the confirmation of the existence of hemo-vasculogenesis.[14,70–73]

There is discussion over which cells appropriately guide human vascular formation. In mice, the primary plexus appears to be formed by sprouting angiogenesis(61) and is guided by specialized "endothelial tip cells," which are themselves guided by retinal astrocytes at the leading point of the growing vasculature.[74,75] These cells express angiogenic markers such as VEGFRs.[75] In contrast, others propose the guider may be "Müller cells," the other macroglial cell of the retina, as opposed to astrocytes.[63] Recent work has found that at 9 weeks of gestation, the inner Müller cells have formed a process for invasion and angiogenesis in the human retina, and by 12 weeks of gestation, they express Notch 1, a protein that plays a role in vascular maturation in several organs.[76,77] VEGF is also localized to the Muller cell process associated inner neuroblastic layer.[9] At 17 weeks of gestation, Müller cells clear extracellular spaces in the inner retina; early blood vessels will grow and mature in these spaces.[78,79] Mice monocytes and stalk endothelial cells express Notch1 and provide guidance at the edges of the vasculature as well as regulate the formation of tip cells which guide angiogenesis and in stalk endothelial cells which control the formation of tip cells during angiogenesis.[74,76]

Insights into vascular maturation, remodeling, and differentiation are provided mostly in mice studies. Vessel propagation from the center to the periphery of the retina is done by VEGF mRNA gradient.[80] VEGF mRNA is expressed in lower quantities in the center of the retina where there is full vascularization and no hypoxia.[81] The periphery of the retina, on the other hand, is hypoxic; here, VEGF mRNA is expressed in high quantities.[80] This differential in the center versus peripheral likely induces a gradient distribution of VEGF protein. Mice possessing only the short VEGF120 isoform, which diffuses freely due to less affinity for the extracellular matrix, experience a flattened VEGF protein gradient.[80] As a result, vascular growth throughout the retina is slowed.[82,83]

The mechanisms of remodeling, the method by which vessels are altered to produce a mature form, is also relatively unknown in humans. In mice, the endothelial cells trailing the tip cells at the forefront of migration remodel and relocate to mature the structures via methods such as selective endothelial cell apoptosis.[62,84] Leukocytes use CD18 to adhere to the vascular network, thereby initiating cell death via Fas ligand (FasL).[84] Inhibiting the action of either FasL or CD18 with antibodies increases vascular density, showing the importance of these mechanisms in pruning the vasculature to allow the mature form to develop.[84] Perhaps paradoxically, the use of intraocular clodronate liposomes, which deplete macrophages, has the opposite effect and reduces vascular density.[85] Immune cells have diverse roles during normal vascular development, required to initiate apoptosis for remodeling yet also seemingly required to maintain or increase vascular density as well, with possible mediators including VEGF and TGF-Beta.[86] In terms of vascular differentiation into arteries, veins, arterioles, and venules, the DLL4/Notch signaling pathway is also needed in the retinal vasculature for artery formation.[74]

After the primary plexus is established, angiogenesis allows for the formation of the deeper plexus (outer plexus).[39,58] Mice and rat models have shown this formation occurring about 1 week postpartum, at about the time the expanding primary plexus reaches the margins of the retina.[87] Angiogenic sprouts emerge from veins, venules, and capillaries near veins.[7] These branches penetrate the retina at a perpendicular angle to the primary plexus and begin their development in the center of the retina after which they move toward the periphery.[9,58] From a molecular standpoint, mRNA VEGF is expressed in the inner nuclear layer (INL), suggesting the necessity for VEGF to allow for sprouting from the primary plexus.[52] Once reaching the inner and outer aspects of the INL, the vascular projections develop longitudinally, growing along the Müller cell processes.[9,56,58] As a result, this movement in relation to the INL creates two capillary beds parallel to the original primary plexus. The deeper plexus develops independently of retinal astrocytes, a distinction from the primary plexus.[56] Apart from this observation, little is known about the cellular and molecular mechanisms that induce and guide deeper plexus angiogenesis in the retina.

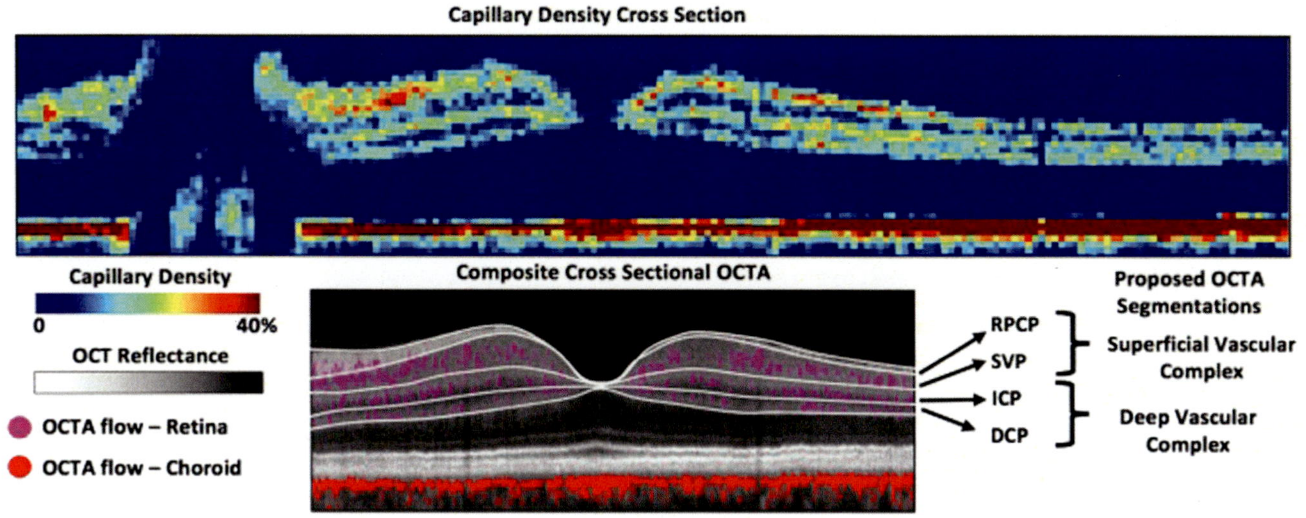

FIGURE 1.3 Optical coherence tomography angiography (OCTa) depicts the capillary densities inside the retina itself. Three layers of concentrated capillary density could be seen in the retina (top layers of upper image): superior vascular complex, intermediate capillary plexus, and deep capillary plexus. One layer of high capillary density is seen in the choriocapillaris (bottom layer). Note the avascular fovea. No changes were made to the figure; the Creative Commons license link: https://creativecommons.org/licenses/by/4.0/. *From Campbell JP, Zhang M, Hwang TS, et al. Detailed vascular anatomy of the human retina by projection-resolved optical coherence tomography angiography.* Scientific Reports. *2017;7. https://doi.org/10.1038/srep42201.*

The development of the fovea

The fovea begins to form at 22 weeks of gestation.[88] The area, developmentally, is characterized by the absence of rods and develops as the temporal arcade vessels advance toward the eventual macula, as confirmed by OCT.[88,89] Maturation completes between 15 and 45 months postpartum.[88] By 26 weeks of gestation, the fovea is a blood vessel-free zone from nasal macula to ora serrata (the junction between the choroid and the ciliary body) Fig. 1.3. Cones, and no rods, are present within this zone.[90] It is the responsibility of antiangiogenic factors to create the signature avascularization of the fovea and macula. Two of the proposed cell signals for this are pigment epithelial-derived factor (PEDF) and brain natriuretic peptide precursor B (NPPB); both of which are found in elevated quantities in the fovea compared to other parts of the fundus which do remain vascular.[91] The absence of vasculature is thought to optimize the optical quality of the central vision by reducing the aberrant scattering of light particles.[92]

References

1. Gupta N, Motlagh M, Singh G. *Anatomy. Head and Neck, Eye Arteries*. StatPearls Publishing; 2022.
2. Kiel JW. *Integrated Systems Physiology: From Molecule to Function to Disease*. 2010.
3. Duroux PE, Rougier J, Moulin J. Anatomie des artères ciliaires longues postérieures. *Lyon Médical*. 1957;89(33):129–145.
4. Hayreh SS, Hayreh SB. Uveal vascular bed in health and disease: lesions produced by occlusion of the uveal vascular bed and acute uveal ischaemic lesions seen clinically. *Eye (Basingstoke)*. 2023;37:2617–2648. https://doi.org/10.1038/s41433-023-02417-y.
5. Sellheyer K. Development of the choroid and related structures. *Eye*. 1990;4(2):255–261. https://doi.org/10.1038/eye.1990.37.
6. Booij JC, Baas DC, Beisekeeva J, Gorgels TGMF, Bergen AAB. The dynamic nature of Bruch's membrane. *Progress in Retinal and Eye Research*. 2010;29(1):1–18. https://doi.org/10.1016/j.preteyeres.2009.08.003.
7. Saint-Geniez M, D'Amore PA. Development and pathology of the hyaloid, choroidal and retinal vasculature. *International Journal of Developmental Biology*. 2004;48(8–9):1045–1058. https://doi.org/10.1387/ijdb.041895ms.
8. Nickla DL, Wallman J. The multifunctional choroid. *Progress in Retinal and Eye Research*. 2010;29(2):144–168. https://doi.org/10.1016/j.preteyeres.2009.12.002.
9. Lutty GA, McLeod DS. Development of the hyaloid, choroidal and retinal vasculatures in the fetal human eye. *Progress in Retinal and Eye Research*. 2018;62:58–76. https://doi.org/10.1016/j.preteyeres.2017.10.001.
10. Balazs EA, Toth LZ, Ozanics V. Cytological studies on the developing vitreous as related to the hyaloid vessel system. *Albrecht von Graefes Archiv für Klinische und Experimentelle Ophthalmologie*. 1980;213(2):71–85. https://doi.org/10.1007/BF00413534.
11. McLeod DS, Hasegawa T, Baba T, et al. From blood islands to blood vessels: morphologic observations and expression of key molecules during hyaloid vascular system development. *Investigative Opthalmology & Visual Science*. 2012;53(13):7912. https://doi.org/10.1167/iovs.12-10140.
12. Kaufman DP, Khattar J, Lappin SL. *Physiology Hemoglobin Fetal*. StatPearls Publishing; 2023.

13. Basilio-De-Oliveira RP, Pannain VLN. Prognostic angiogenic markers (endoglin, VEGF, CD31) and tumor cell proliferation (Ki67) for gastrointestinal stromal tumors. *World Journal of Gastroenterology*. 2015;21(22):6924–6930. https://doi.org/10.3748/wjg.v21.i22.6924.
14. Urbich C, Dimmeler S. Endothelial progenitor cells: characterization and role in vascular biology. *Circulation Research*. 2004;95(4):343–353. https://doi.org/10.1161/01.RES.0000137877.89448.78.
15. Perlingeiro RCR. Endoglin is required for hemangioblast and early hematopoietic development. *Development*. 2007;134(16):3041–3048. https://doi.org/10.1242/dev.002907.
16. Goldie LC, Nix MK, Hirschi KK. Embryonic vasculogenesis and hematopoietic specification. *Organogenesis*. 2008;4(4):257–263. https://doi.org/10.4161/org.4.4.7416.
17. Risau W. Mechanisms of angiogenesis. *Nature*. 1997;386(6626):671–674. https://doi.org/10.1038/386671a0.
18. Chan-Ling T, McLeod DS, Hughes S, et al. Astrocyte-endothelial cell relationships during human retinal vascular development. *Investigative Ophthalmology and Visual Science*. 2004;45(6):2020–2032. https://doi.org/10.1167/iovs.03-1169.
19. Karram K, Chatterjee N, Trotter J. NG2-expressing cells in the nervous system: role of the proteoglycan in migration and glial-neuron interaction. *Journal of Anatomy*. 2005;207(6):735–744. https://doi.org/10.1111/j.1469-7580.2005.00461.x.
20. Patel NA, Fan KC, Al-Khersan H, et al. Persistent tunica vasculosa lentis as an independent risk factor for treatment in retinopathy of prematurity. *Ophthalmology Retina*. 2021;5(2):217–219. https://doi.org/10.1016/j.oret.2020.08.002.
21. Ito M, Yoshioka M. Regression of the hyaloid vessels and pupillary membrane of the mouse. *Anatomy and Embryology*. 1999;200(4):403–411. https://doi.org/10.1007/s004290050289.
22. Gaier ED, Yonekawa Y. Persistent vasa hyaloidea propria/retinae in familial exudative vitreoretinopathy. *Journal of AAPOS*. 2021;25(2):114–116. https://doi.org/10.1016/j.jaapos.2020.11.004.
23. Zhu M, Madigan MC, Van Driel D, et al. The human hyaloid system: cell death and vascular regression. *Experimental Eye Research*. 2000;70(6):767–776. https://doi.org/10.1006/exer.2000.0844.
24. Yoshikawa Y, Yamada T, Tai-Nagara I, et al. Developmental regression of hyaloid vasculature is triggered by neurons. *Journal of Experimental Medicine*. 2016;213(7):1175–1183. https://doi.org/10.1084/jem.20151966.
25. Fuhrmann S. Eye morphogenesis and patterning of the optic vesicle. *Current Topics in Developmental Biology*. 2010;93(C):61–84. https://doi.org/10.1016/B978-0-12-385044-7.00003-5.
26. Lutty GA, Hasegawa T, Baba T, Grebe R, Bhutto I, McLeod DS. Development of the human choriocapillaris. *Eye*. 2010;24(3):408–415. https://doi.org/10.1038/eye.2009.318.
27. Torczynski E. Normal and abnormal ocular development in man. *Progress in Clinical and Biological Research*. 1982;82:35–51.
28. Etchevers HC, Vincent C, Le Douarin NM, Couly GF. The cephalic neural crest provides pericytes and smooth muscle cells to all blood vessels of the face and forebrain. *Development*. 2001;128(7):1059–1068.
29. Sohn EH, Khanna A, Tucker BA, Abràmoff MD, Stone EM, Mullins RF. Structural and biochemical analyses of choroidal thickness in human donor eyes. *Investigative Ophthalmology and Visual Science*. 2014;55(3):1352–1360. https://doi.org/10.1167/iovs.13-13754.
30. Sakamoto T, Sakamoto H, Murphy TL, et al. Vessel Formation by choroidal endothelial cells in vitro is modulated by retinal pigment epithelial cells. *Archives of Ophthalmology*. 1995;113(4):512–520. https://doi.org/10.1001/archopht.1995.01100040134039.
31. Zhao S, Overbeek PA. Elevated TGFβ signaling inhibits ocular vascular development. *Developmental Biology*. 2001;237(1):45–53. https://doi.org/10.1006/dbio.2001.0360.
32. Rousseau B, Larrieu-Lahargue F, Bikfalvi A, Javerzat S. Involvement of fibroblast growth factors in choroidal angiogenesis and retinal vascularization. *Experimental Eye Research*. 2003;77(2):147–156. https://doi.org/10.1016/S0014-4835(03)00127-1.
33. Gogat K, Le Gat L, Van Den Berghe L, et al. VEGF and KDR gene expression during human embryonic and fetal eye development. *Investigative Ophthalmology and Visual Science*. 2004;45(1):7–14. https://doi.org/10.1167/iovs.02-1096.
34. Yi X, Mai LC, Uyama M, Yew DTW. Time-course expression of vascular endothelial growth factor as related to the development of the retinochoroidal vasculature in rats. *Experimental Brain Research*. 1998;118(2):155–160. https://doi.org/10.1007/s002210050267.
35. Oguni M, Tanaka O, Shinohara H, Yoshioka T, Setogawa T. Ultrastructural study on the retinal pigment epithelium of human embryos, with special reference to quantitative study on the development of melanin granules. *Acta Anatomica*. 1991;140(4):335–342. https://doi.org/10.1159/000147079.
36. Hasegawa T, McLeod DS, Bhutto IA, et al. The embryonic human choriocapillaris develops by hemo-vasculogenesis. *Developmental Dynamics*. 2007;236(8):2089–2100. https://doi.org/10.1002/dvdy.21231.
37. Campbell JP, Zhang M, Hwang TS, et al. Detailed vascular anatomy of the human retina by projection-resolved optical coherence tomography angiography. *Scientific Reports*. 2017;7. https://doi.org/10.1038/srep42201.
38. Stone J, van Driel D, Valter K, Rees S, Provis J. The locations of mitochondria in mammalian photoreceptors: relation to retinal vasculature. *Brain Research*. 2008;1189(1):58–69. https://doi.org/10.1016/j.brainres.2007.10.083.
39. Provis JM. Development of the primate retinal vasculature. *Progress in Retinal and Eye Research*. 2001;20(6):799–821. https://doi.org/10.1016/S1350-9462(01)00012-X.
40. Henkind P. Radial peripapillary capillaries of the retina. I. Anatomy: human and comparative. *The British Journal of Ophthalmology*. 1967;51(2):115–123. https://doi.org/10.1136/bjo.51.2.115.
41. Alterman M, Henkind P. Radial peripapillary capillaries of the retina. II. Possible role in Bjerrum scotoma. *The British Journal of Ophthalmology*. 1968;52(1):26–31. https://doi.org/10.1136/bjo.52.1.26.
42. Jia Y, Wei E, Wang X, et al. Optical coherence tomography angiography of optic disc perfusion in glaucoma. *Ophthalmology*. 2014;121(7):1322–1332. https://doi.org/10.1016/j.ophtha.2014.01.021.

43. Watanabe T, Raff MC. Retinal astrocytes are immigrants from the optic nerve. *Nature*. 1988;332(6167):834–837. https://doi.org/10.1038/332834a0.
44. Stone J, Dreher Z. Relationship between astrocytes, ganglion cells and vasculature of the retina. *Journal of Comparative Neurology*. 1987;255(1):35–49. https://doi.org/10.1002/cne.902550104.
45. Chan-Ling T, Halasz P, Stone J. Development of retinal vasculature in the cat: processes and mechanisms. *Current Eye Research*. 1990;9(5):459–478. https://doi.org/10.3109/02713689008999612.
46. Fruttiger M, Calver AR, Krüger WH, et al. PDGF mediates a neuron-astrocyte interaction in the developing retina. *Neuron*. 1996;17(6):1117–1131. https://doi.org/10.1016/S0896-6273(00)80244-5.
47. Chu Y, Hughes S, Chan-Ling T. Differentiation and migration of astrocyte precursor cells and astrocytes in human fetal retina: relevance to optic nerve coloboma. *The FASEB Journal: Official Publication of the Federation of American Societies for Experimental Biology*. 2001;15(11):2013–2015. https://doi.org/10.1096/fj.00-0868fje.
48. Mi H, Barres BA. Purification and characterization of astrocyte precursor cells in the developing rat optic nerve. *Journal of Neuroscience*. 1999;19(3):1049–1061. https://doi.org/10.1523/jneurosci.19-03-01049.1999.
49. Mudhar HS, Pollock RA, Wang C, Stiles CD, Richardson WD. PDGF and its receptors in the developing rodent retina and optic nerve. *Development*. 1993;118(2):539–552.
50. Fruttiger M, Calver AR, Richardson WD. Platelet-derived growth factor is constitutively secreted from neuronal cell bodies but not from axons. *Current Biology*. 2000;10(20):1283–1286. https://doi.org/10.1016/S0960-9822(00)00757-0.
51. Pierce EA, Foley ED, Smith LEH. Regulation of vascular endothelial growth factor by oxygen in a model of retinopathy of prematurity. *Archives of Ophthalmology*. 1996;114(10):1219–1228. https://doi.org/10.1001/archopht.1996.01100140419009.
52. Stone J, Itin A, Alon T, et al. Development of retinal vasculature is mediated by hypoxia-induced vascular endothelial growth factor (VEGF) expression by neuroglia. *Journal of Neuroscience*. 1995;15(7 I):4738–4747. https://doi.org/10.1523/jneurosci.15-07-04738.1995.
53. West H, Richardson WD, Fruttiger M. Stabilization of the retinal vascular network by reciprocal feedback between blood vessels and astrocytes. *Development*. 2005;132(8):1855–1862. https://doi.org/10.1242/dev.01732.
54. Miyawaki T, Uemura A, Dezawa M, et al. Tlx, an orphan nuclear receptor, regulates cell numbers and astrocyte development in the developing retina. *Journal of Neuroscience*. 2004;24(37):8124–8134. https://doi.org/10.1523/JNEUROSCI.2235-04.2004.
55. Uemura A, Kusuhara S, Wiegand SJ, Yu RT, Nishikawa SI. Tlx acts as a proangiogenic switch by regulating extracellular assembly of fibronectin matrices in retinal astrocytes. *Journal of Clinical Investigation*. 2006;116(2):369–377. https://doi.org/10.1172/JCI25964.
56. Fruttiger M. Development of the retinal vasculature. *Angiogenesis*. 2007;10(2):77–88. https://doi.org/10.1007/s10456-007-9065-1.
57. Gariano RF. Cellular mechanisms in retinal vascular development. *Progress in Retinal and Eye Research*. 2003;22(3):295–306. https://doi.org/10.1016/S1350-9462(02)00062-9.
58. Gariano RF, Iruela-Arispe ML, Hendrickson AE. Vascular development in primate retina: comparison of laminar plexus formation in monkey and human. *Investigative Ophthalmology and Visual Science*. 1994;35(9):3442–3455.
59. Gariano RF, Gardner TW. Retinal angiogenesis in development and disease. *Nature*. 2005;438(7070):960–966. https://doi.org/10.1038/nature04482.
60. Ashton N. Retinal angiogenesis in the human embryo. *British Medical Bulletin*. 1970;26(2):103–106. https://doi.org/10.1093/oxfordjournals.bmb.a070758.
61. Flower RW, McLeod DS, Lutty GA, Goldberg B, Wajer SD. Postnatal retinal vascular development of the puppy. *Investigative Ophthalmology and Visual Science*. 1985;26(7):957–968.
62. Hughes S, Chan-Ling T. Roles of endothelial cell migration and apoptosis in vascular remodeling during development of the central nervous system. *Microcirculation*. 2000;7(5):317–333. https://doi.org/10.1111/j.1549-8719.2000.tb00131.x.
63. Fruttiger M. Development of the mouse retinal vasculature: angiogenesis versus vasculogenesis. *Investigative Ophthalmology and Visual Science*. 2002;43(2):522–527.
64. McLeod DS, Hasegawa T, Prow T, Merges C, Lutty G. The initial fetal human retinal vasculature develops by vasculogenesis. *Developmental Dynamics*. 2006;235(12):3336–3347. https://doi.org/10.1002/dvdy.20988.
65. Chen T, Bai H, Shao Y, et al. Stromal cell-derived factor-1/CXCR4 signaling modifies the capillary-like organization of human embryonic stem cell-derived endothelium in vitro. *Stem Cells*. 2007;25(2):392–401. https://doi.org/10.1634/stemcells.2006-0145.
66. Kawakami Y, Ii M, Matsumoto T, et al. SDF-1/CXCR4 Axis in Tie2-lineage cells including endothelial progenitor cells contributes to bone fracture healing. *Journal of Bone and Mineral Research*. 2015;30(1):95–105. https://doi.org/10.1002/jbmr.2318.
67. Wang L, Guo S, Zhang N, et al. The role of SDF-1/CXCR4 in the vasculogenesis and remodeling of cerebral arteriovenous malformation. *Therapeutics and Clinical Risk Management*. 2015;11:1337–1344. https://doi.org/10.2147/TCRM.S87590.
68. Noronha-Matos JB, Costa MA, Magalhães-Cardoso MT, et al. Role of ecto-NTPDases on UDP-sensitive P2Y 6 receptor activation during osteogenic differentiation of primary bone marrow stromal cells from postmenopausal women. *Journal of Cellular Physiology*. 2012;227(6):2694–2709. https://doi.org/10.1002/jcp.23014.
69. Hasegawa T, McLeod DS, Prow T, Merges C, Grebe R, Lutty GA. Vascular precursors in developing human retina. *Investigative Ophthalmology and Visual Science*. 2008;49(5):2178–2192. https://doi.org/10.1167/iovs.07-0632.
70. Dumont DJ, Fong G-, Puri MC, Gradwohl G, Alitalo K, Breitman ML. Vascularization of the mouse embryo: a study of flk-1, tek, tie, and vascular endothelial growth factor expression during development. *Developmental Dynamics*. 1995;203(1):80–92. https://doi.org/10.1002/aja.1002030109.
71. Rehman J, Li J, Orschell CM, March KL. Peripheral blood \endothelial progenitor cells are derived from monocyte/macrophages and secrete angiogenic growth factors. *Circulation*. 2003;107(8):1164–1169. https://doi.org/10.1161/01.CIR.0000058702.69484.A0.

72. Shalaby F, Rossant J, Yamaguchi TP, et al. Failure of blood-island formation and vasculogenesis in Flk-1-deficient mice. *Nature.* 1995;376(6535):62–66. https://doi.org/10.1038/376062a0.
73. Yamashita J, Itoh H, Hirashima M, et al. Flk1-positive cells derived from embryonic stem cells serve as vascular progenitors. *Nature.* 2000;408(6808):92–96. https://doi.org/10.1038/35040568.
74. Hellström M, Phng LK, Hofmann JJ, et al. Dll4 signalling through Notch1 regulates formation of tip cells during angiogenesis. *Nature.* 2007;445(7129):776–780. https://doi.org/10.1038/nature05571.
75. Siemerink MJ, Klaassen I, van Noorden CJF, Schlingemann RO. Endothelial tip cells in ocular angiogenesis: potential target for anti-angiogenesis therapy. *Journal of Histochemistry and Cytochemistry.* 2013;61(2):101–115. https://doi.org/10.1369/0022155412467635.
76. Outtz HH, Tattersall IW, Kofler NM, Steinbach N, Kitajewski J. Notch1 controls macrophage recruitment and Notch signaling is activated at sites of endothelial cell anastomosis during retinal angiogenesis in mice. *Blood.* 2011;118(12):3436–3439. https://doi.org/10.1182/blood-2010-12-327015.
77. Roca C, Adams RH. Regulation of vascular morphogenesis by Notch signaling. *Genes and Development.* 2007;21(20):2511–2524. https://doi.org/10.1101/gad.1589207.
78. McLeod DS, Lutty GA, Wajer SD, Flower RW. Visualization of a developing vasculature. *Microvascular Research.* 1987;33(2):257–269. https://doi.org/10.1016/0026-2862(87)90021-5.
79. Taomoto M, McLeod DS, Merges C, Lutty GA. Localization of adenosine A2a receptor in retinal development and oxygen-induced retinopathy. *Investigative Ophthalmology and Visual Science.* 2000;41(1):230–243.
80. Ruhrberg C, Gerhardt H, Golding M, et al. Spatially restricted patterning cues provided by heparin-binding VEGF-A control blood vessel branching morphogenesis. *Genes and Development.* 2002;16(20):2684–2698. https://doi.org/10.1101/gad.242002.
81. Ford KM, Saint-Geniez M, Walshe T, Zahr A, D'Amore PA. Expression and role of VEGF in the adult retinal pigment epithelium. *Investigative Ophthalmology and Visual Science.* 2011;52(13):9478–9487. https://doi.org/10.1167/iovs.11-8353.
82. Gerhardt H, Golding M, Fruttiger M, et al. VEGF guides angiogenic sprouting utilizing endothelial tip cell filopodia. *Journal of Cell Biology.* 2003;161(6):1163–1177. https://doi.org/10.1083/jcb.200302047.
83. Stalmans I, Ng YS, Rohan R, et al. Arteriolar and venular patterning in retinas of mice selectively expressing VEGF isoforms. *Journal of Clinical Investigation.* 2002;109(3):327–336. https://doi.org/10.1172/JCI0214362.
84. Ishida S, Yamashiro K, Usui T, et al. Leukocytes mediate retinal vascular remodeling during development and vaso-obliteration in disease. *Nat Med.* 2003;9(6):781–788. https://doi.org/10.1038/nm877.
85. Checchin D, Sennlaub F, Levavasseur E, Leduc M, Chemtob S. Potential role of microglia in retinal blood vessel formation. *Investigative Ophthalmology and Visual Science.* 2006;47(8):3595–3602. https://doi.org/10.1167/iovs.05-1522.
86. Vargesson N, Laufer E. Smad7 misexpression during embryonic angiogenesis causes vascular dilation and malformations independently of vascular smooth muscle cell function. *Developmental Biology.* 2001;240(2):499–516. https://doi.org/10.1006/dbio.2001.0481.
87. Engerman RL, Meyer RK. Development of retinal vasculature in rats. *American Journal of Ophthalmology.* 1965;60(4):628–641. https://doi.org/10.1016/0002-9394(65)92251-8.
88. Hendrickson AE, Yuodelis C. The morphological development of the human fovea. *Ophthalmology.* 1984;91(6):603–612. https://doi.org/10.1016/S0161-6420(84)34247-6.
89. Hendrickson A, Possin D, Vajzovic L, Toth CA. Histologic development of the human fovea from midgestation to maturity. *American Journal of Ophthalmology.* 2012;154(5):767–778. https://doi.org/10.1016/j.ajo.2012.05.007.
90. Provis JM, Hendrickson AE. The foveal avascular region of developing human retina. *Archives of Ophthalmology.* 2008;126(4):507–511. https://doi.org/10.1001/archopht.126.4.507.
91. Kozulin P, Natoli R, Bumsted O'Brien KM, Madigan MC, Provis JM. Differential expression of anti-angiogenic factors and guidance genes in the developing macula. *Molecular Vision.* 2009;15:45–59.
92. Chui TYP, Zhong Z, Song H, Burns SA. Foveal avascular zone and its relationship to foveal pit shape. *Optometry and Vision Science.* 2012;89(5):602–610. https://doi.org/10.1097/OPX.0b013e3182504227.

Chapter 2

Animal models

Burak Turgut
Onsekiz Mart University, Faculty of Medicine, Department of Ophthalmology, Canakkale, Turkey

Introduction

Various animal species have been used to research the development, prevention, and treatment of many ocular diseases in experimental studies. Experimental animal models are very important in finding new treatment modalities for various ocular diseases before use in humans. If researchers know which animal model belongs the closest to the anatomy and physiology of the human eye, they will be successful in obtaining significant results in ophthalmologic studies. Currently, some animal models have been used in the research of various choroidal and retinal vascular diseases.[1–5] In this chapter, it has been aimed to provide experimental animal procedures in various retinal and choroidal diseases for researchers.

Choroidal and retinal vascular diseases commonly occur due to the occlusion of the vessels that supplied these tissues and macular edema or subretinal fluid accumulation that arises from the elevation of choroidal intravascular hydrostatic pressure, choroidal and retinal ischemia, choroidal congestion, and the breakdown of blood–retina barriers (BRB). In turn, preretinal neovascularization (NV), tissue edema, and related complications develop. Preretinal NV includes choroidal NV (CNV), retinal NV (RNV), and retinal angiomatous proliferation (RAP). According to the early classification of CNV, Type 1 CNV has been used to define the neovascular vessels which were limited to under retina pigment epithelium (sub-RPE) space and Type 2 CNV has been defined as NV between the neurosensorial retina and the RPE but not in the subretinal space. On the other hand, the last classification includes Type 1 macular NV (MNV) (occult, polypoidal choroidal vasculopathy [PCV] or aneurysmal type 1 NV), Type 2 MNV (classical macular CNV), and Type 3 MNV (retinal angiomatous proliferation [RAP]).[6,7]

Animal models for CNV

Although rodents have disadvantages such as not having a human-like macula and drusen-like accumulation at the RPE level, the most commonly used animal model for this purpose is still mouse or rat. In recent years, especially transgenic mice carrying the AMD genetic mutation have been developed. While monkeys are another model that has found use for this disease, the use of zebrafish, which is highly utilized in diseases such as cataracts and DR, is also increasing in AMD studies.[3,8–11]

The CNV is defined as the NV arising from vessels in choriocapillaris and growing into the sub-RPE and/or subretinal space through the defects in Bruch's membrane (BM). It often occurs in neovascular age-related macular degeneration (n-AMD). However, some diseases such as pathologic myopia, uveitis, central serous chorioretinopathy (CSCR), angioid streaks, choroidal osteoma, hereditary chorioretinal diseases, and iatrogenic disorders may also cause CNV in usually younger patients.[12,13] Idiopathic CNV is used to define the cases without any specific cause. The evolution of CNV begins with a defect in the BM caused by ocular inflammation or trauma, or retinochoroidal degenerative disease. The involutional stage of CNV includes scarring and fibrosis with a reduction in cytokine production.[14–17]

Only the animals having a macula are the primates. Primates (large-brained higher mammals) include humans and humanoids such as monkeys, gorillas, orangutans, chimpanzees, gibbons, lemurs, galagoes, witch lemurs, and lorikeets. The retina, macula, and globe size as well as ocular physiology in monkeys are very similar to those in humans. Thus, the choice of a monkey as a primate model is more appropriate for the studies on drug delivery and antiangiogenic molecules. The primate laser-induced CNV models in monkeys (rhesus and cynomolgus) are used in the preclinical assessment of n-AMD treatment. However, this model has some disadvantages such as the expensiveness, long experiment time, and

Retinal and Choroidal Vascular Diseases of the Eye. https://doi.org/10.1016/B978-0-443-15583-3.00002-0

ethical problems. Rodents (mouse, rat, and rabbits) may be used in experimental models. As a rabbit model can involve the release of various angiogenic factors, the damage to BM and CNV may last for months. It is useful for longitudinal drug studies on efficacy durations and doses. If the study primarily targets the macula, the use of a primate model is more adequate because the cost and ethical issues that could arise from primate use are present.[3,8–11]

The size of a pig eye is near to that of the human eye. Additionally, experimental RNV in these animals is very similar to that in humans. A pig model like a rabbit model has no disadvantages such as the cost and ethical problems like primates, and it is very appropriate to be used in drug delivery experiments. However, the pig has some disadvantages including the lack of performance and longer experiment duration than those laser-induced CNV models in rats and mouse.[3,8–11]

Zebrafish (Danio rerio) model quickly took its place in ophthalmic research because it matures in a short time, lays a large number of eggs, and causes mutations easily. Additionally, it has some advantages such as easy genetic analysis, rapid and easy observable eye development, and human-like ocular morphology. Zebrafish can be easily produced at a low cost. However, the retina of the zebrafish has no macula. Unlike the mouse eye, the zebrafish eye completes its development rapidly and it has a human-like dominant cone cell structure, which is an important advantage for AMD research.[18–24]

A successful animal CNV model requires the disruption of BM by laser and/or mechanical way; the obtaining of vascular endothelial growth factor (VEGF) overexpression in subretinal space, RPE, and release of inflammatory cytokines by subretinal injection of some biomaterials. Currently, there are two categories of animal models of CNV including nongenetical models (laser-induced models and surgically induced models) and genetical models.[3]

The three molecule groups related to the initiation, maintenance, and involution stages of CNV formation are proinflammatory/inflammatory molecules (complement system, cytokines, and chemokines), proangiogenic factors (VEGF, basic fibroblast growth factor [bFGF], and platelet-derived growth factor [PDGF]), and proteolytic molecules (a provisional fibrin matrix).[15,25–27]

Laser-induced CNV models

The mechanical disruption of BM with laser photocoagulation (LFC) can induce CNV in rodents and primates. To CNV induction in rodents with LFC burns, argon green LFC (100 μm spot size, 180–300 mW power, and 0.1 s duration) or diode LFC (100 μm spot size, 150–180 mW power and 0.1 s duration) may use at the positions of 3, 6, 9, and 12 o'clock between the major blood vessels in each eye lacking major vessels and far two discs diameters from the optic disc during a slit lamp biomicroscopy. The observation of an acute vapourization or a subretinal cavitation bubble and hearing a voice of "pop" indicate BM perforation. LFC power of 240–1500 mW may used in primates. The power level, spot size, and duration, which are most commonly used, are 300–700 mW, 50 mm, and 100 ms in nonhuman primates, respectively.[28–32] Some recent studies demonstrated that laser-induced CNV in nonhuman primates was low incidence while some showed that CNV incidence following LFC of the macula increased.[28–33]

Rodent (mouse/rat) models: Laser-induced mouse CNV model created by spot treatments with a krypton laser to create laser photocoagulation (LPC) injuries to BM currently used in preclinical trials for the study of antiangiogenic drugs. This model provides very similar cellular responses that occur in human CNV. As the mouse laser-induced CNV models are small and reproducible models, they are usually used in drug screening studies. In these models, CNV may be created in approximately 80% of lesions. However, the acute insult to BM is different from the long-term chronic development in human AMD.[28,34] In a rat CNV model developed by Frank et al., a krypton laser with green or blue-green wavelength, 500 μm spot size, 50 mW power, and 0.02-s duration was used to create breaks in BM and subsequent CNV.[35] However, a krypton laser in mice with smaller spot size, higher intensity, and shorter duration (blue green or green wavelength, 50 μm, 350–400 mW, 0.05 s) has obtained histological CNV development with 100% frequency.[34,36] Alternatively, a diode laser may also be used for CNV creation (532 nm, 100 μm 50–100 mW, 0.1 s) in rats and mice.[37–43]

The main advantages of this model used in rat or mouse are creation simplicity, cheapness, reproducibility, efficiency, and short experiment time like several weeks. Its disadvantages are a smaller size than that of the human eye, no macula, and necessity of LPC. However, the rat/mouse laser model of CNV is currently the standard animal model.

Primate (monkey) models: The primate model of experimental CNV in monkeys (*Macaca speciosa*), created by Ryan et al., is the first animal model of CNV performed using argon laser disruption of BM (488 and 514 nm, 50–200 μm, 200–950 mW, 0.1–0.5 s). The angiographic and histologic evidence of this method were 40% and 90%, respectively.[28] Criswell et al. reported that CNV with laser parameters including 532 nm, 0.05 s, 75 μm, and 650 mW was obtained in 65% of squirrel monkey laser spot sites.[44]

Intermediate (rabbit/pig) models: El Dirini et al. have developed an intermediate laser-induced CNV model formed in rabbits with subretinal endophotocoagulation. This CNV model has no expensive and ethical problems that the primate model has. However, the rabbit retina has no macula and medullary vascular supply.[45] Some researchers have developed another intermediate laser-induced CNV model in the pig via the formation of defects in BM with a diode laser (75 μm, 400 mW, 0.1 s) for drug delivery studies.[45,46] Saishin et al. reported that this model has obtained 100% histologic CNV evidence.[46] Kiilgaard et al. demonstrated that histopathologic evidence of CNV development in a pig model was 83% with diode laser and 100% with mechanical disruption of BM.[47]

Light-induced CNV model: A cyclic light-induced rat CNV model has been generated in Albino rats that were exposed to 12 h of 3000-lux cyclic light for 1, 3 or 6 months. In this model, sub-RPE NV was observed at 3 months. An advantage of this model is to be a nongenetically model with no laser or mechanical disruption of BM.[47]

Surgically induced CNV models

The CNV could be induced in various animal models by injection of viral vectors containing VEGF, some cells, some synthetic peptides, and materials.

Rodent (rat) models: It has been demonstrated that the injections of a recombinant adenovirus vector having rat VEGF164 by a CMV promoter into the subretinal space caused angiographically and histologically CNV in 80% of the rat eyes.[48] Some authors have obtained CNV development by angiographic and histopathologic evidence in rat eyes with an injection of 2, 5 or 10 μL of an adenoviral vector having VEGF165 or GFP into the subretinal space.[49,50] It has been also reported that successful CNV induction in C57BL6 mice by subretinal injection of RPE and polystyrene microbeads.[51] Additionally, it has been reported that Matrigel injection (a basement membrane extract that stimulates regional vasculogenesis) into the subretinal space or sub-RPE injection induced CNV development in B6 and Ccl2 deficient mice with 31% and 53% incidences, respectively. However, as it has been demonstrated that small CNV development after only subretinal PBS injection, it is possible that the subretinal injection itself can cause CNV development via damage to BM with the trauma.[52]

Rabbit model: Qui et al. reported that CNV incidence was 100% at 1–9 weeks after injection with angiographic and histologic evidence via transvitreal injections of 10 μL of Matrigel and 750 μg of VEGF into the subretinal space of the rabbit eye.[53] The creation of CNV via subretinal injections in rabbits is very expensive. Thus, subretinal injection of vitreous has been developed, and microscopic CNV evidence was obtained in rabbit eyes with induction of RPE proliferation.[54] On the other hand, angiographic and histologic evidence of CNV was detected in 83%–100% of the rabbit eyes, which were injected transscleral 50 μL suspensions of 2.5 μL bFGF impregnated gelatin microspheres into the subretinal space via a defect in BM.[55,56] A study demonstrated that subretinal injections of 12.5–25 μg of linoleic acid hydroperoxide (LHP) into the subretinal space obtained angiographically CNV in approximately a half of rabbit eyes.[57] On the other hand, it has been reported by Ni et al. that an injection of 50 μL mixture consisting endotoxin, growth factor, heparin beads, fibroblast growth factor-2, 100 ng LPS with 50 μg of heparin sepharose beads transvitreally into the subretinal space obtained angiographic and histologic extensive CNV development in 100% of the rabbit eyes.[58] It has also been reported that transvitreal injections of an adenoviral vector (106 IU five times) expressing VEGF-A165 driven by an EF1αHTLV promoter into the subretinal space caused CNV with angiographic and histologic evidence in 85% of the injected rabbit eyes.[59] The rabbit model has some advantages such as large size and long CNV duration while lack of a macula and retinal vasculature similar to the human eye are its main disadvantages.

Kiilgaard et al. demonstrated that the rates of CNV development with surgical debridement of the RPE followed by mechanical break of BM were 100%, compared to xenon LFC (54%) and diode LFC (83%) in the mechanically induced CNV model in pigs.[46] In this model, retinal detachment with subretinal injection of isotonic saline and then transvitreal subretinal debridement of the RPE with a retinal scraper were created. Mechanically induced CNV model provided with the perforation of BM with impact RPE has histologic CNV evidence in 100% of eyes and the easiness to CNV.[46,60–62] It was reported that the mechanical disruption of BM using a retinal perforator provided large CNV with angiographic evidence (present in 92% of eyes), very similar vasculature and CNV configuration to the human retina in the pig model compared to the rabbit model.[62] Additionally, cytokine production in this CNV model was also very similar to that in human CNV. Because pig and rabbit eyes are large, it is very easy to perform drug delivery experiments in these models. Disadvantages in the CNV pig model are similar to mechanically induced CNV in rabbits.

Primate (monkey) models: The first CNV creation in a primate model was performed by Ryan via subretinal injection of collagenase and hyaluronidase into the subretinal space.[28] In another primate CNV model, Cui et al. obtained angiographic evidence of small CNV in 92% of the eyes via the injection of an 80 μL suspension of VEGF-impregnated gelatin microspheres transvitreally through a retinotomy into the subretinal space in monkeys. However, this CNV model has protection difficulties and has no advantages compared to the laser CNV model in the primate.[63]

Genetic (transgenic and knockout mouse) CNV models

VEGF164RPE65 transgenic mouse model: In this model, intrachoroidal vascular anomalies but not CNV developed histologically in all experiments. The importance of this model comes from the fact that in the development of CNV, VEGF produced only by the retina or RPE is not sufficient, and BM damage plays a major role.[64]

Ccr2/Ccl2-deficient mouse model: In this model developed by Ambati et al., the accumulation of C5a and IgG-inducing VEGF production by RPE is obtained by transgenic mice deficient in either Ccl2 (MCP-1) or C–C chemokine receptor-2 (Ccr2) fail to recruit macrophages to the area of the RPE and BM. However, in this model, the microscopical CNV development rate has remained at almost 25% of the mice.[65] CNV has been formed in transgenic mice through the increased expression of VEGF in the retina.[66]

Ccl2/Cx3cr1-deficient mouse model: In this model, a Ccl2/Cx3cr1 deficient double knockout (DKO) transgenic mouse having high levels of N-retinylidene-N-retinylethanomalmine (A2E) and reduced levels of ERp29 in the RPE had histological and ophthalmoscopically RPE changes and drusen-like accumulations, photoreceptor loss, BM thickening, and C3/C4 deposition in the RPE when they were fed a diet low in omega-3 polyunsaturated fatty acids (PUFAs). The frequency and time of histologic CNV development were found as 15% and 6 weeks, respectively.[67]

Senescent SOD1–/– mouse model: Imamura et al. found that histologic CNV evidence was 10% of the senescent *Sod–/–* mice (Cu, Zn-superoxide dismutase [SOD1] deficient mice). In senescent *Sod–/–* mice that CNV developed, 8-hydroxy-2′-2-deoxyguanosine (I-OHdG) in the RP, a marker of oxidative damage to DNA was detected. The CNV was observed in 16-month-old mice.[68]

ApoE overexpression mouse model: AMD-like changes in BM and RPE in hypercholesterolemic mice were shown by Dithmar et al.[69] It has been demonstrated that dry AMD-like changes such as drusen-like and basal laminar-like deposits and ultrastructural CNV in 16–30 months developed in Apolipoprotein E (ApoE4) overexpressing transgenic mice, which were fed a high-fat cholesterol-rich diet. The importance of this model comes from the development of spontaneous CNV formation in the setting of AMD-like lesions.[70]

Cp–/– Heph-/Y knockout mouse model: It has been shown that dry AMD-like changes including RPE abnormalities and photoreceptor degeneration caused by disruption of ceruloplasmin and hephaestin developed in iron overload in transgenic mice. Histological CNV evidence without drusen-like or basal laminar deposit-like lesions in these mice was 100%.[71]

Tet/VMD2/VEGF double transgenic mice model: Oshima et al. demonstrated that histologically confirmed CNV developed with 100% frequency in double transgenic mice (*Tet/VMD2/VEGF*) via subretinal injection of an adenoviral vector expressing *Ang 2*. This model has supported that not only the overexpression of VEGF is necessary for the development of CNV but also the mechanical disruption of BM is needed.[72]

Vldlr TM1Her knockout mice model: In knockout mice with the very low-density lipoprotein receptor gene (Vldlr TM1Her), NV in the outer plexiform layer (OPL) and choroidal anastomosis have been developed by 3 months.[73]

The transgenic mouse line (mcd/mcd) model: This mouse has a mutation in cathepsin D causing photoreceptor outer segment phagocytosis by RPE cells. So, it has shown signs associated with geographic atrophy and dry AMD.[74]

Belly spot and tail (Bst) mutation mouse model: The spontaneous autosomal semidominant mouse mutation Belly spot and tail (Bst) is arisen from the C57/Bl6J inbred strain. It was detected that *Bst* + mice were found to have CNV in the subretinal space.[75]

Animal models for RNV

The RNV arises from retinal vessels, usually extends through the internal limiting membrane (ILM) and grows between the posterior hyaloid face and retina or into the vitreous.[76] It is well known that the main cause for RNV is ischemic retinopathies such as diabetic retinopathy (DR), retinopathy of prematurity (ROP), central retinal vein occlusion (CRVO), and branch retinal vein occlusion (BRVO).[77]

Experimental animal RNV models include transgenic mouse models with overexpression of VEGF, hyperoxia/oxygen-induced models (mouse, rat, cat, dog, and zebrafish), RVO models (rat, pig), and intravitreal injection models of proangiogenic molecules such as dermal fibroblasts, cultured fibroblasts, tumor cells, and implants impregnated with interleukin-1 (mouse, rabbit). RVO in rats may be created by LPC, irradiation, or photodynamic therapy. In the rat and pig RVO models, microvascular changes such as venous dilation, vascular tortuosity, retinal hemorrhage and retinal extracellular edema and subsequently retinal ischemia, and RNV similar to the findings seen in human RVO may be observed hours after the occlusion.[78–94]

Transgenic mouse RNV models

rho/VEGF transgenic mice: It has been shown that the RNV in the transgenic mice with *VEGF* driven by the *rhodopsin* promoter (*rho/VEGF* mice) developed. However, it has been considered that this model is related more to RAP.[66,95] Additionally, Tobe et al. demonstrated that retinal VEGF transgene mRNA existed on the postnatal sixth day and that RNVs reaching the subretinal space were observed postnatal 18th day with no invasion choroidal vessel.[34]

Reverse tetracycline transactivator (rtTA) inducible promoter system coupled with the rhodopsin (rho/rtTA-TRE/VEGF) or interphotoreceptor retinoid-binding protein (IRBP) promoter (IRBP/rtTA-TRE/VEGF) transgenic mice: These models may be used to control the onset of VEGF transgene expression in the retina. In these models, in the case of the presence of doxycycline, more extensive RNV with prominent *VEGF* transgene expression and higher ocular *VEGF* mRNA levels compared to *rho/VEGF* mice occurred. The addition of doxycycline to mice has caused extensive RNV with the activation of the VEGF expression.[3,96]

VEGF165 (hVEGF) overexpressing mice: In this model, VEGF165 (hVEGF) overexpressing transgenic mice has been obtained via microinjection of a DNA construct containing the *hVEGF* gene driven by a truncated mouse *rhodopsin* promoter, and severe RNV could be developed.[8,95]

Müller cell-derived VEGF knockout mice model: In mice that were produced by coupling Müller cell-expressing Cre mice and floxed VEGF mice, Müller VEGF production was disrupted by feeding 2 mg/mL doxycycline from embryonic day 15 to postnatal day 1. However, VEGF disruption in Müller cell failed to prevent the RNV.[97]

Nrl−/−Grk1−/− double knockout mice: It defines the absence of the G-protein coupled receptor kinase 1 gene (*Grk1−/−*) playing a role in the inactivation and normal recovery of cone pigments at phototransduction event in the neural retina leucine zipper knockout mouse (*Nrl−/−*). In this model, the inflammatory-mediated RAP and RNV have been observed at the postnatal first month of age.[98–100]

Zebrafish RNV model: In the retina of adult zebrafish (5−18 months old), NV can be developed in the aquarium environment with air saturation progressively reduced to 10% for 48−72 h and continued for 12 days. The effects of orally administered anti-VEGF agents (sunitinib and ZM323881) on these pathological changes have been demonstrated. This model can be used as a useful model for studies with antiangiogenic agents. In another study, leaky, disorganized blood vessels were also able to form in adult zebrafish that had been in a hypoxic environment for 3−10 days. RNV can also be generated in zebrafish by the transgenic method. Gnn mutant zebrafish elicited a phenotype similar to AMD.[18–24]

Intraocular injection of proangiogenic molecules

The RNV may be induced by intravitreal injection of proangiogenic molecules such as tissue-cultured dermal fibroblasts, tumor cells, and implants impregnated with interleukin-1 in mice and rabbits. Tano et al. demonstrated that RNV with this method in rabbits began with the growth of vessels along the fibrous strand and RNV lasted for 3 months.[80] A modified method generated by Antoszyk et al. includes performing partial posterior vitreous detachment that is induced with repeated injection and aspiration of hyaluronidase before injection of homologous tissue-cultured dermal fibroblasts in rabbits. They reported an absolute RNV frequency of 95% at the 14th day.[93] However, a disadvantage of this method is that it is more traumatic and inflammatory and is necessary for the retinal detachment for the formation of RNV.[95]

The oxygen-induced RNV model

This model is based on the formation of RNV without high serum glucose levels. This model is commonly created with high oxygen exposure and the use of VEGF agents. The use of VEGF is more suitable for large animal models such as rabbits and monkeys. It is also possible to use this model for ROP or proliferative retinopathies due to RVOs.[1–5]

Diabetic retinopathy models

Diabetic retinopathy (DR) is currently one of the leading causes of blindness over the world. While microaneurysms develop as a result of the thickening of the capillary endothelial basement membrane and loss of pericyte, structural changes such as localized enlargement of the venous vessels, coiled retinal vessels, and intraretinal microvascular anomalies (IRMA) occur with the progression of the diabetic process and the effect of developing ischemia, and these changes progress. Although none of these models has been successful in mimicking fully human DR, most of them provide us with early findings of DR. Nonhuman primates and humans have similar eye structures, and both can develop spontaneous diabetes mellitus (DM). However, because of ethical reasons, they could not usually be used in this research. Rodents such as mice and rats are experimental animals that are easily available, relatively inexpensive, and have short reproduction cycles and human-like genetic backgrounds. Even if it is possible to see the early DR findings easily in this animal model, its most important deficiency is that it does not fully reflect the pathological findings in humans and that late findings cannot be created.[101–120,121–147] However, currently, the zebrafish DR model is a promising option.[101,102] It is possible to induce hyperglycemia and subsequently insulin deficiency and Type 1 DM in experimental animals by destruction in β cells of the pancreas with pharmacological agents (injection of streptozotocin [STZ] or alloxan [ALX]) or pancreatectomy, feeding with galactose/sucrose diet and genetic modification (spontaneous models).

Induction of hyperglycemia

Experimental hyperglycemia may be created with the administration of STZ or alloxan (ALX) in rats, a 30% galactose diet (over 36 months) in dogs, administration of STZ-ALX in porcine, pancreatectomy in cats, and administration of STZ (15 years) in primates.[101–103] Although RNV, microaneurysms, and intraretinal vascular abnormalities (IRMA) have not been identified in rats, vascular changes, including capillary and pericyteloss, and thickening at the basement membrane could be developed in rats with STZ or alloxan-induced diabetes after 6 months from the onset of DM.[101–104]

STZ, an antibiotic, is most commonly used in rodents and animals having large eyeballs in the creation of pharmacological hyperglycemia. It has the advantage of time efficiency. In the STZ DR mouse model, it is possible to observe pathological findings related to retinopathy within a week. The methyl nitrosourea in STZ causes cytotoxic effects and pancreatic β cell necrosis. STZ created its effect through the GLUT2 receptors also present in the β cell plasma membranes. Administration of STZ for five consecutive days or a single dose causes hyperglycemia.[105] STZ dosage for multiple intraperitoneal (IP) injections is 40–80 mg/kg. In the DR mouse model created with STZ, there is a wide variety in dose administration (180–200 mg/kg, ip, single dose) (40–100 mg/kg, ip, multiple doses in 2-5 consecutive days), but hyperglycemia within 1–4 weeks in all mice will develop. A single dose IP injection of STZ with 150–200 mg/kg in mice or 30–100 mg/kg in rats induces hyperglycemia. Intravenous injection (IV) of STZ with a dose of 110 mg/kg induces hyperglycemia in rabbits. However, in tree shrews for the induction of hyperglycemia, a single dose of 175 mg/kg is necessary. STZ hyperglycemia often occurs within 1–4 weeks in most animal species.[94,106–136]

In the mouse model after IP STZ injection, the increase in vascular permeability at the end of the first week can be measured by intravenous administration of fluorescein isothiocyanate-dextran. In rat model, IV administration is possible in STZ application (30–65 mg/kg single dose).

In the rabbit model, it is possible to induce DM either pharmacologically or by diet, whereas VEGF-induced new vasculature can be formed. In this model, it is possible to obtain signs of hyperglycemia and various degrees of retinopathy after the 19th week with 100 mg/kg/i.v STZ injections. However, the variety of retinopathies that occur limits the use of the model.[96,109–140]

In the early stages of STZ-diabetic rats, basement membrane thickening, dilatation, occlusion, and vascular leakage at retinal capillaries are observed.[105,137,138]

In the zebrafish model, it is quite easy to induce hyperglycemia by adding high concentrations of glucose to the fish pond. A diet-induced DR model was developed by adding 10% lard, 40% sucrose, and 0.1%–0.5% cholesterol to the standard meals of rabbits for 24 weeks. During this period, microaneurysms can be observed in the retina, and the findings may progress with the prolongation of the feeding period, which shows a slow development consistent with retinopathy in humans. When zebrafish are exposed to fresh water containing 2% glucose, hyperglycemia occurs on the first day, and it eventually becomes permanent on the 30th day. Hyperglycemic retinal anomalies in the zebrafish model are similar to those in humans. The development of hyperglycemia and characteristics of DR with STZ intraperitoneally in the zebrafish model could be created.[104,125]

An alternative to STZ, a pyrimidine derivative, ALX, can also be used in hyperglycemia induction via the β-cell apoptosis in the pancreas by the inhibition of glucokinase in mice, rats, rabbits, and pigs. In rodent models, it has been demonstrated that ALX could be induced to develop hyperglycemia within 1 week of administration however mild DR findings can retard to 3 months in mice.[96,109–128]

Surgical and diet-induced hyperglycemia are other methods for the induction of experimental hyperglycemia. While it is possible to create hyperglycemia in a period of 1–2 weeks, especially in animal models such as cats and monkeys that underwent pancreatectomy as a surgical technique, unfortunately, it may be necessary to wait for years before DR findings can be observed in these animals. In dog models, hyperglycemia often occurs in a month following the surgery. Additionally, hyperglycemia by pancreatectomy may induced with the drugs. Diet-induced hyperglycemia in dogs may be created with a high glucose/galactose diet or a high galactose, fat, and sucrose diet. It has been demonstrated that diet-induced hyperglycemia caused DR development in mice and rats. However, this technique has the disadvantage of time and in this method, the development of DR can take years in dogs and monkeys with much faster in rodents.[96,109–126]

Genetic models of hyperglycemia

Spontaneous hyperglycemia may be created by carrying gene mutations in rodents and zebrafish animal models. These models are relatively easy to work with, and economical to develop and inbreed in controlled environments. With this technique, it is possible to create permanent phenotypes with a high success rate in animal models such as rodents and zebrafish. Experimental animals with spontaneous hyperglycemia caused by endogenous mutation transfer are reproduced by mating and used as DR animal models. The main disadvantages of the technique are the high cost and technology, a long time to DR model creation.[94,107–123]

Mouse models *Transgenic mouse model with high VEGF levels*: Experimental DR model was created by a binary recombinant adeno-associated virus construct producing green fluorescent protein (GFP) and VEGF under the control of the human cytomegalovirus promoter was injected into the subretinal space of C57BL6 mice. In this model, RNVs have gradually developed compared to other transgenic mouse models with high VEGF levels.[148]

db/db mouse model: In the spontaneously created db/db mice model (homozygous for the mutation [Leprdb] db/db mice), hyperglycemia and the signs of T2D begin at 8–10 weeks postnatally. At 18 weeks, microvascular complications begin to appear and the VEGF level rises. The model is suitable for research on DR, obesity, and metabolic disorders.[115,137–140]

The biobreeding (BB) rat model: The biobreeding (BB) rat model was created as a type 1 DM model. In the Zucker diabetic spontaneous type 2 rat model (ZDF), diabetes tendency occurs in obese male rats. Hyperglycemia develops in the 20th week, and this model is ready for DR studies at the 20th postnatal week. The SDT fatty rat model is similar to humans, and it differs from other rats as it shows signs of severe retinopathy. In this model, unlike the normal SDT model, diabetes develops within 5 weeks (4 times shorter after birth).[115,137–140]

Kimba and OIR mice with human VEGF165 gene microinjection or high oxygen exposure: It is possible to model early retinal vascular changes in all these mouse and rat DR models and to conduct VEGF and candidate drug studies for the mechanisms of DM pathology. It is possible to generate retinal neovascularization models in Kimba and OIR mice with human VEGF165 gene microinjection or high oxygen exposure. In Kimba mice, retinal ischemia findings were observed with an increase in retinal new vessels; in OIR mice exposed to 75% oxygen exposure, retinal vasculature regression, ischemia, and neovascularization may occur at postnatal day 12. Retinal neovascularization becomes visible with endothelial cell markers such as fluorescein isothiocyanate-dextran or isolectin-B4 or CD31.[134,136]

Kimba transgenic line (tr029VEGF) mice model: This model in which performed the breeding with the Ins2Akita mice to product an Ins2Akita/VEGF+/genotype has imitated NPDR and mild PDR.[116,138,148]

Diabetic Torii rat (DTR) model: A spontaneously diabetic strain of Sprague–Dawley (SD) rat was recognized in 1988 and named as the spontaneously diabetic Torii (SDT) rat. In this model, besides retinal structural anomalies, complications such as tractional retinal detachment (TRD) and vitreous hemorrhage (VH) similar to the PDR can be demonstrated. In this model, glycosuria and the development of DM are observed at 20 weeks old in males and almost all male rats at 40 weeks, respectively. DP incidence was 33.3% by 65 weeks old in the female rats.[111,145] Additionally, SDT rats develop RNV without retinal ischemia possibly on the basis of the persistent vitreous adhesion.[144]

fa/fa mutation in the leptin receptor gene (Lepr) rat model (noninsulin-dependent DM model): These rodents are obese animals derived from the Zucker rats, and they have a (fa/fa) mutation in the leptin receptor gene (Lepr).[107,108]

Ins2Akita mice model: These animals carry a point mutation in insulin2 that causes abnormal insulin generation by the pancreatic β cells. This model is used for especially the creation of early DR stages in humans.[113,114]

The NOD mouse model: These mice have the autoimmune destruction of the pancreas and they imitate human autoimmune insulin-dependent T1DM.[115,116] Most NOD females have hyperglycemia at 12 weeks old. However, in NOD males, hyperglycemia occurs at 20 weeks old.[116]

Cat model The majority of cats used as DR models use the pancreatectomy technique with or without STZ/ALX injection. Early retinopathies have been reported for several years (5–7 years) in cats that developed hyperglycemia within 2 weeks after surgery, but this model is not widely used due to its cost, long observation period, and lack of markers for molecular studies.[101–106,118]

Dog model It is possible to create a DR model in dogs with STZ and ALX injection or, as a more suitable model, a galactose diet (30% galactose in addition to the normal diet). Studies have shown that early-stage DR findings increase after 3 years in the chemical group and after 5 years in the galactose diet group. For example, in the canine DR model induced with a galactose diet, pathologies such as intraretinal microvascular anomalies, soft exudate formation, and gliosis in the nerve fiber layer were reported at the end of 5 years. while doing so, RNV has been reported after 6 years. Signs appear earlier in young dogs. Lensectomy may be required due to cataracts in these animals during the observation period. The canine DR model advantageously shows human-like retinopathy findings. Unlike mice, rats, and zebrafish, it is a great advantage that dogs have macula, and their eye mass is almost human-like in size. Similar to cats, there are limitations to the use of this model such as cost, lack of molecular markers, and long observation time, as well as emotional resistance to the use of this model.[101–106,120]

Monkey model In monkeys, eye structures are close to humans and they have the macula. Pathologies such as exudates, retinal hemorrhages, IRMA, and microaneurysms can occur in these animals. Type 1 DM can be induced in these animals by STZ injection or pancreatectomy. However, significant DR findings in these animals can occur between 6 and 15 years. Type 2 DM occurs spontaneously in obese rhesus monkeys, but it may take up to 15 years for significant DR findings to occur. Changes in initial retinal morphology findings, absence of advanced retinopathy findings, low birth rate, high cost, long observation period, and ethical concerns are in the research area limits regarding the use of this animal model.[121,122]

Zebrafish model When zebrafish (*Danio rerio*) were exposed to fresh water containing 2% glucose, hyperglycemia occurred on the first day and became permanent at the end of the 30th day. Structural retinal anomalies are seen in the hyperglycemic zebrafish model, similar to humans. Hyperglycemia develops in the zebrafish model in which STZ is given intraperitoneally (i.p), and characteristic DR findings can be created. DR findings, which can easily develop in the zebrafish model, have made this animal model an important alternative today. Zebrafish have similar blood glucose and pancreatic islet cells to those of humans. Thus, IP STZ injection and water with glucose 2% may be used to induce DR in this model.[139,140]

Retinopathy of prematurity/O_2-induced retinopathy models

The retinopathy of prematurity (ROP) is one of the main causes leading to blindness in childhood. O_2-induced retinopathy (OIR) animal models may mimic ROP through high oxygen levels, oxidative stress, and inflammation.[146,147,149–166] Most OIR models provide some features of human ROP because healthy newborns but not premature animals were used in these models. Although mice and rats are the most used animals for the OIR model, beagle puppy and zebrafish have been also used. Animal OIR model has been developed firstly in kittens by Grossniklaus et al.[3] then in various animal species such as rat, mouse, beagle puppy, and zebrafish.[149–166] In this model, it is observed that the formation of RNV between the nonvascularized and vascularized areas in retina.[145] Interestingly, the retinal detachment can be observed in the late stage of OIR in beagle puppies, while it cannot occur in mice or kittens.[146,147,149–166]

Human and animal models have some differences regarding the stages of OIR. Human ROP stages are "early phase" (stages 1–2 ROP), "vascular phase" (stage 3 ROP with plus disease), or "fibrovascular phase" (stages 4–5 ROP). OIR Stage 1 in the rat and mouse conforms to the "early phase" of ROP in humans. OIR Phase 2 in the rat and mouse models conforms to the "vascular phase" in humans. Interestingly, in the beagle OIR model, retinal folds and dragging of retinal vessels seen in stage 4 human ROP may be observed.[146,147,149–166]

Rodent model

The most commonly used OIR models are the mouse and rat models.[141,153] The main advantages of the mouse OIR model are the ease of genetic manipulation, reproducibility of the phenotype, lower costs, and the size of the animals.[141–147,149–166] The main disadvantage of this model is that central retinal vessels are involved in hyperoxia in mice. However, in ROP, the peripheral retinal vessels are affected.[153] In the mouse OIR model, C57BL/6J mouse pups with 1-week-old were exposed to 75% oxygen for 5 days and then returned to room air, and RNV was developed in postnatal 5 days.[130,147] The rats have been also used for this purpose.[141,146,159]

The protocol of the oxygen exposure, both duration and age at which the rat pups are initially exposed to hyperoxia, is different from the mouse OIR protocol. The rat OIR model was generated by exposing the animals to the alternating cycle of 80% and 40% oxygen for 12-h for the first 14 days followed by exposure to the room air.[152] In another protocol, alternating oxygen cycles include 12-h cycles of 50% and 10% oxygen. This protocol has demonstrated a higher incidence and severity of NV compared to above mentioned 80%/40% cycle.[149,152] The rat OIR model has been used in the research of pathogenesis of hypoxia-induced NV and angiostatic molecules.[160−163]

As the rat has central retinal vaso-attenuation and delayed physiologic peripheral RNV in the OIR model, it is the most favorite animal used OIR model. In the newly born pup and dam OIR model, experiments are placed in a controlled oxygen environment and exposed to oxygen levels from 50% to 100% every 24 h for 14 days. Induced hyperoxia causes the elevation of arterial oxygen similar to that in a human preterm infant with severe ROP.[146,147,149,162,163] The appearance of the first delayed physiologic retinal vascular development followed by intravitreal NV at the junction of the vascular and avascular retina at day 18 is similar to type 1 severe ROP. Thus, the rat OIR model imitates very similarly severe ROP in human preterm infants.

Cat model

The kitten model to be used for the OIR model was first described by Ashton et al.[87] In this model, kittens are exposed to 70%−80% of oxygen after birth for 4 days and then returned to normal room air.[152,166]

Dog model

The dog OIR model was generated by McLeod et al.[150] In this model, the newborn pups are exposed to 95%−100% oxygen for 4 days and returned to normal room air until 22−45 days old. Tractional retinal folds and intravitreal vascularized membranes develop after 1 h of hyperoxia. RNV development due to hyperoxia is similar to those of kittens and premature humans.[146,147,149−166]

It has more similarity to the human preterm infant eye than the newborn rodent regarding the sizes of the eyes. The retina of a newborn beagle initially vascularizes via vasculogenesis and then angiogenesis like that occurs in premature human infant retina. However, the beagle OIR model uses very high oxygen to cause OIR, which differs from the pathogenesis of ROP in most premature infants. In the beagle model, newborn pups are placed into 100% oxygen in day 1 postnatal for 4 days and then returned to room air to recreate the phases of OIR.[149,150,152,154,156,158]

Zebrafish model

Cao et al. developed a hypoxia-induced RNV model in adult zebrafish. In this model, nitrogen gas is perfused directly into the water in a sealed aquarium, *Fli*-EGFP-Tg zebrafish are kept in normoxic water, and then the oxygen pressure is gradually diminished to the final 10% air saturation. The fish are exposed to the hypoxic conditions for 3−15 days.[19−25,124−126]

Retinal artery occlusion models

Retinal artery occlusion (RAO) causes significant ischemia and then irreversible injury in approximately 2 h in the human retina. It can be induced by various methods such as embolization, intraocular pressure (IOP) elevation, laser application, vasoconstriction by injection of chemical substances, or occlusion by ligation, clamping, suturing, or applying the occluder. The duration of occlusion lasts for 30−90 min and may be controlled in RAO formed with ligation, clamping, and IOP elevation. In the occlusion by laser, the duration varies between 3 h and 2 weeks before reperfusion. By the occlusion with an occluder or probe, RAO may last between 10 min and 4 h. RAO with the vasoconstriction caused by the injection of endothelin-1 lasts for approximately 5−50 min. RAO by embolization by the injection of leukocyte aggregates and the exposition to hypoxia lasts up to 8 days. Studies using other agents reported occlusions lasting up to 166 s or 20 min.[167−172]

Laser

Experimental RAO laser model has been often created by photo-thrombosis by argon or green LPC (514−577 nm wavelengths) for 2−45 min to induce CRA and Rose Bengal injection (IV) as a photosensitizing agent to increase laser absorption and to optimize the used laser energy. The activation of Rose Bengal by the laser results in intraluminal thrombus formation and CRAO. However, there are also models that are performed by no photosensitizing agent and no photodynamic or similar therapy. In another study, adenosine diphosphate as a platelet-activating agent for induction-targeted delivery of lasers has been used for the same purpose. In the laser CRAO model, the laser beam is applied at one or more retinal arterioles near the optic disc.[167–172]

Intravitreal injection of some substances

In the past, serotonin has been used to create vasospasm in monkeys and it was reported in a study that CRAO developed in 4 of 5 monkeys that applied serotonin. CRAO could also be obtained by injection of Endothelin-1 (ET-1), which is a potent vasoconstrictor drug produced by vascular endothelial cells. In this model, if ET-1 is injected into the posterior vitreous over the optic disc, it causes both RAO and RVO. However, when it is injected in the subconjunctival space, it causes only CRAO. The main advantages of the injection of ET-1 are to be less invasive and there is no necessity for special equipment. Recent studies demonstrated that ET-1 showed similar effects on rabbit and cat eyes following the intravenous, intravitreal, or subconjunctival administration, respectively.[167–172]

IOP elevation

IOP elevation was created most frequently by occlusion or ischemia generated by the cannulation of the anterior chamber with a tube connected to an elevated saline reservoir. Elevated IOP to 90–160 mm Hg will cause RAO in a few hours.[170]

Clamping

On monkeys, the clamping of the CRA at its site of entry into the optic nerve dural sheath may be performed via a lateral orbitotomy.[170]

Ligation

Ligation of CRA may be performed by a ligature or a clip in rodents. However, usually, when ligation of the optic nerve for ligation of CRA, posterior ciliary arteries are also being ligated. In the procedure regarding single CRA, the optic nerve is exposed by the lateral conjunctival dissection and removal of the lateral rectus muscle. Then, CRA is isolated by a longitudinal incision on the dural sheath surrounding the optic nerve. CRA is sutured after the trifurcation of the ophthalmic artery into the CRA and two posterior ciliary arteries. A ligature may be used for the ligation of the CRA in monkeys.[170]

Occluder/probe

A probe or an occluder generated from a steel needle or a glass probe may be used for the creation of experimental BRAO. A puncture is performed through the superior lateral quadrant of the pars plana. Then the instrument is pressed on a branch artery. If the probe is pressing on the optic nerve, the CRAO and posterior ciliary artery occlusion may be developed. The same intervention may be performed by a heat coagulator. The indicator of RAO is the observation of the occluded vessel in direct ophthalmoscopy.[170]

Embolization

It may be created by injecting various substances into the carotid or maxillary artery. It has been demonstrated that BRAO was obtained in approximately half of cases by the injection of fibrin clots, platelet aggregates, or leukocyte aggregates. Ciulla et al. reported that the injection of human atherosclerotic material caused BRAO and CRAO in rabbits.[173] The temporary BRAO with embolization may be produced by air injection BRAO. It has been also reported that drugs containing particles may cause RAO.[167–172]

Intraarterial methods

Transfemoral catheterization, ophthalmic artery balloon catheterization, or embolization with a liquid embolic agent or ophthalmic artery coiling were shown to cause permanent RAO in pigs.[170]

Retinal vein occlusion models

Retinal vein occlusion (RVO) in rats, cats, or rabbits may be induced by mechanical ligation, ET-1 injection, and light coagulation. The number of retinal veins that were intervened determines CRVO or BRVO.[173–186]

BRVO models

Experimental BRVO may be induced by laser photocoagulation, photodynamic coagulation, diathermic cauterization, or intravitreal injection of some substances.[173–186]

Laser photocoagulation or photodynamic coagulation The most commonly used method for inducing RVO in rats is LPC or photodynamic coagulation with a photosensitizer on a slit lamp biomicroscopy using a contact lens. Laser irradiation is applied on selected retinal veins to produce BRVO. Classically, argon LPC [514-nm] burns are placed in approximately 1.0–2.0 optic disc areas from the optic disc, avoiding damage to the retinal arteries for 6–20 min until evidence of venous occlusion is observed. Laser parameters vary depending on the type of laser used, the type of animal, and the use or not of adjuvants. RVO signs such as retinal edema develop in 10–30 min after the application at areas of the occluded veins. Rose Bengal (IV), thrombin (intravitreal), NPe6, erythrosin B, and sodium fluorescein (IP) are commonly used photosensitizers with LPC to facilitate thrombus formation. Rose Bengal and thrombin are administered in doses of 10–50 mg/kg and 50 units, respectively. LPC to retinal vein may be also performed intravitreally with endolaser probe through a sclerotomy until RVO is observed. In Experimental BRVO, an endo illuminator in combination with vitrectomy may be used.[173–186]

Diathermic cauterization In cats, the application to the targeted vein of 20-gauge bipolar diathermy via a pars plana sclerotomy for 5 s with indirect ophthalmoscopy has induced BRVO. In pigs, BRVOs were able to induce the coagulation of major retinal veins one disc diameter away from the optic disc for 5–7 s following a temporal canthotomy, conjunctival incision, and three pars plana sclerotomies using a light source and a bipolar diathermy under direct view with contact fundus lens with aid of an operating microscope.[173–186]

Intravitreal injection of the PD0325901 *N*-[2,3-Dihyroxy-propoxy]-3,4-difluoro-2-[fluoro-4-iodo-phenylamino]-benzamide is a mitogen-activated protein kinase inhibitor used as a rabbit model of BRVO. It is administered as a single intravitreal injection into the midvitreous cavity in a dose of 0.5 or 1.0 mg per eye with a 27-gauge needle at the superior temporal quadrant.[173–186]

CRVO models

Experimental CRVO may be produced by LPC, diathermic cauterization, permanent ligation of the CRV, transient clamping/ligation of the optic nerve, or intravitreal injection of some pharmacologic agents.[173–186]

Laser photocoagulation To induce CRVO, all main branches are irradiated with laser on the slit lamp using a contact lens, 1.0–2.0 optic disc areas from the optic disc, avoiding injury to the retinal arteries. Different types of laser, wavelengths, and photosensitizers mentioned in BRVO induction have been used.[173–186]

Diathermic cauterization Diathermy may be applied on CRV at the inferomedial aspect of the optic nerve, avoiding damage to the ciliary vessels via a lateral orbitotomy in nonhuman primates.[173–186]

Permanent ligation of CRV Mechanical ligation using the operating microscope of the CRV may be used in nonhuman primates to produce CRVO. CRV is identified via the lateral orbital approach and ligated using an 8–0 silk suture. Then, the neoprene introduction into CRV via a cannula or the cutting of CRV is performed.[173–186]

Transient ligation or clamping In rats and pigs, it has been reported that transient ligation/clamping of the optic nerve through a lateral orbital approach for 1–2 h induced CRVO. However, as in this procedure, the ciliary vessels and CRA were also involved, so the creation of an isolated CRVO is not possible.[173–186]

Intravitreal injection of substances In rabbits, CRVO could be successfully created via intravitreally 0.01 mL (5 units) injection with a 27-gauge needle of thrombin solution over the wall of retinal veins under direct view with contact fundus lens with the aid of an operating microscope (Hirosaki model). On the other hand, in the rabbit model, intravitreally 50 or 100 lg injection of a hydrophilic photosensitizer, mono-L-aspartyl chlorin e6 (NPe6) may also be applied through pars plana using a 30-gauge needle. However, in this model, CRVO and CRAO developed together at the end of the first week following injection. Thus, this model fails to obtain pure CRVO. Intravitreally 1000 pmol injection with a 29-gauge needle of ET-1 solution, a potent vasoconstrictor drug, over the optic disc observed by ophthalmoscope-induced CRVO in a rabbit model. The advantage of the ET-1 model is that it is less invasive and does not need special equipment. Additionally, it has been reported that it has similar effects on rabbit and cat eyes following the administration of ET-1 via the intravenous, intravitreal, or subconjunctival routes.[173–186]

Sickle cell retinopathy models

In transgenic mouse models of sickle cell disease (α H β S [β MDD] type), mice retinae were processed by flat-embedding adenosine diphosphatase-reacted retinas in glycol methacrylate. In this model, retinal capillary nonperfusion areas and arterio-venous anastomoses, interestingly, all of RNV, CNV, and RAP were developed due to RVO. In these animals, the frequency of bilateral chorioretinal involvement was 30% over 15 months old experiments. In conclusion, in this mouse model, the retinochoroidal proliferative changes similar to those in human sickle cell retinopathy (SCR) could be obtained.[187–189]

Inflammatory RNV model

In this model, RNVs extending into the choroid were present without significant inflammation, in 38% of the eyes with the immunization of rats with interphotoreceptor retinoid-binding protein peptide on the 45th day.[190]

Radiation retinopathy RNV model

The experimental radiation retinopathy (RR) models in monkeys showed firstly the focal loss of capillary endothelial cells and pericytes after radiation. Later, the merging of the areas of capillary loss and the development of retinal capillary nonperfusion areas, RNV, and rubeosis iris have developed.[191–193]

Animal models for RAP

The MNV3 or RAP may occur in the setting of CNV and/or RNV and when it forms anastomosis between the retinal and choroidal circulation, and it may present the forms as alone intraretinal NV, or intraretinal NV + CNV or CNV alone.[194] MNV3 has been suggested that retinal hypoxia, inflammation, von Hippel–Lindau (VHL)–hypoxia-inducible factor (HIF)–vascular endothelial growth factor (VEGF) pathway, and multiple cell types play a role in the development and progression of MNV3. There are currently some MNV3 models including models of retina-specific VhlKO mice, Vegf overexpression mice, sVEGFR1 knockout mice, Vldlr−/− mice, Cyp27a1−/− mice, endothelial cell-specific Srf knockout mice, and RNV3 (JR5558) mice. The models of retina-specific VhlKO mice, Vegf overexpression, and sVEGFR1 knockout in mouse photoreceptors are related to the hypoxia pathway. Vldlr−/− mice and Cyp27a1−/− mice models are the basis of abnormal lipid metabolism. The endothelial cell-specific Srf knockout mice model is related to the cell migration pathway. Lastly, the RNV3 (JR5558) mice model is based on the gene mutation involved in retinal inflammation.[194]

The PI3K pathway model: The PI3K pathway plays very important roles in angiogenesis.[178,195,196] Blockage of PI3K pathways may cause the downstream signaling for epidermal growth factor receptor (EGFR), vascular endothelial growth factor receptor (VEGFR), and platelet-derived growth factor receptor (PDGFR) and may inhibit the RNV.[195,197,198] Thus, this model may be used for the research on the treatment with PI3K inhibitors of late-stage MNV3.

Rhodopsin promoter VEGF overexpression: In *rhodopsin* promoter/*VEGF* fusion gene transgenic mice (crossbred with C57BL/6 mice), angiographical intraretinal NV extending into the subretinal space has been obtained. Because NV could be developed by overexpression of VEGF, it is possible that NV can arise from both the choroid and retina, chorioretinal anastomosis similar to RAP.[34,66,95]

Spontaneous Bst chromosome 16 mutant: Smith et al. demonstrated that the frequency of histological CNV development in spontaneous tail (Bst) mutation derived from C57BLKS inbred strains of mice was almost 90%. Additionally, RPE abnormalities, retinochoroidal anastomosis, and retinal hamartomatous lesions are due to Bruch's membrane defects. However, this study did not show drusen-like or basal laminar deposit-like lesions. So, the NV obtained in this genetic model is more similar to that seen in RAP.[75]

Vldlr −/− targeted mutant: Heckenlively et al. demonstrated that NVs in the retina and choroidal anastomoses similar to RAP lesions developed in 3 months in the mice homozygous strains of mice with targeted mutations for very low-density lipoprotein receptor gene (VldlrTm1Her).[72]

Pachychoroid disease and central serous chorioretinopathy models

Pachychoroid spectrum diseases include various macular diseases in which pachychoroid defined as choroidal thickening and the dilatation of outer choroidal vessels has occurred. CSCR, polypoidal choroidal vasculopathy, pachychoroid pigment epitheliopathy, and focal choroidal excavation were included in this spectrum.[199] To generate a mouse model of pachychoroid, the vortex vein congestion may be obtained by the suturing of vortex veins in the eyes of wild-type mice. The choroidal congestion is observed as dilatation in vortex veins by fluorescein fundus angiography, and as choroidal thickening in optical coherence tomography.[200] In the pachychoroid spectrum, the most common disease is CSCR. It has been considered that the increased levels of serum catecholamine have an important role in pathogenesis.[201] Recent studies demonstrated that CSCR could be created by intravenous epinephrine in the experimental monkey model. It was postulated that elevated catecholamine levels in patients with CSCR might cause choroidal vasoconstriction by activating the sympathetic nervous system and by the breakdown of tight junctions among RPE cells.[202,203]

In a recent model, using neural crest-specific *Angpt1* knockout mice, it has been shown that Angiopoietin 1, a ligand of the endothelial receptor TEK (also known as Tie2) is essential for choriocapillaris development and vortex vein patterning. In that study, lacking choroidal ANGPT1, neural crest-specific *Angpt1* knockout eyes showed significant vascular attenuation in choriocapillaris and a half percent reduction in number of vortex veins. These eyes had dilated choroidal vessels just following birth. Then RPE dysfunction and CNV similar to that seen in pachychoroid disease developed.[204]

References

1. Dejneka NS, Rex TS, Bennett J. Gene therapy and animal models for retinal disease. *Developments in Ophthalmology*. 2003;37:188–198.
2. Iwata T, Tomarev S. *Animal Models for Eye Diseases and Therapeutics*. Sourcebook of Models for Biomedical Research; 2008:279–287. Totowa, Conn PM.
3. Grossniklaus HE, Kang SJ, Berglin L. Animal models of choroidal and retinal neovascularization. *Progress in Retinal and Eye Research*. 2010;29(6):500–519. https://doi.org/10.1016/j.preteyeres.2010.05.003.
4. Williams DL. Animal models for ophthalmic diseases in human. In: Gelatt KN, ed. *Veterinary Ophthalmology*. 2nd ed. Philadelphia: Lea & Febiger; 1991:1237–1272.
5. Yetemian RM, Craft CM. Retinal neovascular disorders: mouse models for drug development studies. In: LaVail MM, et al., eds. *Retinal Degenerative Diseases, Advances in Experimental Medicine and Biology*. Vol 723. Springer Science+Business Media, LLC; 2012:253–259. https://doi.org/10.1007/978-1-4614-0631-0_33.
6. Piotrowicz A, Paczwa K, Gołebiewska J, Różycki R. Current classification of macular neovascularization in the course of AMD based on the Consensus nomenclature for reporting neovascular age-related macular degeneration data. *Ophthatherapy*. 2021;8(1):5–10.
7. Spaide RF, Jaffe GJ, Sarraf D, et al. Consensus nomenclature for reporting neovascular age-related macular degeneration data: Consensus on neovascular age-related macular degeneration nomenclature study group. *Ophthalmology*. 2020;127(5):616–636.
8. Fabian-Jessing BK, Jakobsen TS, Jensen EG, et al. Animal models of choroidal neovascularization: a systematic review. *Investigative Ophthalmology & Visual Science*. 2022;63(9):11. https://doi.org/10.1167/iovs.63.9.11.
9. Rastoin O, Pagès G, Dufies M. Experimental models in neovascular age related macular degeneration. *International Journal of Molecular Sciences*. 2020;21(13):4627. https://doi.org/10.3390/ijms21134627.
10. Liu CH, Wang Z, Sun Y, Chen J. Animal models of ocular angiogenesis: from development to pathologies. *The FASEB Journal*. 2017;31(11):4665–4681. https://doi.org/10.1096/fj.201700336R.
11. Hollyfield JG, Kondo LK. Animal models for age-related macular degeneration. In: Pang I-H, Clark AF, eds. *Animal Models for Retinal Diseases*. Humana Press, Springer Science+Business Media; 2010:81–98. https://doi.org/10.1007/978-1-60761-541-5_5. Vol 46.
12. Green WR, Wilson DJ. Choroidal neovascularization. *Ophthalmology*. 1986;93(9):1169–1176.
13. Green WR. The uveal tract. In: Spencer WR, ed. *Ophthalmic Pathology: An Atlas and Textbook*. Philadelphia: WB Saunders; 2006:1597–1607.
14. Grossniklaus HE, Green WR. Choroidal neovascularization. *American Journal of Ophthalmology*. 2004;137(3):496–503.
15. Grossniklaus HE, Ling JX, Wallace TM, et al. Macrophage and retinal pigment epithelium expression of angiogenic cytokines in choroidal neovascularization. *Molecular Vision*. 2002;8:119–126.
16. Penfold PL, Madigan MC, Gillies MC, Provis JM. Immunological and aetiological aspects of macular degeneration. *Progress in Retinal and Eye Research*. 2001;20(3):385–414.
17. Kilarski WW, Samolov B, Petersson L, Kvanta A, Gerwins P. Biomechanical regulation of blood vessel growth during tissue vascularization. *Nature Medicine*. 2009;15(6):657–664.
18. Goldsmith P, Harris WA. The zebrafish as a tool for understanding the biology of visual disorders. *Seminars in Cell and Developmental Biology*. 2003;14(1):11–18.
19. Bilotta J, Saszik S. The zebrafish as a model visual system. *International Journal of Developmental Neuroscience*. 2001;19(7):621–629.
20. Dowling J. *The Retina: An Approachable Part of the Brain*. Vol 18. Cambridge, MA, USA: Belknap Press of Harvard University Press; 1987:12–32.

21. Robinson J, Schmitt EA, Harosi FI, et al. Zebrafish ultraviolet visual pigment: absorption spectrum, sequence and localization. *Proceedings of the National Academy of Sciences of the United States of America*. 1993;90(13):6009–6012.
22. Cao R, Jensen LDE, Soll I, et al. Hypoxia-induced retinal angiogenesis in zebrafish as a model to study retinopathy. *PLoS One*. 2008;3(7):e2748.
23. Cao Z, Jensen LD, Rouhi P, et al. Hypoxia-induced retinopathy model in adult zebrafish. *Nature Protocols*. 2010;5(12):1903–1910.
24. Biehlmaier O, Neuhauss SCF, Kohler K. Double cone dystrophy and RPE degeneration in the retina of the zebrafi sh gnn mutant. *Investigative Ophthalmology & Visual Science*. 2003;44(3):1287–1298.
25. Hageman GS, Luthert PJ, Victor Chong NH, Johnson LV, Anderson DH, Mullins RF. An integrated hypothesis that considers drusen as biomarkers of immune-mediated processes at the RPE-Bruch's membrane interface in aging and age-related macular degeneration. *Progress in Retinal and Eye Research*. November 2001;20(6):705–732.
26. Mullins RF, Russell SR, Anderson DH, Hageman GS. Drusen associated with aging and age-related macular degeneration contain proteins common to extracellular deposits associated with atherosclerosis, elastosis, amyloidosis, and dense deposit disease. *The FASEB Journal*. May 2000;14(7):835–846.
27. Schlingemann RO. Role of growth factors and the wound healing response in age-related macular degeneration. *Graefes Archive for Clinical and Experimental Ophthalmology*. January 2004;242(1):91–101.
28. Ryan SJ. Subretinal neovascularization. Natural history of an experimental model. *Archives of Ophthalmology*. November 1982;100(11):1804–1809.
29. Ryan SJ. The development of an experimental model of subretinal neovascularization in disciform macular degeneration. *Transactions of the American Ophthalmological Society*. 1979;77:707–745.
30. Shen M, Yan M, Wang L, Zhang JJ, Liu B, et al. Experimental choroidal neovascularization induced by laser in the eyes of rhesus monkeys. *Sichuan Da Xue Xue Bao Yi Xue Ban*. 2008;39(4):567–569.
31. Shen WY, Lee SY, Yeo I, et al. Predilection of the macular region to high incidence of choroidal neovascularization after intense laser photocoagulation in the monkey. *Archives of Ophthalmology*. 2004;122(3):353–360.
32. Zhang Y, Fortune B, Atchaneeyasakul LO, McFarland T, Mose K, et al. Natural history and histology in a rat model of laser-induced photothrombotic retinal vein occlusion. *Current Eye Research*. 2008;33(4):365376.
33. Gholipour MA, Kanavi MR, Ahmadieh H, Aldavood SJ, Nourinia R, et al. Intravitreal topotecan inhibits laser-induced choroidal neovascularization in a rat model. *Journal of Ophthalmic and Vision Research*. 2015;10(3):295302.
34. Tobe T, Ortega S, Luna JD, et al. Targeted disruption of the FGF2 gene does not prevent choroidal neovascularization in a murine model. *American Journal of Pathology*. November 1998;153(5):1641–1646.
35. Frank RN, Das A, Weber ML. A model of subretinal neovascularization in the pigmented rat. *Current Eye Research*. March 1989;8(3):239–247.
36. Seo MS, Kwak N, Ozaki H, et al. Dramatic inhibition of retinal and choroidal neovascularization by oral administration of a kinase inhibitor. *American Journal of Pathology*. June 1999;154(6):1743–1753.
37. Espinosa-Heidmann DG, Suner I, Hernandez EP, Frazier WD, Csaky KG, Cousins SW. Age as an independent risk factor for severity of experimental choroidal neovascularization. *Investigative Ophthalmology & Visual Science*. May 2002;43(5):1567–1573.
38. Sakurai E, Anand A, Ambati BK, van Rooijen N, Ambati J. Macrophage depletion inhibits experimental choroidal neovascularization. *Investigative Ophthalmology & Visual Science*. August 2003;44(8):3578–3585.
39. Espinosa-Heidmann DG, Suner IJ, Hernandez EP, Monroy D, Csaky KG, Cousins SW. Macrophage depletion diminishes lesion size and severity in experimental choroidal neovascularization. *Investigative Ophthalmology & Visual Science*. August 2003;44(8):3586–3592.
40. Bainbridge JW, Mistry AR, Thrasher AJ, Ali RR. Gene therapy for ocular angiogenesis. *Clinical Science [Lond]*. June 2003;104(6):561–575.
41. Reich SJ, Fosnot J, Kuroki A, et al. Small interfering RNA [siRNA] targeting VEGF effectively inhibits ocular neovascularization in a mouse model. *Molecular Vision*. May 30, 2003;9:210–216.
42. Mori K, Duh E, Gehlbach P, et al. Pigment epithelium-derived factor inhibits retinal and choroidal neovascularization. *Journal of Cellular Physiology*. August 2001;188(2):253–263.
43. Singh SR, Grossniklaus HE, Kang SJ, Edelhauser HF, Ambati BK, Kompella UB. Intravenous transferrin, RGD peptide and dual-targeted nanoparticles enhance anti-VEGF intraceptor gene delivery to laser-induced CNV. *Gene Therapy*. May 2009;16(5):645–659.
44. Criswell MH, Ciulla TA, Hill TE, et al. The squirrel monkey: characterization of a new-world primate model of experimental choroidal neovascularization and comparison with the macaque. *Investigative Ophthalmology & Visual Science*. February 2004;45(2):625–634.
45. elDirini AA, Ogden TE, Ryan SJ. Subretinal endophotocoagulation. A new model of subretinal neovascularization in the rabbit. *Retina*. 1991;11(2):244–249.
46. Saishin Y, Silva RL, Saishin Y, et al. Periocular injection of microspheres containing PKC412 inhibits choroidal neovascularization in a porcine model. *Investigative Ophthalmology & Visual Science*. November 2003;44(11):4989–4993.
47. Kiilgaard JF, Andersen MV, Wiencke AK, et al. A new animal model of choroidal neovascularization. *Acta Ophthalmologica Scandinavica*. December 2005;83(6):697–704.
48. Albert DM, Neekhra A, Wang S, et al. Development of choroidal neovascularization in rats with advanced intense cyclic light-induced retinal degeneration. *Archives of Ophthalmology*. February 2010;128(2):212–222.
49. Spilsbury K, Garrett KL, Shen WY, Constable IJ, Rakoczy PE. Overexpression of vascular endothelial growth factor [VEGF] in the retinal pigment epithelium leads to the development of choroidal neovascularization. *American Journal of Pathology*. July 2000;157(1):135–144.
50. Baffi J, Byrnes G, Chan CC, Csaky KG. Choroidal neovascularization in the rat induced by adenovirus mediated expression of vascular endothelial growth factor. *Investigative Ophthalmology & Visual Science*. October 2000;41(11):3582–3589.

51. Wang F, Rendahl KG, Manning WC, Quiroz D, Coyne M, Miller SS. AAV-mediated expression of vascular endothelial growth factor induces choroidal neovascularization in rat. *Investigative Ophthalmology & Visual Science*. February 2003;44(2):781–790.
52. Schmack I, Berglin L, Nie X, et al. Modulation of choroidal neovascularization by subretinal injection of retinal pigment epithelium and polystyrene microbeads. *Molecular Vision*. 2009;15:146–161.
53. Shen D, Wen R, Tuo J, Bojanowski CM, Chan CC. Exacerbation of retinal degeneration and choroidal neovascularization induced by subretinal injection of Matrigel in CCL2/MCP-1-deficient mice. *Ophthalmic Research*. 2006;38(2):71–73.
54. Qiu G, Stewart JM, Sadda S, et al. A new model of experimental subretinal neovascularization in the rabbit. *Experimental Eye Research*. July 2006;83(1):141–152.
55. Zhu ZR, Goodnight R, Sorgente N, Ogden TE, Ryan SJ. Morphologic observations of retinal pigment epithelial proliferation and neovascularization in the rabbit. *Retina*. 1989;9(4):319–327.
56. Kimura H, Sakamoto T, Hinton DR, et al. A new model of subretinal neovascularization in the rabbit. *Investigative Ophthalmology & Visual Science*. September 1995;36(10):2110–2119.
57. Kimura H, Spee C, Sakamoto T, et al. Cellular response in subretinal neovascularization induced by bFGF-impregnated microspheres. *Investigative Ophthalmology & Visual Science*. February 1999;40(2):524–528.
58. Tamai K, Spaide RF, Ellis EA, Iwabuchi S, Ogura Y, Armstrong D. Lipid hydroperoxide stimulates subretinal choroidal neovascularization in the rabbit. *Experimental Eye Research*. February 2002;74(2):301–308.
59. Ni M, Holland M, Jarstadmarken H, De Vries G. Time-course of experimental choroidal neovascularization in Dutch-Belted rabbit: clinical and histological evaluation. *Experimental Eye Research*. September 2005;81(3):286–297.
60. Julien S, Kreppel F, Beck S, et al. A reproducible and quantifiable model of choroidal neovascularization induced by VEGF A165 after subretinal adenoviral gene transfer in the rabbit. *Molecular Vision*. July 30, 2008;14:1358–1372.
61. Lassota N. Clinical and histological aspects of CNV formation: studies in an animal model. *Acta Ophthalmology*. September 2008;86:1–24.
62. Lassota N, Kiilgaard JF, la Cour M, Scherfig E, Prause JU. Natural history of choroidal neovascularization after surgical induction in an animal model. *Acta Ophthalmology*. August 2008;86(5):495–503.
63. Lassota N, Kiilgaard JF, Prause JU, Qvortrup K, Scherfig E, la Cour M. Surgical induction of choroidal neovascularization in a porcine model. *Graefes Archive for Clinical and Experimental Ophthalmology*. August 2007;245(8):1189–1198.
64. Cui JZ, Kimura H, Spee C, Thumann G, Hinton DR, Ryan SJ. Natural history of choroidal neovascularization induced by vascular endothelial growth factor in the primate. *Graefes Archive for Clinical and Experimental Ophthalmology*. April 2000;238(4):326–333.
65. Rakoczy PE, Brankov M, Fonceca A, Zaknich T, Rae BC, Lai CM. Enhanced recombinant adeno-associated virus-mediated vascular endothelial growth factor expression in the adult mouse retina: a potential model for diabetic retinopathy. *Diabetes*. March 2003;52(3):857–863.
66. Schwesinger C, Yee C, Rohan RM, et al. Intrachoroidal neovascularization in transgenic mice overexpressing vascular endothelial growth factor in the retinal pigment epithelium. *American Journal of Pathology*. March 2001;158(3):1161–1172.
67. Ambati J, Anand A, Fernandez S, et al. An animal model of age-related macular degeneration in senescent Ccl-2- or Ccr-2-deficient mice. *Nature Medicine*. November 2003;9(11):1390–1397.
68. Okamoto N, Tobe T, Hackett SF, Ozaki H, Vinores MA, et al. Transgenic mice with increased expression of vascular endothelial growth factor in the retina: a new model of intraretinal and subretinal neovascularization. *American Journal of Pathology*. 1997;151(1):281–291.
69. Chan CC, Ross RJ, Shen D, Ding X, Majumdar Z, Bojanowski CM, Zhou M, Salem Jr N, Bonner R, Tuo J. Ccl2/Cx3cr1-deficient mice: an animal model for age-related macular degeneration. *Ophthalmic Research*. 2008;40(3–4):124–128.
70. Imamura Y, Noda S, Hashizume K, Shinoda K, Yamaguchi M, Uchiyama S, Shimizu T, Mizushima Y, Shirasawa T, Tsubota K. Drusen, choroidal neovascularization, and retinal pigment epithelium dysfunction in SOD1-deficient mice: a model of age-related macular degeneration. *Proceedings of the National Academy of Sciences of the United States of America*. July 25, 2006;103(30):11282–11287.
71. Dithmar S, Curcio CA, Le NA, Brown S, Grossniklaus HE. Ultrastructural changes in Bruch's membrane of apolipoprotein E-deficient mice. *Investigative Ophthalmology & Visual Science*. July 2000;41(8):2035–2042.
72. Malek G, Johnson LV, Mace BE, Saloupis P, Schmechel DE, Rickman DW, Toth CA, Sullivan PM, Bowes Rickman C. Apolipoprotein E allele-dependent pathogenesis: a model for age-related retinal degeneration. *Proceedings of the National Academy of Sciences of the United States of America*. August 16, 2005;102(33):11900–11905.
73. Hahn P, Qian Y, Dentchev T, Chen L, Beard J, Harris ZL, Dunaief JL. Disruption of ceruloplasmin and hephaestin in mice causes retinal iron overload and retinal degeneration with features of age-related macular degeneration. *Proceedings of the National Academy of Sciences of the United States of America*. September 21, 2004;101(38):13850–13855. https://doi.org/10.1073/pnas.0405146101.
74. Oshima Y, Oshima S, Nambu H, Kachi S, Hackett SF, Melia M, Kaleko M, Connelly S, Esumi N, Zack DJ, Campochiaro PA. Increased expression of VEGF in retinal pigmented epithelial cells is not sufficient to cause choroidal neovascularization. *Journal of Cellular Physiology*. December 2004;201(3):393–400.
75. Heckenlively JR, Hawes NL, Friedlander M, Nusinowitz S, Hurd R, Davisson M, Chang B. Mouse model of subretinal neovascularization with choroidal anastomosis. *Retina*. August 2003;23(4):518–522.
76. Rakoczy PE, Zhang D, Robertson T, Barnett NL, Papadimitriou J, Constable IJ, Lai CM. Progressive age-related changes similar to age-related macular degeneration in a transgenic mouse model. *American Journal of Pathology*. October 2002;161(4):1515–1524.
77. Smith RS, John SW, Zabeleta A, Davisson MT, Hawes NL, Chang B. The bst locus on mouse chromosome 16 is associated with age-related subretinal neovascularization. *Proceedings of the National Academy of Sciences of the United States of America*. February 29, 2000;97(5):2191–2195.

78. Campochiaro PA. Retinal and choroidal neovascularization. *Journal of Cellular Physiology*. September 2000;184(3):301–310.
79. Ashton N. Retinal vascularization in health and disease: proctor award lecture of the association for research in ophthalmology. *American Journal of Ophthalmology*. October 1957;44(4 Pt 2):7–17.
80. Ham DI, Chang K, Chung H. Preretinal neovascularization induced by experimental retinal vein occlusion in albino rats. *Korean Journal of Ophthalmology*. June 1997;11(1):60–64. https://doi.org/10.3341/kjo.1997.11.1.60.
81. Sugita G, Tano Y, Machemer R. Experimental neovascularization of the retina. *International Ophthalmology*. 1980;2(1):33–37.
82. Tano Y, Chandler DB, Machemer R. Retinal neovascularization after intravitreal fibroblast injection. *American Journal of Ophthalmology*. 1981;92(1):103–109.
83. Tano Y, Chandler DB, Machemer R. Vascular casts of experimental retinal neovascularization. *American Journal of Ophthalmology*. 1981;92(1):110–120.
84. Hitchins CA, Grierson I, Hiscott PS. The effects of injections of cultured fibroblasts into the rabbit vitreous. *Graefes Archive for Clinical and Experimental Ophthalmology*. 1985;223(5):237–249.
85. Ophir A, Blumenkranz MS, Clafin AJ. Experimental intraocular proliferation and neovascularization. *American Journal of Ophthalmology*. 1982;94(4):450–457.
86. Hemo I, Maftzir G, BenEzra D. *Retinal Neovascularization. Second International Symposium on Ocular Circulation and Neovascularization*. Baltimore: Wilmer Institute; 1989:36.
87. Hamilton AM, Marshall J, Kohner EM, Bowbyes JA. Retinal new vessel formation following experimental vein occlusion. *Experimental Eye Research*. 1975;20(6):493–497.
88. Jabbour NM, Nork TM, McCormick SA, Brown SC. Extraretinal neovascularization model system in the rhesus monkey. ARVO Abstracts. *Investigative Ophthalmology & Visual Science*. 1989;30:392.
89. Ashton N, Ward B, Serpell G. Effect of oxygen on developing retinal vessels with particular reference to the problem of retrolental fibroplasias. *British Journal of Ophthalmology*. 1954;38(7):397–432.
90. Patz A. Clinical and experimental studies on retinal neovascularization. *American Journal of Ophthalmology*. 1982;94(6):715–743.
91. Ashton N, Ward B, Serpell G. Role of oxygen in the genesis of retrolental fibroplasias. *British Journal of Ophthalmology*. 1953;37(9):513–520.
92. Finkelstein D, Bren S, Patz A, Folkman J, Miller S, et al. Experimental retinal neovascularization induced by intravitreal tumors. *American Journal of Ophthalmology*. 1977;83(5):660–664.
93. Weber ML, Mancini MA, Frank RN. Retinovitreal neovascularization in the royal college of surgeons rat. *Current Eye Research*. 1989;8(1):61–74.
94. Frank RN, Mancini MA. Presumed retinovitreal neovascularization in dystrophic retinas of spontaneously hypertensive rats. *Investigative Ophthalmology & Visual Science*. 1986;27(3):346–355.
95. Antoszyk AN, Gottlieb JL, Casey RC, Hatchell DL, Machemer R. An experimental model of preretinal neovascularization in the rabbit. *Investigative Ophthalmology & Visual Science*. 1991;32(1):46–52.
96. Feit-Leichman RA, Kinouchi R, Takeda M, Fan Z, Mohr S, Kern TS, Chen DF. Vascular damage in a mouse model of diabetic retinopathy: relation to neuronal and glial changes. *Investigative Ophthalmology & Visual Science*. November 2005;46(11):4281–4287.
97. Miller JW. Vascular endothelial growth factor and ocular neovascularization. *American Journal of Pathology*. July 1997;151(1):13–23.
98. Ohno-Matsui K, Hirose A, Yamamoto S, Saikia J, Okamoto N, Gehlbach P, et al. Inducible expression of vascular endothelial growth factor in adult mice causes severe proliferative retinopathy and retinal detachment. *American Journal of Pathology*. 2002;160:711–719.
99. Lai CM, Dunlop SA, May LA, Gorbatov M, Brankov M, Shen WY, Binz N, Lai YK, Graham CE, Barry CJ, Constable IJ, Beazley LD, Rakoczy EP. Generation of transgenic mice with mild and severe retinal neovascularization. *British Journal of Ophthalmology*. July 2005;89(7):911–916.
100. Bai Y, Ma JX, Guo J, Wang J, Zhu M, Chen Y, Le YZ. Müller cell-derived VEGF is a significant contributor to retinal neovascularization. *The Journal of Pathology*. December 2009;219(4):446–454. https://doi.org/10.1002/path.2611.
101. Yetemian RM, Brown BM, Craft CM. Neovascularization, enhanced inflammatory response, and age-related cone dystrophy in the Nrl-/-Grk1-/- mouse retina. *Investigative Ophthalmology & Visual Science*. December 2010;51(12):6196–6206.
102. Zhu X, Brown B, Li A, Mears AJ, Swaroop A, Craft CM. GRK1-dependent phosphorylation of S and M opsins and their binding to cone arrestin during cone phototransduction in the mouse retina. *Journal of Neuroscience*. July 9, 2003;23(14):6152–6160.
103. Zhu X, Brown B, Rife L, Craft CM. Slowed photoresponse recovery and age-related degeneration in cones lacking G protein-coupled receptor kinase 1. *Advances in Experimental Medicine & Biology*. 2006;572:133–139.
104. Adrian MT, Miller CM, Zhu L. Animal models of diabetic retinopathy. In: Pang I-H, Clark AF, eds. *Animal Models for Retinal Diseases, Neuromethods*. Vol 46. Humana Press, Springer Science+Business Media; 2010:113–138.
105. Fletcher EL, Jobling AI, Vessey KA, Luu C, Guymer RH, et al. Animal models of retinal disease. *Progress in Molecular Biology and Translational Science*. 2011;100:211286.
106. Engerman R, Bloodworth JBM. Experimental diabetic retinopathy in dogs. *Archives of Ophthalmology*. 1965;73:205–210.
107. Jiang X, Yang L, Luo Y. Animal models of diabetic retinopathy. *Current Eye Research*. 2015;40(8):761–771.
108. Sosula L, Beaumont P, Hollows FC, Jonson KM. Dilatation and endothelial proliferation of retinal capillaries in streptozotocin-diabetic rats: quantitative electron microscopy. *Investigative Ophthalmology*. November 1972;11(11):926–935.
109. Pitale PM, Gorbatyuk MS. Diabetic retinopathy: from animal models to cellular signaling. *International Journal of Molecular Sciences*. January 27, 2022;23(3):1487. https://doi.org/10.3390/ijms23031487.

110. Sima AA, Chakrabarti S, Garcia-Salinas R, Basu PK. The BB-rat—an authentic model of human diabetic retinopathy. *Current Eye Research.* October 1985;4(10):1087–1092.
111. Peterson RG, Little LA, Neel MA. WKY fatty rat as a model of obesity and non-insulin-dependent diabetes mellitus. *ILAR News (Institute of Laboratory Animal Resources).* January 1990;32(3):13–15.
112. Yokoi N, Hoshino M, Hidaka S, Yoshida E, Beppu M, Hoshikawa R, Sudo K, Kawada A, Takagi S, Seino S. A novel rat model of type 2 diabetes: the zucker fatty diabetes mellitus ZFDM rat. *Journal of Diabetes Research.* 2013;2013:103731. https://doi.org/10.1155/2013/103731.
113. Lu ZY, Bhutto IA, Amemiya T. Retinal changes in Otsuka long-evans Tokushima Fatty rats [spontaneously diabetic rat]—possibility of a new experimental model for diabetic retinopathy. *Japanese Journal of Ophthalmology.* 2003 Jan-Feb;47(1):28–35.
114. Shinohara M, Masuyama T, Shoda T, Takahashi T, Katsuda Y, Komeda K, Kuroki M, Kakehashi A, Kanazawa Y. A new spontaneously diabetic non-obese Torii rat strain with severe ocular complications. *International Journal of Experimental Diabetes Research.* 2000;1(2):89–100.
115. Yoshioka M, Kayo T, Ikeda T, Koizumi A. A novel locus, Mody4, distal to D7Mit189 on chromosome 7 determines early-onset NIDDM in nonobese C57BL/6 [Akita] mutant mice. *Diabetes.* May 1997;46(5):887–894.
116. Barber AJ, Antonetti DA, Kern TS, Reiter CE, Soans RS, Krady JK, Levison SW, Gardner TW, Bronson SK. The Ins2Akita mouse as a model of early retinal complications in diabetes. *Investigative Ophthalmology & Visual Science.* June 2005;46(6):2210–2218.
117. Wicker LS, Todd JA, Peterson LB. Genetic control of autoimmune diabetes in the NOD mouse. *Annual Review of Immunology.* 1995;13:179–200.
118. Makino S, Kunimoto K, Muraoka Y, Mizushima Y, Katagiri K, Tochino Y. Breeding of a non-obese, diabetic strain of mice. *Jikken Dobutsu.* January 1980;29(1):1–13. https://doi.org/10.1538/expanim1978.29.1_1.
119. Hummel KP, Dickie MM, Coleman DL. Diabetes, a new mutation in the mouse. *Science.* September 2, 1966;153(3740):1127–1128.
120. van Eeden PE, Tee LB, Lukehurst S, Lai CM, Rakoczy EP, Beazley LD, Dunlop SA. Early vascular and neuronal changes in a VEGF transgenic mouse model of retinal neovascularization. *Investigative Ophthalmology & Visual Science.* October 2006;47(10):4638–4645.
121. Jo DH, Cho CS, Kim JH, et al. Animal models of diabetic retinopathy: doors to investigate pathogenesis and potential therapeutics. *Journal of Biomedical Science.* 2013;20(38):1–13.
122. Anderson HR, Stitt AW, Gardiner TA, Archer DB. Diabetic rethinopathy: morphometric analysis of basement membrane thickening of capillaries in different retinal layers within arterial and venous environments. *British Journal of Ophthalmology.* 1995;79(12):1120–1123.
123. Kador PF, Blessing K, Randazzo J, et al. Evaluation of the vascular targeting agent combretastatin A-4 prodrug on retinal neovascularization in the galactose-fed dog. *Journal of Ocular Pharmacology and Therapeutics.* 2007;23(2):132–142.
124. Kern TS, Engerman RL. Capillary lesions develop in retina rather than cerebral cortex in diabetes and experimental galactosemia. *Archives of Ophthalmology.* 1996;114(3):306–310.
125. Tso MOM, Kurosawa A, Benhamou E, et al. Microangiopathic retinopathy in experimental diabetic monkeys. *Trans-actions of the American Ophthalmological Society.* 1988;86:389–421.
126. Kim SY, Johnson MA, McLeod DS. Retinopathy in monkeys with spontaneous type 2 diabetes. *Investigative Ophthalmology and Visual Science.* 2004;45(12):4543–4553.
127. Gleeson M, Connaughton V, Arneson LS. Induction of hyperglycaemia in Zebrafish [Danio rerio] leads to morphological changes in the retina. *Acta Diabetologica.* 2007;44(3):157–163.
128. Olsen AS, Sarras MP, Intine RV. Limb regeneration is impaired in an adult zebrafish model of diabetes mellitus. *Wound Repair and Regeneration.* 2010;18(5):532–542.
129. Eleazu CO, Eleazu KC, Chukwuma S, Essien UN. Review of the mechanism of cell death resulting from streptozotocin challenge in experimental animals, its practical use and potential risk to humans. *Journal of Diabetes and Metabolic Disorders.* December 23, 2013;12(1):60. https://doi.org/10.1186/2251-6581-12-60.
130. Sakata N, Yoshimatsu G, Tsuchiya H, Egawa S, Unno M. Animal models of diabetes mellitus for islet transplantation. *Experimental Diabetes Research.* 2012;2012. https://doi.org/10.1155/2012/256707.
131. Gorbatyuk OS, Pitale PM, Saltykova IV, Dorofeeva IB, Zhylkibayev AA, Athar M, Fuchs PA, Samuels BC, Gorbatyuk MS. A novel tree shrew model of diabetic retinopathy. *Frontiers in Endocrinology [Lausanne].* January 3, 2022;12. https://doi.org/10.3389/fendo.2021.799711.
132. Chaudhry ZZ, Morris DL, Moss DR, Sims EK, Chiong Y, Kono T, Evans-Molina C. Streptozotocin is equally diabetogenic whether administered to fed or fasted mice. *Lab Animal.* October 2013;47(4):257–265. https://doi.org/10.1177/0023677213489548.
133. Lai AK, Lo AC. Animal models of diabetic retinopathy: summary and comparison. *Journal of Diabetes Research.* 2013;2013:106594. https://doi.org/10.1155/2013/106594.
134. McLetchie NG. Alloxan diabetes: a discovery, albeit a minor one. *The Journal of the Royal College of Physicians of Edinburgh.* 2002;32(2):134–142.
135. Olivares AM, Althoff K, Chen GF, Wu S, Morrisson MA, DeAngelis MM, Haider N. Animal models of diabetic retinopathy. *Current Diabetes Reports.* August 24, 2017;17(10):93. https://doi.org/10.1007/s11892-017-0913-0.
136. Black HE, Rosenblum IY, Capen CC. Chemically induced [streptozotocin-alloxan] diabetes mellitus in the dog. Biochemical and ultrastructural studies. *American Journal of Pathology.* February 1980;98(2):295–310.
137. Seita M, Noguchi H, Kubota Y, Kawamoto H, Nakaji S, Kobayashi N, Fujiwara T. Development of canine models of type 1 diabetes with partial pancreatectomy and the administration of streptozotocin. *Cell Medicine.* October 21, 2013;6(1–2):25–31. https://doi.org/10.3727/215517913X674289.
138. Engerman RL, Kern TS. Experimental galactosemia produces diabetic-like retinopathy. *Diabetes.* January 1984;33(1):97–100. https://doi.org/10.2337/diab.33.1.97.

139. Chronopoulos A, Roy S, Beglova E, Mansfield K, Wachtman L, Roy S. Hyperhexosemia-induced retinal vascular pathology in a novel primate model of diabetic retinopathy. *Diabetes*. July 2015;64(7):2603–2608.
140. Cohen MP, Hud E, Shea E, Shearman CW. Vitreous fluid of db/db mice exhibits alterations in angiogenic and metabolic factors consistent with early diabetic retinopathy. *Ophthalmic Research*. 2008;40(1):5–9.
141. Han Z, Guo J, Conley SM, Naash MI. Retinal angiogenesis in the Ins2Akita Mouse model of diabetic retinopathy. *Investigative Ophthalmology & Visual Science*. 2013;54(1):574–584.
142. Danis RP, Yang Y. Microvascular retinopathy in the Zucker diabetic fatty rat. *Investigative Ophthalmology & Visual Science*. 1993;34(7):2367–2371.
143. Masuyama T, Katsuda Y, Shinohara M. A novel model of obesity-related diabetes: introgression of the Lepr[fa] allele of the Zucker fatty rat into nonobese Spontaneously Diabetic Torii [SDT] rats. *Experimental Animals*. 2005;54:13–20.
144. Smith LE, Wesolowski E, McLellan A, et al. Oxygen-induced retinopathy in the mouse. *Investigative Ophthalmology & Visual Science*. 1994;35(1):101–111.
145. Connor KM, Krah NM, Dennison RJ, et al. Quantification of oxygen-induced retinopathy in the mouse: a model of vessel loss, vessel regrowth and pathological angiogenesis. *Nature Protocols*. 2009;4(11):1565–1573.
146. Kakehashi A, Saito Y, Mori K, Sugi N, Ono R, Yamagami H, Shinohara M, Tamemoto H, Ishikawa SE, Kawakami M, Kanazawa Y. Characteristics of diabetic retinopathy in SDT rats. *Diabetes/Metabolism Research and Reviews*. 2006;22(6):455–461.
147. Yamada H, Yamada E, Higuchi A, Matsumura M. Retinal neovascularisation without ischaemia in the spontaneously diabetic Torii rat. *Diabetologia*. August 2005;48(8):1663–1668.
148. Miyamura N, Amemiya T. Lens and retinal changes in the WBN/Kob rat [spontaneously diabetic strain]. Electron-microscopic study. *Ophthalmic Research*. 1998;30(4):221–232.
149. Zhang SX, Sima J, Wang JJ, Shao C, Fant J, Ma JX. Systemic and periocular deliveries of plasminogen kringle 5 reduce vascular leakage in rat models of oxygen-induced retinopathy and diabetes. *Current Eye Research*. August 2005;30(8):681–689.
150. Masuzawa K, Jesmin S, Maeda S, Zaedi S, Shimojo N, Miyauchi T, Goto K. Effect of endothelin dual receptor antagonist on VEGF levels in streptozotocin-induced diabetic rat retina. *Experimental Biology and Medicine [Maywood]*. June 2006;231(6):1090–1094.
151. Yanni SE, Penn JS. Animal models of retinopathy of prematurity. In: Pang I-H, Clark AF, eds. *Animal Models for Retinal Diseases, Neuromethods*. Vol 46. Humana Press, Springer Science+Business Media; 2010:99–111.
152. McLeod DS, Brownstein R, Lutty GA. Vaso-obliteration in the canine model of oxygen-induced retinopathy. *Investigative Ophthalmology & Visual Science*. February 1996;37(2):300–311.
153. Penn JS, Tolman BL, Lowery LA. Variable oxygen exposure causes preretinal neovascularization in the newborn rat. *Investigative Ophthalmology & Visual Science*. March 1993;34(3):576–585.
154. Madan A, Penn JS. Animal models of oxygen-induced retinopathy. *Frontiers in Bioscience*. May 1, 2003;8:d1030–d1043. https://doi.org/10.2741/1056.
155. Penn JS, Henry MM, Tolman BL. Exposure to alternating hypoxia and hyperoxia causes severe proliferative retinopathy in the newborn rat. *Pediatric Research*. 1994;36(6):724–731.
156. Lutty GA, McLeod DS, Bhutto I, Wiegand SJ. Effect of VEGF Trap on normal retinal vascular development and oxygen-induced retinopathy in the dog. *Investigative Ophthalmology & Visual Science*. 2011;52(7):4039–4047.
157. Hartnett ME, Penn JS. Mechanisms and management of retinopathy of prematurity. *New England Journal of Medicine*. 2012;367(26):2515–2526.
158. Hartnett ME. Studies on the pathogenesis of avascular retina and neovascularization into the vitreous in peripheral severe retinopathy of prematurity [an American Ophthalmological Society thesis]. *Transactions of the American Ophthalmological Society*. 2010;108:96–119.
159. Pierce EA, Foley ED, Smith LE. Regulation of vascular endothelial growth factor by oxygen in a model of retinopathy of prematurity. *Archives of Ophthalmology*. 1996;114(10):1219–1228.
160. Flower RW, Balke DAS. Retrolental fibroplasia: role of of the prostaglandin cascade in the pathogenesis of oxygen induced retinopathy in the newborn beagle. *Pediatric Research*. 1981;15(9):1293–1302.
161. Aguilar E, Dorrell MI, Friedlander D, Jacobson RA, Johnson A, Marchetti V, Moreno SK, Ritter MR, Friedlander M. Chapter 6. Ocular models of angiogenesis. *Methods Enzymology*. 2008;444:115–158. https://doi.org/10.1016/S0076-6879[08]02806-1.
162. Ritter MR, Banin E, Moreno SK, Aguilar E, Dorrell MI, Friedlander M. Myeloid progenitors differentiate into microglia and promote vascular repair in a model of ischemic retinopathy. *Journal of Clinical Investigation*. December 2006;116(12):3266–3276.
163. Cunningham S, McColm JR, Wade J, Sedowofia K, McIntosh N, et al. A novel model of retinopathy of prematurity simulating preterm oxygen variability in the rat. *Investigative Ophthalmology & Visual Science*. 2000;41(13):4275–4280.
164. Holmes JM, Duffner LA. The effect of postnatal growth retardation on abnormal neovascularization in the oxygen exposed neonatal rat. *Current Eye Research*. April 1996;15(4):403–409. https://doi.org/10.3109/02713689608995831.
165. Holmes JM, Zhang S, Leske DA, Lanier WL. The effect of carbon dioxide on oxygen-induced retinopathy in the neonatal rat. *Current Eye Research*. July 1997;16(7):725–732. https://doi.org/10.1076/ceyr.16.7.725.5054.
166. Penn JS, Rajaratnam VS, Collier RJ, Clark AF. The effect of an angiostatic steroid on neovascularization in a rat model of retinopathy of prematurity. *Investigative Ophthalmology & Visual Science*. January 2001;42(1):283–290.
167. Penn JS, Rajaratnam VS. Inhibition of retinal neovascularization by intravitreal injection of human rPAI-1 in a rat model of retinopathy of prematurity. *Investigative Ophthalmology & Visual Science*. December 2003;44(12):5423–5429. https://doi.org/10.1167/iovs.02-0804.

168. Chan-Ling T, Tout S, Holländer H, Stone J. Vascular changes and their mechanisms in the feline model of retinopathy of prematurity. *Investigative Ophthalmology & Visual Science*. June 1992;33(7):2128–2147.
169. Daugeliene L, Niwa M, Hara A, Matsuno H, Yamamoto T, et al. Transient ischemic injury in the rat retina caused by thrombotic occlusion-thrombolytic reperfusion. *Investigative Ophthalmology & Visual Science*. 2000;41(9):2743–2747.
170. Masuzawa K, Jesmin S, Maeda S, Kaji Y, Oshika T, et al. A model of retinal ischemia-reperfusion injury in rats by subconjunctival injection of Endothelin-1. *Experimental Biology and Medicine*. 2006;231(6):1085–1089.
171. Yu DY, Su EN, Cringle SJ, Alder VA, Yu PK, et al. Effect of betaxolol, timolol and nimodipine on human and pig retinal arterioles. *Experimental Eye Research*. 1998;67(1):73–81.
172. Vestergaard N, Cehofski LJ, Honoré B, Aasbjerg K, Vorum H. Animal models used to simulate retinal artery occlusion: a comprehensive review. *Translational Vision Science & Technology*. August 15, 2019;8(4):23. https://doi.org/10.1167/tvst.8.4.23.
173. Granstam E, Wang L, Bill A. Ocular effects of endothelin-1 in the cat. *Current Eye Research*. 1992;11(4):325–332.
174. Ciulla TA, Moulton R, Oberoi A, Miller JW. Retinal artery occlusion in rabbit eyes using human atheroma. *Current Eye Research*. July 1995;14(7):573–578. https://doi.org/10.3109/02713689508998404.
175. Khayat M, Lois N, Williams M, Stitt AW. Animal models of retinal vein occlusion. *Investigative Ophthalmology & Visual Science*. December 1, 2017;58(14):6175–6192. https://doi.org/10.1167/iovs.17-22788.
176. Fuma S, Nishinaka A, Inoue Y, Tsuruma K, Shimazawa M, Kondo M, Hara H. A pharmacological approach in newly established retinal vein occlusion model. *Science Reports*. March 2, 2017;7:43509. https://doi.org/10.1038/srep43509.
177. Hayashi A, Imai K, Kim HC, et al. Activation of protein tyrosine phosphorylation after retinal branch vein occlusion in cats. *Investigative Ophthalmology & Visual Science*. 1997;38(2):372–380.
178. Takei K, Sato T, Nonoyama T, et al. A new model of transient complete obstruction of retinal vessels induced by endothelin–1 injection into the posterior vitreous body in rabbits. *Graefes Archive for Clinical and Experimental Ophthalmology*. 1993;231(8):476–481.
179. Linner E. Occlusion of the retinal veins in rabbits induced by light coagulation. *Acta Ophthalmology*. 1961;39(4):739–740.
180. Chen W, Wu Y, Zheng M, et al. Establishing an experimental rat model of photodynamically–induced retinal vein occlusion using erythrosine B. *International Journal of Ophthalmology*. 2014;7(2):232–238.
181. Pournaras CJ, Petropoulos IK, Pournaras JA, et al. The rationale of retinal endovascular fibrinolysis in the treatment of retinal vein occlusion: from experimental data to clinical application. *Retina*. 2012;32(8):1566–1573.
182. Kwong Kwong JM, Caprioli J. Animal models of retinal ischemia. In: Pang I–H, Clark AF, eds. *Animal Models for Retinal Diseases, Neuro-methods*. Vol 46. Germany: Humana Press, Springer Science+Business Media; 2016:191–206.
183. Minhas G, Anand A. *Animal models of retinal ischemia in: brain injury–pathogenesis, monitoring, recovery and management. Brain Injury*. 2011:153–174.
184. Minhas G, Morishita R, Anand. Preclinical models to investigate retinal ischemia: advances and drawbacks. *Frontiers in Neurology*. 2012;3:75.
185. Minhas G, Morishita R, Shimamura M, et al. Modeling transient retinal ischemia in mouse by ligation of pterygopalatine artery. *Annals of Neurosciences*. 2015;22(4):222–225.
186. Katz ML, Redmond TM. Effect of Rpe65 knockout on accumulation of lipofuscin fluorophores in the retinal pigment epithelium. *Investigative Ophthalmology & Visual Science*. 2001;42(12):3023–3030.
187. Virdi PS, Hayreh SS. Ocular neovascularization with retinal vascular occlusion: I. Association with experimental retinal vein occlusion. *Archives of Ophthalmology*. 1982;100(2):331–341.
188. Sun M, Hao XF, Xie LK, Jin Q, Wang SH, Xu J. Methods for making animal models of retinal vein occlusion. *Guoji Yanke Zazhi [International Eye Science]*. 2020;20(11):1841–1846.
189. Fabry ME. Transgenic animal models of sickle cell disease. *Experientia*. 1993;49(1):28–36. https://doi.org/10.1007/BF01928785.
190. Promsote W, Powell FL, Veean S, Thounaojam M, Markand S, Saul A, Gutsaeva D, Bartoli M, Smith SB, Ganapathy V, Martin PM. Oral monomethyl fumarate therapy Ameliorates retinopathy in a humanized mouse model of sickle cell disease. *Antioxidants and Redox Signaling*. 2016;25(17):921–935. https://doi.org/10.1089/ars.2016.6638.
191. Lutty GA, McLeod DS, Pachnis A, Costantini F, Fabry ME, Nagel RL. Retinal and choroidal neovascularization in a transgenic mouse model of sickle cell disease. *American Journal of Pathology*. 1994;145(2):490–497.
192. Sakamoto T, Sanui H, Ishibashi T, Kohno T, Takahira K, Inomata H, Kohen L, Ryan SJ. Subretinal neovascularization in the rat induced by IRBP synthetic peptides. *Experimental Eye Research*. February 1994;58(2):155–160. https://doi.org/10.1006/exer.1994.1003.
193. Irvine AR, Wood IS. Radiation retinopathy as an experimental model for ischemic proliferative retinopathy and rubeosis iridis. *American Journal of Ophthalmology*. June 15, 1987;103(6):790–797. https://doi.org/10.1016/s0002-9394[14]74395-8.
194. Ramos MS, Echegaray JJ, Kuhn-Asif S, Wilkinson A, Yuan A, Singh AD, Browne AW. Animal models of radiation retinopathy - from teletherapy to brachytherapy. *Experimental Eye Research*. April 2019;181:240–251.
195. Villaret J, Pouzoulet F, Thomas C, Crisanti P, Delaunay K, Cassoux N, Matet A, Behar-Cohen F. Development of a new animal model of radiation retinopathy using an experimental radiation platform. *Acta Ophthalmology*. 2019;97:S263.
196. Qiang W, Wei R, Chen Y, Chen D. Clinical pathological features and current animal models of type 3 macular neovascularization. *Frontiers in Neuroscience*. 2021;15:734860. https://doi.org/10.3389/fnins.2021.734860.
197. Wert KJ, Mahajan VB, Zhang L, Yan Y, Li Y, Tosi J, Hsu CW, Nagasaki T, Janisch KM, Grant MB, Mahajan M, Bassuk AG, Tsang SH. Neuroretinal hypoxic signaling in a new preclinical murine model for proliferative diabetic retinopathy. *Signal Transduction and Targeted Therapy*. 2016;1:16005. https://doi.org/10.1038/sigtrans.2016.5.

198. Davies EM, Gurung R, Le KQ, Mitchell CA. Effective angiogenesis requires regulation of phosphoinositide signaling. *Advances in Biological Regulation*. January 2019;71:69–78. https://doi.org/10.1016/j.jbior.2018.11.008.
199. Zhang ZY, Bao XL, Cong YY, Fan B, Li GY. Autophagy in age-related macular degeneration: a regulatory mechanism of oxidative stress. *Oxidative Medicine and Cellular Longevity*. 2020;2020. https://doi.org/10.1155/2020/2896036.
200. Gong C, Qiao L, Feng R, Xu Q, Zhang Y, Fang Z, Shen J, Li S. IL-6-induced acetylation of E2F1 aggravates oxidative damage of retinal pigment epithelial cell line. *Experimental Eye Research*. November 2020;200:108219. https://doi.org/10.1016/j.exer.2020.108219.
201. Borooah S, Sim PY, Phatak S, Moraes G, Wu CY, Cheung CMG, Pal B, Bujarborua D. Pachychoroid spectrum disease. *Acta Ophthalmology*. September 2021;99(6):e806–e822.
202. Matsumoto H, Mukai R, Hoshino J, Oda M, Matsuzaki T, Ishizaki Y, Shibasaki K, Akiyama H. Choroidal congestion mouse model: could it serve as a pachychoroid model? *PLoS One*. 2021;16(1):e0246115. https://doi.org/10.1371/journal.pone.0246115.
203. Nicholson B, Noble J, Forooghian F, Meyerle C. Central serous chorioretinopathy: update on pathophysiology and treatment. *Survey Ophthalmology*. 2013;58(2):103–126.
204. Yoshioka H, Katsume Y, Akune H. Studies on experimental central serous chorioretinopathy. Fluorescein angiography and histopathology during the course of spontaneous remission. *Nihon Ganka Gakkai Zasshi*. 1984;88(5):819–828.

Chapter 3

Vascular changes with aging

Dmitrii S. Maltsev and Alexey N. Kulikov
S.M.Kirov Military Medical Academy, Ophthalmology Department, Saint-Petersburg, Russia

Introduction

The evaluation and study of posterior segment microcirculation undoubtedly make a substantial contribution to the diagnosis of retinal disorders. Over a broad spectrum of retinal disorders, we note that the majority are vascular in nature. This includes the four most common retinopathies and maculopathies: diabetic retinopathy and maculopathy, retinal vein occlusion, age-related macular degeneration, and central serous chorioretinopathy. The latter demonstrates circulatory disturbance in choroidal, rather than retinal circulation. Only genetic or inherited retinal disorders such as MacTel2 or retinitis pigmentosa demonstrate primary alteration of the neurosensory retina or retinal pigment epithelium. However, these conditions are followed by secondary vascular changes which may then be used for grading and monitoring.

However, there are several challenges in measuring vascular alteration in posterior segment disorders. These include (1) the dynamic character of blood supply in the posterior segment, including physiological, circadian, and pharmacologically induced changes; (2) the wide range of values for the majority of the vascular indices among healthy individuals; (3) the dependence of some indices on the axial length, which is the main cause of transverse magnification effect; (4) age-related changes in posterior segment vasculature.

If age is not considered as a cofounder, age-related changes in posterior segment vasculature make the evaluation of the majority of vascular parameters meaningless. Moreover, the age of the individual and age-related changes in retinal microcirculation seem to affect the natural course of retinal diseases. Unfortunately, in real clinical practice, it is sometimes impossible to separate age-related changes from the cumulation of subclinical comorbidities, such as undiagnosed cardiovascular disorders. This class of retinal vascular changes is termed as age-related. However, in contrast to the other causes listed above, age-related changes bear the most similarity to pathological abnormalities.

Due to the introduction of new noninvasive high-definition imaging techniques such as OCTA, most data on age-related vascular changes of the retina have been obtained quite recently. Other imaging techniques are still not able to compete with OCTA. Color fundus photography, for example, does not have a high enough resolution to enable detailed analysis. Intravenous contrast angiography cannot be applied routinely across a healthy population because ethical constraints limit the use of invasive techniques. OCT and OCTA also have the ability to visualize the choroid structure, especially the choriocapillaris that can only be observed with this method.

Vessel diameter and arteriovenous ratio

The caliber of retinal arteries and veins in newborn babies, 85.5 and 130.0 mcm, respectively, is substantially lower than in adults.[1] The caliber of retinal arteries and veins doubles over the first 6 years of life, 165.6 and 232.0 mcm, respectively, coming close to that of adults.[2] The arteriovenous ratio in children is higher than in adults due to the substantially thinner arteries. The diameter of arteries increases until 40 years of age, which is followed by a decrease in both veins and arteries. Therefore, arteriolar and venular diameters are lower in the elderly compared to middle-aged adults, independent of arterial pressure and other factors. As there is a similar reduction in the diameter of arteries and veins after 40 years of age (approximately 2 mcm every 10 years), their ratio remains stable. This is one of the reasons why the ratio may be a good indicator of pathological vascular changes. Another argument in favor of AVR as a reliable biomarker for pathological vascular changes is its independence from the transverse magnification effect. The mean diameter of retinal arteries decreases from 204.4 mcm between 43 and 54 years to 196.4 mcm between 75 and 84 years. Similarly, the mean venular

Retinal and Choroidal Vascular Diseases of the Eye. https://doi.org/10.1016/B978-0-443-15583-3.00003-2

diameter decreases from 231.2 mcm between 43 and 54 years to 221.2 mcm between 75 and 84 years. The mean AVR with color fundus photography in 61 years was 0.88.[3]

However, another study in the population ARIC within 49–73 years showed lower AVR in elderly individuals, suggesting a higher rate of age-related arteriolar narrowing.[4] Lower AVR in the elderly also was shown in computational analysis of eye fundus images, which showed a relatively stable diameter of retinal veins (111 ± 6 mcm), while arterial diameter and AVR decreased from 96 ± 6 mcm and 0.870 ± 0.046, respectively, at 2nd decade to 85 ± 4 mcm and 0.761 ± 0.044 at 7th decade[5] (Fig. 3.1).

Meanwhile, in the older population (69–97 years), no correlation was found between AVR and age,[6] indicating no effect of age-related changes on AVR after 70 years. Mean vessel diameter in the outer and inner ETDRS ring, another parameter assessed with color fundus photography, showed a negative correlation with age in both rings and in the whole macula.[7] In contrast to other approaches; however, this parameter shows no difference between arteries and veins.

Age-related changes in retinal vessels include not only changes in the diameter but also changes in the vessel wall. Therefore, the ratio of visible to the actual diameter of retinal vessels is also age sensitive. Indeed, the ratio of retinal vessel wall thickness to the lumen width (normally 0.3 when measured with SLO) shows a positive correlation with age ($r = 0.803$). This ratio increases with age from 0.23 between 20 and 40 years to 0.3 between 40 and 60 years.[8] Changes in retinal wall thickness are clearly associated with changes in vessel elasticity.

Retinal blood flow velocity

Retinal blood flow may be assessed quantitatively by measuring the intensity of erythrocyte motion in consecutive images of the eye fundus with the retinal functional imager. With this approach, hemoglobin plays the role of a natural contrast dye for retinal vessels. With the retinal functional imager, an age-related decrease in retinal vessel density, the thickness of the inner retina, and blood flow velocity were clear and showed interrelation. However, this does not lead to any conclusions about the primary role of one of those changes.[9] Among the healthy population, the mean arterial and venous blood flow velocity was 4.2 ± 0.9 mm/s and 3.3 ± 0.8 mm/s, respectively. Venous blood flow velocity decreases after 40 years (0.32 mm/s per decade, $P < .01$).[10] The analysis of wavefront parameters of blood flow showed substantial age-related changes which, taken together, reflect changes in vascular wall stiffness (including those extraocular such as the aorta). Corrected upstroke time (UTc) was a parameter of the form of the wave significantly correlated with age ($r = 0.497$, $P < .01$). Area declination was a parameter of the form of the wave negatively correlated with age ($r = -0.682$, $P < .01$).[11] Another potential cause of such changes is age-related deterioration of cardiac function, which leads to longer UTc. Analysis of blood flow velocity characteristics in the retina may reveal early changes in circulation before the structural reorganization of vessels. Interestingly, blood flow velocity may be evaluated without special equipment as was described by Grunwald. They based their simulation approach on the blue-field entoptic phenomenon and analyzed healthy

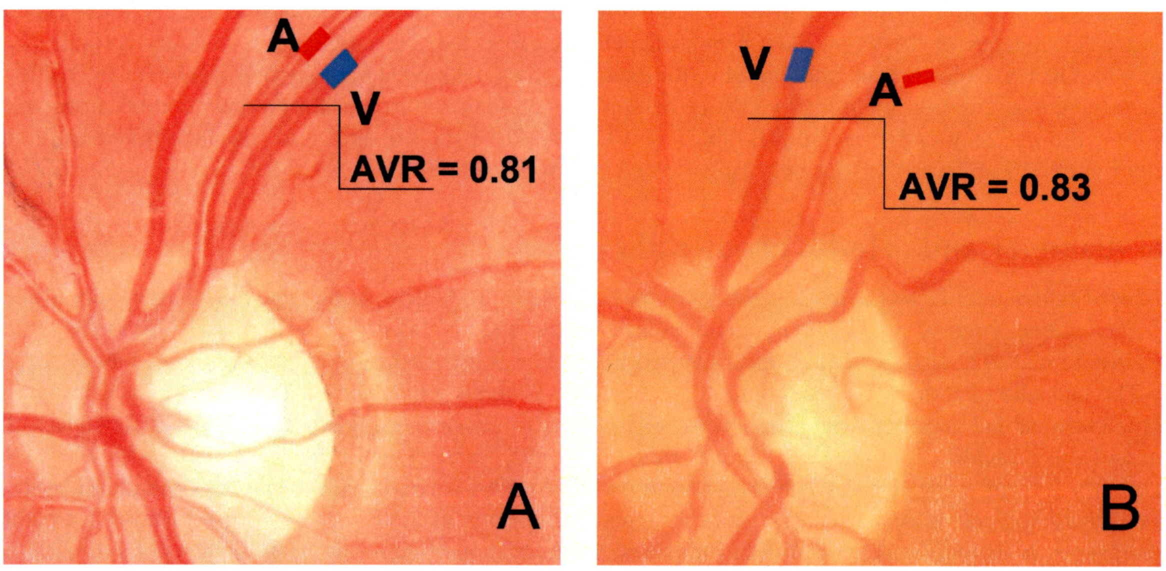

FIGURE 3.1 Arteriovenous ratio measured based on color fundus photography in 20-yo (A) and 80-yo (B) healthy individuals remains stable.

individuals at 49 ± 19 years in whom macular microcirculation was assessed. Subjects were asked to compare the velocity of particle movement in a computer model on a monitor with that observed in the entoptic phenomenon. In a series of repeated experiments, the mean velocity of leukocyte movement was calculated. There was a substantial negative correlation between blood flow velocity and age ($P = .0001$). In elderly persons (aged between 50 and 78 years), blood flow velocity was substantially lower (0.61 ± 0.21 mm/s) than in younger subjects (0.92 ± 0.12 mm/s).[12]

Fractal analysis

Fractal dimension is a global measure of the inner structure of an object, including anatomical objects such as retinal microvasculature. The study of possible models of changes in fractal dimension in young and elderly subjects (10–73 years) based on a log scale showed a nonlinear correlation between fractal dimension and age. Among three polynomial models, linear ($R^2 = 0.127$), quadratic ($R^2 = 0.154$), and cubical ($R^2 = 0.153$), the quadratic regression reflects the dynamic of the fractal dimension substantially better than the linear ($P < .001$). The decrease of fractal dimension accelerates after 50 years. However, the difference between quadratic and cubic models is not significant and is distinguishable mostly before 20 years.[13]

Fractal dimension appears to be a good classifier even between close age groups such as 60–70 and 70–80 years (AUROC = 0.802). However, between younger cohorts, this difference gets smaller. In general, the fractal dimension shows a linear decrease by age in persons from 50 to 90 years.[14] Over a longer timeline (from 3 to 97 years) and in a larger group (850 participants), changes in fractal dimension seem to be nonlinear and show two periods of fast decrease: after 20 years and after 75. This agrees with the absence of changes in fractal dimension (and tortuosity) between 40 and 70 when assessed by OCTA. However, OCTA scans cover only central 36 mm^2 while color fundus photography expands beyond this area.[15]

In both ETDRS rings, fractal dimension showed a weak but statistically significant correlation with age when assessed on fundus color photography.[7] In contrast to other parameters, fractal dimension shows noticeable age-related changes (especially in the inner ring) even in persons older than 25 years, while vessel diameter and tortuosity changes mostly after 45 years.

In fact, the decrease of fractal dimension means rarefication of the vessel pattern, due mostly to the loss of small capillaries, which might be expected under the cumulation of microvascular event risk with aging. At a young age group, the decrease of fractal dimension may be caused by the growth of the eye.

Perivascular capillary-free space

Perivascular capillary-free space is defined as a space without capillaries that can be seen around relatively large retinal vessels (up to third order) with contrast angiography, adaptive optics scanning laser ophthalmoscopy, or OCTA. This space can be observed along both veins and arteries, although around arteries it is 1.5 times wider (approximately 60 vs. 40 mcm round veins). It is considered that this phenomenon is a result of the development of retinal microvasculature at earliest stages which avoids the periarterial space due to the higher partial oxygen pressure. As partial oxygen pressure is higher around arteries than around veins (up to 94% vs. 70%), this space is wider around arteries.[16] The mean width of the capillary-free periarterial space in the elderly population (mean age 64 years) is substantially higher than in the young (mean age 26 years), 78.8 ± 12.3 μm and 67.2 ± 12.8 μm, respectively ($P = .03$) (Fig. 3.2).

The perivenular capillary-free space shows an even higher difference between the elderly and the young, 61.5 ± 11.3 μm and 42.7 ± 5.17 μm, respectively ($P < .001$). Although this parameter may affect vessel density, its relationship with other vascular metrics is not known, and neither has any correlation with FAZ parameters been established.[16] The reason for the expansion of the perivascular capillary-free space is not known. However, the loss of perivascular capillaries may be related to the decrease in oxygen requirements of retinal tissues, secondary to age-related neurodegeneration. However, as OCTA and other imaging modalities show only blood flow but not retinal vessel walls, the thickness of retinal vessel walls may also contribute to the perivascular capillary-free space. In such cases, widening of perivascular capillary-free space (especially that around arteries) may be related to age-related thickening of the walls of the retinal vessels.

Vessel tortuosity

Tortuosity of retinal vessels is defined as the ratio of the length of the vessel between two points to the shortest distance between those points.

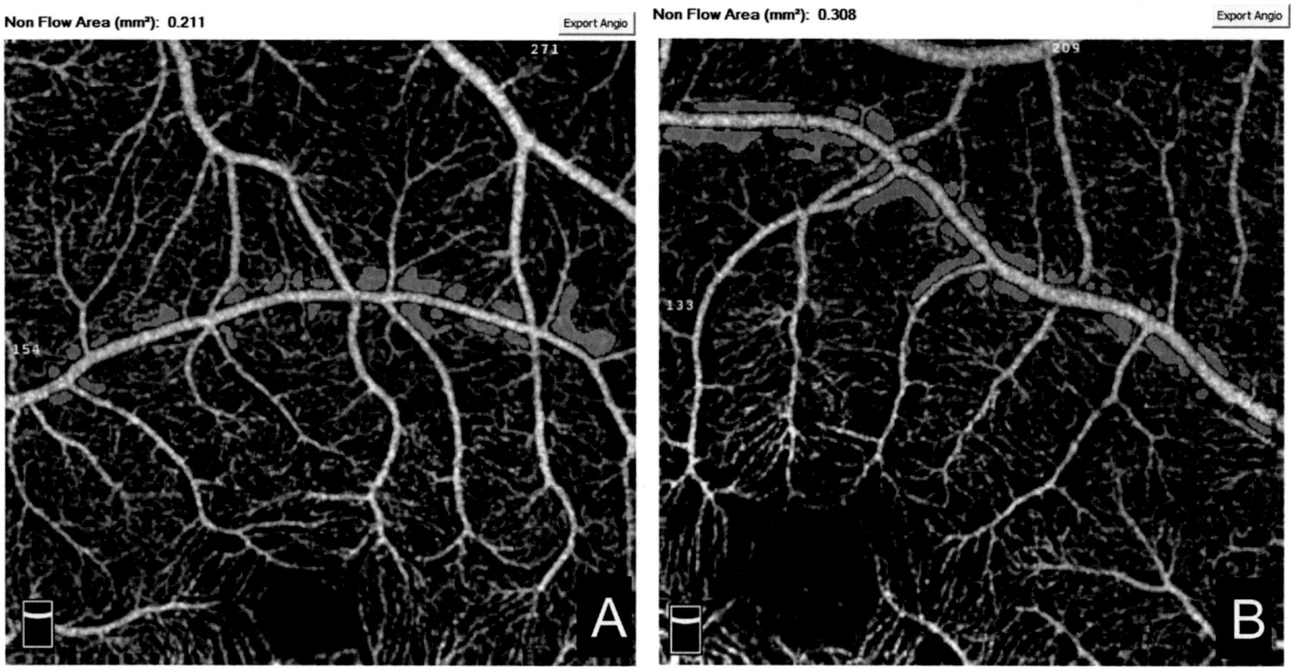

FIGURE 3.2 Periarterial capillary-free space in 19-yo (A) and 67-yo (B) healthy individuals is 0.211 and 0.308 mm^2, respectively. The length of the vessel in both cases is very close 3.2 and 3.4 mm, respectively.

Based on the color fundus photography, the tortuosity of retinal vessels does not show substantial change with aging.[7] However, between subgroups of 26—35 years and older than 46 years some increase of vessel tortuosity can be observed in the outer ETDRS ring, though not statistically significant. Cheung in a large cohort between 40 and 80 years showed a decrease of arteriolar and venular tortuosity with aging.[16] These facts taken together with the changes in tortuosity in numerous retinal vascular disorders lead to the conclusion that vessel tortuosity is appropriate for the estimation of pathological but not age-related changes. However, a decrease in vessel tortuosity with aging could be expected as we know the stiffness of vessel walls increases with aging. At the same time, the decrease of vessel stiffness (even the relative one during the increase of blood pressure) leads to increased tortuosity.

Vessel density

The ratio between vessel area to the entire area of the region analyzed based on color fundus photography showed a substantial correlation with age; however, it is most noticeable in the outer ring compared to the inner and only in the older population (after 35 years).[7]

The SCP and DCP vessel densities were significantly lower in the older age groups. This was noticeable for the whole image, parafovea, and perifovea. The decrease was more prominent after 50 years. Similarly, vessel density within the optic disc decreases with age.[17] In a large population study, higher vessel density in both capillary plexuses was also significantly associated with younger age. It has also been shown that vessel density was stable, or even slightly increased with aging, until age 50, at which point it started to decrease progressively at a rate of approximately 1.3% per decade, based on OCTA data obtained with RTVue-XR.[18] In another study, correlation with age remained statistically significant even at a relatively short timeline from 17 to 64 years.[19]

Different approaches can be used for the estimation of vascular density: perfusion density or skeletonized vessel density, but both show similar trends with aging. Also, vessel density shows similar trends in different segments of the ETDRS map (although the density decreases in the SITN sequence) from 11 to 60 years. In this cohort, skeletonized vessel density increases not significantly until 25 years and then progressively decreases until 60 years. Interestingly, perfusion density between 50 and 60 years remains stable in contrast to skeletonized vessel density[20] (Fig. 3.3).

Analysis of three separate vascular plexuses (when the deep vascular complex divided in the intermediate and deep capillary plexuses) in a cohort between 22 and 76 years, vessel density decreases significantly in all plexuses and segments of the macula. A mean decrease per year in a 9 mm^2 central region was 0.064% in SVP, 0.086% in DVC, 0.055 in ICP%,

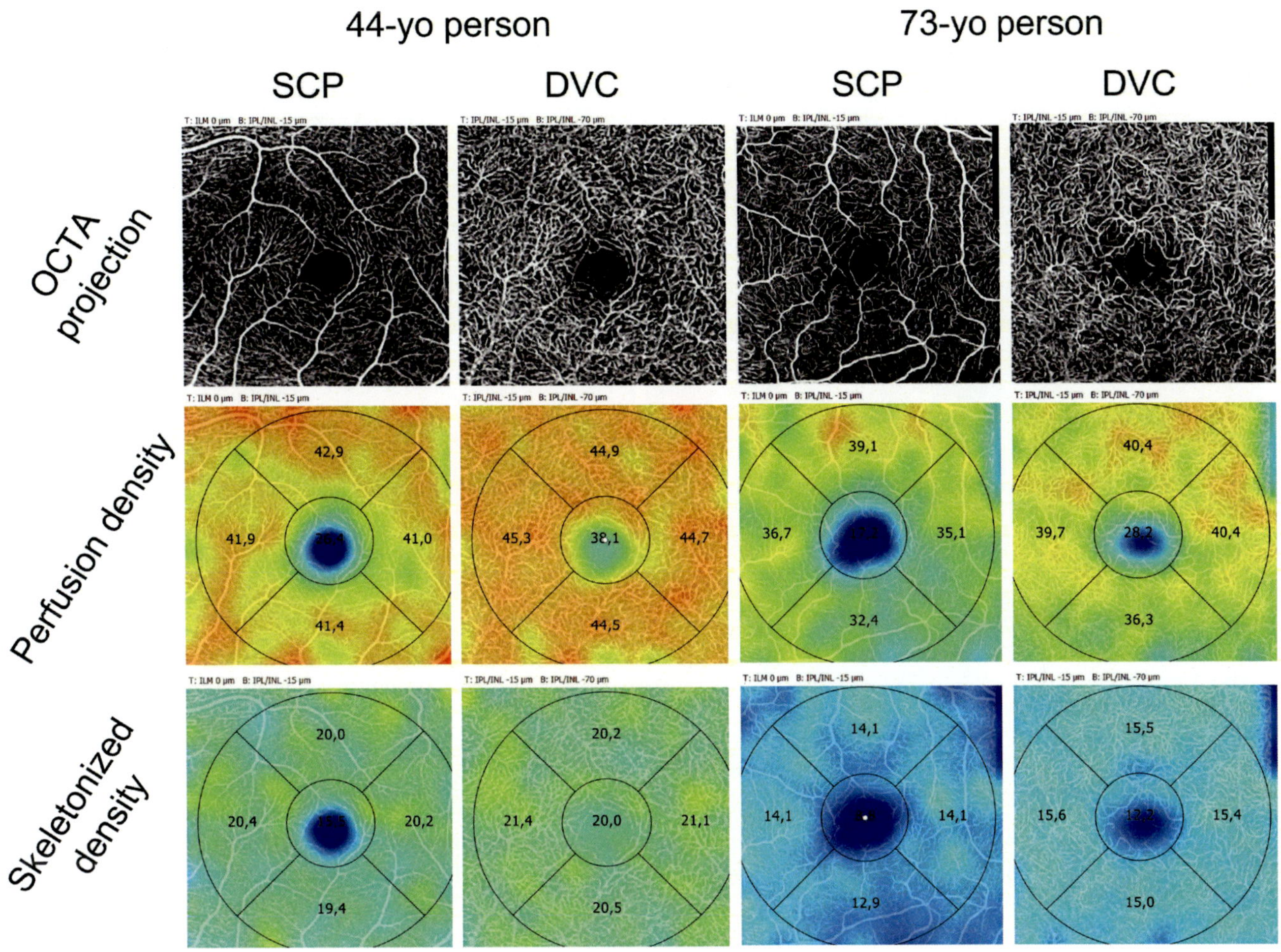

FIGURE 3.3 Age-related difference of vessel density based on perfusion and skeletonized vessel density analysis.

and 0.076% in DCP. Therefore, DCP vessel density shows the most significant decrease with aging ($r = -0.46$) between 20 and 80 years, although it is most noticeable after 30 years. In general, the most notable trend in all plexuses appears after 40 years.[21]

Foveal avascular zone

The correlation of the foveal avascular zone (FAZ) area with age was not statistically significant at the relatively short timeline from 17 to 64 years.[22] Another study showed an increase in the FAZ area in a cohort between 11 and 60 years, from 0.18 to 0.3 mm^2 before 20 years, flowed by a further increase of FAZ area to 0.46 mm^2 until 60 years. Therefore, the increase of FAZ area seems to have a stable but nonlinear character.[20] The perimeter as well as the area of FAZ correlates with age. The mean annual increase of FAZ is about 0.003 mm^2 ($P < .001$). Interestingly, the acircularity index does not correlate with age.[21,23] In general, the FAZ metric suffers from different sources of variability including a wide normal range and age-related changes. The problem of normalization of the FAZ area may be solved through the personalization of this metric. This improved approach for the evaluation of the FAZ area includes measuring the ratio between the FAZ area and the area of the INL-free zone, both of which normally have a high correlation. This ratio has a high potential to compensate for the normal wide range of FAZ area, and there is no doubt that this ratio will remain stable over aging.[24] Several papers have noted that the so-called superficial FAZ does not change with aging while deep FAZ at the level of DVC decreases with time.[25] For FAZ area as a parameter relating to the axial length, the latter should be included in studies of FAZ to explain some differences between reports.[23] In general, the majority of papers indicated an increase of FAZ area with age[26,27] (Fig. 3.4).

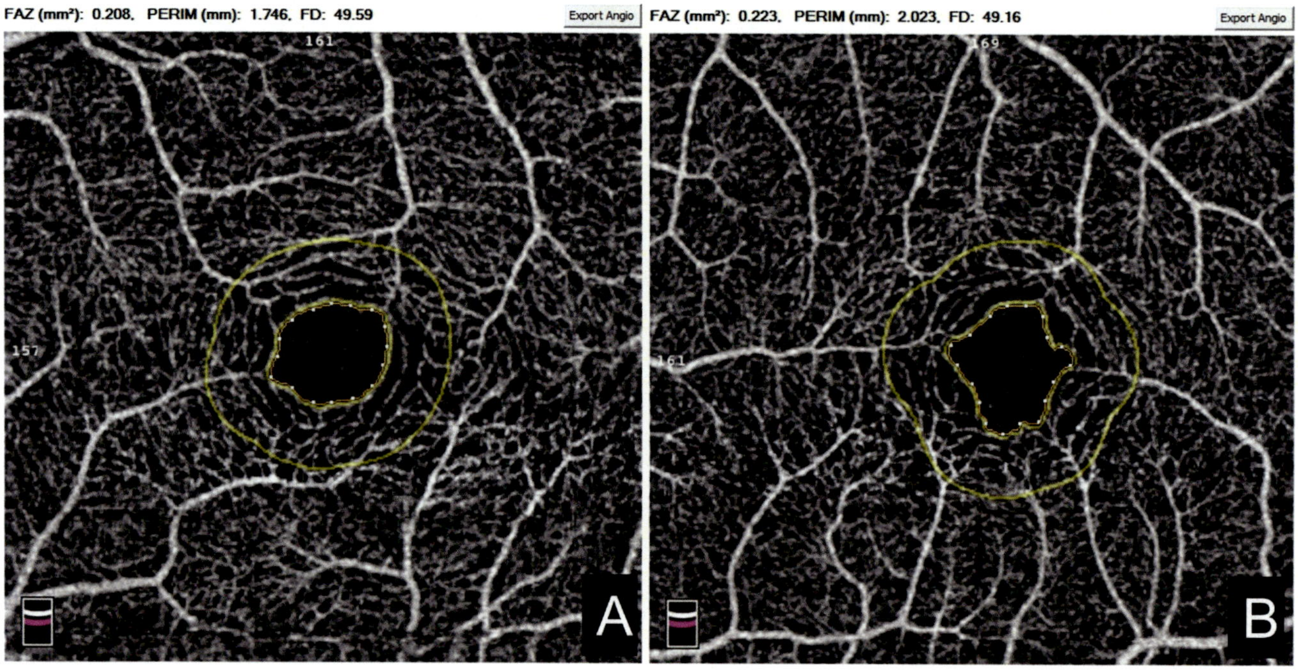

FIGURE 3.4 Representative examples of FAZ metric in 22-yo (A) and 59-yo (B) healthy persons show enlargement of FAZ area (0.208 vs. 0.223 mm^2) and perimeter (1.746 vs. 2.023 mm).

Choriocapillaris

Based on histopathological studies the density and diameter of choriocapillaries linearly decrease over the entire life with a trend toward faster loss of choriocapillaris density after 7th decade. However, persons of older age do not necessarily have a low choriocapillaris density. The main feature of age-related changes in choriocapillaris density is the increased variability of this index. Therefore, in the older age group, some individuals may have choriocapillaris density similar to that of young individuals while others may have substantially lower. The choriocapillaris thins from 9.8 mcm in the 1st decade to 6.5 mcm in the 10th decade.[28] As choriocapillaris density is closely related to the diameter of individual capillaries ($r = 0.64$), age-related changes may simultaneously affect both parameters. This histopathological study also reported age-dependent widening of intercapillary space. The widening of intercapillary space may be partially explained by the thickening of capillary walls (which in the study thickened from 25% in young individuals (four persons aged 21–50 years) to 60% in older people (four persons aged 80–97 years)).

Apart from histopathological studies only OCTA was able to provide quantitative information on the status of the choriocapillaris. Based on OCTA, blood flow deficit increases in a linear fashion from the 3rd to the 9th decade when assessed with a 3-mm OCTA scan. This trend is less remarkable after 50 years if a 6-mm OCTA scan is used for analysis. This difference may result from a better preservation of perfusion in the choriocapillaris outside the perifovea. A comparison of blood flow deficit between the fovea and perifovea also indicates the dominance of foveal changes over a lifetime. In the older age group, there was a substantial variation of normal values rather than merely a general decrease in perfusion.[29]

OCTA with scan averaging technology showed the total flow voids area with the mean size of an individual flow void 0.99 ± 0.20 mm^2 and 567.8 ± 201.5 mcm^2, respectively. Both these indices positively correlated with age, $P = .002$, $R = 0.336$ and $P = .026$, $R = 0.247$, respectively. The annual increase of the total flow void area and the mean size of an individual flow void was 4.20×10^{-3} mm^2 and 3.07 μm^2, respectively.[30] The link between aging and the appearance of large voids was supported by the distribution of voids with the power law[31] Fig. 3.5.

Correlation of the choriocapillaris perfusion area with age was not statistically significant at a relatively short timeline from 17 to 64 years.[22] However, variation of perfusion area increases in older cohort and healthy eyes with low perfusion area are more common in that population Fig. 3.6.

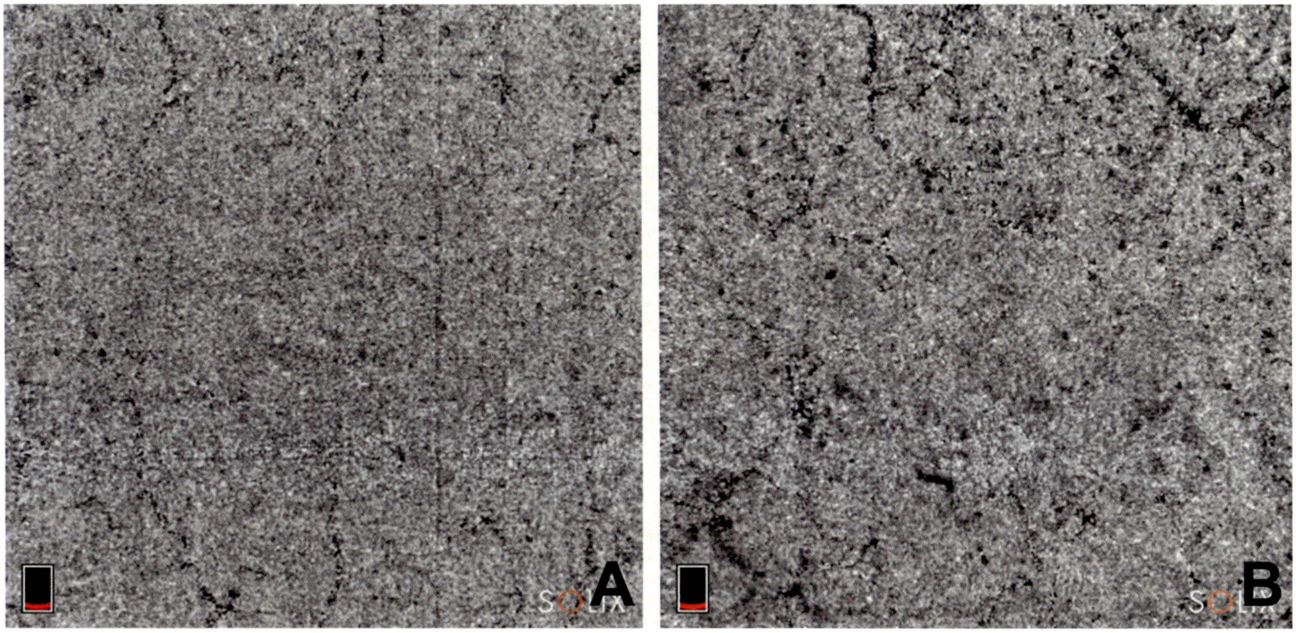

FIGURE 3.5 Status of choriocapillaris perfusion in 21-yo (A) and 59-yo (B) healthy individuals based on image averaging technology (averaging of four 3-mm scans). In the older individual, numerous flow voids are visible.

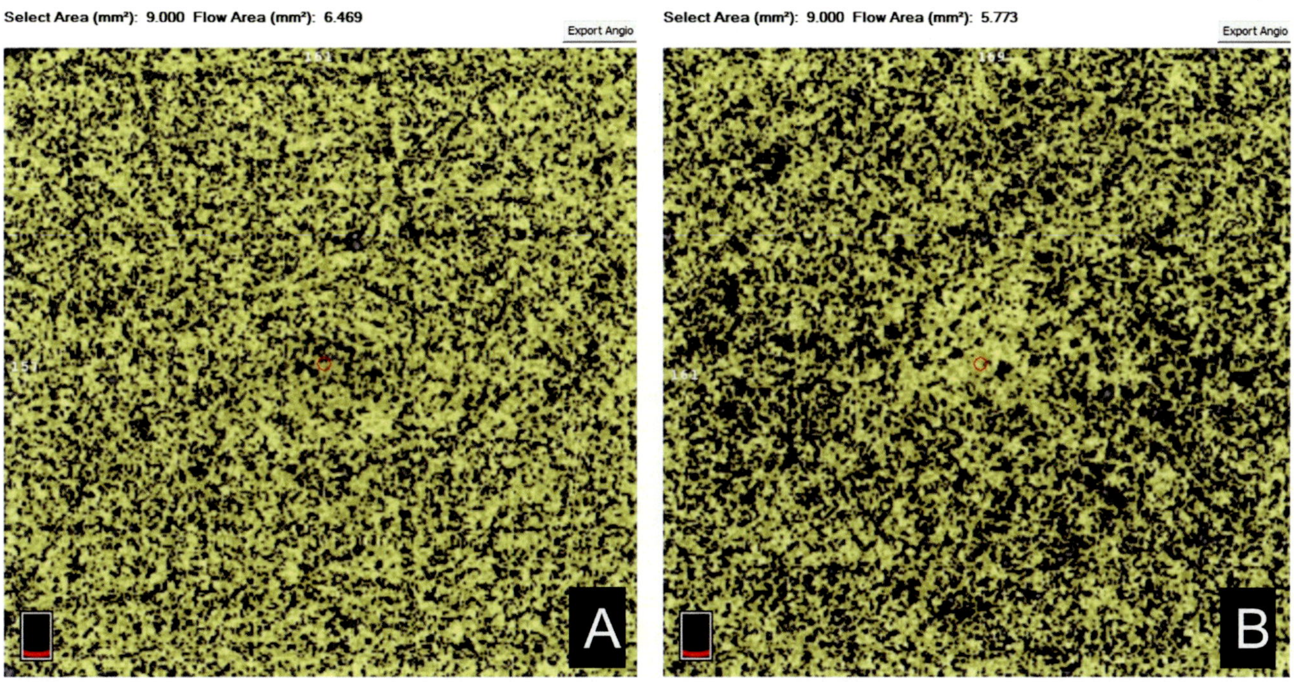

FIGURE 3.6 Choriocapillaris perfusion area measured based on OCTA projection of choriocapillaris slab in 24-yo (A) and 59-yo (B) healthy individuals reveal substantial age-related changes (6.47 vs. 5.77 mm^2).

Choroid

Based on histopathological studies choroidal thickness decreases with age from 193.5 to 84.5 mcm. However significant interindividual differences exist in the older age group.[28] Not only choroidal thickness but choroidal volume, and vascular density decrease with age.[32,33] However, the correlation with age for those indices is moderate with thickness up to −0.53, volume −0.39, and vascular density −0.39.[9,32] Total choroidal area, luminal area, and stromal area decrease with each decade from 20 to 79 years. This decrease can be observed at various distances from the fovea but most notably at eccentric points. Moreover, all these indices decreased significantly after the 5th decade. Moreover, a notable difference is observed in choroidal thickness between the 3rd decade and the following decades. However, between the 6th and 8th decades, this difference is not significant. The maximum difference is therefore observed between the 3rd and 4th and between the 5th and 6th decades. Choroidal vascularity index (CVI) also demonstrates a decrease with age, although in a nonlinear fashion. The most notable decrease was noted after the 5th decade. As with choroidal thickness, peripheral changes in CVI dominate over central changes though they are not statistically significant. All choroidal indices are characterized by an age-related increase in variability.[32] As other studies found that the stromal area remains stable, they concluded that changes in the luminal area were the leading cause of the age-related decrease of CVI.[34] This may explain the weak trend toward a decrease of CVI in other studies which found a decrease in both the stromal and luminal areas. This may also result from the use of auto-segmentation algorithms for choroidal structures and from the analysis of the entire wide-field scan (12 mm), not only in a central B-scan.[33] In general, the volume of the choroid in the macula decreases every decade by 0.547 mm^3.[35] Analysis of choroidal perfusion with laser Doppler flowmetry showed a significant correlation between choroidal blood volume and age ($R = -0.52$, $P = .004$) and between choroidal blood flow and age ($R = -0.54$, $P = .003$) in the foveal region. Choroidal blood flow velocity did not change over age.[12] Age-related decrease of subfoveal choroidal thickness in eyes with refractive error less than 6 D can be described by formula 444/3 × Age.[36] Age-related changes in the ETDRS grid are not homogenous with a slower trend being observed in the nasal quadrant, a faster in the superior quadrant, and the fastest in the temporal and inferior quadrants.[36] Age-related change in the subfoveal choroidal thickness occurs mostly through the thinning of the Sattler layer (choriocapillaris and true Sattler layer), which shows a substantially stronger correlation with age than does the thickness of the large vessels layer ($R^2 = 0.29$ and $R^2 = 0.056$, respectively)[36] Fig. 3.7.

Retinal perfusion

Retinal perfusion is a ratio of retinal blood flow to retinal tissue volume in the same area (usually, central 20°). After 35, the age is negatively associated with retinal perfusion in the macula ($r = -0.26$, $P = .02$); however, on a longer timeline, some increase in perfusion is noticeable until 42 years followed by a progressive decrease. As age-related changes in volumetric characteristics of the retinal tissue are not significant in healthy eyes, the dynamic of retinal blood flow is the factor most responsible for the decreased retinal perfusion.[37]

Another approach to the analysis of blood flow in the posterior eye segment is so-called laser speckle flowgraph in the optic disc region. Mean blur rate as a general characteristic of the blood flow intensity showed a negative correlation with

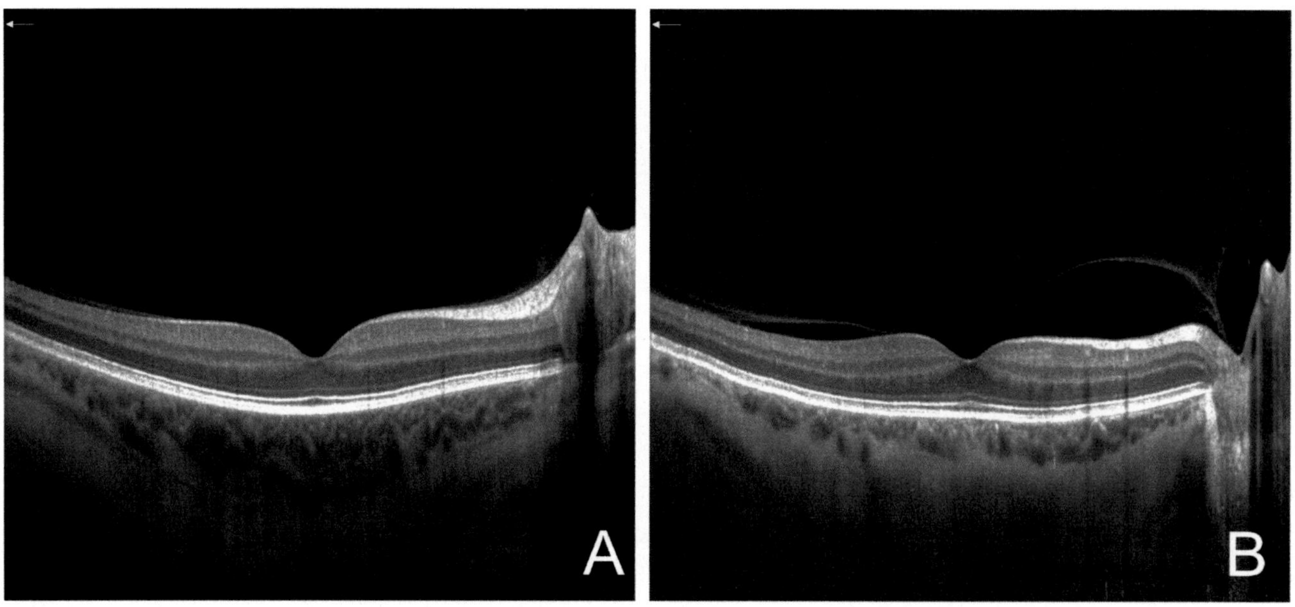

FIGURE 3.7 Enhanced depth OCT imaging of choroid in eyes with a similar axial length of 23.5 mm show thicker choroid in a 21-yo (A) compared to 59-yo (B) healthy individual with no apparent difference in choroidal vascularity.

age for the whole optic disc area and separately for blood vessels and the optic disc. The majority of pulse wave characteristics for the exclusion of RR demonstrated a significant association with age. The strongest correlation was found with the blowout time and failing rate. The correlation with those indices describes age-related changes as related to the decrease of the blood filling, the blood filling the tissues for a shorter period with a faster release.[38,39]

Miscellaneous

Spontaneous venous pulsation on the optic disc is a frequently observed phenomenon in healthy populations (up to 90%) from 20 to 90 years, and this phenomenon is not affected by age ($P = .57$ between young and old population).[40] Retinal vascular reactivity is assessed as changes in retinal vessel diameter and blood flow velocity changes significantly with age (from 22 to 66 years). The rate of reactive arteriolar narrowing also significantly decreases with age ($P = .045$) by 0.16 mcm annually.[41] Similarly, aging leads to a decrease of arterial reactivity to hypercapnia with the choroid remaining unchanged.[42] Small resolved paracentral acute middle maculopathy lesions were described in relation to arterial hypertension, cardiovascular diseases, and retinal vein occlusions.[43–45] Despite the fact that these lesions are tightly related to vascular disorders, they can be observed in a proportion of healthy populations, and this proportion increases with age in the absence of known cardiovascular comorbidity.[45] At the same time, the rate of the occurrence of new lesions appears to be not very high even in high-risk individuals, such as those with retinal vein occlusions. In the healthy eyes of those patients over a mean period of 2 years, no more than 1/3 demonstrated new lesions. In accordance with the results of Doppler fluorometry the blood flow velocity in the optic disc increases with age, although these changes are masked by the general reduction of the vessel density. These changes have a symmetrical pattern that equally affects nasal and temporal segments of the optic disc.[46] Age-related changes of the posterior segment perfusion are also characterized by a significant decrease of the blood flow at night in both the macula and optic disc. These circadian and nocturnal changes in ocular blood flow are not dependent on changes in ocular perfusion pressure.[47]

References

1. Kandasamy Y, Smith R, Wright IMR. Retinal microvasculature measurements in full-term newborn infants. *Microvascular Research*. 2011;82(3):381–384. https://doi.org/10.1016/j.mvr.2011.07.011.
2. Mitchell P, Liew G, Rochtchina E, et al. Evidence of arteriolar narrowing in low-birth-weight children. *Circulation*. 2008;118(5):518–524. https://doi.org/10.1161/CIRCULATIONAHA.107.747329.
3. Wong TY, Klein R, Klein BEK, Meuer SM, Hubbard LD. Retinal vessel diameters and their associations with age and blood pressure. *Investigative Opthalmology & Visual Science*. 2003;44(11):4644. https://doi.org/10.1167/iovs.03-0079.
4. Klein R, Sharrett AR, Klein BE, et al. Are retinal arteriolar abnormalities related to atherosclerosis?: the atherosclerosis risk in communities study. *Arteriosclerosis, Thrombosis, and Vascular Biology*. 2000;20(6):1644–1650. https://doi.org/10.1161/01.atv.20.6.1644.
5. Pose-Reino A, Gomez-Ulla F, Hayik B, et al. Computerized measurement of retinal blood vessel calibre: description, validation and use to determine the influence of ageing and hypertension. *Journal of Hypertension*. 2005;23(4):843–850. https://doi.org/10.1097/01.hjh.0000163154.35577.8e.
6. Wong TY, Klein R, Sharrett AR, et al. The prevalence and risk factors of retinal microvascular abnormalities in older persons: the cardiovascular health study. *Ophthalmology*. 2003;110(4):658–666. https://doi.org/10.1016/S0161-6420(02)01931-0.
7. Feng Z, Wang G, Xia H, et al. Macular vascular geometry changes with sex and age in healthy subjects: a fundus photography study. *Frontiers in Medicine*. 2021;8:778346. https://doi.org/10.3389/fmed.2021.778346.
8. Meixner E, Michelson G. Measurement of retinal wall-to-lumen ratio by adaptive optics retinal camera: a clinical research. *Graefe's Archive for Clinical and Experimental Ophthalmology = Albrecht Von Graefes Archiv Fur Klinische Und Experimentelle Ophthalmologie*. 2015;253(11):1985–1995. https://doi.org/10.1007/s00417-015-3115-y.
9. Wei Y, Jiang H, Shi Y, et al. Age-related alterations in the retinal microvasculature, microcirculation, and microstructure. *Investigative Ophthalmology & Visual Science*. 2017;58(9):3804–3817. https://doi.org/10.1167/iovs.17-21460.
10. Burgansky-Eliash Z, Lowenstein A, Neuderfer M, et al. The correlation between retinal blood flow velocity measured by the retinal function imager and various physiological parameters. *Ophthalmic Surgery, Lasers & Imaging Retina*. 2013;44(1):51–58. https://doi.org/10.3928/23258160-20121221-13.
11. Takizawa Y, Song Y, Tani T, et al. Retinal blood velocity waveform characteristics with aging and arterial stiffening in hypertensive and normotensive subjects. *Translational Vision Science & Technology*. 2021;10(13):25. https://doi.org/10.1167/tvst.10.13.25.
12. Grunwald JE, Piltz J, Patel N, Bose S, Riva CE. Effect of aging on retinal macular microcirculation: a blue field simulation study. *Investigative Ophthalmology & Visual Science*. 1993;34(13):3609–3613.
13. Che Azemin MZ, Ab Hamid F, Aminuddin A, Wang JJ, Kawasaki R, Kumar DK. Age-related rarefaction in retinal vasculature is not linear. *Experimental Eye Research*. 2013;116:355–358.
14. Azemin MZC, Kumar DK, Wong TY, et al. Age-related rarefaction in the fractal dimension of retinal vessel. *Neurobiology of Aging*. 2012;33(1):194. https://doi.org/10.1016/j.neurobiolaging.2010.04.010.

15. Song Y, Cheng W, Li F, et al. Ocular factors of fractal dimension and blood vessel tortuosity derived from OCTA in a healthy Chinese population. *Translational Vision Science & Technology*. 2022;11(5):1. https://doi.org/10.1167/tvst.11.5.1.
16. Arthur E, Alber J, Thompson LI, Sinoff S, Snyder PJ. OCTA reveals remodeling of the peripheral capillary free zones in normal aging. *Scientific Reports*. 2021;11(1):15593. https://doi.org/10.1038/s41598-021-95230-0.
17. Cheung CY-L, Zheng Y, Hsu W, et al. Retinal vascular tortuosity, blood pressure, and cardiovascular risk factors. *Ophthalmology*. 2011;118(5):812−818. https://doi.org/10.1016/j.ophtha.2010.08.045.
18. Abay RN, Akdeniz GŞ, Katipoğlu Z, Kerimoğlu H. Normative data assessment of age-related changes in macular and optic nerve head vessel density using optical coherence tomography angiography. *Photodiagnosis and Photodynamic Therapy*. 2022;37:102624. https://doi.org/10.1016/j.pdpdt.2021.102624.
19. You QS, Chan JCH, Ng ALK, et al. Macular vessel density measured with optical coherence tomography angiography and its associations in a large population-based study. *Investigative Opthalmology & Visual Science*. 2019;60(14):4830. https://doi.org/10.1167/iovs.19-28137.
20. Pujari A, Chawla R, Markan A, et al. Age-related changes in macular vessels and their perfusion densities on optical coherence tomography angiography. *Indian Journal of Ophthalmology*. 2020;68(3):494. https://doi.org/10.4103/ijo.IJO_521_19.
21. Lavia C, Bonnin S, Maule M, Erginay A, Tadayoni R, Gaudric A. Vessel density of superficial, intermediate, and deep capillary plexuses using optical coherence tomography angiography. *Retina*. 2019;39(2):247−258. https://doi.org/10.1097/IAE.0000000000002413.
22. Wang Q, Chan S, Yang JY, et al. Vascular density in retina and choriocapillaris as measured by optical coherence tomography angiography. *American Journal of Ophthalmology*. 2016;168:95−109. https://doi.org/10.1016/j.ajo.2016.05.005.
23. Linderman R, Salmon AE, Strampe M, Russillo M, Khan J, Carroll J. Assessing the accuracy of foveal avascular zone measurements using optical coherence tomography angiography: segmentation and scaling. *Translational Vision Science & Technology*. 2017;6(3):16. https://doi.org/10.1167/tvst.6.3.16.
24. Kulikov AN, Maltsev DS, Burnasheva MA. Improved analysis of foveal avascular zone area with optical coherence tomography angiography. *Graefe's Archive for Clinical and Experimental Ophthalmology*. 2018;256(12):2293−2299. https://doi.org/10.1007/s00417-018-4139-x.
25. Alshareef RA, Al-khersan H, Darwich R, et al. Measurement of normative foveal avascular zone parameters in healthy adults using optical coherence tomography angiography. *Journal of VitreoRetinal Diseases*. 2018;2(4):213−218. https://doi.org/10.1177/2474126418778492.
26. Gómez-Ulla F, Cutrin P, Santos P, et al. Age and gender influence on foveal avascular zone in healthy eyes. *Experimental Eye Research*. 2019;189:107856. https://doi.org/10.1016/j.exer.2019.107856.
27. Kaya M, Ayhan Z, Ozturk AT, Kocak N, Arikan G, Kaynak S. Evaluation of the macular and choroidal perfusion in healthy Turkish population using optical coherence tomography angiography. *Korean Journal of Ophthalmology*. 2021;35(5):360−367. https://doi.org/10.3341/kjo.2020.0122.
28. Ramrattan RS, van der Schaft TL, Mooy CM, de Bruijn WC, Mulder PG, de Jong PT. Morphometric analysis of Bruch's membrane, the choriocapillaris, and the choroid in aging. *Investigative Ophthalmology & Visual Science*. 1994;35(6):2857−2864.
29. Zheng F, Zhang Q, Shi Y, et al. Age-dependent changes in the macular choriocapillaris of normal eyes imaged with swept-source optical coherence tomography angiography. *American Journal of Ophthalmology*. 2019;200:110−122. https://doi.org/10.1016/j.ajo.2018.12.025.
30. Ichioka A, Ooto S, Uji A, et al. Correlations among age, sex, axial length, and subfoveal choroidal thickness in the choriocapillaris structure analyzed using multiple en face image averaging optical coherence tomography angiography. *PLoS One*. 2021;16(11):e0259880. https://doi.org/10.1371/journal.pone.0259880.
31. Spaide RF. Choriocapillaris flow features follow a power law distribution: implications for characterization and mechanisms of disease progression. *American Journal of Ophthalmology*. 2016;170:58−67. https://doi.org/10.1016/j.ajo.2016.07.023.
32. Nivison-Smith L, Khandelwal N, Tong J, Mahajan S, Kalloniatis M, Agrawal R. Normal aging changes in the choroidal angioarchitecture of the macula. *Scientific Reports*. 2020;10(1):10810. https://doi.org/10.1038/s41598-020-67829-2.
33. Zhou H, Dai Y, Shi Y, et al. Age-related changes in choroidal thickness and the volume of vessels and stroma using swept-source OCT and fully automated algorithms. *Ophthalmology Retina*. 2020;4(2):204−215. https://doi.org/10.1016/j.oret.2019.09.012.
34. Koçak N, Subaşõ M, Yeter V. Effects of age and binarising area on choroidal vascularity index in healthy eyes: an optical coherence tomography study. *International Ophthalmology*. 2021;41(3):825−834. https://doi.org/10.1007/s10792-020-01636-6.
35. Barteselli G, Kozak I, El-Emam S, Chhablani J, Cortes MA, Freeman WR. 12-month results of the standardised combination therapy for diabetic macular oedema: intravitreal bevacizumab and navigated retinal photocoagulation. *The British Journal of Ophthalmology*. 2014;98(8):1036−1041. https://doi.org/10.1136/bjophthalmol-2013-304488.
36. Wakatsuki Y, Shinojima A, Kawamura A, Yuzawa M. Correlation of aging and segmental choroidal thickness measurement using swept source optical coherence tomography in healthy eyes. *PLoS One*. 2015;10(12):e0144156. https://doi.org/10.1371/journal.pone.0144156.
37. Lin Y, Jiang H, Liu Y, et al. Age-related alterations in retinal tissue perfusion and volumetric vessel density. *Investigative Ophthalmology & Visual Science*. 2019;60(2):685−693. https://doi.org/10.1167/iovs.18-25864.
38. Luft N, Wozniak PA, Aschinger GC, et al. Ocular blood flow measurements in healthy white subjects using laser speckle flowgraphy. *PLoS One*. 2016;11(12):e0168190. https://doi.org/10.1371/journal.pone.0168190.
39. Liu C, Kobayashi T, Shiba T, Hayashi N, Piknova B. Effects of aging and exercise habits on blood flow profile of the ocular circulation. *PLoS One*. 2022;17(4):e0266684. https://doi.org/10.1371/journal.pone.0266684.
40. Harder B, Jonas JB. Frequency of spontaneous pulsations of the central retinal vein. *The British Journal of Ophthalmology*. 2007;91(3):401−402. https://doi.org/10.1136/bjo.2006.103341.
41. Rose PA, Hudson C. The impact of age on retinal vascular reactivity and the relationship with cardiovascular risk factors. *ARVO Annual Meeting Abstract*. 2007;48:2266.

42. Miyaji A, Ikemura T, Hayashi N. Aging decreases CO2 reactivity in the retinal artery, but not in the ocular choroidal vessels; a cross-sectional study. *Clinical Hemorheology and Microcirculation*. 2018;70(2):231–240. https://doi.org/10.3233/CH-170332.
43. Burnasheva MA, Maltsev DS, Kulikov AN, Sherbakova KA, Barsukov AV. Association of chronic paracentral acute middle maculopathy lesions with hypertension. *Ophthalmology Retina*. 2020;4(5):504–509. https://doi.org/10.1016/j.oret.2019.12.001.
44. Long CP, Chan AX, Bakhoum CY, et al. Prevalence of subclinical retinal ischemia in patients with cardiovascular disease—a hypothesis driven study. *EClinicalMedicine*. 2021;33:100775. https://doi.org/10.1016/j.eclinm.2021.100775.
45. Maltsev DS, Kulikov AN, Burnasheva MA, Chhablani J. Prevalence of resolved paracentral acute middle maculopathy lesions in fellow eyes of patients with unilateral retinal vein occlusion. *Acta Ophthalmologica*. 2020;98(1). https://doi.org/10.1111/aos.14196.
46. Boehm AG, Koeller AU, Pillunat LE. The effect of age on optic nerve head blood flow. *Investigative Opthalmology & Visual Science*. 2005;46(4):1291. https://doi.org/10.1167/iovs.04-0987.
47. Kida T, Liu JHK, Weinreb RN. Effect of aging on nocturnal blood flow in the optic nerve head and macula in healthy human eyes. *Journal of Glaucoma*. 2008;17(5):366–371. https://doi.org/10.1097/IJG.0b013e31815d7838.

Section II

Imaging

Chapter 4

Fundus photography

Xinyuan Zhang[1,2]
[1]*Beijing Tongren Eye Center, Tongren Hospital, Capital Medical University, Beijing, China;* [2]*Beijing Retinal and Choroidal Vascular Disorders Study Group, Beijing, China*

Introduction

Color fundus photography (CFP) has been used in clinical practice for several decades to visually observe the pathophysiological changes of the fundus. Of all the diagnostic imaging modalities, the most basic tool is CFP. Fundus cameras take high-resolution images of the fundus using human eyes' dark adaption and optical imaging capabilities. CFP provides a color or red-free image of the retina with high-resolution (primarily digital) with a wide field, making it a valuable tool to document normal structure and retinal pathology. Recent advances in fundus imaging are advantageous to reduce preventable visual morbidity through timely fundus screening. Digital CFP has evolved into a vital tool for the physical examination and screening of patients, particularly in association with telemedicine and artificial intelligence.

Fundus photography and its invention history

CFP refers to capturing the main structures, which can be visualized in the fundus, including the central area, peripheral retina, optic disc, and macula using fundus cameras consisting of a microscope attached to a flash-enabled camera. The CFP has progressed rapidly over the last decades. In 1839, Daguerre introduced photography using a silver iodide wet plate, laying a foundation for modern CFP. In 1851, Hermann von Helmholtz invented an inspection glass, as a prototype of a modern ophthalmoscope. The early CFP encountered several technical issues that need to be resolved, including insufficient lighting, limited capture field, long exposure, and significant cornel reflexes, which reduced the clarity and details of fundus images. In 1926, Stockholm's Johan Nordenson and Zeiss-Jena Corporation released a commercial device, which was the first modern fundus camera for clinical practitioners. In 1955, Littman invented a flashlight, which could take realistic and clear fundus photographs. Based on the rapid progress of fundus cameras, in 1965, David Alvis and Harold Novotony proposed retinal angiography using the Zeiss fundus camera and digital electronic flash. Meanwhile, nonmydriatic imaging (infrared microlight television technique), electronic illumination control, automatic alignment, and high-resolution digital image capture techniques have been drastically applied in clinical practice. In the 1990s, Japanese researchers applied computer graphic technology to manipulate fundus images, and they also created a stereoscope with (or without) dilated pupils that can extract three-dimensional (3D) stereoscopic fundus images via a charged-couple device (CCD) recording system.

Fundus cameras consist of photographic and observational targeting capabilities (a video recording device for fundus video cameras). The illumination component directs light from the light source to illuminate the fundus; the observation and targeting components search for and observe wards; and the video recording component captures fundus images (or saves them to the computer). Almost all fundoscopic instruments that use ordinary light have imaging optics based on the fundoscopic cameras successfully developed by Zeiss in 1925. Its most notable feature is coaxial illumination. That is, the illumination and recording system employs a coaxial illumination system, in which the light source from the concentrating mirror is imaged at the ring diaphragm and is then partially focused by the projection lens and the edge of the receiving objective lens's long working distance, forming a ring-shaped illumination beam on the fundus through the pupil of the examined eye. The light reflected by the illuminated part of the fundus travels in a parallel pathway through the refractive media of the eye, ejecting from the pupil's center. Light is split into two paths after passing through the eyepiece objective and projective objective: one into the eyepiece and then the eye, and the other into imaging on the detector

Retinal and Choroidal Vascular Diseases of the Eye. https://doi.org/10.1016/B978-0-443-15583-3.00004-4

(e.g., photographic negative, CCD, etc.). The light source and the capture fields (from 35° to montage 220–240°) have been significantly improved.

Types of fundus cameras

Fundus cameras can be distinguished by the light source for illumination as follows:

Halogen light source: Pupil dilation is mainly required. A xenon lamp is added to the lighting system to empower the flash while taking pictures with insufficient light for CFP.

Laser source: The laser light source emits beams in a short time, reflecting horizontally and vertically through a polygonal mirror to perform scanning of the retina. An infrared laser is invisible to the examined eye, and it does not cause reflective contraction of the iris in the nondilated natural pupil eye. The laser source beam can be red, green, blue, or infrared according to the wavelength.

CFP involves two or three different wavelengths while scanning simultaneously, and images are processed and integrated by a computer to form a color image. Red-free refers to green-light CFP using a 590-nm green filter, which primarily focuses on the inner retinal layer, and it is appropriate for recording blood vessels and bleeding lesions. Blue-light CFP uses a blue filter with a 495-nm wavelength, which focuses on the surface of the retina for investigating the retinal nerve fiber layer and macula. Blue light perfectly visualizes the nerve fiber layer defects and macular epiretinal membrane. Infrared-light CFP, with a wavelength of 820-nm light source, can penetrate deeply with the benefit of photographing the retinal pigment epithelium and the choroid.

Types of fundus photography

CFP includes two types of photography: standard field imaging and ultra-widefield imaging (UWI).

Standard field imaging: The standard CFP provides 30°–55° images, including the optic nerve and macula. When taking fundus photos, the internal fixation lamp can be used for macular and optic disc fixation. The former covers the fundus of the posterior pole, and the latter can take pictures of the optic disc center.

UWI: Diabetic retinopathy study (DRS) made the first attempt to capture more retinal area by making a montage of seven standard 30-degree fields, which could capture up to 75° of the retina.[1] Widefield imaging (WUFI) and UWI, referring to images with greater than 50 degrees of the retina, are becoming increasingly popular. UWI, including the Zeiss CLARUS 500 and Optos, catches up to 200 degrees[2] which covers more than 80% of the retinal surface.[3,4] Even the peripheral retina can be photographed through a small pupil. WUFI provides valuable information about the peripheral vasculature and other retinal lesions that traditional imaging systems might miss. The modern WFI (100–200) fundus systems include Pomerantzeff camera (Contact)-image 148° Equator-plus camera (Film camera), Retcam (Clarity Medical Systems, Inc., Pleasanton, CA, USA), contact-based 130° Panoret-1000 camera (Medibell Medical Vision Technologies, Haifa, Israel), contact-based 130° Optos camera (Optos PLC, Dunfermline, UK), noncontact-based 200° Heidelberg Spectralis with the Staurenghi lens (Ocular Staurenghi 230 SLO Retina Lens; Ocular Instruments Inc., Bellevue, WA, USA), noncontact-based for 105° and contact-based for 150° Staurenghi lens for pseudocolor images, Clarus 500 (Carl Zeiss Meditec) (noncontact)—133° in single image and 200° in two images—noncontact imaging—true color imaging and autofluorescence modes (blue, green, and infrared). Ultra-widefield laser color fundus camera CRO produced by Micro-Clear Medical is a single widefield imaging system with an image range of 160°, and automatic image stitching exceeding 240°. The light source of CRO is a solid-state laser with 532 and 785 nm wavelengths. As confocal laser technology provides point-to-point scanning with a higher resolution of 5 μm, more subtle lesions and capillaries can be observed.

Smartphone fundus photography

Some adapters, such as Volk iNview, have facilitated the streamlining of the smartphone-based CFP with greater effectiveness. With the rapid advances and development of camera technology, the posterior pole may be perfectly captured using a cell phone only, an indirect ophthalmoscope-condensing lens, and modern cameras. This simple and convenient process provides more benefits to bedside patients, wheelchair-bound patients, and especially those undergoing offsite examinations.

Advantages of fundus photography for the diagnosis of retinal choroidal diseases

CFP is a noninvasive, time-saving simple procedure and can be used as a routine examination in daily clinical practice. Serial photographs, especially in follow-up conditions, enable physicians to examine the delicate and pathological changes in the nerve fiber, macula, and entire retina. Furthermore, certain internal eye landmarks are more visible in CFP than in other retinal anatomical or functional examinations, especially when combined with different light sources (red-free, infrared-light, blue-light, etc.). In ophthalmic practice, as a noninvasive imaging tool, CFP assists ophthalmologists or physicians from other disciplines to screen specific retinal pathological changes, such as diabetic retinopathy (DR), and to determine treatment guidance for systemic diseases. The conveniently quickly acquired color images could also be clinically valuable for patients and sometimes more worthy than a physician's notes.

Fundus imaging acquisition can be nonmydriatic

A nonmydriatic fundus camera is easily applied, and the image quality is as clear as the mydriatic fundus camera without any inconvenience by mydriatic drugs. A nonmydriatic fundus camera can greatly shorten the diagnostic cycle and reduce inconvenience induced by pupil dilatation drugs. To date, the nonmydriatic camera has become the first option for screening DR. For instance, iCam (Optovue, Fremont, CA), 3nethra (Forus, Bengaluru, India), DRs (CenterVue, Padova, Italy), TRC-NW8FPLUS (Topcon, Tokyo, Japan), and VISUCAM 200 (Zeiss, Oberkochen, Germany) are commercially available nonmydriatic cameras. Some commercially digital nonmydriatic cameras, such as Canon CR-2 PLUS AF, provide both color and fundus auto-fluorescence imaging with a 3.3-mm small pupil size.[3]

Beneficial of CFP for remote diagnosis and guidance of treatment

The community eye clinic (vision centers [VCs]) in India has routine CFP to screen posterior segment diseases.[5] Some studies applied telemedicine for glaucoma to increase access to glaucoma care by improving efficiency in Canada.[6] With the development of telemedicine-based remote digital fundus imaging evaluation techniques, timely screening of retinopathy of prematurity (ROP) is also widely used in the United States, Japan, and Singapore.[7] CFP provides a definitive diagnosis for 47% of patients with acute vision loss in the emergency department.[8] The CFP taken in a developing area can be interpreted by remote sites of experienced consultants through telemedicine. Besides, 3nethra designed by Forus Inc. (Bengaluru, India) intelligently uses a digitally web-based telemedicine system to transmit images to mobile phones for primary screening of eye healthcare.[3]

CFP provides wider fundus fields for the documentation of peripheral retinal and choroidal lesions

Additional details are described in "Types of fundus photography".

CFP equipped with multimodal image analysis software

CFP is generally equipped with a variety of software to improve the resolution of fundus images and to expand the imaging field through Mosaic. Common CFP software includes the large field of view for Mosaic, dereddened imaging software, vascular enhancement imaging software, choroidal enhancement software, nerve fiber enhancement imaging software, cup-to-disc ratio analyzer, and arteriovenous vascular analyzer.

Establishment of electronic health archives: The digital fundus images can easily track the changes of fundus for follow-up. In addition, the establishment of electronic health archives through CFP can provide better health education for patients and improve their awareness of their health conditions, preventing blindness. Digital fundus images could also make it easier for professionals to analyze and manage ophthalmic data.

Diagnosis and screening of a wide range of diseases using CFP

Clinically effective fundus examination benefits the subsequent treatment of retinal and choroidal disorders. A retinal microvascular imaging diagnostic center, which adopted a digital CFP system, was established in Australia.[9] Some prestigious large-scale international epidemiological cohort studies applied nondilated CFP as the main screening tool for various fundus disorders, including the Diabetes Control and Complications Trial (DCCT), the Epidemiology of Diabetes Interventions and Complications (EDIC) study, the United Kingdom Prospective Diabetes Study (UKPDS), the Finnish Diabetic Nephropathy Study (FinnDiane), the Early Treatment Diabetic Retinopathy Study (ETDRS), the Age-Related Eye Disease Study (AREDS), the Early Manifest Glaucoma Trial (EMGT), and a multicenter European epidemiological study (the EUREYE Study).[10–16]

The disadvantages of CFP

Compared with dilated ophthalmoscopy examination by an ophthalmologist, single-field CFP has a sensitivity ranging from 38% to 100% and a specificity ranging from 75% to 100%.[17] Several variables negatively affect fundus image quality, including opacities in the optical media of the eye, such as cataracts, reflections from the cornea or lens, operator-related problems, and difficulty in focusing on the retina or tear film rupture. Furthermore, the following conditions are inappropriate for CFP: vitreous hemorrhage sensitive to light or having light-sensitive substances in their bodies, without ocular lens, unable to fix visual objects, with the ability to penetrate eyeball injury before wound healing, and with pupil diameter of less than 3.5 mm after pupil dilatation.

Normal fundus photography

The fundus of normal adults appears orange-red, and the color depends on the amount of pigmentation and visibility of the choroidal vessels. The optic disk is pale pink with a sharp flat margin. The physiological cup, which lies inside the disc, is typically smaller than six-tenths of the disk's diameter. Macular is an area located temporal to the disc. The fovea is a depression in the macula's center that has a diameter of about 1.5 mm and resembles the optic disk. A tiny pit called the foveola sits in the fovea's center. The central vision corresponds to these regions. The central retinal artery and vein branches are near the surface of the disc into the superior and inferior branches of arterioles and venules. A fundus color image covers the optic disc, macula, retina, blood vessels, choroid, and vitreous body (Fig. 4.1). Generally, children aged ≥3 years can cooperate with CFP. Bright, flake white wet silk-like retinal reflection can be investigated on both sides of the retinal vascular arch in children, and the reflection around the macula resembles a "C" band (Fig. 4.2). Growing with age, posterior vitreous detachment and vitreous opacity can be detected due to vitreous liquefaction. The ring opacity around the optic disc, namely the Weiss ring, can be found in the CFP. In the elderly, due to the reduction of pigment in the retinal pigment epithelium, the atrophy of the choroid and disappearance of a small vasculature network, thick red-brown stripes on the fundus, which is called tessellated retinal fundus arteriosclerosis, and other changes can also be found (Fig. 4.3).

Abnormal fundus photography

Some common abnormal signs, including retinal hemorrhage, exudation, pigmentation, degeneration, retinal hole, bulge, atrophy, and hyperplasia can be found in the CFP images. By recognizing the common abnormal signs in the optic disc, retina, and macula, a preliminary diagnosis can be made, which can be further confirmed using other imaging modalities, such as optical coherence tomography (OCT), fundus fluorescein angiography (FFA), and visual electrophysiology. Furthermore, various abnormal changes on the fundus color map can assist ophthalmologists to find out the prognosis of the disease.

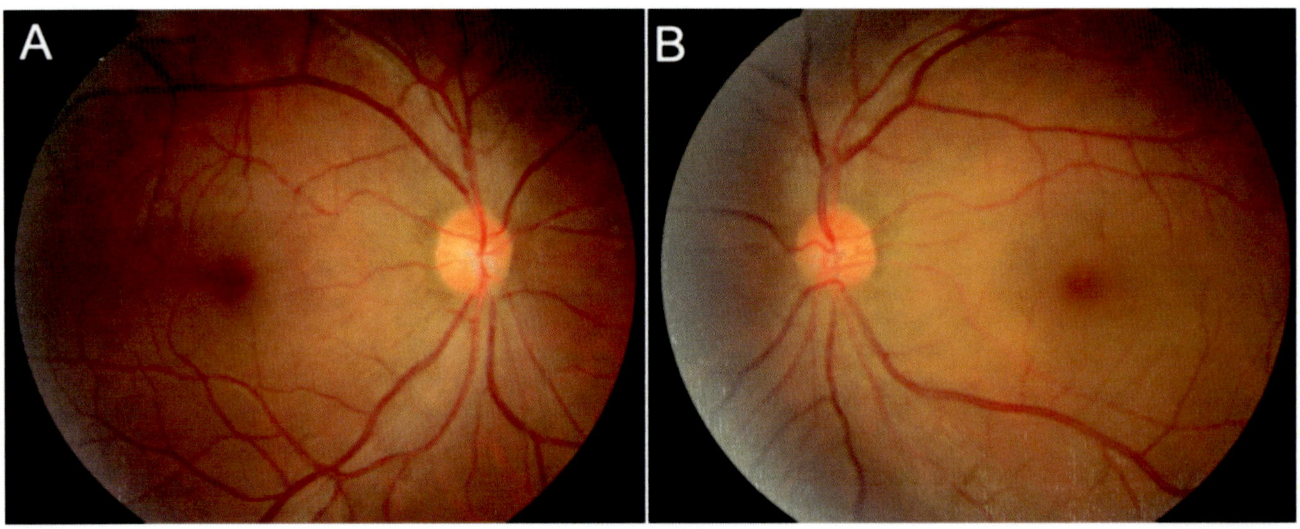

FIGURE 4.1 A 45-degree color fundus photograph (CFP) of a 27-year-old healthy patient. The best corrected visual acuity of both eyes is 20/20. CFP of the right eye (A) and left eye (B) shows an *orange-red* background and a light-red optic disc with a 0.3 cup/disc ratio. The macular fovea reflection is clear, and the blood vessel diameter looks normal. No bleeding and exudation are detected.

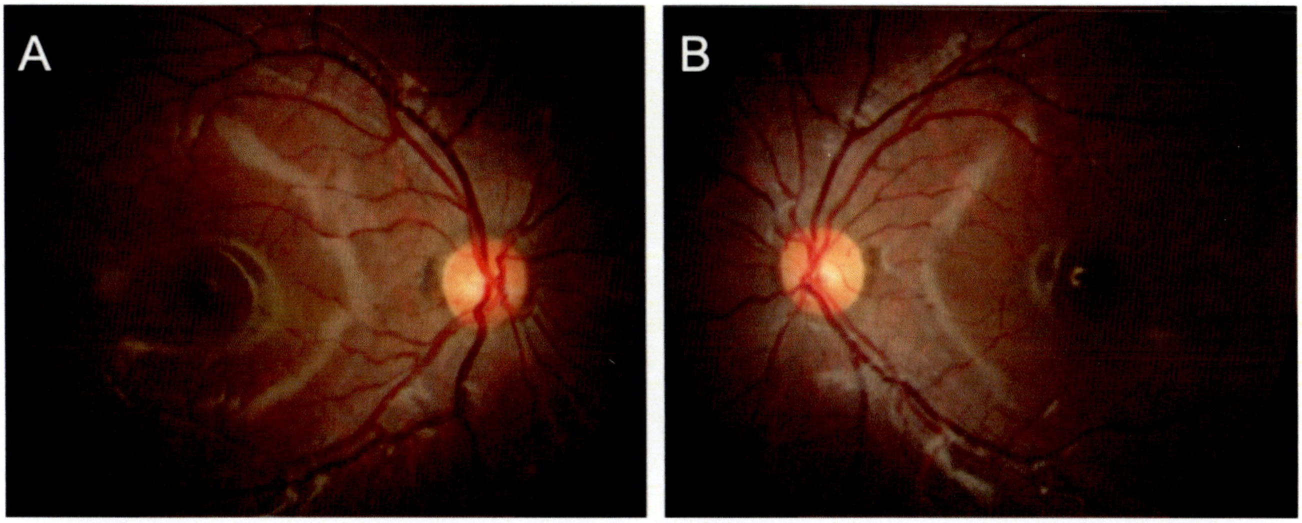

FIGURE 4.2 CFP of an 8-year-old healthy boy with 20/20 visual acuity of both eyes (A and B). The size and color of the optic disc are normal, the boundary is clear, and (C/D) is 0.3. The central fovea reflection is visible, and the bright white wet silk-like retinal reflection can be observed on both sides of the retinal vascular arch, and the reflection around the macula appears as a "C" shape.

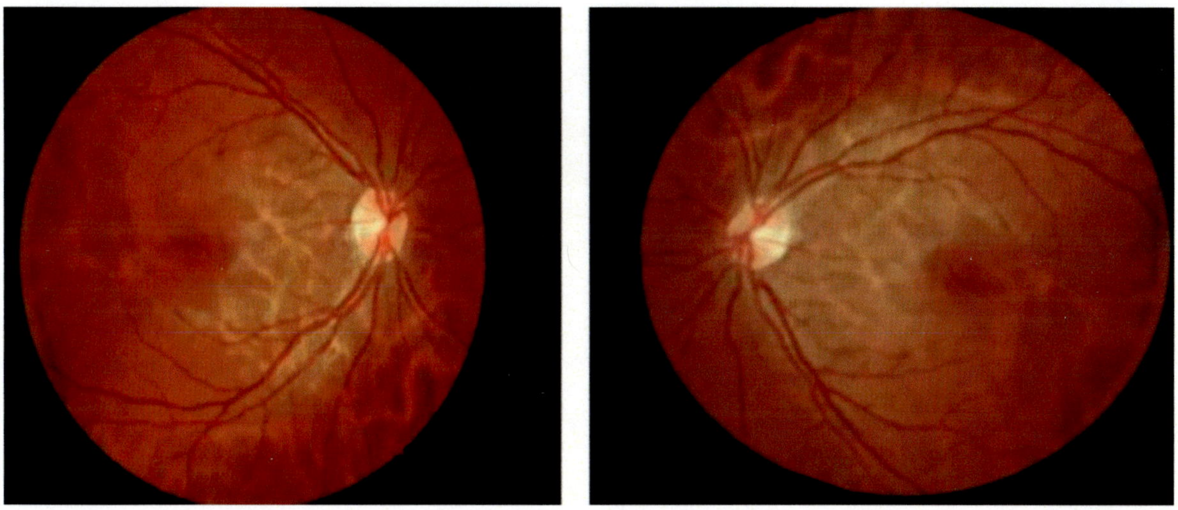

FIGURE 4.3 CFP of a 78-year-old healthy man, the visual acuity of both eyes is 20/20 (A and B). CFP shows that the size and color of the optic disc are normal, the boundary is clear, and (C/D) is 0.3. The ratio of the artery to the vein is about 2:3, and the retinal vascular path is normal. The choroidal vessels are visible, and the retina presents leopard-like changes. The foveal reflection disappeared. No microangioma, hemorrhage, exudation, or neovascularization could be found.

Fig. 4.4 shows retinal exudation: (A—B) hard exudation and (C—D) cotton-wool spot.

In Fig. 4.5, CFP shows retinal hemorrhage.

Although most of the gold diagnostic standards for fundus diseases are not associated with CFP, preliminary screening of patients through CFP is suggested, providing a basis for the diagnosis and treatment of patients.

Retinal vascular disease

Retinal artery occlusion

In Fig. 4.6A, CFP shows the right eye of central retinal artery occlusion. In a 55-year-old woman with retinal edema at the posterior pole, the macular area shows "cherry erythema," and the optic nerve looks pale. The superior temporal artery branches are slender, and the veins are slightly dilated and tortuous. Fig. 4.6B shows branch retinal artery occlusion: a 60-year-old man with branch retinal artery occlusion in his right eye. Retinal edema is found in the occlusion area of the infratemporal branch artery, which looks off-white, and the dividing line with the normal retina can be observed.

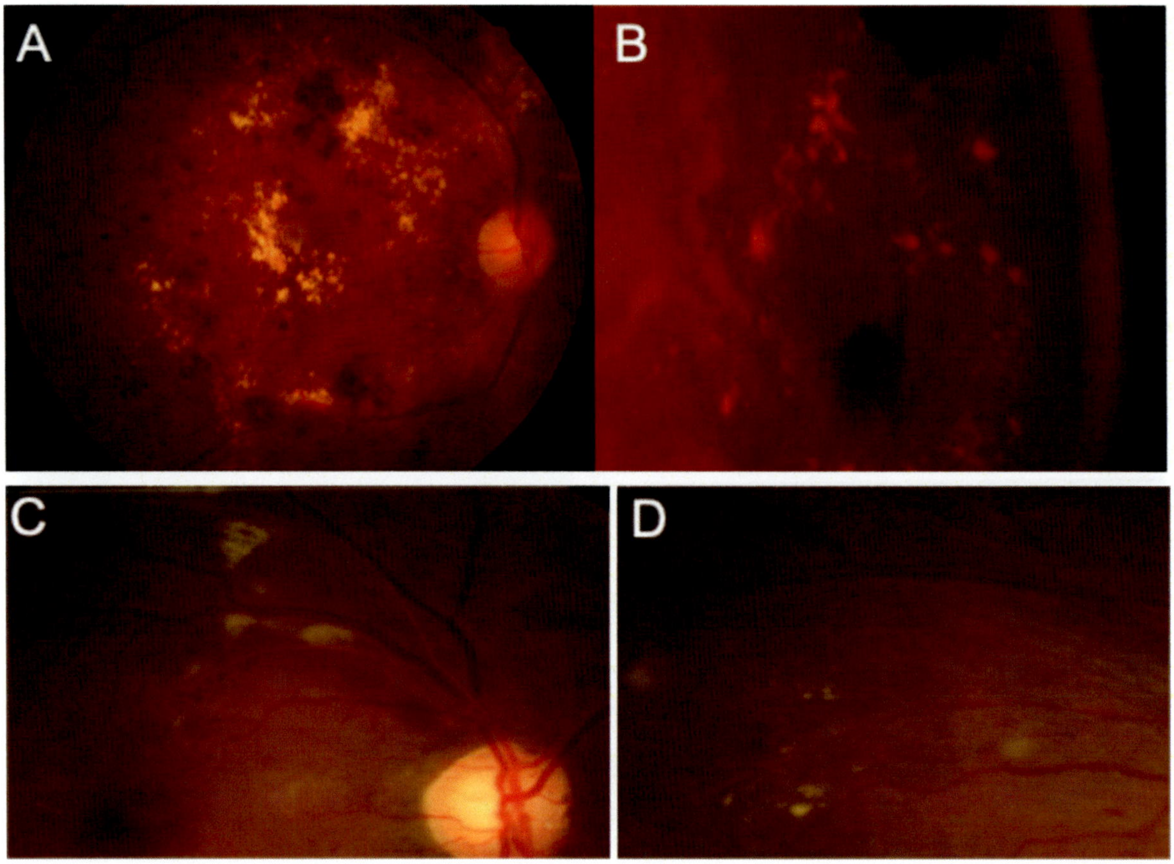

FIGURE 4.4 Retinal exudation: (A–B) hard exudation; (C–D) cotton-wool spot.

Retinal vein occlusion

Fig. 4.7 shows central retinal vein occlusion. CFP of a 65-year-old woman. The image shows a heavy radial flame-like hemorrhage centered on the optic disc in her left eye, and optic disc edema could be found with unclear boundaries. Fig. 4.7B shows branch retinal vein occlusion: CFP of a 56-year-old woman, showing flame bleeding along the temporal branch vein of her right eye.

Diabetic retinopathy

The American Academy of Ophthalmology divides diabetic retinopathy (DR) into five stages of progressive severity: no apparent retinopathy, mild-moderate and severe nonproliferative DR (NRPD), and proliferative DR (PDR). NPDR is characterized by multiple microangiopathies, retinal hemorrhages, venous beading, intraretinal microvascular abnormalities, hard exudative, and cotton wool spots. PDR is characterized by retinal neovascularization, vitreous hemorrhage or retinal detachment. Macular edema can be observed either in NPDR or PDR[18] (Figs. 4.8 and 4.9).

Family exudative vitreoretinopathy

Family exudative vitreoretinopathy (FEVR) is a hereditary retinal vascular dysplasia. Abnormal peripheral retinal vascularization forms a peripheral nonperfusion area, leading to retinal neovascularization, bleeding, and proliferation. The disease is mostly autosomal dominant, which is mainly related to mutations in FZD4 and LRP5 genes[19] (Fig. 4.10).

Coat's disease

Coat's disease is an idiopathic eye disease characterized by retinal vasodilation, extensive retinal exudation, and exudative retinal detachment.[11] The fundus of the disease is characterized by retinal telangiectasia and flaky yellow-white retinal exudation in abnormal vascular areas. Severe retinal exudation can lead to exudative retinal detachment with the spherical bulge (Fig. 4.11).

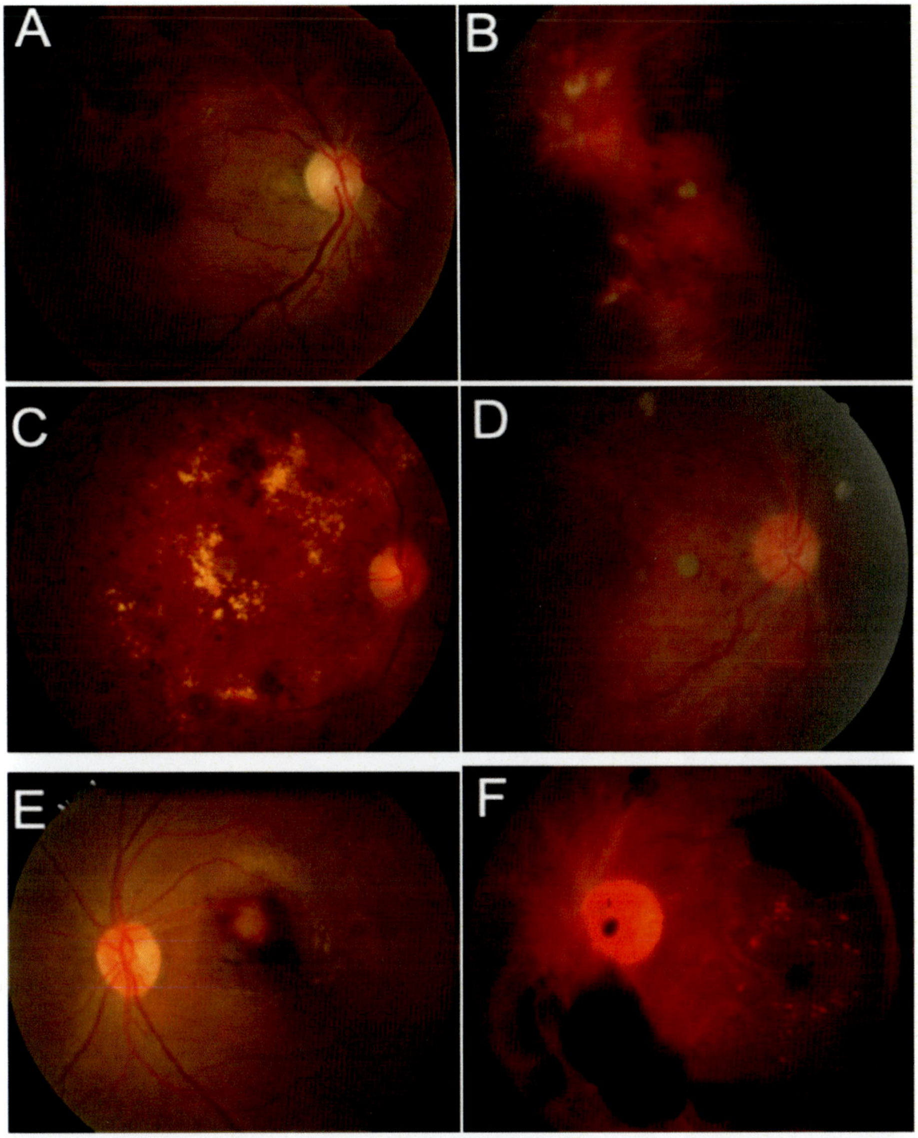

FIGURE 4.5 CFP shows retinal hemorrhage. (A and B) superficial radial-like hemorrhage; (C–E) dot-like deep retinal hemorrhage; (F) preretinal hemorrhage.

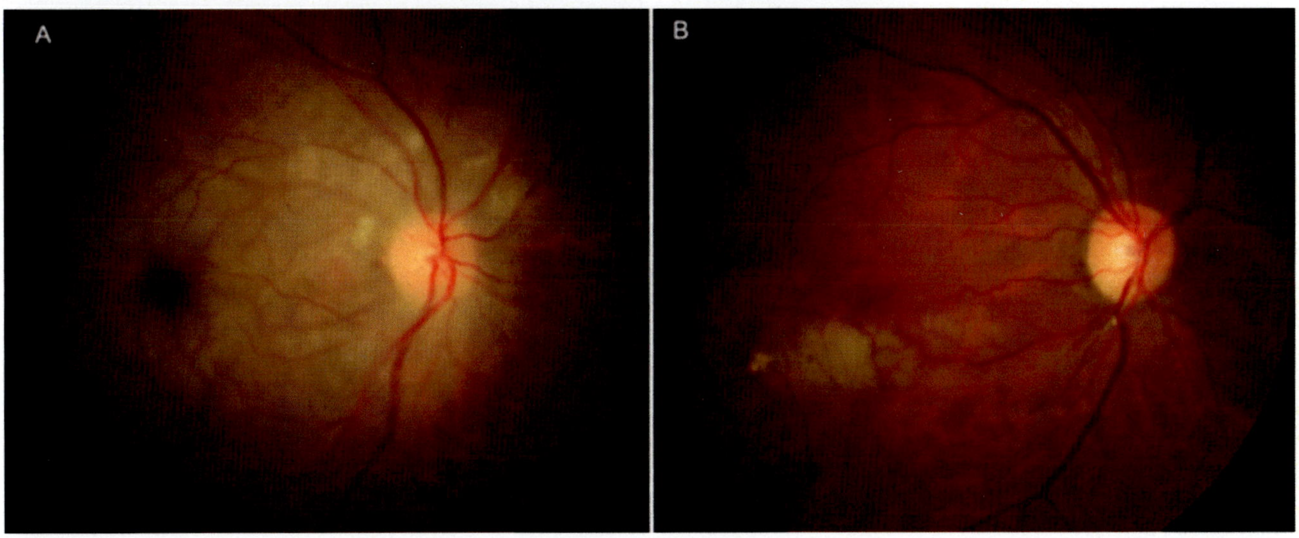

FIGURE 4.6 (A) shows central retinal artery occlusion with pale, thickened retinal infarction and cherry red dot in the right eye of a 55-year-old woman. The optic nerve looks pale. The superior temporal artery branches are slender, and the veins are slightly dilated and tortuous. (B) shows branch retinal artery occlusion in the right eye of a 60-year-old man. Retinal edema is found in the occlusion area of the infratemporal branch artery, which looks off-white, and the dividing line with the normal retina can be observed.

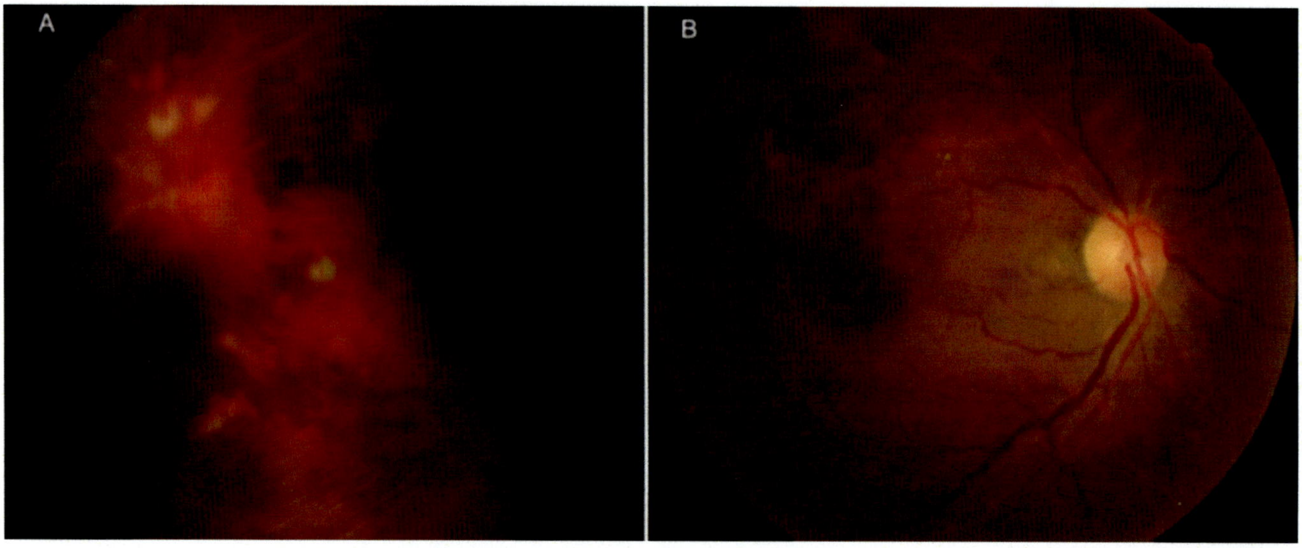

FIGURE 4.7 (A) Central retinal vein occlusion: CFP of a 65-year-old woman. The image shows a heavy radial flame-like hemorrhage centered on the optic disc in her left eye, and optic disc edema could be found with unclear boundaries. (B) Branch retinal vein occlusion: CFP of a 56-year-old woman, showing flame bleeding along the temporal branch vein of her right eye.

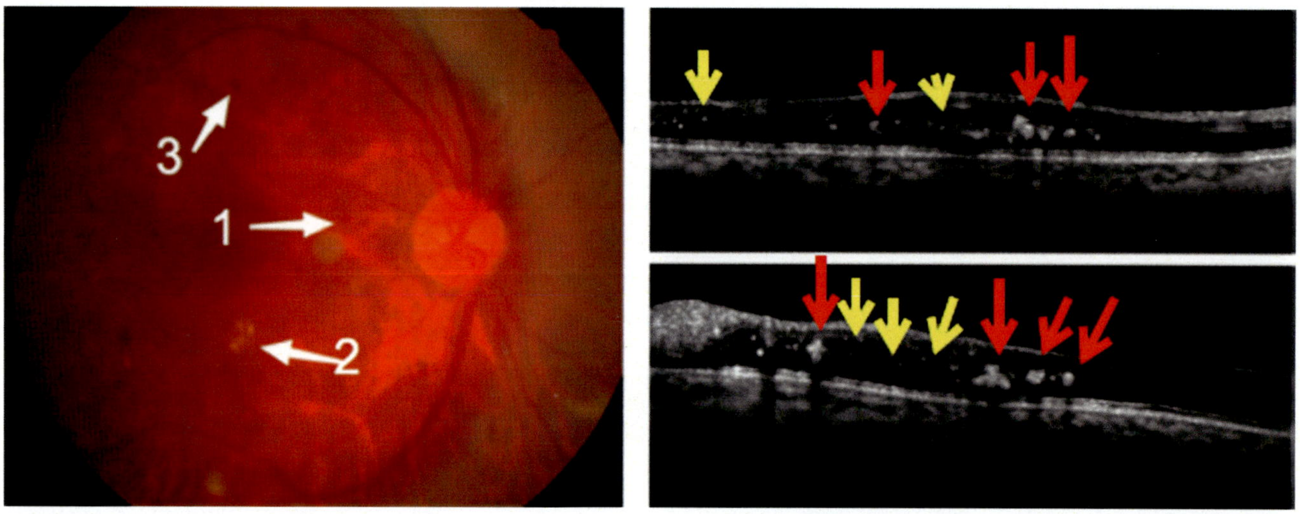

FIGURE 4.8 Nonproliferative diabetic retinopathy. A 68-year-old man, with type 2 diabetes for 10 years. The white arrows on the CFP show retinal microaneurysms (arrow 1), hard exudatives (arrow 2) and dot-like hemorrhages. The red arrows on OCT show the intraretinal hard exudates. Hyper-reflective focis were also shown by yellow arrows.

Retinal microaneurysm

Retinal microaneurysm (RM) is characterized by local dilation of the acquired retinal artery, which is more common in elderly aged over 60 years old[12] (Fig. 4.12).

Macular disease

Central serous chorioretinopathy

Central serous chorioretinopathy (CSC) is one of the most common ocular fundus diseases in the clinic, which is characterized by localized serous detachment at the posterior pole with focal or multifocal retinal pigment epithelium leakage[13] (Fig. 4.13).

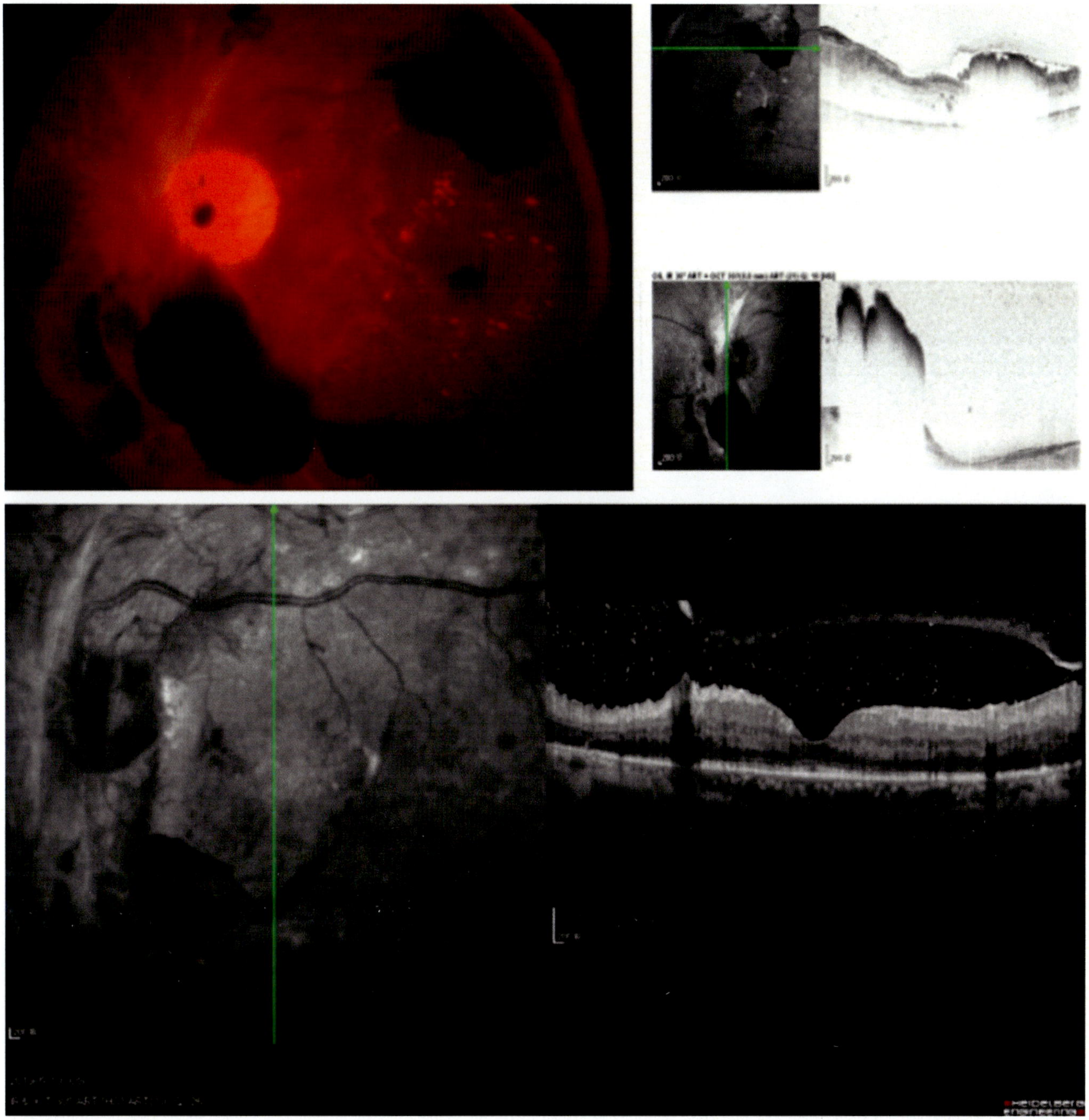

FIGURE 4.9 Proliferative diabetic retinopathy: CFP shows hemorrhage in front of the retina of the left eye, and OCT displays a proliferative membrane in front of the optic disc to drag the retina.

Age-related macular degeneration

Age-related macular degeneration (AMD) is one of the most common causes of irreversible vision loss or even blindness in individuals aged over 50 years old. Drusen and macular geography are the most characteristic manifestations of dry AMD, respectively, showing yellow-white punctate or nodular deposits under the retina. Wet AMD, also known as neovascular AMD, is manifested in both eyes gradually, and subretinal and choroidal neovascularization is the most characteristic manifestation of wet AMD[14] (Fig. 4.14).

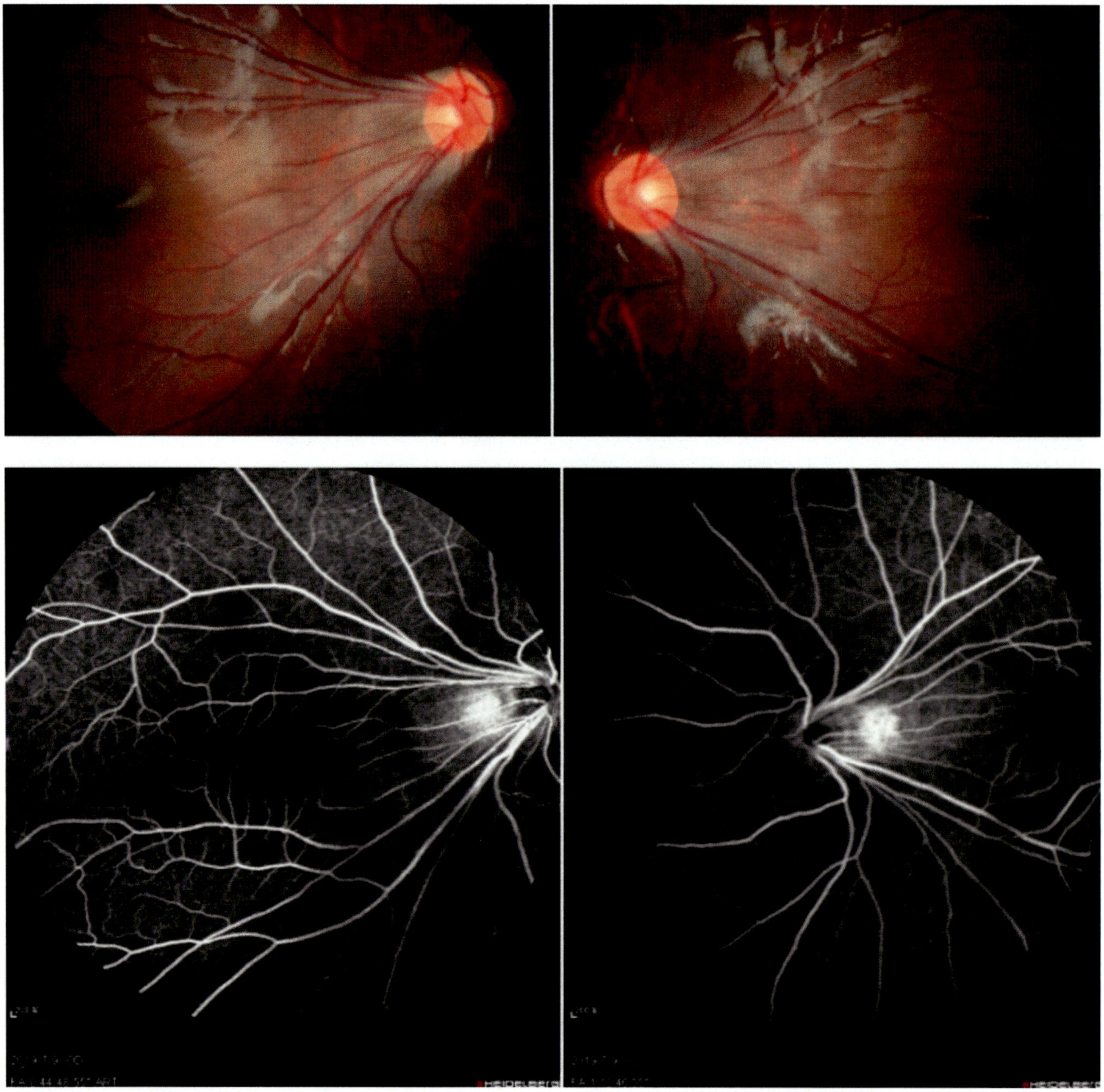

FIGURE 4.10 A 3-year-old boy who suffered from poor vision since his born. CFP (A, right eye; B left eye) shows the optic disc of both eyes dragged, the angle between the superior and inferior temporal vascular arches becomes sharp, and branches of peripheral retinal blood vessels can be observed. FFA image (A, left eye; B left eye) shows increased branches of retinal blood vessels in the lower peripheral part with a nonperfusion area in the peripheral part.

Idiopathic polypoid choroidal vasculopathy

Idiopathic polypoid choroidal vasculopathy (IPCV) is characterized by orange-red nodular lesions under the retina and polyps at the end of the abnormal branching vascular network (BVN) of choroid[15] (Fig. 4.15).

Idiopathic macular hole

Idiopathic macular hole (IMH) refers to the full-thickness defect of the neuroepithelial layer in the macular area caused by no trauma or other definite systemic or ocular causes. Its cause is related to the tangential traction of the inner limiting membrane and the traction of the posterior vitreous cortex[16] (Fig. 4.16).

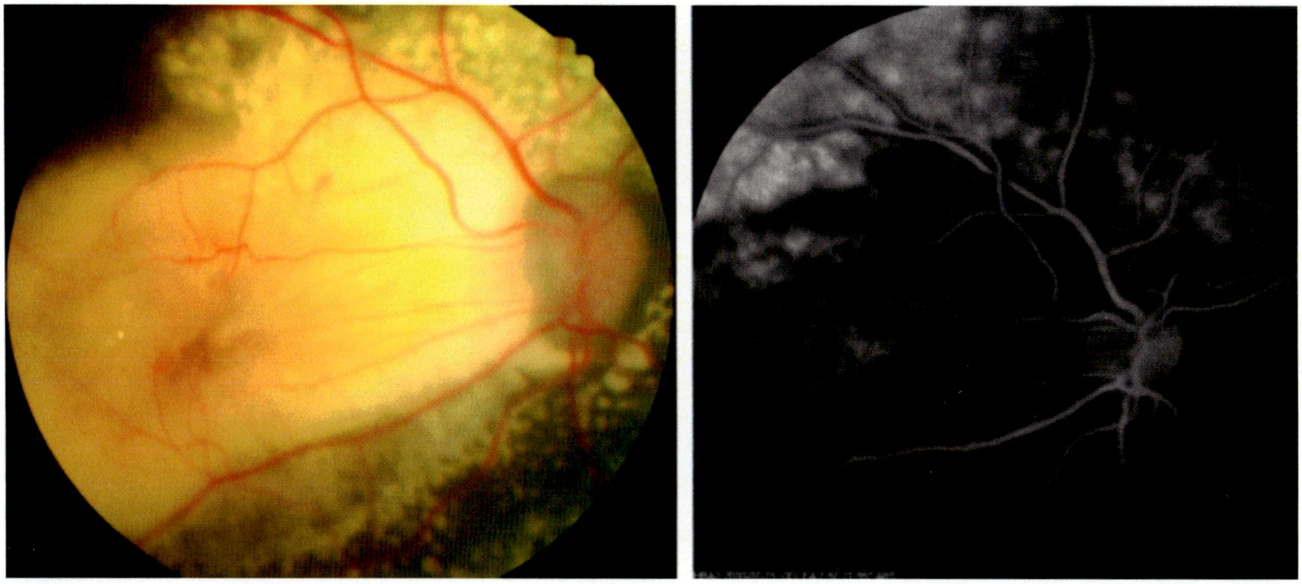

FIGURE 4.11 A 3-year-old boy, with heavy yellow and white hard exudates, enhanced copper-like reflection of retinal artery, and tortuous expansion of vein in his right eye (A). In the FFA of the right eye (B), there was a strong fluorescence leakage in the periphery retina, a military hemangioma-like expansion, and a dot-like strong fluorescence. There was no nonperfusion area in the peripheral retina.

Retinal degeneration diseases

Hereditary macular disease

Stargardt disease

Stargardt disease (STGD) is one of the most common childhood inherited macular dystrophy. It is mostly an autosomal recessive trait, caused by a mutation in the ABCA3 gene on chromosome 1p21-p13(STGD1). STGD is characterized by chronic progressive damage to retinal structure and function, decreased vision, color vision deficiency, and central scotoma (Fig. 4.17).

Retinitis pigmentosa

Retinitis pigmentosa (RP) is a group of hereditary progressive slow-moving degenerative diseases with loss of photoreceptor cells and pigment epithelium. Retinal artery thinning, the sallow color of the optic disc, and osteocyte-like pigmentation can gradually appear in the fundus[20] (Fig. 4.18).

Pathological myopia

Pathological myopia (PM) is the rational change of fundus diseases in patients with high myopia, which is also known as anamorphic myopia. The fundus changes, such as leopard-shaped fundus, myopic arc, choroidal neovascularization in the macular area, posterior scleral staphyloma, macular hole, Fuchs spot, retinal choroidal atrophy in the posterior pole, lacquer crack-like damage, and retinal splitting can be observed[21] (Fig. 4.19).

Retinal detachment

Retinal detachment (RD) can be classified into rhegmatogenous, tractional, exudative, and combined traction-rhegmatogenous detachment (Fig. 4.20).

Fundus congenital and hereditary diseases

Congenital macular coloboma

Congenital macular coloboma (CMC) is also known as macular dysplasia, it is a hereditary and developmental fundus disease characterized by choroidal and retinal defects in the macular region[22] (Fig. 4.21).

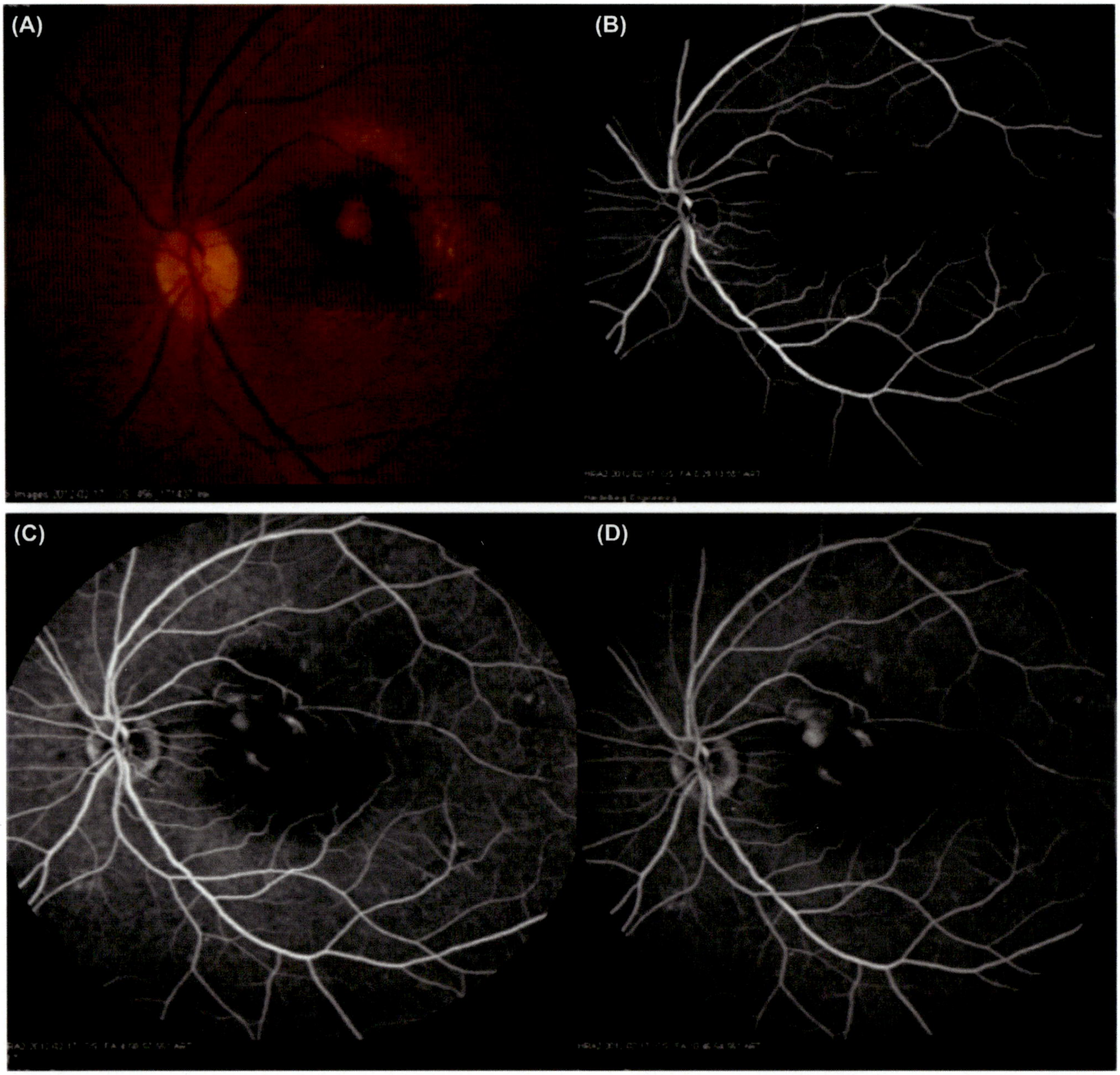

FIGURE 4.12 A 74-year-old woman with a large retinal aneurysm in her left eye. CFP of the left eye shows a large yellow-white hard exudate around the red spherical bulge on the temporal edge of the optic disc of the left eye, involving the fovea of the macula (A). FFA showed the filling and development of the tumor's nutrient arteries and reflux veins (B, C,D).

Choroio retinal coloboma

Choroio retinal coloboma (CRC) is characterized by the congenital absence of part of the retinal pigment epithelium and choroid. CRC may be correlated with other ocular disorders including cataract, microphthalmia, and anophthalmia[23] (Fig. 4.22).

Myelinated retinal nerve fiber layer

Retinal nerve fiber layer (RNFL) is a relatively common and mostly benign congenital anomaly, in which the retinal nerve fibers anterior to the lamina cribrosa have a myelin sheath[24] (Fig. 4.23).

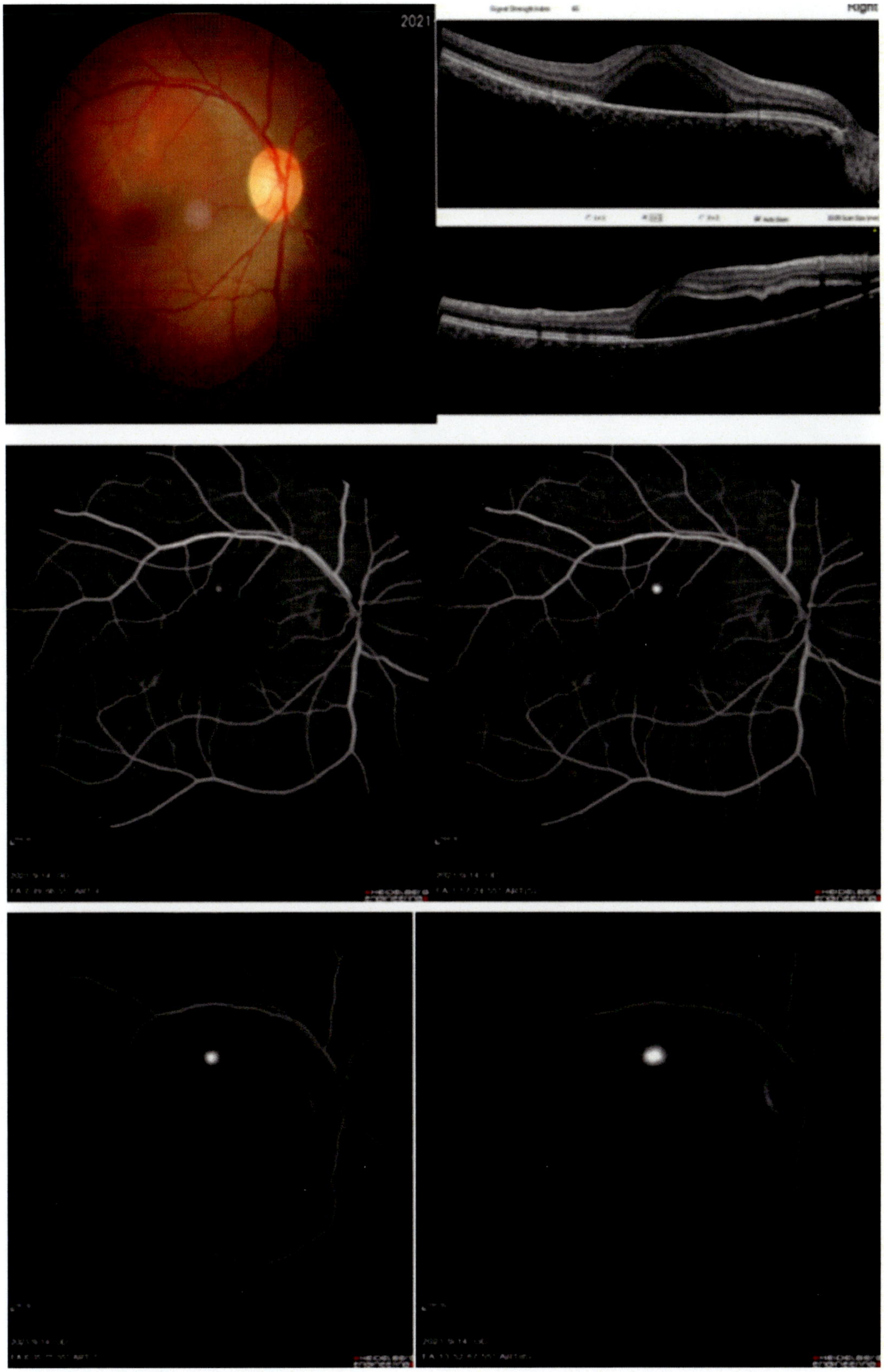

FIGURE 4.13 A 39-year-old man who suffered from blurred vision in his right eye for more than 2 weeks. CFP shows the serous detachment of the neuroepithelial layer of the round retina on the partial temporal side of the macula of the right eye (about 4PD in size). *Yellow* and *white* dots can be observed under the retina of the detachment area. The spot-like high fluorescence in the early stage is shown, and the hyper-fluorescence in the macula is enlarged similar to ink stains.

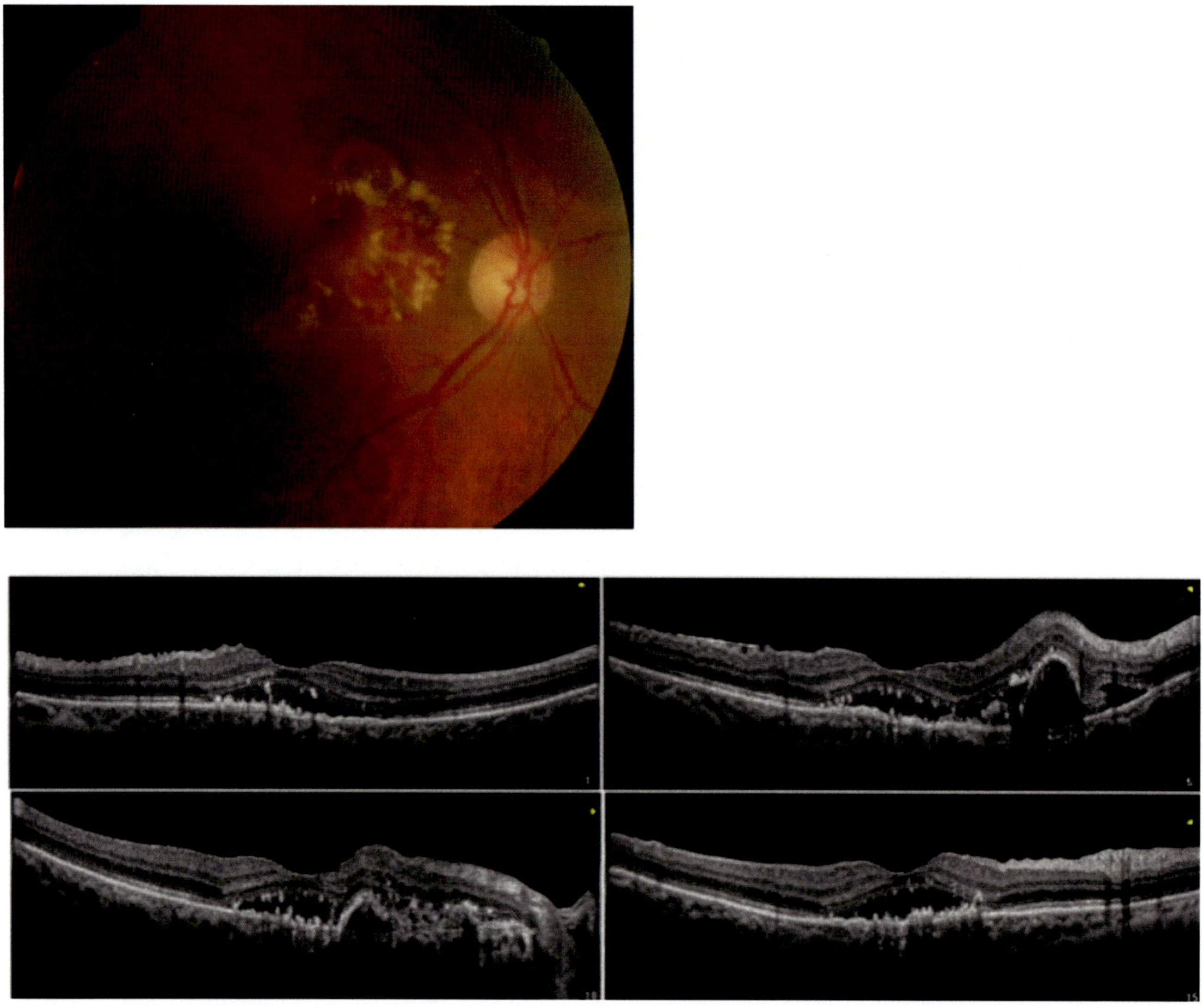

FIGURE 4.14 Wet AMD: A 74-year-old man had decreased vision in his right eye. CFP shows a large number of *yellow-white* hard exudates in the macula of his right eye with patchy hemorrhage. OCT shows that the neuroepithelium of his right eye is separated from the pigment epithelium, and the vitreous wart of the retinal pigment epithelium is localized and raised with the attachment of some highly reflective substances.

Morning glory syndrome

Morning glory syndrome (MGS) is characterized by a funnel-shaped optic nerve, which is a special type of optic nerve defect. PAX6 gene abnormality may be related to its pathogenesis[25] (Fig. 4.24).

Optic nerve disorders

(1) *Ischemic optic neuropathy (ION):* ION can be classified into anterior ION (AION) and posterior ION (PION), in which AION is more common clinically[26] (Fig. 4.25).

(2) *Leber's hereditary optic neuropathy (LHON):* LION, the most common mitochondrially inherited DNA dysfunction disorder that affects young men, is caused by mitochondrial DNA (mtDNA) mutations at 11,778, 3460, and 14,484. Dilated optic disc capillaries, the changes of arterioles, and the nerve fibers around the optic disc swell can be observed in CFP examination, while without FFA leakage in the optic disc[27] (Fig. 4.26).

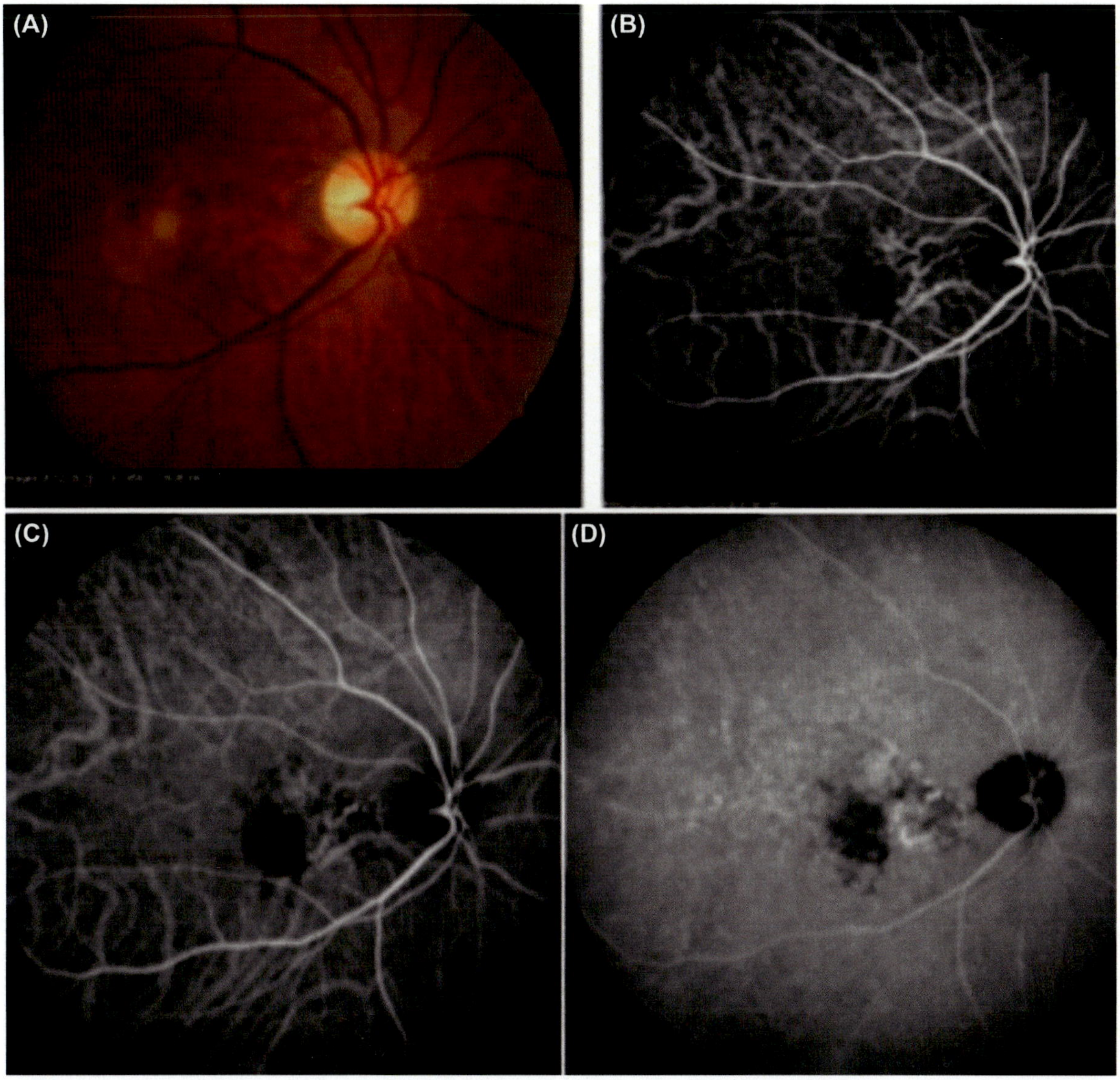

FIGURE 4.15 PCV: A 56-year-old man who suffered from decreased VA in his right eye for 6 months. CFP of the right eye shows a reddish lesion with a clear boundary on the nasal side of the macular, around the lesion, retinal neuroepithelium detachment can be observed (A). Indocyanine green angiography (ICGA) shows the choroidal vascular network with abnormal branches in the macular area, and a polypoid structure with cystic expansion can be observed at the end of these abnormal vessels. The polypoid structure in the macular area shows hyperfluorescence in the late stage of ICGA.

Glaucoma

Glaucoma is a chronic, progressive eye disorder caused by damage to the optic nerve, leading to a visual field defect. Clinical diagnosis of glaucoma should be based on comprehensive judgment of intraocular pressure, visual field, and optic nerve changes. The most common type of glaucoma is open-angle glaucoma, which can be early screened and diagnosed by CFP (the rim shape of the optic disc, cup-to-disc (C/D) ratio, etc.) (Fig. 4.27).

Traumatic fundus changes

See Fig. 4.28.

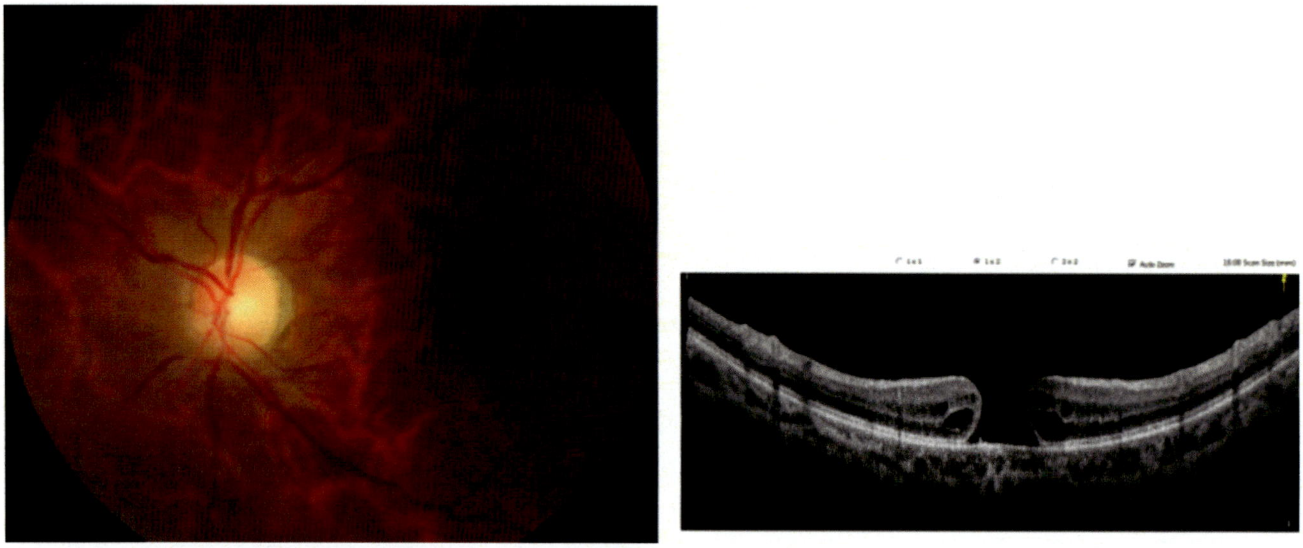

FIGURE 4.16 Idiopathic macular hole: CFP showed that the reflection of the fovea disappeared, and there was a round hole of about 1/4PD in the macular(A). OCT showed that the neuroepithelial layer in the fovea of the macula was a full-thickness defect, with cystic edema between layers in the left eye (B).

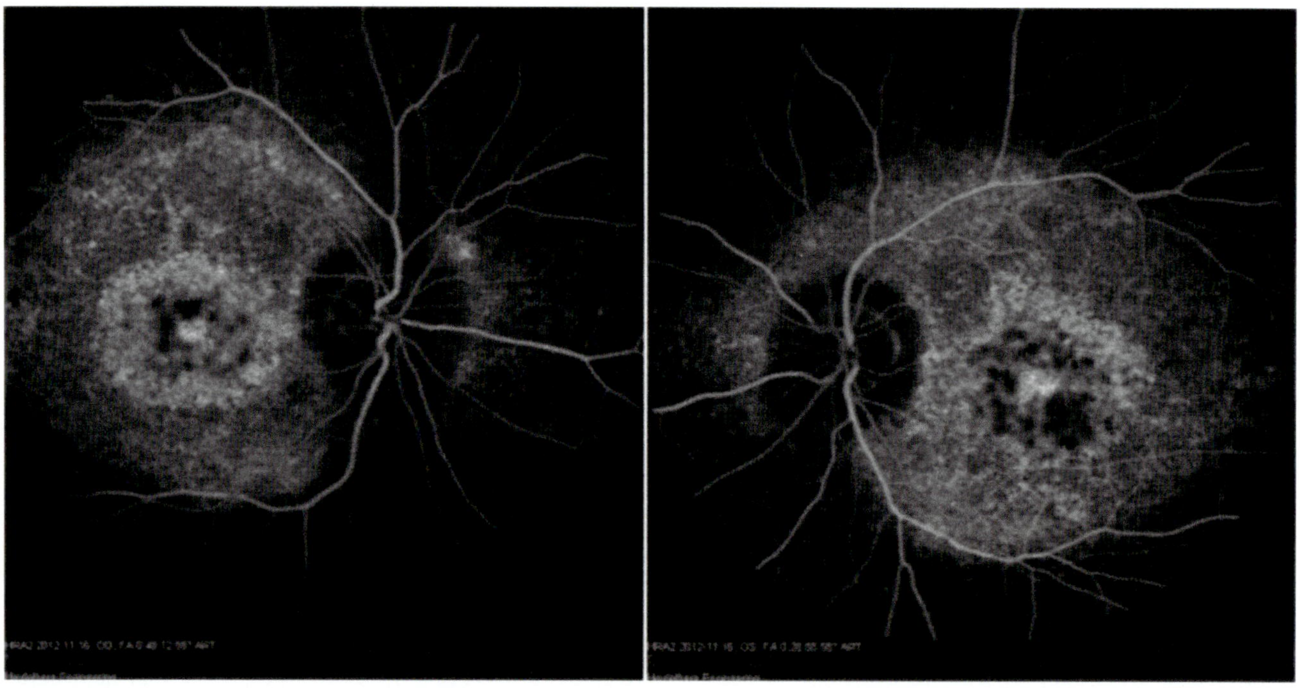

FIGURE 4.17 A 16-year-old male patient with bilateral low vision since childhood. CFP shows symmetric round macular *bluish-gray* atrophy lesions in both eyes, with local pigment disorder and uneven granular appearance. Binocular fundus autofluorescence (FAF) accentuates the classic bull's eye maculopathy. Abnormally increased FAF indicates excessive accumulation of lipofuscin in the retinal pigment epithelium (RPE). Fundus autofluorescence is informative with hyperfluorescence of flecks. A "dark" and "silent" choroid is observed with FFA due to the accumulation of lipofuscin.

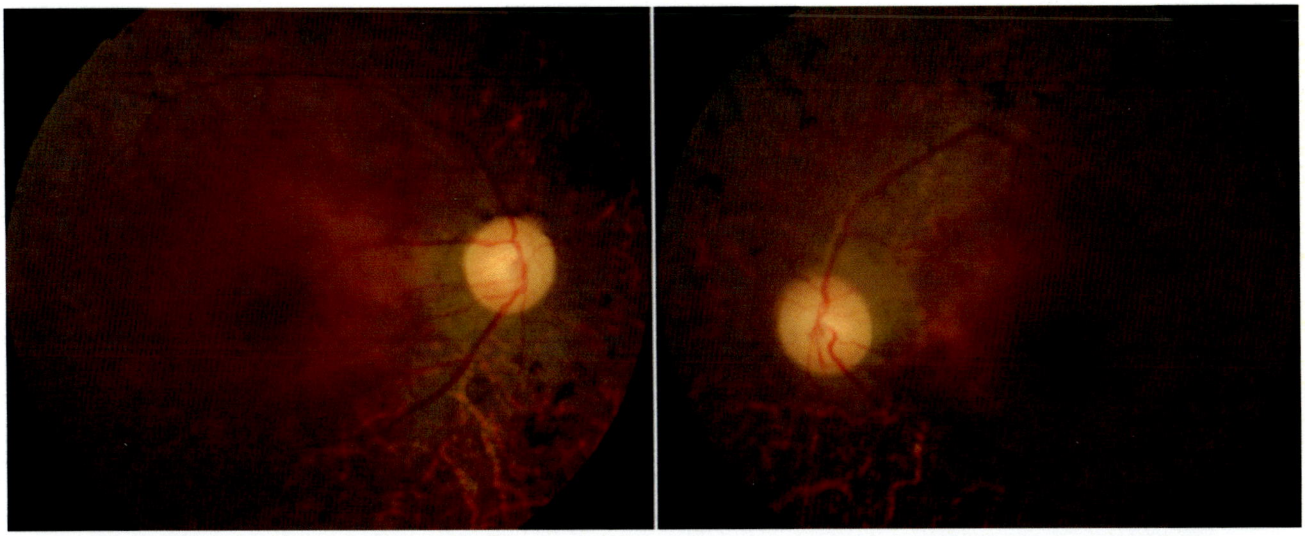

FIGURE 4.18 A 47-year-old man with bilateral night blindness since childhood. CFP shows sallow color of the optic disc, narrowing of the retinal artery, the disappearance of reflection in the fovea, and osteocyte-like pigmentation in the periphery in the right eye (A) and left eye (B).

Eye tumor

Choroidal osteoma is a benign tumor in the choroid, composed of mature bone tissue. It was officially named by Gass in 1978. The majority of the tumors are located near the optic disc, showing a yellow-white or orange-red flat bulge, visible pigmentation, and irregular margin of the tumor, which is similar to pseudopodia extending around, and it can form subretinal space (Fig. 4.29).

The role of artificial intelligence using CFP

Artificial intelligence (AI) is a general term referring to the use of a computer to model intelligent behaviors with minimal human interventions.[28] As a branch of machine learning (ML), deep learning (DL) adopts advanced ML and has achieved great success in computer vision and natural language processing tasks.[29] Ophthalmology is one of the several multidisciplinary domains where AI is advancing quickly.[30] CFP and OCT are the main sources of information and learning objects for AI algorithms in ophthalmology.[31,32] Moreover, the combination of DL with telemedicine showed great advantages in terms of screening and reducing financial pressure on governments and societies without the necessity of reading images from retinal specialists.[33]

CFP is widely used in AI for retinochoroidal diseases. The obtained fundus photographs can be analyzed by AI for a variety of diagnostic analyses, including optic disc, macular localization, vascular segmentation, lesion detection, and lesion grading.[34] CFP is the basis for the validation of AI algorithms in the detection and staging of DR, age-related macular degeneration, and other retinochoroidal diseases.

Diabetic retinopathy

DR is a leading cause of blindness in working-age populations. Approximately 100% of people with type 1 diabetes and more than 60% of people with type 2 diabetes suffer from DR.[35] Prompt treatment and prevention of blindness and visual impairment rely on early identification and screening of DR.[36]

DR is characterized by microaneurysms, hemorrhage, exudation, and neovascularization. These characteristics can be mainly obtained by high-quality fundus images, which is the basis for accurate screening. The limited number of registered

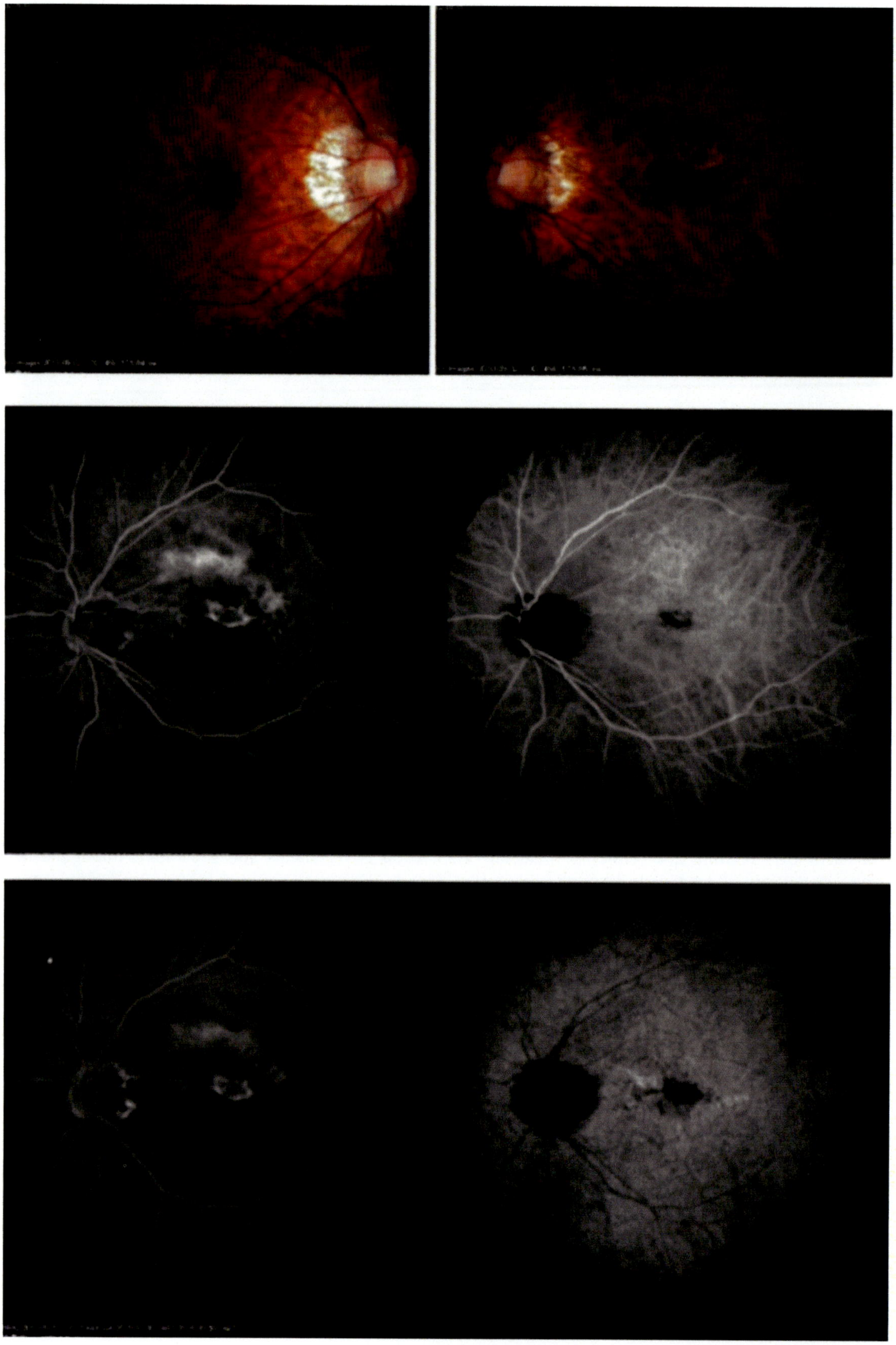

FIGURE 4.19 A 56-year-old man, with a leopard-shaped fundus in his left eye, choroidal atrophy on the temporal side of the optic disc, a gray-white lesion with a blurred boundary above the macula, and hemorrhage is observed. FA shows a high fluorescence leakage of neovascular vessels above the macula and a hypofluorescence in the macula.

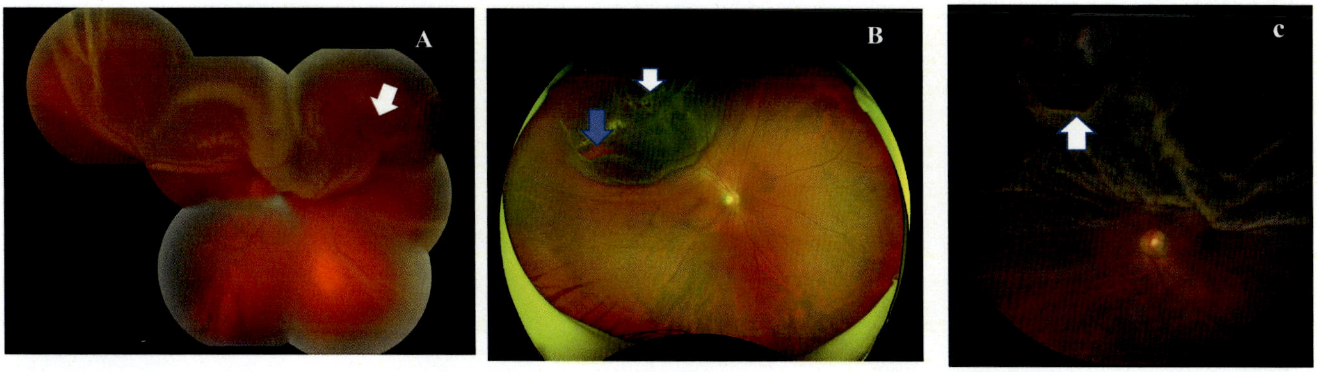

FIGURE 4.20 (A) A montage of seven fields of color fundus photography shows rhegmatogenous retinal detachment with a horseshoe-shaped retinal tear in the superior nasal retina (white arrow). (B) An ultra-widefield image (UWI, Optos Daytona, P200T) shows retinal detachment with lattice degeneration (white arrow) and a tear (blue arrow) in the supratemporal retina. (C) A typical horseshoe-shaped retinal tear (white arrow) with retinal detachment is shown by the ultra-widefield image of CLARUS 500. The advantage of UWI is that a single fundus photo can visualize peripheral retinal abnormalities with a wider field than a traditional CFP.

ophthalmologists, particularly retinal specialists, has a noticeable influence on the effectiveness of screening based on fundus photography. Both ophthalmologists and endocrinologists face noticeable time and financial constraints. DL and ML detect DR more quickly and accurately in comparison with traditional methods.[37] AI identifies characteristic lesions by learning the artificial marker from DR lesions, as well as screening and preliminary diagnosis of diseases using CFP. Google's AI algorithm identified DR lesions based on the characteristics of 128,000 retinal images graded by 54 ophthalmologists on fundus lesions of DR. At present, scholars aim to predict DM based on vascular markers. In summary, deep neural networks have suggested a significantly higher predictive performance for DR screening using CFP.[38,39] The AI-based DR screening model appeared applicable for patients in endocrinology outpatient settings.[40]

AMD

AMD is globally a leading cause of blindness in elderly aging over 65 years old, and it is a common retinal vascular disease in clinical practice. The diagnosis and follow-up of AMD rely on imaging modalities, including CFP. Early and mid-term AMD tends to be asymptomatic and easily missed, while late AMD has a distinct impact on a patient's vision, leading to limited therapeutic options. Identifying early AMD is important for intervention and inhibition of the disease progression. Based on CFP, the John Hopkins team applied a DL algorithm to identify AMD with an accuracy of 88.1%–91.6%, which is significantly consistent with manual interpretation results.[41,42] OCT combined with CFP increases the efficiency of AI algorithms for diagnosing AMD.[43,44]

Other retinochoroidal diseases

In addition to the above-mentioned vascular diseases, AI has shown a promising capability in terms of assisting the diagnosis of other diseases, such as retinal vein occlusion and central serious chorioretinopathy.[44–47] CFP is taking a predominant role in Al analysis with other ophthalmic imaging modalities, including anterior segment photography, corneal topography, anterior or posterior segment OCT, and ultrasound biomicroscopy.[48] In the future, multimodality images, especially cost-saving CFP images taken by nonmydriatic automatic cameras, will be presented and analyzed.

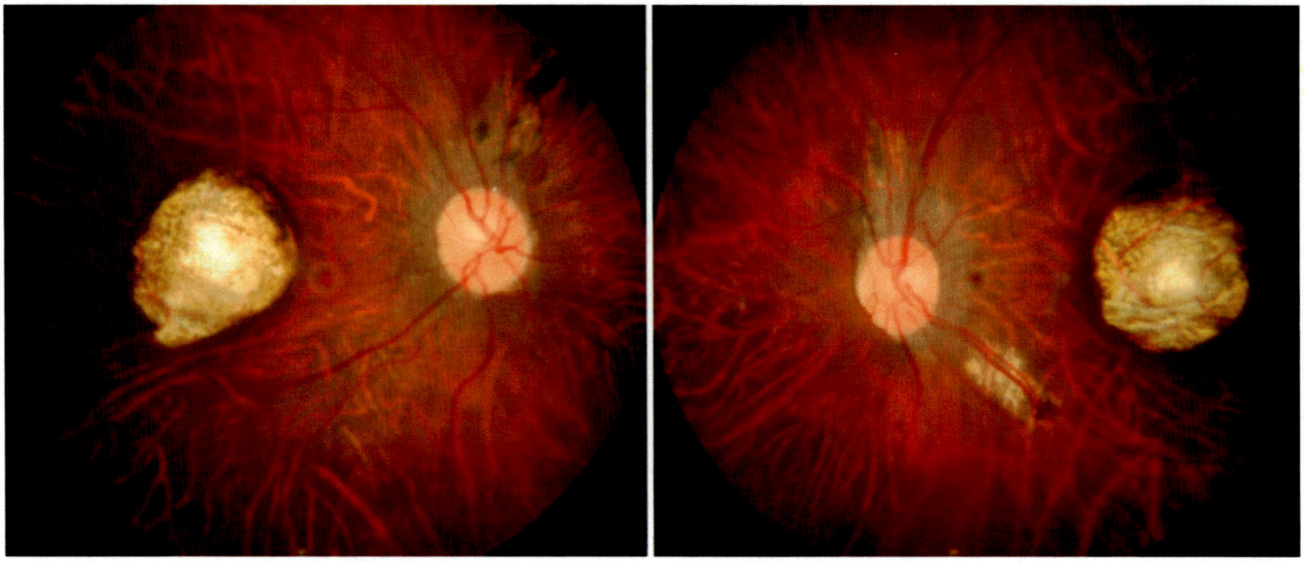

FIGURE 4.21 A 22-year-old woman with low visual acuity since childhood (the visual acuity was 6/200 in both eyes). CFP shows a leopard-like fundus, clear optic disc boundary, and reddish color and round off-white defect area in the macular. *Yellow-white* sclera tissue can be observed in the center of the lesion, residual large choroidal blood vessels and retinal blood vessels can be found in the lesion, and the edges are surrounded by pigments.

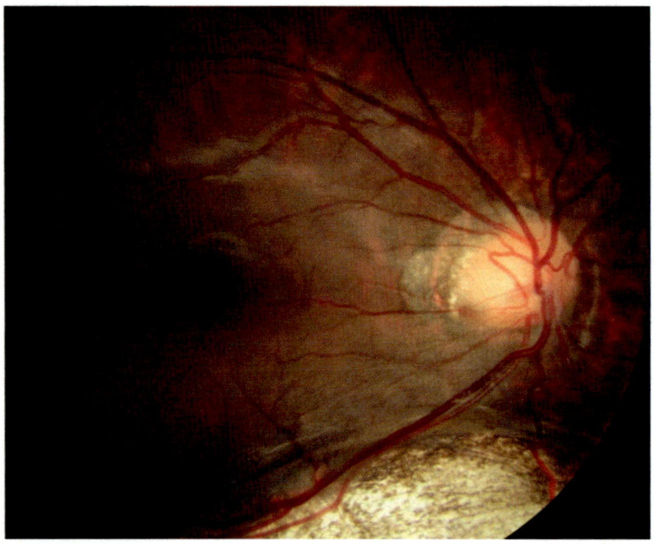

FIGURE 4.22 A 12-year-old boy had a binocular vision of 20/20. CFP shows the optic disc has a clear boundary, and a prominent depigment white zone below the optic disc.

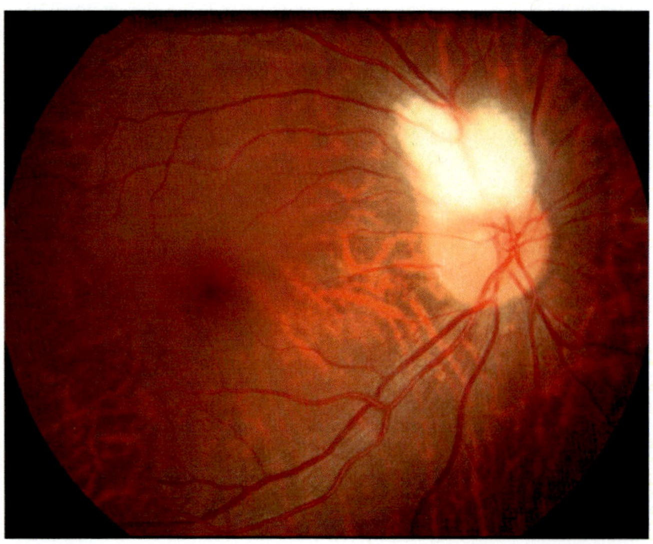

FIGURE 4.23 56-year-old woman with the best corrected visual acuity of 20/20. CFP of the right eye shows that the upper part of the optic disc is covered by thick *yellow-white* retinal myelinated nerve fibers, the vessels underneath cannot be observed.

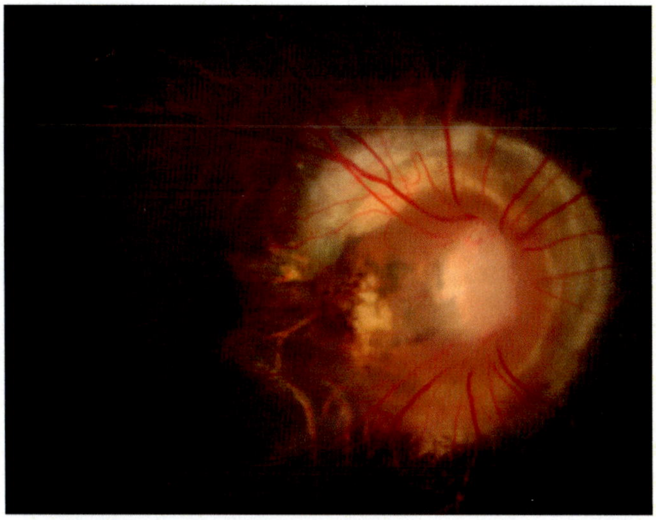

FIGURE 4.24 A 12-year-old girl with a deep bowl-shaped depression in the optic disc of her right eye, which was surrounded by a crested ridge of *yellow* and *white* rings. There is a wide area of peripapillary pigment abnormality, with zones of hyperpigmentation in clumps and hypopigmentation. An increased number of straight retinal vessels arise from the disc margin. An afferent pupillary defect was detected in the same eye.

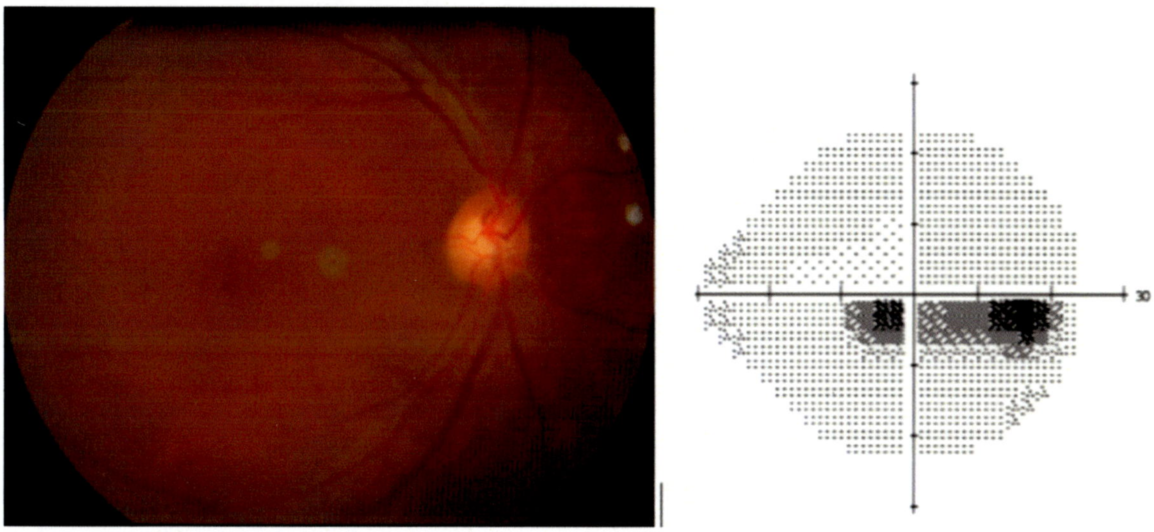

FIGURE 4.25 A 63-year-old woman with a decreased vision for 20 days. A relative afferent pupillary defect (RAPD [+]) was detected in her right eye. CFP shows edema in the lower optic disc with blurred boundaries and small pieces of blood. Cross-impression of arteries and veins can also be observed. Horizontal defects of the lower visual field associated with the physiological blind spot could be detected in the visual field examination.

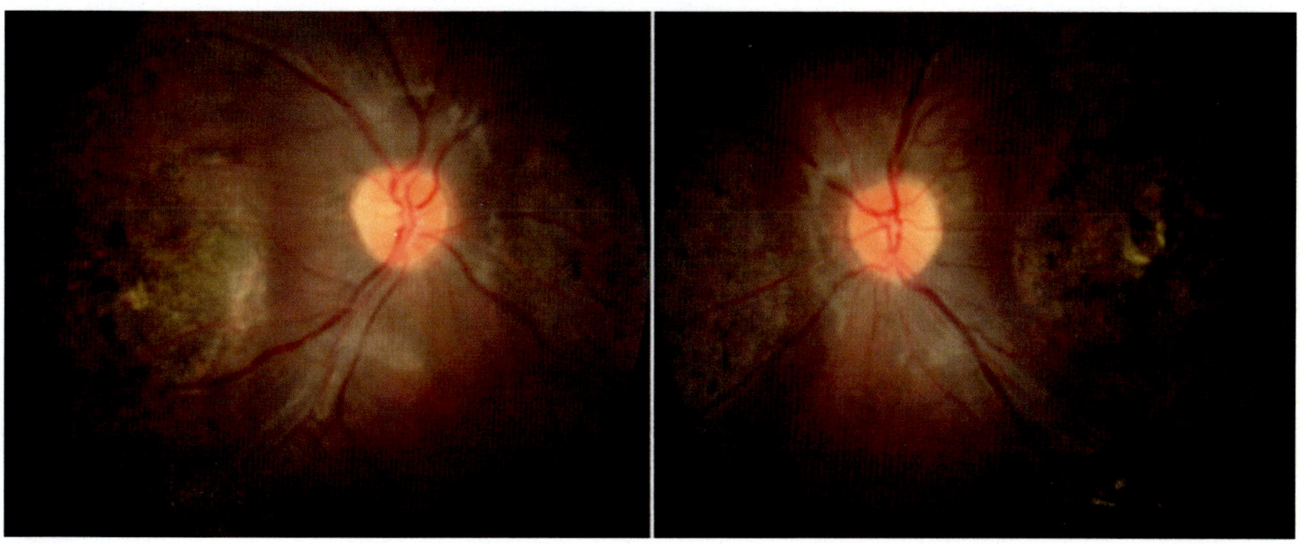

FIGURE 4.26 A 8-year-old boy who suffered from decreased visual acuity for 4 years. CFP shows the temporal side of the optic disc in both eyes is pale, the color of the nasal side is flush, and pigmentation can be found in the macula of both eyes.

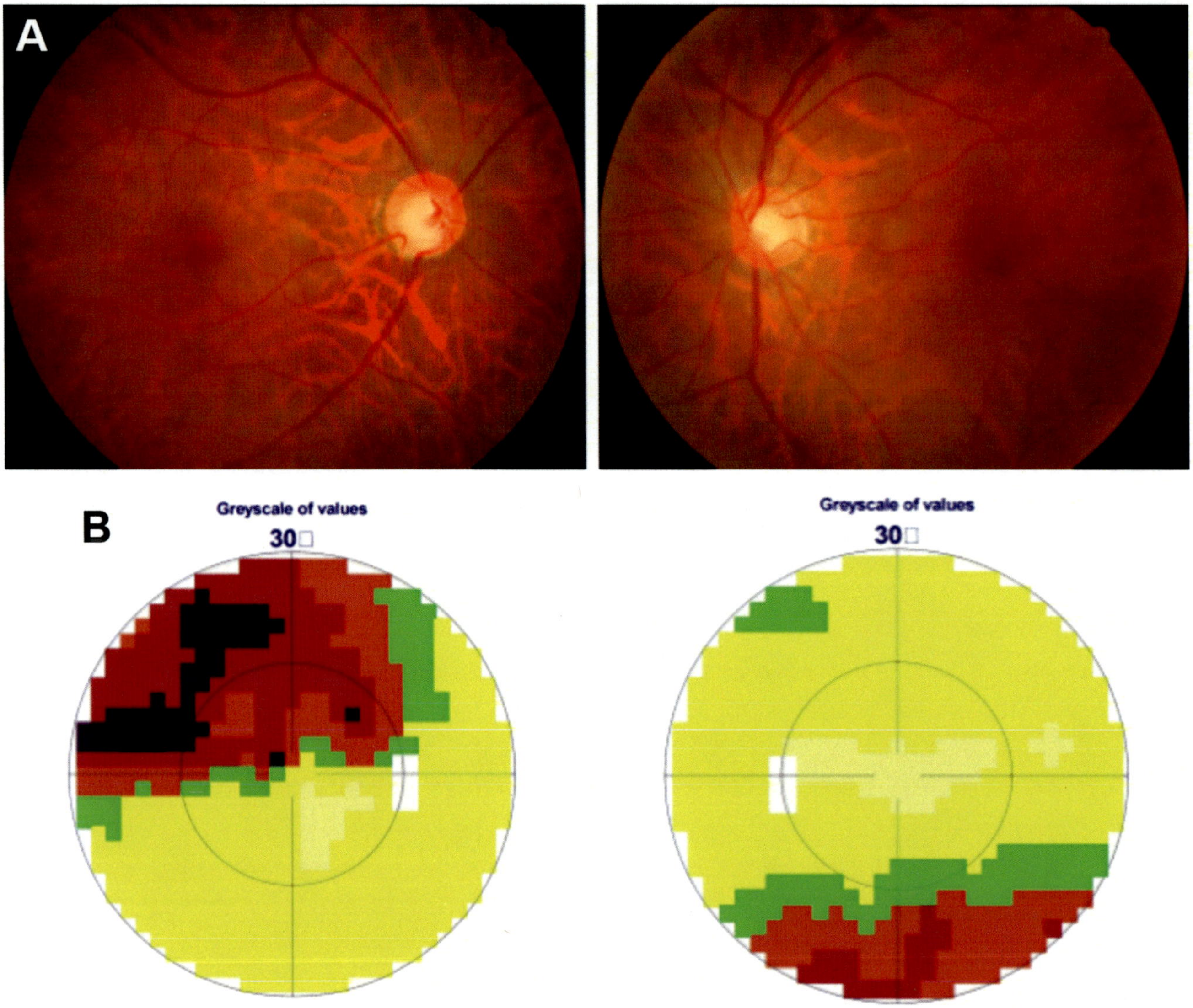

FIGURE 4.27 A 69-year-old man with binocular vision blurred for 16 years. CFP shows that the optic disc becomes lighter, the retinal nerve fiber layer of the lower optic disc shrinks, the C/D is about 0.4, the retinal blood vessels run normally, and no bleeding or exudation is found. Visual field examination shows the typical visual field defect above the right eye and the stepped visual field defect below the left eye.

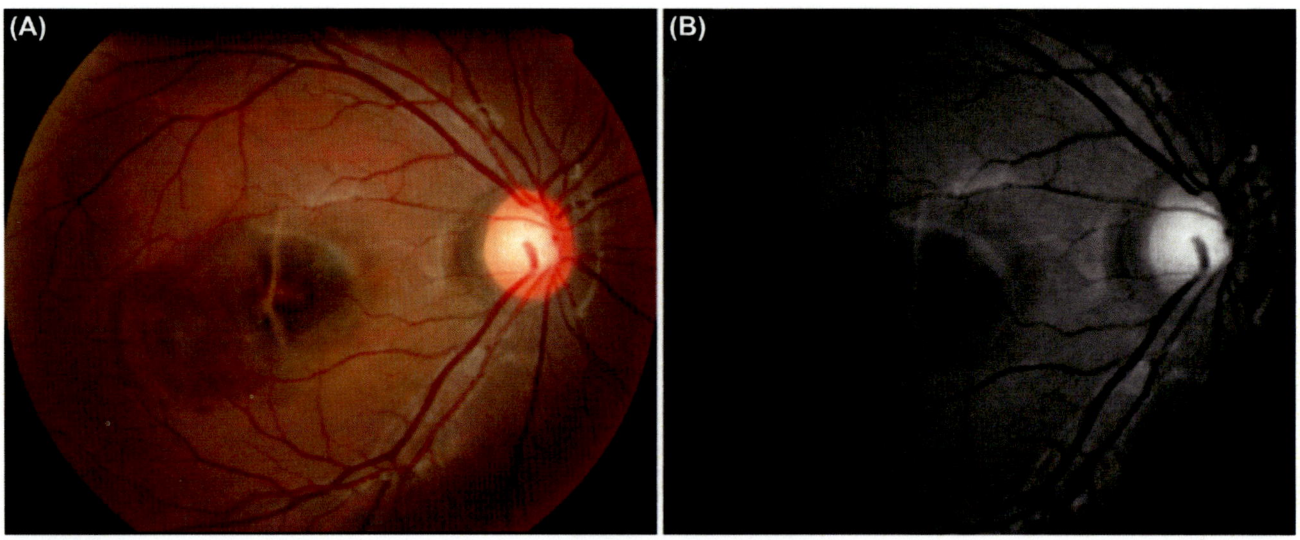

FIGURE 4.28 An 18-year-old man was diagnosed with traumatic optic neuropathy in the right eye after a boxing injury. His visual acuity was 0.01 in the right eye and 0.4 in the left eye 0.4. CFP (A) shows that the choroid in the macula of the right eye was ruptured, indicating a yellow-white crescent-shaped lesion with its concavity facing the optic disc. The infrared fundus photo shows the correspondent lesion (B).

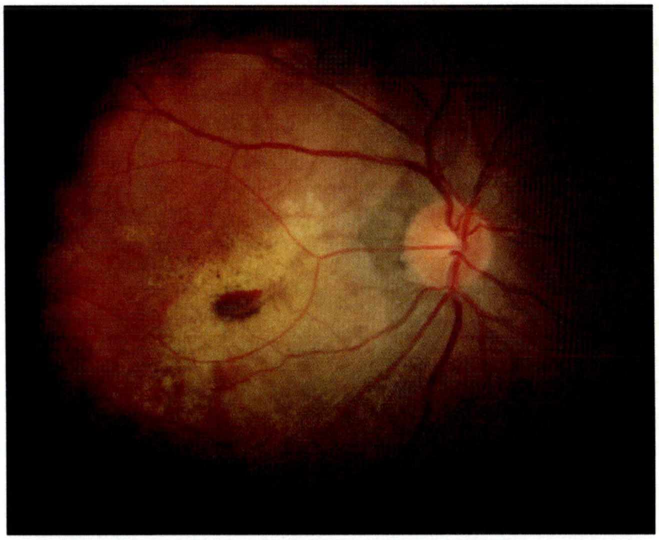

FIGURE 4.29 A 36-year-old female with left eye vision of 0.1. The CFP in her left eye shows an irregular *yellow-white* bulge in the macular area, with *orange-red* changes around it, and a flaky bleeding focus in the center of the lesion. B-mode ultrasound in the left eye showed hyperechoic plaques beside the optic disc, followed by an audio shadow.

References

1. Diabetic retinopathy study. Report Number 6. Design, methods, and baseline results. Report Number 7. A modification of the Airlie house classification of diabetic retinopathy. Prepared by the diabetic retinopathy. *Investigative Ophthalmology & Visual Science*. 1981;21(1 Pt 2).
2. Abràmoff MD, Garvin MK, Sonka M. Retinal imaging and image analysis. *IEEE Reviews in Biomedical Engineering*. 2010;3:169–208. https://doi.org/10.1109/RBME.2010.2084567.
3. Panwar N, Huang P, Lee J, et al. Fundus photography in the 21st century—a review of recent technological advances and their implications for worldwide healthcare. *Telemedicine Journal and E-health*. 2016;22(3):198–208. https://doi.org/10.1089/tmj.2015.0068.
4. Jackman T, Webster J. Milestones, rivalries and controversy, part III. *Transactions of the American Ophthalmological Society*. 2013:9–26.
5. Shekhawat NS, Niziol LM, Sharma SS, et al. The utility of routine fundus photography screening for posterior segment disease: a stepped-wedge, cluster-randomized trial in south India. *Ophthalmology*. 2021;128(7):1060–1069. https://doi.org/10.1016/j.ophtha.2020.11.025.
6. Gan K, Liu Y, Stagg B, Rathi S, Pasquale LR, Damji K. Telemedicine for glaucoma: guidelines and recommendations. *Telemedicine Journal and E-health*. 2020;26(4):551–555. https://doi.org/10.1089/tmj.2020.0009.
7. Fierson WM, Capone A. Telemedicine for evaluation of retinopathy of prematurity. *Pediatrics*. 2015;135(1):e238–e254. https://doi.org/10.1542/peds.2014-0978.
8. Vasseneix C, Bruce BB, Bidot S, Newman NJ, Biousse V. Nonmydriatic fundus photography in patients with acute vision loss. *Telemedicine Journal and E-health*. 2019;25(10):911–916. https://doi.org/10.1089/tmj.2018.0209.
9. Liew G, Mitchell P, Chiha J, et al. Retinal microvascular changes in microvascular angina: findings from the Australian heart eye study. *Microcirculation*. 2019;26(6):e12536. https://doi.org/10.1111/micc.12536.
10. van Leeuwen R, Chakravarthy U, Vingerling JR, et al. Grading of age-related maculopathy for epidemiological studies: is digital imaging as good as 35-mm film? *Ophthalmology*. 2003;110(8):1540–1544. https://doi.org/10.1016/S0161-6420(03)00501-3.
11. Sen M, Shields CL, Honavar SG, Shields JA. Coats disease: an overview of classification, management and outcomes. *Indian Journal of Ophthalmology*. June 2019;67(6):763–771. https://doi.org/10.4103/ijo.IJO_841_19.
12. Pitkänen L, Tommila P, Kaarniranta K, Jääskeläinen JE, Kinnunen K. Retinal arterial macroaneurysms. *Acta Ophthalmologica*. March 2014;92(2):101–104. https://doi.org/10.1111/aos.12210.
13. Pauleikhoff L, Agostini H, Lange C. Central serous chorioretinopathy. *Ophthalmologe*. September 2021;118(9):967–980. https://doi.org/10.1007/s00347-021-01376-7. Chorioretinopathia centralis serosa.
14. Lim LS, Mitchell P, Seddon JM, Holz FG, Wong TY. Age-related macular degeneration. *The Lancet*. 2012;379(9827):1728–1738.
15. Anantharaman G, Sheth J, Bhende M, et al. Polypoidal choroidal vasculopathy: pearls in diagnosis and management. *Indian Journal of Ophthalmology*. 2018;66(7):896.
16. Oh H. Idiopathic macular hole. *Developments in Ophthalmology*. 2014;54:150–158. https://doi.org/10.1159/000360461.
17. Williams GA, Scott IU, Haller JA, Maguire AM, Marcus D, McDonald HR. Single-field fundus photography for diabetic retinopathy screening: a report by the American Academy of Ophthalmology. *Ophthalmology*. 2004;111(5):1055–1062.
18. Zhang X, Lai D, Bao S, Hambly B, Gillies M. Triamcinolone acetonide inhibits p38MAPK activation and neuronal apoptosis in early diabetic retinopathy. *Current Molecular Medicine*. 2013;13(6):946–958.
19. Sõzmaz S, Yonekawa Y, Trese MT. Familial exudative vitreoretinopathy. *Turkish Journal of Ophthalmology*. 2015;45(4):164.

20. Hartong DT, Berson EL, Dryja TP. Retinitis pigmentosa. *The Lancet*. 2006;368(9549):1795–1809.
21. Morgan IG, Ohno-Matsui K, Saw SM. Myopia. *Lancet*. May 5, 2012;379(9827):1739–1748. https://doi.org/10.1016/s0140-6736(12)60272-4.
22. Heckenlively JR, Foxman SG, Parelhoff ES. Retinal dystrophy and macular coloboma. *Documenta Ophthalmologica*. March–April 1988;68(3–4):257–271. https://doi.org/10.1007/bf00156432.
23. Eida H, Ohira A, Amemiya T. Choroidal coloboma in two members of a family. *Ophthalmologica*. 1998;212(3):208–211. https://doi.org/10.1159/000027281.
24. Ramkumar HL, Verma R, Ferreyra HA, Robbins SL. Myelinated retinal nerve fiber layer (RNFL): a comprehensive review. *International Ophthalmology Clinics*. Fall 2018;58(4):147–156. https://doi.org/10.1097/iio.0000000000000239.
25. Lit ES, D'Amico DJ. Retinal manifestations of morning glory disc syndrome. *International Ophthalmology Clinics*. Winter 2001;41(1):131–138. https://doi.org/10.1097/00004397-200101000-00012.
26. Morrow MJ. Ischemic optic neuropathy. *Continuum (Minneap Minn)*. October 2019;25(5):1215–1235. https://doi.org/10.1212/con.0000000000000767.
27. Yu-Wai-Man P, Chinnery PF. Leber hereditary optic neuropathy. In: Adam MP, Everman DB, Mirzaa GM, et al., eds. 1993. GeneReviews(®). University of Washington, Seattle Copyright © 1993–2022, University of Washington, Seattle. Gene Reviews is a registered trademark of the University of Washington, Seattle. All rights reserved.
28. McCarthy J, Minsky ML, Rochester N, Shannon CE. A proposal for the dartmouth summer research project on artificial intelligence, August 31, 1955. *AI Magazine*. 2006;27(4), 12-12.
29. LeCun Y, Bengio Y, Hinton G. Deep learning. *Nature*. 2015;521(7553):436–444. https://doi.org/10.1038/nature14539.
30. Sheng B, Chen X, Li T, et al. An overview of artificial intelligence in diabetic retinopathy and other ocular diseases. *Frontiers in Public Health*. 2022;10:971943. https://doi.org/10.3389/fpubh.2022.971943.
31. Ting DSW, Cheung CY-L, Lim G, et al. Development and validation of a deep learning system for diabetic retinopathy and related eye diseases using retinal images from multiethnic populations with diabetes. *JAMA*. 2017;318(22):2211–2223. https://doi.org/10.1001/jama.2017.18152.
32. Brown JM, Campbell JP, Beers A, et al. Automated diagnosis of plus disease in retinopathy of prematurity using deep convolutional neural networks. *JAMA Ophthalmology*. 2018;136(7):803–810. https://doi.org/10.1001/jamaophthalmol.2018.1934.
33. Ting DSW, Pasquale LR, Peng L, et al. Artificial intelligence and deep learning in ophthalmology. *British Journal of Ophthalmology*. 2019;103(2):167–175. https://doi.org/10.1136/bjophthalmol-2018-313173.
34. Qu M, Ni C, Chen M, et al. Automatic diabetic retinopathy diagnosis using adjustable ophthalmoscope and multi-scale line operator. *Pervasive and Mobile Computing*. 2017;41:490–503. https://doi.org/10.1016/j.pmcj.2017.04.003.
35. Kilpatrick ES, Rigby AS, Atkin SL. The diabetes control and complications trial: the gift that keeps giving. *Nature Reviews Endocrinology*. 2009;5(10):537–545. https://doi.org/10.1038/nrendo.2009.179.
36. Microvascular complications and foot care. *Diabetes Care*. 2021;44(Suppl 1):S151–S167. https://doi.org/10.2337/dc21-S011.
37. Reddy SS, Sethi N, Rajender R, Mahesh G. Withdrawn: extensive analysis of machine learning algorithms to early detection of diabetic retinopathy. *Materials Today: Proceedings*. 2020. https://doi.org/10.1016/j.matpr.2020.10.894.
38. Raman R, Srinivasan S, Virmani S, Sivaprasad S, Rao C, Rajalakshmi R. Fundus photograph-based deep learning algorithms in detecting diabetic retinopathy. *Eye (London)*. 2019;33(1). https://doi.org/10.1038/s41433-018-0269-y.
39. Babenko B, Mitani A, Traynis I, et al. Detection of signs of disease in external photographs of the eyes via deep learning. *Nature Biomedical Engineering*. 2022. https://doi.org/10.1038/s41551-022-00867-5.
40. Keel S, Lee PY, Scheetz J, et al. Feasibility and patient acceptability of a novel artificial intelligence-based screening model for diabetic retinopathy at endocrinology outpatient services: a pilot study. *Scientific Reports*. 2018;8(1):4330. https://doi.org/10.1038/s41598-018-22612-2.
41. Burlina PM, Joshi N, Pekala M, Pacheco KD, Freund DE, Bressler NM. Automated grading of age-related macular degeneration from color fundus images using deep convolutional neural networks. *JAMA Ophthalmology*. 2017;135(11):1170–1176. https://doi.org/10.1001/jamaophthalmol.2017.3782.
42. Grassmann F, Mengelkamp J, Brandl C, et al. A deep learning algorithm for prediction of age-related eye disease study severity scale for age-related macular degeneration from color fundus photography. *Ophthalmology*. 2018;125(9):1410–1420. https://doi.org/10.1016/j.ophtha.2018.02.037.
43. ElTanboly A, Ismail M, Shalaby A, et al. A computer-aided diagnostic system for detecting diabetic retinopathy in optical coherence tomography images. *Medical Physics*. 2017;44(3):914–923. https://doi.org/10.1002/mp.12071.
44. Treder M, Lauermann JL, Eter N. Automated detection of exudative age-related macular degeneration in spectral domain optical coherence tomography using deep learning. *Graefe's Archive for Clinical and Experimental Ophthalmology*. 2018;256(2):259–265. https://doi.org/10.1007/s00417-017-3850-3.
45. Wang R, Zheng L, Xiong C, et al. Retinal optic disc localization using convergence tracking of blood vessels. *Multimedia Tools and Applications*. 2017;76(22):23309–23331. https://doi.org/10.1007/s11042-016-4146-z.
46. Sivaswamy J, Krishnadas S, Chakravarty A, Joshi G, Tabish AS. A comprehensive retinal image dataset for the assessment of glaucoma from the optic nerve head analysis. *JSM Biomedical Imaging Data Papers*. 2015;2(1):1004.
47. van der Heijden AA, Abramoff MD, Verbraak F, van Hecke MV, Liem A, Nijpels G. Validation of automated screening for referable diabetic retinopathy with the IDx-DR device in the Hoorn diabetes care system. *Acta Ophthalmologica*. 2018;96(1):63–68. https://doi.org/10.1111/aos.13613.
48. Sahlsten J, Jaskari J, Kivinen J, et al. Deep learning fundus image analysis for diabetic retinopathy and macular edema grading. *Scientific Reports*. 2019;9(1):10750. https://doi.org/10.1038/s41598-019-47181-w.

Chapter 5

Fluorescein angiography

Federico Ricardi[1], Michele Reibaldi[1], Francesco Bandello[2,3] and Enrico Borrelli[1,2]
[1]*Department of Ophthalmology, University of Turin, Italy;* [2]*School of Medicine, Vita-Salute San Raffaele University, Milan, Italy;* [3]*Division of Head and Neck, Ophthalmology Unit, IRCCS San Raffaele Scientific Institute, Milan, Italy*

Introduction

First introduced in 1961 by Novotny and Alvis,[1] fluorescein angiography (FA) has gained a fundamental role in the diagnostic workflow of retinal diseases. Although some new technologies including optical coherence tomography angiography (OCTA) are emerging and their adoption in everyday clinical practice has risen over time, different guidelines still consider FA as the gold standard for the diagnosis of several retinal disorders.

In this chapter, we will cover the fundamentals of this examination, from the chemical basics to the angiographic aspects that may be appreciated in healthy and pathological eyes.

Basics of fluorescein angiography

Fluorescein is an organic compound that is available as a water-soluble red powder. The salt form of fluorescein (i.e., sodium fluorescein) is used for ophthalmic examinations ($NaC_{20}H_{12}O_5$). Sodium fluorescein is a hydrocarbon with an absorption spectrum between 465 and 490 nm (in the blue-light spectrum) and an emission spectrum with a peak at 520−530 nm (green wavelength). The intensity of fluorescein is also dependent on pH, with activity over the range of 5−9.

In clinical practice, a solution of 1.5 mL with a 10% concentration of sodium fluorescein is usually injected intravenously. Sodium fluorescein is a low-weight molecule (367 Da) that readily diffuses through the fenestrated vessels of the choriocapillaris, while this molecule does not diffuse through the healthy endothelial cells of nonfenestrated retinal vessels and the retinal pigment epithelium (RPE).

Following the injection, most of the sodium fluorescein binds to plasma proteins and does not participate in the fluorescence process. The remaining portion (i.e., 20%−25%) which is not bound to plasma protein is responsible for the fluorescence that is commonly visualized during the angiography procedure. The unbound fluorescein is also free to diffuse in the extravascular tissues, except for larger choroidal vessels, retinal vessels (inner blood−retinal barrier), and cerebral vessels (blood−brain barrier). The fluorescein cannot diffuse in the intraretinal space because of the RPE with its tight junctions (zonulae occludentes), which represent the outer blood−retinal barrier. Due to the diffusion in the systemic extravascular space, skin and mucosal surfaces can be characterized by an orange color after the examination. Fluorescein is completely excreted and eliminated from the body through renal and hepatic metabolism within 24 h.

Seldom (<5%), mild adverse reactions to fluorescein can occur, such as nausea, vomiting, and injection site reactions. Moderately severe reactions (<1%) include hives, shortness of breath, and vasovagal reactions. Conversely, major adverse allergic reactions such as anaphylactic shock, seizure, and cardiovascular collapse are extremely rare (<0.001%). Injection of fluorescein should be contraindicated during pregnancy or in patients with a history of previous severe allergic reactions to fluorescein or other substances.[2−4]

The acquisition of a fluorescein angiogram is possible with the use of a scanning laser ophthalmoscope (SLO). During the latter examination, the light is emitted by a laser source of a specific wavelength, passes through a beam splitter, hits the eye structures, and is absorbed by the fluorescein molecules. A detector then receives the electrons released by the excited fluorescein creating an image. Several commercially available photographic systems are available in the market. Typical lens systems are available for 30 degrees imaging, applicable for excellent macular detail, 55 degrees imaging for wider field imaging, and even widefield lens systems for peripheral angiography. Color and/or red-free fundus photographs may be obtained simultaneously.

Retinal and Choroidal Vascular Diseases of the Eye. https://doi.org/10.1016/B978-0-443-15583-3.00005-6

Phases of fluorescence angiography examination

The arm-retina time is the time between the injection of fluorescein in the antecubital vein and the moment the dye appears in the central retinal artery. It can be prolonged in some patients due to different flow rate perfusion or blood pressure levels. It can be assumed as physiologic in the range of 7−15 s.

Commonly, the fluorescence angiography examination can be divided into different phases which are consistent with the path made by the dye inside the vascular structures of the eye.

Briefly, fluorescein travels from the internal carotid artery to the ophthalmic artery and finally reaches the posterior ciliary and central retinal arteries. At the same time, the short posterior ciliary arteries are perfused, and this finally determines the choroidal phase of fluorescein angiography.

Rapidly after that, the central retinal artery filling characterizes the arterial phase. The arteriovenous phase lasts until fluorescein reaches the retinal veins and is classically subdivided into early, intermediate, and late arteriovenous phases. Finally, the late phase represents the moment when the dye gradually fades away from the eye.

Early phase

- Choroidal flush: the fluorescein appears visible in the choroid about 10 s after venous injection and this fluorescence occurs a few seconds earlier than fluorescein becoming visible within retinal vessels. As stated earlier, choroidal circulation is characterized by fenestrated vessels allowing the movement of fluorescein into the extravascular space with a consequent diffuse and indistinct choroidal fluorescence (Fig. 5.1). Furthermore, the RPE acts as a "filter" and partially obscures the choroidal fluorescence. The latter features do not allow a detailed assessment of the choroidal circulation using FA. The "choroidal flush" phase is important to visualize the possible presence of the cilioretinal artery which originates from the posterior ciliary choroidal arteries.
- Arterial phase: the retinal arterial filling is appreciated 1 or 2 s after the choroidal flush and about 12 s after the dye injection. The hyperfluorescent retinal vessels differ from the hypofluorescent remaining macular area.
- Arterovenous phase: during this phase, there is a complete retinal arterial filling and an initial retinal venous filling. The venous vessels are usually characterized by a greater hyperfluorescence of their walls which is determined by the venous laminar flow occurring near the vascular endothelium. On the contrary, the arterial vessels are completely hyperfluorescent due to their turbulent flow (Fig. 5.2).

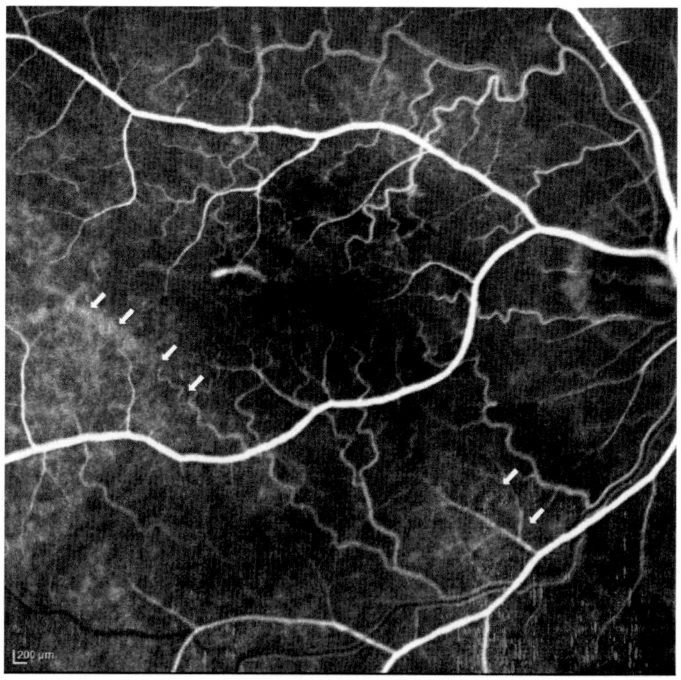

FIGURE 5.1 Choroidal flush: in this angiogram is visible an early and irregularly diffused hyperfluorescence (*white arrows*).

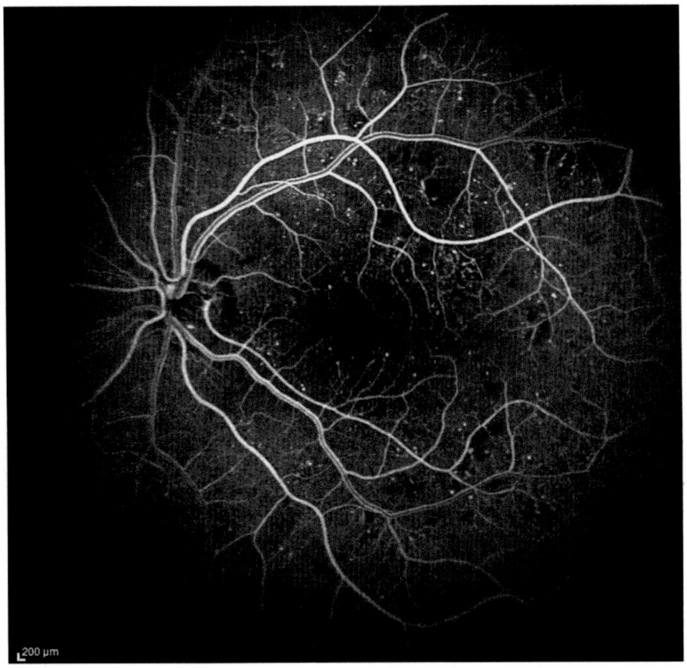

FIGURE 5.2 Arterovenous phase: in this phase is clearly visible the complete retinal filling and the initial venous filling with the laminar flow occurring near the vascular endothelium.

- Venous phase: a complete filling of the veins occurs about 30 s after the fluorescein injection. During the venous phase, the perifoveal capillary network is clearly visualized.

Intermediate phase

It occurs 2–4 min after the fluorescein injection. Veins and arteries are similar in fluorescence. During this phase, the intensity of the fluorescence slowly decreases over time.

Late phase

It is characterized by the progressive elimination of the fluorescein from the retinal and choroidal. This phase occurs approximately 6–8 min after the fluorescein injection. Of note, while the delay in the elimination of fluorescein from the optic disc (i.e., late staining) is considered to be physiological, other hyperfluorescent regions suggest the presence of vascular anomalies.

Fluorescence patterns

Depending on the pathology examined by the angiography, two main patterns of fluorescence could be discerned, as follows: hypofluorescence or hyperfluorescence.

Hypofluorescence

Hypofluorescence may be due to a reduced or absence of fluorescence. This could be secondary to:

- Blocking defect: the preretinal, retinal, or subretinal deposition of abnormal materials, including blood, lipids, exudates, lipofuscin, and pigments (xanthophyll or melanin) may determine an obstruction for the direct visualization of the choroidal and/or retinal fluorescence. In this case, it is useful to determine the anatomical position of the blocking material, which is in turn important for identifying the etiology of the disorder. As an example, a preretinal hemorrhage secondary to proliferative diabetic retinopathy may block the fluorescence of both the retinal and choroidal vasculature, while a subretinal exudate secondary to age-related macular degeneration obscures the choroidal circulation (Fig. 5.3).
- Filling defect: due to a retinal and/or choroidal vascular hypoperfusion. The most common causes of retinal hypoperfusion include retinal vein and arterial occlusions and diabetic retinopathy (Fig. 5.3). Choroidal hypoperfusion can be caused by central retinal artery occlusion, giant cell arteritis (Horton syndrome), or hypertensive choroidopathy.

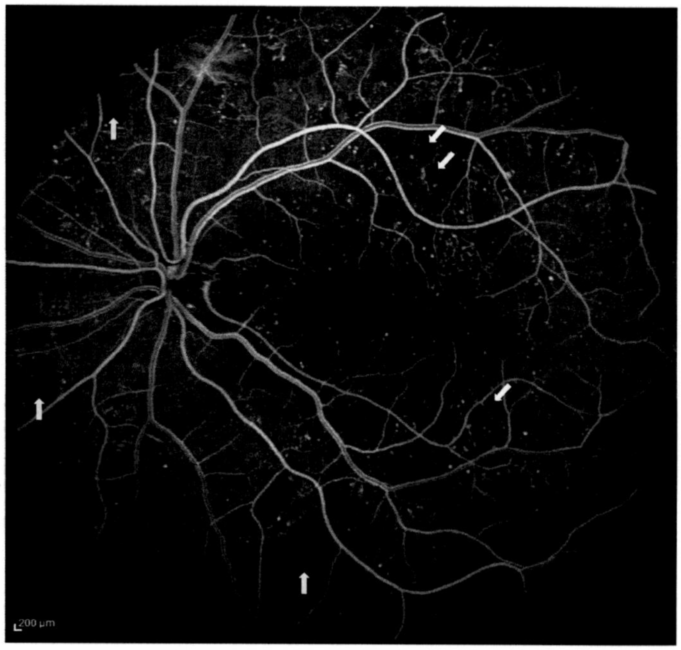

FIGURE 5.3 Blocking effect and hypoperfusion: in this angiogram taken in a patient with diabetic retinopathy are visible areas of hypofluorescence due to blocking effect (e.g., retinal exudates, *white arrows* on the right of the image) and hypoperfusion (e.g., ischemic retinal areas, *yellow arrows* on the left of the image).

Hyperfluorescence

An abnormal hyperfluorescence is mainly caused by:

- Autofluorescence, which is the emission of fluorescent light from ocular structures in the absence of sodium fluorescein. Optic nerve head drusen and astrocytic hamartoma may determine autofluorescence.
- Pseudofluorescence, which occurs when the blue exciter and green barrier filters overlap. The light passes through the system and reflects off highly reflective surfaces as light-colored or white fundus changes (e.g., sclera, exudate, scar tissue, myelinated nerve fibers, foreign body) stimulating the detector with nonfluorescent light. This effect is called pseudofluorescence. Nowadays, filters are well-matched and pseudofluorescence is rare.
- Window effect, which depends on the disruption/atrophy of the outer retina/RPE with a consequent greater visualization of the underlying choroidal fluorescence (Fig. 5.4).
- Leakage is caused by the extravasation of fluorescein. Leakage can be caused by a dysfunction of the inner- or outer-blood retinal barrier (Fig. 5.5). In addition to retinal vasculature and retinal pigment epithelial abnormalities, leakage can be also observed in a variety of retinal disorders associated with the development of new blood vessels (e.g., AMD and proliferative DR).
- Staining refers to a late hyperfluorescence secondary to an accumulation of fluorescein dye which typically increases in intensity throughout the angiogram although the margins remain distinct (Fig. 5.6). Drusen and chorioretinal scars may show staining. Of note, also normal structures such as optic disc and sclera may stain. However, the scleral staining is visible only when there is an impairment or absence of the EPR (window effect) with associated choroidal atrophy.
- Pooling, which refers to a late hyperfluorescence secondary to localized accumulation of fluorescein dye within a distinct anatomical space. The pooling-associated hyperfluorescence has a sharp shape and no shading, as compared to leakage-associated hyperfluorescence. Pooling can occur in all those pathologies associated with an intraretinal, subretinal, or sub-RPE space, including serous neuroretinal detachments associated with central serous chorioretinopathy.

Fig. 5.7 summarizes the main causes of hyperfluorescence and hypofluorescence.

Limits of fluorescein angiography

Despite presenting some risk for moderate and rare complications, the main limitation of FA is the poor anatomical discrimination of the retinal vasculature.[5] The decline in visualization with retinal depth appears to be secondary to light scattering from the retina degrading the angiographic image at deeper levels.[6]

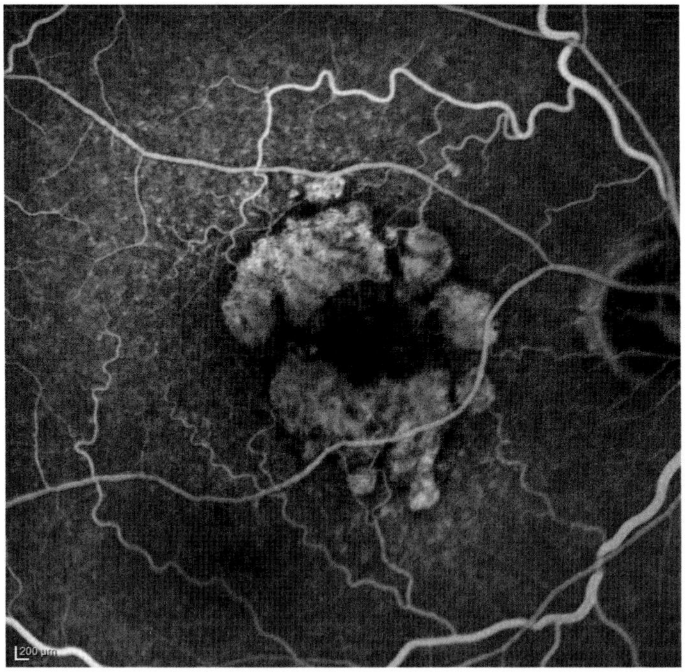

FIGURE 5.4 Window effect: hyperfluorescence due to the absence of retinal pigmented epithelium in geographic atrophy.

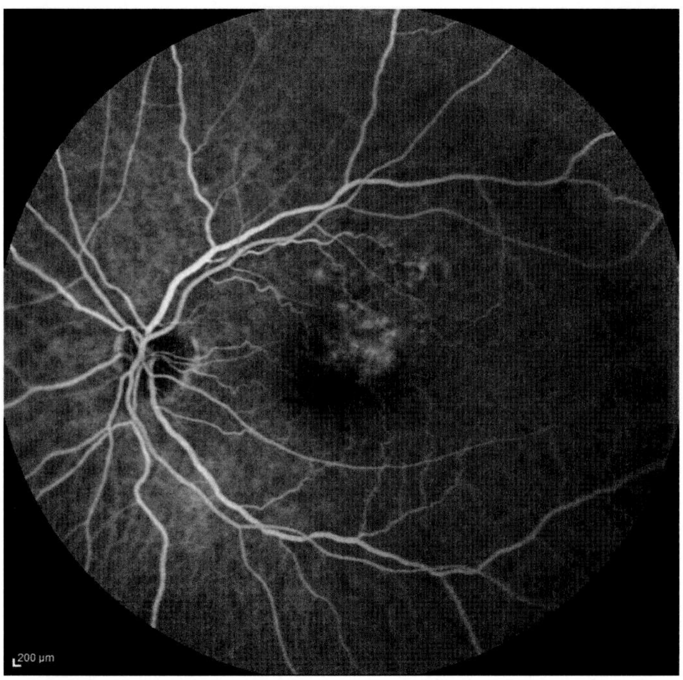

FIGURE 5.5 Leakage: hyperfluorescence due to extravasation of the fluorescein dye.

Widefield fluorescein angiography

Using a single capture, FA images typically cover 30–55 degrees of the central retina. To visualize a larger area of the retina, we may perform a photomontage of traditional angiograms (e.g., seven standard fields [7SF] protocol detecting 75 degrees of the retina) or, alternatively, employ a dedicated lens (e.g., Spectralis ultra-widefield lens [105 degrees]) applied to a confocal scanning laser ophthalmoscope (cSLO). However, the latter techniques are usually characterized by some issues, including the necessity of a complete mydriasis, more advanced technician skills,

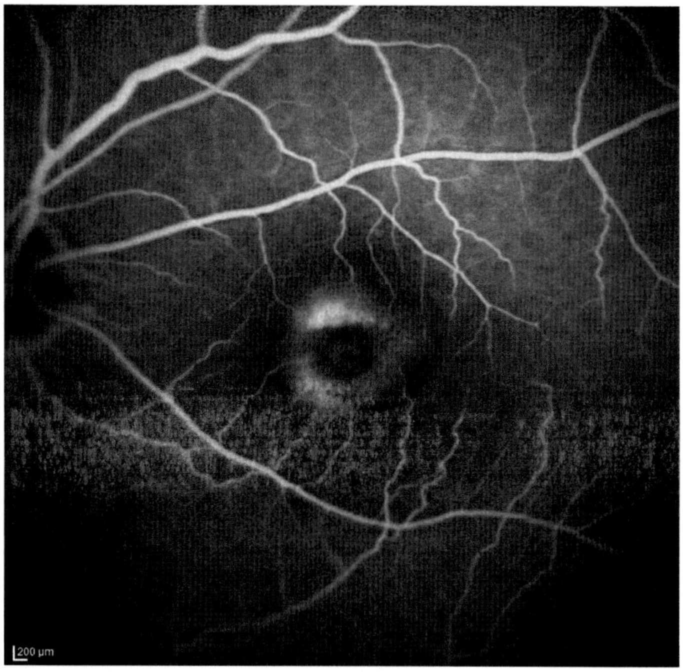

FIGURE 5.6 Staining due to the accumulation of dye in a patient with vitelliform material accumulation.

patient compliance, and a longer time of acquisition. In this scenario, widefield (WF) and ultra-widefield (UWF) FA devices have significantly impacted the diagnosis and management of retinal diseases, as they grant imaging of up to 200 degrees of the posterior pole with a single acquisition. UWF-FA imaging has been shown to be relevant in many retinal disorders including diabetic retinopathy,[7,8] age-related macular degeneration,[9] retinal vein occlusion,[10] uveitis,[11] and retinal dystrophies.[12,13]

Comparison with optical coherence tomography angiography

Although optical coherence tomography angiography (OCTA) is a noninvasive imaging technique which is able to show retinal perfusion, this technique is still characterized by a limited field of view (FOV) as compared to WF and UWF FA. The terms WF and UWF applied to OCTA have been recently defined by a consensus formed by the International Widefield Imaging Study Group as 60−100 degrees and 110−120 degrees of FOV, respectively.[14] Limited FOV in OCTA imaging is mainly secondary to the limits of the currently available scan acquisition hardware. Despite limitations, WF- and UWF-OCTA have been largely employed to characterize several retinal and choroidal disorders.[15−21]

Clinical applications of fluorescein angiography

Drusen

Drusen are deposits of extracellular material, particularly rich in proteins and lipids, which are typically localized below the retinal pigment epithelium (RPE). In the early phases of FA, regions with drusen may be hyperfluorescent in the presence of associated RPE atrophy because of a window defect or, alternatively, large drusen can be hypofluorescent as they mask the choroidal fluorescence. In the late FA phases, according to their composition and the presence of hydrophilic molecules such as phospholipids, drusen can be hyperfluorescent in the presence of dye staining.

Geographic atrophy

The term geographic atrophy (GA) refers to a late stage of AMD characterized by atrophy of the RPE, outer retinal layers, and choriocapillaris. On FA, GA is typically hyperfluorescent because of a window defect. Depending on the amount of choriocapillaris atrophy, major choroidal vessels can be also visualized.

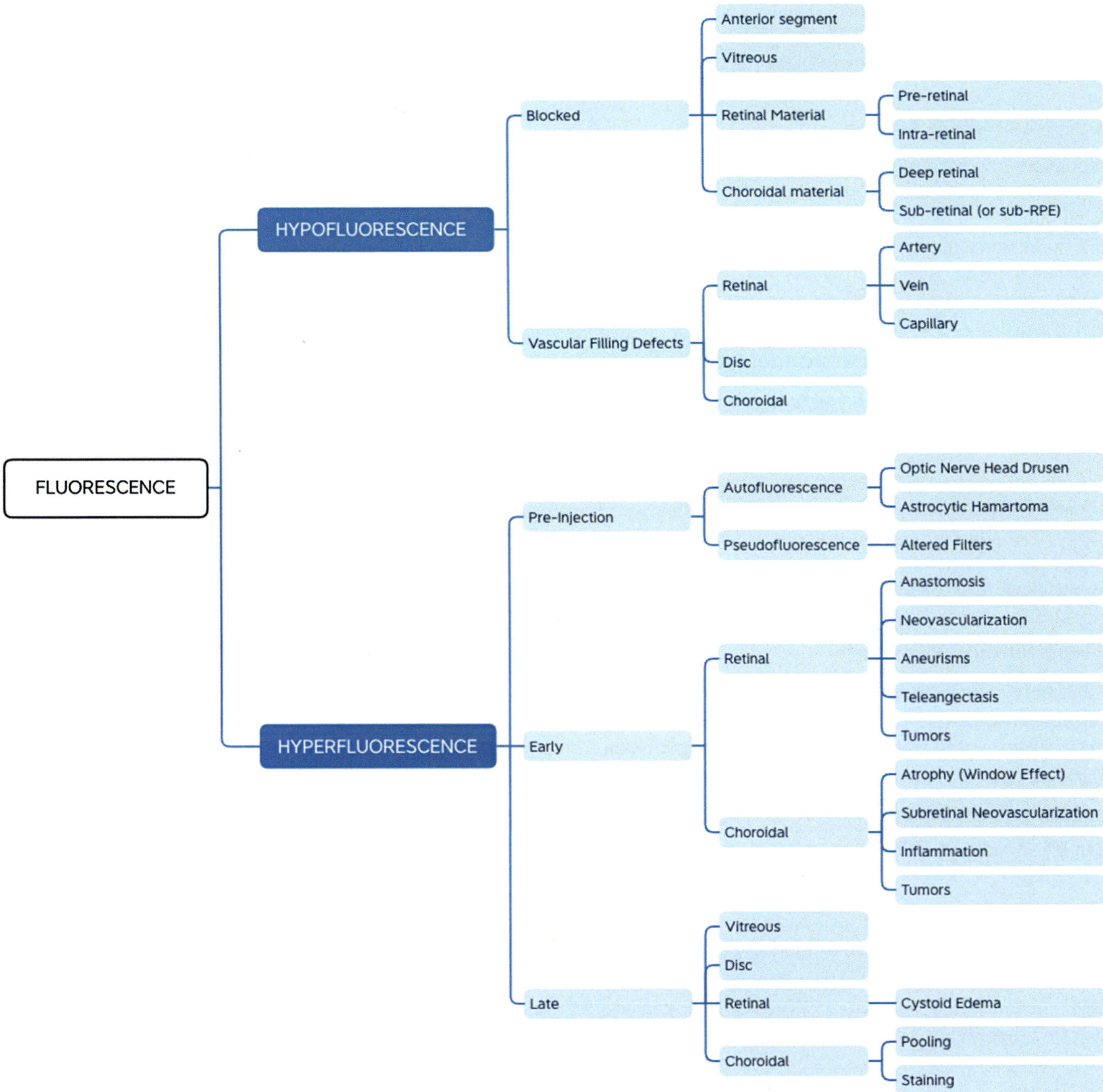

FIGURE 5.7 The main causes of hyperfluorescence and hypofluorescence.

Macular neovascularization

The development of macular neovascularization (MNV) may complicate different macular disorders including AMD. Type 1 MNV is characterized by pathologic angiogenesis occurring between the RPE and its basal lamina and the remainder of Bruch's membrane. Type 2 MNV is characterized by the presence of pathologic vessels in the subretinal space anterior to the RPE.

Type 3 MNV represents a peculiar form of neovascularization that typically originates from the retinal vessels.

On FA, type 1 MNV frequently illustrates a poorly defined or "occult" pattern, which has been described as an area of stippled hyperfluorescence with irregular RPE elevation that may have poorly demarcated boundaries. This type of MNV may be difficult to identify on FA and was often undetected in the initial Macular Photocoagulation Studies (MPS); hence the term "occult" was applied. Type 2 MNV typically exhibits a well-defined or "classic" pattern with FA. It is characterized by vascular leakage which is evident in the intermediate and late phases of the FA examination. Type 3 MNV is usually characterized by intraretinal/subretinal leakage with indistinct margins (i.e., minimally classic or occult pattern).

Cystoid macular edema

Nontractional cystoid macular edema (CME) may characterize different vascular and/or retinal disorders (e.g., diabetic retinopathy) and is usually secondary due to an alteration of the retinal capillary permeability. CME is featured by a cystic accumulation of extracellular fluid mainly localized at the level of the outer plexiform layer (Henle's fiber layer).

In the early phases of the FA exam, a diffuse macular hyperfluorescence is usually visualizable and is thought to be secondary to the relocation of macular pigments. The progressive leakage of dye outside the compromised capillaries determines the formation of cystic hyperfluorescent spaces in the late FA.

Nonproliferative diabetic retinopathy

Diabetic retinopathy (DR) is due to a disruption of pericytes covering the external wall of retinal capillaries, resulting in a damage to the inner retinal blood barrier. Depending on the DR severity, lesions that include microaneurysms, small retinal hemorrhages, hard exudates, cotton wool spots, and macular edema may be visualized.

On FA images, microaneurysms appear as small hyperfluorescent dots, while hard exudates and hemorrhages are typically hypofluorescent because of blocked fluorescence. Hypoperfused retinal areas are hypofluorescent on FA images throughout the examination. Retinal capillaries near areas of nonperfusion can be seen as enlarged and tortuous vessels, which are usually referred to as intraretinal microvascular anomalies (IRMA). In the presence of neovessels, fluorescein dye leakage may be visualized in the late phases.

Hypertensive retinopathy

Chronic systemic hypertension may lead to the sclerotization of retinal arteries and this may eventually cause reduced retinal perfusion. On FA, hypertensive retinopathy may be characterized by hypofluorescent areas secondary to either the presence of cotton wool spots or retinal hypoperfusion.

Central retinal artery occlusion

On FA, a delayed filling of the arterial vessels is usually visualizable.

Retinal vein occlusion

On FA, a delayed filling of retinal veins may be displayed. Also, leakage of dye from vessels may occur during the examination. FA may be also useful in determining the amount of nonperfused retina to discriminate between nonischemic and ischemic CRVO patterns. Of note, in the presence of superficial hemorrhages, it may be challenging to determine whether hypofluorescent areas are secondary to a blockage effect or, alternatively, they represent a sign of retinal hypoperfusion.

Central serous chorioretinopathy

Central serous chorioretinopathy (CSC) is a disease that is commonly characterized by an accumulation of subretinal fluid under the neurosensory retina associated with pachychoroid. In the early phases of FA, a reduction in choroidal fluorescence may be associated with the presence of subretinal fluid. The distribution and pooling of dye in SRF results in diffuse circular hyperfluorescence in the mid and late phases. Serous PED, a common finding in CSC, shows early hyperfluorescence with a progressive increase in the intensity due to the pooling of dye, but the size of hyperfluorescence remains unchanged.

References

1. Novotny HR, Alvis DL. A method of photographing fluorescence in circulating blood in the human retina. *Circulation*. 1961;24:82–86. https://doi.org/10.1161/01.CIR.24.1.82.

2. Halperin LS, Olk RJ, Soubrane G, Coscas G. Safety of fluorescein angiography during pregnancy. *American Journal of Ophthalmology*. 1990;109(5):563–566. https://doi.org/10.1016/S0002-9394(14)70686-5.
3. Yannuzzi LA. Fluorescein angiography complication survey. *Ophthalmology*. 1986;93:33697–33702. https://doi.org/10.1016/S0161-6420.
4. Kornblau IS, El-Annan JF. Adverse reactions to fluorescein angiography: a comprehensive review of the literature. *Survey of Ophthalmology*. 2019;64(5):679–693. https://doi.org/10.1016/j.survophthal.2019.02.004.
5. Mendis KR, Balaratnasingam C, Yu P, et al. Correlation of histologic and clinical images to determine the diagnostic value of fluorescein angiography for studying retinal capillary detail. *Investigative Ophthalmology and Visual Science*. 2010;51(11):5864–5869. https://doi.org/10.1167/iovs.10-5333.
6. Weinhaus RS, Burke JM, Delori FC, Snodderly DM. Comparison of fluorescein angiography with microvascular anatomy of macaque retinas. *Experimental Eye Research*. 1995;61(1):1–16. https://doi.org/10.1016/S0014-4835(95)80053-0.
7. Patel RD, Messner LV, Teitelbaum B, Michel KA, Hariprasad SM. Characterization of ischemic index using ultra-widefield fluorescein angiography in patients with focal and diffuse recalcitrant diabetic macular edema. *American Journal of Ophthalmology*. 2013;155(6):1038–e2. https://doi.org/10.1016/j.ajo.2013.01.007.
8. Wessel MM, Nair N, Aaker GD, Ehrlich JR, D'Amico DJ, Kiss S. Peripheral retinal ischaemia, as evaluated by ultra-widefield fluorescein angiography, is associated with diabetic macular oedema. *British Journal of Ophthalmology*. 2012;96(5):694–698. https://doi.org/10.1136/bjophthalmol-2011-300774.
9. Tan CS, Heussen F, Sadda SR. Peripheral autofluorescence and clinical findings in neovascular and non-neovascular age-related macular degeneration. *Ophthalmology*. 2013;120(6):1271–1277. https://doi.org/10.1016/j.ophtha.2012.12.002.
10. Singer M, Tan CS, Bell D, Sadda SR. Area of peripheral retinal nonperfusion and treatment response in branch and central retinal vein occlusion. *Retina*. 2014;34(9):1736–1742. https://doi.org/10.1097/IAE.0000000000000148.
11. Nicholson BP, Nigam D, Miller D, et al. Comparison of wide-field fluorescein angiography and 9-field montage angiography in uveitis. *American Journal of Ophthalmology*. 2014;157(3):673–677. https://doi.org/10.1016/j.ajo.2013.12.005.
12. Oishi A, Ogino K, Makiyama Y, Nakagawa S, Kurimoto M, Yoshimura N. Wide-field fundus autofluorescence imaging of retinitis pigmentosa. *Ophthalmology*. 2013;120(9):1827–1834. https://doi.org/10.1016/j.ophtha.2013.01.050.
13. Ogura S, Yasukawa T, Kato A, et al. Wide-field fundus autofluorescence imaging to evaluate retinal function in patients with retinitis pigmentosa. *American Journal of Ophthalmology*. 2014;158(5):1093–1098. https://doi.org/10.1016/j.ajo.2014.07.021.
14. Poddar R, Migacz JV, Schwartz DM, Werner JS, Gorczynska I. Challenges and advantages in wide-field optical coherence tomography angiography imaging of the human retinal and choroidal vasculature at 1.7-MHz A-scan rate. *Journal of Biomedical Optics*. 2017;22(10). https://doi.org/10.1117/1.JBO.22.10.106018.
15. Agarwal A, Afridi R, Agrawal R, Do DV, Gupta V, Nguyen QD. Multimodal imaging in retinal vasculitis. *Ocular Immunology and Inflammation*. 2017;25(3):424–433. https://doi.org/10.1080/09273948.2017.1319494.
16. Alibhai AY, De Pretto LR, Moult EM, et al. Quantification of retinal capillary nonperfusion in diabetics using wide-field optical coherence tomography angiography. *Retina*. 2020;40(3):412–420. https://doi.org/10.1097/IAE.0000000000002403.
17. Mastropasqua R, Viggiano P, Borrelli E, et al. In vivo mapping of the choriocapillaris in high myopia: a widefield Swept source optical coherence tomography angiography. *Scientific Reports*. 2019;9(1). https://doi.org/10.1038/s41598-019-55192-w.
18. Matet A, Daruich A, Zografos L. Radiation maculopathy after proton beam therapy for uveal melanoma: optical coherence tomography angiography alterations influencing visual acuity. *Investigative Opthalmology & Visual Science*. 2017;58(10):3851. https://doi.org/10.1167/iovs.17-22324.
19. Sakimoto S, Kawasaki R, Nishida K. Retinal neovascularization-Simulating retinal capillary reperfusion in branch retinal vein occlusion, imaged by wide-field optical coherence tomography angiography. *JAMA Ophthalmology*. 2020;138(2):216–218. https://doi.org/10.1001/jamaophthalmol.2019.5018.
20. Sawada O, Ichiyama Y, Obata S, et al. Comparison between wide-angle OCT angiography and ultra-wide field fluorescein angiography for detecting non-perfusion areas and retinal neovascularization in eyes with diabetic retinopathy. *Graefe's Archive for Clinical and Experimental Ophthalmology*. 2018;256(7):1275–1280. https://doi.org/10.1007/s00417-018-3992-y.
21. You QS, Guo Y, Wang J, et al. Detection of clinically unsuspected retinal neovascularization with wide-field optical coherence tomography angiography. *Retina*. 2020;40(5):891–897. https://doi.org/10.1097/IAE.0000000000002487.

Chapter 6

Widefield indocyanine green angiography

Jiwon Baek

The Catholic University of Korea, College of Medicine, Ophthalmology, Seoul, South Korea; Bucheon St.Mary's Hospital, The Catholic University of Korea, Gyeonggi-do, South Korea

Introduction

While fluorescein angiography (FA) shows a nice spatial and temporal resolution of the retinal vasculature, the resolution of choroidal circulation is limited by FA due to poor transmission. Using indocyanine green (ICG) dye and high-resolution confocal scanning laser ophthalmoscopes (cSLO) on infrared digital fundus cameras, choroidal vasculature through overlying melanin, xanthophyll pigment, serosanguineous fluid, and lipid exudates can be visualized.

ICG is a tricarbocyanine, anionic dye, which is soluble in highly distilled water, and is supplied with a solvent consisting of sterile water at pH 5.5–6.5. The final product contains 5%–9.5% sodium iodine.[1] ICG absorbs light in the near-infrared (NIR) spectrum with the maximum absorption at 790 nm and the maximum emission at 835 nm.[2] These optical properties allow penetration through macular pigment, melanin, and blood. About 98% of ICG is bound to plasma proteins and is excreted by the liver.[3,4]

ICGA was first used by Kogure and Choromokos in 1969 for studying the cerebral circulation in a dog.[5] The first human ICG angiogram was of the carotid artery.[6] In 1972, Flower and Hochheimer performed the first ICG angiography to image the human choroid.[7] Due to the high molecular weight and affinity to plasma proteins, ICG rarely exits from fenestrations in choroidal vessels. These properties make ICG suitable for choroidal vascular network visualization. In the mid-1980s, the resolution of ICGA was improved by using ameliorated filter combinations with sufficient sensitivity for near-infrared wavelengths and the introduction of videoangiography enabled dynamic imaging was enabled.[8] Finally, Yannuzzi et al. described a 1024-line resolution system, which was synthesized with the appropriate flash synchronization and image storage capability, permitting high-resolution, long-duration ICG angiography.[9]

Instruments at present can be divided into two main categories: digital flash fundus cameras and scanning laser ophthalmoscopes (SLOs) according to the camera system and the acquisition modalities. The light source for a digital camera is a white light with an excitation filter (640–780 nm) and a barrier filter (820–900 nm). SLO system uses a laser monochromatic light for excitation (785–790 nm) and a barrier filter at 805 nm. Some SLO systems encompass a confocal aperture that allows the selective acquisition of images of particular tissue layers.[10]

Widefield indocyanine green angiography

Imaging system

The peripheral retina is frequently affected by retinal and choroidal diseases. Widefield imaging and ultra-widefield (UWF) are developed for this reason. Widefield imaging refers to fundus imaging beyond 50 degrees field area and UWF systems can image up to 200 degrees in a single capture. This was a significant advancement considering that conventional fundus imaging modalities are restricted to the central 30–60 degrees. UWF technology is used with a variety of modalities including pseudocolor images, fundus autofluorescence (FAF), FA, and ICGA.[11]

A representative of the UWF imaging system is the Optos fundus camera (Optos PLC, Dunfermline, UK). It is a confocal SLO-based UWF imaging system that produces up to 200 degrees view of the retina, approximately 82% of the

Retinal and Choroidal Vascular Diseases of the Eye. https://doi.org/10.1016/B978-0-443-15583-3.00006-8

surface area.[11] The system utilizes laser (532 and 633 nm lasers for red and green two-color images, 488 nm blue laser for FA, and 805 nm infrared laser for ICGA) and an ellipsoidal mirror to obtain images of the retinal periphery. The system contains optical elements including beam-splitters and a confocal aperture. Oscillating galvanometer mirrors provide rapid two-dimensional raster scanning onto an ellipsoidal mirror and another ellipsoidal mirror is located at the pupillary plane, enabling a wide scanning even without pupillary dilation. The compensated light from the retina is focused through a confocal aperture and various filters and then sent to the image detector.

Other widefield imaging systems at the current include the TRC-50DX ICG (Topcon, Tokyo, Japan), the FF 450plus (Carl Zeiss Meditec, Inc., Dublin, CA, United States), and the VX-10i (Kowa, Tokyo, Japan) using a flash camera, and the Spectralis HRA (Heidelberg Engineering, Heidelberg, Germany), the Mirante (Nidek, Gamagori, Japan), and the Clarus 700 (Carl Zeiss Meditec, Inc., Dublin, CA, United States) using ICG-capable SLO systems.

Injection technique

The concentration and preparation for intravenous injection of ICG for UWF ICGA systems are similar to the conventional systems and vary with the used instrument. For fundus cameras, the standard concentration is 25 mg of ICG dissolved in 5 mL saline and may be increased to 50 mg.[12] For SLOs, the standard dosage is 25 mg of ICG dissolved in 3 mL, and 1 mL of the solution is injected. Fluorescein sodium solution at 10% to 25% concentrations can be used as a solvent for combined FA and ICG angiography. Intravenous injection should be rapid and immediately followed by a 5 mL saline flush. Infracyanine green + glucose-based solvent is available for patients with iodine allergy.

Specific findings on ultra-widefield indocyanine green angiography

Normal eye

Although a number of studies utilized UWF ICGA to evaluate pathological abnormalities in posterior segment disorders, the abnormalities cannot be defined without a clear or complete knowledge of the normal UWF ICGA findings. However, only a few papers have been published regarding UWF ICGA features on healthy eyes.

The characteristic findings vary depending on the time after the dye injection (phase) of the ICGA. The phases of ICGA consist of early (the retinal arteries are not yet filled), mid- (both arteries and veins are filled), and late or recirculation (more than after 10 min) phases (Fig. 6.1).

The early filling phase reflects the subsequent filling of different choroidal layers in order of Haller's layer and Sattler's layer, followed by the choriocapillaris. As seen in Fig. 6.1, UWF ICGA well demonstrates the global distribution of the Haller vessels, whereas the choriocapillaris is less evident as it usually appears as a diffuse haze in the posterior pole on ICGA images.

A watershed zone is present between the medial and lateral posterior ciliary artery, which is seen in the early phase image (Fig. 6.1A). Forty choroidal vortex veins are well visualized in the mid-to-late phase of UWF ICGA (usually four in number, located at the periphery of each quadrant) (Fig. 6.1B and C), and this is one of the distinct advantages of UWF ICGA compared to conventional ICGA. Zones of the retinal area can be divided into the posterior pole, mid-periphery, and periphery regarding the lines of major vascular arcade and an ellipsoid connecting vortex ampullae as seen in Fig. 6.1B. The peripheral choroidal boundary may be defined using UWF ICGA using a method suggested by Verma et al.[13]

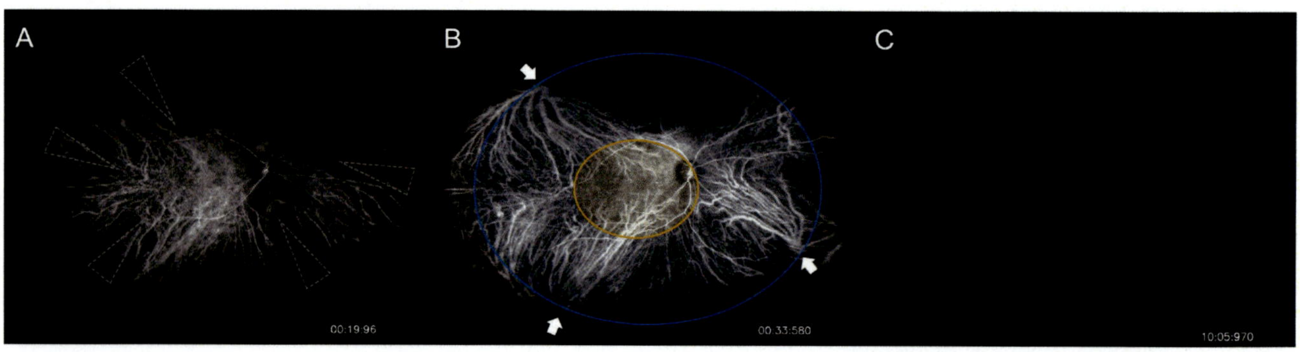

FIGURE 6.1 UWF ICGA in a healthy eye: (A) the early phase, (B) the mid-phase, and (C) the last or recirculation phase of UWF ICGA in a healthy eye. Vortex ampullae are well visualized (white arrows). Gray dash line area demarcates watershed zones in the early phase. Blue (outer) ellipsoid demarcates mid-peripheral retinal area and the yellow (inner) line demarcates the posterior polar area.

Neovascular age-related macular degeneration

There is no published study solely enlightening the features of age-related macular degeneration (AMD) on PubMed by far. AMD is commonly classified as categories and definitions by "Consensus on Neovascular Age-Related Macular Degeneration Nomenclature (CONAN)" study group suggestions. These are type 1 macular neovascularization (MNV; formerly known as type 1 CNV or occult, located under the RPE), type 2 MNV (formerly known as type 2 CNV or classic, located under the neurosensory retina), and type 3 MNV (formerly known as type 3 CNV or retinal angiomatous proliferation, RAP).[14]

Conventional ICGA, with its ability to visualize vascular structures deeper than the RPE, has an advantage in showing type 1 MNV. However, this advantage of ICGA seems to diminish in UWF ICGA regarding its characteristics of weakness in showing choriocapillaris and a relatively small area of the posterior pole in the total UWF ICGA image (Fig. 6.2). In a similar context, UWF ICGA does not seem to improve visualization of the fine structure of the neovascular network of type 2 MNV as seen on the conventional ICGA. Maruyama-Inoue revealed an asymmetrical choroidal venous pattern evident on UWG ICGA in the eyes of patients with AMD as compared with age-matched controls.[15]

Type 3 MNV refers to vascular proliferation that originates from the retinal microvasculature in the deep capillary plexus and extends toward the deeper portions of the retina, establishing anastomosis and leading to sub-RPE neovascularization. In contrast to type 1 and type 2 MNVs, UWF ICGA clearly delineates the vascular structure of the mature type 3 MNV lesions (Fig. 6.3) despite the fact that type 3 MNVs are usually found within the PED.54. Moreover, the characteristics "dye filling of at least one retinal arteriole descending into the deep retinal space to a vascular communication and at least one draining retinal vein." can be observed as seen in Fig. 6.3B.[16] Also, progressive leakage on ICGA in the late phases can help in the differential diagnosis because this feature is not observed in other MNV lesion types (Fig. 6.3C). Type 3 MNV also shows a progressive leakage on ICGA in the late phases of the examination, a feature that is not observed in other MNV lesion types, thus helping in the differential diagnosis.

Aneurysmal type 1 MNV or polypoidal choroidal vasculopathy

Aneurysmal type 1 MNV or polypoidal choroidal vasculopathy (PCV) is a particular subtype in which features of UWF ICGA are well established. It is MNV characterized by a growth of cavernous thin-walled vessels located between the RPE and the Bruch's membrane (branching vascular network [BVN]) ending in one or more aneurysmal structure, which was referred to as "polyps" before.[17] On ICGA, a BVN often arises in the macula, and hyperfluorescent "bulges" become visible shortly after, often characterized by a slow, sometimes pulsatile, filling.[18] ICGA is currently considered to be the most sensitive and specific tool for identifying aneurysmal type 1 MNV.

PCV, or aneurysmal type 1 MNV is known to show two spectrums regarding the choroidal vascular morphology (Fig. 6.4). In terms of choroidal thickness, PCV eyes showed a bimodal distribution at the histogram, with peaks at 170 and 390 μm.[19] And those two groups showed distinctive features on Haller vessel morphology demonstrated by en face OCT images.[20] This difference was confirmed on UWF ICGA qualitatively and quantitatively.[21] Mean choroidal vessel density on UWF ICGA was higher in PCV eyes with thick-choroid compared to PCV eyes with thin-choroid and the vessel density at each quadrant correlated with that of the corresponding Vortex ampulla. This suggest that congestion at the vortex ampulla might be a cause for venous insufficiency which in turn results in both increasing choroidal vessel density of the corresponding quadrant and subfoveal choroidal thickening in PCV, or aneurysmal type 1 MNV, with thick-choroid.

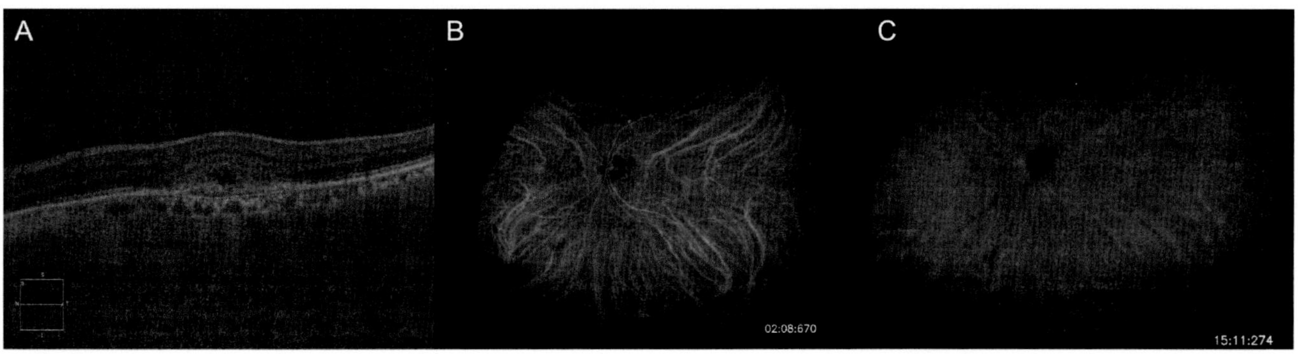

FIGURE 6.2 A UWF ICGA image of a representative type 1 MNV case: (A) OCT, (B) the early phase UWF ICGA, and (C) the late phase UWF ICGA image of a representative type 1 MNV case.

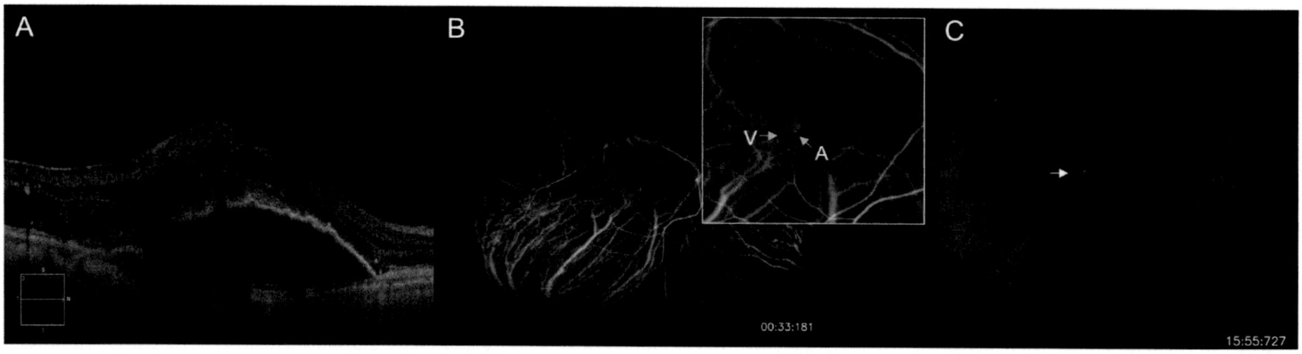

FIGURE 6.3 A representative type 3 MNV case: (A) OCT, (B) the early phase ICGA, and (C) the late phase UWF ICGA of a representative type 3 MNV case. *Yellow arrows* in image B indicate arteriole and venular anastomosis formed in the lesion. White arrow in image C indicates late leakage of the lesion.

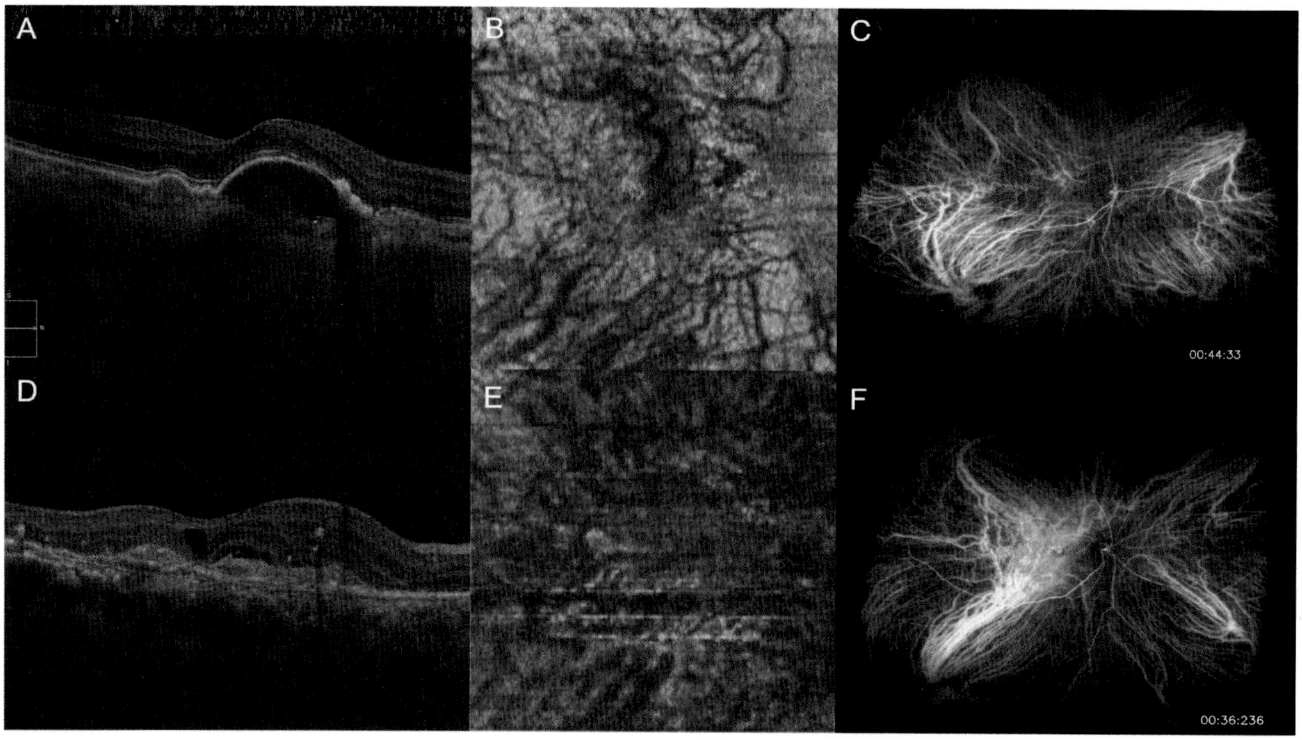

FIGURE 6.4 Representative cases with thin-choroid and thick choroid PCV (aneurysmal type 1 MNV): (A and D) OCTs; (B and E) en face OCTs of Haller layer; (C and F) UWF ICGAs of representative cases with thin-choroid and thick choroid PCV (aneurysmal type 1 MNV). Different patterns between the two eyes in choroidal morphology are demonstrated on OCT b-scan, en face scan, and UWF ICGA.

Pachychoroid disease spectrum

The term "pachychoroid" has been coined relatively recently and may bear some controversy at the moment. It generally refers to a spectrum of clinical conditions characterized by the presence of dilated choroidal vessels with an attenuation of the overlying choriocapillaris.[22] These conditions include central serous chorioretinopathy (CSC), pachychoroid pigment epitheliopathy (PPE), pachychoroid neovasculopathy (PNV), PCV/aneurysmal type 1 neovascularization, focal choroidal excavation (FCE), and peripapillary pachychoroid syndrome (PPS).[23] These diseases show progressive dysfunction of RPE and may eventually develop MNV. Recent studies using UWF ICGA have demonstrated that the manifestations of these choroidal diseases may extend peripherally.

The ICGA of CSC is characterized by multifocal areas of choroidal hyperpermeability in the mid-to-late phase.[24] UWF ICGA in CSC may be useful for identifying peripheral areas of previous or ongoing SRF and choroidal hyperpermeability. UWF ICGA revealed dilated choroidal vessels and choroidal hyperpermeability in areas corresponding to altered AF on UWF autofluorescence (Fig. 6.5E–H).[25] Subsequently, Hirahara et al. quantified the density of the choroidal vasculature

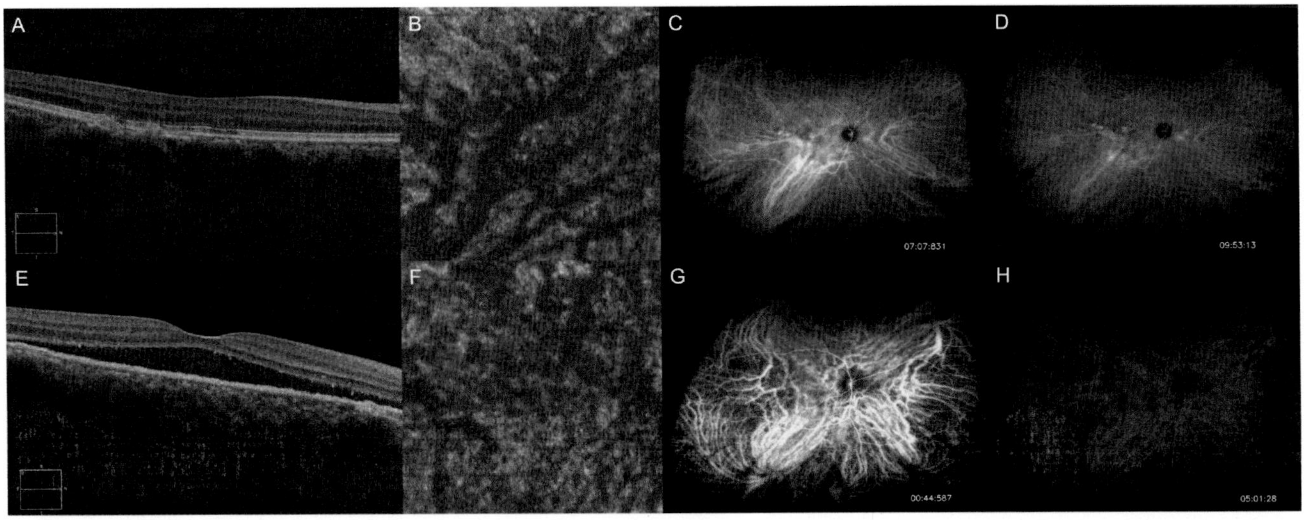

FIGURE 6.5 Representative cases with PPE and CSC: (A and E) OCTs; (B and F) en face OCTs of Haller layer; (C and G) the mid-phase UWF ICGAs; (D and H) the late-phase UWF ICGAs of representative cases with PPE and CSC.

on UWF ICGA using a binarization method and detected increased vascular density and global dilatation of choroidal vessels in eyes with CSC.[26] Quantified vessel densities of CSC shared similarities with PCV eyes with thick-choroid and pachychoroid neovasculopathy (PNV).[21] The increased number and binocular symmetry of engorged vortex veins were observed in CSC and suggested as an anatomical predisposition for CSC.[27] ICGA features of PPE have not been firmly established due to limitations in clinical need for ICGA examination in these patients. Fig. 6.5A–D shows that UWF ICGA shows diffuse Haller vessel dilation in the mid-phase and multiple choroidal vascular hyperpermeability in the late phase as seen in a CSC case.

PNV describes the occurrence of type 1 MNV in eyes with pachychoroid disease and lacks typical findings of drusen-associated AMD.[28] OCT angiography has become the preferred imaging technique because these lesions are often readily detected with OCTA as a well-delineated "tangled network" of flow signal (Fig. 6.6C). Other ICGA findings in the pachychoroid disease spectrum include multiple serous PED, choroidal filling defects, focal or punctate hyperfluorescent areas suggesting possible choroidal ischemia, punctate hyperfluorescent spots, delays in the arterial filling of the choroidal arteries and choriocapillaris, and venous congestion.[29–32] In 2021, Spaide et al. suggested that intervortex venous anastomosis is common in pachychoroid disorders in their study using UWF ICGA.[33] Over a time period of the disease progression, the intervenous anastomotic connections could remodel and enlarge to create large anastomotic vessels in areas that were watershed zones between vortex vein systems as seen in Fig. 6.6D.

Additionally, imbalanced choroidal venous drainage with congestion of specific vortex vein systems may contribute to a state of choroidal venous insufficiency characterized by regional choroidal thickening, choroidal vascular hyperpermeability, and remodeling of venous drainage routes in pachychoroid eyes.[34] Quantitative analysis by Jung et al. again proved this by showing that asymmetric choroidal venous outflow occurs in CSC/PPE eyes on UWF ICGA in each quadrant. Increased brightness levels along inferior quadrants in mid-phase ICGA in this study suggested venous outflow congestion among eyes with CSC or PPE.[35]

These features of UWF ICGA have been proven to be effective in the automated classification of pachychoroid disorders using UWF ICGA on a machine learning (ML) model by Google's AutoML vision.[36] Also, the features were used as predictive values for short-term treatment results with intravitreal aflibercept in PNV.[37]

Other posterior polar diseases

Myopia

UWF ICGA can analyze the characteristic features of choroidal vessels in highly myopic eyes including vortex veins. Vortex venous outflow system drains flow from the corresponding segment of the iris, ciliary body, and choroid. Sometimes, especially in myopic eyes, a vein passing from the choroid through the sclera closely adjacent to the optic nerve head and draining into the venous plexus of the pial sheath of the optic nerve can be observed.[38] These posterior vortex vein plays an important role as routes of choroidal outflow in highly myopic eyes (Fig. 6.7).[39]

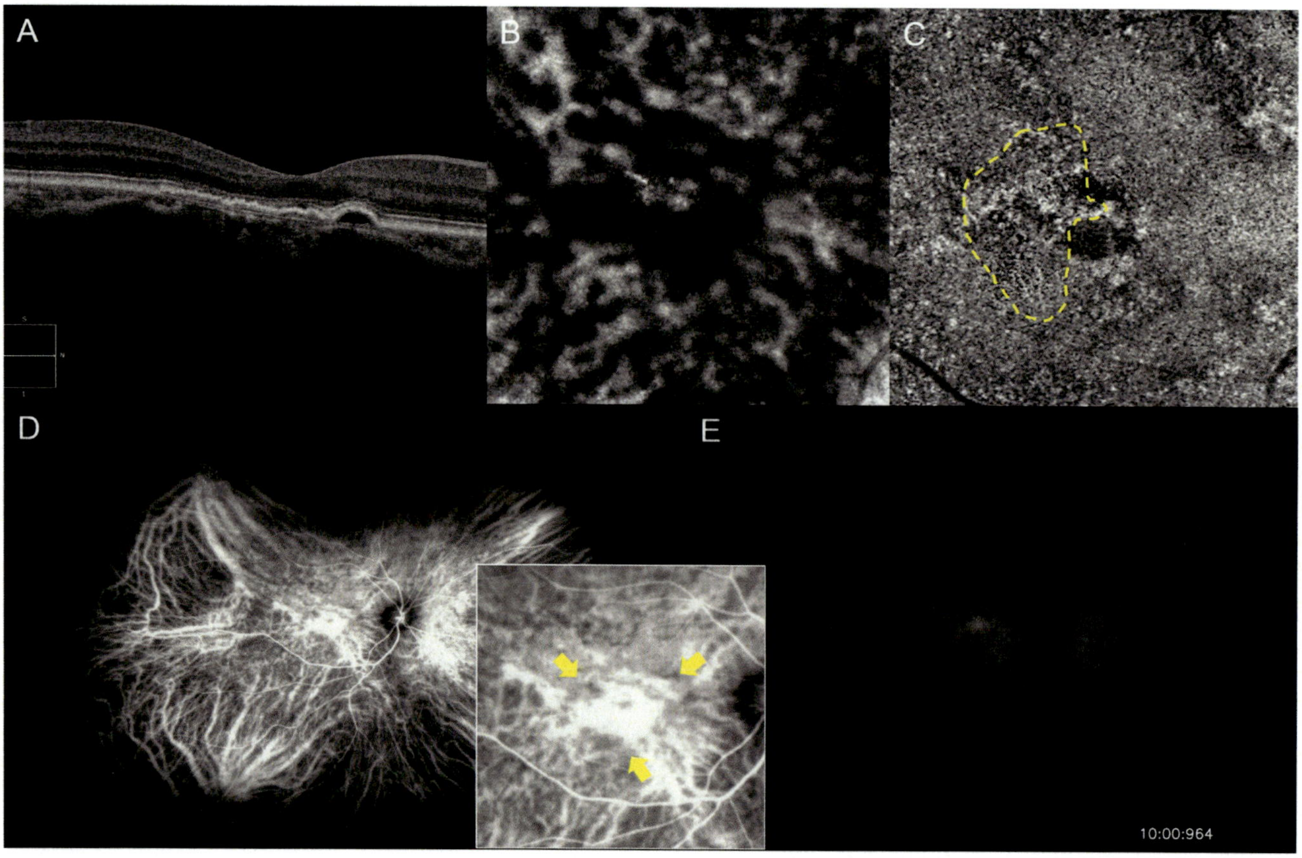

FIGURE 6.6 A representative PNV case: (A) OCT; (B) OCT en face; (C) OCT angiography at choriocapillaris level; (D) the mid-phase UWF ICGA; (E) the late-phase UWF ICGA of a representative PNV case.

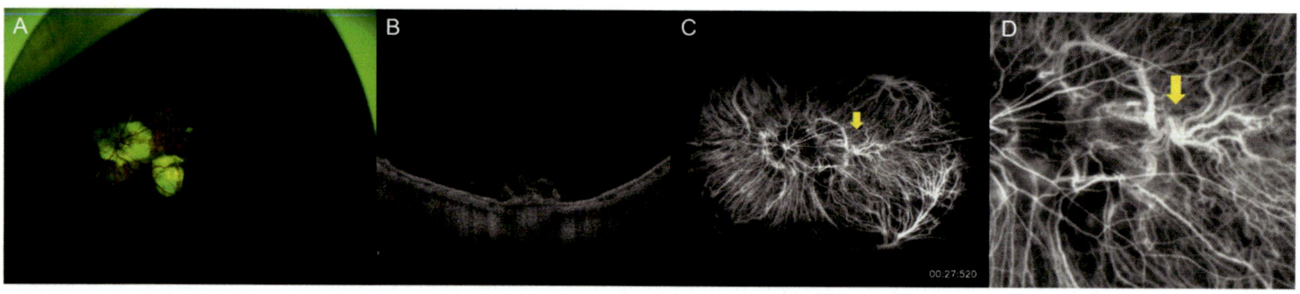

FIGURE 6.7 A representative myopic choroidal neovascularization case: (A) UWF color fundus photography; (B) OCT; (C) UWF ICGA; (D) macular area of UWF ICGA of a representative myopic choroidal neovascularization case with posterior pole vortex outlet (*yellow arrow*).

Uveitis and choroidal inflammations

In Vogt–Koyanagi–Harada (VKH) eyes, diffuse leakage from peripheral choroidal vessels was observed in about 94% of the studied eyes.[40] In addition, vortex obstruction, and choroidal vein dilatation were observed using UWF ICGA in a VKH patient with annular choroidal detachment, suggesting that the pathogenesis of the choroidal detachment might be related to the vortex obstruction (Fig. 6.8).[41]

In sarcoidosis, angiographic changes were found to be most frequent in the inferotemporal periphery area in a UWF FA study.[42] On UWF ICGA, multiple round areas of choroidal hypofluorescence consistent with a choroidal vasculature filling defect and/or blockage from retinal granulomas are visible in sarcoidosis eyes.

In ocular syphilis, UWF ICGA findings include mid-phase diffuse round areas of hypofluorescence and posterior pole hyperfluorescence, and late-phase inferior hypofluorescence and filling defect of a choroidal vessel consistent with nonperfusion.[43]

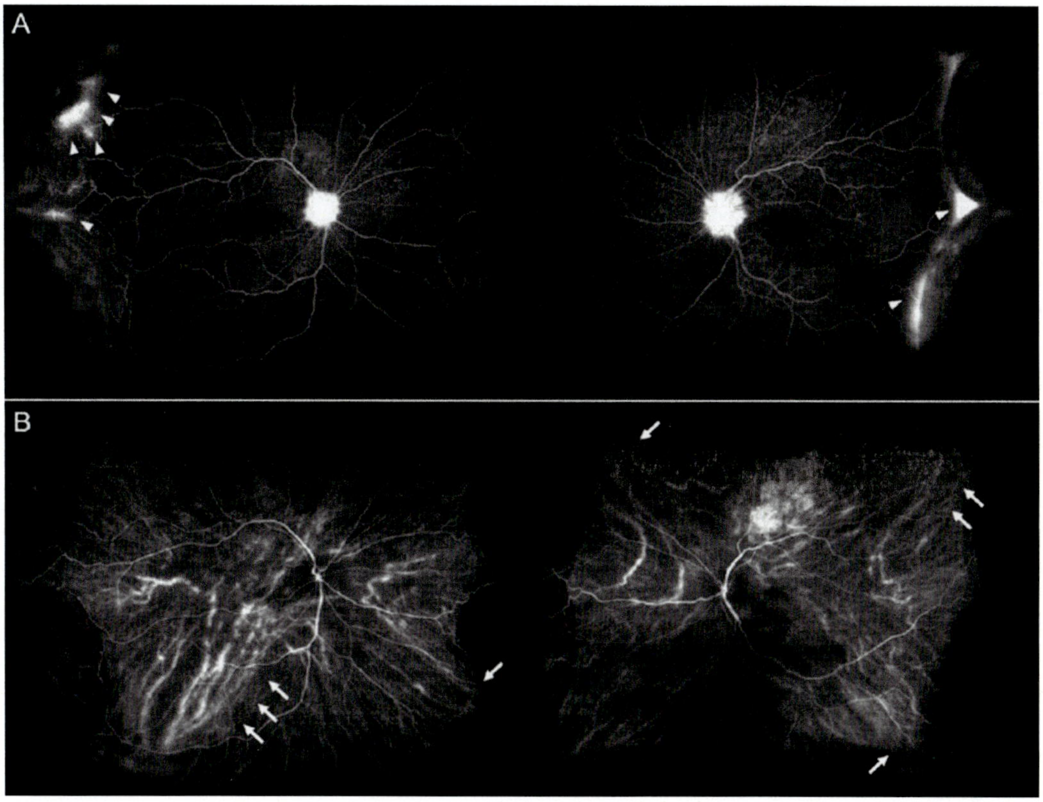

FIGURE 6.8 A VKH patient with annular choroidal detachment: (A) UWF fluorescein angiography (FA); (B) UWF ICGA of a VKH patient with annular choroidal detachment. FA shows multiple dye leakages and dye pooling at the posterior margin of choroidal detachment (arrowheads). ICGA shows dilated choroidal veins (arrows) and obscured vortex ampullae in the late phase.

Multiple evanescent white dot syndrome (MEWDS) is a unilateral inflammatory disease that affects young subjects, more common in women, usually with self-limiting visual impairment.[44] Conventional and UWF ICGA shows a pattern of multiple hypofluorescent areas with a characteristic lobulated shape at the posterior pole and peripheral retina that become visible in the mid-to-late phases and resolve over time (Fig. 6.9).[45] Multiple prominent areas of choroidal hypofluorescence corresponding to white lesions on clinical examination consistent with "holes" in the fluorescence of the choriocapillaris are evident in UWF ICGA of in birdshot chorioretinopathy.[43] In acute zonal occult outer retinopathy (AZOOR), early and late UWF ICGA shows peripapillary zonal areas of hypofluorescence, which correspond to the areas of zonal atrophy on OCT. In the periphery, confluent pinpoint areas of hypofluorescence are also observed.

Lavine et al. studied features of primary vitreoretinal lymphoma on UWF ICGA and showed early and late hypocyanescence in a corresponding pattern to the FA staining, which was more frequent in peripheral areas than in the posterior pole.[46]

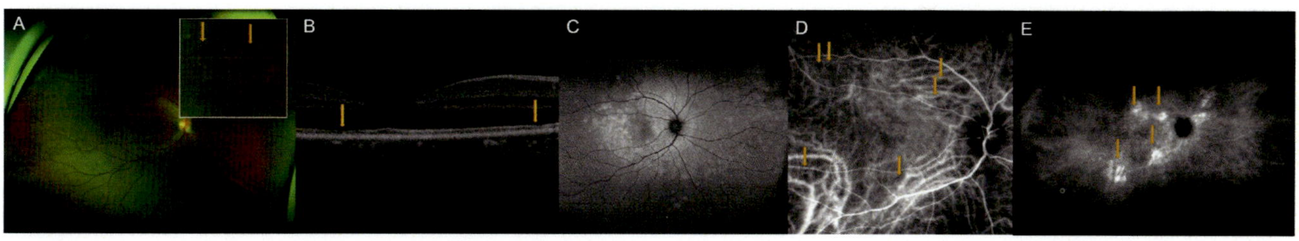

FIGURE 6.9 A representative MEWDS case: (A) UWF color fundus photography; (B) OCT; (C) UWF autofluorescence; (D) macular area of the mid-phase UWF ICGA; (E) the late UWF ICGA of a representative MEWDS case. *Yellow arrows* indicate white dot area.

Systemic conditions

Using UWF ICGA, Kovacs et al. evaluated quantitative measurements of choroidal vasculature in patients with systemic vascular diseases including hypertension, diabetes, obstructive sleep apnea, and congestive heart failure.[47] Patients with obstructive sleep apnea were found to have increased vascular density due to increased mean vessel caliber at peripheral landmarks, which was not present in other systemic vascular diseases. This requires further studies as the number of samples was small as 2–14 patients.

Diabetic choroidopathy is characterized by choriocapillaris dropout, luminal constriction, and large and intermediate blood vessel tortuosity.[48] The qualitative and quantitative UWF ICGA image analysis revealed that choroidal blood vessels were dilated, complex, and hyperpermeable as the retinopathy progressed and these features were more severe in eyes with macular edema.[49]

Conclusion

UWF ICGA allows visualization of up to 200 degrees of the peripheral retinae in a single frame and shows the global status of choroidal vessels, especially large vessels. Obtaining UWF ICGA with the noncontact Optos system was proved to be clinically feasible and comparable with obtaining traditional ICGA.[43] Klufas et al. showed that more than 67.4% of posterior polar diseases including CSC, uveitis, choroidal inflammations, AMD, PCV, and myopia demonstrated peripheral pathologic findings on UWF ICGA. Therefore, UWF ICGA serves as an important diagnostic tool for posterior polar diseases.

In summary, UWF ICGA is particularly useful in pachychoroid spectrum diseases including CSC, in which the pathogenesis is suspected to be largely involved with changes in choroidal circulation. Numerous researchers studied choroidal vasculature using UWF ICGA and revealed many significant abnormalities related to disease pathogenesis.

Further UWF ICGA imaging on additional normal controls and diseases listed above and others will aid in the classification and spectrum of disease with choroidal pathology. The clinical significance of these peripheral findings and the impact on the diagnosis, treatment, and follow-up.

References

1. Bischoff PM, Flower RW. Ten years experience with choroidal angiography using indocyanine green dye: a new routine examination or an epilogue? *Documenta Ophthalmologica*. 1985;60(3):235–291. https://doi.org/10.1007/BF00157827.
2. Flower RW, Hochheimer BF. Indocyanine green dye fluorescence and infrared absorption choroidal angiography performed simultaneously with fluorescein angiography. *Johns Hopkins Medical Journal*. 1976;138(2):33–42.
3. Hunton DB, Bollman JL, Hoffman HN. Studies of hepatic function with indocyanine green. *Gastroenterology*. 1960;39:713–724. https://doi.org/10.1016/s0016-5085(19)35200-x.
4. Ott P. Hepatic elimination of indocyanine green with special reference to distribution kinetics and the influence of plasma protein binding. *Pharmacology & Toxicology*. 1998;83:1–48. https://doi.org/10.1111/j.1600-0773.1998.tb01945.x.
5. Kogure K, Choromokos E. Infrared absorption angiography. *Journal of Applied Physiology*. 1969;26(1):154–157. https://doi.org/10.1152/jappl.1969.26.1.154.
6. Choromokos E, Kogure K, David NJ. Infrared absorption angiography. *Journal of the Biological Photographic Association*. 1969;37(2):100–104.
7. Flower RW, Hochheimer BF. Clinical infrared absorption angiography of the choroid. *American Journal of Ophthalmology*. 1972;73(3):458–459. https://doi.org/10.1016/0002-9394(72)90079-7.
8. Hayashi K, De Laey JJ. Indocyanine green angiography of submacular choroidal vessels in the human eye. *Ophthalmologica*. 1985;190(1):20–29. https://doi.org/10.1159/000309488.
9. Yannuzzi LA, Slakter JS, Sorenson JA, Guyer DR, Orlock DA. Digital indocyanine green videoangiography and choroidal neovascularization. *Retina*. 1992;12(3):191–223. https://doi.org/10.1097/00006982-199212030-00003.
10. Flower RW, Csaky KG, Murphy RP. Disparity between fundus camera and scanning laser ophthalmoscope indocyanine green imaging of retinal pigment epithelium detachments. *Retina*. 1998;18(3):260–268. https://doi.org/10.1097/00006982-199803000-00012.
11. Lee J, Sagong M. Ultra-widefield retina imaging: principles of technology and clinical applications. *Journal of Retina*. 2016;1(1):1–10. https://doi.org/10.21561/jor.2016.1.1.1.
12. Stanga PE, Lim JI, Hamilton P. Indocyanine green angiography in chorioretinal diseases: indications and interpretation: an evidence-based update. *Ophthalmology*. 2003;110(1):15–50. https://doi.org/10.1016/S0161-6420(02)01563-4.
13. Verma A, Maram J, Alagorie AR, et al. Peripheral extent of the choroidal circulation by ultra-widefield indocyanine green angiography in healthy eyes. *British Journal of Ophthalmology*. 2021;105(6):824–828. https://doi.org/10.1136/bjophthalmol-2020-316412.
14. Spaide RF, Jaffe GJ, Sarraf D, et al. Consensus nomenclature for reporting neovascular age-related macular degeneration data: consensus on neovascular age-related macular degeneration nomenclature study group. *Ophthalmology*. 2020;127(5):616–636. https://doi.org/10.1016/j.ophtha.2019.11.004.

15. Maruyama-Inoue M, Yamane S, Satoh H, Sato S, Kadonosono K. Choroidal angioarchitecture according to ultra-widefield indocyanine green angiography in age-related macular degeneration. *Journal of VitreoRetinal Diseases*. 2017;1(6):365−371. https://doi.org/10.1177/2474126417733161.
16. Hartnett ME, Weiter JJ, Staurenghi G, Elsner AE. Deep retinal vascular anomalous complexes in advanced age-related macular degeneration. *Ophthalmology*. 1996;103(12):2042−2053. https://doi.org/10.1016/S0161-6420(96)30389-8.
17. Yannuzzi LA, Sorenson J, Spaide RF, Lipson B. Idiopathic polypoidal choroidal vasculopathy (IPCV). *Retina*. 1990;10(1):1−8. https://doi.org/10.1097/00006982-199001010-00001.
18. Ciardella AP, Donsoff IM, Huang SJ, Costa DL, Yannuzzi LA. Polypoidal choroidal vasculopathy. *Survey of Ophthalmology*. 2004;49(1):25−37. https://doi.org/10.1016/j.survophthal.2003.10.007.
19. Lee WK, Baek J, Dansingani KK, Lee JH, Freund KB. Choroidal morphology in eyes with polypoidal choroidal vasculopathy and normal or subnormal subfoveal choroidal thickness. *Retina*. 2016;36:S73−S82. https://doi.org/10.1097/IAE.0000000000001346.
20. Baek J, Lee JH, Jung BJ, Kook L, Lee WK. Morphologic features of large choroidal vessel layer: age-related macular degeneration, polypoidal choroidal vasculopathy, and central serous chorioretinopathy. *Graefe's Archive for Clinical and Experimental Ophthalmology*. 2018;256(12):2309−2317. https://doi.org/10.1007/s00417-018-4143-1.
21. Lee A, Ra H, Baek J. Choroidal vascular densities of macular disease on ultra-widefield indocyanine green angiography. *Graefe's Archive for Clinical and Experimental Ophthalmology*. 2020;258(9):1921−1929. https://doi.org/10.1007/s00417-020-04772-y.
22. Warrow DJ, Hoang QV, Freund KB. Pachychoroid pigment epitheliopathy. *Retina*. 2013;33(8):1659−1672. https://doi.org/10.1097/IAE.0b013e3182953df4.
23. Cheung CMG, Lee WK, Koizumi H, Dansingani K, Lai TYY, Freund KB. Pachychoroid disease. *Eye (Basingstoke)*. 2019;33(1):14−33. https://doi.org/10.1038/s41433-018-0158-4.
24. Spaide RF, Hall L, Haas A, et al. Indocyanine green videoangiography of older patients with central serous chorioretinopathy. *Retina*. 1996;16(3):203−213. https://doi.org/10.1097/00006982-199616030-00004.
25. Pang CE, Shah VP, Sarraf D, Freund KB. Ultra-widefield imaging with autofluorescence and indocyanine green angiography in central serous chorioretinopathy. *American Journal of Ophthalmology*. 2014;158(2):362−371. https://doi.org/10.1016/j.ajo.2014.04.021.
26. Hirahara S, Yasukawa T, Kominami A, Nozaki M, Ogura Y. Densitometry of choroidal vessels in eyes with and without central serous chorioretinopathy by wide-field indocyanine green angiography. *American Journal of Ophthalmology*. 2016;166:103−111. https://doi.org/10.1016/j.ajo.2016.03.040.
27. Jeong S, Kang W, Noh D, van Hemert J, Sagong M. Choroidal vascular alterations evaluated by ultra-widefield indocyanine green angiography in central serous chorioretinopathy. *Graefe's Archive for Clinical and Experimental Ophthalmology*. 2022;260(6):1887−1898. https://doi.org/10.1007/s00417-021-05461-0.
28. Pang CE, Freund KB. Pachychoroid neovasculopathy. *Retina*. 2015;35(1):1−9. https://doi.org/10.1097/IAE.0000000000000331.
29. Fukuyama H, Iwami H, Araki T, Ishikawa H, Ikeda N, Gomi F. Indocyanine green dye filling time for polypoidal lesions in polypoidal choroidal vasculopathy affects the visibility of the lesions on OCT angiography. *Ophthalmology Retina*. 2018;2(8):803−807. https://doi.org/10.1016/j.oret.2017.11.016.
30. Guyer DR, Yannuzzi LA, Slakter JS, Sorenson JA, Ho A, Orlock D. Digital indocyanine green videoangiography of central serous chorioretinopathy. *Archives of Ophthalmology*. 1994;112(8):1057−1062. https://doi.org/10.1001/archopht.1994.01090200063023.
31. Prünte C, Flammer J. Choroidal capillary and venous congestion in central serous chorioretinopathy. *American Journal of Ophthalmology*. 1996;121(1):26−34. https://doi.org/10.1016/s0002-9394(14)70531-8.
32. Tsujikawa A, Ojima Y, Yamashiro K, et al. Punctate hyperfluorescent spots associated with central serous chorioretinopathy as seen on indocyanine green angiography. *Retina*. 2010;30(5):801−809. https://doi.org/10.1097/IAE.0b013e3181c72068.
33. Spaide RF, Ledesma-Gil G, Gemmy Cheung CM. Intervortex venous anastomosis in pachychoroid-related disorders. *Retina*. 2021;41(5):997−1004. https://doi.org/10.1097/iae.0000000000003004.
34. Bacci T, Oh DJ, Singer M, Sadda SV, Freund KB. Ultra-widefield indocyanine green angiography reveals patterns of choroidal venous insufficiency influencing pachychoroid disease. *Investigative Ophthalmology and Visual Science*. 2022;63(1). https://doi.org/10.1167/iovs.63.1.17.
35. Jung JJ, Yu DJG, Ito K, Rofagha S, Lee SS, Hoang QV. Quantitative assessment of asymmetric choroidal outflow in pachychoroid eyes on ultra-widefield indocyanine green angiography. *Investigative Ophthalmology and Visual Science*. 2020;61(8). ISSN: 15525783. https://doi.org/10.1167/IOVS.61.8.50.
36. Kim IK, Lee K, Park JH, Baek J, Lee WK. Classification of pachychoroid disease on ultrawide-field indocyanine green angiography using auto-machine learning platform. *British Journal of Ophthalmology*. 2021;105(6):856−861. https://doi.org/10.1136/bjophthalmol-2020-316108.
37. Han SY, Lee SH, Lee PK, Ra H, Baek J. Choroidal morphology on ultra-widefield indocyanine green angiography and response to aflibercept in pachychoroid neovasculopathy. *Pharmaceuticals*. 2023;16(1). https://doi.org/10.3390/ph16010073.
38. Hayreh SS. Physiological anatomy of the choroidal vascular bed. *International Ophthalmology*. 1983;6(2):85−93. https://doi.org/10.1007/BF00127636.
39. Moriyama M, Cao K, Ogata S, Ohno-Matsui K. Detection of posterior vortex veins in eyes with pathologic myopia by ultra-widefield indocyanine green angiography. *British Journal of Ophthalmology*. 2017;101(9):1179−1184. https://doi.org/10.1136/bjophthalmol-2016-309877.
40. Kim P, Sun HJ, Ham DI. Ultra-wide-field angiography findings in acute Vogt-Koyanagi-Harada disease. *British Journal of Ophthalmology*. 2019;103(7):942−948. https://doi.org/10.1136/bjophthalmol-2018-312569.

41. Kwak JH, Baek J, Ra H. Multimodal imaging of annular choroidal detachment in a patient with Vogt–Koyanagi–Harada disease. *Ocular Immunology and Inflammation*. 2021;29(5):911–914. https://doi.org/10.1080/09273948.2019.1704023.
42. Suh H, Lee A, Ra H, Lee JH, Baek J. Quantitative analysis of ultra-widefield fluorescein angiography in uveitis associated with sarcoidosis. *Heliyon*. 2022;8(11):e11218. https://doi.org/10.1016/j.heliyon.2022.e11218.
43. Klufas MA, Pang CE, Srinivas S, Sadda S, Freund KB. Feasibility and clinical utility of ultra-widefield indocyanine green angiography. *Retina*. 2015;35:508–520.
44. Jampol LM, Sieving PA, Pugh D, Fishman GA, Gilbert H. Multiple evanescent white dot syndrome: I. Clinical findings. *Archives of Ophthalmology*. 1984;102(5):671–674. https://doi.org/10.1001/archopht.1984.01040030527008.
45. Dell'Omo R, Wong R, Marino M, Konstantopoulou K, Pavesio C. Relationship between different fluorescein and indocyanine green angiography features in multiple evanescent white dot syndrome. *British Journal of Ophthalmology*. 2010;94(1):59–63. https://doi.org/10.1136/bjo.2009.163550.
46. Lavine JA, Singh AD, Sharma S, Baynes K, Lowder CY, Srivastava SK. Ultra-widefield multimodal imaging of primary vitreoretinal lymphoma. *Retina*. 2019;39(10):1861–1871. https://doi.org/10.1097/IAE.0000000000002260.
47. Kovacs KD, Young B, Adelman RA. Quantitative modeling of ultra-widefield choroidal indocyanine green angiography in systemic vascular diseases. *Ophthalmic Surgery Lasers and Imaging Retina*. 2021;52(5):281–287. https://doi.org/10.3928/23258160-20210429-06.
48. Fryczkowski AW, Sato SE, Hodes BL. Changes in the diabetic choroidal vasculature: scanning electron microscopy findings. *Annals of Ophthalmology*. 1988;20(8):299–305.
49. Choi SU, Kim YJ, Lee JY, Lee J, Yoon YH. Qualitative and quantitative evaluation of diabetic choroidopathy using ultra-widefield indocyanine green angiography. *Scientific Reports*. 2023;13(1). https://doi.org/10.1038/s41598-023-29216-5.

Chapter 7

Optical coherence tomography

Jorge Ruiz-Medrano[1,2], Rosa Gutiérrez-Bonet[3] and José M. Ruiz-Moreno[1,2,4,5]
[1]*Ophthalmology Department, Puerta de Hierro-Majadahonda University Hospital, Madrid, Spain;* [2]*Ocular Microsurgery Institute IMO, Madrid, Spain;* [3]*Ophthalmology Department, Gregorio Marañon University Hospital, Madrid, Spain;* [4]*VISSUM, Alicante, Spain;* [5]*Ophthalmology Department, University of Castilla La Mancha (UCLM), Ciudad Real, Spain*

Introduction

Since its original development more than 20 years ago, optical coherence tomography (OCT) has become an almost indispensable tool in the day-to-day analysis of patients with retinal pathology, as it allows the patient to be studied in a fast, reproducible, "in vivo," and noninvasive manner. This technology began its development in the 1980s, with the first images being similar to A-mode ultrasound.[1] In 1992, two-dimensional cross-sectional imaging was developed, with the first in vivo human retinal scans seeing the light of day in 1993.[2]

OCT is based on the principle of low-coherence interferometry. A low-coherence light beam is directed at the target tissue and the light that reflects off the tissue in a scattered manner is combined with a reference light beam producing an interference pattern. This is used to construct an axial A-scan, which is nothing more than a representation of the scattering properties of the tissue that has been hit by the light. From the combination of different A-scans with different points of incidence, we can generate a B-scan, which is the two-dimensional representation of a cross-section of the studied area.[2]

OCT time domain

The first OCT models were called time-domain OCT (TD-OCT). This technology used a single detector and the A-scans were generated by moving a mirror to modify the reference beam and thus study the tissue in different depth planes. The main drawback was the speed of scanning, which together with the lack of resolution and tissue penetration made this technology rapidly obsolete (Fig. 7.1).[2,3]

OCT spectral domain

Then in 2006 appeared the spectral domain OCT (SD-OCT) or Fourier domain OCT that uses multiple detectors instead of multiple reference beams generated from a moving mirror to obtain the A-scans. Its speed is approximately 30,000 A-scans per second, reaching up to 100,000. The speed of the SD-OCT has also allowed us to obtain higher-quality images by reducing image noise, which is responsible for the poorly defined appearance of some B-scans. By acquiring, saving, and combining multiple shots of the same B-scan located in the same position, we obtained an average image, with minimal noise and much higher definition, allowing us to easily differentiate the different layers of the retina (Fig. 7.2).[3] Bruch's membrane is difficult to identify in healthy subjects; however, when there are detachments of the pigmentary epithelium (DEP) of the retina, it is easy to identify (Fig. 7.3). Soon after, software for ocular eye-tracking was developed, which helped us to ensure that images were obtained at the exact same location on different scans.

Enhanced-depth imaging technology

Later, Spaide published the first papers using the new enhanced-depth imaging OCT (EDI-OCT) technique, which provided consistent visualization of the choroid in most eyes and allowed reproducible and accurate thickness measurements.[4,5] One of the consequences of using a Fourier transform to decode the interferometry signal is the appearance of two

Retinal and Choroidal Vascular Diseases of the Eye. https://doi.org/10.1016/B978-0-443-15583-3.00007-X

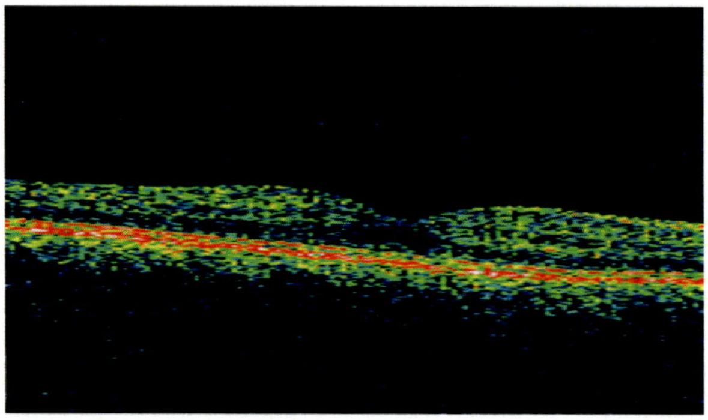

FIGURE 7.1 Time domain OCT.

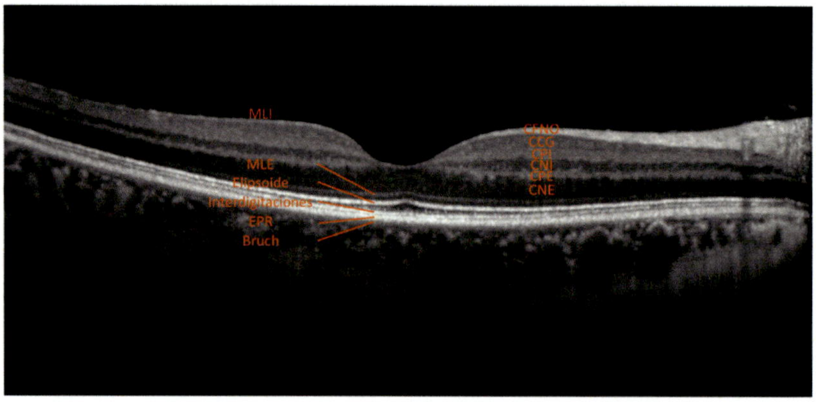

FIGURE 7.2 Spectral-domain OCT of a patient's right eye in which the layers of the normal retina can be clearly identified, from inner to outer: inner limiting membrane (ILM), optic nerve fiber layer (ONNFC), ganglion cell layer (GCC), inner plexiform layer (IPL), inner nuclear layer (INL), outer plexiform layer (EPC), and outer nuclear layer (NEC). The hyperreflective lines of the outer retina can be identified as follows: outer limiting membrane (OLM), photoreceptor ellipsoid, photoreceptor interdigitations, retinal pigment epithelium (RPE), and Bruch's membrane.

conjugate images. However, only one of them is used or displayed in most devices used in daily clinical practice, usually the one showing the retina toward the top of the screen and the choroid below it. If we modify the focus point to place it on the inner sclera, deep structures such as the choroid can be seen with much greater definition. In this way, the conjugated and inverted image appears on the screen, while the classically oriented image appears blank due to the lack of information.

In this way, we gain about 500–800 μm of penetration in the ocular tissues, which in certain patients with thinner choroids, such as high myopes (HM) or patients with age-related macular degeneration (AMD), can allow us to visualize almost the entire sclera and even the orbital fat. But with all these advantages, there is also a drawback, and that is the fact that by obtaining maximum image sensitivity in the choroid, we lose quality in the visualization of the vitreous (Fig. 7.4).

OCT swept-source

The latest techniques for OCT imaging consist of an OCT with a longer wavelength and high penetration: the swept-source OCT (SS-OCT).[6–8] SS-OCT uses a tunable laser operating at 100,000 Hz with a wavelength of about 1 mm. These devices can average images of up to 96 B-scans at the same position, and in case of using the single line scanning mode it generates images averaging up to 1024 A-scans with a length of 12 mm.[9] According to some publications, choroidal thickness can be reliably measured with this device in up to 100% of the patients (Fig. 7.5).[10–12]

OCT "en face"

The latest improvements in OCT techniques allow the acquisition of multiple B-scans generating high-resolution images. En face OCT has recently appeared as a new imaging technique that provides a new way of visualizing the posterior pole.

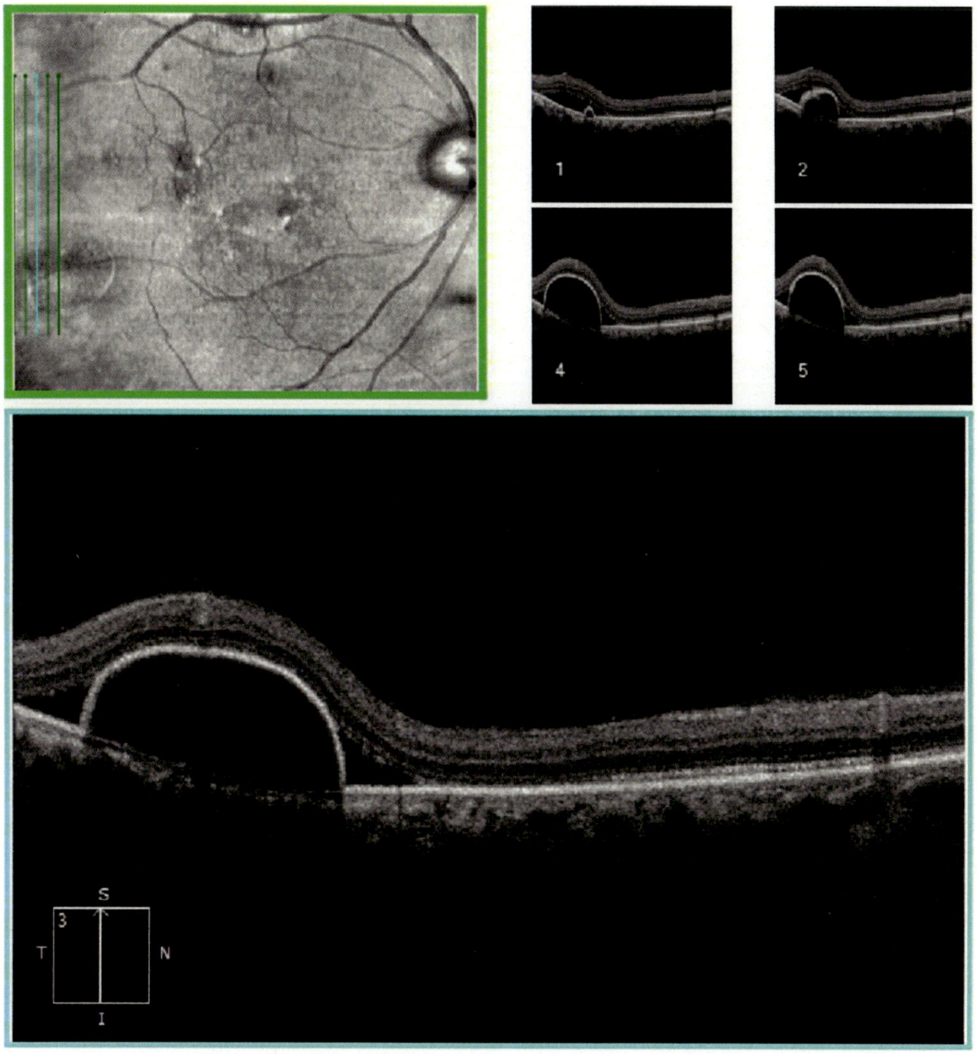

FIGURE 7.3 Spectral-domain optical coherence tomography of a patient showing two serous pigment epithelium detachments, which allow an easy visualization of Bruch's membrane.

While the new C-scans provide us with a coronal plane of the posterior pole, en face technology generates an artificial flattening of the posterior pole tissues (Fig. 7.6) using the selected layer (usually the RPE or Bruch's membrane) as a reference and allowing us to navigate through different layers while modifying the depth of the image by means of these three-dimensional captures.[13] In this way, OCT "en face" allows a new view of the choriocapillaris, Sattler's layer, and Haller's layer (Fig. 7.7) without the need to use contrast as indocyanine green angiography (IVI), even diagnosing the nutrient vessel of a neovascular complex.

OCT B-scans and en face images show choroidal vessels as hyporeflective compared to their surroundings.[14] But not all dark images are actually choroidal vessels, as shadows of retinal vessels can cause misleading images.[15] This technology has been used in pathologies such as central serous chorioretinopathy (CRSC), giving us much more information than the B-scan in relation to its posterior pole involvement in which it has demonstrated the presence of thickened vessels along all layers of the choroid as we will see later.[14] Other studies have also allowed the visualization of more than 90% of the diagnosable polyps by AVI or the vascular ramifications of the neovascular membranes under Bruch's membrane and under these polyps that we will also show.[16] In the case of macular edema, en face images give us important information regarding the surface of the affected retina. It has also been used to confirm previous findings, such as the choriocapillary atrophy found in patients with AMD.[17]

Retinal vessels appear hyperreflective in OCT images, while choroidal vessels are hyporeflective.[12] This difference seems to be related to the velocity of blood cells through them. The blood flow velocity inside choroidal vessels is much

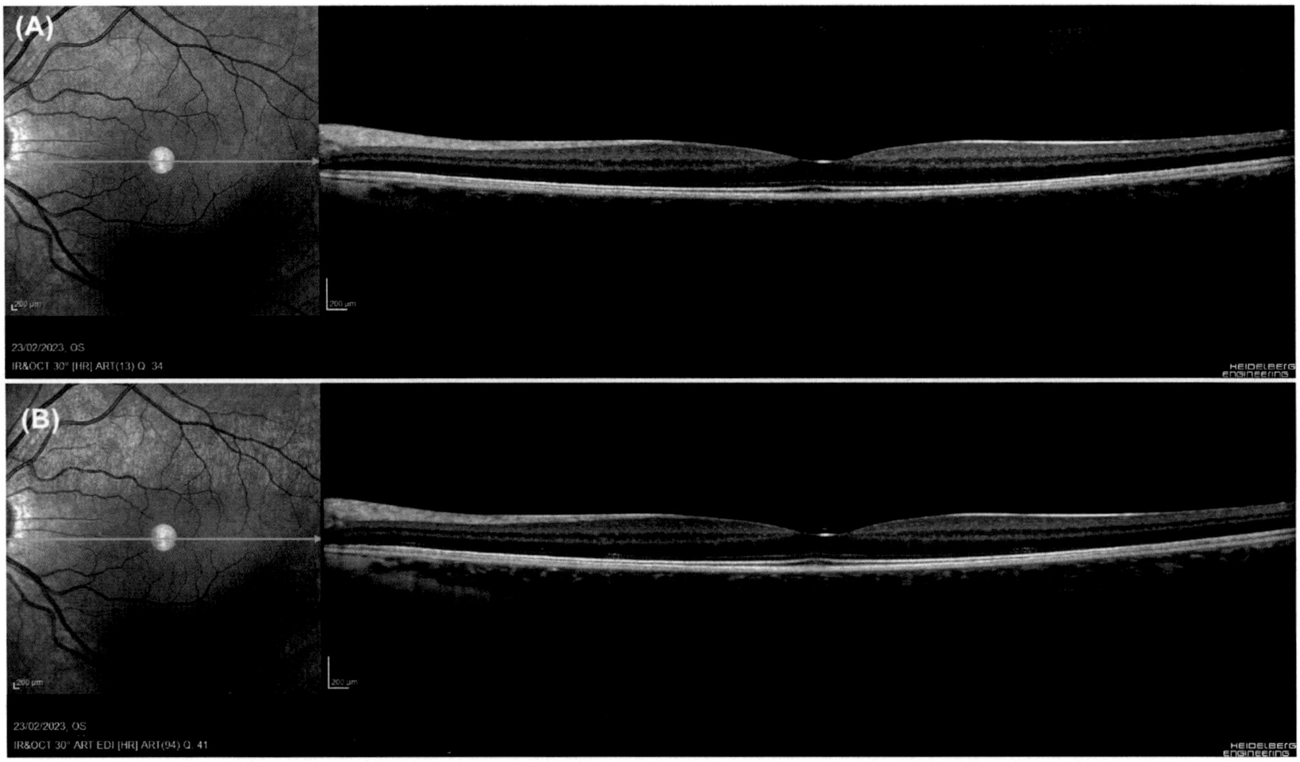

FIGURE 7.4 SD-OCT images of the same patient (MPR) and at the same location. Without EDI technology (*top*) and with EDI technology (*bottom*). The better resolution in deep layers of the second image can be clearly seen.

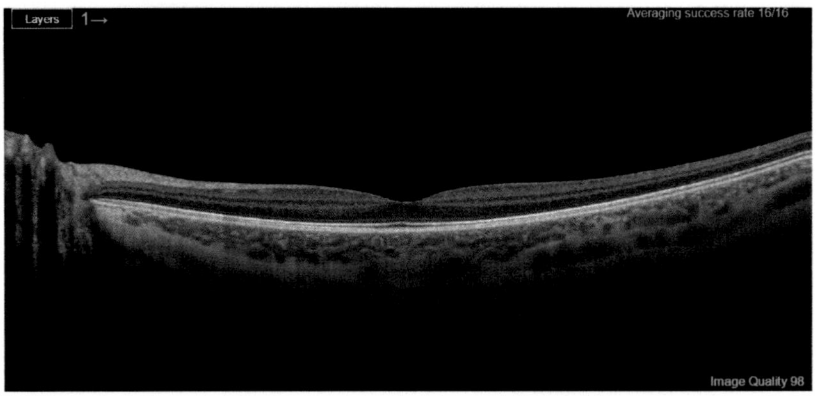

FIGURE 7.5 Swept-source OCT image of the same patient of the previous example (MPR). The higher penetrance of this technology allows us to visualize the sclerochoroidal profile more clearly, which facilitates a more reliable study of the choroid.

higher than that of retinal vessels. OCT imaging is based on interferometry, where an interference fringe is detected to construct the intensity of the signal. The fast speed of the blood cells makes the interference fringe vanish, thus no signal is observed inside the choroidal vessels. On the other hand, blood cell speed is relatively slow in the case of retinal vascular structures, which contributes to the interference signal as a hyperreflective signal.

The role of the choroid in several pathologies has already been proved. They include central serous chorioretinopathy, age-related macular degeneration and polypoidal choroidal vasculopathy, high myopia, posterior uveitis, and choroidal tumors.

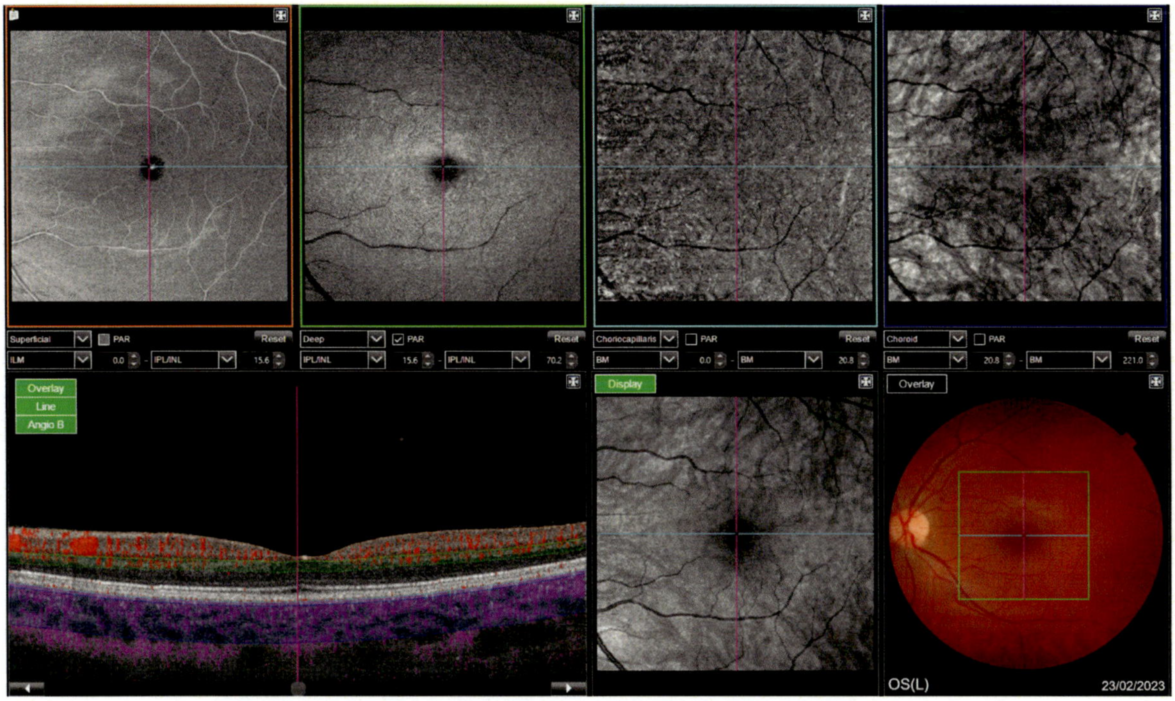

FIGURE 7.6 Swept-source OCT image "en face" showing different choroidal layers.

Choroidal thickness

Age-dependent thinning is a key factor to establish the role of the choroid in retinal conditions, as has been shown in several studies. Margolis et al.[18] studied a group of 54 eyes of 30 52.4-year-old subjects with a mean SE of −1.3 D finding a mean subfoveal choroidal thickness (SFCT) of 287 mm. Chhablani et al. analyzed 211 eyes from 115 healthy 42.8-year-old Indian subjects and found them to have a mean central macular thickness of 216.4 mm, which was thinner than in other reports.[5] Xu reported a mean SFCT of 266 mm in the nondiabetic patient group of his study (1795 subjects).[19] Pediatric choroid has also been studied by SS-OCT and found to be thicker than that of adult patients.[10,20,21]

Our group has conducted several studies on this subject using SS-OCT technology, manually measuring the choroidal thickness of 276 eyes from 154 healthy patients ranging from 3 to 95 years old. Both MCT and SFCT showed a high correlation with age, and both decreased as patients grew older, showing statistically significant differences between age groups. This reduction was found to be progressive until 40 years old, at which point the most significant variation takes place. No statistically significant differences were found between men and women.[22] Following linear regression models, a mean reduction of 10 mm in MCT per decade (one every year) and 8.87 mm of SFCT per decade can be predicted.

This line of investigation came up with the fact that both eyes are not exactly symmetrical.[23] Mean nasal CT was statistically thicker in the right eye than in the left eye (228 vs. 212 mm, $P = .0002$). Chen et al. also found a trend toward a thicker nasal CT (14 mm) in the right eyes.[24] It is difficult to explain the strong and consistent evidence that right eyes have thicker nasal choroids than left eyes in a healthy population. One possibility is a differential in blood flow between the two eyes due to a lack of anatomic symmetry at the aortic arch. Such asymmetry has been suggested to explain the differences in the incidence and prevalence of vascular pathologies between right and left eyes with respect to metastatic bacterial endophthalmitis and retinal artery occlusion. Choroidal circulation is generated from the short posterior ciliary arteries that penetrate through the sclera around the optical nerve in a variable number between 10 and 20. Therefore, the nasal choroid (choroid between the fovea and optical nerve) is directly supplied from these short posterior ciliary arteries. Furthermore, short posterior ciliary arteries are branches of the ophthalmic artery, which is a branch of the internal carotid artery and the common carotid artery. The right common carotid artery is a branch of the brachiocephalic trunk, whereas the left common carotid artery emerges directly from the aorta, and may presumably be responsible for a more proximal and direct blood flow to the right carotid artery.[25] Given that most of the choroidal structure is vascular tissue (the mean vessel density in the

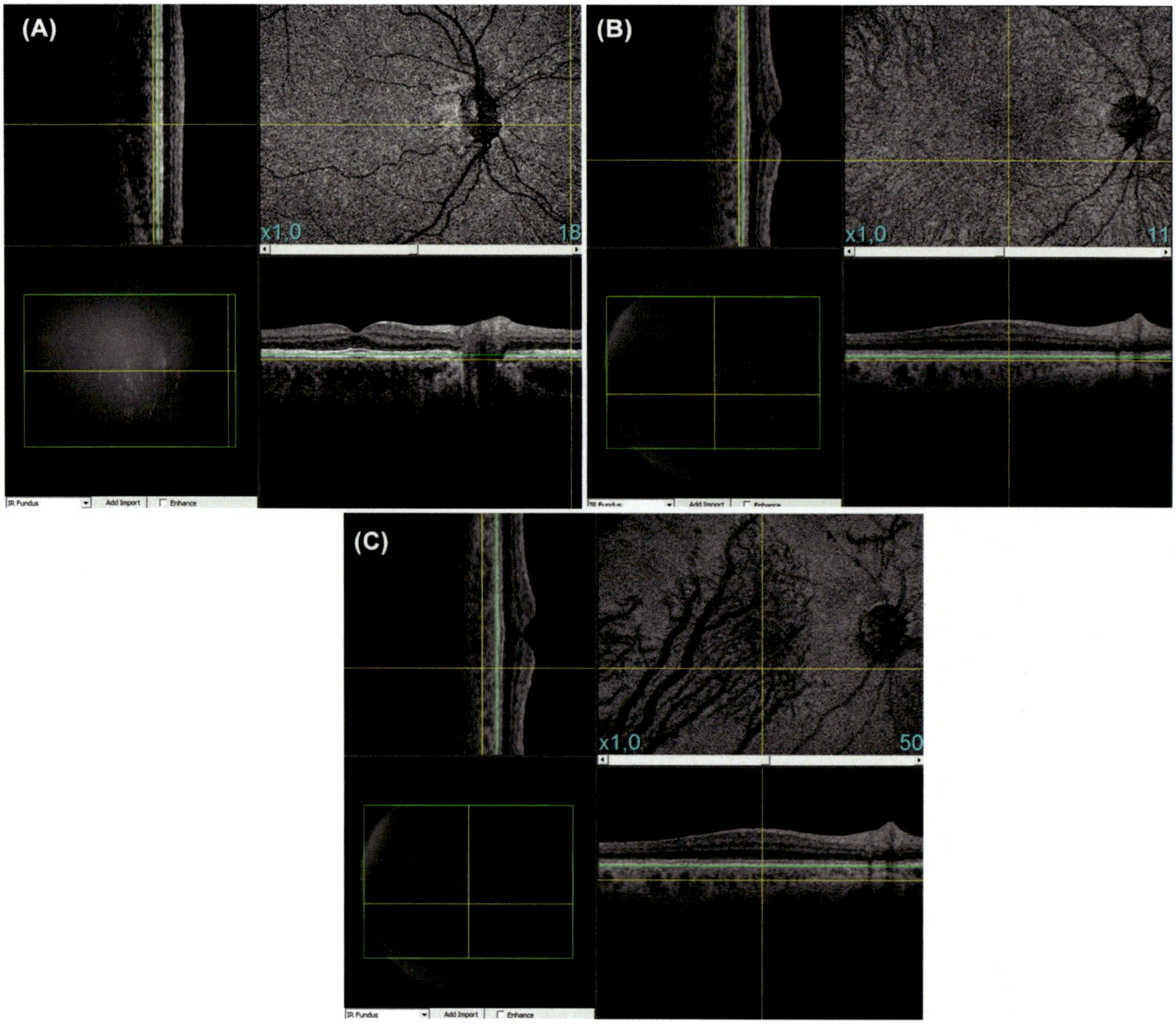

FIGURE 7.7 Swept-source OCT "en face" image of a 50-year-old patient with the C-scan at the level of the choriocapillaris (A), Sattler's layer (B), and Haller's layer (C).

outer choroid is 87%),[17] a supposed higher blood flow in the right versus left short posterior ciliary arteries may explain why the nasal choroid is thicker in the right eye, as stated in our study, and previously reported by Chen et al.[24]

Choroidal morphology

The choroid is normally thickest in the subfoveal region, decreasing slowly and progressively toward the temporal aspect and more steeply toward the optic nerve,[22,24] with a bowl-shaped contour. Adhi et al.[26] described significant alterations in choroidal morphological features in most of the eyes with diabetic retinopathy (DR). They found an irregular temporal choroidoscleral interface inflection point in 8 of 9 eyes with nonproliferative DR (89%), 9 of 10 eyes with proliferative DR (90%), and in 13 of 14 eyes with diabetic macular edema (93%) compared with 0 of 24 in controls. The presence of an irregular temporal choroidoscleral interface inflection contour with focal thinning of the choroid has been detected in diabetic retinopathy and age-related macular degeneration.[26–28] Dolz-Marco et al.[29] reported a bilateral case of focal inferotemporal scleral bulge with a concave choroidal thinning surrounded by normal choroidal tissue. The authors hypothesized that this finding could be related to the inferior oblique muscle leading to inward compression of the choroid, not being able to study the scleral wall, as it was not entirely visible.[29]

Our group performed a study on 276 eyes from 154 patients.[12] A concave or bowl-shaped contour of the horizontal choroidoscleral interface was found in 87.2% of the eyes analyzed and a temporal choroidoscleral interface inflection was identified in 12.8% of the eyes (Fig. 7.4). Mean age of the patients with temporal choroidoscleral interface inflection was 16 ± 19 years versus 36 ± 25 years in the group with bowl-shaped contour ($P = .001$; Student t test for unpaired data).

Reported studies on the anatomy of the posterior aspect of the globe and the vascular and muscular structures and their relation to the sclera show that the anterior and superior limit of the posterior insertion of the inferior oblique muscle coincides with the horizontal meridian within 5.7 mm of the fovea.[30] The mean distance from the inflection point to the fovea was 4427 μm in our series and always located temporal to the fovea. The presence of the inflection point is probably related to the insertion of the inferior oblique muscle. The traction applied to the sclera by the tendon might cause a separation of the scleral wall leading to its deformation and a thickening of the choroid. The fact that this finding is more frequent in younger healthy people, especially those under age 15, could be explained by the reduced scleral stiffness in younger age,[31] that could facilitate the deformation induced by the traction of the inferior oblique muscle tendon.

Most recently, the advent of ultra-widefield OCT offers new possibilities regarding the exploration of the equatorial retina, and in some cases, with proper collaboration from the patient, even further. This technology allows us to visualize from a new perspective pathologies that go beyond the central macula, as can be the case with certain types of posterior staphyloma, that can now be studied as a whole in a quick, noninvasive exploration.

References

1. Huang D, Swanson E, Lin C, Schuman J, Stinson W, Chang W. Optical coherence tomography. *Science*. 1991;254:1178–1181.
2. Drexler W, Fujimoto J. State-of-the-art retinal optical coherence tomography. *Progress in Retinal and Eye Research*. 2008;27:45–88.
3. de Amorim Garcia Filho C, Yehoshua Z, Gregori G, Puliafito CA, Rosenfeld PJ. Optical coherence tomography. In: Ryan SJ, ed. *Retina*. 5th ed. Elsevier Saunders; 2013:82–110.
4. Spaide R, Koizumi H, Pozzoni M. Enhanced depth imaging spectral-domain optical coherence tomography. *American Journal of Ophthalmology*. 2008;146:496–500.
5. Chhablani J, Srinivasa Rao P, Venkata A, et al. Choroidal thickness profile in healthy Indian subjects. *Indian Journal of Ophthalmology*. 2014;62(11):1060–1063.
6. Huber R, Adler D, Srinivasan V, Fujimoto JG. Fourier domain mode locking at 1050 nm for ultra-high-speed optical coherence tomography of the human retina at 236,000 axial scans per second. *Optics Letters*. 2007;32:2049–2051.
7. Unterhuber A, Povazay B, Hermann B, Sattmann H, Chavez-Pirson A, Drexler W. In vivo retinal optical coherence tomography at 1040 nm-enhanced penetration into the choroid. *Optics Express*. 2005;13:3252–3258.
8. Lim H, de Boer J, Park B, Lee E, Yelin R, Yun S. Optical frequency domain imaging with a rapidly swept laser in the 815-870 nm range. *Optics Express*. 2006;14:5937–5944.
9. Ikuno Y, Maruko I, Yasuno Y, et al. Reproducibility of retinal and choroidal thickness measurements in enhanced depth imaging and high-penetration optical coherence tomography. *Investigative Ophthalmology & Visual Science*. 2011;52:5536–5540.
10. Ruiz-Moreno J, Flores-Moreno I, Lugo F, Ruiz-Medrano J, Montero J, Akiba M. Macular choroidal thickness in normal pediatric population measured by swept-source optical coherence tomography. *Investigative Ophthalmology & Visual Science*. 2012;54:353–359. https://doi.org/10.1167/iovs.12-10863.
11. Copete S, Flores-Moreno I, Montero JA, Duker JS, Ruiz-Moreno JM. Direct comparison of spectral-domain and swept-source OCT in the measurement of choroidal thickness in normal eyes. *British Journal of Ophthalmology*. 2014;98:334–338.
12. Ruiz-Medrano J, Flores-Moreno I, Montero JA, Duker JS, Ruiz-Moreno JM. Morphologic features of the choroidoscleral interface in a healthy population using swept-source optical coherence tomography. *American Journal of Ophthalmology*. 2015;160(3):596–601. https://doi.org/10.1016/j.ajo.2015.05.027.
13. Podoleanu AG, Dobre GM, Cucu RG, Rosen RB. Sequential optical coherence tomography and confocal imaging. *Optics Letters*. 2004;29:364–366.
14. Motaghiannezam R, Schwartz D, Fraser S. In vivo human choroidal vascular pattern visualization using high-speed swept-source optical coherence tomography at 1060 nm. *Investigative Ophthalmology & Visual Science*. 2012;53:2337–2348.
15. Mohler KJ, Draxinger W, Klein T, et al. Combined 60° wide-field choroidal thickness maps and high-definition en face vasculature visualization using swept-source megahertz OCT at 1050 nm. *Investigative Ophthalmology & Visual Science*. 2015;56:6284.
16. Alasil T, Ferrara D, Adhi M, et al. En face imaging of the choroid in polypoidal choroidal vasculopathy using swept-source optical coherence tomography. *American Journal of Ophthalmology*. 2015;159:634–643.
17. Sohrab M, Wu K, Fawzi AA. A pilot study of morphometric analysis of choroidal vasculature in vivo, using en face optical coherence tomography. *PLoS One*. 2012;7:e48631.
18. Margolis R, Spaide RF. A pilot study of enhanced depth imaging optical coherence tomography of the choroid in normal eyes. *American Journal of Ophthalmology*. 2009;147(5):811–815. https://doi.org/10.1016/j.ajo.2008.12.008.
19. Koizumi H, Yamagishi T, Yamazaki T, Kawasaki R, Kinoshita S. Subfoveal choroidal thickness in typical age-related macular degeneration and polypoidal choroidal vasculopathy. *Graefe's Archive for Clinical and Experimental Ophthalmology*. 2011;249(8):1123–1128.
20. Park K-A, Oh SY. Choroidal thickness in healthy children. *Retina*. 2013;33(9):1971–1976.

21. Chhablani J, Deshpande R, Sachdeva V, et al. Choroidal thickness profile in healthy Indian children. *Indian Journal of Ophthalmology*. 2015;63(6):474–477.
22. Ruiz-Medrano J, Flores-Moreno I, Peña-García P, Montero JA, Duker JS, Ruiz-Moreno JM. Macular choroidal thickness profile in a healthy population measured by swept-source optical coherence tomography. *Investigative Ophthalmology & Visual Science*. 2014;55(6):3532–3542.
23. Ruiz-Medrano J, Flores-Moreno I, Peña-Garcia P, Montero J, Duker J, Ruiz-Moreno J. Asymmetry in macular choroidal thickness profile between both eyes in a healthy population measured by swept-source optical coherence tomography. *Retina*. 2015;35(10):2067–2073.
24. Chen FK, Yeoh J, Rahman W, Patel PJ, Tufail A, Da Cruz L. Topographic variation and interocular symmetry of macular choroidal thickness using enhanced depth imaging optical coherence tomography. *Investigative Ophthalmology & Visual Science*. 2012;53(2):975–985.
25. Greenwald MJ, Wohl LG, Sell CH. Metastatic bacterial endophthalmitis: a contemporary reappraisal. *Survey of Ophthalmology*. 1986;31(2):81–101.
26. Adhi M, Brewer E, Waheed NK, Duker JS. Analysis of morphological features and vascular layers of choroid in diabetic retinopathy using spectral-domain optical coherence tomography. *JAMA Ophthalmology*. 2013;131(10):1267–1274.
27. Regatieri CV, Branchini L, Carmody J, Fujimoto JG, Duker JS. Choroidal thickness in patients with diabetic retinopathy analyzed by spectral-domain optical coherence tomography. *Retina*. 2012;32(3):563–568.
28. Manjunath V, Goren J, Fujimoto JG, Duker JS. Analysis of choroidal thickness in age-related macular degeneration using spectral-domain optical coherence tomography. *American Journal of Ophthalmology*. 2011;152(4):663–668.
29. Dolz-Marco R, Gallego-Pinazo R, López-Gálvez MI, Díaz-Llopis M, Shields CL. Focal inferotemporal scleral bulge with choroidal thinning: a new observation on high-penetration optical coherence tomography. *Retina*. 2015;35(3):600–602.
30. Siam ALH, El-Mamoun TA, Ali MH. A restudy of the surgical anatomy of the posterior aspect of the globe: an essential topography for exact macular buckling. *Retina*. 2011;31(7):1405–1411.
31. Pallikaris IG, Kymionis GD, Ginis HS, Kounis GA, Tsilimbaris MK. Ocular rigidity in living human eyes. *Investigative Ophthalmology & Visual Science*. 2005;46(2):409–414.

Chapter 8

Optical coherence tomography angiography

Marco Lupidi[1,2], Alessio Muzi[3] and Cesare Mariotti[1]

[1]*Eye Clinic, Department of Experimental and Clinical Medicine, Polytechnic University of Marche, Ancona, Italy;* [2]*Fondazione per la Macula Onlus, Di.N.O.G.Mi., University Eye Clinic, Genova, Italy;* [3]*Department of Ophthalmology, Humanitas Gradenigo, Torino, Italy*

Introduction

Optical coherence tomography angiography (OCT-A) is one of the biggest advances in ophthalmic imaging. In 2012, Jia et al. introduced a new technique able to enhance the OCT visualization of retinal and choroidal perfusion: this was called OCT-angiography.[1] OCT-A allows visualizing retinal and choroidal vasculature in three dimensions, without any dye injection, therefore in a totally noninvasive manner.[1,2] OCT-A principle is based on the possibility of distinguishing, in a steady eye, the movement related to blood cells flowing inside vessels from static tissues. The blood cells' motion can be displayed as a grayscale image by computing the decorrelation of signal amplitude between consecutive B-scans taken at the same retinal location: in this way, a contrast between static and nonstatic tissue is generated. OCT-angiograms are derived from repeated structural B-scans; therefore, they are perfectly coregistered with morphological data. Hence, OCT-A enables a simultaneous assessment of structural and functional retinal or choroidal features (Fig. 8.1).

OCT-A imaging interpretation is generally performed on C-scan or "En-Face" images. To visualize blood flow at different retinal or choroidal layers, C-scans can be moved at different depths: from the inner retinal layers, where the superficial and deep capillary plexuses are located to the deep choroid. Outer retinal layers are normally avascular; therefore, no decorrelation signal could be appreciated at this level. External to the BM is the choriocapillaris (CC): this layer appears as a highly decorrelated, homogeneous vascular texture. Single capillaries may not be distinguished because their maximum diameter is smaller than the lateral or transverse resolutions of the OCT device. Deeper choroidal layers (Sattler's or Haller's layers) are actually difficult to image because of the reflectivity of the structures above, the blood flow intensity of the CC, the decay of the light power and the same wavelength of the light source.

Healthy anatomy

The central retinal artery is responsible for the blood supply of the retina and divides to form two main branches and each of these divides again to form the superior nasal and temporal and the inferior nasal and temporal arteries that supply the four quadrants of the retina. The retinal venous vessels are distributed in a similar pattern. On OCT-A images, arteries are clearly distinguishable from veins by the presence of the surrounding hypointense halo due to the absence of efferent vessels directly out-coming from the walls. The major arterial and venous branches and the successive divisions of the retinal vasculature are present in the retinal nerve fiber layer (RNFL) close to the internal limiting membrane (ILM). As the large arteries extend within the retina toward the periphery, they divide to form successive levels of arteries with progressively smaller diameters until they reach the ora serrata. The retinal arteries branch dichotomously or at right angles from the original vessel. The arterioles coming from the retinal arteries form an extensive capillary network in the inner retina as far as the external border of the inner nuclear layer (INL), either toward the periphery and/or in the macular area. Many macular arterioles, branching from the temporal (superior and inferior) retinal arteries, dive into the retina forming the macular capillary bed with two distinct capillary plexuses: the *superficial capillary plexus (SCP)* and the *deep capillary plexus (DCP)*. The SCP invests with one or more layers of capillaries in the ganglion cell layer (GCL) (Fig. 8.1), while the

Retinal and Choroidal Vascular Diseases of the Eye. https://doi.org/10.1016/B978-0-443-15583-3.00008-1

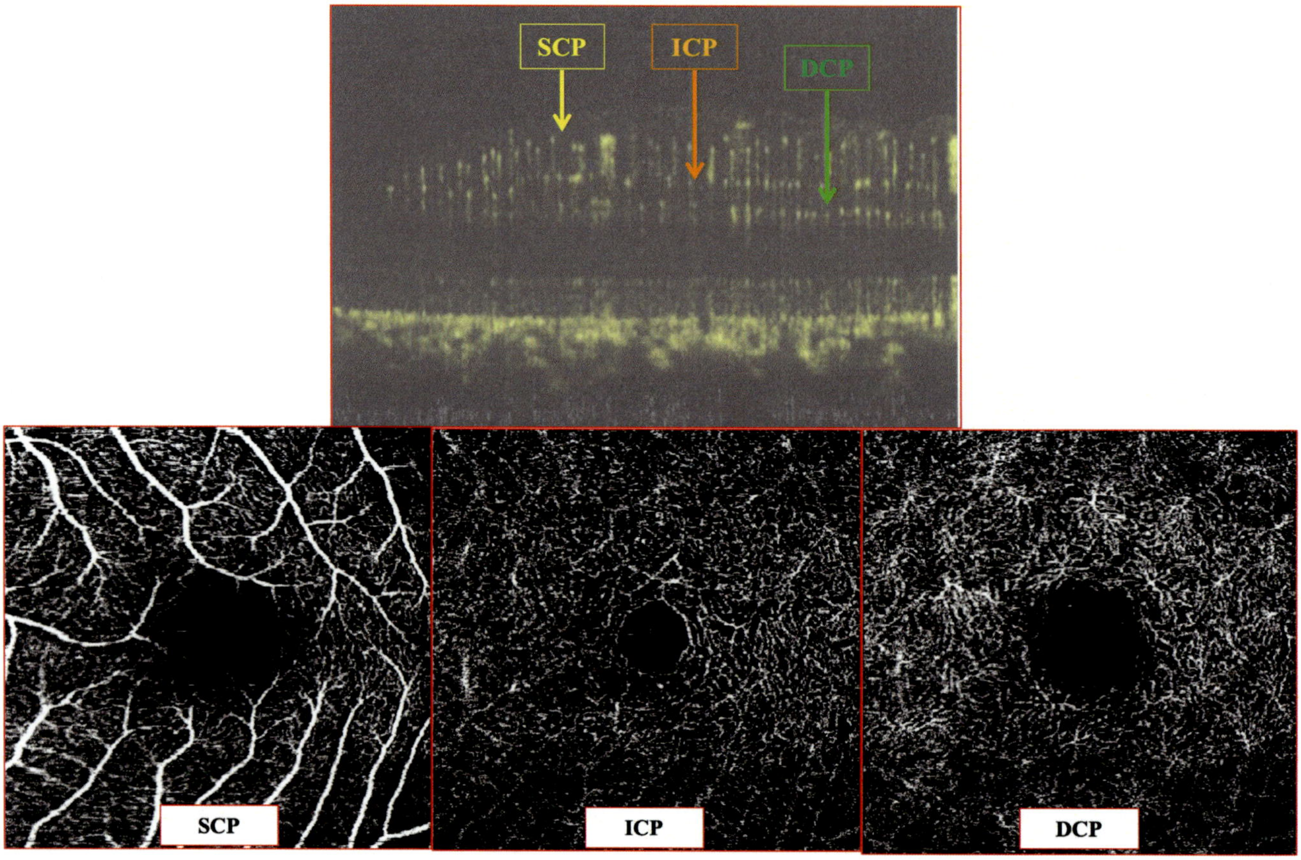

FIGURE 8.1 Retinal OCT-A in health subject. (A) Structural OCT overlapped to decorrelation signal of vascular plexus in OCT-A images. The arrows show the localization of superficial capillary plexus (SCP) (B), the intermediate capillary plexus (ICP) (C), and the deep capillary plexus (DCP) (D).

DCP ordinarily brackets the inner nuclear layer (INL) by a layer of capillaries on either side (Fig. 8.1). The SCP is imaged on OCT-angiograms by starting with the inner border of the GCL to the inner border of the inner plexiform layer (IPL). The DCP is commonly visualized by setting the inner boundary at the outer IPL and the outer boundary at the outer plexiform layer (OPL). *An intermediate capillary plexus* is also described between the inner plexiform layer and the inner nuclear layer. Due to the short distance between the intermediate capillary plexus and the outer component of the deep capillary plexus, they are commonly considered as a unique entity (Fig. 8.1). The SCP appears as a fine capillary network with a hyperintense signal. The perifoveal arcade is well visible at 360 degrees. The deep capillary plexus is shown in the C-scan taken at the level of the inner nuclear layer (INL). A dense capillary network, different from the superficial one, becomes clearly visible and it develops all around the perifoveal area. OCT-A is the first in vivo examination that allows a fine visualization of the DCP. It appears as a very dense, regularly anastomosed network, with sinuous arborization. Arterioles and venules are not distinctly shown. Generally, no vessels extend deeper than the inner nuclear layer. The outer retinal layers and photoreceptors receive oxygen and nutrients from the choriocapillaris.

The choroid is vascularized by two arterial systems: the short posterior ciliary arteries, which supply the posterior choroid and the long posterior ciliary arteries, which supply the anterior portion of the choroid (as well as the iris and ciliary body). Choriocapillaris starts from the Bruch's membrane for a 20-μm distance, toward the choroidal–scleral interface; different C-scans shaped on the BM profile show a relatively homogenous grayish aspect. This aspect seems composed of a large number of tiny dots either hyper- or hypointense. This homogenous pattern could correspond to the very richly anastomosed vascular layer of the choriocapillaris. No vascular channels are clearly detectable at this level probably due to the limited resolution of the OCT-A devices (Fig. 8.2). Sattler's layer (medium choroidal vessels layer) is clearly visible on the OCT-A B-scans with a quite continuous hyperintense signal that is mixed with some hypo-intense structures. The C-scan shows many hypointense, linear (black, tubular) entities resembling the medium vessels network on an almost continuous hyperintense greyish background. The reason why we are not able to appreciate a fine hyperintense vascular network is due to the attenuation of its own decorrelation signal induced by the structures above (RPE and

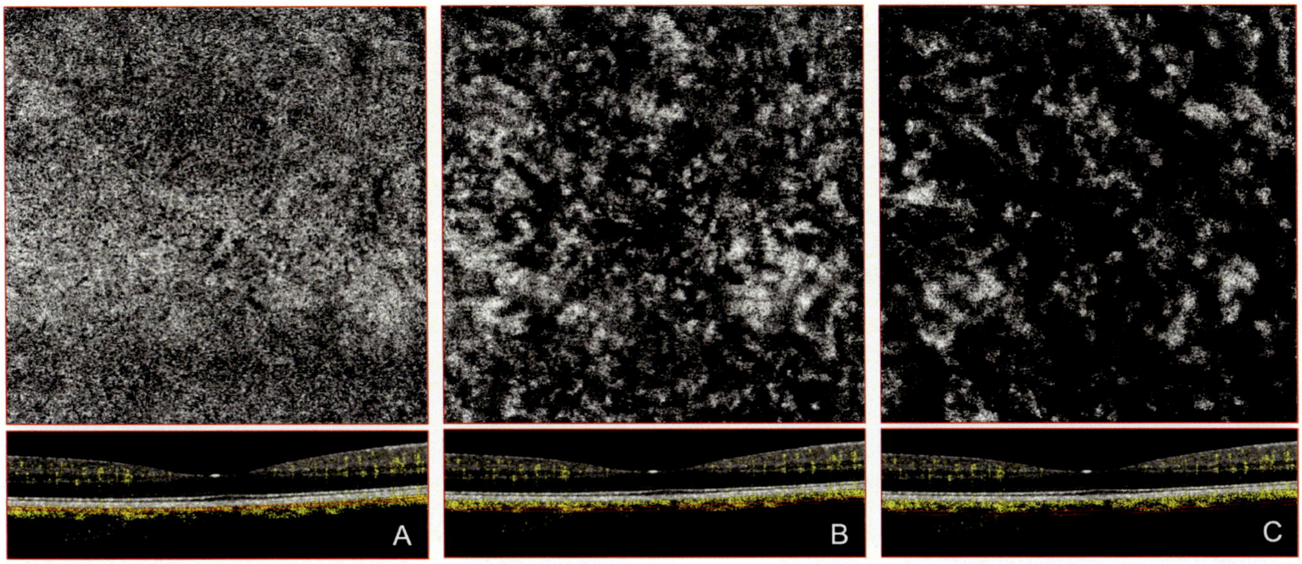

FIGURE 8.2 Choroidal OCT-A in health subject. (A) Choriocapillaris starts 20 μm under the Bruch membrane and is characterized by a relatively homogenous grayish aspect. (B) Sattler's layer (medium choroidal vessels layer) is clearly visible on the OCT-A B-scans with a quite continuous hyperintense signal that is mixed with some hypo-intense structures. (C) Haller's layer and the B-scan section show alternative areas of hypo- (black, tubular) and hyperintense (grayish, diffuse) for the presence of vessels whose caliber is much larger.

Choriocapillaris) that makes it not distinguishable from noise (Fig. 8.2). Deeper we found that Haller's layer and the B-scan section show alternative areas of hypo- (black, tubular) and hyperintense (greyish, diffuse) signal corresponding to these vessels whose caliber is much more larger than the Sattler's layer. The C-scan shows that the signal in this layer is discontinuous with multiple interruptions. In this case also, the aspect is due to the signal attenuation from the structures above (RPE, choriocapillaris, Sattler's layer)[3] (Fig. 8.2).

Neovascular age-related macular degeneration

Neovascular age-related macular degeneration (nAMD) is a degenerative disease that involves the macula, associated with a high risk of blinding. Optical coherence tomography angiography (OCT-A) is one of the biggest advances in ophthalmic imaging. It enables a depth-resolved assessment of retinal and choroidal blood flow, far exceeding the levels of detail commonly obtained with dye angiographies.[4] OCT-A introduction in clinical practice changed significantly the diagnosis and management of n-AMD, making facultative the investigation with conventional imaging such as fluorangiography and indocyanine green angiography.[5] One of the first applications of OCT-A was detecting the presence of choroidal neovascularization (CNV) in age-related macular degeneration (AMD) and establishing its position in relationship to the retinal pigmented epithelium and Bruch's membrane, thereby classifying the CNV as type I (below the RPE) (Fig. 8.3), type II (above the RPE in the subretinal space) (Fig. 8.4), type III (retinal angiomatous proliferation RAP) (Fig. 8.5), or mixed lesions (Fig. 8.6). OCT-angiograms, due to the longer wavelength used by OCT, showed a more distinct CNV vascular pattern than FA since it is less suffering from light scattering or less obscured by overlying subretinal hemorrhages or exudation. A further subtype of CNV might be polypoidal choroidal vasculopathy (PCV), as described by Yannuzzi et al. in 1990.[6] The branching vascular networks, a typical feature of PCV, are now considered to be a variant of type 1 CNV. Kuehlewein et al. showed that OCTA provided the opportunity to directly visualize the occult CNV in AMD and its microvascular details.[7] Through OCTA, they identified two distinct type 1 CNV morphologies: the "medusa" and the "sea-fan" patterns. In the first, vessels branched in all directions from the center of the lesion where a large vessel trunk was identified, while in the second more than 90% of the vessels radiated only from one side of the lesion. OCTA proved effective in detecting neovascular lesions in cases of severe outer retinal impairment or chronic alterations of the RPE, conditions that could compromise the diagnostic sensitivity of dye-based angiography for type 1 CNV. El Ameen et al. reported different type 2 CNV patterns on OCTA: the "medusa-shaped" complex and the "glomerulus-shaped" complex.[8] The medusa-shaped lesion was described as a compact zone of small new blood vessels with minimal hypodense structure inside. The glomerulus-shaped lesion was defined by comparison with the kidney glomerulus, as harboring globular structures of entwined vessels separated by hypodense spaces. Type I and type II may coexist in the same

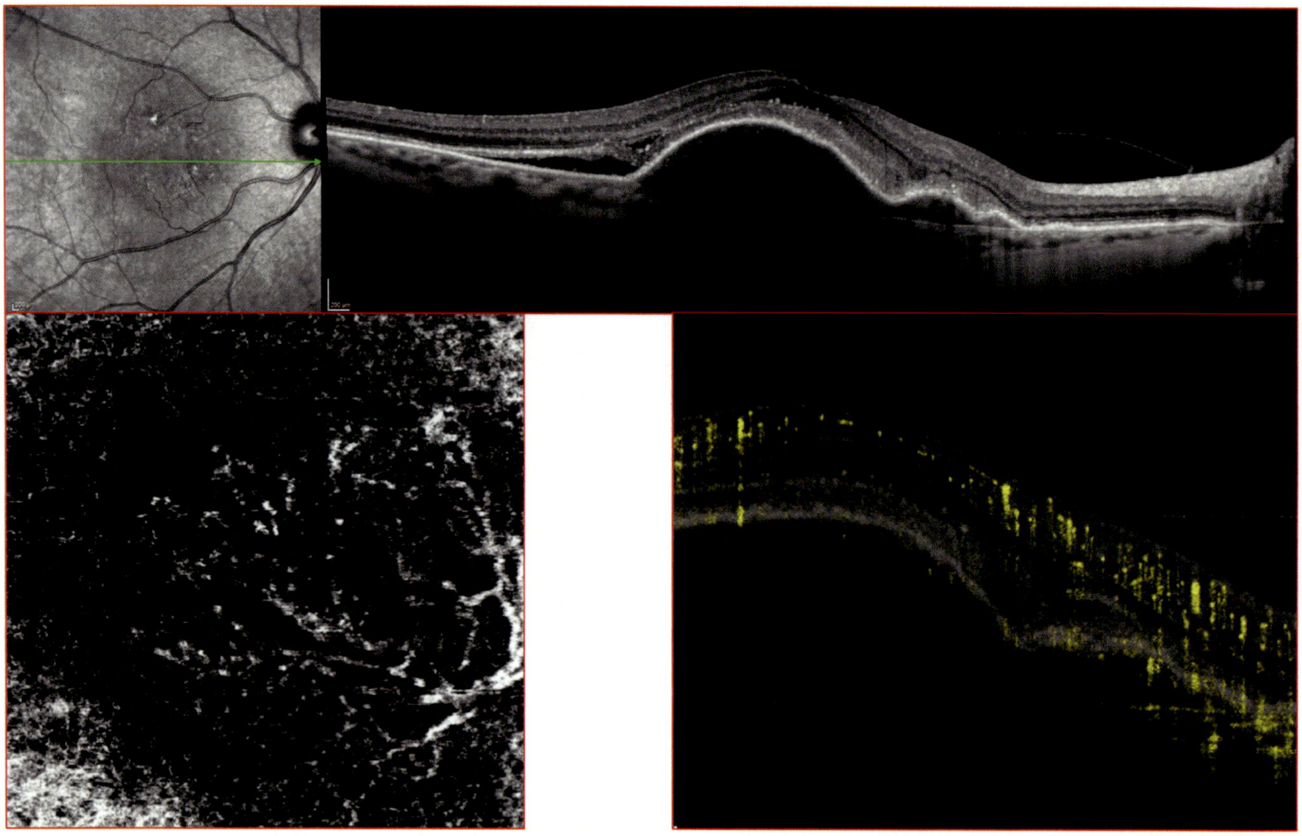

FIGURE 8.3 nAMD with type I CNV. On the top: structural OCT shows the presence of large pigmented epithelial detachment (PED) associated with subretinal fluids (SRF). On the bottom: OCT-A images that evidence the presence of new vessel and the decorrelation signal.

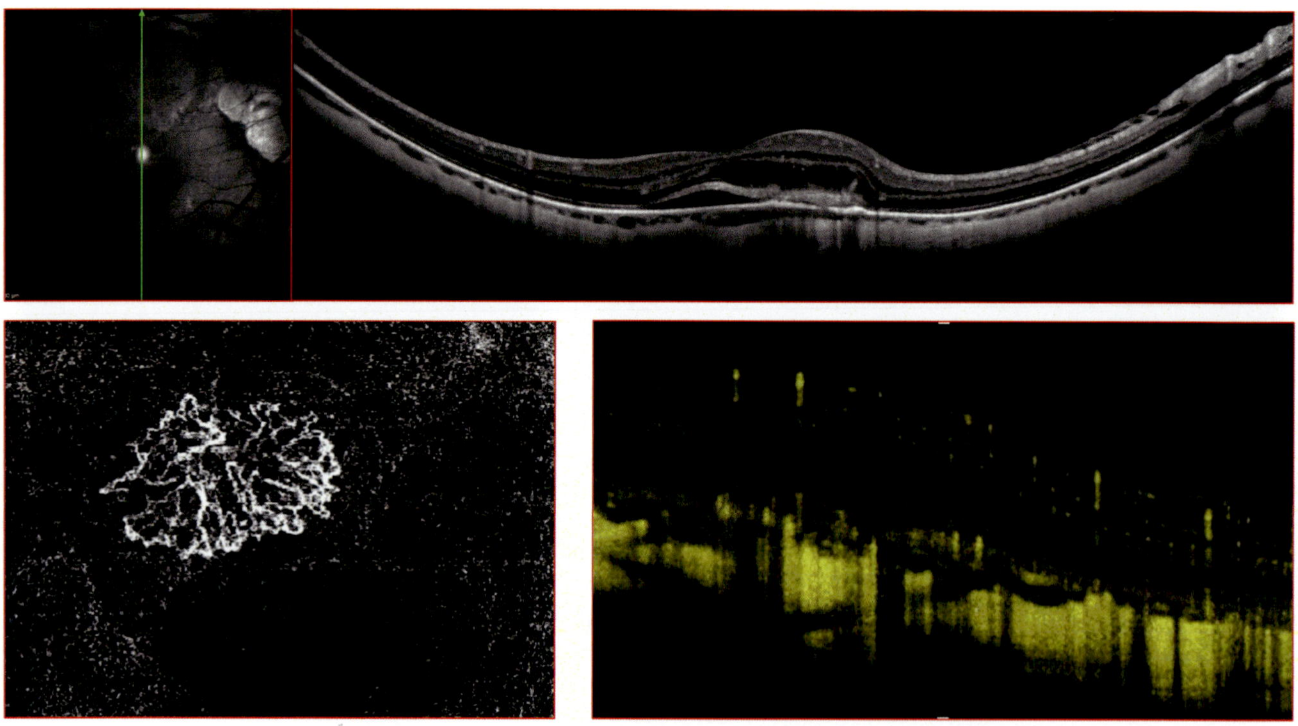

FIGURE 8.4 nAMD with CNV type 2: the NV is situated above the RPE.

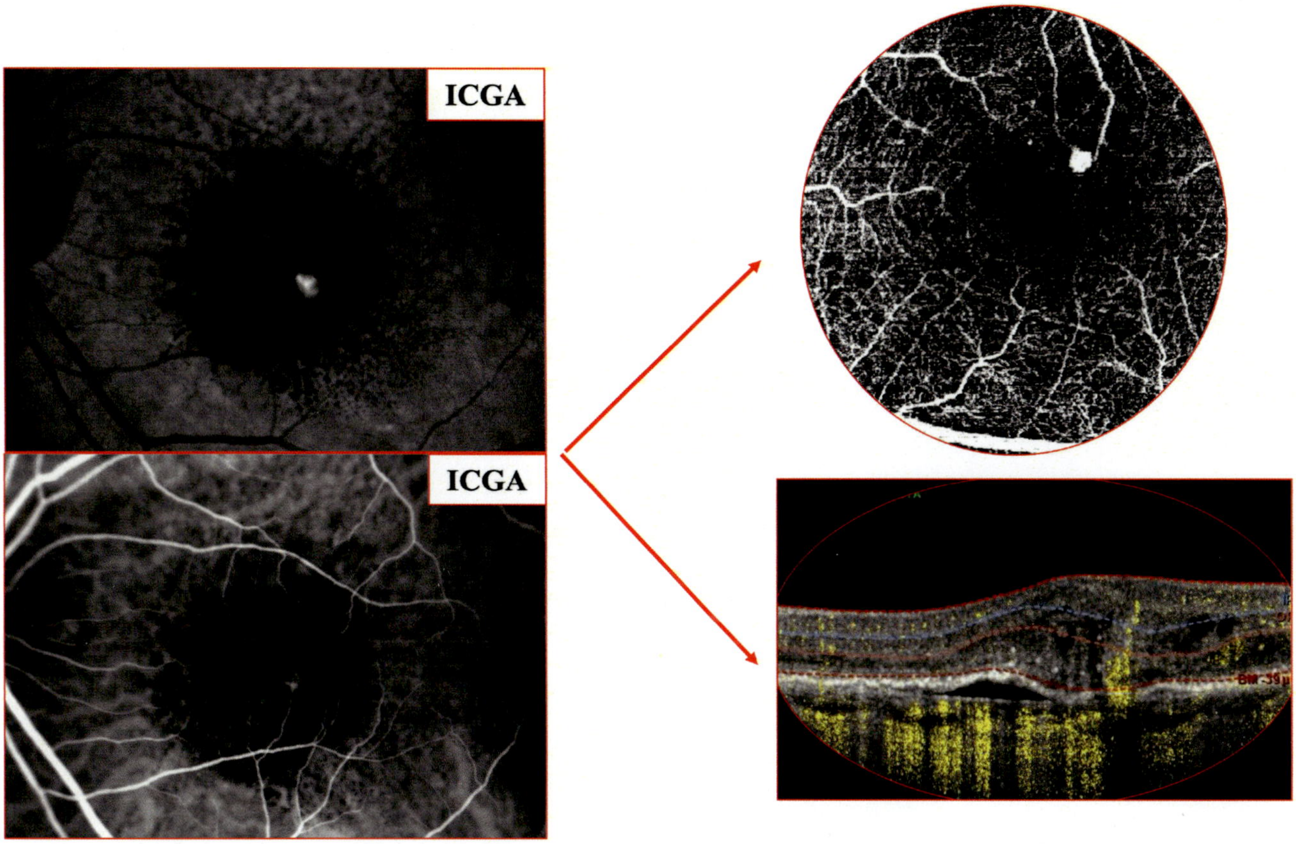

FIGURE 8.5 nAMD type III (retinal angiomatous proliferation RAP). On the left: there are early and late phases of ICGA. On the right: relative OCT-A images, characterized by typical hyperflow "clew-like" lesion located between the deep capillary plexus and the outer retina.

neovascularization: this lesion was once called "minimally classic" CNV. Coscas et al. described on OCT-angiograms the modification of a mixed type I and II CNV during the intravitreal VEGF-trap treatment.[9] Type 3 neovascularization has been commonly referred to as retinal angiomatous proliferation (RAP), and the OCT-A findings are characterized by a typical hyperflow "clew-like" lesion located between the deep capillary plexus and the outer retina.[10] Polypoidal choroidal vasculopathy could be poorly visualized with OCTA, and in most cases the signal within the polyps was clear but not uniform.[11] Srour et al. described the morphological features of polyps on OCTA either as a "hypoflow" round structure or as a "hyperflow" round structure surrounded by a hypointense halo. They also reported that the branching vascular network was always clearly detected on OCTA.[12]

One of the most interesting applications of OCT-A is the possibility to define some quantitative aspects related to neovascular lesions and other abnormalities associated with AMD. The possibility to calculate the extent of a vascular lesion allows us to understand its spatial evolution over time and its structural response to treatment. During follow-up after anti-VEGF injections, Coscas et al. observed a substantial decrease in terms of visibility of tiny branching vessels and anastomoses both in type 1 and type 2 components of a mixed neovascular complex, associated with the persistence of a clear hyperintense signal coming from larger trunks[9] (Fig. 8.7).

Diabetic retinopathy

OCT-A might be a useful noninvasive imaging modality for diabetic patients that can provide helpful information in all stages of diabetic retinopathy (DR).

On OCT-A images is possible to value the presence of findings in the superficial and deep capillary plexuses in diabetic patients before visualization on clinical examination. Remodeling and enlargement of the FAZ and areas of capillary nonperfusion were commonly noted in these eyes even before microaneurysms, which are currently believed to be the first clinical sign of DR.[13]

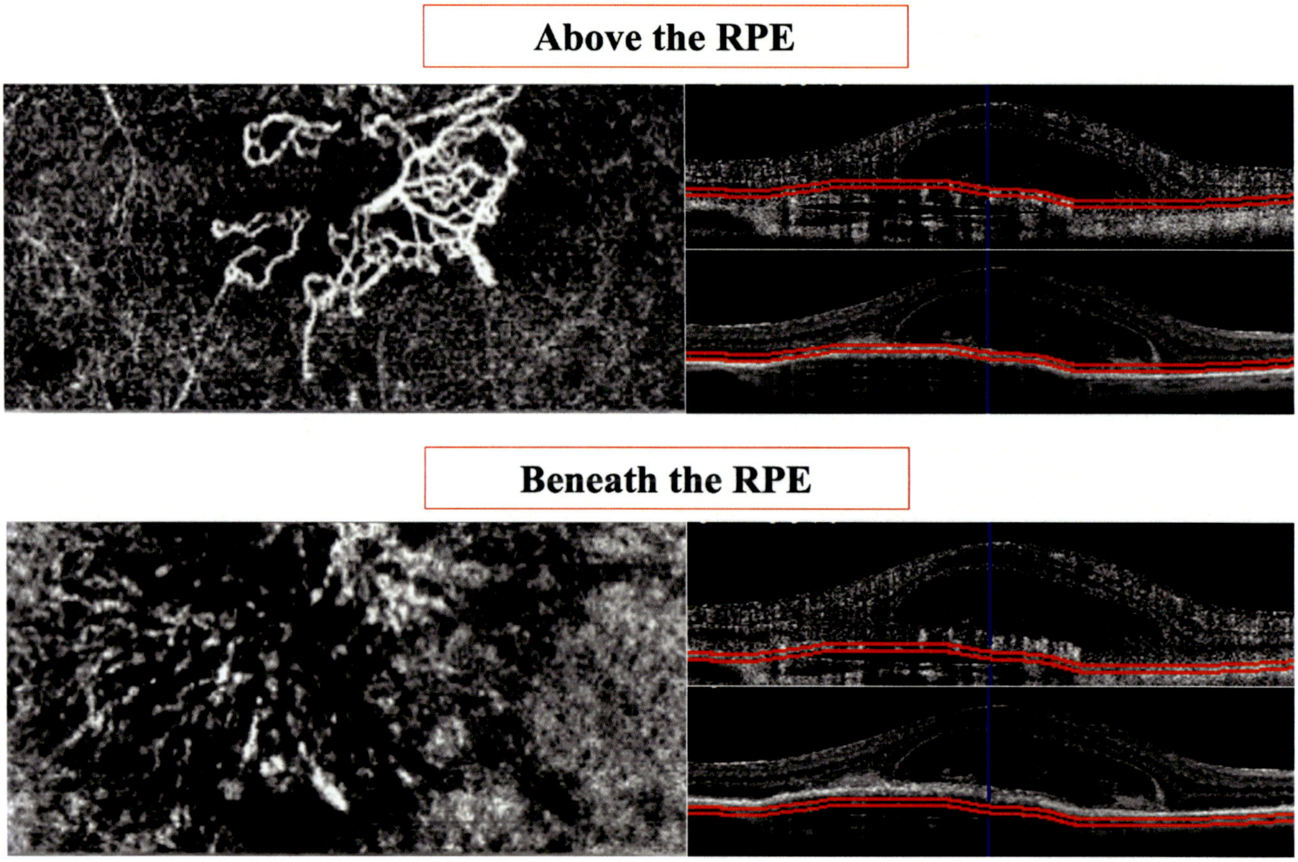

FIGURE 8.6 nAMD mixed lesion. nAMD type I and type II that may be present in the same neovascularization. OCT-A above and beneath the RPE evidence the two lesions.

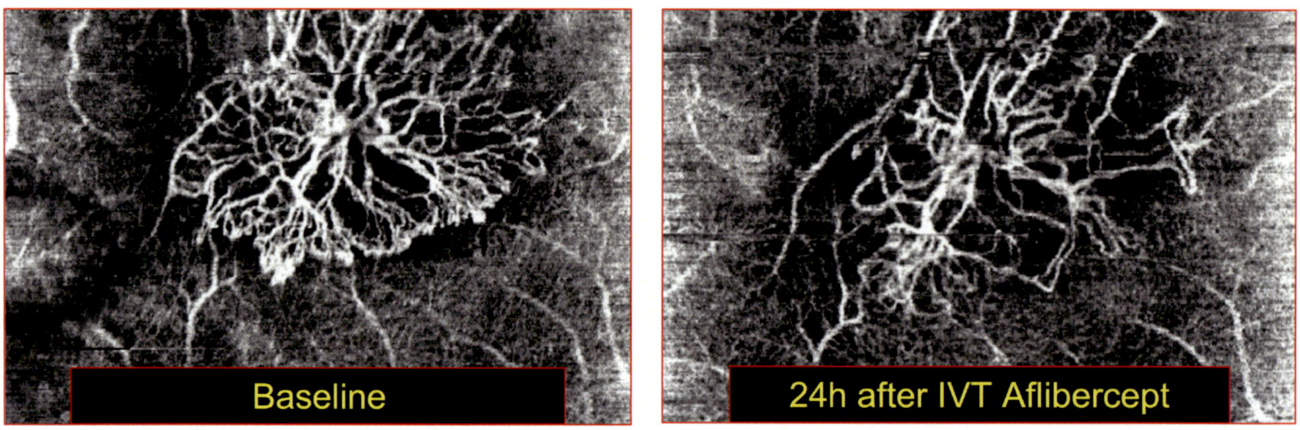

FIGURE 8.7 OCT-A of CNV before (on the left) and after 24 h (on the right) from intravitreal injection with aflibercept. It is evident the decrease in terms of visibility of tiny branching vessels and anastomoses.

Mild nonproliferative diabetic retinopathy is a pathological condition that is mainly characterized by the presence of microaneurysms that are appreciable on OCT-A as roundish or fusiform decorrelated structures. OCT-A may allow identifying some subclinical retinal lesions, such as nonperfused areas that can be hidden on biomicroscopic examinations of the eye fundus The SCP and the perifoveal vascular arcade may appear rarefied and discontinuous. A few, variably sized areas of nonperfusion both at the level of the superficial and deep capillary plexus can be visible: in these areas, the decorrelation signal coming from the vascular network is less intense or absent[3] (Fig. 8.8).

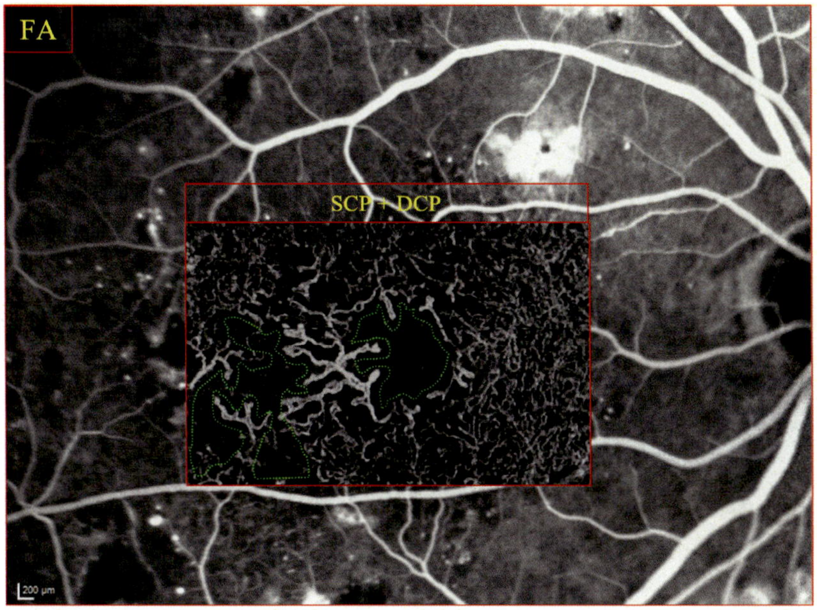

FIGURE 8.8 Overimposion of FA and OCT-A in diabetic patients. The green lines evidence the zone of rarefaction of the capillary density of OCT-A that corresponds to the areas of hypofluorescence on FA.

Severe, nonproliferative diabetic retinopathy is a pathological condition characterized by the presence of more than 20 intraretinal hemorrhages in each of the four quadrants, definite venous beading in two quadrants, prominent intraretinal microvascular abnormalities in one quadrant and no signs of proliferative retinopathy. Intraretinal hemorrhages may block or attenuate the signal coming from decorrelated structures resulting therefore in a sort of masquerading process onto perfused vessels. The venous beading is clearly evaluated on OCT-A as contiguous focal dilations of retinal vessels; in some cases, the blood flow may substantially vary when venous beading occurs, causing the partial visualization of vascular dilations, since they are not homogeneously perfused. Intraretinal microvascular abnormalities (IRMA) are highly decorrelated vascular complexes in the proximity of nonperfused areas. These are characterized by the presence of focal or diffuse dilations and a completely disrupted vascular architecture. They may extend from the inner to the outer retinal vascular layers and sometimes exceed the borders of the inner nuclear layer, as true intraretinal neovascularizations. OCT-A can also highlight the presence of nonperfused areas at different levels of depth. IRMA might be difficult to distinguish from intraretinal neovascularizations (Fig. 8.9).

The main feature of proliferative diabetic retinopathy (PDR) is neovascularization, which is typically located at the vitreoretinal interface. OCT-A also showed the capability of detecting retinal new vessels in PDR as hyperintense vascular structures.[14] The OCT-A is also capable of defining a topographical relationship with nonperfused (or abnormally perfused) areas, and the B-scan angiogram easily localizes the lesion on the retinal surface or into the vitreous cavity (Fig. 8.10).

In the RENOCTA study, Lupidi et al. reported that quantitative OCT-A assessment of retinal new vessels can be a useful noninvasive approach for determining treatment efficacy of laser photocoagulation (Fig. 8.11).[15]

The choroidal changes in diabetic eyes include microaneurysms, dilation of the choriocapillaris, vascular remodeling, increased vascular tortuosity and capillary obstruction, vascular dropout, and choroidal new vessels.[16] Several studies evaluated OCT-A in assessing diabetic-induced choriocapillaris vascular impairment. Agemy et al. demonstrated a statistically significant reduction in terms of vascular density in diabetic versus healthy eyes.[17] Nesper et al. described a novel OCT-A index of retinal and choroidal nonperfusion and the percent area of nonperfusion (PAN). The PAN was significantly correlated with the disease stage when considering the retinal vascular layers, but not any significant correlation was found for the choriocapillaris.[18]

Retinal vein occlusion

The microvascular changes associated with retinal vein occlusions (RVO) in the posterior pole, including nonperfusion areas (NPAs), vascular tortuosity and telangiectasia, disruption of the perifoveal capillary net, and formation of

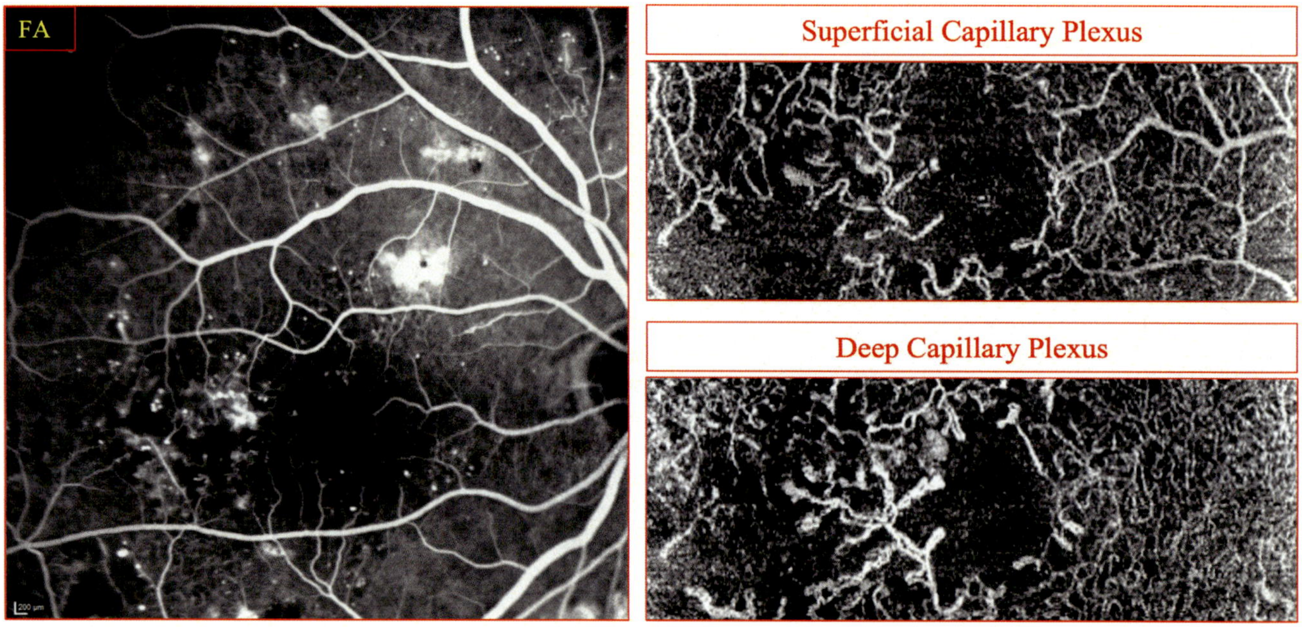

FIGURE 8.9 Intraretinal microvascular abnormalities (IRMA) in severe nonproliferative diabetic retinopathy on fluorescein angiography (FA—left image) and on OCT-angiography (right). On OCT-A images is evident the presence of IRMA that appears as a highly decorrelated vascular complex in proximity of nonperfused areas.

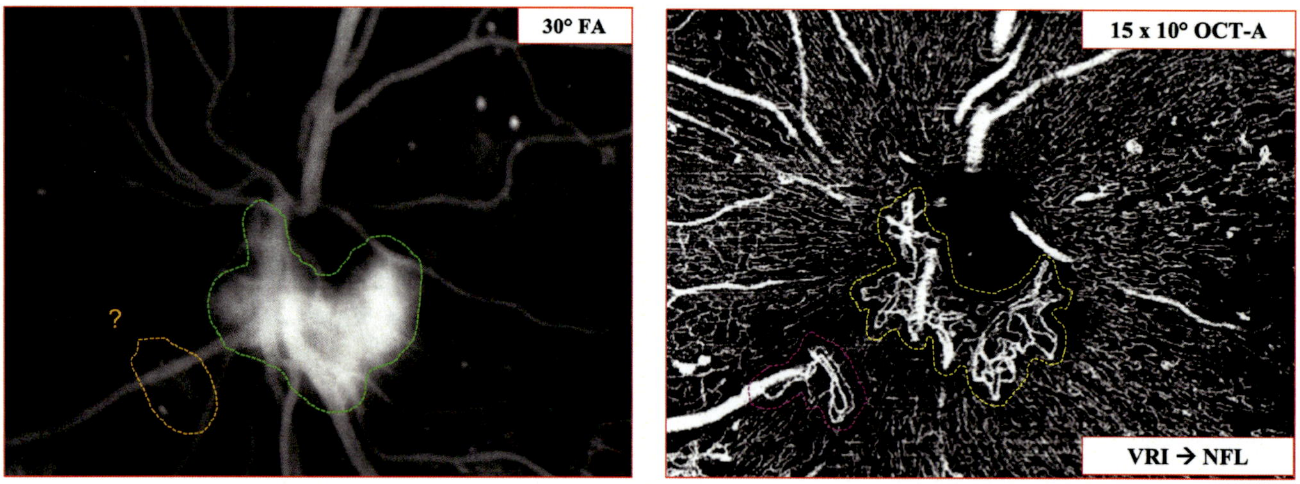

FIGURE 8.10 Proliferative diabetic retinopathy with new-vessels (NV) at disc. OCT-angiography (OCT-A) allowed to identify the presence of NV and establish the true extension (yellow-dashed line). OCT-A helps in detecting also poorly-leaking neovascularizations (purple dashed line).

microaneurysms and collaterals, are all readily visible in OCTA images.[19] The findings can be described as qualitative and quantitative (Fig. 8.12).

NPAs are regions without visible perfused capillaries. In RVOs, they are more extensive in the DCP than in the SCP.[20] Nevertheless, the absence of visible vessels in areas of nonperfusion may not be due to a total absence of flow but may represent areas in which the blood flow has decreased below the device's detection threshold.[21] Vascular tortuosity is seen in both central RVOs and branch RVOs and includes kinking, angulation, and/or spiral twisting of vessels.[22] The perifoveal capillary plexus can be damaged, especially in DCP than in the SCP in RVOs, because DCP drains directly into the larger superficial veins and an increase in intravenous pressure in RVO is directly transmitted to the DCP.[23] Another anomaly visible especially in DCP is the dilation of the capillary plexus and venous dilation. The microaneurysms are principally detected in the border of NPAs, and in collateral vessels and could be a source of persistent leakage and recurrence of edema.[24] Cystoid spaces have no signal and coincide with areas of perfusion abnormalities. The absence of

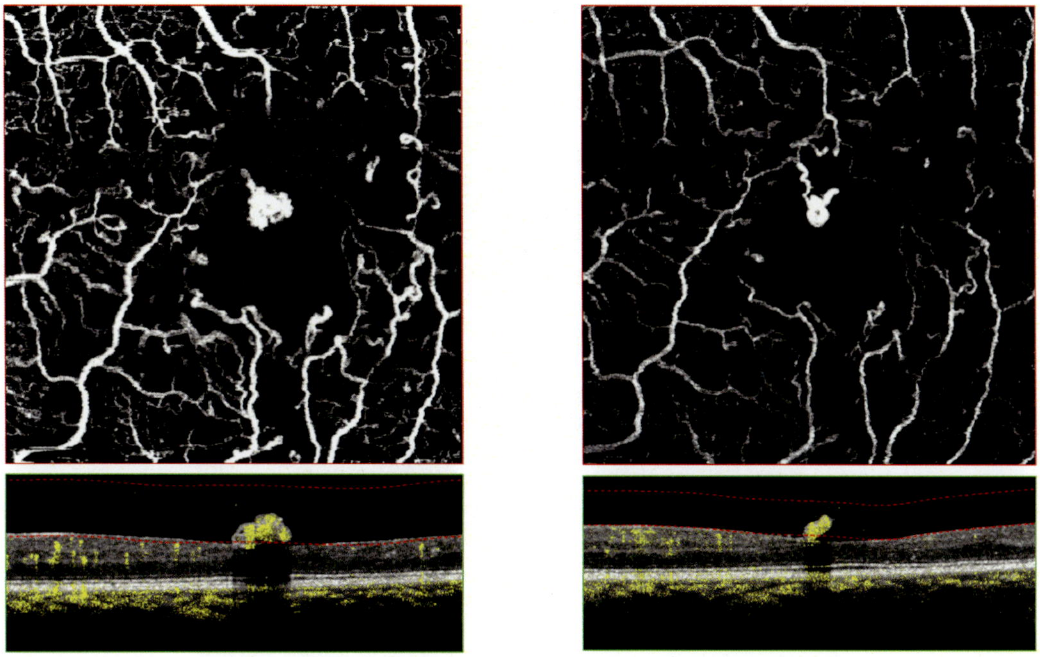

FIGURE 8.11 OCT-A and proliferative diabetic retinopathy images show the regression of NV in a patient after PDR laser treatment.

OCTA signals in the area of cystoid space can be explained by the displacement of capillaries by cysts and the development of cysts in nonperfused areas.[25] The intraretinal hemorrhages may obscure images of one or both intraretinal vascular plexuses, for a shadowing effect, and the level of the hemorrhage can be determined from the degree to which the images are obscured.[20] The development of new vessels can be detected with OCTA with the same efficacy of FA because of the absence of leakage in OCTA. This modality allows to perform quantitative follow-up of new vessels and evaluate the response to treatment.[26] An interesting differential diagnosis is to differentiate optic disc venous collaterals (OVCs) from neovascularization of disc (NVD). OCTA shows optic disc collaterals at the level of the superficial peripapillary plexus whereas neovascular vessels are visible above the retina at the level of the vitreous. OVCs are loopy vessels whereas new vessels are a mesh of fine vessels.[19]

As a concern, in the quantitative findings on OCT-A images in RVOs, the most significative parameters are as follows: the measure of nonperfusion area (that correlates with VA and macular sensitivity in microperimetry in eyes with RVO) and the measure of FAZ.[19]

A major limitation of early OCTA technology was the limited field of view, but the introduction of wide-field OCTA allowed the acquisition of the largest images useful in pathology such as DR and RVO.[27,28]

Retina artery occlusion

Retinal artery occlusion is an emergency in ophthalmology that brings to the rapid and irreversible loss of vision. The collapse of capillary flow leads to axoplasmic stasis, intracellular edema, and ischemic necrosis of the inner retinal layers. Over the course of several weeks compare inner retinal atrophy and persistent retinal vessel narrowing.[29]

OCT-A in both the acute and chronic phases has demonstrated capillary nonperfusion in affected areas, involving both the SCP and DCP. These areas of nonperfusion on OCT-A images match perfectly with ischemic fluorangiography changes.[30]

Quantitatively, it was also found that vessel density parameters in the SCP and DCP of RAO eyes were reduced compared to fellow eyes and that of normal controls. While the size of FAZ was not found to be enlarged in RAO eyes, the acquisition of OCT-A images could be difficult because of the poor capacity of fixation and the segmentation of the layer may be nonsharply for the presence of retinal edema or atrophy.

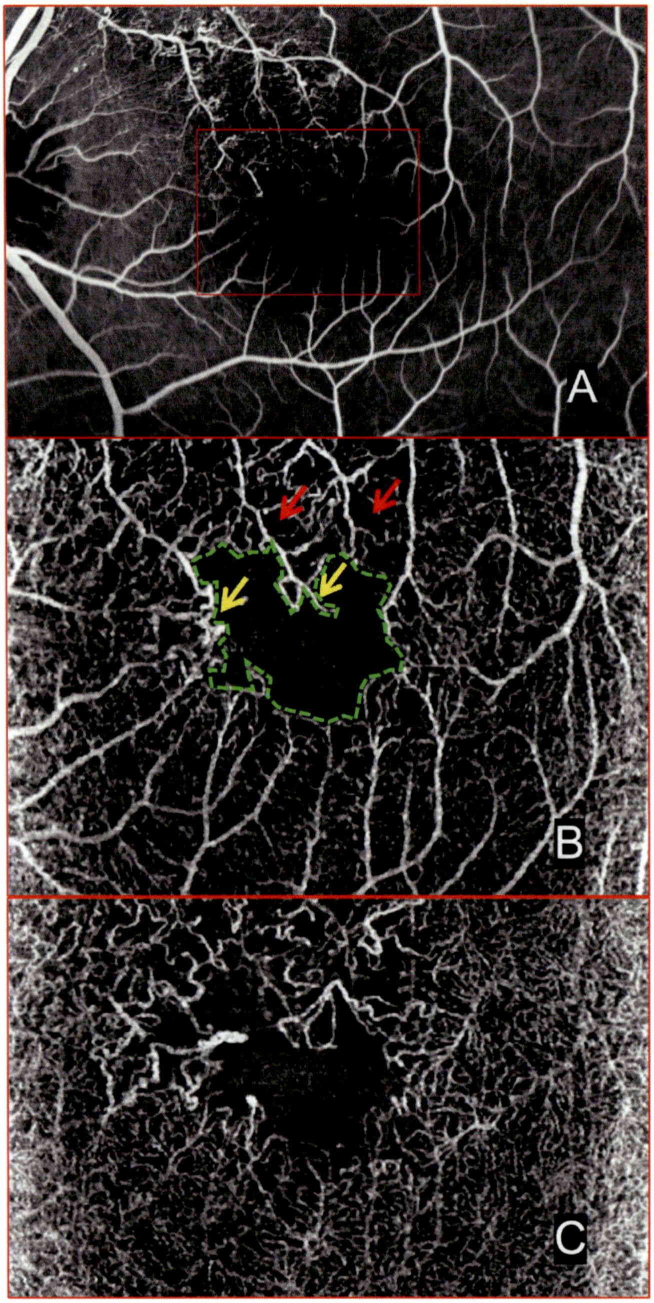

FIGURE 8.12 Branch retinal vein occlusion. (A) Fluorescein angiography shows the presence of a BRVO involving the macular area. (B) Superficial capillary plexus: low arrows evidence the presence of focal vascular dilatations, the dashed line indicates the presence of FAZ enlargment, and the top arrows the areas of impaired perfusion. (C) The same alterations are more extensive in the DCP than in the SCP.

Myopia

Myopia, particularly high myopia, is characterized by excessive axial elongation and an increased risk of developing a series of complications, such as degenerative retinopathy, lacquer cracks, choroidal neovascularization, and posterior staphylomas.[31] OCT-A mainly contributes to the diagnosis and follow-up of CNV. Myopic CNVs are usually type 2 neovascularizations; in many cases, myopic CNVs appeared as irregular in shape, with poorly defined margins, and without a visible core. Moreover, OCT-A showed that in most cases active myopic CNVs appeared to be mainly interlacing, while in inactive CNVs the neovascular network appeared to be mainly tangled (Fig. 8.13).

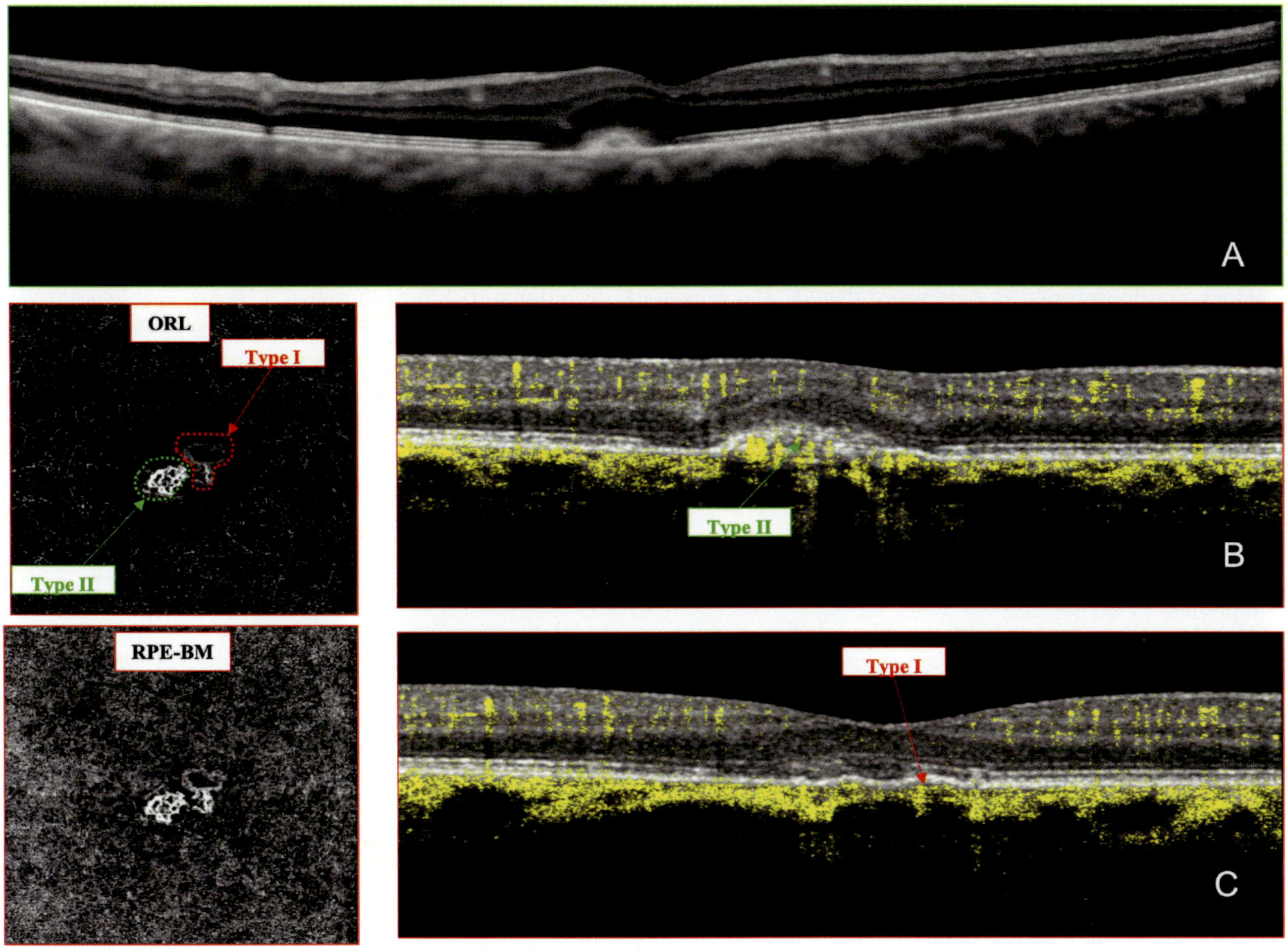

FIGURE 8.13 Myopic eye with CNV. (A) Structural OCT evidence the presence of a hyperreflective lesion in the foveal area. (B) OCT-A images with segmentation at ORL (outside retinal layer) evidence the presence of CNV II. (C) Segmentation at the RPE-BM complex: OCT-A evidence a component of CNV type I.

These data suggest that OCT-A could be considered as a potentially useful tool in characterizing CNV by its morphology and detecting CNV activity in eyes with pathological myopia.[32] Also, Bruyère identified two types of CNV complicating high myopia: small, disorganized vascular loops, suggesting an immature neovascular network and a larger, highly structured, interlacing network, suggesting a mature lesion.[33] They hypothesized as in case of nAMD[34] a process of vascular remodeling, suggesting the existence of a three-stage life cycle of neovascular membranes in general and complicating high myopia in particular: immature, mature, and fibrotic neovascularizations.[33]

Although the gold standard for identifying myopic CNV is fluorography, Querques et al. found high sensitivity (90.48%) and specificity (93.75%) for neovascular detection with OCT-A in pathological myopia,[32] according to Myata et al. who found a successful detection in 94.1% of cases.[35]

Another interesting challenge is the differential diagnosis between macular hemorrhages due to myopic macular neovascularization (m-MNV) and idiopathic macular hemorrhage (IMH) in patients with high myopia (HM) as the therapy and prognosis are different. Querques found a extremely good accordance between OCTA and fluorescein angiography. Moreover, OCT-A is not influenced as fluorescein angiography by the presence of extensive retinal hemorrhage that could block the underneath fluorescence, partially or totally masking the neovascularization and causing failure of the diagnostic process.[36]

Central serous chorioretinopathy

One of the first applications of OCT-A was detecting choroidal neovascularizations (CNV) in central serous chorioretinopathy (CSCR) within those flat irregular pigmented epithelium detachments that are commonly observed in cases of long-standing disease. Several studies reported that no clear OCT-A signs were associated with the leaking point[37] and that choroidal flow might be either normal or impaired.[37,38]

The choriocapillaris (CC) imaged in patients with CSCR usually shows areas "low-flow signals" "flow rarefaction areas" and "flow-voids" at the CC layer on OCT-angiograms also in some fellow eye without CSCR.[39,40] Teussink et al. stated that on OCT-A, chronic CSCR showed irregular CC flow patterns, corresponding to ICGA abnormalities. These results suggest focal CC ischemia with surrounding hyperperfusion that may lead to subretinal fluid leakage.[41]

One of the first applications of OCT-A in CSCR cases was detecting type I CNVs inside those shallow pigmented epithelium detachments (PED) that are often seen in long-standing diseases.[40] The CNV corresponds to the small undulating RPE detachment on B-scan (Fig. 8.14).

Quaranta-El Maftouhi et al. reported that OCT-A allows the detection of CNV in chronic CSCR not visible with other imaging techniques.[38] Bousquet et al. in a group of 81 CSCR patients evaluated the rate of choroidal CNV detected by OCT-A in flat irregular pigment epithelium detachment (FIPED) in chronic central serous chorioretinopathy. They concluded that one-third of FIPEDs in chronic CSCR contained CNVs.[42] This approach might allow an appropriate treatment and a better visual outcome. Lupidi et al. reported an increased sensitivity of OCT-A to detach CNV in FIPDs in patients who performed a hand grip stress, compared to a rest condition[43] (Fig. 8.15).

Choriocapillaris alterations underneath CNVs might be difficult to visualize due to the shadowgraphic artifact that a CNV lesion itself can cast onto the layers below.[44] These projection artifacts could make difficult the differentiation between normal physiological vessels and pathological ones.[7] However, modern OCT-A software now provide "artifacts removal" tools that subtract projection artifact from superficial vessels but might sometimes cause some loss of signal of pathological blood vessels.

Uveitis and inflammatory eye desease

The clinical application of OCT-A has been studied also in inflammatory disorders, such as sarcoidosis, ocular tuberculosis, and multifocal choroiditis.

Sarcoidosis is a multisystemic disease characterized by cellular immunity activity with the formation of noncaseating granulomas in various organ systems. The most common manifestations of sarcoidosis involving the posterior segment include vitritis, snow-balls, peripheral multifocal choroiditis, and retinal vasculitis, mainly in the form of segmental periphlebitis and macular edema. Other manifestations include solitary choroidal or optic disc granuloma, retinal macroaneurysms, and exudative retinal detachment. Hypoperfused/nonperfused areas, capillary abnormalities, and capillary

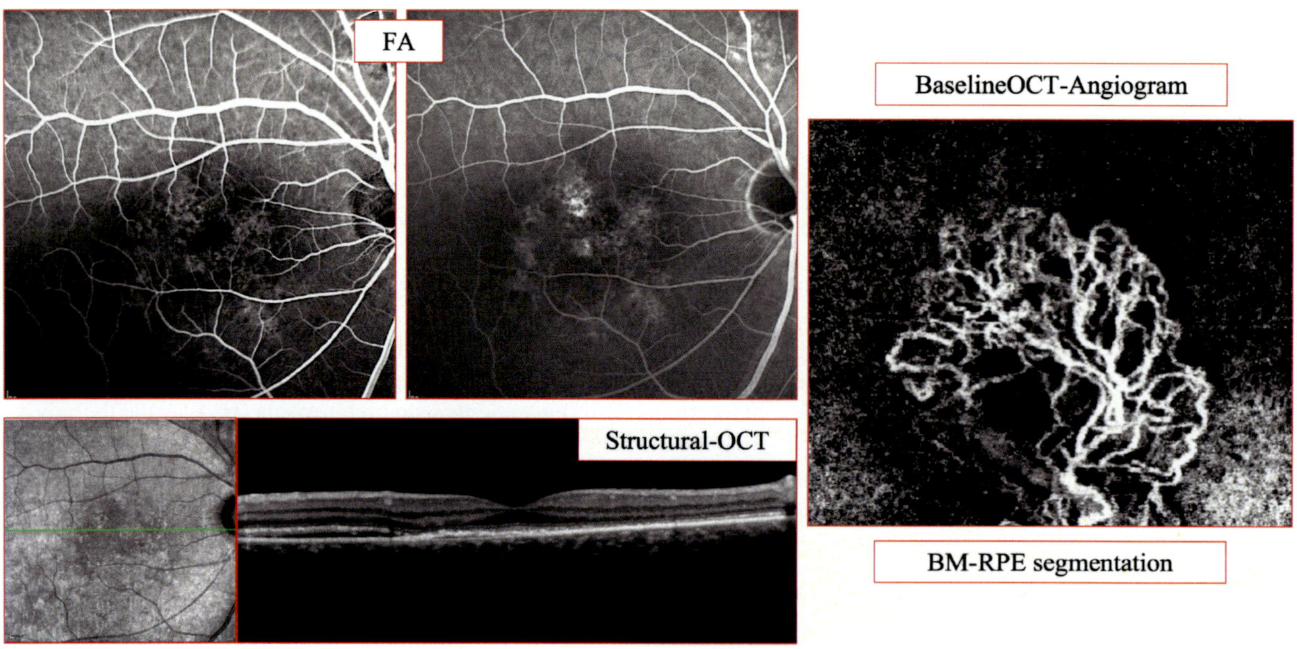

FIGURE 8.14 Patient affected by CSCR with PED. The CNV can be visible with the CNV under the RPE detachment on B-scan.

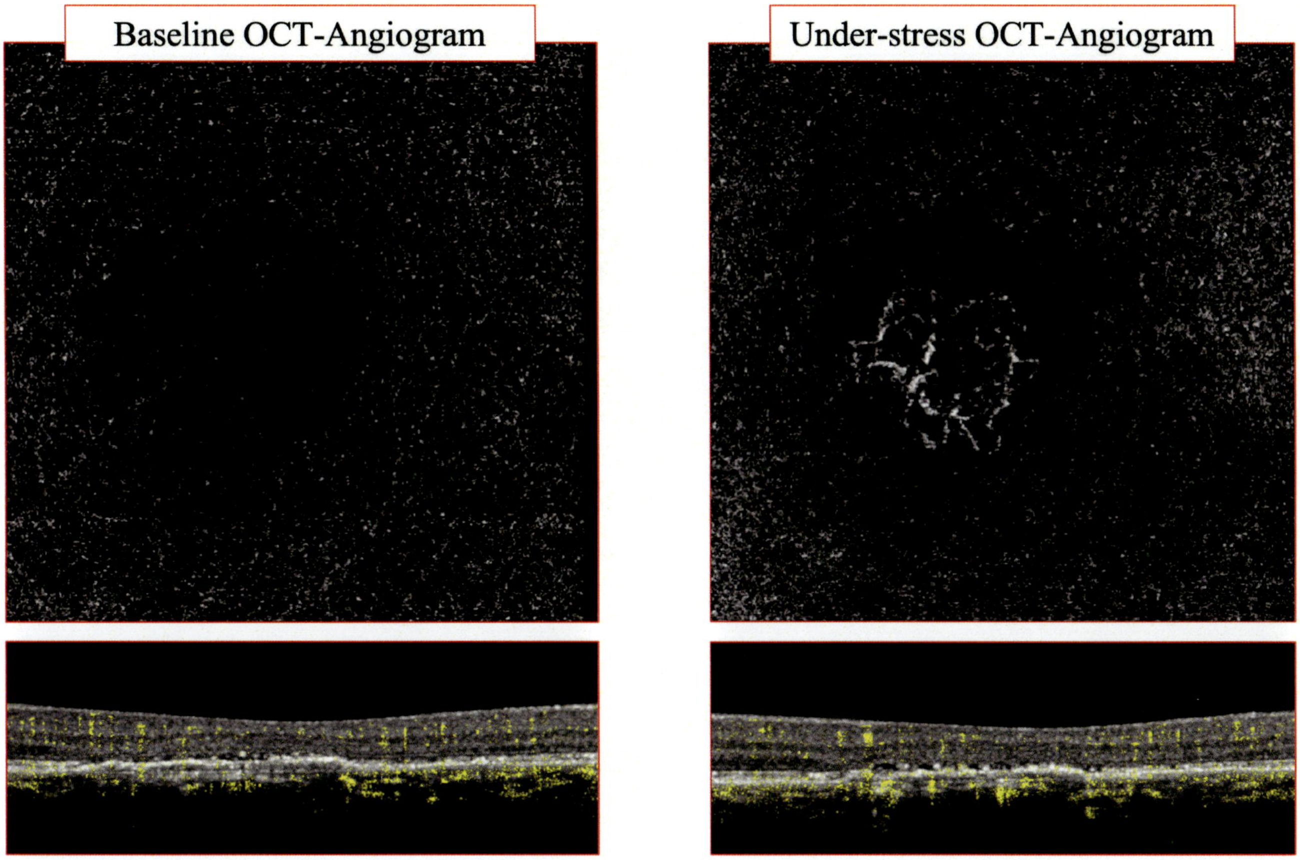

FIGURE 8.15 The influence of blood pressure in CSCR patients. On the left: OCT-A was performed on a patient in rest condition. On the right: OCT-A of the same patient after isometric exercise. It is evident how with an increase in the blood flow, CNV signal is more evident.

network disorganization were the most common OCT-A findings in this study. The DCP appeared more severely compromised than the SCP in terms of hypoperfused/nonperfused areas, disorganization of the capillary network, and cystoid spaces. OCT-A also revealed the presence of flow void areas at the level of the choriocapillaris plexus and could be probably explained by the presence of low-flow regions due to active granulomas or to chronic tissue damage secondary to previously active granulomas.[45]

Intraocular tuberculosis (TB) is characterized by a varied clinical presentation including granulomatous anterior uveitis, intermediate uveitis, posterior uveitis, or panuveitis. Posterior uveitis is the most common presentation and may manifest as choroidal tubercles, granulomas, occlusive retinal vasculitis, and serpiginous-like choroiditis or multifocal serpiginoid choroiditis.[46] As described in the preceding sections, on ICGA, choriocapillaris involvement in tubercular serpiginous-like choroiditis appears as hypofluorescent areas suggestive of nonperfusion. The availability of OCTA has provided additional tools to image the choroidal circulation in vivo with very high resolution. The findings of choriocapillaris low void on en-face OCTA correlated very well with ICGA as well as EDI-OC. As the lesions heal, a subset of patients may develop choriocapillaris atrophy, which results in the visualization of medium-to-large choroidal vessels in the choriocapillaris zone on en face OCTA. The development of CNV could be found with OCT-A.[47]

Inflammatory choroidal neovascularization (iCNV) is a sight-threatening complication of retinochoroidal inflammatory and infectious conditions. The presence of choroidal neovascularization (CNV) is considered the most common cause of poor visual acuity (VA) at presentation in idiopathic choroiditis such as multifocal choroiditis (MFC). The distinction between inflammatory lesions and CNV is mandatory for the therapeutic approach, but sometimes it may be tricky. CNV in MFC is virtually always of type 2 and is clearly visible with fluorography in the absence of large hemorrhage. OCT-A could contribute to the clinical decision indeed the absence of any decorrelation signal in correspondence with macular lesions, which were very suspicious of CNV on CMI, allowed us to better understand the inflammatory nature of the subretinal lesion responsible for the patient's central visual loss[48] (Fig. 8.16).

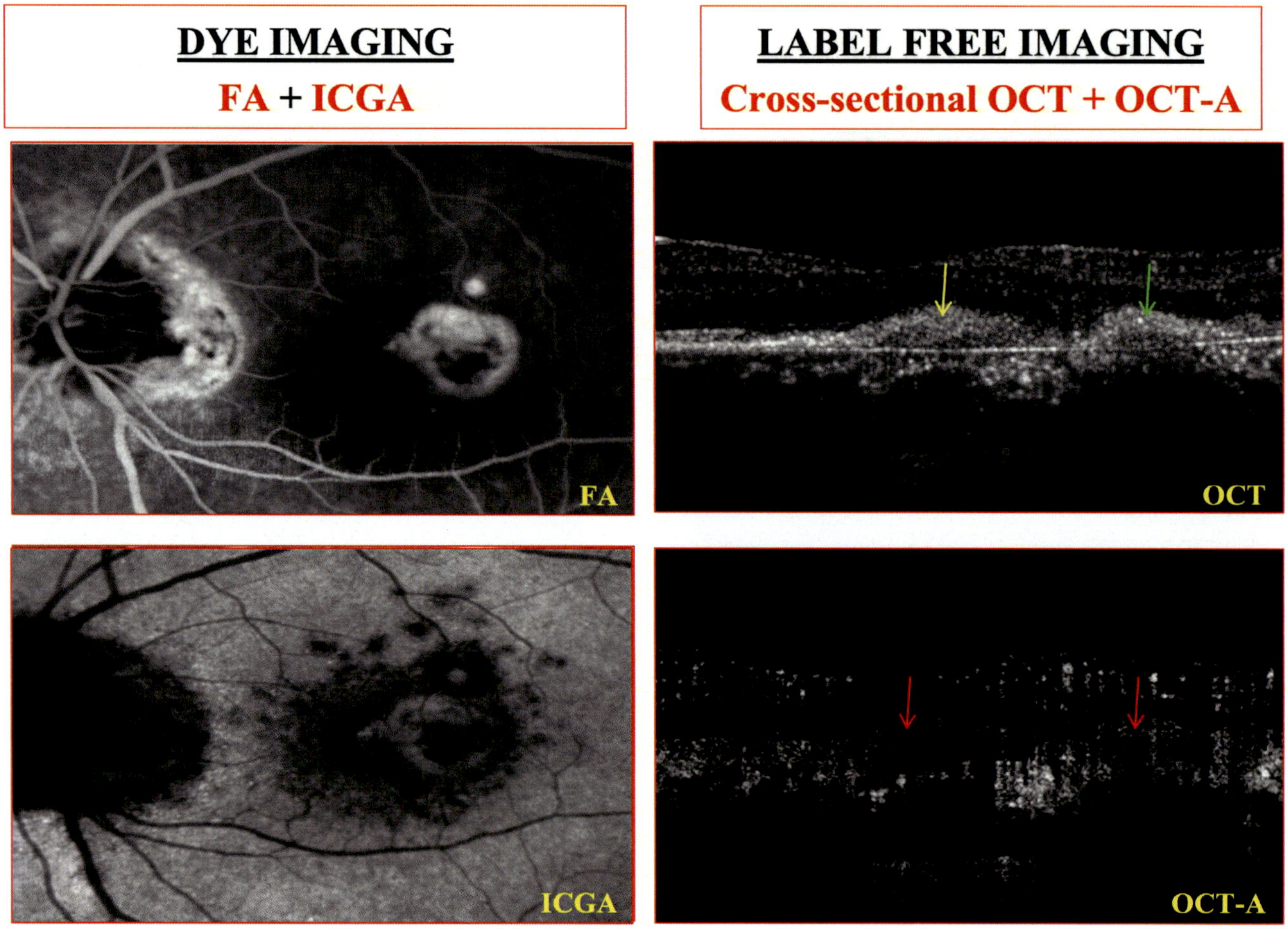

FIGURE 8.16 Multifocal choroiditis (MFC). On the left, a large stained lesion with uncertain foveal leakage on FA; lesion that appears as a fibrotic scar with some areas of RPE atrophy in the early ICGA phases, and then as a CC and RPE impairment in stromal and inversion phases, respectively. On the right, we have subretinal hyperreflective material accumulation on structural OCT and an absence of any decorrelation signal on OCT-A; so from one side the CNV is uncertain and on the other side there is no evidence of neovascular pattern.

Macular telangiectasia

Macular telangiectasia (MacTel), also known as idiopathic macular telangiectasia, is a rare group of neurodegenerative diseases caused by loss of Müller cells with abnormal anastomoses of foveal and perifoveal capillaries. It results in the degeneration of the outer retinal layer and even the development of a macular hole in the advanced stage.[49] Macular telangiectasia type 1 is an aneurysmal type of macular telangiectasia, considered a central variant of Coats disease. Macular telangiectasia type 2 is a nonaneurysmal perifoveal capillary telangiectasia, associated with atrophy of neurosensory retina, presenting in the nonproliferative or proliferative forms. MacTel type 2 is bilateral and the most frequent in the population. In MacTel 2, an increased intervascular network with diminished capillaries and abnormal vessel anastomosis is noted on OCTA. The quantified superficial and deep retinal capillary density and branches are significantly reduced compared to healthy eyes.[50,51] This finding suggests that the abnormal anastomosis of vessels contributes to hypoperfusion of the retina, which may play a role in the disease progression. Zeimer et al. have noted vascular alterations mainly characterized by reduction of the capillary network in the inner retinal vascular plexus and dilatation and telangiectasis in the deep plexus, associated with vascular proliferation of the outer retina.[52] Zhang et al. report better visualization of retinal vessel proliferation in MacTel 2 eyes when using OCTA rather than fluorescein angiography.[53] Optical coherence tomography angiography shows capillary proliferation in the outer retina corresponding to areas of EZ loss, which could be a useful marker to monitor the efficacy of possible treatments in MacTel2 disease (Fig. 8.17).

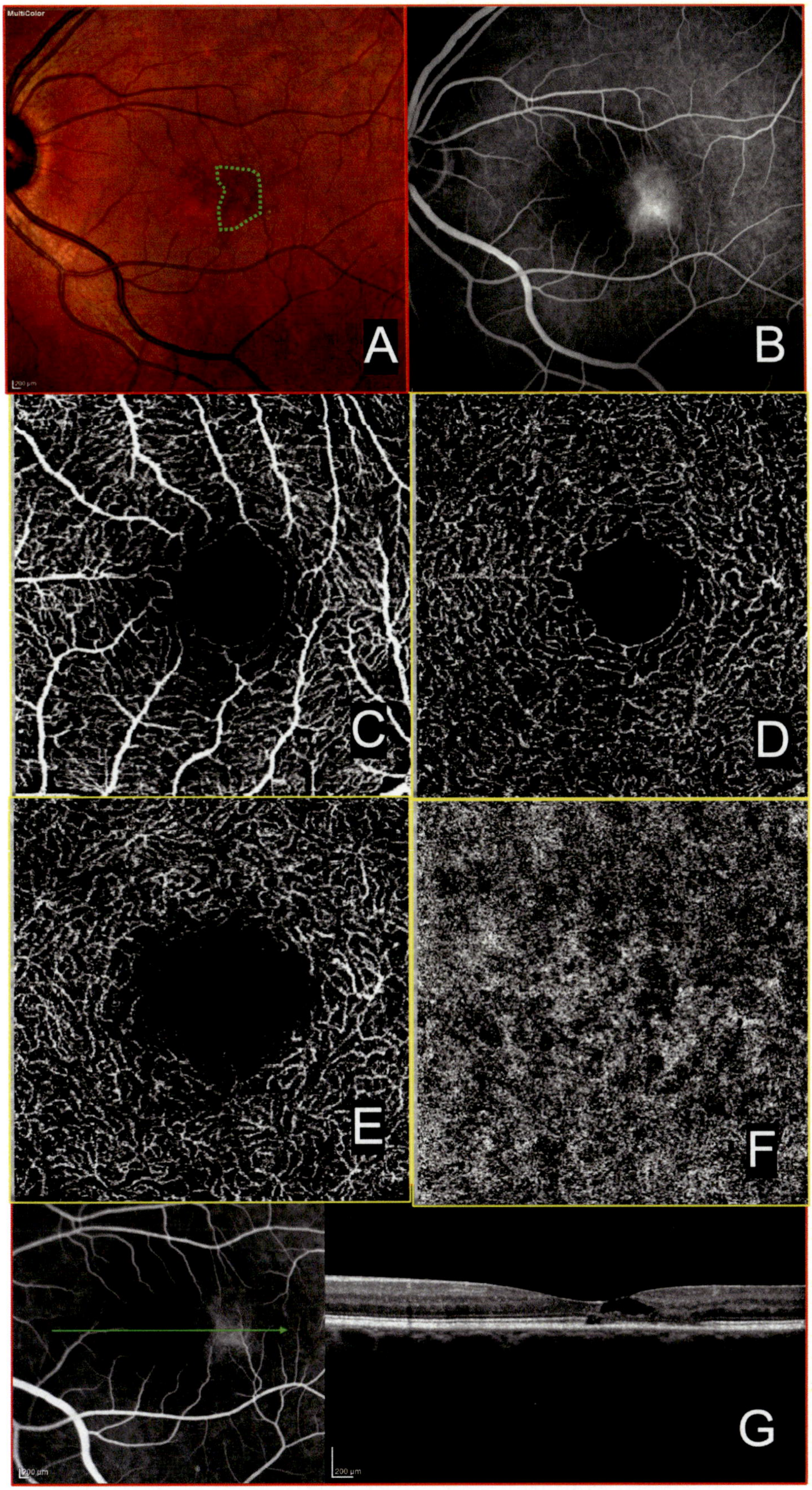

FIGURE 8.17 Patient with MACTEL type 2. (A) Multicolor image evidence of macular temporal changes. (B) FA shows temporal foveal telangiectatic vessels that leak in the later stage. (C−F) Vascular alterations mainly characterized by reduction of the capillary network in the inner retinal vascular plexus and dilatation and telangiectasis in the deep plexus, associated with vascular proliferation of the outer retina. (G) Structural OCT with degeneration of the outer retinal layer and development of a pseudomacular hole.

References

1. Jia Y, Tan O, Tokayer J, et al. Split-spectrum amplitude-decorrelation angiography with optical coherence tomography. *Optics Express*. 2012;20(4):4710–4725. https://doi.org/10.1364/OE.20.004710.
2. Potsaid B, Baumann B, Huang D, et al. Ultrahigh speed 1050nm swept source/fourier domain OCT retinal and anterior segment imaging at 100,000 to 400,000 axial scans per second. *Optics Express*. 2010;18(19):20029–20048. https://doi.org/10.1364/OE.18.020029.
3. Coscas G, Lupidi M, Coscas F, Chhablani J, Cagini C. Optical coherence tomography angiography in healthy subjects and diabetic patients. *Ophthalmologica*. 2018;239(2–3):61–73. https://doi.org/10.1159/000485323.
4. Lupidi M, Cerquaglia A, Chhablani J, et al. Optical coherence tomography angiography in age-related macular degeneration: the game changer. *European Journal of Ophthalmology*. 2018;28(4):349–357. https://doi.org/10.1177/1120672118766807.
5. Lupidi M, Schiavon S, Cerquaglia A, et al. Real-world outcomes of anti-VEGF therapy in treatment-naïve neovascular age-related macular degeneration diagnosed on OCT angiography: the reveal study. *Acta Ophthalmologica*. 2022;100(4):e936–e942. https://doi.org/10.1111/aos.15005.
6. Yannuzzi LA, Sorenson J, Spaide RF, Lipson B. Idiopathic polypoidal choroidal vasculopathy (IPCV). *Retina*. 1990;10(1):1–8. https://doi.org/10.1097/00006982-199001010-00001.
7. Kuehlewein L, Bansal M, Lenis TL, et al. Optical coherence tomography angiography of type 1 neovascularization in age-related macular degeneration. *American Journal of Ophthalmology*. 2015;160(4):739–748. https://doi.org/10.1016/j.ajo.2015.06.030.
8. El Ameen A, Cohen SY, Semoun O, et al. Type 2 neovascularization secondary to age-related macular degeneration imaged by optical coherence tomography angiography. *Retina*. 2015;35(11):2212–2218. https://doi.org/10.1097/IAE.0000000000000773.
9. Coscas G, Lupidi M, Coscas F, Français C, Cagini C, Souied EH. Optical coherence tomography angiography during follow-up: qualitative and quantitative analysis of mixed type I and II choroidal neovascularization after vascular endothelial growth factor trap therapy. *Ophthalmic Research*. 2015;54(2):57–63. https://doi.org/10.1159/000433547.
10. Mastropasqua L, Toto L, Borrelli E, Carpineto P, Di Antonio L, Mastropasqua R. Optical coherence tomography angiography assessment of vascular effects occurring after aflibercept intravitreal injections in treatment-naive patients with wet age-related macular degeneration. *Retina*. 2017;37(2):247–256. https://doi.org/10.1097/iae.0000000000001145.
11. Inoue M, Balaratnasingam C, Freund KB. Optical coherence tomography angiography of polypoidal choroidal vasculopathy and polypoidal choroidal neovascularization. *Retina*. 2015;35(11):2265–2274. https://doi.org/10.1097/IAE.0000000000000777.
12. Srour M, Querques G, Semoun O, et al. Optical coherence tomography angiography characteristics of polypoidal choroidal vasculopathy. *British Journal of Ophthalmology*. 2016;100(11):1489–1493. https://doi.org/10.1136/bjophthalmol-2015-307892.
13. Hwang TS, Gao SS, Liu L, et al. Automated quantification of capillary nonperfusion using optical coherence tomography angiography in diabetic retinopathy. *JAMA Ophthalmology*. 2016;134(4):367–373. https://doi.org/10.1001/jamaophthalmol.2015.5658.
14. Ishibazawa A, Nagaoka T, Yokota H, et al. Characteristics of retinal neovascularization in proliferative diabetic retinopathy imaged by optical coherence tomography angiography. *Investigative Ophthalmology and Visual Science*. 2016;57(14):6247–6255. https://doi.org/10.1167/iovs.16-20210.
15. Lupidi M, Gujar R, Cerquaglia A, et al. OCT-Angiography as a reliable prognostic tool in laser-treated proliferative diabetic retinopathy: the RENOCTA Study. *European Journal of Ophthalmology*. 2021;31(5):2511–2519. https://doi.org/10.1177/1120672120963451.
16. Foo VHX, Gupta P, Nguyen QD, et al. Decrease in Choroidal Vascularity Index of Haller's layer in diabetic eyes precedes retinopathy. *BMJ Open Diabetes Research & Care*. 2020;8(1). https://doi.org/10.1136/bmjdrc-2020-001295.
17. Agemy SA, Scripsema NK, Shah CM, et al. Retinal vascular perfusion density mapping using optical coherence tomography angiography in normals and diabetic retinopathy patients. *Retina*. 2015;35(11):2353–2363. https://doi.org/10.1097/IAE.0000000000000862.
18. Nesper PL, Roberts PK, Onishi AC, et al. Quantifying microvascular abnormalities with increasing severity of diabetic retinopathy using optical coherence tomography angiography. *Investigative Ophthalmology and Visual Science*. 2017;58(6):BIO307–BIO315. https://doi.org/10.1167/IOVS.17-21787.
19. Tsai G, Banaee T, Conti FF, Singh RP. Optical coherence tomography angiography in eyes with retinal vein occlusion. *Journal of Ophthalmic and Vision Research*. 2018;13(3):315–332. https://doi.org/10.4103/jovr.jovr_264_17.
20. Kashani AH, Lee SY, Moshfeghi A, Durbin MK, Puliafito CA. Optical coherence tomography angiography of retinal venous occlusion. *Retina*. 2015;35(11):2323–2331. https://doi.org/10.1097/IAE.0000000000000811.
21. Tokayer J, Jia Y, Dhalla AH, Huang D. Blood flow velocity quantification using splitspectrum amplitude-decorrelation angiography with optical coherence tomography. *Biomedical Optics Express*. 2013;4(10):1909–1924. https://doi.org/10.1364/BOE.4.001909.
22. Rispoli M, Savastano MC, Lumbroso B. Capillary network anomalies in branch retinal vein occlusion on optical coherence tomography angiography. *Retina*. 2015;35(11):2332–2338. https://doi.org/10.1097/IAE.0000000000000845.
23. Paques M, Tadayoni R, Sercombe R, et al. Structural and hemodynamic analysis of the mouse retinal microcirculation. *Investigative Ophthalmology and Visual Science*. 2003;44(11):4960–4967. https://doi.org/10.1167/iovs.02-0738.
24. Suzuki N, Hirano Y, Tomiyasu T, et al. Retinal hemodynamics seen on optical coherence tomography angiography before and after treatment of retinal vein occlusion. *Investigative Ophthalmology and Visual Science*. 2016;57(13):5681–5687. https://doi.org/10.1167/iovs-16-20648.
25. Couturier A, Mané V, Bonnin S, et al. Capillary plexus anomalies in diabetic retinopathy on optical coherence tomography angiography. *Retina*. 2015;35(11):2384–2391. https://doi.org/10.1097/iae.0000000000000859.
26. Sogawa K, Nagaoka T, Ishibazawa A, Takahashi A, Tani T, Yoshida A. En-face optical coherence tomography angiography of neovascularization elsewhere in hemicentral retinal vein occlusion. *International Medical Case Reports Journal*. 2015;8:263–266. https://doi.org/10.2147/IMCRJ.S93702.

27. Tang W, Liu W, Guo J, et al. Wide-field swept-source OCT angiography of the periarterial capillary-free zone before and after anti-VEGF therapy for branch retinal vein occlusion. *Eye and Vision*. 2022;9(1). https://doi.org/10.1186/s40662-022-00297-z.
28. Yang J, Zhang B, Wang E, Xia S, Chen Y. Ultra-wide field swept-source optical coherence tomography angiography in patients with diabetes without clinically detectable retinopathy. *BMC Ophthalmology*. 2021;21(1). https://doi.org/10.1186/s12886-021-01933-3.
29. Mendis KR, Balaratnasingam C, Yu P, et al. Correlation of histologic and clinical images to determine the diagnostic value of fluorescein angiography for studying retinal capillary detail. *Investigative Ophthalmology and Visual Science*. 2010;51(11):5864–5869. https://doi.org/10.1167/iovs.10-5333.
30. Yang S, Liu X, Li H, Xu J, Wang F. Optical coherence tomography angiography characteristics of acute retinal arterial occlusion. *BMC Ophthalmology*. 2019;19(1). https://doi.org/10.1186/s12886-019-1152-8.
31. Ruiz-Medrano J, Montero JA, Flores-Moreno I, Arias L, García-Layana A, Ruiz-Moreno JM. Myopic maculopathy: current status and proposal for a new classification and grading system (ATN). *Progress in Retinal and Eye Research*. 2019;69:80–115. https://doi.org/10.1016/j.preteyeres.2018.10.005.
32. Querques L, Giuffrè C, Corvi F, et al. Optical coherence tomography angiography of myopic choroidal neovascularisation. *British Journal of Ophthalmology*. 2017;101(5):609–615. https://doi.org/10.1136/bjophthalmol-2016-309162.
33. Bruyère E, Miere A, Cohen SY, et al. Neovascularization secondary to high myopia imaged by optical coherence tomography angiography. *Retina*. 2017;37(11):2095–2101. https://doi.org/10.1097/IAE.0000000000001456.
34. Spaide RF. Optical coherence tomography angiography signs of vascular abnormalization with antiangiogenic therapy for choroidal neovascularization. *American Journal of Ophthalmology*. 2015;160(1):6–16. https://doi.org/10.1016/j.ajo.2015.04.012.
35. Miyata M, Ooto S, Hata M, et al. Detection of myopic choroidal neovascularization using optical coherence tomography angiography. *American Journal of Ophthalmology*. 2016;165:108–114. https://doi.org/10.1016/j.ajo.2016.03.009.
36. Battista M, Sacconi R, Borrelli E, et al. Discerning between macular hemorrhages due to macular neovascularization or due to spontaneous Bruch's membrane rupture in high myopia: a comparative analysis between OCTA and fluorescein angiography. *Ophthalmology and Therapy*. 2022;11(2):821–831. https://doi.org/10.1007/s40123-022-00484-0.
37. Feucht N, Maier M, Lohmann CP, Reznicek L. OCT angiography findings in acute central serous chorioretinopathy. *Ophthalmic Surgery Lasers and Imaging Retina*. 2016;47(4):322–327. https://doi.org/10.3928/23258160-20160324-03.
38. Quaranta-El Maftouhi M, El Maftouhi A, Eandi CM. Chronic central serous chorioretinopathy imaged by optical coherence tomographic angiography. *American Journal of Ophthalmology*. 2015;160(3):581–587.e1. https://doi.org/10.1016/j.ajo.2015.06.016. Epub 2015 Jun 29. PMID: 26133250.
39. Nelis P, Clemens CR, Alten F, Eter N. OCT angiography reveals changes in foveal vessel architecture and foveal flow in central serous chorioretinopathy. *Acta Ophthalmologica*. 2017;95(8):e802–e803. https://doi.org/10.1111/aos.13299.
40. De Carlo TE, Rosenblatt A, Goldstein M, Baumal CR, Loewenstein A, Duker JS. Vascularization of irregular retinal pigment epithelial detachments in chronic central serous chorioretinopathy evaluated with OCT angiography. *Ophthalmic Surgery Lasers and Imaging Retina*. 2016;47(2):128–133. https://doi.org/10.3928/23258160-20160126-05.
41. Teussink MM, Breukink MB, Van Grinsven MJJP, et al. Oct angiography compared to fluorescein and indocyanine green angiography in chronic central serous chorioretinopathy. *Investigative Ophthalmology and Visual Science*. 2015;56(9):5229–5237. https://doi.org/10.1167/iovs.15-17140.
42. Bousquet E, Bonnin S, Mrejen S, Krivosic V, Tadayoni R, Gaudric A. Optical coherence tomography angiography of flat irregular pigment epithelium detachment in chronic central serous chorioretinopathy. *Retina*. 2018;38(3):629–638. https://doi.org/10.1097/iae.0000000000001580.
43. Lupidi M, Fruttini D, Eandi CM, et al. Chronic neovascular central serous chorioretinopathy: a stress/rest optical coherence tomography angiography study. *American Journal of Ophthalmology*. 2020;211:63–75. https://doi.org/10.1016/j.ajo.2019.10.033.
44. Moult E, Choi WJ, Waheed NK, et al. Ultrahigh-speed swept-source OCT angiography in exudative AMD. *Ophthalmic Surgery, Lasers and Imaging Retina*. 2014;45(6):496–505. https://doi.org/10.3928/23258160-20141118-03.
45. Cerquaglia A, Iaccheri B, Fiore T, et al. New insights on ocular sarcoidosis: an optical coherence tomography angiography study. *Ocular Immunology and Inflammation*. 2019;27(7):1057–1066. https://doi.org/10.1080/09273948.2018.1497665.
46. Gupta V, Gupta A, Rao NA. Intraocular tuberculosis-an update. *Survey of Ophthalmology*. 2007;52(6):561–587. https://doi.org/10.1016/j.survophthal.2007.08.015.
47. Agarwal A, Mahajan S, Khairallah M, Mahendradas P, Gupta A, Gupta V. Multimodal imaging in ocular tuberculosis. *Ocular Immunology and Inflammation*. 2017;25(1):134–145. https://doi.org/10.1080/09273948.2016.1231332.
48. Cerquaglia A, Lupidi M, Fiore T, Iaccheri B, Perri P, Cagini C. Deep inside multifocal choroiditis: an optical coherence tomography angiography approach. *International Ophthalmology*. 2017;37(4):1047–1051. https://doi.org/10.1007/s10792-016-0300-x.
49. Zhang L, Van Dijk EHC, Borrelli E, Fragiotta S, Breazzano MP. OCT and OCT angiography update: clinical application to age-related macular degeneration, central serous chorioretinopathy, macular telangiectasia, and diabetic retinopathy. *Diagnostics*. 2023;13(2). https://doi.org/10.3390/diagnostics13020232.
50. Chidambara L, Gadde SGK, Yadav NK, et al. Characteristics and quantification of vascular changes in macular telangiectasia type 2 on optical coherence tomography angiography. *British Journal of Ophthalmology*. 2016;100(11):1482–1488. https://doi.org/10.1136/bjophthalmol-2015-307941.
51. Pauleikhoff D, Gunnemann F, Book M, Rothaus K. Progression of vascular changes in macular telangiectasia type 2: comparison between SD-OCT and OCT angiography. *Graefe's Archive for Clinical and Experimental Ophthalmology*. 2019;257(7):1381–1392. https://doi.org/10.1007/s00417-019-04323-0.

52. Zeimer M, Gutfleisch M, Heimes B, Spital G, Lommatzsch A, Pauleikhoff D. Association between changes in macular vasculature in optical coherence tomography- and fluorescein- angiography and distribution of macular pigment in type 2 idiopathic macular telangiectasia. *Retina*. 2015;35(11):2307–2316. https://doi.org/10.1097/iae.0000000000000868.
53. Thorell MR, Zhang Q, Huang Y, et al. Swept-source OCT angiography of macular telangiectasia type 2. *Ophthalmic Surgery, Lasers and Imaging Retina*. 2014;45(5):369–380. https://doi.org/10.3928/23258160-20140909-06. PMID: 25291783.

Chapter 9

Oximetry

Jakob Grauslund[1,2,3,4], **Anna Stage Vergmann**[1,2] and **Søren Leer Blindbæk**[1,2]
[1]*Department of Ophthalmology, Odense University Hospital, Odense, Denmark;* [2]*Department of Clinical Research, University of Southern Denmark, Odense, Denmark;* [3]*Steno Diabetes Center Odense, Odense University Hospital, Odense, Denmark;* [4]*Department of Ophthalmology, Vestfold Hospital Trust, Tønsberg, Norway*

Introduction

The retina is the only part of the human vascular system available for direct, noninvasive imaging. As such, retinal inspection does not only allow for the detection of ocular disease but can often be used to detect systemic diseases such as diabetes, hypertension, and hematological disease by direct inspection of retinal lesions like microaneurysms, hemorrhages, and hard and soft exudates.[1–3] Additional information can often be obtained by semiautomated analysis of the retinal vascular tree, as it has consistently been demonstrated that the retinal vascular caliber is indicative and often predictive of ocular and systemic disease, like in diabetes,[4] hypertension,[5] and cerebrovascular disease.[6]

In alignment with these observations, a logical next step in the use of noninvasive retinal examination would be the evaluation of the retinal metabolism, which begins with the measure of the oxygen saturation of the retinal arterioles and venules. This would not only provide information on the concentration of oxygen entering and leaving the retina but would also give valuable information on the retinal oxygen consumption, as measured by the difference between the retinal arterial and venular oxygen saturation. In fact, these observations have been made for more than half a century, as pioneered by Hickam and coworkers.[7,8] It did, however, take some decades for the technology to evolve to a stage that would allow retinal oximetry to be used as a clinically applicable semiautomated method to evaluate retinal metabolism.[9] Within the last 15 years, retinal oximetry has expanded substantially from highly reproducibly[10] studies of normal physiology[11,12] to numerous studies of human disease.[9]

The aim of this chapter is to introduce the technology of retinal oximetry and to present this entity as a potential marker of ocular and systemic disease along with current barriers to clinical implementation.

Technical background

Retinal oximetry is a noninvasive method for determining the percentwise oxygen saturation in retinal vessels by spectrophotometry. The method depends on quantifying differences between oxyhemoglobin and deoxyhemoglobin light absorption spectra at different wavelengths.[11]

A retinal oximeter consists of a fundus camera, a beam splitter, and a digital camera. A monochromatic image is captured and separated by the beam splitter into two (or more) optical channels. A different narrow bandwidth filter is situated in each channel, through which only light of a specific wavelength can pass. One is chosen at an isosbestic point of the hemoglobin, whereas the other is chosen such that a large difference exists between the absorption coefficients of oxygenated and deoxygenated hemoglobin.[11,13,14] The optical density is a measure of light absorbance of blood and is calculated as the logarithmic contrast of a vessel versus the surrounding retinal tissue. It has been demonstrated that the ratio of optical densities at specific wavelengths has an inverse and almost linear relationship to the oxygen saturation of hemoglobin.[11,15] In this way, the oxygenation of hemoglobin can be estimated by analyzing the contrast inside and just outside a retinal vessel at two wavelengths of light.

Currently, there are two commercially available devices: the Oxymap retinal oximeter (Oxymap, Reykjavik, Iceland)[16] and the VesselMap system (iMEDOS Health GmbH, Jena, Germany).[11] Both systems include semiautomated software for

Retinal and Choroidal Vascular Diseases of the Eye. https://doi.org/10.1016/B978-0-443-15583-3.00009-3

image analysis. While the specific wavelengths applied by the two systems differ slightly, the general methods for image capture and image analysis described above adhere to both systems.

Retinal oximetry has been demonstrated as a reliable tool for estimating oxygen saturations in retinal arterioles and venules. Thus, the standard deviation of repeated measures using the same instrument on the same subject is approximately 3% and 5% oxygen saturation for retinal arterioles and venules, respectively. Furthermore, the method has been demonstrated to be sensitive to changes in oxygen saturation by respiration of 100% oxygen in comparison to normoxic conditions.[13,16] However, comparison of measurement across instruments is difficult as oxygen saturation readings are subjected to different and somewhat arbitrary calibrations of the oximeters.

Retinal oximetry in diabetes

For a number of reasons, most studies in retinal oximetry so far have been performed in patients with diabetic retinopathy (DR).[9,17] First, DR is an ischemic retinal disease, and metabolic changes would be expected. Second, DR is a progressing disease with well-defined stages, which would allow for retinal oximetry measurements within the entire spectrum of the disease. Third, DR has two end-stage complications (diabetic macular edema [DME] and proliferative DR [PDR]), and measurements of retinal oximetry might differ between those. Fourth, retinal oximetry may serve as a marker of treatment response for DME and PDR. Fifth, and perhaps most importantly, screening is indicated for DR,[18] and retinal oximetry might serve as a noninvasive method, which could potentially be included in a screening program.

In 2017, the literature on retinal oximetry was evaluated in a systematic review by Rilvén et al., who identified six studies with retinal oximetry measurements at various stages of DR.[17] It was universally found that patients with PDR had a higher retinal venular oxygen saturation than healthy controls or diabetes patients without DR. Likewise, increased retinal arteriolar oxygen saturation was reported in some, but not all, studies. As speculated by Hardarson and Stefansson,[19] a higher retinal venular oxygen saturation in advanced DR might be explained by a number of reasons, including (1) poor oxygen distribution to the retinal tissue, (2) increased oxygen supply in diabetes, and (3) a lower oxygen consumption within the ischemic retinal tissue.

A few recent studies have examined retinal oximetry along the entire spectrum of DR. In Denmark, Weisner et al. reported gradual increments of retinal venular oxygen saturation in patients with increasing levels of nonproliferative DR (51.3% vs. 53.3% vs. 54.2% vs. 59.8% for DR-levels 0−3, $P = .02$).[20] In contrast, the retinal arteriolar oxygen saturation remained unchanged across the entire spectrum. While the Danish study included a mixture of patients with type 1 and type 2 diabetes, Veiby et al. exclusively examined patients with type 1 diabetes ($n = 185$) and nondiabetes controls ($n = 94$) in a cross-sectional study from Norway.[21] The authors reported that the retinal arteriolar saturation, as well as venular oxygen saturation, was higher in patients with moderate or severe DR than in healthy controls as well as in patients without DR. While the majority of retinal oximetry studies in diabetes have been cross-sectional, a number of longitudinal studies have also provided information on the dynamic changes in retinal oximetry. In a 3-year prospective study of 214 patients with diabetes, Hardarson et al. presented an annual 1.74%-point increment and a 0.99%-point decrement in the retinal venular oxygen saturation and the arteriovenous oxygen saturation difference, respectively.[14]

In one of the first studies to establish retinal oximetry as a potential decision-supporting tool in clinical medicine, Blindbæk et al. measured the retinal metabolism before and after intravitreal and focal/grid laser treatment in a treatment-naïve cohort of patients with macular-involving DME. While the retinal oximetry did not change after initial loading with intravitreal aflibercept,[22] the retinal arteriolar oxygen saturation was independently able to predict the need for postloading treatment.[23] For patients with the highest baseline values of retinal arteriolar oxygen saturation, a higher number of intravitreal treatments were needed within the first year of treatment (5.0 vs. 3.6 injections, $P = .002$).

In treatment-naïve patients with PDR, Torp et al. evaluated retinal oximetry before and 3 months after panretinal laser treatment.[24] When comparing the retinal metabolism according to disease activity after treatment, patients with progression of PDR despite treatment developed a statistically significantly higher retinal venular oxygen saturation between baseline and follow-up (+4.1% vs. −1.8% for PDR-progression and nonprogression, respectively, $P = .02$). In fact, the increase in retinal venular oxygen saturation is independently associated with a 30% higher risk of PDR-progression (OR 1.30 per 1%-point increment in retinal venular oxygen saturation, 95% CI 1.05−1.62, $P = .02$). Combined with the results from the observational studies, this might indicate a progressing increment in the retinal venular oxygen saturation in reflection to the increasing severity of DR and that this trend is only broken when PDR is successfully treated. In contrast, for progressing PDR despite treatment, the retinal venular oxygen saturation would continue to rise.

Finally, Vergmann et al. evaluated potential differences in retinal oximetry in a 6-month, randomized trial with full panretinal laser compared to individualized treatment with the treatment load individualized according to disease actively (i.e., less laser in patients with new vessels present in fewer retinal quadrants).[25] While the retinal arteriolar oxygen saturation did not differ between groups, eyes with less extensive panretinal laser treatment developed a lower retinal venular oxygen saturation from baseline to follow-up. As speculated by the authors, this could indicate that less extensive laser may sufficiently improve the retinal metabolic environment in parallel with PDR regression.

In conclusion, there has been substantial evidence that in particular, the retinal venular oxygen saturation increases in parallel to increasing severity of DR and that it may also serve as a marker of treatment response. Thereby, it may act as a decision-supporting imaging entity. However, validation studies are still needed, and it has still not been elucidated, if retinal oximetry would provide sufficient value to be included in DR screening.

Retinal vascular occlusion

Intuitively, severe disturbances in retinal vascular oxygen saturation would be expected in patients with retinal artery (RAO) and vein occlusion (RVO), given the metabolic disturbances caused by these diseases.

RAO is an acute occlusive disease causing severe and irreversible loss of vision and visual field. Retinal oximetry has only been studied sparely in RAO. In the acute phase of a patient with central RAO, Hardarson et al. reported substantial lowering of the retinal arterial oxygen saturation in the affected eye (71% vs. 95% in the unaffected fellow eye) of a 69-year-old woman with a history of polymyalgia rheumatica.[26] Even though the eye was still blind 1 month later, the retinal arteriolar oxygen saturation had returned to 100%. The retinal venular oxygen saturation was neither affected at the onset nor follow-up. This indicates severe hypoxia at the onset of central RAO followed by adaption of the retinal metabolism despite irreversible loss of visual function.

More data have been presented on RVO, which is a thrombotic retinal disease affecting the central or branch retinal vein, thereby causing loss of vision or visual field due to retinal ischemia or macular edema. In a systematic review, Rilvén et al.[17] identified four studies measuring retinal oximetry in patients with central RVO.[27–29] In central RVO, it was a universal finding among studies that the retinal venular oxygen saturation was decreased in affected eyes, as compared to the healthy fellow eyes. Likewise, the retinal arteriolar oxygen saturation was unaffected, which led to an increased retinal arteriovenous oxygen saturation difference. As the retinal venular oxygen saturation is a measurement of remaining oxygen after passage through the retinal microvasculature, a lower retinal venular oxygen saturation might reflect a higher retinal oxygen extraction caused by a slower retinal blood flow and a correspondingly higher microvascular oxygen passage.[27] Traustason et al. demonstrated in 17 eyes with central RVO that 6 months of intravitreal ranibizumab treatment did indeed increase the retinal venular oxygen saturation (from 35.5% at baseline to 43.5% at follow-up, $P = .012$), but that this still remained lower than in the healthy control eyes (59%).[29]

Studies of retinal oximetry have also been performed in branch RVO using different study designs and outcomes. Osaka et al. reported a case series of 50 eyes with resolved branch RVO, categorized into nonischemic ($n = 32$) and ischemic ($n = 18$) branch RVO.[30] In the affected retinal hemisphere, a higher retinal arteriolar oxygen saturation was found in ischemic versus nonischemic branch RVO (106.9% vs. 101.3%, $P = .044$). As the retinal venular oxygen saturation did not differ between groups, this also corresponded to a higher retinal arteriovenous oxygen saturation difference in ischemic branch RVO (51.9% vs. 43.4%, $P = .028$). Both of these findings were also confirmed in multivariable linear regression models and were supported by Lin et al., who included 26 nonischemic and 24 ischemic eyes with branch RVO as well as fellow nonaffected eyes and those of 100 healthy controls.[31] In contrast to eyes with nonischemic branch RVO, those with ischemic branch RVO had a higher retinal arteriolar oxygen saturation in the occluded vessels (106.0% vs. 94.0% for unaffected vessels vs. 86.1% for contralateral eye vessels, $P < .001$), the central retinal artery (100.7% vs. 96.0% for the contralateral eye, $P = .011$), and the entire retinal arterial oxygen saturation (96.6% vs. 92.5% for the contralateral eye, $P = .021$). As compared to healthy controls, a higher central retinal artery oxygen saturation was found in eyes with ischemic (101.3% vs. 94.7%, $P < .001$) and nonischemic branch RVO (97.4% vs. 94.4%, $P = .035$).

In contrast, in a smaller study of 22 eyes with branch RVO and fellow unaffected eyes, Yang et al. also demonstrated changes in the retinal venular oxygen saturation.[32] In ischemic branch RVO, the authors reported a higher retinal arteriolar oxygen saturation (107.5% vs. 93.1%, $P = .005$) and a lower retinal venular oxygen saturation (46.4% vs. 55.7%, 0.02). In eyes with nonischemic branch RVO, only differences in the retinal venular oxygen saturation were statistically significant (39.8% vs. 58.2%, $P = .03$). In a 12-month prospective case series of 73 eyes with branch RVO, Jeppesen and Bek reported a higher retinal arteriolar (100.9% vs. 96.5%, $P = .01$) and a lower retinal venular oxygen saturation (46.5% vs. 57.3%, $P = .02$) at baseline in patients with hemiretinal vein occlusion as compared to macular RVO, but the retinal vascular oxygen saturation was not able to predict the clinical outcome after intravitreal angiostatic treatment.[33]

In conclusion, studies of branch RVO seem to indicate that the retinal venular oxygen saturation is not as affected as in central RVO. This might reflect less affection of retinal tissue and a lower ischemic drive in branch RVO. Interestingly, ischemic branch RVO associates with a higher retinal arteriolar oxygen saturation than in central RVO or nonischemic branch RVO. The reasons for this are not evident and may to some extent reflect differences in study populations, duration of disease, and underlying cardiovascular risk factors.

Ocular degenerative disease

Age-related macular degeneration

Retinal oximetry has only been examined in a few studies of age-related macular degeneration (AMD). In an Icelandic case-control study, Geirsdottir et al. compared retinal oxygen saturation between 46 eyes with treatment-naïve neovascular AMD and a younger group of 120 health-control eyes.[34] In cases, the retinal venular oxygen correlated with age with a 0.45% increment for each year of age, compared to a 0.13% decrease per year in controls. At age 80, this added up to a 2.7%-point higher retinal venular oxygen saturation in cases and, correspondingly, a 4.2% lower arteriovenous oxygen saturation difference, indicating a lower retinal oxygen extraction in eyes with untreated neovascular AMD.

In a twofold study from Denmark, Jakobsen et al. compared retinal oximetry between health controls ($n = 43$) and eyes with dry ($n = 30$) or neovascular AMD ($n = 76$).[35] In both groups of AMD, the retinal arteriolar oxygen saturation was higher than among controls (dry AMD: 95.2%, neovascular AMD: 94.3%, controls: 92.6%, $P = .04$), but in contrast to the study from Iceland,[34] the retinal venular oxygen saturation did not differ between groups. Jakobsen et al. then followed 57 of the treatment-naïve eyes with neovascular AMD prospectively during a clinical loading phase with three monthly injections of aflibercept. After loading, patients were stratified according to their structural and functional response, and development in retinal oximetry was measured. In a multiple logistic regression model, each 1%-point increase in retinal venular oxygen saturation after loading is independently associated with a 17% higher risk of negative structural treatment response. Most recently, Donicova et al. reported a reduced flicker response of venous oxygen saturation in eyes with early and late AMD ($n = 60$) as compared to controls ($n = 23$).[36] As flicker stimulation would normally lead to higher retinal oxygen consumption, these results may indicate that the regulation of oxygen supply and consumption is impaired in AMD.

In conclusion, there is some evidence that a higher retinal venular oxygen saturation is associated with neovascular AMD and may also be negatively correlated to the short-term clinical structural outcome. At present, it is not known, whether this is a consequence of a lower retinal oxygen demand or an impaired ability of oxygen extraction from the vascular system to the retinal tissue. Long-term studies are clearly indicated to address this.

Glaucoma

Glaucoma is an optic neuropathy with progressive loss of retinal ganglion cells and corresponding visual field defects. As this may lead to a decreased retinal metabolic demand, it has been speculated that glaucoma would lead to a higher retinal venular oxygen saturation and a correspondingly decreasing retinal arteriovenous oxygen saturation difference. This has in fact been demonstrated in a number of studies[37–42], but not in others.[43,44] Matters are, in fact, complicated, as the retinal arteriovenous oxygen saturation difference is not only proportional to the metabolic demand of the retinal tissue but also inversely correlated with the retinal blood flow.[45] As it is well known that the retinal blood flow is decreased in glaucoma,[46] a higher retinal arteriovenous oxygen saturation difference would physiologically be expected. When this difference was in fact decreased in most studies, this would indicate that a reduced retinal metabolic demand (due to retinal tissue atrophy) was the underlying cause of the observations, as speculated by Shughoury et al.[47]

Retinal oximetry has also been measured secondary to glaucoma treatment. As compared to healthy controls, 12 patients with acute primary angle closure developed a lower retinal venular oxygen saturation (50.0% vs. 59.5%, $P = .004$) and a higher retinal arteriovenous oxygen saturation difference (47.6% vs. 33.5%, $P = .04$) 1 month after trabeculectomy.[48] This was in consistence with a case series of 30 patients with glaucoma in whom the retinal venular oxygen saturation increased from 54.9% to 57.4% ($P = .01$) 10 days after glaucoma surgery.[49] Results from studies of topical treatment have been less consistent, and most studies have been in animals or cross-over human studies testing the retinal oxygen saturation in various topical treatment regimens.[47]

Retinal oximetry as a marker of systemic disease

In recent years, there has been growing attention to retinal oximetry in systemic disease. This is based on the fundamental observation that the retinal vasculature is the only part of human circulation, which is directly available for noninvasive observation. Initially, this led to a number of studies with semiautomated, structural measurements of the retinal vascular tree in various diseases like diabetes, hypertension, and cardiovascular disease,[5,6,50,51] but in recent years studies on retinal oximetry in systemic disease have also emerged.

Neurodegenerative diseases

As the retina is embryologically derived from the brain, a particular focus has been given to retinal oximetry in neurodegenerative disease. Most notably, this includes Alzheimer's disease, which is an irreversible neurodegenerative disease and a leading cause of dementia. The diagnosis of Alzheimer's disease can be challenging, is often clinically delayed, and, hence, would be important with early noninvasive identification of patients at risk or at the early stages of the disease.

In 18 patients with Alzheimer's disease and 18 age- and gender-matched healthy controls, Einarsdottir et al. found a higher retinal arteriolar (93.7% vs. 90.5%, $P = .019$) but not venular (51.3% vs. 49.7%, $P = .46$) oxygen saturation in patients than in controls.[52] These findings were expanded in a study by Olafsdottir et al.[53] In an age-matched case-control study of 42 patients with mild cognitive impairment and 42 healthy individuals, the authors reported a higher retinal arteriolar (93.1% vs. 91.1%, $P = .01$) and venular (59.6% vs. 54.9%, $P = .001$) oxygen saturation in cases, indicating an affected retinal oxygen metabolism even in the early stages of cerebral neurodegeneration. This may reflect that systemic changes in cerebral oxygen metabolism can be measured noninvasively in the eye, even though we are not aware of any studies, which has yet compared systemic and retinal oxygen extraction to confirm this hypothesis.

Multiple sclerosis is another degenerative cerebral disease, which often affects young adults. It is a demyelinating disease causing axonal degeneration,[54] and optic neuritis is often an initial finding.[55] Einarsdottir et al. measured retinal oximetry in a case-control study of eight patients with multiple sclerosis and a history of optic neuritis and 22 healthy age- and gender-matched unaffected controls.[56] Although the retinal arterial oxygen saturation did not differ between groups, a higher retinal venular oxygen saturation (70.7% vs. 66.2%, $P = .021$) and a correspondingly lower retinal arteriovenous oxygen saturation differences (26.6% vs. 30.5%) were found in cases, reflecting a lower retinal oxygen consumption in affected eyes. In patients, the retinal vascular oxygen saturation did not differ between the affected and the unaffected fellow eyes.

Results have also been reported in the acute phase in a case series of 32 patients with optic neuritis due to multiple sclerosis by Svrcinová et al.[57] In that study, retinal oximetry was performed within 3 months of the onset of symptoms, a higher retinal arteriovenous oxygen saturation difference (34.2% vs. 31.3%, $P = .044$) was found in the affected eye, which may reflect metabolic and inflammatory changes in the acute phase of optic neuritis. In a longitudinal study from the same group, it was confirmed that in 6 months, this was followed by a lower retinal arteriovenous oxygen saturation difference, which may reflect a lower retinal oxygen consumption in the chronic phase, while this was only partly reflected in retinal nerve fiber layer thickness reduction.[58]

Rheumatic diseases

Sarcoidosis is another systemic inflammatory disease, with potential affection of the central nervous system and the eye. Kindt et al. measured retinal oximetry in 201 eyes of 103 biopsy-verified patients with sarcoidosis categorized according to organ affection.[59] While the retinal arteriolar and venular oxygen saturation did not differ between groups, differences were found in the retinal arteriovenous oxygen saturation between patients with ($n = 29$ eyes) and without ($n = 109$ eyes) ocular and central nervous system affection (42.1% vs. 37.7%, $P = .003$). This was in alignment with the observations in multiple sclerosis-induced optic neuritis[57] and might reflect that ocular inflammation is associated with higher retinal oxygen consumption.

Lowater et al. measured retinal oximetry in patients with giant cell arteritis ($n = 20$) and polymyalgia rheumatica ($n = 20$).[60] Although this has traditionally been considered two separate disease entities, patients in both groups had a higher retinal arterial caliber, as compared to a group of 12 healthy controls (124.4 vs. 124.8 vs. 115.3 μm, $P = .023$ and $P = .049$). This did, however, not reflect any differences in retinal vascular oxygen saturation between the three groups.

Cardiovascular disease

Retinal oximetry has also been evaluated in systemic hypoxic and cardiovascular diseases. In a case-control study of 11 patients with severe chronic obstructive pulmonary disease and 11 age- and gender-matched healthy controls, Eliasdottir et al. reported lower retinal arterial (87.2% vs. 93.4%, $P = .02$) and venular (45.0% vs. 55.2%, $P = .01$) oxygen saturation in patients, of which only the retinal arteriolar oxygen saturation was reversed (87.2%−89.5%, $P = .02$) after administration of prescribed supplementary oxygen.[61] These findings were confirmed in a study of 16 patients with a cyanotic cardiac defect (Eisenmenger syndrome), who had a statistically significantly lower retinal arterial oxygen saturation than 17 health controls (81% vs. 93%, $P < .001$).[62]

Dinesen et al. evaluated the retinal oxygen saturation in patients undergoing coronary artery bypass surgery for cardiovascular disease. As compared to 39 healthy controls, 38 recently coronary artery bypass grafting patients had higher

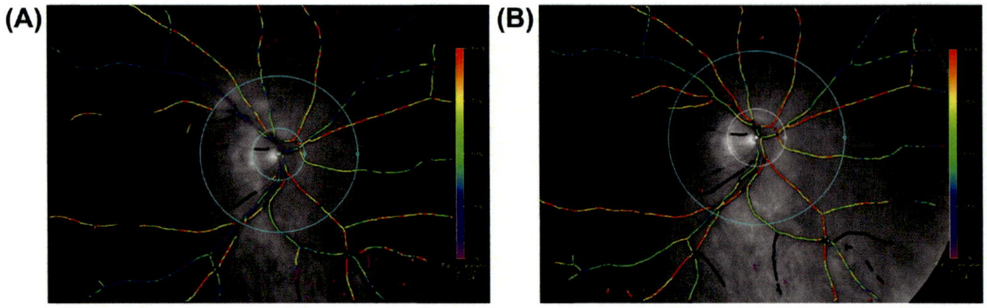

FIGURE 9.1 Retinal oximetry before and after coronary artery bypass grafting. Retinal oximetry in a patient with coronary heart disease before (A) and after (B) coronary artery bypass grafting. In the right part of both pictures, an indicator of the retinal vascular oxygen saturation is shown with dark red indicating high saturation and purple indicating a low saturation. The retinal arteriolar and venular oxygen saturation increased from 89.0% to 94.5% and 46.0% to 70.0%, respectively, which emphasizes the ability of retinal oximetry to detect changes in the systemic circulation.

retinal arterial (93.1% vs. 90.5%, $P = .001$) and venular (57.4% vs. 53.5%, $P = .048$) oxygen saturations, which was also statistically significant in multivariable linear regression models.[63] Although preoperative values were not known for the entire group of patients, a substudy of nine patients demonstrated increments in retinal arteriolar (90.8%−94.9%, $P = .04$) as well as venular (50.8%−69.7%, $P = .01$) oxygen saturation, as measured a few days before and after surgery (Fig. 9.1).[64] This might potentially be explained by a higher cardiac output after coronary revascularization, which may lead to optimized retinal circulation.[65] This would lead to an increased retinal blood flow, which could decrease the oxygen extraction as a result of the increased linear blood flow velocity.[66] A similar mechanism might be present after cytoreductive treatment in blood dyscrasia caused by myeloproliferative neoplasms. In a case series of five patients, Willerslev et al. reported a 8.1% increased retinal blood flow velocity as well as a 7.2% higher retinal arteriolar oxygen saturation 12 weeks after cytoreductive treatment.[67]

Conclusions

Retinal oximetry enables noninvasive measurement of the retinal oxygen consumption including evaluation of the retinal arteriolar and venular oxygen saturation as well as the arteriovenous oxygen saturation difference, which reflects oxygen consumption of the retinal tissue.

In this chapter, we have summarized the major findings of clinical retinal oximetry studies in ocular and systemic disease. Most studies have been performed on retinal vascular diseases. In particular, DR has been a major field of clinical studies of retinal oximetry, and a decrement in the retinal venular oxygen saturation has consistently been demonstrated in these patients. Likewise, the retinal arteriolar oxygen saturation has been altered in some, but not all, studies. As expected, findings in RAO indicate severe retinal hypoxia at the onset of the disease. While the retinal metabolism is restored at a later stage, this is unlikely to be followed by clinical improvement. In RVO, different patterns can be observed between studies of central and branch RVO. For the former, a decreased retinal venular oxygen saturation could reflect a higher retinal oxygen extraction from the retinal vascular system, which could potentially be explained by a slower retinal blood flow. In branch RVO, the retinal venular oxygen saturation is not affected to the same degree as in central RVO. However, higher retinal arteriolar oxygen saturations and arteriovenous oxygen saturation differences have been demonstrated in ischemic than in nonischemic branch RVO, which may reflect underlying differences in the retinal metabolism between these entities.

In degenerative retinal diseases like AMD and glaucoma, studies have indicated higher retinal venular oxygen saturations, which could reflect a lower retinal oxygen consumption. Likewise, prospective studies have demonstrated that retinal oximetry may relate to prospective clinical outcomes, though more studies are needed to confirm this. In recent years, retinal oximetry has also been tested as a potential marker of systemic disease. In neurodegenerative disease, higher retinal arteriolar oxygen saturations may reflect early changes in the retinal metabolism of cerebral disease, and if these can also be observed at the preclinical stage, retinal oximetry parameters may act as an early warning of upcoming systemic disease, which could allow for timely treatment before irreversible loss of cognitive function. In studies of inflammatory systemic disease, a higher retinal oxygen consumption has been measured in the acute phases of optic neuritis and ocular sarcoidosis, followed by a lower retinal oxygen consumption in the inactive phases of the diseases. Finally, studies of systemic hypoxia and cardiovascular disease have demonstrated a close relation between systemic and ocular hypoxia, and subsequent improvement in retinal metabolism was found after coronary revascularization.

In conclusion, retinal oximetry does not only reflect ocular morbidity but can also be closely related to systemic disease. However, a number of factors need to be addressed before clinical implementation. First, most studies have been cross-sectional and should be confirmed by longitudinal studies to address cause and effect. Second, interpretations across different studies have often been complicated due to differences in age, gender, disease definition, and heterogeneous study populations. To exemplify, a number of excellent studies have been performed in Iceland, but these studies represent a relatively small population, which may have different genetic traits than a number of other populations. Third, there are no universally accepted thresholds for retinal oximetry measurements. To act as a reliable diagnostic tool, it would be important to define cutoff values to distinguish healthy and diseased eyes. To validate such testing, it would be important to know the sensitivity, specificity, and area under the curve in comparison to gold-standard testing.

We would encourage upcoming studies in retinal oximetry to validate this as a valid and reliable endpoint to detect and prognosticate ocular and systemic disease. Based on the clinical findings so far, solid proof-of-concept has been demonstrated that retinal oximetry may be a valuable tool to detect upcoming disease, diagnose present conditions, and predict upcoming clinical outcomes.

Abbreviations

AMD age-related macular degeneration
DME diabetic macular edema
DR diabetic retinopathy
PDR proliferative diabetic retinopathy
RAO retinal artery occlusion
RVO retinal vein occlusion

References

1. van den Born B-J, Hulsman CAA, Hoekstra JBL, Schlingemann RO, van Montfrans GA. Value of routine funduscopy in patients with hypertension: systematic review. *BMJ*. 2005;331(7508).
2. Delestre F, Blanche P, Bouayed E, et al. Ophthalmic involvement of chronic lymphocytic leukemia: a systematic review of 123 cases. *Survey of Ophthalmology*. 2021;66(1):124–131. https://doi.org/10.1016/j.survophthal.2020.05.001.
3. Yau JWY, Rogers SL, Kawasaki R, et al. Global prevalence and major risk factors of diabetic retinopathy. *Diabetes Care*. 2012;35(3):556–564. https://doi.org/10.2337/dc11-1909.
4. Broe R, Rasmussen ML, Frydkjaer-Olsen U, et al. Retinal vessel calibers predict long-term microvascular complications in type 1 diabetes: the Danish cohort of pediatric diabetes 1987 (DCPD1987). *Diabetes*. 2014;63(11):3906–3914. https://doi.org/10.2337/db14-0227.
5. Ding J, Wai KL, McGeechan K, et al. Review: retinal vascular caliber and the development of hypertension: a meta-analysis of individual participant data. *Journal of Hypertension*. 2014;32(2):207–215. https://doi.org/10.1097/HJH.0b013e32836586f4.
6. Rim TH, Teo AWJ, Yang HHS, Cheung CY, Wong TY. Retinal vascular signs and cerebrovascular diseases. *Journal of Neuro-Ophthalmology*. 2020;40(1):44–59. https://doi.org/10.1097/WNO.0000000000000888.
7. Hickam JB, Ross JC. An improved method for photographic measurement of retinal venous blood oxygen. *Technical Documentary Report. SAM-TDR. USAF School of Aerospace Medicine*. 1962;62.
8. Hickam JB, Frayser R, Ross JC. A study of retinal venous blood oxygen saturation in human subjects by photographic means. *Circulation*. 1963;27:375–385. https://doi.org/10.1161/01.CIR.27.3.375.
9. Stefánsson E, Olafsdottir OB, Eliasdottir TS, et al. Retinal oximetry: metabolic imaging for diseases of the retina and brain. *Progress in Retinal and Eye Research*. 2019;70:1–22. https://doi.org/10.1016/j.preteyeres.2019.04.001.
10. Palsson O, Geirsdottir A, Hardarson SH, Olafsdottir OB, Kristjansdottir JV, Stefánsson E. Retinal oximetry images must be standardized: a methodological analysis. *Investigative Ophthalmology and Visual Science*. 2012;53(4):1729–1733. https://doi.org/10.1167/iovs.11-8621.
11. Beach JM, Schwenzer KJ, Srinivas S, Kim D, Tiedeman JS. Oximetry of retinal vessels by dual-wavelength imaging: calibration and influence of pigmentation. *Journal of Applied Physiology*. 1999;86(2):748–758. https://doi.org/10.1152/jappl.1999.86.2.748.
12. Geirsdottir A, Palsson O, Hardarson SH, Olafsdottir OB, Kristjansdottir JV, Stefánsson E. Retinal vessel oxygen saturation in healthy individuals. *Investigative Ophthalmology and Visual Science*. 2012;53(9):5433–5442. https://doi.org/10.1167/iovs.12-9912.
13. Hammer M, Vilser W, Riemer T, Schweitzer D. Retinal vessel oximetry-calibration, compensation for vessel diameter and fundus pigmentation, and reproducibility. *Journal of Biomedical Optics*. 2008;13(5):054015. https://doi.org/10.1117/1.2976032.
14. Hardarson SH, Stefansson E, Bek T. Retinal oxygen saturation changes progressively over time in diabetic retinopathy. *PLoS One*. 2021;16(5). https://doi.org/10.1371/journal.pone.0251607.
15. Harris A, Dinn RB, Kagemann L, Rechtman E. A review of methods for human retinal oximetry. *Ophthalmic Surgery Lasers and Imaging*. 2003;34(2):152–164. https://doi.org/10.3928/1542-8877-20030301-16.
16. Hardarson SH, Harris A, Karlsson RA, et al. Automatic retinal oximetry. *Investigative Ophthalmology and Visual Science*. 2006;47(11):5011–5016. https://doi.org/10.1167/iovs.06-0039.
17. Rilvén S, Torp TL, Grauslund J. Retinal oximetry in patients with ischaemic retinal diseases. *Acta Ophthalmologica*. 2017;95(2):119–127. https://doi.org/10.1111/aos.13229.

18. Stefánsson E, Bek T, Porta M, Larsen N, Kristinsson JK, Agardh E. Screening and prevention of diabetic blindness. *Acta Ophthalmologica Scandinavica*. 2000;78(4):374–385. https://doi.org/10.1034/j.1600-0420.2000.078004374.x.
19. Hardarson SH, Stefánsson E. Retinal oxygen saturation is altered in diabetic retinopathy. *British Journal of Ophthalmology*. 2012;96(4):560–563. https://doi.org/10.1136/bjophthalmol-2011-300640.
20. Weisner G, Blindbæk SL, Tang FY, et al. Non-invasive structural and metabolic retinal markers of disease activity in non-proliferative diabetic retinopathy. *Acta Ophthalmologica*. 2021;99(7):790–796. https://doi.org/10.1111/aos.14761.
21. Veiby NCBB, Simeunovic A, Heier M, et al. Venular oxygen saturation is increased in young patients with type 1 diabetes and mild nonproliferative diabetic retinopathy. *Acta Ophthalmologica*. 2020;98(8):800–807. https://doi.org/10.1111/aos.14462.
22. Blindbæk SL, Peto T, Grauslund J. Correlation between diabetic retinopathy severity and oxygen metabolism in patients with diabetic macular edema during treatment with intravitreal aflibercept. *Ophthalmic Research*. 2020;63(2):106–113. https://doi.org/10.1159/000503930.
23. Blindbæk SL, Peto T, Grauslund J. Retinal arteriolar oxygen saturation predicts the need for intravitreal aflibercept in patients with diabetic macular oedema. *BMJ Open Ophthalmology*. 2020;5(1):e000382. https://doi.org/10.1136/bmjophth-2019-000382.
24. Torp TL, Kawasaki R, Wong TY, Peto T, Grauslund J. Changes in retinal venular oxygen saturation predict activity of proliferative diabetic retinopathy 3 months after panretinal photocoagulation. *British Journal of Ophthalmology*. 2018;102(3):383–387. https://doi.org/10.1136/bjophthalmol-2017-310576.
25. Vergmann AS, Torp TL, Kawasaki R, et al. Retinal vascular oxygen saturation in response to a less extensive laser treatment in proliferative diabetic retinopathy. *Acta Ophthalmologica*. 2021;99(7):783–789. https://doi.org/10.1111/aos.14727.
26. Hardarson SH, Elfarsson A, Agnarsson BA, Stefánsson E. Retinal oximetry in central retinal artery occlusion. *Acta Ophthalmologica*. 2013;91(2):189–190. https://doi.org/10.1111/j.1755-3768.2012.02393.x.
27. Eliasdottir TS, Bragason D, Hardarson SH, Kristjansdottir G, Stefánsson E. Venous oxygen saturation is reduced and variable in central retinal vein occlusion. *Graefe's Archive for Clinical and Experimental Ophthalmology*. 2015;253(10):1653–1661. https://doi.org/10.1007/s00417-014-2849-2.
28. Hardarson SH, Stefansson E. Oxygen saturation in central retinal vein occlusion. *American Journal of Ophthalmology*. 2010;150(6):871–875. https://doi.org/10.1016/j.ajo.2010.06.020.
29. Traustason S, La Cour M, Larsen M. Retinal vascular oximetry during ranibizumab treatment of central retinal vein occlusion. *British Journal of Ophthalmology*. 2014;98(9):1208–1211. https://doi.org/10.1136/bjophthalmol-2013-304580.
30. Osaka R, Nakano Y, Takasago Y, et al. Retinal oximetry in branch retinal vein occlusion. *Acta Ophthalmologica*. 2019;97(6):e896–e901. https://doi.org/10.1111/aos.14070.
31. Lin LL, Dong YM, Zong Y, et al. Study of retinal vessel oxygen saturation in ischemic and non-ischemic branch retinal vein occlusion. *International Journal of Ophthalmology*. 2016;9(1):99–107. https://doi.org/10.18240/ijo.2016.01.17.
32. Yang JY, You B, Wang Q, Chan SY, Jonas JB, Wei WB. Retinal vessel oxygen saturation in healthy subjects and early branch retinal vein occlusion. *International Journal of Ophthalmology*. 2017;10(2):267–270. https://doi.org/10.18240/ijo.2017.02.14.
33. Jeppesen SK, Bek T. Lack of predictive value of retinal oxygen saturation for visual outcome after angiostatic treatment of branch retinal vein occlusion. *Acta Ophthalmologica*. 2022;100(3):e737–e742. https://doi.org/10.1111/aos.14988.
34. Geirsdottir A, Hardarson SH, Olafsdottir OB, Stefánsson E. Retinal oxygen metabolism in exudative age-related macular degeneration. *Acta Ophthalmologica*. 2014;92(1):27–33. https://doi.org/10.1111/aos.12294.
35. Jakobsen DB, Torp TL, Stefansson E, Peto T, Grauslund J. Retinal metabolic and structural alterations in response to aflibercept treatment in neovascular age-related macular degeneration. *Acta Ophthalmologica*. 2019;97(5):525–531. https://doi.org/10.1111/aos.13996.
36. Donicova E, Ramm L, Augsten R, Hammer M. The flicker response of venous oxygen saturation is significantly reduced in the early and late stages of age-related macular degeneration. *Graefe's Archive for Clinical and Experimental Ophthalmology*. 2020;258(1):31–37. https://doi.org/10.1007/s00417-019-04533-6.
37. Mordant DJ, Al-Abboud I, Muyo G, Gorman A, Harvey AR, McNaught AI. Oxygen saturation measurements of the retinal vasculature in treated asymmetrical primary open-angle glaucoma using hyperspectral imaging. *Eye*. 2014;28(10):1190–1200. https://doi.org/10.1038/eye.2014.169.
38. Olafsdottir OB, Hardarson SH, Gottfredsdottir MS, Harris A, Stefánsson E. Retinal oximetry in primary open-angle glaucoma. *Investigative Ophthalmology and Visual Science*. 2011;52(9):6409–6413. https://doi.org/10.1167/iovs.10-6985.
39. Pinto LA, Willekens K, Van Keer K, et al. Ocular blood flow in glaucoma – the Leuven eye study. *Acta Ophthalmologica*. 2016;94(6):592–598. https://doi.org/10.1111/aos.12962.
40. Ramm L, Jentsch S, Peters S, Augsten R, Hammer M. Investigation of blood flow regulation and oxygen saturation of the retinal vessels in primary open-angle glaucoma. *Graefe's Archive for Clinical and Experimental Ophthalmology*. 2014;252(11):1803–1810. https://doi.org/10.1007/s00417-014-2766-4.
41. Van Keer K, Pinto LA, Willekens K, Stalmans I, Vandewalle E. Correlation between peripapillary choroidal thickness and retinal vessel oxygen saturation in young healthy individuals and glaucoma patients. *Investigative Ophthalmology and Visual Science*. 2015;56(6):3758–3762. https://doi.org/10.1167/iovs.14-16225.
42. Vandewalle E, Pinto LA, Olafsdottir OB, et al. Oximetry in glaucoma: correlation of metabolic change with structural and functional damage. *Acta Ophthalmologica*. 2014;92(2):105–110. https://doi.org/10.1111/aos.12011.
43. Michelson G, Scibor M. Intravascular oxygen saturation in retinal vessels in normal subjects and open-angle glaucoma subjects. *Acta Ophthalmologica Scandinavica*. 2006;84(3):289–295. https://doi.org/10.1111/j.1600-0420.2005.00631.x.
44. Olafsdottir OB, Vandewalle E, Pinto LA, et al. Retinal oxygen metabolism in healthy subjects and glaucoma patients. *Br J Ophthalmol*. 2014;98(3):329–333. https://doi.org/10.1136/bjophthalmol-2013-303162.

45. Hickam JB, Frayser R. Spectrophotometric determination of blood oxygen. *The Journal of Biological Chemistry*. 1949;180(1):457–465.
46. Hamard P, Hamard H, Dufaux J, Quesnot S. Optic nerve head blood flow using a laser Doppler velocimeter and haemorheology in primary open angle glaucoma and normal pressure glaucoma. *British Journal of Ophthalmology*. 1994;78(6):449–453. https://doi.org/10.1136/bjo.78.6.449.
47. Shughoury A, Mathew S, Arciero J, et al. Retinal oximetry in glaucoma: investigations and findings reviewed. *Acta Ophthalmologica*. 2020;98(6):559–571. https://doi.org/10.1111/aos.14397.
48. Wang J, Gao X, Du S, et al. Aqueous humor concentration of VEGF and retinal oxygen saturation after unilateral acute primary angle closure. *Acta Ophthalmologica*. 2016;94(4):380–385. https://doi.org/10.1111/aos.12664.
49. Nitta E, Hirooka K, Shimazaki T, et al. Retinal oxygen saturation before and after glaucoma surgery. *Acta Ophthalmologica*. 2017;95(5):e350–e353. https://doi.org/10.1111/aos.13274.
50. Blindbæk SL, Torp TL, Lundberg K, et al. Noninvasive retinal markers in diabetic retinopathy: advancing from bench towards bedside. *Journal of Diabetes Research*. 2017;2017. https://doi.org/10.1155/2017/2562759.
51. Sabanayagam C, Lye WK, Klein R, et al. Retinal microvascular calibre and risk of diabetes mellitus: a systematic review and participant-level meta-analysis. *Diabetologia*. 2015;58(11):2476–2485. https://doi.org/10.1007/s00125-015-3717-2.
52. Einarsdottir AB, Hardarson SH, Kristjansdottir JV, Bragason DT, Snaedal J, Stefánsson E. Retinal oximetry imaging in Alzheimer's disease. *Journal of Alzheimer's Disease*. 2015;49(1):79–83. https://doi.org/10.3233/JAD-150457.
53. Olafsdottir OB, Saevarsdottir HS, Hardarson SH, et al. Retinal oxygen metabolism in patients with mild cognitive impairment. *Alzheimer's and Dementia: Diagnosis, Assessment and Disease Monitoring*. 2018;10:340–345. https://doi.org/10.1016/j.dadm.2018.03.002.
54. Compston A, Coles A. Multiple sclerosis. *Lancet*. 2008;372(9648):1502–1517. https://doi.org/10.1016/S0140-6736(08)61620-7.
55. Kale N. Optic neuritis as an early sign of multiple sclerosis. *Eye and Brain*. 2016;8:195–202. https://doi.org/10.2147/EB.S54131.
56. Einarsdottir AB, Olafsdottir OB, Hjaltason H, Hardarson SH. Retinal oximetry is affected in multiple sclerosis. *Acta Ophthalmologica*. 2018;96(5):528–530. https://doi.org/10.1111/aos.13682.
57. Svrcinova T, Mares J, Chrapek O, et al. Changes in oxygen saturation and the retinal nerve fibre layer in patients with optic neuritis—a pilot study. *Acta Ophthalmologica*. 2018;96(3).
58. Svrčinová T, Hok P, Šínová I, et al. Changes in oxygen saturation and the retinal nerve fibre layer in patients with optic neuritis associated with multiple sclerosis in a 6-month follow-up. *Acta Ophthalmologica*. 2020;98(8):841–847. https://doi.org/10.1111/aos.14463.
59. Kindt A, Byg K-E, Wied J, Ellingsen T, Davidsen JR, Grauslund J. Altered retinal oxygen metabolism in patients with combined ocular and central nervous system sarcoidosis. *Rheumatology*. 2021;60(7):3301–3306. https://doi.org/10.1093/rheumatology/keaa781.
60. Lowater SJ, Ellingsen TJ, Pedersen JK, Wied J, Grauslund J, Byg KE. Non-invasive metabolic and structural retinal markers in patients with giant cell arteritis and polymyalgia rheumatica: a cross-sectional study. *Metabolites*. 2022;12(9). https://doi.org/10.3390/metabo12090872.
61. Eliasdottir TS, Bragason D, Hardarson SH, et al. Retinal oximetry measures systemic hypoxia in central nervous system vessels in chronic obstructive pulmonary disease. *PLoS One*. 2017;12(3):e0174026. https://doi.org/10.1371/journal.pone.0174026.
62. Traustason S, Jensen AS, Arvidsson HS, Munch IC, Søndergaard L, Larsen M. Retinal oxygen saturation in patients with systemic hypoxemia. *Investigative Ophthalmology and Visual Science*. 2011;52(8):5064–5067. https://doi.org/10.1167/iovs.11-7275.
63. Dinesen S, Jensen PS, Bloksgaard M, et al. Coronary artery bypass surgery independently associates with retinal vascular oxygen saturation. *Acta Ophthalmologica*. 2020;98(7):709–715. https://doi.org/10.1111/aos.14444.
64. Dinesen S, Jensen PS, Bloksgaard M, et al. Retinal vascular oxygen saturation increases after cardiac surgery. *Acta Ophthalmologica*. 2019;97(6):e941–e942. https://doi.org/10.1111/aos.14084.
65. Van Keer K, Van Keer J, Breda JB, et al. Retinal oxygen saturation as a non-invasive estimate for mixed venous oxygen saturation and cardiac output. *Acta Ophthalmol*. 2018;97:e308–e312.
66. Jeppesen SK, Bek T. The retinal oxygen saturation measured by dual wavelength oximetry in larger retinal vessels is influenced by the linear velocity of the blood. *Current Eye Research*. 2019;44(1):46–52. https://doi.org/10.1080/02713683.2018.1524015.
67. Willerslev A, Hansen MM, Klefter ON, et al. Non-invasive imaging of retinal blood flow in myeloproliferative neoplasms. *Acta Ophthalmologica*. 2017;95(2):146–152. https://doi.org/10.1111/aos.13249.

Chapter 10

Widefield imaging of the peripheral retina

Niveditha Pattathil[1,2], Samantha Orr[1,2], Jay Chhablani[5] and Netan Choudhry[1,2,3,4]

[1]*Octane Imaging Lab, Toronto, ON, Canada;* [2]*Vitreous Retina Macula Specialists of Toronto, Toronto, ON, Canada;* [3]*Department of Ophthalmology & Vision Sciences, University of Toronto, Toronto, ON, Canada;* [4]*Cleveland Clinic Canada, Toronto, ON, Canada;* [5]*Department of Ophthalmology, University of Pittsburgh School of Medicine, Pittsburgh, PA, United States*

Introduction

Since the first photographs of the retina were captured in the late 1800s, retinal imaging has progressed significantly. One major improvement has been the increased field of view of retinal imaging. The first retinal photograph imaging the peripheral retina beyond the equator was captured in 1977 by Oleg Pomeranzeff, who used a widefield contact lens to gain up to a 148 degrees field of view. The most common method of obtaining widefield imaging over the years has been through the use of image montaging, as with the standardized Early Treatment Diabetic Retinopathy Study (ETDRS) 7-standard field.

With improving imaging capabilities, the International Widefield Imaging Study Group created a consensus on terminology, published in 2019.[1] Widefield (WF) imaging was defined as an image that captures the retina beyond the arcade vessels up to the posterior border of the vortex vein ampulla—the midperiphery of the retina. This correlates with an approximate field of view of 60—100 degrees. However, this field-of-view measurement is difficult to standardize across devices, making anatomical definitions preferable. The term ultra-widefield (UWF) refers to images that capture the far-periphery of the retina beyond the anterior border of the vortex vein ampulla and correlates with an approximate field of view of 110—220 degrees.

Imaging the mid- and far-periphery of the retina has provided new insights into peripheral pathologies, aiding with their diagnosis, informing their management, and generating prognostic information. This field is rapidly evolving. Widefield and UWF terminology applies to many retinal imaging modalities, the most common of which will be introduced below. In this chapter, multimodal widefield imaging of different pathologies will be reviewed, with an emphasis on the advantage and clinical utility of each modality. This discussion will be supplemented with multimodal case studies.

Imaging modalities

Pseudocolor or multicolor imaging is similar to color fundus photography in that a color image of the retina is generated. However, the method by which this image is captured and generated is different and allows for additional insights. Pseudocolor images, such as those captured by Optos (Optos TLC, Edinburgh, UK) devices are generated using two monochromatic images of different wavelengths: one from a green laser, one from a red laser. From these scans, a "pseudocolor" image is generated. When the monochromatic images are reviewed separately, additional details may be detected. The green-separated image presents a detailed view of the neurosensory retina down to the RPE, while the red-separated image has an enhanced depth of focus to provide higher detail of the choroid below the RPE. Multicolor images, such as those acquired with the Heidelberg Spectralis device (HRA-OCT, Heidelberg Engineering, Heidelberg, Germany) function similarly. Three monochromatic images are acquired with lasers of different wavelengths, each with a different depth of penetration. These infrared, green, and blue monochromatic images can be viewed separately or combined to generate a multicolor image with improved structural detail.

Retinal and Choroidal Vascular Diseases of the Eye. https://doi.org/10.1016/B978-0-443-15583-3.00010-X

Near-infrared retinal imaging employs light of 804–895 nm, typically using a scanning laser ophthalmoscope to limit light scatter artifacts. Similarly to the red laser used in pseudocolor imaging, the longer wavelength allows for greater depth of penetration into retinal tissue. It can therefore provide detailed visualization of the outer retina, RPE, and choroid, details of which can be difficult to appreciate on clinical examination. The longer wavelength has additional advantages in image acquisition, including improved ability to image through media opacities such as cataracts and improved comfort for the patient, as near-infrared wavelengths may cause less discomfort than bright light of lower wavelengths.[2]

Fluorescein angiography (FA) employs the use of fluorescein dye, typically injected intravenously, to image retinal vasculature by capturing the fluorescence emitted by the dye following excitation with laser over a period of several minutes. Along with providing information on retinal blood flow, it can show leakage, neovascularization, and tissue staining—providing insight into pathology. The fluorescent properties of indocyanine green dye can be used to image the vasculature of the choroid because this dye has a high affinity for proteins in the choroidal vessels.[2]

Unlike FA, autofluorescence imaging is based on the innate fluorescent properties of macular pigments, most notably lipofuscin, which can be detected upon excitation with blue or green laser. Lipofuscin is a waste byproduct found in retinal pigment epithelium (RPE) cells. The amount and distribution of this pigment can provide insight into the health and status of the RPE layer. By extension, pathologies that affect production or clearance of lipofuscin, via cell atrophy or cellular distress can be monitored over time with this imaging modality.[2]

Optical coherence tomography (OCT) has become indispensable in the field of ophthalmology since its development in the 1990s. OCT allows for the construction of a 3-dimensional view of the retina by measuring light reflected off retinal tissues. Since its inception, the field of view for this modality has increased, and some OCT technology now allows for ultra-widefield OCT.[2] Some devices allow for targeted OCT volume, line, or radial scans in the retinal periphery. In these situations, the International Widefield Imaging Study Group recommends that terminology must identify the scan size (mm), retinal location, and type.[1]

OCT angiography represents an evolution of OCT technology by providing information about retinal blood flow based on detection of red blood cell movement upon repeated scanning of the same location. As a consequence of this approach, turbulent or slow blood flow may not be detected by OCTA. Unlike FA, vascular leakage and tissue staining are not seen. However, OCTA can provide depth-resolved details, allowing separate evaluation of superficial and deep capillary plexuses, choroidal vessels, and detection of abnormal vasculature, such as choroidal neovascular membranes. Another advantage of OCTA is its noninvasive nature—no intravenous dye is required—making this test safe and easily repeatable.[2] If the OCTA is not a montaged view of the entire WF or UWF area, this should be denoted with the term "asymmetric" and should include the scan size (mm).[1]

Retinal vascular disease

Diabetic retinopathy

Diabetic retinopathy (DR) is a common microvascular complication of diabetes mellitus (DM), which can lead to significant loss of vision, particularly when not managed appropriately. Retinal imaging plays an important role in the diagnosis and monitoring of DR, and since this pathology affects the entire retina and not just the posterior pole, WF and UWF imaging have proven valuable. The standard 7-field ETDRS montaged color fundus photograph has been used as the basis for the staging of DR severity since 1991.[3] More recently, the staging of DR and risk of progression have been assessed using WF and UWF color photography. Compared to staging based on standard ETDRS color photos, this approach has yielded similar results.[3,4] In approximately 10% of cases, the wider field-of-view increased the stage of DR severity by at least two steps, detecting additional lesions that were not visualized in the ETDRS fields.[3]

Some DR lesions are not easily detected in color photography, and multimodal imaging can be valuable in detecting additional lesions to develop a more complete and accurate assessment of diagnosis, staging, and prognosis.[5] Angiographic imaging modalities such as FA and OCTA are particularly useful in the detection of vascular lesions such as microaneurysms, intraretinal microvascular abnormalities (IRMA), and neovascularization (NV) in the retinal periphery.[6,7] These findings are critical for the accurate staging of DR: the presence of IRMA denotes severe nonproliferative DR (NPDR), which carries an increased risk of disease progression, whereas the presence of NV marks the development of proliferative DR (PDR).[6,7] The gold standard remains FA, and UWF-FA has proven effective in identifying peripheral vascular lesions. Given previously discussed advantages of OCTA, many studies have compared the performance of peripheral OCTA to that of FA and have found that OCTA performs similarly in detecting IRMA, NV, and microaneurysms and in distinguishing between these lesions.[5,8] OCTA outperforms UWF color photography in the detection of NV, with one study reporting 100% versus 32% sensitivity, making OCTA a useful adjunct in DR screening and diagnosis.[7]

Assessing or measuring areas of nonperfused retina is another biomarker that can aid in prognostication of DR. Nonperfusion can be assessed using UWF-FA or UWF-OCTA with the use of ischemic index (ISI), ratio of nonperfusion (RNP) measures, or nonperfusion area (NPA) measurements.[9] In the measurement of NPA, projection distortion adjustments are required to avoid overestimation of area in the retinal periphery.[10] Distortion correction algorithms are now incorporated in Optos software for the measurement of NPA on UWF-FA.[11] The baseline NPA in a patient with DR has been correlated to the risk of disease severity progression. OCTA has been used to assess for choriocapillaris changes in the setting of diabetic retinopathy—also termed diabetic choroidopathy.[12,13] Choroidal vascularity indices (CVI) show a steep decrease in PDR, which could be related to choroidal depletion of vascular shrinkage. However, CVI changes in the macula were more distinct than changes in the periphery, and thus macular changes may be a more valuable biomarker.[12,13]

Multimodal WF and UWF imaging in patients with diabetes, but no diabetic retinopathy has led to the detection of preclinical retinal changes even in the absence of DR pathology and insight into the development of retinopathy: this is termed diabetes without retinopathy (DWR).[9,14] Ultra-widefield OCTA has revealed retinal capillary dilation and tortuosity in DWR that was more commonly seen in the periphery than the central retina. Ratio of nonperfusion (RNP) indicated capillary dropout in the central and periphery in these patients, further supporting that microvascular changes are occurring. This measure of RNP in the periphery has been used to differentiate DWR from NPDR and PDR.[9] One study reported that either nonperfusion, capillary dilation, or tortuosity were seen on UWF-OCT in 58% of eyes in DWR.[14] These findings also align with central retinal findings that report a higher perfused capillary density (PCD) in DWR, possibly representing capillary dilation, with progressively decreasing PCD in NPDR and PDR.[15]

UWF-OCT can also provide clinically valuable insights, with the ability to document vitreous traction and detect tractional retinoschisis or retinal detachments.[16]

Here, we present a case of a 61-year-old female with proliferative diabetic retinopathy in both eyes. Multimodal imaging of the left eye is shown and visual acuity at the time of imaging was 20/40. The montaged pseudocolor image (Optos California, Optos PLC, Edinburgh, UK) shows dot-blot hemorrhages, exudates, and neovascularization (Fig. 10.1A). On autofluorescence imaging (Optos California), discrete areas of hypoautofluorescence are seen, which colocalize with dot-blot hemorrhages on pseudocolor imaging, and colocalize with neovascularization seen on FA and OCTA (Fig. 10.1B). Ultra-widefield montaged fluorescein angiography (Optos California) shows nonperfusion in the periphery and several areas of vascular leakage and neovascularization (Fig. 10.1C). These angiographic findings are corroborated with montaged widefield OCTA (Plex Elite, Zeiss), which also demonstrates peripheral nonperfusion and neovascularization (Fig. 10.1D).

Retinal vein occlusions

Central or branch retinal vein occlusions (RVO) are another common retinal vascular pathology in which imaging of the retinal periphery can provide valuable clinical information. On UWF color photography, pathologic findings such as intraretinal hemorrhages can be detected. Based upon such findings and others on color photography, deep learning algorithms have been able to learn to detect branch and central RVO.[17,18] Distinguishing ischemic from nonischemic central RVO on UWF-pseudocolor photographs can present a challenge, particularly in less severe ischemic cases with no iris neovascularization. In these cases, examination of the red-separated image may be valuable and has been reported to significantly improve the accurate detection of ischemic central RVO without iris neovascularization.[19]

OCTA and FA of the retinal periphery play an important role in characterizing and prognosticating this disease. RVO can be classified as ischemic or nonischemic based on the amount of nonperfusion present, measured as ISI, or a quantitative measurement of area on FA or OCTA.[11] The UWF component is particularly important in RVO as nonperfusion typically begins in the far-periphery of the retina and progresses toward the posterior pole. Comparisons between FA and OCTA measurements in RVO suggest that OCTA is a reasonable alternative for assessing NPA or ISI.

RVO can lead to neovascularization in both the retina and the iris, which can result in complications such as vitreous hemorrhage or neovascular glaucoma. Unsurprisingly, the risk of neovascularization has been linked to the extent of nonperfusion. One study reported that all cases with retinal NV had an ISI of at least 45% and an average ISI of 75%. NPA and ISI have also been linked to the risk of iris NV.

An additional OCTA measurement relevant to the assessment of BRVO is the periarterial capillary free zone (paCFZ).[20] In the superficial capillary plexus, there is a physiological avascular area surrounding retinal arteries that can become enlarged in the setting of pathology. PaCFZ, as well as the ratio of paCFZ to the artery area (P/A ratio), are measured at first and second-order arteries, necessitating a larger field of view than a traditional 3 × 3 or 6 × 6 OCTA of the macula. Research has suggested that paCFZ and P/A ratios correlate with baseline visual acuity and central retinal thickness in patients with BRVO and macular edema. These OCTA measurements also decreased over time with anti-VEGF therapy.[20] Improvements in the assessment and prognostication of RVO can play an important role in guiding treatment and managing patient expectations.

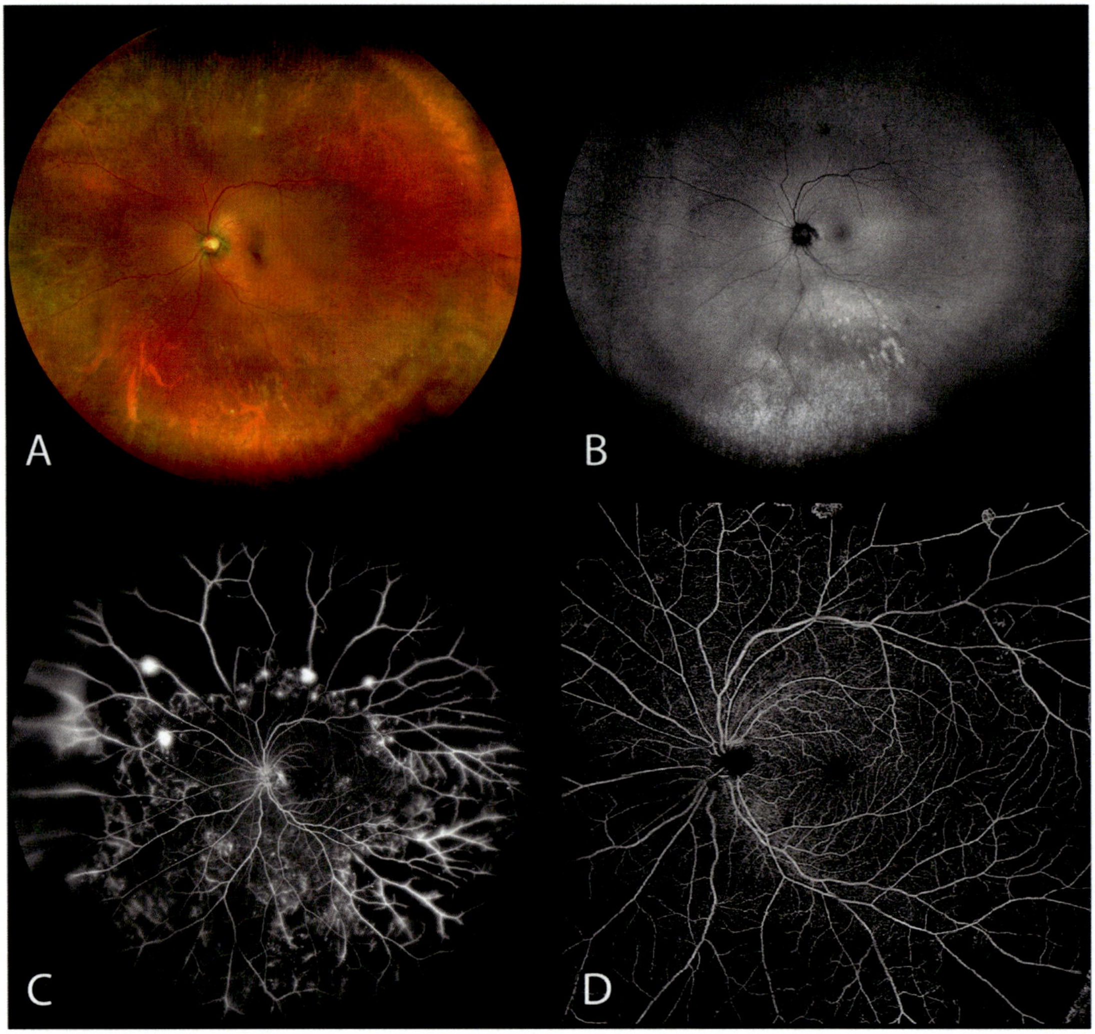

FIGURE 10.1 Multimodal imaging of 61 year old woman with proliferative diabetic retinopathy in both eyes.

Age-related macular degeneration

Age-related macular degeneration (AMD) is typically considered to be a pathology of the posterior pole. With WF and UWF imaging available in many AMD patients, pathologic findings have been detected in the retinal periphery, although their clinical significance has not yet been elucidated.[21] On pseudocolor imaging, peripheral drusen have been found in the mid- and far-peripheral retina in many AMD patients.[22,23] Senile reticular pigmentary changes, RPE hypopigmentation, geographic atrophy, and neovascularization are other reported peripheral findings in AMD patients. Peripheral reticular pseudodrusen were best detected with blue-light autofluorescence or UWF pseudocolor images.[23] Abnormal green-light autofluorescence findings such as peripheral nummular hypoautofluorescence and granular hyperautofluorescence were reported in 86% of eyes. The nummular hypoautofluorescent pattern may correspond to atrophy or cobblestone degeneration.[23]

Peripheral retinal degenerations

There is a clear value of WF and UWF imaging for the diagnosis, monitoring, and understanding of peripheral retinal degenerations.[16,21,24,25] Peripheral retinal degenerations include intraretinal pathologies such as degenerative retinoschisis and peripheral cystoid degeneration, vitreoretinal degenerations such as lattice degeneration and retinal tufts, and chorioretinal degenerations including paving-stone degeneration and retinal drusen.

Color photographs of the retinal periphery allow for clear documentation of pathologies that may or may not be seen easily on clinical examination.[16,24–26] Using this technology, the ability to monitor for changes or complications is enhanced. With UWF pseudocolor imaging, the additional review of color-separated images can play an important role in the detection and diagnosis of peripheral pathologies, based upon the principles of depth of focus and location of pathology. For example, degenerative retinoschisis involves splitting of the neurosensory retina, typically at the level of the outer plexiform layer (OPL).[27] The green-separated monochromatic image can enhance the view of this pathology because it provides clearer details of the neurosensory retina. This highlights the importance of reviewing not only the pseudocolor image that has been generated but also the color-separated images, as additional insights can be gained.

Many peripheral retinal degenerations are also visible on autofluorescence imaging. However, the clinical utility of these images is less established.[28] In paving-stone degeneration, a nummular pattern is seen on autofluorescence. In both white-without-pressure and dark-without-pressure, hypoautofluorescence is seen.

Peripheral OCT may be the most clinically valuable modality for peripheral retinal degenerations. Presence of pathologies including retinal holes, retinal tufts, and peripheral drusen can be confirmed using OCT. In some studies, OCT scans have been correlated with histologic findings.[16] In lattice degeneration, retinal thinning has been reported as the most common OCT finding, which correlates with histological findings of retinal thinning and vitreoretinal attachment. OCT findings in other peripheral retinal degenerations including vitreous traction, retinal splitting, and vitreous condensation were also supported by histological studies.[16] Many studies have reported that clinical details detected on OCT imaging have supported or altered management recommendations. For example, the presence or absence of vitreous traction in pathologies, including retinal holes, retinal tufts, and lattice degeneration, can be the deciding factor in the recommendation for prophylactic laser treatment.[24–26]

In the cases where the prophylactic laser is performed, UWF imaging can document the response to treatment and monitor for any further changes. In one study of lattice degeneration in myopic eyes, prophylactic laser treatment led to blunting of vitreous traction and appeared to halt the progression of vitreous traction.[29] The laser-treated retina is identifiable on OCT by scarring of the ellipsoid zone.

This case presents the multimodal imaging of a 45-year-old female with degenerative retinoschisis. No prophylactic treatment was recommended due to an absence of risk factors such as inner retinal holes. Multimodal imaging of the right eye reveals superotemporal retinoschisis. On montaged pseudocolor imaging, a smooth dome-shaped elevation is seen with overlying darkened vasculature (Fig. 10.2A). A subtle reticular pattern and yellow-white hyperreflective foci at the anterior border (Fig. 10.2A). On the corresponding green-separated image, these findings are emphasized (Fig. 10.2B). On autofluorescence imaging, the area of schisis is isoautofluorescent with a hypoautofluorescent posterior border (Fig. 10.2C). Peripheral OCT corresponding to lines on pseudocolor image shows splitting of the neurosensory retina (Fig. 10.2D and E).

Retinal detachment

Rhegmatogenous and tractional retinal detachments typically involve the peripheral retina. These pathologies can lead to significant loss of vision; comprehensive and timely assessment is key to appropriate management. Color or pseudocolor imaging can detect the presence and extent of detachment and OCT can confirm the presence of a neurosensory retinal detachment.[21] A device with integrated OCT capabilities with fundus imaging, such as the Optos Silverstone, allows for easily targeted OCT of the area of interest. UWF-OCT can provide valuable information regarding vitreous attachments or traction, which can inform risk of progression and management decisions.[24] Localization of the retinal break or tear with modalities including pseudocolor and OCT can also guide treatment.[24,25] On autofluorescence imaging, acute retinal detachments often display hypoautofluorescence over the area of detachment.[30] Hyperautofluorescence at the leading edge of the detachment can suggest the presence of subretinal fluid, a finding which can be verified with the use of OCT.[21]

Using UWF-FA, retinal vascular changes in the setting of retinal detachments have been described. Peripheral capillary nonperfusion has been reported along with vascular leakage from small peripheral vessels. The clinical relevance of these findings has not yet been elucidated.[31]

Posttreatment or postoperatively, imaging of the periphery can provide insight into response to treatment. Thirty minutes after laser retinopexy, WF-OCT of the treated area showed disruption of retinal lamination, ONL splitting, and RPE disruption with localized thickening and columnar hyperreflectivity.[32] These retinal changes, suggestive of coagulative necrosis, were confined to the discrete laser spot, with normal surrounding retina. In the following weeks, the treated area became progressively thinner and atrophic. Within 24 h of cryotherapy treatment, thinning and disruption of retinal layers with hyperreflectivity and loss of RPE layer.[32] Unlike postlaser retinopexy OCT, the retina postcryotherapy was detached from RPE remnants. This persisted through day 4, at which time choroidal separation from the sclera was also noted. These findings were resolved by day 6.

Chronicity of retinal detachments can be difficult to determine and holds clinical relevance. The presence of a demarcation line on clinical examination or fundus imaging can indicate chronicity.[32] This demarcation line correlates with

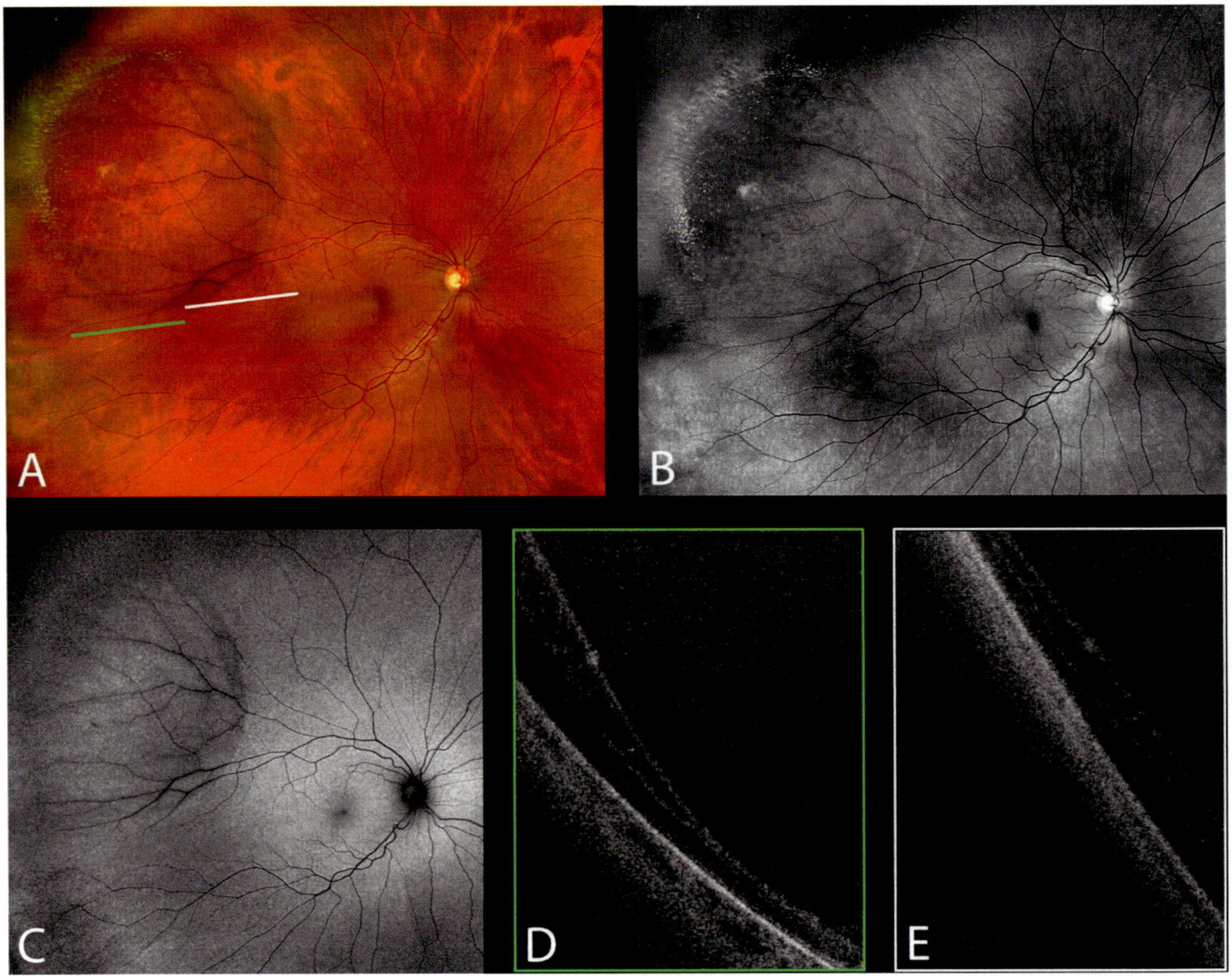

FIGURE 10.2 Multimodal imaging of 45 year old woman with degenerative retinoschisis.

hyperreflective RPE proliferation on OCT.[32] Autofluorescence imaging of chronic detachments may appear isoautofluorescent, in contrast to hypoautofluorescence and other findings in acute retinal detachments as described above.[27,30]

Here, we present a case of a 60-year-old female referred for urgent assessment after a change in vision who was diagnosed with a foveal splitting retinal detachment. No obvious breaks were seen, and the patient was referred to an on-call specialist for pneumatic retinopexy. Multimodal widefield imaging included pseudocolor, autofluorescence, and OCT. Montaged pseudocolor image shows a temporal retinal detachment with a corrugated appearance, no retinal tears seen (Fig. 10.3A). Autofluorescence imaging reveals hypoautofluorescence in the area of detached retina (Fig. 10.3B). On OCT line scan through the fovea, extending through the optic nerve out to the temporal periphery, a foveal-splitting retinal detachment can be seen, with involvement of the temporal foveal area (Fig. 10.3C).

Uveitis

Infectious and noninfectious uveitis can present a diagnostic challenge. Some pathognomic characteristics of different types of uveitis can be detected with imaging of the retinal periphery. One category of noninfectious uveitis, termed white dot syndromes (WDS), can appear similar on clinical examination, and multimodal imaging thus plays an important role in distinguishing these pathologies—which have distinct management recommendations and risk of complications.[33,34]

Autofluorescence imaging can reveal the characteristics of WDS lesions.[33–35] Multiple evanescent white dot syndrome (MEWDS) lesions are hyperautofluorescent and correspond to RPE hypertrophy on OCT. By contrast, in acute posterior multifocal placoid pigment epitheliopathy (APMPPE), lesions are hypoautofluorescent on FAF due to the dysfunctional and damaged RPE, seen as RPE hyperreflectivity on OCT imaging. Lesions in multifocal choroiditis (MFC) and punctate inner choroidopathy (PIC) are hypoautofluorescent with hyperautofluorescent margins. On OCT imaging, these lesions

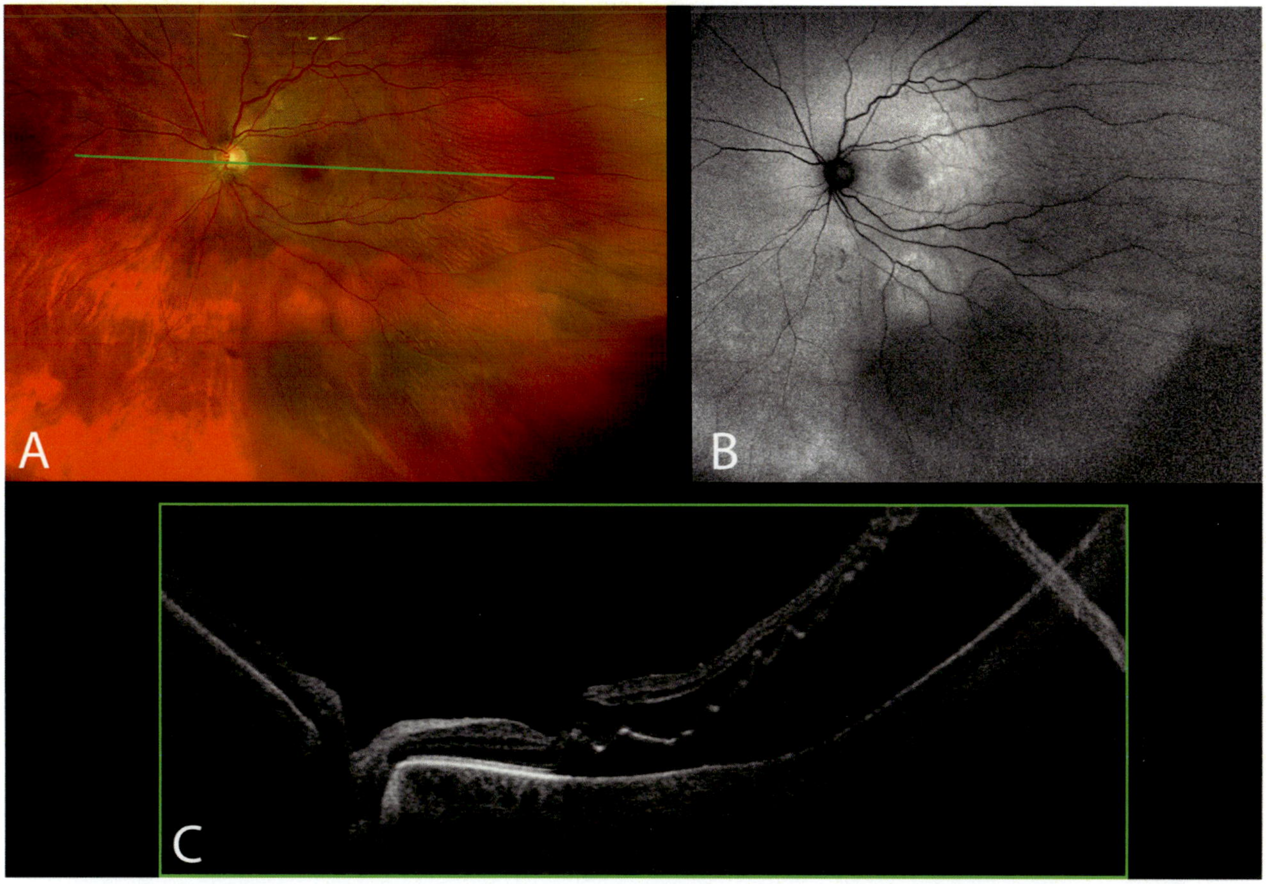

FIGURE 10.3 Multimodal imaging of a 60 year old woman with foveal splitting retinal detachment.

show elevation of the RPE with surrounding outer retinal disruption. In serpiginous choroiditis, autofluorescence imaging can provide insight into the activity of lesions: lesions begin as hyperautofluorescent and become progressively more hypoautofluorescent with resolution over time. Again, this correlates with findings on OCT, which show a progression from initial outer retinal thickening to decreased thickness with RPE atrophy. In acute zonal occult outer retinopathy (AZOOR) lesions, autofluorescence imaging reveals a characteristic pattern useful for identifying this pathology. The center of the lesion, "zone 3," is hypoautofluorescent and corresponds with RPE and choroidal atrophy on OCT. Surrounding this is "zone 2," with speckled hyperautofluorescence corresponding to subretinal hyperreflective material on OCT. The outer region of the lesion, "zone 1," shows normal autofluorescence.

Angiographic tests, including FA, ICGA, and OCTA can also assist with diagnosis and provide insight into the pathophysiology of these conditions.[33–35] Whether the pathology is primarily retinal or choroidal can sometimes be elucidated using these modalities. The multimodal approach allows for lesion findings to be cross-referenced between modalities to draw more in-depth conclusions. MEWDS lesions show a "wreath-like" pattern of hyperfluorescence on FA, which appears early and persists through the later phase of the test whereas APMPPE lesions become hyperfluorescent only in the late phase of the test. On ICGA imaging, MEWDS lesions show late hypofluorescence and APMPPE shows hypofluorescent spots in all phases of the test.

To compare these findings to the third angiographic modality, OCTA, MEWDS typically reveals no flow voids in the choriocapillaris whereas APMPPE lesions are colocalized with flow deficits in the choriocapillaris.[33–35] These findings support the hypothesis that APMPPE is a disease primarily of the choriocapillaris and that MEWDS pathology begins in the outer retina and RPE.[33–36] Other WDS show colocalization of hypofluorescence on ICGA and flow reduction in the choriocapillaris on OCT, including MFC, PIC, and serpiginous choroiditis. FA of the retinal periphery is particularly useful to detect and document associated inflammatory vasculitis and possible macular leakage. OCTA also plays an important role in the detection of secondary choroidal neovascularization, a potentially vision-threatening complication of some WDS, including MFC, PIC, APMPPE, and birdshot chorioretinopathy.

We present multimodal imaging from the left eye of a 27-year-old female with multifocal choroiditis (MFC) OU. Visual acuity was 20/20 OU and management consisted of intravitreal bevacizumab and oral prednisone taper. On a montaged pseudocolor image, multiple lesions are scattered throughout the midperiphery of the retina, with a higher distribution in the nasal region (Fig. 10.4A). On autofluorescence imaging, areas of hypoautofluorescence are colocalized with lesions seen on pseudocolor imaging (Fig. 10.4B). Some lesions also have a hyperautofluorescent border (Fig. 10.4B).

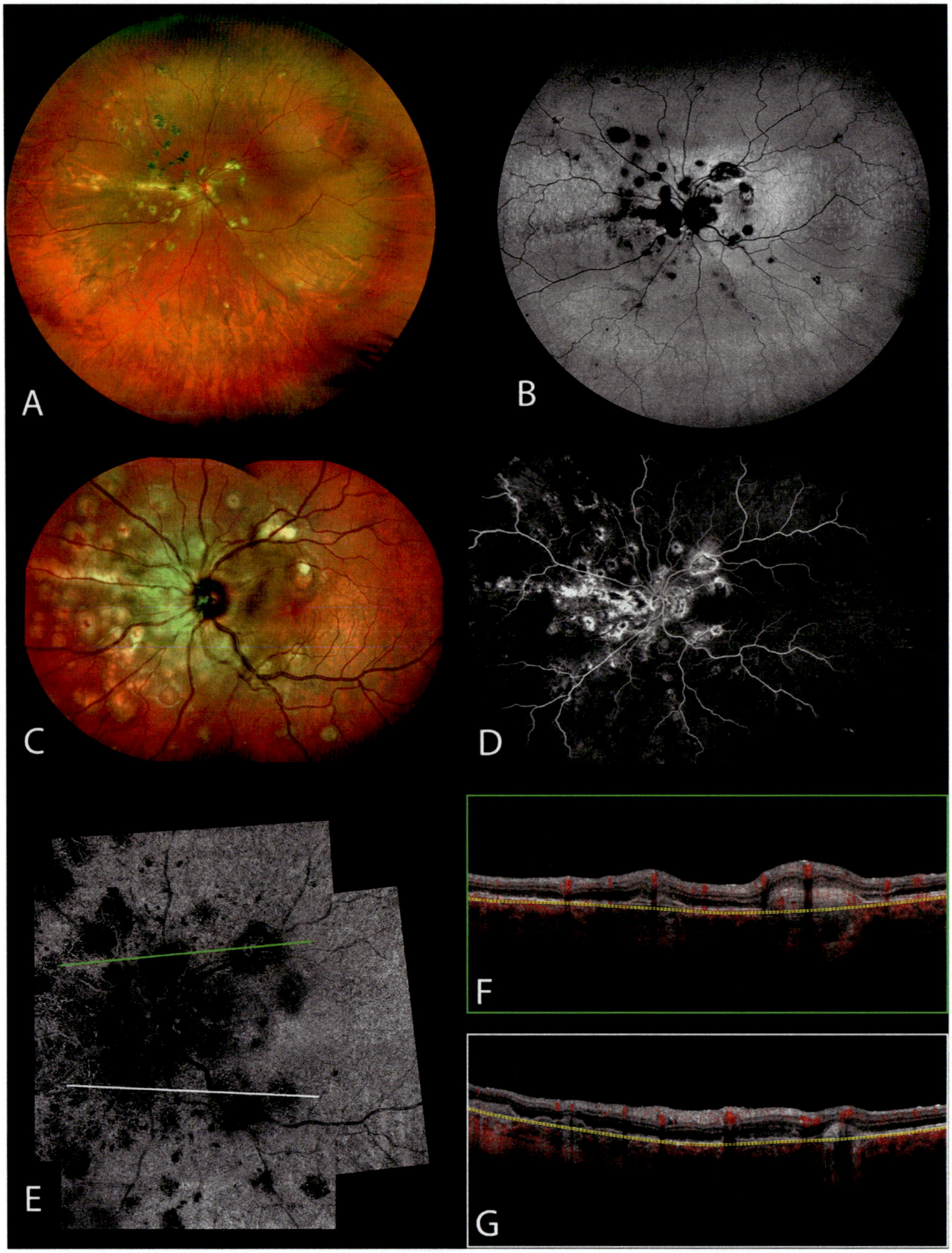

FIGURE 10.4 Multimodal imaging of a 27 year old woman with multifocal choroiditis.

Montaged multispectral imaging reveals scattered lesions corresponding with those seen on other modalities (Fig. 10.4C). OCT angiography reveals areas of flow deficit in the choriocapillaris which colocalize with lesions and colocalize with RPE elevation on OCT line scans (Fig. 10.4E–G). Fluorescein angiography reveals hyperfluorescence in the area of some lesions and a lack of fluorescence with surrounding hyperfluorescence in the area of other lesions (Fig. 10.4D).

Peripheral retinal and choroidal tumors

The use of multimodal WF and UWF imaging can provide valuable clinical insight into masses in the retinal periphery. They can play an important role in the distinction between benign and malignant pathologies as well as the characterization of the retinal or choroidal mass. One diagnosis not to be missed is choroidal melanoma. "Pseudomelanoma" is the term given to any mass or pathology that can mimic or be mistaken for melanoma.

Choroidal nevi are one of the most common pseudomelanomas. Characteristics suspicious for melanoma or for increased risk of conversion from nevus to melanoma can be assessed with the assistance of WF and UWF multimodal imaging. Shields and colleagues have identified these predictive imaging biomarkers in choroidal nevi: size (>2 mm depth, >5 mm diameter), presence of subretinal fluid, hyperautofluorescent orange pigment, and acoustic hollowness on ultrasound.[37] Depth was measured with ultrasound and diameter was measured with photography.[37] UWF-pseudocolor imaging of choroidal masses has also been used to measure size with good correlation to ultrasound measurements, although care must be taken to consider issues of projection distortion.[21] Although OCT was not a validated tool for measuring choroidal lesion size, estimates of depth and diameter can be generated. WF and UWF-OCT imaging is particularly valuable in the detection of subretinal fluid associated with choroidal masses. Finally, pigmentary changes that appear hyperautofluorescent on autofluorescence imaging could potentially be visualized on color or pseudocolor imaging as well.

The use of color-separated images may also be valuable in the characterization of peripheral masses, with malignant choroidal lesions reportedly appearing darker on the red-separated image.[21] Compared to benign nevi, choroidal melanoma lesions purportedly appear less intense on autofluorescence imaging. Peripheral exudative hemorrhagic chorioretinopathy (PEHCR) is another common pseudomelanoma typically located in the retinal periphery.[38] OCT of this pathology can provide clarification, as the presence of retinal exudation and RPE atrophy suggest a diagnosis of PEHCR.[38] Another mass that can be mistaken for choroidal melanoma is choroidal hemangioma. Autofluorescence imaging can reveal a granular pattern that is hypoautofluorescent or isoautofluorescent with or without hyperautofluorescent pigmentary changes.[39]

Angiography with FA, ICGA, or OCTA can show the vascular features of masses, although OCT can be employed to monitor for secondary retinal effects of tumors. FA and OCTA of retinoblastoma tumors reveal a complex branching pattern of intrinsic vessels. Pathology studies have suggested that increased tumor vascular density may correlate with the risk of metastasis and poor outcomes, highlighting the clinical value of this in vivo angiographic analysis.[40]

Here, we present the multimodal imaging of a 40-year-old male with a choroidal melanoma in the left eye. Widefield pseudocolor image of the left eye reveals a superotemporal raised mass (Fig. 10.5B). Montaged color photograph reveals pigmentary changes, particularly around the circumference of the mass (Fig. 10.5A). Autofluorescence imaging of the mass reveals hypoautofluorescence over the elevated area, with a patch of hyperautofluorescence inferiorly, consistent with the presence of subretinal fluid (Fig. 10.5D). Fluorescein angiography reveals staining in the area of the mass with a lack of fluorescence colocalizing with pigmentary changes on the color image (Fig. 10.5C). OCT line scan through the mass demonstrates a choroidal mass and confirms the presence of subretinal fluid over the inferior border of the mass (Fig. 10.5E). The patient was referred to the ocular oncology service for plaque radiation therapy.

Retinal dystrophies

Retinal dystrophies are a collection of genetically and clinically diverse degenerative illnesses of the retina. These progressive diseases commonly include presentations such as color blindness, night blindness, anomalies in peripheral vision, and eventual progression to total blindness.[41] WF and UWF imaging can enable the discovery of peripheral retinal abnormalities that would otherwise go undetected, making it an effective tool for the management and diagnosis of retinal dystrophies.[42]

UWF imaging is valuable for the evaluation of retinal integrity in inherited retinal dystrophies, including Stargardt disease, retinitis pigmentosa (RP), cone dystrophies (CD), and Best vitelliform dystrophy.[43] UWF with color or pseudocolor and fundus autofluorescence images allow for better visualization of abnormal vessels, aneurysms, and early and progressive retinal changes in the periphery. These imaging modalities correlate with visual field testing or electroretinograms and facilitate the diagnosis of these retinal dystrophies.

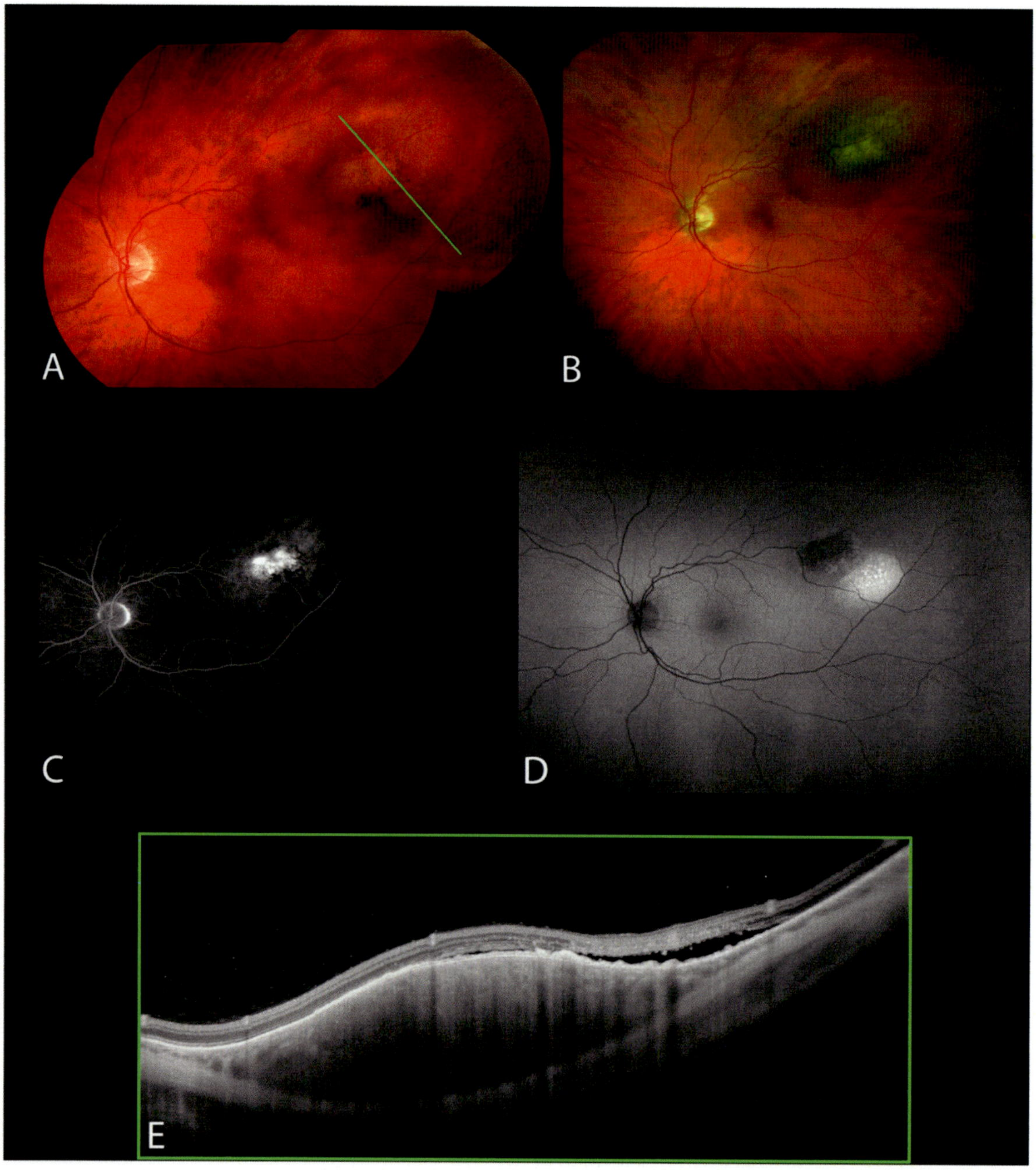

FIGURE 10.5 Multimodal imaging of a 40 year old man with choroidal melanoma in the left eye.

UWF FAF is helpful in identifying Stargardt disease. A study by Klufas and colleagues stratified ABCA4 Stargardt disease in three types using UWF FAF: type I presents with lesions confined to the macula with no peripheral findings; type II has macular atrophy with flecks only in the periphery, and type III has macular atrophy with varying degrees of peripheral atrophy.[44] More specifically, they concluded that although type I could be assessed with FAF of the macula alone, types II and III have peripheral retinal findings that would not be visualized without UWF imaging. These findings include increased hyper- and hypoautofluorescent changes secondary to the flecks extending to the equator in type II and peripheral atrophy as well as mottled areas of increased hyper- and hypoautofluorescent changes in type III. The study also suggested that more extensive UWF FAF patterns found in Type III are associated with more severe panretinal ERG dysfunction.

In certain cases of CD, UWF FAF reveals a hyperautofluorescent ring in the retinal periphery.[45] The area of this peripheral hyperautofluorescent ring inversely correlates with residual visual function on both Goldmann perimetry and

ERG responses in cases of CD.[46] UWF FAF demonstrates a ring of hypo- and hyperautofluorescence and hypoautofluorescent flecks in the midperiphery.[47] The study by Oishi and colleagues also suggested that WF FAF imaging is clinically relevant as it can evaluate the entire retina and find aforementioned retinal abnormalities that have been shown to have a strong correlation visual function of patients measured with Goldmann perimetry or ERG.[46]

Autofluorescence imaging in patients with RP shows perifoveal hyperautofluorescence which is deemed as the landmark between the spared central fovea and the diseased retinal.[43] However, UWF FAF has a greater predictive value for estimating residual retinal sensitivity than autofluorescence of the posterior pole in patients with RP. The area of hypoautofluorescence on UWF autofluorescence imaging and Goldmann functional scotoma are well correlated with 91% agreement.[43,48–50]

In Best vitelliform dystrophy, extrafoveal lesions can be assessed and monitored with UWF imaging.[43] Up to 30% of patients may present with multifocal lesions both central and extrafoveal. Extrafoveal lesions extending beyond the vascular arcades have similar imaging characteristics to macular lesions, which can be visualized on pseudocolor imaging, green autofluorescence (dishomogeneous hyperfluorescence ring), and OCT (subretinal hyporeflective lesion); however, they are typically smaller in size than foveal lesions.[43] There have not been any specific studies addressing the clinical significance of retinal periphery lesions found in best vitelliform dystrophy, however continuous assessment of lesions may provide a better clinical understanding of disease progression.[43]

Pathologic myopia

High myopia, or pathologic myopia, is typically defined as a refractive error of greater than or equal to negative 6 diopters.[40] Although this refractive error can be corrected to optimize visual acuity, these patients are left at increased risk of complications including retinal detachments, maculopathy, and optic neuropathy. As many abnormalities associated with myopia are present in the retinal periphery, UWF imaging can be a valuable adjunct in screening for these complications.[51] UWF imaging has also proven useful in the assessment of posterior staphylomas, as this allows visualization of the full curvature of the retina.[52,53]

Ultra-widefield pseudocolor imaging can detect and identify peripheral lesions such as retinal tears and lattice degeneration. Studies have thus far reported low sensitivity in detecting retinal breaks (33%–57%) and lattice degeneration (66%) in myopic patients, due in part to incomplete visualization of the retina.[51] This highlights the importance of obtaining imaging of the entire UWF view of the retina and making efforts to reduce or eliminate imaging artifacts such as eyelashes. As these pathologies carry an increased risk of retinal detachment, it is imperative that they are detected and treated prophylactically, if appropriate. UWF-OCT of these areas of pathology can inform management decisions with additional information regarding vitreous attachments and traction as well as presence of subretinal fluid.[51]

The presence of posterior staphyloma has been linked to poorer visual outcomes in myopic patients. Although they can be detected on clinical examination, studies utilizing UWF fundus images in combination with 3D-MRI were able to further characterize and classify these staphylomas into six types.[52,53] Given the limited availability of MRI, this is not a viable screening approach. However, most staphylomas can now be documented with SS-OCT 16 × 14 mm scans, a modality more amenable to screening purposes. The presence of choroidal thinning on OCT may also denote the staphyloma edge.[52] UWF-OCT of dome-shaped macula have proven useful in determining that they are distinct in their pathophysiology from posterior staphylomas.[54]

On UWF-FA, areas of nonperfusion in the far-periphery of the retina are common in those with pathologic myopia (83%–5% in emmetropic eyes) although the clinical significance of these areas of nonperfusion remains unclear.[51]

References

1. Choudhry N, Duker JS, Freund KB, et al. Classification and guidelines for widefield imaging: recommendations from the international widefield imaging study group. *Ophthalmology Retina*. 2019;3(10):843–849. https://doi.org/10.1016/j.oret.2019.05.007.
2. Fogel-Levin M, Sadda SVR, Rosenfeld PJ, et al. Advanced retinal imaging and applications for clinical practice: a consensus review. *Survey of Ophthalmology*. 2022;67:1373–1390. https://doi.org/10.1016/j.survophthal.2022.02.004.
3. Aiello LP, Odia I, Glassman AR, et al. Comparison of early treatment diabetic retinopathy study standard 7-field imaging with ultrawide-field imaging for determining severity of diabetic retinopathy. *Journal of the American Medical Association Ophthalmology*. 2019;137(1):65–73. https://doi.org/10.1001/jamaophthalmol.2018.4982.
4. Silva PS, Cavallerano JD, Haddad NMN, et al. Peripheral lesions identified on ultrawide field imaging predict increased risk of diabetic retinopathy progression over 4 years. *Ophthalmology*. 2015;122(5):949–956. https://doi.org/10.1016/j.ophtha.2015.01.008.

5. Cui Y, Zhu Y, Wang JC, et al. Comparison of widefield swept-source optical coherence tomography angiography with ultra-widefield color fundus photography and fluorescein angiography for detection of lesions in diabetic retinopathy. *British Journal of Ophthalmology*. 2021;105(4):577. https://doi.org/10.1136/BJOPHTHALMOL-2020-316245.
6. Arya M, Sorour O, Chaudhri J, et al. Distinguishing intraretinal microvascular abnormalities from retinal neovascularization using optical coherence tomography angiography. *International Journal of Retina and Vitreous*. 2019:1–10. https://doi.org/10.1097/IAE.0000000000002671.
7. Pichi F, Smith SD, Abboud EB, et al. Wide-field optical coherence tomography angiography for the detection of proliferative diabetic retinopathy. *Graefe's Archive for Clinical and Experimental Ophthalmology*. 2020;258:1901–1909. https://doi.org/10.1007/s00417-020-04773-x.
8. Khalid H, Schwartz R, Nicholson L, et al. Widefield optical coherence tomography angiography for early detection and objective evaluation of proliferative diabetic retinopathy. *British Journal of Ophthalmology*. 2021;105(1):118–123. https://doi.org/10.1136/bjophthalmol-2019-315365.
9. Wang FP, Saraf SS, Zhang Q, Wang RK, Rezaei KA. Ultra-widefield protocol enhances automated classification of diabetic retinopathy severity with OCT angiography. *Ophthalmology Retina*. 2020;4(4):415–424. https://doi.org/10.1016/j.oret.2019.10.018.
10. Or C, Sabrosa AS, Sorour O, Arya M, Waheed N. Use of OCTA, FA, and ultra-widefield imaging in quantifying retinal ischemia: a review. *Asia-Pacific Journal of Ophthalmology*. 2018;7(1):46–51. https://doi.org/10.22608/APO.201812.
11. Kadomoto S, Muraoka Y, Uji A, et al. Nonperfusion area quantification in branch retinal vein occlusion: a widefield optical coherence tomography angiography study. *Retina*. 2021;41:1210–1218. https://doi.org/10.1097/IAE.0000000000002999.
12. Choi W, Waheed NK, Moult EM, et al. Ultrahigh speed OCT angiography of retinal and choriocapillaris alterations in diabetic patients with and without retinopathy using swept source optical coherence tomography. *Retina*. 2017;37(1):11–21. https://doi.org/10.1097/IAE.0000000000001250.
13. Nicolini N, Tombolini B, Barresi C, et al. Assessment of diabetic choroidopathy using ultra-widefield optical coherence tomography. *Translational Vision Science & Technology*. 2022;11(3):35. https://doi.org/10.1167/tvst.11.3.35.
14. Yang J, Zhang B, Wang E, Xia S, Chen Y. Ultra-wide field swept-source optical coherence tomography angiography in patients with diabetes without clinically detectable retinopathy. *BMC Ophthalmology*. 2021;21(1):1–8. https://doi.org/10.1186/s12886-021-01933-3.
15. Rosen RB, Andrade Romo JS, Krawitz BD, et al. Earliest evidence of preclinical diabetic retinopathy revealed using optical coherence tomography angiography perfused capillary density. *American Journal of Ophthalmology*. 2019;203:103–115. https://doi.org/10.1016/J.AJO.2019.01.012.
16. Stanga PE, Pastor-Idoate S, Reinstein U, et al. Navigated single-capture 3D and cross-sectional wide-field OCT of the mid and peripheral retina and vitreoretinal interface. *European Journal of Ophthalmology*. 2022;32(3):1642–1651. https://doi.org/10.1177/11206721211026100.
17. Nagasato D, Tabuchi H, Masumoto H, et al. Automated detection of a nonperfusion area caused by retinal vein occlusion in optical coherence tomography angiography images using deep learning. *PLoS One*. 2019;14(11):1–14. https://doi.org/10.1371/journal.pone.0223965.
18. Nagasato D, Tabuchi H, Ohsugi H, et al. Deep neural network-based method for detecting central retinal vein occlusion using ultrawide-field fundus ophthalmoscopy. *Journal of Ophthalmology*. 2018;2018. https://doi.org/10.1155/2018/1875431.
19. Sato A, Asaoka R, Tanaka S, et al. Using ultra-widefield red channel images to improve the detection of ischemic central retinal vein occlusion. *PLoS One*. 2021;16(11 November):1–7. https://doi.org/10.1371/journal.pone.0260383.
20. Tang W, Liu W, Guo J, et al. Wide-field swept-source OCT angiography of the periarterial capillary-free zone before and after anti-VEGF therapy for branch retinal vein occlusion. *Eye and Vision*. 2022;9(1):1–10. https://doi.org/10.1186/s40662-022-00297-z.
21. Kumar V. Ultra-wide field retinal imaging: a wider clinical perspective. *Indian Journal of Ophthalmology*. 2021;69(4):824–835. https://doi.org/10.4103/ijo.IJO.
22. Forshaw TRJ, Minör ÅS, Subhi Y, Sørensen TL. Peripheral retinal lesions in eyes with age-related macular degeneration using ultra-widefield imaging: a systematic review with meta-analyses. *Ophthalmology Retina*. 2019;3(9):734–743. https://doi.org/10.1016/j.oret.2019.04.014.
23. Domalpally A, Clemons TE, Danis RP, et al. Peripheral retinal changes associated with age-related macular degeneration in the age-related eye disease study 2: age-related eye disease study 2 report number 12 by the age-related eye disease study 2 Optos PEripheral RetinA (OPERA) study research Group. *Ophthalmology*. 2017;124(4):479–487. https://doi.org/10.1016/j.ophtha.2016.12.004.
24. Sodhi SK, Golding J, Trimboli C, Choudhry N. Feasibility of peripheral OCT imaging using a novel integrated SLO ultra-widefield imaging swept-source OCT device. *International Ophthalmology*. 2021;41(8):2805–2815. https://doi.org/10.1007/S10792-021-01837-7/FIGURES/3.
25. Kurobe R, Hirano Y, Ogura S, Yasukawa T, Ogura Y. Ultra-widefield swept-source optical coherence tomography findings of peripheral retinal degenerations and breaks. *Clinical Ophthalmology*. 2021;15:4739–4745. https://doi.org/10.2147/OPTH.S350080.
26. Kovacs KD, Mahrous MA, Gonzalez L, et al. Feasibility and clinical utility of ultra-widefield–navigated swept-source optical coherence tomography imaging. *Journal of VitreoRetinal Diseases*. 2021;5(5):396–404. https://doi.org/10.1177/2474126421997335.
27. Ness S, Subramanian ML, Chen X, Siegel NH. Diagnosis and management of degenerative retinoschisis and related complications. *Survey of Ophthalmology*. 2022;67:892–907. https://doi.org/10.1016/j.survophthal.2021.12.004.
28. Cheung R, Ly A, Katalinic P, et al. Visualization of peripheral retinal degenerations and anomalies with ocular imaging. *Seminars in Ophthalmology*. 2022;37(5):554–582. https://doi.org/10.1080/08820538.2022.2039222.
29. Tsai CY, Hung KC, Wang SW, Chen MS, Ho TC. Spectral-domain optical coherence tomography of peripheral lattice degeneration of myopic eyes before and after laser photocoagulation. *Journal of the Formosan Medical Association*. 2019;118:679–685. https://doi.org/10.1016/J.JFMA.2018.08.005.
30. Navaratnam J, Salvanos P, Vavvas DG, Bragadottir R. Ultra-widefield autofluorescence imaging findings in retinoschisis, rhegmatogenous retinal detachment and combined retinoschisis retinal detachment. *Acta Ophthalmologica*. 2021;99:195–200. https://doi.org/10.1111/aos.14521.
31. Tripathy K, Chawla R, Wadekar BR, Venkatesh P, Sharma YR. Evaluation of rhegmatogenous retinal detachments using Optos ultrawide field fundus fluorescein angiography and comparison with ETDRS 7 field overlay. *Journal of Current Ophthalmology*. 2018;30(3):263–267. https://doi.org/10.1016/j.joco.2018.06.006.

32. Lee WW, Muni RH. Single-capture ultra-widefield guided swept-source optical coherence tomography in the management of rhegmatogenous retinal detachment and associated peripheral vitreoretinal pathology. *British Journal of Ophthalmology*. 2022;0:1−7. https://doi.org/10.1136/bjophthalmol-2021-320149.
33. Testi I, Modugno RL, Pavesio C. Multimodal imaging supporting the pathophysiology of white dot syndromes. *Journal of Ophthalmic Inflammation and Infection*. 2021;11(32):1−7. https://doi.org/10.1186/S12348-021-00261-3.
34. Pradas M, Rodriguez-Merchante MP, Estébanez N, et al. Navigating the white dot syndromes with optical coherence tomography (OCT) and OCT angiography (OCT-A). *Ocular Immunology and Inflammation*. 2022;30(3):1−11. https://doi.org/10.1080/09273948.2022.2046798.
35. Pichi F, Sarraf D, Morara M, Mazumdar S, Neri P, Gupta V. Pearls and pitfalls of optical coherence tomography angiography in the multimodal evaluation of uveitis. *Journal of Ophthalmic Inflammation and Infection*. 2017;7(20):1−12. https://doi.org/10.1186/S12348-017-0138-Z.
36. Pichi F, Srvivastava SK, Chexal S, et al. En face optical coherence tomography and optical coherence tomography angiography of multiple evanescent white dot syndrome: new insights into pathogenesis. *Retina*. 2016;36(12):S178−S188. https://doi.org/10.1097/IAE.0000000000001255.
37. Shields CL, Dalvin LA, Ancona-Lezama D, et al. Choroidal nevus imaging features in 3,806 cases and risk factors for transformation into melanoma in 2,355 cases: the 2020 Taylor R. Smith and victor T. Curtin lecture. *Retina*. 2019;39(10):1840−1851. https://doi.org/10.1097/IAE.0000000000002440.
38. Shields CL, Salazar PF, Mashayekhi A, Shields JA. Peripheral exudative hemorrhagic chorioretinopathy simulating choroidal melanoma in 173 eyes. *Ophthalmology*. 2009;116(3):529−535. https://doi.org/10.1016/j.ophtha.2008.10.015.
39. Heimann H, Jmor F, Damato B. Imaging of retinal and choroidal vascular tumours. *Eye*. 2013;27(2):208−216. https://doi.org/10.1038/eye.2012.251.
40. House R, Hsu ST, Thomas A, et al. Vascular findings in a small retinoblastoma tumor using OCT angiography. *Ophthalmology Retina*. 2019;3(2):194−195. https://doi.org/10.1016/j.oret.2018.09.018.Vascular.
41. Chawla H, Vohra V. Retinal dystrophies. In: *StatPearls*. StatPearls Publishing; 2023. http://www.ncbi.nlm.nih.gov/books/NBK564379/.
42. Hood DC, Ramachandran R, Holopigian K, Lazow M, Birch DG, Greenstein VC. Method for deriving visual field boundaries from OCT scans of patients with retinitis pigmentosa. *Biomedical Optics Express*. 2011;2(5):1106−1114. https://doi.org/10.1364/BOE.2.001106.
43. Cicinelli MV, Marchese A, Bordato A, Manitto MP, Bandello F, Battaglia Parodi M. Reviewing the role of ultra-widefield imaging in inherited retinal dystrophies. *Ophthalmology Therapy*. 2020;9(2):249−263. https://doi.org/10.1007/s40123-020-00241-1.
44. Klufas MA, Tsui I, Sadda SR, Hosseini H, Schwartz SD. Ultrawidefield autofluoresence in ABCA4 stargardt disease. *Retina*. 2018;38(2):403−415. https://doi.org/10.1097/IAE.0000000000001567.
45. Oishi A, Miyata M, Numa S, Otsuka Y, Oishi M, Tsujikawa A. Wide-field fundus autofluorescence imaging in patients with hereditary retinal degeneration: a literature review. *International Journal of Retina and Vitreous*. 2019;5(Suppl 1):23. https://doi.org/10.1186/s40942-019-0173-z.
46. Oishi M, Oishi A, Ogino K, et al. Wide-field fundus autofluorescence abnormalities and visual function in patients with cone and cone-rod dystrophies. *Investigative Ophthalmology & Visual Science*. 2014;55(6):3572−3577. https://doi.org/10.1167/iovs.14-13912.
47. Kothari N, Pineles S, Sarraf D, et al. Clinic-based ultra-wide field retinal imaging in a pediatric population. *International Journal of Retina and Vitreous*. 2019;5(Suppl 1):21. https://doi.org/10.1186/s40942-019-0171-1.
48. Lee J, Asano S, Inoue T, et al. Investigating the usefulness of fundus autofluorescence in retinitis pigmentosa. *Ophthalmology Retina*. 2018;2(10):1062−1070. https://doi.org/10.1016/j.oret.2018.03.007.
49. Ogura S, Yasukawa T, Kato A, et al. Wide-field fundus autofluorescence imaging to evaluate retinal function in patients with retinitis pigmentosa. *American Journal of Ophthalmology*. 2014;158(5):1093−1098. https://doi.org/10.1016/j.ajo.2014.07.021.
50. Hariri AH, Gui W, Datoo O'Keefe GA, Ip MS, Sadda SR, Gorin MB. Ultra-widefield fundus autofluorescence imaging of patients with retinitis pigmentosa: a standardized grading system in different genotypes. *Ophthalmology Retina*. 2018;2(7):735−745. https://doi.org/10.1016/j.oret.2017.10.018.
51. Ludwig CA, Moon J, Garg I, Miller JB. Ultra-widefield imaging for evaluation of the myopic eye. *Seminars in Ophthalmology*. 2021;36(4):185−190. https://doi.org/10.1080/08820538.2021.1887904.
52. Shinohara K, Shimada N, Moriyama M, et al. Posterior staphylomas in pathologic myopia imaged by widefield optical coherence tomography. *Investigative Ophthalmology & Visual Science*. 2017;58(9):3750−3758.
53. Nakao N, Igarashi-Yokoi T, Takahashi H, Xie S, Shinohara K, Ohno-Matsui K. Quantitative evaluations of posterior staphylomas in highly myopic eyes by ultra-widefield optical coherence tomography. *Investigative Ophthalmology & Visual Science*. 2022;63(8):1−8. https://doi.org/10.1167/IOVS.63.8.20.
54. Saito R, Shinohara K, Tanaka N, Takahashi H, Yoshida T, Ohno-Matsui K. Association between dome-shaped macula and posterior staphyloma in highly myopic eyes investigated by ultra-widefield optical coherence tomography. *International Journal of Retina and Vitreous*. 2021;41(3):646−652. https://doi.org/10.1097/IAE.0000000000002889.

Chapter 11

Upcoming invasive and noninvasive imaging modalities

Julia Fallon, Sofia Ahsanuddin, Nilesh Raval, Toco Chui and Richard Rosen
New York Eye and Ear Infirmary of Mount Sinai, Department of Ophthalmology, New York, NY, United States

Laser Doppler holography

In vivo imaging of retinal and choroidal blood flow is achievable using laser Doppler holography.[1] In this imaging technique, near-infrared light along with fast CMOS (complementary metal-oxide-semiconductor) cameras are used to obtain dynamics of vascular flow. This technique can map the variations in flow between retinal arterial and venous networks corresponding with the systolic and diastolic phases of the heart.[2] It offers enhanced resolution of temporal indices of blood flow compared to the snapshot of optical coherence tomography angiography (OCT-A). Using spatio-temporal filtering, Puyo and colleagues through early work have demonstrated the ability to differentiate the retinal and choroidal vasculatures.[2] Compared to indocyanine green (ICG) and OCT-A, these images also reveal more details of choroidal vascular network dynamics than previously available, as highlighted in Fig. 11.1.[3] Efforts to minimize the effects of global eye movement on measurements, especially in conditions of reduced blood flow, such as pathologic vascular states, may have important diagnostic value.[4]

Laser Doppler holography holds immense promise in providing quantitative measurements of retinal blood flow throughout the cardiac cycle offering unprecedented insight into both normal physiology and systemic hemodynamic abnormalities associated with retinal diseases. Recent studies on healthy eyes have consistently demonstrated continuous anterograde flow throughout the cardiac cycle. In contrast, diseased eyes affected by conditions such as glaucoma or central retinal vein occlusion exhibit significant reductions and even reversals in arterial flow within the retinal vasculature.[5] Ongoing efforts are focused on identifying quantitative biomarkers that can effectively grade pathologies and assess the impact of treatment. Considering the prevalence of impaired and pathologic blood flow in various retinal conditions, the application of laser Doppler holography has the potential to enhance diagnosis and effective monitoring of treatment for many disease states.

Adaptive optics

Adaptive optics (AO) is a noninvasive imaging technique that measures wavefront distortion errors and corrects them with a compensating mechanism, most commonly a deformable mirror.[6] It has found its way into common use in telescopes for removing atmospheric aberrations, enabling diffraction-limited imaging.[7–9] A wavefront sensor measures the optical distortions and provides this information to the deformable mirror that provides compensatory corrections. Wavefront distortions typically present as either low- or high-order aberrations. Low-order aberrations may be compensated for by using tip-tilt correction, which uses a rapidly moving tip-tilt mirror with a large correcting power that functions in two dimensions.[9] Higher order aberrations are compensated for by a deformable mirror, which can correct distortion errors with multiple degrees of freedom.[6] The number of actuators, or devices that physically move the deformable mirror, determines the number of degrees of freedom the mirror is capable of correcting.

AO has recently found its way into ophthalmic imaging through incorporation into scanning laser ophthalmoscopes (SLOs), which provide ultra-high-resolution in vivo imaging of retinal photoreceptors and vasculature.[6,10] SLO uses a confocal laser with horizontal and vertical scanning mirrors to scan the retina. However, due to optical aberrations created

Retinal and Choroidal Vascular Diseases of the Eye. https://doi.org/10.1016/B978-0-443-15583-3.00011-1

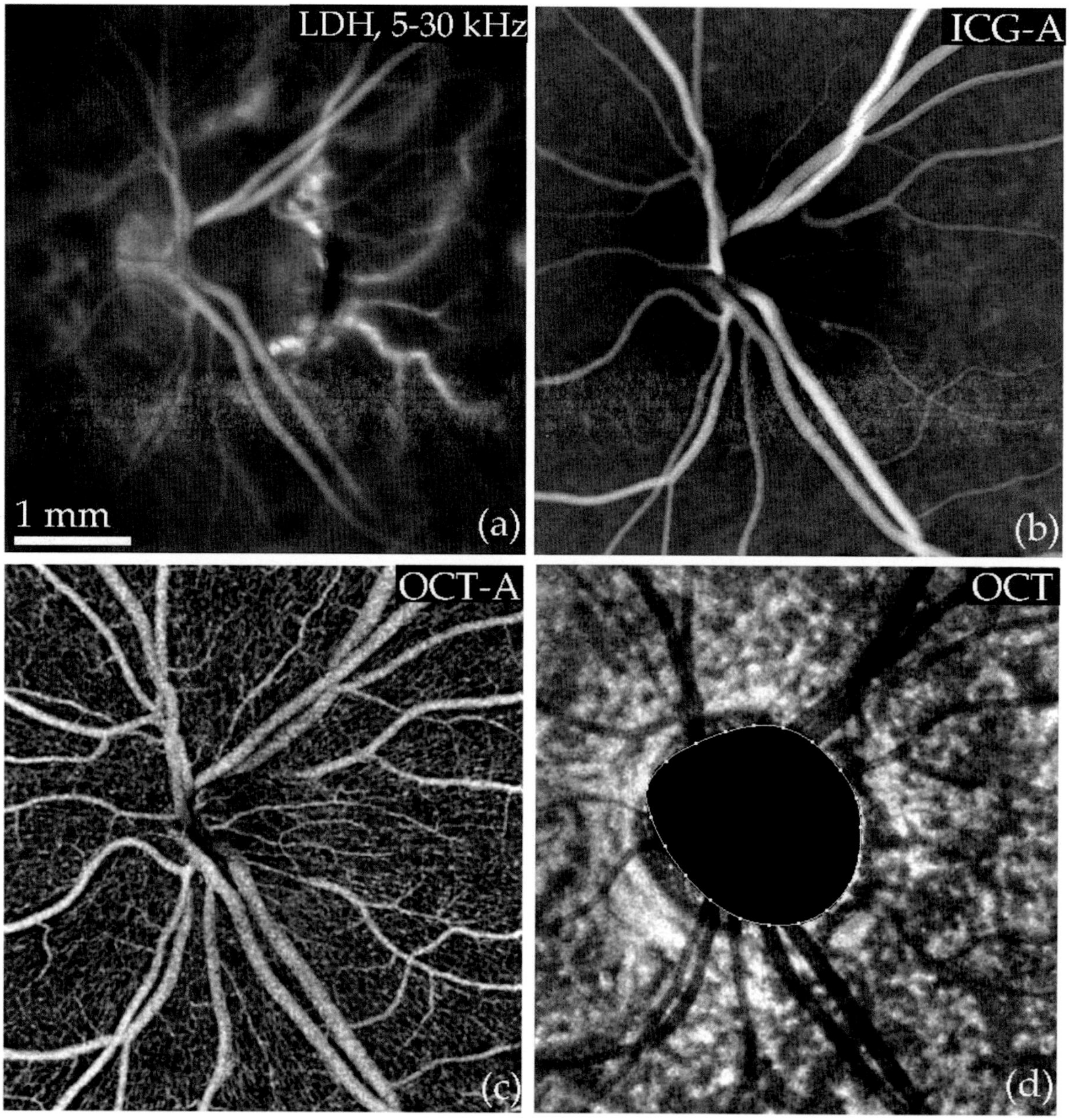

FIGURE 11.1 Angiographic images of the optic nerve head (ONH) with different imaging methodologies. (A) Laser Doppler holography (LDH) reveals the retinal vasculature and deeper choroidal arteries originating from para optic posterior ciliary arteries (SPCAs). (B) A late ICG-A (Heidelberg—Spectralis) reveals the retinal vessels. (C and D) OCT-A image of the retinal layer and OCT image of the choroid without the choriocapillaris (Optovue, Avanti with AngioVue), respectively. *Reproduced with permission by Dr. Leo Puyo and Dr. Michael Atlan from Puyo L, Paques M, Fink M, Sahel JA, Atlan M. Choroidal vasculature imaging with laser Doppler holography.* Biomedical Optics Express *January 31, 2019;10(2):995–1012.*

by anterior segment structures such as the cornea, anterior chamber, and lens, along with natural eye movements, the incoming wavefront is distorted.[6,10] The deformable mirror corrects these aberrations in real time and allows for high-resolution in vivo retinal imaging. The increased resolution from adaptive optics scanning light ophthalmoscopy (AOSLO) allows for the assessment of the spatial distribution and density of single cells, for example, foveal cone cells.[6,10,11] Migacz et al. demonstrated the ability to monitor changes in a single hyalocyte of the vitreous cortex using AOSLO.[12] As demonstrated in Fig. 11.2, the subtle morphological and spatial changes in a hyalocyte can be monitored over time using AOSLO. Although AOSLO is not currently widely used in the clinical setting, it has the potential to be used for phenotype tracking in inherited retinal diseases such as cone dystrophies or retinitis pigmentosa.[10,11]

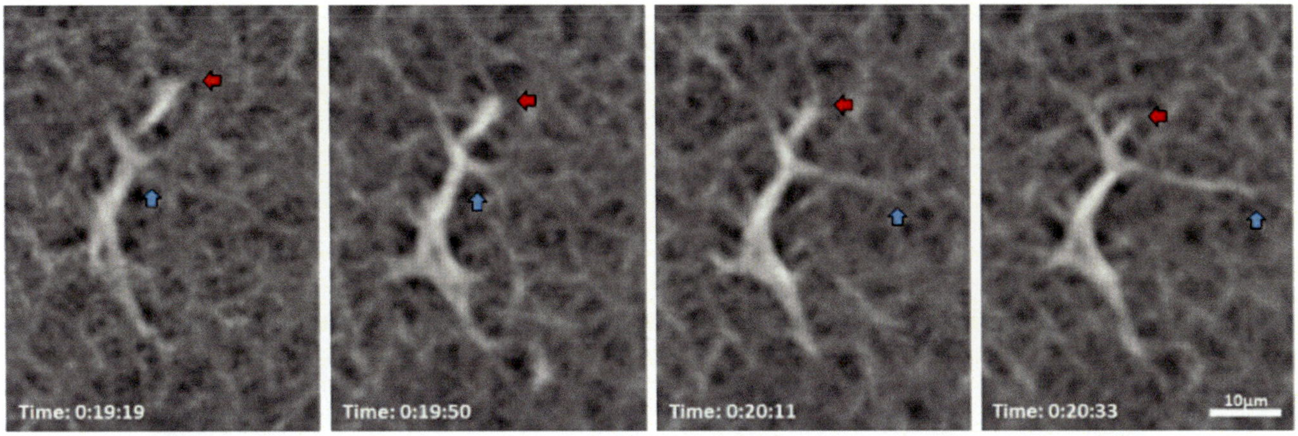

FIGURE 11.2 Hyalocytes with AOSLO images demonstrate a hyalocyte with extension (cyan arrows) and retraction (red arrows) of processes over 74 s in a 32-year-old healthy male subject. Time of acquisition of each image is displayed in hours:minutes:seconds. *Reproduced with permission by Dr. Richard Rosen and Dr. Toco Y. P. Chui from Migacz JV, Otero-Marquez O, Zhou R, Rickford K, Murillo B, Zhou DB, Castanos MV, Sredar N, Dubra A, Rosen RB, Chui TYP. Imaging of vitreous cortex hyalocyte dynamics using nonconfocal quadrant-detection adaptive optics scanning light ophthalmoscopy in human subjects.* Biomedical Optics Express *March 1, 2022;13(3):1755–1773.*

Furthermore, focal vascular lesions, such as microaneurysms and intraretinal microvascular abnormalities, can be imaged over time to track changes in these lesions at high resolution. Pinhas et al. describe the use of AOSLO in conjunction with fluorescein angiography showing the ability to identify microscopic detail of the retinal capillary beds.[13] In sickle cell retinopathy, AOSLO was able to capture unique histologic aspects of the disease including stacks of cells in rouleaux formation and even single-sickled red blood cells, as shown in Fig. 11.3.[14]

AO in conjunction with forms of invasive and noninvasive angiography has shown useful in imaging of the choriocapillaris.[15,16] Kurokawa et al. describe using AO-OCT-A to identify individual capillaries of the choroid, a task previously difficult due to its highly interconnected and dense nature.[15] Jung et al. describe utilizing AO-indocyanine green angiography (ICG-A) to measure the nature and interaction of the retinal pigment epithelium and choriocapillaris, which could hold major implications in monitoring disease states as well as the dynamics of the blood–brain barrier.[16]

Optoretinography

Optoretinography is an emerging technology that gives us the ability to analyze retinal function at a cellular level. Detecting the physical changes a photoreceptor undergoes during the phototransduction cascade, optoretinography is able to capture and measure the functional capacity of the photoreceptor.[17,18] In contrast to an electroretinogram, the optoretinogram measures the cones' optical properties and physical changes when exposed to light rather than electrical changes. Most described optoretinograms in the literature are constructed using AO imaging in conjunction with optical coherence tomography (OCT), an SLO or flood illumination ophthalmoscopes.[19–23]

Lassoued et al. utilized AO and phase-sensitive OCT to measure cone photoreceptor function in retinitis pigmentosa patients.[19] With this imaging technique, they can analyze both the structure and function (phototransduction ability) of a single cone, facilitating differentiation of long, medium, and short wavelength cones. Single cone functional resolution will enable earlier detection of disease and monitoring of treatment response, potentially useful for gene therapy.

Utilizing a different method, Jonnal et al. described AO-enabled flood illumination for measuring scintillations of cones that correspond to phototransduction as measured by ERG.[20] ERG, however, measures electrical changes rather than the physical changes detected by optoretinography. Optoretinography, in contrast, has the sensitivity as well as spatial and temporal resolution to identify changes at the cellular level, as demonstrated in Fig. 11.4.

Recently, Cooper et al. described the use of AOSLO, similar to the work described above by Lassoued et al. with AO OCT., to help differentiate between different types of cones.[21] This work shows cone response as a function of stimulus intensity. With the ability to measure retinal function on a cellular level, the optoretinogram promises to enhance the understanding and treatment of retinal disease, including pathophysiology, early detection, discovery of new therapies, and monitoring response to treatment.

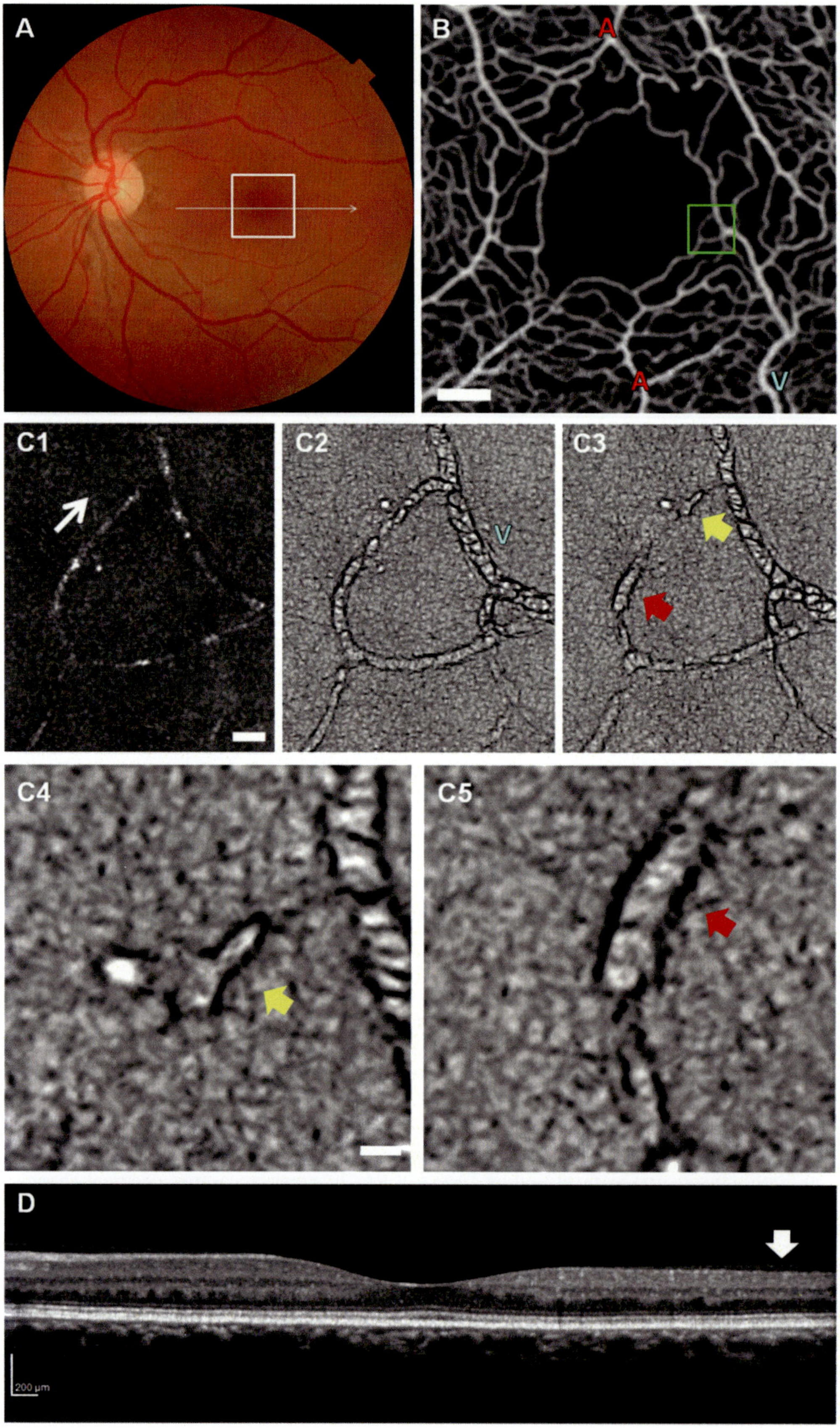
A
B
A
A
V
C1
C2
V
C3
C4
C5
D
200 μm

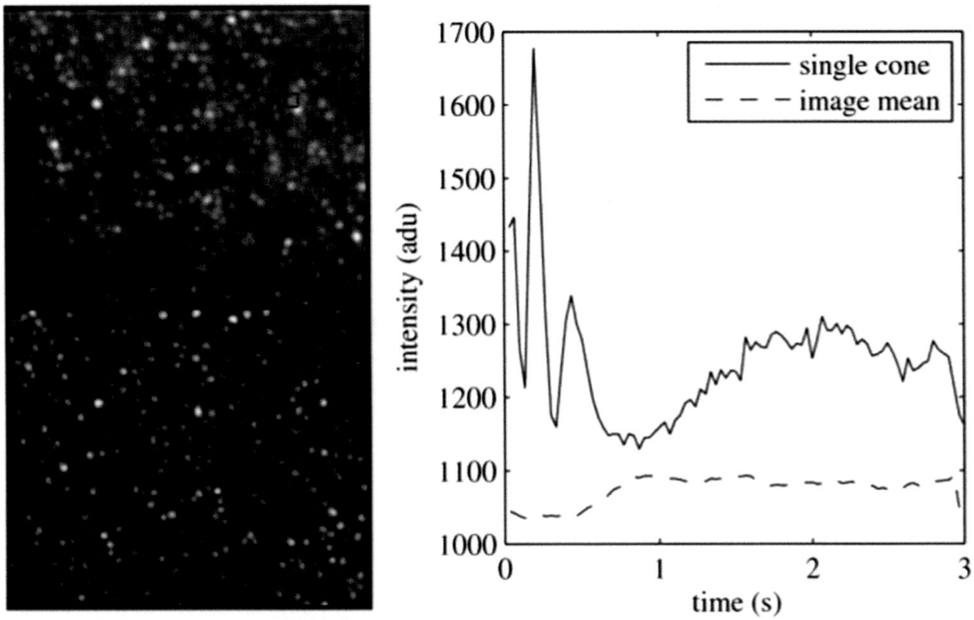

FIGURE 11.4 Cone imaging on a cellular level. Cone imaging results using the same 670 nm laser diode (Lc = 144 μm) to stimulate and image the retina over a 3-s duration. The *top left* image represents the registering and coadding of 90 images acquired at 30 Hz of the same patch of retina. The *bottom left* image is the time-RMS across the same data set. The plot on the right traces the scintillation of a single cone (*solid blue line*) whose location in the cone mosaic (*left*) is indicated by the *blue boxes*. Also shown is the average intensity of the entire image (*dashed line*). *Reproduced with permission by Dr. Ravi S. Jonnal from Jonnal RS, Rha J, Zhang Y, Cense B, Gao W, Miller DT. In vivo functional imaging of human cone photoreceptors.* Opticals in Express *November 26, 2007;15(24):16141–16160.*

Flavoprotein fluorescence imaging

Mitochondria are integral to cellular energy generation and are particularly sensitive to oxidative stress. Mitochondrial dysfunction has been previously implicated in the pathophysiology of many retinal disease states including diabetic retinopathy and age-related macular degeneration.[24] Flavoproteins are enzymes within the electron transport chain of mitochondria that emit a green fluorescence in an oxidized state when stimulated by blue light. Flavoprotein fluorescence (FPF), therefore, can serve as a marker of mitochondrial stress.[25] The spectral emission and excitation properties of flavoproteins are distinct from other retinal fluorophores such as lipofuscin, allowing researchers to isolate mitochondrial FPF using specialized band filters.[25] FPF imaging detects oxidative stress at a subcellular level, providing important insights into the pathophysiology and treatment efficacy in a variety of common retinal diseases.

Multiple studies have looked at FPF in diabetic patients using two different parameters: FPF intensity and FPF heterogeneity.[26–28] The former reflects the average pixel intensity over a 17 degrees × 21.5 degrees region centered at a select region of interest (ROI) and the latter captures the variation of signal intensity of all the pixels at the same location. Elner et al. examined FPF at the macula of healthy controls compared to diabetic patients with and without retinopathy.[26] They found that control patients had significantly lower levels of FPF compared to diabetic patients without retinopathy, who in turn had significantly lower levels of fluorescence compared to diabetic patients with retinopathy. Using cell cultures, this study also showed that antioxidants can block FPF following exposure of a cell to an oxidative stressor which induces apoptosis. Field et al. published similar results showing that even when controlling for age, diabetics exhibited

FIGURE 11.3 AO in sickle cell retinopathy. (A) Standard color fundus photograph. Line represents horizontal OCT scan through fovea. *White box* localizes the parafoveal area imaged with OCT angiography (OCT-A). (B) Averaged OCT-A of 10 scans. Red A's label the associated arterioles, and blue V labels the associated venule. *Green box* localizes the region of interest (ROI) imaged with adaptive optics scanning light ophthalmoscopy (AOSLO). Scale bar represents 200 μm. (C1) Single frame from a confocal, and (C2, C3) two frames from a nonconfocal AOSLO imaging video. *White arrow* represents the direction of blood flow in the capillary segment of interest. *Red arrow* points to a red blood cell (RBC) rouleau. *Yellow arrow* points to a single-sickled RBC. Blue V labels the associated venule from (B). Scale bar represents 200 μm. (C4, C5) Zoomed-in views from (C3) highlighting morphological features of the torpedo-shaped single-sickled RBC (*yellow arrow*) and the RBC rouleau (*red arrow*). Scale bar represents 10 μm. (D) Macular OCT. *White arrow* indicates temporal macular thinning. *Reproduced with permission by Dr. Richard Rosen and Dr. Toco Y. P. Chui from Pinhas A, Migacz JV, Zhou DB, Castanos Toral MV, Otero-Marquez O, Israel S, Sun V, Gillette PN, Sredar N, Dubra A, Glassberg J, Rosen RB, Chui TYP. Insights into sickle cell disease through the retinal microvasculature: adaptive optics scanning light ophthalmoscopy correlates of clinical OCT Angiography.* Ophthalmology Science *July 12, 2022;2(4):100196.*

greater levels of FPF when compared to controls.[27] Andrade Romo et al. took this one step further and analyzed how FPF could have value in the management of patients with diabetic retinopathy. Measuring FPF in patients with diabetic macular edema before and after injection of intravitreal antivascular endothelial growth factor (anti-VEGF) agents, he found that FPF levels improved before significant change in central macular thickness (CMT). This finding suggests that FPF can be utilized as a marker of early response to treatment before structural change.[28] Fig. 11.5 shows pre- and postinjection FPF maps highlighting the stark difference in FPF before and after administration of anti-VEGF.

In addition to diabetic retinopathy, FPF imaging has also been used to quantify mitochondrial oxidative stress in age-related macular degeneration, glaucoma, and central serous chorioretinopathy (CSR).[29–32] Muste et al. examined FPF in 456 eyes and found that intermediate age-related macular degeneration, geographic atrophy, and neovascular age-related macular degeneration all had significantly increased FPF intensity compared to healthy age-matched controls.[29] Field et al. showed similar results with increased levels of FPF in a small group of dry, nonexudative age-related macular degeneration patients.[30] Zhou et al. more recently demonstrated that FPF can be used as a marker of mitochondrial dysfunction in the optic disks of primary open-angle glaucoma eyes (Fig. 11.6).[31] Moreover, Ahsanuddin et al. showed that FPF intensity is highly sensitive in detecting mitochondrial dysfunction in a variety of retinal pathologies including exudative AMD, retinal vein occlusion, diabetic retinopathy, and CSR (Fig. 11.7).[32] This correlation with disease status suggests that FPF could serve as a biomarker for disease activity and response to treatment.

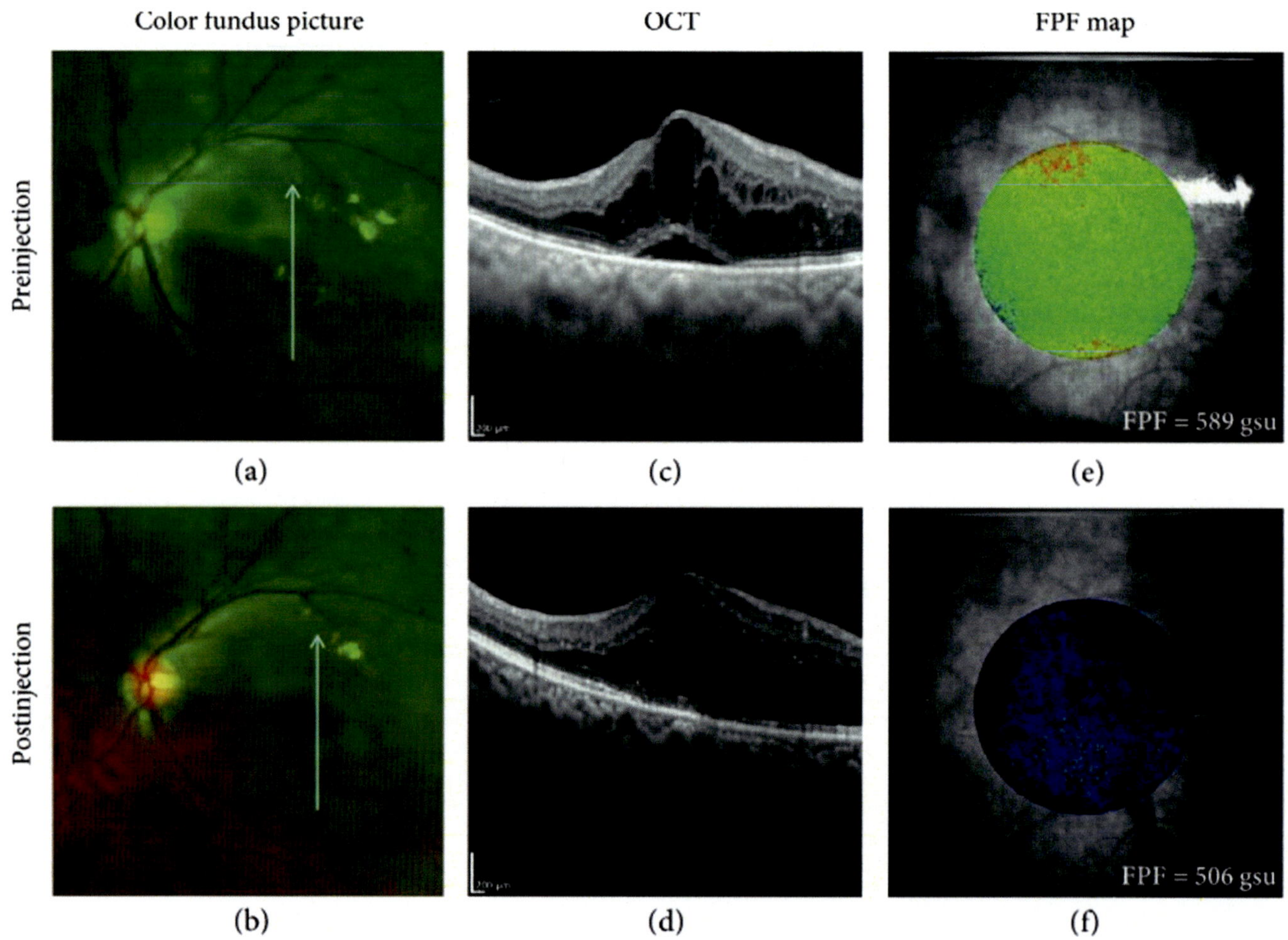

FIGURE 11.5 Flavoprotein fluorescence before and after anti-VEGF. (A, B) Color fundus photos illustrating macular edema, hard exudates, and hemorrhages in a patient with diabetic retinopathy. (C) OCT CMT before anti-VEGF administration (574 μm); (D) OCT CMT after anti-VEGF administration (541 μm). (E) FPF map preinjection shows an increase in *green-yellow* coloring, suggesting greater mitochondrial dysfunction (oxidative stress) compared to the (F) *blue-green* postinjection FPF map. *CMT*, central macular thickness; *FPF*, flavoprotein fluorescence; *GSU*, grayscale units; *OCT*, optical coherence tomography. *Reproduced with permission by Dr. Richard Rosen from Andrade Romo JS, Lynch G, Liu K, Kim D, Jansen M, Field MG, Elner VM, Rosen RB. Flavoprotein fluorescence correlation with visual acuity response in patients receiving anti-VEGF injection for diabetic macular Edema.* Oxidative Medicine and Cellular Longevity *August 9, 2018;2018:3567306.*

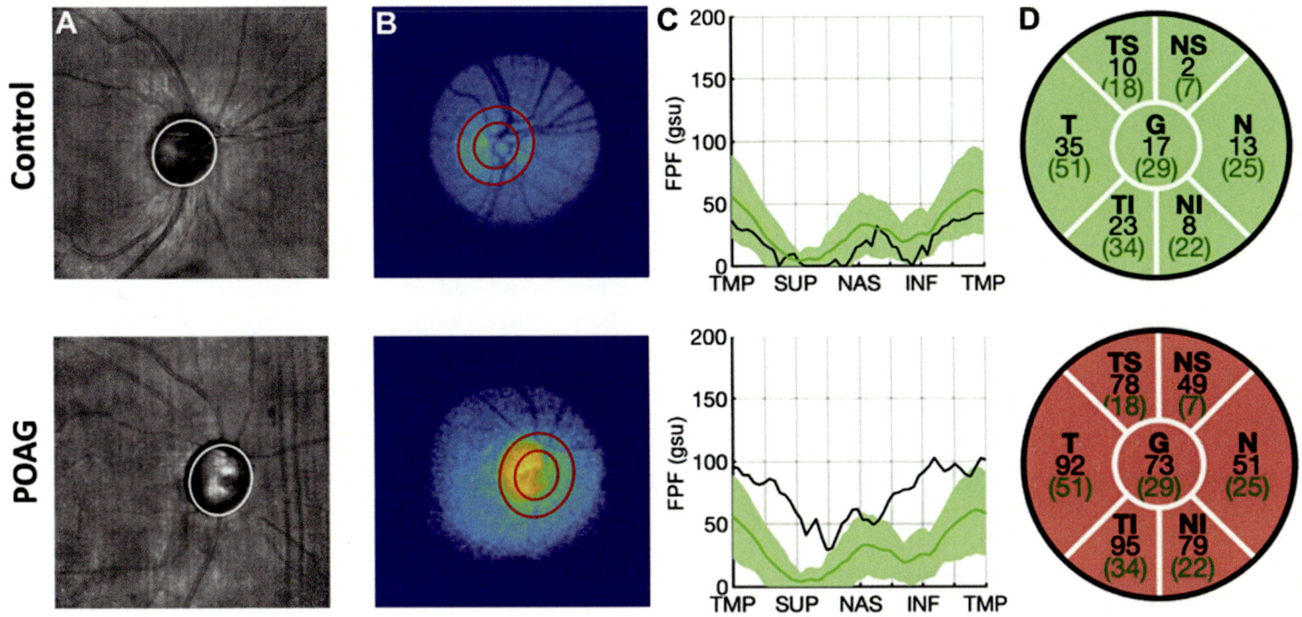

FIGURE 11.6 Flavoprotein fluorescence (FPF) measurements at the optic nerve head (ONH) from representative primary open-angle glaucoma (POAG) and control eyes. (A) 23 × 23 degrees infrared image of the optic nerve head (ONH) with the white ellipse indicating the algorithmically identified ONH, (B) FPF heatmap with the red circles highlighting the central circular FPF capture field spanning 13 degrees in diameter. FPF measurement at the ONH is obtained between the inner and outer red ellipses generated at ratiometric distances 0.5 and 1.1 times, respectively, of the white ONH ellipse perimeter. This annulus captures the highest concentration of FPF signal. (C) FPF sector maps, and (D) FPF intensity profile. Higher levels of FPF can be observed in the POAG eye compared with the healthy control eye on both the FPF sector map and profile. *gsu*, gray scale unit; *INF*, inferior; *N*, nasal; *NI*, nasal inferior; *NS*, nasal superior; *SUP*, superior; *T*, temporal; *TI*, temporal inferior; *TMP*, temporal; *TS*, temporal superior. *Reproduced with permission by Dr. Richard Rosen from Zhou DB, Castanos MV, Geyman L, et al. Mitochondrial dysfunction in primary open-angle glaucoma characterized by flavoprotein fluorescence at the optic nerve head.* Ophthalmologica Glaucoma *2022;5(4):413–420.*

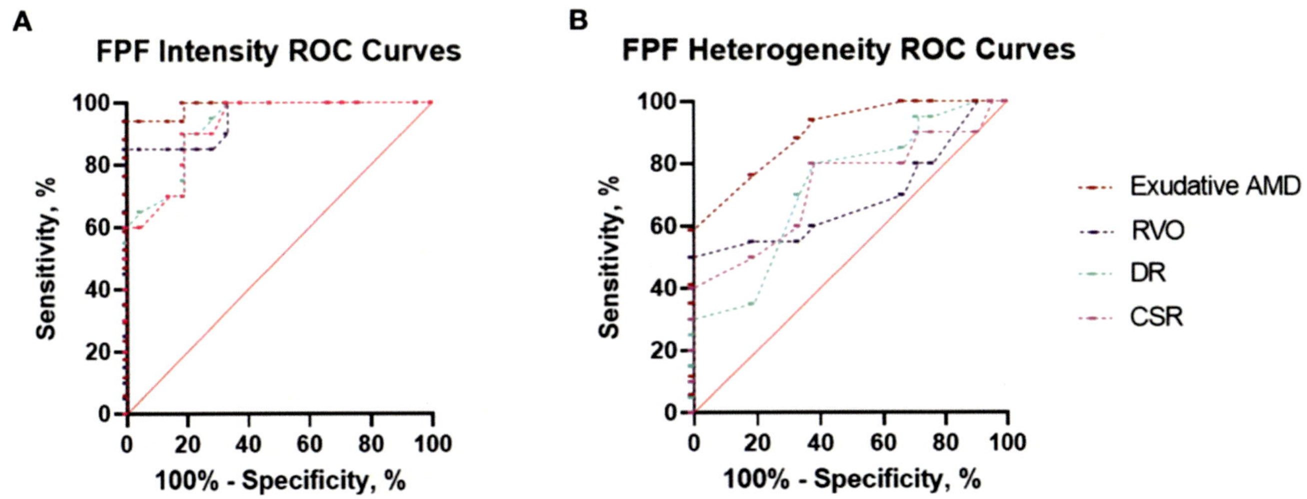

FIGURE 11.7 FPF in pathology. Area under the receiver operating characteristic (AUROC) curve analyses demonstrating the diagnostic ability of (A) FPF intensity and (B) FPF heterogeneity to discriminate between eyes with retinal pathology (exudative AMD, RVO, DR, and CSR) and healthy control eyes. *Reproduced with permission by Dr. Richard Rosen and Dr. Toco Y. P. Chui from Ahsanuddin S, Rios H, Otero-Marquez O, et al. Flavoprotein fluorescence elevation is a marker of mitochondrial oxidative stress in patients with retinal disease.* Frontiers in Ophthalmology *February 2023;3(4).*

Doppler optical coherence tomography

Since the advent of OCT in the 1990s, a myriad of technical extensions have been developed to improve its functional utility for quantifying in vivo blood flow dynamics in the posterior pole of the eye.[33] Given that many ocular disease states harbor vascular flow abnormalities,[34–36] quantitative blood flow measurement is an appealing strategy for further enhancing noninvasive visualization of tissue structure and dynamics in the living human eye. Simultaneous imaging of both structural abnormalities and blood flow could potentially provide critical information for the early detection and diagnosis of ocular diseases.[37]

One such functional extension of OCT is Doppler OCT, which is a technique that employs the Doppler effect to measure flow parameters of individual retinal arteries and veins. First introduced in 1997, Doppler OCT was first applied using a spectrogram method, which is based on the motion that backscattered light from a moving particle can produce a frequency shift.[38] Subsequent iterations of Doppler OCT focused on refining the phase-resolved method and Fourier-domain technique, the latter of which enables quantification of transverse flow and provides high spatial resolution, image acquisition speed, and velocity sensitivity.

Because phase-resolved D-OCT techniques are sensitive to the orientation and pulsatile nature of blood flow,[37] Doppler angle measurements can provide accurate characterization of blood flow parameters. A recently developed automated Doppler OCT measurement technique enables instantaneous Doppler angle measurement and retinal blood flow over 2 s, encompassing approximately two cardiac cycles (Fig. 11.8).[39,40] The flow velocity is derived from the Doppler shift caused by the moving blood, as indicated by the equation in Fig. 11.7C, where $\Delta\Phi$ is the phase difference at the same depth location between adjacent profiles after Fourier transform, z is the depth location, n is the refractive index of blood, λ is the center wavelength, t is the time interval between the adjacent profiles, and θ is the Doppler angle between the flow vector and the incident probe beam. To capture retinal blood flow measurements, a color fundus photo is taken and a vessel that is located around one disk diameter from the optic nerve head is selected (Fig. 11.9).[39] The scan location is perpendicular to the selected vessel, and the transverse scan length is 0.9 mm with 1024 A-lines. Doppler blood flow measurements were performed where the Doppler angle (θ) is significantly less than 90 degrees (i.e., between 78 and 83 degrees).

Imaging vitreous cortex hyalocytes (VCHs) using clinical en face OCT

Recent technological advances in OCT and AOSLO have enabled in vivo visualization of macrophages resident in the vitreous above the human retina, so-called hyalocytes.[12] Hyalocytes play essential roles in immune modulation, tissue repair, and neuroprotection.[41,42] Under normal physiological circumstances, hyalocytes synthesize vitreous hyaluronan and

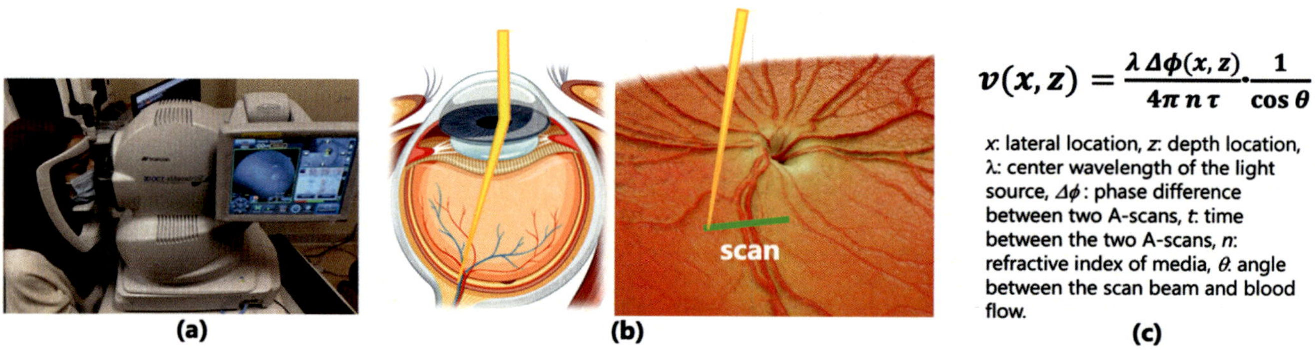

FIGURE 11.8 Prototype Maestro D-OCT flowmeter and schematic images of RBF calculation. (A) Prototype device set-up. (B) Oblique incident angle against the target vessels is achieved by shifting the scan beam away from the pupil center. OCT beam is scanned across the vessel(s) repeatedly. Doppler angle estimating routine is implemented followed by blood flow measurement scans for 2 s. (C) Pixel-by-pixel intensity is calculated from the obtained phase signal and the Doppler angle with the formula shown here. *Photo courtesy of Masahiro Akiba and Kana Minamide, Topcon Corporation.*

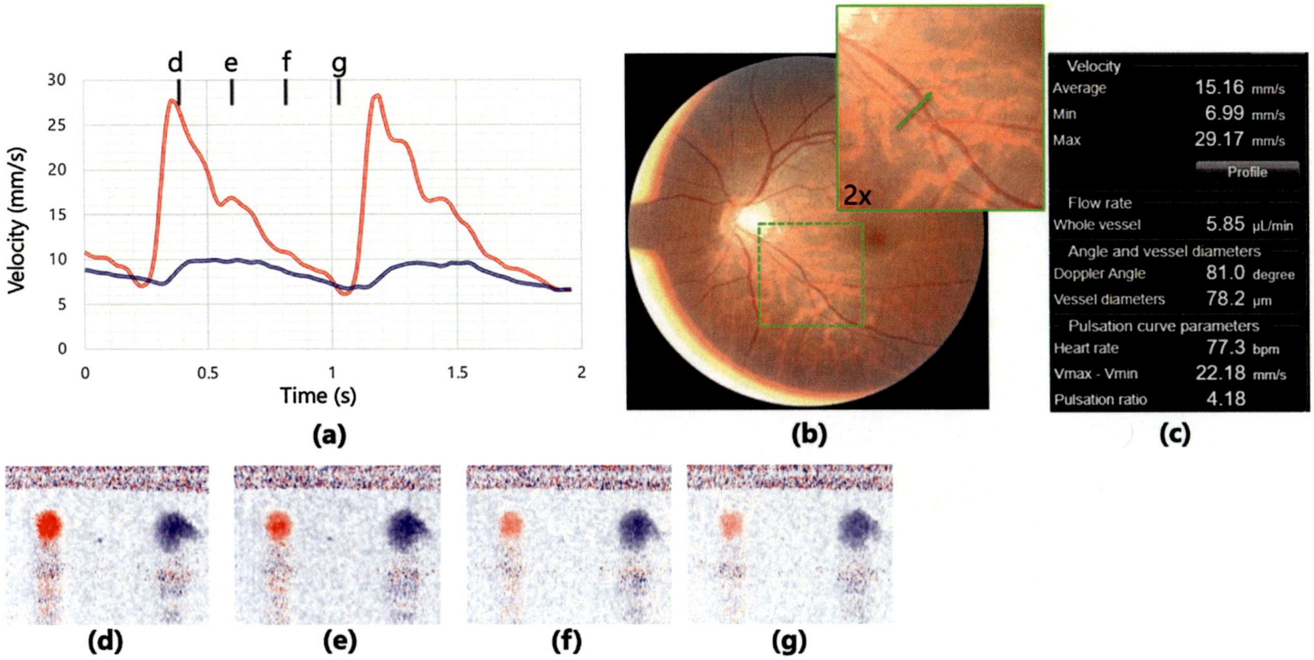

FIGURE 11.9 Schematic images of RBF measurement. (A) Acquired flow profile from a select retinal artery and vein located 2–3.5 mm away from the optic nerve head (ONH). Phase images are shown at different time locations in a single cardiac cycle (D–G). (B) Color fundus photo with *green arrows* indicating the location of scan location. (C) Blood flow parameters for selected artery.

other extracellular matrix components necessary for maintaining vitreous transparency. In pathological states, hyalocytes have been implicated in premacular/epiretinal membrane formation and proliferative vitreoretinopathy. While the precise identity of these cells, detected in vivo on OCT, cannot be defined without immunostaining, the morphological characteristics and spatial distribution of these cells on OCT sufficiently resemble that of hyalocytes as opposed to other monocyte-derived macrophages or microglia.

Prior studies have demonstrated that hyalocytes can be visualized at the vitreoretinal interface and can assume a variety of morphological configurations depending on whether they serve a neuroprotective or neurotoxic role. In the "activated" state, hyalocytes adopt an ameboid morphology characterized by a spherical cell body and retracted cell processes. In a "quiescent" state, hyalocyte somata become more elongated and spindle-like with ramified, filopodia-like processes. In this section, we will demonstrate the differences in morphology and spatial distribution of these cells in health[41] and a variety of retinal pathologies, including paracentral acute middle maculopathy,[42] pigmented paravenous chorioretinal atrophy,[43] acute macular neuroretinopathy,[44] and partial posterior vitreous detachment in a patient with retinal vaso-occlusive disease.[45] Hyalocytes can serve as a potential biomarker for assessing the progression of disease and for determining treatment efficacy, as shown here.

Hyalocytes in healthy subjects

See Figs. 11.10–11.13.

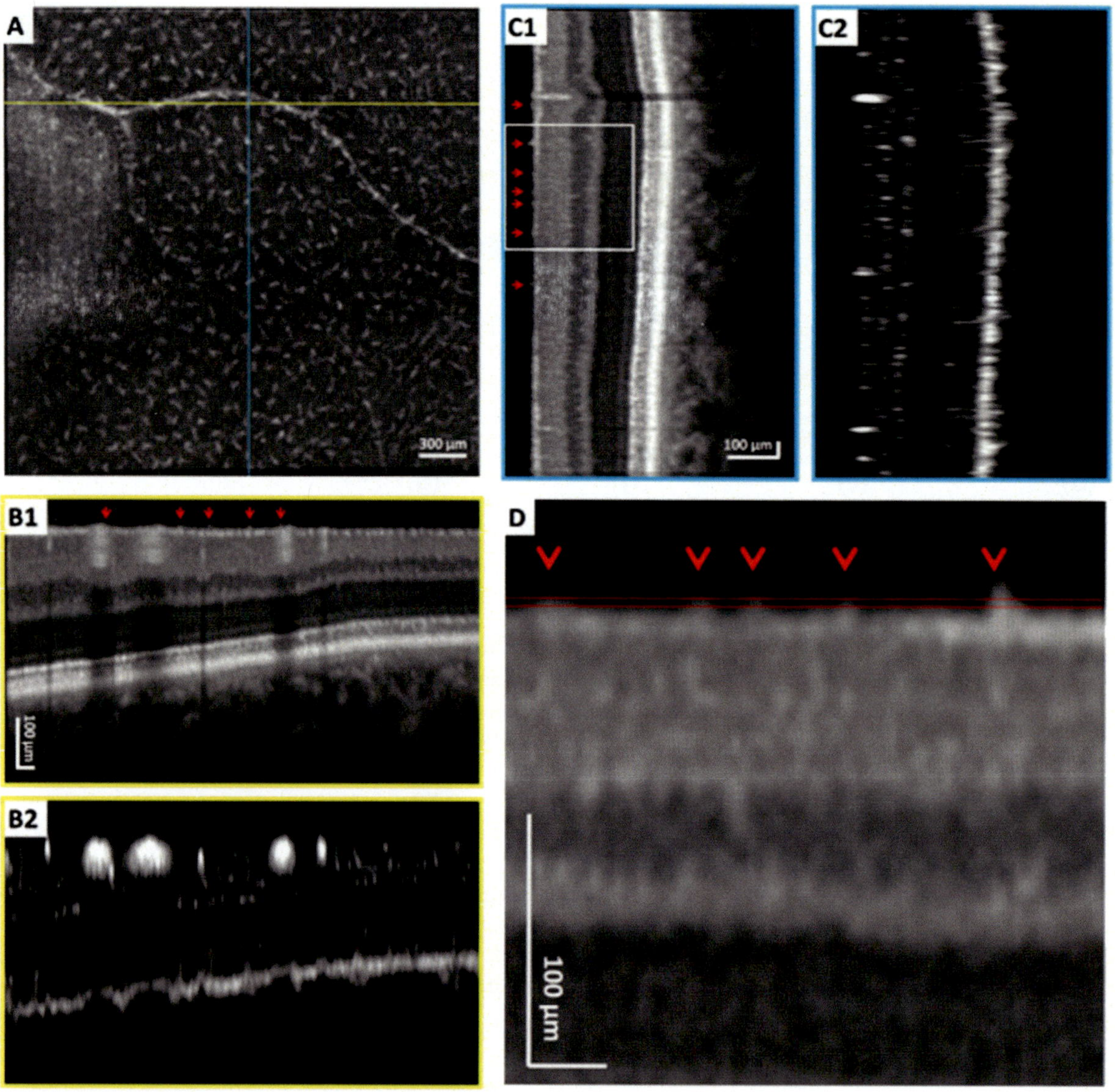

FIGURE 11.10 3 × 3-mm en face OCT-R imaging of the temporal retina in a healthy subject using clinical OCT. (A) The 3-µm OCT-R slab located above the ILM demonstrates regular spatial separation of hyalocytes. The fovea is located to the left of the image. (B1, B2) Horizontal OCT and OCT-A B-scan at the position indicated by the *yellow horizontal line* in **A**. (C1, C2) Vertical OCT and OCT-A B-scan at the position indicated by the *blue vertical line* in **A**. (D) Magnified OCT B-scan (rotated horizontally) at the region indicated by the *white box* in **C1**. The parallel *red lines* indicate the axial depth of the 3-µm OCT-R slab in **A**. Individual hyalocytes are indicated by the *red arrows* on the surface of the ILM. *Reproduced with permission by Dr. Richard Rosen and Dr. Toco Y. P. Chui from Castanos MV, Zhou DB, Linderman RE, et al. Imaging of macrophage-like cells in living human retina using clinical OCT.* Investigative Ophthalmology & Visual Science *2020;61(6):48.*

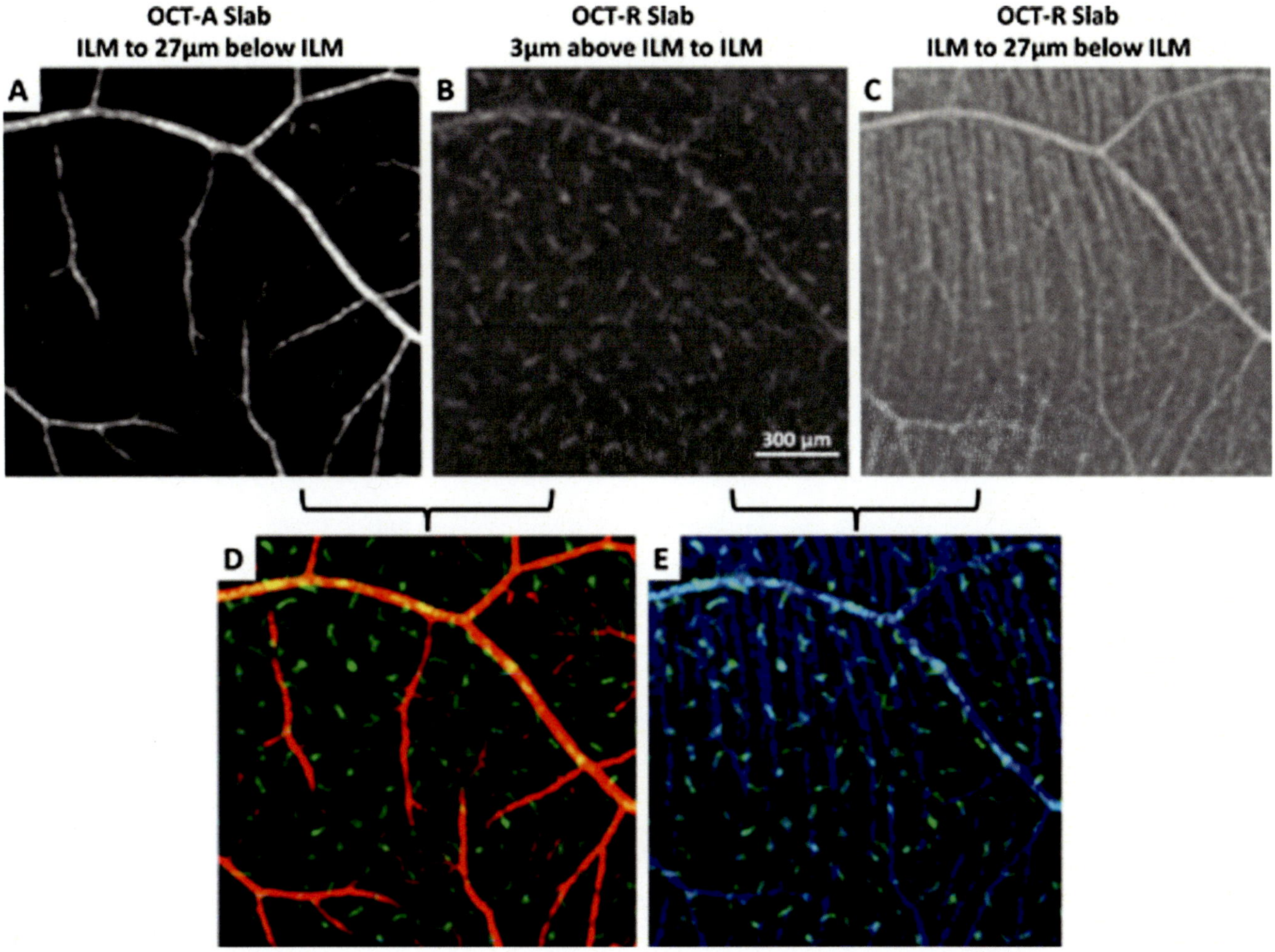

FIGURE 11.11 Hyalocytes in healthy subjects. (A) Superficial retinal vascular slab (ILM—27 μm below the ILM), (B) OCT-R visualizing hyalocytes 3 μm above the ILM, and (C) OCT-R of the RNFL (ILM—27 μm below the ILM) at the temporal retina in a healthy subject. (D, E) Color overlays of hyalocytes (*green*) superimposed upon the superficial retinal vascular network (*red*), and retinal nerve fiber bundles (*blue*). The fovea is located to the left of all images. *Reproduced with permission by Dr. Richard Rosen and Dr. Toco Y. P. Chui from Castanos MV, Zhou DB, Linderman RE, et al. Imaging of macrophage-like cells in living human retina using clinical OCT.* Investigative Ophthalmology & Visual Science *2020;61(6):48.*

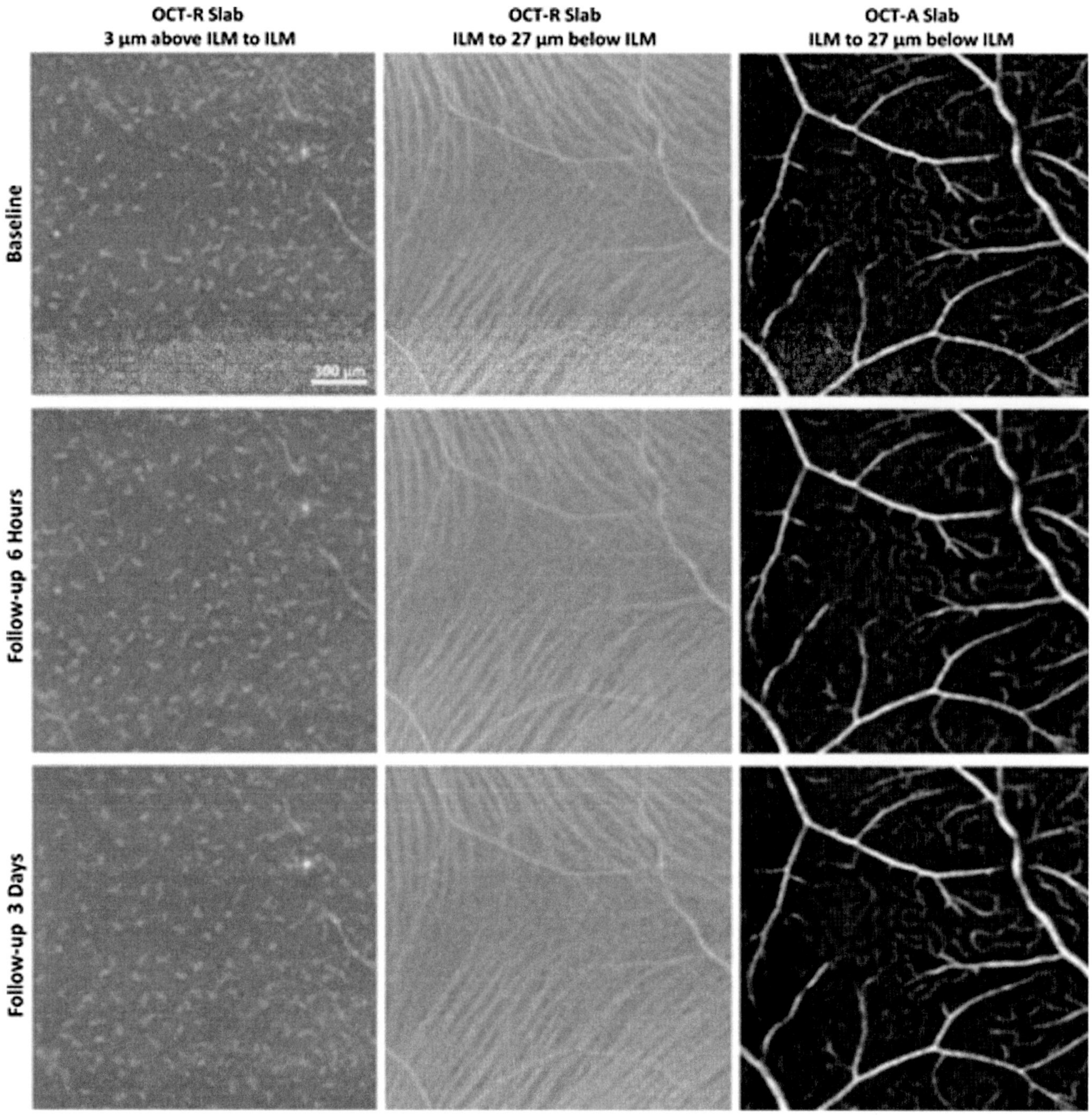

FIGURE 11.12 In vivo imaging of hyalocytes at (A) baseline, (B) 6 h, and (C) 3 days in a healthy subject. *Left column:* 3-µm OCT-R slabs located above the ILM surface. *Middle column:* 27-µm OCT-R slabs located below the ILM surface demonstrating the RNFL bundles at the temporal raphe. *Right column*: 27-µm OCT-A slabs located below the ILM surface. The fovea is located to the left of all images. *Reproduced with permission by Dr. Richard Rosen and Dr. Toco Y. P. Chui from Castanos MV, Zhou DB, Linderman RE, et al. Imaging of macrophage-like cells in living human retina using clinical OCT.* Investigative Ophthalmology & Visual Science *2020;61(6):48.*

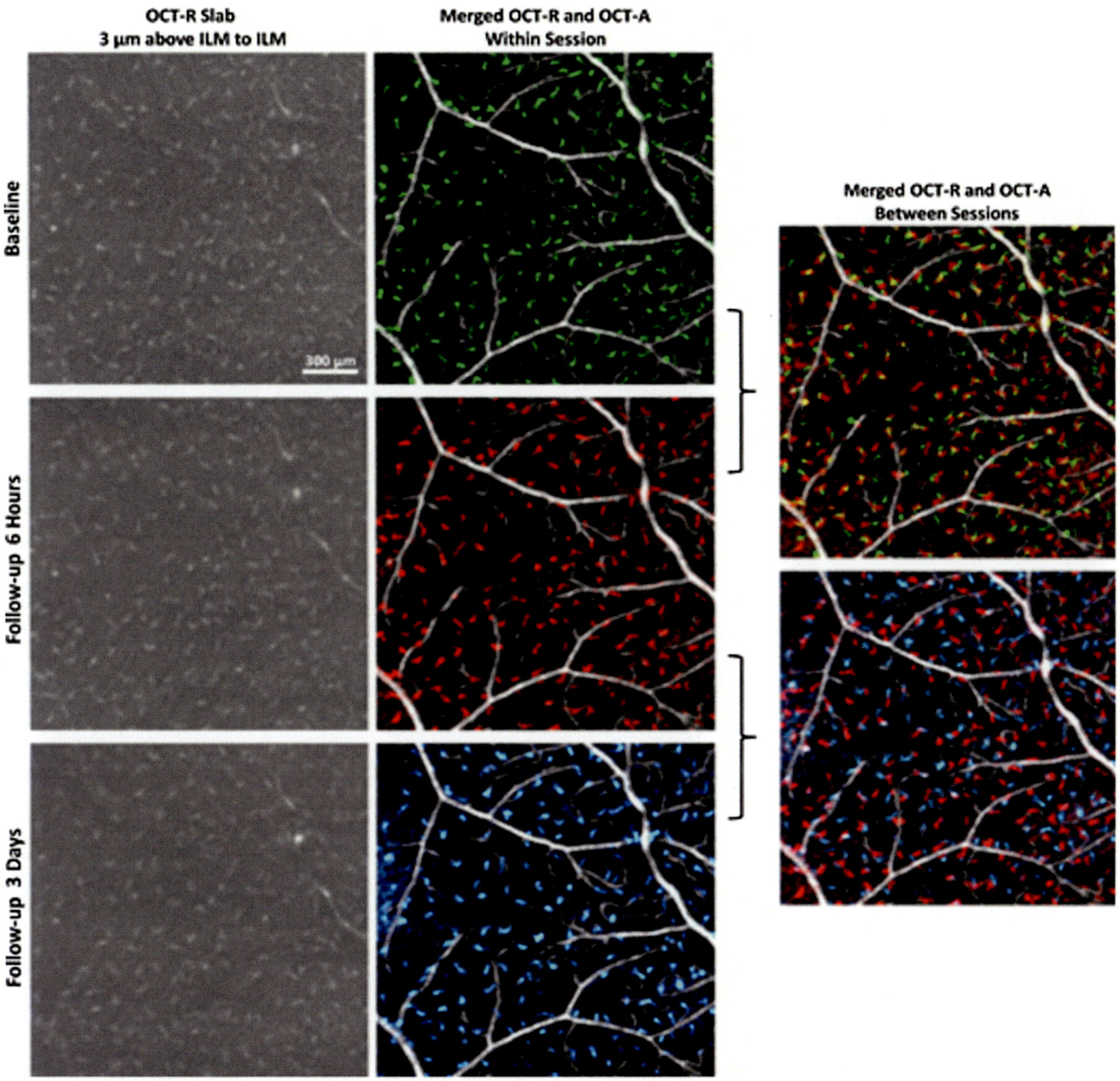

FIGURE 11.13 In vivo imaging of hyalocytes over time in a healthy control. *Left column:* OCT-R slabs located 3-µm above the ILM surface at the temporal retina. *Middle column:* OCT-R (baseline location of hyalocytes shown in *green*, 6 h follow-up location of hyalocytes shown in *red*, and 3-day follow-up location of hyalocytes shown in *blue*) and OCT-A overlaid. *Right column*: Merged OCT-R and OCT-A images from two imaging sessions demonstrating differences in hyalocyte locations at 6 h and 3 days later, suggesting cell translocation over time. The fovea is located to the left of all images. *Reproduced with permission by Dr. Richard Rosen and Dr. Toco Y. P. Chui from Castanos MV, Zhou DB, Linderman RE, et al. Imaging of macrophage-like cells in living human retina using clinical OCT.* Investigative Ophthalmology & Visual Science *2020;61(6):48.*

Hyalocytes in paracentral acute middle maculopathy (PAMM)

See Fig. 11.14.

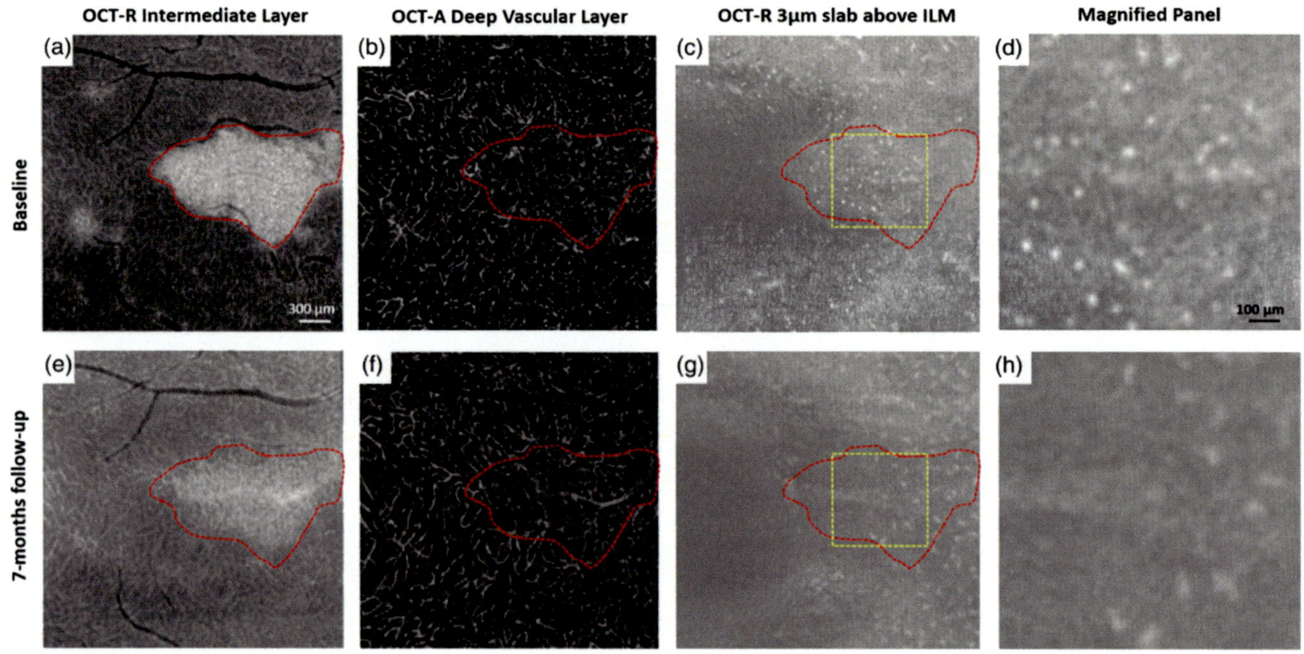

FIGURE 11.14 En face OCT-R and OCT-A of a PAMM patient at baseline and at follow-up 7 months later. (A) On initial presentation, OCT-R imaging of the intermediate layer revealed a significant hyperreflective ischemic zone. (B) OCT-A deep vascular layer demonstrated decreased perfusion in the ischemic zone. (C) Hyalocytes located within the ischemic zone appear round with a spherical morphological configuration. (D) Magnified panel shows a better visualization of the hyalocytes in an "activated" state. (E) At 7 months follow-up, OCT-R imaging of the intermediate layer showed a decrease in the hyperreflectivity of the ischemic area corresponding to increased perfusion in the ischemic region as shown in the OCT-A deep vascular layer (F). (G) Hyalocytes appear to adopt a spindle-like configuration, with a more uniform distribution across the retinal surface. (H) Magnified panel shows a better visualization of the hyalocytes in a "uiescent" state. *Reproduced with permission by Dr. Richard Rosen and Dr. Toco Y. P. Chui from Otero-Marquez O, Ahsanuddin S, Spellman L, Pinhas A, Chui TY, Rosen RB. Imaging vitreous cortex hyalocyte response to paracentral acute middle maculopathy (PAMM) using en face OCT [published online ahead of print, 2022 Nov 14].* European Journal of Ophthalmology *2022;11206721221138303.*

Hyalocytes in asymmetric pigmented paravenous chorioretinal atrophy (PPCRA)

See Fig. 11.15.

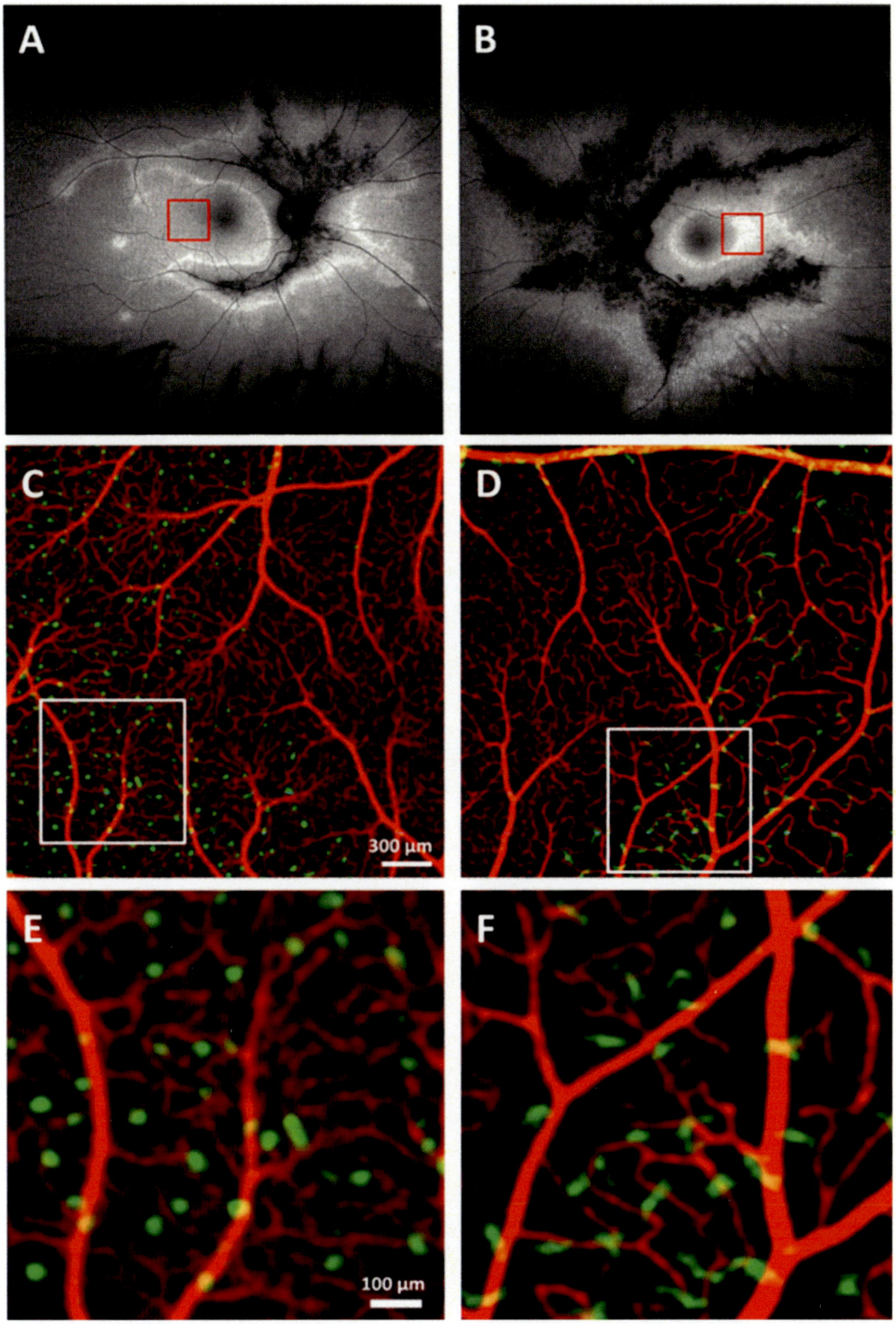

FIGURE 11.15 Hyalocytes in PPCRA. Fundus autofluorescence (FAF) of a patient with PPCRA. (A) OD and (B) OS. (C, D) Color overlay of 3 × 3 mm OCT-A (*in red*) of full retinal vascular layer ILM to OPL + 9 μm and OCT-R (*in green*) of a 3 μm slab anterior to the ILM at 9 degrees temporal to the fovea OU corresponding to the red squares marked in A and B, respectively. Decreased capillary density can be seen in the temporal half of the image in OS. (E, F) Magnified color overlays of the *white boxes* in C and D show hyalocytes located 3 μm above the ILM in green and retinal capillaries in red, respectively. The morphology of the hyalocytes differs between the patients' eyes; activated, ameboid-shaped hyalocytes with fewer, retracted processes can be visualized in OD. More elongated, ramified hyalocytes consistent with the morphological characteristics seen in a "quiescent" state can be visualized in OS. *Reproduced with permission by Dr. Richard Rosen and Dr. Toco Y. P. Chui from Fallon J, Ahsanuddin S, Otero-Marquez O, et al. Posterior vitreous cortex hyalocytes visualization in asymmetric pigmented paravenous chorioretinal atrophy (PPCRA) using en face OCT.* American Journal of Ophthalmology Case Reports *2023;30:101846.*

Hyalocytes in acute macular neuroretinopathy (AMN)

See Fig. 11.16.

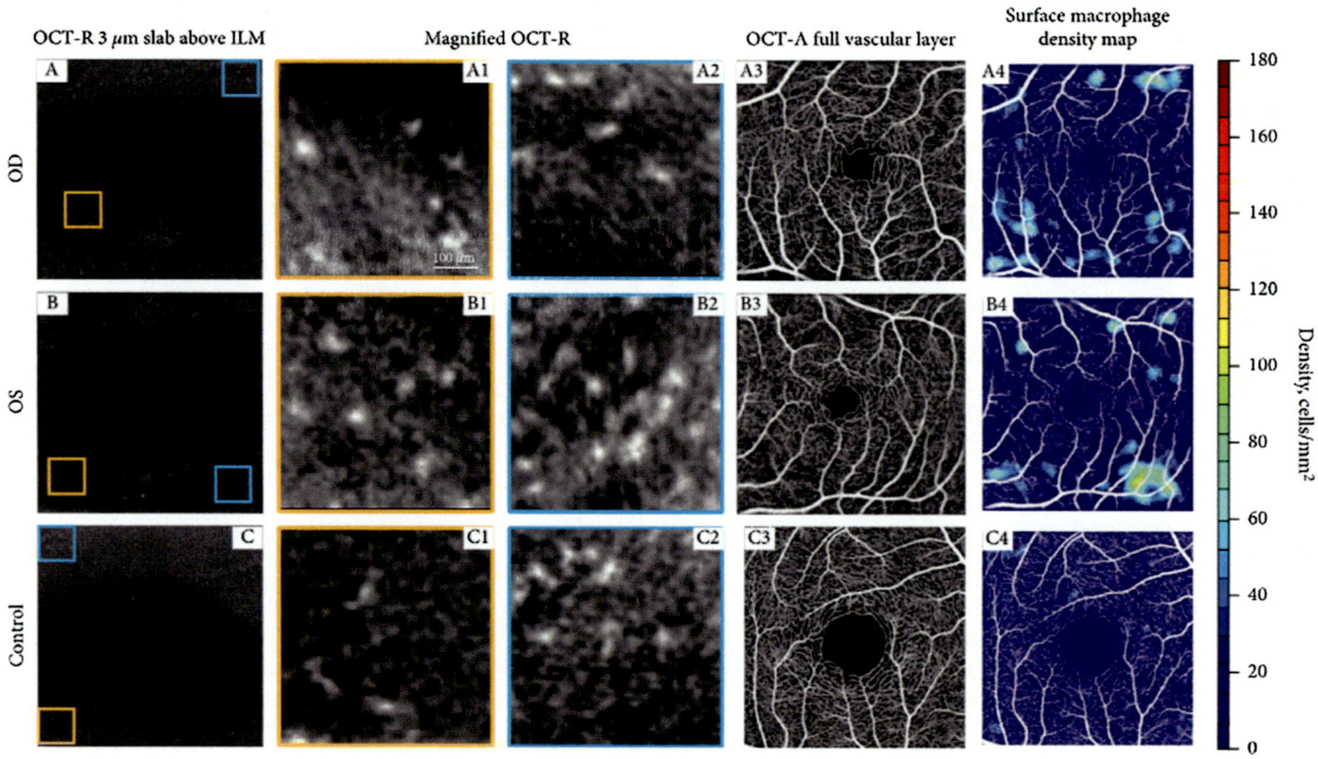

FIGURE 11.16 Hyalocyte spatial distribution and morphology in a 19-year-old male patient with AMN compared to a 35-year-old healthy control. (A—C) OCT-R slab 3 μm above the ILM surface demonstrates a higher density of hyalocytes in eyes with AMN. *Orange and blue squares* indicate two different regions of interest (ROIs) with a magnified view of surface macrophages (A1—C2). (A1—C2) Magnified view of hyalocytes in AMN compared to a healthy control. In the eyes with AMN, the cells appear round with fewer, shortened processes, consistent with an "activated" state. In the healthy control, there are fewer cells and they appear slender with spindle-like configuration consistent with a "quiescent" state. (A3, B3, C3) OCT-A full vascular layer of OU in our patient with AMN and a healthy control. (A4, B4, C4) Overlay of hyalocyte density map onto the OCT-A. These findings suggest that hyalocyte density increases in acute ischemic conditions such as AMN. *Cell density:* AMN eyes, OD 9.26 cells/mm^2, OS 10.38 cells/mm^2. Healthy control eye: 2.39 cells/mm^2. *Reproduced with permission by Dr. Richard Rosen from Schatz MJ, Otero-Marquez O, Rosen RB, Parikh D. Multimodal imaging and macular hyalocyte count in a patient with acute macular neuroretinopathy.* Case Reports in Ophthalmology Medicine *2022;2022:2855191.*

Hyalocytes in partial posterior vitreous detachment (PVD) and vaso-occlusive disease

See Figs. 11.17 and 11.8.

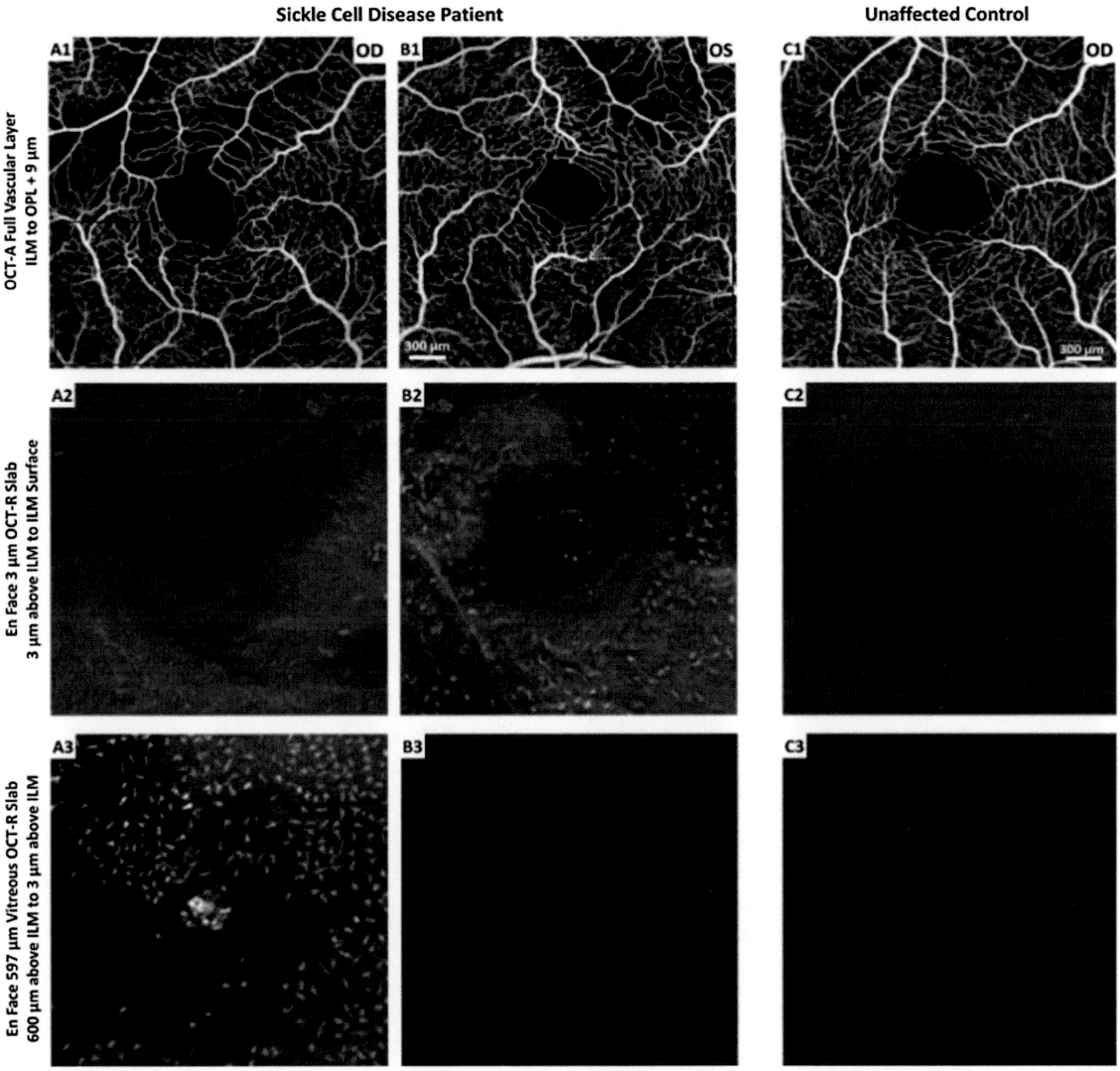

FIGURE 11.17 Hyalocytes in vaso-occlusive disease. (A1, B1, C1) Full retinal vascular layer OCT-A and coronal plane en face OCT-R images of a sickle cell disease patient (OU) (A1, B1) compared to a healthy control (C1). Full retinal vascular layer centered on the fovea shows focal capillary nonperfusion in the superotemporal quadrant of both eyes with a larger foveal avascular zone in OD (A1, B1). Magnified regions of the *red boxes* are shown in Fig. 11.9. (A2 and B2) Coronal plane en face OCT-R slab 3 μm immediately above the ILM demonstrates few hyalocytes in OD compared to OS, which shows numerous cells scattered above the ILM. (A3) Coronal plane en face 597 μm vitreous OCT-R slab above the ILM centered at the fovea of OD reveals a layer of hyalocytes suspended in the vitreous. (B3) Few hyalocytes are observed in the vitreous of OS. (C2 and C3) Minimal hyper-reflective foci are visualized on the coronal plane en face 3 μm OCT-R slab above the ILM and the en face 597 μm vitreous OCT-R slab above the ILM centered at the fovea of OD in the unaffected control. *Reproduced with permission by Dr. Richard Rosen and Dr. Toco Y. P. Chui from Ahsanuddin S, Rios HA, Glassberg JA, Chui TYP, Sebag J, Rosen RB. 3-D OCT imaging of hyalocytes in partial posterior vitreous detachment and vaso-occlusive retinal disease.* American Journal of Ophthalmology Case Reports *2023;30:101836.*

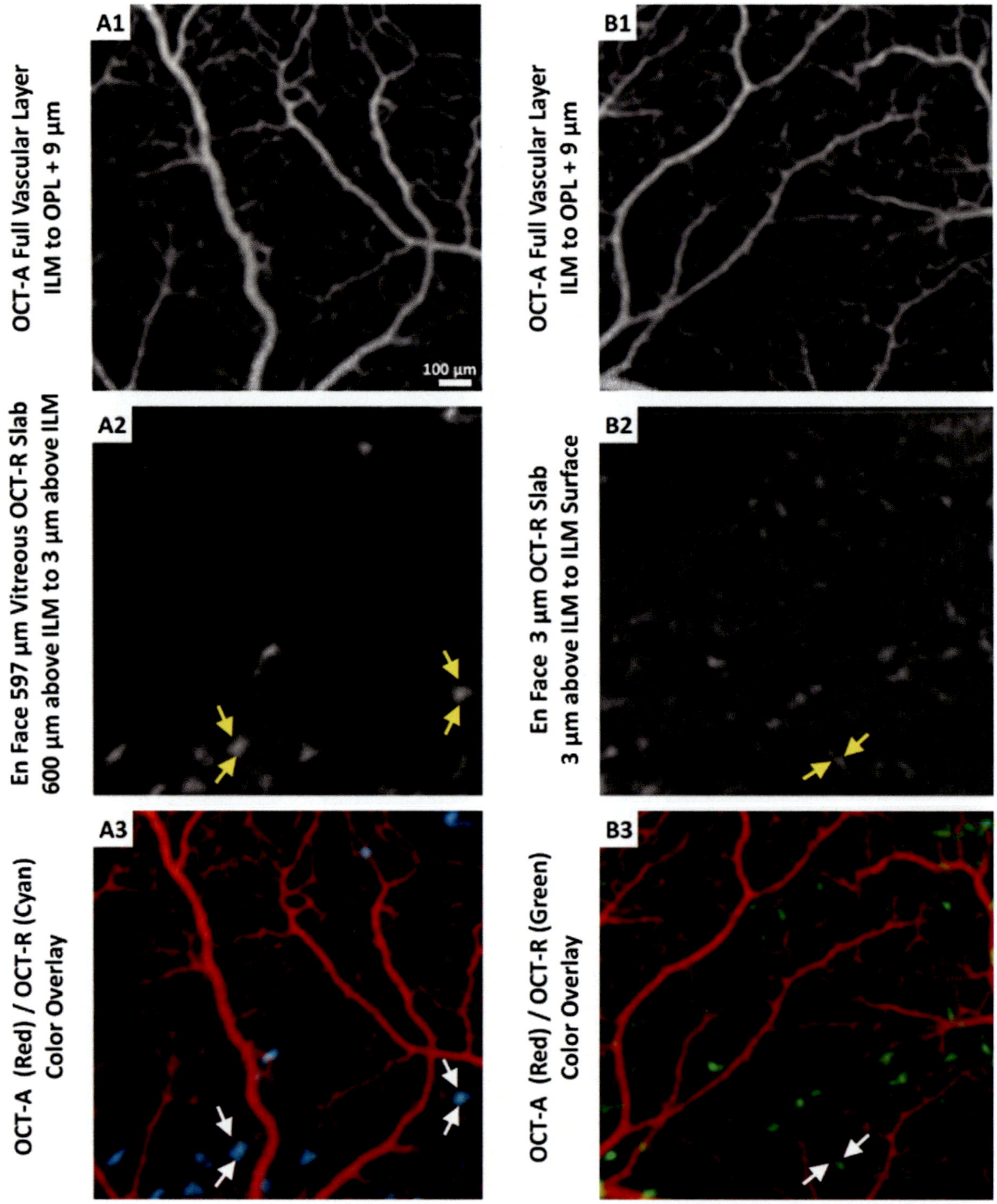

FIGURE 11.18 Magnified view of hyalocytes in vaso-occlusive disease. Magnified superotemporal regions from Fig. 11.16 of both eyes showing the full vascular layer OCT-A (A1, B1), the en face slabs where the hyalocytes are most densely located (A2, B2) and a color overlay of the two images (A3, B3). (A1 and B1) Capillary nonperfusion in the superotemporal quadrant of both eyes. (A2) Coronal plane en face 597 µm vitreous OCT-R slab located anterior to the ILM demonstrates the presence of hyalocytes with varying morphologies in OD (*yellow arrows* indicate plump, ameboid morphologies indicating an "activated" state). (B2) Coronal plane en face 3 µm OCT-R slab immediately anterior to the ILM shows the presence of hyalocytes with varying morphologies anterior to the ILM surface in the OS (*yellow arrows* indicate slender, elongated morphology indicating a more "quiescent" state). (A3 and B3) Color overlay of hyalocytes in the vitreous OCT-R slab (*cyan*) and anterior to the ILM surface OCT-R slab (*green*) superimposed on the corresponding full vascular layer of the OCT-A scans (*red*) in the OD and OS, respectively (*white arrows* indicate the location of the same hyalocytes). *Reproduced with permission by Dr. Richard Rosen and Dr. Toco Y. P. Chui from Ahsanuddin S, Rios HA, Glassberg JA, Chui TYP, Sebag J, Rosen RB. 3-D OCT imaging of hyalocytes in partial posterior vitreous detachment and vaso-occlusive retinal disease.* American Journal of Ophthalmology Case Reports *2023;30:101836.*

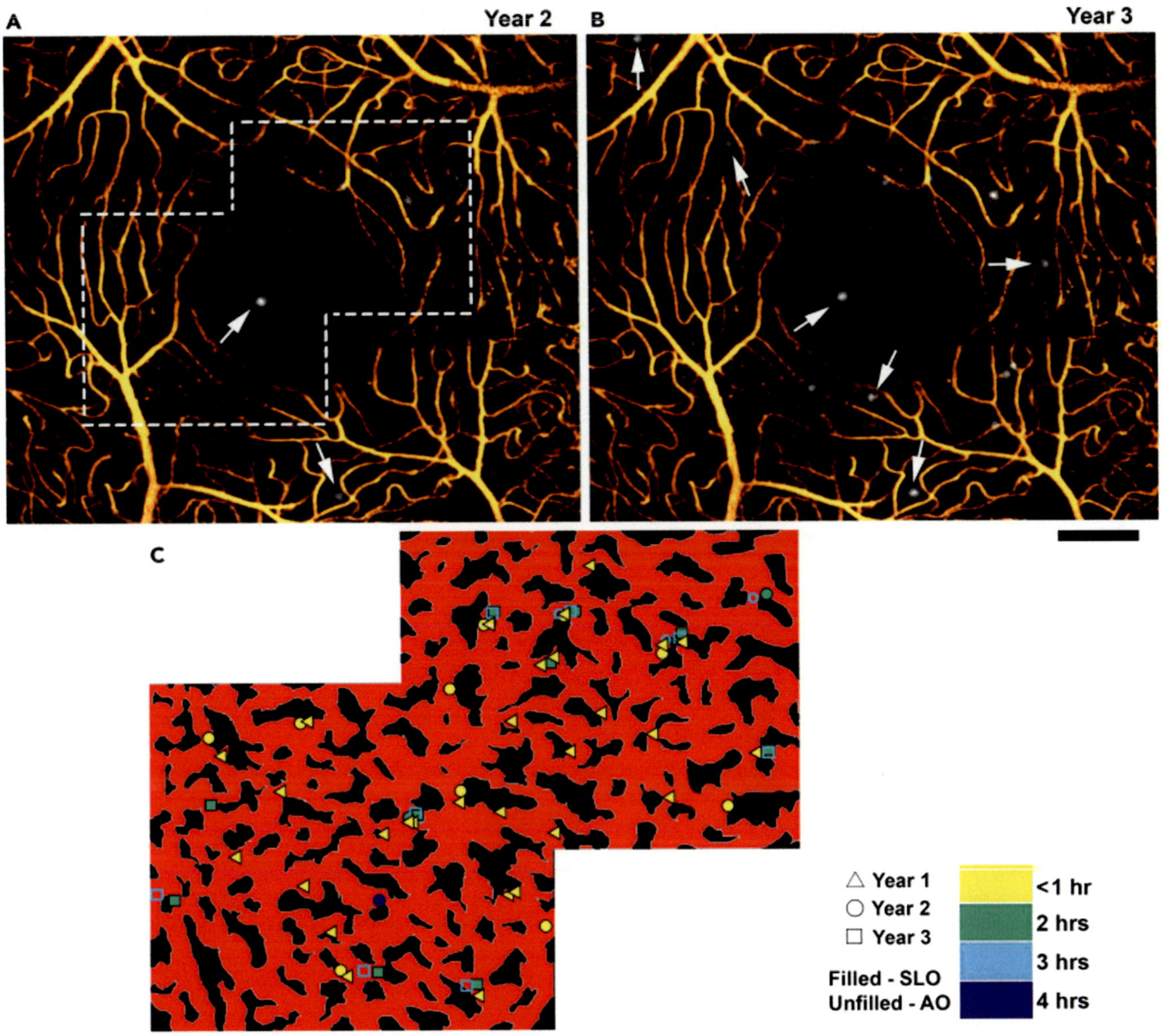

FIGURE 11.19 Erythrocyte-mediated angiography. Erythrocyte stasis occurred within the short vessel segments of the choroidal microvasculature. Only standard EMA was performed in year 1, and both standard EMA and AO-EMA were performed in years 2 and 3. (A and B) Higher resolution AO images of erythrocyte in stasis (white circular dots imaged using AO-ICG; [A]: year 2, [B]: year 3) coregistered to AO images of the inner retinal microvasculature (*yellow*, imaged using AO nonconfocal split detection). (C) In this segmentation map of the choriocapillaris, the images of the retinal vessels were acquired at a focal plane that was anterior to the erythrocytes in stasis. Erythrocyte stasis marked from multiple timepoints associated with three separate EMA procedures across 3 years. Cells recorded from different years and procedures are categorized by different markers (filled markers: imaged using SLO; unfilled markers: imaged using AO). The colormap indicates the time points at which these cells in stasis were captured by imaging. All of the cells in stasis that could be identified within this region were located within the choriocapillaris vessel segments. Scale bar: 200 μm. *Reproduced with permission by Dr. Osamah Saeedi from Li J, Wang D, Pottenburgh J, Bower AJ, Asanad S, Lai EW, Simon C, Im L, Huryn LA, Tao Y, Tam J, Saeedi OJ. Visualization of erythrocyte stasis in the living human eye in health and disease.* iScience *December 7, 2022;26(1):105755.*

Erythrocyte-mediated angiography

Erythrocyte-mediated angiography is an emerging invasive technique to measure retinal blood flow on a cellular level with a wide field of view. The technique requires intravenous injection of ICG-tagged erythrocytes, which are then imaged through fluorescein angiogram with laser diodes specific to ICG.[46–48] These angiograms can then be interpreted, with erythrocytes being tracked by expert reviewers. Manual tracking can be time consuming, although accurate.[49] Recent literature has explored more automated means for interpreting these angiograms.[47,48] Saeedi et al. have recently described an automated and objective way to measure blood flow described as "automated velocimetry," which showed a high correlation with trained manual tracking.[48] The ability to measure single erythrocyte movement and velocity; thus, overall blood flow holds large implications for further study of retinal vasculature as well as its role in disease states. The ability to note a lack of movement, termed "stasis," holds similar value. Using EMA, Li et al. describe the phenomenon of red blood cell stasis within the choriocapillaris, as illustrated in Fig. 11.19.[50] The presence or absence of red blood cell movement was found to be altered in glaucomatous eyes compared to normal, suggesting future utility in understanding, and monitoring common disease states. Although invasive in nature, the amount of ICG used is significantly lower than in conventional ICG angiography, thus suggesting likely less risk for adverse events.[48]

References

1. Puyo L, Paques M, Fink M, Sahel JA, Atlan M. In vivo laser Doppler holography of the human retina. *Biomedical Optics Express*. August 6, 2018;9(9):4113–4129. https://doi.org/10.1364/BOE.9.004113.
2. Puyo L, Paques M, Fink M, Sahel JA, Atlan M. Waveform analysis of human retinal and choroidal blood flow with laser Doppler holography. *Biomedical Optics Express*. September 5, 2019;10(10):4942–4963.
3. Puyo L, Paques M, Fink M, Sahel JA, Atlan M. Choroidal vasculature imaging with laser Doppler holography. *Biomedical Optics Express*. January 31, 2019;10(2):995–1012.
4. Puyo L, Paques M, Atlan M. Spatio-temporal filtering in laser Doppler holography for retinal blood flow imaging. *Biomedical Optics Express*. May 26, 2020;11(6):3274–3287.
5. Puyo L, Paques M, Atlan M. Retinal blood flow reversal quantitatively monitored in out-of-plane vessels with laser Doppler holography. *Scientific Reports*. September 8, 2021;11(1):17828.
6. Roorda A, Romero-Borja F, Donnelly III WJ, Queener H, Hebert TJ, Campbell MCW. Adaptive optics scanning laser ophthalmoscopy. *Optics Express*. 2002;10:405–412.
7. Webb RH, Hughes GW. Scanning laser ophthalmoscope. *IEEE Transactions on Biomedical Engineering*. 1981;28(7):488–492.
8. Sharp P, Manivannan A. The scanning laser ophthalmoscope. *Physics in Medicine and Biology*. 1997;42:951.
9. Sengupta AR, Jensen-Clem R. In: *Research Notes of the AAS*. vol. 4. 2020:244.
10. Duncan JL, Zhang Y, Gandhi J, et al. High-resolution imaging with adaptive optics in patients with inherited retinal degeneration. *Investigative Ophthalmology & Visual Science*. 2007;48(7):3283–3291.
11. Chui TY, Song H, Burns SA. Individual variations in human cone photoreceptor packing density: variations with refractive error. *Investigative Ophthalmology & Visual Science*. 2008;49(10):4679–4687.
12. Migacz JV, Otero-Marquez O, Zhou R, et al. Imaging of vitreous cortex hyalocyte dynamics using non-confocal quadrant-detection adaptive optics scanning light ophthalmoscopy in human subjects. *Biomedical Optics Express*. March 1, 2022;13(3):1755–1773.
13. Pinhas A, Dubow M, Shah N, et al. In vivo imaging of human retinal microvasculature using adaptive optics scanning light ophthalmoscope fluorescein angiography. *Biomedical Optics Express*. July 12, 2013;4(8):1305–1317.
14. Pinhas A, Migacz JV, Zhou DB, et al. Insights into sickle cell disease through the retinal microvasculature: adaptive optics scanning light ophthalmoscopy correlates of clinical OCT angiography. *Ophthalmology Science*. July 12, 2022;2(4):100196.
15. Kurokawa K, Liu Z, Miller DT. Adaptive optics optical coherence tomography angiography for morphometric analysis of choriocapillaris [Invited]. *Biomedical Optics Express*. February 24, 2017;8(3):1803–1822.
16. Jung H, Liu T, Liu J, Huryn LA, Tam J. Combining multimodal adaptive optics imaging and angiography improves visualization of human eyes with cellular-level resolution. *Communications Biology*. November 14, 2018;1:189.
17. Roorda A. Optoretinography is coming of age. *Proceedings of the National Academy of Sciences of the U S A*. December 21, 2021;118(51):e2119737118.
18. Jonnal RS. Toward a clinical optoretinogram: a review of noninvasive, optical tests of retinal neural function. *Annals of Translational Medicine*. August 2021;9(15):1270. https://doi.org/10.21037/atm-20-6440.
19. Lassoued A, Zhang F, Kurokawa K, et al. Cone photoreceptor dysfunction in retinitis pigmentosa revealed by optoretinography. *Proceedings of the National Academy of Sciences of the U S A*. November 23, 2021;118(47):e2107444118.
20. Jonnal RS, Rha J, Zhang Y, Cense B, Gao W, Miller DT. In vivo functional imaging of human cone photoreceptors. *Optics Express*. November 26, 2007;15(24):16141–16160.
21. Cooper RF, Brainard DH, Morgan JIW. Optoretinography of individual human cone photoreceptors. *Optics Express*. December 21, 2020;28(26):39326–39339.
22. Rha J, Schroeder B, Godara P, Carroll J. Variable optical activation of human cone photoreceptors visualized using a short coherence light source. *Optics Letters*. December 15, 2009;34(24):3782–3784.
23. Kim TH, Wang B, Lu Y, Son T, Yao X. Functional optical coherence tomography enables in vivo optoretinography of photoreceptor dysfunction due to retinal degeneration. *Biomedical Optics Express*. August 27, 2020;11(9):5306–5320.
24. Barot M, Gokulgandhi MR, Mitra AK. Mitochondrial dysfunction in retinal diseases. *Current Eye Research*. December 2011;36(12):1069–1077.
25. Chen AX, Conti TF, Hom GL, et al. Functional imaging of mitochondria in retinal diseases using flavoprotein fluorescence. *Eye*. January 2021;35(1):74–92.

26. Elner SG, Elner VM, Field MG, Park S, Heckenlively JR, Petty HR. Retinal flavoprotein autofluorescence as a measure of retinal health. *Transactions of the American Ophthalmological Society*. 2008;106:215–222. discussion 222–4.
27. Field MG, Elner VM, Puro DG, et al. Rapid, noninvasive detection of diabetes-induced retinal metabolic stress. *Archives of Ophthalmology*. July 2008;126(7):934–938.
28. Andrade Romo JS, Lynch G, Liu K, et al. Flavoprotein fluorescence correlation with visual acuity response in patients receiving anti-VEGF injection for diabetic macular edema. *Oxidative Medicine and Cellular Longevity*. August 9, 2018;2018:3567306.
29. Muste JC, Russell MW, Chen AX, et al. Functional imaging of mitochondria in age-related macular degeneration using flavoprotein fluorescence. *Ophthalmic Surgery Lasers Imaging Retina*. January 2023;54(1):24–31.
30. Field MG, Comer GM, Kawaji T, Petty HR, Elner VM. Noninvasive imaging of mitochondrial dysfunction in dry age-related macular degeneration. *Ophthalmic Surgery, Lasers and Imaging*. September–October, 2012;43(5):358–365.
31. Zhou DB, Castanos MV, Geyman L, et al. Mitochondrial dysfunction in primary open-angle glaucoma characterized by flavoprotein fluorescence at the optic nerve head. *Ophthalmology Glaucoma*. 2022;5(4):413–420.
32. Ahsanuddin S, Rios H, Otero-Marquez O, et al. Flavoprotein fluorescence elevation is a marker of mitochondrial oxidative stress in patients with retinal disease. *Frontiers in Ophthalmology*. February 2023;3:4.
33. Leitgeb RA, Werkmeister RM, Blatter C, Schmetterer L. Doppler optical coherence tomography. *Progress in Retinal and Eye Research*. 2014;41(100):26–43.
34. Grunwald JE, Brucker AJ, Grunwald SE, Riva CE. Retinal hemodynamics in proliferative diabetic retinopathy. A laser Doppler velocimetry study. *Investigative Ophthalmology & Visual Science*. 1993;34(1):66–71.
35. Piltz-Seymour JR. Laser Doppler flowmetry of the optic nerve head in glaucoma. *Survey of Ophthalmology*. 1999;43 Suppl. 1:S191–S198.
36. Jackson TL, Danis RP, Goldbaum M, et al. Retinal vascular abnormalities in neovascular age-related macular degeneration. *Retina*. 2014;34(3):568–575.
37. Li Y, Chen J, Chen Z. Advances in Doppler optical coherence tomography and angiography. *Translation Biophotonics*. 2019;1(1–2):e201900005.
38. Chen Z, Milner TE, Srinivas S, et al. Noninvasive imaging of in vivo blood flow velocity using optical Doppler tomography. *Optics Letters*. 1997;22(14):1119–1121.
39. Sakai J, Minamide KJ, Nakamura S, et al. Retinal arteriole pulse waveform analysis using a fully-automated Doppler optical coherence tomography flowmeter: a pilot study. *Translation Vision Science & Technology*. 2019;8(3):13.
40. Abe T, Yoshioka T, Song Y, et al. Glaucoma diagnostic performance of retinal blood flow measurement with Doppler optical coherence tomography. *Translation Vision Science & Technology*. 2022;11(10):11.
41. Castanos MV, Zhou DB, Linderman RE, et al. Imaging of macrophage-like cells in living human retina using clinical OCT. *Investigative Ophthalmology & Visual Science*. 2020;61(6):48.
42. Otero-Marquez O, Ahsanuddin S, Spellman L, Pinhas A, Chui TY, Rosen RB. Imaging vitreous cortex hyalocyte response to paracentral acute middle maculopathy (PAMM) using en face OCT [published online ahead of print, 2022 Nov 14]. *European Journal of Ophthalmology*. 2022;33:11206721221138303.
43. Fallon J, Ahsanuddin S, Otero-Marquez O, et al. Posterior vitreous cortex hyalocytes visualization in asymmetric pigmented paravenous chorioretinal atrophy (PPCRA) using en face OCT. *American Journal of Ophthalmology Case Reports*. 2023;30:101846.
44. Schatz MJ, Otero-Marquez O, Rosen RB, Parikh D. Multimodal imaging and macular hyalocyte count in a patient with acute macular neuroretinopathy. *Case Reports in Ophthalmology Medicine*. 2022;2022:2855191.
45. Ahsanuddin S, Rios HA, Glassberg JA, Chui TYP, Sebag J, Rosen RB. 3-D OCT imaging of hyalocytes in partial posterior vitreous detachment and vaso-occlusive retinal disease. *American Journal of Ophthalmoloy Case Reports*. 2023;30:101836.
46. Flower R, Peiretti E, Magnani M, et al. Observation of erythrocyte dynamics in the retinal capillaries and choriocapillaris using ICG-loaded erythrocyte ghost cells. *Investigative Ophthalmology & Visual Science*. December 2008;49(12):5510–5516.
47. Flower RW, Kling R. Observation and characterization of microvascular vasomotion using erythrocyte mediated ICG angiography (EM-ICG-A). *Microvascular Research*. September 2017;113:78–87.
48. Wang D, Haytham A, Mayo L, Tao Y, Saeedi O. Automated retinal microvascular velocimetry based on erythrocyte mediated angiography. *Biomedical Optics Express*. June 27, 2019;10(7):3681–3697.
49. Saeedi O, Tracey B, Renner C, et al. Determination of absolute erythrocyte velocity and flow in the human retinal microvasculature by direct visualization of icg-labelled erythrocytes. *Investigative Ophthalmology & Visual Science*. 2018;59:3950.
50. Li J, Wang D, Pottenburgh J, et al. Visualization of erythrocyte stasis in the living human eye in health and disease. *iScience*. December 7, 2022;26(1):105755.

Further reading

1. Pandiyan VP, Maloney-Bertelli A, Kuchenbecker JA, et al. The optoretinogram reveals the primary steps of phototransduction in the living human eye. *Science Advances*. September 9, 2020;6(37):eabc1124.

Section III

Retinal vascular diseases

Chapter 12

Central retinal vein occlusion

Sara Labay-Tejado[1], Covadonga Menendez-Acebal[1], Carolina Bernal-Morales[1,2], Socorro Alforja[1,2] and Javier Zarranz-Ventura[1,2,3]

[1]*Hospital Clínic de Barcelona, Barcelona, Spain;* [2]*Institut d'Investigacions Biomediques August Pi i Sunyer (IDIBAPS), Barcelona, Spain;* [3]*Universitat de Barcelona, Barcelona, Spain*

Overview bullets

- Central retinal vein occlusion (CRVO) is a retinal vascular disease that occurs at or near the lamina cribosa of the optic nerve.
- Early classification between nonischemic and ischemic CRVO is essential.
- Ischemic CRVO has more risk of severe complications (i.e., neovascularization) and requires close monitoring.
- Macular edema is the most common complication and causes visual loss.
- The treatment for macular edema usually starts with anti-VEGF intravitreal injections. Steroids could also be considered, especially in nonresponders to anti-VEGF.
- Panretinal photocoagulation is indicated in patients with active neovascularization.

Pathogenesis

Retinal vein occlusion (RVO) is the second most common retinal vascular disease after diabetic retinopathy and usually causes moderate to severe visual loss in older individuals.[1–4] Among the different types of RVO, which are classified according to the site of occlusion, central retinal vein occlusion (CRVO) typically occurs at or near the lamina cribosa of the optic nerve.[2,5,6] In addition, hemi-CRVO refers to the occlusion of only one trunk of the central retinal vein and it has historically been considered as a distinct entity.[7] In all cases of CRVO, a subdivision is made between nonischemic and ischemic forms, which differ in their pathogenesis, clinical features, management, and prognosis.[2,8]

The exact pathogenesis of CRVO remains unclear. The intrinsic condition may be due to the combination of three vascular factors known as Virchow's triad: venous stasis, degenerative changes of the vessel wall, and blood hypercoagulability.[1,2]

The pathogenesis of both ischemic and nonischemic CRVO involves ischemia and inflammation to some degree. Ischemia activates the hypoxia-inducible factor (HIF) transcription that regulates vascular endothelial growth factor (VEGF) expression. VEGF is expressed by vascular endothelial and retinal glial cells and participates in endothelial proliferation, angiogenesis, and vascular permeability.[9,10] Other factors regulated by HIF are erythropoietin (EPO) and placental growth factor (PIGF). EPO grants neuroprotection to the retina against ischemic injury. Higher levels of EPO are associated with higher levels of VEGF. PIGF is homologous to VEGF and binds exclusively with a higher affinity to VEGF-1 receptor (VEGFR-1).[10]

CRVO-related macular edema (ME) is a common complication, caused by the increased vascular permeability related to inner blood—retinal barrier breakdown and associated osmotic and hydrostatic pressure changes, which contribute to fluid accumulation within the retina. The severity of ME seems to be directly related to the levels of VEGF and also the grade of retinal ischemia although it also appears in nonischemic CRVO. These findings suggest that ME pathogenesis is not completely explained by retinal ischemia.[9,10]

Inflammation also plays an important role in CRVO. Multiple cytokines and vasoactive proteins are released in the extracellular space (i.e., blood components), such as VEGF, which induces the expression of both receptors 1 (VEGFR-1) and 2 (VEGFR-2). The VEGFR expression in monocytes/macrophages (VEGFR-1) and vascular endothelial cells

Retinal and Choroidal Vascular Diseases of the Eye. https://doi.org/10.1016/B978-0-443-15583-3.00012-3

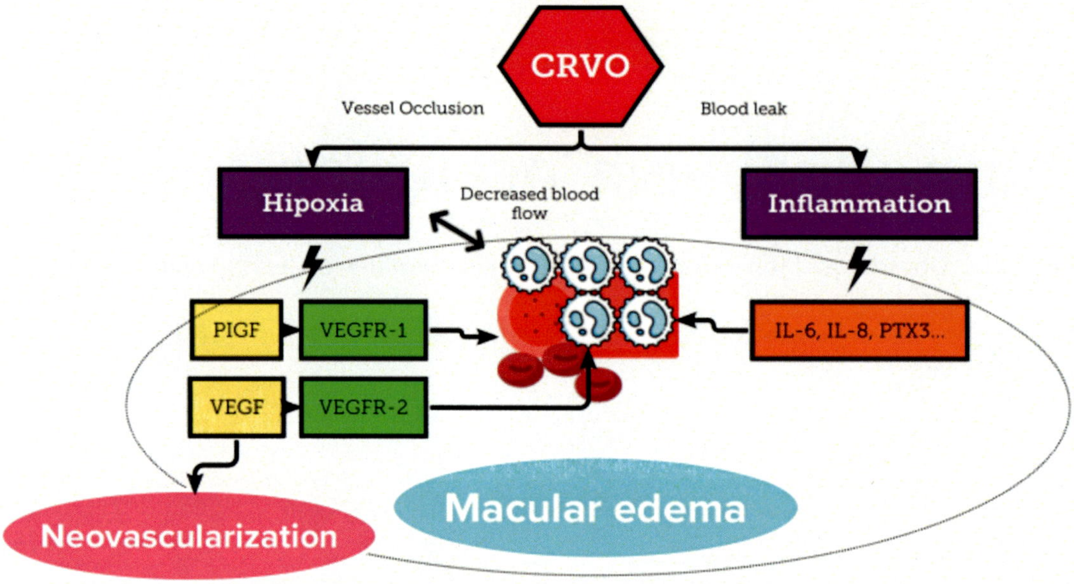

FIGURE 12.1 Pathogenesis of macular edema and neovascularization in central retinal vein occlusion (CRVO). *Adapted from Noma H, Yasuda K, Shimura M. Cytokines and pathogenesis of central retinal Vein occlusion.* Journal of Clinical Medicine. *2020;9(11):1–16.*

(VEGFR-2) induces leukocyte chemotaxis, proinflammatory factor activation, and increased vascular permeability in the case of VEGFR-2, causing ME.[9–11] These proinflammatory factors include interleukin 6 (IL-6), interleukin 8 (IL-8), platelet-derived growth factor (PDGF), pentraxin 3 (PTX-3), and chemokine monocyte chemoattractant protein (MCP-1), among others. IL-6 plays a role in protecting retinal ganglion cells; meanwhile, overexpression of PTX-3 manifests retinal damage and necrosis. MCP-1 phosphorylates tight junction proteins, and it is associated with retinal detachment in CRVO.[9,10] Leukocyte chemotaxis and adhesion to the vascular endothelium decrease blood flow worsening retinal ischemia and therefore perpetuating inflammation. This hypothesis indicates that inflammation seems to gain importance in the pathogenesis of CRVO as it turns into a chronic condition. Therefore, corticosteroids may reduce the risk of CRVO becoming ischemic by blocking this positive feedback.[9,10] Chronic ischemia induces VEGF overexpression promoting neovascularization with its corresponding complications: neovascular glaucoma, retinal detachment, and vitreous hemorrhage (Fig. 12.1).[9,10]

Risk factors

The incidence of RVO increases with age, which is the most important risk factor. In fact, more than 50%–70% of CRVO cases appear in patients over 60 years.[2,12–15] Other known risk factors are systemic arterial hypertension, hyperlipidemia, stroke, and sleep apnea.[3,6,9] Usually, the central vein is compressed by an arteriosclerotic central retinal artery near the lamina cribosa leading to venous stasis and a thrombotic process.[2,16] Some studies found a gender predilection toward men, while others did not report significant differences. The same occurs with laterality where the *IRIS* registry cohort observed a slightly left-eye onset preference in CRVO.[12]

The role of diabetes mellitus has historically been controversial. Recently, a meta-analysis of 37 studies found a statistically significant association with CRVO but not with BRVO.[17] Glaucoma is one of the classic risk factors associated with CRVO. Patients with glaucoma generally have narrower retinal veins than normal subjects, and it is also hypothesized that elevated intraocular pressure may compress vessel walls causing intimal layer proliferation and risk of collapse. Narrow angle and shallow anterior chamber in angle-closure glaucoma have also been associated with RVO.[9,18] Thrombophilia is a relevant factor, especially in young people, in patients with a family history of thrombosis or bilateral involvement. Factor V Leiden, prothrombin mutation, lupus anticoagulant, antiphospholipid antibody, and plasminogen activator inhibitor (PAI-1) have been related to RVO.[1,2,19] Hyperhomocysteinemia (HH) is characterized by the C677T allele with a point mutation at position 677 (677 C $>$ T) of the MTHFR gene in chromosome 1 and it is considered an independent factor for atherosclerosis and thrombosis. An association between HH and RVO has also been described.[1,9] Finally, the Coronavirus Disease-19 has also been proposed as a risk factor for CRVO, secondary to disseminated intravascular coagulation.[20,21]

Clinical features

CRVO usually presents as a sudden, painless, unilateral vision loss, and the clinical features differ between nonischemic and ischemic forms.[16,22] Nonischemic CRVO, the most common type of CRVO (75%), usually presents as a low or mild visual impairment with blurry vision and VA above 20/200.[16,23] Symptoms are more marked after waking up in the morning due to venous stasis and usually improve throughout the day.[16] Ischemic CRVO's (25%) symptomatology is more aggressive, with acute and severe visual loss, which usually is less than 20/200.[16,23] In both cases, the patient may describe previous episodes of amaurosis fugax before the permanent visual drop onset.[16]

Confrontation visual field testing often shows a campimetric defect, more frequently seen in ischemic CRVO than in nonischemic forms, being a peripheral inferior nasal visual field scotoma the most frequent presentation.[9,16] Ischemic CRVO usually manifests a relative afferent pupillary defect (RAPD), less frequently seen in nonischemic forms.[9,16]

Diagnosis of CRVO is directed by fundus examination.[24] The increased intravenous pressure results in venous dilatation and tortuosity, retinal flame-shaped (superficial) dot or blot (ischemic) hemorrhages throughout all four retinal quadrants and most numerous in the periphery, cotton wool spots, hard exudates, and in some cases optical nerve edema and/or ME.[2,3] Cotton wool spots are produced by nerve axonal ischemia, and hard exudates appear as yellow lipid deposits in the limits between the healthy and the affected retina (Table 12.1).[2,24]

CRVO in conjunction with cilioretinal artery occlusion (CLRAO) is a particular case of combined retinal vascular disease. Cilioretinal artery (CLRA) is a branch of the posterior ciliary artery (PCA) or the choroidal vascular plexus and may irrigate part of the macula and one or more retinal sectors. The most accepted theory to explain the pathogenesis of this combined retinal vascular disease is that the sudden occlusion of the central retinal vein elevates the intraluminal pressure in the capillary bed and results in a hemodynamic block of the CLRA.[25] The patient may describe purple blur during the episodes of amaurosis fugax before the permanent blurry vision, centrocecal visual field defects could appear, and the papillomacular bundle would appear pale in these cases.[16,25]

Visual prognosis and complications

The most relevant prognostic factor for visual acuity recovery is the CRVO subtype, with poorer prognosis for ischemic CRVO compared to nonischemic subtypes.[2] Several reports have described a 10%–33% risk conversion from nonischemic to ischaemic types, predominantly in the first month of follow-up.[2,9] The influence of age has also been suggested, with young patients associating a faster recovery and less associated complications, arguably related to a relatively thicker choroid that presumably provides resistance to ischemic injury.[22] Older age, renal disease, diabetes mellitus, grade of retinal hemorrhages and optic disc shallowing, and poor baseline VA have been defined as risk factors for poor visual prognosis in young individuals.[26,27]

Collateral circulation over an atrophic optic nerve head (retina–retina, retina–choroid anastomosis), peripheric capillary stenosis, and macular microaneurysms are signs of chronicity.[3,28] The most sight-threatening effects of CRVO are ME and NV, particularly in ischemic forms. Chronic ME can lead to cystoid macular degeneration, retinal pigment epithelium atrophy, premacular fibrosis, or epiretinal membrane worsening visual prognosis.[16] NV is the result of sustained

TABLE 12.1 Differences in clinical presentation and findings between nonischemic and ischemic central retinal vein occlusion.

	Nonischemic CRVO	Ischemic CRVO
Visual acuity	>20/100	≤20/100
Visual field defect	Rare	Common
RAPD	Rare	Present
Fundus signs	Less: hemorrhages, disc/macular edema, venous tortuosity, cotton-wool spots	More: hemorrhages, disc/macular edema, venous tortuosity, cotton-wool spots
	Retinal capillary nonperfusion <10 disc areas	Retinal capillary nonperfusion >10 disc areas

CRVO, central retinal vein occlusion.

TABLE 12.2 Differences in fundoscopy between acute and chronic CRVO.

Acute CRVO	Chronic CRVO
Initial macular edema	Cystoid macular edema Retinal pigment epithelium atrophy Epiretinal membrane
Optic disc edema	Optic disc atrophy
Venous dilatation and atrophy	Collateral vessels
Hard exudates, cotton-wool spots Retinal hemorrhages	Neovascularization and its complications: vitreous hemorrhages, retinal detachment, neovascular glaucoma

retinal ischemia and more than 60% of patients with ischemic CRVO may develop it. Abnormal new vessel growth in the iris (rubeosis iridis) and the iridocorneal angle may result in neovascular glaucoma. Retinal neovascularization is less frequent and could lead to vitreous hemorrhages and/or retinal detachment (Table 12.2).[3,8,24]

Imaging findings

Ultra-widefield fundus imaging, fundus autofluorescence

Ultra-widefield fundus imaging (UWF) has become essential in retinal vascular diseases for diagnosis, progression monitoring, and management purposes.[29] As a digital type of image, the information can also be evaluated in different color channels (red and green). According to recent reports, the red channel UWF fundus images have more sensitivity to detect ischemic CRVO than color fundus images, especially in nonrubeotic eyes.[30] Fundus autofluorescence (FAF) imaging allows a noninvasive mapping of endogenous fluorophores in the ocular fundus. Atrophy of retinal pigment epithelium is shown as areas of hypopigmentation or decreased autofluorescence due to the reduction in lipofuscin density.[31] A perivenular fern-like FAF hypoautofluorescence pattern has been described in patients with a recent-onset CRVO.[32]

Fundus fluorescein angiography and indocyanine green angiography

Fundus fluorescein angiography (FFA) is a dynamic and invasive technique that can identify the site of occlusion and assess the reduction of flow velocity and the state of retinal perfusion. The definition of ischemic CRVO is a nonperfused area equivalent to a minimum of 10 optic disc areas (DA) in a standard 55 degrees fundus picture technology.[2,24] As suggested by a recent report using UWF, it seems that the distribution of retinal nonperfusion areas is more determinant than its total area extension. According to this study, the involvement of 10 DA of nonperfusion in the posterior pole seemed to highly increase the risk of neovascularization when compared to the same extent of affected areas in the periphery.[33] Retinal nonperfusion can also be estimated by the ischemic index, directed to estimate capillary nonperfused areas (CNP) compared to the total retinal area observed. This parameter has been positively correlated with the central macular thickness (CMT) by some authors, suggesting that targeted retinal photocoagulation may potentially reduce the frequency of anti-VEGF injections.[29,34] FFA can also detect vascular proliferations and help to distinguish collateral vessels from neovascularization.[3,35] Finally, another important role of FFA is to identify and localize veno-venous anastomosis. In cases of no leakage through collateral vessel walls, this feature could be considered as a good prognosis marker, suggesting that reperfusion is effective.[35] Interestingly, some authors have observed a decreased microvascular density in fellow eyes with adaptive-optics scanning light ophthalmoscope fluorescein angiography and suggested that this could be a predictor of future occlusion in those eyes.[36,37] Dynamic indocyanine green angiography (ICGA) is a useful test to evaluate the hemodynamic alterations in CRVO, particularly in the early course of the disease when a pulsatile filling of retinal arteries can be detected due to the retrograde transmission of elevated venous pressure from the site of occlusion.[38]

Optical coherence tomography

Macular optical coherence tomography (OCT) is the most used imaging technique in CRVO for the detection of ME. Quantitative parameters such as CMT, defined as the mean thickness between the inner and outer retinal boarders in the central 1-mm area, macular volume, or average retinal thickness, are routinely used in retinal clinics worldwide on a daily basis. CMT is the most relevant parameter to evaluate activity, progression, and response to treatment in ME. Intraretinal fluid (IRF), subretinal fluid (SRF), or hyperreflective foci (HRF) can also be distinguished and measured in ME associated with CRVO.[24] Qualitative analysis of OCT images includes the description of initial signs of retinal ischemia, such as the presence of a prominent hyperreflective band in the middle retinal layers defined as paracentral acute middle maculopathy in early phases of vein occlusions, as well as the disorganization of the retinal inner layers (DRIL) or the ellipsoid zone (EZ) and external limiting membrane (ELM) disruption, being all these prognostic markers of impaired visual outcome.[24,39] Chronic CRVO changes observed in macular OCT scans include retinal pigment epithelium atrophy or epiretinal membrane formation.[16,24]

Optical coherence tomography angiography

Recently, optical coherence tomography angiography (OCT-A) has been introduced as a noninvasive imaging technique to assess retinal and choroidal vascularization. Microvascular changes in CRVO are visible in OCTA images in the superficial and deep capillary plexus of the retina, including decreased vessel density (VD), vascular tortuosity, microaneurysms, nonperfused areas, capillary dilation and telangiectasias, disruption of the foveal perivascular network, and collateral vessel formation.[40,41] OCT-A parameters such as the foveal avascular zone (FAZ) and vessel density (VD) seem to be correlated with visual function, being the FAZ the best predictor according to some series.[39,42] Some studies have been directed to evaluate the performance of OCT-A compared to FFA to detect areas of retinal nonperfusion, with both techniques being effective for this purpose.[40,43] With regard to nonperfusion areas quantification, a Delphi approach conducted by international experts reached a consensus to define ischemic CRVO in OCTA as a 30% decreased flow area compared to the absolute imaged area.[44] However, cases with concomitant cystoid ME could cause problems of accurate segmentation and shadowing due to artifacts, which may hinder the correct detection of capillaries and may overestimate the areas of nonperfusion.[41]

Clinical trial updates

The optimal treatment of the complications related to CRVO, mainly ME and NV, has evolved in the last decades and includes classic laser therapy, intravitreal injections of anti-VEGF and corticosteroids, as well as the surgical management of selected cases.

Macular laser photocoagulation and panretinal photocoagulation

Classically, the use of laser was widely extended to treat ME (macular laser) and as a preventive measure and to treat NV secondary to CRVO panretinal photocoagulation (PRP). However, the randomized *Central Retinal Vein Occlusion (CRVO)* study showed that ME diminished with macular laser but did not have a statistically significant effect on visual acuity (VA).[45] Regarding NV, a large, randomized, multicentric study by the *Central Vein Occlusion Study Group* concluded that prophylactic PRP is not effective in preventing iris or angle NV, and showed that once any of them appear, eyes with no previous laser treatment respond better to PRP and thus the prevention of neovascular glaucoma (NVG) is more effective.[46]

Intravitreal anti-VEGF

The *CRUISE* trial, a randomized, multicenter study including 392 participants who had ME due to CRVO showed that ranibizumab was better than placebo in terms of visual function.[47] Other later studies replicated these findings, the *CRYSTAL* study revealing a steady gain in VA at months 12 and 24 regardless of macular ischemia at baseline.[48] In addition, the *RELATE* trial showed no further benefit with ranibizumab at a higher dose (2.0 mg), or adding grid laser photocoagulation to the anti-VEGF injections. Interestingly, *McAllister* et al. found a significant reduction in the number of ranibizumab injections needed in patients who received supplementary laser-induced chorioretinal anastomosis.[49,50]

Regarding aflibercept, another two randomized clinical trials, the *COPERNICUS* and the *GALILEO* studies, showed that aflibercept was superior to sham injections in terms of VA change and CRT improvement at 6 months and was maintained in patients that either continued with aflibercept monthly or changed from sham to monthly aflibercept from the 6th month onward to the 52nd week.[51] However, the change in central subfield thickness (CST) was only weak to moderately associated with the best corrected visual acuity (BCVA).[52] Additionally, the phase IV *CENTERA* trial suggested that the treat-and-extend (TAE) protocol of administration of aflibercept adjusted in 2-week intervals after disease stability showed clinically relevant improvements both functionally and anatomically.[53]

Several clinical trials have been directed to compare the efficacy of different anti-VEGF drugs in CRVO. In general, no great differences between drugs have been found although many trials suggest that aflibercept may achieve greater anatomical outcomes and reduce the number of injections required compared to the other compounds.[54,55]

In the *CRAVE* study, no functional or anatomical differences were found between monthly bevacizumab and ranibizumab for 6 months.[56] The *SCORE2* randomized trial comparing aflibercept and bevacizumab showed that aflibercept produced better anatomical results with noninferior VA outcomes compared to bevacizumab at 6 months.[57,58] In addition, they presented that greater macular thickness fluctuations translated into worse VA letter score (VALS) at 12 months, being bevacizumab the one that displayed a higher variation and thus worse VALS results.[59] Moreover, in anti-VEGF treatment responders, no significant differences in VA change were observed between monthly and treat-and-extend regimens for either drug.[60] A specific subanalysis was conducted to evaluate the effect of switching treatment after poor initial response to bevacizumab or aflibercept to another anti-VEGF or to dexamethasone implant; however, no superior results were observed for this switch strategy.[61]

The *LEAVO* study, a multicentric, noninferiority trial described that aflibercept was both noninferior and nonsuperior to ranibizumab when the noninferiority margin was marked to −5 letters and described a reduced number of injections favoring aflibercept compared to ranibizumab (10 vs. 11.8 injections) or bevacizumab (11.5 injections) at 100 weeks. This study was unable to demonstrate that bevacizumab was noninferior to ranibizumab or aflibercept at the intention-to-treat population, and therefore, the possibility that bevacizumab was worse than ranibizumab and aflibercept by five visual acuity letters could not be ruled out. However, they also carried out a compelling economic analysis and concluded that bevacizumab was the most cost-effective alternative with no great differences observed in quality of life-adjusted years (QALYs) among the study cohorts.[62]

Intravitreal corticosteroids

By the first decade of 2000, the use of intravitreal triamcinolone was being widely studied. Promising results started to appear, although soon the VA gain that had been described was considered to be nonlasting.[63,64] Additionally, a specific report of the *SCORE trial* showed that it was also not useful in preventing neovascular events or retinal capillary non-perfusion.[65] Recently, a randomized clinical trial suggests that it improves anatomic results and lowers the number of injections needed when added to the anti-VEGF bevacizumab as opposed to both treatments separately.[66]

Intravitreal dexamethasone implant is licensed for the treatment of ME in RVO. The multicenter, masked, randomized *GENEVA study* showed the efficacy of both 0.7 and 0.35 mg of dexamethasone implants in the treatment of ME due to CRVO compared to sham injections. Improvement in VA persisted longer than the CRT reduction observed in OCT, suggesting that other factors besides CRT reduction could play a role in eyes treated with dexamethasone implant.[67] The main adverse events were ocular hypertension (typically transient in most of the cases), pain, and anterior chamber cells, with no significant difference in the rate of cataract development between treatment groups.[67–69] The *COMRADE* trial was directed to evaluate the efficacy of intravitreal dexamethasone implants compared to anti-VEGF, and reported greater VA gains and a better safety profile for ranibizumab at 6 and 12 months; however, these differences may be explained by the strict treatment criteria that prevented reinjections of the dexamethasone implant before 6 months.[70,71] In the real world, the effectiveness of the dexamethasone implant in CRVO has been demonstrated by multiple studies in terms of visual and anatomical results, with a reduced burden of retreatment compared to anti-VEGF drugs at the expense of side effects such as IOP elevation or cataract progression in phakic eyes.[72,73]

Future treatments: New molecules and surgical approaches

New options are being studied for the treatment of CRVO and its complications. The use of new anti-VEGF molecules such as *brolucizumab* or *faricimab* for the treatment of CRVO-related ME is still being investigated in different clinical trials, such as the *RAVEN* (brolucizumab) or *COMINO (faricimab)* trials, respectively. The *FALCON* study and other groups found similar results of the fusion protein *conbercept* (Lumitin) compared to other anti-VEGFs.[74,75] Hara et al. tried injecting 0.05 mL of *human Activated Protein C* in 10 eyes of 9 patients with ischemic CRVO or hemi-CRVO, with 50% of the eyes achieving complete resolution of the ME with only one injection.[76] Additionally, a phase I–II trial by Ikeda et al. brought significant improvement in BCVA at week 12 with intravitreal injections of a new neuroprotectant drug, named *KUS121*.[77] The use of surgical robots for intraluminal delivery of substances has also been investigated. Willekens et al. completed a phase I trial using robot-assisted retinal vein cannulation with ocriplasmin infusion in four patients, demonstrating that this approach was feasible and safe with promising results.[78] Finally, topical drugs are being investigated to evaluate its ability to reduce the development of neovascularization in CRVO. An example of this is *aganirsen*, a new topical treatment that is being investigated in the *STRONG* study, a phase II/III, randomized, placebo-controlled, multicenter ongoing study.[79]

Management strategies

The management of CRVO encompasses the immediate inhibition of vascular leakage with anti-VEGF drugs, with steroids considered in suitable patients with systemic cardiovascular risk, laser treatment for neovascular complications, surgery for selected cases, as well as a basic systemic workup to recognize systemic concomitants.[24] Early classification between nonischemic and ischemic CRVO is essential, as the latter usually entails a higher rate of complications and thus requires more exhaustive monitoring and treatment.

As CRVO is associated with an elevated risk of atherosclerosis and its risk factors, cardiovascular health should be promptly assessed and managed. Systemic hypertension and hyperlipemia must be treated and smoking cessation is imperative.[80,81] Oral contraceptives should probably be avoided as they have been found to be the most common risk factor associated with RVO in young women, although there is no consensus on this topic.[82] Currently, there is no evidence to advise the use of antithrombotic drugs for the management or prevention of CRVO.[80,81]

Observation alone after a newly diagnosed CRVO with no ME or NV could be a valid initial option, a situation which would correspond in most of the cases to a nonischemic CRVO. Should complications such as ME or NV appear, prompt management must be initiated the following treatments, making individualized decisions for each patient:

- **Intravitreal anti-VEGF**: Ranibizumab and aflibercept are two licensed anti-VEGF used for this purpose, although bevacizumab (Avastin) is also widely used off-label due to the multiple pieces of research supporting its clinical and cost-effectiveness. Repeated injections of either drug are usually needed, with over one-third of the patients needing only three, yet more than 56% still requiring injections 5 years after initiating the therapy—which is usually implemented using loading doses of monthly intravitreal injections for 3 months followed by a treat-and-extend (TAE) or pro-re-nata (PRN) protocol-until the ME is solved (or, in a high number of cases, stabilized with the longest period between injections without recurrence).[83,84] Early treatment is advised, as delaying intravitreal treatment up to 6 months has shown to reduce potential VA gain.[80] Nonresponse to anti-VEGF therapy should be considered if there is no gain in VA or anatomical improvement via OCT central subfield thickness after a number of six injections.[80]
- **Intravitreal corticosteroids**:
  - *Triamcinolone*: it has been shown to cause a temporary increase in VA that is soon reverted, at the expense of considerable side effects (i.e., sterile endophthalmitis). Due to its off-label use and its safety profile, it is currently not extended in routine clinical practice and it has been relegated by other approved steroid implants.
  - *Dexamethasone implant* (Ozurdex): It is licensed for the management of ME associated with CRVO and has demonstrated efficacy in visual and particularly anatomical outcomes. Due to its side effects (mainly IOP elevation and cataract development), it is considered a second-line treatment after confirmed nonresponse to anti-VEGF treatment or in the first line in suitable naïve selected cases (i.e., systemic cardiovascular risk).[24,83,85]
- **Panretinal photocoagulation (PRP)**: Laser PRP causes necrosis of the treated area, reducing the total area of ischemic tissue and diminishing the production of inflammatory factors and VEGF, which in turn reduces the risk of neovascularization.[67] In CRVO, it is often applied in the 360-degree retinal periphery, causing a loss in peripheral vision as a side effect. In ischemic CRVOs, it is widely used for the prevention and treatment of neovascular glaucoma (NVG) with iris or angle neovascularization (NVI, NVA). However, it has been proven not effective as a prevention method and should not be used before NV appear.[82,86]
- **Vitrectomy surgery with endolaser photocoagulation**: In selected cases with associated vitreous hemorrhage preventing adequate laser PRP treatment, a surgical procedure −generally a pars plana vitrectomy (PPV)− may be required with intraoperative 360-degree laser PRP treatment.

Patients with either type of CRVO must be followed up after diagnosis even if no complications are found in the first visit. In ischemic CRVO, monthly visits with a careful search for iris, angle, retinal or disc NV, and ME are recommended over the first 6 months and then every few months for the next 2−3 years.[82] In nonischemic CRVO, the first follow-up visit can be delayed to 3 months and continued for around 1 year if no complications appear.[82]

Clinical cases

Case 1: RE nonischemic CRVO converted to ischemic CRVO

A 70-year-old male presents with RE central retinal vein occlusion (CRVO) and visual acuity of 20/80. Ultrawidefield (UWF) fundus retinography at the time of diagnosis (A) shows multiple flame-shaped, dot and blot hemorrhages in the four quadrants of the retina. Fundus fluorescein angiography (B) at the time of diagnosis reveals mild peripheral ischemic changes and capillary leakage in the far periphery. OCT (C) reveals macular edema with multiple intraretinal fluid cysts and subfoveal neurosensorial detachment. A diagnosis of nonischemic CRVO was made, and a course of intravitreal aflibercept was initiated. During follow-up, 6 months after CRVO onset, the patient comes to casualty for severe RE visual acuity loss, recorded as 20/400. UWF fundus retinography (D) shows ischemic changes with ghost vessels in the temporal periphery, FFA (E) reveals massive ischemic changes with severe peripheral dropout and OCT (F) shows temporal thinning with hyperreflective band in the middle retinal layers, all consistent with a conversion to ischemic CRVO.

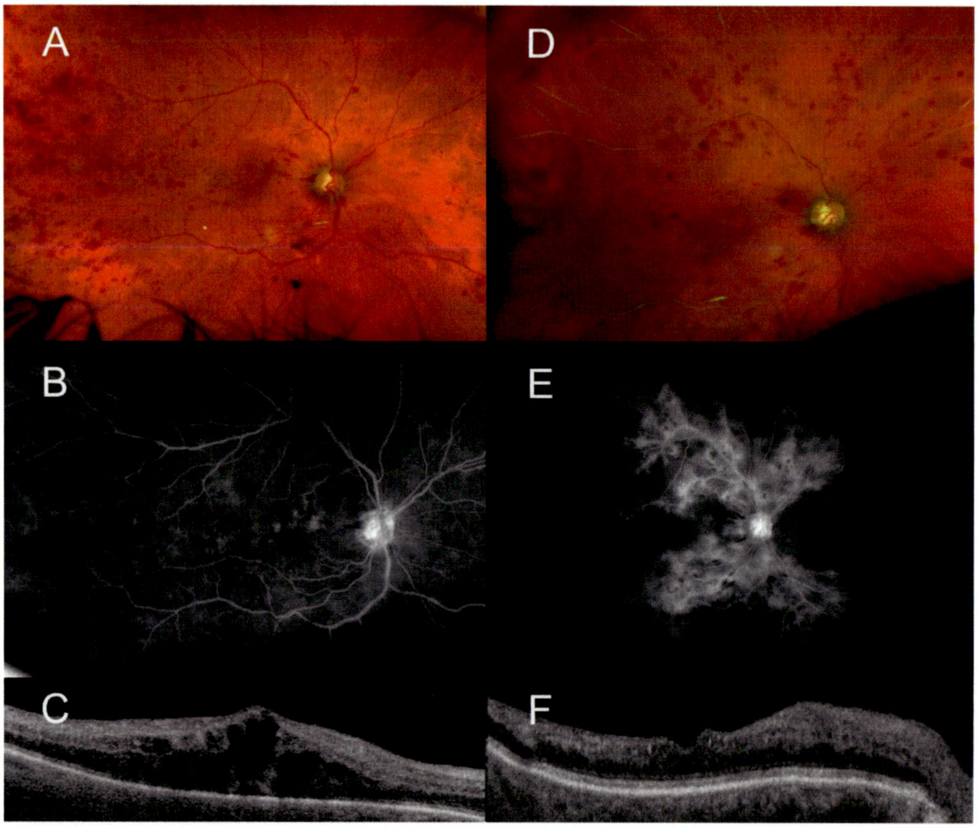

Case 2: LE ischemic CRVO associated with heart transplant

A 52-year-old man was diagnosed with LE CRVO in the emergency room, 1 month after a heart transplant for complex congenital heart disease. Visual acuity was LE 20/63 and RE 20/20. UWF fundus retinography (A) showed abundant flame and dot hemorrhages in all four quadrants, venous dilation, and vascular tortuosity. LE OCT (B) revealed hyperreflective band-like lesions in the middle retinal layers suggesting paracentral acute middle maculopathy (PAMM). One month later, LE visual acuity was 20/100, OCT (C) revealed macular edema with intraretinal fluid and subfoveal neurosensorial detachment, and intravitreal therapy with monthly ranibizumab injections was started. Fundus fluorescein angiography (FFA) performed after the loading dose showed peripheral vascular dropout in the early arteriovenous phase (D) and mild peripheral vascular leakage with no central edema in the late venous phase (E). Following antiangiogenic treatment (F) for relapsing macular edema (G), retinal hemorrhages markedly improved, and several cotton-wool spots in nasal retina and temporal vascular arcades were observed. After 2 years of follow-up, FFA (H) revealed small areas of temporal macular retinal ischemia, and at late phase (I), petaloid macular edema and areas of peripheral vascular closure with vascular leakage and signs of early neovascularization and panretinal laser photocoagulation were indicated. Four years after CRVO onset (J), LE visual acuity is 20/32 and OCT (K) shows no macular edema and temporal macular thinning.

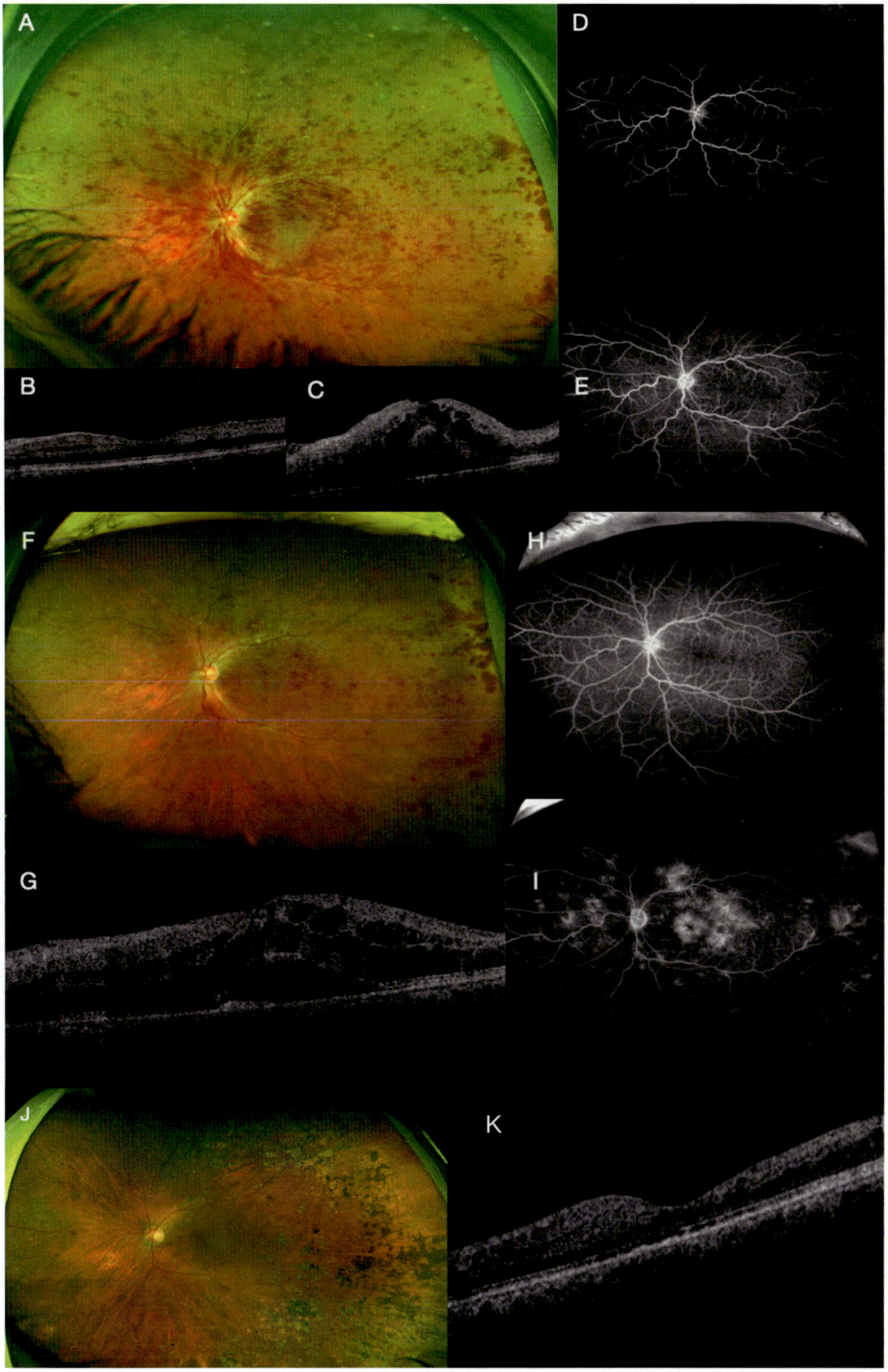

Case 3: LE CRVO associated with prothrombin mutation

A 47-year-old woman was referred for LE CRVO, 1 year after being diagnosed and treated with intravitreal anti-VEGF injections in another center. Visual acuity was LE 20/25 and RE 20/20. She presented high blood pressure and a systemic workup revealed a heterozygous prothrombin G20210A mutation. UWF fundus retinography (A) showed vascular tortuosity and venous dilation, hemorrhages in four retinal quadrants, and cotton-wool spots in the posterior pole. OCT (B) presented isolated diffuse intraretinal cysts and a minimal flat subfoveal neurosensorial detachment. Two months later, LE macular edema relapsed, VA decreased to 20/63, and OCT (C) revealed greater intraretinal and subretinal fluid. A course of intravitreal aflibercept was restarted, with improvement in the signs of occlusion in both retinography (D) and OCT (E). OCT angiography showed no apparent signs of vascular closure in the superficial (F) and deep capillary plexus (G). After completing three monthly doses of aflibercept, we observed an improvement in venous tortuosity and the number of hemorrhages in the UWF fundus retinography (H), and fluorescein angiography (I–J) revealed areas of ischemia in the temporal periphery with no macular edema.

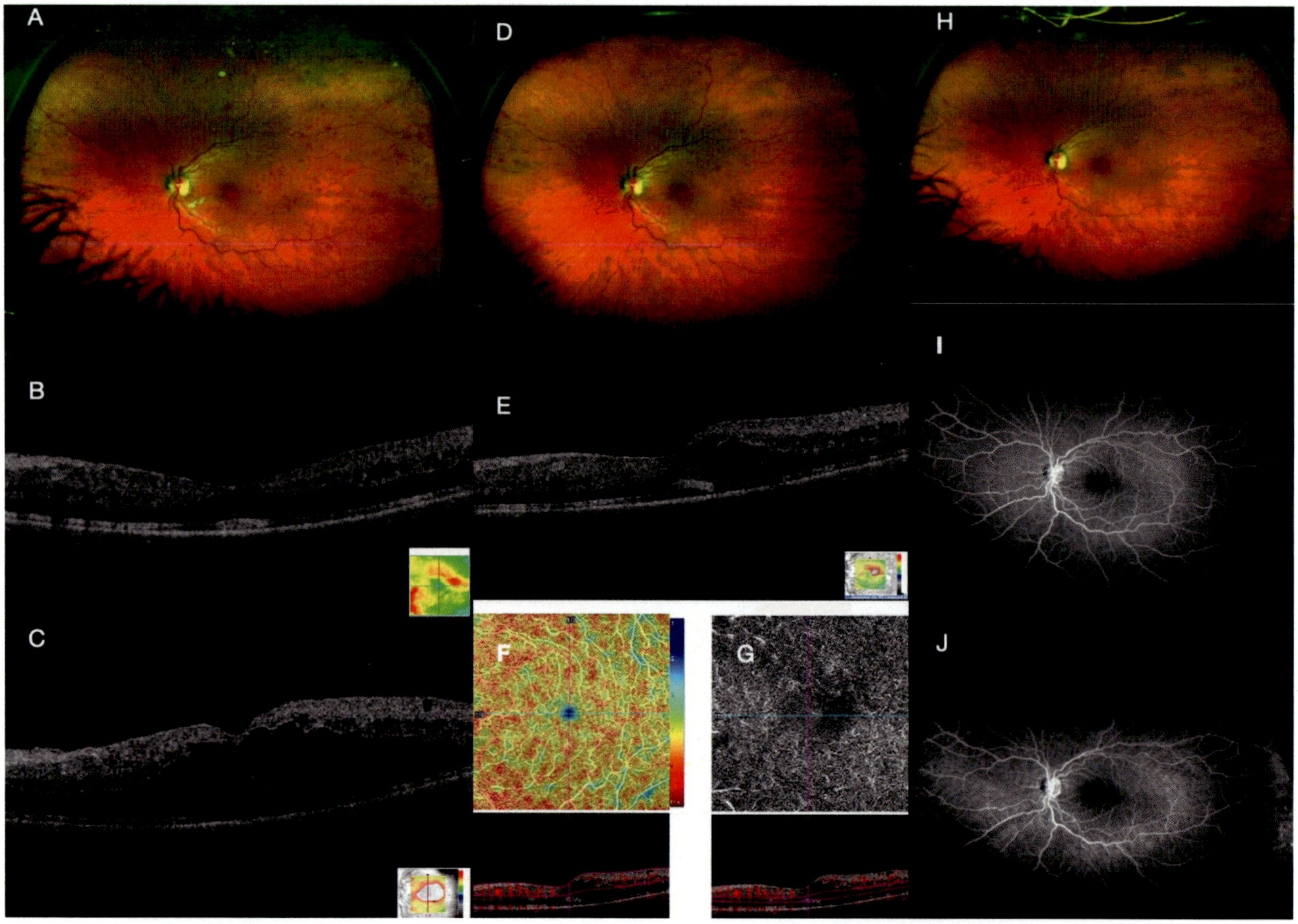

References

1. Vieira MJ, Campos A, do Carmo A, Arruda H, Martins J, Sousa JP. Thrombophilic risk factors for retinal vein occlusion. *Scientific Reports*. 2019;9(1):5–10.
2. Rehak M, Wiedemann P. Retinal vein thrombosis: pathogenesis and management. *Journal of Thrombosis and Haemostasis*. 2010;8(9):1886–1894.
3. Ip M, Hendrick A. Retinal vein occlusion review. *Asia-Pacific Journal of Ophthalmology*. 2018;7(1):40–45.
4. Cugati S, Wang JJ, Knudtson MD, et al. Retinal vein occlusion and vascular mortality. Pooled data analysis of 2 population-based cohorts. *Ophthalmology*. 2007;114(3):520–524.

5. Song P, Xu Y, Zha M, Zhang Y, Rudan I. Global epidemiology of retinal vein occlusion: a systematic review and meta-analysis of prevalence, incidence, and risk factors. *Journal of Global Health*. 2019;9(1).
6. Ponto KA, Elbaz H, Peto T, et al. Prevalence and risk factors of retinal vein occlusion: the Gutenberg Health Study. *Journal of Thrombosis and Haemostasis*. 2015;13(7):1254–1263.
7. Hayreh SS, Hayreh MS. Hemi-central retinal vein occlusion: pathogenesis, clinical features, and natural history. *Archives of Ophthalmology*. September 1980;98(9):1600–1609.
8. Hayreh SS. Classification of central retinal vein occlusion. *Ophthalmology*. 1983;90(5):458–474.
9. Tang Y, Cheng Y, Wang S, Wang Y, Liu P, Wu H. Review: the development of risk factors and cytokines in retinal vein occlusion. *Frontiers of Medicine*. 2022;9(June):1–18.
10. Noma H, Yasuda K, Shimura M. Cytokines and pathogenesis of central retinal Vein occlusion. *Journal of Clinical Medicine*. 2020;9(11):1–16.
11. Wang B, Zhang X, Chen H, Koh A, Zhao C, Chen Y. A review of intraocular biomolecules in retinal vein occlusion: toward potential biomarkers for companion diagnostics. *Frontiers in Pharmacology*. 2022;13(April):1–13.
12. Li Y, Hall NE, Pershing S, et al. Age, gender, and laterality of retinal vascular occlusion: a retrospective study from the IRIS registry. *Ophthalmology Retina*. 2022;6(2):161–171.
13. Klein R, Moss SE, Meuer SM, Klein BEK. The 15-year cumulative incidence of retinal vein occlusion. *Archives of Ophthalmology*. 2008;126(4):513–518.
14. Baseline and early natural history report: the central vein occlusion study. *Archives of Ophthalmology*. August 1993;111(8):1087–1095.
15. Yannuzzi LA, Sorenson JA, Sobel RS, et al. Risk factors for central retinal vein occlusion. *Archives of Ophthalmology*. May 1996;114(5):545–554.
16. Hayreh SS. Retinal vein occlusion. *Indian Journal of Ophthalmology*. 1994;42(3):109–132.
17. Wang Y, Wu S, Wen F, Cao Q. Diabetes mellitus as a risk factor for retinal vein occlusion: a meta-analysis. *Medicine*. 2020;99(9).
18. Yin X, Li J, Zhang B, Lu P. Association of glaucoma with risk of retinal vein occlusion: a meta-analysis. *Acta Ophthalmologica*. 2019;97(7):652–659.
19. Bucciarelli P, Passamonti SM, Gianniello F, Artoni A, Martinelli I. Thrombophilic and cardiovascular risk factors for retinal vein occlusion. *European Journal of Internal Medicine*. 2017;44:44–48.
20. Ullah I, Sohail A, Shah MUFA, et al. Central Retinal Vein Occlusion in patients with COVID-19 infection: a systematic review. *Annals of Medicine and Surgery*. 2021;71(September):102898.
21. Invernizzi A, Pellegrini M, Messenio D, et al. Impending central retinal vein occlusion in a patient with Coronavirus disease 2019 (COVID-19). *Ocular Immunology and Inflammation*. 2020;28(8):1290–1292.
22. Zhang XT, Zhong YF, Xue YQ, et al. Clinical features of central retinal vein occlusion in young patients. *Ophthalmology and Therapy*. 2022;11(4):1409–1422.
23. Hayreh SS, Podhajsky PA, Zimmerman MB. Natural history of visual outcome in central retinal vein occlusion. *Ophthalmology*. January 2011;118(1):119–133.e2.
24. Schmidt-Erfurth U, Garcia-Arumi J, Gerendas BS, et al. Guidelines for the management of retinal vein occlusion by the European society of retina specialists (EURETINA). *Ophthalmologica*. 2019;242(3):123–162.
25. Tesfaw AK, Batra NN, Phan CT. Clinical features and treatment outcome of a concurrent central retinal vein occlusion and cilioretinal artery occlusion. *Case Reports in Ophthalmology*. 2022;13(3):770–776.
26. Koh YY, Lai CC, Wu WC, et al. Baseline clinical features predict visual outcome in young patients with central retinal vein occlusion. *Graefe's Archive for Clinical and Experimental Ophthalmology*. 2020;258(7):1367–1377.
27. Eah KS, Kim YN, Park YJ, et al. Central retinal vein occlusion in young patients: clinical characteristics and prognostic factors. *Retina*. 2021;41(3):630–637.
28. Hayreh SS, Zimmerman MB. Fundus changes in central retinal vein occlusion. *Retina*. 2015;35(1):29–42.
29. Patel SN, Shi A, Wibbelsman TD, Klufas MA. Ultra-widefield retinal imaging: an update on recent advances. *Therapeutic Advances in Ophthalmology*. January 2020;12(6).
30. Sato A, Asaoka R, Tanaka S, et al. Using ultra-widefield red channel images to improve the detection of ischemic central retinal vein occlusion. *PLoS One*. 2021;16(11 November):1–7.
31. Schmitz-Valckenberg S, Pfau M, Fleckenstein M, et al. Fundus autofluorescence imaging. *Progress in Retinal and Eye Research*. 2021;81(August 2020):100893.
32. Domain S, Coherence O. Perivenular whitening in central vein occlusion described by fundus auto fluorescence and spectral domain optical coherence tomography. *Retina*. 2012:1438–1439.
33. Nicholson L, Vazquez-Alfageme C, Patrao NV, et al. Retinal nonperfusion in the posterior Pole is associated with increased risk of neovascularization in central retinal vein occlusion. *American Journal of Ophthalmology*. 2017;182:118–125.
34. Kumar V, Surve A, Kumawat D, et al. Ultra-wide field retinal imaging: a wider clinical perspective. *Indian Journal of Ophthalmology*. 2021;69(4):824.
35. Turczyńska MJ, Krajewski P, Brydak-Godowska JE. Wide-field fluorescein angiography in the diagnosis and management of retinal vein occlusion: a retrospective single-center study. *Medical Science Monitor*. 2021;27:1–10.
36. Pinhas A, Dubow M, Shah N, et al. Fellow eye changes in patients with nonischemic central retinal vein occlusion. *Retina*. 2015;35(10):2028–2036.
37. Fan W, Fleming A, Van Hemert J, et al. Retinal vascular bed area in eyes with retinal vein occlusion on ultra-widefield fluorescein angiography: WAVE study. *Retina*. 2022;42(10):1883–1888.

38. Paques M, Garmyn V, Catier A, Naoun K, Vicaut E, Gaudric A. Analysis of retinal and choroidal circulation during central retinal vein occlusion using indocyanine green videoangiography. *Archives of Ophthalmology.* 2001;119(12):1781–1787.
39. Hajdu D, Told R, Angeli O, et al. Identification of microvascular and morphological alterations in eyes with central retinal non-perfusion. *PLoS One.* 2020;15(11 November):1–13.
40. Glacet-Bernard A, Miere A, Houmane B, Tilleul J, Souied E. Nonperfusion assessment in retinal vein occlusion: comparison between ultra-widefield fluorescein angiography and widefield optical coherence tomography angiography. *Retina.* 2021;41(6):1202–1209.
41. Tsai G, Banaee T, Conti FF, Singh RP. Optical coherence tomography angiography in eyes with retinal vein occlusion. *Journal of Ophthalmic and Vision Research.* 2018;13(3):315–332.
42. Moussa M, Leila M, Bessa AS, et al. Grading of macular perfusion in retinal vein occlusion using en-face swept-source optical coherence tomography angiography: a retrospective observational case series. *BMC Ophthalmology.* 2019;19(1):1–13.
43. Nobre Cardoso J, Keane PA, Sim DA, et al. Systematic evaluation of optical coherence tomography angiography in retinal vein occlusion. *American Journal of Ophthalmology.* 2016;163:93–107.
44. Munk MR, Kashani AH, Tadayoni R, et al. Recommendations for OCT angiography reporting in retinal vascular disease: a Delphi approach by international experts. *Ophthalmology Retina.* 2022;6(9):753–761.
45. Clarkson JG, Chuang E, Gass D, Pedroso M, Cubillas T, Duria ES. Evaluation of grid pattern photocoagulation for macular edema in central vein occlusion. Central vein occlusion study group M report. *Ophthalmology.* 1995;102(10):1425–1433.
46. Central Vein Occlusion Study Group. A randomized clinical trial of early panretinal photocoagulation for ischemic central vein occlusion. The Central vein occlusion study group N report. *Ophthalmology.* 1995;102(10):1434–1444.
47. Thach AB, Yau L, Hoang C, Tuomi L. Time to clinically significant visual acuity gains after ranibizumab treatment for retinal vein occlusion: BRAVO and CRUISE trials. *Ophthalmology.* May 2014;121(5):1059–1066.
48. Larsen M, Waldstein SM, Priglinger S, et al. Sustained benefits from ranibizumab for central retinal vein occlusion with macular edema: 24-month results of the CRYSTAL study. *Ophthalmology Retina.* February 2018;2(2):134–142.
49. Campochiaro PA, Hafiz G, Mir TA, et al. Scatter photocoagulation does not reduce macular edema or treatment burden in patients with retinal vein occlusion: the RELATE trial. *Ophthalmology.* July 2015;122(7):1426–1437.
50. McAllister IL, Smithies LA, Chen FK, Mackey DA, Sanfilippo PG. Two-year efficacy of ranibizumab plus laser-induced chorioretinal anastomosis vs ranibizumab monotherapy for central retinal vein occlusion: a randomized clinical trial. *JAMA Ophthalmology.* December 2018;136(12):1391–1397.
51. Pielen A, Clark WL, Boyer DS, et al. Integrated results from the COPERNICUS and GALILEO studies. *Clinical Ophthalmology.* August 2017;11:1533–1540.
52. Gu SZ, Nanegrungsunk O, Bressler SB, et al. Correlation between change in central subfield thickness and change in visual acuity in macular edema due to retinal vein occlusion: post hoc analysis of COPERNICUS, GALILEO, and VIBRANT. *Graefes Archive for Clinical and Experimental Ophthalmology.* December 2022;260(12):3799–3807.
53. Korobelnik JF, Larsen M, Eter N, et al. Efficacy and safety of intravitreal aflibercept treat-and-extend for macular edema in central retinal vein occlusion: the CENTERA study. *American Journal of Ophthalmology.* July 2021;227:106–115.
54. Lotfy A, Solaiman KAM, Abdelrahman A, Samir A. Efficacy and frequency of intravitreal aflibercept versus bevacizumab for macular edema secondary to central retinal vein occlusion. *Retina.* 2018;38(9):1795–1800.
55. Casselholm De Salles M, Amrén U, Kvanta A, Epstein DL. Injection frequency of aflibercept versus ranibizumab in a treat-and-extend regimen for central retinal vein occlusion: a randomized clinical trial. *Retina.* 2019;39(7):1370–1376.
56. Rajagopal R, Shah GK, Blinder KJ, et al. Bevacizumab versus ranibizumab in the treatment of macular edema due to retinal vein occlusion: 6-month results of the CRAVE study. *Ophthalmic Surgery, Lasers and Imaging Retina.* September 2015;46(8):844–850.
57. Etheridge T, Blodi B, Oden N, et al. Spectral domain OCT predictors of visual acuity in the study of comparative treatments for retinal vein occlusion 2: SCORE 2 report 15. *Ophthalmology Retina.* October 2021;5(10):991–998.
58. Scott IU, Van Veldhuisen PC, Ip MS, et al. Effect of bevacizumab vs aflibercept on visual acuity among patients with macular edema due to central retinal vein occlusion: the SCORE2 randomized clinical trial. *Journal of the American Medical Association.* May 2017;317(20):2072–2087.
59. Scott IU, Oden NL, VanVeldhuisen PC, Ip MS, Blodi BA. SCORE2 Report 17: macular thickness fluctuations in anti-VEGF-treated patients with central or hemiretinal vein occlusion. *Graefes Archive for Clinical and Experimental Ophthalmology.* May 2022;260(5):1491–1500.
60. Scott IU, VanVeldhuisen PC, Ip MS, et al. Comparison of monthly vs treat-and-extend regimens for individuals with macular edema who respond well to anti-vascular endothelial growth factor medications: secondary outcomes from the SCORE2 randomized clinical trial. *JAMA Ophthalmology.* April 2018;136(4):337–345.
61. Ip MS, Oden NL, Scott IU, et al. Month 12 outcomes after treatment change at month 6 among poor responders to aflibercept or bevacizumab in eyes with macular edema secondary to central or hemiretinal vein occlusion. *JAMA Ophthalmology.* March 2019;137(3):281.
62. Hykin P, Prevost AT, Sivaprasad S, et al. Intravitreal ranibizumab versus aflibercept versus bevacizumab for macular oedema due to central retinal vein occlusion: the LEAVO non-inferiority three-arm RCT. *Health Technology Assessment.* June 2021;25(38):1–196.
63. Jonas JB, Akkoyun I, Kamppeter B, Kreissig I, Degenring RF. Intravitreal triamcinolone acetonide for treatment of central retinal vein occlusion. *European Journal of Ophthalmology.* 2005;15(6):751–758.
64. Ramezani A, Entezari M, Moradian S, Tabatabaei H, Kadkhodaei S. Intravitreal triamcinolone for acute central retinal vein occlusion; a randomized clinical trial. *Graefes Archive for Clinical and Experimental Ophthalmology.* December 2006;244(12):1601–1606.

65. Chan CK, Ip MS, Vanveldhuisen PC, et al. SCORE study report #11: incidences of neovascular events in eyes with retinal vein occlusion. *Ophthalmology*. July 2011;118(7):1364–1372.
66. Vinković M, Bosnar D, Tedeschi Reiner E, De Salvo G, Matić S. Combined treatment with bevacizumab and triamcinolone acetonide for macular edema due to retinal vein occlusion. *Acta Clinica Croatica*. 2020;59(4):569–575.
67. Haller JA, Bandello F, Belfort R, et al. Randomized, sham-controlled trial of dexamethasone intravitreal implant in patients with macular edema due to retinal vein occlusion. *Ophthalmology*. 2010;117(6).
68. Rajesh B, Zarranz-Ventura J, Fung AT, et al. Safety of 6000 intravitreal dexamethasone implants. *British Journal of Ophthalmology*. January 2012;104(1):1–8.
69. Zarranz-Ventura J, Sala-Puigdollers A, Velazquez-Villoria D, et al. Long-term probability of intraocular pressure elevation with the intravitreal dexamethasone implant in the real-world. *PLoS One*. 2019;14(1):1–13.
70. Feltgen N, Hattenbach LO, Bertelmann T, et al. Comparison of ranibizumab versus dexamethasone for macular oedema following retinal vein occlusion: 1-year results of the COMRADE extension study. *Acta Ophthalmologica*. 2018;96(8):e933–e941.
71. Pielen A, Feltgen N, Hattenbach LO, et al. Ranibizumab pro Re nata versus dexamethasone in the management of ischemic retinal vein occlusion: post-hoc analysis from the COMRADE trials. *Current Eye Research*. May 2020;45(5):604–614.
72. Garweg JG, Zandi S. Retinal vein occlusion and the use of a dexamethasone intravitreal implant (Ozurdex) in its treatment. *Graefe's Archive for Clinical and Experimental Ophthalmology*. 2016;254(7):1257–1265. https://doi.org/10.1007/s00417-016-3350-x.
73. Soliman MK, Zarranz-Ventura J, Chakravarthy U, et al. United Kingdom database study of intravitreal dexamethasone implant (Ozurdex) for macular edema related to retinal vein occlusion. *Retina*. December 14, 2022;43(4):679–687.
74. Sun Z, Zhou H, Lin B, et al. Efficacy and safety of intravitreal conbercept injections in macular edema secondary to retinal vein occlusion. *Retina*. September 2017;37(9):1723–1730.
75. Song W, Jiao W, Li F, Ma A, Zhao B. Evaluation of microvascular structure changes after conbercept treatment on macular edema secondary to retinal vein occlusion. *BioMed Research International*. 2020;2020.
76. Hara C, Kamei M, Sakaguchi H, et al. Activated protein C for ischemic central retinal vein occlusion: one-year results. *Ophthalmology Retina*. 2019;3(1):93–94.
77. Ikeda HO, Muraoka Y, Hata M, et al. Safety and effectiveness of a novel neuroprotectant, KUS121, in patients with non-arteritic central retinal artery occlusion: an open-label, non-randomized, first-inhumans, phase 1/2 trial. *PLoS One*. 2020;15(2):1–16.
78. Willekens K, Gijbels A, Smits J, et al. Phase I trial on robot assisted retinal vein cannulation with ocriplasmin infusion for central retinal vein occlusion. *Acta Ophthalmologica*. February 2021;99(1):90–96.
79. Lorenz K, Scheller Y, Bell K, et al. A prospective, randomised, placebo-controlled, double-masked, three-armed, multicentre phase II/III trial for the study of a topical treatment of Ischaemic central retinal vein occlusion to prevent neovascular glaucoma - the STRONG study: study protocol for a randomised controlled trial. *Trials*. March 2017;18(128):1–12.
80. Nicholson L, Talks SJ, Amoaku W, Talks K, Sivaprasad S. Retinal vein occlusion (RVO) guideline: executive summary. *Eye*. March 2022;36(5):909–912.
81. Ageno W, Beyer-Westendorf J, Garcia DA, Lazo-Langner A, McBane RD, Paciaroni M. Guidance for the management of venous thrombosis in unusual sites. *Journal of Thrombosis and Thrombolysis*. January 2016;41(1):129–143.
82. Salmon JF. Oclusiones venosas retininanas. In: *Kanski Oftalmología Clínica Un Enfoque Sistemático*. 9th ed. Barcelona: Elsevier España; 2020:514–525.
83. Scott IU, Campochiaro PA, Newman NJ, Biousse V. Retinal vascular occlusions. *Lancet (London, England)*. December 2020;396(10266):1927–1940.
84. Hunt A, Nguyen V, Bhandari S, et al. Central retinal vein occlusion 36-month outcomes with anti-VEGF: the fight retinal blindness! Registry. *Ophthalmology Retina*. November 9, 2022;7(4):1–8. https://doi.org/10.1016/j.oret.2022.11.001.
85. Yuan Q, Gao Y, Liu Y, Xu H, Wang T, Zhang M. Efficacy of single-dose intravitreal dexamethasone implantation for retinal vein occlusion patients with refractory macular edema: a systematic review and meta-analysis. *Frontiers in Pharmacology*. 2022;13(September):1–10.
86. Hayreh SS. Photocoagulation for retinal vein occlusion. *Progress in Retinal and Eye Research*. 2021;85:1–21.

Chapter 13

Branch retinal vein occlusion

Laurentino Biccas Neto[1], Fernando M. Penha[2], Michel Pacques[3] and Ricardo Leitão Guerra[4]
[1]*Ocular Vitória, Vitória, ES, Brazil;* [2]*Universidade Regional de Blumenau, Blumenau, SC, Brazil;* [3]*Paris Eye Imaging Group, Centre Hospitalier National d'Ophtalmologie des Quinze-Vingts, INSERM-DHOS Clinical Investigation Center, Paris, France;* [4]*Leitão Guerra Oftalmologia, Salvador, BA, Brazil*

Overview

- Introduction
- Classification
- Epidemiology
- Pathogenesis: general overview of disease mechanisms in branch retinal vein occlusion (BRVO)
- Clinical features: symptoms, clinical findings, risk factors, prognosis
- Imaging findings: color pictures, angiography, OCT, OCT-A, wide-field imaging
- Management strategies: laser photocoagulation, pharmacological treatment (steroids, anti-VEGF, etc.), surgical treatment (vitrectomy)
- Recent clinical trials updates: review of recent publications in the field
- New perspectives: adaptive optics in the diagnosis of BRVO

Introduction

Retinal vein occlusions (RVOs) are common retinal disorders, second only to diabetic retinopathy, and are important causes of visual loss.[1] RVOs comprise central (CRVO) and branch (BRVO) presentations if the central retinal vein or any of its branches are affected, respectively. Additionally, hemi-RVO, which affects the anterior part of a trunk from the central retinal vein, has been considered as a subtype either of CRVO or BRVO, although it should better be considered a separate entity.[2]

Classification

Since the first description by Leber,[1] RVOs can be classified according to their anatomic location either as major when one of the major venous branches draining one of the retinal quadrants is involved, or as macular when smaller venules within the macula are affected (Fig. 13.1).

Noteworthily, the superotemporal quadrant is by far the most frequent location in BRVOs, probably due to the increased presence of arteriovenous crossings or because extramacular, mostly nasal BRVOs, may not cause visual symptoms and stay undetected[3] (Figs. 13.2 and 13.3). Moreover, retinal capillary perfusion assessed by fluorescein angiography may be used to divide BRVOs into ischemic (more than five disc-diameter area of capillary nonperfusion) and nonischemic types, although this is of limited value for BRVO as opposed for CRVO.

Epidemiology and risk factors

The incidence of BRVO is three times more common than of CRVO. BRVO occurrence increases with age, especially around the 7th decade, and may affect men and women equally. Hypertension is a major risk factor, and all patients with

Retinal and Choroidal Vascular Diseases of the Eye. https://doi.org/10.1016/B978-0-443-15583-3.00013-5

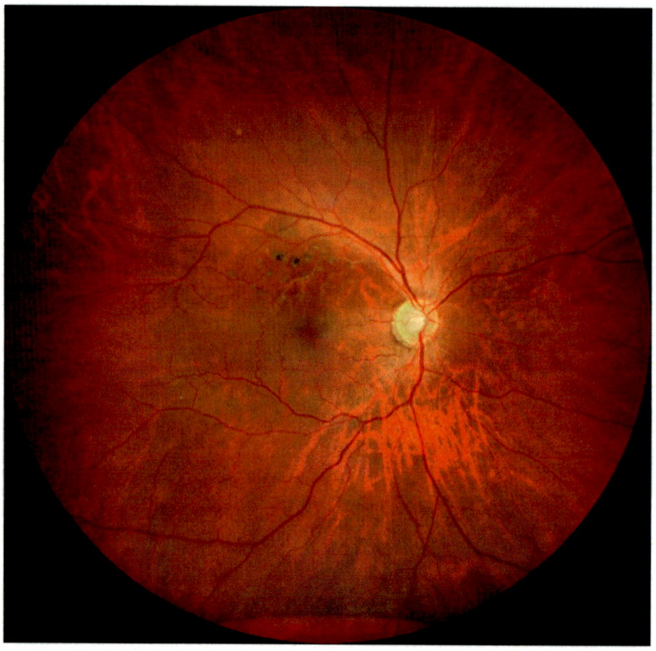

FIGURE 13.1 Superior macular BRVO in the OD treated by grid laser photocoagulation. Laser scars in the parafoveal superior region.

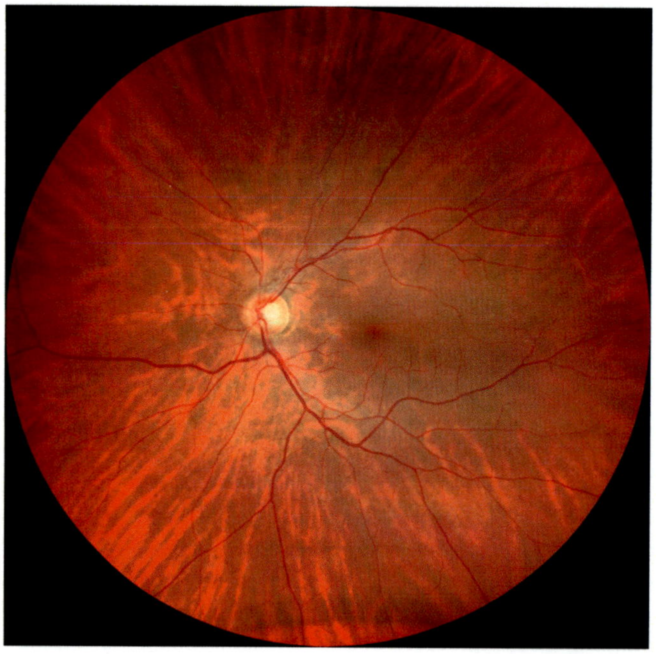

FIGURE 13.2 Asymptomatic BRVO affecting the superior extramacular area of the OS (BCVA: 20/20). Superficial retinal hemorrhages and ghost vessels can be seen superiorly.

RVO should be medically evaluated. Glaucoma is another important risk factor for RVOs in general, and it increases by 2.5 the risk for BRVO.

Usually, extensive diagnostic workups are noncontributory. Nevertheless, in the absence of cardiovascular disease or glaucoma, a search for other causative or predisposing systemic conditions should be considered, mainly in patients younger than 50 years old. These patients may have benefits for an additional workup for other causes of occlusion, such as thrombophilia, inflammatory conditions, drug abuse, or infectious diseases. However, an extensive hematological evaluation is controversial. Hayreh states that RVO occlusion in young patients (less than 45 years old) is not common but

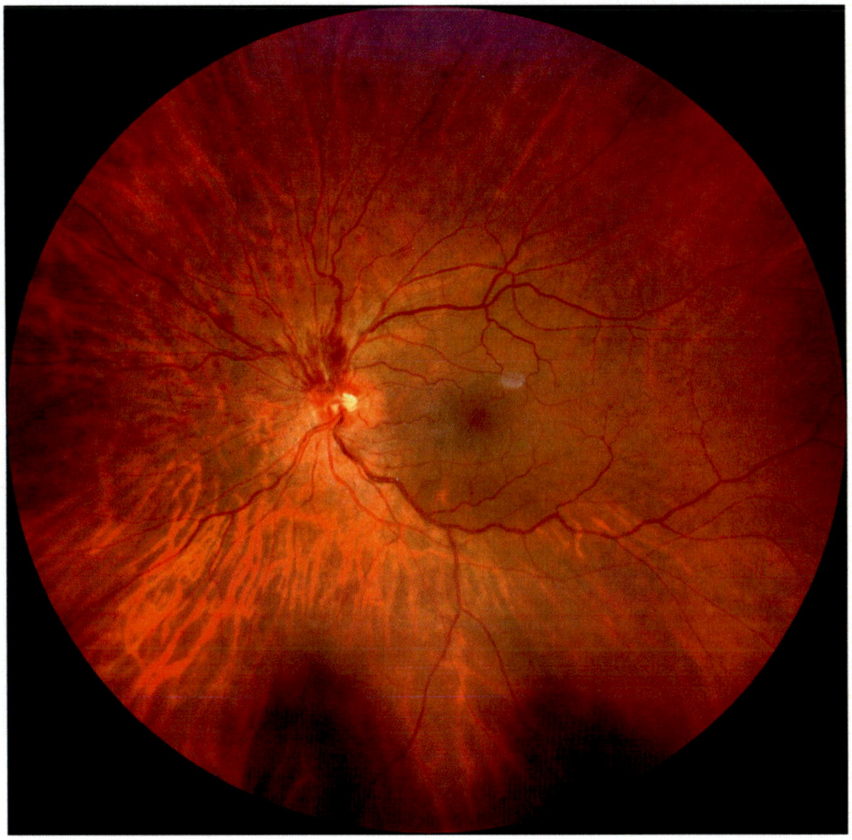

FIGURE 13.3 Asymptomatic BRVO affecting the superior and nasal extramacular areas of the OS (BCVA: 20/20). Retinal hemorrhages in the nasal and superior region but no macular edema.

occurs and represents 5% of all BRVO cases.[4] Usually, an inexpensive hematological evaluation is usually sufficient, unless, of course, there are indications to the contrary. It is important to state that venous thrombosis is crucial for the development of RVO, including BRVO. Since the discovery of antithrombin deficiency in 1965, many thrombophilic conditions have been described so far, which have then allowed to currently detect an inherited or acquired predisposition in approximately 60%–70% of patients with thromboembolic disorders.[5] Thus, there may be other factors not yet described that may contribute to and could explain this thrombotic event in young patients.

Pathogenesis

Characteristically, BRVOs most commonly occur at an arteriovenous crossing, where thickening of the arterial wall and glial proliferation compress the adjacent vein within a common adventitial sheath (Fig. 13.4). It may also occur in a different setting, however, in this situation, other underlying diseases should be investigated, such as retinochoroiditis or retinal vasculitis.

Despite the clinical observation that the vein occlusion occurs at an arterial compression over the vein, what can actually be seen in most of the cases (in 97.6% of the eyes in some reports[6]), it has not been documented so far in histopathological studies. This phenomenon may occur due to a deeper position of the vein, rather than as a result of true artery compression from the neighboring artery.[7]

Branch retinal vein occlusion is a multifactorial disease, and the three classical components of Virchow's triad play a role in the process: venous stasis and vascular injury at arteriovenous crossings lead to a hypercoagulable state and consequently occlusion at the site.

The disease can manifest as ischemic or nonischemic. Ischemic lesions typically appear to have extensive hemorrhagic infarcts of the retina, with retinal edema, leading to congestion and dilation of the veins. In a later stage of the disease, the artery serving the area of occlusion may become narrowed and sheathed. The affected retina will clear out the hemorrhage over time, in some cases resulting in irreversible fibrotic scars.[6]

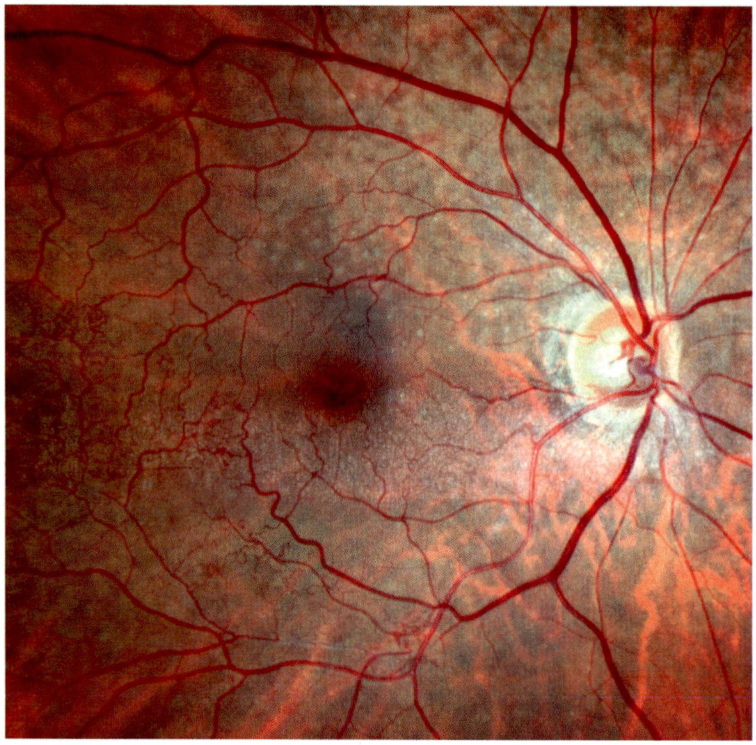

FIGURE 13.4 BRVO in the OD, with typical occlusion at an AV crossing. An occlusion site can be seen at an A–V crossing inferiorly.

Clinical features

Clinical findings

The vein obstructs at the level of arteriovenous crossing and may occur in a main branch or in a macular subbranch, resulting in more sectoral disease and may be ischemic or nonischemic[7]—ischemic BRVO is classically defined as an occlusion with >5-disc diameters of nonperfusion on fluorescein angiography (FA). Clinical features, prognosis, and management of these two subtypes of BRVO are different.[4,8] There is also a special type of BRVO that may occur at the edge of the disk resulting in a hemi-central retinal vein occlusion (HCRVO) with clinical features and prognosis similar to CRVO. Major branch occlusions can progress to retinal ischemia, neovascularization, and macular edema.

BRVO is characterized by the obstruction and retrograde dilation of retinal veins, usually associated in the acute phase with intraretinal hemorrhages, intraretinal fluid, cotton-wool spots, hard exudates, and macular edema. It is usually unilateral in 90% of the cases.[9,10] The most affected quadrant is the superotemporal (63%), and a clinical finding of nasal obstruction is very unusual.[11,12] There may be vitreous hemorrhage, especially in the acute phase (Figs. 13.5 and 13.6).

The most common symptom of the disease is vision loss, usually acute and nonresolving. However, in some cases, it may present initially with transient vision loss. Vision may deteriorate further in the evolution from nonischemic to ischemic BRVO.

Prognosis

In acute disease, the visual impact of BRVO is directly associated with the presence or absence of macular or foveal involvement. The presence of retinal hemorrhages may lead to visual field loss, but the foveal involvement along with macular edema will affect visual acuity.

Before the availability of anti-VEGF agents, the Branch Vein Occlusion Study (BVOS) found that the incidence of neovascularization from the retina or optic nerve was 36% in eyes with extensive retinal ischemia, defined as an area of nonperfusion in the fluorescein angiogram of at least five disc-diameters in size. In the absence of laser photocoagulation, 60%–90% of those cases could develop vitreous hemorrhage.[13,14]

Permanent vision loss may occur over time, which could be related to macular ischemia, retinal atrophy, cystoid macular edema (CME), hard exudates in the central fovea, pigmentary macular changes, subretinal fibrosis, and epiretinal membrane formation. Other less common complications could include vitreous hemorrhage and tractional retinal detachment.

Neovascular glaucoma is a rare complication of BRVO, as marked ischemia of at least half of the retina is considered necessary to cause it.[15]

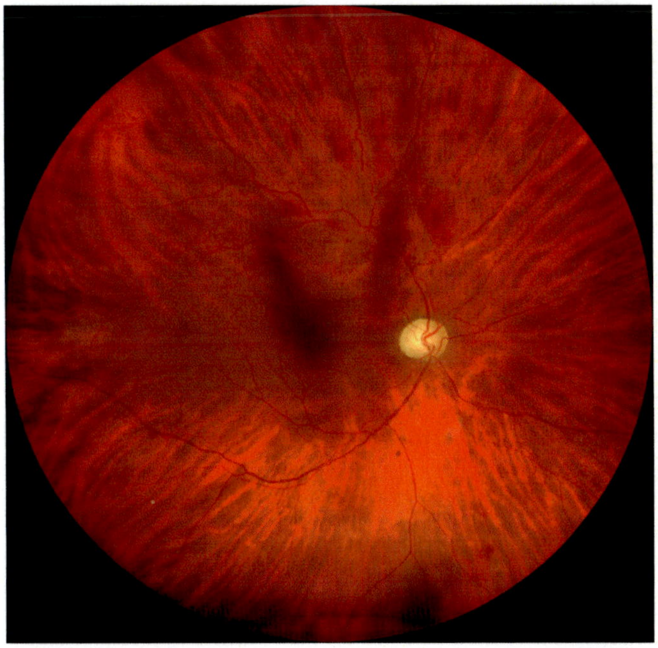

FIGURE 13.5 Color fundus picture of a discrete vitreous hemorrhage in the OD of a patient with extensive BRVO superiorly. Vitreous hemorrhage sometimes may preclude a fair retinal assessment.

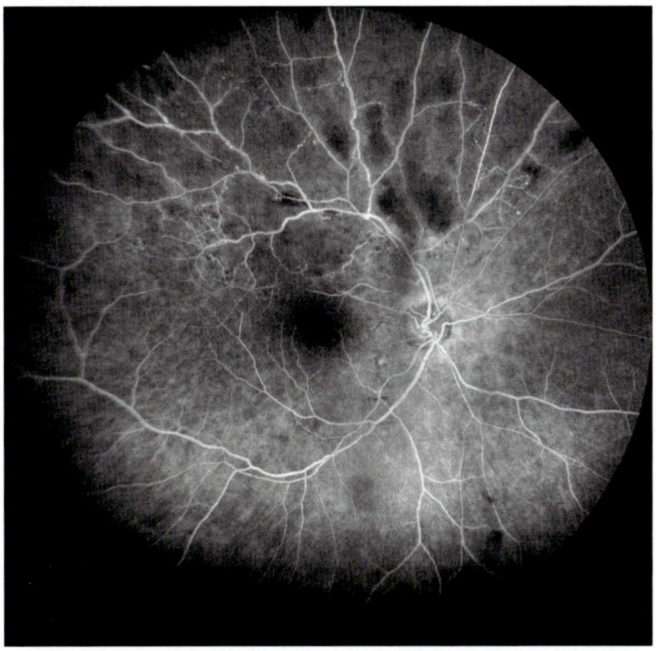

FIGURE 13.6 Mid-transit FA of the same eye. The FA shows extensive areas of impaired retinal perfusion, capillary telangiectasia, and aneurysms.

Imaging findings

Fluorescein angiography

FA is useful for diagnosing BRVO and differentiating ischemic and nonischemic forms. The diagnosis of BRVO is confirmed by the presence of an increased dye transit time of more than 5 seconds, compared to a normal retina. FA also provides information on inflammation of the capillary bed, including ischemia and hyperpermeability.

Macular edema, in FA, presents with hyperfluorescence and delayed leakage within the macular area in radially oriented cystic cavities. FA is crucial for evaluating macular perfusion and has a direct impact on treatment decisions. Macular edema associated with macular ischemia may or may not be associated with fluorescein leakage, which has therapeutic consequences. For patients with macular nonperfusion, the macular laser grid photocoagulation is usually contraindicated.[13,14]

With FA, it is also possible to define the extent of the areas of poor capillary perfusion, as well as to evaluate the areas of collateral circulation and optociliary shunt vessels. Intraretinal hemorrhages are the rule and may preclude proper assessment of perfusion in the acute phase (Figs. 13.7 and 13.8).

Peripheral analysis with wide-field angiography from new devices such as the California or Silverstone (Optos), Mirante (Nidek Co, Ltd), Clarus (Carl Zeiss Meditech), and HRA (Heidelberg Engineering), to name a few, allows for assessing the area of poor perfusion with greater precision (Fig. 13.9).

Indocyanine green angiography (ICGA)

More recently, it has been shown that capillary macroaneurysms, or telangiectatic capillaries (TelCaps) are a common late complication of BRVO. These may be missed by OCTA or FA and indocyanine green angiography (ICGA) is the technique of choice for their diagnosis[16] (Fig. 13.10). Targeted photocoagulation of the TelCaps may contribute to resolve macular edema in these cases.[17]

Optical coherence tomography

Optical coherence tomography (OCT) assesses the macular anatomy noninvasively and is the most useful imaging test for BRVO diagnosis, appraisal, documentation, and follow-up. The rapid evolution of the method, with spectral domain and swept-source technologies, allows for a microstructural evaluation of the retina with almost histological resolution. The central retinal thickness (CRT) or central macular thickness (CMT), defined as the average thickness of the retina between its inner and outer edges in the central area (1 mm), is the most used feature of OCT as a primary endpoint in most clinical trials, being used to assess disease activity and progression, as well as response to treatment. Increased macular thickness generally correlates with worse vision, while its decrease correlates with functional gains.

As in other retinal diseases, OCT can be used for diagnosis, staging, and monitoring response to treatment of macular edema in BRVO. There has been great interest in finding prognostic biomarkers in noninvasive imaging.[18] Subretinal fluid (SRF) and intraretinal fluid (IRF) are signs of acute macular edema, but they can be present in the chronic phases as well,

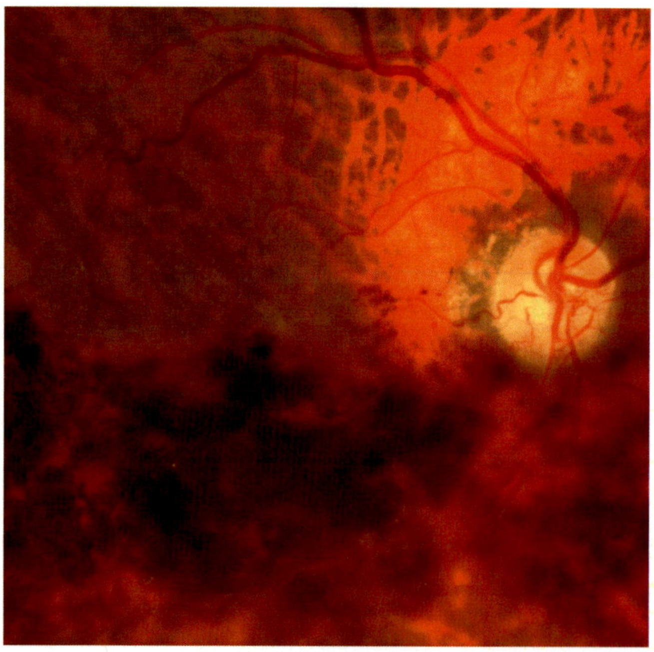

FIGURE 13.7 Retinography of a BRVO with overlaying vitreous hemorrhage. Superficial retinal hemorrhages may preclude proper assessment of capillary perfusion.

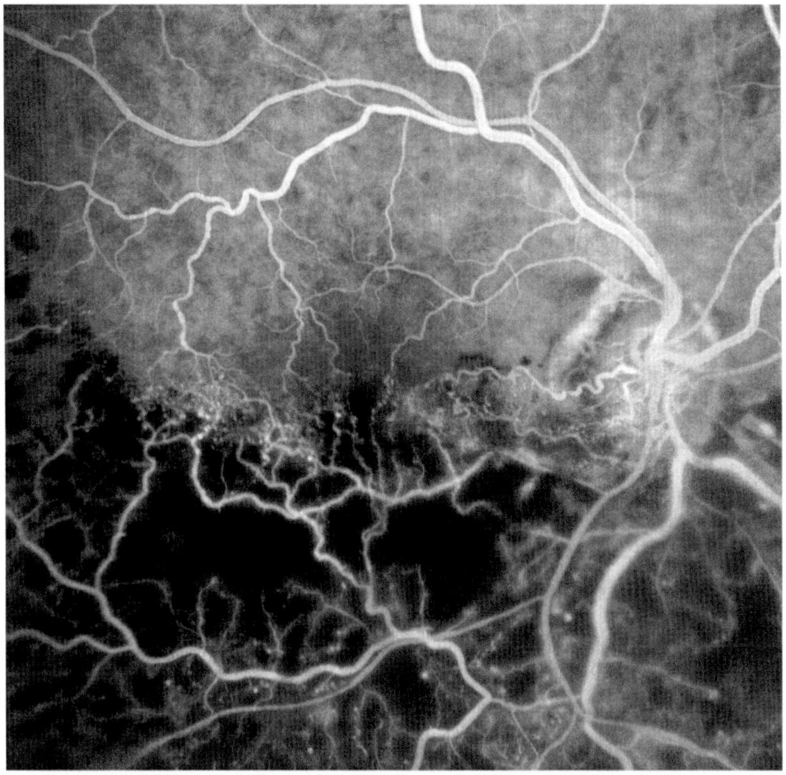

FIGURE 13.8 Fluorescein angiography of a BRVO with overlaying vitreous hemorrhage. Mid-transit phase of the FA showing a discrete vitreous hemorrhage precluding proper assessment of the retinal perfusion.

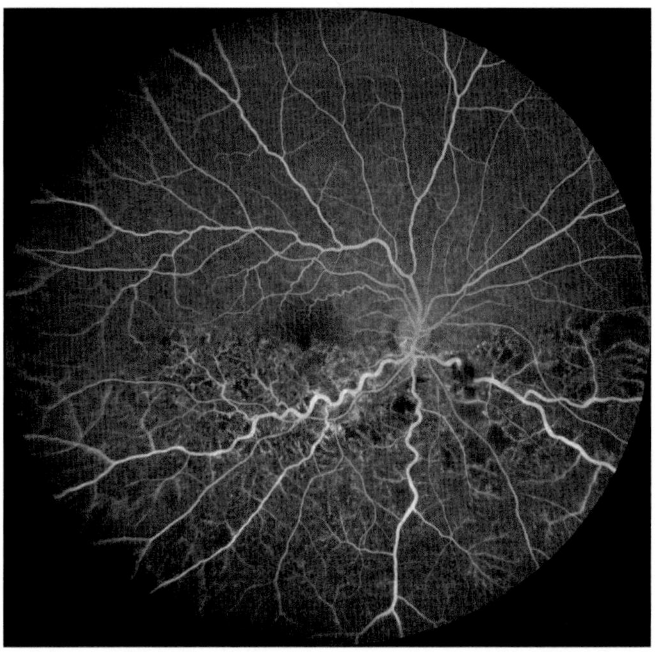

FIGURE 13.9 Widefield FA performed with the Zeiss Clarus. Mid-transit FA demonstrating increased tortuosity of the inferior retinal vein and many areas of capillary dropout and telangiectasia in a case of hemispherical RVO.

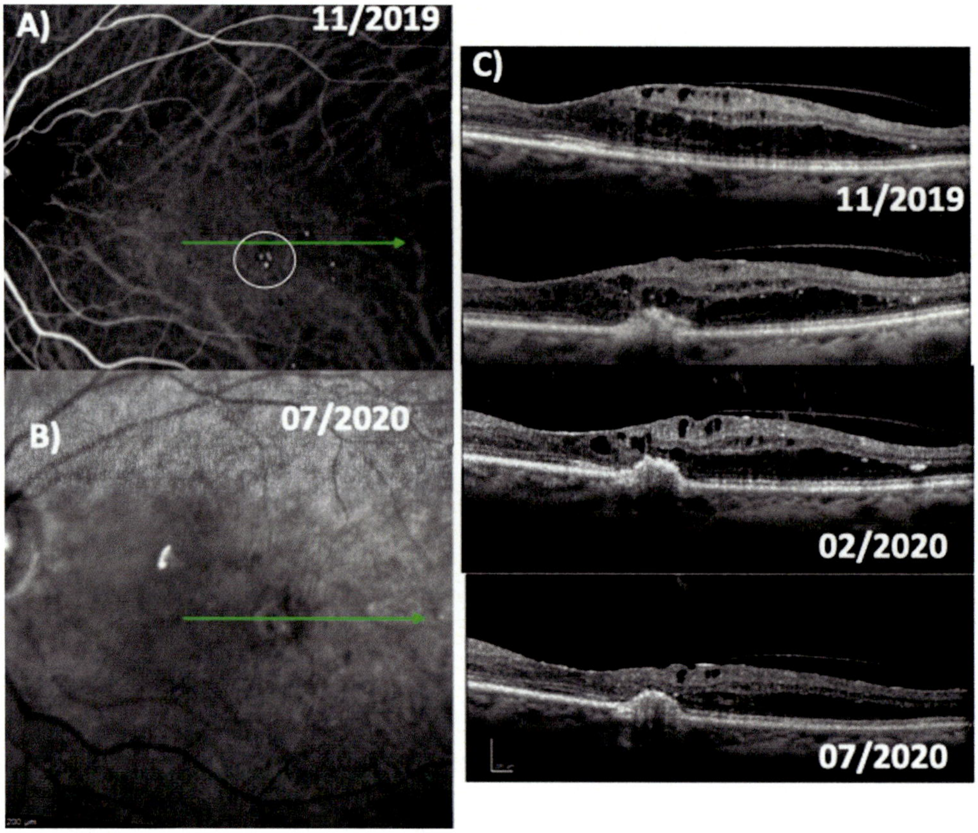

FIGURE 13.10 Clinical case of TelCaps in BRVO. (A) Image of the initial ICGA where TelCaps are observed as an aggregate of lesions in the middle of the circle; (B) IR image 8 months later showing scar formation at the same site; (C) succession of OCT scans at the same region over time showing scarring of the RPE-Bruchs membrane complex at the site of the TelCaps.

especially with cystoid spaces larger than 600 μm.[19] The inner and outer nuclear layers may be affected (Figs. 13.11 and 13.12). Disorganization of the inner layers of the retina (DRIL), disorganization of the line of the inner and outer segments of the photoreceptors (ellipsoid zone), or the outer limiting membrane show damage to the structure of the retina that can lead to irreversible loss of cells and associated visual loss. The presence of retinal hyperreflective spots (correlating to the hard exudates seen clinically), as well as vitreomacular abnormalities (e.g., traction and epiretinal membrane), can be noted at different stages of the disease. The presence of a prominent hyperreflective line located in the outer plexiform layer (imaging sign known as "medium limiting membrane—MLM") and the occurrence of paracentral acute middle maculopathy (PAMM) are considered signs of ischemia due to obstruction of the retinal capillary plexuses. PAMM is a hyperreflective band-like lesion in the inner nuclear layer and outer plexiform layer on spectral domain optical coherence tomography (SD-OCT).[20] Higher reflectivity of the inner retinal layers may also be a surrogate for intracellular edema and ischemia (Fig. 13.13).

Studies like CRYSTAL and BRIGHTER have analyzed a series of biomarkers looking for the structure–function correlation. The only predictive features among the 15 potential morphological biomarkers were CRT, SRF, HRF, and IRF with CRT being the most predictive, closely linked to the presence of IRF.[21]

Optical coherence tomography angiography

OCT angiography (OCTA) in BRVO usually shows a decrease in the density of retinal vessels in the affected eye compared to the contralateral eye, as seen in vascular density maps. In the acute phase of occlusion, vascular abnormalities such as loops, vascular tortuosity, telangiectasia, and structures like microaneurysms in the superficial and deep plexuses can be observed. However, it is very important to be aware of signal-blocking artifacts in acute cases that present with superficial retinal hemorrhages, as they cast projection artifacts (shadows) that can be confused with areas of lack of flow. For this reason, it is always important to pair OCTA with the structural en face image to interpret the signal distribution in the area in question.

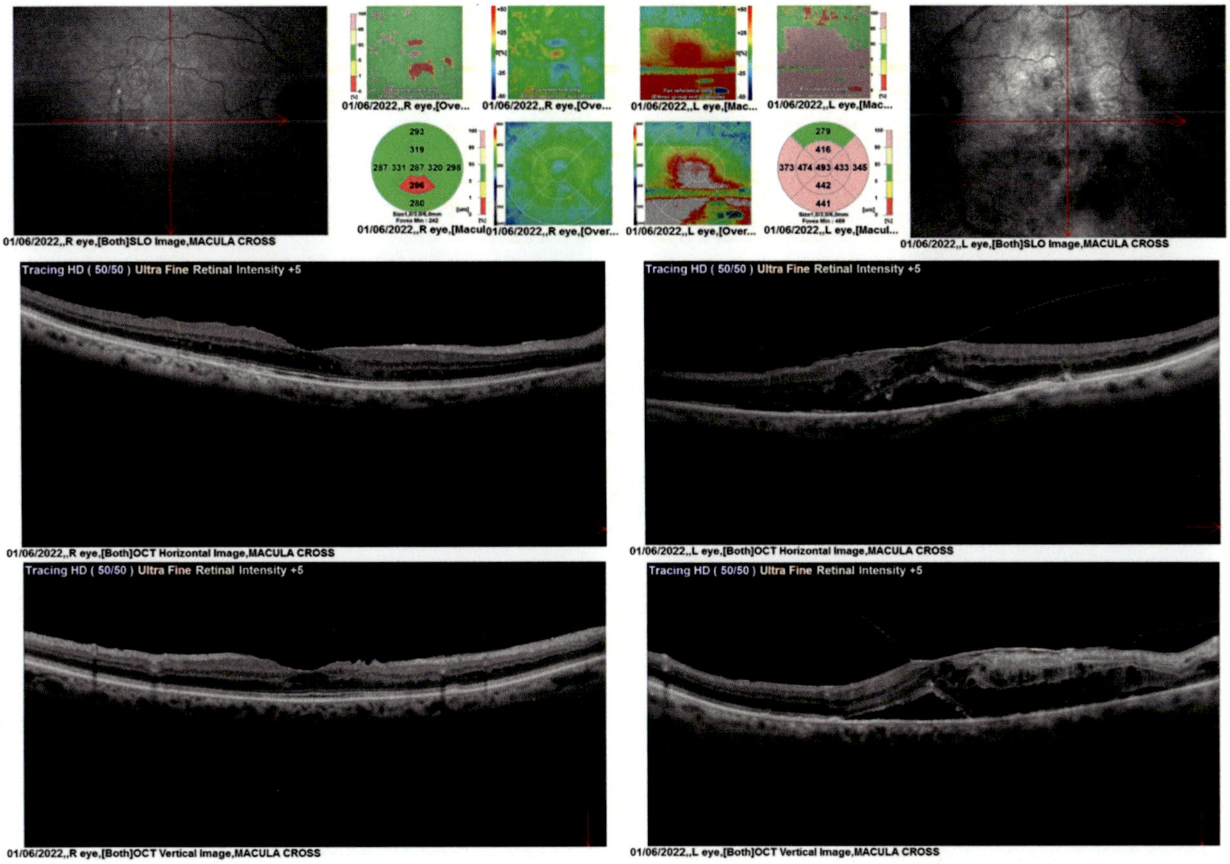

FIGURE 13.11 OCT of an inferior BRVO in the OS. OCT shows posterior vitreous attachment to the foveal area, a discrete epiretinal membrane, serous foveal detachment, and intraretinal fluid in both inner and outer nuclear layers.

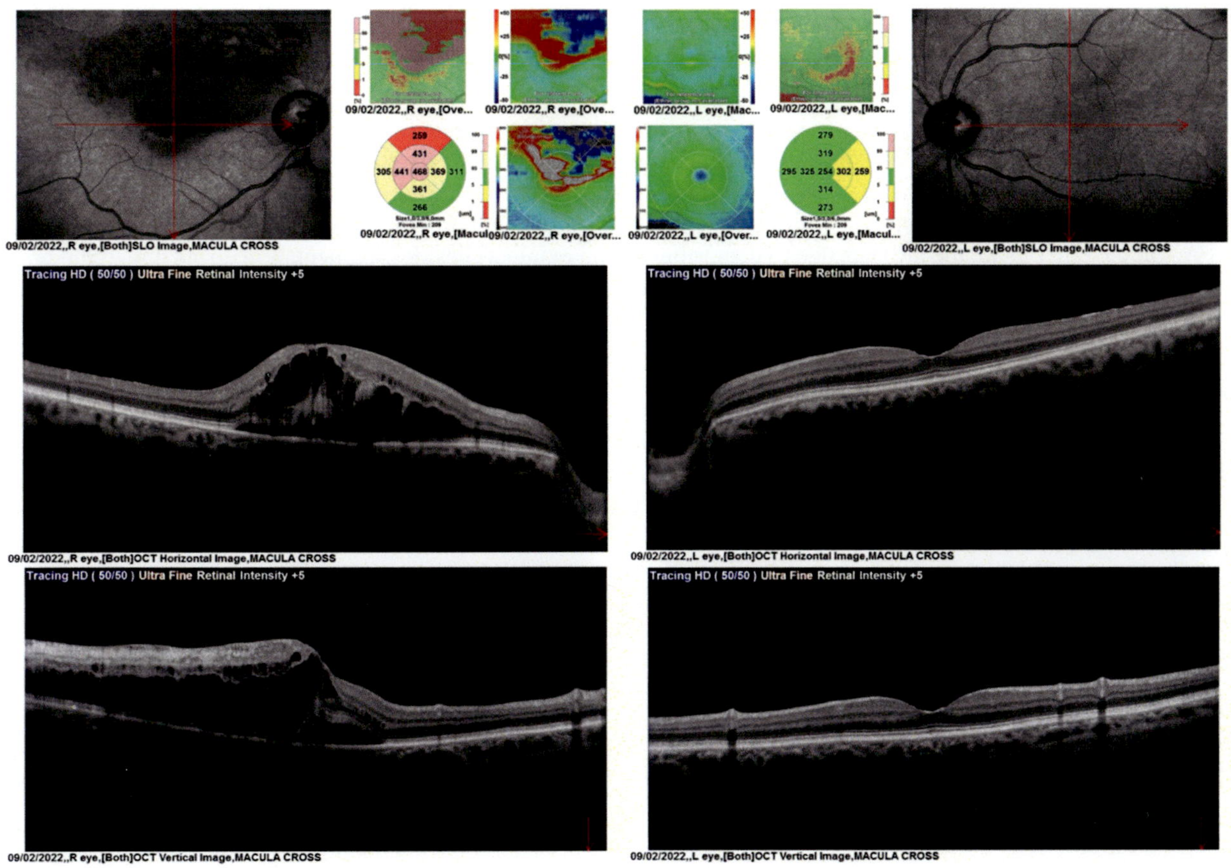

FIGURE 13.12 BRVO in the superior macula of the OD. Cystic IRF may accumulate in both inner and outer nuclear layers, usually the latter being more affected as seen in this vertical OCT scan.

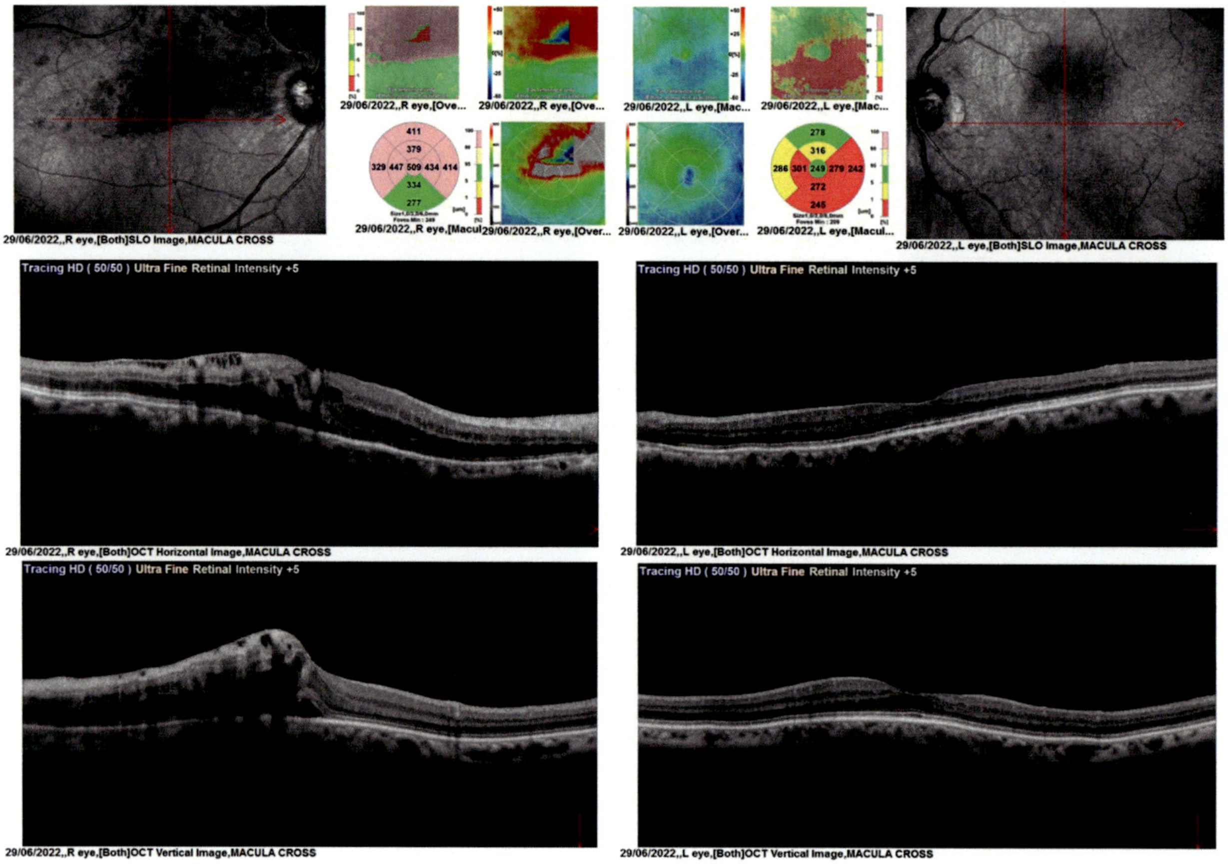

FIGURE 13.13 BRVO in the superior macula of the OD. Higher reflectivity of the retinal inner layers seen in OCT, along with moderate macular edema and serous foveal detachment in the acute phase in the OD of a patient with BRVO of the temporal superior branch.

In the chronic phase, poor capillary perfusion, increased foveal avascular zone, and telangiectasias are observed.[22] Studies comparing OCTA with FA have shown that the former can detect the extent of retinal ischemia in most cases, but not all, in addition to enabling quantitative measurements through maps and vascular density indices. Another advantage would be the stratified observation of the different retinal and choriocapillaris plexuses separately, until then impossible with FA. On the other hand, the artifacts due to hemorrhages and exudates besides segmentation errors due to the presence of macular edema and tissue disorganization may compromise the usefulness of OCTA in a significant number of cases, especially in the acute phase of RVO.[22]

Adaptive optics

Using adaptive optics (AO) imaging, the notion of arterial thickening as the cause of arteriovenous nicking has been challenged. Indeed, high-resolution imaging by adaptive optics showed that there is no evidence of parietal thickening of arteries at arteriovenous crossings with nicking compared to those without nicking[23] (Fig. 13.14). Hence, this stresses the role of the arteriovenous interface as the mediator of venous changes leading to BRVO.

Management strategies

Laser photocoagulation

Branch Vein Occlusion Study (BVOS) was a landmark for the treatment of BRVO in the 80s. It showed that laser grid photocoagulation may be indicated in patients with macular edema and visual acuity of 20/40 or less. Patients treated with laser were more likely to gain two lines of visual acuity (65% of treated patients) than untreated eyes (37% of the untreated patients). After a 3-year follow-up, more treated eyes presented with visual acuity of 20/40 or better (60% vs. 34%, respectively), with a mean visual acuity improvement of 1.3 ETDRS lines (vs. 0.2 line in untreated eyes). However, before

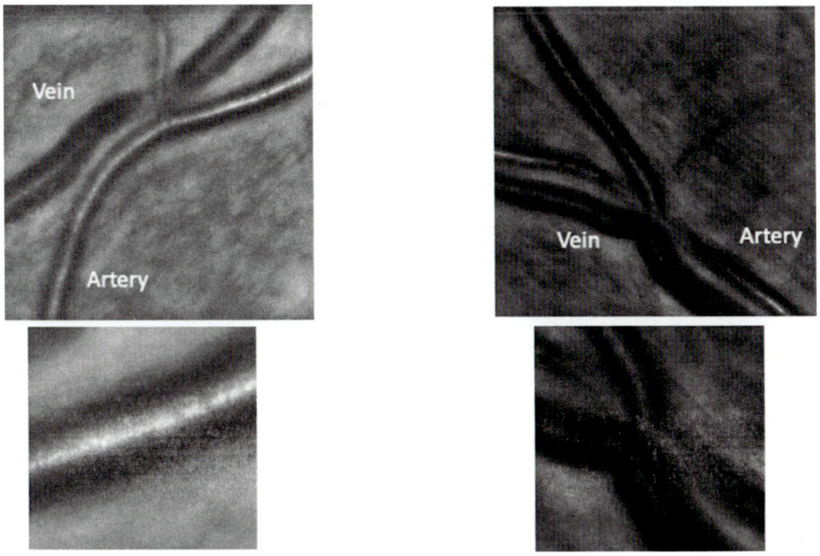

FIGURE 13.14 Two cases of arteriovenous nickings with adaptative optics imaging. Note the absence of parietal thickening of the artery and the focal opacification of the vein.

treatment is applied, the BVOS recommended at least 3 months to allow for the maximum spontaneous resolution of intraretinal blood and edema.[13,14] Another important lesson from the BVOS is the assessment of macular perfusion with fluorescein angiography. FA is necessary before laser treatment, as using laser on macular nonperfusion areas could potentially result in a decrease in visual acuity and should therefore be avoided.[13,14]

In the last 2 decades, the use of intravitreal agents for macular edema has diminished the role of grid photocoagulation for BRVO. The Standard Care versus Corticosteroids for Retinal Vein Occlusion study (SCORE) compared the effect of intravitreal triamcinolone with grid photocoagulation for eyes with BRVO and macular edema and found that there was no difference in visual acuity between the two at 12 months.[24] Moreover, considering the rates of adverse events of intravitreal triamcinolone (particularly elevated intraocular pressure and cataracts), it favored grid photocoagulation. Nonetheless, grid photocoagulation for macular edema has its own complications, including retinal atrophy due to the enlargement of laser scars over time, scotomas, and changes in contrast sensitivity, all of which must also be taken into consideration (Fig. 13.15).

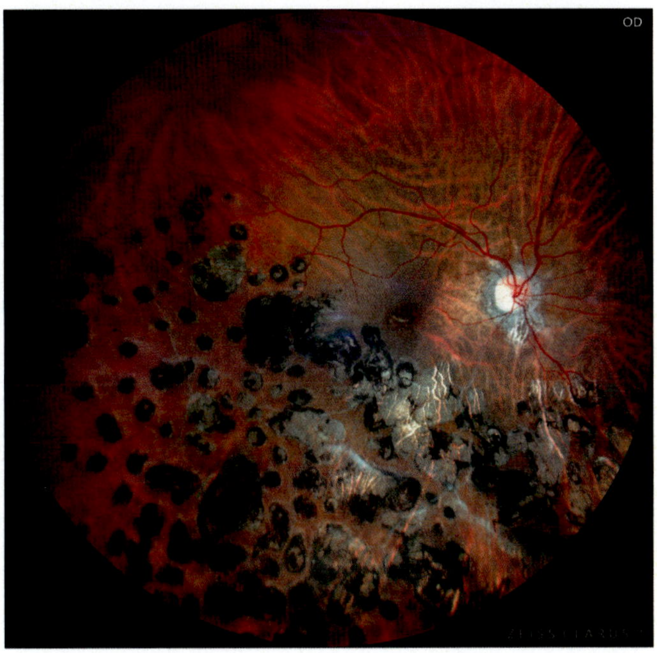

FIGURE 13.15 Chronic case of BRVO in the inferotemporal region of the OLHO DIREITO. Laser scars may enlarge with time, causing extensive retinal atrophy, as seen in this case of severe BRVO in the inferotemporal area of the OD, with widespread venous sheathing. There is also a very relevant compromise of the arterial supply as well, seen by ghost vessels in the area.

Besides treating macular edema, the laser could be applied in BRVO patients for the control of secondary neovascularization. The BVOS showed new vessel regression in nonperfused areas after laser treatment—laser should be applied when new vessels were present, but ischemia alone was not an indication for treatment. Close follow-up is highly recommended especially in patients with large nonperfused areas, considered to be at higher risk of neovascularization.

Some general guidelines for the laser treatment in BRVO[7]:

- Medium white burns 200–500 μm in diameter should be applied one burn width apart covering the entire involved ischemic segment and sparing the area within two disk diameters of the fovea;
- Power should be titrated being increased in 50 mW increments starting at 200 mW until the desired burn intensity is achieved;
- Treatment duration may be variable from 0.1 s near the center to up to 0.2 s in the periphery;
- A Goldmann—or wide angle—contact lens may be used; the adjacent unaffected retina should not be treated;
- Topical anesthesia is mostly sufficient, and local anesthesia is rarely necessary;
- Neovascularization and areas of vitreoretinal traction should be spared and not photocoagulated directly. [IB5][GU6]

Scanning lasers may also be employed, streamlining the procedure, and making a more consistent pattern throughout the occluded area (Fig. 13.12). Infrared (810 nm), red (670 nm), green (532 nm), and yellow (577 nm) wavelengths can be used with different power settings, usually titrated by the reaction of discrete tissue whitening within the treated retina (Figs. 13.16 and 13.17).

Steroids

Pharmacological treatment is currently the gold standard for RVOs. Numerous cytokines are involved in the pathophysiology of the BRVO, such as TNF-alpha, interleukins 6 and 8, ICAM-1, TGF-β, and -2 and VEGF.[25] Therefore, intravitreal steroids, due to their wide range of action, may also be useful by stabilizing the blood–retinal barrier, decreasing vascular permeability, as well as acting upon perivascular and intraluminal inflammation. However, steroids are not devoid of side effects, mainly cataracts and an increase in intraocular pressure. Noteworthy, the SCORE study found no difference in visual acuity and central foveal thickness at 12 months between patients treated with 1 and 4 mg triamcinolone and grid photocoagulation, but there were higher rates of adverse events, particularly elevated intraocular pressure (IOP) and cataracts in the group treated with 4 mg of triamcinolone.[24] Therefore, intravitreal triamcinolone, especially at

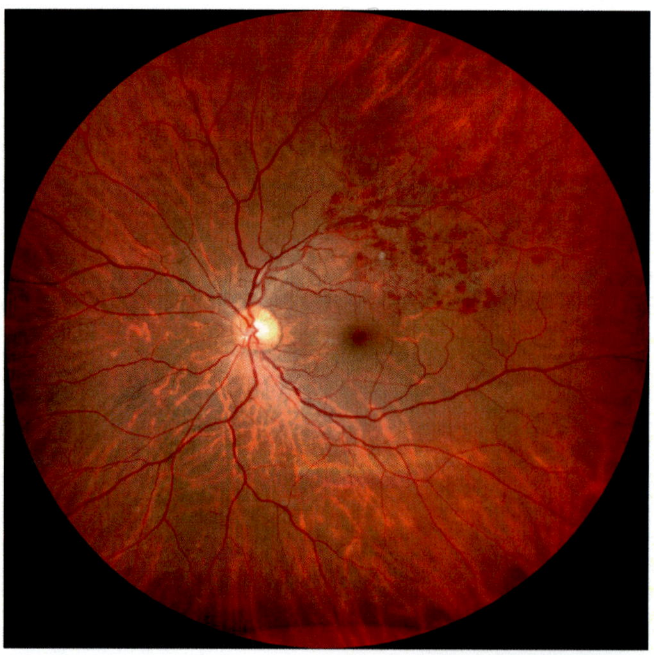

FIGURE 13.16 Color picture of a BRVO in the superotemporal area of the OS with macular edema. Superficial and deep retinal hemorrhages are widely spread in this acute BRVO.

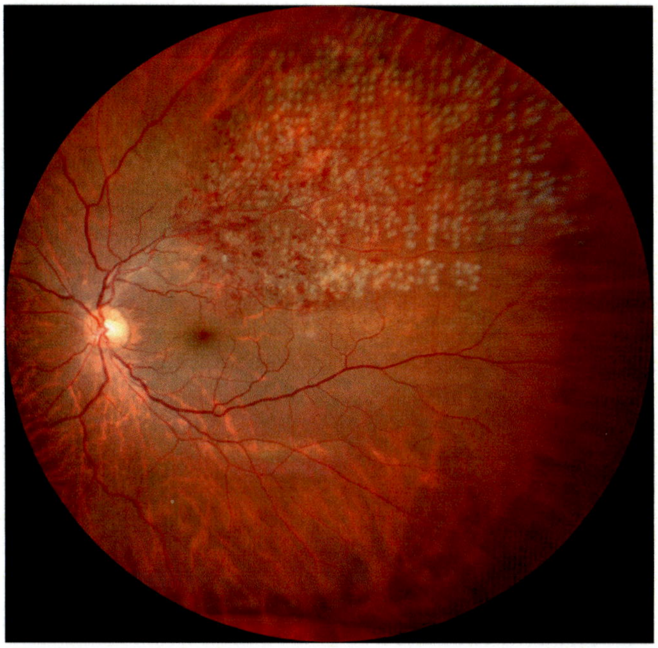

FIGURE 13.17 Color picture of the OS immediately after the pattern scanning laser applied to the occluded area. Multiple laser scars in square patterns can be seen in the temporosuperior region.

the 1 mg dose, could be considered as an alternative at least for pseudophakic patients without previous history of glaucoma.

On the other hand, long-action intravitreal dexamethasone implants for BRVO were evaluated in the global evaluation of implantable dexamethasone in retinal vein occlusion with macular edema (GENEVA), favoring the use of the dexamethasone implant—there was a 15 letter VA gain at 180 days in 41% of treated patients with the 0.7 mg implant, 40% in the 0.35 mg group, and 23% in the sham group.[26] Monitoring side effects, the 12-month GENEVA extension study combined BRVO and CRVO data. There was greater cataract progression in 29% of eyes retreated with dexamethasone implants compared with 6% of eyes that received a single implant. IOP-lowering medications were started in 26% of eyes in the initial dexamethasone treatment groups and in an additional 10% of those who had a second implant.[27] Intravitreal dexamethasone implants are also effective for the treatment of macular edema due to BRVO and could be an option for pseudophakic patients without a previous history of uncontrolled glaucoma. Another important observation is that this implant is contraindicated for aphakic patients or those who had complicated cataract surgery with posterior capsule rupture. In these situations, the implant may migrate to the anterior chamber and cause major complications in the anterior segment such as corneal damage requiring immediate explantation.

Antiangiogenic agents

Intravitreal anti-VEGF therapies are now considered first-line treatment for macular edema secondary to BRVO.[28] There are several anti-VEGFs available for clinical use. At the date that this chapter was written, ranibizumab and aflibercept were approved for use in macular edema due to RVOs. Bevacizumab remains an off-label alternative; and other anti-VEGF agents are under investigation, such as brolucizumab and faricimab.

Bevacizumab

Bevacizumab is an effective anti-VEGF agent against macular edema due to BRVO. It remains an off-label drug, but with a widespread use due to its cost-effectiveness. Its efficacy was demonstrated in several uncontrolled and some controlled studies.[29–31] The comparative trials with ranibizumab and aflibercept will be covered in the next paragraphs.

Ranibizumab

Ranibizumab has been the anti-VEGF agent most widely investigated in RCTs, first in the Branch Retinal Vein Occlusion: Evaluation of Efficacy and Safety (BRAVO) study,[32,33] then in HORIZON,[34] an open-label, single-arm extension trial for

BRAVO and other phase III ranibizumab studies and subsequently the extended follow-up of patients with macular edema due to bRanch rETinal vein occlusion or centrAl retinal veIn occlusioN previously treated with intravitreal ranibizumab (RETAIN) study.[35]

BRAVO was a phase III RCT that evaluated patients with center-involved macular edema secondary to BRVO. Patients were randomized into three groups: 0.3 mg ranibizumab, 0.5 mg ranibizumab, and sham.[33] Overall, patients who received ranibizumab had greater visual gains (16.6 letters in 0.3 mg and 18.3 letters in 0.5 mg) and greater central foveal thickness (CFT) reductions compared with patients who received sham treatment (7.3 letters) at 6 months and after crossover to ranibizumab, the sham group experienced visual gains but not to the same level of improvement as other groups which were treated with ranibizumab early on.[32] This reinforced the importance of initiating treatment early.

BRIGHTER trial was designed to compare ranibizumab intravitreal injections with macular laser photocoagulation. Ranibizumab arm had a 14.8 letter gain versus a 6 letter gain for the laser arm at 6 months, the primary endpoint. At 24-month ranibizumab continued to have a better outcome when compared with laser as monotherapy. Overall, the BRIGHTER study results suggested that the addition of laser did not lead to better visual outcomes and did not reduce the number of ranibizumab injections (11.4 vs. 11.3; $P = .4259$).[36]

There were two small studies that compared bevacizumab versus ranibizumab for RVO, including BRVO and CRVO patients. The MARVEL study evaluated 75 participants comparing intravitreal injections of bevacizumab versus ranibizumab. The 6-month findings suggested no significant difference in visual outcomes between the drugs using the PRN (pro re nata) strategy with an average of three to four injections needed in the first 6 months.[29] The Comparison of AntiVEGF Agents in the Treatment of Macular Edema from Retinal Vein Occlusion Trial (CRAVE) randomized 98 patients to ranibizumab or bevacizumab every month for 6 months and reported no difference in functional and anatomical outcomes between treatment arms in the short term.[30] A larger RCT addressed this issue, evaluating 277 patients with macular edema secondary to RVO. Bevacizumab group improved 15.3 letters versus 15.5 letters in ranibizumab group at 6 months with monthly treatment. This study showed that, based on change of visual acuity, bevacizumab is noninferior to ranibizumab for the treatment of macular edema secondary to RVO with injections every 4 weeks.[37]

Ranibizumab was also compared against dexamethasone implants in the COMRADE-B trial and in a study by Bandello and colleagues, but the trials comprised different treatment regimens. COMRADE-B, with 215 eyes, reported greater visual acuity gains in patients treated with ranibizumab compared with dexamethasone over a 6-month study period,[38] whereas Bandello and colleagues[39] reported no significant difference in visual outcomes between the arms, though acknowledged their study was underpowered. The COMRADE extension study was a phase IV open-label study, that followed 92 patients with BRVO, from the original COMRADE-B trial.[40] Patients were monitored monthly and received either 0.5 mg ranibizumab or 0.7 mcg implant of dexamethasone as needed. Of the 52 BRVO patients treated with ranibizumab in the extension study, 50 patients (96.2%) received at least one reinjection (maximum: nine), while 16 of 40 patients (40.0%) treated with DEX received one retreatment. The mean (±standard deviation (SD)) cumulative number of injections from after the initial treatment phase (month 3 of core studies) until the end of the extension was 4.46 ± 2.09 for ranibizumab and 0.40 ± 0.50 for DEX. The drawbacks were the adverse events against DEX implants. The incidence of increases in IOP of ≥21 mmHg was investigated over 12 months in a post hoc analysis. Such increases were reported more frequently for DEX (29 patients (72.5%)) than ranibizumab (7 patients (13.5%)), $P < .001$.

Aflibercept

The VIBRANT study was a randomized, double-masked phase III trial that evaluated whether aflibercept was superior to grid laser for macular edema secondary to BRVO—a total of 181 eyes were included with center-involved macular edema.[38,41] Patients received intravitreal aflibercept every 4 weeks until week 20 compared with laser photocoagulation. After week 24, patients in the aflibercept arm received the drug every 8 weeks until week 52. At 24 weeks, the laser-treated arm was allowed rescue therapy with aflibercept. Monthly aflibercept led to greater vision gains and central retinal thickness (CRT) reduction at 6 months compared with the macular laser arm.[41] The CRT and BCVA improved at 52 weeks when aflibercept was added in the laser arm, however, the improvement was not as great as in those eyes treated initially with aflibercept.[38]

The Study of Comparative Treatments for Retinal Vein Occlusion 2 (SCORE2) study compared bevacizumab with aflibercept in 362 patients with ME resulting from CRVO or hemi-CRVO.[31] This study confirmed the noninferiority of bevacizumab compared to aflibercept regarding visual outcomes after six monthly injections. The 24-month data evaluated 236 patients, 65.2% from the original cohort. The mean visual acuity improved from baseline to 12 months by 21.6 letters in the aflibercept group compared to 21.9 letters in the bevacizumab group (difference, −0.3; 99% CI, −5.6 to 4.9), then worsened from those values by a mean of 7.6 letters in the aflibercept group and 7.5 letters in the bevacizumab group. The

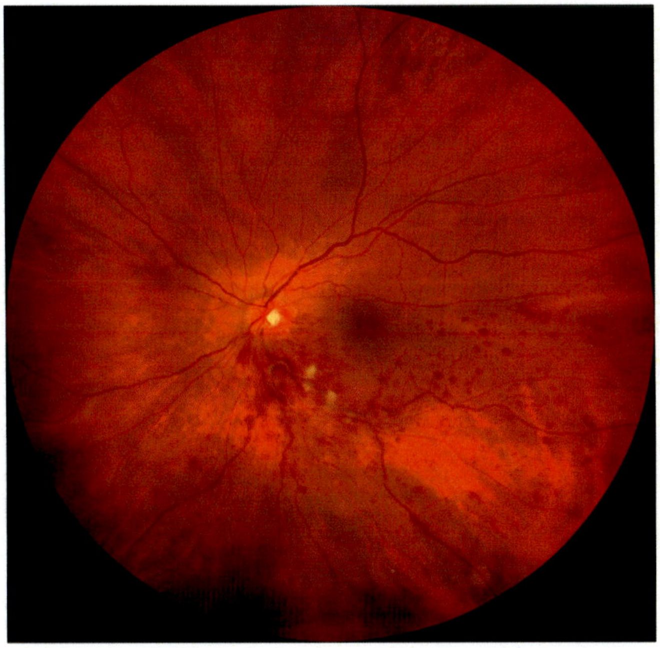

FIGURE 13.18 Color picture of BRVO in the temporoinferior region of the OS. Retinal hemorrhages and cotton-wool spots (focal retinal ischemia) are seen in the area of the BRVO.

result at 24 months remained the same with comparable outcomes between the two drugs; however, it is important to note the drop in visual acuity in the second year when the treatment was at the investigator's discretion.[42] Aflibercep is one of the mainstays of therapy today for BRVO (Figs. 13.18–13.20).

Real-world data

There is no doubt that anti-VEGFs are now the gold standard treatment for macular edema secondary to retinal vein occlusion. However, all these data came from controlled scenarios inside the randomized clinical trials. Their performance in a real-world scenario must therefore be verified with real-world data.

The OCEAN study employed real-world data (RWD) from 744 patients treated for RVOs in Germany.[43] From those patients, 27% had BRVO, 16% CRVO, and the remaining unspecified RVO. The mean baseline BCVA was 55.9 for BRVO, which improved rapidly within the first 3 months, reaching 64.3 letters at 12 months and 64.7 at 24 months. Patients received a median of 4 (5) injections over 12 (24) months, with 100% of patients receiving injections at baseline, 70% at Month 1 and 81% at Month 2. Overall, 40% of patients demonstrated a ≥15 letter increase within the first 3 months. Patients with worse initial BCVA (<50 letters) showed greater improvement than patients with better baseline BCVA.

The largest analysis of RWD regarding anti-VEGF therapy for RVOs is from the Vestrum Health Retina Database, comprising naïve patients from 2013 to 2019. Six-month and 1-year visual acuity outcomes were analyzed. At 6 months, in 8876 eyes with BRVO and macular edema, after a mean of 4.5 anti-VEGF injections, VA increased by 9.4 letters (95% confidence interval (CI) for change in VA +8.94 to +9.78, $P < .001$) from a baseline of 55.1 letters. At 1 year, results were comparable—7.4 injections, +8.1 letters (95% CI + 7.55 to +8.57, $P < .001$). At 6 months and 1 year, visual gains increased with the number of anti-VEGF injections. Patient eyes with baseline VA of 20/40 or better tended to lose VA at 1 year.[44] This data support the fact that in RWD the results are very different from the RCT, correlating directly with less frequent dosing in routine clinical practice, outside the controlled ambient of clinical trials.

Another interesting study, not focused on RWD, evaluated the long-term outcomes in patients treating macular edema secondary to BRVO. After the initial years of these trials, patients were treated according to the investigator's discretion, some kind of RWD. For BRVOs, the baseline VA was 55.4 (95% CI 51.7–59.1) letters. After 2 years, BCVA improved in BRVO by 9.1 letters to 64.5 (95% CI 58.2–70.7, $P < .01$) letters. After 3 years, it improved from baseline to 67.3 (95% CI 61.4–73.1, $P < .01$) letters. BCVA improved from baseline at 5 years for BRVOs by 16.2–71.6 (95% CI: 59.0–84.3, $P = .01$) letters. When drugs were compared, not all treatment types had follow-up periods beyond 2 years. Ranibizumab had 28 cohorts at baseline (2262 participants) and eight cohorts (280 participants) with data at 3 years. BCVA for

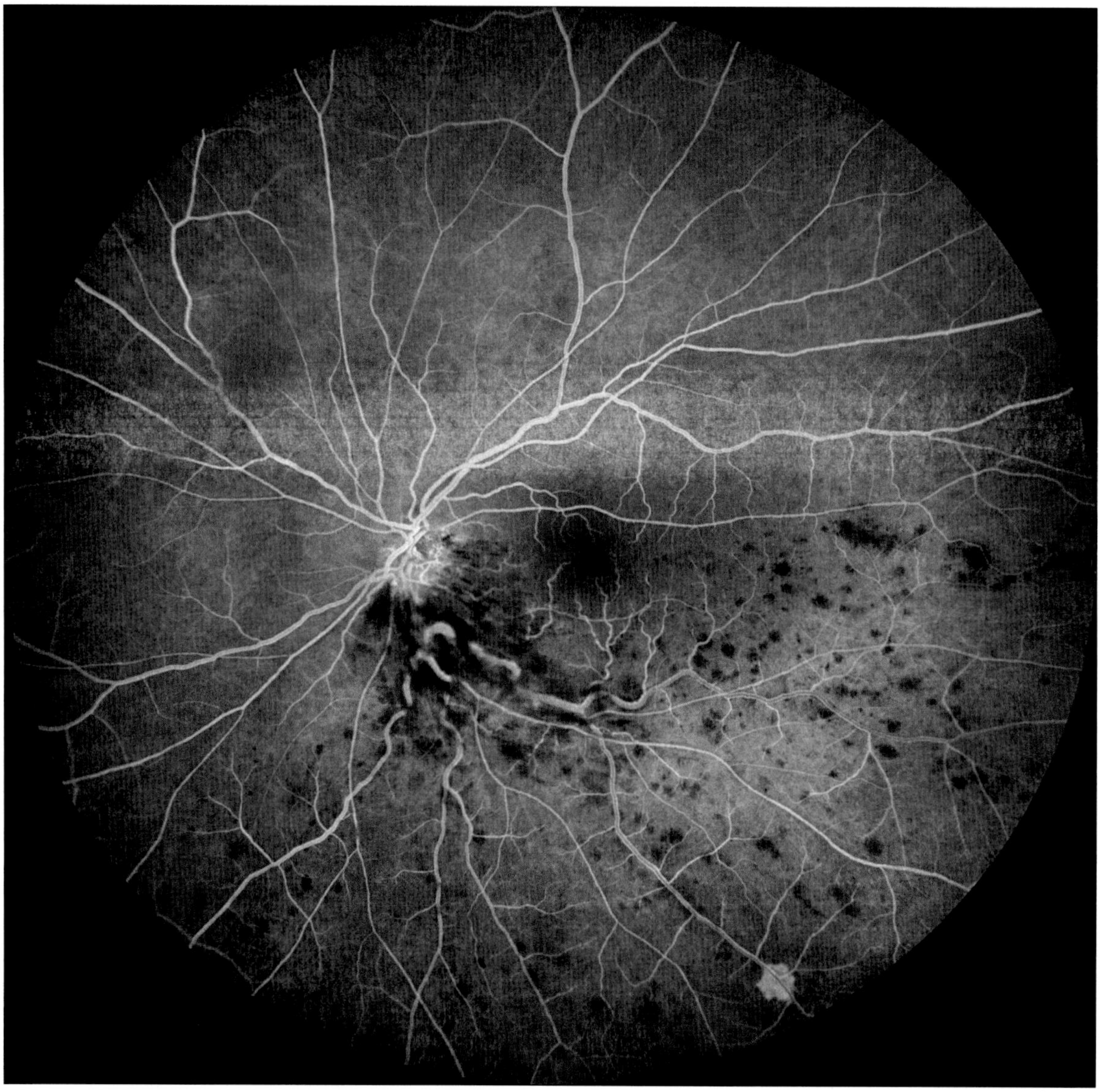

FIGURE 13.19 Fluorescein angiogram helps assessment of retinal perfusion. Fair retinal perfusion could still be seen in the mid-transit phase of the FA.

ranibizumab improved by 11.2 letters from 53.4 at baseline to 3 years (95% CI 49.1−57.6, $P < .01$). Dexamethasone had eight cohorts at baseline with 1165 participants and five cohorts (703 participants) at 3 years. BCVA for dexamethasone improved by 6.3 letters from 50.1 at baseline at 3 years (95% CI 41.6−58.6, $P = .55$). Bevacizumab had eight cohorts at baseline with 582 participants and 1 cohort (57 participants) at 3 years. BCVA with bevacizumab decreased by 2.5 letters from 55.1 letters baseline at 3 years (95% CI 40.7−69.2) to 52.6 at 3 years. Two of the studies with "ranibizumab only" cohorts had 5-year follow-up: mean BCVA decreased from 3 years to 61.8 (95% CI 19.2−104.3) letters ($P = .077$), though this still was an improvement from baseline.[45]

Surgical approach

In the anti-VEGF area, vitrectomy may be indicated for eyes that develop vitreous hemorrhage, retinal detachment, or epiretinal membrane. Some authors also advocate for the importance of vitrectomy and internal limiting membrane peeling for chronic macular edema, with the same rationale for DME. In theory, it could promote oxygenation of the retina and relief of any tractional factor that may exist thus leading to improvement of visual acuity.[46]

Arteriovenous adventitial sheathotomy also deserves to be mentioned. It comprises the surgical separation of the retinal artery from the underlying vein at the site of the AV crossing that causes the BRVO. It had a very appealing rationale of

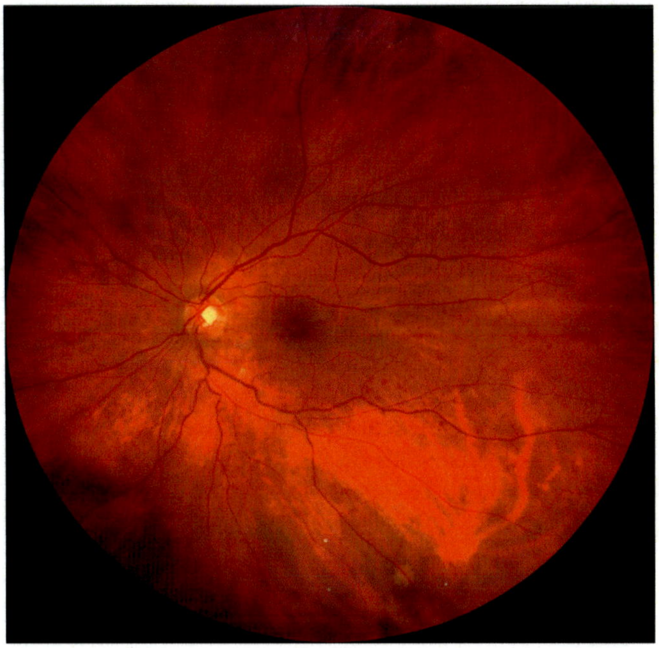

FIGURE 13.20 Fundus picture showing improvement of the occlusion after three monthly intravitreal injections of aflibercept. Retinography showing good resolution of the retinal hemorrhages.

relieving the pressure over the vein at the AV crossing, to improve the blood flow and resolve the occlusion.[47]Unfortunately, the inclusion criteria and the postoperative results of all published studies were very heterogeneous and hardly comparable. To date, it cannot be concluded that surgical sheathotomy is more beneficial than other therapies for BRVO or even the natural course of the disease.

Clinical trial updates

The RAPTOR and RAVEN studies were designed to evaluate the safety and efficacy of brolucizumab versus aflibercept in patients with visual impairment due to macular edema secondary to BRVO. Both studies included six initial monthly injections. They were both halted in the interest of public safety due to the high incidence of intraocular inflammation, including retinal vasculitis and retinal vascular occlusion, which were also observed in patients following a four-week regimen of brolucizumab during the MERLIN trial for wet AMD.

The BALATON (NCT04740905) is a randomized, multicenter, double-masked, global Phase III study evaluating the efficacy and safety of faricimab compared to aflibercept. For the first 20 weeks, patients are randomized 1:1 to receive six monthly injections of either faricimab (6.0 mg) or aflibercept (2.0 mg). From weeks 24 to 72, all patients receive faricimab (6.0 mg) up to every 4 months—according to a personalized treatment interval dosing regimen—using a treat-and-extend approach. BALATON study enrolled 553 people with branch retinal vein occlusion and the primary endpoint was the change in best-corrected visual acuity (BCVA) from baseline at 24 weeks. Secondary endpoints include change in central subfield thickness from baseline over time up to week 24. Results have not been published yet.

KSI-301 is an antibody-biopolymer conjugate structure designed to decrease clearance from the eye and increase the duration of the effect. Developed by Kodiak Sciences is now in a phase IIb/III study under the BEACON trial. This is a Phase 3, prospective, randomized, double-masked, two-arm, multicenter noninferiority study evaluating the efficacy and safety of repeated intravitreal dosing of KSI-301 5 mg in 568 participants with visual impairment due to treatment-naïve macular edema secondary to RVO (either branch or central type), compared to aflibercept. The primary endpoint will be assessed at Week 24; additional secondary endpoints for efficacy will be assessed at Week 24 and Week 48. Results are expected on July 20, 2023.[48]

Another candidate is ONS-5010/Lytenava, an investigational ophthalmic formulation of bevacizumab. It has the potential to mitigate the risks associated with the use of off-label Avastin, and once approved, it will be the first and only FDA-approved, on-label, responsibly priced bevacizumab for ophthalmic indications.

EYP-1901 (vorolanib) by EyePoint Pharmaceuticals, is a sustain-release tyrosine kinase inhibitor that acts as a long sustain-release anti-VEGF therapy. It is under a phase 2 clinical trial for wet AMD but has the potential to be used in RVO patients as well.

In addition to these emerging drugs, there are a few early-stage retinal vein occlusion drugs worth mentioning. ANXV by Annexin Pharmaceuticals is a human recombinant protein (Annexin A5), which is intended primarily for acute treatment of patients with cardiovascular diseases, where vascular damage and inflammation are involved, such as RVO. ANXV is now under a phase II trial (NCT05532735) and promised to change the way we treat RVO targeting retinal vein reperfusion. With the same idea targeting the major cause of retinal vein obstruction, autologous Bone Marrow CD34+ Stem Cells by The Emmes Company are being tested. The goal of this phase I/II prospective, randomized, sham-controlled, double-masked clinical trial (NCT03981549) is to determine whether intravitreal autologous CD34+ stem cell therapy is safe, feasible, and potentially beneficial in eyes with vision loss from RVO. CD34+ stem cells in human bone marrow are mobilized into the circulation in response to tissue ischemia for tissue revascularization and repair. As local delivery of CD34+ stem cells benefits ischemic tissue, intravitreal delivery of CD34+ stem cells may benefit vision and retinal ischemia in eyes with RVO. A pilot clinical trial has shown no major safety or feasibility concerns using intravitreal autologous CD34+ bone marrow stem cells. In this proposed expanded phase I/II study, 20 participants (20 eyes) with persistent vision loss from CRVO will be enrolled and followed up for 1 year.

References

1. Leber T. Die Krankheite der Netzhaut und des Sehnerven. In: *Graefe-Saemisch. Handbuch der Gesamten Augenheikunde*. Leipzig: Verlag von Wilhelm Engelmann; 1877.
2. Appiah AP, Trempe CL. Differences in contributory factors among hemicentral, central, and branch retinal vein occlusions. *Ophthalmology*. 1989;96(3):364–366. https://doi.org/10.1016/S0161-6420(89)32884-3.
3. Jaulim A, Ahmed B, Khanam T, Chatziralli IP. Branch retinal vein occlusion. *Retina*. 2013;33(5):901–910. https://doi.org/10.1097/iae.0b013e3182870c15.
4. Hayreh SS. Prevalent misconceptions about acute retinal vascular occlusive disorders. *Progress in Retinal and Eye Research*. 2005;24:493–519.
5. Salvagno GL, Pavan C, Lippi G. Rare thrombophilic conditions. *Annals of Translational Medicine*. 2018;6(17). https://doi.org/10.21037/atm.2018.08.12.
6. Rabinowicz IM, Litman S, Michaelson IC. Branch venous thrombosis—a pathological report. *Transactions of the Ophthalmological Societies of the United Kingdom*. 1969;88:191–210.
7. Joussen AM, Gardner TW, Ryan SJ. *Retinal Vascular Disease: Branch Retinal Vein Occlusion*. Springer; 2007:467–506.
8. Hayreh SS, Zimmerman MB, Podhajsky P. Incidence of various types of retinal vein occlusion and their recurrence and demographic characteristics. *American Journal of Ophthalmology*. 1994;117(4):429–441. https://doi.org/10.1016/S0002-9394(14)70001-7.
9. David R, Zangwill L, Badarna M, Yassur Y. Epidemiology of retinal vein occlusion and its association with glaucoma and increased intraocular pressure. *Ophthalmologica*. 1988;197(2):69–74. https://doi.org/10.1159/000309923.
10. Frucht J, Shapiro A, Merin S. Intraocular pressure in retinal vein occulsion. *British Journal of Ophthalmology*. 1984;68(1):26–28. https://doi.org/10.1136/bjo.68.1.26.
11. Blankenship GW, Okun E. Retinal tributary vein occlusion: history and management by photocoagulation. *Archives of Ophthalmology*. 1973;89(5):363–368. https://doi.org/10.1001/archopht.1973.01000040365002.
12. Gutman FA, Zegarra H. The natural course of temporal retinal branch vein occlusion. *Transactions of the American Academy of Ophthalmology and Otolaryngology*. 1974;78(2):OP178–OP192.
13. Branch Vein Occlusion Study Group. Argon laser photocoagulation for macular edema in branch vein occlusion. *American Journal of Ophthalmology*. 1985;99:218–219.
14. Branch Vein Occlusion Study Group. Argon laser scatter photocoagulation for prevention of neovascularization and vitreous hemorrhage in branch vein occlusion. A randomized clinical trial. *Arch Ophthalmol*. 1986;104:34–41.
15. Hayreh SS, Rojas P, Podhajsky P, Montague P, Woolson RF. Ocular neovascularization with retinal vascular occlusion-III. *Ophthalmology*. 1983;90(5):488–506. https://doi.org/10.1016/s0161-6420(83)34542-5.
16. Bourhis A, Girmens JF, Boni S, et al. Imaging of macroaneurysms occurring during retinal vein occlusion and diabetic retinopathy by indocyanine green angiography and high resolution optical coherence tomography. *Graefe's Archive for Clinical and Experimental Ophthalmology*. 2010;248(2):161–166. https://doi.org/10.1007/s00417-009-1175-6.
17. Paques M, Philippakis E, Bonnet C, et al. Indocyanine-green-guided targeted laser photocoagulation of capillary macroaneurysms in macular oedema: a pilot study. *British Journal of Ophthalmology*. 2017;101(2):170–174. https://doi.org/10.1136/bjophthalmol-2015-308142.
18. Wang Q, Chan SY, Yan Y, et al. Optical coherence tomography angiography in retinal vein occlusions. *Graefe's Archive for Clinical and Experimental Ophthalmology*. 2018;256(9):1615–1622. https://doi.org/10.1007/s00417-018-4038-1.
19. Hoeh AE, Ruppenstein M, Ach T, Dithmar S. OCT patterns of macular edema and response to bevacizumab therapy in retinal vein occlusion. *Graefe's Archive for Clinical and Experimental Ophthalmology*. 2010;248(11). ISSN: 0721832X:1567–1572. https://doi.org/10.1007/s00417-010-1419-5.

20. Sarraf D, Rahimy E, Fawzi AA, et al. Paracentral acute middle maculopathy a new variant of acute macular neuroretinopathy associated with retinal capillary ischemia. *JAMA Ophthalmology*. 2013;131(10):1275–1287. https://doi.org/10.1001/jamaophthalmol.2013.4056.
21. Schmidt-Erfurth U, Garcia-Arumi J, Gerendas BS, et al. Guidelines for the management of retinal vein occlusion by the European Society of retina Specialists (EURETINA). *Ophthalmologica*. 2019;242(3):123–162. https://doi.org/10.1159/000502041.
22. Ciulla T, Pollack JS, Williams DF. Visual acuity outcomes and anti-VEGF therapy intensity in macular oedema due to retinal vein occlusion: a real-world analysis of 15 613 patient eyes. *British Journal of Ophthalmology*. 2021;105(12):1696–1704. https://doi.org/10.1136/bjophthalmol-2020-317337.
23. Paques M, Brolly A, Benesty J, et al. Venous nicking without arteriovenous contact the role of the arteriolar microenvironment in arteriovenous nickings. *JAMA Ophthalmology*. 2015;133(8):947–950. https://doi.org/10.1001/jamaophthalmol.2015.1132.
24. Scott IU, Ip MS, VanVeldhuisen PC, et al. A randomized trial comparing the efficacy and safety of intravitreal triamcinolone with standard care to treat vision loss associated with macular edema secondary to branch retinal vein occlusion: the standard care vs corticosteroid for retinal vein occlusion (SCORE) study report 6. *Archives of Ophthalmology*. 2009;127(9):1115–1128. https://doi.org/10.1001/archophthalmol.2009.233.
25. Rezar-Dreindl S, Eibenberger K, Pollreisz A, et al. Effect of intravitreal dexamethasone implant on intra-ocular cytokines and chemokines in eyes with retinal vein occlusion. *Acta Ophthalmologica*. 2017;95(2):e119–e127. https://doi.org/10.1111/aos.13152.
26. Haller JA, Bandello F, Belfort R, et al. Randomized, sham-controlled trial of dexamethasone intravitreal implant in patients with macular edema due to retinal vein occlusion. *Ophthalmology*. 2010;117(6):1134–e3. https://doi.org/10.1016/j.ophtha.2010.03.032.
27. Haller JA, Bandello F, Belfort R, et al. Dexamethasone intravitreal implant in patients with macular edema related to branch or central retinal vein occlusion: twelve-month study results. *Ophthalmology*. 2011;118(12):2453–2460. https://doi.org/10.1016/j.ophtha.2011.05.014.
28. *Clinical Guidelines: Retinal Vein Occlusion (RVO) Guidelines*. London, UK: The Royal College of Ophthalmologists; 2015.
29. Vogl WD, Waldstein SM, Gerendas BS, Schlegl T, Langs G, Schmidt-Erfurth U. Analyzing and predicting visual acuity outcomes of anti- VEGF therapy by a longitudinal mixed effects model of imaging and clinical data. *Investigative Ophthalmology and Visual Science*. 2017;58(10):4173–4181. https://doi.org/10.1167/iovs.17-21878.
30. Rajagopal R, Shah GK, Blinder KJ, et al. Bevacizumab versus ranibizumab in the treatment of macular edema due to retinal vein occlusion: 6-month results of the CRAVE study. *Ophthalmic Surgery Lasers and Imaging Retina*. 2015;46(8):844–850. https://doi.org/10.3928/23258160-20150909-09.
31. Scott IU, Van Veldhuisen PC, Ip MS, et al. Effect of bevacizumab vs aflibercept on visual acuity among patients with macular edema due to central retinal vein occlusion the SCORE2 randomized clinical trial. *JAMA - Journal of the American Medical Association*. 2017;317(20):2072–2087. https://doi.org/10.1001/jama.2017.4568.
32. Brown DM, Campochiaro PA, Bhisitkul RB, et al. Sustained benefits from ranibizumab for macular edema following branch retinal vein occlusion: 12-month outcomes of a phase III study. *Ophthalmology*. 2011;118(8). ISSN: 15494713:1594–1602. https://doi.org/10.1016/j.ophtha.2011.02.022.
33. Campochiaro PA, Heier JS, Feiner L, et al. Ranibizumab for macular edema following branch retinal vein occlusion. Six-month primary end point results of a phase III study. *Ophthalmology*. 2010;117(6):1102–e1. https://doi.org/10.1016/j.ophtha.2010.02.021.
34. Heier JS, Campochiaro PA, Yau L, et al. Ranibizumab for macular edema due to retinal vein occlusions: long-term follow-up in the HORIZON trial. *Ophthalmology*. 2012;119(4):802–809. https://doi.org/10.1016/j.ophtha.2011.12.005.
35. Campochiaro PA, Sophie R, Pearlman J, et al. Long-term outcomes in patients with retinal vein occlusion treated with ranibizumab: the RETAIN study. *Ophthalmology*. 2014;121(1):209–219. https://doi.org/10.1016/j.ophtha.2013.08.038.
36. Tadayoni R, Waldstein SM, Boscia F, et al. Sustained benefits of ranibizumab with or without laser in branch retinal vein occlusion: 24-month results of the BRIGHTER study. *Ophthalmology*. 2017;124(12):1778–1787. https://doi.org/10.1016/j.ophtha.2017.06.027.
37. Vader MJC, Schauwvlieghe ASME, Verbraak FD, et al. Comparing the efficacy of bevacizumab and ranibizumab in patients with retinal vein occlusion: the bevacizumab to ranibizumab in retinal vein occlusions (BRVO) study, a randomized trial. *Ophthalmology Retina*. 2020;4(6):576–587. https://doi.org/10.1016/j.oret.2019.12.019.
38. Bandello F, Augustin A, Tufail A, Leaback R. A 12-month, multicenter, parallel group comparison of dexamethasone intravitreal implant versus ranibizumab in branch retinal vein occlusion. *European Journal of Ophthalmology*. 2018;28(6):697–705. https://doi.org/10.1177/1120672117750058.
39. Feltgen N, Hattenbach LO, Bertelmann T, et al. Comparison of ranibizumab versus dexamethasone for macular oedema following retinal vein occlusion: 1-year results of the COMRADE extension study. *Acta Ophthalmologica*. 2018;96(8):e933–e941. https://doi.org/10.1111/aos.13770.
40. Campochiaro PA, Clark WL, Boyer DS, et al. Intravitreal aflibercept for macular edema following branch retinal vein occlusion: the 24-week results of the VIBRANT study. *Ophthalmology*. 2015;122(3):538–544. https://doi.org/10.1016/j.ophtha.2014.08.031.
41. Clark WL, Boyer DS, Heier JS, et al. Intravitreal aflibercept for macular edema following branch retinal vein occlusion 52-week results of the VIBRANT study. *Ophthalmology*. 2016;123(2):330–336. https://doi.org/10.1016/j.ophtha.2015.09.035.
42. Scott IU, Oden NL, VanVeldhuisen PC, Ip MS, Blodi BA, Chan CK. Month 24 outcomes after treatment initiation with anti-vascular endothelial growth factor therapy for macular edema due to central retinal or hemiretinal vein occlusion: SCORE2 report 10: a secondary analysis of the SCORE2 randomized clinical trial. *JAMA Ophthalmology*. 2019;137(12):1389–1398. https://doi.org/10.1001/jamaophthalmol.2019.3947.
43. Callizo J, Ziemssen F, Bertelmann T, et al. Real-world data: ranibizumab treatment for retinal vein occlusion in the OCEAN Study. *Clinical Ophthalmology*. 2019;13:2167–2179. https://doi.org/10.2147/OPTH.S209253.
44. Narayanan R, Panchal B, Das T, Chhablani J, Jalali S, Ali MH. A randomised, double-masked, controlled study of the efficacy and safety of intravitreal bevacizumab versus ranibizumab in the treatment of macular oedema due to branch retinal vein occlusion: MARVEL Report No. 1. *British Journal of Ophthalmology*. 2015;99(7):954–959. https://doi.org/10.1136/bjophthalmol-2014-306543.

45. Hunter A, Williams M. Long-term outcomes for patients treated for macular oedema secondary to retinal vein occlusion: a systematic review. *BMJ Open Ophthalmology*. 2022;7(1):e001010. https://doi.org/10.1136/bmjophth-2022-001010.
46. Cicinelli MV, Chatziralli I, Touhami S, et al. Epiretinal membrane peeling in eyes with retinal vein occlusion: visual and morphologic outcomes. *Ophthalmology and Therapy*. 2022;11(2):661–675. https://doi.org/10.1007/s40123-022-00461-7.
47. Osterloh MD, Charles S. Surgical decompression of branch retinal vein occlusions. *Archives of Ophthalmology*. 1988;106(10):1469–1471. https://doi.org/10.1001/archopht.1988.01060140633037.
48. Stern HD, Hussain RM. KSI-301: an investigational anti-VEGF biopolymer conjugate for retinal diseases. *Expert Opinion on Investigational Drugs*. 2022;31(5):443–449. https://doi.org/10.1080/13543784.2022.2052042.

Chapter 14

Central retinal artery occlusion

Ramesh Venkatesh[1], Rubble Mangla[1], Nikitha Gurram Reddy[2], Vishma Prabhu[1], Isha Acharya[1] and Ashit Handa[1]
[1]*Narayana Nethralaya, Department of Retina and Vitreous, Bangalore, Karnataka, India;* [2]*Anant Bajaj Retina Institute, L.V. Prasad Eye Hospital, Kallam Anji Reddy Campus, Hyderabad, Telangana, India*

Introduction

Central retinal artery occlusion (CRAO) is the sudden blockage of the central retinal artery, which leads to retinal hypoperfusion, retinal infarction, rapidly progressive cellular damage, and painless vision loss. The ability of retinal tissue to survive depends on the extent of vascular collateralization and the length of retinal ischemia. Avoiding irreversible retinal damage and blindness requires prompt diagnosis and treatment to dislodge or dissolve the offending embolus or thrombus. Over the years, voluminous literature has accumulated on its various aspects, rife with grave controversies and misunderstandings, especially regarding its management. In this chapter, we will discuss with our readers the various aspects of this catastrophic retinal vascular disease, from epidemiology to treatment.

History

Since 1859, when von Graefe first described CRAO as an embolic disease caused by endocarditis, it has been recognized as a clinical entity.[1] Following that, Schweigger described it on ophthalmoscopy in 1864.[2] Mauthner proposed in 1868 that "spasmodic contractions" could cause retinal artery occlusion.[3] Hayreh SS had studied various aspects of CRAO extensively since 1955, including anatomical studies on the central retinal artery,[4,5] experimental studies,[6–11] and multiple clinical studies on its various features.[12–24]

Epidemiology

In the United States, the incidence of CRAO is approximately 1 per 100,000 people and 1 per 10,000 ophthalmic outpatient visits, with bilateral involvement occurring in less than 2% of cases.[3] Although a few cases have been reported in patients younger than 30 years, the average age at presentation is the early sixth to seventh decade of life. The probable etiology of occlusion varies with age at presentation.[25] Men have a slightly higher incidence than women. The life expectancy of CRAO patients is 5.5 years, compared to 15.4 years for age-matched non-CRAO patients. Similar to other thromboembolic diseases, cardiovascular risk factors such as hypertension, smoking, hyperlipidemia, diabetes, hypercoagulable states, and male gender are associated with CRAO. Approximately one-third of patients with CRAO have significant carotid artery stenosis.[26]

Vascular anatomy

The central retinal artery (CRA) is a branch of the ophthalmic artery, the initial branch of the internal carotid artery.[5] The CRA supplies the surface layer of the optic disc. From this point, the artery divides into two major branches (superior and inferior), which then divide into temporal and nasal branches that supply blood to the four quadrants of the retina.[21] The outer retina is supplied by the choriocapillaris of the choroid, which branches from the short posterior ciliary artery (SPCA). As both CRA and SPCA are derived from the OA, both must be active to maintain retinal function. This is distinguished by the presence of a cilioretinal artery (CLRA). A study of 1000 consecutive patients utilizing fundus

Retinal and Choroidal Vascular Diseases of the Eye. https://doi.org/10.1016/B978-0-443-15583-3.00014-7

fluorescein angiography revealed that 49.5% of patients had a CLRA.[27] It supplies the papillomacular bundle, which contains the maximum number of photoreceptors required for central vision, when present. In these patients with acute CRAO, the macula may still be perfused. This indicates that it is possible to maintain good vision. Nevertheless, this is not always the case. In 35 of 260 eyes with CRAO, a thorough examination revealed the presence of the cilioretinal artery. Sixty percent of these had an initial VA of 6/30 or worse.[15] Due to the variable size of the cilioretinal artery and the area it supplies, such poor outcomes are observed. Infarcts caused by CRAO will deprive the retina of blood flow, thereby reducing the thickness of the inner retinal layers. Depending on how much of the retina it supplies, a cilioretinal artery, if present, will maintain the thickness of the retina to varying degrees.[15] It is debatable exactly where CRAO occurs. Anatomical studies reveal that the narrowest portion of the CRA lumen is where it pierces the dural sheath of the optic nerve and not the lamina cribrosa and that this was the most common site of CRAO.[28]

Etiology

Two potential causes of CRAO exist:

1. *Emboli:* CRAO is typically described as a thrombo-embolic disorder, with the embolus originating from atheroma in the carotid arteries or cardiac valves. Cholesterol, calcium, and platelet-fibrin are the three major types of emboli. Typically, both cholesterol and platelet-fibrin emboli originate from atherosclerosis in the carotid arteries. Calcium emboli typically arise from cardiac valves. Calcium emboli appear white, cholesterol emboli (Hollenhorst plaques) appear orange, and platelet-fibrin emboli appear dull white on clinical fundus examination.
2. *Spasm of ocular circulation:* Spasmodic contractions of the ocular arterial circulation in an already compromised ipsilateral carotid artery, ophthalmic artery, or central retinal artery due to the presence of in situ thrombosis may result in a sudden decrease in the ocular blood flow and retinal perfusion, which may also result in CRAO. Thrombi may be caused by atherosclerosis, collagen vascular disease, inflammatory conditions, or hypercoagulable states. Polycythemia vera, sickle cell anemia, multiple myeloma, systemic lupus erythematosus, factor V Leiden, prothrombin III mutation, hyperhomocysteinemia, polyarteritis nodosa, giant cell (temporal) arteritis, antiphospholipid syndrome, activated protein C resistance, Behçet disease, and syphilis are among the predisposing conditions.

Pathology

Gross pathology

Immediate occlusion of the CRA causes retinal layer opacification due to inner layer edema. Following a sudden decrease in retinal perfusion, extracellular edema of the ganglion cell nuclei causes the retina to lose its transparency. As a result of the thickened nerve fiber layer and ganglion cells in the macula, the opacity is densest in the posterior pole. The severity of the occlusion is proportional to the density of the retinal opacification, the thickness of the retina, and the inability to see the choroidal vasculature beneath the retina. In addition, the foveola takes on a cherry-red hue due to a combination of two factors: (1) the foveolar retina remains transparent because it is nourished by the choriocapillaris, and (2) the intact retinal pigment epithelium and choroid beneath the fovea are outlined by the opaque surrounding retina. and death of the nuclei of ganglion cells (Fig. 14.1). After a few weeks, the opacification disappears, but the retina remains thin and atrophic; despite the possibility of arteriolar constriction and optic atrophy, the retina may appear deceptively normal.[29]

Once the CRA is occluded, retinal opacification takes between 15 min and several hours to become visible. The ability of the retina to recover is dependent upon the removal of the offending embolus or thrombus and, more importantly, the retinal ischemic tolerance time.[10] In old atherosclerotic hypertensive rhesus monkeys, electrophysiological, histopathological, and morphometric studies revealed that a CRAO of 97 min caused no detectable damage. Visual-evoked potential measurements revealed a variable degree of partial retinal recovery between 105 and 240 min.[10] It was discovered that the relationship between the length of CRAO and the extent of retinal damage varied substantially between individuals. At 240 min, all eyes exhibited total or nearly total optic nerve atrophy and nerve fiber damage, resulting in massive irreversible retinal damage.[10] This indicates that the intervention window is finite and inversely proportional to the degree of recovery. The precise retinal tolerance time at which irreversible damage occurs is unknown, but it appears to be less than 4 hours. However, a 2018 meta-analysis by Tobalem et al. suggests that retinal ischemia most likely occurs within 12–15 min and that earlier experimental evidence suggesting a window of 90–120 min may be flawed and not applicable to the general population, as the majority of experiments supporting this theory were conducted on animals in a controlled environment of hypothermia and under barbiturate anesthesia, which has additional neuroprotective effects. This may explain why the majority of rescue efforts in acute CRAO are ineffective.[30] In 4–6 weeks, the retinal opacification

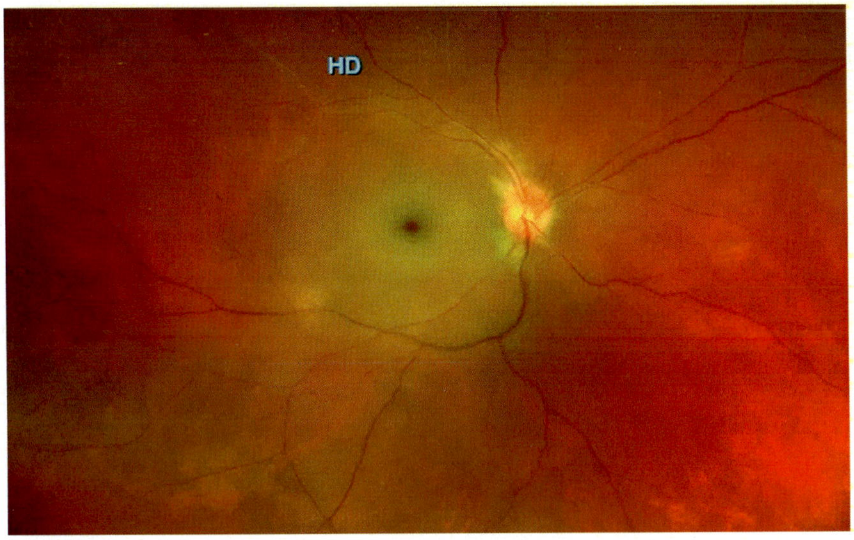

FIGURE 14.1 Case of central retinal artery occlusion with dense posterior pole opacification and cherry red spot at the fovea.

resolves. The resulting pathology reflects a catastrophic injury to the inner retinal layers, as indicated by the diminution of retinal arterioles and the pallor of the optic nerve. Pigmentary changes are generally absent because the retinal pigment epithelium is unaffected; however, vesicular, pearly, and occasionally pigmented patches may appear in the retina. In both arteries and veins, the impaired blood column has a "box-carring" appearance. Wolf and Davies observed this result by experimentally occluding both ophthalmic arteries and veins with a clamp.

Histopathology

All tissues and organs are able to withstand variable durations of ischemia without functional impairments or signs of damage. Once a critical duration of ischemia, which varies by cell type and organ, has been exceeded, cell damage and/or death occurs. Both the magnitude and duration of ischemia impact the extent of cell dysfunction, injury, and/or death. Several intracellular biochemical and metabolic changes occur during prolonged ischemia, leading to cell swelling and rupture and cell death via necrotic, necroptotic, apoptotic, and autophagic mechanisms.[31] Although prompt reperfusion restores the delivery of oxygen and substrates, reperfusion appears to have deleterious effects due to an increase in the production of reactive oxygen species and proinflammatory neutrophilic infiltration of ischemic tissues. This is referred to as reperfusion injury.[32,33] Thus, following acute CRAO, retinal cell injury and/or death can occur due to severe retinal ischemia and reperfusion injury.

Classification of CRAO

In 2005, Hayreh and Zimmerman classified CRAO into four subclasses.[15] Classification of CRAO is essential for comprehending visual outcome variations.

1. *Nonarteritic CRAO (NA-CRAO):* This exhibits the classic clinical presentation of permanent CRAO with retinal infarction, cherry red spot, and retinal arterial changes, absent or poor residual retinal circulation on fluorescein fundus angiography, and no evidence of giant cell arteritis.
2. *NA-CRAO with cilioretinal artery sparing:* This has the classic clinical picture of permanent NA-CRAO, except for the presence of a patent cilioretinal artery, which can have a significant impact on visual outcome and retinal circulation.
3. *Transient NA-CRAO:* In this condition, occlusion of the CRA can range from a few minutes to many hours, depending on the occlusion's cause (Fig. 14.2). Depending solely on the duration of transient CRAO, the visual outcome of this type of CRAO can be completely distinct from that of any other type.
4. *Arteritic CRAO:* In this condition, giant cell arteritis is responsible for the permanent development of CRAO. Almost always, associated arteritic anterior ischemic optic neuropathy is present in these eyes. Clinically, these eyes exhibit the classic fundus findings of CRAO, with the exception of optic disk edema caused by arteritic anterior ischemic optic neuropathy.

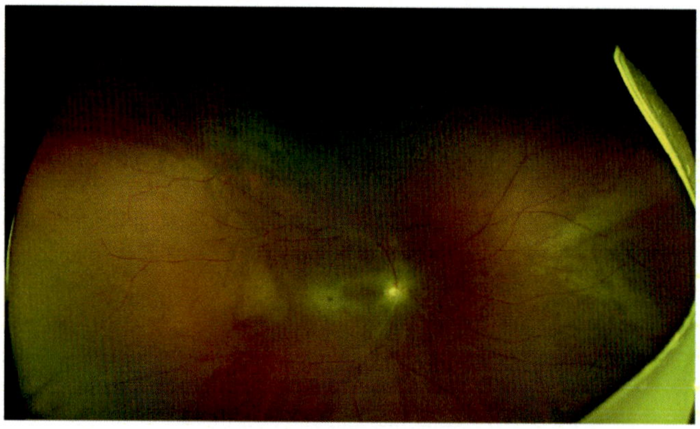

FIGURE 14.2 A case of ophthalmic artery occlusion typically characterized by choroidal and retinal opacification, pale optic disc, and absent cherry red spot.

Clinical features

Ocular history

Overall, 90% of patients with CRAO present with acute, unilateral, persistent, painless vision loss, typically in the range of counting fingers to light perception. The clinician should consider an ophthalmic artery occlusion if the visual acuity is worse than light perception. The presenting visual acuity in these patients is influenced by a number of variables. These include the severity of occlusion and degree of acute retinal ischemia, type and cause of CRAO, site of occlusion in the CRA, presence of residual retinal circulation, and the presence of a cilioretinal artery and its perfusion to the papillomacular bundle.[10,34–37] Some patients with CRAO report a history of amaurosis fugax, which may result from transient CRAO. Amaurosis fugax is a transient vision loss that lasts seconds to minutes but may last up to 2 hours. The vision typically returns to normal following an episode of amaurosis fugax.

Medical history and examination

The past medical history should add any condition that could predispose to embolus formation (e.g., atrial fibrillation, endocarditis, atherosclerotic disease, hypercoagulable state). To determine potential causes, a physical examination should be conducted with a focus on the cardiovascular system. Important are the radial pulse rate and cardiac rhythm, as irregular rhythms suggest atrial fibrillation, which is a risk factor for embolic phenomena. Given the association between CRAO and hypertension, it is crucial to measure blood pressure. In older patients, the clinician should inquire about temporal arteritis symptoms (e.g., headache, jaw claudication, scalp tenderness, proximal muscle and joint aches, anorexia, weight loss, and fever). To exclude the possibility of temporal arteritis, scalp tenderness and an examination for temporal artery nodules should be performed. Young patients should undergo a thorough examination to identify potential autoimmune connective tissue diseases that may predispose them to vasculitis. Additional risk factors include prolonged direct pressure to the globe during drug-induced stupor and improper positioning during face-down surgery. Additionally, the physician should inquire about illicit drug use.

Ocular findings

Anterior segment findings

In general, intraocular pressures, anterior chamber examinations, and extraocular eye movements are within normal limits. When CRAO is caused by ocular ischemic syndrome, the affected eye may develop iris and angle neovascularization; otherwise, CRAO eyes do not develop any neovascularization.[13,14]

Posterior segment findings

For an accurate diagnosis of CRAO, a thorough fundoscopic examination is essential, and a dilated exam should be performed on any patient without contraindications to mydriatic medications who presents with symptoms suggestive of CRAO. The fundoscopic findings in CRAO vary based on the time elapsed since the event and the CRAO subtype. The retina will appear diffusely pale with a cherry-red central spot upon fundoscopy. This spot is the result of the visible preservation of choroidal circulation beneath the thin fovea. The absence of a cherry-red spot in the presence of retinal opacification may indicate an ophthalmic artery occlusion (Fig. 14.2). Additionally, narrowing, "boxcarring," or segmentation of the arterioles may be valued (Fig. 14.3). Arteriolar emboli are occasionally visible. Early findings performed within 7 days of CRAO showed the following findings: posterior pole retinal opacity (58%), cherry-red spot (90%), retinal arterial attenuation (32%), retinal trucking (19%), and optic disk edema (22%) and pallor (39%).[17] Fundoscopic findings at later stages revealed optic atrophy (91%), retinal arterial attenuation (58%), cilioretinal collaterals (18%), and macular retinal pigment epithelial alterations (11%) (Fig. 14.4).[17] CRAO with sparing of the cilioretinal artery was related to a decreased incidence of all macular and optic disk abnormalities (Fig. 14.5). Fundoscopic findings for transient CRAO varied greatly. The presence of optic disc edema in conjunction with typical CRAO findings may indicate an arteritic form of CRAO. In 20% of CRAO patients, the fundus examination revealed the presence of intraarterial emboli.[17] This is essential, as the morphology of the emboli can help determine the cause of CRAO. For instance, the presence of small, yellow, refractile plaques, also known as "Hollenhorst plaques," suggests the presence of cholesterol emboli, whereas single, white, nonscintillating plaques located in the proximal retinal vasculature are caused by calcific emboli and fibrin-platelet emboli appear as small pale bodies (Fig. 14.6).[38,39]

FIGURE 14.3 "Box-carring" effect seen in CRAO.

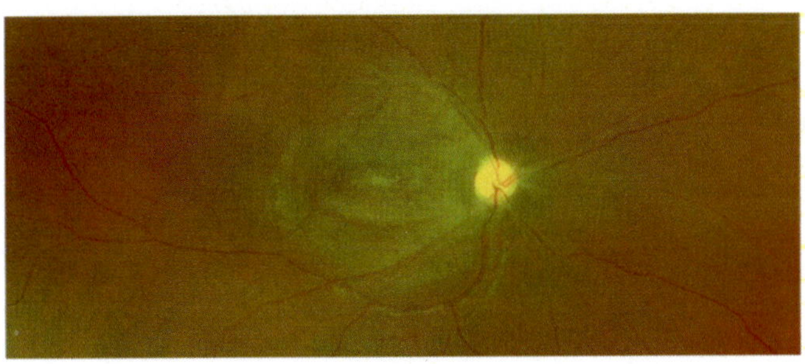

FIGURE 14.4 Clinical fundus image of a case of resolved CRAO.

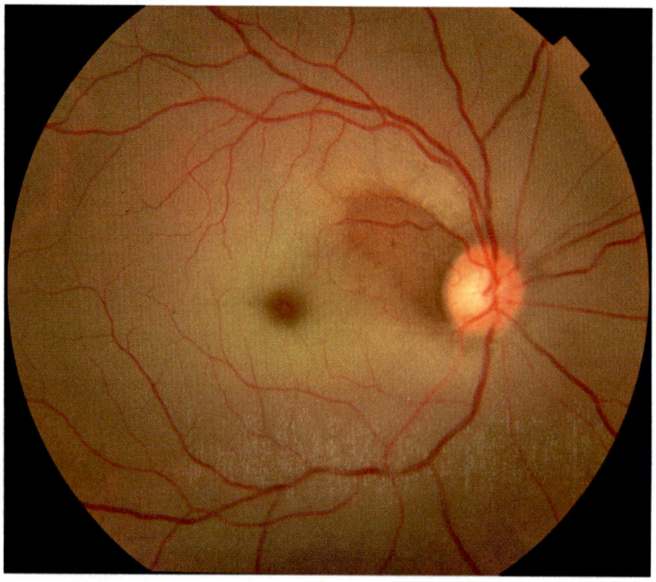

FIGURE 14.5 CRAO with cilioretinal artery sparing.

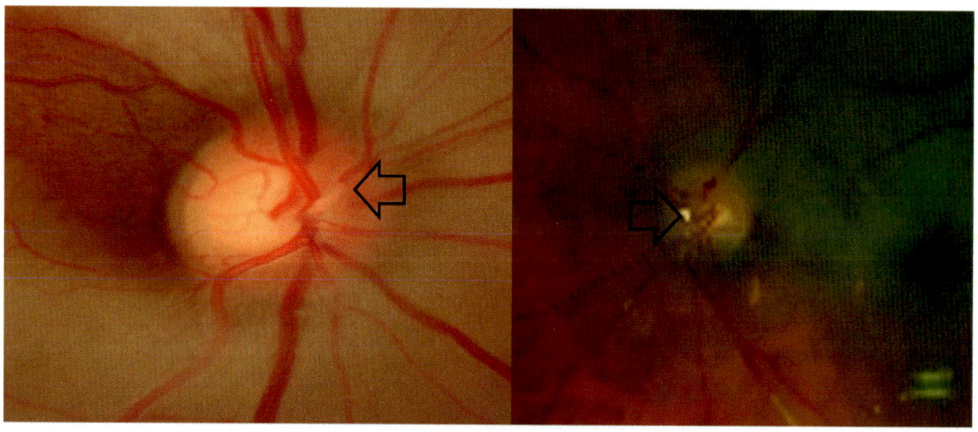

FIGURE 14.6 Various emboli seen in CRAO.

Contralateral eye examination

There may be hints of underlying pathologies, such as hypertensive retinopathy, arteriole changes, or previous vaso-occlusive diseases, in the contralateral eye. Additional ophthalmologic examinations can suggest the cause of CRAO. For instance, increased intraocular pressures, the presence of preretinal arterial loops, and drusen on the optic nerve head may predispose to a CRAO because these conditions reduce the mean arterial perfusion across the optic nerve head. If hypertensive or sickle cell retinopathy is detected during ophthalmoscopy, this may indicate the presence of small vessel disease.

Differential diagnosis

Purtscher's or Purtscher's-like retinopathy is characterized by retinal opacification patches on fundus examination and a history of head or chest trauma and sudden onset of severe vision loss. Other conditions, such as commotio retinae and metabolic storage disorders such as Tay–Sachs disease, and Niemann–Pick disease, may present with a cherry red spot on fundoscopic examination, but they should be easy to differentiate based on clinical presentation.[40]

Physical investigations

CRAO is comparable to a cerebral vascular accident (stroke) or a myocardial infarction. Patients with CRAO are at increased risk for stroke. During the first three and a half years following a CRAO, the risk of having a stroke is up to 10 times greater than in the general population, according to studies.[32,41] This risk may persist for 10 years following the CRAO event. Consequently, the evaluation for CRAO closely resembles the evaluation for stroke or transient ischemic attacks. See Table 14.1.

Ocular investigations

Visual field testing

Modern automated static threshold perimetry, as opposed to manual kinetic perimetry performed with a Goldmann perimeter, does not provide complete information about the peripheral visual fields, as automated perimetry does not provide information beyond the central 30 degrees or less of the field, whereas kinetic perimetry covers the peripheral visual field up to about 90 degrees. On a visual field examination, five distinctive visual field defects are identified: peripheral constriction only (8%), paracentral scotoma (3%), central and cecocentral scotoma (19%), temporal island (59%) and no visual field (10%).[42] Severe defects are linked to severe stages of CRAO, poor BCVA at baseline, delayed retinal arterial perfusion, and severe retinal morphologic changes on optical coherence tomography (OCT). Even without treatment, the visual field improved in 39% of all cases during the follow-up periods. Visual field improvement was significantly associated with mild CRAO stages, good best-corrected visual acuity at baseline, mild retinal morphologic changes, and mild initial field defects.

TABLE 14.1 Investigations for CRAO.

Tests to be advised:	Reason:
(1) Baseline tests	
(A) Hematological tests	
Complete blood counts including peripheral blood smear	To rule out blood dyscrasias and hyperviscosity states
Erythrocyte sedimentation rate and C-reactive protein	To rule out temporal or giant cell arteritis in patients >50 years with CRAO and abnormal optic disc
Coagulation profile (bleeding time and clotting time)	To rule out coagulation disorders
Blood sugar levels and lipid profile including triglycerides and cholesterol levels	To rule out atherosclerotic diseases
Blood cultures	To rule out bacterial endocarditis or septic emboli
Serum homocysteine levels	To rule out hypercoagulable states, especially in young patients with CRAO
Vasculitis profile (Mantoux test and chest X-ray)	To rule out hypercoagulable states, especially in young patients with CRAO and associated signs of retinal vasculitis
(B) Carotid doppler	To rule out atheromatous plaques causing stenosis in the ipsilateral carotid arteries
(C) 2D echocardiography	For identifying the source of emboli
(2) Advanced tests:	
(A) Hematological tests:	
Factor V Leiden, prothrombin mutation, homocysteine levels, fibrinogen, antiphospholipid antibodies, prothrombin time/activated partial thromboplastin time [PT/aPTT], serum protein electrophoresis	To rule out coagulation disorders
(B) MR or CT angiography of the brain and neck	To rule out vascular occlusive disease

Optical coherence tomography

CRAO primarily affects the inner retinal layers, and the most prominent findings are at the posterior pole.[43] OCT has been an invaluable, noninvasive imaging tool in identifying a variety of inner and outer retinal structural features in CRAO.[14,35,44,45] In routine clinical practice, OCT findings such as presence of inner retinal layer hyperreflectivity, thickening, and loss of inner retinal layer stratification are the common findings used for the diagnosis of an acute CRAO on OCT.[44] All cells of the inner retinal layers suffer from acute swelling in acute CRAO, resulting in hyperreflectivity and thickening of the entire inner retinal layers on OCT. The chronic stage of the disease is identified by the presence of a thinner inner retina and preservation of inner retinal layer stratification.[46] The severity of an acute CRAO can be graded based on OCT findings into mild, moderate, and severe occlusion (and Fig. 14.7).[35]

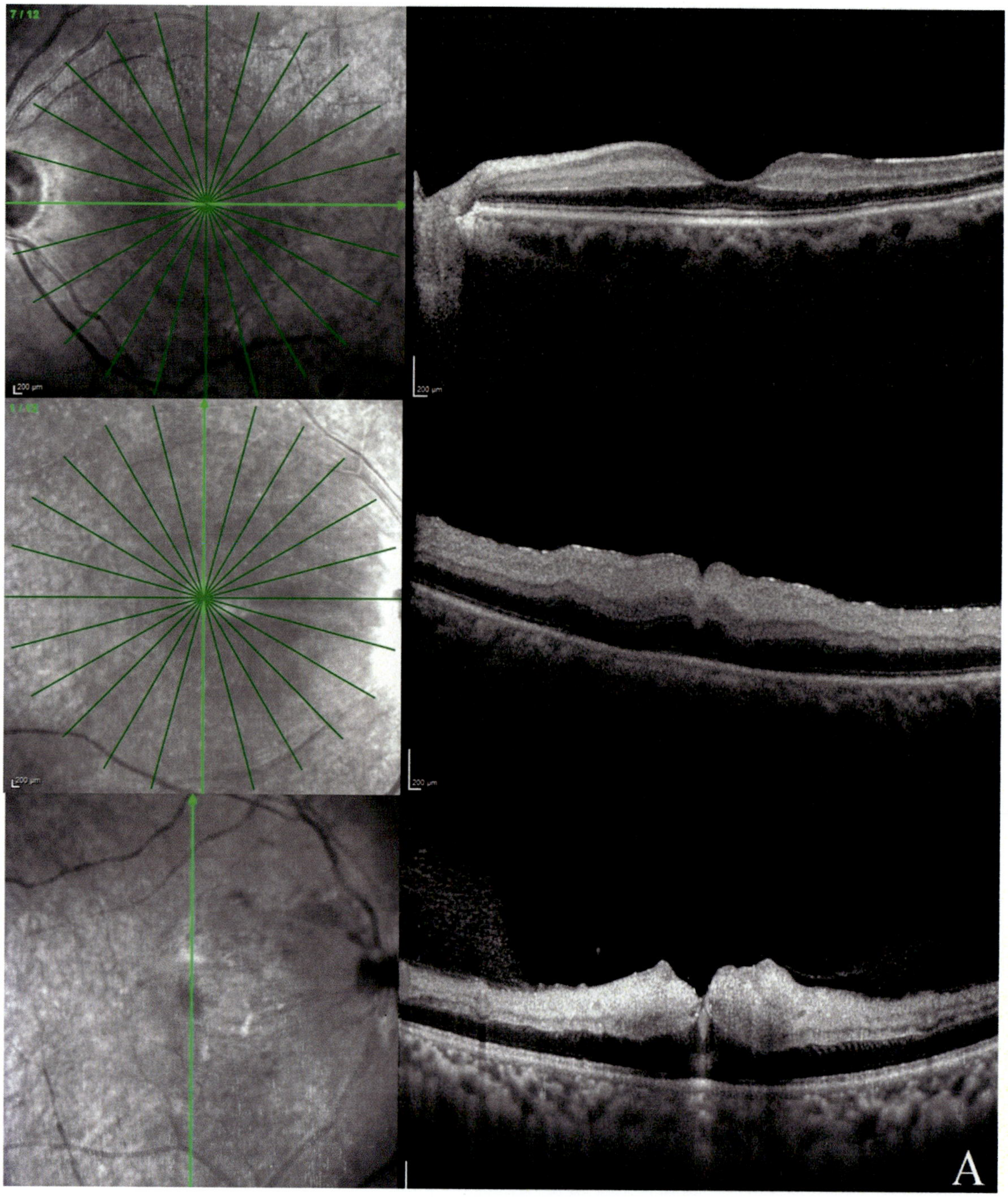

FIGURE 14.7 OCT-based classification of CRAO.

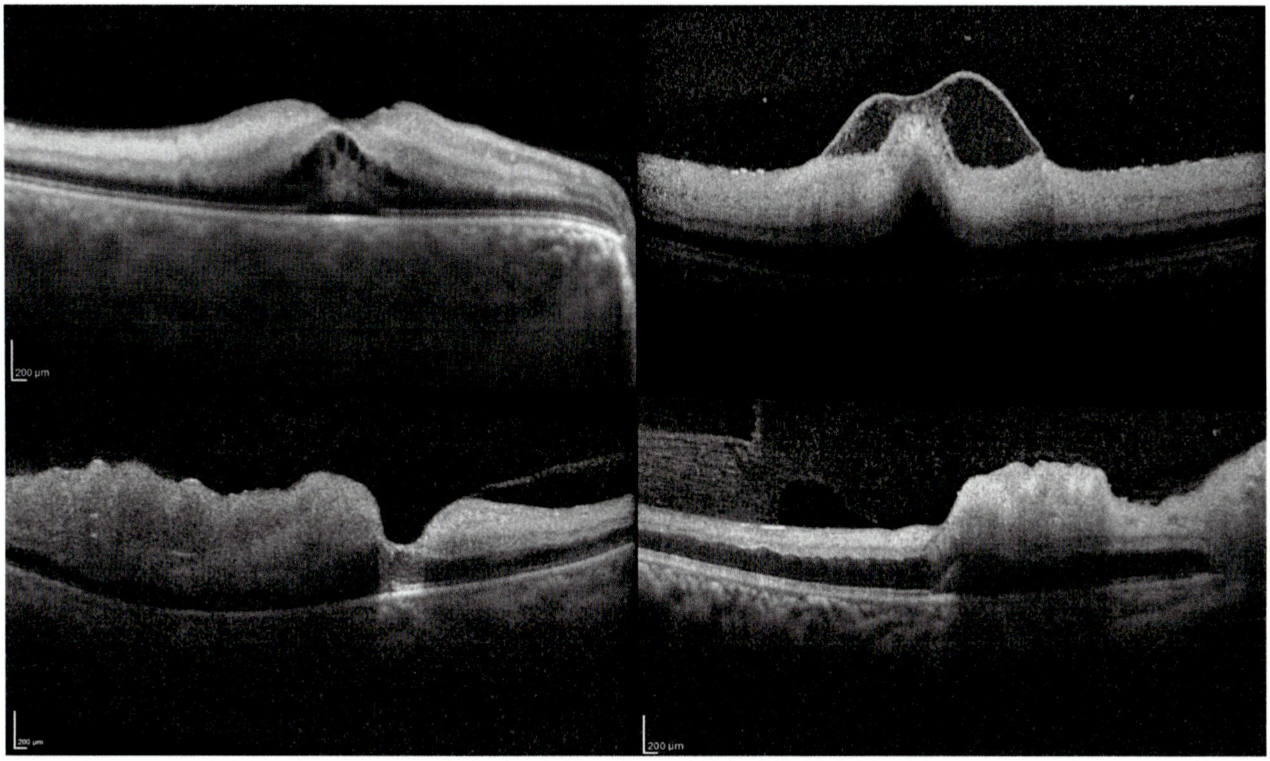

FIGURE 14.8 Other OCT findings in CRAO.

The descriptive characteristics that are observed on OCT include (Figs. 14.8−14.10):

i. Posterior vitreous opacities;
ii. Inner retinal layer features such as inner limiting membrane detachment (ILMD), hyperreflectivity, thickening and loss of individual layer stratification, hyperreflective foci, hyperreflective fluid-filled cavities;
iii. Empty vessel sign—On OCT, the blood vessel appears as a vertically oval cavity in the inner retina with hyperreflective upper and lower borders, minimal back shadowing, and a hyporeflective lumen. The lumen's hyporeflectivity indicates that the vessel is empty.[14]
iv. Middle retinal layer opacification and prominent middle limiting membrane (p-MLM) sign;
v. Neurosensory detachment and
vi. Prominent retinal pigment epithelium (RPE) at the fovea.

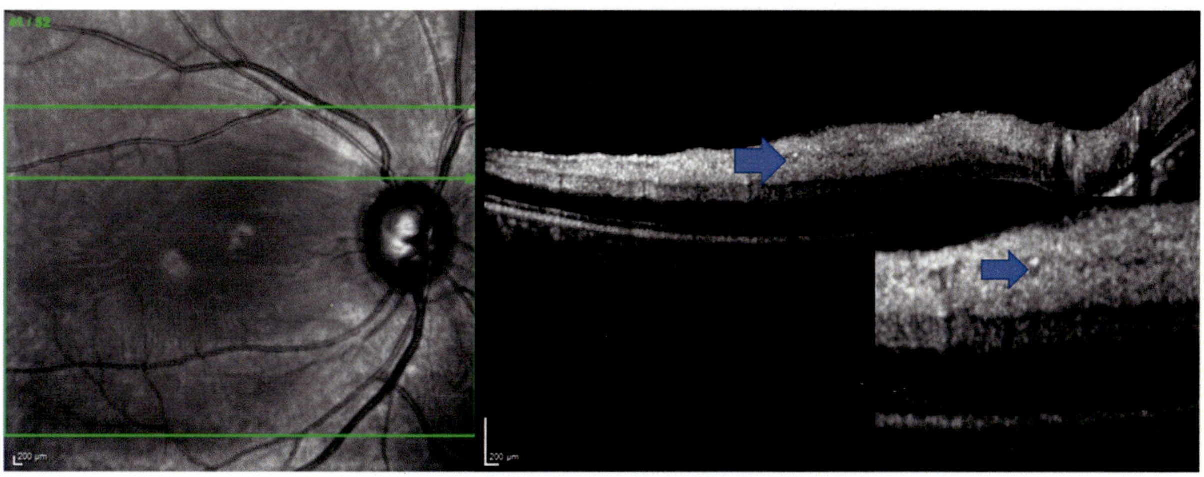

FIGURE 14.9 Empty vessel sign in CRAO.

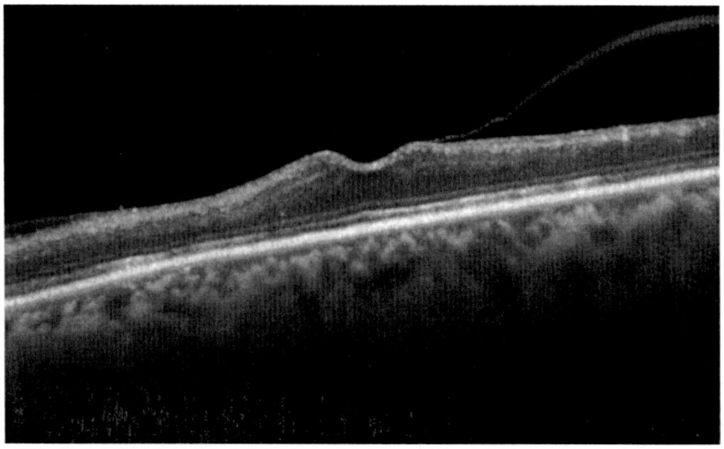

FIGURE 14.10 Inner retinal thinning and atrophy on OCT in a case of chronic CRAO.

The middle retinal layer opacification and p-MLM sign are the first changes to appear in eyes with mild acute cases of CRAO.[47,48] These findings are usually absent in higher grades of acute CRAO. In mild acute CRAO cases, OCT findings such as inner retinal fluid-filled hyporeflective cavities, neurosensory detachment, ILMD, inner retinal hyperreflective foci, and posterior vitreous hyperreflective opacities are not seen. These findings are more noticeable in severe grades of acute CRAO. The middle retinal layer is primarily made up of two synaptic plexuses, a dense deep microvascular capillary plexus, and the cell bodies of bipolar cells, horizontal cells, and amacrine cells.[49] In comparison to the superficial capillary plexus, which is located in the retinal nerve fiber layer, the deep capillary plexus is a denser capillary network and is predominantly supplied by the central retinal artery via the superficial vascular plexus.[49] When the central retinal artery suddenly gets hypo-perfused, the perfusion of the deep capillary plexus gets primarily affected along with the superficial capillary plexus, resulting in cellular swelling and extracellular edema predominantly in the middle retinal layers, which causes its opacification.[50] As a result, with a milder grade of CRAO, the opacification is limited to the middle retinal layers, with no significant retinal thickening or loss of retinal layer stratification. While the severity of the occlusion increases, the superficial capillary plexus becomes involved as well, resulting in opacification of the inner retinal layers, retinal thickening, and the absence of inner retinal layer opacification.[47] Cell body death occurs over time, resulting in retinal layer thinning and preservation of retinal layer stratification. The retinal opacification fades out depending on the severity of the occlusion, with the opacification in the middle retinal layers being the last to resolve.[47]

The bipolar cell synapses are located in the MLM, which is located at the inner boundary of the outer plexiform layer.[51] In mild and moderate grades of acute CRAO, there is acute swelling predominantly of bipolar cell synapses, resulting in a hyperreflective line on OCT suggestive of a p-MLM sign. The p-MLM sign is an indicator of less severe ischemic damage and may be a factor associated with a better visual acuity outcome.

Other signs seen on OCT such as inner retinal fluid-filled hyporeflective cavities, neurosensory detachment, ILMD, inner retinal hyperreflective foci, and posterior vitreous hyperreflective opacities can be explained based on the concept of reperfusion injury seen with acute CRAO. Although prompt retinal reperfusion restores oxygen and substrate delivery, reperfusion appears to have negative consequences due to an increase in reactive oxygen species generation and proinflammatory neutrophilic infiltration of the ischemic tissues.[32,33] As a result of severe retinal ischemia and reperfusion injury, retinal cell injury and/or death can occur following acute CRAO. The retinal tissue undergoes necrosis and apoptosis, resulting in intraretinal degeneration, permanent necrosis, and atrophy resulting in the OCT signs mentioned above. The tissue injury caused by retinal ischemia and reperfusion increases proportionally to the magnitude of ischemia caused by CRAO.

Fundus fluorescein angiography

In all cases of CRAO, fundus fluorescein angiography (FFA) must be performed to help distinguish the various types of CRAO and determine the extent of fundus circulatory abnormalities.

FFA patterns in various kinds of CRAO

1. NA-CRAO eyes almost always exhibit sluggish filling of the retinal vasculature and variable amounts of residual retinal circulation; complete absence of retinal vascular filling is uncommon in these eyes. Due to the variable amount of residual retinal circulation, the majority of ophthalmologists incorrectly believe that the CRA is not completely occluded in CRAO.
2. Retinal circulation is normal or nearly normal in NA-CRAO of transient variety.
3. In CRAO with sparing of the cilioretinal artery, FFA is the only way to outline definitively the region supplied by the patent cilioretinal artery, which is crucial for determining the visual outcome.
4. The key diagnostic feature of arteritic CRAO is the presence of associated posterior ciliary artery occlusion, as giant cell arteritis causes thrombosis of the common trunk of the posterior ciliary artery and the CRA arising from the ophthalmic artery. Failing to diagnose arteritic CRAO and, consequently, giant cell arteritis can result in catastrophic loss of vision in both eyes; therefore, FFA is crucial. Therefore, it must be performed on all CRAO patients aged 50 and older.

" No reflow" phenomenon

During the acute phase of CRAO, the macular region is primarily affected by whitish retinal opacification and swelling. If circulation is restored in the CRA, the retinal capillaries in the central, thickest portion of the macula do not refill due to compression by the surrounding swollen superficial retinal tissue, resulting in the "no reflow phenomenon" and permanent ganglion cell death in the nonperfused retina (Fig. 14.11).[10] The area of central retinal capillary nonfiling can vary from eye to eye based on the severity of macular retinal swelling. Consequently, the size of the permanent central scotoma is variable.

Retrograde filling of the retinal vein in CRAO

A retrograde filling of a major branch retinal vein on FFA in eyes with CRAO is a less commonly mentioned finding in literature (Fig. 14.12).[26] This finding is noted only in cases of CRAO with a patent CLRA. This phenomenon can be explained as follows.

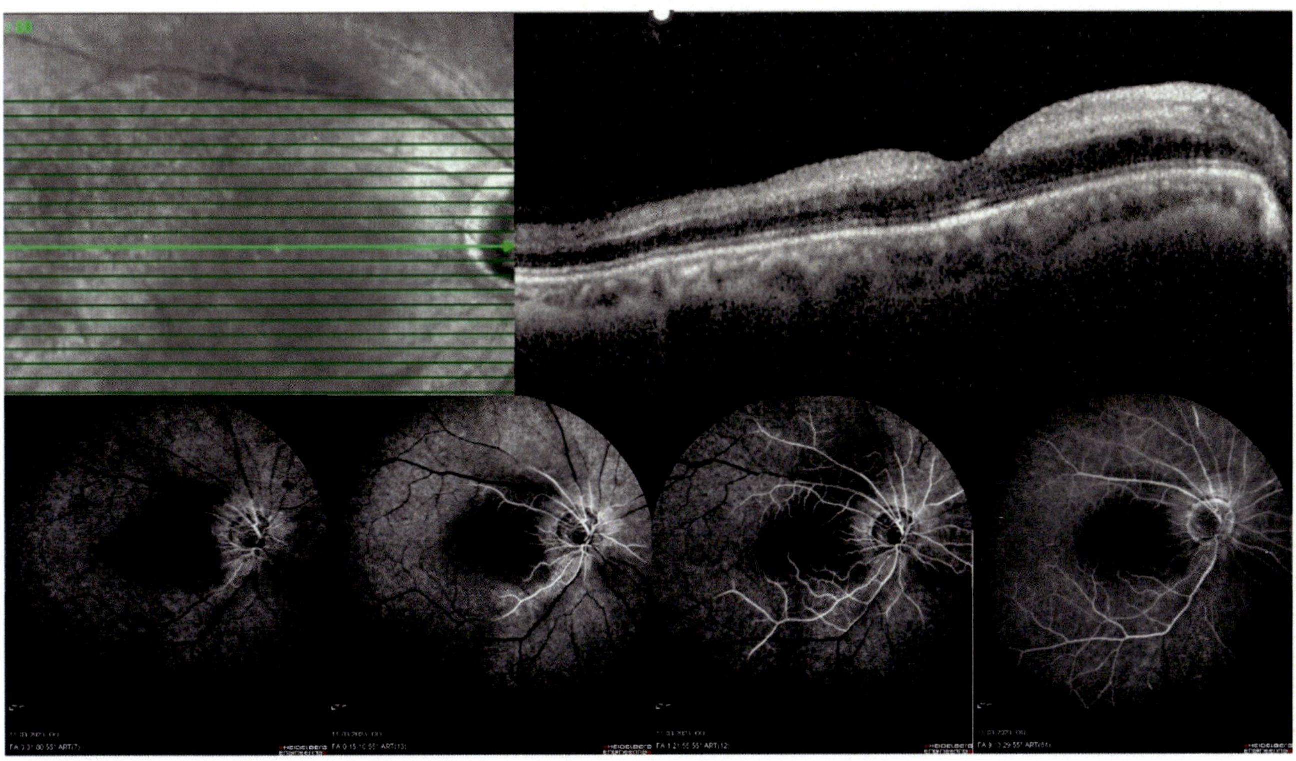

FIGURE 14.11 "No reflow" phenomenon in CRAO on FFA.

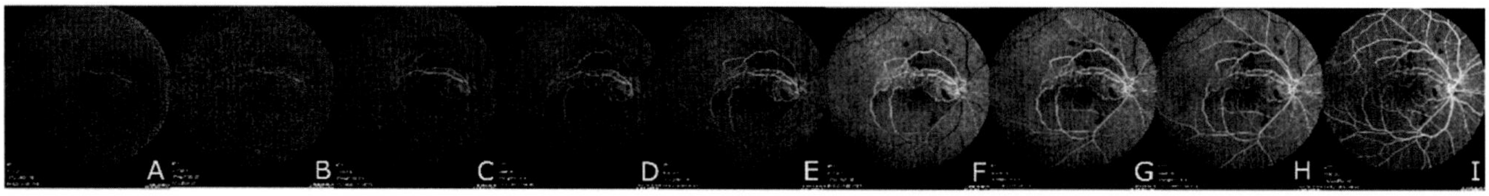

FIGURE 14.12 Retrograde filling of the retinal vein in a case of acute CRAO with a patent cilioretinal artery.

In normal retinal circulation, the blood from the retinal arteries and arterioles travels through the capillaries to the retinal venules before filling the retinal veins and eventually emptying from the eye by the central retinal vein.[52,53] The retinal artery fills from the optic disc to the retinal periphery. The proximal-most part of the retinal artery fills first as blood enters a major branch of the central retinal artery. This segment of the retinal artery supplies the retinal arterioles, which in turn supply the retinal capillary plexuses in the superficial and deeper layers of the inner retina in that region. Finally, blood from the retinal capillaries enters postcapillary venules and then the major branch of the accompanying retinal vein before draining into the central retinal vein. As the retinal artery fills up further, another segment of retinal capillaries fills up, eventually filling up that segment's retinal vein. Hence, in a normal circulation, the blood fills the retinal veins from the optic disc to the retinal periphery, while the blood drains from the retinal vein toward the optic disc into the central retinal vein.[52]

In the setting of a CRAO with the presence of a patent CLRA, the CLRA supplies the retinal capillaries in the region of the papillomacular bundle and macula. The retinal capillaries in the perfused segment of the CLRA fill into the tributaries of an adjacent major branch retinal vein. The retinal vein fills from the distal segment of the vein to the proximal portion of the vein in a retrograde fashion. Once the blood enters the retinal vein, it gets drained into the central retinal vein by flowing toward the optic disc (Fig. 14.13). In an eye with CRAO without a patent CLRA, the retinal capillaries do not get filled by the retinal artery and hence no filling of the dye into the retinal vein in a retrograde fashion occurs.

Arm-to-retina time in CRAO

Gong et al. examined the association between central visual impairment and FFA characteristics in patients with CRAO and identified three angiographic patterns, namely poor perfusion, exudative, and mixed, based on the arm-retina time and retinal vessel and optic disc leakage.[54] Vision impairment caused by exudative CRAO was less severe than that caused by poor perfusion and mixed types.

Mid-phase hyperfluorescent spots

Acute CRAO was characterized by the presence of a single or multiple hyperfluorescent pinpoint spots during the middle phase of the fundus fluorescein angiogram. These are known as mid-phase hyperfluorescent spots (MPHS). These MPHS were characterized by their predominant posterior pole location, prevalence in moderate to severe grades of acute CRAO, and emergence during the middle phase of the FFA with late staining and subtle adjacent perivascular leakage. These MPHS persisted until the severity of occlusion existed and vanished during the CRAO resolving stages. There were not any statistically significant associations between the presence of MPHS and the age of presentation, duration of symptoms, severity of CRAO, or arm-retina time delay. The exact nature of these MPHSs is not known. One may consider them to be true emboli (fibrin-platelet or cholesterol) released from the thrombus responsible for the obstruction of the retinal artery. This may be unlikely, however, in the absence of nonperfusion distal to these hyperfluorescent spots. Red blood cell (RBC) aggregates in the retinal circulation are an alternative explanation for these hyperfluorescent spots. Following an acute occlusion of the CRA, the blood flow distal to the occlusion becomes stagnant or reduced. Blood flow is affected more frequently at higher occlusion grades. This results in the clustering and aggregation of RBCs and the formation of rouleaux.[55] RBC aggregation would also depend on the surface properties of RBCs and fibrinogen concentrations.[56] As indicated by FFA, the formation of rouleaux damages the endothelium of the capillaries, resulting in the breakdown of the blood-retinal barrier and leakage from the capillaries into the intraretinal space. Rouleaux formation is more prevalent in the retinal capillary circulation as a result of the smaller vessel diameter.[33] This also suggests that the spots are more likely to be observed at the retinal level than at the choroidal level because only the retinal arteriolar and capillary anatomy is suitable for such a staining. This may not be possible in the choroidal circulation due to the larger caliber and high flow characteristics of the choroidal vessels.[57] MPHS is typically observed in patients with a higher severity grade of CRAO and at the posterior pole, where retinal capillary density is highest and capillary vessel diameter is smallest. In the milder forms

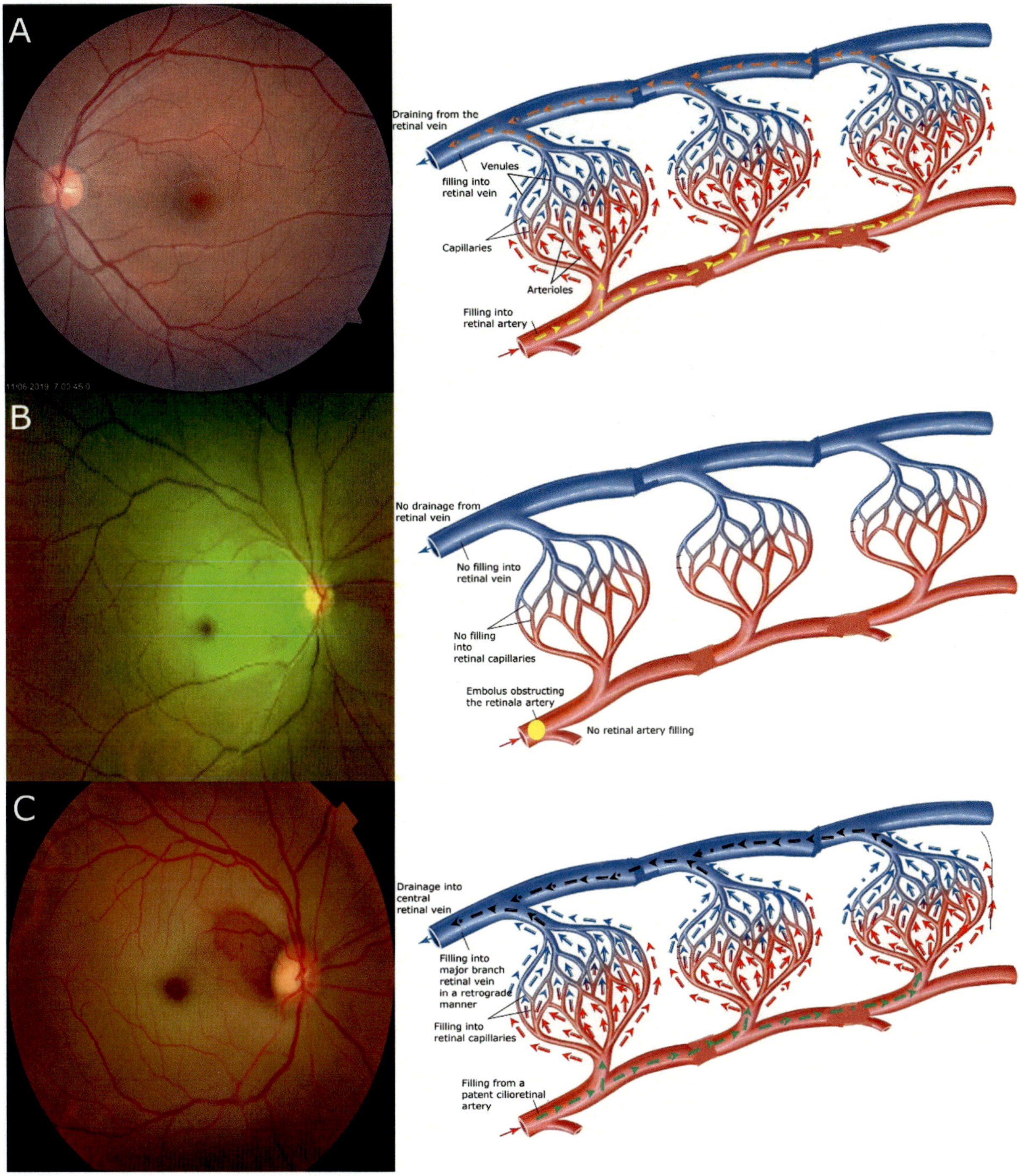

FIGURE 14.13 Diagrammatic image explaining the finding of retrograde flow of the dye into a retinal vein in a case of CRAO with patent cilioretinal artery.

of acute CRAO, there does not appear to be any obstruction of blood flow that would result in RBC aggregation and rouleaux formation. In the most severe stages of acute CRAO, widespread retinal whitening and thickening may obscure the MPHS beneath the retina on FFA. Thus, MPHSs were observed more frequently on FFA in patients with moderate occlusion severity than in those with milder and more severe disease. The spots resolve with the resolution of CRAO (Fig. 14.14).

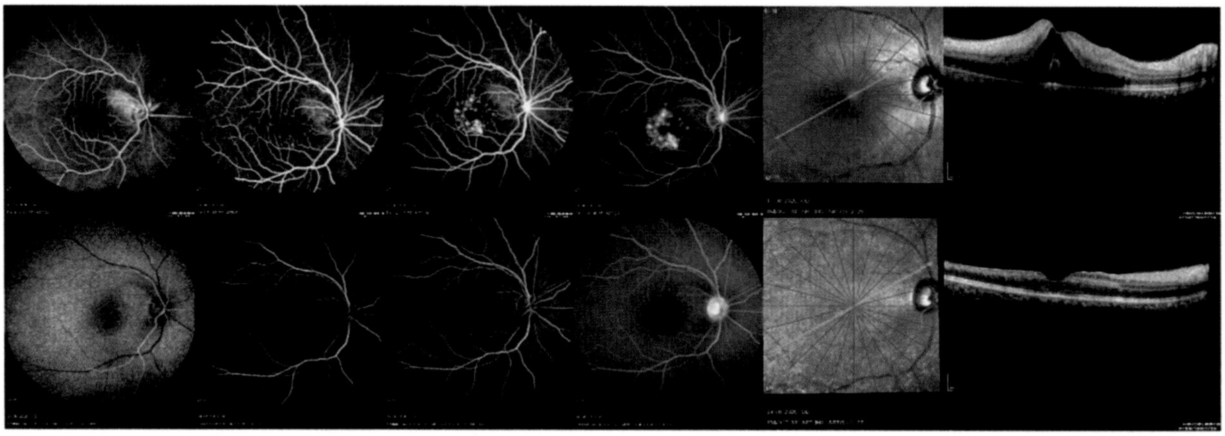

FIGURE 14.14 Mid-phase pin-point hyperfluorescent spots on FFA in an acute and resolving stage of CRAO.

Indocyanine green angiography

In eyes with CRAO, indocyanine green angiography is not routinely performed. Venkatesh et al. analyzed eight indocyanine green angiography (ICGA) studies in five patients with acute retinal artery occlusions in a recently published study.[58] According to the authors, ICGA provides valuable and useful information regarding choroidal perfusion in patients with acute retinal artery occlusions. As a result of the occlusion, rouleaux formation manifests as hypercyanescent spots in the retinal circulation during the late phase of ICGA.

OCT-angiography

Optical coherent tomography angiography (OCTA) is a novel noninvasive technique that obviates the need for dye injection when assessing retinal microvasculature. In CRAO, OCTA reveals decreased vascular perfusion in the superficial and deep retinal capillary plexuses, which corresponds to poor perfusion on FFA. In patients with the variant sparing the cilioretinal artery, the deep capillary plexus maintains perfusion. A recent case report by Bhanushali et al. demonstrated the restoration of retinal perfusion on OCTA following anterior chamber paracentesis in a patient with acute CRAO.[59] Unlike FFA, OCTA is unable to demonstrate a transit time delay. However, OCTA findings in retinal artery occlusions in eyes with severe inner retinal layer hyperreflectivity, thickening, and loss of retinal layer stratification must be interpreted with extreme caution.

Fundus autofluorescence

In the acute phase, fundus autofluorescence in ischemic areas is decreased because of retinal edema blocking the normal RPE (Fig. 14.15). Eventually, this could return to normal baseline or may be associated with increased autofluorescence owing to a window defect created by the thinned-out inner retinal layers.[60]

Electroretinography

Several authors have examined the relationship between various electroretinography (ERG) components, visual acuity, and the severity of retinal artery occlusion.[61,62] Three ERG parameters, notably the amplitude of the B-wave in scotopic 3.0, the B/A ratio, and the amplitude of the photopic negative response (PhNR), were found to be reliable predictors of visual prognosis. In mild-to-severe ischemia, the PhNR amplitudes indicating ganglion cell function showed a more pronounced decrease than the B-wave amplitudes. In terms of visual function and outcome, baseline visual acuity and visual field defects were only correlated with B/A ratio (both P .001), whereas visual acuity and visual field improvements were correlated with B-wave amplitude in dark-adapted 3.0 ($P = .004$ and .006), B/A ratio ($P = .023$ and .008), and PhNR amplitude (P .001 and .004).[61] Visual function and prognosis are correlated with B-wave amplitude in dark-adapted 3.0, B/A ratio, and PhNR amplitude changes in eyes with central retinal artery occlusion.

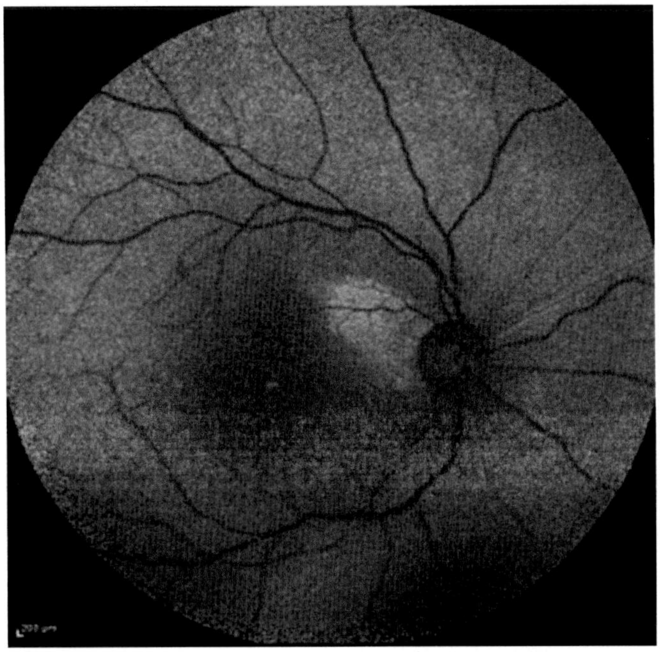

FIGURE 14.15 Fundus autofluorescence in a case of acute CRAO with patent cilioretinal artery.

Management

CRAO is an ophthalmic emergency characterized by catastrophic visual impairment. Numerous measures have been proposed to enhance visual outcomes. Typically, one or a combination of the following conventional treatments for acute CRAO have been advocated and claimed to be effective.[26] These include: (1) ocular massage to dislodge the embolus in the CRA; (2) a reduction of intraocular pressure (IOP) by paracentesis, massage of the eyeball, administration of acetazolamide, etc., to improve blood flow; (3) vasodilation of the CRA; (4) inhalations of 95% oxygen and 5% carbon dioxide; (5) rebreathing of expired CO_2 in a bag; and (6) retrobulbar vasodilators. With the exception of ocular massage, which occasionally dislodges the embolus, there is no evidence that any of the other treatments are effective. Other advocated treatments include (1) intravenous administration of a thrombolytic agent; (2) isovolumic hemodilution; (3) hyperbaric oxygen; (4) administration of Pentoxifylline to reduce red blood cell rigidity; (5) intravenous administration of systemic steroids to reduce vascular endothelial edema after CRAO; (6) neodymium:yttrium–aluminum-garnet laser arteriotomy and embolectomy; (7) cannulation of the supraorbital artery and retrograde injection of antispasmodic papaverine; and (8) pars plana vitrectomy with posterior vitreous detachment to create sudden hypotony and subsequent dislodgement of the embolus.

Local intraarterial thrombolysis is currently the most popular of these treatments.[18] Super-selective administration of a thrombolytic agent is utilized to deliver intraarterial fibrinolysis directly to the ophthalmic artery in this mode of treatment. Regarding intraarterial thrombolytic therapy, a beneficial effect is only possible if this treatment is initiated aggressively in significantly less than 4 h after the onset of CRAO and if the occlusion is caused by a fibrin embolus and not a cholesterol or calcific embolus. In conclusion, CRAO is conventionally described as "a disease with no remedy but multiple treatments."

References

1. Graefe A. Ueber Embolie der Arteria centralis retinae als Ursache plötzlicher Erblindung. *Albrecht von Graefes Archiv für Ophthalmologie*. 1859;5(1):136–157. https://doi.org/10.1007/bf02720764.
2. Schweigger K. *Vorlesungen über den Gebrauch des Augenspiegels*. 1864. Berlin Mylius'sche Verlags-Buchhandlung).
3. Graham RH, Ebrahim S. *Central Retinal Artery Occlusion*. 2023.
4. Hayreh SS. *A Study of the Central Atery of the Retina in Human Beings*. 1958.
5. Singh S, Dass R. The central artery of the retina I. Origin and course. *British Journal of Ophthalmology*. 1960;44(4):193–212. https://doi.org/10.1136/bjo.44.4.193.

6. Hayreh SS, Weingeist TA. Experimental occlusion of the central artery of the retina. I. Ophthalmoscopic and fluorescein fundus angiographic studies. *British Journal of Ophthalmology*. 1980;64(12):896–912. https://doi.org/10.1136/bjo.64.12.896.
7. Hayreh SS, Kolder HE, Weingeist TA. Central retinal artery occlusion and retinal tolerance time. *Ophthalmology*. 1980;87(1):75–78. https://doi.org/10.1016/S0161-6420(80)35283-4.
8. Jonas JB, Hayreh SS. Optic disk morphology in experimental central retinal artery occlusion in rhesus monkeys. *American Journal of Ophthalmology*. 1999;127(5):523–530. https://doi.org/10.1016/S0002-9394(99)00030-6.
9. Hayreh SS, Jonas JB. Optic disk and retinal nerve fiber layer damage after transient central retinal artery occlusion: an experimental study in rhesus monkeys. *American Journal of Ophthalmology*. 2000;129(6):786–795. https://doi.org/10.1016/S0002-9394(00)00384-6.
10. Hayreh SS, Zimmerman MB, Kimura A, Sanon A. Central retinal artery occlusion. Retinal survival time. *Experimental Eye Research*. 2004;78(3):723–736. https://doi.org/10.1016/S0014-4835(03)00214-8.
11. Kwon YH, Rickman DW, Baruah S, et al. Vitreous and retinal amino acid concentrations in experimental central retinal artery occlusion in the primate. *Eye*. 2005;19(4):455–463. https://doi.org/10.1038/sj.eye.6701546.
12. Hayreh SS, Pratesi F. *Central Retinal Vascular Occlusion- Morphology and Pathogenesis International Symposium on Ocular Angiology (1974: Firenze)*. 1977:15–40.
13. Hayreh SS, Podhajsky P. Ocular neovascularization with retinal vascular occlusion: II. Occurrence in central and branch retinal artery occlusion. *Archives of Ophthalmology*. 1982;100(10):1585–1596. https://doi.org/10.1001/archopht.1982.01030040563002.
14. Mizener JB, Podhajsky P, Hayreh SS. Ocular ischemic syndrome. *Ophthalmology*. 1997;104(5):859–864. https://doi.org/10.1016/S0161-6420(97)30221-8.
15. Hayreh SS, Zimmerman MB. Central retinal artery occlusion: visual outcome. *American Journal of Ophthalmology*. 2005;140(3). https://doi.org/10.1016/j.ajo.2005.03.038.
16. Hayreh SS. Prevalent misconceptions about acute retinal vascular occlusive disorders. *Progress in Retinal and Eye Research*. 2005;24(4):493–519. https://doi.org/10.1016/j.preteyeres.2004.12.001.
17. Hayreh SS, Zimmerman MB. Fundus changes in central retinal artery occlusion. *Retina*. 2007;27(3):276–289. https://doi.org/10.1097/01.iae.0000238095.97104.9b.
18. Hayreh SS. Intra-arterial thrombolysis for central retinal artery occlusion. *British Journal of Ophthalmology*. 2008;92(5):585–587. https://doi.org/10.1136/bjo.2007.136481.
19. Hayreh SS. Transient central retinal artery occlusion following viperine snake bite. *Archives of Ophthalmology*. 2008;126(6):870–871.
20. Hayreh SS, Podhajsky PA, Zimmerman MB. Retinal artery occlusion. Associated systemic and ophthalmic abnormalities. *Ophthalmology*. 2009;116(10):1928–1936. https://doi.org/10.1016/j.ophtha.2009.03.006.
21. Hayreh SS. Acute retinal arterial occlusive disorders. *Progress in Retinal and Eye Research*. 2011;30(5):359–394. https://doi.org/10.1016/j.preteyeres.2011.05.001.
22. Hayreh SS, Zimmerman MB. Ocular arterial occlusive disorders and carotid artery disease. *Ophthalmology Retina*. 2017;1(1):12–18. https://doi.org/10.1016/j.oret.2016.08.003.
23. Hayreh SS. Do patients with retinal artery occlusion need urgent neurologic evaluation? *American Journal of Ophthalmology*. 2018;196:53–56. https://doi.org/10.1016/j.ajo.2018.08.015.
24. Hayreh S. *Ocular Vascular Occlusive Disorders*. Heidelberg: Springer-Verlag; 2015:31–53.
25. Ratra D, Dhupper M. Retinal arterial occlusions in the young: systemic associations in Indian population. *Indian Journal of Ophthalmology*. 2012;60(2):95–100. https://doi.org/10.4103/0301-4738.94049.
26. Hayreh SS. Central retinal artery occlusion. *Indian Journal of Ophthalmology*. 2018;66(12). https://doi.org/10.4103/ijo.ijo_1446_18.
27. Justice J, Lehmann RP. Cilioretinal arteries: a study based on review of stereo fundus photographs and fluorescein angiographic findings. *Archives of Ophthalmology*. 1976;94(8):1355–1358. https://doi.org/10.1001/archopht.1976.03910040227015.
28. Hayreh SS. Pathogenesis of occlusion of the central retinal vessels. *American Journal of Ophthalmology*. 1971;72(5):998–1011. https://doi.org/10.1016/0002-9394(71)91706-5.
29. Sir D-E. Obstruction of the retinal arteries. In: *Textbook of Ophthalmology*. Vol 36. St Louis: The C. V. Mosby Company; 1941:2561–2563.
30. Tobalem S, Schutz JS, Chronopoulos A. Central retinal artery occlusion-rethinking retinal survival time. *BMC Ophthalmology*. 2018;18(1). https://doi.org/10.1186/s12886-018-0768-4.
31. Bulkley GB. Free radical-mediated reperfusion injury: a selective review. *British Journal of Cancer*. 1987;55(8):66–73.
32. Jennings RB, Reimer KA. The cell biology of acute myocardial ischemia. *Annual Review of Medicine*. 1991;42:225–246. https://doi.org/10.1146/annurev.me.42.020191.001301.
33. Yellon DM, Hausenloy DJ. Myocardial reperfusion injury. *New England Journal of Medicine*. 2007;357(11):1121–1135. https://doi.org/10.1056/nejmra071667.
34. Yuzurihara D, Iijima H. Visual outcome in central retinal and branch retinal artery occlusion. *Japanese Journal of Ophthalmology*. 2004;48(5):490–492. https://doi.org/10.1007/s10384-004-0102-y.
35. Furashova O, Matthé E. Retinal changes in different grades of retinal artery occlusion: an optical coherence tomography study. *Investigative Ophthalmology & Visual Science*. 2017;58(12). https://doi.org/10.1167/iovs.17-22411.
36. Ahn SJ, Woo SJ, Park KH, Jung C, Hong JH, Han MK. Retinal and choroidal changes and visual outcome in central retinal artery occlusion: an optical coherence tomography study. *American Journal of Ophthalmology*. 2015;159(4):667–676. https://doi.org/10.1016/j.ajo.2015.01.001.

37. Kim YH, Park KH, Woo SJ. Clinical manifestations and visual prognosis of cilioretinal artery sparing central retinal artery occlusion. *Korean Journal of Ophthalmology*. 2020;34(1):27–34. https://doi.org/10.3341/kjo.2019.0099.
38. Russell R. The source of retinal emboli. *The Lancet*. 1968;292(7572):789–792. https://doi.org/10.1016/s0140-6736(68)92453-7.
39. Ros MA, Magargal LE, Uram M. Branch retinal-artery obstruction: a review of 201 eyes. *Annals of Ophthalmology*. 1989;21(3):103–107.
40. Tripathy K, Patel BC. Cherry red spot. In: *StatPearls*. Treasure Island (FL): StatPearls Publishing; 2023.
41. Rim TH, Han J, Choi YS, et al. Retinal artery occlusion and the risk of stroke development: twelve-year nationwide cohort study. *Stroke*. 2016;47(2):376–382. https://doi.org/10.1161/STROKEAHA.115.010828.
42. Chang YS, Jan RL, Weng SF, et al. Retinal artery occlusion and the 3-year risk of stroke in Taiwan: a nationwide population-based study. *American Journal of Ophthalmology*. 2012;154(4):645–652.e1. https://doi.org/10.1016/j.ajo.2012.03.046.
43. Kim HM, Park YJ, Park KH, Woo SJ, Vavvas DG. Visual field defects and changes in central retinal artery occlusion. *Plos One*. 2019;14(1). https://doi.org/10.1371/journal.pone.0209118.
44. Venkatesh R, Jayadev C, Sridharan A, et al. Internal limiting membrane detachment in acute central retinal artery occlusion: a novel prognostic sign seen on OCT. *International Journal of Retina and Vitreous*. 2021;7(1). https://doi.org/10.1186/s40942-021-00323-7.
45. Kanakamedla A, Shandil A, Simakurthy S, Konana VK, Gudimetla J, Kumar RM. Empty vessel in central retinal artery occlusion: a sign denovo. *Indian Journal of Ophthalmology - Case Reports*. 2021;1(2). https://doi.org/10.4103/ijo.ijo_2018_20.
46. Falkenberry SM, Ip MS, Blodi BA, Gunther JB. Optical coherence tomography findings in central retinal artery occlusion. *Ophthalmic Surgery Lasers and Imaging*. 2006;37(6):502–505. https://doi.org/10.3928/15428877-20061101-12.
47. Wenzel DA, Poli S, Casagrande M, et al. Inner retinal layer hyperreflectivity is an early biomarker for acute central retinal artery occlusion. *Frontiers in Medicine*. 2022;9. https://doi.org/10.3389/fmed.2022.854288.
48. Schnieder M, Fischer-Wedi CV, Bemme S, Kortleben M-L, Feltgen N, Liman J. The retrobulbar spot sign and prominent middle limiting membrane as prognostic markers in non-arteritic retinal artery occlusion. *Journal of Clinical Medicine*. 2021;10(2). https://doi.org/10.3390/jcm10020338.
49. Campbell JP, Zhang M, Hwang TS, et al. Detailed vascular anatomy of the human retina by projection-resolved optical coherence tomography angiography. *Scientific Reports*. 2017;7. https://doi.org/10.1038/srep42201.
50. Varma DD, Cugati S, Lee AW, Chen CS. A review of central retinal artery occlusion: clinical presentation and management. *Eye*. 2013;27(6):688–697. https://doi.org/10.1038/eye.2013.25.
51. *Webvision: The Organization of the Retina and Visual System*. 1995.
52. Kuwabara T, Cogan DG. Studies of retinal vascular patterns: Part I. Normal architecture. *Archives of Ophthalmology*. 1960;64(6):904–911. https://doi.org/10.1001/archopht.1960.01840010906012.
53. Gong H, Song Q, Wang L. Manifestations of central retinal artery occlusion revealed by fundus fluorescein angiography are associated with the degree of visual loss. *Experimental and Therapeutic Medicine*. 2016;11(6):2420–2424. https://doi.org/10.3892/etm.2016.3175.
54. Chien S, Jan K-ming. Ultrastructural basis of the mechanism of rouleaux formation. *Microvascular Research*. 1973;5(2):155–166. https://doi.org/10.1016/0026-2862(73)90068-x.
55. Weng X, Cloutier G, Beaulieu R, Roederer GO. Influence of acute-phase proteins on erythrocyte aggregation. *American Journal of Physiology-Heart and Circulatory Physiology*. 1996;271(6):H2346–H2352. https://doi.org/10.1152/ajpheart.1996.271.6.h2346.
56. Baskurt OK, Meiselman HJ. RBC aggregation: more important than RBC adhesion to endothelial cells as a determinant of in vivo blood flow in health and disease. *Microcirculation*. 2008;15(7):585–590. https://doi.org/10.1080/10739680802107447.
57. Delaey C, van de Voorde J. Regulatory mechanisms in the retinal and choroidal circulation. *Ophthalmic Research*. 2000;32(6):249–256. https://doi.org/10.1159/000055622.
58. Venkatesh R, Reddy NG, Prabhu V, et al. Indocyanine green angiography imaging findings in artery occlusions. *European Journal of Ophthalmology*. 2022;32(4):2395–2403. https://doi.org/10.1177/11206721211037832.
59. Bhanushali DR, Yadav NK, Dabir S, Chidambara L, Srinivasan P, Shetty R. Spectral domain optical coherence tomography angiography features in a patient of central retinal arterial occlusion before and after paracentesis. *Retina*. 2016;36(5):e36–e38. https://doi.org/10.1097/IAE.0000000000000908.
60. Mathew R, Papavasileiou E, Sivaprasad S. Autofluorescence and high-definition optical coherence tomography of retinal artery occlusions. *Clinical Ophthalmology*. 2010;4(1):1159–1163. https://doi.org/10.2147/OPTH.S13592.
61. Kim HM, Park KH, Woo SJ. Correlation of electroretinography components with visual function and prognosis of central retinal artery occlusion. *Scientific Reports*. 2020;10(1). https://doi.org/10.1038/s41598-020-68957-5.
62. Matsumoto CS, Shinoda K, Nakatsuka K. High correlation of scotopic and photopic electroretinogram components with severity of central retinal artery occlusion. *Clinical Ophthalmology*. 2011;5(1):115–121. https://doi.org/10.2147/OPTH.S16053.

Chapter 15

Ocular ischemic syndrome

Şengül Özdek[1], Burak Acar[1] and Ece Ozdemir Zeydanli[2]
[1]*Gazi University School of Medicine, Ankara, Turkey;* [2]*Ankara Retina Clinic, Ankara, Turkey*

Introduction

Ocular ischemic syndrome (OIS) is a clinical picture that occurs as a result of impaired ocular perfusion due to stenosis or occlusion of the carotid arteries. In 1963, Hedges[1] reported marked venous engorgement and tortuosity in a patient with carotid artery stenosis. In the same year, Kearns and Hollenhorst[2] associated a similar clinical picture with carotid artery stenosis and named the condition *venous stasis retinopathy*. Later, different names were used, but the most descriptive name was given by Brown and Magargal[3] as *ocular ischemic syndrome*. Although it has been known for many years, it is easily missed or misdiagnosed in our daily practice due to its nonspecific and subtle findings. It may mimic some other ocular conditions such as diabetic retinopathy or central retinal vein occlusion, but the visual expectancy is much lower and the mortality rate is much higher in OIS.[4]

Systemic associations

Since atherosclerosis is the major cause of OIS, it has a significant relationship with systemic diseases. Approximately 70% of patients with OIS have systemic arterial hypertension, 60% have diabetes mellitus, and 40% have ischemic heart disease. One-third of patients have a previous cerebrovascular accident.[4–6] These rates are significantly higher when compared to the aged-matched population.[4] Patients have an average stroke rate of 4% per year, regardless of whether they receive treatment or not.[4] The 5-year mortality rate in OIS patients is around 40%, and the main cause of death is cardiovascular diseases.[4]

The clinician should keep in mind that ocular findings may be the first sign of carotid artery occlusion. Considering that cardiovascular disease and stroke are the main causes of death in these patients, appropriate systemic evaluation can save these patients' lives.[4]

Epidemiology and pathogenesis

The average age of patients is 65 years, and it very rarely occurs before 50 years old.[2,3,5] Both eyes are involved in 25% of patients, and men are affected twice as often as women, in parallel with the higher incidence of atherosclerotic disease in men.[2,3,5] Although the incidence of OIS is calculated as 7.5 cases per million people per year, this rate is thought to be much higher as the diagnosis is often easily overlooked.[7]

Atherosclerosis is the main cause of OIS. However, other diseases such as Behçet's disease, giant cell arteritis, fibrovascular dysplasia, Takayasu arteritis, dissecting aneurysm of the carotid artery, and complications after radiotherapy for nasopharyngeal carcinoma or after intravitreal anti-VEGF injections may be involved in the etiology.[8–14]

OIS mostly occurs with occlusions of 80% or more of the ipsilateral common or internal carotid arteries, and the affected carotid artery is completely occluded in half of the patients.[3,15,16] However, besides the degree of stenosis, the presence or absence of collateral vessels, variations in the anastomotic channels that compensate for the occluded vessels, and the chronicity of the carotid artery disease are also implicated in the pathogenesis. Patients with well-developed collateral circulation may not develop OIS even with total occlusion of the carotid artery. Contrarily, those with poor collateral circulation between the internal and external carotid systems or between the two internal carotid arteries may develop OIS even with a stenosis of less than 50%.

Retinal and Choroidal Vascular Diseases of the Eye. https://doi.org/10.1016/B978-0-443-15583-3.00015-9

Also, patients with OIS exhibit reduced blood flow in the retrobulbar arteries and reversal of blood flow in the ophthalmic artery. This causes a steal phenomenon that shunts blood flow away from the eye to the low-resistance intracranial circuit causing further reduction of blood flow and more pronounced ocular ischemia.[17]

Symptoms

Decreased vision

The most common symptom of OIS is decreased vision, reported in over 90% of the affected eyes.[3] It typically develops gradually, over weeks to months, due to chronic retinal ischemia as well as optic nerve damage related to secondary glaucoma. However, a smaller percentage can present with sudden-onset visual loss due to embolic occlusion of a central retinal artery or when intraocular pressure exceeds the perfusion pressure in eyes with neovascular glaucoma. In these cases, cherry red spots can be observed due to acute inner-layer retinal ischemia.[3,18]

Amaurosis fugax

Temporary vision loss can be seen in 10%–15% of patients.[3,6] It may develop as a result of transient embolization in the central retinal artery or due to vasospasm.[3] Patients typically describe a dark or black shade spreading across the visual field that lasts for a few seconds to minutes. More rarely, it may develop due to choroidal hypoperfusion.[19,20] Increased retinal metabolic demand or decreased perfusion pressure may trigger a temporary vision loss.[21] This has been reported after exposure to bright light, eating a meal, or postural change.[21,22]

Prolonged light recovery

Prolonged recovery of visual function has been reported after exposure to bright light, particularly in patients with advanced carotid artery stenosis, associated with severe macular ischemia.[22]

Pain

Up to 40% of patients with OIS have ocular pain.[3,21] It may develop due to neovascular glaucoma as well as due to ischemia of the ocular region or/and ipsilateral meninges.[3] Ischemic pain may be defined as ocular angina and described as a dull, constant ache in the affected eye, over the orbit, temple, and upper face that begins gradually over hours to days. It is typically relieved by lying down and worsens in the upright position. In elderly patients with ocular pain, giant cell arteritis should be in the differential diagnosis.

Visual field changes—Scotomas

Patients with OIS may have normal visual fields (23%) or have central scotoma (27%), nasal defects (23%), and more rarely, centrocaecal defects (5%) or only central or temporal islands (5%).[6]

Signs

Anterior segment

Although posterior segment findings of ocular ischemia are usually more common, anterior segment findings may be the only manifestation of the disease in some cases. Anterior segment examination may show conjunctival injection and dilated episcleral vessels, indicative of collateral blood flow from the external carotid artery in cases of internal carotid artery occlusion. Depending on the amount of ischemia of the pupillary sphincter, the pupillary light reaction may be sluggish or a fixed pupil can be observed.[3,23]

Rubeosis iridis can be seen in up to 70% of the cases at the pupillary border or at the iridocorneal angle (Fig. 15.1A).[3] A gonioscopy should be performed in a suspected case of OIS before dilation as angle neovascularization may develop in the absence of iris neovascularization. Despite the high frequency of rubeosis, neovascular glaucoma occurs only in one-third of cases and normal tension or even hypotonia may occur in some cases. This is because severe carotid stenosis impairs ciliary body perfusion and decreases aqueous humor formation.[3,21] A low-grade anterior chamber reaction, not exceeding grade 2, may occur. The flare is more prominent than the cellular reaction. Rarely, small keratic precipitates can be seen in cases with rubeosis iridis. Hyphema and ectropion uvea can be observed in cases with severe rubeosis iridis.[6,7] If neovascular glaucoma is not properly controlled, descemet folds and corneal edema may develop, which can progress to bullous keratopathy in advanced stages. In patients with unilateral advanced carotid artery disease, asymmetric cataract development may be seen.[3,6,7]

Posterior segment

The characteristic posterior segment finding is dot-blot configuration retinal hemorrhages located at the midperipheral retina, seen in more than 80% of patients with OIS[3] (Figs. 15.1B and 15.2B). These hemorrhages are very rarely numerous, almost never confluent, and mostly located in deep retinal layers. Some of these hemorrhages can be seen at the nerve fiber layer in a flame-shaped form. These hemorrhages are thought to be due to leakage from the small retinal vessels with ischemic endothelial damage or as a result of rupture of capillary microaneurysms. Microaneurysms are observed as frequently as retinal hemorrhages in OIS. They may be located both in the macula or at midperiphery and appear more prominent on fundus fluorescein angiography (FFA) (Fig. 15.2C). Also, diffuse macular capillary telangiectasia may develop and together with microaneurysms contribute to macular edema in these cases. The retinal arteries are narrowed. In contrast, retinal veins are dilated but not tortuous.[3] Appearance of retinal veins is important to distinguish it from other retinal vascular diseases. The venous dilation may be accompanied by beading; however, it is not as extensive as severe diabetic retinopathy. Also, unlike OIS, that is, caused by inflow obstruction, retinal veins are both dilated and tortuous in a central retinal vein occlusion in which outflow is compromised. Nonetheless, both retinal arteries and veins may be narrowed in some cases of OIS.

As previously noted, a cherry red spot can be seen due to ischemia of the inner retinal layers when the central retinal artery is occluded by an embolus or, more frequently, when the perfusion pressure falls below the intraocular pressure, especially in cases with neovascular glaucoma. New vessel development on the optic disc can be observed in one-third of cases.[3] Optic disc neovascularization is more frequent than new vessels elsewhere. These pathological vessels may bleed in 4% of patients and, rarely, progress to severe fibrovascular proliferation.[24]

Choroidal ischemia and chorioretinal atrophy may occur due to compromised choroidal circulation.[25,26] Choroidal neovascular membranes may rarely develop in these ischemic areas.[6,27] Other posterior segment findings include cotton wool spots due to local ischemia, spontaneous retinal arterial pulsations, and anterior and posterior ischemic optic neuropathy.[21,28]

Rarely, orbital infarction syndrome can be observed, which is the ischemia of all the intraorbital and intraocular structures.[29] It is characterized by orbital pain, anterior and posterior segment ischemia with intraocular inflammation and hypotony, ophthalmoplegia, ptosis, and corneal hypesthesia.

Ancillary tests

Fluorescein angiography

Prolonged choroidal filling time, although not the most common, is the most specific angiographic finding of OIS. The normal choroidal filling normally begins 10−12 s after dye injection and is completed within 5 s after the first appearance of the dye in the choroidal arteries. In OIS, choroidal filling may be delayed up to 45 s after the dye is injected.[3] Patchy or delayed choroidal filling is reported in about 60% of patients.[3]

Prolonged arteriovenous transit time is found in 95% of patients and is the most sensitive finding, but since it is also seen in other retinal diseases, its specificity is low.[3] In advanced ischemia, retinal veins may not be filled during the entire study.

Staining of the retinal vessels is observed in 85% of the cases and represents the second most common angiographic finding in OIS patients[3] (Figs. 15.1C and 15.2C). Although both arteries and veins are affected, it is usually more prominent in arteries. The ischemic damage to the endothelial cells and disruption of the inner blood−retina barrier causes vessel staining.[30,31] Macular edema and leakage can be observed in 15% of patients in the late phases of FFA.[3,32] Even if the optic disc appearance is normal, late staining can be found (Fig. 15.1C and D). Occasionally, retinal capillary nonperfusion can be observed in the midperipheral retina (Figs. 15.1C, D and 15.2C).[3]

Indocyanine green angiography

Indocyanine green angiography (ICGA) shows the choroidal structures better, therefore ischemic changes in choroidal circulation appear more prominent.[33] As in FFA, arm-to-choroid circulation time (normally 10−12 s) and intrachoroidal circulation time (normally 5–6 s) are delayed due to choroidal hypoperfusion. Slow filling of the watershed zones including both posterior (i.e., between the optic disc and the macula) and the peripheral zones are also characteristics of OIS.[33]

Optical coherence tomography

Optical coherence tomography (OCT) provides quick and noninvasive detection of macular edema that may occur during the course of OIS. Also, a centrally located, highly reflective band can be detected within or above the outer plexiform layer, the so-called "middle-limiting membrane sign," indicative of acute retinal ischemia[34] (Fig. 15.1E). Segmental hyperreflectivity of the inner retinal layers may be seen between the outer plexiform layer and the ganglion cell layer.[34] Enhanced depth imaging on OCT is a valuable tool for measuring choroidal thickness in these cases, as studies have shown a decrease in subfoveal choroidal thickness and volume in eyes with OIS and symptomatic carotid artery stenosis. There appears to be a negative correlation between the degree of carotid artery stenosis and choroidal thickness.[25,26,35–38] This decrease in choroidal thickness is thought to be due to impaired choroidal circulation because improvements in choroidal thickness and choroidal vascular flow are observed after carotid artery stenting.[35]

Optical coherence tomography angiography

Optical coherence tomography angiography (OCTA) allows us to examine the vascular structure of the retina and the choroid with a noninvasive method. With multilayered OCTA images, we can see the origin and anatomy of newly formed vessels, the presence of collateral shunt vessels, and vascular changes after intravitreal anti-VEGF injections.[39] Although FFA gives information about the activity of neovascularization indirectly by showing leakage, OCTA technology enables us to directly examine the morphology of the vascular network. In most of the patients with OIS, nonperfusion areas can be observed in all layers. Also, the foveal avascular zone (FAZ) may be enlarged, and vascular density may be decreased.[34,38,40] However, it should be noted that when inner retinal ischemia is present, outer retinal findings may be overlooked due to artifacts. In cases with OIS, a reduction in the FAZ and an increase in vascular density in all layers can be observed after carotid artery stenting.[41] Studies on OCTA in patients with OIS are still limited. As it has great potential in demonstrating choroidal capillaries and detecting retinal vascular abnormalities, it is highly likely that it will have a great place in the diagnosis and follow-up of cases with OIS.

Electroretinography

Electroretinography (ERG) is rarely used to establish the diagnosis of OIS. The b-wave of the ERG correlates with the activity of inner layer retinal function while the a-wave correlates with the activity of outer layer retina function. As both layers are affected by ischemia in OIS, a diminution or absence of both a and b waves can be observed. This is in contrast to central retinal artery occlusions, where only the inner retina is ischemic and the amplitude of b-wave is reduced.[42]

Imaging methods for the evaluation of carotid occlusive disease

Carotid duplex ultrasound

It is the most commonly used noninvasive test because it uses B-mode ultrasound and Doppler ultrasound simultaneously and gives information about both the anatomical features of the vessels and the flow rates.[43] It can detect both carotid artery stenosis and occlusion with a high sensitivity and a specificity.[44]

Color Doppler ultrasound

It gives information about retrobulbar and carotid circulation by measuring the flow velocity in the orbital artery. In people with carotid artery stenosis and occlusion, deterioration in these values is more pronounced than in a normal age-matched population.[45]

Magnetic resonance angiography and computed tomographic angiography

Over the years, MRA and CTA have gradually replaced conventional carotid arteriography. These non- or minimally invasive techniques are now used as the second-line investigation tools after carotid ultrasonography in cases with suspected carotid stenosis. For stenosis over 70%, both MRA and CTA have very high sensitivity and specificity.[44,46] Although duplex ultrasonography is as accurate in detecting complete occlusions, MRA and CTA have a better discriminatory power in diagnosing 70%–99% stenosis.

Carotid arteriography

Conventional carotid arteriography has long been the gold standard for the evaluation of carotid stenosis.[47] However, since there is a risk of transient ischemic attack, cerebral infarction, and even death, noninvasive tests are now used primarily in diagnosis and follow-up.[48] Conventional angiography may be reserved for cases where the results of ultrasonography and CTA/MRA are contradictory or inconclusive.

Differential diagnosis

The differential diagnosis of OIS mainly includes diabetic retinopathy and mild–moderate central retinal vein occlusion (CRVO) (Table 15.1). The most distinctive difference of OIS from these two entities is decreased retinal perfusion pressure. As a result, light digital pressure on the eyelid may induce retinal arterial pulsations in the affected eyes.[21] Another distinctive feature of OIS is delayed and patchy choroidal filling observed in FFA. Also, the presence of dilated retinal veins but the absence of tortuosity helps to differentiate OIS from CRVO.

In diabetic retinopathy, intraretinal hemorrhages are more numerous and found in the posterior pole, whereas in OIS they are less numerous and located in the midperiphery. The hard exudates that we see frequently in diabetic retinopathy are not observed in cases with OIS. Nevertheless, OIS can coexist with diabetic retinopathy. Patients with significant asymmetry of retinopathy should be investigated for carotid artery disease.[9]

Hyperviscosity syndromes as well as new-onset uveitis over the age of 50 should also be considered in differential diagnosis of OIS.

TABLE 15.1 Key differentiating features of ocular ischemic syndrome, diabetic retinopathy, and central retinal vein occlusion.

	OIS	DRP	CRVO
Age	50–80s	Variable	50–80s
Laterality	80% unilateral	Bilateral	Usually unilateral
Posterior segment			
Retinal perfusion pressure	Decreased	Normal	Normal
Retinal veins	Dilated but not tortuous	Dilated and beaded	Dilated and tortuous
Optic disk	Normal	Papillopathy (rare)	Swollen
Hemorrhage	Dot-blot (midperiphery)	Dot-blot and flame-shaped (posterior pole and midperiphery)	Flame-shaped (all quadrants)
Microaneurysms	Common (midperiphery)	Common (posterior pole)	Uncommon
Hard exudates	No	Common	Rare
Fluorescein angiography			
Choroidal filling	Delayed, patchy	Normal	Normal
Arteriovenous transit time	Prolonged	Usually normal	Prolonged
Retinal vessel staining	Arteries > Veins	Usually absent	Veins > arteries

CRVO, central retinal vein occlusion; *DRP*, diabetic retinopathy; *OIS*, ocular ischemic syndrome.

Treatment

As atherosclerosis is the main cause in the development of OIS, controlling diseases such as systemic arterial hypertension and diabetes mellitus is the cornerstone of the treatment. Therefore, OIS patients should be treated and followed by a multidisciplinary team of specialists, including primary care physicians, neurologists, cardiologists, interventional radiologists, and vascular surgeons as required.

Ocular treatment

Two main focuses of ocular therapy are retinal ischemia and neovascular glaucoma. Half of patients require one or more topical treatments to control the intraocular pressure. Topical β-adrenergic blockers or α-2-agonists are preferred together, along with carbonic anhydrase inhibitors. Prostaglandin analogs should be avoided due to their proinflammatory effects. About 10% of patients are refractory to medical treatment. In these patients, trabeculectomy, aqueous shunt implants or diode laser cyclophotocoagulation are the preferred treatment options.[6,49] In affected eyes, lowering the intraocular pressure as much as possible is essential for effective ocular perfusion.[6] After treatment of carotid artery stenosis, increased intraocular pressure may develop due to ocular hyperperfusion because of improved oxygen supply and aqueous humor production.[50,51]

After the development of neovascularization in the anterior or posterior segment, the application of panretinal photocoagulation (PRP) may prevent progression to neovascular glaucoma.[52–54] However, PRP has been shown to be effective in regressing iris neovascularization in only one-third of the patients.[54] In these cases, intravitreal bevacizumab can help to regress iris and disk neovascularization.[55–58] It has been reported that intravitreal bevacizumab can improve macular perfusion and it is effective for the treatment of macular edema.[55,57] If there is no response to intravitreal anti-VEGF therapy, intravitreal triamcinolone is another option available.[59]

Carotid artery surgical management

Iris neovascularization indicates severe ocular ischemia, and it is unlikely to reverse ischemia at this stage. Most favorable outcomes could be achieved when treatment is initiated before the development of neovascular glaucoma.[60] Currently, there are carotid artery endarterectomy, carotid artery stenting, and extracranial-intracranial arterial bypass surgery for the treatment of carotid artery occlusion.[17,61] Carotid artery stenting is an effective alternative in cases at high risk of developing complications after carotid artery endarterectomy.[61]

Prognosis

Despite treatment, 60% of patients have worsening of vision and those with visual acuity less than counting fingers increased from 22% to 60% after 1 year follow-up.[54] At the end of follow-up, one-third of patients have a bare light perception or less.[54] Having low vision, having rubeosis iridis, and developing rubeosis iridis in the first 3 months are poor prognostic indicators for visual acuity.[6,54] Although diode laser cyclophotocoagulation and PRP are ineffective in improving vision, they reduce the risk of neovascular development.[54] Carotid artery surgeries in advanced stages do not improve vision; however, they may be useful in selected cases with asymptomatic severe carotid artery disease.[6,54,62]

Case presentations

Case 1

A 70-year-old female patient with a known diagnosis of diabetes mellitus and hypertension for 10 years was referred to our clinic for neovascular glaucoma in the right eye. The visual acuity was hand motions in the right eye and 0.8 in the left eye. Ocular examination revealed rubeosis iridis (Fig. 15.1A) and dot-blot configuration retinal hemorrhages in the mid-peripheral retina (Fig. 15.1B) in the right eye. FFA of the patient showed ischemia in the inferotemporal quadrant (Fig. 15.1C and D). OCT of the patient showed a prominent middle limiting membrane sign (*) (Fig. 15.1E). Computed tomography angiography of the patient revealed total occlusion of the right carotid artery (Fig. 15.1F). Retinal photocoagulation was applied to the ischemic areas in the inferotemporal quadrant seen in FFA. However, ischemia became evident in all four quadrants 6 months later (Fig. 15.1G).

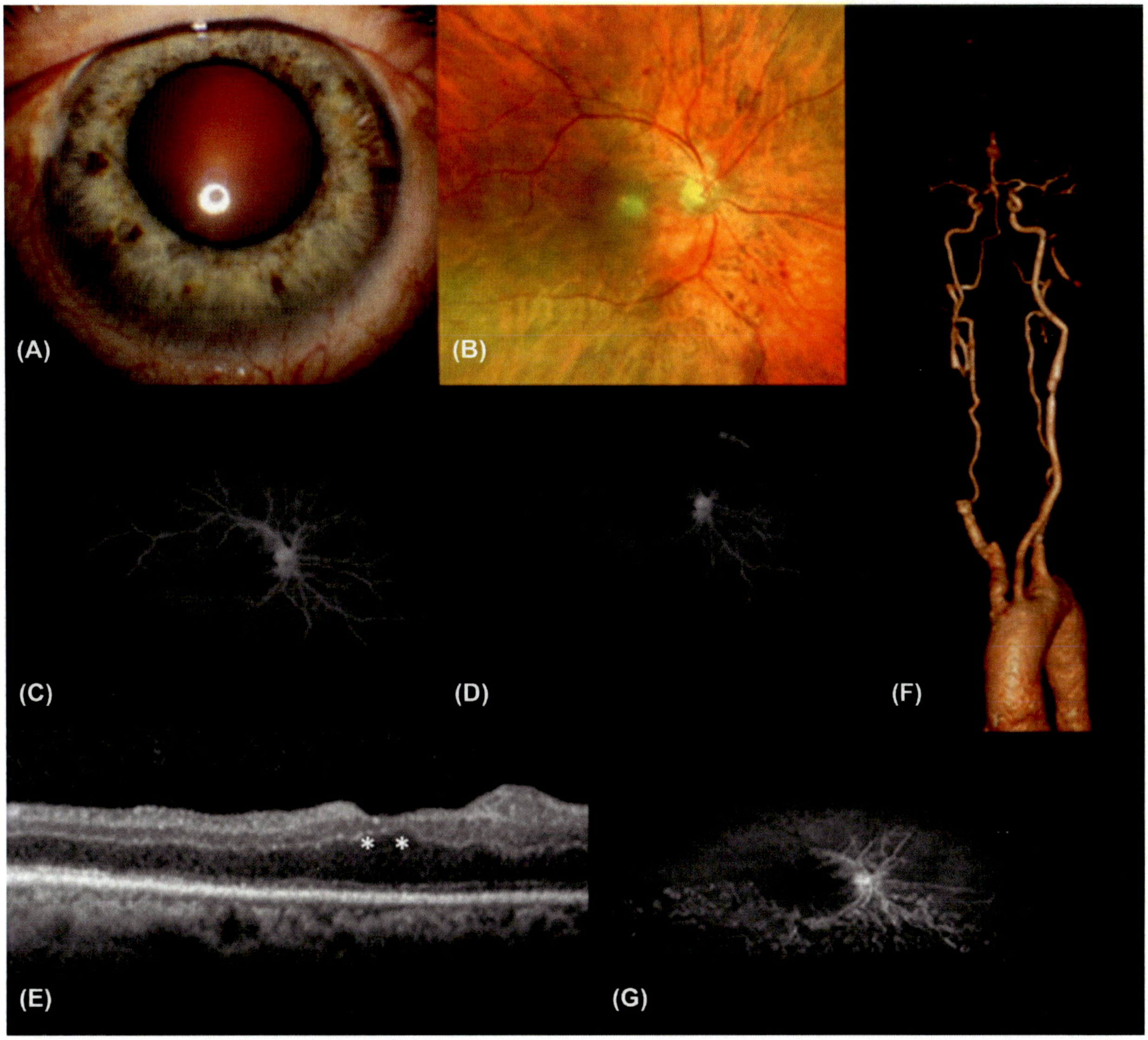

FIGURE 15.1 Ocular and radiological findings of case 1. *Case Courtesy of H. Baran Özdemir, MD, Gazi University School of Medicine.*

Case 2

A 71-year-old male patient presented with the complaint of a gradual visual decrease in both eyes for the past several months. He had a history of uncontrolled diabetes and hypertension for 5 years and a cerebrovascular infarct 3 months ago, after which he developed left facial palsy and a persistent epithelial defect in the left eye. The visual acuity was 0.7 in the right and 0.05 in the left eye. The intraocular pressure was 16 mmHg in the right and 35 mmHg in the left eye. Anterior segment examination revealed neovascularization at the anterior chamber angle in both eyes and ciliary injection, irregular corneal epithelium with central corneal haze, and a fixed dilated pupil in the left eye (Fig. 15.2A and D). Fundus examination revealed numerous dot blot configuration retinal hemorrhages in the midperipheral retina in the right eye (Fig. 15.2B). Although the fundus view was not clear in the left eye due to the corneal pathology, a few dot-blot hemorrhages were noted (Fig. 15.2E). FFA showed delayed arteriovenous transit time, diffuse microaneurysms, and peripheral ischemia in the right eye (Fig. 15.2C); however, the view was suboptimal in the left eye. Computed tomography angiography of the patient revealed a 90% stenosis in the right and a 75% stenosis in the left carotid artery. He underwent left carotid artery stenting and later bilateral carotid endarterectomy. Panretinal photocoagulation stabilized the ocular ischemic findings in the right eye. However, in the left eye, the epithelial defect persisted despite treatment with bandage contact lenses and intense topical medication. Although rubeosis iridis regressed after regular anti-VEGF injections, intraocular

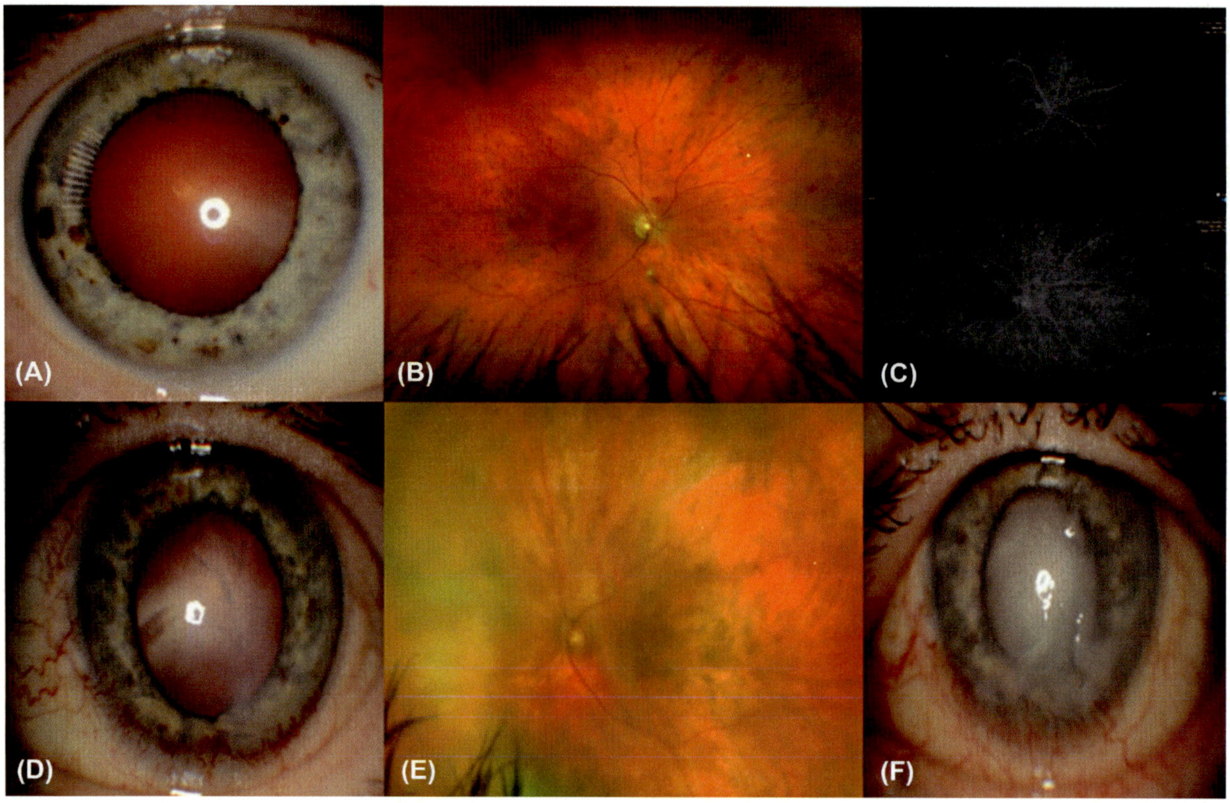

FIGURE 15.2 Ocular findings of case 2 with OIS.

pressure remained high despite maximum antiglaucoma medication. Systemic conditions and severe ocular surface problems did not allow a glaucoma surgery at the time. The corneal pathology in the left eye progressed and resulted in severe vascularization as well as a dense white cataract developed, and the patient had no light perception at the final follow-up (Fig. 15.2F). Unfortunately, the patient died following an intensive care unit hospitalization after bypass surgery (Fig. 15.2).

References

1. Hedges TR. Ophthalmoscopic findings in internal carotid artery occlusion. *American Journal of Ophthalmology*. 1963;55:1007–1012.
2. Kearns TP, Hollenhorst RW. Venous-stasis retinopathy of occlusive disease of the carotid artery. *Proceedings of the Staff Meetings. Mayo Clinic*. 1963;38:304–312.
3. Brown GC, Magargal LE. The ocular ischemic syndrome. Clinical, fluorescein angiographic and carotid angiographic features. *International Ophthalmology*. 1988;11(4):239–251.
4. Sivalingam A, Brown GC, Magargal LE, Menduke H. The ocular ischemic syndrome. II. Mortality and systemic morbidity. *International Ophthalmology*. 1989;13(3):187–191.
5. Luo J, Yan Z, Jia Y, Luo R. Clinical analysis of 42 cases of ocular ischemic syndrome. *Journal of Ophthalmology*. 2018;2018:2606147. https://doi.org/10.1155/2018/2606147.
6. Mizener JB, Podhajsky P, Hayreh SS. Ocular ischemic syndrome. *Ophthalmology*. 1997;104(5):859–864.
7. Sturrock GD, Mueller HR. Chronic ocular ischaemia. *The British Journal of Ophthalmology*. 1984;68(10):716–723.
8. Casson RJ, Fleming FK, Shaikh A, James B. Bilateral ocular ischemic syndrome secondary to giant cell arteritis. *Archives of Ophthalmology (Chicago, IL: 1960)*. 2001;119(2):306–307.
9. Duker JS, Belmont JB. Ocular ischemic syndrome secondary to carotid artery dissection. *American Journal of Ophthalmology*. 1988;106(6):750–752.
10. Effeney DJ, Krupski WC, Stoney RJ, Ehrenfeld WK. Fibromuscular dysplasia of the carotid artery. *The Australian and New Zealand Journal of Surgery*. 1983;53(6):527–531.

11. Huang Z-L, Lin K-H, Lee Y-C, Sheu M-M, Tsai R-K. Acute vision loss after intravitreal injection of bevacizumab (avastin) associated with ocular ischemic syndrome. *Ophthalmologica. Journal international d'ophtalmologie. International journal of ophthalmology. Zeitschrift fur Augenheilkunde*. 2010;224(2):86–89. https://doi.org/10.1159/000235726.
12. Hwang JM, Girkin CA, Perry JD, Lai JC, Miller NR, Hellmann DB. Bilateral ocular ischemic syndrome secondary to giant cell arteritis progressing despite corticosteroid treatment. *American Journal of Ophthalmology*. 1999;127(1):102–104.
13. Peter J, David S, Danda D, Peter JV, Horo S, Joseph G. Ocular manifestations of Takayasu arteritis: a cross-sectional study. *Retina*. 2011;31(6):1170–1178. https://doi.org/10.1097/IAE.0b013e3181fe540b.
14. Tang Y, Luo D, Peng W, Huang F, Peng Y. Ocular ischemic syndrome secondary to carotid artery occlusion as a late complication of radiotherapy of nasopharyngeal carcinoma. *Journal of Neuro-Ophthalmology: The Official Journal of the North American Neuro-Ophthalmology Society*. 2010;30(4):315–320. https://doi.org/10.1097/WNO.0b013e3181dee914.
15. Smith VH. Pressure changes in the ophthalmic artery after carotid occlusion: an experimental study in the rabbit. *The British Journal of Ophthalmology*. 1961;45(1):1–26.
16. Kobayashi S, Hollenhorst RW, Sundt TM. Retinal arterial pressure before and after surgery for carotid artery stenosis. *Stroke*. 1971;2(6):569–575.
17. Kearns TP, Siekert RG, Sundt TM. The ocular aspects of carotid artery bypass surgery. *Transactions of the American Ophthalmological Society*. 1978;76:247–265.
18. Konuk O, Ozdek S, Onal B, Tiftikçioğlu Y, Gürelik G, Hasanreisoğlu B. Ocular ischemic syndrome presenting as central retinal artery occlusion in scleroderma. *Retina*. 2006;26(1):102–104.
19. Winterkorn JM, Teman AJ. Recurrent attacks of amaurosis fugax treated with calcium channel blocker. *Annals of Neurology*. 1991;30(3):423–425.
20. Hayreh SS, Zimmerman MB. Amaurosis fugax in ocular vascular occlusive disorders: prevalence and pathogeneses. *Retina*. 2014;34(1):115–122. https://doi.org/10.1097/IAE.0b013e31829234b5.
21. Mendrinos E, Machinis TG, Pournaras CJ. Ocular ischemic syndrome. *Survey of Ophthalmology*. 2010;55(1):2–34. https://doi.org/10.1016/j.survophthal.2009.02.024.
22. Donnan GA, Sharbrough FW, Whisnant JP. Carotid occlusive disease. Effect of bright light on visual evoked response. *Archives of Neurology*. 1982;39(11):687–689.
23. Legocki AT, Davis JA, Francis CE. Anisocoria due to ocular ischemic syndrome. *Journal of Neuro-Ophthalmology: The Official Journal of the North American Neuro-Ophthalmology Society*. 2022. https://doi.org/10.1097/WNO.0000000000001539.
24. Brown GC, Brown MM, Magargal LE. The ocular ischemic syndrome and neovascularization. *Transactions - Pennsylvania Academy of Ophthalmology and Otolaryngology*. 1986;38(1):302–306.
25. Kwon DH, Kim YC, Kang KT. Clinical significance of choroidal thickness in eyes with ocular ischemic syndrome. *Korean Journal of Ophthalmology*. 2022;36(1):66–73. https://doi.org/10.3341/kjo.2021.0155.
26. Kang HM, Choi JH, Koh HJ, Lee SC. Significant changes of the choroid in patients with ocular ischemic syndrome and symptomatic carotid artery stenosis. *PLoS One*. 2019;14(10):e0224210. https://doi.org/10.1371/journal.pone.0224210.
27. Flaxel C, Gregor Z, Broadway D. An uncommon presentation of the ocular ischaemia syndrome. *Eye (London, England)*. 1998;12(Pt 1):154–156.
28. Brown GC. Anterior ischemic optic neuropathy occurring in association with carotid artery obstruction. *Journal of Clinical Neuro-Ophthalmology*. 1986;6(1):39–42.
29. Bogousslavsky J, Pedrazzi PL, Borruat FX, Regli F. Isolated complete orbital infarction: a common carotid artery occlusion syndrome. *European Neurology*. 1991;31(2):72–76.
30. Michelson PE, Knox DL, Green WR. Ischemic ocular inflammation. A clinicopathologic case report. *Archives of Ophthalmology (Chicago, IL: 1960)*. 1971;86(3):274–280.
31. Kahn M, Green WR, Knox DL, Miller NR. Ocular features of carotid occlusive disease. *Retina*. 1986;6(4):239–252.
32. Brown GC. Macular edema in association with severe carotid artery obstruction. *American Journal of Ophthalmology*. 1986;102(4):442–448.
33. Utsugi N, Takahashi K, Kishi S. Choroidal vascular occlusion in internal carotid artery obstruction. *Retina*. 2004;24(6):915–919.
34. Feucht N, Zapp D, Reznicek L, Lohmann CP, Maier M, Mayer CS. Multimodal imaging in acute retinal ischemia: spectral domain OCT, OCT-angiography and fundus autofluorescence. *International Journal of Ophthalmology*. 2018;11(9):1521–1527. https://doi.org/10.18240/ijo.2018.09.15.
35. Durusoy GK, Gumus G, Onay M, Altay CM, Binboga AB. Early choroidal structure and choroidal vascularity index change after carotid stenting. *Photodiagnosis and Photodynamic Therapy*. 2022;38:102748. https://doi.org/10.1016/j.pdpdt.2022.102748.
36. Kim DY, Joe SG, Lee JY, Kim J-G, Yang SJ. Choroidal thickness in eyes with unilateral ocular ischemic syndrome. *Journal of Ophthalmology*. 2015;2015:620372. https://doi.org/10.1155/2015/620372.
37. Wang H, Wang Y-L, Li H-Y. Subfoveal choroidal thickness and volume in severe internal carotid artery stenosis patients. *International Journal of Ophthalmology*. 2017;10(12):1870–1876. https://doi.org/10.18240/ijo.2017.12.13.
38. Wong CL, Ang M, Tan ACS. Clinical applications of optical coherence angiography imaging in ocular vascular diseases. *Applied Sciences*. 2019;9(12):2577. https://doi.org/10.3390/app9122577.
39. Lupidi M, Fiore T, Cerquaglia A, Coscas G, Cagini C. Depth-resolved imaging of papillary vitreoretinal neovascularization: OCT-angiography assessment in ocular ischemic syndrome. *Retina*. 2017;37(4):e42–e44. https://doi.org/10.1097/IAE.0000000000001422.
40. İncekalan TK, Taktakoğlu D, Şimdivar GHN, Öztürk İ. Optical cohorence tomography angiography findings in carotid artery stenosis. *International Ophthalmology*. 2022;42(8):2501–2509. https://doi.org/10.1007/s10792-022-02297-3.

41. Saito K, Akiyama H, Mukai R. Alteration of optical coherence tomography angiography in a patient with ocular ischemic syndrome. *Retinal Cases & Brief Reports*. 2021;15(5):588–592. https://doi.org/10.1097/ICB.0000000000000857.
42. Coleman K, Fitzgerald D, Eustace P, Bouchier-Hayes D. Electroretinography, retinal ischaemia and carotid artery disease. *European Journal of Vascular Surgery*. 1990;4(6):569–573.
43. Taylor DC, Strandness DE. Carotid artery duplex scanning. *Journal of Clinical Ultrasound*. 1987;15(9):635–644.
44. Nederkoorn PJ, van der Graaf Y, Hunink MGM. Duplex ultrasound and magnetic resonance angiography compared with digital subtraction angiography in carotid artery stenosis: a systematic review. *Stroke*. 2003;34(5):1324–1332.
45. Hu HH, Sheng WY, Yen MY, Lai ST, Teng MM. Color Doppler imaging of orbital arteries for detection of carotid occlusive disease. *Stroke*. 1993;24(8):1196–1203.
46. Koelemay MJW, Nederkoorn PJ, Reitsma JB, Majoie CB. Systematic review of computed tomographic angiography for assessment of carotid artery disease. *Stroke*. 2004;35(10):2306–2312.
47. Marlow C. MRC European Carotid Surgery Trial: interim results for symptomatic patients with severe (70–99%) or with mild (0–29%) carotid stenosis. European Carotid Surgery Trialists' Collaborative Group. *Lancet (London, England)*. 1991;337(8752):1235–1243.
48. Davies KN, Humphrey PR. Complications of cerebral angiography in patients with symptomatic carotid territory ischaemia screened by carotid ultrasound. *Journal of Neurology, Neurosurgery, and Psychiatry*. 1993;56(9):967–972.
49. Sivak-Callcott JA, O'Day DM, Gass JD, Tsai JC. Evidence-based recommendations for the diagnosis and treatment of neovascular glaucoma. *Ophthalmology*. 2001;108(10):1767–1776.
50. Lee KM, Kim EJ, Heo SH, Jin KH. Paradoxical development of neovascular glaucoma following carotid angioplasty and stenting. *Interventional neuroradiology: Journal of Peritherapeutic Neuroradiology, Surgical Procedures and Related Neurosciences*. 2016;22(5):540–543. https://doi.org/10.1177/1591019916653938.
51. Ganaie HA, Gupta V, Parthasarathy R, Londhe S, Anand S. Blindness following carotid artery stenting due to ocular hyperperfusion—report and review of literature. *Neurology India*. 2020;68(4):897–899. https://doi.org/10.4103/0028-3886.293455.
52. Johnston ME, Gonder JR, Canny CL. Successful treatment of the ocular ischemic syndrome with panretinal photocoagulation and cerebrovascular surgery. *Canadian Journal of Ophthalmology*. 1988;23(3):114–119.
53. Carter JE. Panretinal photocoagulation for progressive ocular neovascularization secondary to occlusion of the common carotid artery. *Annals of Ophthalmology*. 1984;16(6):572–576.
54. Sivalingam A, Brown GC, Magargal LE. The ocular ischemic syndrome. III. Visual prognosis and the effect of treatment. *International Ophthalmology*. 1991;15(1):15–20.
55. Choi EY, Lee SC, Kim M. Optical coherence tomography angiography before and after bevacizumab injection in ocular ischemic syndrome. *Retina*. 2018;38(3):e23–e25. https://doi.org/10.1097/IAE.0000000000001981.
56. Lee SJ, Lee JJ, Kim SY, Kim SD. Intravitreal bevacizumab (Avastin) treatment of neovascular glaucoma in ocular ischemic syndrome. *Korean Journal of Ophthalmology: KJO*. 2009;23(2):132–134. https://doi.org/10.3341/kjo.2009.23.2.132.
57. Amselem L, Montero J, Diaz-Llopis M, et al. Intravitreal bevacizumab (Avastin) injection in ocular ischemic syndrome. *American Journal of Ophthalmology*. 2007;144(1):122–124.
58. Asif H, Si Z, Quan S, et al. Neovascular glaucoma from ocular ischemic syndrome treated with serial monthly intravitreal bevacizumab and panretinal photocoagulation: a case report. *Case Reports in Ophthalmological Medicine*. 2022;2022:4959522. https://doi.org/10.1155/2022/4959522.
59. Klais CM, Spaide RF. Intravitreal triamcinolone acetonide injection in ocular ischemic syndrome. *Retina*. 2004;24(3):459–461.
60. Ala-Kauhaluoma M, Nuotio K, Ijäs P, et al. Ocular signs of carotid stenosis in ipsi- and contralateral eyes before and after carotid endarterectomy: a prospective study. *Acta Ophthalmologica*. 2022;100(4):e1015–e1023. https://doi.org/10.1111/aos.15019.
61. Bonati LH, Lyrer P, Ederle J, Featherstone R, Brown MM. Percutaneous transluminal balloon angioplasty and stenting for carotid artery stenosis. *The Cochrane Database of Systematic Reviews*. 2012:9CD000515. https://doi.org/10.1002/14651858.CD000515.pub4.
62. Thompson JE. Carotid endarterectomy for asymptomatic stenosis: benefits confirmed by randomized trial. *Journal of Vascular Surgery*. 1993;17(4):798–800.

Chapter 16

Diabetic retinopathy and choroidopathy: Pathophysiology, imaging findings, and treatment updates

Sukhum Silpa-archa, Mongkol Tadarati, Peranut Chotcomwongse and Paisan Ruamviboonsuk
Department of Ophthalmology, Rajavithi Hospital, College of Medicine, Rangsit University, Bangkok, Thailand

Diabetic retinopathy (DR) remains a common complication of diabetes and a leading cause of preventable blindness in the working-age population. As strongly associated with highly elevated blood sugar, DR can mainly deteriorate vision through diabetic macular edema (DME) and neovascular complications. According to the International Diabetes Federation, the global population with diabetes mellitus (DM) was expected to be 700 million by 2045.[1] The number of adults worldwide with DR, vision-threatening DR, and clinically significant macular edema is expected to rise to 161 million, 45 million, and 29 million, respectively, by 2045.[2] Diabetic retinopathy ranges in severity from nonproliferative diabetic retinopathy (NPDR) to more severely proliferative diabetic retinopathy (PDR), which causes abnormal vessel growth and complicated retinal detachment.

Hyperglycemia is thought to play an important role in the pathogenesis of retinal microvascular damage in DR, causing tissue damage via several mechanisms. In addition, increasing laboratory and clinical evidence revealed that, besides microvascular changes, inflammation and retinal neurodegeneration may be implicated in the early stages of DR as independent pathogenesis pathways.[3–5] Regarding the structural changes in DR, the technological advancement in retinal imaging has discovered retinal changes before the clinical onset of DR found by routine fundus examination or fundus photography. These encompass ultra-widefield (UWF) imaging,[6,7] optical coherence tomography (OCT),[8–10] optical coherence tomography angiography (OCTA),[11,12] and fundus fluorescein angiography (FFA).[13,14] Besides, functional changes in the electroretinogram have been reported before the onset of clinically overt DR.[15,16] The standard treatments of DR include antivascular endothelial growth factor (anti-VEGF) for DME, and laser panretinal photocoagulation (PRP) with or without anti-VEGF for PDR.[17,18] However, a significant proportion of patients do not respond well to the treatment, indicating that factors other than vascular endothelial growth factors (VEGF) are involved.[4,5]

A certain proportion of patients with DR may have complications related to the proliferative stage despite recent advancements in retinal imaging for early disease detection. The standard treatments for these sight-threatening complications, such as long-standing vitreous hemorrhage or traction retinal detachment are vitrectomy with additional intraoperative procedures as necessary.[19–21]

Pathophysiology

The pathophysiology of DR is highly interconnected, dynamic, and complex in nature, which can be elaborated in aspects of biochemical and molecular changes, and alterations in cells and tissues. Hyperglycemia causes tissue damage through four major mechanisms: (1) the polyol pathway; (2) advanced glycation end products (AGEs) accumulation; (3) the protein kinase C (PKC) pathway; and (4) the hexosamine pathway (Fig. 16.1). All of these mechanisms interact in various ways[22,23] and have all been linked to mitochondrial dysfunction and endoplasmic reticulum stress, promoting reactive oxygen species (ROS) accumulation that, in turn, promotes cellular damage and contributing to diabetic complications development and progression.[24,25]

Retinal and Choroidal Vascular Diseases of the Eye. https://doi.org/10.1016/B978-0-443-15583-3.00016-0

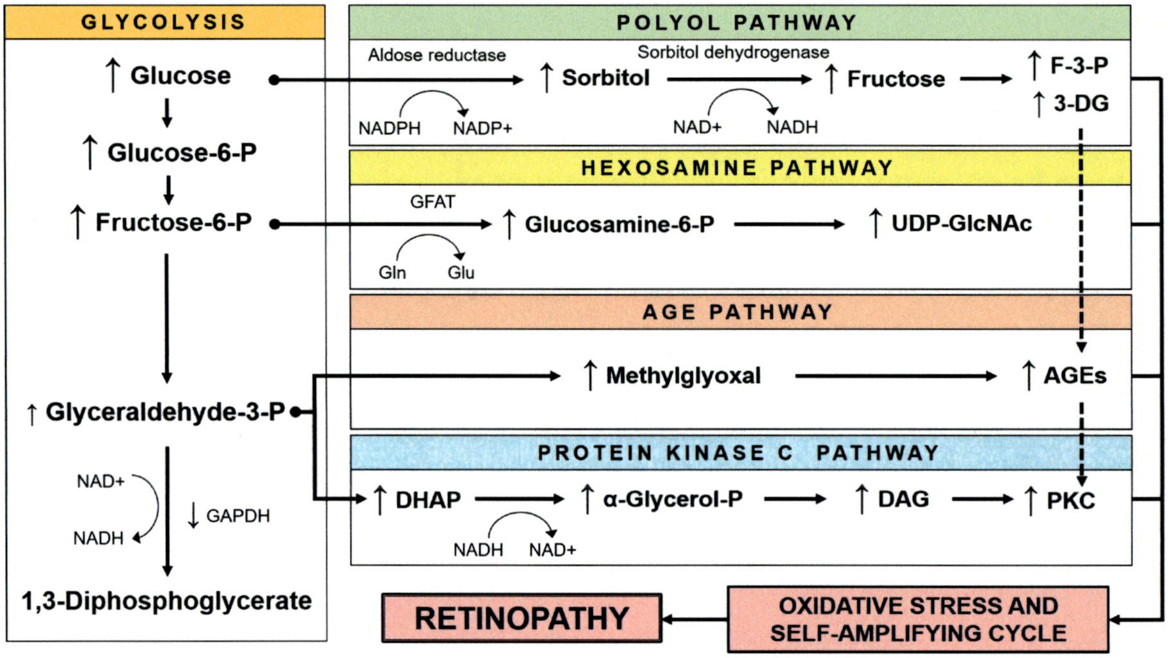

FIGURE 16.1 The major biochemical pathways of DR demonstrate oxidative stress as a unifying mechanism leading to DR. These pathways include the polyol pathway, the hexosamine pathway, the formation of advanced glycation end products (AGEs), and protein kinase C (PKC) activation pathway. All of these mechanisms interact in various ways as the interconnection between the polyol, AGEs, and PKC pathway was demonstrated by the dotted line. *3-DG*, 3-deoxyglucosone; *DAG*, diacylglycerol; *DHAP*, dihydroxyacetone phosphate; *F-3-P*, fructose-3-phosphate; *GAPDH*, glyceraldehyde 3-phosphate dehydrogenase; *GFAT*, glutamine:fructose-6-phosphate amidotransferase; *P*, phosphate; *UDP-GlcNAc*, uridine diphosphate *N*-acetylglucosamine.

Biochemical and molecular mechanism in the pathogenesis of diabetic retinopathy

Polyol pathway

Excess glucose is metabolized through the polyol pathway, which produces sorbitol via an aldose reductase (AR)-mediated pathway.[26] The aldose reductase pathway is inactive in normoglycemic conditions because glucose is a poor substrate for aldose reductase. When intracellular glucose levels rise, this pathway becomes active once the other enzymatic pathways of glucose metabolism become saturated. The first and rate-limiting enzyme in the pathway, aldose reductase, converts glucose to sorbitol with the help of NADPH. Sorbitol accumulates inside cells and causes osmotic damage because it is impermeable to cellular membranes.[27,28] Sorbitol is then converted to fructose by sorbitol dehydrogenase, and then to fructose-3-phosphate and deoxyglucosone. Both are strong glycolyzing agents that result in the formation of AGEs, which cause retinopathy.[26,29] Furthermore, upregulation of the polyol pathway reduces the availability of NADPH, increasing the sensitivity of affected cells to oxidative stress and depleting the reduced glutathione pool, which leads to elevated ROS levels.[30]

Advanced glycation end products accumulation

The formation of AGEs is the result of the nonenzymatic process of glycation, which occurs when reactive or reducing sugars react with proteins, lipids, or nucleic acids. A growing body of evidence shows that the interaction of AGEs with their receptor, namely the receptor for AGE (RAGE), causes the generation of oxidative stress, which in turn causes a variety of cells to encounter proliferative, inflammatory, thrombotic, and fibrotic reactions.[31–33] AGE formation leads to the cross-linking of proteins and subsequent changes in the physicochemical properties of tissues, and in diabetes, this includes basement membranes, cellular receptors, and blood vessel wall components. The accumulation of AGEs has been also correlated to pericyte loss.[34,35]

Protein kinase C pathway

Hyperglycemia also increases glucose flux via the glycolysis pathway. As a result, the production of diacylglycerol increases, activating the protein kinase C (PKC) pathway. PKC is a widely distributed enzyme that seems to promote the development of many diabetes complications without involving the aldose reductase pathway. PKC is involved in a wide range of biochemical signaling pathways, and as a result, it impacts a number of molecular processes when upregulated in DR, including modulating enzymatic activities such as mitogen-activated protein (MAP) kinase pathway, cytosolic phospholipase A2, Na+, K+-ATPase, and gene expressions.[36–38] This results in vascular abnormalities including changes in retinal and renal blood flow, contractility, permeability, proliferation, and basement membrane.[36]

Hexosamine pathway

With hyperglycemia, some extra glucose is diverted into the hexosamine pathway causing numerous changes in gene and protein levels that all contribute to the pathogenesis of DR.[39,40] Another study suggests that increased glucose flux via the hexosamine pathway may induce apoptosis in retinal neurons.[41] In the hexosamine pathway, fructose-6-phosphate is transformed to *N*-acetylglucosamine-6-phosphate by glutamine:fructose-6-phosphate amidotransferase (GFAT). *N*-Acetylglucosamine-6-phosphate is then converted to *N*-acetylglucosamine-1,6-phosphate, and uridine diphosphate *N*-acetylglucosamine (UDP-GlcNAc).[42–44] UDP-GlcNAc is a substrate for proteoglycan synthesis and for O-linked glycosylation of certain proteins. Increased UDP-GlcNAc production may alter gene expression and cellular function via excessive protein glycosylation, ultimately causing cell apoptosis in retinal neurons and endothelial cells.[42] Clinically, increased GFAT activity has been found to be closely related to HbA1c levels in diabetic patients.[45]

Alterations in cells and tissues in diabetic retinopathy

Pericytes are the first vascular cells affected by diabetes and their loss is the earliest and most specific sign of DR. This discovery was made while examining trypsin-digested retinal vasculature flat mounts from diabetic human subjects.[3,46] Pericyte loss alters vascular intercellular contacts and compromises the inner blood–retina barrier. This has severe consequences because the vascular walls weaken causing clinically visible venous dilation and beading, endothelial cell proliferation, and microaneurysm development.[47–50] While pericyte loss is the first histological sign of DR, microaneurysm is the first clinically visible sign of DR.[51] There are two leading theories that attempt to explain how hyperglycemia leads to pericyte degeneration. One theory points to the involvement of the aldose reductase pathway,[52–54] while the other implicates platelet-derived growth factor-beta (PDGF-$_\beta$).[49,55] Nevertheless, it's important to note that the precise mechanism behind this process remains largely unknown.

Basement membrane thickening is the most prominent histological feature of diabetic microangiopathy and can be seen under electron microscopy (EM). The biochemical mechanism that causes basement membrane thickening is unknown, but studies suggest that aldose reductase and the sorbitol pathway play a role.[56,57] Other morphological changes in diabetes from EM also included "Swiss cheese-like" vacuolization and deposition of fibrillar collagen within the homogenous structure of the BM.[58,59] Thickening of the basement membrane, in conjunction with systemic and local hypertension, hyperglycemia, the formation of advanced glycation end products, and hypoxia, may disrupt the tight link between pericytes and endothelial cells, resulting in pericyte apoptosis, while endothelium, deprived of proliferation control, can give rise to new vessels.[60] Various angiogenic factors stimulate or modulate DR in different ways. The VEGF, however, seems to be primarily in charge of the ischemia-driven angiogenesis and permeability response in PDR, according to in vivo and in vitro studies.[61–63] The VEGF is a key mediator in the breakdown of the inner blood–retina barrier. This potent permeabilizing factor rapidly induced phosphorylation of occludin and zonula occludens-1 (ZO-1). Both proteins play critical roles in the formation and function of endothelial cell tight junctions.[64,65] The angiopoietin family is involved in the regulation of vascular maturation and neovascularization (NV). Angiopoietin (Ang)-1 is a strong agonist of the Tie-2 receptor, and Ang-2 acts as an agonist or antagonist in a context-dependent manner.[66] While Ang-1 induces Tie-2 activation and phosphorylation, thereby promoting vessel stabilization and maturation under physiological conditions, high levels of VEGF and Ang-2 were seen among patients with preproliferative and proliferative retinopathy.[67] The role of the VEGF and angiopoietin families in DR has prompted a number of drug-development projects. Dual inhibition of VEGF and Ang-2 has been shown to promote vascular stability and decrease vascular leakage. Moreover, dual VEGF and Ang-2 inhibition resulted in antiinflammatory effects in preclinical models, including decreased inflammatory cell infiltration of the retina.[68]

In addition to DR, choroidal abnormalities in DM are common in patients with diabetes and suggest a significant role in the pathogenesis of DR given that the choroid is largely responsible for the nutrition and oxygenation of the outer retinal layers.[69] Before the development of OCT technologies that elaborate more detailed images of the choroidal anatomy and topography,[8–10,70] the studies for diabetic choroidopathy were based on indocyanine angiography (ICGA)[71–73] and histologic findings[74–76] at the beginning. The choroidal abnormalities in diabetic eyes include microaneurysms, luminal narrowing and marked basement membrane thickening of the choriocapillaris and other small choroidal blood vessels, choriocapillaris degeneration, vascular remodeling with increased vascular tortuosity, leakage of proteinaceous fluid into the choroidal stroma, capillary dropout, and choroidal neovascularization with subretinal fibrovascular membranes.[74–77]

Clinical features and classification

As the most common retinal vascular disorder, microvascular changes in DR were pathologically characterized by increased vascular permeability and closure of the retinal vessels. These lead to the clinically visible retinal microvascular changes as used for the current measure for the stages of DR. Diabetic retinopathy is classified into NPDR and PDR. When the proper intervention is not made, DR progresses gradually from mild NPDR to more severe stages, including moderate NPDR, severe NPDR, and PDR. Evaluation and staging of DR are critical for providing appropriate care and treatment to avoid severe vision loss.[78] Retinal vascular abnormalities in NPDR are characterized by microaneurysms, intraretinal hemorrhages, retinal edema, capillary closure, venous dilation, and cotton wool spots.[79]

Microaneurysms appear as grape-like or spindle-shaped dilations of retinal capillaries under confocal laser scanning microscopy.[80] Microaneurysms are visible as tiny intraretinal red dots in the inner retina during a dilated fundus examination. They appear as punctate hyperfluorescent dots with variable amounts of fluorescein leakage on FFA, but they can also be nonperfused. Intraretinal hemorrhages in NPDR have a variable appearance and are reversible with proper glycemic control. Dot-blot hemorrhages, which are typically small and have clearly defined borders, are more common than flame-shaped hemorrhages, which are larger and have indistinct borders. Besides microaneurysms, chronic hyperglycemia results in microvascular changes including capillary occlusion, arteriolar narrowing, venous beading, and Intraretinal microvascular abnormalities (IRMA) (Fig. 16.2). Venous beading is characterized by localized increases in a venous caliber that resemble a string of beads.[81,82] Cotton-wool spots, which are small white patches of nerve fiber layer infarctions, commonly develop and disappear over weeks or months. IRMAs are shunt vessels that appear as dilated and tortuous retinal vasculature adjacent to areas of capillary nonperfusion. They are slightly larger in caliber, have a broader arrangement, and are always contained within the intraretinal layers when compared to the extraretinal NV in PDR. NV tends to grow along the posterior hyaloid interface, particularly near the optic disc and periphery (Fig. 16.2). On FFA, IRMA has little or no leakage and can be distinguished from NV, which frequently has late leakage. Nonetheless, FA leakage can occur in settings other than neovascularization elsewhere (NVE), such as dilated capillaries and vascular abnormalities. As a result, leakage in FA alone may not be sufficient for distinguishing NVE from IRMA, necessitating the use of other imaging modalities, such as spectral domain-OCT, to supplement IRMA and NVE evaluation.[83] Although the idea that IRMA is a direct precursor lesion of NVE is debatable, the severity of IRMA has been shown to be a risk factor

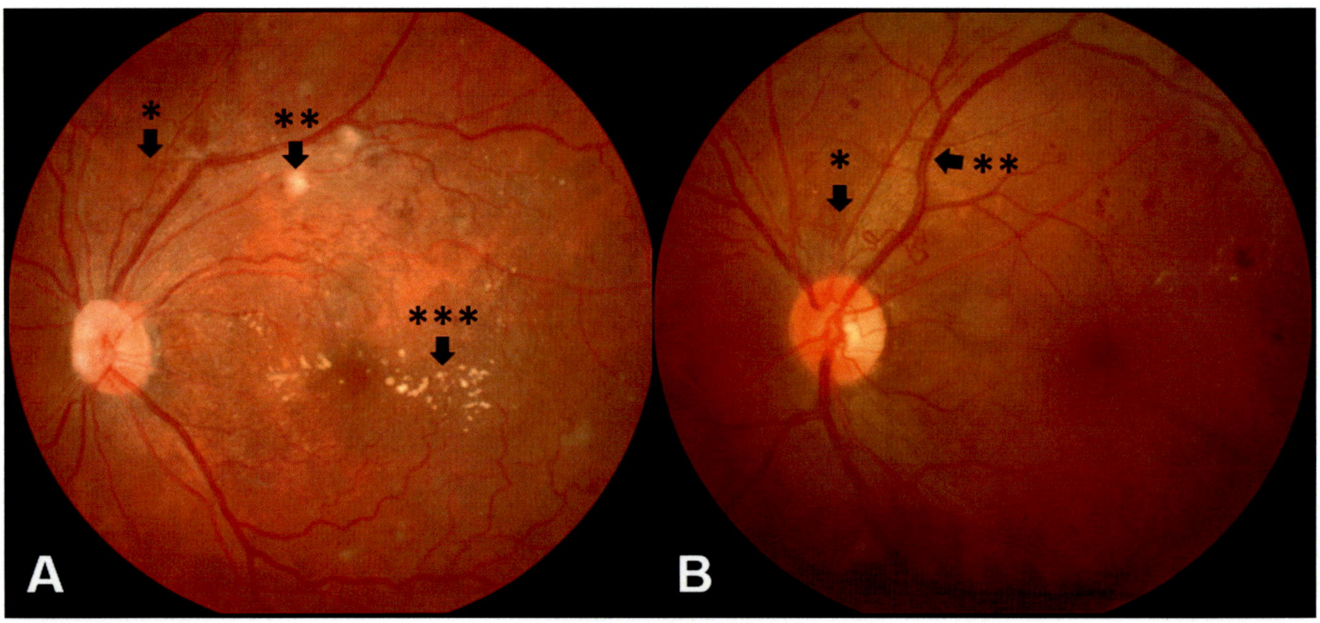

FIGURE 16.2 Findings in diabetic retinopathy. (A) Color photograph of nonproliferative diabetic retinopathy reveals blot hemorrhages (*), cotton wool spot (**), and a group of hard exudates (***) representing macular edema. (B) Color photograph of proliferative diabetic retinopathy shows neovascularization at the disc (*) and venous beading (**).

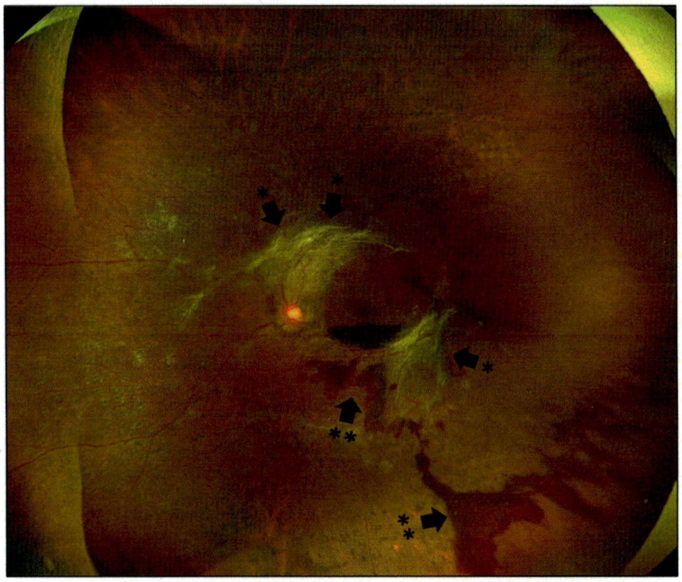

FIGURE 16.3 Ultra-widefield fundus image of proliferative diabetic retinopathy demonstrates tractional retinal detachment involving macula (*) and preretinal hemorrhages (**).

for PDR progression.[83,84] Initially, NV is typically primarily vascular, but as time goes on, it becomes increasingly more fibrous. Following a contraction of the fibrous tissue, tractional retinal detachment and vitreous hemorrhage can occur, which are the two most common complications associated with severe visual loss in PDR (Fig. 16.3).

Diabetic macular edema, caused by retinal vascular hyperpermeability, is defined as macular thickening resulting from diabetes and is a common cause of vision loss in diabetic patients. It may manifest in eyes with a variety of underlying retinopathies, ranging from mild NPDR to PDR. It can happen in regions of retinal ischemia or capillary nonperfusion in addition to areas of clearly visible retinal vascular incompetence, as seen by FFA. Hard exudates appear on ophthalmoscopy as sharply defined yellow-white deposits within the retina. They can appear as small collections in the posterior pole or as a circinate ring in association with or without macular thickening.

Classification

The classification of DR was established as Diabetic Retinopathy Severity Scale (DRSS) by an adaptation of the Airlie House classification of diabetic retinopathy, and this was extended for grading severity of retinopathy in the Diabetic Retinopathy Study (DRS) and the Early Treatment Diabetic Retinopathy Study (ETDRS).[85]

This classification grades retinopathy from no retinopathy (level 10) to advanced proliferative retinopathy (level 95) using modified Airlie House seven-field 30-degree nonsimultaneous stereo color fundus photographs by trained readers. Given that its complexity prevents it from being used in practice, an international classification with greater simplicity, which is based on color photographs, was put forward in 2003 and widely used for determining the degree of retinopathy, including DME (Table 16.1).[17,86] A 4-2-1 rule that includes a particular set of severe intraretinal hemorrhages, venous beading, and IRMA simplifies the definition of severe NPDR (any of these) and very severe NPDR (any two of these). The ETDRS study demonstrated that, within a year, half of the patients with severe NPDR will progress to PDR while patients with very severe NPDR have a 75% chance of developing PDR.[87]

The term "clinically significant macular edema (CSME)" was first used by the ETDRS[88] to refer to DME, which was derived from the observation that the center of the macula was commonly threatened by or involved in retinal thickening or hard exudation. Later, the international classification of DME further classified the severity of DME according to whether retinal thickening and hard exudates affected or threatened the center of the macula, which reflects the significance of foveal involvement in prognosis and management (Table 16.1).[86] Such a scale allows DME to be classified as mild, moderate, or severe DME, or as center-involved or noncenter-involved DME.

Currently, with the advancement of multimodal imaging and artificial intelligence (AI), a newer DR classification should be developed for a more accurate assessment of the progression or regression of DR, the effectiveness of anti-VEGF therapy, and the visual prognosis with greater accuracy.[89,90]

TABLE 16.1 International clinical diabetic retinopathy and diabetic macular edema disease severity scales.

Disease severity level		Findings of dilated fundus examination
Diabetic retinopathy	No apparent retinopathy	No abnormalities
	Mild NPDR	Microaneurysm only
	Moderate NPDR	More than just microaneurysms but less than severe NPDR
	Severe NPDR	Any of the following (4-2-1 rule) and no signs of PDR: • More than 20 intraretinal hemorrhages in each of 4 quadrants • definite venous beading in 2 or more quadrants Prominent IRMA in 1 or more quadrants
	PDR	One or more of the following: • Neovascularization • Vitreous/preretinal hemorrhage
Diabetic macular edema	No apparent DME	No apparent retinal thickening or hard exudates in posterior pole
	Mild DME	Some retinal thickening or hard exudates in posterior pole but distant from the center of the macula
	Moderate DME	Retinal thickening or hard exudates approaching the center of the macula but not involving the center
	Severe DME	Retinal thickening or hard exudates involving the center of the macula

DME, Diabetic macular edema; *IRMA*, intraretinal microvascular abnormalities; *NPDR*, nonproliferative diabetic retinopathy; *PDR*, proliferative diabetic retinopathy.

Multimodal imaging in diabetic retinopathy

Novel imaging modality has now allowed more precise assessment of the retinal and choroidal vascular changes in DR.[91,92] Optical coherence tomography angiography has been used to visualize microvascular changes in DR, such as enlarged foveal avascular zone (FAZ), capillary dropout and increasing vessel tortuosity.[93] Also, choroidal vascular changes in DR such as choriocapillaris flow impairment have been demonstrated using newer swept-source OCTA systems.[94]

Ultrawide field imaging

Current noncontact UWF systems can yield retinal images covering up to a 200-degree field of view or approximately 82% of the total retinal surface area compared to that of 30% in the ETDRS 7-field standards (Fig. 16.3).[95,96] The Diabetic Retinopathy Clinical Research Network (DRCR.net) protocol AA reported that imaging by the ETDRS 7-field and UWF imaging systems had moderate to substantial agreement when determining the severity of DR within the 7 standard fields.[97] As detected by UWF imaging, predominantly peripheral DR lesions (PPL), which represent any DR lesion type that was predominantly present in any peripheral field outside the ETDRS 7 standard fields, provide important information that is required to fully assess the risk of DR progression. When compared to eyes without PPL, eyes with PPL had a 3.2-fold increased risk of 2-step or more DR progression and a 4.7-fold increased risk for progression to PDR over 4 years.[6]

Optical coherence tomography

Optical coherence tomography has become the gold standard for treatment decision-making in eyes with DME in both clinical trials and clinical practice.[98] The morphologic patterns of DME on OCT can be described including sponge-like retinal swelling (diffuse retinal thickening), cystoid macular edema, serous retinal detachment, posterior hyaloid traction, and tractional retinal detachment (Fig. 16.4).[99,100] OCT may predict a worse visual prognosis in eyes with DME that show intraretinal cysts, ellipsoid layer disruption, disorganization of retinal inner layers, and the presence of foveal hard exudates.[92]

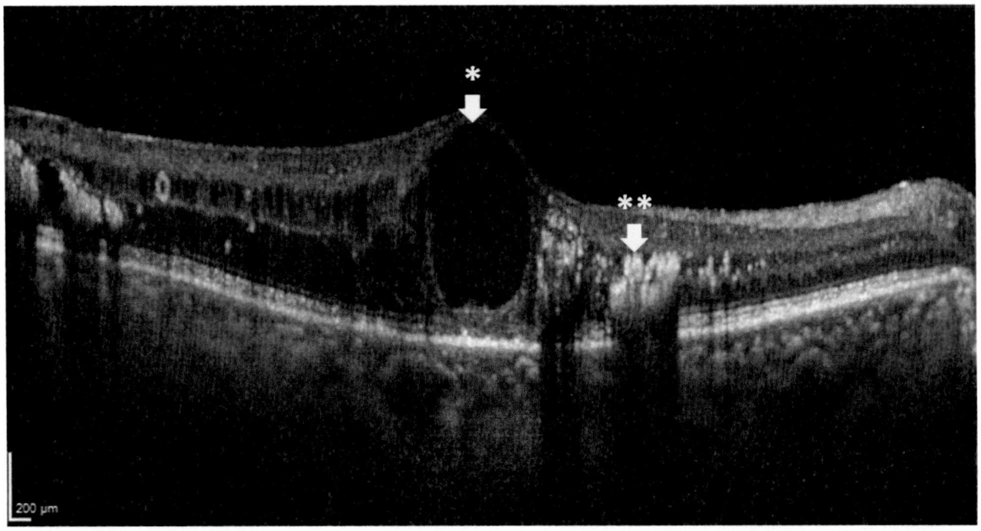

FIGURE 16.4 Spectral-domain optical coherence tomography B-scan of the right eye with center-involved diabetic macular edema. The cystoid macular edema (*) with hard exudates (**) and hyperreflective foci were demonstrated. The shadow below hard exudates was also noted.

To assess the choroidal structure, OCT with enhanced depth imaging (EDI) software and swept-source OCT (SS-OCT) has been widely used. Subfoveal choroidal thickness (SFCT) and choroidal vascularity index on EDI-OCT can be used as quantitative metrics for the evaluation of choroidal structure.[9,101] A meta-analysis and subgroup analyses adjusted for axial length revealed that SFCT was significantly higher in DR eyes, particularly in severe NPDR and PDR.[9] The increase in SFCT in a later stage of DR may be due to elevated VEGF resulting in various choroidal alterations including choroidal vasodilation, the elevation of the choroidal blood flow and vascular permeability, and choroidal neovascularization.[102] These effects can contribute to both stromal and luminal components of the choroid and subsequently increase the choroidal thickness.[103] Choroidal vascularity index (CVI) was defined as the ratio of the luminal area to the total choroidal area in the subfoveal choroidal area with a width of 1000–1500 μm, centered at the fovea. It was decreased in DR patients, especially in mild-to-moderate NPDR.[9] As such, the decrease in CVI indicates that the increase in SFCT is primarily due to the stromal component. This may imply that in DR, choroidal vascular constriction and low blood flow reduce the luminal component of the choroid thereby decreasing the choroidal vascularity index, while proliferative changes primarily increase the stromal component and result in an increased SFCT.[9]

Optical coherence tomography angiography

Optical coherence tomography angiography is a relatively new and noninvasive modality. It provides depth-resolved images of blood flow in the retina and choroid by utilizing motion contrast imaging, which in this case is primarily the movement of red blood cells in the intravascular space.[104] Qualitative characteristics of DR are better visualized with OCTA, including distortion or enlargement of the FAZ, parafoveal microvasculature changes, focal areas of capillary nonperfusion, microaneurysms, cystic edema, and preretinal NV (Fig. 16.5).[92,105] OCTA better defined areas of capillary nonperfusion, which were otherwise obscured by fluorescein leakage on FFA. It was found that some of the foci of focal leakage during the early phases of FFA, which were previously believed to be microaneurysms, were actually small tufts of NV that extended above the internal limiting membrane.[106]

Before clinically detectable DR, OCTA can identify foveal microvascular changes. Compared to healthy eyes, diabetic eyes had more FAZ changes and capillary nonperfusion. OCTA could therefore be used as a screening tool to identify diabetic eyes that are at risk of developing retinopathy.[12] Moreover, OCTA may predict responsiveness to anti-VEGF treatment. Poor responders tended to show greater damage and more microaneurysms in the deep capillary plexus (DCP) and a larger FAZ area.[107]

Quantitative characteristics of DR on OCTA including vessel density, average vessel caliber, vessel tortuosity, capillary density index, and fractal dimension (FD). The density of retinal capillaries was significantly lower in diabetic eyes compared to normal eyes.[105,108] In addition, trend analysis revealed a significant decrease in vessel density values as retinopathy progressed. As a result, OCTA may provide an objective method for monitoring DR progression.[108] In DME,

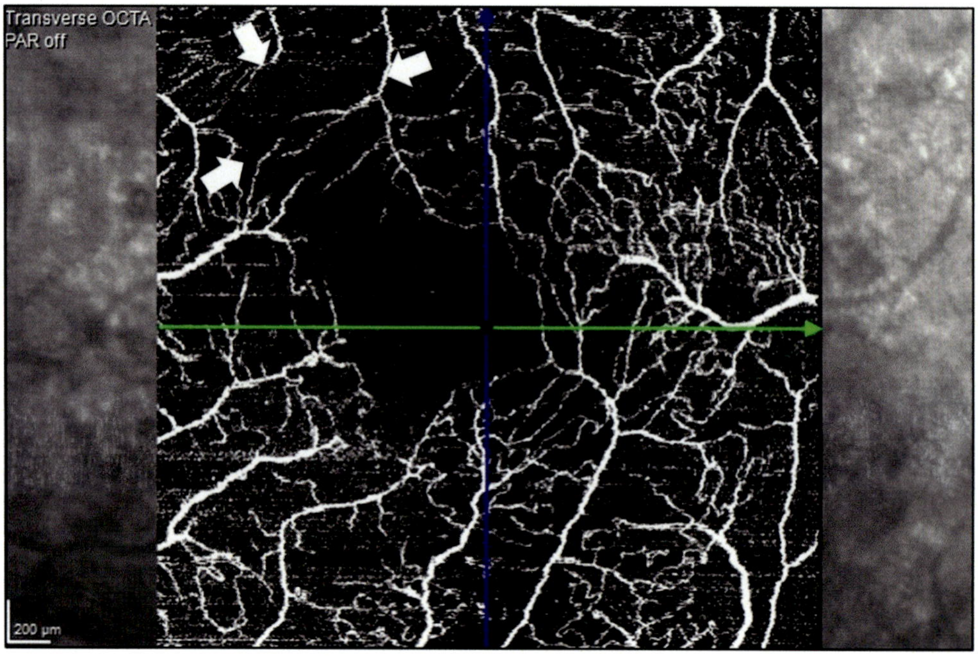

FIGURE 16.5 3 × 3-mm en face optical coherence tomography angiography image of the left eye with mild nonproliferative diabetic retinopathy. There is an enlarged and irregular foveal avascular zone, as well as some capillary nonperfusion areas (*arrow*) in the superficial vascular complex slab.

parafoveal vessel density in the superficial layer at baseline was an independent predictor of visual improvement after loading ranibizumab treatment.[105] Furthermore, greater vessel caliber and vessel tortuosity were observed on OCTA of DME eyes compared to normal eyes.[105]

Ting and colleagues used the capillary density index and FD to describe vascular density and morphologic features in a prospective observational study.[93] The capillary density index was calculated by the custom semiautomated algorithm and was defined as the ratio of the luminal area to the total area region within the 1.5-mm-radius circle centering on the subfoveal region. FD is a metric ranging between 1 and 2 with higher values reflecting the increased complexity of a vascular network can be calculated from OCTA scans to distinguish eyes with retinal capillary branching complexity from those without.[93,109] The study showed that the capillary density index decreased with a worsening stage of DR, while FD was positively correlated with increasing DR staging levels.[93]

SS-OCTA system can visualize changes in DR on UWF montage images. In a prospective cross-sectional study, Yang and colleagues found that diabetic eyes without any obvious signs already had microvascular impairments such as capillary dropout area, capillary nonperfusion area, IRMA, and NV. These impairments were found more frequently in peripheral regions than in macular regions. As a result, UWF OCTA images can be used to detect microvascular changes in eyes with preclinical DR and may be useful for DR screening.[110]

The inability of OCTA to demonstrate leakage, blood flow velocity, or transit time, which can be obtained from FFA, is one of its current limitations. The current use of OCTA in general clinical practice may still be constrained by projection or motion image artifacts, a small field of view (much less than the 200 degrees achievable by UWF FFA), increased image acquisition time, a lack of standardized interoperability across OCTA devices, difficulty in interpretation, cost, and a lack of widespread availability of OCTA devices.[92,96]

Fundus fluorescein angiography

Fundus fluorescein angiography has been the gold standard for retinal vascular structure assessment. It can be used to visualize microaneurysms, IRMA, NV, enlarged and irregular FAZ, and, especially, vascular perfusion and leakage in DR. On FFA, a capillary nonperfusion area can be seen, most commonly in the midperipheral region of NPDR eyes (Fig. 16.6).[96] Visualization of the peripheral retina is essential to evaluate nonperfusion areas, vascular leakage, microvascular abnormalities, and NV, which is crucial for screening, diagnosis, monitoring, treatment, and prognosis of DR (Fig. 16.7).[111] DRCR.net protocol AA reported that 70% of the nonperfusion area in diabetic eyes was located outside the posterior pole. The increased nonperfusion area is associated with PPL, suggesting that UWF-FFA may be able to

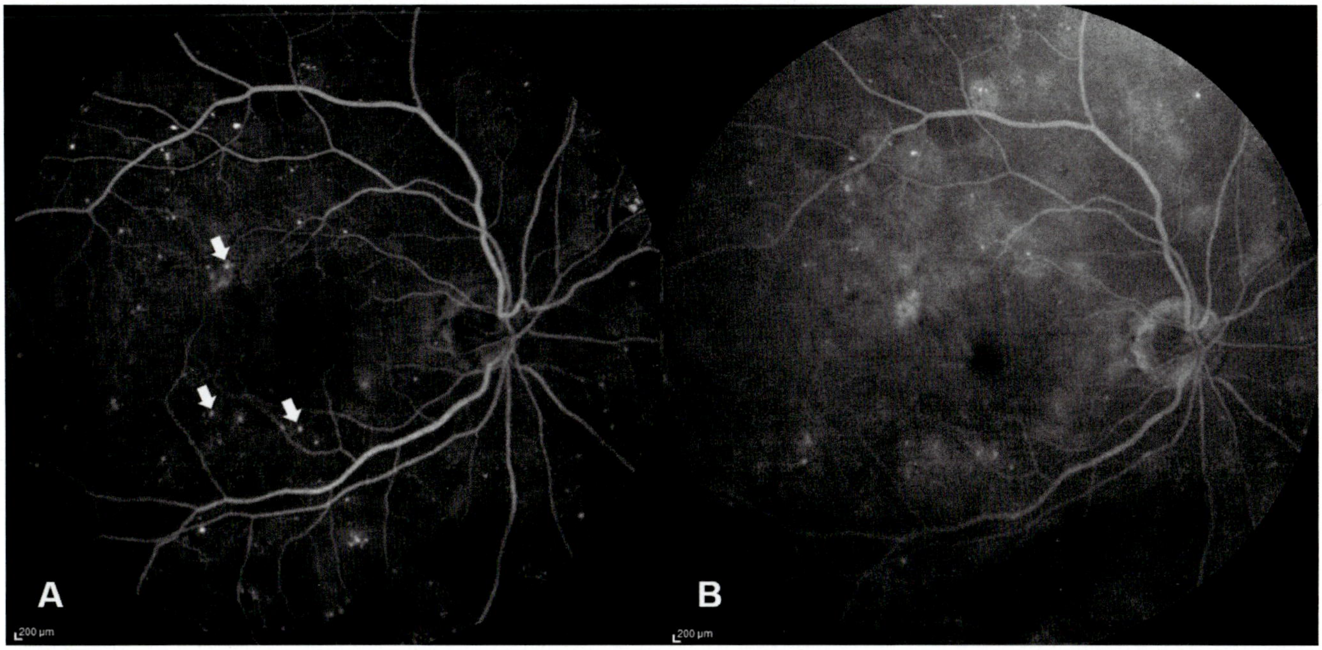

FIGURE 16.6 Fundus fluorescein angiography (FFA) of the right eye with severe nonproliferative diabetic retinopathy. (A) Numerous microaneurysms (*arrow*) and small areas of capillary nonperfusion were demonstrated in an early phase of FFA. (B) Late-phase FFA shows profound leakage of microaneurysms.

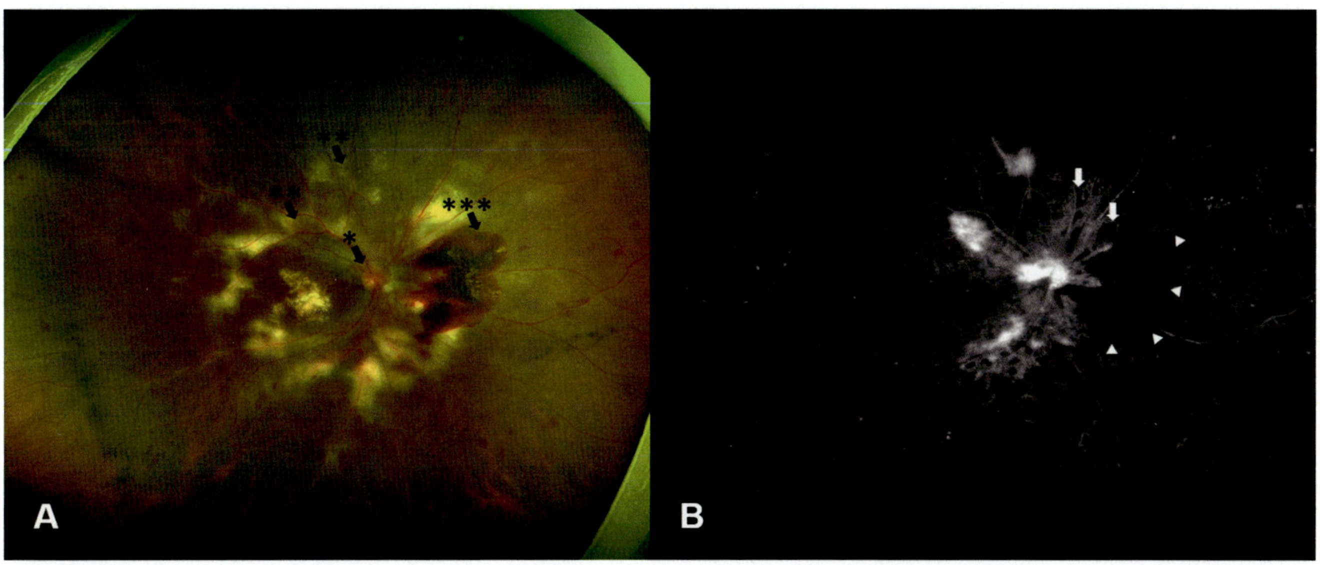

FIGURE 16.7 Ultra-widefield imaging of the right eye with proliferative diabetic retinopathy. (A) Neovascularization at the disc (*), neovascularization elsewhere (**), preretinal hemorrhage (***), and hard exudates were demonstrated at the posterior pole of the fundus. Dot and blot hemorrhages were also well visualized at the posterior pole to midperiphery of the fundus. (B) An ultra-widefield fundus fluorescein angiogram shows numerous microaneurysms and areas of capillary nonperfusion (*arrow*). The leakage of neovascularization was profound. An area of fluorescence blockage near the optic disc due to preretinal hemorrhage was also noted (*arrow head*).

predict future DR worsening better than UWF-photography alone.[97,112] Silva and coworkers conducted a prospective, multicenter, cohort study with data for 508 eyes with NPDR and gradable nonperfusion on UWF-FFA at baseline. This 4-year longitudinal study has demonstrated that both greater baseline retinal nonperfusion and predominantly peripheral lesions on UWF-FFA are associated with a higher risk of DR worsening.[113]

Nonetheless, the limitation of FFA includes a time-consuming, invasive imaging technique that requires intravenous fluorescein dye access and can have adverse effects. Nausea and more serious allergic reactions are among the side effects. In patients with renal disease or during pregnancy, FFA may be contraindicated. FFA had also a higher number of nongradable images. In addition, one eye must be chosen as the initial transit eye causing an imbalance in the amount of information obtained for each eye. To address FFA's limitations, it has been demonstrated that noninvasive OCTA can be a comparable imaging modality to FFA for evaluating macular vascular changes or diabetic NV.[114,115]

Application of DR imaging in clinical practice

Although a number of novel technologies have been developed for better evaluation of various retinal diseases, it may not be necessary to use them for all DR cases in general clinical practice. According to the Diabetic Retinopathy Preferred Practice Pattern developed by the American Academy of Ophthalmology, OCT and FFA could be the mainstay tools for screening and management of DR in clinical practice.[17] OCT should be used to detect, quantify, and monitor DME, as well as to identify areas of vitreomacular traction. It should also be used to investigate other causes of macular swelling and to screen patients with no or minimal DR. FFA for DR cases aids in guiding laser treatment for CSME, identifying suspected but clinically obscured retinal NV, occasionally ruling out other causes of macular swelling, and identifying large areas of capillary nonperfusion, but it should never be used to screen a patient who has no or minimal DR.

Future imaging studies: Artificial intelligence

Recently, deep learning (DL), a subset of AI, has been found to be an effective tool for assisting the screening for referable DR. Based on retrospective external validations on new datasets of retinal photographs of patients with diabetes, most DL models for DR screening were found to achieve robust performance with sensitivity and specificity of more than 95%.[116–127] From studies on prospective validation of DL models, which were published in a much fewer number when compared with the retrospective validations, the sensitivity and specificity were lower at approximately 90% and 85%, respectively.[128–138] These diagnostic parameters of DL are still acceptable and on par with trained graders or ophthalmologist experts.[126,136] In the real-world application of DL for DR screening, there are critical considerations beyond diagnostic accuracy. These include the seamless integration of DL into the DR screening workflow, compatibility with existing hospital management systems, and the acceptance of both patients and healthcare personnel. These factors are essential for ensuring the long-term sustainability of the implementation.[139] The approval of DL models by a corresponding regulatory body before deployment is also necessary.[140] In a bigger picture, the deployment of AI in healthcare, not only for DR screening, should follow the concept of AI governance framework to ensure fairness, transparency, trustworthiness, and accountability.[141]

Not only for DR screening, there have been attempts to apply DL for early detection or monitoring of DME by automated classification or segmentation on OCT images with an acceptable performance from internal validation.[142,143] Additional validation of these DL models on new, external datasets is essential before deployment in real-world clinical practices.

Recent clinical trial updates for diabetic retinopathy management

Glycemic and other metabolic control

As hyperglycemia is directly related to DR, well-glycemic control should be exercised to avoid this visually threatening condition. This has been approved by the two large randomized controlled trials for both patients with type 1[144,145] and type 2[146] diabetes that tight glycemic control could reduce the incidence and progression of DR. However, a number of patients suffer from hypoglycemic episodes due to tight glycemic control, which necessitates the close supervision of a primary care physician.[147,148] In view of other metabolic control in diabetic patients, combination therapy for dyslipidemia slowed the progression of DR in diabetic patients,[147] whereas strong pieces of evidence have identified the efficacy of intensive blood pressure control in reducing the incidence of DR,[149,150] but there is considerable debate about the efficacy of intensive blood pressure control in reducing the progression of DR.[147,151,152]

Diabetic macular edema

Anti-VEGF and other antiangiogenic medications

Anti-VEGF therapy, another paradigm shift in ophthalmology, has now replaced focal laser photocoagulation in many retinal vascular diseases, including DR and DME. Many previous studies found that a monthly or pro re nata (PRN) schedule of DME treatment with ranibizumab improved visual outcomes significantly more than a conventional focal/grid laser.[98,153–156] VIVID and VISTA trials also found that 2-monthly interval aflibercept following 5-month loading was superior above focal/grid laser.[157] According to the DRCR.net protocol T 2-year follow-up, the efficacy of ranibizumab and aflibercept for treating patients with an initial visual acuity (VA) of 20/32-20/400 showed no significant difference, but both were more effective than bevacizumab when the initial VA was worse than 20/40.[158] For center-involved DME with good initial VA (20/20-20/25), DRCR.net protocol V reported no difference in final BCVA at year 2 after receiving aflibercept, focal/grid laser, and observation.[159] The potential roles of focal/grid laser treatment for DME include lowering the number of anti-VEGF injections and treating noncenter-involved DME.[159]

Bevacizumab is a full-length, recombinant, humanized antibody that recognizes all VEGF-A isoforms. Initially, FDA approved for colon cancer treatment, it has been widely used as an off-label intravitreal agent for various ophthalmic conditions due to its cost-effectiveness. Although the benefit of bevacizumab for DME has been established in the past[160–162] and present,[158] a number of patients have not responded adequately to prior bevacizumab therapy, necessitating a switch to the approved ranibizumab or aflibercept.[163] Nevertheless, DRCR.net recently published protocol AC, a randomized clinical trial, to indicate the difference between bevacizumab and aflibercept-monotherapy as the first medicine. Results showed no significant difference in visual outcomes and central subfield thickness between two groups over a 2-year period.[164]

Given the limitations of anti-VEGF therapy, such as the burdensome need for frequent intravitreal injections, new anti-VEGF agents, dosage, and drug delivery systems aimed at sustaining efficacy are being developed.[165,166] Brolucizumab is a single-chain antibody fragment targeting all isoforms of VEGF-A. KITE and KESTREL trials are 2 Phase III, 100-week, randomized clinical trials comparing the efficacy of brolucizumab to aflibercept in DME patients.[167] Patients in the brolucizumab arm received five loading injections every 6 weeks followed by a 12-week interval, while ones in the aflibercept arm received five loading injections every 4 weeks followed by an 8-week interval. At weeks 32 and 52, brolucizumab was found to be noninferior to aflibercept in terms of visual outcome. Furthermore, fewer patients receiving brolucizumab experienced persistent subretinal and/or intraretinal fluid, and half of the subjects were able to continue treatment at every 12-week dosing interval until week 52. However, the most serious issue with brolucizumab is intraocular inflammation (IOI). The incidences of IOI for brolucizumab 3 mg, 6 mg, and aflibercept 2 mg in KESTREL were 4.7%, 3.7%, and 0.5%, respectively. However, in KITE, the incidences of IOI were 1.7%, which was comparable to the rates for both brolucizumab and aflibercept.[167] This is consistent with the HAWK and HARRIER data for patients with wet AMD.[168] To minimize the risk of IOI-related visual loss after brolucizumab treatment, experts advised early diagnosis, prompt and timely intervention, and intensive treatment.[169]

Faricimab, the first bispecific antibody designed for intraocular use, independently inhibits VEGF-A and Ang-2 with high affinity and specificity. Ang-2 inhibition by faricimab results in endothelial stability and prevents pericyte loss.[170] YOSEMITE and RHINE are two parallel phase 3 clinical trials evaluating the efficacy and safety of faricimab in DME. These studies also introduced personalized treatment interval (PTI) protocol for extending follow-up intervals. The result showed that the faricimab 6 mg PTI group was noninferior to aflibercept 2 mg bimonthly at 1 year. Moreover, at week 52, almost three forth of the faricimab PTI patients were able to achieve every-12-week or every-16-week dosing in YOSEMITE and RHINE. The incidence of adverse events was similar among all treatment groups.[171] The US Food and Drug Administration (FDA) approved faricimab (VABYSMO) in 2022 for the treatment of DME with a modified extension of the follow-up period following the resolution of edema based on the central subfield thickness of the macula as determined by OCT.[172] Later, the US FDA approved brolucizumab (BEOVU) for the treatment of DME, but it also provided a warning about retinal vasculitis and retinal vascular occlusion, which frequently occur in association with intraocular inflammation.

Port delivery system

A port delivery system (PDS), a novel surgically implanted refill-exchange drug delivery system, was recently introduced. This treatment strategy required surgical implantation into the sclera and a 6-month refill-exchange of ranibizumab.[173] The PAGODA trial is a phase III randomized clinical trial studying the efficacy and safety of PDS evaluating PDS with ranibizumab in DME patients.[174] PAVILION trial is a phase III randomized clinical trial evaluating the effectiveness and safety of PDS in DR without center-involved DME.[175] Given that Susvimo was previously approved by the US FDA as

PDS for the treatment of neovascular age-related macular degeneration on the basis of promising findings from the multicenter Phase III Archway study, it was expected that PDS in DR and DME would yield results that were comparable.[176] However, due to PDS's drawbacks, such as the potential for surgical complications, ensuing endophthalmitis,[173,176] and, most recently, septum dislodgements,[177] caution should be taken.

Intravitreal corticosteroids

Corticosteroids have been used in DR treatment in various forms, including injectable suspension, eye drops, and solid slow-release devices.[178–183] Among these, FDA-approved intravitreal sustained-release implants such as Ozurdex (sustained-release dexamethasone intravitreal implant)[184] and Iluvien (fluocinolone acetonide intravitreal insert),[178] as well as intravitreal triamcinolone injection, have been used as a second-line treatment for DME after anti-VEGF therapy. They may be indicated when first-line anti-VEGF therapy is contraindicated, ineffective, or has an incomplete response.[185,186] Any advantages should, however, be weighed against IOP elevation and cataract progression as the common side effects of the corticosteroids.[187]

Proliferative diabetic retinopathy and nonproliferative diabetic retinopathy

PDR management strategies are currently integrated and include PRP, vitrectomy, and anti-VEGF agents. PRP is still the mainstay treatment for PDR due to its long-lasting effects and high cost-effectiveness once off therapy, although intravitreal anti-VEGF injections have recently emerged as a safe and effective treatment alternative for PDR.[188] According to the DRS and ETDRS, PRP laser treatment for severe NPDR and PDR was recommended. The goal is to avoid severe vision loss (VA worse than 5/200).[189,190] The DRS also suggested that prompt photocoagulation treatment is highly required for eyes with high-risk characteristics.[189] Recent evidence of advancements in laser therapy includes the use of pattern scan laser photocoagulators, subthreshold laser, and fundus camera-based photocoagulation system that is integrated with retinal eye-tracking technology.[191] With the Pascal (OptiMedica, Santa Clara, California), shorter-pulsed patterned scanning laser photocoagulation, the procedure is less painful, has fewer side effects, and is more efficient.[192,193] In contrast to the conventional PRP technique that was first described in the DRS, the advantage of UWF imaging led to the practice of UWF-image-guided retinal laser photocoagulation, also known as targeted retinal photocoagulation, in which the Pascal photocoagulation was applied selectively to areas of nonperfusion as shown on UWF FFA.[194] This may reduce the adverse effects of PRP, such as loss of peripheral vision, night blindness, color vision changes, exacerbation of macular edema, uveal effusion, and tractional retinal detachment. In a pilot randomized trial, Muqit and colleagues compared targeted retinal photocoagulation to conventional PRP in patients with treatment-naive PDR. There was no difference in the rates of PDR regression and ocular adverse events between the two techniques after a 12-week follow-up. However, the study was probably underpowered to identify a difference between the two groups because it only included 30 eyes.[195]

Vitrectomy should be considered for patients with PDR who have macular-threatening tractional retinal detachment, long-standing nonclearing vitreous hemorrhage, or glaucoma secondary to vitreous hemorrhage. The recommended indications for PPV in DR are listed in Table 16.2. Currently, a trend toward earlier diabetic retinopathy intervention has been linked to the development and widespread use of smaller gauge instrumentation and intraoperative wide-angle viewing systems.[196]

The vitrectomy can be supplemented with anti-VEGF, making the procedure much simpler. Both preoperative and intraoperative applications are possible. According to meta-analyses, anti-VEGF treatment before vitrectomy reduces surgery time, retinal breaks, and intraoperative bleeding,[197–199] while intraoperative anti-VEGF treatment reduces the incidence of early postoperative vitreous hemorrhage.[198] However, due to the rapid progression of preexisting TRD after anti-VEGF injection in some patients, caution should be exercised.[200]

Furthermore, anti-VEGF demonstrated efficacy as an alternative treatment for PDR and in reducing DR severity. Anti-VEGF agents such as ranibizumab and aflibercept were not inferior to PRP laser treatment in PDR treatment. Additionally, when compared to the laser treatment group, the outcomes of anti-VEGF therapy resulted in better visual function preservation, fewer vitrectomies, and fewer new cases of DME-related vision loss.[201–203] Although these findings are likely to ignite a paradigm shift in the treatment of PDR, choosing anti-VEGF over PRP should be approached with caution. The limitations of anti-VEGF injections for PDR treatment included the increased number of treatments and visits, patient compliance,[204] injection anxiety,[205] and endophthalmitis.[206] The anti-VEGF injection alone should be considered for patients with reliable follow-up.[17,204] Large randomized controlled trials have also shown that anti-VEGF preventive therapy can reduce the severity of DR.[155,157,207–209] Nevertheless, it is necessary to carefully assess the cost- and risk-effectiveness of using anti-VEGF in NPDR.

TABLE 16.2 Indications for vitrectomy in diabetic retinopathy.[19–21]

- Vitreous hemorrhage does not resolve within 1 month
- Tractional retinal detachment involving or threatening the fovea
- Combined rhegmatogenous and tractional retinal detachment
- Dense premacular subhyaloid hemorrhage
- DME with the presence of VMT and/or ERM
- Macular hole
- Macular tractional retinoschisis
- Progressive fibrovascular proliferation
- Anterior segment NV with an associated media opacity

DME, diabetic macular edema; *ERM*, epimacular membrane; *NV*, neovascularization; *VMT*, vitreomacular traction.

Conclusion

DME and PDR are still the most frequent ocular complications in diabetic patients and they need to be promptly treated to prevent vision loss. These complications are the results of the retinal microvascular damage that hyperglycemia initially caused. In clinical practice, ocular imaging, including UWF imaging and OCT, along with or without AI-based imaging processing tools, may be the mainstay tools for screening and managing DR. Although studies have shown a promising role for the treatment including PRP, intravitreal medications, and vitrectomy, the key to achieving this goal is patient compliance with regular follow-up examinations and systemic blood glucose and other comorbidity control. Clinical trials examining the optimal integration of these treatment modalities and novel drugs to reduce treatment burden while achieving optimal clinical efficacy are required.

References

1. Saeedi P, Petersohn I, Salpea P, et al. Global and regional diabetes prevalence estimates for 2019 and projections for 2030 and 2045: results from the international diabetes federation diabetes atlas, 9th edition. *Diabetes Research and Clinical Practice*. 2019;157:107843. https://doi.org/10.1016/j.diabres.2019.107843.
2. Teo ZL, Tham YC, Yu M, et al. Global prevalence of diabetic retinopathy and projection of burden through 2045: systematic review and meta-analysis. *Ophthalmology*. 2021;128(11):1580–1591. https://doi.org/10.1016/j.ophtha.2021.04.027.
3. Ren J, Zhang S, Pan Y, et al. Diabetic retinopathy: involved cells, biomarkers, and treatments. *Frontiers in Pharmacology*. 2022;13. https://doi.org/10.3389/fphar.2022.953691.
4. Tang L, Xu GT, Zhang JF. Inflammation in diabetic retinopathy: possible roles in pathogenesis and potential implications for therapy. *Neural Regeneration Research*. 2023;18(5):976–982. https://doi.org/10.4103/1673-5374.355743.
5. Wang W, Lo ACY. Diabetic retinopathy: pathophysiology and treatments. *International Journal of Molecular Sciences*. 2018;19(6). https://doi.org/10.3390/ijms19061816.
6. Silva PS, Cavallerano JD, Haddad NMN, et al. Peripheral lesions identified on ultrawide field imaging predict increased risk of diabetic retinopathy progression over 4 years. *Ophthalmology*. 2015;122(5):949–956. https://doi.org/10.1016/j.ophtha.2015.01.008.
7. Wessel MM, Aaker GD, Parlitsis G, Cho M, D'Amico DJ, Kiss S. Ultra-wide-field angiography improves the detection and classification of diabetic retinopathy. *Retina*. 2012;32(4):785–791. https://doi.org/10.1097/IAE.0b013e3182278b64.
8. Damian I, Roman G, Nicoară SD. Analysis of the choroid and its relationship with the outer retina in patients with diabetes mellitus using binarization techniques based on spectral-domain optical coherence tomography. *Journal of Clinical Medicine*. 2021;10(2):1–18. https://doi.org/10.3390/jcm10020210.
9. Jiang J, Liu J, Yang J, Jiang B. Optical coherence tomography evaluation of choroidal structure changes in diabetic retinopathy patients: a systematic review and meta-analysis. *Frontiers in Medicine*. 2022;9. https://doi.org/10.3389/fmed.2022.986209.
10. Kung EWT, Chan VTT, Tang Z, Yang D, Sun Z, Wang YM, et al.. Alterations in the choroidal sublayers in relationship to severity and progression of diabetic retinopathy: a swept-source OCT study. *Ophthalmology Science*. 2022;2:100130.
11. Arias JD, Arango FJ, Parra MM, et al. 2021. Early microvascular changes in patients with prediabetes evaluated by optical coherence tomography angiography. Therapeutic Advances in Ophthalmology 2021;13.
12. De Carlo TE, Chin AT, Bonini Filho MA, et al. Detection of microvascular changes in eyes of patients with diabetes but not clinical diabetic retinopathy using optical coherence tomography angiography. *Retina*. 2015;35(11):2364–2370. https://doi.org/10.1097/IAE.0000000000000882.
13. Bursell SE, Clermont AC, Kinsley BT, Simonson DC, Aiello LM, Wolpert HA. Retinal blood flow changes in patients with insulin- dependent diabetes mellitus and no diabetic retinopathy: a video fluorescein angiography study. *Investigative Ophthalmology and Visual Science*. 1996;37(5):886–897.

14. Yamana Y, Ohnishi Y, Taniguchi Y, Ikeda M. Early signs of diabetic retinopathy by fluorescein angiography. *Japanese Journal of Ophthalmology*. 1983;27(1):218–227.
15. Bresnick GH, Korth K, Groo A, Palta M. Electroretinographic oscillatory potentials predict progression of diabetic retinopathy: preliminary report. *Archives of Ophthalmology*. 1984;102(9):1307–1311. https://doi.org/10.1001/archopht.1984.01040031057023.
16. Luu CD, Szental JA, Lee SY, Lavanya R, Wong TY. Correlation between retinal oscillatory potentials and retinal vascular caliber in type 2 diabetes. *Investigative Ophthalmology and Visual Science*. 2010;51(1):482–486. https://doi.org/10.1167/iovs.09-4069.
17. Flaxel CJ, Adelman RA, Bailey ST, et al. Diabetic retinopathy preferred practice pattern. *Ophthalmology*. 2020;127(1):P66–P145. https://doi.org/10.1016/j.ophtha.2019.09.025.
18. Lin K-Y, Hsih W-H, Lin Y-B, Wen C-Y, Chang T-J. Update in the epidemiology, risk factors, screening, and treatment of diabetic retinopathy. *Journal of Diabetes Investigation*. 2021;12(8):1322–1325. https://doi.org/10.1111/jdi.13480.
19. Early vitrectomy for severe vitreous hemorrhage in diabetic retinopathy: two-year results of a randomized trial diabetic retinopathy vitrectomy study report 2 the diabetic retinopathy vitrectomy study research group. *Archives of Ophthalmology*. 1985;103(11):1644–1652. https://doi.org/10.1001/archopht.1985.01050110038020.
20. El Annan J, Carvounis PE. Current management of vitreous hemorrhage due to proliferative diabetic retinopathy. *International Ophthalmology Clinics*. 2014;54(2):141–153. https://doi.org/10.1097/IIO.0000000000000027.
21. Sharma T, Fong A, Lai TY, Lee V, Das S, Lam D. Surgical treatment for diabetic vitreoretinal diseases: a review. *Clinical and Experimental Ophthalmology*. 2016;44(4):340–354. https://doi.org/10.1111/ceo.12752.
22. Hao HH, Shao ZM, Tang DQ, et al. Preventive effects of rutin on the development of experimental diabetic nephropathy in rats. *Life Sciences*. 2012;91(19–20):959–967. https://doi.org/10.1016/j.lfs.2012.09.003.
23. Obrosova IG, Kador PF. Aldose reductase/polyol inhibitors for diabetic retinopathy. *Current Pharmaceutical Biotechnology*. 2011;12(3):373–385. https://doi.org/10.2174/138920111794480642.
24. Brownlee M. The pathobiology of diabetic complications: a unifying mechanism. *Diabetes*. 2005;54(6):1615–1625. https://doi.org/10.2337/diabetes.54.6.1615.
25. Fiorentino TV, Prioletta A, Zuo P, Folli F. Hyperglycemia-induced oxidative stress and its role in diabetes mellitus related cardiovascular diseases. *Current Pharmaceutical Design*. 2013;19(32):5695–5703. https://doi.org/10.2174/1381612811319320005.
26. Szwergold BS, Kappler F, Brown TR. Identification of fructose 3-phosphate in the lens of diabetic rats. *Science*. 1990;247(4941):451–454. https://doi.org/10.1126/science.2300805.
27. Gabbay KH, Hyperglycemia, polyol metabolism, and complications of diabetes mellitus Annual Review of Medicine 1975;26:521–536.
28. Steinmetz PR, Balko C, Gabbay KH. The sorbitol pathway and the complications of diabetes. *New England Journal of Medicine*. 1973;288(16):831–836. https://doi.org/10.1056/NEJM197304192881609.
29. Zong H, Ward M, Stitt AW, AGEs, RAGE, and diabetic retinopathy. Current Diabetes Reports. 2011;11(4):244–252.
30. Whitehead M, Wickremasinghe S, Osborne A, Van Wijngaarden P, Martin KR. Diabetic retinopathy: a complex pathophysiology requiring novel therapeutic strategies. *Expert Opinion on Biological Therapy*. 2018;18(12):1257–1270. https://doi.org/10.1080/14712598.2018.1545836.
31. Indyk D, Bronowicka-Szydełko A, Gamian A, Kuzan A. Advanced glycation end products and their receptors in serum of patients with type 2 diabetes. *Scientific Reports*. 2021;11(1). https://doi.org/10.1038/s41598-021-92630-0.
32. Singh R, Barden A, Mori T, Beilin L. Advanced glycation end-products: a review. *Diabetologia*. 2001;44(2):129–146. https://doi.org/10.1007/s001250051591.
33. Yamagishi SI, Nakamura N, Suematsu M, Kaseda K, Matsui T. Advanced glycation end products: a molecular target for vascular complications in diabetes. *Molecular Medicine*. 2015;21:S32–S40. https://doi.org/10.2119/molmed.2015.00067.
34. Chen Bh, Jiang Dy, Tang LS. Advanced glycation end-products induce apoptosis involving the signaling pathways of oxidative stress in bovine retinal pericytes. *Life Sciences*. 2006;79(11):1040–1048. https://doi.org/10.1016/j.lfs.2006.03.020.
35. Kim J, Kim KM, Kim CS, et al. Puerarin inhibits the retinal pericyte apoptosis induced by advanced glycation end products in vitro and in vivo by inhibiting NADPH oxidase-related oxidative stress. *Free Radical Biology and Medicine*. 2012;53(2):357–365. https://doi.org/10.1016/j.freeradbiomed.2012.04.030.
36. Ishii H, Koya D, King GL. Protein kinase C activation and its role in the development of vascular complications in diabetes mellitus. *Journal of Molecular Medicine*. 1998;76(1):21–31. https://doi.org/10.1007/s109-1998-8101-y.
37. Ueda Y, Hirai SI, Osada SI, Suzuki A, Mizuno K, Ohno S. Protein kinase C δ activates the MEK-ERK pathway in a manner independent of Ras and dependent on Raf. *Journal of Biological Chemistry*. 1996;271(38):23512–23519. https://doi.org/10.1074/jbc.271.38.23512.
38. Xia P, Inoguchi T, Kern TS, Engerman RL, Oates PJ, King GL. Characterization of the mechanism for the chronic activation of diacylglycerol-protein kinase C pathway in diabetes and hypergalactosemia. *Diabetes*. 1994;43(9):1122–1129. https://doi.org/10.2337/diab.43.9.1122.
39. Kolm-Litty V, Sauer U, Nerlich A, Lehmann R, Schleicher ED. High glucose-induced transforming growth factor β1 production is mediated by the hexosamine pathway in porcine glomerular mesangial cells. *Journal of Clinical Investigation*. 1998;101(1):160–169. https://doi.org/10.1172/JCI119875.
40. Wells L, Vosseller K, Hart GW. Glycosylation of nucleocytoplasmic proteins: signal transduction and O-GlcNAc. *Science*. 2001;291(5512):2376–2378. https://doi.org/10.1126/science.1058714.
41. Nakamura M, Barber AJ, Antonetti DA, et al. Excessive hexosamines block the neuroprotective effect of insulin and induce apoptosis in retinal neurons. *Journal of Biological Chemistry*. 2001;276(47):43748–43755. https://doi.org/10.1074/jbc.M108594200.

42. Giacco F, Brownlee M. Oxidative stress and diabetic complications. *Circulation Research*. 2010;107(9):1058–1070. https://doi.org/10.1161/CIRCRESAHA.110.223545.
43. Nerlich AG, Sauer U, Kolm-Litty V, Wagner E, Koch M, Schleicher ED. Expression of glutamine:fructose-6-phosphate amidotransferase in human tissues: evidence for high variability and distinct regulation in diabetes. *Diabetes*. 1998;47(2):170–178. https://doi.org/10.2337/diab.47.2.170.
44. Schleicher ED, Weigert C. Role of the hexosamine biosynthetic pathway in diabetic nephropathy. *Kidney International, Supplement*. 2000;58(77):S13–S18. https://doi.org/10.1046/j.1523-1755.2000.07703.x.
45. Yki-Jarvinen H, Daniels MC, Virkamaki A, Makimattila S, DeFronzo RA, McClain D. Increased glutamine:fructose-6-phosphate amidotransferase activity in skeletal muscle of patients with NIDDM. *Diabetes*. 1996;45(3):302–307. https://doi.org/10.2337/diabetes.45.3.302.
46. Cogan DG, Toussaint D, Kuwabara T. Retinal vascular patterns: IV. Diabetic retinopathy. *Archives of Ophthalmology*. 1961;66(3):366–378. https://doi.org/10.1001/archopht.1961.00960010368014.
47. Aiello LP, Cavallerano J, Bursell SE. Diabetic eye disease. *Endocrinology and Metabolism Clinics of North America*. 1996;25(2):271–291. https://doi.org/10.1016/S0889-8529(05)70325-X.
48. Hammes HP, Lin J, Renner O, et al. Pericytes and the pathogenesis of diabetic retinopathy. *Diabetes*. 2002;51(10):3107–3112. https://doi.org/10.2337/diabetes.51.10.3107.
49. Lindahl P, Johansson BR, Levéen P, Betsholtz C. Pericyte loss and microaneurysm formation in PDGF-B-deficient mice. *Science*. 1997;277(5323):242–245. https://doi.org/10.1126/science.277.5323.242.
50. Wilkinson-Berka JL, Babic S, De Gooyer T, et al. Inhibition of platelet-derived growth factor promotes pericyte loss and angiogenesis in ischemic retinopathy. *American Journal of Pathology*. 2004;164(4):1263–1273. https://doi.org/10.1016/S0002-9440(10)63214-2.
51. Friedenwald JS. Diabetic retinopathy*. *American Journal of Ophthalmology*. 1950;33(8):1187–1199. https://doi.org/10.1016/0002-9394(50)90988-3.
52. Buzney SM, Frank RN, Varma SD, Tanishima T, Gabbay KH. Aldose reductase in retinal mural cells. *Investigative Ophthalmology Visual Science*. 1977;16(5):392–396.
53. Kern TS, Engerman RL. Distribution of aldose reductase in ocular tissues. *Experimental Eye Research*. 1981;33(2):175–182. https://doi.org/10.1016/S0014-4835(81)80066-8.
54. Ludvigson MA, Sorenson RL. Immunohistochemical localization of aldose reductase. II. Rat eye and kidney. *Diabetes*. 1980;29(6):450–459. https://doi.org/10.2337/diab.29.6.450.
55. Hellstrom M, Kal n M, Lindahl P, Abramsson A, Betsholtz C. Role of PDGF-B and PDGFR-beta in recruitment of vascular smooth muscle cells and pericytes during embryonic blood vessel formation in the mouse. *Development*. 1999;126(14):3047–3055. https://doi.org/10.1242/dev.126.14.3047.
56. Das A, Frank RN, Zhang NL, Samadani E. Increases in collagen type IV and laminin in galactose-induced retinal capillary basement membrane thickening-Prevention by an aldose reductase inhibitor. *Experimental Eye Research*. 1990;50(3):269–280. https://doi.org/10.1016/0014-4835(90)90211-C.
57. Robison WG, Kador PF, Kinoshita JH. Retinal capillaries: basement membrane thickening by galactosemia prevented with aldose reductase inhibitor. *Science*. 1983;221(4616):1177–1179. https://doi.org/10.1126/science.6612330.
58. Frank RN, Keirn RJ, Kennedy A, Frank KW. Galactose-induced retinal capillary basement membrane thickening: prevention by sorbinil. *Investigative Ophthalmology and Visual Science*. 1983;24(11):1519–1524.
59. Roy S, Sato T, Paryani G, Kao R. Downregulation of fibronectin overexpression reduces basement membrane thickening and vascular lesions in retinas of galactose-fed rats. *Diabetes*. 2003;52(5):1229–1234. https://doi.org/10.2337/diabetes.52.5.1229.
60. Beltramo E, Porta M. Pericyte loss in diabetic retinopathy: mechanisms and consequences. *Current Medicinal Chemistry*. 2013;20(26):3218–3225. https://doi.org/10.2174/09298673113209990022.
61. Aiello LP, Arrigg PG, Shah ST, et al. Vascular endothelial growth factor in ocular fluid of patients with diabetic retinopathy and other retinal disorders. *New England Journal of Medicine*. 1994;331(22):1480–1487. https://doi.org/10.1056/NEJM199412013312203.
62. Miller JW, Adamis AP, Shima DT, et al. Vascular endothelial growth factor/vascular permeability factor is temporally and spatially correlated with ocular angiogenesis in a primate model. *American Journal of Pathology*. 1994;145(3):574–584.
63. Pierce IA, Avery RL, Foley ED, Aiello LP, Smith LEH. Vascular endothelial growth factor/vascular permeability factor expression in a mouse model of retinal neovascularization. *Proceedings of the National Academy of Sciences of the United States of America*. 1995;92(3):905–909. https://doi.org/10.1073/pnas.92.3.905.
64. Antonetti DA, Barber AJ, Hollinger LA, Wolpert EB, Gardner TW. Vascular endothelial growth factor induces rapid phosphorylation of tight junction proteins occludin and zonula occluden 1. A potential mechanism for vascular permeability in diabetic retinopathy and tumors. *Journal of Biological Chemistry*. 1999;274(33):23463–23467. https://doi.org/10.1074/jbc.274.33.23463.
65. Harhaj NS, Felinski EA, Wolpert EB, Sundstrom JM, Gardner TW, Antonetti DA. VEGF activation of protein kinase C stimulates occludin phosphorylation and contributes to endothelial permeability. *Investigative Ophthalmology and Visual Science*. 2006;47(11):5106–5115. https://doi.org/10.1167/iovs.06-0322.
66. Maisonpierre PC, Suri C, Jones PF, et al. Angiopoietin-2, a natural antagonist for Tie2 that disrupts in vivo angiogenesis. *Science*. 1997;277(5322):55–60. https://doi.org/10.1126/science.277.5322.55.
67. Lip PL, Chatterjee S, Caine GJ, et al. Plasma vascular endothelial growth factor, angiopoietin-2, and soluble and angiopoietin receptor tie-2 in diabetic retinopathy: effects of laser photocoagulcution and angiotensin receptor blockade. *British Journal of Ophthalmology*. 2004;88(12):1543–1546. https://doi.org/10.1136/bjo.2004.048587.

68. Eter N, Singh RP, Abreu F, Asik K, Basu K, Baumal C, et al.. YOSEMITE and RHINE: phase 3 randomized clinical trials of faricimab for diabetic macular edema: study design and rationale. *Ophthalmol Sci.* 2022;1:100111.
69. Mrejen S, Spaide RF. Optical coherence tomography: imaging of the choroid and beyond. *Survey of Ophthalmology.* 2013;58(5):387−429. https://doi.org/10.1016/j.survophthal.2012.12.001.
70. Iwase T, Yamamoto K, Kobayashi M, Ra E, Murotani K, Terasaki H. What ocular and systemic variables affect choroidal circulation in healthy eyes. *Medicine.* 2016;95(43):e5102. https://doi.org/10.1097/md.0000000000005102.
71. Saracco JB, Gastaud P, Ridings B, Ubaud CA. LA choroidopathie diabetique. *Journal Francais d'Ophtalmologie.* 1982;5(4):231−236.
72. Shiragami C, Shiraga F, Matsuo T, Tsuchida Y, Ohtsuki H. Risk factors for diabetic choroidopathy in patients with diabetic retinopathy. *Graefe's Archive for Clinical and Experimental Ophthalmology.* 2002;240(6):436−442. https://doi.org/10.1007/s00417-002-0451-5.
73. Weinberger D, Kramer M, Priel E, Gaton DD, Axer-Siegel R, Yassur Y. Indocyanine green angiographic findings in nonproliferative diabetic retinopathy. *American Journal of Ophthalmology.* 1998;126(2):238−247. https://doi.org/10.1016/S0002-9394(98)00148-2.
74. Cao J, McLeod DS, Merges CA, Lutty GA. Choriocapillaris degeneration and related pathologic changes in human diabetic eyes. *Archives of Ophthalmology.* 1998;116(5):589−597. https://doi.org/10.1001/archopht.116.5.589.
75. Hidayat AA, Fine BS. Diabetic choroidopathy: light and electron microscopic observations of seven cases. *Ophthalmology.* 1985;92(4):512−522. https://doi.org/10.1016/S0161-6420(85)34013-7.
76. McLeod DS, Lutty GA. High-resolution histologic analysis of the human choroidal vasculature. *Investigative Ophthalmology and Visual Science.* 1994;35(11):3799−3811.
77. Melancia D, Vicente A, Cunha JP, Abegão Pinto L, Ferreira J. Diabetic choroidopathy: a review of the current literature. *Graefe's Archive for Clinical and Experimental Ophthalmology.* 2016;254(8):1453−1461. https://doi.org/10.1007/s00417-016-3360-8.
78. Ferris FL. How effective are treatments for diabetic retinopathy? *JAMA: The Journal of the American Medical Association.* 1993;269(10):1290−1291. https://doi.org/10.1001/jama.1993.03500100088034.
79. An D, Chandrasekera E, Yu D-Y, Balaratnasingam C. Non-proliferative diabetic retinopathy is characterized by nonuniform alterations of peripapillary capillary networks. *Investigative Opthalmology & Visual Science.* 2020;61(4):39. https://doi.org/10.1167/iovs.61.4.39.
80. Moore J, Bagley S, Ireland G, McLeod D, Boulton ME. Three dimensional analysis of microaneurysms in the human diabetic retina. *Journal of Anatomy.* 1999;194(1):89−100. https://doi.org/10.1017/S0021878298004403.
81. Chen C, Shah CP. Review of therapeutic advances in diabetic retinopathy. *Therapeutic Advances in Endocrinology and Metabolism.* 2011;2(1):39−53. https://doi.org/10.1177/2042018810396750.
82. Cheung CY, Ikram MK, Klein R, Wong TY. The clinical implications of recent studies on the structure and function of the retinal microvasculature in diabetes. *Diabetologia.* 2015;58(5):871−885. https://doi.org/10.1007/s00125-015-3511-1.
83. Lee CS, Lee AY, Sim DA, et al. Reevaluating the definition of intraretinal microvascular abnormalities and neovascularization elsewhere in diabetic retinopathy using optical coherence tomography and fluorescein angiography. *American Journal of Ophthalmology.* 2015;159(1):101−110. https://doi.org/10.1016/j.ajo.2014.09.041.
84. DaCosta J, Bhatia D, Crothers O, Talks J. Utilisation of optical coherence tomography and optical coherence tomography angiography to assess retinal neovascularisation in diabetic retinopathy. *Eye.* 2022;36(4):827−834. https://doi.org/10.1038/s41433-021-01544-8.
85. Grading diabetic retinopathy from stereoscopic color fundus photographs—an extension of the modified Airlie House classification: ETDRS report number 10. *Ophthalmology.* 1991;98(5):786−806. https://doi.org/10.1016/S0161-6420(13)38012-9.
86. Wilkinson CP, Ferris FL, Klein RE, et al. Proposed international clinical diabetic retinopathy and diabetic macular edema disease severity scales. *Ophthalmology.* 2003;110(9):1677−1682. https://doi.org/10.1016/S0161-6420(03)00475-5.
87. Early photocoagulation for diabetic retinopathy: ETDRS report number 9. *Ophthalmology.* 1991;98(5):766−785. https://doi.org/10.1016/S0161-6420(13)38011-7.
88. Photocoagulation for Diabetic Macular Edema. Early treatment diabetic retinopathy study report number 1. Early treatment diabetic retinopathy study research group. *Arch Ophthalmol.* 1985;103(12):1796−1806.
89. Jampol LM, Tadayoni R, Ip M. Need for a new classification of diabetic retinopathy. *Retina.* 2021;41(3):459−460. https://doi.org/10.1097/iae.0000000000003070.
90. Tan T-E, Wong TY. Diabetic retinopathy: looking forward to 2030. *Frontiers in Endocrinology.* 2023;13. https://doi.org/10.3389/fendo.2022.1077669.
91. Tan B, Lim NA, Tan R, et al. Combining retinal and choroidal microvascular metrics improves discriminative power for diabetic retinopathy. *British Journal of Ophthalmology.* 2022;107:993−999. https://doi.org/10.1136/bjophthalmol-2021-319739.
92. Tran K, Pakzad-Vaezi K. Multimodal imaging of diabetic retinopathy. *Current Opinion in Ophthalmology.* 2018;29(6):566−575. https://doi.org/10.1097/ICU.0000000000000524.
93. Ting DSW, Tan GSW, Agrawal R, et al. Optical coherence tomographic angiography in type 2 diabetes and diabetic retinopathy. *JAMA Ophthalmology.* 2017;135(4):306−312. https://doi.org/10.1001/jamaophthalmol.2016.5877.
94. Gendelman I, Alibhai AY, Moult EM, et al. Topographic analysis of macular choriocapillaris flow deficits in diabetic retinopathy using swept-source optical coherence tomography angiography. *International Journal of Retina and Vitreous.* 2020;6(1). https://doi.org/10.1186/s40942-020-00209-0.
95. Horie S, Ohno-Matsui K. Progress of imaging in diabetic retinopathy—from the past to the present. *Diagnostics.* 2022;12(7):1684. https://doi.org/10.3390/diagnostics12071684.

96. Nanegrungsunk O, Patikulsila D, Sadda SR. Ophthalmic imaging in diabetic retinopathy: a review. *Clinical and Experimental Ophthalmology*. 2022;50(9):1082–1096. https://doi.org/10.1111/ceo.14170.
97. Aiello LP, Odia I, Glassman AR, et al. Comparison of early treatment diabetic retinopathy study standard 7-field imaging with ultrawide-field imaging for determining severity of diabetic retinopathy. *JAMA Ophthalmology*. 2019;137(1):65–73. https://doi.org/10.1001/jamaophthalmol.2018.4982.
98. Elman MJ, Aiello LP, Beck RW, et al. Randomized trial evaluating ranibizumab plus prompt or deferred laser or triamcinolone plus prompt laser for diabetic macular edema. *Ophthalmology*. 2010;117(6):1064. https://doi.org/10.1016/j.ophtha.2010.02.031.
99. Kim BY, Smith SD, Kaiser PK. Optical coherence tomographic patterns of diabetic macular edema. *American Journal of Ophthalmology*. 2006;142(3):405. https://doi.org/10.1016/j.ajo.2006.04.023.
100. Otani T, Kishi S, Maruyama Y. Patterns of diabetic macular edema with optical coherence tomography. *American Journal of Ophthalmology*. 1999;127(6):688–693. https://doi.org/10.1016/S0002-9394(99)00033-1.
101. Tan KA, Laude A, Yip V, Loo E, Wong EP, Agrawal R. Choroidal vascularity index – a novel optical coherence tomography parameter for disease monitoring in diabetes mellitus? *Acta Ophthalmologica*. 2016;94(7):e612–e616. https://doi.org/10.1111/aos.13044.
102. Gupta P, Thakku SG, Sabanayagam C, et al. Characterisation of choroidal morphological and vascular features in diabetes and diabetic retinopathy. *British Journal of Ophthalmology*. 2017;101(8):1038–1044. https://doi.org/10.1136/bjophthalmol-2016-309366.
103. Laíns I, Figueira J, Santos AR, et al. Choroidal thickness in diabetic retinopathy: the influence of antiangiogenic therapy. *Retina*. 2014;34(6):1199–1207. https://doi.org/10.1097/IAE.0000000000000053.
104. Spaide RF, Fujimoto JG, Waheed NK, Sadda SR, Staurenghi G. Optical coherence tomography angiography. *Progress in Retinal and Eye Research*. 2018;64:1–55. https://doi.org/10.1016/j.preteyeres.2017.11.003.
105. Hsieh YT, Alam MN, Le D, et al. OCT angiography biomarkers for predicting visual outcomes after ranibizumab treatment for diabetic macular edema. *Ophthalmology Retina*. 2019;3(10):826–834. https://doi.org/10.1016/j.oret.2019.04.027.
106. Hwang TS, Jia Y, Gao SS, et al. Optical coherence tomography angiography features of diabetic retinopathy. *Retina*. 2015;35(11):2371–2376. https://doi.org/10.1097/IAE.0000000000000716.
107. Lee J, Moon BG, Cho AR, Yoon YH. Optical coherence tomography angiography of DME and its association with anti-VEGF treatment response. *Ophthalmology*. 2016;123(11):2368–2375. https://doi.org/10.1016/j.ophtha.2016.07.010.
108. Agemy SA, Scripsema NK, Shah CM, et al. Retinal vascular perfusion density mapping using optical coherence tomography angiography in normals and diabetic retinopathy patients. *Retina*. 2015;35(11):2353–2363. https://doi.org/10.1097/IAE.0000000000000862.
109. Zahid S, Dolz-Marco R, Freund KB, et al. Fractal dimensional analysis of optical coherence tomography angiography in eyes with diabetic retinopathy. *Investigative Ophthalmology and Visual Science*. 2016;57(11):4940–4947. https://doi.org/10.1167/iovs.16-19656.
110. Yang J, Zhang B, Wang E, Xia S, Chen Y. Ultra-wide field swept-source optical coherence tomography angiography in patients with diabetes without clinically detectable retinopathy. *BMC Ophthalmology*. 2021;21(1). https://doi.org/10.1186/s12886-021-01933-3.
111. Rabiolo A, Parravano M, Querques L, et al. Ultra-wide-field fluorescein angiography in diabetic retinopathy: a narrative review. *Clinical Ophthalmology*. 2017;11:803–807. https://doi.org/10.2147/OPTH.S133637.
112. Silva PS, Liu D, Glassman AR, et al. Assessment of fluorescein angiography nonperfusion in eyes with diabetic retinopathy using ultrawide field retinal imaging. *Retina*. 2022;42(7):1302–1310. https://doi.org/10.1097/IAE.0000000000003479.
113. Silva PS, Marcus DM, Liu D, et al. Association of ultra-widefield fluorescein angiography-identified retinal nonperfusion and the risk of diabetic retinopathy worsening over time. *JAMA Ophthalmology*. 2022;140(10):936–945. https://doi.org/10.1001/jamaophthalmol.2022.3130.
114. Al-khersan H, Russell JF, Lazzarini TA, et al. Comparison between graders in detection of diabetic neovascularization with swept source optical coherence tomography angiography and fluorescein angiography. *American Journal of Ophthalmology*. 2021;224:292–300. https://doi.org/10.1016/j.ajo.2020.11.020.
115. Soares M, Neves C, Marques IP, et al. Comparison of diabetic retinopathy classification using fluorescein angiography and optical coherence tomography angiography. *British Journal of Ophthalmology*. 2017;101(1):62–68. https://doi.org/10.1136/bjophthalmol-2016-309424.
116. Abràmoff MD, Folk JC, Han DP, et al. Automated analysis of retinal images for detection of referable diabetic retinopathy. *JAMA Ophthalmology*. 2013;131(3):351–357. https://doi.org/10.1001/jamaophthalmol.2013.1743.
117. Abràmoff MD, Lou Y, Erginay A, et al. Improved automated detection of diabetic retinopathy on a publicly available dataset through integration of deep learning. *Investigative Ophthalmology and Visual Science*. 2016;57(13):5200–5206. https://doi.org/10.1167/iovs.16-19964.
118. Bhaskaranand M, Ramachandra C, Bhat S, et al. Automated diabetic retinopathy screening and monitoring using retinal fundus image analysis. *Journal of Diabetes Science and Technology*. 2016;10(2):254–261. https://doi.org/10.1177/1932296816628546.
119. Bhaskaranand M, Ramachandra C, Bhat S, et al. The value of automated diabetic retinopathy screening with the EyeArt system: a study of more than 100,000 consecutive encounters from people with diabetes. *Diabetes Technology and Therapeutics*. 2019;21(11):635–643. https://doi.org/10.1089/dia.2019.0164.
120. Grzybowski A, Brona P. Analysis and comparison of two artificial intelligence diabetic retinopathy screening algorithms in a pilot study: IDx-DR and retinalyze. *Journal of Clinical Medicine*. 2021;10(11):2352. https://doi.org/10.3390/jcm10112352.
121. Gulshan V, Peng L, Coram M, et al. Development and validation of a deep learning algorithm for detection of diabetic retinopathy in retinal fundus photographs. *Journal of the American Medical Association*. 2016;316(22):2402–2410. https://doi.org/10.1001/jama.2016.17216.
122. Hansen MB, Abràmoff MD, Folk JC, et al. Results of automated retinal image analysis for detection of diabetic retinopathy from the Nakuru study, Kenya. *PLoS One*. 2015;10(10):e0139148. https://doi.org/10.1371/journal.pone.0139148.

123. Larsen M, Godt J, Larsen N, et al. Automated detection of fundus photographic red lesions in diabetic retinopathy. *Investigative Ophthalmology and Visual Science.* 2003;44(2):761−766. https://doi.org/10.1167/iovs.02-0418.
124. Larsen N, Godt J, Grunkin M, Lund-Andersen H, Larsen M. Automated detection of diabetic retinopathy in a fundus photographic screening population. *Investigative Ophthalmology and Visual Science.* 2003;44(2):767−771. https://doi.org/10.1167/iovs.02-0417.
125. Oliveira CM, Cristóvão LM, Ribeiro ML, Abreu JRF. Improved automated screening of diabetic retinopathy. *Ophthalmologica.* 2011;226(4):191−197. https://doi.org/10.1159/000330285.
126. Raumviboonsuk P, Krause J, Chotcomwongse P, et al. Deep learning versus human graders for classifying diabetic retinopathy severity in a nationwide screening program. *Npj Digital Medicine.* 2019;2(1). https://doi.org/10.1038/s41746-019-0099-8.
127. Ting DSW, Cheung CYL, Lim G, et al. Development and validation of a deep learning system for diabetic retinopathy and related eye diseases using retinal images from multiethnic populations with diabetes. *Journal of the American Medical Association.* 2017;318(22):2211−2223. https://doi.org/10.1001/jama.2017.18152.
128. Abràmoff MD, Lavin PT, Birch M, Shah N, Folk JC. Pivotal trial of an autonomous AI-based diagnostic system for detection of diabetic retinopathy in primary care offices. *Npj Digital Medicine.* 2018;1(1). https://doi.org/10.1038/s41746-018-0040-6.
129. Bawankar P, Shanbhag N, K. SS, et al. Sensitivity and specificity of automated analysis of single-field non-mydriatic fundus photographs by Bosch DR Algorithm—comparison with mydriatic fundus photography (ETDRS) for screening in undiagnosed diabetic retinopathy. *PLoS One.* 2017;12(12):e0189854. https://doi.org/10.1371/journal.pone.0189854.
130. Bellemo V, Lim ZW, Lim G, et al. Artificial intelligence using deep learning to screen for referable and vision-threatening diabetic retinopathy in Africa: a clinical validation study. *The Lancet Digital Health.* 2019;1(1):e35−e44. https://doi.org/10.1016/S2589-7500(19)30004-4.
131. Heydon P, Egan C, Bolter L, et al. Prospective evaluation of an artificial intelligence-enabled algorithm for automated diabetic retinopathy screening of 30,000 patients. *British Journal of Ophthalmology.* 2021;105(5):723−728. https://doi.org/10.1136/bjophthalmol-2020-316594.
132. Ipp E, Liljenquist D, Bode B, et al. Pivotal evaluation of an artificial intelligence system for autonomous detection of referrable and vision-threatening diabetic retinopathy. *JAMA Network Open.* 2021;4(11):e2134254. https://doi.org/10.1001/jamanetworkopen.2021.34254.
133. Kim TN, Aaberg MT, Li P Comparison of automated and expert human grading of diabetic retinopathy using smartphone-based retinal photography. Eye (Basingstoke). 2021;235(1):334−342.
134. Natarajan S, Jain A, Krishnan R, Rogye A, Sivaprasad S. Diagnostic accuracy of community-based diabetic retinopathy screening with an offline artificial intelligence system on a smartphone. *JAMA Ophthalmology.* 2019;137(10):1182−1188. https://doi.org/10.1001/jamaophthalmol.2019.2923.
135. Rajalakshmi R, Subashini R, Anjana RM, Mohan V. Automated diabetic retinopathy detection in smartphone-based fundus photography using artificial intelligence. *Eye.* 2018;32(6):1138−1144. https://doi.org/10.1038/s41433-018-0064-9.
136. Ruamviboonsuk P, Tiwari R, Sayres R, et al. Real-time diabetic retinopathy screening by deep learning in a multisite national screening programme: a prospective interventional cohort study. *The Lancet Digital Health.* 2022;4(4):e235−e244. https://doi.org/10.1016/S2589-7500(22)00017-6.
137. Zhang Y, Shi J, Peng Y, et al. Artificial intelligence-enabled screening for diabetic retinopathy: a real-world, multicenter and prospective study. *BMJ Open Diabetes Research and Care.* 2020;8(1). https://doi.org/10.1136/bmjdrc-2020-001596.
138. van der Heijden AA, Abramoff MD, Verbraak F, van Hecke MV, Liem A, Nijpels G. Validation of automated screening for referable diabetic retinopathy with the IDx-DR device in the Hoorn Diabetes Care System. *Acta Ophthalmologica.* 2018;96(1):63−68. https://doi.org/10.1111/aos.13613.
139. Yuan A, Lee AY. Artificial intelligence deployment in diabetic retinopathy: the last step of the translation continuum. *The Lancet Digital Health.* 2022;4(4):e208−e209. https://doi.org/10.1016/S2589-7500(22)00027-9.
140. Tabuchi H. Understanding required to consider AI applications to the field of ophthalmology. *Taiwan Journal of Ophthalmology.* 2022;12(2):123−129. https://doi.org/10.4103/tjo.tjo_8_22.
141. Reddy S, Allan S, Coghlan S, Cooper P. A governance model for the application of AI in health care. *Journal of the American Medical Informatics Association.* 2020;27(3):491−497. https://doi.org/10.1093/jamia/ocz192.
142. Kaothanthong N, Limwattanayingyong J, Silpa-archa S, et al. The classification of common macular diseases using deep learning on optical coherence tomography images with and without prior automated segmentation. *Diagnostics.* 2023;13(2):189. https://doi.org/10.3390/diagnostics13020189.
143. Schlegl T, Waldstein SM, Bogunovic H, et al. Fully automated detection and quantification of macular fluid in OCT using deep learning. *Ophthalmology.* 2018;125(4):549−558. https://doi.org/10.1016/j.ophtha.2017.10.031.
144. Schifferdecker E, Schmidt K, Boehm BO, Schatz H. Long-term compliance of intensified insulin therapy. *Diabetes Research and Clinical Practice.* 1994;23(1):17−23. https://doi.org/10.1016/0168-8227(94)90122-8.
145. White NH, Sun W, Cleary PA, et al. Prolonged effect of intensive therapy on the risk of retinopathy complications in patients with type 1 diabetes mellitus: 10 years after the diabetes control and complications trial. *Archives of Ophthalmology.* 2008;126(12):1707−1715. https://doi.org/10.1001/archopht.126.12.1707.
146. Turner R. Intensive blood-glucose control with sulphonylureas or insulin compared with conventional treatment and risk of complications in patients with type 2 diabetes (UKPDS 33). *Lancet.* 1998;352(9131):837−853. https://doi.org/10.1016/S0140-6736(98)07019-6.
147. Chew EY, Ambrosius WT, Davis MD, et al. Effects of medical therapies on retinopathy progression in type 2 diabetes. *New England Journal of Medicine.* 2010;363(3):233−244. https://doi.org/10.1056/NEJMoa1001288.
148. Egger M, Davey Smith G, Stettler C, Diem P. Risk of adverse effects of intensified treatment in insulin-dependent diabetes mellitus: a meta-analysis. *Diabetic Medicine.* 1997;14(11):919−928. https://doi.org/10.1002/(SICI)1096-9136(199711)14:11<919::AID-DIA456>3.0.CO;2-A.

149. Do DV, Wang X, Vedula SS, et al. Blood pressure control for diabetic retinopathy. *Sao Paulo Medical Journal.* 2015;133(3):278–279. https://doi.org/10.1590/1516-3180.20151333T1.
150. Zhou JB, Song ZH, Bai L, Zhu XR, Li HB, Yang JK. Could intensive blood pressure control really reduce diabetic retinopathy outcomes? Evidence from meta-analysis and trial sequential analysis from randomized controlled trials. *Diabetes Therapy.* 2018;9(5):2015–2027. https://doi.org/10.1007/s13300-018-0497-y.
151. Beulens JWJ, Patel A, Vingerling JR, et al. Effects of blood pressure lowering and intensive glucose control on the incidence and progression of retinopathy in patients with type 2 diabetes mellitus: a randomised controlled trial. *Diabetologia.* 2009;52(10):2027–2036. https://doi.org/10.1007/s00125-009-1457-x.
152. Group UKPDS. Tight blood pressure control and risk of macrovascular and microvascular complications in type 2 diabetes: UKPDS 38. UK Prospective Diabetes Study Group. *BMJ.* 1998;317(7160).
153. Massin P, Bandello F, Garweg JG, et al. Safety and efficacy of ranibizumab in diabetic macular edema (RESOLVE study): a 12-month, randomized, controlled, double-masked, multicenter phase II study. *Diabetes Care.* 2010;33(11):2399–2405. https://doi.org/10.2337/dc10-0493.
154. Mitchell P, Bandello F, Schmidt-Erfurth U, et al. The RESTORE study: ranibizumab monotherapy or combined with laser versus laser monotherapy for diabetic macular edema. *Ophthalmology.* 2011;118(4):615–625. https://doi.org/10.1016/j.ophtha.2011.01.031.
155. Nguyen QD, Brown DM, Marcus DM, et al. Ranibizumab for diabetic macular edema: results from 2 phase iii randomized trials: RISE and RIDE. *Ophthalmology.* 2012;119(4):789–801. https://doi.org/10.1016/j.ophtha.2011.12.039.
156. Nguyen QD, Shah SM, Khwaja AA, et al. Two-year outcomes of the ranibizumab for edema of the mAcula in diabetes (READ-2) study. *Ophthalmology.* 2010;117(11):2146–2151. https://doi.org/10.1016/j.ophtha.2010.08.016.
157. Korobelnik JF, Do DV, Schmidt-Erfurth U, et al. Intravitreal aflibercept for diabetic macular edema. *Ophthalmology.* 2014;121(11):2247–2254. https://doi.org/10.1016/j.ophtha.2014.05.006.
158. Wells JA, Glassman AR, Ayala AR, et al. Aflibercept, bevacizumab, or ranibizumab for diabetic macular edema two-year results from a comparative effectiveness randomized clinical trial. *Ophthalmology.* 2016;123(6):1351–1359. https://doi.org/10.1016/j.ophtha.2016.02.022.
159. Baker CW, Glassman AR, Beaulieu WT, et al. Effect of initial management with aflibercept vs laser photocoagulation vs observation on vision loss among patients with diabetic macular edema involving the center of the macula and good visual acuity: a randomized clinical trial. *Journal of the American Medical Association.* 2019;321(19):1880–1894. https://doi.org/10.1001/jama.2019.5790.
160. Goyal S, Lavalley M, Subramanian ML. Meta-analysis and review on the effect of bevacizumab in diabetic macular edema. *Graefe's Archive for Clinical and Experimental Ophthalmology.* 2011;249(1):15–27. https://doi.org/10.1007/s00417-010-1452-4.
161. Michaelides M, Kaines A, Hamilton RD, et al. A prospective randomized trial of intravitreal bevacizumab or laser therapy in the management of diabetic macular edema (BOLT study). 12-Month data: report 2. *Ophthalmology.* 2010;117(6):1078–e2. https://doi.org/10.1016/j.ophtha.2010.03.045.
162. Scott IU, Edwards AR, Beck RW, Bressler NM, Chan CK, et al.. A phase II randomized clinical trial of intravitreal bevacizumab for diabetic macular edema. *Ophthalmology.* 2007;114(10):1860–1867.
163. Madjedi K, Pereira A, Ballios BG, et al. Switching between anti-VEGF agents in the management of refractory diabetic macular edema: a systematic review. *Survey of Ophthalmology.* 2022;67(5):1364–1372. https://doi.org/10.1016/j.survophthal.2022.04.001.
164. Jhaveri CD, Glassman AR, Ferris FL, et al. Aflibercept monotherapy or bevacizumab first for diabetic macular edema. *New England Journal of Medicine.* 2022;387(8):692–703. https://doi.org/10.1056/NEJMoa2204225.
165. Liberski S, Wichrowska M, Kociecki J. Aflibercept versus faricimab in the treatment of neovascular age-related macular degeneration and diabetic macular edema: a review. *International Journal of Molecular Sciences.* 2022;23(16). https://doi.org/10.3390/ijms23169424.
166. Randomized. Active-controlled phase 2/3 study of the efficacy and safety of high-dose aflibercept in patients with diabetic macular edema. *Double-Masked.* 2022.
167. Brown DM, Emanuelli A, Bandello F, et al. KESTREL and KITE: 52-week results from two phase III pivotal trials of brolucizumab for diabetic macular edema. *American Journal of Ophthalmology.* 2022;238:157–172. https://doi.org/10.1016/j.ajo.2022.01.004.
168. Singer M, Albini TA, Seres A, et al. Clinical characteristics and outcomes of eyes with intraocular inflammation after brolucizumab: post hoc analysis of HAWK and HARRIER. *Ophthalmology Retina.* 2022;6(2):97–108. https://doi.org/10.1016/j.oret.2021.05.003.
169. Baumal CR, Bodaghi B, Singer M, et al. Expert opinion on management of intraocular inflammation, retinal vasculitis, and vascular occlusion after brolucizumab treatment. *Ophthalmology Retina.* 2021;5(6):519–527. https://doi.org/10.1016/j.oret.2020.09.020.
170. Le NT, Kroeger ZA, Lin WV, Khanani AM, Weng CY. Novel treatments for diabetic macular edema and proliferative diabetic retinopathy. *Current Diabetes Reports.* 2021;21(10). https://doi.org/10.1007/s11892-021-01412-5.
171. Wykoff CC, Abreu F, Adamis AP, et al. Efficacy, durability, and safety of intravitreal faricimab with extended dosing up to every 16 weeks in patients with diabetic macular oedema (YOSEMITE and RHINE): two randomised, double-masked, phase 3 trials. *The Lancet.* 2022;399(10326):741–755. https://doi.org/10.1016/S0140-6736(22)00018-6.
172. Shirley M. Faricimab: first approval. *Drugs.* 2022;82(7):825–830. https://doi.org/10.1007/s40265-022-01713-3.
173. Ranade SV, Wieland MR, Tam T, et al. The Port Delivery System with ranibizumab: a new paradigm for long-acting retinal drug delivery. *Drug Delivery.* 2022;29(1):1326–1334. https://doi.org/10.1080/10717544.2022.2069301.
174. Phase, III Multicenter, Randomized. Active-comparator study of the efficacy, safety, and pharmacokinetics of the port delivery system with ranibizumab in patients with diabetic macular edema (pagoda). *Visual Assessor-Masked.* 2022.
175. Phase, III Multicenter. *Randomized Study of the Efficacy, Safety, and Pharmacokinetics of the Port Delivery System with Ranibizumab in Patients with Diabetic Retinopathy.* 2022.

176. Holekamp NM, Campochiaro PA, Chang MA, et al. Archway randomized phase 3 trial of the port delivery system with ranibizumab for neovascular age-related macular degeneration. *Ophthalmology*. 2022;129(3):295−307. https://doi.org/10.1016/j.ophtha.2021.09.016.
177. Timmons K, Heckmann LC, Ren Y, Gunderson I, Makkouk F, Berger BB. Ranibizumab injection (susvimo) implant septum dislodgement in a patient with neovascular age-related macular degeneration. *JAMA Ophthalmology*. 2022;140(8):832. https://doi.org/10.1001/jamaophthalmol.2022.2495.
178. Campochiaro PA, Brown DM, Pearson A, et al. Long-term benefit of sustained-delivery fluocinolone acetonide vitreous inserts for diabetic macular edema. *Ophthalmology*. 2011;118(4):626. https://doi.org/10.1016/j.ophtha.2010.12.028.
179. Gillies M, Sutter F, Simpson J, Larsson J, Ali H, Zhu M. Intravitreal triamcinolone for refractory diabetic macular EdemaTwo-year results of a double-masked, placebo-controlled, randomized clinical trial. *Ophthalmology*. 2006;113(9):1533−1538. https://doi.org/10.1016/j.ophtha.2006.02.065.
180. Jonas JB, Kreissig I, Söflzer A, Degenring RF. Intravitreal injection of triamcinolone for diffuse diabetic macular edema. *Archives of Ophthalmology*. 2003;121(1):57−61. https://doi.org/10.1001/archopht.121.1.57.
181. Kuppermann BD, Blumenkranz MS, Haller JA, et al. Randomized controlled study of an intravitreous dexamethasone drug delivery system in patients with persistent macular edema. *Archives of Ophthalmology*. 2007;125(3):309−317. https://doi.org/10.1001/archopht.125.3.309.
182. Nakano Goto S, Yamamoto T, Kirii E, Abe S, Yamashita H. Treatment of diffuse diabetic macular oedema using steroid eye drops. *Acta Ophthalmologica*. 2012;90(7):628−632. https://doi.org/10.1111/j.1755-3768.2010.02066.x.
183. Nakano S, Yamamoto T, Kirii E, Abe S, Yamashita H. Steroid eye drop treatment (difluprednate ophthalmic emulsion) is effective in reducing refractory diabetic macular edema. *Graefe's Archive for Clinical and Experimental Ophthalmology*. 2010;248(6):805−810. https://doi.org/10.1007/s00417-010-1316-y.
184. Boyer DS, Yoon YH, Belfort R, et al. Three-year, randomized, sham-controlled trial of dexamethasone intravitreal implant in patients with diabetic macular edema. *Ophthalmology*. 2014;121(10):1904−1914. https://doi.org/10.1016/j.ophtha.2014.04.024.
185. Khan Z, Kuriakose RK, Khan M, Chin EK, Almeida DRP. Efficacy of the intravitreal sustained-release dexamethasone implant for diabetic macular edema refractory to anti-vascular endothelial growth factor therapy: meta-analysis and clinical implications. *Ophthalmic Surgery Lasers and Imaging Retina*. 2017;48(2):160−166. https://doi.org/10.3928/23258160-20170130-10.
186. Pacella F, Romano MR, Turchetti P, et al. An eighteen-month follow-up study on the effects of intravitreal dexamethasone implant in diabetic macular edema refractory to anti-VEGF therapy. *International Journal of Ophthalmology*. 2016;9(10):1427−1432. https://doi.org/10.18240/ijo.2016.10.10.
187. Rittiphairoj T, Mir TA, Li T, Virgili G. Intravitreal steroids for macular edema in diabetes. *Cochrane Database of Systematic Reviews*. 2020;2020(11). https://doi.org/10.1002/14651858.CD005656.pub3.
188. El Rami H, Barham R, Sun JK, Silva PS. Evidence-based treatment of diabetic retinopathy. *Seminars in Ophthalmology*. 2017;32(1):67−74. https://doi.org/10.1080/08820538.2016.1228397.
189. Clinical application of diabetic retinopathy study (DRS) findings, DRS report number 8. The diabetic retinopathy study research group. *Ophthalmology*. 1981;88(7):583−600.
190. Early treatment diabetic retinopathy study design and baseline patient characteristics: ETDRS report number 7. *Ophthalmology*. 1991;98(5):741−756. https://doi.org/10.1016/S0161-6420(13)38009-9.
191. Moutray T, Evans JR, Lois N, Armstrong DJ, Peto T, Azuara-Blanco A. Different lasers and techniques for proliferative diabetic retinopathy. *Cochrane Database of Systematic Reviews*. 2018;2018(3). https://doi.org/10.1002/14651858.CD012314.pub2.
192. Muqit MMK, Sanghvi C, McLauchlan R, et al. Study of clinical applications and safety for Pascal laser photocoagulation in retinal vascular disorders. *Acta Ophthalmologica*. 2012;90(2):155−161. https://doi.org/10.1111/j.1755-3768.2009.01854.x.
193. Sanghvi C, McLauchlan R, Delgado C, et al. Initial experience with the Pascal photocoagulator: a pilot study of 75 procedures. *British Journal of Ophthalmology*. 2008;92(8):1061−1064. https://doi.org/10.1136/bjo.2008.139568.
194. Muqit MMK, Marcellino GR, Henson DB, et al. Optos-guided pattern scan laser (Pascal)-targeted retinal photocoagulation in proliferative diabetic retinopathy. *Acta Ophthalmologica*. 2013;91(3):251−258. https://doi.org/10.1111/j.1755-3768.2011.02307.x.
195. Muqit MMK, Young LB, McKenzie R, et al. Pilot randomised clinical trial of Pascal TargETEd Retinal versus variable fluence PANretinal 20 ms laser in diabetic retinopathy: PETER PAN study. *British Journal of Ophthalmology*. 2013;97(2):220−227. https://doi.org/10.1136/bjophthalmol-2012-302189.
196. Mansour SE, Browning DJ, Wong K, Flynn HW, Bhavsar AR. The evolving treatment of diabetic retinopathy. *Clinical Ophthalmology*. 2020;14:653−678. https://doi.org/10.2147/OPTH.S236637.
197. Simunovic MP, Maberley DAL. ANTI-vascular endothelial growth factor therapy for proliferative diabetic retinopathy. *Retina*. 2015;35(10):1931−1942. https://doi.org/10.1097/IAE.0000000000000723.
198. Smith JM, Steel DHW. Anti-vascular endothelial growth factor for prevention of postoperative vitreous cavity haemorrhage after vitrectomy for proliferative diabetic retinopathy. *Cochrane Database of Systematic Reviews*. 2015;2015(8). https://doi.org/10.1002/14651858.CD008214.pub3.
199. Zhao XY, Xia S, Chen YX. Antivascular endothelial growth factor agents pretreatment before vitrectomy for complicated proliferative diabetic retinopathy: a meta-analysis of randomised controlled trials. *British Journal of Ophthalmology*. 2018;102(8):1077−1085. https://doi.org/10.1136/bjophthalmol-2017-311344.
200. Oshima Y, Shima C, Wakabayashi T, et al. Microincision vitrectomy surgery and intravitreal bevacizumab as a surgical adjunct to treat diabetic traction retinal detachment. *Ophthalmology*. 2009;116(5):927−938. https://doi.org/10.1016/j.ophtha.2008.11.005.

201. Gross JG, Glassman AR, Jampol LM, et al. Panretinal photocoagulation vs intravitreous ranibizumab for proliferative diabetic retinopathy: a randomized clinical trial. *Journal of the American Medical Association*. 2015;314(20):2137–2146. https://doi.org/10.1001/jama.2015.15217.
202. Lang GE, Stahl A, Voegeler J, et al. Efficacy and safety of ranibizumab with or without panretinal laser photocoagulation versus laser photocoagulation alone in proliferative diabetic retinopathy – the PRIDE study. *Acta Ophthalmologica*. 2020;98(5):e530–e539. https://doi.org/10.1111/aos.14312.
203. Sivaprasad S, Prevost AT, Vasconcelos JC, et al. Clinical efficacy of intravitreal aflibercept versus panretinal photocoagulation for best corrected visual acuity in patients with proliferative diabetic retinopathy at 52 weeks (CLARITY): a multicentre, single-blinded, randomised, controlled, phase 2b, non-inferiority trial. *The Lancet*. 2017;389(10085):2193–2203. https://doi.org/10.1016/S0140-6736(17)31193-5.
204. Gross JG, Glassman AR, Liu D, et al. Five-year outcomes of panretinal photocoagulation vs intravitreous ranibizumab for proliferative diabetic retinopathy: a randomized clinical trial. *JAMA Ophthalmology*. 2018;136(10):1138–1148. https://doi.org/10.1001/jamaophthalmol.2018.3255.
205. Senra H, Balaskas K, Mahmoodi N, Aslam T. Experience of anti-VEGF treatment and clinical levels of depression and anxiety in patients with wet age-related macular degeneration. *American Journal of Ophthalmology*. 2017;177:213–224. https://doi.org/10.1016/j.ajo.2017.03.005.
206. Falavarjani KG, Nguyen QD. Adverse events and complications associated with intravitreal injection of anti-VEGF agents: a review of literature. *Eye (Basingstoke)*. 2013;27(7):787–794. https://doi.org/10.1038/eye.2013.107.
207. Bressler SB, Liu D, Glassman AR, et al. Change in diabetic retinopathy through 2 years secondary analysis of a randomized clinical trial comparing aflibercept, bevacizumab, and ranibizumab. *JAMA Ophthalmology*. 2017;135(6):558–568. https://doi.org/10.1001/jamaophthalmol.2017.0821.
208. Brown DM, Wykoff CC, Boyer D, et al. Evaluation of intravitreal aflibercept for the treatment of severe nonproliferative diabetic retinopathy: results from the PANORAMA randomized clinical trial. *JAMA Ophthalmology*. 2021;139(9):946–955. https://doi.org/10.1001/jamaophthalmol.2021.2809.
209. Maturi RK, Glassman AR, Josic K, et al. Effect of intravitreous anti-vascular endothelial growth factor vs sham treatment for prevention of vision-threatening complications of diabetic retinopathy: the protocol W randomized clinical trial. *JAMA Ophthalmology*. 2021:E1–E12. https://doi.org/10.1001/jamaophthalmol.2021.0606.

Chapter 17

Paracentral acute middle maculopathy and acute macular neuroretinopathy: Distinct but related entities

Elodie Bousquet[1,2], Ahmad Santina[1] and David Sarraf[1]
[1]*Retinal Disorders and Ophthalmic Genetics Division, Stein Eye Institute, University of California Los Angeles, Los Angeles, CA, United States;*
[2]*Paris Cité University, Hôpital Lariboisière, AP-HP, Ophthalmology, Paris, France*

Pathogenesis

Retinal capillary plexuses

The retina is sustained by a dual blood supply: the outer retina (i.e., photoreceptor layers) is predominantly supplied by the choriocapillaris while the inner retina (i.e., the nerve fiber layer and the ganglion cell layer) and the middle retina (i.e., the inner nuclear layer and inner plexiform layer) are supplied by the retinal capillary system (RCP).

In the parafovea region, the retinal capillary plexuses (RCPs) are organized in a predominantly vertical system of three capillary layers of varying depths: the superficial retinal capillary plexus (SCP), intermediate retinal capillary plexus (ICP), and the deep retinal capillary plexus (DCP).

Advanced multimodal imaging with optical coherence tomography (OCT)-angiography and animal models have given new insight into the organization of the RCP. Major arterial inflows occur at the level of SCP and ICP while venous outflow predominantly tracks through the DCP.[1,2] Indeed, the DCP receives only arteriolar supply from the overlying ICP, explaining why the DCP may be at greater risk of lower perfusion pressure.[1] Macular blood supply is limited to minimize interference of light passage to the photoreceptors, putting the parafovea at particular risk of ischemic damage.[3]

Pathogenesis of paracentral acute middle maculopathy and the ischemic cascade

In 2013, the term paracentral acute middle maculopathy (PAMM) was introduced for the first time and was defined as a paracentral hyperreflective band at the level of the middle retina with OCT representing infarction of the inner nuclear layer (INL).[4] PAMM may represent the mildest form of macular ischemia in eyes with central retinal vein occlusion (CRVO), incomplete central retinal arterial occlusion (CRAO), and other forms of retinal vasculopathy.[1] The mildest forms of CRVO and CRAO are associated with perivenular PAMM, but these disorders can progress to more severe grades of macular ischemia including diffuse globular PAMM on en face OCT corresponding to a continuous hyperreflective band at the level of the INL on cross-sectional OCT.[5] In the most severe cases of occlusion, diffuse whitening of the middle and inner retina can occur.[1,5–7] This progression can take place within hours or days. The ischemic cascade refers to this progression of ischemia with time and severity. In short, ischemia starts at the level of the DCP (perivenular fern-like PAMM), progresses laterally to diffusely involve the INL (globular PAMM), and then ascends anteriorly to involve the inner retina at the level of SCP.[1,5–8] Typically, patients will present at one of these three stages of severity; identification of the evolution through two or all three stages in the same patient is unusual but can occur.

The ischemic cascade is evidence of the predominant vertical organization of the retinal capillary plexus. This has been confirmed in animal histological studies and more recently in autopsy eyes using confocal microscopy and OCT

Retinal and Choroidal Vascular Diseases of the Eye. https://doi.org/10.1016/B978-0-443-15583-3.00017-2

angiography.[1,2] The DCP is the major level of venous outflow and does not receive major arterial inflow.[2] As a result, the DCP may be subject to lower perfusion pressure or lower oxygen saturation during vascular obstruction.[2] This may be further aggravated by the increased metabolic demand of the middle retina, especially the synaptic outer plexiform layer replete with mitochondria.[9,10] As such, the middle retina may be at greatest risk of ischemic injury, especially at the venous end of the circulation, and may represent the first level of ischemic injury in the ischemic cascade or the only level of infarction in mild occlusive retinal vascular disorders.

Paracentral acute middle maculopathy

Clinical presentation

PAMM can be caused by a broad range of retinal vascular disorders especially retinal vein and retinal artery obstruction. Other less common etiologies include sickle cell disease, diabetes mellitus, hypertension, retinal vasculitis, hypotensive shock, postsurgery, and postvaccination.[1] Clinical examination and multimodal retinal imaging may provide clues to the underlying etiologic condition.

Clinical symptoms include decreased visual acuity and paracentral scotoma, which can be permanent. Visual acuity will depend on the extent of the lesions and the underlying etiology.

PAMM can be challenging to identify with ophthalmoscopy or with color fundus photography and is often only captured with OCT or near-infrared reflectance (NIR) imaging. When detectable, PAMM may be evident as perivenular retinal whitening identified with ophthalmoscopy or with color fundus photography. Unlike cotton wool spots, these lesions are deeper and grayer when present.[1,4] Multimodal retinal imaging is critical to properly capture and evaluate PAMM.

Advanced retinal imaging

Optical coherence tomography

While PAMM is poorly captured with fundus photography, fluorescein angiography (FA), and fundus auto-fluorescence (FAF), these lesions are readily identified with OCT analysis as band-like hyperreflectivity at the level of the inner nuclear layer (INL) with or without extension into the adjacent inner or outer plexiform layers (IPL, OPL), representing INL infarction. In the chronic phase, PAMM leaves a legacy of INL (and OPL) thinning, and more recently, this pathoanatomical alteration is referred to as retinal ischemic perivascular lesions (RIPL).[11] RIPL is easily identified on OCT as a focal area of irregularity in the middle retina and it is interesting that RIPL can be a biomarker of cardiovascular disease and atrial fibrillation.[11]

En face optical coherence tomography

There are three distinct patterns of PAMM with en face OCT according to Sridhar and colleagues:[12]

1) An arteriolar pattern with a band-like area of distribution along a major arteriole.
2) A fern-like pattern in which the PAMM lesions precisely colocalize with the veins.
3) A globular pattern in which the PAMM lesion has a large focal area of involvement.

The fern-like perivenular pattern of PAMM is most interesting.[5,6,12] A spectrum of perivenular PAMM is possible with en face OCT including mild cases with thin PAMM lesions noted around the veins or more severe cases of macular ischemia with diffuse lesions around the veins. In fact, progression from perivenular PAMM to a diffuse PAMM pattern or progression that includes hyperreflectivity and infarction of the inner retina can be noted in the same patient after hours or days of follow-up.[5,7,8] This progression is referred to as the ischemic cascade.

OCT angiography

The introduction of OCT angiography (OCTA) into the field of ophthalmology represents a breakthrough in retinal imaging. OCTA detects motion contrast and provides depth resolution and is a tool that, for the first time, can noninvasively image the superficial, intermediate, and deep retinal capillary plexus. Because of the development of OCTA, the etiology of PAMM lesions can be illustrated and is the result of hypoperfusion through the DCP.[13] This is in contrast to cotton wool spots which develop because of axoplasmic stasis due to hypoperfusion of the SCP.

Fluorescein angiography

FA does not adequately illustrate the intermediate and deep retinal capillary plexus, and therefore FA is a very poor modality to capture PAMM.[1,14]

But in certain cases of PAMM due to CRVO or CRAO, FA can be valuable to demonstrate delayed arteriovenous retinal transit times.[1] Because of the lack of depth resolution with FA, OCT may provide a better tool to grade macular ischemia in eyes with CRVO.[15]

Fundus autofluorescence

FAF is also not typically helpful but in some cases can capture PAMM due to hypoautofluorescent blockage. FAF can rarely demonstrate the perivenular pattern of PAMM.

Near-infrared reflectance

Near-infrared reflectance is an excellent tool that can readily detect PAMM as hyporeflective well-demarcated paracentral lesions.[1] While the sensitivity of NIR to capture PAMM is excellent, the specificity is not ideal as other entities can have similar features including AMN.

Etiologies of PAMM

Retinal vascular occlusions

The most common disorders causing PAMM are retinal vascular occlusions especially CRVO (Fig. 17.1). Rahimy et al. showed that 5% of CRVO cases show evidence of PAMM.[16] It is important to note that perivenular PAMM can also be the only ocular finding of incomplete CRAO (Fig. 17.2), partial branch retinal artery occlusion (RAO), and cilioretinal artery insufficiency/obstruction.[17–19] It is essential therefore to elicit other ocular signs to differentiate CRVO and CRAO, as recently outlined by Zhao et al.,[7] as the management can be remarkably different for each condition (Fig. 17.3).

PAMM and giant cell arteritis

There have been recent reports of PAMM in association with giant cell arteritis.[20,21] In fact, PAMM can be the only ocular sign of giant cell arteritis (GCA) and therefore the detection of this OCT signal can prevent vision loss or even blindness (Fig. 17.4).

PAMM can also be a marker of arteritic anterior ischemic optic neuropathy. The cause of PAMM in these cases may be incomplete or occult CRAO. In a retrospective study including 96 patients with GCA, Mairot et al.[22] reported ocular involvement in 52 cases (54.2%). Of these ocular cases, PAMM was identified in 16 eyes (30.8%) and was the only ocular finding in seven eyes (13.4%). As expected in GCA, the most frequent ophthalmologic finding was A-AION but the second most frequent ocular feature was PAMM.[22]

Other retinal vascular diseases and systemic associations

Other associations with PAMM include hypertension, diabetes, sickle cell disease, autoimmune diseases such as lupus, hypercoagulable states, hypovolemia or hypotensive shock, migraine disorder, febrile illness, following various surgeries or procedures (including systemic vascular or cardiovascular surgery, eye surgery, coronary angiography, and cosmetic facial or filler injection), eye compression injury causing global ocular ischemia, Purtscher's retinopathy, systemic medication use, excessive coffee use, postupper respiratory infection, and post-H1N1 vaccine.[1] Recently, PAMM has been reported in association with COVID-19 disease and even after COVID-19 vaccination.[23]

The detection of PAMM can also be an important clue to the diagnosis of very serious systemic disorders such as internal carotid dissection or following internal carotid artery stenting[24] and the etiology of PAMM in these cases may be ocular ischemia syndrome or CRAO.

Management strategies

It is critical to first determine the etiology of PAMM (Fig. 17.3).

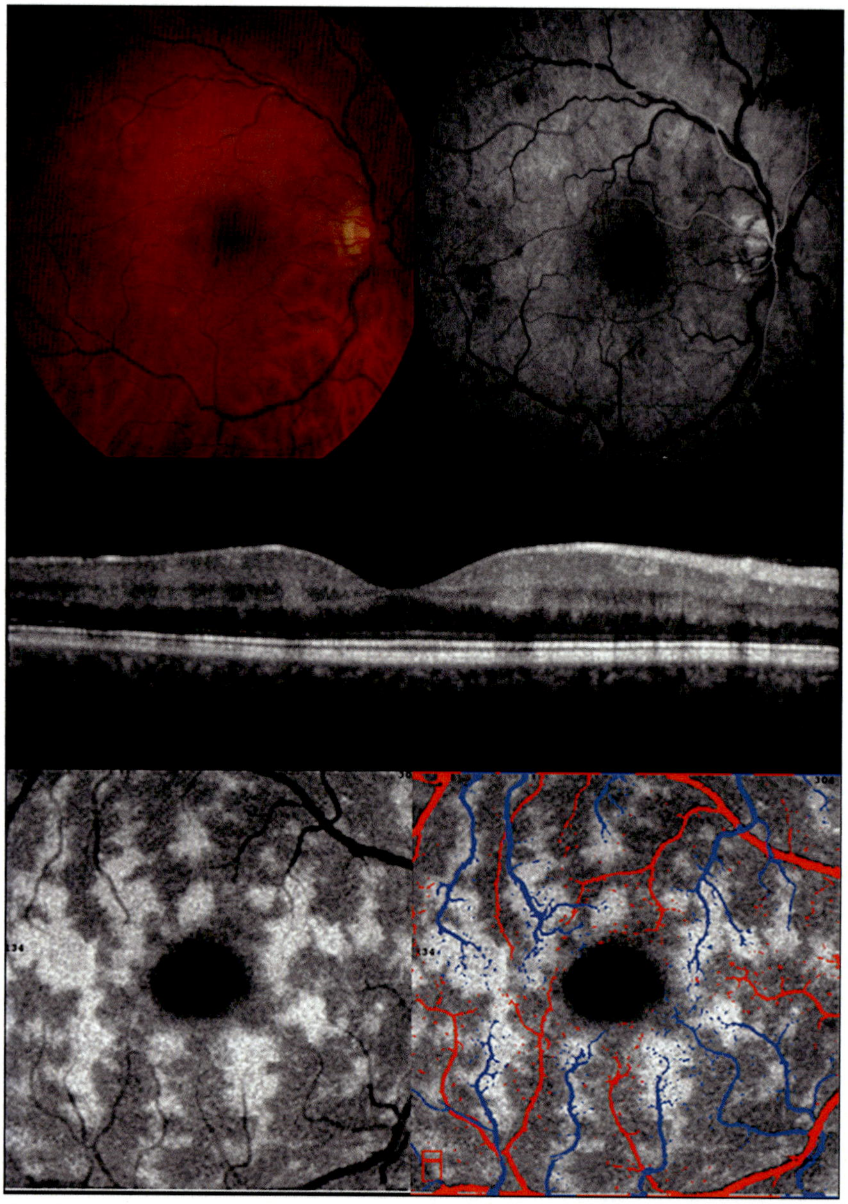

FIGURE 17.1 **Color fundus photography (*top left*) and fluorescein angiography (*top right*) of a 36-year-old male patient illustrate findings consistent with the diagnosis of central retinal vein occlusion.** Cross-sectional optical coherence tomography (OCT) demonstrates areas of inner nuclear layer (INL) hyperreflectivity consistent with paracentral acute middle maculopathy (PAMM). En face OCT (*bottom left*) segmented at the level of the INL and with vascular overlay mapping (*bottom right*) display the perivenular distribution of fern-like PAMM with periarterial sparing. *Courtesy of Falavarjani KG, Phasukkijwatana N, Freund KB et al. En face optical coherence tomography analysis to assess the spectrum of perivenular ischemia and paracentral acute middle maculopathy in retinal vein occlusion.* American Journal of Ophthalmology. *2017;177:131–138.*

A diagnosis of CRVO should elicit a work-up to rule out glaucoma and high blood pressure. Focal lesions may be the result of a branch retinal artery occlusion and careful retinal examination combined with systemic imaging is critical to rule out an embolus to the retinal arterial system. Perivenular PAMM can be the result of CRVO or CRAO. The diagnosis of CRAO will require a systemic embolic workup and baby aspirin therapy may be considered as cases of perivenular PAMM may be less commonly due to incomplete CRAO, which can rarely progress to cause more severe vision loss due to complete CRAO. As already mentioned, PAMM can be the only ocular sign of GCA, and therefore it is imperative to assess for this potentially blinding disease as systemic steroids can prevent severe vision loss in the affected or fellow eye.

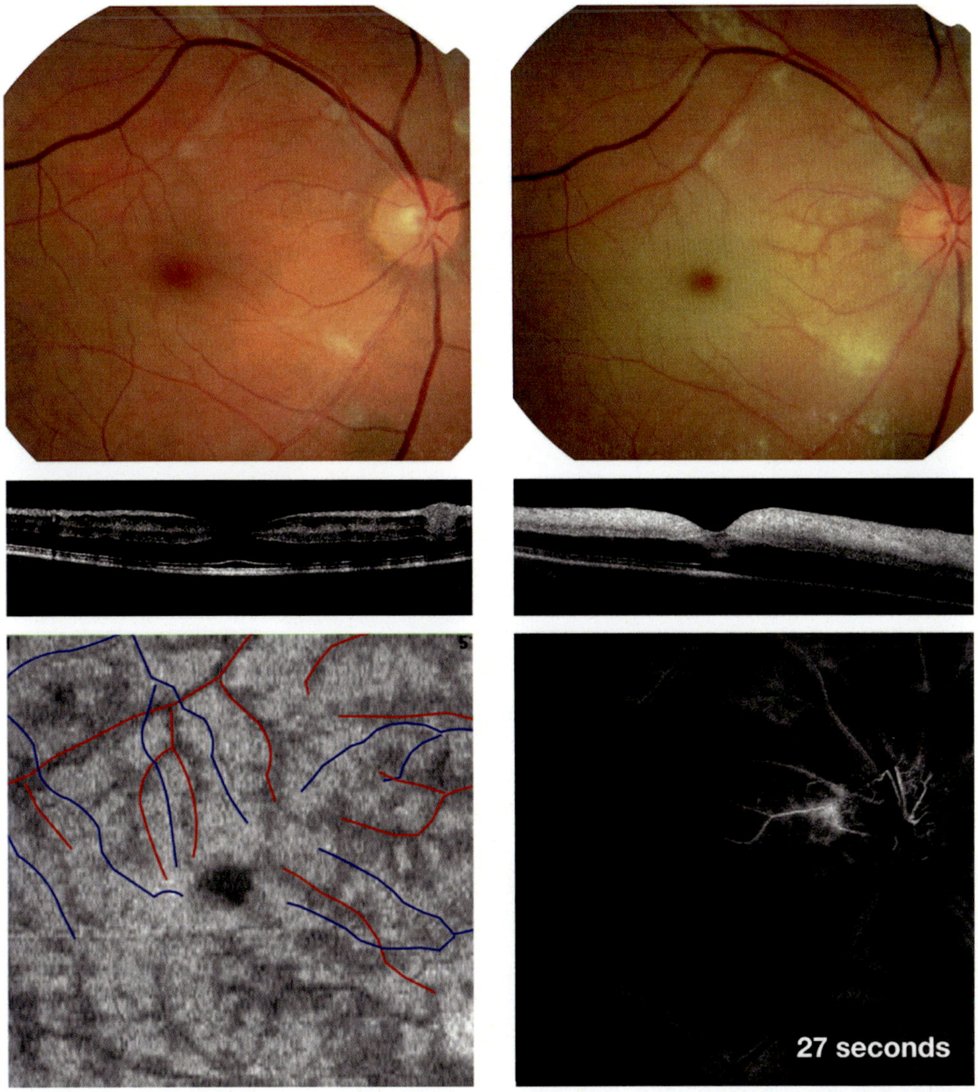

FIGURE 17.2 **Multimodal imaging of a 60-year-old hypertensive man with partial central retinal artery occlusion (CRAO) and perivenular paracentral acute middle maculopathy (PAMM) at baseline and complete CRAO with diffuse middle and inner retinal ischemia with optical coherence tomography (OCT) at follow-up.** This progression is consistent with the ischemic cascade. Fundus photography illustrates perifoveal retinal whitening (which colocalizes with PAMM on OCT) in the central macular region and cotton-wool spots in the region of the vascular arcades (*top left*) consistent with a partial CRAO. Note the presence of PAMM (hyperreflective bands at the level of inner nuclear layer (INL)) in a skip pattern on the cross-sectional OCT (*middle left*). En face OCT with vascular overlay mapping illustrates a precise perivenular fern-like pattern with periarterial sparing (*bottom left*). Five days later, the patient's vision declined to counting fingers at 1 foot. Fundus photography displays a cherry-red spot and arterial attenuation (*top right*) and progression to a complete CRAO. SD-OCT illustrates diffuse middle and inner retinal ischemia (*middle right*). Fluorescein angiography confirms disproportionate arterial delay with cilioretinal artery sparing (frame shown at 27 s) (*bottom left*). *Courtesy of Bakhoum MF, Freund KB, Dolz-Marco R et al. Paracentral acute middle maculopathy and the ischemic cascade associated with retinal vascular occlusion.* American Journal of Ophthalmology. *2018;195:143–153. https://doi.org/10.1016/j.ajo.2018.07.031*

Acute macular neuroretinopathy

Pathogenesis of acute macular neuroretinopathy

Acute macular neuroretinopathy (AMN), originally described by Bos and Deutman,[25] is characterized on color fundus photography by a dark-reddish lesion pointing toward the fovea in young women using oral contraceptives. Advanced retinal imaging, especially OCT, provides much greater insight into the pathoanatomy of this disorder, which occurs at the level of the middle and outer retina.

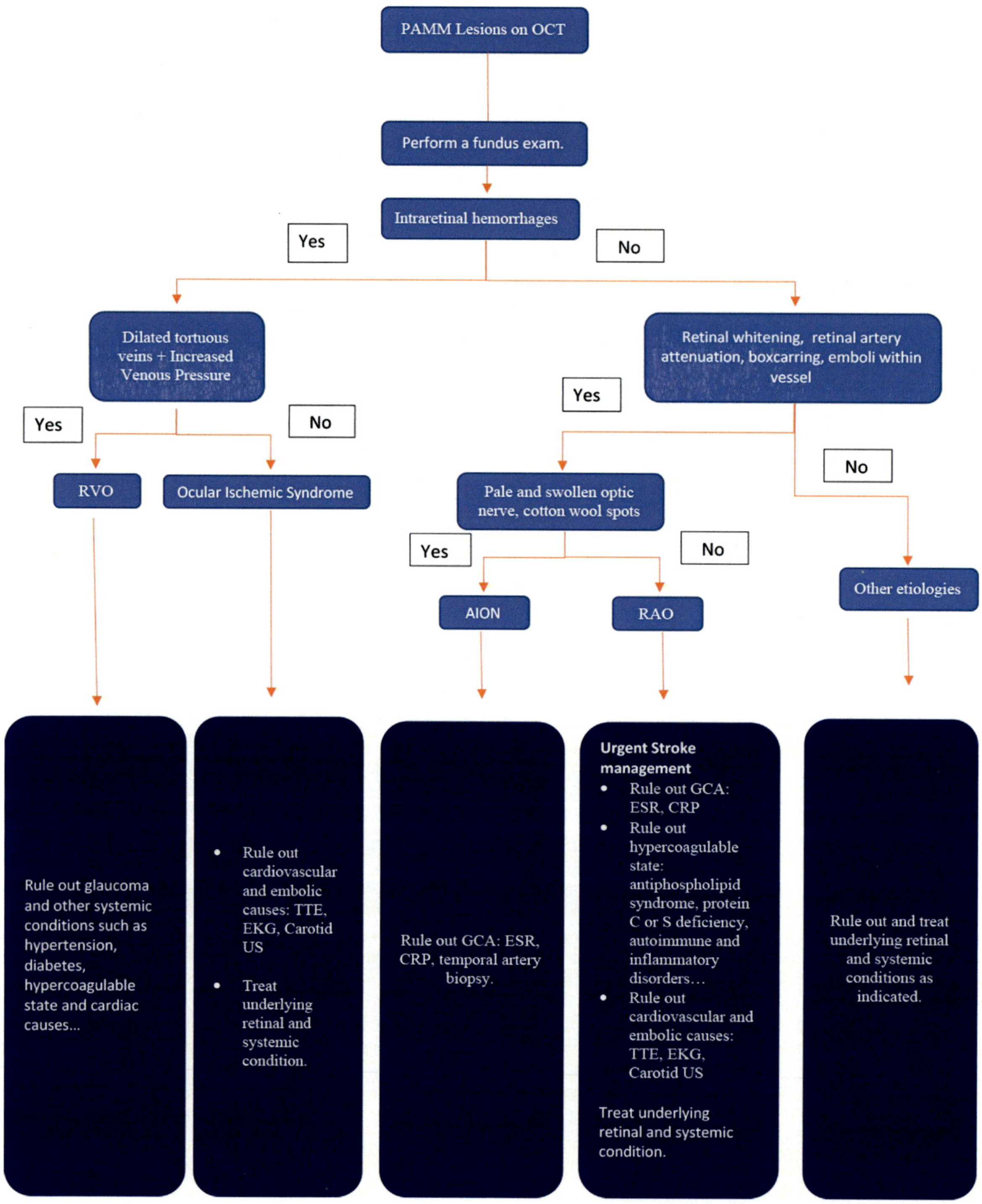

FIGURE 17.3 Algorithm for the evaluation of paracentral acute middle maculopathy (PAMM).

The exact pathophysiology, however, of AMN is unknown. While PAMM and AMN lesions typically are independent findings, they can occur coincidently and these cases have provided clues regarding the etiology of AMN (Fig. 17.5).[26] OCT and OCTA studies indicate that impaired perfusion of the DCP, at the most distal position of the collecting venules, leads to contiguous disruption of the HFL and the outer retina.[10,26,27] AMN displays ASHH with OCT, that is, an angular sign of Henle fiber layer hyperreflectivity.[10] This may result from DCP hypoperfusion, which can cause ischemia and disruption of the contiguous outer plexiform layer (OPL) and Henle fiber layer (HFL) leading to associated EZ loss.

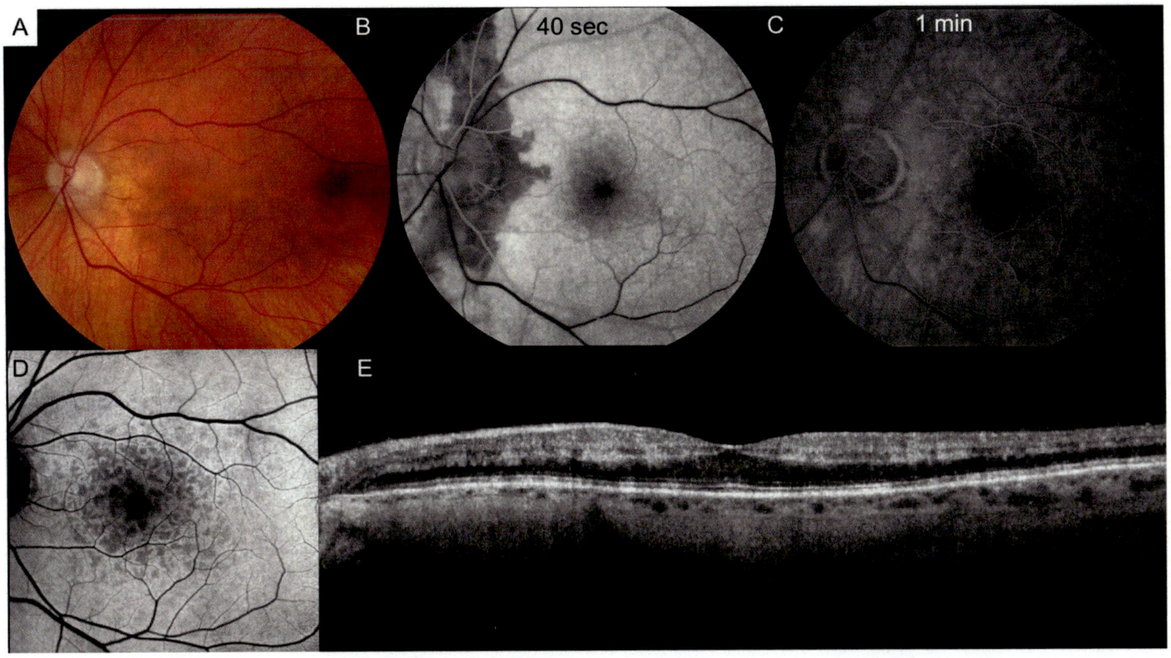

FIGURE 17.4 Multimodal retinal imaging of the left eye of a patient in his eighth decade with paracentral acute middle maculopathy (PAMM) and giant cell arteritis. (A) Color fundus photography of the left eye shows no signs of retinal vascular obstruction or optic neuropathy. (B) and (C) Fluorescein angiography shows a hypofluorescent watershed zone corresponding to the medial and lateral posterior ciliary arteries and a delayed arteriovenous transit time. (D) Fundus autofluorescence illustrates numerous round hypo-autofluorescent lesions around the fovea which colocalize with multiple PAMM lesions on cross-sectional optical coherence tomography (OCT) (E).

Therefore, DCP impairment leading to INL infarction causes PAMM. DCP impairment leading to OPL and HFL disruption causes AMN.[26]

Clinical presentation

AMN can be unilateral or bilateral. In a review of 101 cases, Bhavsar et al. reported bilateral involvement in 54% of cases.[28] AMN usually occurs in young women. In Bhavsar et al. review,[28] 84% of patients were female and the mean age was 30 years old.

Patients complain mainly of a paracentral scotoma or blurred or decreased vision. The lesions are deep in the retina, typically wedge-shaped or flower-like in shape, pointing to the fovea and located in the paracentral region of the macula. The petaloid configuration may be due to the pathoanatomical HFL involvement. These lesions are better captured with red-free photography but best identified with near-infrared reflectance imaging as a hyporeflective paracentral lesion.

Advanced retinal imaging

Optical coherence tomography

AMN displays retrograde ASHH with OCT[10] because of the pathoanatomical disruption of the HFL. Acutely, however, only the involvement of the OPL may be appreciated. Subsequent retrograde hyperreflective involvement of the HFL results in ASHH. Associated EZ loss is also typical. Resolution of ASHH in AMN is often followed by OPL/HFL and ONL thinning and varying degrees of EZ/IZ disruption (Fig. 17.5).[10,29]

OCT angiography

Projection-resolved OCTA demonstrates a reduced DCP flow signal that colocalizes with the AMN lesion on OCT.[30,31] Recently, using nonlinear transformation to correct the projection of AMN lesions in the DCP for HFL length and orientation, Cabral et al.[27] observed a high prevalence of either flow deficits in the capillary vortices or complete flow defects within the centroid of AMN lesions, supporting a distal ischemic insult at the level of the capillary vortex or collecting venule in AMN.

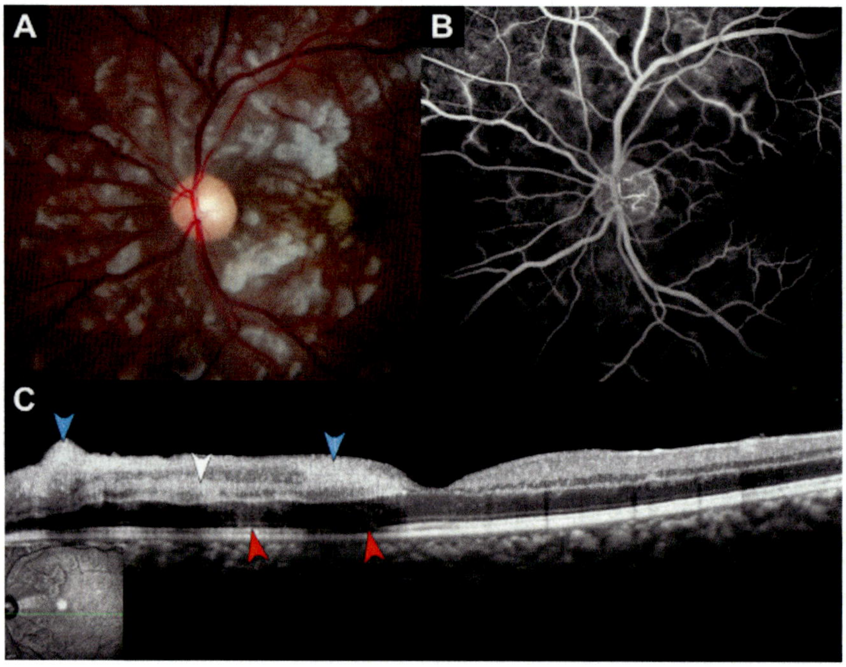

FIGURE 17.5 Coincident paracentral acute middle maculopathy (PAMM) and acute macular neuroretinopathy (AMN) in the left eye of a patient with Purtscher's retinopathy. (A) Color fundus photography shows a peripapillary distribution of superficial and deep retinal whitening consistent with Purtscher's retinopathy OS. Note the presence of cotton wool spots in the region of the arcades and deep retinal whitening (i.e., Purtscher flecken) around the fovea that corresponds to PAMM on the cross-sectional optical coherence tomography (OCT). (B) Late-phase fluorescein angiography image OS shows blocked choroidal fluorescence, areas of capillary nonperfusion, and late segmental retinal venous staining and leakage. (C) Cross-sectional OCT OS shows foci of inner retinal hyperreflectivity (*blue arrowheads*) corresponding to cotton wool spots. Diffuse PAMM lesions (*white arrowhead*) are visible in the middle retina (i.e., the inner nuclear layer) of the nasal macula and vertically colocalize with more subtle AMN lesions (*red arrowheads*) in the outer retina. The perifoveal AMN lesion shows a distinctive Z-shaped configuration and radiates along the Henle fiber layer consistent with retrograde ASHH (angular sign of Henle fiber layer hyperreflectivity). The inset (C) near-infrared reflectance image with the green line indicates the location of the OCT B-scan. *Courtesy of Iovino C, Au A, Ramtohul P et al. Coincident PAMM and AMN and insights into a common pathophysiology.* American Journal of Ophthalmology. *2022;236:136–146.*

Choriocapillaris flow deficits in AMN, as detected by prior OCTA studies, may be better explained by shadowing artifact of the hyperreflective AMN lesions,[32] although inner choroidal ischemia can also lead to outer retinal disruption and even anterograde ASHH in disorders such as acute posterior multifocal placoid pigment epitheliopathy[10] (APMPPE) and GCA.[33]

Fluorescein and indocyanine angiography

Fluorescein and indocyanine angiography (FA and ICGA) are usually unremarkable in patients with AMN although a subtle hypofluorescent aspect of the lesion on FA has been described.[28]

Fundus autofluorescence

FAF is also not helpful to diagnose AMN although a subtle hypoautofluorescence of the lesion can be captured in rare cases.[34]

Near-infrared reflectance

Near-infrared reflectance is a valuable modality that can readily detect AMN as a hyporeflective dark or gray, petaloid or wedge-shaped, paracentral lesion that correlates with the patient's scotoma and colocalizes with EZ/IZ disruption.[10,28,35]

Visual field

The most common visual field deficit is a paracentral scotoma that corresponds to the shape and location of the perifoveal lesions.[28] Although AMN is a self-limited disorder often with spontaneous recovery of visual acuity, persistent visual field deficits are reported in more than 50% of patients.[28]

Etiologies of AMN

Although several associated risk factors are reported, including the use of oral contraceptives, vasoactive substances (e.g., epinephrine compounds as in cough medicines), infectious diseases (e.g., Dengue fever), and flu-like illnesses and more recently COVID-19 infection or vaccination, systemic and laboratory workup are largely unremarkable.[28]

In recent studies of coincident PAMM and AMN, AMN versus ASHH lesions were identified in various retinal vascular disorders including CRVO, Purtscher's retinopathy, idiopathic retinal vasculitis, and even CRAO.[26,36] The appropriate systemic workup should be initiated according to the underlying diagnosis. While these cases may be accurately characterized as ASHH, the fundamental outer retinal findings are essentially the same as AMN. Therefore, AMN may be the preferred designation when there is no evidence of an underlying retinal disorder while ASHH may be secondary to a primary retinal condition such as CRVO or CRAO.

Management strategies

AMN is self-limited. Visual acuity and visual field defects improve slowly over time, although the paracentral scotoma may persist. No treatment is indicated or has proven to be effective. The appropriate systemic workup should be initiated if there is an underlying retinal disorder such as CRVO or CRAO, as was described above for PAMM.

Conclusion

PAMM is a common OCT finding that is most commonly the result of retinal vascular occlusion such as CRVO or CRAO. PAMM can also be associated with grave systemic conditions such as GCA and carotid dissection so detection with OCT can be impactful. PAMM refers to a paracentral hyperreflective band at the level of the middle retina with optical coherence tomography (OCT) that represents infarction of the inner nuclear layer (INL). It is critical to first determine the etiology of PAMM, in particular, perivenular PAMM caused by an incomplete CRAO, which should lead to an urgent stroke evaluation to rule out cardiovascular disease and to exclude giant cell arteritis in elderly patients.

Aacute macular neuroretinopathy (AMN) is a rare condition characterized by dark wedge-shaped or petaloid paracentral lesions, corresponding to retrograde hyperreflectivity of the Henle fiber layer (referred to as ASHH) and associated with ellipsoid zone (EZ) loss on OCT. Recent OCTA studies suggest that distal DCP impairment at the level of the capillary vortices or collecting venules leading to OPL and HFL disruption causes AMN.

References

1. Scharf J, Freund KB, Sadda SV, Sarraf D. Paracentral acute middle maculopathy and the organization of the retinal capillary plexuses. *Progress in Retinal and Eye Research*. 2021;81. https://doi.org/10.1016/j.preteyeres.2020.100884.
2. An D, Yu P, Freund KB, Yu D-Y, Balaratnasingam C. Three-dimensional characterization of the normal human parafoveal microvasculature using structural criteria and high-resolution confocal microscopy. *Investigative Opthalmology & Visual Science*. 2020;61(10). https://doi.org/10.1167/iovs.61.10.3.
3. Yu DY, Yu PK, Cringle SJ, Kang MH, Su EN. Functional and morphological characteristics of the retinal and choroidal vasculature. *Progress in Retinal and Eye Research*. 2014;40:53–93. https://doi.org/10.1016/j.preteyeres.2014.02.001.
4. Sarraf D, Rahimy E, Fawzi AA, et al. Paracentral acute middle maculopathy a new variant of acute macular neuroretinopathy associated with retinal capillary ischemia. *JAMA Ophthalmology*. 2013;131(10):1275–1287. https://doi.org/10.1001/jamaophthalmol.2013.4056.
5. Bakhoum MF, Freund KB, Dolz-Marco R, et al. Paracentral acute middle maculopathy and the ischemic cascade associated with retinal vascular occlusion. *American Journal of Ophthalmology*. 2018;195:143–153. https://doi.org/10.1016/j.ajo.2018.07.031.
6. Ghasemi Falavarjani K, Phasukkijwatana N, Freund KB, et al. En face optical coherence tomography analysis to assess the spectrum of perivenular ischemia and paracentral acute middle maculopathy in retinal vein occlusion. *American Journal of Ophthalmology*. 2017;177:131–138. https://doi.org/10.1016/j.ajo.2017.02.015.
7. Zhao PY, Johnson MW, McDonald HR, Sarraf D. Paracentral acute middle maculopathy and the ischemic cascade: toward interventional management. *American Journal of Ophthalmology*. 2022;234:15–19. https://doi.org/10.1016/j.ajo.2021.07.030.
8. Abtahi S-H, Nourinia R, Mazloumi M. Retinal ischemic cascade: new insights into the pathophysiology and imaging findings. *Survey of Ophthalmology*. 2022;2022(22):173.
9. Stone J, van Driel D, Valter K, Rees S, Provis J. The locations of mitochondria in mammalian photoreceptors: relation to retinal vasculature. *Brain Research*. 2008;1189(1):58–69. https://doi.org/10.1016/j.brainres.2007.10.083.
10. Ramtohul P, Cabral D, Sadda S. The OCT angular sign of Henle fiber layer (HFL) hyperreflectivity (ASHH) and the pathoanatomy of the HFL in macular disease. *Progress in Retinal and Eye Research*. 2023;95:101135.

11. Long CP, Chan AX, Bakhoum CY, et al. Prevalence of subclinical retinal ischemia in patients with cardiovascular disease — a hypothesis driven study. *EClinicalMedicine*. 2021;33. https://doi.org/10.1016/j.eclinm.2021.100775.
12. Sridhar J, Shahlaee A, Rahimy E, et al. Optical coherence tomography angiography and en face optical coherence tomography features of paracentral acute middle maculopathy. *American Journal of Ophthalmology*. 2015;160(6):1259—1268.e2. https://doi.org/10.1016/j.ajo.2015.09.016.
13. Nemiroff J, Kuehlewein L, Rahimy E, et al. Assessing deep retinal capillary ischemia in paracentral acute middle maculopathy by optical coherence tomography angiography. *American Journal of Ophthalmology*. 2016;162:121—132.e1. https://doi.org/10.1016/j.ajo.2015.10.026.
14. Weinhaus RS, Burke JM, Delori FC, Snodderly DM. Comparison of fluorescein angiography with microvascular anatomy of macaque retinas. *Experimental Eye Research*. 1995;61(1):1—16. https://doi.org/10.1016/S0014-4835(95)80053-0.
15. Au A, Hilely A, Scharf J, et al. Relationship between nerve fiber layer hemorrhages and outcomes in central retinal vein occlusion. *Investigative Opthalmology & Visual Science*. 2020;61(5). https://doi.org/10.1167/iovs.61.5.54.
16. Rahimy E, Sarraf D, Dollin ML, Pitcher JD, Ho AC. Paracentral acute middle maculopathy in nonischemic central retinal vein occlusion. *American Journal of Ophthalmology*. 2014;158(2):372—380.e1. https://doi.org/10.1016/j.ajo.2014.04.024.
17. Yu S, Pang CE, Gong Y, et al. The spectrum of superficial and deep capillary ischemia in retinal artery occlusion. *American Journal of Ophthalmology*. 2015;159(1):53—63.e2. https://doi.org/10.1016/j.ajo.2014.09.027.
18. Iyer PG, Swaminathan SS, Trivizki O, et al. Widefield en face optical coherence tomography monitoring of the peri-venular fern-like pattern of paracentral acute middle maculopathy. *American Journal of Ophthalmology Case Reports*. 2021;22. https://doi.org/10.1016/j.ajoc.2021.101047.
19. Pichi F, Fragiotta S, Freund KB, et al. Cilioretinal artery hypoperfusion and its association with paracentral acute middle maculopathy. *British Journal of Ophthalmology*. 2019;103(8):1137—1145. https://doi.org/10.1136/bjophthalmol-2018-312774.
20. Ahuja AS, El-Dairi MA, Hadziahmetovic M, Gospe SM. Paracentral acute middle maculopathy as a manifestation of giant cell arteritis. *J Neuro Ophthalmol: The Official Journal of the North American Neuro-Ophthalmology Society*. 2021;41(2):e153—e156. https://doi.org/10.1097/WNO.0000000000001170.
21. Pellegrini F, Brocca D, Calandriello L, Cuna A, Leonardi F, Lee AG. Paracentral acute middle maculopathy as the presenting manifestation of giant cell arteritis. *Journal of Neuro-Ophthalmology*. 2022. https://doi.org/10.1097/wno.0000000000001483.
22. Mairot K, Sené T, Lecler A, et al. Paracentral acute middle maculopathy in giant cell arteritis. *Retina*. 2022;42(3):476—484. https://doi.org/10.1097/iae.0000000000003339.
23. Dutta Majumder P, Agarwal A. Acute macular neuroretinopathy and paracentral acute middle maculopathy during SARS-CoV-2 infection and vaccination. *Vaccines*. 2023;11(2). https://doi.org/10.3390/vaccines11020474.
24. Alsberge JB, McDonald HR. Paracentral acute middle maculopathy following internal carotid artery stenting. *American Journal of Ophthalmology Case Reports*. 2022;28. https://doi.org/10.1016/j.ajoc.2022.101704.
25. Bos PJM, Deutman AF. Acute macular neuroretinopathy. *American Journal of Ophthalmology*. 1975;80(4):573—584. https://doi.org/10.1016/0002-9394(75)90387-6.
26. Iovino C, Au A, Ramtohul P, et al. Coincident PAMM and AMN and insights into a common pathophysiology. *American Journal of Ophthalmology*. 2022;236:136—146. https://doi.org/10.1016/j.ajo.2021.07.004.
27. Cabral D, Ramtohul P, Zatreanu L, et al. Deep capillary plexus features in acute macular neuroretinopathy: novel insights based on the anatomy of Henle fiber layer. *Investigative Opthalmology & Visual Science*. 2022;63(13). https://doi.org/10.1167/iovs.63.13.4.
28. Bhavsar KV, Lin S, Rahimy E, et al. Acute macular neuroretinopathy: a comprehensive review of the literature. *Survey of Ophthalmology*. 2016;61(5):538—565. https://doi.org/10.1016/j.survophthal.2016.03.003.
29. Fawzi AA, Pappuru RR, Sarraf D, et al. Acute macular neuroretinopathy: long-term insights revealed by multimodal imaging. *Retina*. 2012;32(8):1500—1513. https://doi.org/10.1097/IAE.0b013e318263d0c3.
30. Chu S, Nesper PL, Soetikno BT, Bakri SJ, Fawzi AA. Projection-resolved OCT angiography of microvascular changes in paracentral acute middle maculopathy and acute macular neuroretinopathy. *Investigative Ophthalmology and Visual Science*. 2018;59(7):2913—2922. https://doi.org/10.1167/iovs.18-24112.
31. Nemiroff J, Sarraf D, Davila JP, Rodger D. Optical coherence tomography angiography of acute macular neuroretinopathy reveals deep capillary ischemia. *Retinal Cases and Brief Reports*. 2018;12:S12—S15.
32. Thanos A, Faia LJ, Yonekawa Y, Randhawa S. Optical coherence tomographic angiography in acute macular neuroretinopathy. *JAMA Ophthalmology*. 2016;134(11):1310—1314. https://doi.org/10.1001/jamaophthalmol.2016.3513.
33. Garrity ST, Holz EJ, Sarraf D. Amalric triangular syndrome associated with outer nuclear layer infarction. *Ophthalmic Surgery Lasers and Imaging Retina*. 2017;48(8):668—670. https://doi.org/10.3928/23258160-20170802-10.
34. Yeh S, Hwang TS, Weleber RG, Watzke RC, Francis PJ. Acute Macular Outer Retinopathy (AMOR): a reappraisal of acute macular neuroretinopathy using multimodality diagnostic testing. *Archives of Ophthalmology*. 2011;129(3):365—368. https://doi.org/10.1001/archophthalmol.2011.22.
35. Casalino G, Arrigo A, Romano F, Munk MR, Bandello F, Parodi MB. Acute macular neuroretinopathy: pathogenetic insights from optical coherence tomography angiography. *British Journal of Ophthalmology*. 2019;103(3):410—414. https://doi.org/10.1136/bjophthalmol-2018-312197.
36. Sarrasin E, Malclès A, Sarraf D. Acute macular neuroretinopathy secondary to central retinal artery occlusion. *American Journal of Ophthalmology Case Reports*. 2023;29. https://doi.org/10.1016/j.ajoc.2023.101793.

Section IV

Choroidal vascular diseases

Chapter 18

Pachychoroid syndrome

Claudio Iovino

Eye Clinic, Multidisciplinary Department of Medical, Surgical and Dental Sciences, University of Campania Luigi Vanvitelli, Naples, Italy

Introduction

The term pachychoroid (Greek: παχύ-, pachy-, thick) refers to a choroidal phenotype shared by several disorders, which unifies them into a spectrum.

Warrow and colleagues in 2013 reported a series of eyes showing retinal pigment epithelium (RPE) alterations associated with choroidal thickening in the absence of subretinal fluid and coined for the first time the term pachychoroid pigment epitheliopathy (PPE) to describe these findings.[1]

The typical choroidal features of the pachychoroid disease spectrum (PDS) on multimodal imaging evaluation include reduced fundus tessellation, focal or diffuse choroidal thickening, dilated Haller's layer vessels (termed "pachyvessels") with thinning of the overlying inner choroid, and RPE abnormalities.[2]

Continuous progress in choroidal imaging techniques, including enhanced depth imaging optical coherence tomography (EDI-OCT), swept-source optical coherence tomography (SS-OCT), en face OCT, and optical coherence tomography angiography (OCTA) have facilitated an in-depth analysis of choroidal vasculature.[3] The diffusion of these novel imaging modalities has allowed a new and more precise definition of the pachychoroid phenotype to emphasize the morphology of pachyvessels over absolute choroidal thickness.

Although the etiology of pachychoroid disorders remains controversial, choroidal abnormalities are believed to play an important pathogenic role in the development of the following clinical manifestations that reside within the PDS: central serous chorioretinopathy (CSC), PPE, pachychoroid neovasculopathy (PNV), polypoidal choroidal vasculopathy (PCV)/aneurysmal type 1 neovascularization (AT1N), focal choroidal excavation (FCE), and peripapillary pachychoroid syndrome (PPS).

Regarding the pathogenesis of the PDS, recent evidence suggests a sequence wherein choroidal dysfunction may be followed by RPE alterations and ultimately, in some cases neovascularization. The latter is thought to be caused by the loss of the inner choroid, which produces an ischemic environment.[4,5]

More specifically, choroidal thickening is thought to be secondary to the expansion of the pachyvessels, which may lead to compressive damage in the overlying tissues, such as choriocapillaris, Bruch's membrane, and RPE.

Nevertheless, it is not well established whether the occurrence of the pachyvessels is a primary or secondary event to the inner choroidal attenuation.[2]

Moreover, it is still unclear why some eyes with typical pachychoroid disease phenotype show no evidence of RPE damage and subretinal fluid while others present progressive chorioretinal damage, neovascularization, and eventually atrophy.

Older literature also proposes an aberrant steroid metabolism as an upstream factor, with mineralocorticoid receptors expressed in the choroid. Stimulation of these receptors increases choroidal thickness and congestion and the effect is reversed by mineralocorticoid receptor antagonists.[6]

A correct diagnosis of pachychoroid disorders is important, especially the differentiation between the presence of an active neovascularization or a purely exudative stage. A complete multimodal imaging evaluation is essential to distinguish the different entities within the PDS.

Several treatment strategies have been proposed for pachychoroid disorders, depending on the presence of intra/subretinal fluid and eventually a neovascularization.[5,7]

Retinal and Choroidal Vascular Diseases of the Eye. https://doi.org/10.1016/B978-0-443-15583-3.00018-4

The aim of the present chapter is to briefly discuss the main clinical and imaging features of pachychoroid diseases. The disease's natural course and types of management will also be discussed. We will mainly focus on CSC, PPE, PNV, and FCE, whereas PCV and PPS will be discussed in dedicated chapters.

Central serous chorioretinopathy

CSC is the most common disorder of the PDS. The disorder usually occurs unilaterally in males aged between 30 and 50 as a self-limiting disease that tends to resolve with a good visual prognosis. Acute CSC is the most common form of presentation and manifests as localized neurosensory retinal detachment and/or pigment epithelium detachment (PED) that tends to spontaneously regress in 4–6 months (Fig. 18.1A and B).

However, about 5%–10% of patients develop recurrent or persistent subretinal fluid. These patients classified as having the chronic form of CSC may present serous detachments that gravitate inferiorly, resulting in pigment epithelial atrophic tracts.[8]

Patients with chronic CSC usually complain of decreased vision, relative scotoma, and micropsia.

The venous choroidal congestion associated with delayed arterial perfusion has been recognized for more than 20 years as a cardinal sign of CSC. Indeed, an abnormal regulation of the choroidal blood circulation could be one of the major driving mechanisms leading to "pachychoroid" features.

Multimodal imaging evaluation

Color picture features of patients with acute CSC include the presence of a localized neurosensory retinal detachment and/or PED. The chronic forms are characterized by retinal pigmentary changes with or without subretinal fluid.

Fundus autofluorescence (FAF) may show no abnormalities in the acute forms. A hypoautofluorescence can be detected at the initial presentation corresponding to the leakage site on fluorescein angiography (FA), and a faint hypoautofluorescence corresponding to the area of subretinal fluid.[9]

Serous PEDs appear as focal areas of hyperautofluorescence. A granular pattern of hyperautofluorescenceis also typical of acute CSC.

The chronic forms of CSC typically present descending gravitational tracts of a variable hypoautofluorescence, areas of diffuse hyperautofluorescence corresponding to outer retinal atrophy, and a granular hypoautofluorescence (Fig. 18.2A and B).[10]

A variety of patterns can be seen on FA in acute CSC including the inkblot pattern and the smokestack pattern. In chronic cases, FA often shows multiple leakage areas and diffuse RPE window defects with patchy hyperfluorescence because of RPE atrophy (Fig. 18.3A and B).

Indocyanine green angiography (ICGA) is a useful imaging modality for PDS e specifically for CSC. The benefit of ICGA over FA alone is the ability to demonstrate choroidal abnormalities that cannot be detected with FA. The principal

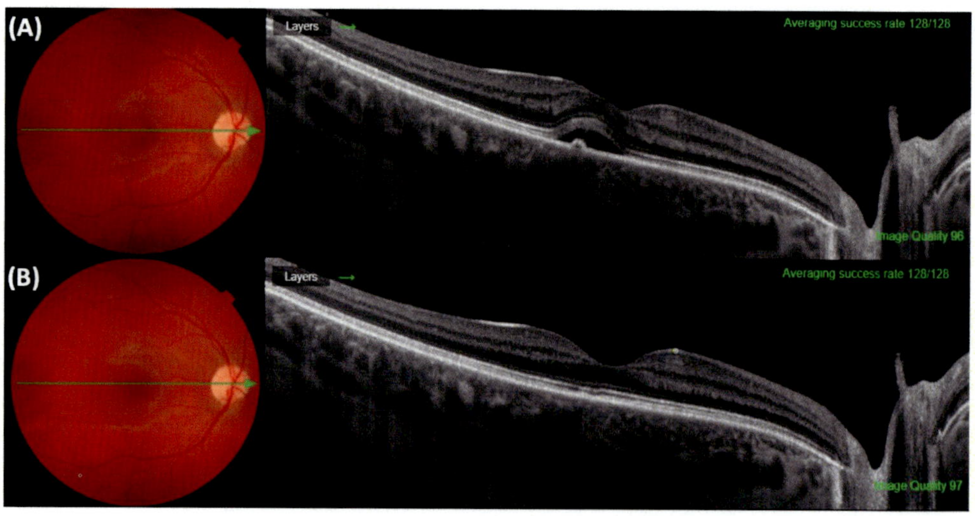

FIGURE 18.1 Acute central serous chorioretinopathy case of a 34-year-old man. (A) Baseline color picture with corresponding swept-source optical coherence tomography (SS-OCT) showing a subfoveal neurosensory retinal detachment. Note the dilated choroidal Haller layer vessels. (B) Three months follow-up color picture with corresponding SS-OCT demonstrating the spontaneous reabsorption of the subretinal fluid.

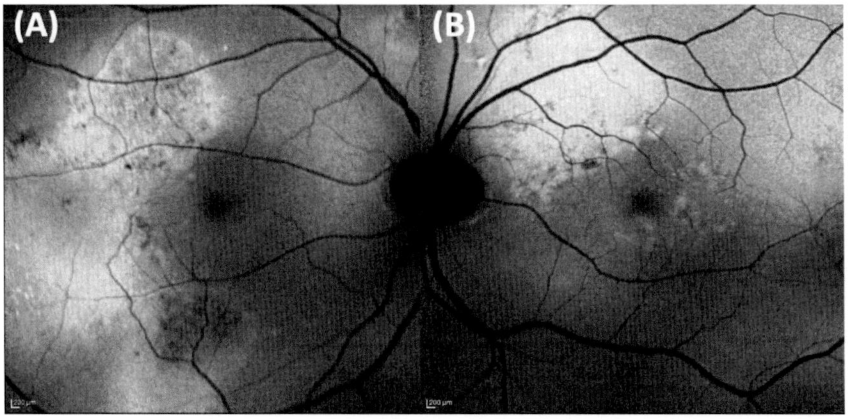

FIGURE 18.2 Fundus autofluorescence (FAF) of a 55-year-old man with an 8-year history of chronic central serous chorioretinopathy. (A) Right eye FAF shows areas of hyper-autofluorescence at the posterior pole with the characteristic descending gravitational tracts. (B) Left eye FAF shows a hyperautofluorescence along the superotemporal vascular arcade and a mottled hyperautofluorescence in the macular area.

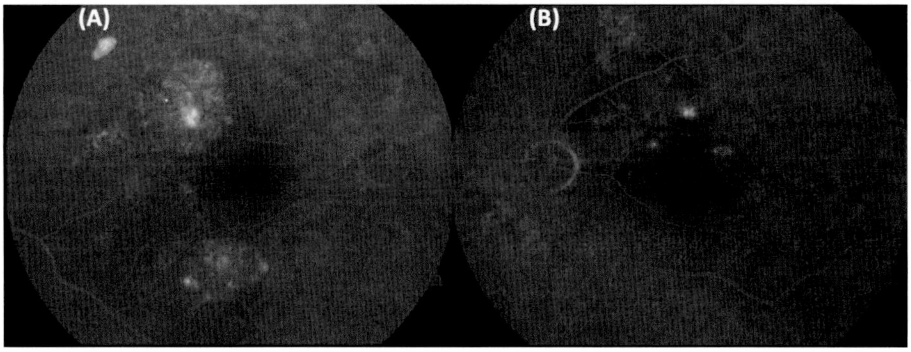

FIGURE 18.3 Fluorescein angiography (FA) of the same patient of Fig. 18.2. (A) Right eye and (B) left eye FA showing multiple hyperfluorescent areas corresponding to the outer retina and RPE atrophy and few leakage points.

choroidal vascular abnormalities of acute forms detectable by ICGA are choroidal filling delay, venous dilations, focal or diffuse choroidal hyperfluorescence, and hypofluorescent points/areas.

The chronic forms are characterized by a more extensive and diffuse retinal pigment epitheliopathy, which is identified by ICGA as multifocal areas of choroidal vascular hyperpermeability, not necessarily having clinical and FA counterparts (Fig. 18.4A–C).[11]

Technologies such as EDI-OCT or SS-OCT frequently can demonstrate a thickened choroid (pachychoroid) underlying the neurosensory retinal detachment (Fig. 18.5). Although the choroid seems to be diffusely enlarged in CSC patients, they also tend to show focal thickening of large vessels. OCT signs of acute and chronic CSC are summarized in Table 18.1.

Several studies investigated the role of OCTA in assessing CSC patients.[12–15] It has been reported that no clear OCTA signs were associated with the leaking point and that choroidal flow might be either normal or impaired. Nevertheless, OCTA appeared very useful to detect any choroidal neovascularization complicating chronic CSC (Fig. 18.6A–D), with a sensitivity as good as FA and ICGA.[16,17]

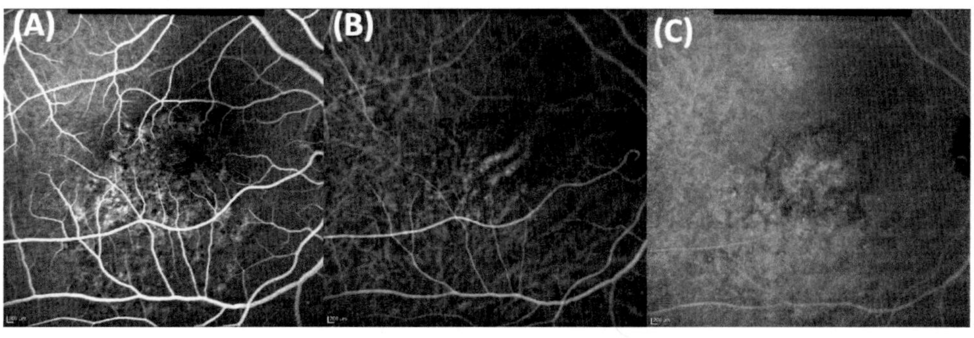

FIGURE 18.4 Fluorescein angiography (FA) and indocyanine green angiography (ICGA) of a 58-year-old man with chronic central serous chorioretinopathy. (A) Early-phase FA showing granular hyperfluorescence in the macular area. (B) Early-phase ICGA showing choroidal filling delay with venous dilation, and (C) late-phase ICGA confirming a macular area of hyperpermeability. *Courtesy of E. Peiretti.*

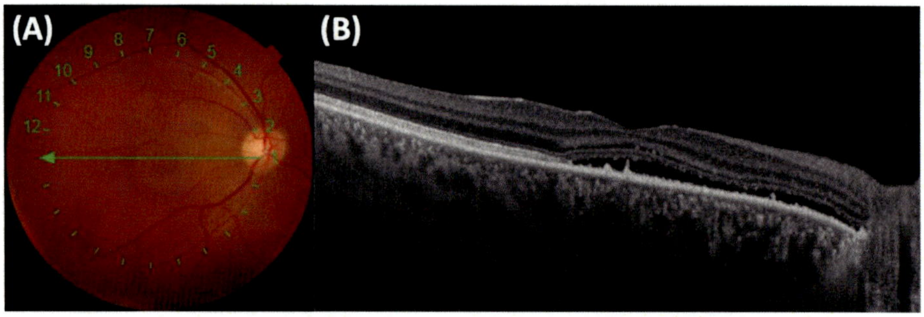

FIGURE 18.5 Representative case of a 44-year-old man with a 3-year history of treatment naïve chronic central serous chorioretinopathy. (A) Color picture with corresponding swept-source optical coherence tomography (SS-OCT) showing a subfoveal neurosensory retinal detachment. Note the thicker choroid and the hyperreflective spots in the choroidal stroma.

TABLE 18.1 Key optical coherence tomography signs in acute and chronic central serous chorioretinopathy.

	Acute Central Serous Chorioretinopathy	*Chronic Central Serous Chorioretinopathy*
Choroid	• Increased thickness • Increased Haller's layer thickness • Reduced Sattler's and choriocapillaris	• Increased thickness • Increased Haller's layer thickness • Reduced Sattler's and choriocapillaris • Thickened vessel walls • Hyperreflective spots in the stroma
Retinal pigment epithelium	• Serous PED • RPE phagosomes • Excrescences • Local epitheliopathy	• Serous PED • Flat PED • Excrescences • Local epitheliopathy • Diffuse epitheliopathy • RPE apertures
Neurosensory retina	• Neurosensory detachment • Elongation of photoreceptor outer segments	• Neurosensory detachment • Elongation of photoreceptor outer segments • Loss of the ellipsoid band • Disappearance of outer segments + thinning of ONL • Hyperreflective intraretinal spots • Intraretinal cysts • Outer retinal tubulations • Subretinal hyperreflective material

ONL, outer nuclear layer; *PED*, pigment epithelium detachment; *RPE*, retinal pigment epithelium.

Management

The hallmark feature of CSC is the presence of subretinal fluid, which is also the main target of any treatment option. Nevertheless, as previously mentioned, the acute form is self-limiting and the neurosensory retinal detachment usually disappears spontaneously. On the other hand, in chronic CSC, the subretinal fluid may persist for months, or years, leading to RPE atrophy. Several treatment options including conventional laser, subthreshold micropulse laser, photodynamic therapy (PDT), intravitreal antivascular endothelial growth factor (anti-VEGF), oral agents (mineralocorticoid antagonist, antibiotic, carbonic anhydrase inhibitor, etc.), and other minor therapies (transpupillary thermotherapy, patching, acupuncture etc.) have been proposed.[18]

Based on efficacy and safety data from the prospective multicenter–randomized controlled PLACE trial, half-dose (or half-fluence), PDT should be considered the treatment of choice for chronic CSC (Fig. 18.7A–D).[19]

Evidence also suggests that CSC cases with choroidal neovascularization should be treated using intravitreal injections of anti-VEGF compounds and/or half-dose or half-fluence PDT.

Subtreshold micropulse laser remains a good treatment option in cases where PDT cannot be performed (Fig. 18.8A and B).[18]

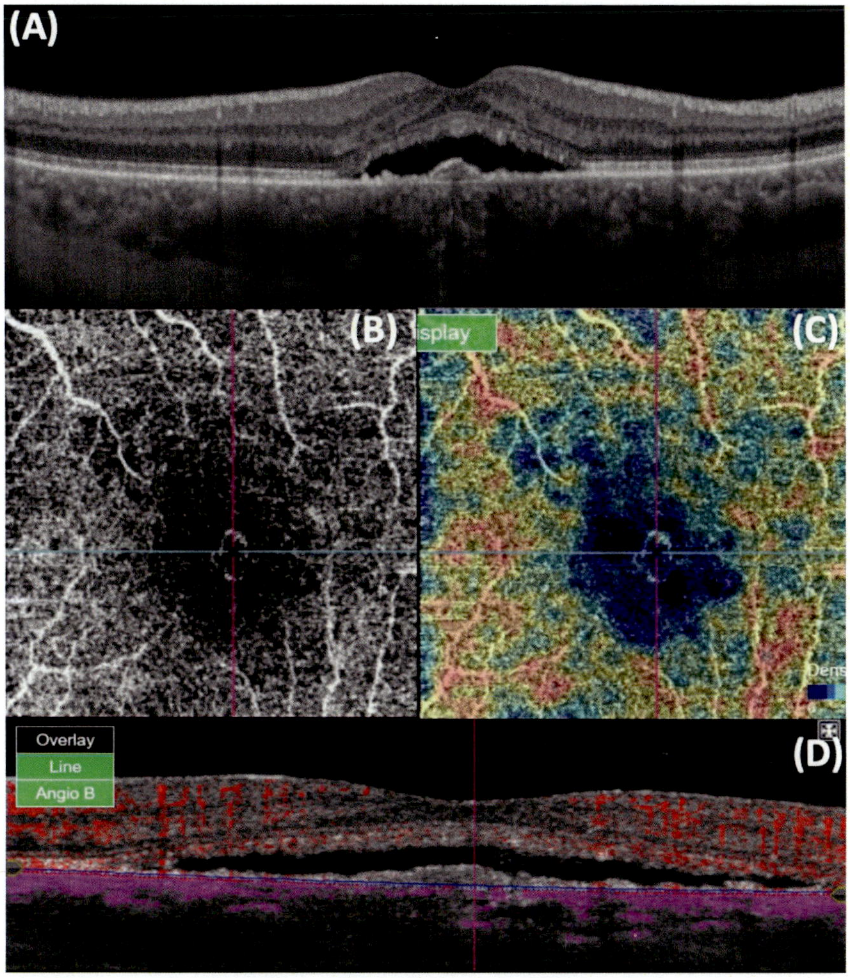

FIGURE 18.6 Representative case of a 54-year-old man with a 2-year history of treatment naïve chronic central serous chorioretinopathy complicated with type 1 neovascularization. (A) Swept-source optical coherence tomography (SS-OCT) showing a subfoveal neurosensory retinal detachment with underlying flat pigment epithelium detachment. Note the thicker choroid with dilated Haller choroidal vessels. (B) Macular 3 × 3 optical coherence tomography angiography (OCTA) scan with corresponding flow density map (C) showing a small vascular network in the foveal region. (D) B scan OCTA showing the choriocapillaris segmentation slab.

Pachychoroid pigment epitheliopathy

PPE is a relatively new disease concept characterized by dilation of choroidal vessels, increased choroidal thickness, and choroidal vascular hyperpermeability, associated with RPE abnormalities.[1] It may be isolated or may occur in the fellow eyes of patients with other pachychoroid spectrum disorders.

Patients with PPE are usually asymptomatic and can be misdiagnosed with age-related macular degeneration (AMD), pattern dystrophy, or inflammatory chorioretinopathies.

However, the clinical appearance of the pigment epitheliopathy includes mottling of the RPE, irregular areas of RPE elevation termed "drusenoid RPE lesions" or "pachydrusen," and an absence of soft drusen seen in eyes with AMD.[20,21]

PPE is also considered as a form *fruste* of CSC,[1] since the subretinal fluid is absent. These eyes may progress to CSC over time[22] or even develop a PCV/AT1N.[23]

Multimodal imaging evaluation

Eyes with PPE usually exhibit decreased tessellation and increased pigmentation in the macular area at fundus examination. Specifically, the RPE abnormalities including RPE elevations, small PED, or isolated pachydrusen may resemble AMD or pattern dystrophy-like retinal changes.

FAF reflects the RPE abnormalities as hyper- and hypofluorescent changes with a mottled or reticular pattern.

Choroidal ischemia is seen as delayed arterial filling of the choroid on FA and ICGA. Choroidal hyperpermeability is the main ICGA feature of PPE, and the hypercyanescent areas are located in the areas of RPE alterations (Fig. 18.9A).[24]

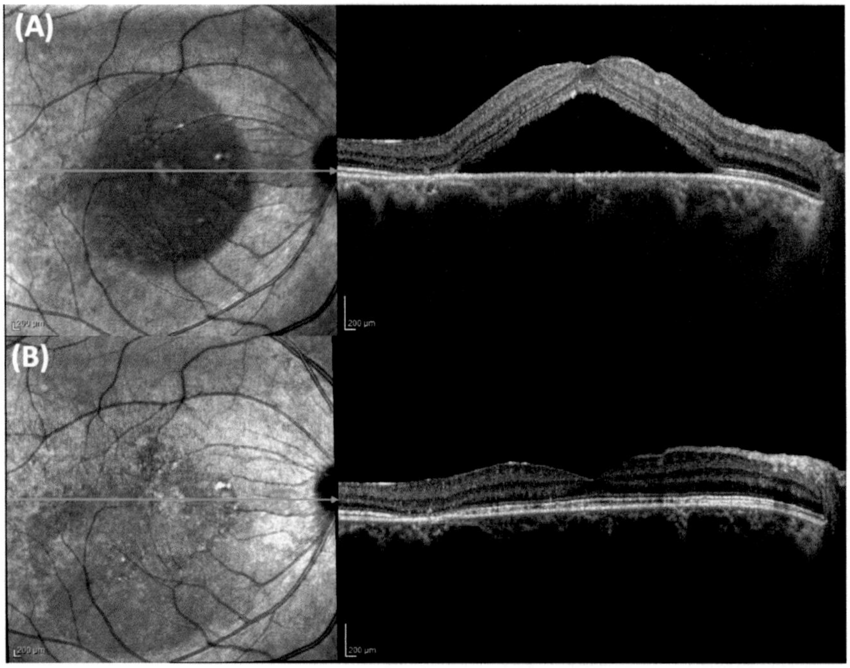

FIGURE 18.7 Representative case of a 44-year-old man with chronic central serous chorioretinopathy receiving photodynamic therapy (PDT). (A) Baseline near-infrared (NIR) image with corresponding enhanced depth imaging (EDI) spectral domain optical coherence tomography (SD-OCT) B scan displays a significant amount of subretinal fluid and a thicker choroid. (B) Three months after full dose half-fluence PDT, NIR with corresponding tracked EDI SD-OCT scan shows a complete reabsorption of the subretinal fluid.

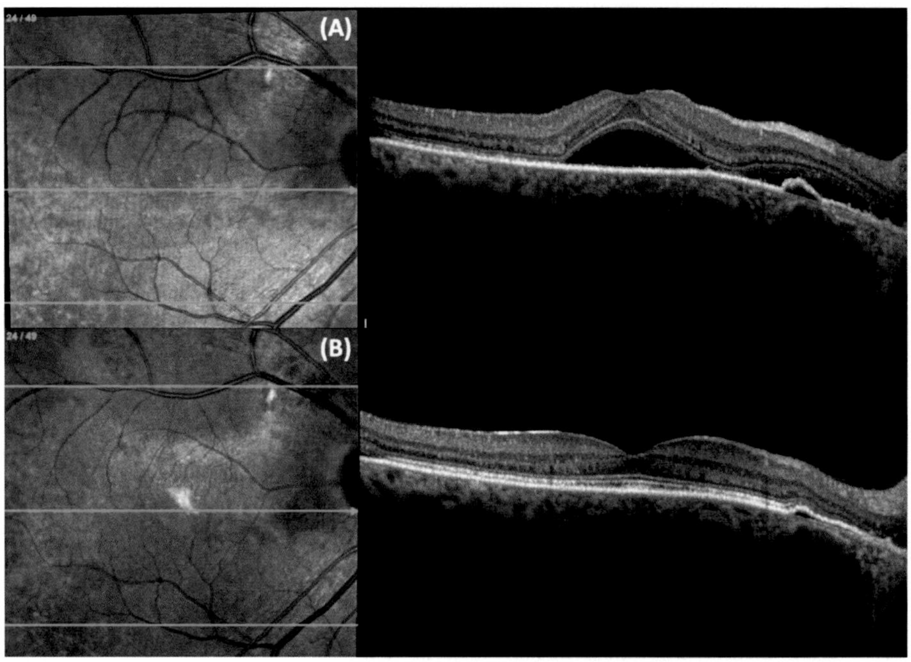

FIGURE 18.8 Representative case of a 35-year-old man with chronic central serous chorioretinopathy receiving navigated *yellow* subthreshold micropulse laser (YSML). (A) Baseline near-infrared (NIR) image with corresponding enhanced depth imaging (EDI) spectral domain optical coherence tomography (SD-OCT) B scan displays a macular subretinal fluid, with a small pigment epithelium detachment. (B) Three months after YSML treatment, NIR with corresponding tracked EDI SD-OCT scan shows a complete reabsorption of the subretinal fluid.

OCT evaluation is mandatory to investigate all the typical features of PDS. The hallmark OCT sign of PPE is the presence of the pachyvessels (dilated Haller layer vessels), better visualized with SS-OCT or EDI-OCT, that compress on the overlying Sattler and choriocapillaris layers (Fig. 18.9B).[22] Moreover, pigment epitheliopathy can lead to photoreceptors atrophy with a consequent outer nuclear layer thinning.[25] En face OCT segmented below the Bruch membrane can also nicely show the distribution and morphology of pachyvessels in PPE.

In the same correspondent retinal area, above the pachyvessels, different RPE and outer retina changes may be shown on B scan evaluation, including:

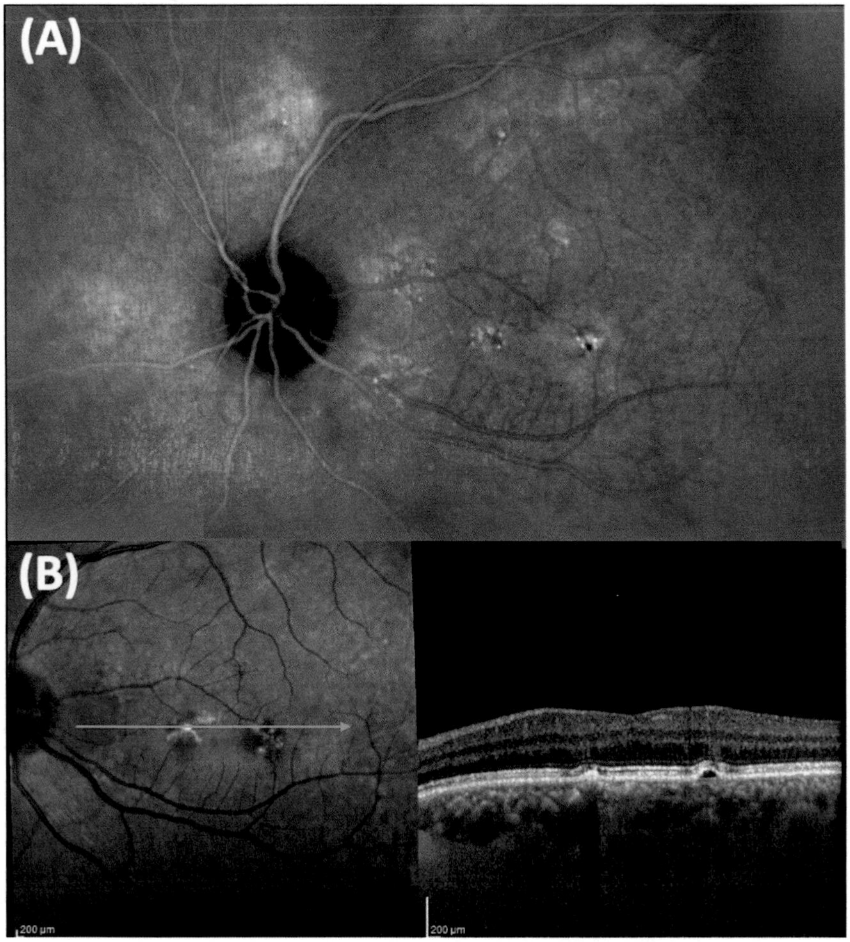

FIGURE 18.9 Pachychoroid pigment epitheliopathy in a 44-year-old asymptomatic patient. (A) Late-phase indocyanine *green* angiography shows multiple areas of choroidal hyperpermeability. (B) Near-infrared image with corresponding spectral domain optical coherence tomography B scan displays small retinal pigment epithelium elevations with underlying dilated choroidal vessels. *Courtesy of S. Demirel.*

- Small RPE elevation or shallow irregular PED (Fig. 18.10)
- Serous PED without any evidence of subretinal fluid
- Isolated or multiple drusen-like hyperreflective deposits under the RPE also called "pachydrusen"
- Very small fissure appearance between RPE cells, also named "microbreak," often associated with slight detachments of the RPE[22]
- Focal disruption of the ellipsoid and/or the external limiting membrane
- Outer nuclear layer thinning

OCTA can show a high prevalence of focal inner choroidal flow signal attenuation areas in eyes with PPE, and most of these zones are related anatomically to pachyvessels. It has been postulated that these flow void areas represent zones of inner choroidal ischemia.[26,27]

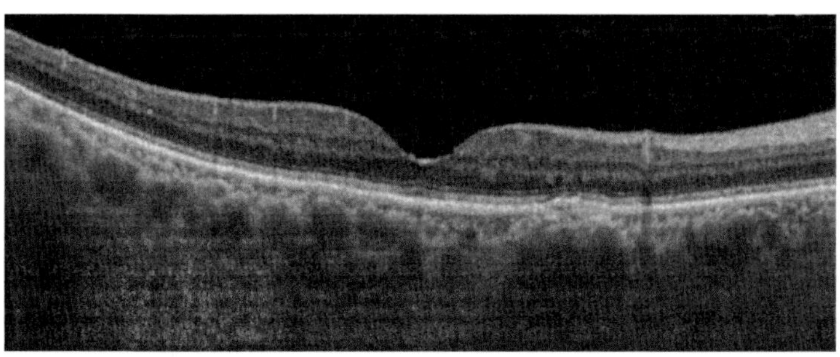

FIGURE 18.10 Swept-source optical coherence tomography in a case of pachychoroid pigment epitheliopathy. Note the flat retinal pigment epithelium elevation with an underlying pachyvessels.

Management

Considering the absence of subretinal fluid, most cases of PPE do not need any kind of treatment.

Nevertheless, observation is required due to the possibility of progression to CSC, PNV, or even PCV.

Pachychoroid neovasculopathy

Differentiating PDS disorders from AMD can sometimes be challenging as the two conditions may have very similar features on FA and ICGA. Indeed, PNVs also share some features typical for occult neovascularization in AMD, although drusen are usually absent.[5] Pang and Freund described the development of type 1 neovascularization in three eyes over background changes consistent with PPE and introduced the term "pachychoroid neovasculopathy" to describe this entity.[28]

This kind of type 1 neovascularization usually occurs overlying the focal areas of choroidal thickening, and it was named PNV to distinguish the pachychoroid-driven pathophysiologic mechanism of this disease process from other causes of type 1 neovascularization.[5] PNV can convert to PCV for the onset of polyps in about 13% of cases, and younger age and sustained choroidal thickening were identified as risk factors for conversion.[29]

Multimodal imaging evaluation

Color photographs of eyes with PNV show reduced tessellation indicative of thickened choroid, associated with a reddish appearance in the area of the neovascular lesion. No signs of AMD are usually present. Subretinal fibrin may be found in association with the subretinal fluid occurring in PNV, but hemorrhages and lipid exudation are less common.[20]

FAF usually fails to exhibit zonal areas of hyper/hypoautofluorescence or gravitational tracts. A mottled hyperautofluorescence surrounding the neovascular tissue where the subretinal fluid accumulation is located may be shown.

The presence of PNV can be confirmed by the detection of a late leakage on FA, with a corresponding late staining "plaque" on ICGA. Aneurysmal (polypoidal) lesions may be also identified with ICGA. Moreover, similar CSC angiographic signs may be detected, as a result of diffuse RPE disturbance and choroidal hyperpermeability.

Characteristic OCT lesions include the double layer sign (a shallow separation of the RPE from the Bruch's membrane) overlying pachyvessels, small peaked PEDs with heterogeneously hyperreflective material, and subretinal and/or intraretinal fluid as exudative manifestation (Fig. 18.11A–D). Subretinal hemorrhages can be recognized as high-density subretinal hyperreflective material.

PNV may also be displayed noninvasively by means of OCTA as a tangled network of signals between the RPE and Bruch's membrane (Fig. 18.12A and B). OCTA showed a superior detection rate for identifying PNV over dye angiography (both FA and ICGA).[30]

Management

Eyes with PNV respond favorably to intravitreal anti-VEGF therapy with a significantly lower dependence on repeated injections if compared to typical AMD cases. This may be related to a lower intraocular VEGF concentration in PNV eyes compared to AMD.[31] Nevertheless, some eyes with PNV can be refractory to intravitreal therapy alone. Combined anti-VEGF + PDT therapy appears useful in these cases, with good anatomical and functional results.[32]

Tanaka et al. also demonstrated PDT monotherapy to be a valuable treatment option in eyes with PNV, with 89% of patients requiring no additional intravitreal injections.[33]

Focal choroidal excavation

The term FCE was first coined by Margolis et al. in 2011 to describe a new idiopathic entity characterized by one or more focal areas of choroidal excavation without evidence of posterior staphyloma or scleral ectasia.[34] This finding is usually present unilaterally and in one single retinal location (typically macular area); though exceptions to these patterns have been reported.[35] The disease is more frequently diagnosed in the fourth to fifth decade, shows no gender predilection, and tends to be asymptomatic. Patients usually are myopic. The exact pathogenetic mechanism leading to FCE formation is not yet fully understood, but it has been speculated that it may be formed by RPE retraction due to focal scarring of choroidal tissue.

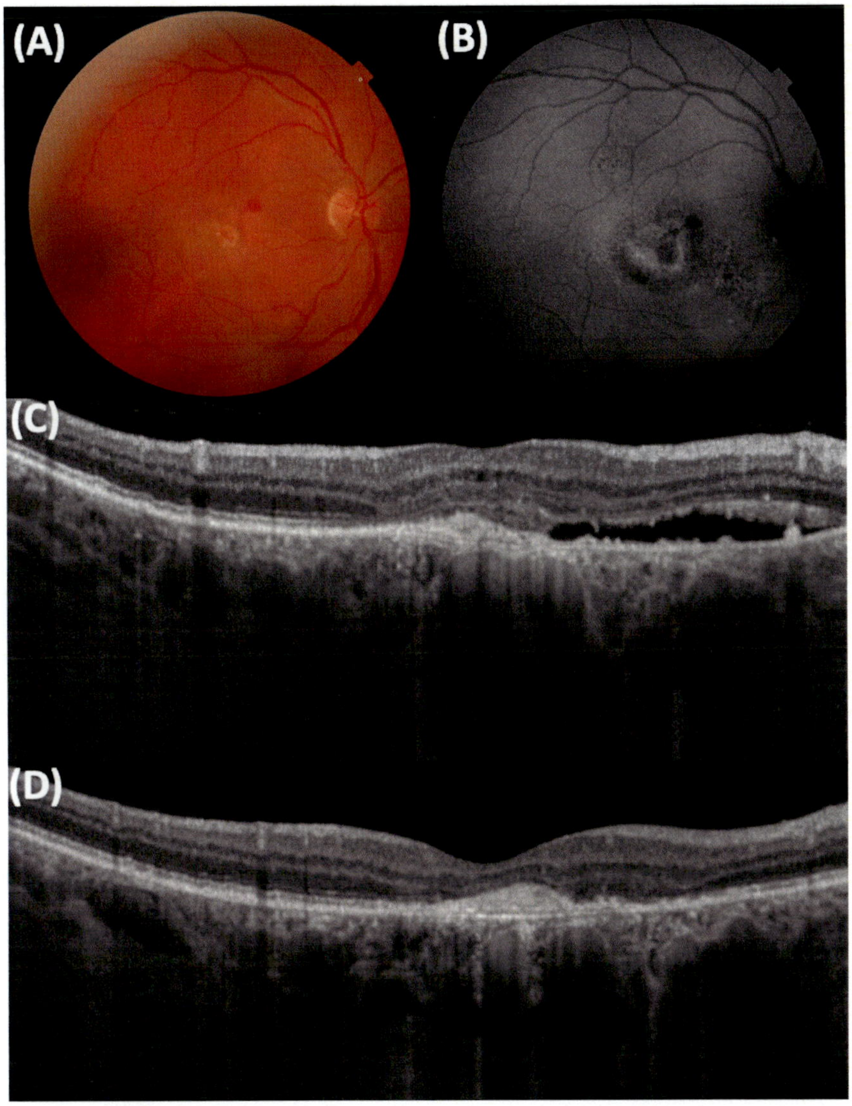

FIGURE 18.11 Multimodal imaging in a case of pachychoroid neovasculopathy. (A) Color picture shows reduced tessellation, with subfoveal yellowish material and a perifoveal hemorrhage. (B) Fundus autofluorescence displays a mottled hyperautofluorescence surrounding the neovascular tissue with a hypofluorescent spot corresponding to intraretinal hemorrhage. (C) Swept-source optical coherence tomography (SS-OCT) shows some hyperreflective subretinal material, and subretinal fluid as exudative manifestation. Note the underlying pachyvessels. (D) Tracked SS-OCT after three intravitreal antivascular endothelial growth factor shows reabsorption of the subretinal fluid.

Multimodal imaging evaluation

Color fundus picture may be normal or show some nonspecific pigmentary changes or yellowish spots in an area or reduced fundus tessellation.

FAF may show either hypoautofluorescence or hyperautofluorescence in the area where the FCE is located.

FA shows varying degrees of hypofluorescence or hyperfluorescence as a result of RPE changes. Hyperfluorescence (window defect) related to an RPE attenuation/atrophy is frequent.

Different patterns of hypocyanescence with abnormal staining on late phases are shown on ICGA, suggesting choriocapillaris loss. Other pachychoroid typical features on ICGA have been described in eyes with FCE such as choroidal vascular hyperpermeability.[36]

OCT is the best imaging modality to visualize FCE. It shows the RPE band following the contour of the choroidal excavation. OCT also allows us to distinguish the FCE into two forms[34]:

- Conforming FCE: there is no separation between the photoreceptor tips and the RPE, and the outer nuclear layer usually appears thicker compared with the retina areas not affected by excavation.
- Nonconforming FCE: the photoreceptor tips appear to be detached from the underlying RPE with intervening hyporeflective spaces representing subretinal fluid

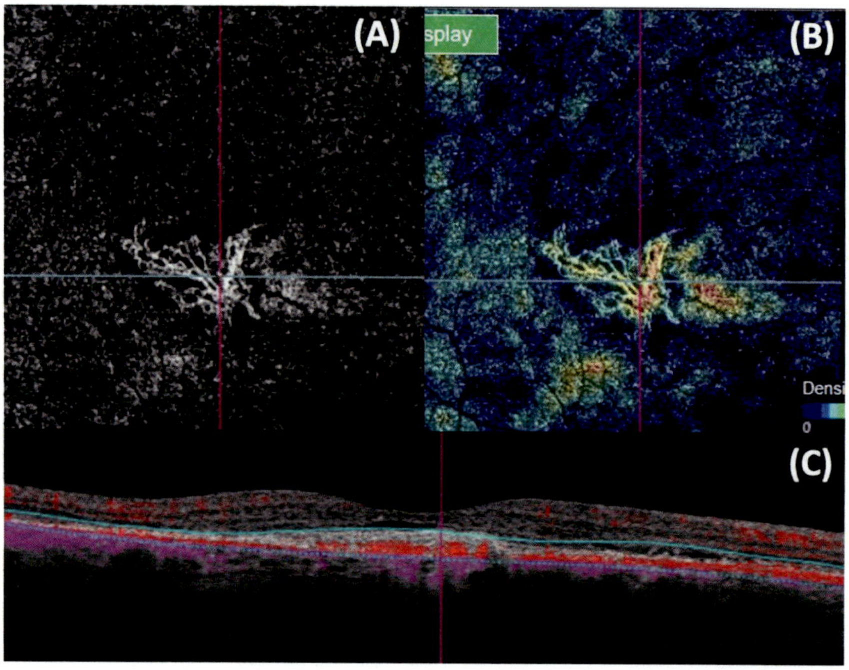

FIGURE 18.12 Baseline optical coherence tomography angiography (OCTA) of the same case of Fig. 18.11 (pachychoroid neovasculopathy). (A) Macular 3 × 3 OCTA scan with corresponding flow density map (B) clearly shows a foveal neovascular tissue with a perilesional dark halo. (C) OCTA B scan shows the segmentation slab and the *red* flow within the subretinal hyperreflective material.

The alterations at the level of RPE or choroid may lead to the development of choroidal neovascularization, which may be either type 1 or type 2. The neovascular tissue usually located within the boundary of the FCE.

OCTA is a good noninvasive, alternative imaging technique for the diagnosis of neovascularization and should be considered as an essential imaging tool in FCE diagnosis and follow-up. Demirel et al. concluded that OCTA seems to be a superior technique to ICGA with regard to revealing neovascularization over the areas that show pachychoroid characteristics.[37]

Management

FCE usually does not require any kind of treatment unless a choroidal neovascularization occurs. Patients with an active exudation related to the presence of neovascular tissue show a satisfactory response to anti-VEGF monotherapy with good visual improvement, even though SRF persistence is described.[38]

Conclusions

The clinical knowledge about all disorders within the PDS has significantly improved in the last few years. Recent advances in multimodal imaging have enabled a better understanding of all the choroidal features of the pachychoroid phenotype.

Although all diseases within the PDS share some common features, it is important to distinguish each single entity for the different clinical courses and treatments needed as previously discussed. It is still unclear why some eyes with thick choroids and pachyvessels show no evidence of RPE alterations or subretinal fluid (uncomplicated pachychoroid).

Future studies are necessary to better understand the abnormal choroidal blood flow and venous congestion typical of the PDS and to better clarify several areas of uncertainty.

References

1. Warrow DJ, Hoang QV, Freund KB. Pachychoroid pigment epitheliopathy. *Retina*. 2013;33(8):1659–1672. https://doi.org/10.1097/IAE.0b013e3182953df4.
2. Mazzeo TJMM, Leber HM, da Silva AG, et al. Pachychoroid disease spectrum: review article. *Graefe's Archive for Clinical and Experimental Ophthalmology*. 2022;260(3):723–735. https://doi.org/10.1007/s00417-021-05450-3.

3. Borrelli E, Sarraf D, Freund KB, Sadda SR. OCT angiography and evaluation of the choroid and choroidal vascular disorders. *Progress in Retinal and Eye Research*. 2018;67:30–55. https://doi.org/10.1016/j.preteyeres.2018.07.002.
4. Dansingani KK, Gal-Or O, Sadda SR, Yannuzzi LA, Freund KB. Understanding aneurysmal type 1 neovascularization (polypoidal choroidal vasculopathy): a lesson in the taxonomy of 'expanded spectra' – a review. *Clinical and Experimental Ophthalmology*. 2018;46(2):189–200. https://doi.org/10.1111/ceo.13114.
5. Pang CE, Freund KB. *Retina*. 2015.
6. Zhao M, Célérier I, Bousquet E, et al. Mineralocorticoid receptor is involved in rat and human ocular chorioretinopathy. *Journal of Clinical Investigation*. 2012;122(7):2672–2679. https://doi.org/10.1172/JCI61427.
7. Xu, Garg, Lee K, Sakurada Y, Amphornphruet, Phasukkijwatana, et al. Long-term visual and anatomic outcomes of patients with peripapillary pachychoroid syndrome. *British Journal of Ophthalmology*. 2022;29.
8. Fung AT, Yang Y, Kam AW. Central serous chorioretinopathy: a review. *Clinical and Experimental Ophthalmology*. 2023;51(3):243–270. https://doi.org/10.1111/ceo.14201.
9. Iacono P, Battaglia PM, Papayannis A, La Spina C, Varano M, Bandello F. Acute central serous chorioretinopathy: a correlation study between fundus autofluorescence and spectral-domain OCT. *Graefe's Archive for Clinical and Experimental Ophthalmology*. 2015;253(11):1889–1897. https://doi.org/10.1007/s00417-014-2899-5.
10. Zola M, Chatziralli I, Menon D, Schwartz R, Hykin P, Sivaprasad S. Evolution of fundus autofluorescence patterns over time in patients with chronic central serous chorioretinopathy. *Acta Ophthalmologica*. 2018;96(7):e835–e839. https://doi.org/10.1111/aos.13742.
11. Peiretti E, Iovino C. Indocyanine green angiography. *Central Serous Chorioretinopathy*. 2019:97–113. https://doi.org/10.1016/B978-0-12-816800-4.00009-7.
12. Cennamo G, Montorio D, Comune C, Clemente L, Iovino C, Carandente R. Study of vessel density by optical coherence tomography angiography in patients with central serous chorioretinopathy after low-fluence photodynamic therapy. *Photodiagnosis Photodynamic Therapy*. 2020;30.
13. Sahoo NK, Mishra SB, Iovino C, et al. Optical coherence tomography angiography findings in cystoid macular degeneration associated with central serous chorioretinopathy. *British Journal of Ophthalmology*. 2019;103(11):1615–1618. https://doi.org/10.1136/bjophthalmol-2018-313048.
14. Mandadi S, Singh SR, Sahoo NK, Mishra SB, Sacconi R, Iovino C. Optical coherence tomography angiography findings in fellow eyes of choroidal neovascularisation associated with central serous chorioretinopathy. *British Journal of Ophthalmology*. 2019;105:1280–1285.
15. Peiretti E, Iovino C, Sacconi R, Caminiti G, Querques G. Optical coherence tomography angiography characteristics of polypoidal choroidal vasculopathy secondary to chronic central serous chorioretinopathy. *Retina*. 2019;39(9):1693–1700. https://doi.org/10.1097/iae.0000000000002234.
16. Bonini Filho MA, De Carlo TE, Ferrara D, et al. Association of choroidal neovascularization and central serous chorioretinopathy with optical coherence tomography angiography. *JAMA Ophthalmology*. 2015;133(8):899–906. https://doi.org/10.1001/jamaophthalmol.2015.1320.
17. Bansal R, Dogra M, Mulkutkar S, et al. Optical coherence tomography angiography versus fluorescein angiography in diagnosing choroidal neovascularization in chronic central serous chorioretinopathy. *Indian Journal of Ophthalmology*. 2019;67(7):1095–1100. https://doi.org/10.4103/ijo.IJO_1238_18.
18. Rijssen TJ, Dijk, Yzer, et al. Central serous chorioretinopathy: towards an evidence-based treatment guideline. *Progress in Retinal and Eye Research*. 2019;73.
19. van Dijk EHC, Fauser S, Breukink MB, et al. Half-dose photodynamic therapy versus high-density subthreshold micropulse laser treatment in patients with chronic central serous chorioretinopathy: the PLACE trial. *Ophthalmology*. 2018;125(10):1547–1555. https://doi.org/10.1016/j.ophtha.2018.04.021.
20. Cheung CMG, Lee WK, Koizumi H, Dansingani K, Lai TYY, Freund KB. Pachychoroid disease. *Eye*. 2019;33(1):14–33. https://doi.org/10.1038/s41433-018-0158-4.
21. Spaide RF. Disease expression in nonexudative age-related macular degeneration varies with choroidal thickness. *Retina*. 2018;38(4):708–716. https://doi.org/10.1097/IAE.0000000000001689.
22. Karacorlu M, Ersoz MG, Arf S, Hocaoglu M, Sayman Muslubas I. Long-term follow-up of pachychoroid pigment epitheliopathy and lesion characteristics. *Graefe's Archive for Clinical and Experimental Ophthalmology*. 2018;256(12):2319–2326. https://doi.org/10.1007/s00417-018-4144-0.
23. Tang J, Han X, Tang R, et al. Case series: pachychoroid pigment epitheliopathy transformed to polypoidal choroidal vasculopathy after long-term follow-up. *BMC Ophthalmology*. 2022;22(1). https://doi.org/10.1186/s12886-022-02487-8.
24. Giray Ersoz M, Arf S, Hocaoglu M, Muslubas IS, Karacorlu M. Indocyanine green angiography of pachychoroid pigment epitheliopathy. *Retina*. 2017;38(9):1668–1674. https://doi.org/10.1097/IAE.0000000000001773.
25. Ersoz MG, Karacorlu M, Arf S, Hocaoglu M, Muslubas IS. Outer nuclear layer thinning in pachychoroid pigment epitheliopathy. *Retina*. 2018;38(5):957–961. https://doi.org/10.1097/IAE.0000000000001655.
26. Gal-Or O, Dansingani KK, Sebrow D, Dolz-Marco R, Bailey Freund K. Inner choroidal flow signal attenuation in pachychoroid disease: optical coherence tomography angiography. *Retina*. 2018;38(10):1984–1992. https://doi.org/10.1097/IAE.0000000000002051.
27. Demirel S, Değirmenci MFK, Batöoğlu F, Özmert E. Evaluation of the choroidal features in pachychoroid spectrum diseases by optical coherence tomography and optical coherence tomography angiography. *European Journal of Ophthalmology*. 2021;31(1):184–193. https://doi.org/10.1177/1120672119887095.
28. Pang CE, Freund KB. Pachychoroid pigment epitheliopathy may masquerade as acute retinal pigment epitheliitis. *Investigative Ophthalmology and Visual Science*. 2014;55(8). https://doi.org/10.1167/iovs.14-14959.

29. Siedlecki J, Klaas JE, Keidel LF, et al. Progression of pachychoroid neovasculopathy into aneurysmal type 1 choroidal neovascularization or polypoidal choroidal vasculopathy. *Ophthalmology Retina*. 2022;6(9):807−813. https://doi.org/10.1016/j.oret.2022.04.004.
30. Su Y, Zhang X, Gan Y, Zeng Y, Wen F. Detection of pachychoroid neovasculopathy with optical coherence tomography angiography versus dye angiography imaging. *Photodiagnosis and Photodynamic Therapy*. 2022;40. https://doi.org/10.1016/j.pdpdt.2022.103126.
31. Hata M, Yamashiro K, Ooto S, et al. Intraocular vascular endothelial growth factor levels in pachychoroid neovasculopathy and neovascular age-related macular degeneration. *Investigative Ophthalmology and Visual Science*. 2017;58(1):292−298. https://doi.org/10.1167/iovs.16-20967.
32. Miki A, Kusuhara S, Otsuji T, et al. Photodynamic therapy combined with anti-vascular endothelial growth factor therapy for pachychoroid neovasculopathy. *PLoS One*. 2021;16(3). https://doi.org/10.1371/journal.pone.0248760.
33. Tanaka K, Mori R, Wakatsuki Y, Onoe H, Kawamura A, Nakashizuka H. Two-thirds dose photodynamic therapy for pachychoroid neovasculopathy. *Journal of Clinical Medicine*. 2021;10(10). https://doi.org/10.3390/jcm10102168.
34. Margolis R, Mukkamala SK, Jampol LM, et al. The expanded spectrum of focal choroidal excavation. *Archives of Ophthalmology*. 2011;129(10):1320−1325. https://doi.org/10.1001/archophthalmol.2011.148.
35. Braimah IZ, Rapole S, Dumpala S, Chhablani J. Focal choroidal excavation in retinal dystrophies. *Seminars in Ophthalmology*. 2018;33(2):161−166. https://doi.org/10.1080/08820538.2016.1182634.
36. Shinojima A, Kawamura A, Mori R, Yuzawa M. Morphologic features of focal choroidal excavation on spectral domain optical coherence tomography with simultaneous angiography. *Retina*. 2014;34(7):1407−1414. https://doi.org/10.1097/iae.0000000000000108.
37. Demirel S, Beğar PG, Yanõk Ö, Batõoğlu F, Özmert E. Visualization of type-1 macular neovascularization secondary to pachychoroid spectrum diseases: a comparative study for sensitivity and specificity of indocyanine green angiography and optical coherence tomography angiography. *Diagnostics*. 2022;12(6). https://doi.org/10.3390/diagnostics12061368.
38. Szabelska P, Medrzycka J, Brydak-Godowska J, Różycki R, Gołebiewska J. Focal choroidal excavation: epidemiology, clinical characteristics and multimodal imaging findings. *Diagnostics*. 2023;13(4). https://doi.org/10.3390/diagnostics13040602.

Chapter 19

Venous overload choroidopathy

Arman Zarnegar[1], Joshua Ong[2], Sumit Randhir Singh[3] and Jay Chhablani[1]
[1]Department of Ophthalmology, University of Pittsburgh School of Medicine, Pittsburgh, PA, United States; [2]Michigan Medicine, University of Michigan, Ann Arbor, MI, United States; [3]Sri Sai Eye Hospital, Patna, Bihar, India

Introduction

Several pathologic conditions have recently been proposed to be caused by venous overload choroidopathy. Venous overload choroidopathy is increased pressure within the choroidal venous system due to abnormal outflow that can result in remodeling and anastomosis. Downstream, there can be negative effects on the retinal pigment epithelium (RPE) and neurosensory retina that cause visual impairment. Herein, the relevant anatomy and physiology of choroidal vasculature and proposed mechanisms of venous overload choroidopathy are reviewed.

Overview of the choroid

The choroid is a tetra-layered subretinal structure that hosts arteries, capillaries, and veins that perfuse and drain blood from the outer retina. On the external border of the choroid lies the suprachoroid (a transitional zone), then the sclera, through which short posterior ciliary arteries enter, and vortex veins exit the globe. The outermost layers of the choroid, Haller and Sattler layers, contain large- and medium-diameter vessels, respectively. Sattler layer lies adjacent to the choriocapillaris, a 7—10 μm thick structure through which highly anastomosed choroidal capillaries course.[1] Bruch membrane separates the choroid from the RPE and neurosensory retina. Appropriate choroidal function is paramount to maintaining photoreceptor homeostasis.

Oxygenated blood enters choroidal circulation from the posterior ciliary arteries, branches of the ophthalmic artery. The posterior ciliary arteries branch into the medial and lateral posterior ciliary arteries, which give rise to short posterior ciliary arteries. The short posterior ciliary arteries pass through the sclera to ultimately reach the choriocapillaris. They run in a segmental fashion, as can be observed in Haller and Sattler layers. There is little anastomosis present between these layers.

Once blood arrives at the choriocapillaris, nutrients, waste, and oxygen, pass through fenestrations in the capillary walls and diffuse to and from the photoreceptors. Fluid extrusion from the retina toward the choroid is likely primarily dependent on choroidal vascular flow rather than on an RPE pump mechanism.[2] Bruch membrane and the RPE act as barriers to prevent undesired molecules from passing out of circulation. Blood from the choriocapillaris next drains into venules, followed by larger veins, culminating with the vortex veins.

The majority of the population possesses four to five vortex veins, though some studies indicate the number to be around eight (Fig. 19.1).[3—5] Their purpose is to drain venous blood from the choroidal circulation to either the cavernous sinus or the pterygoid plexus.[6] Vortex veins exit the eye posterior to the equator to reach these sites by passing through the sclera. There are watershed areas between vortex veins that divide the eye into quadrants. Vortex venous pressure is generally higher than intraocular pressure; it may not be highly dependent upon extreme changes and fluctuations of intraocular pressure due to a Starling resistor mechanism of the choroid. Choroidal venous pressure is likely regulated at the level of the ampulla of the vortex vein and/or at the site of scleral outflow.[2]

Retinal and Choroidal Vascular Diseases of the Eye. https://doi.org/10.1016/B978-0-443-15583-3.00019-6

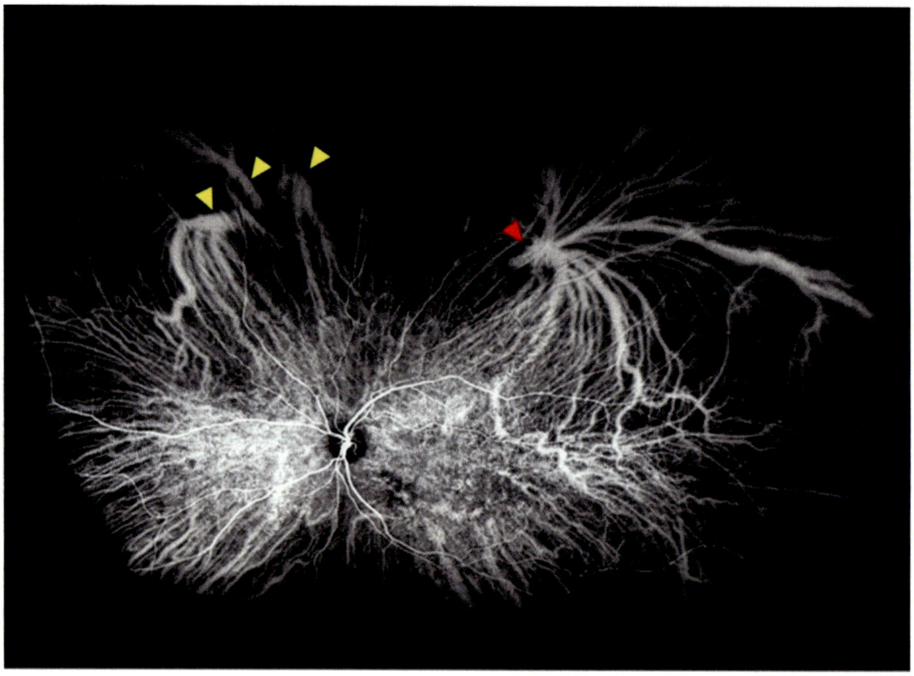

FIGURE 19.1 Indocyanine green angiography of the left eye. *Reprinted with permission from Funatsu R, Terasaki H, Shiihara H, et al. Quantitative evaluations of vortex vein ampullae by adjusted 3D reverse projection model of ultra-widefield fundus images.* Scientific Reports. *2021;11:8916. https://doi.org/10.1038/s41598-021-88265-w under Creative Commons Attribution Non-Commercial license (https://creativecommons.org/licenses/by/4.0/).*

Causes of choroidal venous overload

Systemic causes

Systemic causes of venous overload choroidopathy are related to systemic venous congestion and autonomic dysfunction. Venous congestion may arise secondary to cardiopulmonary failure, such as right heart failure or pulmonary arterial hypertension. A case report by Zhou et al. describes a patient with a history of pulmonary arterial hypertension treated with low-dose (5 mg daily) sildenafil who presented with bilateral diminishing central vision and metamorphopsia.[7] He was found on optical coherence tomography (OCT) angiography to have changes consistent with central serous chorioretinopathy (CSCR), and dilated, tortuous vortex veins were observed on indocyanine green angiography (ICGA). Oral torasemide, and spironolactone, as well as brinzolamide eye drops, were added to his regimen. A similar case of a patient with right heart failure and pulmonary hypertension presenting with CSCR was described by Li et al.[8] This patient also exhibited improvement in choroidal thickness following the treatment of systemic conditions. On return visit, OCT angiography revealed subretinal fluid (SRF) resolution. Cases of sildenafil-induced CSCR have been described in the literature and may also contribute to venous overload via nitric oxide-mediated dilatory effects on choroidal vasculature.[9,10]

Hemodynamic changes, as seen in exercise or hypertension, may result in increased choroidal flow and vessel engorgement when choroidal regulatory mechanisms fail. In a resting, nonhypertensive state, fluctuations in blood pressure are regulated at the level of the choroid at least in part by the sympathetic nervous system. Increased choroidal blood flow triggers vasoconstriction to achieve constant blood flow.[11] Piccolino et al. demonstrated that patients who engaged in vigorous physical activity had up to a 5.58 increased probability of having CSCR.[12] Their group also describes a phenomenon where patients with CSCR achieve elevated blood pressure during physical stress compared to controls.[13] Dysregulation of choroidal flow increases flow throughout the system, most notably in the vortex veins.

Vortex vein occlusion and varix

Vortex vein occlusion has been extensively studied in animal models such as nonhuman primates. On occlusion of all vortex veins in rhesus monkeys, fundus fluorescein angiography (FA) revealed uniformly delayed filling of the entire choroidal bed due to backflow. Elevated intraocular pressure was noted as well. The occlusion of three of four vortex veins demonstrated the filling of a single quadrant during the late phase. There was a general lack of cross-talk between vortex venous systems in the control group.[14] In their studies with nonhuman primate models, Nishikawa et al. and Chen et al.

found decreased choroidal filling on ICGA in regions of vortex vein occlusion. Chen et al. discovered that delayed filling persisted for only 1 week following the ligation of two vortex veins, indicating compensation occurred. They similarly found increased values for the choroidal vascularity index, a measure of choroidal vessel density.[15,16]

While vortex vein occlusions can occur idiopathically, physical compression of the vortex veins is a known complication of procedures such as scleral buckling surgeries.[17,18] Takahashi and Kishi reported vortex vein occlusion and choroidal congestion in quadrants local to the buckle. Furthermore, novel drainage routes were formed via vortex vein anastomoses observed at the 3-month follow-up in affected eyes.[19] Suprachoroidal buckling, although a relatively rare procedure, can disrupt vortex venous circulation and potentially cause vortex vein varices.[20]

Vortex vein varicosities are usually benign dilatations of the vortex vein ampulla(e) (Fig. 19.2). Given their presentation on fundoscopy as elevated, dark-colored masses, they can be uveal tumor masqueraders.[21] Application of pressure to the globe is sufficient to cause blood drainage and reduce the vortex varix, although when withdrawn, the varix will reform. The example of the vortex varix lends further support to a choroidal flow model modulated by in- and outflow at the sclera.

Central serous chorioretinopathy and uveal effusion syndrome

Central serous chorioretinopathy (CSCR) is a common ocular disease that may result in decreased vision and, ultimately, vision loss. It is the fourth most common vision-threatening retinal disorder.[22–24] While the disease commonly self-resolves within 3 months, up to 31% of patients are estimated to develop a chronic form of the disease.[25,26] Four population-based studies regarding CSCR have been conducted thus far.[26–29] Three studies were conducted in Asia: Taiwan, South Korea, and Japan, respectively with reported incidence rates between 21 and 34 per 100,000 person-years. One study examined CSCR incidence in Minnesota, USA citing a mean annual age-adjusted incidence of 9.9 per 100,000 men and 1.7 per 100,000 women. All studies agreed that CSCR has a higher incidence rate among males than females, most often occurring between ages 35 and 39. Similarly, a greater incidence of CSCR among Asian populations than in Caucasians was reported.

First described by von Graefe in 1866 as a central retinitis of an infectious etiology, CSCR has since been renamed several times as our understanding of its pathophysiology has been shaped through vigorous research. Theories regarding CSCR's pathogenesis have undergone extensive revision since first described. The advent of FA in 1961 allowed for imaging active leaks in CSCR patients. Studies using FA in the 1960s revealed serous pigment epithelial detachment (PED), thus inspiring the initial concept of some underlying RPE defect and/or choroidal dysfunction.[30–32]

One of the many roles of the RPE is to regulate fluid movement. Given the early FA findings in CSCR, it was proposed that RPE damage altered fluid movement dynamics, resulting in serous retinal detachment.[33] Due to a general lack of

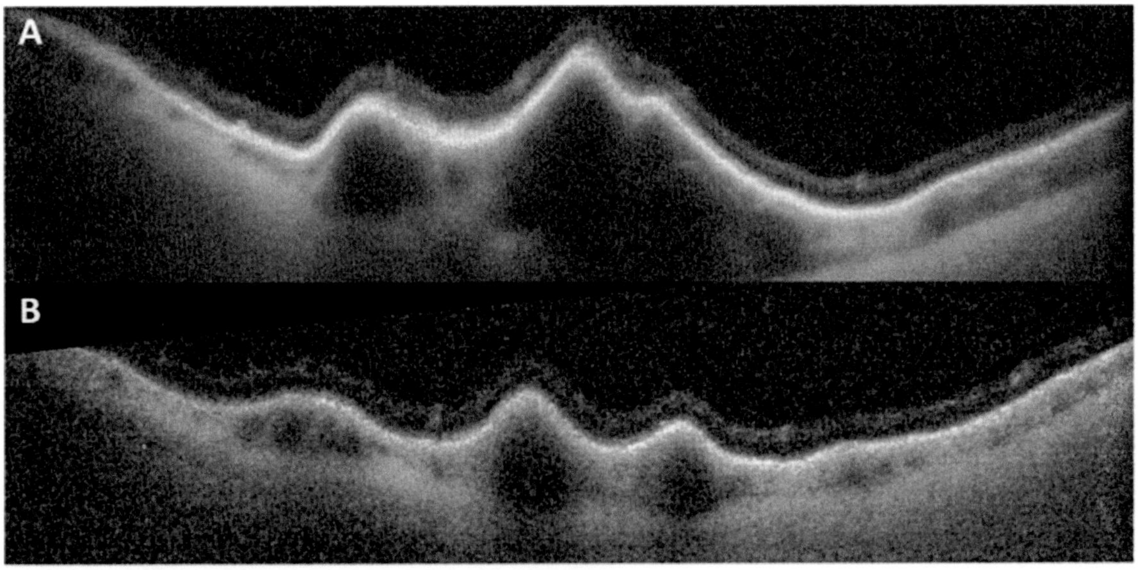

FIGURE 19.2 Optical coherence tomography image of 96-year-old man with a vortex varix of the left eye reveals choroidal congestion at presentation (A) and relative improvement at 9-month follow-up (B). *Reprinted with permission from Weidmayer SL, Demirci H. The spontaneous resolution of a vortex vein varix: case report.* BMC Ophthalmology. *2021;21:101. https://doi.org/10.1186/s12886-021-01861-2 under a Creative Commons Attribution 4.0 International License (http://creativecommons.org/licenses/by/4.0/).*

effective animal models, support for the theory of focal RPE disruption is limited. Furthermore, a study by Negi and Marmor[34] demonstrated that focal RPE damage in rabbit eyes produced fluid flow from the retina to the choroid and increased the rate of subretinal bleb resorption. Further studies may elucidate the role of the RPE in CSCR.

Choroidal dysfunction has recently received support as the likely primary cause of serous retinal detachment and PED. Gass initially proposed that choroidal vessel hyperpermeability increases hydrostatic pressure in the choroid, causing SRF accumulation and RPE disruption.[30] This was not confirmed until Hayashi et al. later demonstrated that choriocapillaris leakage observed with indocyanine green angiography (ICG) colocalized with FFA leaks.[35] Several studies during the 1990s found evidence of choroidal vascular hyperpermeability using FFA and ICG, notably observing midphase staining.[36,37] Prünte and Flammer found evidence of delayed arterial filling with venous engorgement and ischemic areas of the choroid.[36] Venous overload choroidopathy may underlie the choroidal changes observed in CSCR.

Clinically, CSCR is characterized by neurosensory retinal detachment secondary to SRF accumulation in the macular region.[38] This lesion is often accompanied by pigment epithelial detachment. Eyes with CSCR exhibit overall choroidal thickening, largely resulting from Haller layer engorgement with accompanying choriocapillaris thinning. This phenomenon may be due to ischemia of the choriocapillaris.[39] Congestion and dilatation of at least one vortex vein were noted in a majority of eyes with CSCR.[40] Further studies revealed asymmetric vortex venous dilation was more frequently observed in patients with diseases of the pachychoroid spectrum, such as CSCR, than in those with healthy eyes.[41] When the asymmetric vortex vein is a dominant vein responsible for draining the macular region, there may not be adequate compensatory mechanisms. The resultant increase in hydrostatic pressure causes fluid exudation from the choriocapillaris into the subretinal space.[42] While vortex vein asymmetry may contribute to venous overload choroidopathy in CSCR, recent studies have also identified a relationship between CSCR, hyperopia, and increased scleral thickness.[43–46]

As discussed above, there are two likely sites where choroidal blood flow is modulated. One of which includes where vortex veins penetrate the sclera. Hyperopia, a potential risk factor for CSCR, results from a shortened axial length and is associated with rigid sclera. Scleral thickening, especially in the posterior and equatorial regions, may disrupt vortex venous outflow and increase resistance, causing backflow, stasis, and congestion.[43] There may also be decreased diffusion across a thickened scleral barrier, leading to fluid accumulation in the choroid and suprachoroid.[47] Disruption at the sclera likely helps drive the venous overload, which is seen in diseases such as CSCR and uveal effusion syndrome.

An interesting phenomenon appreciated in eyes with CSCR is the formation of vortex venous anastomoses. Appreciable via modalities such as en face optical coherence tomography, anastomoses form between different vortex venous systems in up to 90% of eyes with CSCR, according to a study by Matsumoto et al.[48] Similar anastomoses were also observed in pachychoroid diseases like polypoidal choroidal vasculopathy and pachychoroid neovasculopathy.[49] One explanation for this occurrence is that anastomoses form to relieve the pressure caused by venous overload in instances of vortex vein asymmetry (Fig. 19.3).

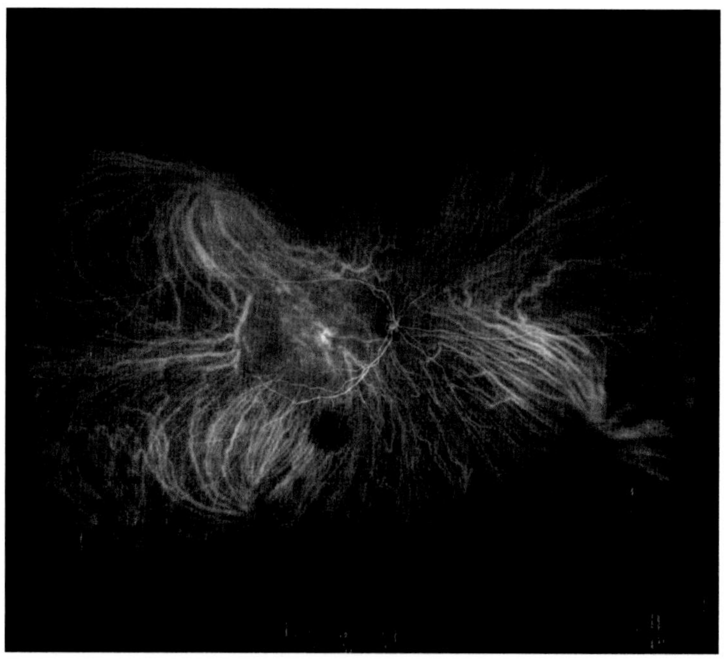

FIGURE 19.3 Indocyanine green angiography of a 69-year-old male with CSCR revealing asymmetric vortex vein outflow and vortex venous anastomoses.

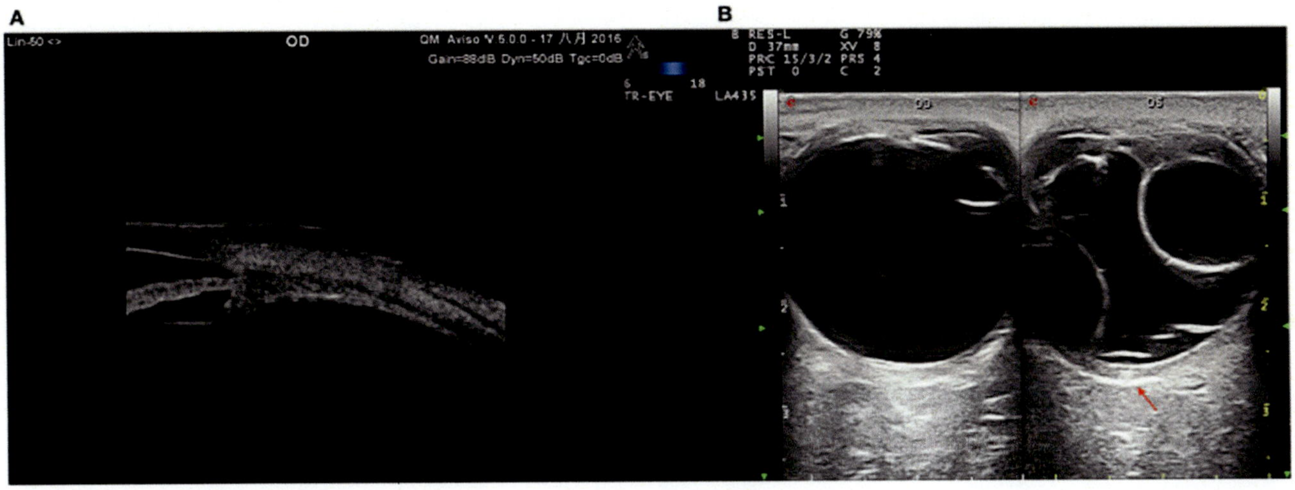

FIGURE 19.4 UES (type 2) captured with ultrasound biomicroscopy that shows ciliary body effusion (A) and scleral thickening with bullous choroidal detachment (B). *Reprinted with permission from Zhou N, et al. Uveal effusion syndrome: clinical characteristics, outcome of surgical treatment, and histopathological examination of the sclera.* Frontiers in Medicine (Lausanne). *2022;9:785444. https://doi.org/10.3389/fmed.2022.785444. PMID: 35755052; PMCID: PMC9218343 under Creative Commons Attribution License (CC BY).*

Uveal effusion syndrome (UES) is a condition in which suprachoroidal and subretinal fluid collection causes serous retina, choroid, and ciliary body detachment.[50] UES can be idiopathic but has been associated with hypermetropia, similar to CSCR.[51] Patients with hyperopia tend to have choroidal thickening, a common feature of both UES and CSCR. In UES, thickened sclerae impair transscleral outflow and mechanical compression of vortex veins.[52] Elevated hydrostatic pressure in the choroidal vasculature may lead to fluid leakage characteristic of the disease. It is unsurprising that scleral thinning surgery yields a rapid resolution of UES (Fig. 19.4).[53]

Intracranial causes of venous overload choroidopathy

Disorders that result in increased hydrostatic pressure within intracranial blood vessels, such as carotid cavernous fistula (CCF), are known to cause venous overload choroidopathy. Interestingly, the pathogenesis of spaceflight-associated neuro-ocular syndrome (SANS), a neuro-ophthalmic phenomenon unique to the spaceflight microenvironment, has also been proposed to be partly due to venous overload.

CCF is an aberrant shunt from the carotid artery to the cavernous sinus, which may arise spontaneously or, more commonly, following trauma.[54] CCF presents with chemosis, proptosis and orbital pain, visual impairment, and episcleral vessels. Signs of glaucoma, painful headaches, or severe proptosis warrant surgical intervention. Depending on the etiology, CCF may resolve without treatment, however.[54] Given that the ophthalmic vein drains directly into the cavernous sinus, backflow causes elevated hydrostatic pressure transmitted to the choroidal circulation. Pachychoroid features (such as increased subfoveal thickness and choroidal vascularity index), PEDs, and choroidal folds are common features seen on imaging.[55–57]

Another potentially perplexing ophthalmic finding from venous overload was observed during spaceflight termed SANS. SANS is a collection of ocular findings encountered by astronauts following long-duration spaceflight, including optic disc edema, choroidal folds, optic nerve sheath distension, posterior globe flattening, and hyperopic refractive error[58,59] (Fig. 19.5). It has been noted that this unique phenomenon is a large barrier to future spaceflight and is rated as a "red" risk by the National Aeronautics and Space Administration for prolonged spaceflight missions.[60] Several hypotheses exist, including cephalad fluid shift, ocular glymphatic system, genetic susceptibility, vitamin B status, choroidal congestion, and upward brain shift.[58,59] During microgravity, there is a reduction in hydrostatic pressure leading to a physiologic cephalad fluid shift. On Earth, lymphatic, cerebrospinal fluid, and venous drainage are processes that are assisted by gravity.[61] During weightlessness and subsequent cephalad fluid shift, cerebrospinal fluid and cerebral venous blood accumulate, causing elevated intracranial pressure. It is hypothesized that vortex venous congestion ensues, resulting in choroidal thickening.[58,62] Return to earth alleviates the choroidal swelling, albeit after several weeks.[63] However, several findings of SANS, including refractive error and posterior globe flattening, may persist for years after returning to Earth.[64,65]

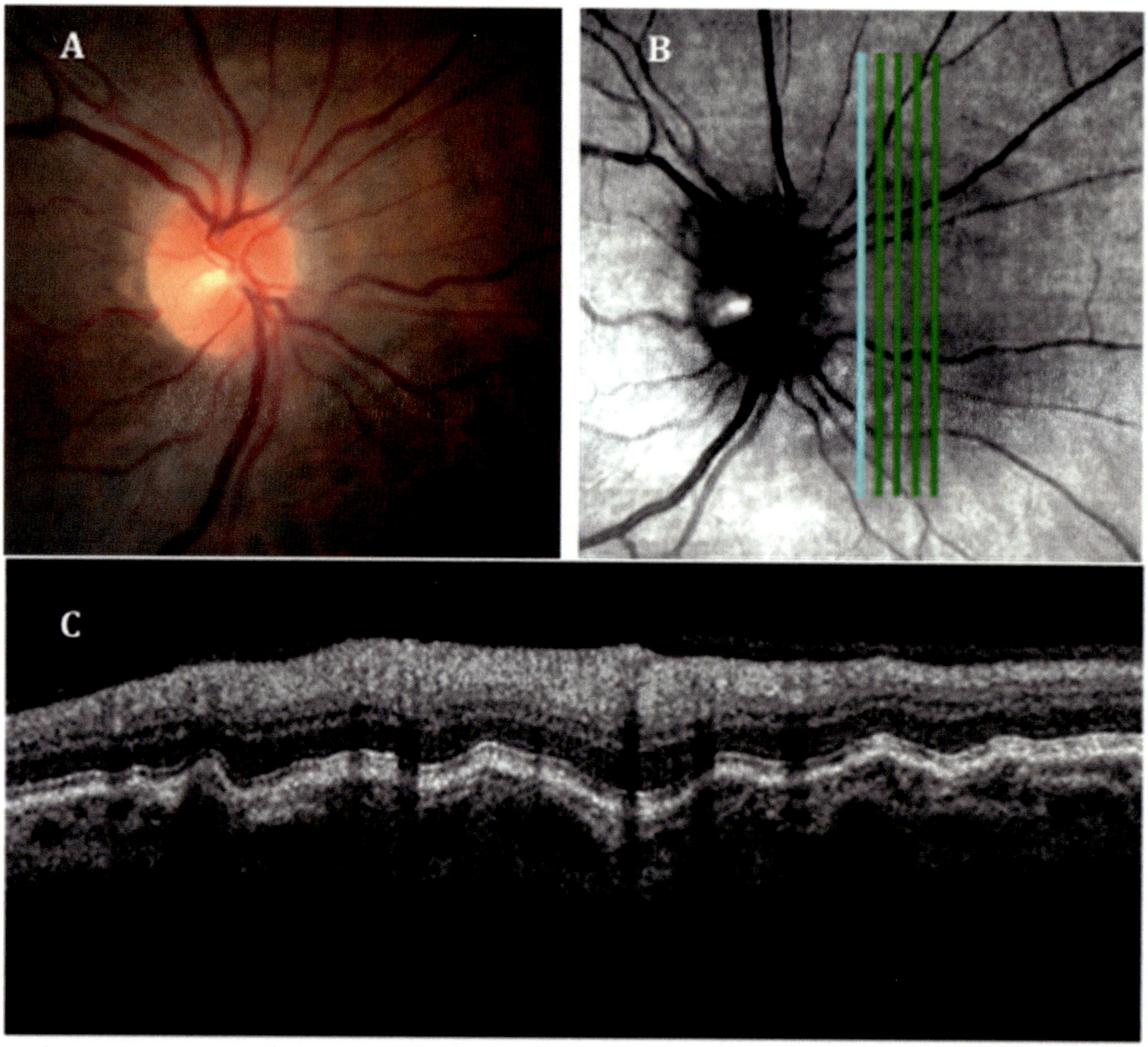

FIGURE 19.5 Postflight multimodal imaging showcasing signs of spaceflight-associated neuro-ocular syndrome (SANS). Fundus photography demonstrates an edematous optic disc with choroidal folds (A). Scanning laser ophthalmoscopy from OCT (B) and OCT B-scans show choroidal folding. *Reprinted with permission from Lee AG, Mader TH, Gibson CR, et al. Spaceflight associated neuro-ocular syndrome (SANS) and the neuro-ophthalmologic effects of microgravity: a review and an update.* NPJ Microgravity. *2020;6:7. https://doi.org/10.1038/s41526-020-0097-9 under Creative Commons Attribution Non-Commercial license (http://creativecommons.org/licenses/by/4.0/).*

Conclusion

Venous overload choroidopathy results from impaired vortex venous outflow, which causes a range of chorioretinal pathologies such as vortex venous dilation and choroidal hyperpermeability. Venous overload choroidopathy likely underlies a host of choroidal disorders, including systemic, regional, and intracranial etiologies. Systemically, congestion of the venous system secondary to heart failure or pulmonary hypertension increases systemic venous hydrostatic pressure that affects the choroidal vasculature. Autoregulation of choroidal blood flow may be partly dysfunctional due to autonomic dysregulation. As may be seen in hypertensive patients with CSCR, increases in cardiac output, such as during vigorous exercise, led to unchecked increases in choroidal flow. Locally, diseases, including spontaneous vortex venous occlusion, CSCR, and UES, demonstrate features consistent with a venous overload model. Vortex venous asymmetry has been observed in a considerable proportion of eyes with CSCR, suggesting variation in vortex venous drainage patterns similarly cause venous overload. In CSCR and UES, a link has been established between scleral thickening and disease, which may contribute to vortex venous outflow abnormalities. Lastly, intracranial disorders such as CCF and SANS lead to venous overload choroidopathy via complex mechanisms, including elevated intracranial pressure.

Disclosures

JC: Allergan, Salutaris, Biogen, Erasca. All else: None.

References

1. Lejoyeux R, Benillouche J, Ong J, et al. Choriocapillaris: fundamentals and advancements. *Progress in Retinal and Eye Research*. 2022;87:100997.
2. Spaide RF, Gemmy Cheung CM, Matsumoto H, et al. Venous overload choroidopathy: a hypothetical framework for central serous chorioretinopathy and allied disorders. *Progress in Retinal and Eye Research*. 2022;86:100973.
3. Verma A, Maram J, Alagorie AR, et al. Distribution and location of vortex vein ampullae in healthy human eyes as assessed by ultra-widefield indocyanine green angiography. *Ophthalmology Retina*. 2020;4(5):530–534.
4. Kutoglu T, Yalcin B, Kocabiyik N, Ozan H. Vortex veins: anatomic investigations on human eyes. *Clinical Anatomy*. 2005;18(4):269–273.
5. Funatsu R, Terasaki H, Shiihara H, et al. Quantitative evaluations of vortex vein ampullae by adjusted 3D reverse projection model of ultra-widefield fundus images. *Scientific Reports*. 2021;11(1):8916.
6. Brismar J. Orbital phlebography. II. Anatomy of superior ophthalmic vein and its tributaries. *Acta Radiologica: Diagnosis*. 1974;15(5):481–496.
7. Zhou X, Zhang J, Gu L, Zhou H. Case report: bilateral central serous chorioretinopathy-like abnormalities in a man with pulmonary arterial hypertension. *Frontiers of Medicine*. 2022;9:983548.
8. Li XQ, Pryds A, Carlsen J, Larsen M. Multifocal central serous chorioretinopathy with photoreceptor-retinal pigment epithelium diastasis in heritable pulmonary arterial hypertension. *Retinal Cases & Brief Reports*. 2015;9(1):83–87.
9. Mohammadpour M, Khodaparast M, Khorrami-Nejad M. Central serous chorioretinopathy following ingestion of sildenafil citrate. *Clinical Optometry*. 2019;11:73–75.
10. Roy R, Panigrahi PK, Saurabh K, Das D, Lobo A. Central serous chorioretinopathy following oral tadalafil intake. *Clinical and Experimental Optometry*. 2014;97(5):473–474.
11. Tittl M, Maar N, Polska E, Weigert G, Stur M, Schmetterer L. Choroidal hemodynamic changes during isometric exercise in patients with inactive central serous chorioretinopathy. *Investigative Ophthalmology & Visual Science*. 2005;46(12):4717–4721.
12. Piccolino FC, Fruttini D, Eandi C, et al. Vigorous physical activity as a risk factor for central serous chorioretinopathy. *American Journal of Ophthalmology*. 2022;244:30–37.
13. Cardillo Piccolino F, Lupidi M, Cagini C, et al. Choroidal vascular reactivity in central serous chorioretinopathy. *Investigative Ophthalmology & Visual Science*. 2018;59(10):3897–3905.
14. Hayreh SS, Baines JA. Occlusion of the vortex veins. An experimental study. *British Journal of Ophthalmology*. 1973;57(4):217–238.
15. Chen LL, Wang Q, Yu WH, Chen YX. Choroid changes in vortex vein-occluded monkeys. *International Journal of Ophthalmology*. 2018;11(10):1588–1593.
16. Nishikawa M, Matsunaga H, Takahashi K, Matsumura M. Indocyanine green angiography in experimental choroidal circulatory disturbance. *Ophthalmic Research*. 2009;41(1):53–58.
17. Doi N, Uemura A, Nakao K. Complications associated with vortex vein damage in scleral buckling surgery for rhegmatogenous retinal detachment. *Japanese Journal of Ophthalmology*. 1999;43(3):232–238.
18. Mantel I, Schalenbourg A, Zografos L. Evanescent vaso-occlusive choroidal pseudo-tumor with acute painful onset: a presumed vortex vein occlusion. *Graefes Archive for Clinical and Experimental Ophthalmology*. 2014;252(5):753–759.
19. Takahashi K, Kishi S. Remodeling of choroidal venous drainage after vortex vein occlusion following scleral buckling for retinal detachment. *American Journal of Ophthalmology*. 2000;129(2):191–198.
20. Milani P, Mazzola M, Bergamini F. Suprachoroidal haemhorrage and vortex vein varix: a potential association. *European Journal of Ophthalmology*. 2022;32(1):NP130–N133.
21. Ferry AP. Lesions mistaken for malignant melanoma of the posterior uvea. A clinicopathologic analysis of 100 cases with ophthalmoscopically visible lesions. *Archives of Ophthalmology*. 1964;72:463–469.
22. Wang M, Munch IC, Hasler PW, Prünte C, Larsen M. Central serous chorioretinopathy. *Acta Ophthalmologica*. 2008;86(2):126–145.
23. Sahoo NK, Ong J, Selvam A, et al. Longitudinal follow-up and outcome analysis in central serous chorioretinopathy. *Eye*. 2022;37(4):732–738.
24. Sahoo NK, Ong J, Selvam A, et al. Ten-year follow-up and sequential evaluation of multifocal retinal pigment epithelium abnormalities in central serous chorioretinopathy. *Graefe's Archive for Clinical and Experimental Ophthalmology*. 2023;261(7):1883–1891.
25. Gilbert CM, Owens SL, Smith PD, Fine SL. Long-term follow-up of central serous chorioretinopathy. *British Journal of Ophthalmology*. 1984;68(11):815–820.
26. Kitzmann AS, Pulido JS, Diehl NN, Hodge DO, Burke JP. The incidence of central serous chorioretinopathy in Olmsted County, Minnesota, 1980-2002. *Ophthalmology*. 2008;115(1):169–173.
27. Tsai DC, Chen SJ, Huang CC, et al. Epidemiology of idiopathic central serous chorioretinopathy in Taiwan, 2001-2006: a population-based study. *PLoS One*. 2013;8(6):e66858.

28. Rim TH, Kim HS, Kwak J, Lee JS, Kim DW, Kim SS. Association of corticosteroid use with incidence of central serous chorioretinopathy in South Korea. *JAMA Ophthalmology*. 2018;136(10):1164–1169.
29. Kido A, Miyake M, Tamura H, et al. Incidence of central serous chorioretinopathy (2011-2018): a nationwide population-based cohort study of Japan. *British Journal of Ophthalmology*. 2022;106(12):1748–1753.
30. Gass JD. Pathogenesis of disciform detachment of the neuroepithelium. *American Journal of Ophthalmology*. 1967;63(3):Suppl:1–139.
31. Gemenetzi M, De Salvo G, Lotery AJ. Central serous chorioretinopathy: an update on pathogenesis and treatment. *Eye (Lond)*. 2010;24(12):1743–1756.
32. Maumenee AE. Macular diseases: clinical manifestations. *Transactions of the American Academy of Ophthalmology & Otolaryngology*. 1965;69:605–613.
33. Bouzas EA, Karadimas P, Pournaras CJ. Central serous chorioretinopathy and glucocorticoids. *Survey of Ophthalmology*. 2002;47(5):431–448.
34. Negi A, Marmor MF. Experimental serous retinal detachment and focal pigment epithelial damage. *Archives of Ophthalmology*. 1984;102(3):445–449.
35. Hayashi K, Hasegawa Y, Tokoro T. Indocyanine green angiography of central serous chorioretinopathy. *International Ophthalmology*. 1986;9(1):37–41.
36. Prünte C, Flammer J. Choroidal capillary and venous congestion in central serous chorioretinopathy. *American Journal of Ophthalmology*. 1996;121(1):26–34.
37. Spaide RF, Hall L, Haas A, et al. Indocyanine green videoangiography of older patients with central serous chorioretinopathy. *Retina*. 1996;16(3):203–213.
38. Zarnegar A, Ong J, Matsyaraja T, Arora S, Chhablani J. Pathomechanisms in central serous chorioretinopathy: a recent update. *International Journal of Retina and Vitreous*. 2023;9(1):3.
39. Wong RL, Singh SR, Rasheed MA, et al. En-face choroidal vascularity in central serous chorioretinopathy. *European Journal of Ophthalmology*. 2021;31(2):536–542.
40. Pang CE, Shah VP, Sarraf D, Freund KB. Ultra-widefield imaging with autofluorescence and indocyanine green angiography in central serous chorioretinopathy. *American Journal of Ophthalmology*. 2014;158(2):362–371.e2.
41. Mori K, Kanno J, Gehlbach PL, Yoneya S. Montage images of spectral-domain optical coherence tomography in eyes with idiopathic macular holes. *Ophthalmology*. 2012;119(12):2600–2608.
42. Hiroe T, Kishi S. Dilatation of asymmetric vortex vein in central serous chorioretinopathy. *Ophthalmology Retina*. 2018;2(2):152–161.
43. Spaide RF, Fisher YL, Ngo WK, Barbazetto I. Regional scleral thickness as a risk factor for central serous chorioretinopathy. *Retina*. 2022;42(7):1231–1237.
44. Fernández-Vigo JI, Moreno-Morillo FJ, Shi H, et al. Assessment of the anterior scleral thickness in central serous chorioretinopathy patients by optical coherence tomography. *Japanese Journal of Ophthalmology*. 2021;65(6):769–776.
45. Imanaga N, Terao N, Nakamine S, et al. Scleral thickness in central serous chorioretinopathy. *Ophthalmology Retina*. 2021;5(3):285–291.
46. Lee YJ, Lee JY, Lee S. A pilot study of scleral thickness in central serous chorioretinopathy using anterior segment optical coherence tomography. *Scientific Reports*. 2021;11(1):5872.
47. Spaide RF, Ryan EH. Loculation of fluid in the posterior choroid in eyes with central serous chorioretinopathy. *American Journal of Ophthalmology*. 2015;160(6):1211–1216.
48. Matsumoto H, Hoshino J, Mukai R, et al. Vortex vein anastomosis at the watershed in pachychoroid spectrum diseases. *Ophthalmology Retina*. 2020;4(9):938–945.
49. Matsumoto H, Hoshino J, Arai Y, et al. Quantitative measures of vortex veins in the posterior pole in eyes with pachychoroid spectrum diseases. *Scientific Reports*. 2020;10(1):19505.
50. Gass JD, Jallow S. Idiopathic serous detachment of the choroid, ciliary body, and retina (uveal effusion syndrome). *Ophthalmology*. 1982;89(9):1018–1032.
51. Pautler SE, Browning DJ. Isolated posterior uveal effusion: expanding the spectrum of the uveal effusion syndrome. *Clinical Ophthalmology*. 2015;9:43–49.
52. Uyama M, Takahashi K, Kozaki J, et al. Uveal effusion syndrome: clinical features, surgical treatment, histologic examination of the sclera, and pathophysiology. *Ophthalmology*. 2000;107(3):441–449.
53. Zhou N, Yang L, Xu X, Wei W. Uveal effusion syndrome: clinical characteristics, outcome of surgical treatment, and histopathological examination of the sclera. *Frontiers of Medicine*. 2022;9:785444.
54. Miller NR. Diagnosis and management of dural carotid-cavernous sinus fistulas. *Neurosurgical Focus*. 2007;23(5):E13.
55. Inam O, Arat YO, Yavas GF, Arat A. Retinal and choroidal optical coherence tomography findings of carotid cavernous fistula. *American Journal of Ophthalmology*. 2019;206:264–273.
56. Fuzzard DRW, Gin TJ, Hardy TG, Harper CA. Subretinal fluid and choroidal venous congestion secondary to a presumed dural cavernous sinus fistula. *Clinical and Experimental Ophthalmology*. 2020;48(4):517–519.
57. González Martín-Moro J, Sales-Sanz M, Oblanca-Llamazares N, Bustos-García A, Méndez-Cendón JC, Fandiño E. Choroidal thickening in a case of carotid cavernous fistula. *Orbit*. 2018;37(4):306–308.
58. Ong J, Mader TH, Gibson CR, Mason SS, Lee AG. Spaceflight associated neuro-ocular syndrome (SANS): an update on potential microgravity-based pathophysiology and mitigation development. *Eye (Lond)*. 2023.

59. Ong J, Tarver W, Brunstetter T, et al. Spaceflight associated neuro-ocular syndrome: proposed pathogenesis, terrestrial analogues, and emerging countermeasures. *British Journal of Ophthalmology*. 2023;107(7):895–900.
60. Patel ZS, Brunstetter TJ, Tarver WJ, et al. Red risks for a journey to the red planet: the highest priority human health risks for a mission to Mars. *NPJ Microgravity*. 2020;6(1):33.
61. Yang JW, Song QY, Zhang MX, et al. Spaceflight-associated neuro-ocular syndrome: a review of potential pathogenesis and intervention. *International Journal of Ophthalmology*. 2022;15(2):336–341.
62. Lee AG, Mader TH, Gibson CR, et al. Spaceflight associated neuro-ocular syndrome (SANS) and the neuro-ophthalmologic effects of microgravity: a review and an update. *NPJ Microgravity*. 2020;6:7.
63. Macias BR, Patel NB, Gibson CR, et al. Association of long-duration spaceflight with anterior and posterior ocular structure changes in astronauts and their recovery. *JAMA Ophthalmol*. 2020;138(5):553–559.
64. Ong J, Tavakkoli A, Strangman G, et al. Neuro-ophthalmic imaging and visual assessment technology for spaceflight associated neuro-ocular syndrome (SANS). *Survey of Ophthalmology*. 2022;67(5):1443–1466.
65. Mader TH, Gibson CR, Barratt MR, et al. Persistent globe flattening in astronauts following long-duration spaceflight. *Neuro-Ophthalmology*. 2020;45(1):29–35.

Chapter 20

Peripapillary pachychoroid syndrome

Reza Mirshahi[1], Kimia Rahimi Ardali[2] and Khalil Ghasemi Falavarjani[1]
[1]*Eye Research Center, Department of Ophthalmology, Iran University of Medical Sciences, Tehran, Iran;* [2]*Student Research Committee, Qazvin University of Medical Sciences, Qazvin, Iran*

The term "pachychoroid" is used for a group of related diseases in which distinctive structural and functional changes in the choroid are the key pathologic mechanisms in associated retinal disorders. The pachychoroid disease spectrum (PDS) encompasses different overlapping disorders such as polypoidal choroidal vasculopathy (PCV), central serous chorioretinopathy (CSC), pachychoroid pigment epitheliopathy (PPE), focal choroidal excavation, peripapillary pachychoroid, and pachychoroid neovasculopathy (PNV). The common pathologic findings in these entities include choroidal hyperpermeability, increased choroidal thickness, presence of "pachyvessels," and choroidal congestion in outer choroid associated with rarefaction of choriocapillaris.[1]

Recent advances in multimodal imaging have shed light on previously unknown features and revealed new findings in various chorioretinal disorders. The "peripapillary pachychoroid syndrome" or PPS was introduced in 2018 by Phasukkijwatana et al.[1] as a new entity in the pachychoroid spectrum. It had been previously reported (and possibly misinterpreted) as isolated posterior uveal effusion syndrome (EUS) in some case reports,[2,3] however, by means of multimodal imaging, Dr. Sarraf's team demonstrated distinctive choroidal features of the pachychoroid spectrum. In the following chapter, we aim to review different aspects of this newly described syndrome.

Clinical features

In the first case series of 16 patients published by Phasukkijwatana et al.,[1] visual acuity ranged from 20/250 to 20/20, with most patients having the best corrected visual acuity of 20/30 or better. The majority of cases were male with a mean age of 71 years and had hypertension as a significant medical history. The axial length of patients ranged from 20.00 to 24.59 mm and nearly all were hyperopic with a mean spherical equivalent of +2.50 diopters. Choroidal folds along vascular arcade were usually present, and the small crowded disc was reported as a relatively common finding (up to one-third of the patients). Chronic bilateral disc edema was also reported for two patients. Other published studies reported fairly the same demographics and presentations (Table 20.1).

Peripapillary retinal pigment epithelium (RPE) mottling and associated gravitational tracts pronounced around the optic disc are also diagnostic clues. Intraretinal fluid (IRF) originating from the temporal optic disc margin (peripapillary fluid pocket [PFP])[4] and cystoid macular edema was virtually evident in optical coherence tomography (OCT) of all patients in the first reported series. It has been speculated that the peripapillary choroidal thickening in PPS, might exert forces on the disc margin, which leads to lamina cribrosa defects and subsequent extension of IRF into the retina (similar to acquired optic disc pit in glaucoma).[1,13] Serous pigment epithelial detachment (PED) and subretinal fluid (SRF) have been reported in 42% and 74% of eyes, respectively. There is also a report of peripapillary pachychoroid neovasculopathy diagnosed by OCT and OCT angiography (OCTA).[8]

Etiology

Similar to PDS, the exact etiology of PPS is unknown. Although the symptoms of PDS are caused by retinal involvement, the true condition seems to lie in the choroid as longitudinal imaging has shown that choroidal hyperpermeability precedes retinal changes.[14] However, the pathogenic cascade of the choroidal disease is yet to be determined. There is a body of evidence pointing toward the circulating steroid and its receptors in the choroid. It is corroborated not only by reports of

Retinal and Choroidal Vascular Diseases of the Eye. https://doi.org/10.1016/B978-0-443-15583-3.00020-2

TABLE 20.1 Summary of major case series reporting the outcomes of peripapillary pachychoroid syndrome.

Study	Year	Number of cases	Visual acuity	Imaging modalities	Associated finding	Special findings	Management	Outcome
Phasukkijwatana et al.[1]	2018	31 eyes of 16 patients	20/250 to 20/20, most patients (65%, 20/31 eyes) 20/30 or better	OCT, FAF, FA, ICGA	Monoclonal gammopathy of uncertain significance, Raynaud disease, Chiari I malformation, HTN, sick sinus syndrome, breast cancer status post-mastectomy and radiation, OSN, hyperlipidemia, prostrate cancer, transient ischemic attack, anxiety, DM type 2, multiple myeloma, melanoma, osteoarthritis, osteopenia, CKD, noninvasive bladder tumors, polymyalgia rheumatica (on prednisolone)	Optic disc edema (13%), choroidal folds (77%), short axial lengths (39% less than 23 mm), and hyperopia (86%)	Intravitreal anti-VEGF (3 eyes), PDT (1 eye), topical dorzolamide (3 eyes), oral acetazolamide (1 patient), observation (1 patient)	No response to anti-VEGF and oral acetazolamide, fluid resolved after PDT, fluid resolved after topical dorzolamide in 2 eyes, spontaneous resolution with observation
Xu et al.[4]	2020	56 eyes of 35 patients	Baseline VA: 20/36 (IQR 20/25–20/63)	OCT, FA, ICGA	N/A	Peripapillary fluid pocket	Observation [13 eyes (23%)], anti-VEGF (23 eyes (41%)), PDT (8 eyes (14%)), in rest of eyes: focal macular laser, oral or topical carbonic anhydrate inhibitors, or topical or intravitreal steroid	Final VA: 20/32 (IQR 20/25–20/53), observation group: VA remained stable, anti-VEGF: SCT and SRT decreased but VA remained stable, PDT: trend toward decreased SRT, no change in SCT, VA remained stable
Iovino et al.[5]	2021	25 eyes of 23 patients	Baseline VA: 0.59 ± 0.44 LogMAR (20/80)	OCT, FAF, FA, ICGA	N/A	Choroidal folds (24%)	PDT full-dose half-fluence 14 eyes (56%) half-dose full-fluence 3 eyes (12%) half-dose half-fluence 7 eyes (28%) full-dose full-fluence 1 eye (4%)	Successfully treated in 64%, VA: 0.51 ± 0.45 LogMAR (20/63) at month 3 ($P = .021$) and 0.65 ± 0.39 LogMAR at month 12 ($P = .982$)

Pothof et al.[6]	2022	17 eyes of 11 patients	Baseline VA: 20/33	OCT, FAF, FA, ICGA	N/A	N/A	Topical prednisolone (pred forte (PF) 10 mg/mL) TDS for 4 weeks, which tapered off after reduction of intraretinal fluid	All patients experienced an initial reduction in intraretinal fluid which persisted on tapering off in 35%. VA improved in 9 (53%) eyes and remained stable in 4 (24%) eyes.
Govetto et al.[7]	2019	2 eyes of 1 patient	20/20 and 20/40	OCT, OCTA FAF, FA, NIR, ICGA	HTN, pulmonary HTN	Flow defects at the choriocapillaris level in OCTA	Observation	Spontaneous resolution of subretinal fluid
Montero Hernández et al.[8]	2020	2 eyes of 1 patient	20/20 20/20	OCT, FAF, ICGA, OCTA	Glaucoma	Hyperopia, peripapillary pachychoroid neovasculopathy	Intravitreal aflibercept	Successful control of subretinal fluid
Manayath et al.[9]	2021	2 eyes of 1 patient	RE: 20/200 LE: 20/40	OCT, FA, ICGA			Low-fluence PDT	At 6 months: VA improved to 20/120 in the right eye and remained stable in the left eye with complete anatomical resolution of fluid in both eyes
Bouzika et al.[10]	2022	3 eyes of 2 patients	20/63 to 20/40	OCT, FA, ICGA	HTN CAD	N/A	PDT in one case and topical treatment (dorzolamide drops) with a switch of acetazolamide tablets to eplerenone tablets 25 mg bid and dexamethasone eye drops bid in the second case	CME and subretinal fluid resolved and VA improved 2 months after PDT, complete fluid absorption in the RE and only a small amount of IRF in the LE of the second case
Iovino et al.[11]	2022	2 eyes of 1 patient	20/20 and 20/32	OCT, FA, FAF OCTA	N/A	Flow voids at the level of the choriocapillaris in OCTA	Navigated SML	VA improved to 20/20 with complete resolution of subretinal and persistent peripapillary intraretinal cysts
Hubschman et al.[12]	2022	1 patient		OCT, FAF FA	HTN OSA	Severe CME	Observation	Spontaneous resolution after 3 months

CAD, coronary artery disease; *CKD*, chronic kidney disease; *CME*, cystoid macular edema; *DM*, diabetes mellitus; *FA*, fluorescein angiography; *FAF*, fundus autofluorescence; *HTN*, hypertension; *ICGA*, indocyanine green angiography; *NIR*, near-infrared; *OCT*, optical coherence tomography; *OCTA*, optical coherence tomography angiography; *OSA*, obstructive sleep apnea; *PAH*, pulmonary arterial hypertension; *PDT*, photodynamic therapy; *SCT*, subfoveal choroidal thickness, *SML*, subthreshold micropulse laser; *SRT*, subfoveal retinal thickness; *VA*, visual acuity; *VEGF*, vascular endothelial growth factor.

CSC exacerbations after the introduction of exogenous corticosteroids but also by the evidence of the decrease in choroidal congestion and thickness following consumption of mineralocorticoid receptor antagonists such as spironolactone and eplerenone.[15]

Recently, venous overload has been proposed as an alternative explanation for PDS. The hydrostatic pressure in choriocapillaris is determined by feeding arterioles and outflow venules. In contrast to other venular systems, these draining venules do not coalesce to form larger veins. Instead, multiple parallel venules take course toward the vortex ampullae independently. In addition, the choroid lacks a lymphatic system to egress the fluid. Therefore, the excessive fluid passes through the sclera to be drained from the choroid.[16] There is evidence in favor of increased scleral thickness in CSC patients.[17] The increased scleral thickness not only attenuates the permeating capability of the sclera, but also compresses the vortex-draining veins leading to choroidal congestion. Moreover, the presence of intervortex venous anastomoses in PDS supports the theory of venous outflow obstruction in this spectrum.[16]

Similar to CSC, choroidal hyperpermeability and intervortex venous anastomoses have also been shown in PPS.[1,18] However, the characteristic difference is the peripapillary location of these abnormalities in PPS in comparison to the macula in CSC. At the edge of the optic disc, the intermediary tissue of Kuhnt and the layer of Jacoby border the termination of the retina and choroid, respectively.[19] The insulation of these tissues is impaired due to the expansion of choroid in PPS, leading to ingression fluid into the retina. In addition, the venous outflow problems similar to CSC are also considered a leading cause of PPS.[16]

Scrutiny of ICGA images of PDS has shown that the areas of pachychoroid and hyperpermeability are regional.[15] We have encountered CSC patients with increasing choroidal thickness toward the SRF location. Therefore, it can be speculated that variations can happen in the location of the thickest choroid and that PPS can merely be a form of asymmetric CSC. Nasal manifestations of retinal changes with foveal sparing during the early course of the disease lead to chronic changes such as IRF and CME in symptomatic patients. Future studies with the aid of artificial intelligence may be beneficial in the subcategorization of PDS.[20]

Diagnosis and multimodal imaging

The diagnosis of PPS is based on multimodal imaging (Fig. 20.1). The proposed criteria in the literature are mainly based on the choroidal thickness profile. The peripapillary choroidal thickening (more than the expected thickness for age) along with localized IRF and/or SRF and evidence of choroidal hyperpermeability seem sufficient for the diagnosis of PPS.

The studies on normal profile for macular choroidal thickness demonstrated that the choroid is thickest subfovealy followed by temporal and nasal parts.[21] However, in PPS, this pattern is altered; choroidal thickness is increased in the nasal macula compared to subfovea and shows thinning toward the temporal macula. Interestingly, the subfoveal choroidal thickness in eyes with PPS is comparable with CSC eyes.[1]

Juxtapapillary RPE alterations evident by granular hyperfluorescence in fluorescein angiography (FA) and mottled autofluorescence are suggestive of PPS. Moreover, ring-like diffuse leakage and hyperfluorescence around the optic disc in FA and peripapillary choroidal vessel dilations and hyperpermeability in indocyanine green angiography (ICGA) are other angiographic findings.

As discussed previously, IRF and CME are common in OCT of eyes with PPS. Atrophy and attenuation of the external limiting membrane (ELM), ellipsoid zone, and RPE are seen later in the course of the chronic disease. The ensuing retinal thinning and RPE atrophy result in peripapillary choroidal signal hypertransmission. IRF may be present in the nasal side of the optic disc. Bacillary layer detachment was also reported in PPS.[22]

Similar to CSC, OCTA is an invaluable tool for the diagnosis of peripapillary pachychoroid neovasculopathy.[8]

Differential diagnosis

Several retinochoroidal diseases may simulate the PPS.

Uveal effusion syndrome

The first case reports of PPS were introduced as isolated posterior pole UES.[2,3] Although the short axial length, hyperopia, and presence of choroidal folds in patients with PPS are similar to UES, there are considerable differences between the two entities. Gravitation tract and serous PED are common findings in PPS that have not been reported in UES. In addition, the diffuse leopard pattern of UES is absent in PPS and localized ICGA findings (pachyvessels and choroidal hyperpermeability) of peripapillary pachychoroid syndrome are compatible with PDS.

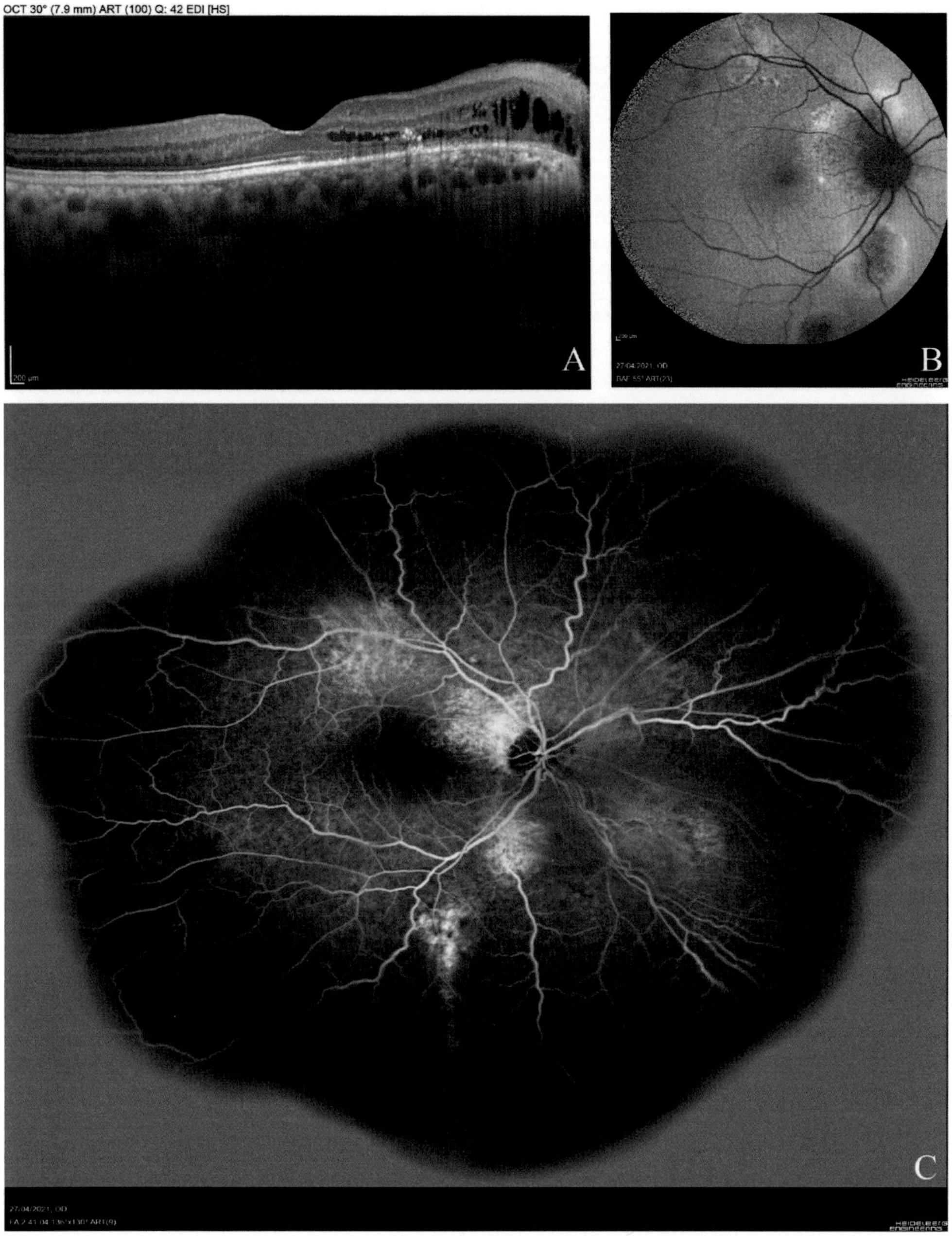

FIGURE 20.1 Spectral-domain optical coherence tomography (OCT) with enhanced depth imaging (EDI) showing pachychoroid, pachyvessels, and attenuation of choriocapillaris along with peripapillary fluid pockets and cystoid macular edema extending from the optic disc (A). Fundus autofluorescence imaging of the same eye showing gutter sign and peripapillary areas of altered autofluorescence (B). Fluorescein angiography showing multiple areas of hyperfluorescence in the peripapillary area and superior to the fovea. Speckled hyperfluorescent gravitational tract extends from the optic disc inferiorly (C).

The UES is believed to be caused by reduced permeability of the sclera and in view of the recent hypothesis regarding the impediment venous outflow, there might be some overlap between the two syndromes. The older age of PPS patients might affect scleral thickness and its egressing function.[1,16]

Chronic central serous chorioretinopathy

As discussed previously, there is considerable overlap between CSC and PPS. Multimodal imaging of both entities is similar with the findings being more pronounced in the juxtapapillary region in PPS. The distinctive feature is the peripapillary location of the highest choroidal thickness in PPS compared with the subfoveal location of pachychoroid in CSC. Moreover, choroidal folds, old age, mild disc leakage in FA associated with a small cup-to-disk ratio, IRF, and cystoid macular edema are more prevalent in PPS.[1]

Maculopathy associated with optic disc excavation

Maculopathy secondary to primary or acquired optic disc pit and other disc excavations may simulate the PPS. Differential findings are preserved choroidal profile, normal disc appearance, and absence of choroidal hyperpermeability and leakage.

Peripapillary choroidal neovascularization

Choroidal neovascularization around the optic disc may simulate the PPS. Hemorrhage is absent in PPS, and multimodal imaging especially OCTA is essential for ruling out the CNV. Of note, CNV has been reported in the presence of PPS.[8]

Management

Various treatment options have been suggested for the management of symptomatic PDS. Considering the association between exogenous steroids and the development and exacerbation of CSC, the first line of management would be cessation of corticosteroids if the medical condition allows. It has been shown that SRF is resolved in the majority of eyes with CSC after discontinuation of steroids.[23] Similarly, in hypothetical cases of PPS associated with exogenous corticosteroid or herbal medicine affecting the steroid receptors (e.g., cordyceps and ginseng), stopping the offending drug is strongly advisable.[15]

SRF usually resolves spontaneously in most cases of acute CSC, although the recurrent rate can reach up to 50%.[24] Therefore, it has been suggested that the treatment can be deferred for up to 3 months after the initial onset of the disease in CSC unless the visual demands of the patient prohibit inaction. Although PPS usually has chronic manifestations, it can show a wax and waning course at least in a minority of cases.[7,12] One case (of 16 cases) in the original case series of Phasukkijwatana et al.[1] and 23% of eyes (13 from 56 eyes) in another large case series[4] showed spontaneous resolution of IRF and SRF with stable visual acuity over time. Therefore, 3–6 months of follow-up without any treatment can be a viable option[12] in fovea-sparing cases.

An old treatment option for CSC is thermal laser photocoagulation of extrafoveal leakage points on FA. However, there are concerns regarding the development of secondary choroidal neovascularization (CNV) at the treatment site. Also, subfoveal leakage points are not amenable to this treatment modality. Recently, subthreshold micropulse laser (SML) photocoagulation has been introduced for delivering controlled less-thermal energy for stimulation of RPE cells with applications even in the subfoveal leakage points. Although micropulse laser seems to be less effective in chronic cases of CSC with diffuse leakage, its efficacy has been shown in randomized clinical trials.[25] In a recent case report, a 65-year-old man with PPS was successfully treated using navigated 577-nm SML targeted at the leakage points temporal to the optic nerve.[11]

Verteporfin photodynamic therapy (PDT) is currently considered the gold standard for the treatment of chronic cases of CSC. The ultimate goal of treatment in PDS is targeting the underlying pathology, that is, choroidal hyperpermeability. Considering the potential visual complications (transient visual loss, secondary CNV, RPE atrophy, etc.) associated with conventional standard dose PDT, half-dose and half-fluence PDT were safely used for the treatment of chronic CSC with favorable outcomes.[15] In a recent case series by Iovino et al.,[5] PDT was used for the successful treatment of 16 of 25 eyes (64%) with PPS. Significant improvement in visual acuity was achieved by a significant reduction of SRF during a 3-month period, which remained stable 12 months later.

While there is no standard recommendation for topical medications for the treatment of CSC, they have been used in PPS. Although oral acetazolamide has been ineffective in the treatment of PPS, topical dorzolamide resulted in fluid resorption in two of three eyes in the original case series reported by Phasukkijwatana et al.[1] The exact mechanism of carbonic anhydrase inhibitors in the resolution of SRF and CME is unknown; however, it seems that they exert their effect through redirecting the fluid flow across the blood–retinal–barrier.[26] Pothof et al.[6] investigated the efficacy of topical prednisolone in the treatment of 17 eyes with PPS. They observed a reduction of IRF in all cases following 1 month of 3 times daily medication with sustained improvement following tapering to once daily dose in 35% of patients. Considering the beneficial effects of topical steroids in PPS, it can be postulated that some degree of underlying inflammatory processes may be present in patients with PPS.[6,12]

In rare cases of PPS complicated with neovasculopathy, intravitreal injection of anti-vascular endothelial growth factor (VEGF) can be tried with successful outcomes.[8]

Although the number of studies is limited for long-term outcomes of PPS, it seems that PPS patients experience a reduction in choroidal and retinal thickness with various treatments and even spontaneously in some cases. Moreover, the overall visual prognosis is favorable despite visual decline in some cases.[4]

References

1. Phasukkijwatana N, Freund KB, Dolz-Marco R, et al. Peripapillary pachychoroid syndrome. *Retina*. 2018;38(9):1652−1667. https://doi.org/10.1097/IAE.0000000000001907.
2. Liu Q, Hemarat K, Kayser DL, Stewart JM. A case of posterior uveal effusion syndrome masquerading as uveitis. *Retinal Cases and Brief Reports*. 2017;11(Suppl 1):S124−S127. https://doi.org/10.1097/ICB.0000000000000463.
3. Pautler SE, Browning DJ. Isolated posterior uveal effusion: expanding the spectrum of the uveal effusion syndrome. *Clinical Ophthalmology*. 2015;9:43−49. https://doi.org/10.2147/OPTH.S75448.
4. Xu D, Garg E, Lee K, et al. Long-term visual and anatomic outcomes of patients with peripapillary pachychoroid syndrome. *British Journal of Ophthalmology*. 2022;106(4):576−581. https://doi.org/10.1136/bjophthalmol-2019-315550.
5. Iovino C, Peiretti E, Tatti F, et al. Photodynamic therapy as a treatment option for peripapillary pachychoroid syndrome: a pilot study. *Eye (London, England)*. 2022;36(4):716−723. https://doi.org/10.1038/s41433-021-01515-z.
6. Pothof AB, Fernández-Avellaneda P, Behar-Cohen F, Ciriano JPM, Yzer S. Potential treatment for peripapillary pachychoroid syndrome. *Retinal Cases and Brief Reports*. 2022.
7. Govetto A, Sarraf D, Scialdone A. "HIDE and seek" neurosensory retinal detachments in peripapillary pachychoroid syndrome associated with pulmonary arterial hypertension. *Retinal Cases & Brief Reports*. 2022;16(1):118−125. https://doi.org/10.1097/ICB.0000000000000942.
8. Montero Hernández J, Remolí Sargues L, Monferrer Adsuara C, Castro Navarro V, Navarro Palop C, Cervera Taulet E. Peripapillary pachychoroid neovasculopathy: a novel entity. *European Journal of Ophthalmology*. 2022;32(1):NP149−NP153. https://doi.org/10.1177/1120672120953071.
9. Manayath GJ, Verghese S, Ranjan R, Narendran V. Photodynamic therapy for peripapillary pachychoroid syndrome-a case report. *Digital Journal of Ophthalmology: DJO*. 2022;28(1):7−11. https://doi.org/10.5693/djo.02.2021.11.001.
10. Bouzika P, Georgalas I, Sotirianakou M-E, et al. Peripapillary pachychoroid syndrome (PPS): diagnosing and treating a rare entity. *Case Reports in Ophthalmological Medicine*. 2022;2022:9124630. https://doi.org/10.1155/2022/9124630.
11. Iovino C, Di Iorio V, Paolercio L, Giordano C, Testa F, Simonelli F. Navigated 577-nm subthreshold micropulse retinal laser treatment for peripapillary pachychoroid syndrome. *American Journal of Ophthalmology Case Reports*. 2022;28:101757. https://doi.org/10.1016/j.ajoc.2022.101757.
12. Hubschman S, Hou K, Sarraf D, Tsui I. An unusual presentation of peripapillary pachychoroid syndrome. *American Journal of Ophthalmology Case Reports*. 2022;25:101338. https://doi.org/10.1016/j.ajoc.2022.101338.
13. Lee JH, Park HL, Baek J. Alterations of the lamina cribrosa are associated with peripapillary retinoschisis in glaucoma and pachychoroid spectrum disease. *Ophthalmology*. 2016;123(10):2066−2076. https://doi.org/10.1016/j.ophtha.2016.06.033.
14. Guyer DR, Yannuzzi LA, Slakter JS, Sorenson JA, Ho A, Orlock D. Digital indocyanine green videoangiography of central serous chorioretinopathy. *Archives of Ophthalmology (Chicago, Ill.: 1960)*. 1994;112(8):1057−1062.
15. Cheung CMG, Lee WK, Koizumi H, Dansingani K, Lai TYY, Freund KB. Pachychoroid disease. *Eye (London, England)*. 2019;33(1):14−33. https://doi.org/10.1038/s41433-018-0158-4.
16. Spaide RF, Gemmy Cheung CM, Matsumoto H, et al. Venous overload choroidopathy: a hypothetical framework for central serous chorioretinopathy and allied disorders. *Progress in Retinal and Eye Research*. 2022;86:100973. https://doi.org/10.1016/j.preteyeres.2021.100973.
17. Oh J-H, Oh J, Togloom A, Kim S-W, Huh K. Biometric characteristics of eyes with central serous chorioretinopathy. *Investigative Ophthalmology & Visual Science*. 2014;55(3):1502−1508. https://doi.org/10.1167/iovs.13-13183.
18. Spaide RF, Ledesma-Gil G. Choriocapillaris vascular parameters in normal eyes and those with pachychoroid with and without disease. *Retina*. 2021;41(4):679−685. https://doi.org/10.1097/IAE.0000000000002988.
19. Hogan MJ. Histology of the human eye. In: *An Atlas and Textbook*. WB saunders; 1971.
20. Mirshahi R, Naseripour M, Shojaei A, et al. Differentiating a pachychoroid and healthy choroid using an unsupervised machine learning approach. *Scientific Reports*. 2022;12(1):16323. https://doi.org/10.1038/s41598-022-20749-9.
21. Margolis R, Spaide RF. A pilot study of enhanced depth imaging optical coherence tomography of the choroid in normal eyes. *American Journal of Ophthalmology*. 2009;147(5):811−815. https://doi.org/10.1016/j.ajo.2008.12.008.
22. Ramtohul P, Comet A, Denis D. Bacillary layer detachment in peripapillary pachychoroid syndrome. *Ophthalmology Retina*. 2020;4(6):587. https://doi.org/10.1016/j.oret.2020.01.017.
23. Sharma T, Shah N, Rao M, et al. Visual outcome after discontinuation of corticosteroids in atypical severe central serous chorioretinopathy. *Ophthalmology*. 2004;111(9):1708−1714.
24. Nicholson B, Noble J, Forooghian F, Meyerle C. Central serous chorioretinopathy: update on pathophysiology and treatment. *Survey of Ophthalmology*. 2013;58(2):103−126. https://doi.org/10.1016/j.survophthal.2012.07.004.
25. Kretz FTA, Beger I, Koch F, Nowomiejska K, Auffarth GU, Koss MJ. Randomized clinical trial to compare micropulse photocoagulation versus half-dose verteporfin photodynamic therapy in the treatment of central serous chorioretinopathy. *Ophthalmic Surgery, Lasers & Imaging Retina*. 2015;46(8):837−843. https://doi.org/10.3928/23258160-20150909-08.
26. Liew G, Ho I-V, Ong S, Gopinath B, Mitchell P. Efficacy of topical carbonic anhydrase inhibitors in reducing duration of chronic central serous chorioretinopathy. *Translational Vision Science & Technology*. 2020;9(13):6. https://doi.org/10.1167/tvst.9.13.6.

Section V

Vascular abnormalities

Chapter 21

Retinal artery aneurysmal abnormalities

Lisa Checchin[1,2], Ilaria Zucchiatti[1,2], Riccardo Sacconi[1,2], Francesco Bandello[1,2] and Giuseppe Querques[1,2]
[1]*School of Medicine, Vita-Salute San Raffaele University, Milan, Italy;* [2]*Division of Head and Neck, Ophthalmology Unit, IRCCS San Raffaele Scientific Institute, Milan, Italy*

Macroaneurysm of retinal arteries

Definition

Macroaneurysm of a retinal artery (RAM) is an acquired focal dilatation of the wall of a retinal artery that is 200—400 μm in diameter[1] and differs from other aneurysmal lesions such Coat's disease or diabetic retinopathy (DR) in either site and etiology. RAM has been encoded as a clinical entity in its own right by Robertson in 1973. It occurs in the first three orders of the bifurcation of the central retinal artery with a higher frequency of the second order (56%) followed by the third order (36%) and finally the first order (8%),[2] while the optic disc is rarely involved.[3,4]

Epidemiology

The prevalence is around 1 of 1500—4500 subjects,[5,6] and they are mostly single, unilateral, and temporal to the optic disc with almost equal distribution between the superotemporal and inferotemporal arcades[7]; 7%—10% of cases may be bilateral, while 20% of cases are multiple, either on the same vessel or on different arteries.[8—10] Typically, they occur in women over 60 years old although cases of younger patients have been described.[3,11,12]

The main risk factors in RAM formation are age, hypertension, and atherosclerosis. In fact, up to 75% of patients have systemic hypertension and many case reports reveal an association with vascular diseases such as myocardial infarction, stroke, branch retinal vein occlusion (BRVO), or branch retinal artery occlusion (BRAO) at the site of RAM or distally to it.[7,13—16] In addition, RAM has also been reported in patients with diabetes mellitus,[14,17] polyarteritis nodosa,[3,8] rheumatoid arthritis,[3,7] elevated hematocrits,[13] elevated serum cholesterol,[7,14,18] macular hole (MH), multifocal chorioretinitis,[19] Raynaud's disease,[3] and arterovenous malformations type 1.[20,21] Lewis et al. reported that within 5 years, patients with RAM may die from stroke, myocardial infarction, and resection of abdominal aortic aneurysm. Therefore, RAM may be a rare sign of hypertension or systemic vascular disease that could compromise the patient's life expectancy and should consequently encourage the ophthalmologist to meticulously assess the patient's blood pressure and cardiovascular risk,[22] even if a higher death rate was not confirmed in further studies.[13]

Pathogenesis

RAMs are thought to be due to local arterial wall damage mechanisms associated with intrinsic vascular wall weakness present in specific locations. In fact, they have a predilection for arterial bifurcations and arteriovenous crossings, which are frequent sites of emboli or intrinsic vascular fragility.[22] In the latter location, the arterial vessel may be encouraged to form aneurysms under the pull of hydrostatic pressure because it shares the same adventitious tonaca with the vein; therefore, the arterial wall is more delicate and may be less capable to resist to further damages, as in case of increased intraluminal pressure or embolic occlusion. Moreover, through this strict contact, any venous disease can provoke local damage to the arterial wall by inducing focal arterial thrombosis and endothelial harm, contributing to the formation of RAM. By inducing an increase in retinal vessel hydrostatic pressure, systemic hypertension may be the cause of both BRVO and RAM formation.[13] So, in addition to sharing the same risk factor, this pathogenetic mechanism may be a plausible

Retinal and Choroidal Vascular Diseases of the Eye. https://doi.org/10.1016/B978-0-443-15583-3.00021-4

explanation for the association between RAM and retinal vascular occlusions reported in the literature. However, both systemic hypertension and atherosclerosis and aging may have the same final findings, characterized by the replacement of muscle tonaca with collagen tissue and stiffening of arterial vessels resulting in loss of elastic return and autoregulatory tone and vascular dilatation.

It has also been reported that local arterial wall weakness or turbulent flow can cause an aneurysm in a dilated collateral vessel.[7]

Furthermore, retinal vessels around the RAM may be altered too by vascular remodeling which occurs after embolic occlusions.[13]

Histologic examination revealed focal thickening of the vessel wall caused by the replacement of smooth muscle cells with infiltration of fibrin and hyaline material (PAS-positive), foamy macrophages, and hemosiderin. This process induces retinal artery wall damage and subsequent protrusion of an aneurysmal sac in response to increased intraluminal pressure and loss of muscle tonaca.[23] Moreover, the progression and rupture of the RAM may be consequent to the development of splits in the vessel wall.[24]

A recent study suggested that pericyte function, vascular inflammatory response, and extracellular matrix remodeling may be influenced by the action of angiotensin II and be predisposing factors in the formation of arterial dilatation and blood-retinal barrier breakdown.[25]

Clinical features

Clinically and histologically RAM is characterized by an active process of development, widening with or without blood-retinal barrier breakdown and gradual thrombosis, fibrosis, and spontaneous involution after a variable time period[13]; thus, RAM has mostly benign course with spontaneous regression and may be an occasional finding during ophthalmoscopy in case of macular sparing. However, the aneurysmal dilatation may undergo rupture before reaching a large dimension.[26] Symptomatic RAM normally causes sudden central or generalized visual loss, scotomas, or floating spots due to hemorrhage or exudation involving the posterior pole.[3,13]

Sporadically, the aneurysmal abnormality may develop on small ramifications of the major arterial tree and, in that case, it tends to achieve a vast size and cause severe exudation.[9] Therefore, patients may complain sudden vision loss due to vitreous hemorrhage or retinal macular bleeding or gradual visual impairment secondary to exudationThey may be completely asymptomatic when the macular region remains unscathed.[13] At fundus examination RAM appears as focal arterial dilatations surrounded by a circinate pattern of hard exudates or associated with yellow lipid deposits in the macular region, regardless the site of the lesion.[3,7] Perforated RAM may be identified as a hemorrhage lying on an artery with a white lesion in the center.[8,23] The aneurysm is pulsatile in 10%–60% of the cases[8,27,28] and its expansion may increase up to 50% in size with each pulse; such characteristic was suggested to be an indicator of imminent vascular rupture and severe bleeding.[11,27] However, other studies denied this finding.[13] At fundus exam, RAM is often associated with signs of systemic vascular disease such as narrow, irregular, and copper wire arteries and arteriovenous crossing modifications as well as signs suggestive of BRVO or an intraluminal embolic plaque.

RAM has two kinds of classification.

First, based on their anatomical conformation, it can be distinguished into three main types:

- saccular, the most frequent (68,75%), representing a focal dilatation of the vessel wall,
- fusiform, characterized by a circumferential enlargement of the artery wall,
- indefinable, in which anatomical structure is incomprehensible.[7]

Secondly, according to their barrier disfunction, it is divided into:

- exudative, in which exudation is the largest constituent, has a caliber greater than 1-disc and is the reason of loss of visual acuity (VA),
- hemorrhagic, in which hemorrhage expands more than the 1-disc diameter, is larger than exudation and causes vision impairment,
- quiescent, in which hemorrhage or exudate is absent or extends for less than 1-disc diameter and avoids the macular region; thus, it does not cause visual loss.[7,13]

Recently the earlier, subclinical condition of quiescent RAM, characterized by arterial dilatation without hemorrhage or signs of exudation, has been studied and it has been shown to tend to enlarge (67%) and progress to rupture (50%).[24]

For either saccular or fusiform dilatations, the predominant feature was hemorrhagic with almost 75% of cases in the former and 90% of cases in the latter.[7] Moreover, hemorrhagic aneurysms are associated to subjects with higher systolic blood pressure whereas the other form is more common in retinal vein occlusions, reflecting its pathogenesis.[13]

Exudation is most often associated with fusiform RAM but tends to decrease as the aneurysmal dilatation occurs away from the optic disc; however, arteries at this site have a larger caliber and higher flow than peripheral vessels, so these elements can increase transmural stress and induce vessel rupture, which would explain why hemorrhage is still the most frequent feature in this aneurysm as well.

The saccular type is more frequently associated with hemorrhages because it has a thin and stretched aneurysmal wall subjected to high perfusion pressure in the first three orders of arterioles, thus it is easier to perforate.[7]

Aneurysmal rupture causes a hemorrhage at any level of the retinal layers: intravitreal (20%), preretinal (5%–40%), subinner limiting membrane (ILM) (65%), intraretinal (80%–100%), subretinal (60%–90%), or multilayers (30%).[2,10] Intraretinal and subretinal blood accumulation are associated with posterior or posterolateral perforation of the vascular dilatation, whereas preretinal and vitreous bleeding are subsequent to an anterior rupture toward the vitreous.[7,9]

A grayish-yellowish saddle-shaped plaque, with RAM in the saddle, may be identified at the site of resorbed subretinal hemorrhage, resembling a disciform macular scar. In the case of RAM regression, a pathognomonic Z-shaped kink may appear due to traction on adventitial tonaca subsequent to fibrotic modifications in the muscular tonaca of the vascular wall. Therefore, the appearance of the Z-shaped kink is the result of a conflict between the fibrous traction on the one hand and the tensile force of the adventitial attachments to the ILM of the retina on the other. Other common findings are also an arteriolar stenosis proximal to the aneurysmal dilatation or an arterial sheathing distal to it.[9]

Visual outcome

Most of the RAMs have a favorable visual outcome also without treatment.[19]

Visual prognosis is influenced by the intraretinal position of RAM in relation to the affected retinal artery, its proximity to the macula, the onset and duration of macular exudation, and the site of bleeding into the retinal layers.[2,7] Initial low VA is more common in hemorrhagic RAMs but, generally, the visual outcome is worse in case of macular edema or exudates which endure for many months, while it is better in case of hemorrhagic or nonpulsatile aneurysms.[7,8] However, also the site of bleeding is an important prognostic factor: in preretinal and intraretinal hemorrhages patients recover good vision when the blood is reabsorbed, whereas in subretinal bleeding, as well as macular exudation, results in irreversible photoreceptors damage due to retinal fibrosis, disruption of the pigment epithelium or macular arterial occlusion, with subsequent permanent morphological changes and poor visual outcome.[3,7,19] Preretinal and intraretinal hemorrhages usually reabsorb without complication within 6–10 weeks.[3,8] Subretinal blood accumulation has been shown to induce retinal degeneration within 2 weeks of its occurrence[19]; indeed, as blood is toxic for photoreceptors, its accumulation leads to iron and fibrin-related toxicity and oxidative stress and hinders the supply of oxygen and nutrients from the choroid to the retina through a mechanical barrier effect.[2,7]

It was reported that complete resorption of the hemorrhage occurs 2–3 months before the best-corrected visual acuity (BCVA) is achieved.[9] However, it was recently revealed that intraretinal hemorrhage can be an unfavorable prognostic factor when associated with subretinal blood accumulation, even after treatment with a macular displacement.[29]

Furthermore, it was suggested that the site of RAM within the retinal layers could be a predictive factor for spontaneous rupture, as it was seen that dilations in the inner retina were associated with both exudation and hemorrhages, in contrast to the deeper ones that were only exudative.[30]

Occasionally, the poor visual outcome may occur in preretinal hemorrhages because of the onset of epiretinal membrane (MER), while chronic macular edema may also result in MER, cystoid macular edema (CME), and MH formation.[3]

Imaging

Diagnosis and follow-up of RAMs are possible through multimodal imaging, particularly optical coherence tomography (OCT), OCT angiography (OCTA), fluorescein angiography (FA), and indocyanine green angiography (ICGA) (Figs. 21.1 and 21.2). As explained in the previous paragraph, retinal aneurysms have been divided into fusiform and saccular through angiographic techniques because they have different behaviors. In saccular RAM dye saturation is minimal in the early phase of FA and more evident in subsequent frames whereas fusiform one fills with dye already in the arterial stage and engages most of the vascular circumference.

However, retinal bleeding may conceal the RAM through a shadow effect, rendering FA useless for diagnostic purposes; in this case, ICGA is necessary as the absorption and emission spectra of its dye are in the near-infrared range,

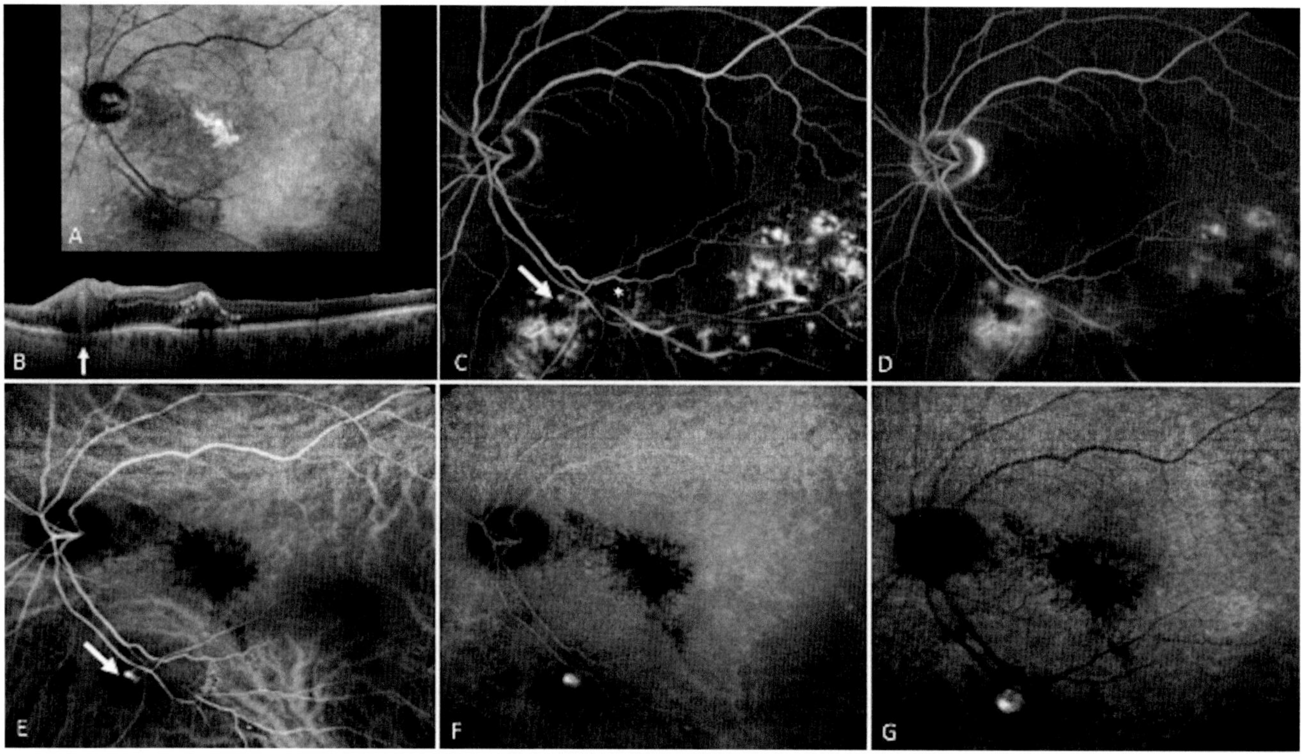

FIGURE 21.1 IR scan reveals a hyperreflective spot surrounded by hyporeflective areas along the inferotemporal vascular arcade with marked hyperreflective macular material referring to hard exudates (A). OCT (B) evidences a large, round hyperreflective lesion within retinal hyperreflective material inducing a shadow effect (*arrow*) associated with macular subretinal fluid, perilesional retinal edema, and hard exudates. FA early frames (C) show an unfilled venous segment (*) starting from an arteriovenous crossing, referring to a branch retinal vein occlusion, associated to a hyperfluorescent dilatation (*arrow*) arising from the artery, referring to RAM, surrounded by fluorescence block resulting from retinal hemorrhage. FA also shows areas of retinal ischemia, capillary dilatations, and compensation circles in the inferotemporal sector with leakage in the late phase of dye perfusion (D). ICGA, early (E), intermediate (F), and late (G) frames, reveals a hyperfluorescent vascular dilation, starting from the artery (*arrow* E), with late leakage, confirming the diagnosis of RAM, associated with macular and perilesional hypofluorescent areas, due to hard exudates. A 78-year-old woman with hypercholesterolemia and without any cardiovascular disease reports visual loss in her left eye, starting from the past month.

allowing a better view of the structures underlying the hemorrhage.[7] Furthermore, ICGA images may be clearer to understand than FA ones, as the dye of the latter may leak more than indocyanine.[19] ICGA reveals dye impregnation in the RAM in premature frames with pooling in later stages.[26,28]

FA shows also other abnormalities associated with RAM: arteriolar stenosis proximal to the lesion, arteriolar sheathing distal to the aneurysm that may result in occlusion,[13] disruption of retinal pigment epithelium (RPE) and photoreceptors and the Z-shaped kink in case of resolved RAM, late hyperfluorescence due to CME and a shadow effect induced by hard exudates, and leakage from the aneurysm directly or from small inefficient vessels surrounding it[3]; such permeability to dye may persist even in involuted RAM causing the retinal damage responsible for permanent vision loss.

On the other hand, it has been identified microangiopathy adjacent to the RAM characterized by intraretinal microvascular abnormalities such as enlargement of the periarterial or perianeurysmal capillary-free zone, small zones of capillary dilatation and capillary nonperfusion, and development of intraarterial collateral vessels.[22] These modifications are responsible for fluorescein leakage in the retinal area surrounding the RAM. Some authors have suggested that embolic processes or mechanical arterial wall tension result in decreased local blood flow and hypoxia and consequently induce reorganization of small vessel networks, representing microangiopathy.[7,22]

OCT, OCTA, and the infrared (IR) scan are other useful tools to diagnose the presence, the site within the retinal layers, and the evolution of RAM and its complications such as retinal bleeding, fluid accumulations, atrophy, and fibrosis; with respect to dye examinations, these methods have the advantage of being quick and noninvasive.

On OCT, RAM appears as an abnormal dilatated vessel localized in the inner retinal layers characterized by hyporeflective lumen and hyperreflective wall; the lesion alters the normal retinal organization through the elevation of the ILM and retinal ganglion cell layers and causes a shadow effect on the deeper retinal layers. However, a case of RAM localized in the deeper retinal layers and not altering the profile of the innermost retinal layers was reported.[31]

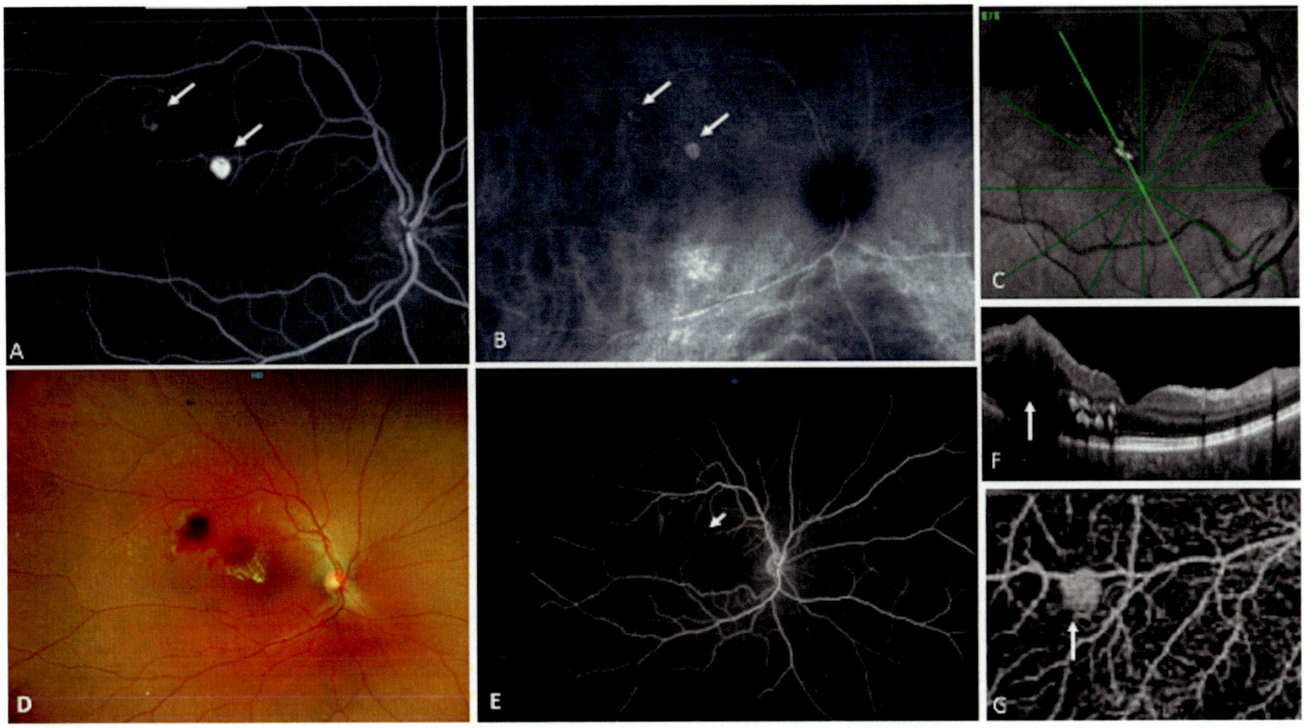

FIGURE 21.2 RAM is diagnosed as an occasional finding in a 34-year-old man with good general health during a follow-up eye examination. UWF FA (A) shows two hyperfluorescent spots near the superotemporal vascular arcades, referring to focal dilatation of the wall of the retinal arteries, that are confirmed on ICGA UWF (B) (*arrows*). However, a few weeks later, the patient reports sudden visual loss in his right eye. Infrared (C) and UWF color fundus image (D) reveals retinal hemorrhages at the site of the lesions with hard exudates expanding toward the fovea, secondary to the rupture of the RAM. UWF FA (E) shows fluorescence blockage from retinal hemorrhages with weak leakage (*arrow*), without any other abnormalities. OCT (F) passing through the RAM (*arrow*) shows a round hyperfluorescent alteration with shadow effect, perilesional retinal edema, and hyperreflective foci, referring to hard exudate, reaching the fovea. Superficial capillary plexus on OCTA (G) clearly shows the retinal microaneurysm (*arrow*).

It can be associated with modifications in the adjacent retina both in the perianeurysmal or macular regions. Such additional features identified are: hyperreflective dots in the outer plexiform layer (OPL) suggestive of intraretinal lipids, retinal edema and hyporeflective intraretinal cysts in the outer retinal layers, subretinal fluid (SRF) accumulation and hemorrhages, which appear as hyperreflective areas creating a shadowing effect.[1] Moreover, intraretinal hemorrhage is sited in the OPL while subretinal bleeding is often associated to photoreceptor damage, with retinal thinning and disruption of the ellipsoid zone.[29]

Other authors have reported a further OCT finding that is macular serous pigment epithelium detachment (PED); this feature has been associated with massive hemorrhagic leakage from RAM sparing the macula.[32]

Blood flow is evidenced by OCTA just below the inner plexiform layer (IPL) or in the deeper retinal layers.[33]

Additionally, OCT and OCTA could be useful methods for assessing the response to RAM treatment by analyzing the intraluminal blood flow after laser photocoagulation: the patented aneurysm was shown to have a hyporreflective lumen, which converted to a homogeneous strong hyperreflectivity after laser treatment, suggesting the formation of a thrombus[30]; other studies with OCTA confirmed the same findings, as the disappearance of intraluminal blood flow.[31,33]

Differential diagnosis

RAM is frequently misdiagnosed. Symptomatic RAMs may resemble other vision-threatening retinal diseases that usually occur in elderly subjects such as age-related macular degeneration (AMD) or diabetic maculopathy.[34] However, the absence of choroidal neovascularization (CNV) on FA and the presence of peculiar features associated with RAMs may help to differentiate the illnesses.[14]

Massive subretinal hemorrhage can seem a neoplasm, such as choroidal melanoma or cavernous hemangioma. RAM may also simulate hypertensive retinopathy or BRVO by flame-shaped retinal hemorrhages.

RAM must be differentiated from other hereditary, idiopathic, or secondary ocular diseases characterized by capillary or venous dilatations such as Coats' disease and von Hippel–Lindau disease, radiation retinopathy, or retinal arteritis through their respective clinical, epidemiologic, and multimodal-imaging features.[19]

Therapy

To date, no standard protocol has been approved for the treatment of RAMs; therefore, an individual therapeutic approach is necessary. Many treatment options are reported in the literature including observation, direct or indirect laser photocoagulation, Nd:YAG laser hyaloidotomy, injection of antivascular endothelial growth factor (anti-VEGF), pars plana vitrectomy with ILM-peeling, recombinant tissue plasminogen activator (rtPA) injection, and pneumatic displacement. However, a prospective randomized trial of superiority between all treatment modalities has not yet been conducted.[35,36] Each therapeutic approach showed a favorable response, despite related side effects.

All patients should first undergo a comprehensive investigation for systemic vascular disease, especially hypertension, to set up the appropriate treatment.

Regarding RAM management, observation is indicated for cases with macular sparing.

Laser photocoagulation is a safe and efficient treatment option suggested in cases of macular exudation and leaking microvascular abnormalities around the aneurysmatic dilatation.[35]

Direct photocoagulation is executed at the RAM site to induce regression of the lesion, while indirect laser photocoagulation is performed in the retina around the aneurysm to stop or keep the exudation away from the macular region, reducing oxygen consumption and blood flow within the RAM.[35,36] One study analyzed combined indirect and direct treatments in two separate periods 4 weeks apart, aiming to prevent vast exudation and extensive hemorrhage due to the rupture of the RAM following direct photocoagulation, applying a ring of fibrotic scar. However, this treatment did not lead to a visual advantage than observation alone after 3 years.[35,37]

The main complications of photocoagulation are vitreous hemorrhage, BRAO, late widening of the laser scar, MER, retinal traction, subretinal fibrosis, secondary CNV, increased exudation, and recurrence of the lesion.[36,38]

To obtain the same morpho-functional benefits and to decrease the number of complications related to photocoagulation, subthreshold laser treatment (STLT) has been suggested as an effective therapy; in fact, resolution of exudation with subsequent improvement in VA without visible retinal scarring and complications has been demonstrated, probably due to reduced protein denaturation resulting from decreased duration of laser exposure and localized thermal effects on the RPE.[36,39,40]

Despite this, laser treatment is contraindicated when the RAM is near the fovea or the optic disc; in this case, the use of anti-VEGF, which has no focal destructive results, is recommended.

Intravitreal anti-VEGF injection has been shown to be an efficient treatment for RAMs with hemorrhages and exudation as it may induce a rapid and safe functional improvement with little complications. It has also been revealed that combined laser and anti-VEGF therapy can be effective in decreasing macular exudation.[36]

However, other studies denied these findings observing recurrence of macular edema or no advantage in BCVA between treated and untreated subjects.[37,38]

The mechanisms of RAM involution after anti-VEGF are not fully elucidated although some authors suggested the blockade of the angiogenic and vasopermeability effect and reduction of nitric oxide, resulting in vasoconstriction and thus limitation of macular edema. In addition, an alteration of balance between coagulation and fibrinolysis processes has been proposed, inducing the resolution of various retinal hemorrhages.[38]

Reported complications are lens injury, endophthalmitis, and retinal detachment.[35,38]

Pars plana vitrectomy (PPV) should be considered in case of submacular hemorrhages that prevent the laser treatment or in case of vitreous hemorrhages that have persisted for more than 3 months or that mask the macula. In the former, also the combination of rtPA with submacular surgery or pneumatic displacement has been shown to be effective. In fact, in the case of subfoveal and subretinal blood, rapid blood removal by pneumatic displacement, rtPA injection, or subretinal drainage is advisable.[35]

Complications are retinal tear, progression of cataracts, retinal detachment, endophthalmitis, MH, massive subretinal hemorrhage, and recurrent vitreous bleeding.[36]

Subhyaloid hemorrhage may be treated with Nd: YAG laser treatment to improve visual function without a more invasive vitreoretinal method allowing rapid passage of the blood from the subhyaloid area into the vitreous space through a defect in the posterior hyaloid membrane.

Side effects are vitreous, intraretinal, subretinal hemorrhage, choroidal bleeding, MH, retinal detachment, and MER.[36,41]

Perifoveal vascular anomalous complex

Definition

Perifoveal exudative vascular anomalous complex (PEVAC) is an uncommon macular disease first described by Querques et al. in 2011 and refers to an idiopathic, large, unilateral, isolated, perifoveal aneurysmal dilatation surrounded by exudation and rarefaction of capillary vessels, which occurs in healthy subjects, and usually does not respond to the treatment with anti- VEGF.[42]

Recently, it has been introduced a new definition of this entity, named perifoveal vascular anomalous complex (PVAC); it is divided into two stages: nonexudative perifoveal vascular anomalous complex (nePVAC), characterized by an isolated aneurysmal dilatation and localized microvascular alterations without signs of exudation or visual decline, and exudative perifoveal vascular anomalous complex (ePVAC) where exudation is present.[43] It is still unclear whether nePVAC stands for a preclinical preexudative phase of the lesion rather than a moment in the evolution of PVAC itself, as it is not a steady pathology and may present a series of modifications as both appearance and spontaneous resolution of exudation may occur.[44]

Epidemiology

Originally, PVAC was described as a monofocal and unilateral disease affecting elderly individuals around 60–70 years old[45–47]; however, multifocal, bilateral, or both cases have been reported.[48] Nevertheless, a multifocal PVAC with more than three lesions has never been described.[49]

PVAC was first identified in healthy subjects without other retinal disorders. Later, it has been described in patients with coincident findings such as AMD (40%–50%), MER (17%), advanced lamellar hole stage (LH), and myopic macular degeneration (DMM, 13%–17%); it has also been diagnosed in patients with well-controlled diabetes and hypertension when the retina is not (or only mildly) involved.[43,45,50–55] An association between PVAC and pachychoroidal pigmentary epitheliopathy (PPE, 17%) has been also reported, characterized by improved visual outcomes after anti-VEGF treatment. However, for the authors, it remains unclear whether the choroid is involved in its pathogenesis or whether this was just a coincidental finding.[56]

Furthermore, a study reported that up to 40% of the patients had a stroke in the past and 70% of cases have systemic hypertension, suggesting that PVAC may represent a sign of an underlying cardiovascular disease.[49]

Pathogenesis

The exact mechanism of PVAC pathogenesis is not yet understood. However, a gradual retinal endothelial cell loss has been suggested as the leading cause of the vasogenic cellular mechanism leading to the formation of PEVAC.[42]

Another study hypothesized that the pathogenic mechanism might be related to the degeneration of the retinal vessel due to age or systemic cardiovascular disease which induces apoptosis and the decline of retinal pericytes, specialized contractile cells whose function is to maintain stationary vascular tone and perfusion pressure in capillaries; this may induce proliferation of the endothelium into microaneurysms while subsequent expansion may be consequent to an imbalance between wall resistance and tension. In addition, microvascular regression has been found in the perilesional area in 90% of subjects, which is known to be the result of compromised angiogenesis sustained by loss of angiogenic stimuli, disruption of the endothelium–pericyte connection, angiogenesis inhibitors, and endothelial dysfunction; therefore, aging and pericyte loss may explain both the onset of the aneurysmal changes and perilesional microvascular rarefaction.[49]

Clinical features

Clinically PVAC causes visual impairment or metamorphopsia consequent to CME although some cases without exudation are asymptomatic.[43,47]

At fundus examination it appears as single large aneurismal change localized near the fovea surrounded by little hemorrhages, intraretinal exudation, and hard exudates in the absence of other retinal pathologies.[42] However, the lesion may be present in eyes affected by pathologic myopia, PPE, or AMD.[47,51,56]

Multimodal imaging

Multimodal imaging is essential for a correct diagnosis and is based on OCT, OCTA, FA, and ICGA.[53]

On OCT PVAC appears as a round lesion with a variably reflective lumen surrounded by a hyperreflective wall sited between the OPL and the inner nuclear layer (INL) although, in some cases, it may be identified in the IPL and in the ganglion cell complex (GCC). Dilatation is surrounded by intraretinal cystic spaces or hard exudates and is also rarely accompanied by SRF or PED.[47,56] However, exudation may occasionally be missing.

Most PVAC aneurysmal abnormalities are disclosed in an area located less than 500 μm from the center of the fovea, but some lesions may be identified between 500 and 1500 μm from the fovea.[47]

In a recent work, it has been shown that PVAC was between 100 and 220 μm in diameter and that although aneurysmal dilatation typically originates from capillaries, occasionally the lesions were venous and connected to capillaries.[49,53]

On FA, PVAC presents as a sharp and isolated hyperfluorescent aneurysmal dilation with variable leakage at late stages; whereas, on ICGA, it shows as a hyperfluorescent lesion without leakage in the tardive frames of the study. Both examinations reveal no other retinal or choroidal alterations at the posterior pole or in the retinal periphery combined with PVAC[47] (Fig. 21.3).

OCTA evidences an isolated large aneurysmal lesion with detectable flow mainly in the deep capillary plexus (DCP, 33% of cases); furthermore, blood flow may be noticeable in the superficial capillary plexus (SCP, 17% of cases), both DCP and SCP (33% of cases) or in the DCP and the avascular slab (17%). The lesion is accompanied by a shadow effect in the choriocapillaris segmentation; furthermore, OCTA revealed no signs of CNV or abnormal flow in the choriocapillaris or other macular alterations near the lesion (Fig. 21.4). Finally, the rarefaction of retinal capillaries in the perilesional area is considered a hallmark of the PVAC spectrum.[43,47]

Visual outcome

Most cases have a stable clinical course, with no significant improvement or worsening in the visual function and CME.[56] The edema is usually unresponsive to intravitreal injections of anti–VEGF although few cases of complete resolution of

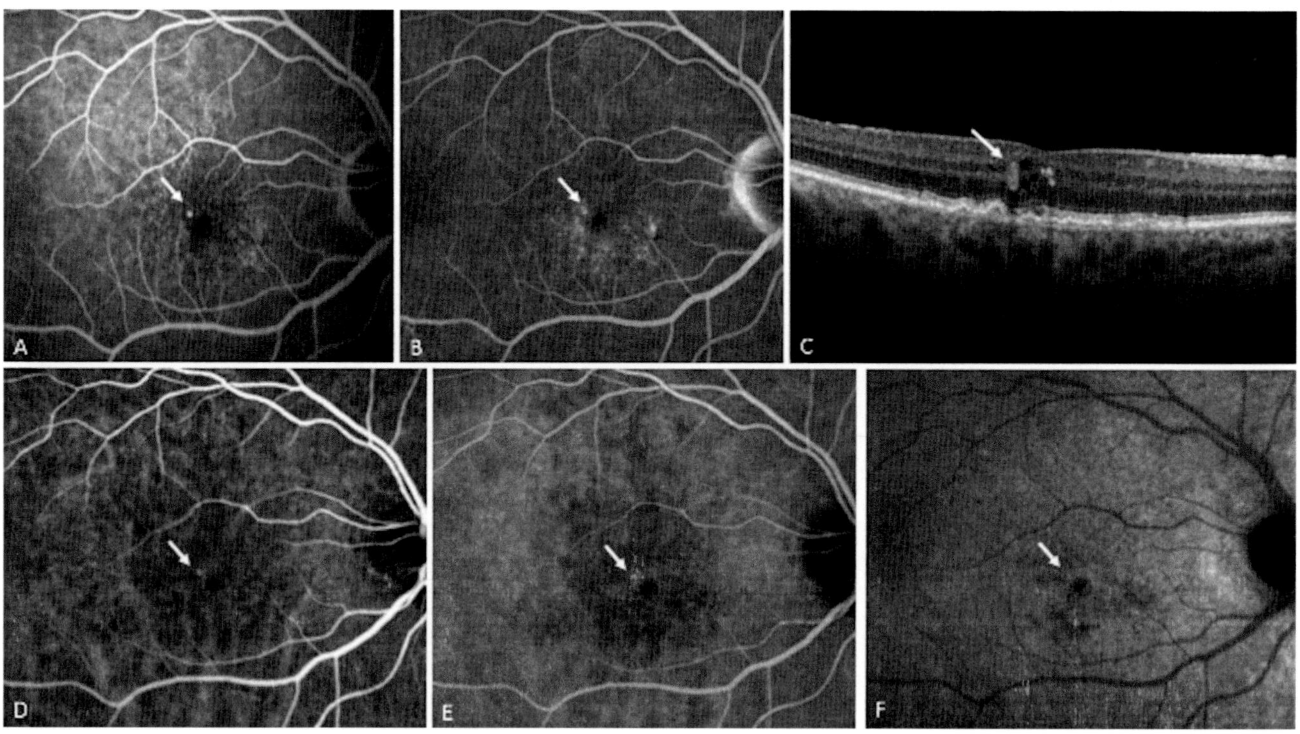

FIGURE 21.3 PVAC in a 79-year-old man with wet AMD in his left eye and intermediate AMD in his right. FA reveals a sharp hyperfluorescent spot in the early phase (A) with weak leakage in late frames (B) (*arrow*), associated with diffuse and smaller hyperfluorescent spots compatible with drusen. ICGA (D–F) confirms the presence of an early sharp hyperfluorescent spot (D) without late leakage (F) (*arrow*), without sign of macular neovascularization. OCT (C) shows a reflective dilatation with hyperreflective lumen and shadow effect on deeper retinal layers (*arrow* C) surrounded by intraretinal cysts and hard exudates, referring to PVAC.

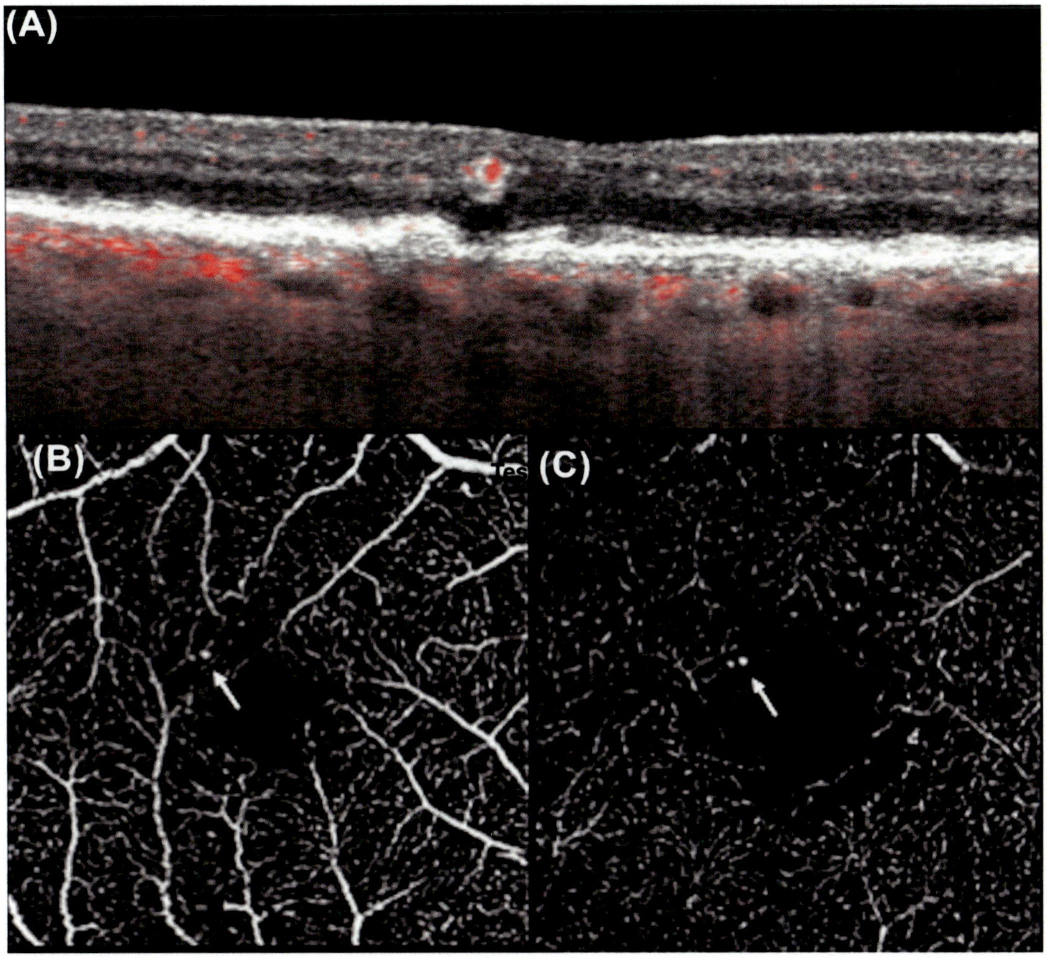

FIGURE 21.4 OCTA of the same patient of Fig. 21.3 affected by PVAC. The patient underwent an OCTA exam (A–C) that reveals blood flow within the lesion (A). OCTA shows perifoveal vessel dilatation on SCP (B) and DCP (C) with perilesional capillary rarefaction, more evident in the DCP, confirming the diagnosis of PVAC.

the exudation with subsequent visual improvement have been reported.[53,57] Occasionally, spontaneous resolution of the CME and the lesion may occur.[45–47,53]

Follow-up is also necessary when a nePVAC lesion is disclosed since 75% of cases will develop exudation after a mean of 15 months; OCTA may be a useful tool for this purpose because the intralesional flow can change over time, therefore increasing blood flow may suggest the progression to exudative complications while closure of the nePVAC lesion may indicate resolution of the process.[43]

Differential diagnosis

PVAC may be misdiagnosed as other diseases, such as macular telangiectasia type 1 (MacTel 1), nascent type 3 macular neovascularization or pseudophakic edema,[57] RAM, retinal capillary macroaneurysm (RCM), or other PEVAC-resembling lesions.

The latter is microangiopathy with clinical and vascular features very similar to PVAC, so they can be differentiated only by etiology; the underlying retinal vascular diseases that induce the formation of the alterations are usually diabetic or hypertensive retinopathy, retinal vein occlusion, inflammatory diseases, prepapillary arterial loops, ocular ischemic syndrome, or MacTel 1.[49,53] The amount and location of lesions of the two diseases aid their differentiation; in fact, most PVAC lesions are identified within 500 μm of the center of the fovea and tend to be single and unilateral, whereas only half of the PEVAC-resembling ones are so central. Alterations of the latter are also more often multifocal and bilateral.[49]

Nascent type 3 macular neovascularization is characterized by intraretinal hyperreflective spots that tend to grow toward the RPE and show noticeable flow in the deep vascular complex (DVC); exudation appears when the lesion exceeds the RPE layer. In contrast, nePVAC does not reveal expansion to the sub-RPE space and thus tends to remain stationary in the intraretinal layers.[43] Furthermore, nascent type 3 macular neovascularization has a good response to anti-VEGF treatment, unlike PVAC.[47]

RAM is a dilation of a large artery at the first three bifurcations of the central retinal artery; therefore, it is located rather distant from the fovea, unlike PVAC. Moreover, retinal aneurysm is associated with good visual outcomes after treatment in most cases.[58]

RCM is a new entity described in patients without any other retinal vascular disease, which is usually persistent, solitary, and larger than PVAC (greater than 200 in diameter) and with an incomplete response to anti-VEGF. It originates from capillary vessels and tends to progress over time and may be discovered in the retinal periphery, unlike PVAC.[53] However, RCM shares multiple similarities to PVAC lesions.

MacTel 1 is an idiopathic macular telangiectasia that occurs in young patients. Like PVAC, rarefaction of the capillary network around the lesion is present; however, MacTel 1 is characterized by multiple capillary, venular, and arteriolar dilatations sited at the level of the SCP and DCP in the temporal juxtafoveal zone and associated with severe exudation; moreover, in this disease, anti-VEGF treatment is also typically incisive.[53,58]

Treatment

There are currently no guidelines for the proper management of PVAC, as it is a rare disease and therefore only a few cases have been reported in the literature.

Observation and careful follow-up are suggested for nePVAC; however, the best treatment for ePVAC has not yet been established.[59]

Intravitreal injections of anti-VEGF have given incomplete or no results in most cases[42,46,47]; nevertheless, some studies have reported persistent regression of the exudation after one to four injections, and in one case even after 13 injections, without recurrence after a long follow-up.[45,53,57] It is still unclear whether the resolution of CME is due to a delayed effect of anti-VEGF or to the natural progress of the pathology.[45]

Recently a case treated with intravitreal injection of triamcinolone acetonide has been reported but it has been proved ineffective too[60]; on the other hand, another patient with PVAC and edema in the lamellar hole-associated epiretinal proliferation (LHEP) has been reported to be partially responsive to three injections of anti-VEGF.[52]

In contrast, it was reported that laser treatment applied directly to the lesion resulted in anatomical and functional improvement with resolution of the edema; however, because of the closeness of the dilatation to the fovea, it may be associated with serious side effects such as permanent paracentral scotoma or visual field loss due to scar itself or its enlargement, secondary hemorrhage, or secondary CNV.[60] This is why new laser approaches have been tried out.

The same results have been obtained from many studies using different laser procedures and instruments, changing wavelength and parameters of the laser therapy: focal thermal photocoagulation, micropulse laser (MPL), subthreshold micropulse one (SMPL), and eye-tracking with angiography-guided focal photocoagulation.[59]

However, edema recurrence or laser inefficacy was detected in a few reports in which focal thermal laser photocoagulation was applied.[57,61]

MPL is also based on the mechanism of photocoagulation but is repeatable and minimally damaging and may reduce the local inflammatory response by deactivating retinal microglia.[58]

Indeed eye-tracking photocoagulation has the main advantage of avoiding laser burns at unwanted sites in addition to being less painful.[62]

SMPL is a noninvasive therapy used in different macular diseases due to the intervals between short pulses that promote cooling of the retinal tissue.[60]

Recently, to reduce the magnitude of the side effects and improve the efficacy of the laser therapy, Arruabarena et al. suggested a combined treatment with three intravitreal anti-VEGF followed by selective focal laser photocoagulation on the ePVAC lesions about 2 weeks after the last injection; the authors hypothesized that this regimen might reduce the surrounding edema to better visualize the PVAC and consequently limit the amount of the laser to be used.[59]

Finally, a case of PVAC that regressed after 1 month of a topical nonsteroidal antiinflammatory drug was reported. However, the authors acknowledged that it was also possible that the favorable response was due to the natural course of the disease.[63]

Retinal capillary macroaneurysm

Definition and epidemiology

RCM is a rare entity introduced by Spaide et al. in 2018, and it refers to a large, single, unilateral, and persistent capillary dilatation, greater than 200 μm in diameter, with surrounding CME, which occurs in healthy subjects and has an incomplete response to anti-VEGF injections.[64]

The average age of onset is 64–70 years, and it tends to increase in caliber over time. Some cases are associated with stable DR or retinal vein occlusion (RVO).[64,65]

Pathogenesis

The mechanism leading to the formation of RCM is not yet understood. It has been hypothesized that aneurysmal enlargement is due to both reduced wall strength consequent to decreased pericyte coverage and increased expression of matrix-metalloproteinase (MMP)-9, which limits the structural integrity of the basement membrane, and enhanced wall tension, as dictated by Laplace's law.

This may also explain the lack of response to the leakage from RCM to anti-VEGF treatment.[64]

Clinical features

Patients tend to be asymptomatic until they develop exudation; in this case, they refer to metamorphopsia and visual loss. On fundus examination, RCM appears as small reddish circular spots surrounded by CME or hard exudates[64,65] and evolves into a whitish nodular lesion after photocoagulation.[64]

Imaging

Multimodal imaging based on FA, ICGA, and OCT is helpful for the diagnosis of RCM.[65]

On OCT RCM is situated in the inner retinal layers and has a hyperreflective wall and hyporeflective lumen and evolves into a uniformly hyperreflective lesion of reduced diameter after laser treatment[65,66]; however, a case of RCM expanding to the outer nuclear layer (ONL) has been described. The lesion may be associated with intraretinal cysts in the INL and in Henle's fiber layer and rarely to SRF.[64,66]

On FA, the vascular abnormality appears hyperfluorescent with late leakage.

Differential diagnosis

RCM must be differentiated from other aneurysmal dilations such as microaneurysms, RAM, and PVAC.

RAM arises from arterioles and is generally larger than RCM. Microaneurysms, by definition, have a smaller caliber, involve the capillaries, and are usually consequent to retinal vascular diseases such as DR and RVO. RCM, similarly to microaneurysms, arise from capillaries, but they differ from the previous for the absence of other concomitant retinal vascular diseases and because they are recalcitrant to anti-VEGF injections.

Differentiating RCM from PVAC is a controversial debate because the two entities are very similar. Originally, PEVAC lesions were described as isolated aneurysms, similar to large microaneurysms, located in the area around the fovea (both in the area less than 500 mm from the center of the fovea, and also in the area between 500 and 1500 mm from the fovea), associated to small hemorrhages, intraretinal exudation, and hard exudates. RCM also can potentially be found in the macular zone but also anywhere else in the retina although; so far, only macular cases of RCM have been described in the literature.[47,64,67]

Treatment

RCMs have recently been introduced in the literature, so there is currently no guideline for their treatment.

It has been revealed that intravitreal injection of anti-VEGF is ineffective, whereas laser photocoagulation may induce the regression of lesions. A combined therapy (laser and anti-VEGF) in the case of RCM associated with other retinal diseases has been shown to be effective. Functional and anatomical improvement without side effects has also been achieved with navigated focal laser photocoagulation, which is more indicated in the case of RCMs near the fovea.[64,65,68]

References

1. Goldenberg D, Soiberman U, Loewenstein A, Goldstein M. Heidelberg spectral-domain optical coherence tomographic findings in retinal artery macroaneurysm. *Retina*. 2012;32:990–995. https://doi.org/10.1097/IAE.0B013E318229B233.
2. Sakaguchi S, Muraoka Y, Kadomoto S, et al. Three-dimensional locations of ruptured retinal arterial macroaneurysms and their associations with the visual prognosis. *Scientific Reports*. 2022;12. https://doi.org/10.1038/S41598-021-04500-4.
3. Rabb MF, Gagliano DA, Teske MP. Retinal arterial macroaneurysms. *Survey of Ophthalmology*. 1988;33:73–96. https://doi.org/10.1016/0039-6257(88)90160-9.
4. Venkatesh R, Gurav P. Optic nerve head retinal artery macroaneurysm: report of a case. *SAGE Open Medical Case Reports*. 2019;7:2050313X1986947. https://doi.org/10.1177/2050313X19869470.
5. Nangia V, Jonas JB, Khare A, Sinha A, Lambat S. Prevalence of retinal macroaneurysms. The Central India Eye and Medical Study. *Acta Ophthalmologica*. 2013;91. https://doi.org/10.1111/j.1755-3768.2012.02465.x.
6. Xu L, Wang Y, Jonas JB. Frequency of retinal macroaneurysms in adult Chinese: the Beijing Eye Study. *British Journal of Ophthalmology*. 2007;91:840–841. https://doi.org/10.1136/BJO.2006.107342.
7. Moosavi RA, Fong KCS, Chopdar A. Retinal artery macroaneurysms: clinical and fluorescein angiographic features in 34 patients. *Eye*. 2006;20:1011–1020. https://doi.org/10.1038/sj.eye.6702068.
8. François J. Acquired macroaneurysms of the retinal arteries. *International Ophthalmology*. 1979;1:153–161. https://doi.org/10.1007/BF00137493.
9. Abdel-Khalek MN, Richardson J. Retinal macroaneurysm: natural history and guidelines for treatment. *British Journal of Ophthalmology*. 1986;70:2. https://doi.org/10.1136/BJO.70.1.2.
10. Hughes EL, Dooley IJ, Kennelly KP, Doyle F, Siah WF, Connell P. Angiographic features and disease outcomes of symptomatic retinal arterial macroaneurysms. *Graefe's Archive for Clinical and Experimental Ophthalmology*. 2016;254:2203–2207. https://doi.org/10.1007/s00417-016-3388-9.
11. Shults WT, Swan KC. Pulsatile aneurysms of the retinal arterial tree. *American Journal of Ophthalmology*. 1974;77:304–309. https://doi.org/10.1016/0002-9394(74)90734-X.
12. Pierro L, Battista M, Arrigo A, Manitto MP, Bandello F. Multimodal imaging in pediatric arterial macroaneurysm: a case report. *European Journal of Ophthalmology*. 2021;31:NP58–NP62. https://doi.org/10.1177/1120672119899033.
13. Lavin MJ, Marsh RJ, Peart S, Rehman A. Retinal arterial macroaneurysms: a retrospective study of 40 patients. *British Journal of Ophthalmology*. 1987;71:817–825. https://doi.org/10.1136/bjo.71.11.817.
14. Cleary PE, Kohner EM, Hamilton AM, Bird AC. Retinal macroaneurysms. *British Journal of Ophthalmology*. 1975;59:355–361. https://doi.org/10.1136/BJO.59.7.355.
15. Nadel AJ, Gupta KK. Macroaneurysms of the retinal arteries. *Archives of Ophthalmology*. 1976;94:1092–1096. https://doi.org/10.1001/archopht.1976.03910040012000.
16. Kakihara S, Kurenuma T, Murata T. Branch retinal artery occlusion with prepapillary macroaneurysm enlargement. *Ophthalmology Retina*. 2021;5:1124. https://doi.org/10.1016/J.ORET.2021.08.003.
17. Robertson DM. Macroaneurysms of the retinal arteries. *Transactions of the American Academy of Ophthalmology & Otolaryngology*. 1973;77(1):55–67. PMID: 4730083.
18. Palestine AG, Robertson DM, Goldstein BG. Macroaneurysms of the retinal arteries. *American Journal of Ophthalmology*. 1982;93:164–171. https://doi.org/10.1016/0002-9394(82)90411-1.
19. Pitkänen L, Tommila P, Kaarniranta K, Jääskeläinen JE, Kinnunen K. Retinal arterial macroaneurysms. *Acta Ophthalmologica*. 2014;92:101–104. https://doi.org/10.1111/AOS.12210.
20. Noble CW, di Nicola M, Hermanson ME, Williams BK. Retinal arterial macroaneurysm in the setting of type 1 arteriovenous malformation. *Retina*. 2021;41:e15–e16. https://doi.org/10.1097/IAE.0000000000003024.
21. Sebrow DB, Cunha de Souza E, Belúcio Neto J, et al. Macroaneurysms associated with congenital retinal macrovessels. *Retinal Cases & Brief Reports*. 2020;14:61–65. https://doi.org/10.1097/ICB.0000000000000619.
22. Lewis RA, Norton EWD, Gass JDM. Acquired arterial macroaneurysms of the retina. *British Journal of Ophthalmology*. 1976;60:21–30. https://doi.org/10.1136/BJO.60.1.21.
23. Fichte C, Streeten BW, Friedman AH. A histopathologic study of retinal arterial aneurysms. *American Journal of Ophthalmology*. 1978;85:509–513. https://doi.org/10.1016/S0002-9394(14)75249-3.
24. Sakaguchi S, Muraoka Y, Nishigori N, et al. Detection and characteristics of unruptured retinal arterial macroaneurysms. *Retina*. 2022;42:1909–1914. https://doi.org/10.1097/IAE.0000000000003548.
25. Chen H, Zhao XY, Chen YX, Deng TT. Angiotensin II is a crucial factor in retinal aneurysm formation. *Experimental Eye Research*. 2021;213. https://doi.org/10.1016/J.EXER.2021.108810.
26. Ohno-Matsui K, Hayano M, Futagami S, Tokoro T, Ohno S. Spontaneous involution of a large retinal arterial macroaneurysm. *Acta Ophthalmologica Scandinavica*. 2000;78:114–117. https://doi.org/10.1034/j.1600-0420.2000.078001114.x.
27. Abdul-Rahman A, Morgan W, Yu DY. Near infra-red reflectance videography in the evaluation of retinal artery macroaneurysm pulsatility. *America Journal of Ophthalmology Case Reports*. 2022;27. https://doi.org/10.1016/J.AJOC.2022.101664.
28. Cahuzac A, Scemama C, Mauget-Faÿsse M, Sahel JA, Wolff B. Retinal arterial macroaneurysms: clinical, angiographic, and tomographic description and therapeutic management of a series of 14 cases. *European Journal of Ophthalmology*. 2015;26:36–43. https://doi.org/10.5301/ejo.5000641.

29. Doi S, Kimura S, Morizane Y, et al. Adverse effect of macular intraretinal hemorrhage on the prognosis of submacular hemorrhage due to retinal arterial macroaneurysm rupture. *Retina*. 2020;40:989–997. https://doi.org/10.1097/IAE.0000000000002460.
30. Kyoung Lee E, Woo J, Ahn J, Park H. Morphologic characteristics of retinal arterial macroaneurysm and its regression pattern on spectral-domain optical coherence tomography. *Retina*. 2011;31. https://doi.org/10.1097/IAE.0b013e3182111711. PMID: 21716167.
31. Miura M, Muramatsu D, Hong YJ, et al. Noninvasive vascular imaging of ruptured retinal arterial macroaneurysms by Doppler optical coherence tomography. *BMC Ophthalmology*. 2015;15. https://doi.org/10.1186/s12886-015-0077-0.
32. Takahashi K, Kishi S. Serous macular detachment associated with retinal arterial macroaneurysm. *Japanese Journal of Ophthalmology*. 2006;50:460–464. https://doi.org/10.1007/s10384-006-0347-0.
33. Alnawaiseh M, Schubert F, Nelis P, Wirths G, Rosentreter A, Eter N. Optical coherence tomography (OCT) angiography findings in retinal arterial macroaneurysms. *BMC Ophthalmology*. 2016;16. https://doi.org/10.1186/s12886-016-0293-2.
34. Harold F, Spalter M. Retinal macroaneurysms; a new masquerade syndrome. *Transactions of the American Ophthalmological Society*. 1982;80:113–130. PMID: 7182954.
35. Koinzer S, Heckmann J, Tode J, Roider J. Long-term, therapy-related visual outcome of 49 cases with retinal arterial macroaneurysm: a case series and literature review. *British Journal of Ophthalmology*. 2015;99:1345–1353. https://doi.org/10.1136/BJOPHTHALMOL-2014-305884.
36. Meng Y, Xu Y, Li L, He L, Yi Z, Chen C. Retinal arterial macroaneurysm rupture by Valsalva maneuver: a case report and literature review. *BMC Ophthalmology*. 2022;22:461. https://doi.org/10.1186/s12886-022-02662-x.
37. Mansour AM, Foster RE, Gallego-Pinazo R, et al. Intravitreal anti-vascular endothelial growth factor injections for exudative retinal arterial macroaneurysms. *Retina*. 2019;39:1133–1141. https://doi.org/10.1097/IAE.0000000000002131.
38. Cho HJ, Rhee TK, Kim HS, et al. Intravitreal bevacizumab for symptomatic retinal arterial macroaneurysm. *American Journal of Ophthalmology*. 2013;155:898–904. https://doi.org/10.1016/j.ajo.2012.12.003.
39. Battaglia Parodi M, Iacono P, Ravalico G, Bandello F. Subthreshold laser treatment for retinal arterial macroaneurysm. *British Journal of Ophthalmology*. 2011;95:534–538. https://doi.org/10.1136/BJO.2009.175547.
40. Battaglia Parodi M, Iacono P, Pierro L, Papayannis A, Kontadakis S, Bandello FM. Subthreshold laser treatment versus threshold laser treatment for symptomatic retinal arterial macroaneurysm. *Investigative Ophthalmology & Visual Science*. 2012;53:1783–1786. https://doi.org/10.1167/IOVS.11-8772.
41. Rennie CA, Newman DK, Snead MP, Flanagan DWN. YAG laser treatment for premacular subhyaloid haemorrhage. *Eye*. 2001;15. https://doi.org/10.1038/eye.2001.166.
42. Querques G, Kuhn D, Massamba N, Leveziel N, Querques L, Souied EH. Perifoveal exudative vascular anomalous complex. *Journal Français d'Ophtalmologie*. 2011;34:559.e1–559.e4. https://doi.org/10.1016/j.jfo.2011.03.002.
43. Sacconi R, Borrelli E, Sadda S, et al. Nonexudative perifoveal vascular anomalous complex: the subclinical stage of perifoveal exudative vascular anomalous complex? *American Journal of Ophthalmology*. 2020;218:59–67. https://doi.org/10.1016/J.AJO.2020.04.025.
44. Verhoekx JSN, Smid LM, Vermeer KA, Martinez Ciriano JP, Yzer S. Comment on: nonexudative perifoveal vascular anomalous complex: the subclinical stage of perifoveal exudative vascular anomalous complex? *American Journal of Ophthalmology*. 2021;223:450–451. https://doi.org/10.1016/j.ajo.2020.10.010.
45. Kim JH, Kim JW, Kim CG, Lee DW. Characteristics of perifoveal exudative vascular anomalous complex in Korean patients. *Seminars in Ophthalmology*. 2019;34:353–358. https://doi.org/10.1080/08820538.2019.1626450.
46. Verhoekx JSN, Smid LM, Vermeer KA, Martinez Ciriano JP, Yzer S. Anatomical changes on sequential multimodal imaging in perifoveal exudative vascular anomalous complex. *Retina*. 2021;41:162–169. https://doi.org/10.1097/IAE.0000000000002809.
47. Sacconi R, Freund KB, Yannuzzi LA, et al. The expanded spectrum of perifoveal exudative vascular anomalous complex. *American Journal of Ophthalmology*. 2017;184:137–146. https://doi.org/10.1016/J.AJO.2017.10.009.
48. Fernández-Vigo JI, Burgos-Blasco B, Dolz-Marco R, et al. Atypical perifoveal exudative vascular anomalous complex (PEVAC) with multifocal and bilateral presentation. *American Journal of Ophthalmology Case Reports*. 2020;18. https://doi.org/10.1016/j.ajoc.2020.100717.
49. Smid LM, Verhoekx JSN, Martinez Ciriano JP, Vermeer KA, Yzer S. Multimodal imaging comparison of perifoveal exudative vascular anomalous complex and resembling lesions. *Acta Ophthalmology*. 2021;99:553–558. https://doi.org/10.1111/AOS.14650.
50. Ayachit AG, Sacconi R, Ayachit GS, Joshi S, Querques G. Perifoveal exudative vascular anomalous complex-like lesion as a complication of prepapillary arterial loops. *Ophthalmic Surgery, Lasers and Imaging Retina*. 2018;49:974–978. https://doi.org/10.3928/23258160-20181203-11.
51. Sacconi R, Borrelli E, Bandello F, Querques G. Perifoveal exudative vascular anomalous complex in a highly myopic eye. *Therapeutic Advances in Ophthalmology*. 2020;12:251584142094793. https://doi.org/10.1177/2515841420947930.
52. Siedlecki J, Vounotrypidis E, Vogt D, Wolf A, Priglinger SG, Schumann RG. Lamellar hole-associated epiretinal proliferation presenting with perifoveal exudative vascular anomalous complex. *American Journal of Ophthalmology Case Reports*. 2019;14:112–116. https://doi.org/10.1016/J.AJOC.2019.03.008.
53. Torrell-Belzach N, Miere A, Souied E, Cohen SY. Long-term resolution of perifoveal exudative vascular anomalous complex after intravitreal injections of anti-vascular endothelial growth factor. *Case Reports Ophthalmology*. 2022;13:936. https://doi.org/10.1159/000526991.
54. Venkatesh R, Yadav NK, Bavaharan B, Prabhu V, Sinha S. Multimodal imaging in perifoveal exudative vascular anomalous complex with co-existent diabetic retinopathy. *Clinical and Experimental Optometry*. 2019;102:528–532. https://doi.org/10.1111/cxo.12868.
55. Zhang Z, Xu L, Wu Z, Zhang J. Case report: perifoveal exudative vascular anomalous complex in a Chinese patient with diabetes mellitus. *Optometry and Vision Science*. 2019;96:531–535. https://doi.org/10.1097/OPX.0000000000001401.

56. Herda N, Semoun O, Srour M, et al. Atypical case of perifoveal exudative vascular anomalous complex associated with pachychoroid pigment epitheliopathy. *American Journal of Ophthalmology Case Reports*. 2022;28:101691. https://doi.org/10.1016/J.AJOC.2022.101691.
57. Mrejen S, Le HM, Nghiem-Buffet S, Tabary S, Quentel G, Cohen SY. Insights into perifoveal exudative vascular anomalous complex. *Retina*. 2020;40:80−86. https://doi.org/10.1097/IAE.0000000000002435.
58. Fu M, Hu P, Zhang G, et al. Case report: a case of perifoveal exudative vascular anomalous complex with a good prognosis. *Frontiers of Medicine*. 2021;8. https://doi.org/10.3389/fmed.2021.757313.
59. Arruabarrena C, Liaño G, Hernáez-Leonato JM, Martínez-Sánchez M, Cañas-Martín J, Teus MA. Selective focal laser therapy as the primary treatment for exudative perifoveal vascular anomalous complex (ePVAC): case series and literature review. *Ophthalmic Surgery, Lasers and Imaging Retina*. 2023;54:43−49. https://doi.org/10.3928/23258160-20221219-01.
60. Kang YK, Park HS. Multiple-session subthreshold micropulse laser therapy for exudative perifoveal vascular anomalous complex: a case report. *Retinal Cases & Brief Reports*. 2021;17(3). https://doi.org/10.1097/ICB.0000000000001181.
61. Venkatesh R, Pereira A, Sridharan A. 'Hematocrit sign' in PEVAC lesion. *Asia Pacific Journal of Ophthalmology (Philadelphia)*. 2021;10:224−225. https://doi.org/10.1097/APO.0000000000000385.
62. Remolí-Sargues L, Montero-Hernández J, Monferrer Adsuara C, Castro Navarro V, Cervera-Taulet E. Perifoveal exudative vascular anomalous complex treated with navigated retina laser. *Journal Français d'Ophtalmologie*. 2022;45:e9−e11. https://doi.org/10.1016/j.jfo.2021.03.011.
63. Tombolini B, Cavalleri M, Sacconi R, et al. Progressive resolution of exudation from perifoveal vascular anomalous complex: a possible role of diclofenac therapy? *American Journal of Ophthalmology Case Reports*. 2022;26:101472. https://doi.org/10.1016/J.AJOC.2022.101472.
64. Spaide RF, Barquet LA. Retinal capillary macroaneurysms. *Retina*. 2019;39:1889−1895. https://doi.org/10.1097/IAE.0000000000002406.
65. Amoroso F, Pedinielli A, Colantuono D, et al. Selective photocoagulation of capillary macroaneurysms by navigated focal laser. *Ophthalmic Surgery, Lasers and Imaging Retina*. 2021;52:366−373. https://doi.org/10.3928/23258160-20210628-02.
66. Rehmani AS, Banaee T, Makkouk F. Subretinal leakage of a retinal capillary macroaneurysm - a case report. *BMC Ophthalmology*. 2021;21. https://doi.org/10.1186/S12886-021-01984-6.
67. Spaide RF. Reply. *Retina*. 2019;39:E49−E50. https://doi.org/10.1097/IAE.0000000000002568.
68. Nadig RR, Kashyap H, Loganathan MP, Bhende M. Effect of combined laser and anti-vascular endothelial growth factor in a retinal capillary macroaneurysm. *Indian Journal of Ophthalmology*. 2022;70:2712−2713. https://doi.org/10.4103/IJO.IJO_581_22.

Chapter 22

Macular telangiectasia

Apoorva Ayachit, Guruprasad Ayachit and Shrinivas Joshi
Department of Vitreoretina, M. M. Joshi Eye Institute, Hubballi, Karnataka, India

Overview bullets

- Macular telangiectasia
- Epidemiology and classification systems
- Fundoscopic features
- Imaging—fluorescein angiography, autofluorescence, confocal blue reflectance, macular pigment optical density, optical coherence tomography, optical coherence tomography angiography
- Clinical, OCT, and OCTA correlates
- Vascular patterns on OCTA
- Quantitative metrics on OCTA
- Multifocal ERG
- Microperimetry
- Artificial intelligence, deep learning

Introduction

Gass and Oyakawa identified parafoveal telangiectatic retinal vessels that were not associated with any other disease and named the condition, juxta foveolar retinal telangiectasia. They classified the condition into four groups.[1] Gass and Blodi later classified it into *group 1A,* unilateral congenital telangiectatic capillaries temporal to the fovea; *group 1B,* unilateral, small area of incompetent capillaries next to the foveal avascular zone; *group 2A,* bilateral idiopathic, acquired para foveolar telangiectasis with right-angle venules, crystalline deposits retinal pigment epithelial plaques, subretinal neovascularization; *group 2B,* juvenile familial idiopathic juxtafoveal telangiectasia (IJT); *group 3A,* occlusive IJT with obliteration of perifoveal capillaries; and *group 3B,* occlusive IJT associated with central nervous system vasculopathy.[2]

Macular telangiectasia type 2 (Mac Tel) is a disease of an idiopathic nature wherein, characteristic macular alterations with progressive macular atrophy and subsequent neovascularization are noted. Based on data from Caucasian studies, the prevalence has been seen to be between 0.004% and 0.022% (Australian study)[3] and up to 0.1% (seen in Beaver Dam study) in the age group of 40−80 years.[4]

The cause of Mac Tel is unclear, but the understanding of the disease and its progression has advanced significantly due to improvements in retinal imaging. Yannuzzi classified Mac Tel into either nonproliferative or proliferative stages based on the presence of subretinal neovascularization (SRNV) on clinical examination, FA, and OCT.[5]

Clinical classification of Mac Tel

Mac Tel type 2 is a bilateral and often symmetric disease. However, it may be more advanced in one eye than the other.[6] The changes begin in the temporal parafovea and then occupy a characteristic oval region within one disk diameter from the foveal center, with a horizontal diameter greater than the vertical diameter. Temporal changes are followed by foveal and nasal involvement. Eyes with advanced disease may manifest only temporal changes and the nasal side may be unaffected. The increase in severity of the disease may not necessarily imply there is a progression from the temporal to the nasal fovea.

Retinal and Choroidal Vascular Diseases of the Eye. https://doi.org/10.1016/B978-0-443-15583-3.00022-6

Fundoscopic features of Mac Tel

1. *Blunting of foveolar reflex*: This may be the earliest finding. Although this is not uncommon in the older population, it may be significant in cases of involvement of the other eye in later stages.
2. *Retinal graying* (loss of transparency) in the parafoveolar area is a hallmark of the disease. Theories postulated for graying include mild extra/intracellular edema that leads to light scatter. Muller cells are involved in ion homeostasis and act as light-transmitting fibers; hence, their dysfunction may account for a grayish appearance (Fig. 22.1).[7] Parafoveal graying may decrease in intensity after continuous light exposure from an indirect ophthalmoscope. The graying reappears after dark adaptation for 15 min. This could be a photochemical reaction due to chromophore release from abnormal Muller cells.[8]
3. *Crystalline deposits*: These are seen in the inner retina and arranged along the nerve fiber layers. Crystalline deposits are frequently bilateral and can be seen in all stages of the disease.[9]
4. *Blunted, dilated venules*: These are associated with ectatic capillaries.[2] Ectatic capillaries are difficult to discern clinically; however, blunted, slightly dilated venules can be made out that taper toward the foveola and abruptly dip into the deeper layers of the retina.[5] These venules appear dilated because of increased blood flow in the abnormal capillaries or also because they are seen in association with retinochoroidal anastomosis (RCA). The draining venule is said to be draining the choroid because of the RCA and hence appears more dilated.[10]
5. Pseudo holes, lamellar holes, and full-thickness atrophic holes may be seen. Lamellar holes are due to cavitary changes in the inner retina. Macular thinning and photoreceptor loss may cause outer lamellar holes. Full-thickness holes typically do not successfully close with vitreoretinal surgery.
6. *Retinal pigment/pigmented plaques*: These start as small foci of hyperplasia, progress to become larger, and migrate intraretinally2[2] (Fig. 22.1). The contact of capillaries with RPE due to the collapse of the retinal layers causes intraretinal migration of pigment surrounding the retinal capillaries.[11]
7. *Neovascular complexes*: These lesions cause sudden vision loss and can complicate the course of the disease at any stage.[2] The lesions are preceded by blunted venules and pigmented plaques. Active CNVM lesions may present with edema, exudates, and subretinal or intraretinal hemorrhage.[5] These are distinct from neovascular age-related macular degeneration in that, they originate from the retinal vasculature, gain access to subretinal space, and may develop chorioretinal shunts in some eyes.

Fluorescein angiography (FA) has been considered the gold standard for Mac Tel 2. In the earlier stages of the disease, diffuse hyperfluorescence in the late phase in the temporal fovea is seen. This is postulated to be due to loss of macular pigment that results in a relative window defect. The hyperfluorescence from ectatic capillaries is more due to the staining of capillary walls rather than leakage. FA also demonstrates right-angled vessels and intraretinal and/or subretinal

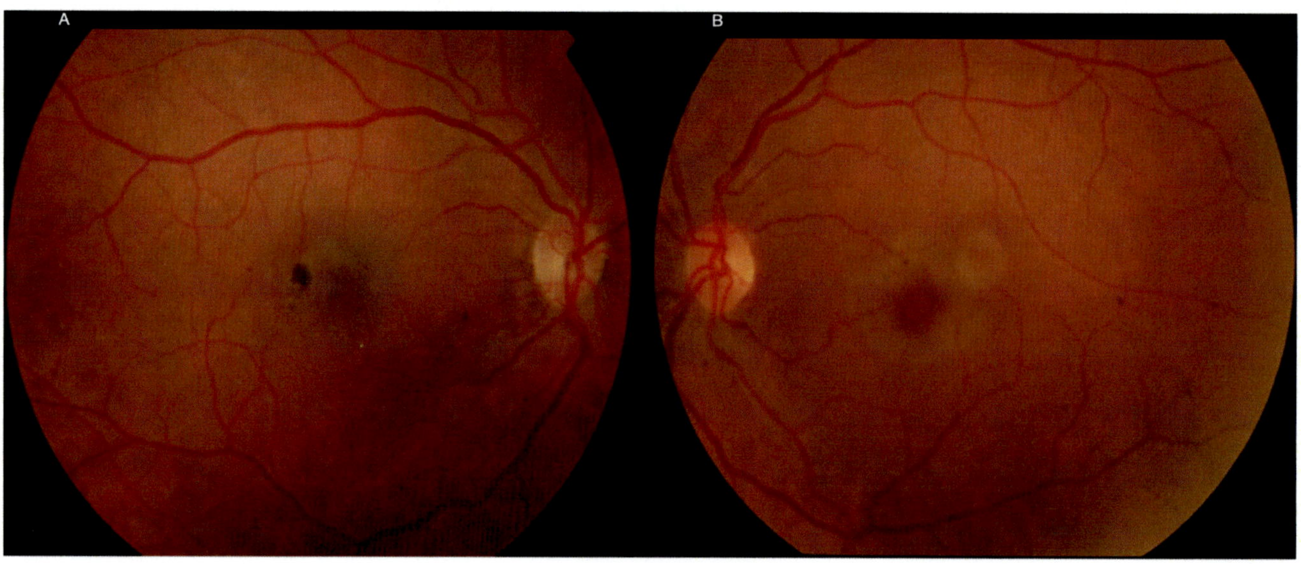

FIGURE 22.1 A) Area of perifoveal graying, with pigment in the temporal parafovea. (B) Perifoveal graying in early Mac Tel.

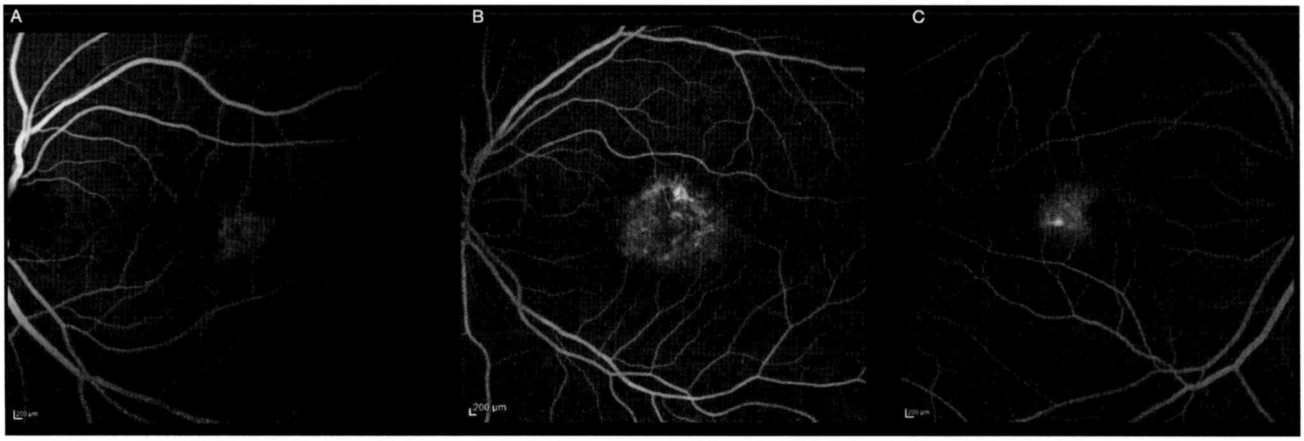

FIGURE 22.2 A) FFA showing only temporal parafoveal hyperfluorescence. (B) Circumferential parafoveal hyperfluorescence. (C) Temporal parafoveal hyperfluorescent background with a central brighter hyperfluorescence suggestive of SRNV.

anastomosis.[2] The presence of subretinal neovascular membrane (SRNVM) is seen as an early focal hyperfluorescence that gets merged with the background diffuse hyperfluorescence in later frames (Fig. 22.2).

Autofluorescence—Short-wave AF signal is derived from lipofuscin in the retinal pigment epithelium. It also depends on the presence and number of absorbing pigments or structures. Macular pigment is present maximally at the fovea. The pigment absorbs blue light; therefore, autofluorescence recorded using 488 nm (blue light) shows a hypo-AF signal due to absorption by macular pigment. Loss of this hypo-AF at the fovea is the earliest feature of Mac Tel 2. Wong et al. categorized disease progression based on clinical, angiographic, OCT, AF, and microperimetry findings. Category 0 eyes had normal clinical and imaging findings. Category 1 eyes had increased foveal AF as the only imaging abnormality. Category 2 eyes had mild-to-moderate foveal hyperautofluorescence along with other clinical and angiographic findings of Mac Tel. Category 3 had a moderate to marked increase in foveal AF and foveal atrophy on OCT. Category 4 had a mixed pattern of hypo- and hyperautofluorescence with clinically evident pigment clumping. Loss of macular pigment, retinal atrophy, pigment, and a changing composition of fluorophores in the RPE underly these changes in the AF (Fig. 22.3).[12]

Loss of macular pigment, retinal atrophy, pigment, and a changing composition of fluorophores in the RPE underlie these changes in the AF.[12]

Confocal blue reflectance—Imaging with a blue laser at 488 nm known as the confocal blue reflectance (CBR) modality demonstrates a well-defined, generally oval, parafoveal area of increased reflectance that corresponds to a larger area than seen clinical graying or angiographically. The intensity of reflectance is increased by dark adaptation. This is because CBR shows the area of macular pigment loss and explains why altered reflectance appears earlier than angiographically visible vascular alterations. Atrophy seen in advanced stages of Mac Tel results in a reduction of the reflectivity.[13] Fruttiger noted that all typical signs of Mac Tel appear to be located within an area defined by increased blue reflectance seen on CBR and suggested the term "Mac Tel area."[13] Further deficits on OCT occur later and expand to fill this Mac Tel area. There is distortion in the SRNVM stage and contraction of the membrane. However, any changes in OCT, FA, and AF do not extend beyond the Mac Tel area defined by CBR.[13] Early detection of CBR may be relevant in genetic studies. Mac Tel area on CBR suggests abnormal pigment is in an oval shape surrounding the macula even though there are only temporal parafoveal findings initially. This susceptible area may be a metabolic attribute of this area related to abnormal interactions between Muller cells and photoreceptor cells (Fig. 22.4).[13]

Macular pigment optical density (MPOD) is lost in a characteristic pattern. The loss is more pronounced temporal to the fovea than nasal. The loss of pigment is confined to within 6.5 degrees in the horizontal and 5 degrees in the vertical axis. The loss of macular pigment is divided into three classes: class 1—wedge-shaped loss of pigment in the temporal parafovea; class 2 shows the enlarged area with foveal center involvement; class 3 is the loss of pigment in an oval-shaped area centered on the foveola.[14]

OCT and OCTA have made possible, the huge strides achieved in the understanding of Mac Tel pathophysiology, disease progression, and the impact of structural abnormalities on visual function. OCT utilizes a light source, a charge-coupled device (CCD) camera, and a spectrometer (spectral domain OCT or SD-OCT). Alternatively, it sweeps through a range of optical frequencies (swept-source OCT or SS-OCT). The light source is of a broad bandwidth that enables micrometer (micron) resolution. SS-OCT devices use wavelengths greater than 1000 nanometer (nm) and operate at speeds

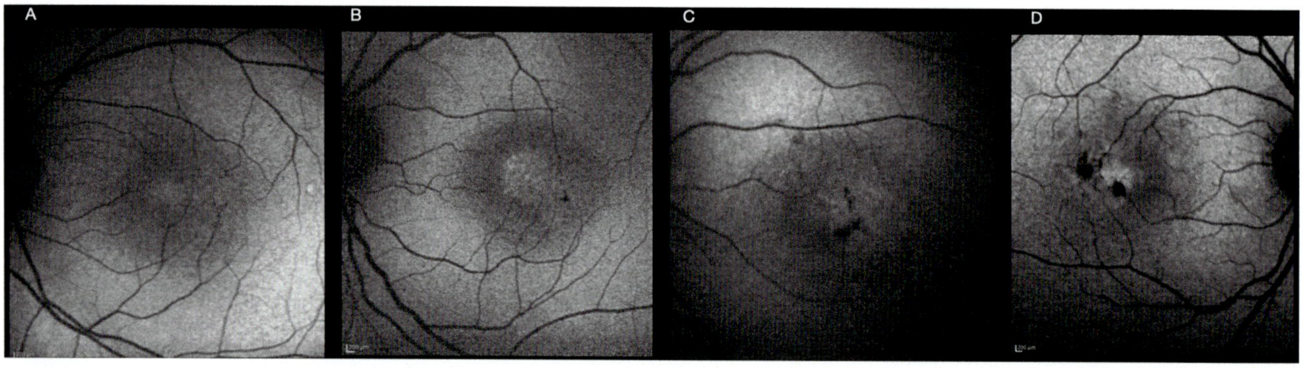

FIGURE 22.3 A) Loss of foveal hypoautofluorescence. (B) Loss of foveal hypo-AF with a speckle of temporal hypo-AF. (C and D) Mixed hypo-AF and hyper-AF signals.

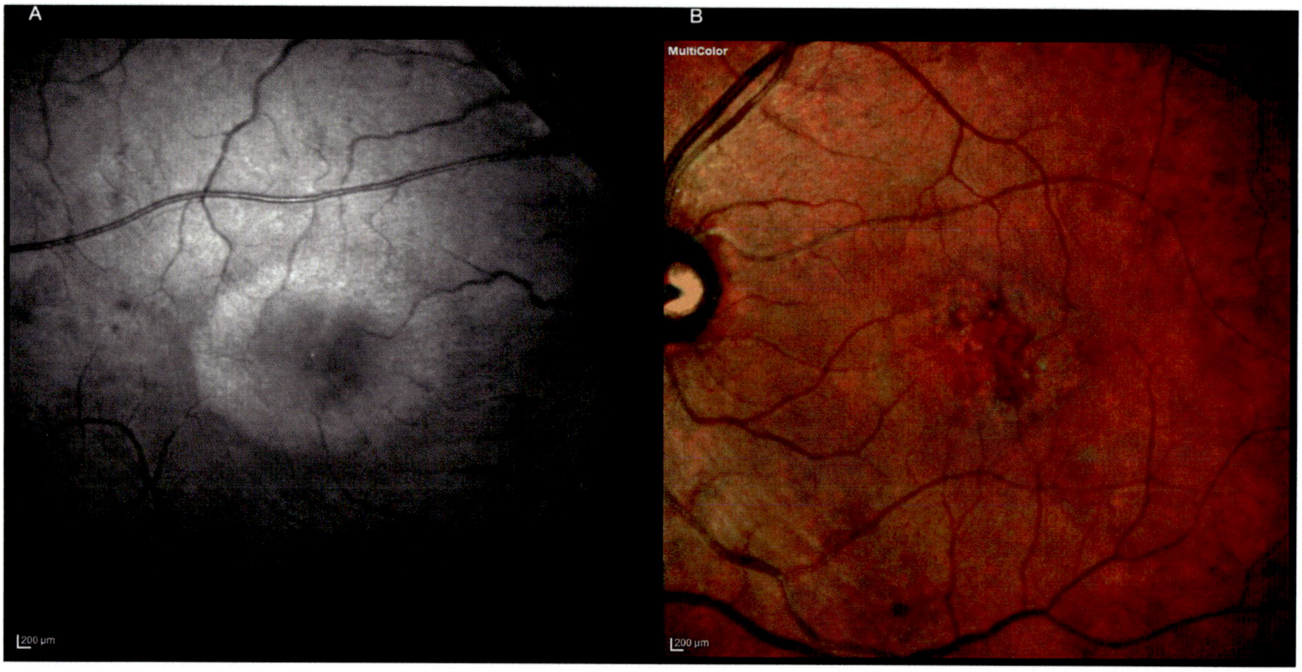

FIGURE 22.4 A) Confocal blue reflectance depicted by an oval area of hyperreflectance indicating "Mac Tel area." (B) Multicolor image of a patient with advanced Mac Tel-reddish plaque indicating pigment and more severe changes temporally than nasally.

at or above 100 kiloHertz (kHz).[15] Spectral domain OCT (SD-OCT) uses a wavelength source of 800–900 nm with scan speeds that range from 40,000 to 100,000 Hz.

OCT features—Important OCT findings in Mac Tel include hyporeflective spaces in the inner and outer retina, internal limiting membrane (ILM) drape, ellipsoid zone (EZ) and external limiting membrane (ELM) loss, macular thinning, neurosensory detachment, lamellar macular hole, full-thickness macular hole, and pigment clumps.[16] Sallo et al. in a multimodal imaging study of Mac Tel proposed an OCT grading, which proposed four grades. Grade 1 was a discontinuity in the EZ, with small retinal cysts confined to the temporal parafovea. Grade 2 was a larger break in the EZ that reached the foveal center, central outer retinal cysts, and inner cysts involving the center and the nasal side as well. Collapse (subsidence) of the outer plexiform layers (OPLs) and the layers internal to it toward the choroid is also a distinct feature. Grades of subsidence have been described as 0 if subsidence is less than one-third depth of the outer nuclear layer (ONL). Grade 1 designates extension from 1/3rd to 2/3rd the thickness and Grade 2 greater than 2/3rd of the entire thickness. In Grade 3 of OCT staging hyperreflective intraretinal lesions due to pigment plaques are visible. Grade 4 has, in addition to Grades 1–3, signs of subretinal neovascularization (SRNV) evidenced by hyperreflective lesion in the subretinal space with subretinal and/or intraretinal fluid. Increased macular thickening is associated with SRNVM (Fig. 22.5).[17,18]

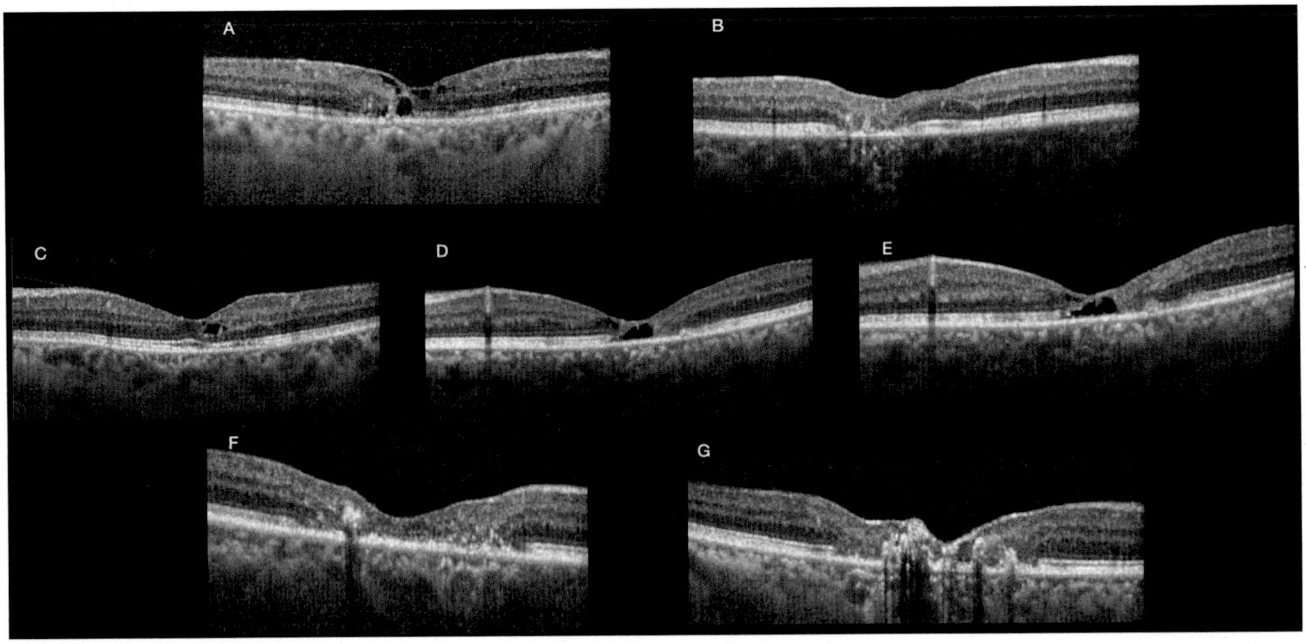

FIGURE 22.5 A) Outer foveal loss, inner retinal cavitary changes. ILM drape. (B) Temporal parafoveal subsidence indicated by the collapse of inner retinal layers toward the choroid. (C, D, E) Longitudinal OCT changes of Mac Tel in the left eye. (C) Changes limited to temporal parafovea, (D) degenerative space involving fovea, and (E) late involvement of nasal fovea. (F) Intraretinal migration of pigment. (G) Transretinal pigment migration associated with back shadowing and outer retinal hyperreflectivity in nasal fovea.

Chew et al. have recently proposed a simplified classification based on OCT.[19] Ellipsoid zone and its integrity are of interest because Mac Tel eyes show loss of EZ in proportion to the progression and severity. EZ and ELM loss start in the temporal parafovea and progressively involve the foveal and nasal parafovea (Fig. 22.5). The EZ and ELM band loss correspond to loss of retinal function as evidenced by microperimetry (MP).[20,21] Structural damage to the outer retinal photoreceptor layers is associated with loss of retinal sensitivity.[22]

Optical coherence tomography angiography (OCTA) This noninvasively images retinal and choroidal microvasculature without the use of intravenous dye injection. OCTA provides better imaging through segmentation in three dimensions (3D). OCTA works on a similar principle as OCT, utilizing two OCT signals: one of which is backscattered from static structural tissue (steady-signal), and the other is backscattered from moving red blood corpuscles (RBCs) in vessels.[23] Changes in OCT signals in subsequent scans are caused by moving particles that generate angiographic contrast and detail.

Considering Mac Tel affects all the vascular layers of the retina and results in neovascularization of the retina and choroid, it is an ideal pathology for imaging by OCTA.

Thorell et al. first published high-quality OCTA images of Mac Tel using the optical coherence tomography microangiography (OMAG) algorithm with scans performed on the Zeiss 1050 nm SS-OCT protype.[24] In early nonproliferative Mac Tel, dilated vessels were seen in the DCP in the temporal fovea corresponding to the disruption of the EZ. There are no visible alterations in the SCP suggesting that the earliest changes are in the DCP.[25] In intermediate nonproliferative Mac Tel, the aneurysmal changes extended into the SCP and outer retina, and the changes corresponded to the loss of EZ in temporal, subfoveal, and nasal fovea (Fig. 22.6).

In proliferative Mac Tel, anastomoses appear in the juxtafoveal capillaries that extend into the outer retina where the EZ is disrupted forming a subretinal fibrovascular plaque.[26] Another of the earlier descriptions of OCTA in Mac Tel was by Spaide et al. in 2015.[27] They utilized the RTvue XR AVANTI (Optovue Inc) to obtain SSADA images in 14 eyes of seven patients. OCTA analysis showed loss of capillaries in both the superficial and deep plexus (more in the deep vascular plexus). The vessels in the outer plexus showed widespread thinning on OCT in advanced Mac Tel stages, along with decreased flow. In addition, there was the loss of perifoveal capillaries in the inner plexus and the invasion of deeper layers and subretinal space with new vessels (Fig. 22.7). The authors reiterated the theories of Muller cell abnormalities and depletion. As Muller cells participate in inflammatory and proliferative processes such as removal and metabolization of

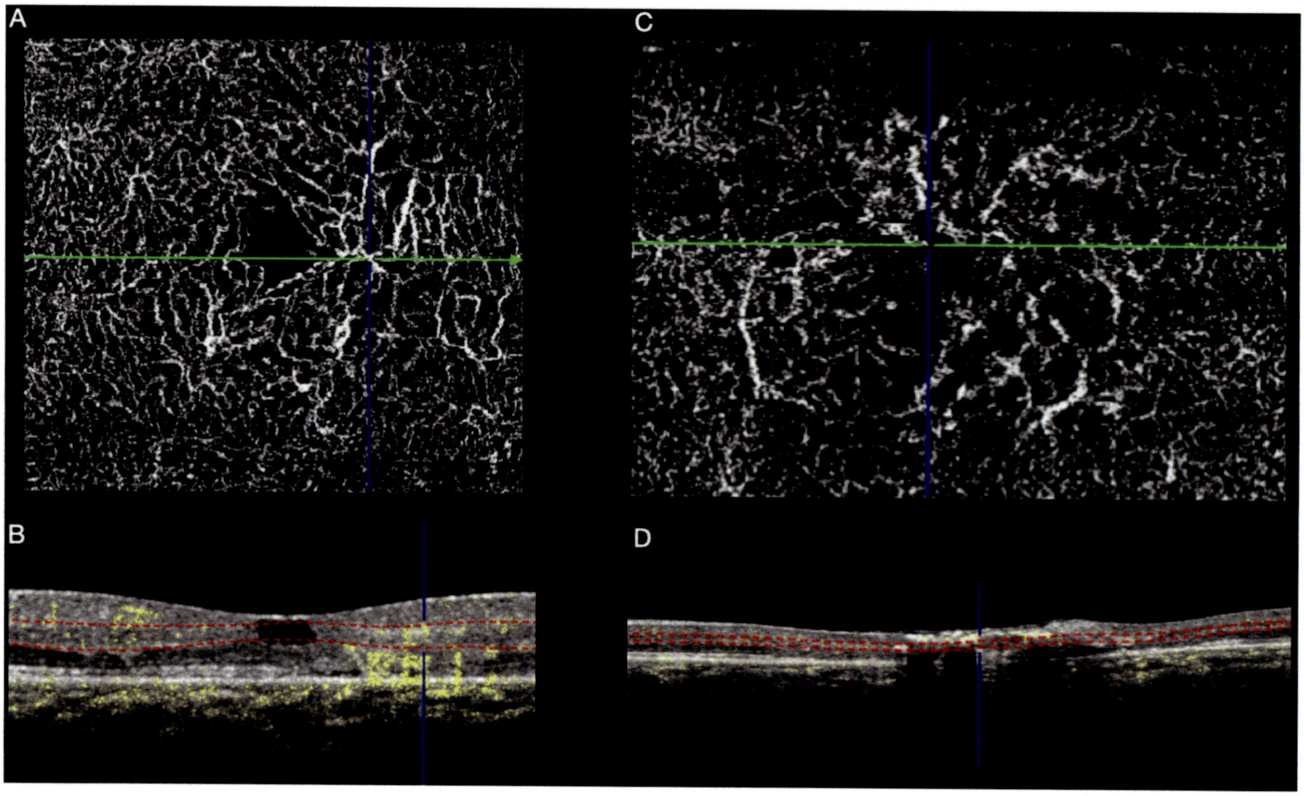

FIGURE 22.6 A) Abnormal capillaries in temporal parafovea seen in DVC, (B) flow signal prominent due to abnormal capillaries in temporal parafovea in early Mac Tel. (C) Circumferential parafoveal changes in advanced Mac Tel. (D) Advanced Mac Tel evidenced by macular thinning and pigment with back shadowing.

neurotransmitters, synaptogenesis, and neuroprotection,[28] the loss of Muller cells causes both neuronal and vascular abnormalities of photoreceptor loss, telangiectasia, and eventual subretinal and intraretinal neovascularization.[29]

Clinical, OCT, and OCTA correlates

Right-angled venules and their OCT correlate

Tzaridis studied the OCT correlation of RAVs with regard to their origin, course, and relation to SRNV. A comparison of OCT and OCTA revealed hyperreflective lesions in the former whose size corresponded to the caliber of dilated vessels. RAVs were noted to dive into deeper layers on OCTA and form anastomoses visualized in the deep and/or superficial retinal layers. All SRNV eyes had RAVs, which could not only be tracked from superficial to outer retinal layers but also seen to be directly connecting to the SRNV complexes. The authors also found subclinical RAVs on OCTA in 69% of stage 1 eyes and 77% of stage 2 eyes. The finding of RAVs in clinical stage 1 and stage 2 eyes were in contradiction to the classical Gass classification that classifies RAVs as stage 3.[30]

Retinochoroidal anastomosis—OCT and OCTA correlate

The detection of RCA in earlier stages of Mac Tel has changed the understanding of disease progression[31] wherein RCA was classified as a feature of advanced Mac Tel.[32] Spaide et al. described RCA on OCT and volume-rendered OCTA in earlier stages of the disease.[10] The anastomotic vessels of RCA connected the deep vessels at the macula with the choriocapillaris at the regions of subsidence. Predictors of RCA were the presence of pigment and Grade 2 subsidence. It is interesting that RCAs were seen even in the nasal parafovea. FA showed reduced fluorescence in the upstream venous segment draining from the right-angled vein. Right-angled veins were associated with RCA, which descended directly through the RPE layer. The most reliable indicator of RCA was a right-angled vein, which is purported to be a compensatory change in the draining vein rather than the cause for RCA. The presence of multiple RCA without development SRNV, coursing perpendicularly (hitherto termed stages 4 and 5), and their presence even in the nasal macula

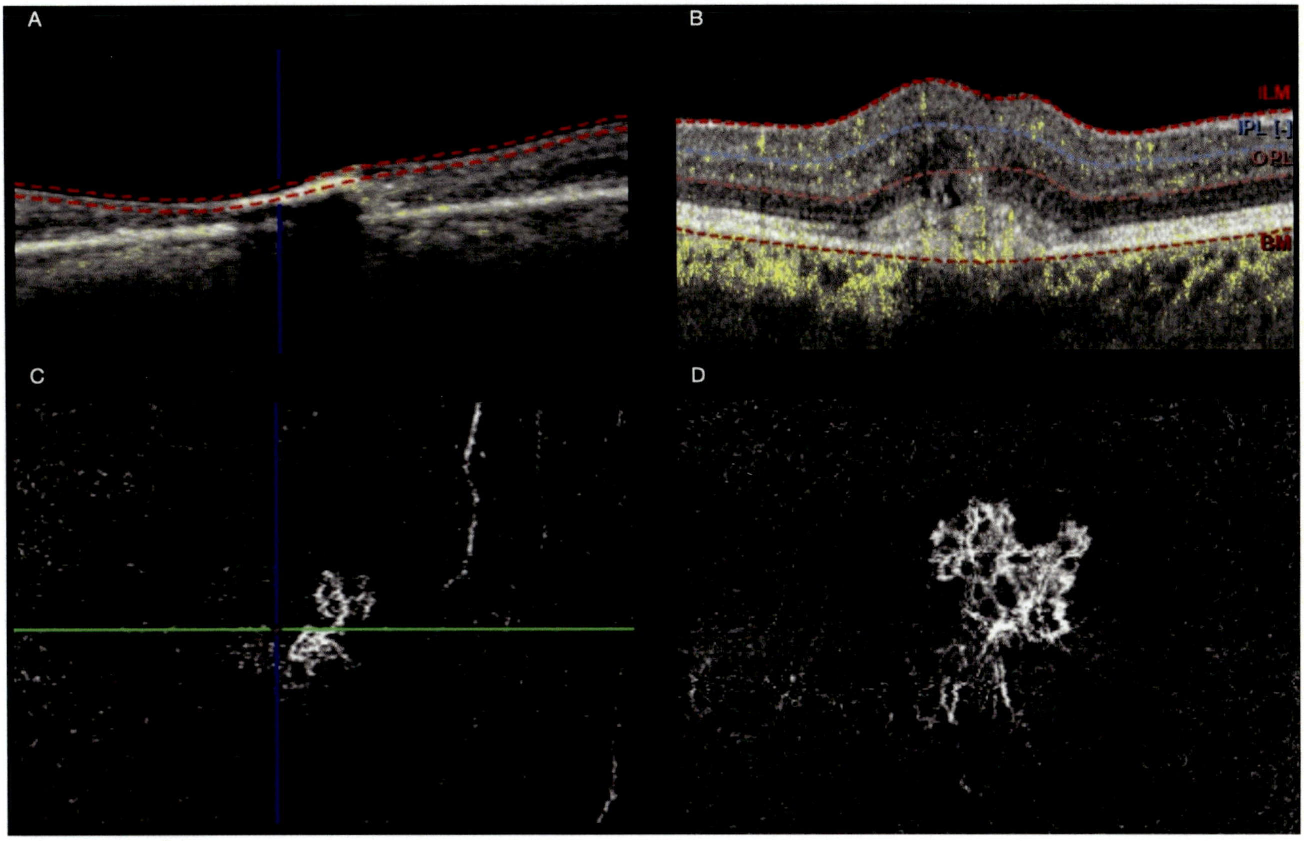

FIGURE 22.7 (A) Epiretinal neovascularization at the vitreoretinal interface on line scan. (B) Network seen on VR interface. (C) Mac Tel CNVM with thickening. (D) Network seen in outer retina.

are in contradiction to previously existing theories that RCA is seen only in end-stage Mac Tel.[32] As an extension of this study, Brezzeano et al. found RCAs were increasingly seen in association with focal hyperpigmentation, EZ defects, and hyperreflective outer retinal lesions.

Retinal pigment—OCT and OCTA correlate

Krivosic et al. found the OCT features[33] of EZ loss and outer retinal hyperreflectivity (ORH) with no clinically visible pigmentation at baseline could evolve into visible pigmentation. They noted that ORH either developed or increased in a large proportion of eyes with EZ loss at baseline and was a predecessor for pigment to appear clinically. In the eyes that developed clinically visible pigment, ORH reached more than half the retinal thickness. In this study, outer retinal neovascularization (ORN) was described as having a flow signal in the ONL on OCTA (Fig. 22.8).

This was different from SRNV, which is characterized by increased retinal thickness with cystoid spaces. ORN showed a distinct additional abnormal capillary network, connected to, but external to the DCP. They noted that this capillary proliferation progresses throughout the course of Mac Tel. In their series, ORN developed or increased in 36% of cases, only in eyes with baseline EZ loss (Fig. 22.8). Thus, ORH and ORN appeared long before the occurrence of clinically obvious pigment and SRNV, respectively (Fig. 22.9).

Mueller also found similar findings in their study on ORN and ORH. They proposed an interesting corollary wherein ONV associated with ORH could be a target for treatment with anti-VEGF and probably limit the progression of EZ loss because ORNV was associated with EZ loss, which may cause hypoxia and invasion of neovascularization from DVC to the outer retina. As ORaH was seen without extensively pigmented plaque lesions, it can be postulated that even unpigmented/lightly pigmented RPE cells may surround NVs and still cause clinically evident hyperpigmentation that is not enough to cause back shadowing.[34] Venkatesh et al. in their study described ORN as “preproliferative Mac Tel,” which developed into proliferative Mac Tel after a mean of 58 $\pm$ 31.9 months.[35]

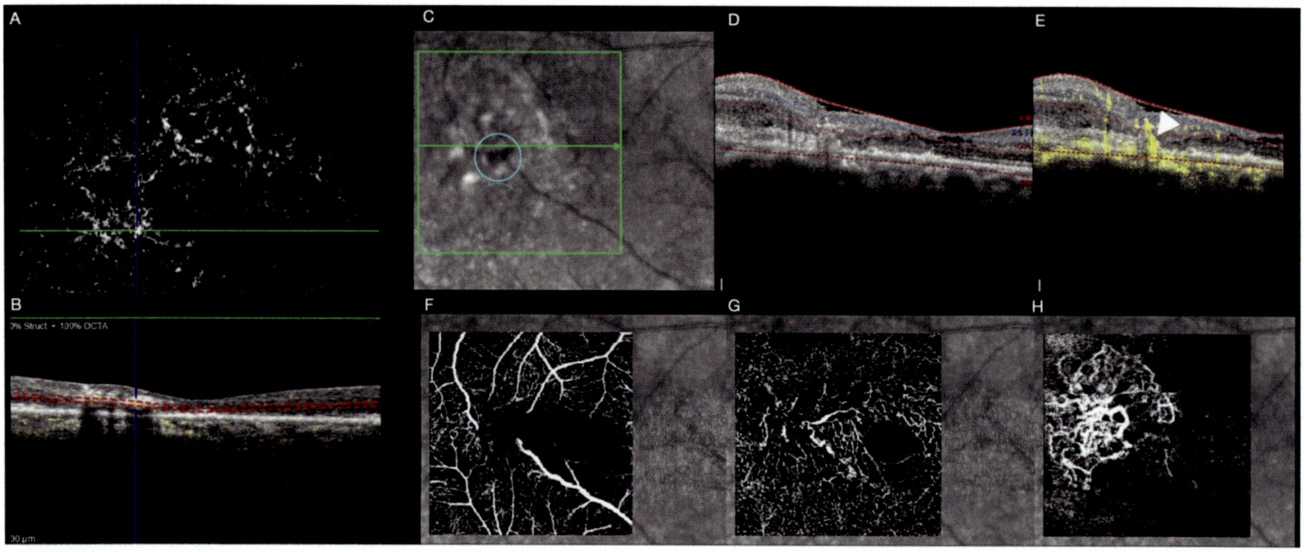

FIGURE 22.8 A) Network of outer retinal neovascularization (ORN). (B) Corresponding ORH showing a flow signal. (C) Enface OCT showing pigment (blue circle) with a large-caliber draining venule. (D) ORH with macular thickening in the temporal macula. (E) ORH with flow signal showing communication from intraretinal to sub-RPE space-suggesting ORCA (white arrowhead). (F) SVC showing the large-caliber draining venule. (G) Abnormal bunching of DVC capillaries. (H) Large neovascular complex in the outer retinal slab.

Epiretinal neovascularization (ERN) was described by Ayachit et al. in seven eyes of four patients with Mac Tel.[36] ERN was defined as a well-defined vascular structure on enface OCTA at the vitreoretinal interface. Notably, all eyes had severe retinal thinning with pigments in the superficial retina. ( ) There was the absence of degenerative spaces intraretinally in these eyes. It was hypothesized that like SRNVM, ERNM could be a distinct endpoint suggesting the proliferative stage of Mac Tel as they are seen in eyes with pigment deposition (stage 4 clinical). Ayachit et al. also described sub-ILM neovascularization wherein a distinct membrane was seen in the superficial retina, just below the ILM.[37]

Tzaridis et al. evaluated the utility of OCTA for monitoring activity, progression, and response to therapy of SRNV in Mac Tel.[38] In this retrospective study, SRNVs were reviewed over a mean period of 13.4 months in eyes that received anti-VEGF treatments on a PRN basis. Retreatment was instituted when clinical and B-scan OCT signs of activity were noted. Interestingly, in this study, the growth of neovascular vessels on OCTA preceded signs of neovascular activity on OCT. The authors noted that, despite the PRN treatment protocol, NV complexes showed progressive growth over time. New vessels preferentially grew and extended within areas of decreased choriocapillaris perfusion. The authors observed changes in vessels in all retinal layers with a shifting toward the outer retina. Considering OCTA documented the growth of NV despite a lack of activity on OCT, the authors suggested an OCTA-based treatment regimen for retreatments in Mac Tel.

Patterns of vessels on OCTA

Paueliokoff et al. sought to investigate the different appearances of vascular changes in Mac Tel and to describe their possible progression. The patterns in different layers were analyzed with OCTA and correlated with the OCT disease severity scale based on the location and extent of EZ loss.[39] The vascular parameters analyzed were (1) flow density, (2) total vessel length, (3) number of vessel branches, (4) number of vessel segments, and (5) fractal dimension.

Significant differences in vascular parameters were seen among all grades of OCT disease particularly in nasal and temporal as well as in the total field of the ETDRS grid. The sequence of vascular changes, that is, the temporal involvement of the DCP and later the SCP in first and later nasal areas was evidenced by mathematical analysis in this study. The morphological progression of the disease was found to correlate significantly with progressive vascular changes, especially in parameters such as the number of branches, vessel segments, and fractal dimension.

Quantitative OCTA metrics in Mac Tel

Park et al. studied the retinal vascular changes on OCTA and assessed their correlation with visual acuity.[40] The foveal VD of the superficial plexus and the deep capillary plexus was significantly lower in the Mac Tel 2 group than in the control

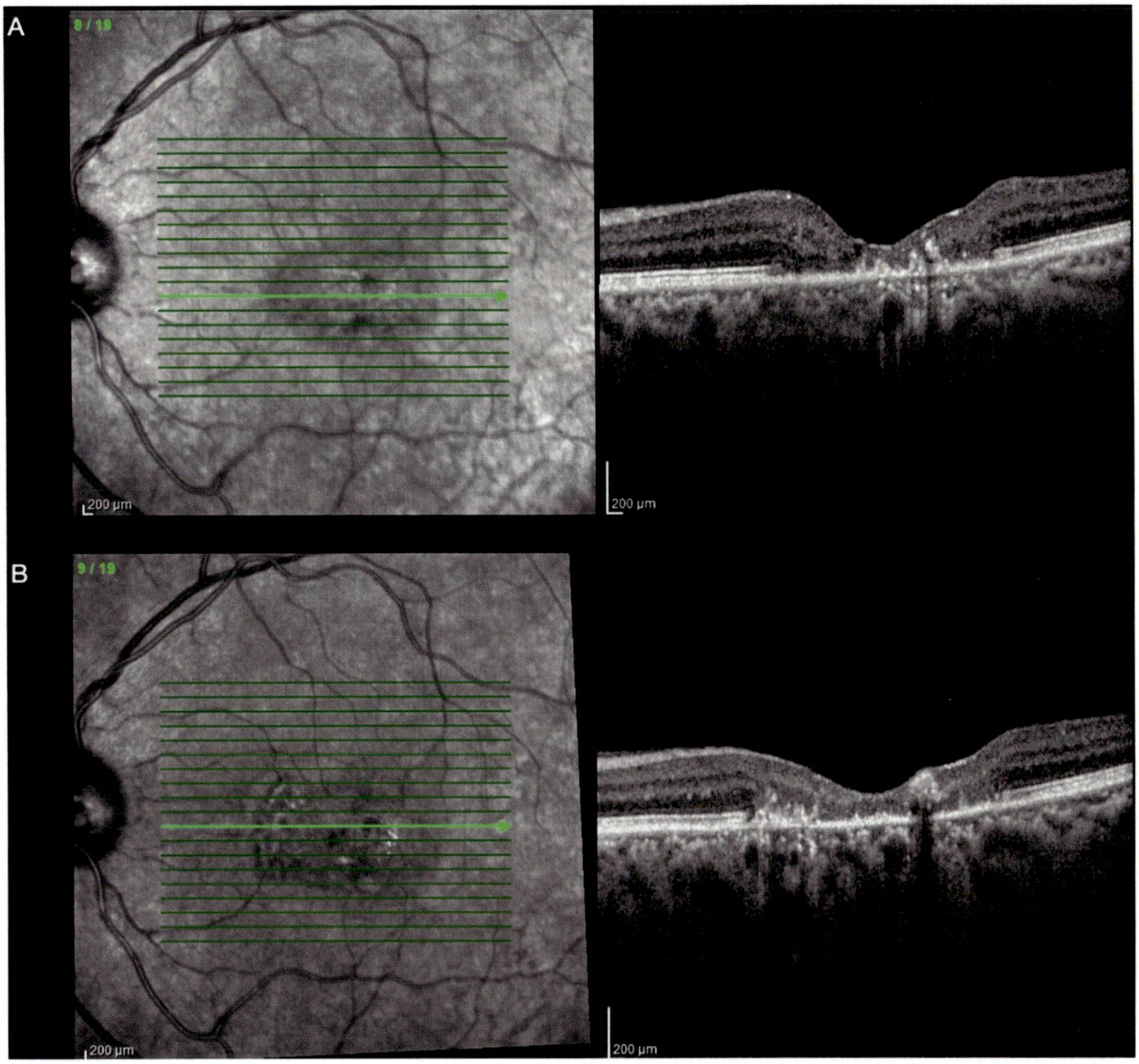

FIGURE 22.9 A) EZ and IZ loss at baseline with ORH in a patient with Mac Tel. (B) Four years later, intraretinal migration of pigment corresponding to visible pigmentation (evidenced by the IR panel).

group. The parafoveal vascular density in the superficial plexus did not differ significantly between the two groups. However, parafoveal vascular density parameters for the deep capillary plexus were significantly lower in the Mac Tel 2 group than in the control group (superior/temporal/inferior/nasal: $P = .001$, $P = .006$, $P = .027$, $P = .035$, and $P = .017$, respectively). The enlarged FAZ of both superficial and deep plexus was negatively correlated with visual acuity. Unlike other studies. EZ disruption in these eyes was not correlated with VA. This is in accordance with Heeren et al. who also reported that EZ loss correlated significantly with change in scotomas but not with BCVA loss. One possible explanation for this is that foveal vessel rarefaction and foveal cavitations with atrophy are related to FAZ enlargement.

Interdigitation zone, EZ, OCTA, and adaptive optics

Ong et al. characterized interdigitation zone (IZ) abnormalities and their correlation to EZ disruption and vascular changes on OCTA. In the advanced stages of Mac Tel without neovascularization, DCP telangiectasia was localized to the extent of EZ loss. However, earlier Mac Tel stages showed DCP telangiectasias showed IZ loss with no evidence of EZ loss.[41] IZ loss area was significantly higher in both moderate and advanced Mac Tel; whereas, both EZ loss areas were higher in only advanced Mac Tel. They noted that when IZ and EZ loss coexisted, EZ loss tended to be in the center of a larger zone of IZ loss. On adaptive optics scanning laser ophthalmoscopy (AOSLO), areas of IZ disruption colocalize with hyporeflective lesions even in areas of preserved EZ. This study postulated that areas of IZ disruption could be indicative of "photoreceptors at risk" and that they precede EZ disruption.

Multifocal ERG

Mali Okada et al., in their study, showed significant asymmetry between temporal and nasal responses on multifocal ERG, which is consistent with the progression of Mac Tel based on imaging. Peripheral mf ERG responses were unaffected as Mac Tel is a localized disturbance of macular function. There was preservation of the early N1 in R1 but selective reduction of the P1. They further stated that an increase in the size of the EZ break area was significantly associated with decreasing central hexagon P1 amplitudes. The primary pathology in Mac Tel is thus postphotoreceptoral.[42]

Microperimetry

Microperimetry as a functional test has been studied in relation to Mac Tel. As microperimetry can test for sensitivity and assess extrafoveal visual loss, it has been used as an endpoint in clinical trials of Mac Tel as a secondary outcome. However, challenges are test duration and the requirement of good fixation. However, after the initial learning curve, high test—retest reliability has been found. Kihara et al. developed a deep-learning network that could directly estimate retinal function from superpositions of high-resolution OCT scans and microperimetry results. The predictions were used to create high-density enface sensitivity maps of the macula. In their study, they found that activation maps generated showed that the important structures identified were in the outer retina, especially the EZ.[43]

Automated segmentation—artificial intelligence and deep learning

Loo et al. studied the comparison between a fully automatic, deep-learning-based segmentation algorithm and an expert-dependent semiautomated segmentation algorithm (previously validated) in measuring the primary outcome of a clinical trial for Mac Tel type 2.[44] Deep OCT atrophy detection (DOCTAD) was used as the automatic segmentation method to assess EZ defects on OCT images of phase 2 clinical trial of ciliary neurotrophic factor (CNTF). The quantitative analysis in both automated and semiautomated segmentation methods showed a good agreement at baseline and month 24. The authors of this study proposed that the automated method's masked design reduces the probability of developing a system that tends to study a singular data set vis a vis a learning-based algorithm that can deal with a diverse set of pathological and anatomic features. It is also interesting to note that ultimately both segmentation methods resulted in the same clinical outcome, that is, the slowing of the progression of EZ loss with the CNTF implant. This study also showed how clinical endpoints are better metrics than performance metrics. The laborious expert-dependent semiautomated was seen to be potentially replaceable by the automated method.

Treatments in Mac Tel

Multiple studies have established the efficacy of anti-VEGF treatments in the proliferative phase of Mac Tel. Improvement of visual acuity and reduction in CMT has been noted in eyes treated with ranibizumab, bevacizumab, and aflibercept. Numerous studies have shown that anti-VEGF agents cause an improvement in BCVA and a reduction in mean CMT. The follow-up duration in these studies ranged from 12 to 76 months. The consensus seems to be that although the macular thickness reduces with an improvement in vision along with the need for reduced injections compared to neovascular AMD, the neurodegenerative nature of Mac Tel ultimately causes macular atrophy.

Anti-VEGF agents, focal and grid laser, subthreshold laser, PDT, and intravitreal steroids have been tried in nonproliferative stage of Mac Tel. The laser caused increased pigmentation, subretinal hemorrhage, scarring, and fibrovascular tissue. Use of PDT, micropulse laser, and intravitreal steroid have shown mixed results. The fluid in nonproliferative Mac Tel is more degenerative in contrast to the fluid due to active neovascular Mac Tel. This could be why the effects of various pharmacotherapeutic and laser treatments on fluid reduction are temporary.

An intravitreal implant (NT-501) delivering transfected RPE cells programmed to overexpress CNTF has completed phase 2 and phase 3 trials and has shown beneficial effects in terms of reduced area of neurodegeneration and EZ disruption on enface OCT at 24 months from baseline. Other outcome measures like reading speed and retinal sensitivity on microperimetry were in favor of the implant over the sham group. The phase 3 multicenter trial was completed in September 2022, and the results are awaited.

Other research on mechanisms of macular telangiectasia type 2

Genetic abnormalities associated with a deficiency in the amino acid serine have been identified in Mac Tel. Insufficient serine due to defects in serine metabolism leads to the accumulation of toxic deoxysphingolipids. Although these findings have revealed crucial areas for further research, there is no evidence to show that serine supplementation would be beneficial in Mac Tel.

References

1. Aung KZ, Wickremasinghe SS, Makeyeva G, Robman L, Guymer RH. The prevalence estimates of macular telangiectasia type 2: the Melbourne collaborative cohort study. *Retina*. 2010;30(3):473–478. https://doi.org/10.1097/IAE.0b013e3181bd2c71.
2. Ayachit A, Fatima F, Salvi R. Subinternal limiting membrane neovascularization in macular telangiectasia. *Ophthalmology Retina*. 2022;6(1):42. https://doi.org/10.1016/j.oret.2021.10.010.
3. Ayachit AG, Reddy LU, Joshi S, Ayachit GS. Epiretinal neovascularization: a novel OCT angiography finding in macular telangiectasia type 2. *Ophthalmology Retina*. 2019;3(6):516–522. https://doi.org/10.1016/j.oret.2019.01.022.
4. *Cellular Signaling and Factors Involved in Müller cell Gliosis: Neuroprotective and Detrimental Effects*. 1966. PubMed.
5. Charbel Issa P, Berendschot TTJM, Staurenghi G, Holz FG, Scholl HPN. Confocal blue reflectance imaging in type 2 idiopathic macular telangiectasia. *Investigative Opthalmology & Visual Science*. 2008;49(3):1172. https://doi.org/10.1167/iovs.07-0636.
6. Chew E, Peto T, Clemons T, et al. A New Classification for Macular Telangiectasia type 2 based on multi-modal imaging. *Investigative Ophthalmology & Visual Science*. 2019;60(9).
7. Clemons TE, Gillies MC, Chew EY, et al. Baseline characteristics of participants in the natural history study of macular telangiectasia (MacTel) MacTel project report No. 2. *Ophthalmic Epidemiology*. 2010;17(1):66–73. https://doi.org/10.3109/09286580903450361.
8. Drexler W, Fujimoto JG. State-of-the-art retinal optical coherence tomography. *Progress in Retinal and Eye Research*. 2008;27(1):45–88. https://doi.org/10.1016/j.preteyeres.2007.07.005.
9. Engelbrecht NE, Aaberg TM, Sung J, Lewis ML. Neovascular membranes associated with idiopathic juxtafoveolar telangiectasis. *Archives of Ophthalmology*. 2002;120(3):320–324. https://doi.org/10.1001/archopht.120.3.320.
10. Franze K, Grosche J, Skatchkov SN, et al. Müller cells are living optical fibers in the vertebrate retina. *Proceedings of the National Academy of Sciences of the United States of America*. 2007;104(20):8287–8292. https://doi.org/10.1073/pnas.0611180104.
11. Gass D, Oyakawa RT. Idiopathic juxtafoveolar retinal telangiectasis. *Archives of Ophthalmology*. 1982;100(5):769–780. https://doi.org/10.1001/archopht.1982.01030030773010.
12. Gass JDM. Chorioretinal anastomosis probably occurs infrequently in type 2A idiopathic juxtafoveolar retinal telangiectasis. *Archives of Ophthalmology*. 2003;121(9):1345. https://doi.org/10.1001/archopht.121.9.1345-a.
13. Gass JDM, Blodi BA. Idiopathic juxtafoveolar retinal telangiectasis: update of classification and follow-up study. *Ophthalmology*. 1993;100(10):1536–1546. https://doi.org/10.1016/S0161-6420(93)31447-8.
14. Alex D, Giridhar A, Gopalakrishnan M, et al. Early spectral-domain optical coherence tomography biomarkers to confirm fellow eye changes in asymmetric type 2 macular telangiectasia: a case control study (India Macular Telangiectasia Report 1). *Retina*. 2021;41(3):471–479.
15. Jaissle GB, May CA, Van De Pavert SA, et al. Bone spicule pigment formation in retinitis pigmentosa: insights from a mouse model. *Graefe's Archive for Clinical and Experimental Ophthalmology*. 2010;248(8):1063–1070. https://doi.org/10.1007/s00417-009-1253-9.
16. Jindal A, Choudhury H, Pathengay A, Flynn HW. A novel clinical sign in macular telangiectasia type 2. *Ophthalmic Surgery Lasers and Imaging Retina*. 2015;46(1):134–136. https://doi.org/10.3928/23258160-20150101-26.
17. Kihara Y, Heeren TFC, Lee CS, et al. Estimating retinal sensitivity using optical coherence tomography with deep-learning algorithms in macular telangiectasia type 2. *JAMA Network Open*. 2019;2(2). https://doi.org/10.1001/jamanetworkopen.2018.8029.
18. Kim YH, Chung YR, Oh J, et al. Optical coherence tomographic features of macular telangiectasia type 2: Korean Macular Telangiectasia Type 2 Study—report No. 1. *Scientific Reports*. 2020;10(1). https://doi.org/10.1038/s41598-020-73803-9.
19. Klein R, Blodi BA, Meuer SM, Myers CE, Chew EY, Klein BEK. The prevalence of macular telangiectasia type 2 in the beaver dam eye study. *American Journal of Ophthalmology*. 2010;150(1):55. https://doi.org/10.1016/j.ajo.2010.02.013.
20. Krivosic V, Lavia C, Aubineau A, Tadayoni R, Gaudric A. OCT of outer retinal hyperreflectivity, neovascularization, and pigment in macular telangiectasia type 2. *Ophthalmology Retina*. 2021;5(6):562–570. https://doi.org/10.1016/j.oret.2020.09.012.
21. Ledolter AA, Holder GE, Ristl R, Schmidt-Erfurth U, Ritter M. Electrophysiological findings show generalised post-photoreceptoral deficiency in macular telangiectasia type 2. *British Journal of Ophthalmology*. 2018;102(1):114–119. https://doi.org/10.1136/bjophthalmol-2017-310228.
22. Ledolter AA, Ristl R, Palmowski-Wolfe AM, et al. Macular Telangiectasia type 2: multimodal assessment of retinal function and microstructure. *Acta Ophthalmologica*. 2022;100(6):e1240–e1252. https://doi.org/10.1111/aos.15072.
23. Loo J, Fang L, Cunefare D, Jaffe GJ, Farsiu S. Deep longitudinal transfer learning-based automatic segmentation of photoreceptor ellipsoid zone defects on optical coherence tomography images of macular telangiectasia type 2. *Biomedical Optics Express*. 2018;9(6):2681–2698. https://doi.org/10.1364/BOE.9.002681.

24. Micevych PS, Lee HE, Fawzi AA, Vavvas DG. Overlap between telangiectasia and photoreceptor loss increases with progression of macular telangiectasia type 2. *PLoS One*. 2019;14(10):e0224393. https://doi.org/10.1371/journal.pone.0224393.
25. Mueller S, Gunnemann F, Rothaus K, et al. Incidence and phenotypical variation of outer retina-Associated hyperreflectivity in macular telangiectasia type 2. *British Journal of Ophthalmology*. 2021;105(4):573–576. https://doi.org/10.1136/bjophthalmol-2020-317997.
26. Mukherjee D, Lad EM, Vann RR, et al. Correlation between macular integrity assessment and optical coherence tomography imaging of ellipsoid zone in macular telangiectasia type 2. *Investigative Ophthalmology & Visual Science*. 2017;58(6):BIO291–BIO299. https://doi.org/10.1167/IOVS.17-21834.
27. Okada M, Robson AG, Egan CA, et al. Electrophysiological characterization of macular telangiectasia type 2 and structure–function correlation. *Retina*. June 2017. https://doi.org/10.1097/IAE.0000000000001746.
28. Park YG, Park YH. Quantitative analysis of retinal microvascular changes in macular telangiectasia type 2 using optical coherence tomography angiography. *PLoS One*. 2020;15(4). https://doi.org/10.1371/journal.pone.0232255.
29. Pauleikhoff D, Gunnemann F, Book M, Rothaus K. Progression of vascular changes in macular telangiectasia type 2: comparison between SD-OCT and OCT angiography. *Graefe's Archive for Clinical and Experimental Ophthalmology*. 2019;257(7):1381–1392. https://doi.org/10.1007/s00417-019-04323-0.
30. Roisman L, Rosenfeld PJ. Optical coherence tomography angiography of macular telangiectasia type 2. *Developments in Ophthalmology*. 2016;56:146–158. https://doi.org/10.1159/000442807.
31. Sallo FB, Leung I, Zeimer M, et al. Abnormal retinal reflectivity to short-wavelength light in type 2 idiopathic macular telangiectasia. *Retina*. 2018;38(1):S79–S88. https://doi.org/10.1097/iae.0000000000001728.
32. Sallo FB, Leung I, Chung M, et al. Retinal crystals in type 2 idiopathic macular telangiectasia. *Ophthalmology*. 2011;118(12):2461–2467. https://doi.org/10.1016/j.ophtha.2011.05.022.
33. Spaide RF, Klancnik JM, Cooney MJ. Retinal vascular layers in macular telangiectasia type 2 imaged by optical coherence tomographic angiography. *JAMA Ophthalmology*. 2015;133(1):66–73. https://doi.org/10.1001/jamaophthalmol.2014.3950.
34. Spaide RF, Yannuzzi LA, Maloca PM. Retinal-choroidal anastomosis in macular telangiectasia type 2. *Retina*. 2018;38(10):1920–1929. https://doi.org/10.1097/IAE.0000000000002289.
35. Thorell MR, Zhang Q, Huang Y, et al. Swept-source OCT angiography of macular telangiectasia type 2. *Ophthalmic Surgery Lasers and Imaging Retina*. 2014;45(5):369–380. https://doi.org/10.3928/23258160-20140909-06.
36. Tzaridis S, Hess K, Friedlander M, Holz FG. Optical coherence tomography-angiography for monitoring neovascularisations in macular telangiectasia type 2. *British Journal of Ophthalmology*. 2021;105(5):735–740. https://doi.org/10.1136/bjophthalmol-2020-316021.
37. Unterlauft JD, Eichler W, Kuhne K, et al. Pigment epithelium-derived factor released by Müller Glial cells exerts neuroprotective effects on retinal ganglion cells. *Neurochemical Research*. 2012;37(7):1524–1533. https://doi.org/10.1007/s11064-012-0747-8.
38. Venkatesh R, Reddy NG, Mishra P, et al. The preproliferative stage in type 2 macular telangiectasia (MacTel type 2). *Graefe's Archive for Clinical and Experimental Ophthalmology*. 2022;260(1):121–132. https://doi.org/10.1007/s00417-021-05371-1.
39. Wang RK, Jacques SL, Ma Z, Hurst S, Hanson SR, Gruber A. Three dimensional optical angiography. *Optics Express*. 2007;15(7):4083–4097. https://doi.org/10.1364/OE.15.004083.
40. Wong WT, Forooghian F, Majumdar Z, Bonner RF, Cunningham D, Chew EY. Fundus autofluorescence in type 2 idiopathic macular telangiectasia: correlation with optical coherence tomography and microperimetry. *American Journal of Ophthalmology*. 2009;148(4):573–583. https://doi.org/10.1016/j.ajo.2009.04.030.
41. Wu L. When is macular edema not macular edema? An update on macular telangiectasia type 2. *Taiwan Journal of Ophthalmology*. 2015;5(4):149–155. https://doi.org/10.1016/j.tjo.2015.09.001.
42. Yannuzzi LA, Bardal AMC, Freund KB, Chen KJ, Eandi CM, Blodi B. Idiopathic macular telangiectasia. *Archives of Ophthalmology*. 2006;124(4):450–460. https://doi.org/10.1001/archopht.124.4.450.
43. Zeimer MB, Spital G, Heimes B, Lommatzsch A, Pauleikhoff D. Macular telangiectasia-changes in macular pigment optical density during a 5-year follow-up. *Retina*. 2014;34(5):920–928. https://doi.org/10.1097/IAE.0000000000000023.
44. Zhang Q, Wang RK, Chen CL, et al. Swept source optical coherence tomography angiography of neovascular macular telangiectasia type 2. *Retina*. 2015;35(11):2285–2299. https://doi.org/10.1097/IAE.0000000000000840.

Chapter 23

Coats disease

Kristine Wang, Hana A. Mansour, Taku Wakabayashi, Meera D. Sivalingam and Yoshihiro Yonekawa
Wills Eye Hospital, Mid Atlantic Retina, Thomas Jefferson University, Philadelphia, PA, United States

Overview bullets

- Coats disease is a rare, idiopathic, congenital retinal vascular disorder, typically seen unilaterally.
- Coats disease is characterized by telangiectasia, aneurysmal dilations, peripheral nonperfusion, intraretinal and subretinal exudation, and exudative retinal detachment.
- Bilateral involvement is more frequent than previously thought, albeit very asymmetric.
- Patients with advanced disease commonly present with decreased visual acuity, strabismus, and xanthocoria.
- A classification system developed by Shields et al. divides the clinical spectrum into five major stages.
- Widefield fluorescein angiography is valuable for visualizing peripheral vascular abnormalities and guiding treatment.
- Laser photocoagulation is the main treatment for Coats disease.
- Intravitreal injections of antivascular endothelial growth factor (VEGF) agents and corticosteroids are useful adjunctive treatments in some patients, especially those with extensive exudation.
- Surgical drainage is considered for bullous exudative retinal detachment.
- In recent years, improvements in the diagnosis and management of Coats disease have resulted in better visual and anatomic outcomes, including fewer enucleations.
- Poor visual outcomes (<20/200) are common among patients who present with advanced-stage disease with subretinal fibrotic nodules, which tend to be patients who present at a younger age.

Introduction

Coats disease is a congenital, noninherited, idiopathic retinal vascular condition characterized by telangiectatic vessels, aneurysmal dilatations, and peripheral capillary nonperfusion, that can lead to exudation and ultimately to exudative retinal detachment.[1–3] It is typically a unilateral disease and has a relative male preponderance.[4] Intraretinal and subretinal lipid exudation and exudative retinal detachments may result in significant vision loss.[5]

Coats disease was first described by George Coats in 1908 as a retinal vascular disease with massive subretinal exudation that typically appeared unilaterally in young male patients.[4] In 1912, Theodor Leber described a similar disease with many miliary aneurysms but lacking the massive exudation reported by Coats.[6] In 1956, Algernon Reese presented evidence that the condition described by Leber was the same clinical entity as Coats disease, albeit a less severe form, thereby emphasizing the wide spectrum of clinical presentations that can be seen in Coats patients—ranging from asymptomatic retinal telangiectasia to total, bullous retinal detachment with phthisis bulbi that may eventually require enucleation.[5,7] Further research has clarified that Coats disease should be defined as idiopathic retinal telangiectasia with exudation but without noticeable vitreoretinal traction.[2] The underlying etiology of why patients with Coats disease harbor abnormal retinal vasculature is still unknown. Coats disease should be distinguished from other Coats-like exudative retinopathies that may accompany other diagnoses, such as in the setting of retinitis pigmentosa,[8] facioscapulohumeral muscular dystrophy,[9,10] and Leri–Weill dyschondrosteosis,[11] that may cause exudation similar to Coats disease.

Recent advances in retinal imaging, such as widefield fundus imaging and fluorescein angiography (FA), optical coherence tomography (OCT), and OCT angiography (OCTA), have significantly improved our understanding of the pathophysiological disease course as well as the response to treatments of Coats disease.[12–16] Some recent progress in our understanding of Coats disease includes mild but frequent fellow eye involvement and the identification of various Coats-like

Retinal and Choroidal Vascular Diseases of the Eye. https://doi.org/10.1016/B978-0-443-15583-3.00023-8

exudative retinopathies associated with systemic conditions.[9,16–20] In addition, although laser photocoagulation is a mainstay of the treatment, antivascular endothelial growth factor (VEGF) injections and steroid injections are available, and surgery is also indicated depending on the severity of exudation and exudative retinal detachments. Combining these treatments has likely improved visual outcomes and allowed greater globe salvage.[21,22]

In this chapter, we summarize the recent advances in diagnosis, imaging, and treatment of Coats disease.

Demographics and epidemiology

Coats disease is a fairly rare disease with an estimated incidence of 0.09 per 100,000 of the population.[23] However, most retina specialists may encounter Coats disease as it is one of the more common diseases that we manage in our pediatric retina clinic.[24] Coats disease is typically seen in young patients, with most patients diagnosed within the first 2 decades of their lives. One study that included 351 eyes from a tertiary referral center reported a median age of 6 years at diagnosis, while a population-based study reported a slightly higher median age of 8 years.[22,23] Patients can present as neonates also.[2] However, Coats disease can be diagnosed in adulthood, with the oldest age of 79 years.[25–30] Adult-onset Coats disease is typically milder and has a slower progression and smaller areas of involvement of the retina.[25,30] It is unclear whether these patients truly have adult "onset" Coats disease, or whether these vascular abnormalities were present congenitally but not detected earlier.

Idiopathic Coats disease shows no racial predisposition or any systemic association, but it is more common in males, with one study of 351 patients reporting that 84% of cases were male.[2,22] While patients typically present unilaterally, recent literature has documented vascular abnormalities in the fellow eye, although asymmetric with very mild disease in the fellow eye.[2,3,16–19]

Genetics

The molecular mechanism underlying the formation of telangiectatic vessels and aneurysmal dilatations in Coats disease remains unknown, and no genetic association has been found in patients with Coats disease. Nevertheless, a few genetic associations have been implicated. Black et al. hypothesized that a somatic mutation in the *NDP* gene may result in Coats disease by causing a deficiency in the protein norrin, which plays an important role in retinal vascular formation. They reported a female patient with Coats disease whose son had Norrie disease. Genetic analysis revealed that both individuals carried a missense mutation within the *NDP* gene on chromosome Xp11.2. When analyzing the enucleated eyes of nine male Coats patients, one of the retinal tissue samples contained a somatic mutation in the *NDP* gene.[31] However, further studies have not verified this gene association.

Robitaille et al. examined the role of *FZD4*, a gene associated with familial exudative vitreoretinopathy (FEVR), in Coats disease. While 11 *FZD4* mutations were found in 68 cases of FEVR, no *FZD4* mutations were found in any of the 16 cases of Coats disease, suggesting that *FZD4* mutations are not a cause of Coats disease.[32]

An association between the *CRB1*gene and retinitis pigmentosa (RP) with Coats-like retinopathies has been reported.[33,34] Den Hollander et al. showed that 7 out of 16 RP patients with this complication had *CRB1* mutations.[33] Likewise, Cremers et al. found that 55% of their patients with RP and Coats-like exudative vasculopathy had *CRB1* mutations.[34] However, no mutations in the *CRB1* gene were found in 18 patients with classic Coats disease, suggesting that *CRB1* mutations are not causative of classic Coats disease.[33]

Pathogenesis

The pathogenesis of Coats disease is not fully understood; however, it is associated with abnormal vascular permeability.[35] The telangiectatic and aneurysmal vessels in Coats disease are considered congenital or secondary to a dysfunctional blood–retinal barrier with abnormal endothelial permeability.[35,36] Consequently, leakage of lipoprotein-rich fluid from abnormal vessels creates intraretinal and subretinal exudates and retinal edema, which can disrupt retinal structures. Extensive leakage may also lead to exudative, nonrhegmatogenous retinal detachment.[35,37,38]

Early histopathological studies, often derived from enucleations for advanced Coats disease or misdiagnosis of retinoblastoma, have revealed several vascular abnormalities including saccular and fusiform aneurysms, vessels with thickened hyaline walls containing periodic acid-Schiff positive materials, loss of endothelial cells and pericytes, telangiectatic vessels with occasional thrombi, and capillary dropout.[35,36,38] Collectively, these findings indicate a loss of vascular integrity (the blood–retinal barrier) that allows subsequent fluid and lipid leakage into and under the retina. Mononuclear cells and polymorphonuclear leukocytes, such as monocytes and eosinophils, can be found within and surrounding the vessels as well as in the subretinal space.[35] Accumulations of eosinophilic exudates thicken the outer retina and may also invade the inner retina in some cases. Additional findings include retinal infiltration by lipid and/or pigment-laden macrophages (foam cells) and ghost cells, especially in the inner retina, and gliosis.[35,36,38] In eyes with exudative retinal detachment, the proteinaceous subretinal exudates contain foamy histiocytes and cholesterol clefts. In up to 20% of eyes, a submacular fibrous nodule with occasional bone formation and calcification is also present and is a common cause of vision loss in Coats disease.[38–40]

Clinical features

Clinical presentation

The common presenting symptoms of Coats disease are decreased vision, strabismus, and xanthocoria.[2] These signs and symptoms could also be seen in retinoblastoma; thus, correctly distinguishing Coats disease from retinoblastoma is important, as retinoblastoma is life threatening.[41] Coats disease is clinically unilateral in 90%–100% of cases,[2,22,42,43] although bilateral involvement with asymptomatic abnormal findings in the fellow-eye has been reported in 18.8% of cases.[16–19]

In a study of 158 eyes with Coats disease, the most common first reported symptoms were decreased visual acuity in 43% of eyes, followed by strabismus in 23%, and leukocoria/xanthocoria in 20%. Other less common presenting symptoms included eye pain in 3%, heterochromia in 1%, and nystagmus in 1%. Of note, 8% of eyes were asymptomatic and were diagnosed on routine examination. Visual acuity on presentation was 20/50 or better in 12% of eyes, 20/60–20/100 in 11%, 20/200 to counting fingers in 18%, and hand motion to no light perception in 58%.[2]

Fundus features

The hallmark, classic fundus features used in the clinical diagnosis of Coats disease include telangiectasias, "light bulb" aneurysmal dilatations, peripheral nonperfusion, and intraretinal and subretinal exudation.[2,44,45] These abnormal vessels are predominantly seen in the peripheral retina, extending posteriorly, but they are rarely found in the macula, with one study reporting only 1% of 158 eyes demonstrating telangiectasias in the macula.[2] Although the leaking vessels are found in the periphery, exudation tends to migrate and accumulate in the macula, which can lead to the formation of a subfoveal nodule, appearing as a yellow, bulging, spheroidal lesion on fundus examination.[2,5,39] The presence of a subfoveal nodule is a predictor of worse visual outcomes due to the subsequent development of macular fibrosis.[39,40] Daruich et al. found that while all cases with a subfoveal nodule later developed macular fibrosis, this finding was only seen in 15.8% of patients without a subfoveal nodule.[39]

Some advanced cases of Coats disease may also demonstrate partial or total exudative retinal detachment, retinal hemorrhage, retinal macrocysts, and vasoproliferative tumors.[2,46] Retinal macrocysts correlate with a longer duration of retinal detachment and worse visual prognosis.[2] Rare but reported findings include vitreous hemorrhage and lamellar hole.[2,46] In the most advanced cases, neovascular glaucoma or angle-closure glaucoma may result from either iris neovascularization or a bullous retinal detachment displacing the lens-iris diaphragm anteriorly.[47,48] Coats disease is progressive, although a few cases have shown spontaneous regression.[49,50]

While the anterior segment examination usually reveals no abnormal findings in 90% of cases, some advanced eyes show cataracts, cholesterolosis of the anterior chamber, corneal edema, megalocornea, and shallow anterior chamber.[2] Although leukocoria/xanthocoria is a common presenting symptom in children, it is typically not how adults with Coats disease present.[25,30]

Classification

The most widely used classification of Coats disease was developed by Shields and colleagues in 2000.[5] Stage 1 is characterized by retinal telangiectasia alone. Stage 2 includes intraretinal exudation. Stage 2A is characterized by extrafoveal exudation, while stage 2B has foveal exudation.[5] A recent classification has suggested further dividing stage 2B into 2B1 (absence of subfoveal nodule) and 2B2 (presence of subfoveal nodule) due to the prognostic value of the subfoveal nodule.[39] Stage 3 is the most prevalent stage at presentation and is characterized by exudative retinal detachment. Stage 3A includes cases with subtotal detachment. If the detachment is extrafoveal, it is classified as stage 3A1, but if it involves the fovea, it is classified as stage 3A2. A total retinal detachment is classified as stage 3B. When a total exudative retinal detachment is accompanied by glaucoma, it is classified as stage 4. Stage 5 encompasses advanced end-stage disease with phthisis bulbi.[5]

Imaging findings

Wide-field fundus imaging and fluorescein angiography

The hallmark features of telangiectasias and aneurysmal dilatations in Coats disease can be diagnosed on funduscopic exam; however, FA has been the gold standard for highlighting areas of nonperfusion and subtle vascular changes and for guiding laser treatment toward the leaking vessels.[13,51] Vascular changes that can be evaluated using FA include telangiectasias, peripheral nonperfusion, light-bulb aneurysms, vessel beading, vascular occlusion, capillary dropout, vascular leakage, arteriovenous shunts, and vascular sheathing.[3,22,45,52] The aneurysms, which can be seen in the arteries, capillaries, and veins, are usually found in areas of avascularity and capillary drop out.[3,45] FA shows greater involvement of arteries than veins, especially in mild Coats disease.[45] Telangiectasias start to show hyperfluorescence at different time points, but they are all filled by the venous phase and demonstrate leakage into the surrounding retina.[2] FA is useful for identifying the degree of leakage from abnormal vessels to allow targeting of the vessels that are most responsible for exudates before laser treatment.[51] Exudates initially appear hypofluorescent but demonstrate mild staining in later phases.[2]

Given the predominantly pediatric population associated with Coats disease, an examination under anesthesia (EUA) is warranted for fundus imaging if a child is too young for in-office imaging.[2,53] During EUA, a contact wide-angle retinal camera with handheld devices, such as RetCam (Clarity Medical Systems, Pleasanton, CA, USA) and ICON (Phoenix Technology group, Pleasanton, CA, USA), can be used for widefield retinal imaging.[51,54,55] Widefield FA is particularly useful in managing Coats disease, as it allows the identification of all peripheral vascular pathologies, which is essential for successful treatment, response monitoring, and identification of new areas of pathology.[13] It is also useful for identifying more subtle peripheral vascular pathology in fellow eyes.[16]

In children who can tolerate intravenous access, intravenous fluorescein with a noncontact wide-angle camera, such as Optos California and 200-TX Optomap (Optos, Dumfernline, MA, USA), can be used in the outpatient setting, thereby reducing the need for EUA.[13] For younger children who cannot tolerate intravenous access but can cooperate for fundus imaging, oral fluorescein can be administered in the outpatient setting.[56,57]

Optical coherence tomography

Spectral domain-OCT (SD-OCT) and swept-source OCT (SS-OCT) are noninvasive imaging modalities that are useful in Coats disease evaluation and management.[14,58] OCT can visualize the presence and extent of intraretinal and subretinal exudation, macular edema, subretinal fluid, subfoveal nodule, ellipsoid zone and external limiting membrane disruption, and vitreoretinal interface abnormalities. Many of these OCT findings correlate with worse visual prognosis.[12] Intraretinal exudation can be seen in all retinal layers but is most commonly found in the outer nuclear layer. Histopathological findings such as cholesterol crystals and macrophages may appear on SD-OCT as hyperreflective linear structures and hyperreflective dots, respectively.[59] OCT is useful for monitoring disease progression and response to treatment, but it does not allow for the direct assessment of the retinal vasculature.[14]

Optical coherence tomography angiography

OCTA allows for imaging of both the superficial and deep retinal vasculature, thereby providing additional insight beyond what FA can supply since the deep capillary plexus cannot be visualized using FA.[14,60] Another advantage of OCTA compared to FA is that it does not require injections of intravenous dye. However, OCTA cannot currently serve as a substitute for FA due to its limitations, which include the inability to visualize leakage and lesions in the periphery; the presence of possible artifacts due to poor visual acuity or fixation; and potential blockage by hemorrhage.[14,61] Vascular abnormalities visualized by OCTA in Coats disease include aberrant superficial vessels in the foveal avascular zone; rarefaction of both the superficial and deep vascular plexuses; and vascular loops, aneurysmal dilations, and areas of nonperfusion in the superficial capillary plexus.[14,19,62,63]

Differential diagnosis

Retinoblastoma

Advanced Coats disease should be differentiated from retinoblastoma because Coats disease misdiagnosed as retinoblastoma can lead to unnecessary enucleation, while retinoblastoma misdiagnosed as Coats disease can have serious adverse systemic and ocular consequences.[50] In an early study of 150 eyes with Coats disease, almost half were initially misdiagnosed as a different condition, with the most common misdiagnosis being retinoblastoma in 43 eyes.[2]

Both Coats disease and retinoblastoma may present with strabismus, decreased visual acuity, pain from secondary glaucoma, retinal detachment, and leukocoria[41,44,64]; however, retinoblastoma presents bilaterally in 40% of cases, has no predilection for sex, and is typically diagnosed at a younger age than Coats disease, with a median diagnosis age of 1 year for bilateral cases and 2 years for unilateral cases.[50,65] Additionally, while all cases of Coats disease are sporadic, retinoblastoma may be hereditary.[2,65] The pupillary reflex in Coats disease is usually yellow (xanthocoria), whereas it is white to gray (leukocoria) in retinoblastoma. Retinoblastoma manifests as a white mass, while Coats disease manifests as lipoproteinaceous exudates. Retinal vessels in retinoblastoma are uniformly dilated and often tortuous, whereas they are irregularly dilated in Coats disease. The retinal vessels in Coats disease remain visible throughout the fundus, in contrast to the vessels in retinoblastoma, which usually disappear into the tumor.[66] When a diagnosis cannot be reached based on these clinical features, ancillary testing including ultrasonography, CT, and MRI may be used.[44]

Familial exudative vitreoretinopathy

Coats disease can simulate familial exudative vitreoretinopathy (FEVR) due to its exudative nature. Both diseases have exudation and peripheral avascularity. However, FEVR often has a positive family history, demonstrates bilaterality, and typically exhibits temporal dragging of the vascular arcades and macular heterotopia.[67,68] Fluorescein angiography findings are also different. Coats disease presents with the aforementioned telangiectasia and aneurysms, while often FEVR presents with anomalous peripheral vasculature with arborization, typically without aneurysms. Genetic testing can also be helpful, but the sensitivity may be limited depending on the genes on the panel because there are many potential genes that are associated with FEVR, in addition to classic FEVR genes, which are *FZD4*, *LRP5*, *NDP*, and *TSPAN12*.[69]

Systemic diseases with coats-like retinopathies

Coats disease has no association with systemic diseases.[2,70] However, some systemic diseases, including facioscapulohumeral muscular dystrophy,[9,10,71] Turner syndrome,[72] Alport syndrome,[73] Cornelia de Lange syndrome,[74] aplastic anemia,[75] 13q deletion syndrome,[76] Senior–Loken syndrome (familial renal-retinal dystrophy),[77] epidermal nevus syndrome,[78] Hallermann–Streiff syndrome,[79] renal transplantation,[80] multiple glomus tumors,[81] pericentric inversion of chromosome 3,[82] telangiectasia of the nasal mucosa,[83] congenital plasminogen deficiency type I,[84] focal segmental glomerulosclerosis,[85] LeriWeill dyschondrosteosis,[11] and Parry–Romberg syndrome,[20] exhibit bilateral features of Coats-like exudative retinopathies. Patients presenting with Coats-like retinopathies should therefore be referred for systemic evaluation to rule out these possible conditions.

Coats plus syndrome

Coats' plus syndrome presents as bilateral Coats retinal findings with additional signs, including intracranial calcification, dystrophic nails, sparse hair, skeletal abnormalities, occasional seizures, postnatal growth failure, leukodystrophy, and movement disorder.[86–88] This syndrome is in the spectrum of dyskeratosis congenita, a family of disorders of telomere homeostasis.[89]

Other ocular conditions associated with coats-like retinopathies

Other ocular conditions that may manifest Coats-like features include persistent fetal vasculature,[50] retinopathy of prematurity,[50,90] toxocariasis,[37] Norrie disease,[31] retinal capillary hemangioma,[91] congenital cataract,[92] uveal melanoma,[93] incontinentia pigmenti,[94] branch retinal vein occlusion,[95] ocular toxoplasmosis,[96] retinitis pigmentosa,[97–103] and morning glory optic disc anomaly.[104] Macular telangiectasia type 1 is considered similar to or possibly on the spectrum of Coats disease, as it manifests as telangiectasia, aneurysms, and exudation at the macula and is mainly a unilateral disease that mostly affects men.[105–107]

Management strategies

The aim of treatment in Coats disease is to ablate the abnormal vasculature to prevent further leakage, promote resolution of the exudates, and preserve vision.[108] Laser photocoagulation is the mainstay of treatment.[109–113] Widefield FA is useful for identifying the full extent of the vascular abnormalities and for guiding treatment.[13,51] Children can be treated at the time of EUA after FA, or in clinic for older children. No strict guidelines have been established for the treatment of Coats disease. Therefore, depending on the severity of disease and the practice patterns of the physician, cryotherapy and adjunct use of periocular or intraocular corticosteroids and intravitreal anti-VEGF injections are also considered.[5,42,114] For very mild, subtle cases with trace telangiectasia but lacking any exudation or leakage (stage 1), close observation with periodic follow-up is acceptable, especially if the patients are adults or older children in whom disease may be more indolent and easily monitored5[5]; however in our practice, we prefer to treat all telangiectasias and aneurysms with laser photocoagulation with the aim of preventing potential future exudation.[24,109] For more advanced disease with total exudative retinal detachment, surgical intervention is a reasonable approach for salvaging the globe.[5,110] Patients with end-stage disease (stage 5) and a comfortable eye can be observed, but enucleation may be indicated in those with painful eyes with no light perception.[5]

Ablative therapies: Laser photocoagulation and cryotherapy

Laser photocoagulation is the first-line treatment for Coats disease (Fig. 23.1).[109–111,113,115] A focal laser is applied to ablate the leaking aneurysms, followed by "painting" of the telangiectatic vessels using long-duration burns with continuous or repeat mode laser. Adequate photocoagulation of the leaking vessels until their color changes to whitish is important, as is avoidance of the creation of any iatrogenic retinal breaks with extensive burns. Scatter laser should then be added to the peripheral nonperfusion area.[24,111,114] Lasers that are recommended for the treatment of Coats disease are green (532 nm)[112] or yellow laser (577 nm),[111,116] as these are well absorbed by hemoglobin within the vessels.[117,118] The advantage of a yellow laser over a green laser is the lower scatter levels, which reduces the power required, and easier delivery through exudates.[118,119]

Regular reevaluation of the patient is recommended after treatment, as response to treatment is gradual and complete ablation of abnormal vessels may require multiple treatment sessions.[5,116] Beginning at 4–8 weeks after the telangiectatic vessels and aneurysms are successfully obliterated, the exudates will decrease as they are absorbed, leading to resolution of the exudative retinal detachment.[116] Because the goal of treatment is to ablate the leaking vessels, rather than to perform retinopexy, which must be applied on an attached retina, laser is effective even in cases of exudative retinal detachment.[111–113,116,120] However, if the laser treatment is administered to a highly elevated retinal detachment or if the retina is immediately posterior to the lens, cataracts may develop.[111,112]

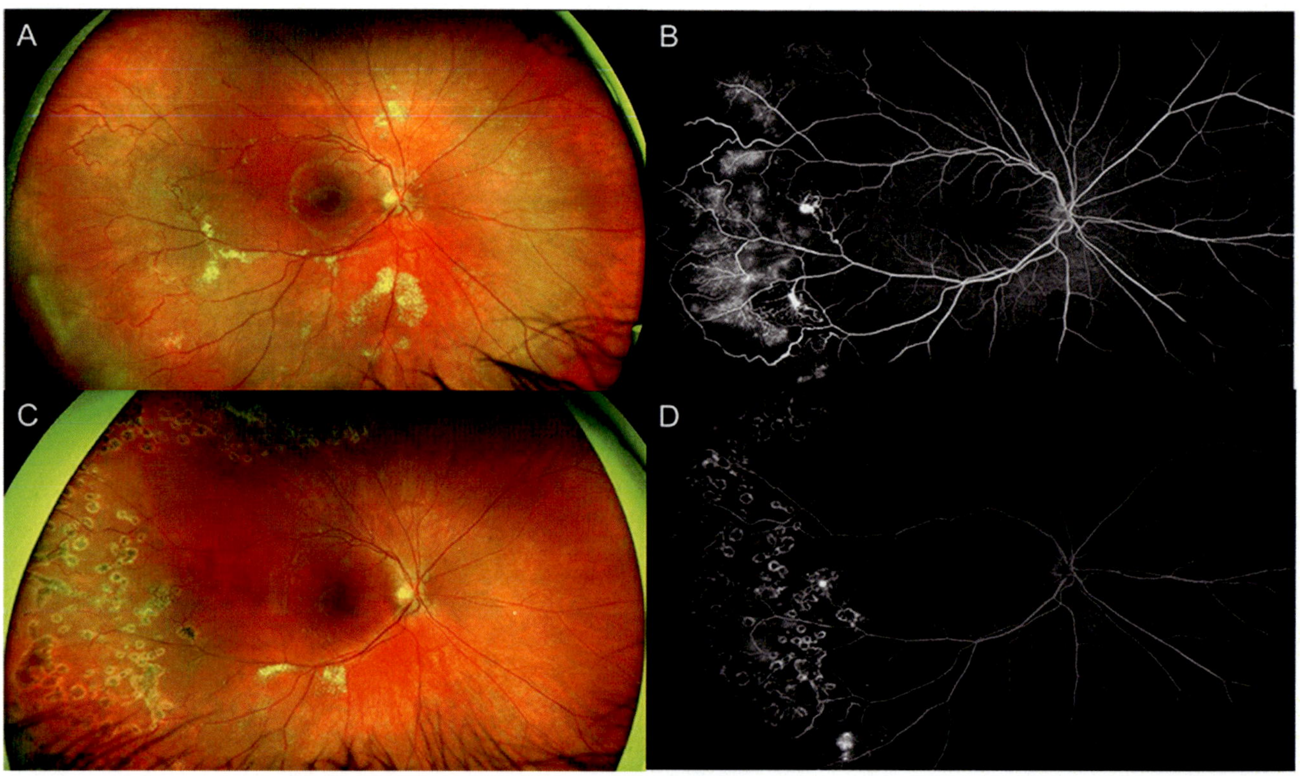

FIGURE 23.1 Ultra-widefield fundus imaging and fluorescein angiography of a 7-year-old with Coats disease before and after laser photocoagulation. (A) Color fundus image shows yellowish exudates secondary to leaky retinal vessels. (B) Fluorescein angiography highlights the aneurysmal dilatations, telangiectasia, and peripheral nonperfusion. (C, D) sixteen months after the first laser treatment. Laser treatment was effective in reducing exudation.

Cryotherapy is also an effective method of vascular ablation in Coats disease.[113] A double or triple freeze–thaw technique has traditionally been performed when telangiectasias are present in the far periphery and are associated with thick exudation or exudative detachment that cannot be effectively lasered.[5,113,115] However, cryotherapy induces an inflammatory response,[121] initially exacerbating exudation. Epiretinal membrane formation and tractional retinal detachment can ensue as well.[122–124]

Pharmacotherapies

Intravitreal anti-VEGF can be used as an adjunctive treatment to ablative therapies. Elevated VEGF levels have been shown in eyes with Coats disease, with greater concentrations with increasing disease severity.[125–127] One study comparing the intraocular VEGF levels in four eyes with Coats disease to control eyes with rhegmatogenous retinal detachment found that the mean level in eyes with Coats disease was 2394.5 pg/mL compared to 15.3 pg/mL in control eyes.[126] The elevated VEGF levels may be secondary to the retinal ischemia associated with Coats disease.[128] Upregulated VEGF levels may potentiate the vascular permeability already present due to the morphological abnormalities of the vessels, leading to increased exudation, retinal edema, and exudative retinal detachment.[126,128] After one bevacizumab injection, the intraocular VEGF levels in an eye with stage 2B Coats disease decreased from 1247 to 20.4 pg/mL in 1 month along with a reduction in macular edema and an improvement in visual acuity.[126] Several other case reports and case series have corroborated the efficacy of anti-VEGF treatment in reducing macular edema, subretinal fluid, and exudation when used alone[129–131] or alongside triamcinolone injections[132] or ablative therapies.[132–138] Flattening the retina reduces the energy required to ablate abnormal vessels, thereby decreasing the inflammation caused by the ablation.[133]

Anti-VEGF injections may allow for more effective ablative procedures, especially in severe cases, by potentially decreasing the amount of subretinal fluid present.[131,133,135,136] A case-control series comparing 10 patients who received intravitreal bevacizumab and ablative therapy with 10 matched patients who received only ablative therapy revealed that anti-VEGF neither reduced the number of treatment sessions required nor the time to resolution. However, two patients in the control group with the most advanced stage of disease in the study (stage 3B) failed to reach resolution, while the matched patients in the bevacizumab group with similar disease severity successfully achieved resolution.[135] This finding suggests laser photocoagulation allows definitive treatment of the source of the exudation, while anti-VEGF may help the process in more severely affected eyes.

The long-term effects of anti-VEGF injections in children are currently unknown.[21] One study reported that 50% of their patients ($n = 4$) developed vitreoretinal fibrosis after anti-VEGF injections and ablative therapies.[139] Although a pooled data analysis of all published reports of Coats disease did not indicate an association between anti-VEGF injections and an increased risk of vitreoretinal fibrosis or tractional retinal detachment (TRD), patients should be carefully monitored after anti-VEGF injections.[124]

Corticosteroids, including intravitreal and subtenon triamcinolone injections and intravitreal dexamethasone implants, can also be used as an adjunctive treatment to laser ablation, especially in cases with extensive subretinal fluid.[140–144] As an antiinflammatory drug, triamcinolone decreases vascular permeability and has been reported to reduce macular edema, subretinal fluid, exudates, and the extent of telangiectatic vessels in Coats disease.[128,140,145–147] Sufficient absorption of the fluid in response to triamcinolone helps the delivery of laser and/or cryotherapy.[145,147] However, there are common complications of intravitreal steroid injections, including elevated intraocular pressure and cataract formation.[140,148,149]

Surgery

Although nonsurgical methods can be effective for resolving exudative retinal detachments,[131,133,147,150] surgical drainage is used in cases of total bullous retinal detachment abutting the lens, when laser and/or cryotherapy cannot reach the leaking vasculature without lens complications, or when the disease is refractory to other treatments.[110,140] The goal of surgical intervention is to prevent further exudation, secondary glaucoma, progression to phthisis bulbi, and ultimately, to preserve the globe.[50,113] Visual prognosis is very limited in these very advanced eyes.[151] Several methods are available for the surgical treatment of bullous exudative detachments in Coats disease, but the fundamental goal is to drain the fluid externally and perform ablative therapy to treat the abnormal vessels.[113] Silodor et al. found that in 13 cases with total bullous exudative retinal detachment, four out of six untreated Coats eyes developed painful glaucoma and required enucleation, whereas none of the seven eyes managed with subretinal fluid drainage and cryotherapy developed secondary glaucoma, indicating a benefit of surgical intervention for globe salvage in cases of advanced Coats disease.[47]

Various surgical techniques for Coats disease have been reported, including transscleral external drainage with a balanced salt solution infusion into the anterior chamber with or without pars plana vitrectomy, followed by laser photocoagulation and gas or silicone oil tamponade.[110,152–158] Laser photocoagulation with a nonvitrectomy approach can be effective if the subretinal drainage is adequately performed.[159] However, in cases that require vitrectomy, the combination of vitrectomy and perfluorocarbon liquid may allow efficient subretinal fluid drainage, as perfluorocarbon assists in the egress of subretinal fluid through an external sclerotomy site.[160] Successful drainage results in the flattening of the retina, thereby allowing subsequent lasers to more effectively ablate the leaky vessels to reduce exudation.[5,50,110] External drainage is preferred over internal drainage retinotomy to avoid the risk of severe proliferative vitreoretinopathy and persistent rhegmatogenous retinal detachment.[113,161]

Although the advancements in the treatment of Coats disease have decreased the need for enucleation in recent years, enucleation may still be necessary in cases of blind, painful eyes with neovascular or severe angle closure glaucoma.[5,22,131] Predictive factors for enucleation include presence of neovascularization of the iris, retinal detachment on ultrasound, elevation of detachment in millimeters, open or closed funnel detachment, and the extent of light bulb aneurysms determined by FA.[162] Transscleral diode laser cyclophotocoagulation and ocular hypotensive drops may be able to control secondary glaucoma and prevent enucleation in some cases.[21,163]

Outcomes and prognosis

Since Coats disease is typically unilateral, patients usually have a good overall visual prognosis when considering overall visual function.[1] However, poor visual outcomes of 20/200 or worse are common in the affected eye, even though 76% of eyes achieve anatomical stabilization or improvement after treatment. In a series of 124 eyes, a final visual acuity of 20/50 or better was reported in 14% of eyes, 20/60–20/100 in 6%, 20/200 to counting fingers in 24%, hand motion to no light perception in 40%, and 16% of eyes were enucleated. The poor visual outcomes (≤20/200) were attributed to persistent retinal detachment, foveal exudation, and/or subfoveal fibrosis.[5]

The visual outcome of the affected eye depends on the stage of the disease at presentation. In a study that included 160 eyes, Shields et al. found that more advanced stages were associated with more frequent poor presenting and final visual acuity (<20/200), as well as a greater need for enucleation and less frequent resolution of subretinal fluid, foveal exudation, and leaking telangiectasias after treatment. Poor visual outcomes (<20/200) were observed in 0% of stage 1 eyes, 12% of stage 2A, 36% of stage 2B, 31% of stage 3A1, 45% of stage 3A2, 88% of stage 3B, and 100% of stage 4 and 5 eyes. However, visual acuity loss (≤3 lines) and gain (≥2 lines) did not differ between stages, suggesting the possibility that visual outcomes may be improved by early detection of Coats disease before it reaches the advanced stages.[151]

The age of diagnosis may also serve as a prognostic factor. In a study that included 55 eyes, stage 3 or worse disease was more common in younger patients.[23] Similarly, Daruich et al. found that younger age was associated with more severe disease stages at presentation, a higher incidence of enucleation and structural complications including total tractional retinal detachment and macular fibrosis or atrophy, and worse final visual acuity.[164] A cohort of 351 eyes found similar associations of younger age with more advanced disease, worse presenting visual acuity, less disease resolution, greater ultimate need for enucleation, and poorer visual outcomes (<20/200).[42] Those who present at a young age typically have advanced disease, because the retinal detachment or extensive exudation causes more signs and symptoms that draw attention from their caretakers. Therefore, young age is confounded by the higher stages of disease, and age itself is not a risk factor per se.

In recent years, advances in the diagnosis and management of Coats disease have led to improvements in outcomes. A study of 39 patients treated at a single center over a 20-year period reported differences in the ocular characteristics and outcomes of patients diagnosed in decade 1 (1995–2005) versus decade 2 (2006–15). Patients who were diagnosed in decade 2 were less likely to show advanced stages of disease compared to eyes diagnosed during decade 1, suggesting earlier detection in recent years. In addition, while visual acuity tended to decline through follow-up in eyes from decade 1, eyes from decade 2 remained stable. Furthermore, a trend toward better final visual acuity was noted in decade 2, which could be explained by the decreased severity of disease at presentation, a decrease in management with observation alone, and a greater average number of procedures received per patient, among other factors.[21] A series assessing the outcomes over 5 decades from the 1970s through the 2010s revealed more advanced disease in the 1980s and improved outcomes in the most recent decade, including greater resolution of subretinal fluid and fewer enucleations needed, as the usage of laser photocoagulation and corticosteroid and anti-VEGF injections has increased.[22] Historically, 78% of stage 4 Coats disease eyes were enucleated.[5] However, with the adjunct use of anti-VEGF to conventional treatment, Gaillard et al. reported in 2014 that all 6 of the stage 4 eyes in their study were salvaged.[131]

Conclusion

Coats disease was first reported over 110 years ago.[4] Since then, significant progress has been made in understanding its clinical features and management strategies. Advances in imaging technologies have identified various clinical presentations, such as peripheral nonperfusion, telangiectasias, "light bulb" aneurysmal dilatations, intraretinal and subretinal exudation, macular edema, and exudative retinal detachment, in Coats disease. Management with laser and several adjunctive treatments has allowed us to improve the visual outcomes and globe salvage rate in patients with Coats disease.[21,22] However, the etiology remains unclear. In addition, recurrence has been reported in up to 33% of eyes even after successful treatment, with no difference in recurrence by stage; therefore, long-term follow-up is crucial for patients with any stage of Coats disease.[151,165] Further research is needed to provide a better understanding of the pathogenesis of the disease and to further improve visual outcomes.

Disclosures

Consultant for Alcon (YY), EyePoint (YY), Pykus (YY), Versant Health (YY).

References

1. Shields JA, Shields CL. Review: Coats disease—the 2001 LuEsther T. Mertz lecture. *Retina*. 2002;22(1):80–91. https://doi.org/10.1097/00006982-200202000-00014.
2. Shields JA, Shields CL, Honavar SG, Demirci H. Clinical variations and complications of Coats disease in 150 cases: the 2000 Sanford Gifford memorial lecture. *American Journal of Ophthalmology*. 2001;131(5):561–571. https://doi.org/10.1016/S0002-9394(00)00883-7.
3. Egerer I, Tasman W, Tomer TL. Coats disease. *Archives of Ophthalmology*. 1974;92(2):109–112. https://doi.org/10.1001/archopht.1974.01010010115006.
4. Coats G. Forms of retinal disease with massive exudation. *Roy London Ophthalmology*. 1908;17:440–525.
5. Shields JA, Shields CL, Honavar SG, Demirci H, Cater J. Classification and management of Coats disease: the 2000 proctor lecture. *American Journal of Ophthalmology*. 2001;131(5):572–583. https://doi.org/10.1016/S0002-9394(01)00896-0.
6. Leber T. Uber eine durch Vorkommen multipler Miliaraneurysmen charakterisierte Form von Retinaldegeneration. *Albrecht von Graefes Archiv für Ophthalmologie*. 1912;81:1–14.
7. Reese AB. Telangiectasis of the retina and Coats' disease*. *American Journal of Ophthalmology*. 1956;42(1):1–8. https://doi.org/10.1016/0002-9394(56)90002-2.
8. Moinuddin O, Sathrasala S, Jayasundera KT, et al. Coats-like exudative vitreoretinopathy in retinitis pigmentosa: ocular manifestations and treatment outcomes. *Ophthalmology Retina*. 2021;5(1):86–96. https://doi.org/10.1016/j.oret.2020.03.026.
9. Desai UR, Sabates FN. Long-term follow-up of facioscapulohumeral muscular dystrophy and Coats' disease. *American Journal of Ophthalmology*. 1990;110(5):568–569. https://doi.org/10.1016/S0002-9394(14)77885-7.
10. Shields CL, Zahler J, Falk N, et al. Neovascular glaucoma from advanced Coats disease as the initial manifestation of facioscapulohumeral dystrophy in a 2-year-old child. *Archives of Ophthalmology*. 2007;125(6):840–842. https://doi.org/10.1001/archopht.125.6.840.
11. Soares RR, Stem MS, Yonekawa Y. Exudative retinal telangiectasia associated with leri-Weill dyschondrosteosis. *JAMA Ophthalmology*. 2021;139(8):914–915. https://doi.org/10.1001/jamaophthalmol.2021.1856.
12. Gupta MP, Dow E, Jeng-Miller KW, et al. Spectral domain optical coherence tomography findings in Coats disease. *Retina*. 2019;39(6):1177–1185. https://doi.org/10.1097/IAE.0000000000002120.
13. Kang KB, Wessel MM, Tong J, D'Amico DJ, Paul Chan RV. Ultra-widefield imaging for the management of pediatric retinal diseases. *Journal of Pediatric Ophthalmology and Strabismus*. 2013;50(5):282–288. https://doi.org/10.3928/01913913-20130528-04.
14. Hautz W, Gołebiewska J, Kocyła-Karczmarewicz B. Optical coherence tomography and optical coherence tomography angiography in monitoring Coats' disease. *Journal of Ophthalmology*. 2017;2017:7849243. https://doi.org/10.1155/2017/7849243.
15. Henry CR, Berrocal AM, Hess DJ, Murray TG. Intraoperative spectral-domain optical coherence tomography in Coats' disease. *Ophthalmic Surgery, Lasers & Imaging*. 2012;43:e80–e84. https://doi.org/10.3928/15428877-20120719-01.
16. Jeng-Miller KW, Soomro T, Scott NL, et al. Longitudinal examination of fellow-eye vascular anomalies in Coats' disease with widefield fluorescein angiography: a multicenter study. *Ophthalmic Surgery, Lasers and Imaging Retina*. 2019;50(4):221–227. https://doi.org/10.3928/23258160-20190401-04.
17. Blair MP, Ulrich JN, Elizabeth Hartnett M, Shapiro MJ. Peripheral retinal nonperfusion in fellow eyes in Coats disease. *Retina*. 2013;33(8):1694–1699. https://doi.org/10.1097/IAE.0b013e318285cb86.
18. Shane TS, Berrocal AM, Hess DJ. Bilateral fluorescein angiographic findings in unilateral Coats' disease. *Ophthalmic Surgery, Lasers and Imaging Retina*. 2011;42:e15–e17. https://doi.org/10.3928/15428877-20110203-03.
19. Muakkassa NW, De Carlo TE, Choudhry N, Duker JS, Baumal CR. Optical coherence tomography angiography findings in Coats' disease. *Ophthalmic Surgery, Lasers and Imaging Retina*. 2016;47(7):632–635. https://doi.org/10.3928/23258160-20160707-04.
20. Agarwal A, Pilania RK, Anjani G, Choudhary H, Gupta A, Gupta V. Retinal vasculitis with Coats-like response in a young girl with parry-romberg syndrome. *Journal of Clinical Rheumatology*. 2021;27(8 S):S543–S545. https://doi.org/10.1097/RHU.0000000000001052.
21. Ong SS, Buckley EG, McCuen BW, et al. Comparison of visual outcomes in Coats' disease: a 20-year experience. *Ophthalmology*. 2017;124(9):1368–1376. https://doi.org/10.1016/j.ophtha.2017.03.051.

22. Shields CL, Udyaver S, Dalvin L, et al. Coats disease in 351 eyes: analysis of features and outcomes over 45 years (by decade) at a single center. *Indian Journal of Ophthalmology*. 2019;67(6):772−783. https://doi.org/10.4103/ijo.IJO_449_19.
23. Morris B, Foot B, Mulvihill A. A population-based study of Coats disease in the United Kingdom I: epidemiology and clinical features at diagnosis. *Eye*. 2010;24(12):1797−1801. https://doi.org/10.1038/eye.2010.126.
24. Soares RR, Fine HF, Yonekawa Y. Diagnosis and management of Coats' disease. *Ophthalmic Surgery, Lasers and Imaging Retina*. 2021;52(12):630−635. https://doi.org/10.3928/23258160-20211128-01.
25. Smithen LM, Brown GC, Brucker AJ, Yannuzzi LA, Klais CM, Spaide RF. Coats' disease diagnosed in adulthood. *Ophthalmology*. 2005;112(6):1072−1078. https://doi.org/10.1016/j.ophtha.2004.12.038.
26. Mandura RA, Alqahtani AS. Coats' disease diagnosed during adulthood. *Cureus*. 2021;13(7):e16303. https://doi.org/10.7759/cureus.16303.
27. Sivakami M, Periyanayagi M, Chandrasekar B. Coats' disease diagnosed in adults: a rare case report. *TNOA Journal of Ophthalmic Science and Research*. 2017;55(4):304−306. https://doi.org/10.4103/tjosr.tjosr_23_17.
28. Mahajan S, Patel O, Sharma R, Kumar V. A rare case of unilateral adult onset Coats' disease. *Delhi Journal of Ophthalmology*. 2018;28:67−68. https://doi.org/10.7869/djo.341.
29. Dhawan B, Goel A, Singh R, Vig V. Adult onset Coats disease in a 56-year-old male: an atypical presentation. *Medical Journal of Dr. D.Y. Patil University*. 2015;8(3):350−353. https://doi.org/10.4103/0975-2870.157084.
30. Rishi E, Rishi P, Appukuttan B, Uparkar M, Sharma T, Gopal L. Coats' disease of adult-onset in 48 eyes. *Indian Journal of Ophthalmology*. 2016;64(7):518−523. https://doi.org/10.4103/0301-4738.190141.
31. Black GCM, Perveen R, Bonshek R, et al. Coats' disease of the retina (unilateral retinal telangiectasis) caused by somatic mutation in the NDP gene: a role for norrin in retinal angiogenesis. *Human Molecular Genetics*. 1999;8(11):2031−2035. https://doi.org/10.1093/hmg/8.11.2031.
32. Robitaille JM, Zheng B, Wallace K, et al. The role of Frizzled-4 mutations in familial exudative vitreoretinopathy and Coats disease. *British Journal of Ophthalmology*. 2011;95(4):574−579. https://doi.org/10.1136/bjo.2010.190116.
33. Den Hollander AI, Davis J, Van Der Velde-Visser SD, et al. CRB1 mutation spectrum in inherited retinal dystrophies. *Human Mutation*. 2004;24(5):355−369. https://doi.org/10.1002/humu.20093.
34. Cremers FPM, Maugeri A, den Hollander AI, Hoyng CB. *The expanding roles of ABCA4 and CRB1 in inherited blindness. Retinal Dystrophies: Functional Genomics to Gene Therapy*. 2008:68−84. https://doi.org/10.1002/0470092645.ch6.
35. Tripathi R, Ashton N. Electron microscopical study of Coat's disease. *British Journal of Ophthalmology*. 1971;55(5):289−301. https://doi.org/10.1136/bjo.55.5.289.
36. Egbert PR, Chan CC, Winter FC. Flat preparations of the retinal vessels in Coats' disease. *Journal of Pediatric Ophthalmology*. 1976;13(6):336−339.
37. Fernandes BF, Odashiro AN, Maloney S, Zajdenweber ME, Lopes AG, Burnier MN. Clinical-histopathological correlation in a case of Coats' disease. *Diagnostic Pathology*. 2006;1:24. https://doi.org/10.1186/1746-1596-1-24.
38. Chang M, McLean IW, Merritt JC. Coats' disease: a study of 62 histologically confirmed cases. *Journal of Pediatric Ophthalmology & Strabismus*. 1984;21(5):163−168. https://doi.org/10.3928/0191-3913-19840901-03.
39. Daruich AL, Moulin AP, Tran HV, Matet A, Munier FL. Subfoveal nodule in Coats' disease: toward an updated classification predicting visual prognosis. *Retina*. 2017;37(8):1591−1598. https://doi.org/10.1097/IAE.0000000000001399.
40. Jumper JM, Pomerleau D, McDonald HR, Johnson RN, Fu AD, Cunningham ET. Macular fibrosis in Coats disease. *Retina*. 2010;30(4 Suppl):S9−S14. https://doi.org/10.1097/IAE.0b013e3181cfd3e7.
41. Smirniotopoulos JG, Bargallo N, Mafee MF. Differential diagnosis of leukokoria: radiologic-pathologic correlation. *Radiographics*. 1994;14(5):1059−1079. https://doi.org/10.1148/radiographics.14.5.7991814.
42. Dalvin LA, Udyaver S, Lim LAS, et al. Coats disease: clinical features and outcomes by age category in 351 cases. *Journal of Pediatric Ophthalmology and Strabismus*. 2019;56(5):288−296. https://doi.org/10.3928/01913913-20190716-01.
43. Spitznas M, Joussen F, Wessing A, Meyer-Schwickerath G. Coats' disease - an epidemiologie and fluorescein angiographic study. *Albrecht von Graefes Archiv für Klinische und Experimentelle Ophthalmologie*. 1975;195(4):241−250. https://doi.org/10.1007/BF00414937.
44. Shields JA, Shields CL. Differentiation of Coats' disease and retinoblastoma. *Journal of Pediatric Ophthalmology & Strabismus*. 2001;38(5):262−266. https://doi.org/10.3928/0191-3913-20010901-05.
45. Tarkkanen A, Laatikainen L. Coat's disease: clinical, angiographic, histopathological findings and clinical management. *British Journal of Ophthalmology*. 1983;67(11):766−776. https://doi.org/10.1136/bjo.67.11.766.
46. Ioannidis AS, Liasis A, Sheldrick J, Snead M, Nischal KK. Lamellar macular hole as the presenting feature in a child with Coats' disease. *Journal of Pediatric Ophthalmology & Strabismus*. 2005;42(6):378−379. https://doi.org/10.3928/01913913-20051101-09.
47. Silodor SW, Augsburger JJ, Shields JA, Tasman W. Natural history and management of advanced Coats' disease. *Ophthalmic Surgery*. 1988;19(2):89−93.
48. Ridley ME, Shields JA, Brown GC, Tasman W. Coats' disease: evaluation of management. *Ophthalmology*. 1982;89(12):1381−1387. https://doi.org/10.1016/S0161-6420(82)34634-5.
49. Deutsch TA, Rabb MF, Jampol LM. Spontaneous regression of retinal lesions in Coats' disease. *Canadian Journal of Ophthalmology*. 1982;17(4):169−172.
50. Haik BG. Advanced Coats' disease. *Transactions of the American Ophthalmological Society*. 1991;89:371−475.
51. Suzani M, Moore AT. Intraoperative fluorescein angiography-guided treatment in children with early Coats' disease. *Ophthalmology*. 2015;122(6):1195−1202. https://doi.org/10.1016/j.ophtha.2015.02.002.

52. Theodossiadis GP. Some clinical, fluorescein-angiographic, and therapeutic aspects of Coats' disease. *Journal of Pediatric Ophthalmology & Strabismus*. 1979;16(4):257–262. https://doi.org/10.3928/0191-3913-19790701-12.
53. Mulvihill A, Morris B. A population-based study of Coats disease in the United Kingdom II: investigation, treatment, and outcomes. *Eye*. 2010;24(12):1802–1807. https://doi.org/10.1038/eye.2010.127.
54. Calvo CM, Hartnett ME. The utility of ultra-widefield fluorescein angiography in pediatric retinal diseases. *International Journal of Retina and Vitreous*. 2018;4:21. https://doi.org/10.1186/s40942-018-0122-2.
55. Koozekanani DD, Connor TB, Wirostko WJ. RetCam II fluorescein angiography to guide treatment and diagnosis of Coats disease. *Ophthalmic Surgery, Lasers and Imaging Retina*. 2010;41(3):1–3. https://doi.org/10.3928/15428877-20100215-86.
56. Ali SMA, Khan I, Khurram D, Kozak I. Ultra-widefield angiography with oral fluorescein in pediatric patients with retinal disease. *JAMA Ophthalmology*. 2018;136(5):593–594. https://doi.org/10.1001/jamaophthalmol.2018.0462.
57. Manoharan N, Pecen PE, Cherof AM, Oliver SCN, Palestine AG. Comparison of oral versus intravenous fluorescein widefield angiography in ambulatory pediatric patients. *Journal of VitreoRetinal Diseases*. 2017;1(3):191–196. https://doi.org/10.1177/2474126417705256.
58. Cirafici P, Musolino M, Saccheggiani M, Traverso CE, Nicolò M. Multimodal imaging findings and treatment with dexamethasone implant in three cases of idiopathic macular telangiectasia type 1. *Case Reports in Ophthalmology*. 2021;12(1):92–97. https://doi.org/10.1159/000509850.
59. Ong SS, Cummings TJ, Vajzovic L, Mruthyunjaya P, Toth CA. Comparison of optical coherence tomography with fundus photographs, fluorescein angiography, and histopathologic analysis in assessing Coats disease. *JAMA Ophthalmology*. 2019;137(2):176–183. https://doi.org/10.1001/jamaophthalmol.2018.5654.
60. Spaide RF, Klancnik JM, Cooney MJ. Retinal vascular layers imaged by fluorescein angiography and optical coherence tomography angiography. *JAMA Ophthalmology*. 2015;133(1):45–50. https://doi.org/10.1001/jamaophthalmol.2014.3616.
61. de Carlo TE, Romano A, Waheed NK, Duker JS. A review of optical coherence tomography angiography (OCTA). *International Journal of Retina and Vitreous*. 2015;1:5. https://doi.org/10.1186/s40942-015-0005-8.
62. Yonekawa Y, Todorich B, Trese MT. Optical coherence tomography angiography findings in Coats' disease. *Ophthalmology*. 2016;123(9):1964. https://doi.org/10.1016/j.ophtha.2016.05.004.
63. Stanga PE, Papayannis A, Tsamis E, et al. Swept-source optical coherence tomography angiography of paediatric macular diseases. *Developments in Ophthalmology*. 2016;56:166–173. https://doi.org/10.1159/000442809.
64. Galluzzi P, Hadijstilianou T, Venturi C. Leukocoria: clinical and neuroradiological findings. *Rivista di Neuroradiologia*. 2004;17(1):61–78. https://doi.org/10.1177/197140090401700109.
65. Aerts I, Lumbroso-Le Rouic L, Gauthier-Villars M, Brisse H, Doz F, Desjardins L. Retinoblastoma. *Orphanet Journal of Rare Diseases*. 2006;1(1). https://doi.org/10.1186/1750-1172-1-31.
66. Shields JA, Shields CL. Ophthalmoscopic differentiation of Coats' disease from retinoblastoma. *Journal of American Association for Pediatric Ophthalmology and Strabismus*. 2011;15(1):e7. https://doi.org/10.1016/j.jaapos.2011.01.029.
67. Joussen AM, Gordes RS, Heußen FA, Müller B. Retinal exsudative Gefäßerkrankungen im Kindesalter: morbus Coats und familiär exsudative Vitreoretinopathie (FEVR). *Klinische Monatsblatter fur Augenheilkunde*. 2013;230(9):902–913. https://doi.org/10.1055/s-0033-1350689.
68. Shields JA, Shields CL, Parsons HM. Review: differential diagnosis of retinoblastoma. *Retina*. 1991;11(2):232–243. https://doi.org/10.1097/00006982-199111020-00009.
69. Kondo H. Complex genetics of familial exudative vitreoretinopathy and related pediatric retinal detachments. *Taiwan Journal of Ophthalmology*. 2015;5(2):56–62. https://doi.org/10.1016/j.tjo.2015.04.002.
70. Collin A, Lecler A. Coats plus syndrome. *JAMA Neurology*. 2019;76(4):501. https://doi.org/10.1001/jamaneurol.2018.4610.
71. Small RG. Coats' disease and muscular dystrophy. *Transactions - American Academy of Ophthalmology and Otolaryngology*. 1968;72(2):225–231.
72. Cameron JD, Yanoff M, Frayer WC. Coats' disease and Turner's syndrome. *American Journal of Ophthalmology*. 1974;78(5):852–854. https://doi.org/10.1016/0002-9394(74)90310-9.
73. Kondra L, Cangemi FE, Pitta CG. Alport's syndrome and retinal telangiectasia. *Annals of Ophthalmology*. 1983;15(6):550–551.
74. Folk JC, Genovese FN, Biglan AW. Coats' disease in a patient with Cornelia de Lange syndrome. *American Journal of Ophthalmology*. 1981;91(5):607–610. https://doi.org/10.1016/0002-9394(81)90059-3.
75. Kajtár P, Méhes K. Bilateral Coats retinopathy associated with aplastic anaemia and mild dyskeratotic signs. *American Journal of Medical Genetics*. 1994;49(4):374–377. https://doi.org/10.1002/ajmg.1320490404.
76. Genkova P, Toncheva D, Tzoneva M, Konstantinov I. Deletion of 13q12.1 in a child with Coats disease. *Acta paediatrica Hungarica*. 1986;27(2):141–143.
77. Schuman JS, Lieberman KV, Friedman AH, Berger M, Schoeneman MJ. Senior-Loken syndrome (familial renal-retinal dystrophy) and Coats' disease. *American Journal of Ophthalmology*. 1985;100(6):822–827. https://doi.org/10.1016/S0002-9394(14)73374-4.
78. Burch JV, Leveille AS, Morse PH. Ichthyosis hystrix (epidermal nevus syndrome) and Coats' disease. *American Journal of Ophthalmology*. 1980;89(1):25–30. https://doi.org/10.1016/0002-9394(80)90225-1.
79. Newell SW, Hall BD, Anderson CW, Lim ES. Hallermann-streiff syndrome with Coats disease. *Journal of Pediatric Ophthalmology & Strabismus*. 1994;31(2):123–125. https://doi.org/10.3928/0191-3913-19940301-16.
80. Berger M, Lieberman KV, Schoeneman MJ, Schuman JS, Friedman AH. Coats' disease in a renal transplant recipient. *Nephrology Dialysis Transplantation*. 1987;2(2):120–123.
81. Bhushan M, Kumar S, Griffiths CEM. Multiple glomus tumours, Coats' disease and basic fibroblast growth factor. *British Journal of Dermatology*. 1997;137(3):454–456. https://doi.org/10.1046/j.1365-2133.1997.18691938.x.

82. Skuta GL, France TD, Stevens TS, Laxova R. Apparent Coats' disease and pericentric inversion of chromosome 3. *American Journal of Ophthalmology*. 1987;104(1):84–86. https://doi.org/10.1016/0002-9394(87)90299-6.
83. Gärtner J, Draf W. Leber's miliary aneurysms associated with telangiectasia of the nasal mucosa. *American Journal of Ophthalmology*. 1975;79(1):56–58. https://doi.org/10.1016/0002-9394(75)90456-0.
84. Patrassi GM, Sartori MT, Piermarocchi S, Viero M, Boeri G, Girolami A. Unusual thrombotic-like retinopathy (Coats' disease) associated with congenital plasminogen deficiency type I. *Journal of Internal Medicine*. 1993;234(6):619–623. https://doi.org/10.1111/j.1365-2796.1993.tb01024.x.
85. Reynolds BC, Lemmers RJLF, Tolmie J, Howatson AG, Hughes DA. Focal segmental glomerulosclerosis, Coats'-like retinopathy, sensorineural deafness and chromosome 4 duplication: a new association. *Pediatric Nephrology*. 2010;25(8):1551–1554. https://doi.org/10.1007/s00467-010-1474-5.
86. Briggs TA, Abdel-Salam GMH, Balicki M, et al. Cerebroretinal microangiopathy with calcifications and cysts (CRMCC). *American Journal of Medical Genetics Part A*. 2008;146A(2):182–190. https://doi.org/10.1002/ajmg.a.32080.
87. Crow YJ, McMenamin J, Haenggeli CA, et al. Coats' plus: a progressive familial syndrome of bilateral Coats' disease, characteristic cerebral calcification, leukoencephalopathy, slow pre- and post-natal linear growth and defects of bone marrow and integument. *Neuropediatrics*. 2004;35(1):10–19. https://doi.org/10.1055/s-2003-43552.
88. Tolmie JL, Browne BH, McGettrick PM, Stephenson JBP. A familial syndrome with Coats' reaction retinal angiomas, hair and nail defects and intracranial calcification. *Eye*. 1988;2(3):297–303. https://doi.org/10.1038/eye.1988.56.
89. Han E, Patel NA, Yannuzzi NA, et al. A unique case of Coats plus syndrome and dyskeratosis congenita in a patient with CTC1 mutations. *Ophthalmic Genetics*. 2020;41(4):363–367. https://doi.org/10.1080/13816810.2020.1772315.
90. Peng J, Zhang Q, Chen C, Huang Q, Li Y, Zhao P. Early onset Coats' disease initially treated as unilateral ROP at 39 weeks postmenstrual age: a case report. *BMC Ophthalmology*. 2017;17:145. https://doi.org/10.1186/s12886-017-0536-x.
91. Shanmugam PM, Ramanjulu R. Vascular tumors of the choroid and retina. *Indian Journal of Ophthalmology*. 2015;63(2):133–140. https://doi.org/10.4103/0301-4738.154387.
92. Haider S, Qureshi W, Ali A. Leukocoria in children. *Journal of Pediatric Ophthalmology and Strabismus*. 2008;45(3):179–180. https://doi.org/10.3928/01913913-20080501-13.
93. Gupta N, Terrell W, Schoenfield L, Kirsch C, Cebulla CM. Uveal melanoma mimicking advanced Coats' disease in a young patient. *Ocular Oncology and Pathology*. 2016;2(3):140–143. https://doi.org/10.1159/000441526.
94. Sanghi G, Dogra M, Ray M, Gupta A. Predominant exudative retinopathy in incontinentia pigmenti and clinical course after peripheral laser photocoagulation. *Indian Journal of Ophthalmology*. 2011;59(3):255–256. https://doi.org/10.4103/0301-4738.81022.
95. Luckie AP, Hamilton AMP. Adult Coats' disease in branch retinal vein occlusion. *Australian and New Zealand Journal of Ophthalmology*. 1994;22(3):203–206. https://doi.org/10.1111/j.1442-9071.1994.tb01717.x.
96. Frezzotti R, Berengo A, Guerra R, Cavallini F. Toxoplasmic Coats' retinitis. A parasitologically proved case. *American Journal of Ophthalmology*. 1965;59(6):1099–1102. https://doi.org/10.1016/0002-9394(65)93431-8.
97. Khan JA, Ide CH, Strickland MP. Coats'-type retinitis pigmentosa. *Survey of Ophthalmology*. 1988;32(5):317–332. https://doi.org/10.1016/0039-6257(88)90094-X.
98. Fogle JA, Welch RB, Green WR. Retinitis pigmentosa and exudative vasculopathy. *Archives of Ophthalmology*. 1978;96(4):696–702. https://doi.org/10.1001/archopht.1978.03910050386018.
99. Lanier JD, McCrary JA, Justice J. Autosomal recessive retinitis pigmentosa and Coats disease: a presumed familial incidence. *Archives of Ophthalmology*. 1976;94(10):1737–1742. https://doi.org/10.1001/archopht.1976.03910040511009.
100. Spallone A, Carlevaro G, Ridling P. Autosomal dominant retinitis pigmentosa and Coats'-like disease. *International Ophthalmology*. 1985;8(3):147–151. https://doi.org/10.1007/BF00136491.
101. Arrigg PG, Lahav M, Hutchins RK, Weiter JJ. Pigmentary retinal degeneration and Coats' disease: a case study. *Ophthalmic Surgery*. 1988;19(6):432–436.
102. Kim RY, Kearney JJ. Coats-type retinitis pigmentosa in a 4-year-old child. *American Journal of Ophthalmology*. 1997;124(6):846–848. https://doi.org/10.1016/S0002-9394(14)71707-6.
103. Kan E, Yilmaz T, Aydemir O, Güler M, Kurt J. Coats-like retinitis pigmentosa: reports of three cases. *Clinical Ophthalmology*. 2007;1(2):193–198.
104. Kremer I, Cohen S, Izhak RB, Ben-Sira I. An unusual case of congenital unilateral Coats's disease associated with morning glory optic disc anomaly. *British Journal of Ophthalmology*. 1985;69(1):32–37. https://doi.org/10.1136/bjo.69.1.32.
105. Yannuzzi LA, Bardal AMC, Freund KB, Chen KJ, Eandi CM, Blodi B. Idiopathic macular telangiectasia. *Archives of Ophthalmology*. 2006;124(4):450–460. https://doi.org/10.1001/archopht.124.4.450.
106. Gass D, Oyakawa RT. Idiopathic Juxtafoveolar retinal telangiectasis. *Archives of Ophthalmology*. 1982;100(5):769–780. https://doi.org/10.1001/archopht.1982.01030030773010.
107. Gass JDM, Blodi BA. Idiopathic Juxtafoveolar retinal telangiectasis: update of classification and follow-up study. *Ophthalmology*. 1993;100(10):1536–1546. https://doi.org/10.1016/S0161-6420(93)31447-8.
108. Sen M, Shields CL, Honavar S, Shields J. Coats disease: an overview of classification, management and outcomes. *Indian Journal of Ophthalmology*. 2019;67(6):763–771. https://doi.org/10.4103/ijo.IJO_841_19.
109. Grosso A, Pellegrini M, Cereda MG, Panico C, Staurenghi G, Sigler EJ. Pearls and pitfalls in diagnosis and management of Coats disease. *Retina*. 2015;35(4):614–623. https://doi.org/10.1097/IAE.0000000000000485.

110. Mrejen S, Metge F, Denion E, Dureau P, Edelson C, Caputo G. Management of retinal detachment in Coats disease: study of 15 cases. *Retina*. 2008;28(3):S26–S32. https://doi.org/10.1097/IAE.0b013e31816b3158.
111. Levinson JD, Hubbard GB. 577-NM yellow laser photocoagulation for Coats disease. *Retina*. 2016;36(7):1388–1394. https://doi.org/10.1097/IAE.0000000000000874.
112. Shapiro MJ, Chow CC, Karth PA, Kiernan DF, Blair MP. Effects of green diode laser in the treatment of pediatric Coats disease. *American Journal of Ophthalmology*. 2011;151(4):725–731. https://doi.org/10.1016/j.ajo.2010.10.024.
113. Sigler EJ, Randolph JC, Calzada JI, Wilson MW, Haik BG. Current management of Coats disease. *Survey of Ophthalmology*. 2014;59(1):30–46. https://doi.org/10.1016/j.survophthal.2013.03.007.
114. Kusaka S. Surgical management of Coats disease. *Asia-Pacific Journal of Ophthalmology*. 2018;7(3):156–159. https://doi.org/10.22608/APO.201867.
115. Ghorbanian S, Jaulim A, Chatziralli IP. Diagnosis and treatment of Coats' disease: a review of the literature. *Ophthalmologica*. 2012;227(4):175–182. https://doi.org/10.1159/000336906.
116. Nucci P, Bandello F, Serafino M, Wilson ME. Selective photocoagulation in Coats' disease: ten-year follow-up. *European Journal of Ophthalmology*. 2002;12(6):501–505. https://doi.org/10.1177/112067210201200609.
117. Swartz M. Histology of macular photocoagulation. *Ophthalmology*. 1986;93(7):959–963. https://doi.org/10.1016/S0161-6420.
118. Lock JHJ, Fong KCS. An update on retinal laser therapy. *Clinical and Experimental Optometry*. 2011;94(1):43–51. https://doi.org/10.1111/j.1444-0938.2010.00529.x.
119. Castillejos-Rios D, Devenyi R, Moffat K, Yu E. Dye yellow vs. argon green laser in panretinal photocoagulation for proliferative diabetic retinopathy: a comparison of minimum power requirements. *Canadian Journal of Ophthalmology*. 1992;27(5):243–244.
120. Schefler AC, Berrocal AM, Murray TG. Advanced Coats' disease: management with repetitive aggressive laser ablation therapy. *Retina*. 2008;28(3):S38–S41. https://doi.org/10.1097/IAE.0b013e318163cd7c.
121. Allegaert K, Devlieger H, Casteels I. Reduced inflammatory response after laser photocoagulation compared with cryoablation for threshold retinopathy of prematurity. *Journal of Pediatric Ophthalmology and Strabismus*. 2005;42(5):264–266. https://doi.org/10.3928/0191-3913-20050901-04.
122. Elnahry AG, Sallam EM, Guirguis KJ, Talbet JH, Abdel-Kader AA. Vitrectomy for a secondary epiretinal membrane following treatment of adult-onset Coats' disease. *American Journal of Ophthalmology Case Reports*. 2019;15:100508. https://doi.org/10.1016/j.ajoc.2019.100508.
123. Glaser BM, Vidaurri-Leal J, Michels RG, Campochiaro PA. Cryotherapy during surgery for giant retinal tears and intravitreal dispersion of viable retinal pigment epithelial cells. *Ophthalmology*. 1993;100(4):466–470. https://doi.org/10.1016/S0161-6420(93)31620-9.
124. Adeniran JF, Duff SM, Mimouni M, Lambert N, Ramasubramanian A. Treatment of Coats' disease: an analysis of pooled results. *International Journal of Ophthalmology*. 2019;12(4):668–674. https://doi.org/10.18240/ijo.2019.04.23.
125. Sun Y, Jain A, Moshfeghi DM. Elevated vascular endothelial growth factor levels in Coats disease: rapid response to pegaptanib sodium. *Graefe's Archive for Clinical and Experimental Ophthalmology*. 2007;245(9):1387–1388. https://doi.org/10.1007/s00417-007-0559-8.
126. He YG, Wang H, Zhao B, Lee J, Bahl D, McCluskey J. Elevated vascular endothelial growth factor level in Coats' disease and possible therapeutic role of bevacizumab. *Graefe's Archive for Clinical and Experimental Ophthalmology*. 2010;248(10):1519–1521. https://doi.org/10.1007/s00417-010-1366-1.
127. Zhao Q, Peng XY, Chen FH, et al. Vascular endothelial growth factor in Coats' disease. *Acta Ophthalmologica*. 2014;92(3):e225–e228. https://doi.org/10.1111/aos.12158.
128. Jun JH, Kim YC, Kim KS. Resolution of severe macular edema in adult Coats' disease with intravitreal triamcinolone and bevacizumab injection. *Korean Journal of Ophthalmology: KJO*. 2008;22(3):190–193. https://doi.org/10.3341/kjo.2008.22.3.190.
129. Entezari M, Ramezani A, Safavizadeh L, Bassirnia N. Resolution of macular edema in Coats' disease with intravitreal bevacizumab. *Indian Journal of Ophthalmology*. 2010;58(1):80–82. https://doi.org/10.4103/0301-4738.58482.
130. Álvarez-Rivera LG, Abraham-Marín ML, Flores-Orta HJ, Mayorquín-Ruiz M, Cortés-Luna CF. Enfermedad de Coats tratada con bevacizumab (Avastin). *Archivos de la Sociedad Española de Oftalmología*. 2008;83(5):329–331. https://doi.org/10.4321/s0365-66912008000500010.
131. Gaillard MC, Mataftsi A, Balmer A, Houghton S, Munier F. Ranibizumab in the management of advanced Coats disease stages 3b and 4: long-term outcomes. *Retina*. 2014;34(11):2275–2281. https://doi.org/10.1097/IAE.0000000000000248.
132. Çakir M, Çekiç O, Yilmaz OF. Combined intravitreal bevacizumab and triamcinolone injection in a child with Coats disease. *Journal of AAPOS*. 2008;12(3):309–311. https://doi.org/10.1016/j.jaapos.2008.01.009.
133. Lin CJ, Hwang JF, Chen YT, Chen SN. The effect of intravitreal bevacizumab in the treatment of Coats disease in children. *Retina*. 2010;30(4):617–622. https://doi.org/10.1097/IAE.0b013e3181c2e0b7.
134. Patel NA, Berrocal AM, Murray TG, Villegas VM. Advanced Coats' disease treated with intravitreal brolucizumab combined with laser photocoagulation. *American Journal of Ophthalmology Case Reports*. 2020;19:100815. https://doi.org/10.1016/j.ajoc.2020.100815.
135. Ray R, Barañano DE, Hubbard GB. Treatment of Coats' disease with intravitreal bevacizumab. *British Journal of Ophthalmology*. 2013;97(3):272–277. https://doi.org/10.1136/bjophthalmol-2012-302250.
136. Cackett P, Wong D, Cheung CMG. Combined intravitreal bevacizumab and argon laser treatment for Coats' disease. *Acta Ophthalmologica*. 2010;88(2):e48–e49. https://doi.org/10.1111/j.1755-3768.2009.01514.x.
137. Kodama A, Sugioka K, Kusaka S, Matsumoto C, Shimomura Y. Combined treatment for Coats' disease: retinal laser photocoagulation combined with intravitreal bevacizumab injection was effective in two cases. *BMC Ophthalmology*. 2014;14:36. https://doi.org/10.1186/1471-2415-14-36.
138. Villegas VM, Gold AS, Berrocal AM, Murray TG. Advanced Coats' disease treated with intravitreal bevacizumab combined with laser vascular ablation. *Clinical Ophthalmology*. 2014;8:973–976. https://doi.org/10.2147/OPTH.S62816.

139. Ramasubramanian A, Shields CL. Bevacizumab for Coats' disease with exudative retinal detachment and risk of vitreoretinal traction. *British Journal of Ophthalmology*. 2012;96(3):356–359. https://doi.org/10.1136/bjophthalmol-2011-300141.
140. Bergstrom CS, Hubbard GB. Combination intravitreal triamcinolone injection and cryotherapy for exudative retinal detachments in severe Coats disease. *Retina*. 2008;28(3):S33–S37. https://doi.org/10.1097/IAE.0b013e318159ecad.
141. Sein J, Tzu JH, Murray TG, Berrocal AM. Treatment of Coats' disease with combination therapy of intravitreal bevacizumab, laser photocoagulation, and sub-tenon corticosteroids. *Ophthalmic Surgery, Lasers and Imaging Retina*. 2016;47(5):443–449. https://doi.org/10.3928/23258160-20160419-07.
142. Alsakran WA, Nowilaty SR, Ghazi NG, et al. Adjunctive intravitreal triamcinolone acetonide for exudative retinal detachment in Coats disease. *Journal of VitreoRetinal Diseases*. 2022;6(1):54–62. https://doi.org/10.1177/24741264211018957.
143. Saatci AO, Doruk HC, Yaman A. Intravitreal dexamethasone implant (Ozurdex) in Coats' disease. *Case Reports in Ophthalmology*. 2013;4(3):122–128. https://doi.org/10.1159/000355363.
144. Martínez-Castillo S, Gallego-Pinazo R, Dolz-Marco R, Marín-Lambíes C, Díaz-Llopis M. Adult Coats' disease successfully managed with the dexamethasone intravitreal implant (Ozurdex) combined with retinal photocoagulation. *Case Reports in Ophthalmology*. 2012;3(1):123–127. https://doi.org/10.1159/000337481.
145. Othman IS, Moussa M, Bouhaimed M. Management of lipid exudates in Coats disease by adjuvant intravitreal triamcinolone: effects and complications. *British Journal of Ophthalmology*. 2010;94(5):606–610. https://doi.org/10.1136/bjo.2009.168013.
146. Jarin RR, Teoh SCB, Lim TH. Resolution of severe macular oedema in adult Coat's syndrome with high-dose intravitreal triamcinolone acetonide. *Eye*. 2006;20(2):163–165. https://doi.org/10.1038/sj.eye.6701828.
147. Ghazi NG, Al Shamsi H, Larsson J, Abboud E. Intravitreal triamcinolone in Coats' disease. *Ophthalmology*. 2012;119(3):648–649. https://doi.org/10.1016/j.ophtha.2011.09.059.
148. Çekiç O, Chang S, Tseng JJ, Akar Y, Barile GR, Schiff WM. Cataract progression after intravitreal triamcinolone injection. *American Journal of Ophthalmology*. 2005;139(6):993. https://doi.org/10.1016/j.ajo.2005.01.022.
149. Thompson JT. Cataract formation and other complications of intravitreal triamcinolone for macular edema. *American Journal of Ophthalmology*. 2006;141(4):629–e1. https://doi.org/10.1016/j.ajo.2005.11.050.
150. Zhao T, Wang K, Ma Y, Jiang YR. Resolution of total retinal detachment in Coats' disease with intravitreal injection of bevacizumab. *Graefe's Archive for Clinical and Experimental Ophthalmology*. 2011;249(11):1745–1746. https://doi.org/10.1007/s00417-010-1563-y.
151. Shields CL, Udyaver S, Dalvin LA, et al. Visual acuity outcomes in Coats disease by classification stage in 160 patients. *British Journal of Ophthalmology*. 2020;104(3):422–431. https://doi.org/10.1136/bjophthalmol-2019-314363.
152. Stanga PE, Jaberansari H, Bindra MS, Gil-Martinez M, Biswas S. Transcleral drainage of subretinal fluid, anti-vascular endothelial growth factor, and wide-field imaging-guided laser in Coats exudative retinal detachment. *Retina*. 2016;36(1):156–162. https://doi.org/10.1097/IAE.0000000000000669.
153. Adam RS, Kertes PJ, Lam WC. Observations on the management of Coats' disease: less is more. *British Journal of Ophthalmology*. 2007;91(3):303–306. https://doi.org/10.1136/bjo.2006.103382.
154. Yoshizumi MO, Kreiger AE, Lewis H, Foxman B, Hakakha BA. Vitrectomy techniques in late-stage Coats'-like exudative retinal detachment. *Documenta Ophthalmologica*. 1995;90(4):387–394. https://doi.org/10.1007/BF01268124.
155. Muftuoglu G, Gulkilik G. Pars plana vitrectomy in advanced Coats' disease. *Case Reports in Ophthalmology*. 2011;2(1):15–22. https://doi.org/10.1159/000323616.
156. Kranias G, Krebs TP. Advanced Coats' disease successfully managed with vitreo-retinal surgery. *Eye*. 2002;16(4):500–501. https://doi.org/10.1038/sj.eye.6700019.
157. Suesskind D, Altpeter E, Schrader M, Bartz-Schmidt KU, Aisenbrey S. Pars plana vitrectomy for treatment of advanced Coats' disease - presentation of a modified surgical technique and long-term follow-up. *Graefe's Archive for Clinical and Experimental Ophthalmology*. 2014;252(6):873–879. https://doi.org/10.1007/s00417-013-2512-3.
158. Ucgul AY, Ozdek S, Ertop M, Atalay HT. External drainage alone versus external drainage with vitrectomy in advanced Coats disease. *American Journal of Ophthalmology*. 2021;222:6–14. https://doi.org/10.1016/j.ajo.2020.09.006.
159. Cai X, Zhao P, Zhang Q, Jin H. Treatment of stage 3 Coats' disease by endolaser photocoagulation via a two-port pars plana nonvitrectomy approach. *Graefe's Archive for Clinical and Experimental Ophthalmology*. 2015;253(7):999–1004. https://doi.org/10.1007/s00417-015-2984-4.
160. Devenyi RG. The use of perfluorocarbon liquid in the surgical repair of a bullous retinal detachment secondary to Coats' disease. *Canadian Journal of Ophthalmology*. 1998;33(7):385–386.
161. Imaizumi A, Kusaka S, Takaesu S, Sawaguchi S, Shimomura Y. Subretinal fluid drainage and vitrectomy are helpful in diagnosing and treating eyes with advanced Coats' disease. *Case Reports in Ophthalmology*. 2016;7(1):223–229. https://doi.org/10.1159/000445795.
162. Udyaver S, Dalvin LA, Lim L-AS, et al. Predictors of enucleation in Coats disease: analysis of 259 eyes of 259 patients at a single center. *Journal of American Association for Pediatric Ophthalmology and Strabismus*. 2019;23(5):266. https://doi.org/10.1016/j.jaapos.2019.05.015.
163. De Silva DJ, Brookes JL. Cyclodiode treatment of neovascular glaucoma secondary to Coats' disease. *British Journal of Ophthalmology*. 2007;91(5):690–691. https://doi.org/10.1136/bjo.2006.097121.
164. Daruich A, Matet A, Munier FL. Younger age at presentation in children with Coats disease is associated with more advanced stage and worse visual prognosis a retrospective study. *Retina*. 2018;38(11):2239–2246. https://doi.org/10.1097/IAE.0000000000001866.
165. Shienbaum G, Tasman WS. Coats disease. *Retina*. 2006;26(4):422–424. https://doi.org/10.1097/01.iae.0000238545.44132.a2.

Chapter 24

Arteriovenous malformations

Javier Muñoz-Solano[2], Roberto Gallego-Pinazo[2] and J. Fernando Arévalo[1]
[1]*The Johns Hopkins University School of Medicine, Edmund F. and Virginia B. Ball, Department of Ophthalmology, Johns Hopkins Bayview Medical Center, Retina Division, Wilmer Eye Institute, Baltimore, MD, United States;* [2]*Unit of Macula and Clinical Trials, Oftalvist Clinic, Valencia, Spain*

Overview bullets: Congenital retinal macrovessel

- Congenital retinal macrovessel is an abnormal retinal vessel, usually a vein, that crosses the median raphe of the macula.
- They are usually asymptomatic and stable over time.
- In fluorescein angiography, premature filling and (Fig. 24.1D) delayed emptying are observed.
- OCT reveals a highly back-scattering hyper-reflective vessel with a shadow effect along its path.
- In 24% of the cases, they are associated with cerebral venous malformations. Therefore, it is recommended to perform a brain MRI with contrast.
- Differential diagnosis include racemose angioma, capillary hemangioma, cavernous hemangioma, retinoblastoma, choroidal melanoma, and retinal vein branch obstruction.

Primary vascular malformations of the retina are still a challenge for most retinal clinicians. They conform to a heterogenous group of alterations in the normal arterial or venous retinal circulation. These morphological changes are decisive in their evolution and clinical manifestations, as well as the location and extension of the vascular malformation. Most of these conditions result from abnormal embryogenesis and could be associated with systemic diseases. Their etiopathogenesis remains unclear. These vascular conditions are usually benign and characterized by mature and competent vessels, but in some cases, a disturbance of homeostasis can cause different alterations, such as retinal hemorrhages. The aim of this chapter is to assess the main characteristics of congenital retinal macrovessels (CRM), preretinal vascular loops, familial retinal arteriolar tortuosity (fRAT), and retinal arteriovenous malformations (AVMs), their pathogenesis, main features in multimodal imaging, and their ocular and systemic associations, which may be life threatening in some occasions. This knowledge is essential as they have different management, prognosis, associations, and a wide differential diagnosis. For this purpose, multimodal imaging has a crucial role to understand these pathologies.

Preretinal vascular loop

- Preretinal vascular loop is a congenital vascular anomaly, usually arterial, that originates from a main branch of the central retinal artery on the optic disk.
- Most are found along the inferotemporal arcade and tend to project into the vitreous cavity.
- They are usually asymptomatic and stable over time.
- Fluorescein angiography can be (Fig. 24.2) useful to differentiate whether they are arterial or venous.
- Differential diagnosis include persistent hyaloid artery, Bergmeister's papilla, optociliary shunt vessels, optic disc neovessels, and IRVAN and Wyburn-Mason syndrome.

Familial retinal arteriolar tortuosity

- Familial retinal arteriolar tortuosity (fRAT) is used to term the tortuosity of the arterioles of second and third order in the macular and peripapillary area, of probable autosomal dominant inheritance.

Retinal and Choroidal Vascular Diseases of the Eye. https://doi.org/10.1016/B978-0-443-15583-3.00024-X

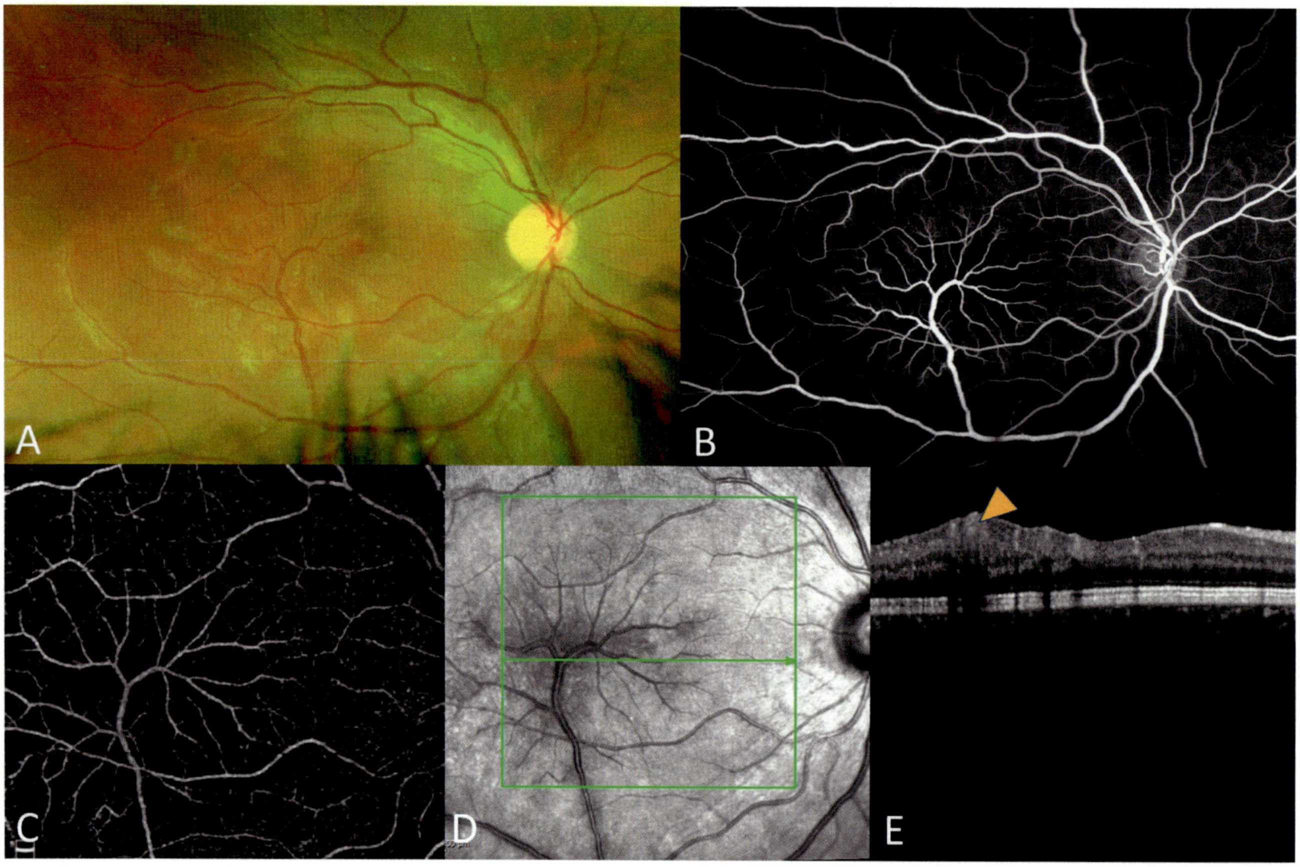

FIGURE 24.1 Widefield color (A) and fluorescein angiography (B) images show a congenital retinal macrovessel arising from the inferior temporal major vein, crossing the medial raphe to the superior retina. (C) En face optical coherence tomography angiography of the superior capillary plexus of the retina, with spared round-shaped foveal avascular zone and the adjacent macrovessel. (D) Infrared image of the congenital retinal macrovessel with a *green line* across, which represents the B-scan optical coherence tomography (E) showing highly back-scattering hyperreflective vessel *(arrowhead)* with shadow effect.

- It is a rare and normally asymptomatic vascular malformation.
- Diagnosis is clinical. Fluorescein (Fig. 24.3) angiography may be helpful in ruling out leaks, staining, or hypoperfusion.
- Differential diagnosis include healthy hyperopic eyes, neurofibromatosis type 1, and tortuosity secondary to an epiretinal membrane.

Retinal arteriovenous malformations

- Retinal arteriovenous malformations (AVM) range from a single arteriovenous communication to a complex anastomotic system.
- They are generally low-grade, asymptomatic and unilateral. Visual acuity correlates directly with the degree of retinal AVMs and the presence of systemic AVMs.
- In fluorescein angiography a characteristic rapid filling without capillary loss immediately after the arterial phase is observed.
- Differential diagnosis include peripapillary telangiectatic microangiopathy, Leber's hereditary optic neuropathy, capillary and cavernous hemangiomas, and retinal angiomas.

Congenital retinal macrovessel

Introduction

In 1869, Mauthner described isolated enlarged vessels crossing the media raphe at the central macula. However, Brown et al. were the first who used the term *congenital retinal macrovessel* (CRM) for an abnormal unilateral retinal vessel, usually a vein, which traversed through the macula and had large tributaries on both sides of the raphe along its path.[1–3].

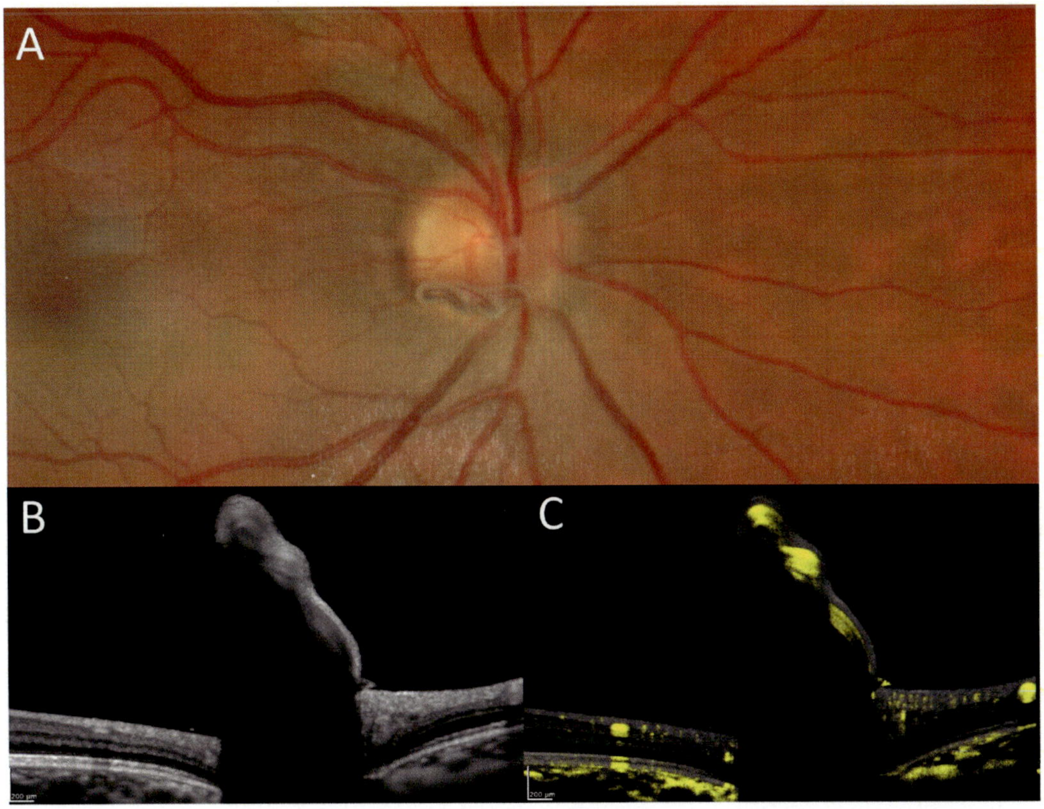

FIGURE 24.2 (A) color fundus image of a small prepapillary arterial loop, which supplies the inferior papillary arterial system. A fibroglial sheath is encasing a portion of the loop. B-scan optical coherence tomography (OCT) (B) and angiography (OCTA) (C) show the projection into the vitreous cavity of the prepapillary arterial loop.

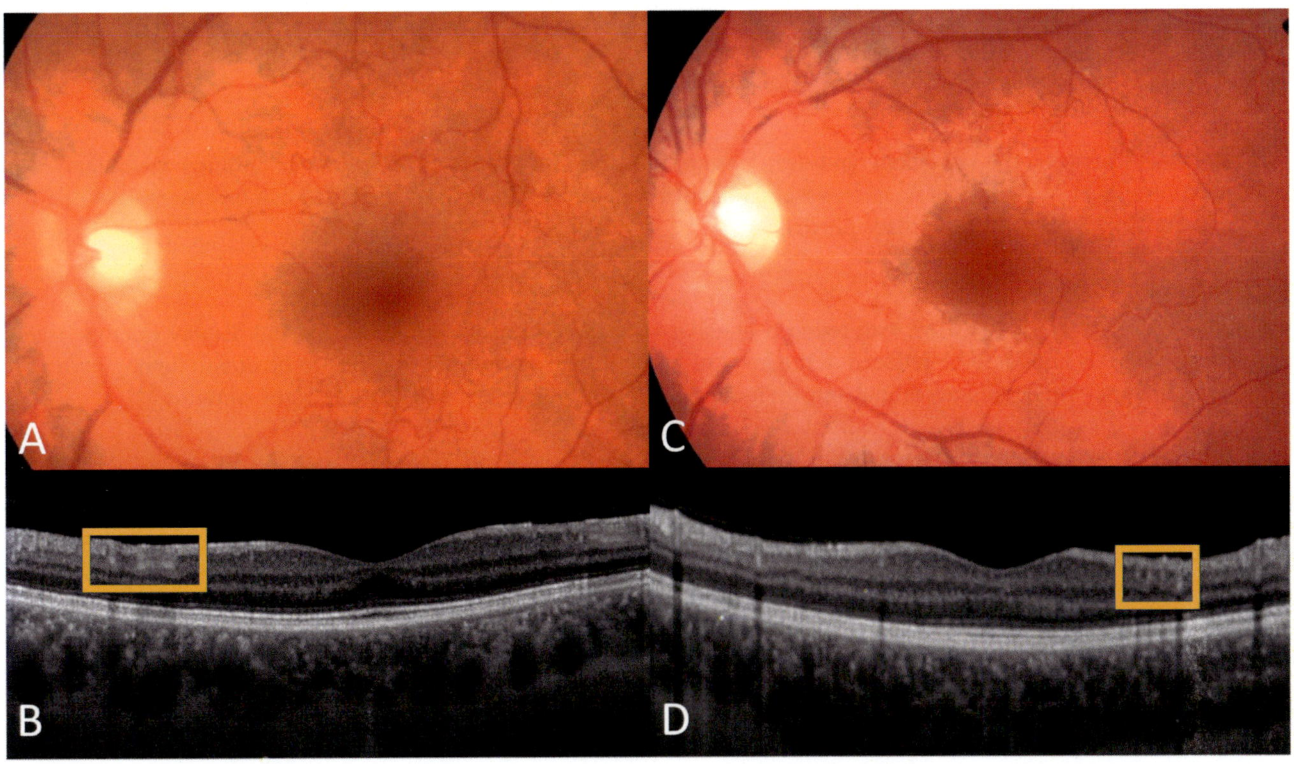

FIGURE 24.3 Color fundus image of a 46 y-o healthy women (A) shows an evident tortuosity of the arterioles of second and third order in the macular and peripapillary area. B-scan optical coherence tomography (B) showing multiple vessels at superficial plexus level (*box*) because the scan is made through tortuous vessels. The same findings could be observed in the images of her healthy 21 y-o daughter (C and D).

CRM is a rare condition. Its prevalence has been estimated at approximately 1 in 200,000 in the United States.[4] Women represent 65% of the patients, being the mean age of diagnosis at 44.0 ± 20.9 years.[3] Nevertheless, this may be an underestimation because CRMs are normally identified during a routine retinal examination due to their lack of symptoms.[3,5]

CRM has an excellent visual prognosis and is stable over time.[1]

Etiopathogenesis

CRM is the result of abnormal embryogenesis, but the etiopathogenesis remains unclear. Around weeks 15th and 16th of gestation, mesenchymal cells appear near the hyaloid artery, in the optic disc, and gradually invade the nerve fiber layer of the retina. Later, these mesenchymal cells begin to differentiate into endothelial cells, forming cords that will become capillaries. Progressively, these capillaries keep differentiating, resulting in arteries and veins.[1,6,7]

By the fifth month of gestation, the macula is surrounded by primitive capillaries.[7,8] There is controversy regarding foveal avascularization. Some authors argue that vascular proliferation never reaches the center of the fovea[8] and others state that it is due to the high concentration of oxygen, which downregulates the production of vascular stimulating factors in the retinal cells.[7]

It is possible for a vessel to enlarge abnormally and reach atypical positions in the retina during the restructuring phases. However, it is not possible to determine neither the exact moment in which it occurs nor the stimulus that generates it.[1] According to Volk[9] and Ashton,[7] throughout this process, CRM does not limit the development of the neural component of the macula.

Multimodal imaging (Fig. 24.1)

Fluorescein angiography

Fluoroangiographic findings of CRM consist of an early filling and delayed emptying and may be accompanied by microvascular alterations, such as enlargement of the foveal avascular zone or aneurysmatic changes.[10]

OCT

The B-scan OCT image revealed a highly back-scattering hyper-reflective vessel with a shadow effect along its path. Other findings have also been reported, such as cystoid macular edema, serous macular detachment, and adhesions between the vitreous and the aberrant macrovessel.[11]

OCTA

CRM is located between the superficial and deep vascular layers of the retina, from the inner limiting membrane to the outer plexiform layer.[12] With OCTA, microvascular capillary abnormalities around the CRM can be observed in one-third of patients, mainly in the deep capillary plexus.[3]

Microperimetry

Shah et al. reported a case of CRM with reduced macular sensitivity measured by liquid crystal display microperimetry. Retinal sensitivity is known to be reduced at the level of larger retinal vessels. This seems to be due to hemoglobin in the retinal vasculature, which reduces retinal sensitivity, especially in the presence of a white light stimulus.[13]

Ocular and systemic associations

In most patients, CRM is a stable condition over time and asymptomatic, with no effects on visual acuity.[5] CRM can cause amblyopia in children due to the alteration in macular development caused by the foveolar compromise by the CRM or its branches.[14] In contrast, decreased visual acuity in adult patients is rare and may be due to a scotoma generated by the alteration of foveal morphology, foveal ectopia, decreased retinal sensitivity, macular ischemia, retinal artery branch occlusion, cavernous hemangioma, retinal detachment, juxtafoveal macroaneurysms, telangiectatic changes, macular hemorrhage, or vitreous hemorrhage, among others.[15]

Other less common causes of bleeding associated with this type of vascular alterations are those produced by sudden increases in intraocular venous pressure, such as Valsalva retinopathy.[16]

Pichi et al.[3] observed that 24% of patients with CRM have associated venous malformations in the brain, visible with MRI. Most of these abnormalities were ipsilateral to the CRM and were mainly located in the frontal lobe in 75% of the cases. For all these reasons, it is currently recommended to perform a cerebral MRI with contrast for the screening of cerebral venous malformations associated with CRM.

Differential diagnosis

Physiological findings can be misdiagnosed as vascular malformations, such as cilioretinal arteries, which start from the papillomacular bundle, have a normal diameter and lack of tortuosity. They can cross the central middle raphe without foveal compromise.

The finding of enlarged isolated retinal vessels can be observed in different diseases such as racemose angioma of the retina, capillary hemangioma of the retina, cavernous hemangioma, retinoblastoma, and malignant melanoma of the choroid or retinal branch vein obstruction.[1]

Preretinal vascular loop

Introduction

In 1871, Liebreich first described the preretinal vascular loop,[17] as a congenital vascular anomaly that originates from a main branch of the central retinal artery at the optic disk.[18] Hirschberg[19] and Goldstein and Wexler[18] demonstrated that preretinal vascular loops have a different origin than embryonic hyaloid vasculature. The incidence of preretinal vascular loops is estimated to be 1 per 9000 ordinary ophthalmology consultations.[20,21] Nevertheless, this may be an underestimation of its real incidence considering its lack of symptoms.[21]

Preretinal vascular loops are usually arterial and unilateral. Approximately 9%–17% of cases are bilateral.[22] Oxilia reported that 45% of arterial loops occurred along the inferotemporal branch and 20% along the superotemporal.[23] Preretinal vascular loops can adopt different morphologies since their spiral turns can be single or multiple. In different studies, it has been observed that this vascular malformation tends to project toward the vitreous cavity although there are occasions in which it is placed in the plane of the optic nerve and peripapillary retina.[22]

Etiopathogenesis

The embryonic origin of preretinal vascular loops remains unclear. They are probably derived from a bud of mesenchymal cells at the optic disc.[24] One hypothesis regarding the origin of this malformation is that during the normal development of the retinal vascular circulation, the retinal vessels evolve abnormally toward the vitreous cavity through the Cloquet canal, caused by less resistance in this area, and are subsequently redirected to the retina originating the prepapillary vascular loop. The stimulus that causes this abnormal vascular development is not yet known. During the growth phase, the preretinal vascular loop begins to acquire its final morphology arising from different loops, being able to acquire even a glial veil.[21] Shakin et al.[25] reported a histological study that determines that the preretinal arterial loop communicated with the retinal arterial system seemed to be independent of the hyaloid vasculature because it did not have an internal elastic lamina in its structure. In addition, the connective tissue of the prepapillary arterial loop is composed of a lower concentration of hyaluronic acid compared to the vitreous.[26]

Multimodal imaging (Fig. 24.2)

Fluorescein angiography

Although the diagnosis of preretinal vascular loops is clinical, knowing about their angiographic characteristics can be useful, especially to differentiate if they are arterial or venous. Preretinal arterial loops fill in the same phase as the retinal arterial vessels. In the case of a prepapillary venous loop, the filling begins 1–2 s after arterial filling with laminar flow.[21]

Justice, J. reported that 75% of eyes with preretinal vascular loops were associated with the presence of cilioretinal arteries in FA.[27]

OCT

OCT can be very useful for differential diagnosis. Preretinal vascular loops are associated with posterior shadowing at the level of the optic disc in OCT B-scans. In addition, a posterior hyaloid attached to the vascular loop might be observed, leading occasionally to certain complications such as vitreous hemorrhages.[28]

OCTA

The preretinal vascular loop can be identified in en face OCTA images segmenting the superficial retina and the vitreous interface.[29] Recent studies with OCTA in patients with arterial occlusion associated with vascular loop showed nonflow areas in the superficial capillary plexus.[30]

Ocular and systemic associations

Preretinal vascular loops are asymptomatic and generally stable over time malformations and are not associated with other systemic conditions.[21,22,31] One of the associated complications is branch retinal arterial occlusion. It occurs in 4% of eyes with arterial loops. The main cause of occlusion is believed to be the obstruction of the vascular loop by its own torsion.[22] Atherosclerotic changes may also occur, not necessarily being related, since it has been observed in young adults without any atherosclerotic features. Alike, an embolic cause has been described.[31] Vitreous hemorrhage could occur when the posterior hyaloid is adhered to the vascular loop due to the traction.[22] Finally, this congenital vascular anomaly is not associated with other systemic conditions.[21]

Differential diagnosis

Occasionally, patients with preretinal vascular loops are misdiagnosed with persistent hyaloid arteries. Persistent hyaloid artery is a fetal remnant single vessel that originated in the optic disc and projected toward the vitreous cavity through the posterior lens capsule, without returning to the optic disc or retina. To observe the entire course of the persistent hyaloid artery is uncommon. However, we can see different stages as Bergmeister's papilla if a remnant of the artery is located at the optic disc, or as Mittendorf's dot if it is located at the posterior lens capsule.[21]

Optociliary shunt vessels develop after chronic retinal vein occlusions, being the majority venous, with thinner caliber than vascular loops. Optic disc preretinal neovascularization shows leakage on FA due to their abnormal endothelium.[32] Wyburn-Mason syndrome is characterized by extensive arteriovenous communications without leakage on FA.[28] IRVAN syndrome is a rare entity characterized by the presence of multiple aneurysmal dilatations along retinal arterioles and optic nerve arterioles, which can be confused with a vascular loop.[28]

Familial retinal arteriolar tortuosity

Introduction

Familial retinal arteriolar tortuosity (fRAT) is characterized by progressive tortuosity of second- and third-order arterioles in the macular and peripapillary region. It is a rare vascular malformation, with most probably autosomal dominant inheritance that develops at an early age.[33] Generally, fRAT is asymptomatic, however, it has occasionally been associated with spontaneous retinal hemorrhages.[34]

Etiopathogenesis

The etiology and pathophysiology of fRAT are not yet known, and it has been found that most of them remain stable over time. However, the presence of this condition in successive genealogical lines and in both sexes suggests an autosomal dominant mode of inheritance. It is suspected that they may be due to point mutations, although other causes, such as isolated environmental phenocopies of other etiologies, cannot be excluded.[33] Vascular tortuosity occurs in mature vessels attached to the irrigating tissue to compensate for their increase in size. It is hypothesized that patients with fRAT have a poorer tolerance to elevated intraluminal pressures, which in certain situations can lead to complications. However, there is not a specific mechanism or other vascular abnormalities that can explain this condition.[35]

Recent studies suggest that a heterozygous mutation in COL4A1 may be the cause of this malformation in some cases. This mutation seems to be involved in the production of basement membrane proteins.[36] There are also other systemic conditions produced by this mutation that occur with retinal arteriolar tortuosity, such as hereditary angiopathy with nephropathy, aneurysms, and muscle cramp syndrome (HANAC).[37]

Multimodal imaging (Fig. 24.3)

Fluorescein angiography

Although fRAT is predominantly a clinical diagnosis, fluorescein angiography can be helpful in ruling out the presence of leaks, staining, or hypoperfusion, which are not typical of fRAT.[33,37] This tool can demonstrate that fRAT maintains intact the integrity and functionality of the internal blood–retinal barrier.[33]

OCTA

OCTA can be applied to objectively demonstrate vascular tortuosity in a noninvasive manner.[38] With recent advances such as swept-source OCTA, the speed and resolution of image acquisition are improved, allowing highly detailed images of the retinal and choroidal vasculature to be obtained.

Ocular and systemic associations

fRATs are usually asymptomatic, however, they can occasionally be associated with complications, such as intraretinal, preretinal, or even subretinal hemorrhages.[39] These retinal hemorrhages can occur spontaneously or during physical exertion and can be asymptomatic if they are not centrally involved. With the resorption of hemorrhages, previous visual acuity is restored.[33–35,40] The frequency of bleeding in fRAT is uncertain and shows a wide variation.[40] In addition, small RPE defects and pigment alterations in the fovea have been observed after resorption of recurrent retinal hemorrhages, which can lead to an irreversible reduction in vision.[33]

Several authors have speculated and searched for associations of fRAT with other conditions but they seem arbitrary,[33] such as sixth nerve palsy[41] or teleangiectasia of the bulbar conjunctiva.[42] However, the few associations described in the literature seem arbitrary.[33]

According to some studies, fRAT may be associated with abnormalities of blood coagulation[33,41,43] and pregnancy does not appear to be a risk factor.[33]

The management of fRAT consists of periodic photographic monitoring to analyze the progression and record of family history. In the case of retinal hemorrhages, it is normally sufficient to observe until their resolution. Recommending avoiding exertion or Valsalva would probably help to reduce the frequency of bleeding.[33]

Differential diagnosis

Generalized vascular tortuosity can be found in healthy hyperopic eyes. It can also be an idiopathic finding. Isolated tortuosity of the small arterioles at the posterior pole, with normal caliber and branching, is a pathognomonic finding of fRAT. Tortuosity secondary to an epiretinal membrane affects both arterial and venous vessels and is usually unilateral. Corkscrew-shaped vessels have also been described in type 1 neurofibromatosis,[44] but only affect the venules. Finally, fRAT should be included in the differential diagnosis of all patients with unexplained retinal hemorrhages.

Retinal arteriovenous malformations

Introduction

In 1874, Magnus first described retinal arteriovenous malformations.[45] Several terms have been used to describe these congenital malformations, including cirsoid aneurysm, racemose aneurysm, or arteriovenous communications of the retina (AVCR). Retinal arteriovenous malformations (AVMs) are a nonhereditary congenital vascular disorder with a low incidence.[45,46] They are usually asymptomatic and unilateral. Bilateral cases of AVM are rare, 4% of the known cases, in which 2% of the cases are with cerebral AVM.[46] Men and women are equally affected.[47]

The spectrum of AVM is wide, it can range from a single arteriovenous communication to a complex anastomotic system.[48] Tortuosity is related to the level of arteriovenous communication, and it is greater between arterial and venous branches than between an arteriole and a venule.[46]

AVMs are usually located in the papillomacular bundle (41%) and the superotemporal arch (41%) and, occasionally, in the nasal areas of the optic disc (6%).[49] Spontaneous regression has occasionally been seen in cases of retinal AVM of different degrees.[50]

Etiopathogenesis

The pathogenesis of retinal AVMs remains unknown. The alteration in vascular development is believed to occur in the capillary retraction stage in small AVMs and during the differentiation of primitive capillaries into arteries and veins in larger AVMs.[7,45]

As the vascular development of the central nervous system[51] precedes retinal vascular development, the association of high-grade retinal AVM with brain AVM is feasible.[46] This is because a defect in the differentiation of the primordial cerebral vascular channels into their arterial, venous, and capillary components could explain this variety of vascular malformations.[45,52]

The development of retinal AVMs appears to be associated with a local decrease in resistance to blood flow that can lead to a decrease in blood pressure and an increase in venous pressure. The retina around the AVM, therefore, may be ischemic due to the stealing phenomenon. Increased hydrostatic pressure and blood flow in the retinal veins can cause "arterialization" and atherosclerosis of the veins, leading to vein thrombosis.[53] Associated retinal or vitreous hemorrhages can occur from areas of capillary ischemia or increased hydrostatic pressure.[53]

Archer et al.[45] classified retinal AVMs into three groups according to the caliber of the communicating vessels, the presence of capillary plexuses joining the vessels, and the degree of visual impairment.[54]

Group 1

It is characterized by the existence of an abnormal arteriolar or capillary plexus between the main communicating vessels. They can be isolated or multiple and are generally located in one sector or quadrant of the retina, mainly affecting the macular area. They are usually accidental findings on routine examination.[45] This group of AVMs is generally asymptomatic and rarely decompensates. They are not associated with other ocular abnormalities. These retinal AVMs are unlikely to be associated with major cerebrovascular malformations.[45] Therefore, group 1 retinal AVMs do not require any treatment or to perform carotid angiograms to look for cerebral vascular malformations.[45]

Group 2

It includes direct arteriovenous communications without interposition of capillary or arteriolar plexus. These arteriovenous communications can be single or multiple, and their size is variable. The larger the caliber of the anastomoses, the greater the alteration in the adjacent retinal microvasculature.[45,50,55] In fluorescein angiography, AVMs have a rapid filling, and the dye transit is usually complete within 2–3 s after the appearance of the dye in the main arteries of the retina.[45]

Group 2 retinal AVM tends to remain stable over time; however, decompensation can occur.[45,56,57] Decompensation can manifest itself in highly variable ways, from minimal subtle fluid leaks from affected vessels that can only be detected on fluorescein angiography, to external edema, intraretinal exudates, and hemorrhages.[57,58] Due to the high oxygen concentration maintained in the retina by the high flow of saturated blood, neovascularization does not develop.[45]

When the spontaneous resolution of these malformations does not occur, photocoagulation can be used to reduce the caliber of the afferent artery and decompress the abnormal communication.[45,50,57]

Group 3

Group 3 retinal AVM was previously described on the basis of its clinical appearance as "snake-shaped contoured vessels."5[59] It is characterized by its great extension, complexity of the vascular network, and secondary retinal complications that can seriously affect visual acuity. This malformation is made up of large-caliber vessels that have frequently become so adherent and intertwined that it complicates the differentiation of their arterial and venous components. The caliber of the vessels can exceed 500–600 μm in diameter.[45] Numerous changes occur in these vessels, mainly in the veins as they are exposed to high pressure. They develop fibromuscular medial layers with almost acellular fibrohyaline adventitious coatings.[60] In addition, the retina can suffer advanced cystic degeneration.[61]

Visual acuity is usually very affected, even being leading to a blind eye from birth.[45] Due to the significant hemodynamic stress to which the retinal AVM vessels are subjected, their walls can become sclerosed. Wyburn-Mason in 1943[62] also reported a high incidence (80%) of brain symptoms associated with group 3 retinal AVMs.[52]

The treatment of this degree of retinal AVM is much more complicated because, on many occasions, the ocular lesions are unapproachable either because of their extension or because they have caused blindness from birth. The main therapeutic challenge lies in the approach to the cerebrovascular malformations that are usually associated.[45]

Multimodal imaging (Fig. 24.4)

Fluoresceine angiography

Retinal AVMs have a characteristic rapid filling without capillary loss on fluorescein angiography immediately after the arterial phase. Peripheral nonperfused areas can also be detected. Generally, arterial laminar flow is observed, instead of normal laminar venous flow.[49]

OCTA

OCTA allows us to see the enlarged tortuous (Fig. 24.4) vessels in the most superficial retinal layers. SS-OCTA also reveals echolucent changes in the inner layers of the retina and disruption of the outer layers of the retina below the altered vessels.[63]

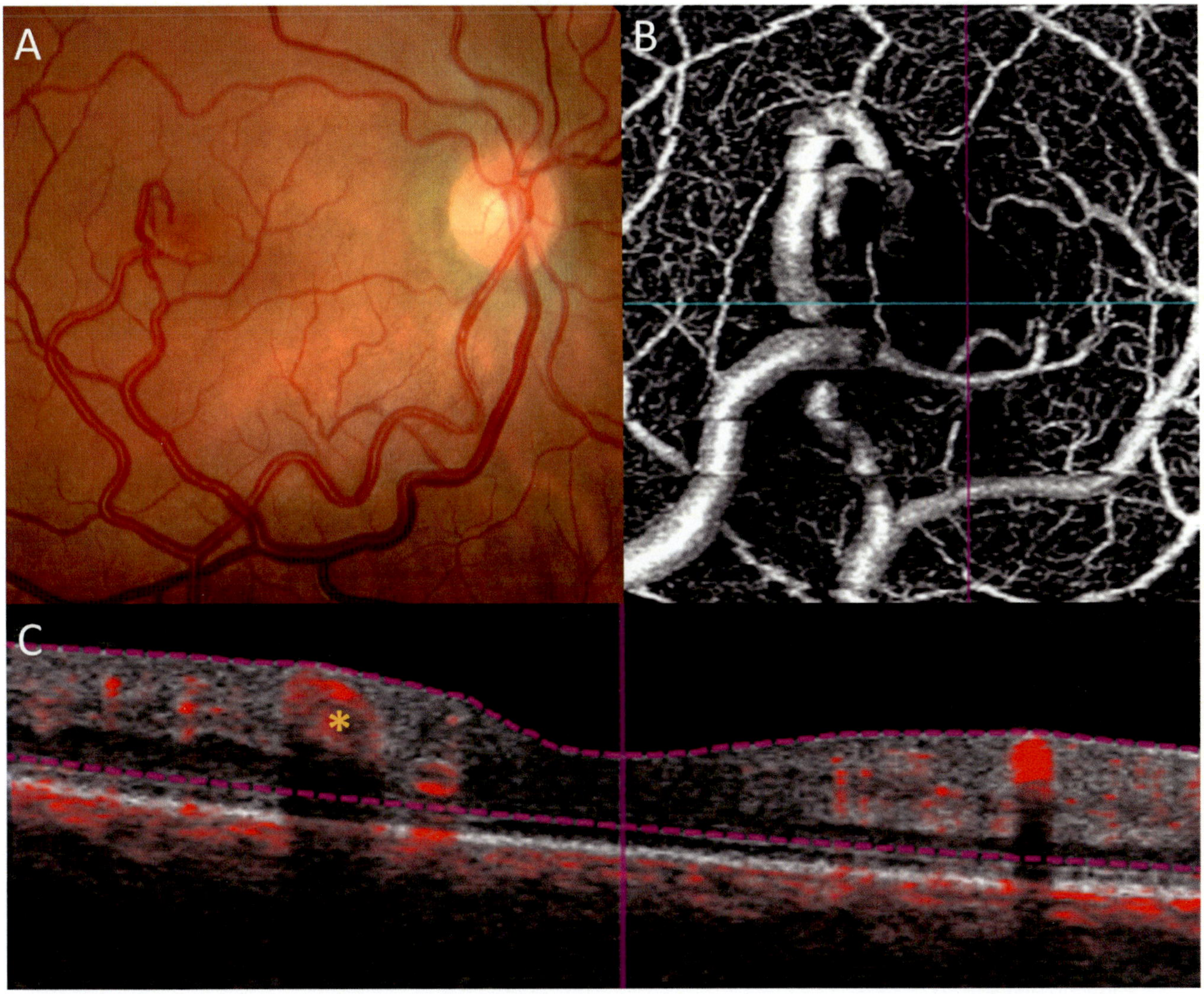

FIGURE 24.4 (A) color image showing retinal arteriovenous malformation (AVM) arising from the inferior temporal major arterial and vein and extending to the foveal area. (B) En face optical coherence tomography angiography of the superior capillary plexus of the retina with a distorted foveal avascular zone shape. (C) B-scan optical coherence tomography angiography (OCTA) shows a spared foveal morphology with an increase in the size of the vessels involved in the retinal AVM (*asterisk*), at the level of superficial plexus. The moderate caliber of the altered vasculature allows the classification as a grade 2 retinal AVM.

Ocular and systemic associations

Visual acuity is directly correlated with the degree of retinal AVM and the presence of systemic AVM. Low-grade retinal AVMs usually maintain good visual acuity and only 10% had visual acuity lower than counting fingers, while in group 3 retinal AVMs it occurs in 50% of cases. Causes of decreased vision in retinal AVMs include its macular location and ocular complications like macular edema, macular hemorrhage, aneurysms,[54] vascular occlusion,[64] neovascular glaucoma, or vitreous hemorrhage.[65]

Bonnet et al. described in 1937 the association between retinal AVM and systemic AVM.[66] Systemic AVMs include central nervous system (25%), orbit (8%), buccal area nasopharyngeal (7%), eyelid (2%), skin elsewhere (1%), lung (1%), and spine (1%). AVMs in the head usually affect the ipsilateral eye. Brain AVMs are generally located in the parietal lobes and are equally distributed between the right and left cerebral hemispheres.[67] Pulmonary AVMs are usually located in the lower lobes.[68]

Differential diagnosis

A wide spectrum of pathologies is included in the differential diagnosis of primary retinal AVMs, such as peripapillary telangiectatic microangiopathy, Leber's hereditary optic neuropathy (LHON),[69] capillary and cavernous hemangiomas,[70] retinal angioma and other retinal tumors,[46,65] arterioarterial and venovenous communications, and CRM.[1]

Congenital retinal AVM should be suspected when clinical and multimodal imaging findings are compatible, brain and orbital computed tomography do not present other findings, and genetic screening is negative.[65]

Secondary retinal AVMs have a smaller caliber, adjacent ischemia, and ocular and systemic defects typical of the primary systemic condition from which they derive.[46]

Peripapillary telangiectatic microangiopathy is described in LHON.[71] Optic neuropathy is preceded by micro-angiopathy. These vessels do not leak on fluorescein angiography.[69]

Capillary hemangiomas of the retina can occur sporadically or in the context of Von Hippel–Lindau disease.[72] They are usually associated with exudation and arteriovenous communications.[65]

Cavernous hemangioma is a rare unilateral congenital anomaly with a "bunch of grapes" appearance. This malformation is located between two retinal veins and has different histological characteristics.[70]

Acquired optic disc venous collaterals secondary to retinal venous occlusions or optic disc meningioma must also be distinguished from retinal AVMs.[65]

References

1. Brown GC, Donoso LA, Magargal LA, Goldberg RE, Sarin LK. Congenital retinal macrovessels. *Archives of Opthalmology*. 1982;100(1):1430–1436.
2. Shah VA, Chalam KV. Congenital retinal macrovessel causes reduced retinal sensitivity at the macula. *European Journal of Ophthalmology*. 2004;14:341–344.
3. Pichi F, Freund KB, Ciardella A, et al. Congenital retinal macrovessel and the association of retinal venous malformations with venous malformations of the brain. *JAMA Ophthalmology*. 2018;136(4):372–378.
4. Petropoulos IK, Petkou D, Theoulakis PE, Kordelou A, Pournaras CJ, Katsimpris JM. Congenital retinal macrovessels: description of three cases and review of the literature. *Klinische Monatsblätter für Augenheilkunde*. 2008;225(5):469–472.
5. Goel N, Kumar V, Seth A, Ghosh B. Branch retinal artery occlusion associated with congenital retinal macrovessel. *Oman Journal of Ophthalmology*. 2014;7(2):96–97.
6. Sanfilippo CJ, Sarraf D. Congenital macrovessel associated with cystoid macular edema and an ipsilateral intracranial venous malformation. *Retinal Cases and Brief Reports*. 2015;9(4):357–359.
7. Ashton N. Retinal angiogenesis in the human embryo. *British Medical Bulletin*. 1970;26(2):103–106.
8. Engerman RL. Development of the macular circulation. *Investigative Ophthalmology and Visual Science*. 1976;15(10):835–840.
9. Volk D. Visual function studies in a case of large aberrant vessels in the macula. *AMA Archives of Ophthalmology*. 1956;55(1):119–122.
10. Goel N, Kumar V, Ghosh B. Congenital retinal macrovessel associated with vitreous hemorrhage. *Journal of American Association for Pediatric Ophthalmology and Strabismus*. 2017;21(1):83–85.
11. Ipek SC, Kavukcu S, Men S, Saatci AO. Multimodal imaging features of a spontaneously resolved unilateral congenital macrovessel-related macular edema in a 13-year-old boy. *GMS Ophthalmology Cases*. 2020;10:40.
12. Strampe MR, Wirostko WJ, Carroll J. A case of congenital retinal macrovessel in an otherwise normal eye. *American Journal of Ophthalmology Case Reports*. 2017;8:18–21.
13. Poloschek CM, Sutter EE. The fine structure of multifocal ERG topographies. *Journal of Vision*. 2002;2(8):577–587.
14. Bhatia HK, Sharma S, Laxminarayana P. Ophthalmology and clinical research report. *ClinMed International Library*. 2015:2–5.
15. Chalam KV, Gupta SK, Vinjamaram S, Shah VA. Congenital anomalous retinal artery associated with a leaking macroaneurysm congenital. *Archives of Ophthalmology*. 2003;121:409–410.
16. de Crecchio G, Alfieri MC, Cennamo G, Forte R. Congenital macular macrovessels. *Graefe's Archive for Clinical and Experimental Ophthalmology*. 2006;244(9):1183–1187.
17. Liebreich R. Demonstration of diseases of the eye. Persistent hyaloid artery and vein. *Transactiones Pathological Society*. 1871;22:222.
18. Goldstein I, Wexler E. The preretinal artery: an anatomic study. *Archives of Ophthalmology*. 1929;1:324.
19. Hirschberg J. Ein Fall von prapapillarer Gef- Ssschlinge der Netzhautschlagader. *Zentralbl Prakt Augenheilkdk*. 1885;9:205.
20. Romano PE. Prepapillary vascular loops. *Clinical and Experimental Ophthalmology*. 2001;29(2):90–91.
21. Degenhart W, Brown GC, Augsburger JJ, Magargal L. Prepapillary vascular loops. *Ophthalmology*. 1981;88(11):1126–1131.
22. Brucker AJ, Michels RG, Fine SL. Congenital retinal arterial loops and vitreous hemorrhage. *American Journal of Ophthalmology*. 1977;84(2):220–223.
23. Oxilia E. Anomalie vascolari della retina:ansa arteriosa prepapillare. *Annali di Ottalmologia e Clinica Oculistica*. 1946;72:408.
24. Mann I. Developmental abnormalities of the eye. *Journal of American Medical Association*. 1957:133–136, 2nd ed.
25. Shakin EP, Shields JA, Augsburger JJ, Brown GC. Clinicopathologic correlation of a prepapillary vascular loop. *Retina*. 1988;8:55–58.
26. Makino S, Ohkubo Y, Tampo H. Prepapillary vascular loop associated with persistent hyperplastic primary vitreous. *Case Reports in Ophthalmological Medicine*. 2013;2013:1–2.
27. Justice J. Cilioretinal arteries. *Archives of Opthalmology*. 1976;94(8):1355.
28. Karti O, Top Karti D, Ayhan Z, Celebisoy N, Saatci AO. Bilateral congenital prepapillary retinal arterial loop. *Neuro Ophthalmology*. September 2, 2020;44(5):344–348.

29. Gaier ED, Miller JB. Retinal loop vessel captured with optical coherence tomography-angiography. *Ophthalmology*. 2017;124(8):1174.
30. Zermeno-Arce AJ, Ledesma-Gil G, Moreno-Mendoza R, Graue-Wiechers F. Arterial occlusion associated with a prepapillary vascular loop. Anatomical changes on the long term follow up. *Archivos de la Sociedad Española de Oftalmología*. 2019;94(8):405–408.
31. Bisland T. Vascular loops in the vitreous. *Archives of Opthalmology*. 1953;49:514.
32. Hsieh YT, Yang CM. The clinical study of congenital looped/coiled peripapillary retinal vessels. *Eye*. 2005;19(8):906–909.
33. Sutter FKP, Helbig H. Familial retinal arteriolar tortuosity: a review. *Survey of Ophthalmology*. 2003;48(3):245–255.
34. Beyer EM. Familiare tortuositas der kleinen netzhautarterien mit makulablutung. *Klin Monatsbl Augenheilkd*. 1958;132:532–539.
35. James Bartlett W, Price J. Familial retinal arteriolar tortuosity with retinal hemorrhage. *American Journal of Ophthalmology*. 1983;95(4):556–558.
36. Zenteno JC, Crespí J, Buentello-Volante B, et al. Next generation sequencing uncovers a missense mutation in COL4A1 as the cause of familial retinal arteriolar tortuosity. *Graefes Archive for Clinical and Experimental Ophthalmology*. 2014;252(11):1789–1794.
37. Saraf SS, Tyring AJ, Chen CL, et al. Familial retinal arteriolar tortuosity and quantification of vascular tortuosity using swept-source optical coherence tomography angiography. *American Journal of Ophthalmology Case Reports*. 2019;14(March):74–78.
38. Reif R, Qin J, An L, Zhi Z, Dziennis S, Wang R. Quantifying optical microangiography images obtained from a spectral domain optical coherence tomography system. *International Journal of Biomedical Imaging*. 2012;2012.
39. Dausch D. [Recurring macular hemorrhages with tortuosity of the paramacular arterioles in both eyes]. *Bericht Uber die Zusammenkunft. Deutsche Ophthalmologische Gesellschaft*. 1977;74:820–825.
40. Wells CG, Kalina RE. Progressive inherited retinal arteriolar tortuosity with spontaneous retinal hemorrhages. *Ophthalmology*. 1985;92(8):1015–1024.
41. Tesinsky P, Lorenz A. [Familial tortuosity of small retinal arteries connected with hemorrhage (study of 12 patients from 1 family)]. *Cesk Oftalmol*. 1975;31:172–175.
42. Takeda M. Congenital retinal vascular anomaly part 4:Bilateral retinal arteriolar tortuosity with retinal haemorrhages. *Folia Ophthalmologica Japonica*. 1985;36:850–855.
43. Stefani F, Greite J-H, Schramm W. [Recurring retinal haemorrhages with tortuosity of the retinal arteriols]. *Klin Monatsbl Augenheilkd*. 1975;167:608–612.
44. Muci-Mendoza R, Ramella M, Fuenmayor-Rivera D. Corkscrew retinal vessels in neurofibromatosis type 1: report of 12 cases. *British Journal of Ophthalmology*. 2002;86(3):282–284.
45. Archer DB, Deutman A, Ernest JT, Krill AE. Arteriovenous communications of the retina. *American Journal of Ophthalmology*. 1973;75(2):224–241.
46. Mansour AM, Walsh JB, Henkind P. Arteriovenous anastomoses of the retina. *Ophthalmology*. 1987;94(1):35–40.
47. Meinhold JR. Arteriovenous communications of the retina. *Journal of the American Optometric Association*. 1996;67:279–282.
48. Pauleikhoff D, Wessing A. Arteriovenous communications of the retina during a 17-year follow-up. *Retina*. 1991;11:433–436.
49. Sharma P, Sridhar J, Rayess N, Maguire JI. Optical coherence tomography angiography (OCT-A) of type 2 retinal arteriovenous malformation. *Canadian Journal of Ophthalmology*. 2015;50:e93–e96. Elsevier.
50. Wolff B, Tick S, Cohen SY. Photocoagulation therapy of leaking archer's type 2 retinal arteriovenous communication. *Retinal Cases and Brief Reports*. 2013;7(1):95–97.
51. Streeter G. The developmental alterations in the vascular system of the brain of the human embryo. *Contribution to Embryology*. 1918;8:5–38.
52. Singh AD, Rundle PA, Rennie I. Retinal vascular tumors. *Ophthalmol Clinics of North America*. 2005;18(1):167–176.
53. Mansour M, Wellls CG, Jampol LM, Kalina RE. Ocular complications of arteriovenous communications of the retina. *Archives of Ophtalmology*. 1989;107:232–236.
54. Tilanus MD, Hoyng C, Deutman AF, Cruysberg JRM, Aandekerk AL. Congenital arteriovenous communications and the development of two types of leaking retinal macroaneurysms. *American Journal of Ophthalmology*. 1991;112(1):31–33.
55. Elizalde J, Vasquez LM. Spontaneous regression in a case of racemose haemangioma archer's type 2. *Retin Cases Brief Reports*. 2011;5(4):294–296.
56. Bonamour G. Intére^t de la fluorographie pour la compréhension de certains aspects de "chorioretinites" exudatives centrales. *Proceedings of the Inernational Symposium of Fluorescein Angiography*. 1971;112.
57. Baurmann H, Meyer F, Oberhoff P. Komplikationen bei der arteriovenösen Anastomose der Netzhaut. *Klin Monatsbl Augenheilkd*. 1968;153:562.
58. Hoyt WF, Beeston D. *The Ocular Fundus in Neurologic Disease*. St. Louis: C. V. Mosby; 1966:90.
59. Biró I. Racemose retinal aneurysm. *Ophthalmologica*. 1951;121:201–207.
60. Cameron ME, Greer CH. Congenital arteriovenous aneurysm of the retina. A post mortem report. *The British Journal of Ophthalmology*. 1968;52(10):768–772.
61. Krug EF, Samuels B. Venous angioma of the retina, optic nerve, chiasm and brain: case report with postmortem observations. *Archives of Opthamology*. 1932;8:871.
62. Wyburn-mason R. Arteriovenous aneurysm of mid-brain and retina, facial nævi and mental changes. *Brain*. 1943;66(3):163–203.
63. Chawla R, Pujari A, Mittal K, Rakheja V, Markan A. Retinal arteriovenous malformation assessment using swept-source OCT angiography. *Ophthalmic Surgery, Lasers and Imaging Retina*. March 1, 2018;49(3):216–219.
64. Shah GK, Shields JA, Lanning RC. Branch retinal vein obstruction secondary to retinal arteriovenous communication. *American Journal of Opthamology*. 1998;126(3):446–448.
65. Chatterjee S, Hope-Ross M, Clark IH, Chell PB. A rare and unusual congenital arteriovenous communication of the retina [1]. *Eye*. 2004;18(1):74–75.

66. Bonnet P, Dechaume J, Blanc E. L'anéurysme circöide de la rétine (aneurysme racémeux ses relations avec l'anéurysme circöide de la face et avec l'anéurysme circöide du cerveau). *Journal de Medecine de Lyon*. 1937;18:165.
67. Perret G, Nishioka H. Report on the cooperative study of intracranial aneurysms and subarachnoid hemorrhage. Section VI. Arteriovenous malformations. An analysis of 545 cases of cranio-cerebral arteriovenous malformations and fistulae reported to the cooperative study. *Journal of Neurosurgy*. 1966;25(4):467–490.
68. Dines D, Arms R, Bernatz P, Gomes M. Pulmonary arteriovenous fistulas. *Mayo Clinic Proceedings*. 1974;49:460–465.
69. Nikoskelainen E. Ophthalmoscopic findings in leber's hereditary optic neuropathy. *Archives of Ophthalmology*. 1983;101(7):1059.
70. Horiuchi T, Gass JDM, David NJ. Arteriovenous malformation in the retina of a monkey. *American Journal of Ophthalmology*. 1976;82(6):896–904.
71. Nikoskelainen E, Hoyt WF, Nummelin K. Fundus findings in leber's hereditary optic neuroretinopathy. *Ophthalmic Genetics*. 1985;5(1–2):125–129.
72. Webster AR, Maher ER, Bird AC, Gregor ZJ, Moore AT. A clinical and molecular genetic analysis of solitary ocular angioma. *Ophthalmology*. 1999;106(3):623–629.

Chapter 25

Choroidal macrovessel

Kushal Agrawal[1] **and Vishal Raval**[2]
[1]*Jupiter Hospital, Thane, Maharashtra, India;* [2]*Anant Bajaj Retina Institute, LV Prasad Eye Institute, KAR Campus, Hyderabad, Telangana, India*

Definition

Choroidal macrovessel appears ophthalmoscopically as a single, hypopigmented, tortuous, and dilated vessel extending from the macula to the equator in the horizontal orientation in most cases.[1] Choroidal macrovessel is a rare condition with few reported cases in the literature. It has been found more commonly in white race and female sex of middle to older age groups.[1] It has been reported at a very young age of 12 years and in an older age of 80 years.[2,3]

No systemic risk factor has been proven to be associated with it. However, hyperlipidemia, hypertension, and hypothyroidism have been associated with reported cases of the choroidal macrovessel.[1] Similarly, various ocular conditions such as glaucoma, age-related macular degeneration, nonproliferative diabetic retinopathy, retinal detachment, and vein occlusion have been seen in an eye with choroidal macrovessel. However, no definitive relationship with any ocular condition has been identified.[4]

Description

Choroidal macrovessel was first reported in 2011 by Lima et al.[1] Normal dimension of choroidal vessels varies between 20 and 100 μm and can extend up to 140 ± 40 μm.[1] Choroidal macrovessel is larger than that and can have a mean diameter of 250 μm (median 250, range 200–300 μm) at the ampulla.

Choroidal macrovessel is unilateral in the majority of the cases but can be bilateral in up to 23% of the cases.[4] Choroidal macrovessel is located temporal to macula and it travels toward equator, but rarely it can be observed at the nasal edge of the macula or can have a vertical or an oblique orientation too.[4,5]

Choroidal blood supply

Ocular circulation is constituted mainly by two sets of arteries named central retina artery and posterior ciliary artery (PCA). Both the above-mentioned arteries are branches of the ophthalmic artery. Central retinal artery is the main blood supply of the retina, which supplies up to the inner retina. The posterior ciliary artery can be one to five in numbers, and these numbers vary individually. The posterior ciliary arteries are named as medial (MPCA) and lateral (LPCA), and in some patients superior posterior ciliary artery (Fig. 25.1). Hayreh et al. in their experimental study have shown that medial, lateral, or superior posterior ciliary arteries do not anastomose with each other and behave like end arteries. The posterior ciliary artery supplies blood to the optic nerve head choroid up to the equator, retinal pigment epithelium (RPE), outer 130 μm of the retina, medial and lateral segments of the ciliary body, and iris. The cilioretinal artery belongs to the posterior ciliary artery and when present the entire thickness of the retina in that region is supplied by the posterior ciliary artery system.[6]

The main posterior ciliary artery is further divided into short and long posterior ciliary arteries. Short posterior ciliary arteries are 10–20 in number and are further divided into paraoptic and distal branches. Paraoptic short posterior ciliary arteries enter the sclera close to the optic nerve and supply blood in the paraoptic area, while the distal short posterior ciliary artery enters the sclera a short distance away from the optic nerve on the medial and lateral sides and runs radially toward the equator (Fig. 25.1).[6]

Retinal and Choroidal Vascular Diseases of the Eye. https://doi.org/10.1016/B978-0-443-15583-3.00025-1

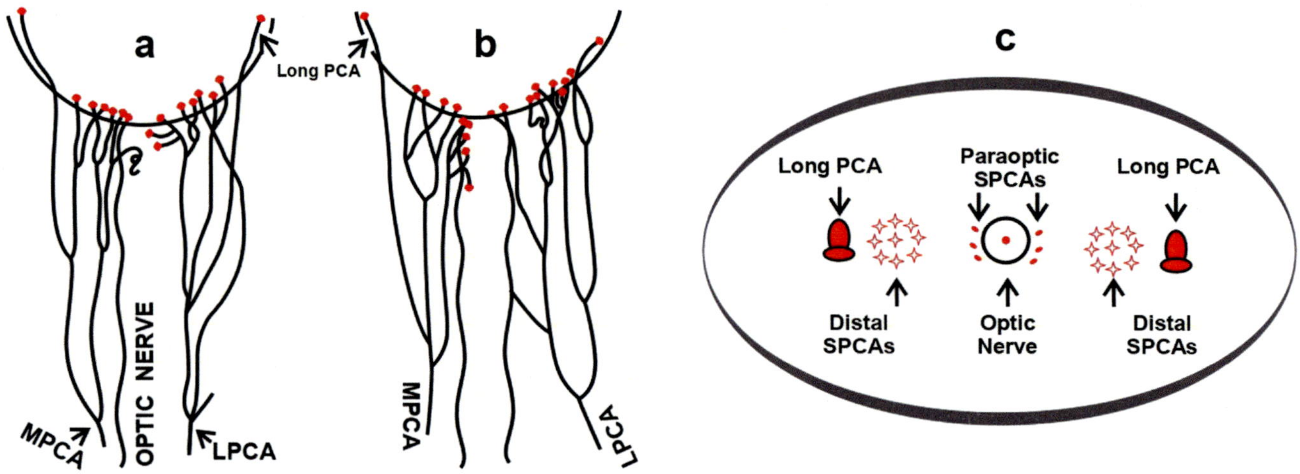

FIGURE 25.1 Origin of medial (MPCA) and lateral (LPCA) posterior ciliary artery from the ophthalmic artery which further divides into short and long posterior ciliary arteries (A and B), while from the back part of the eye the origin of branches of short (SPCAs) and long posterior ciliary arteries (long PCA) which further divide into paraoptic (paraoptic SPCAs) and distal (distal SPCAs) branches (C).

Long posterior ciliary artery (branch of posterior ciliary artery) has two branches: one medial and another lateral branch. A long posterior ciliary artery enters the eyeball in the horizontal plane on the medial and lateral sides. Furthermore, it runs radially in the horizontal meridian toward the iris. Siam et al. studied 12 cadaveric eyes and found that lateral long PCA enters the sclera 3 mm temporal to the optic nerve sheath.[7] Thus, distal branches of short posterior ciliary arteries (SPCA) and long posterior ciliary artery (LPCA) do have radial course and it runs toward the equator.[6]

Hypothetical correlation of choroidal macrovessel with posterior ciliary artery

Previous studies have proposed different origins of the choroidal macrovessel, which could be a dilated choroidal vein, congenital choroidal macrovessel, arteriovenous choroidal vessel anastomosis, retinochoroidal malformation, or a dilated posterior ciliary artery (PCA).[1,4,8,9] Bowen et al. studied the anatomical origin of two eyes with choroidal macrovessel using three-dimensional reconstruction of choroidal macrovessel with 3D-slicer.[10] They noticed that choroidal macrovessel does not have any retinal vasculature involvement so it is unlikely that it could be retinochoroidal malformation or congenital retinal macrovessel. They noticed certain characteristics of the choroidal macrovessel, which correlates with the anatomy and imaging characteristics of the posterior ciliary artery, which is described in the next paragraph.

They noticed certain characteristics of the choroidal macrovessel like: (1) early hyperfluorescence on indocyanine green angiography (ICGA); (2) the entry side of the choroidal macrovessel corresponded to the scleral entry site of the posterior ciliary artery clinically and on 3D reconstructed swept-source OCT; (3) the pattern of the choroidal macrovessel is spoke like with radial distribution; (4) the most reported choroidal macrovessel faded before reaching the equator. All of these characteristics correspond to the anatomy of the posterior ciliary artery. Considering this, they made the hypothesis that the choroidal macrovessel could be either a short or long posterior ciliary artery.[10]

Diagnostic signs

Optical coherence tomography illustrates the hyporeflective area of focal enlargement at the level of the choroid corresponding to the lumen of the choroidal macrovessel. It is not associated with the thickening of the surrounding choroidal tissue. It can cause indentation on the RPE layer with disruption of RPE and ellipsoid layers. Subretinal fluid in the area of the choroidal macrovessel has also been observed.[10] Fundus autofluorescence in such cases shows hyperautofluorescence and/or hypoautofluorescence due to various stages of RPE abnormalities.[4]

OCT-angiography (OCT-A) shows the level of macrovessel at the level of choroid and choriocapillaris.[4] Most of the reported studies demonstrated high flow within the lumen of the choroidal macrovessel on OCT-A.[4,10] Gallo et al. studied that the internal flow within the lumen of the choroidal macrovessel was outside the detectable threshold on OCT-A. They mentioned that this could be related to the high velocity of blood flow within the lumen of the vessel due to an interference fringe washout phenomenon.[4] Kataoka et al. reported high blood flow within the lumen of the choroidal macrovessel using laser speckle flowgraphy (LSG).[11] As per study by Bowen et al., and Moreira-Neto et al. OCT-A had shown the absence of decorrelation signals within the lumen of the blood vessel. Decorrelation signals emerge predominantly from the movement of erythrocytes and represent the blood flow and vascular structure.[8,12] The absence of decorrelation signals could be related to either the thick endothelium, which is not allowing penetration of the laser light of OCT A, or blood flow is too low or too high and outside detectable normal limits. A low blood flow is very unlikely as the early phase of ICGA filling of the choroidal macrovessel is appreciated.[8,10] So, all of the above studies concluded that the choroidal macrovessel is a high-flow vessel.

Otero et al. and some of the other authors demonstrated a low flow within the vessel with a possible explanation that choroidal macrovessel is an arteriole that is possibly a terminal vessel with no established connection with other vessels.[3,8,13] However, further clinical-pathological studies will be needed to prove this hypothesis.

Fundus fluorescein angiography (FFA) shows increased fluorescence of the choroidal macrovessel in the early phase with a lack of leakage in the late phase.[10] Similarly, on ICG angiography, early filling of the macrovessel is seen and it suggests an arterial origin but no leakage is usually appreciated.[10] This finding also suggests that choroidal macrovessel is not an abnormal vessel that could have leaked on FFA or ICG. An early filling of the choroidal macrovessel on FFA and ICGA suggests that it could be either an artery or a vein that carries arterialized blood flow.[4] Ultrasound B-scan through the choroidal macrovessel shows a nodular lesion with low internal reflectivity. Sometimes it is difficult to detect choroidal macrovessel on ultrasound considering their relatively small caliber (Fig. 25.2: an illustrative case).[4]

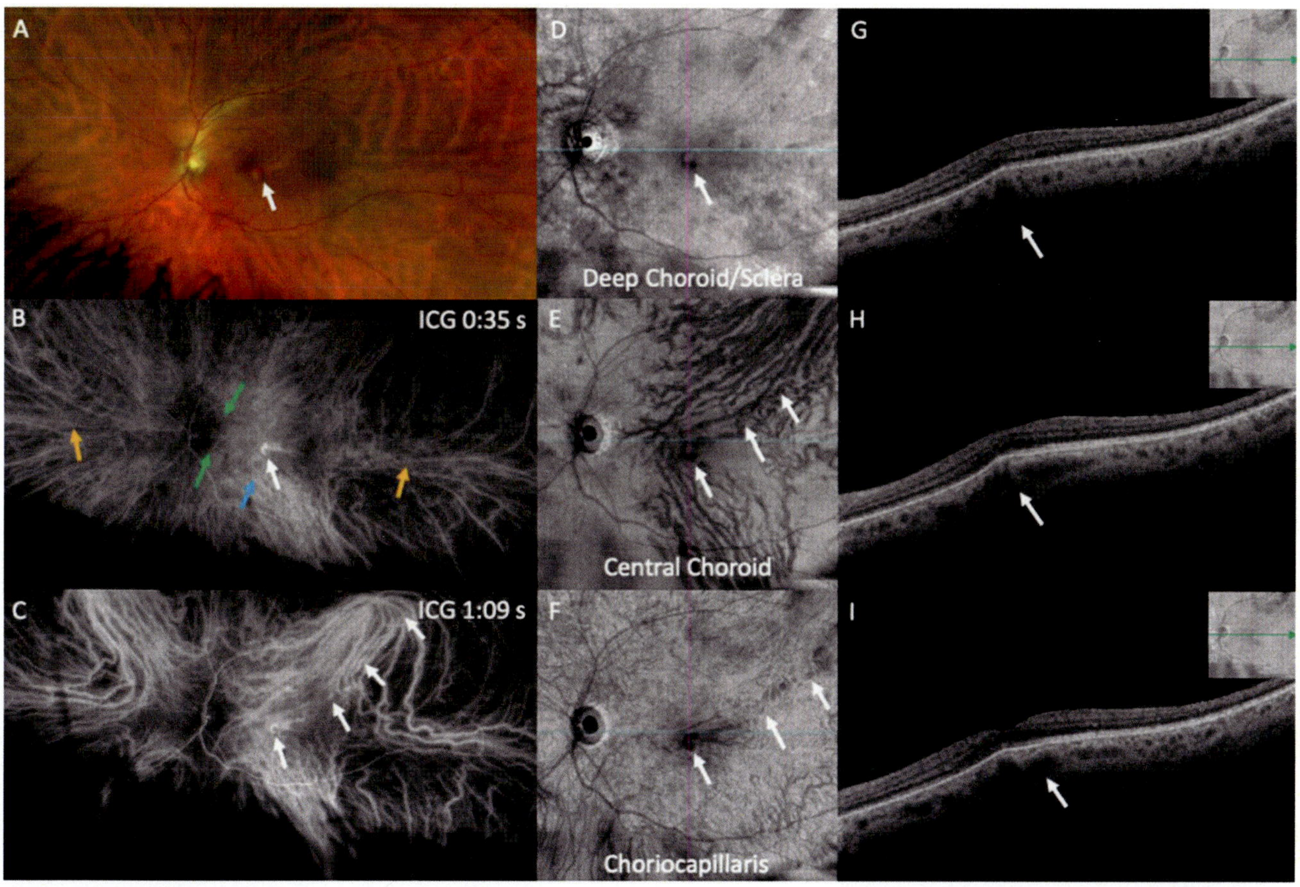

FIGURE 25.2 Left eye macrovessel from case 1: Optos fundus photo shows an enlarged vessel in the subfoveal choroid which extends temporally (A).[10] ICG angiography at 35 s shows lateral and medial long ciliary artery (3 and 9 o' clock hours), distal SPCA (6 o' clock hours), paraoptic SPCA (12 and 7 o' clock hours), and an enlarged distal SPCA representing a macrovessel (white arrow) (B). ICG at 1:09 min shows hypercyanescent macrovessel tracking superotemporally (C). Structural en face OCT at deep choroid/sclera (D), central choroid layer (E), and choriocapillaris highlights the enlarged vessel (white arrow) in the choroid which extends temporally (F). OCT illustrates the location of the enlarged artery entering into the choroid (white arrow) (G). Macrovessel results in choroidal thickening at the subfoveal region (H–I). Abbreviations: *ICG*, indocyanine green; *OCT*, optical coherence tomography; *SPCA*, short posterior ciliary artery.

Differential diagnosis

Choroidal macrovessel should be differentiated from subretinal tracts formed by nematodes. Different nematodes such as *Oedemagena tarandi* (ophthalmo-myiasis), *Alaria americanus*, gnathostomiasis, and nematode-associated diffuse unilateral subacute neuroretinitis (DUSN) are some infections that can form subretinal tracts.[14] Other differential diagnoses could be pachychoroid diseases, choroidal vascular neoplasms such as choroidal haemangioma, vortex vein varix, retinochoroidal anastomosis, and inflammatory choroidopathies.[8]

Choroidal macrovessel is different from the pachychoroid spectrum because pachychoroid diseases are associated with diffuse thickening of the choroid seen on OCT; while macrovessel is associated with focal thickening of the choroid at the site of the choroidal macrovessel with normal surrounding area of the choroid on OCT. Pachychoroid disease spectrum such as central serous chorioretinopathy (CSCR), pachychoroid pigment epitheliopathy, and pachychoroid neovasculopathy can cause vision problems due to pigment epithelial detachment (PED) or choroidal neovascular membrane (CNVM).[10] However, choroidal macrovessel has not been associated with such vision-threatening complications.

Moreover, choroidal macrovessel can be confused with choroidal tumors such as choroidal hemangioma. Choroidal hemangioma shows classic early filling of the lesion and late washout on ICG angiography, echodense mass with high internal reflectivity on ultrasound, and diffuse choroidal thickening on OCT, which can help it in differentiating from the choroidal macrovessel.[15]

Vortex vein varix is a dilated ampullae of the vortex vein, which is venous in origin and located at the equator in the area of the vortex vein. In contrast to it, choroidal macrovessel is more likely to be arterial in origin and the most common location is at the macula.[16]

Complications

Patients with the choroidal macrovessel are usually asymptomatic. However, metamorphopsia and blurry vision have been reported.[1] This could be related to changes in the outer retinal layers, RPE layer, or due to the presence of the subretinal fluid. Choroidal macrovessel can cause compression of the choriocapillaris, which can lead to the above manifestations.[8] However, there are no reported cases of CSCR, choroidal neovascular membrane (CNVM), PED, or aneurysm due to choroidal macrovessel.[8] This suggests that choroidal macrovessel is not a pathological condition but a physiologically dilated vessel.[1]

Treatment options

Any vision-threatening treatable manifestations such as CSCR, CNVM, PED, or aneurysm have not been associated with the choroidal macrovessel. Considering this, no treatment is required and it can be observed on a yearly follow-up if the patient is stable.[1]

Literature review

Not more than 20 cases of the choroidal macrovessel until date are reported as per the literature search. This has been mentioned in Table 25.1. The mean age was found to be 59.3 years (median, 64 years, range, 12–80 years). Out of the reported patients, the majority (64.3%) were asymptomatic, whereas others (35.7%) were symptomatic. Reported symptoms were blurry vision, metamorphopsia, and/or floaters. Almost all reported cases of the choroidal macrovessel originated at or near the macula. The course of the choroidal macrovessel after origin was either temporal (81.9%) or superior (18.1%) from the macula in a radial orientation. None of the patients showed a nasal course of the vessel. Thus, most choroidal macrovessels had radial orientation from the macula to the temporal or superior equator.

OCT demonstrated retinal pigment epithelium (RPE) elevation in the area of the choroidal macrovessel in all (100%) patients along with outer retinal layer changes (69.6%) and subretinal fluid (21.8%). OCT-A revealed a level of macrovessel in the choroidal area and reflectivity similar to the adjacent choroidal vessels in almost all cases. Moreover, OCT-A revealed choroidal macrovessel to be a high-flow vessel (69.2%) in most patients and low flow (31.8%) in others. High flow within the choroidal macrovessel on OCTA was correlated with its choroidal location, arterial origin, and an early filling seen on ICGA. FFA and ICG revealed early filling and hyperfluorescence of the choroidal macrovessel in almost all cases, demonstrating its choroidal origin. None of the reported patients (0%) had leakage on FFA or ICGA in the late phase, suggesting it was not an abnormal vessel.

TABLE 25.1 Reported cases of choroidal macrovessel with their clinical characteristics.

Cases	Authors	Year published	Age at diagnosis	Symptoms	Posterior location (direction)	Anterior extent	OCT-RPE elevation	OCT-irregular EZ	OCT-SRF	OCTA	FFA	ICG early filling	Leakage on FFA or ICG
1	Lima et al.[14]	2011	42	ND	Macula (temporal)	Macula-equator	+	−	−	ND	Hyper	Normal	No
2	Ehlers et al.[9]	2014	76	Asymptomatic	Macula (temporal)	Equator	+	+	−	ND	Hyper	Hyper	No
3	Choudhry et al.[17]	2016	42	Metamorphopsia	Macula (temporal)	Macula-equator	+	+	−	ND	ND	ND	ND
4	Pichi et al.[18]	2016	ND	ND	Macula (temporal)	Equator	+	+	−	ND	ND	ND	ND
5	Kovach et al.[19]	2016	70	Asymptomatic	Macula (temporal)	Macula-equator	+	+	−	ND	ND	ND	ND
6	Mahroo et al.[20]	2017	79	ND	Macula (temporal)	Macula-equator	+	+	−	ND	ND	Hyper	No
7	Hampton et al.[3]	2017	80	ND	Macula (temporal)	Macula-equator	+	+	−	Low flow	ND	ND	ND
8	Dalvin et al.[1]	2018	55	Asymptomatic	Macula (temporal)	Macula-equator	+	+	+	ND	Hyper	Hyper	No
9	Dalvin et al.[1]	2018	68	Blurry vision	Macula (temporal)	Equator	+	+	+	ND	Hyper	Hyper	No
10	Turgut et al.[2]	2018	12	ND	Macula (temporal)	Equator	+	+	ND	ND	ND	ND	ND
11	Casalino et al.[13]	2019	65	Asymptomatic	Macula (temporal)	Macula to equator	+	+	+	Low flow	Hyper	Hyper	No
12	Mori et al.[21]	2020	79	Metamorphopsia	Macula (temporal)	Macula to equator	+	+	−	ND	Hyper	Hyper	No
13	Katakoa et al.[11]	2020	39	Blurry vision	ND	ND	+	+	−	High flow	Hyper	Hyper	No
14	Otero et al.[8]	2020	69	Asymptomatic	Macula (temporal)	Macula to equator	+	−	−	Low flow	ND	ND	ND
15	Moreira et al.[12]	2021	64	Asymptomatic	Macular (temporal)	Macula-equator	+	−	−	High flow	ND	Hyper	No

Continued

TABLE 25.1 Reported cases of choroidal macrovessel with their clinical characteristics.—cont'd

Cases	Authors	Year published	Age at diagnosis	Symptoms	Posterior location (direction)	Anterior extent	OCT-RPE elevation	OCT-irregular EZ	OCT-SRF	OCTA	FFA	ICG early filling	Leakage on FFA or ICG
16	Moreira et al.[12]	2021	71	Asymptomatic	Macular (superiorp)	Macula-equator	+	–	–	High flow	ND	ND	ND
17	Gallo et al.[4]	2021	ND	ND	Macula (temporal)	Macula-equator	+	+	+	High flow	ND	Hyper	No
18	Gallo et al.[4]	2021	27	ND	Macula (superiro)	Macula-equator	+	–	–	High flow	ND	ND	ND
19	Gallo et al.[4]	2021	ND	ND	Macula (temporal)	Macula-equator	+	–	–	High flow	ND	ND	ND
20	Bowen et al.[10]	2022	57	ND	Macula (supero-temporal)	Macula-equator	+	+	–	High flow	ND	Hyper	No
21	Bowen et al.[10]	2022	61	ND	Macula (superior)	Macula-equator	+	–	–	High flow	ND	Hyper	No
22	Mopuru et al.[22]	2022	72	Asymptomatic	Macula (temporal)	Macula-equator	+	+	–	High flow	ND	Hyper	No
23	Fogel-levin et al.[5]	2022	58	Asymptomatic	Extramacular (superior)	Macula-equator	+	+	+	ND	ND	Hyper	No
24	Quijada et al.[23]	2022	59	Metamorphopsia	Macular (temporal)	Macula-equator	+	+	–	Low flow	Hyper	Hyper	No

EZ, ellipsoid zone; *FFA*, fundus fluorescein angiography; *ICG*, indocyanine green angiography; *ND*, no data; *OCT*, optical coherence tomography; *OCTA*, optical coherence tomography angiography; *RPE*, retinal pigment epithelium; *SRF*, subretinal fluid.

Conclusion

Choroidal macrovessel is probably a dilated physiological vessel that is usually asymptomatic in most patients. It can be confused with various nematodes, choroidal hemangioma, or pachychoroid diseases. Knowing its characteristic signs can help diagnose and differentiate this condition from other vision-threatening disorders.

References

1. Dalvin LA, Pennington JD, Mashayekhi A, Shields CL. Multimodal imaging of macular choroidal macrovessel. *Retinal Cases and Brief Reports*. July 2018:1.
2. Turgut B, Kobat GS. The detection of an occult choroidal macrovessel with enhanced depth and en-face imagings in optical coherence tomography. *International Journal of Retina and Vitreous*. 2018;27:77–81.
3. Hampton BM, Kovach JL. Choroidal macrovessel imaged with OCT angiography. *Ophthalmology Retina*. 2017 Nov-Dec;1(6):562.
4. Gallo B, de Silva SR, Mahroo OA, et al. Choroidal macrovessels: multimodal imaging findings and review of the literature. *British Journal of Ophthalmology*. April 2022;106(4):568–575.
5. Fogel-Levin M, Wong A, Sadda SR, Freund KB, Sarraf D. Rare case of extramacular choroidal macrovessel. *Canadian Journal of Ophthalmology*. June 2022;57(3):e89–e91.
6. Hayreh SS. The posterior ciliary artery. In: *In Ocular Vascular Occlusive Disorders*. Cham: Springer International Publishing; 2015:65–98.
7. Siam AL, El-Mamoun TA, Ali MH. A restudy of the surgical anatomy of the posterior aspect of the globe: an essential topography for exact macular buckling. *Retina*. 2011;31(7):1405–1411.
8. Otero-Marquez O, Ledesma-Gil G, Alauddin S, Smith RT. Non-invasive imaging of a choroidal macrovessel. *American Journal of Ophthalmology Case Reports*. August 12, 2020;20:100871.
9. Ehlers JP, Rayess H, Spaide RF. Isolated choroidal macrovessel: a tracklike choroidal lesion. *Canadian Journal of Ophthalmology*. December 2014;49(6):e158–e160.
10. Bowen RC, Raval V, Soto H, Singh AD. Choroidal macrovessel: systematic review and analysis of anatomic origin. *Survey of Ophthalmology*. 2022 Mar-Apr;67(2):570–578.
11. Kataoka K, Kase S, Noda K, Ishida S. Laser speckle flowgraphy findings in a patient with choroidal macrovessel. *Ophthalmology Retina*. November 2020;4(11):1123–1124.
12. Moreira-Neto CA, Lima LH, Zett C, Pereira R, Moreira Jr C. En-face OCT and OCT angiography analysis of macular choroidal macrovessel. *American Journal of Ophthalmology Case Reports*. January 11, 2021;21:101012.
13. Casalino G, De Simone L, Scialdone A, Pavesio C. Choroidal macrovessel: optical coherence tomography angiography and imaging features. *Indian Journal of Ophthalmology*. July 2019;67(7):1168–1170.
14. Lima LH, Laud K, Chang LK, Yannuzzi LA. Choroidal macrovessel. *British Journal of Ophthalmology*. September 2011;95(9):1333–1334.
15. Mashayekhi A, Shields CL. Circumscribed choroidal hemangioma. *Current Opinion in Ophthalmology*. June 2003;14(3):142–149.
16. Gündüz K, Shields CL, Shields JA. Varix of the vortex vein ampulla simulating choroidal melanoma: report of four cases. *Retina*. 1998;18(4):343–347.
17. Choudhry N, Rao RC. Enhanced depth imaging features of a choroidal macrovessel. *Retinal Cases and Brief Reports*. 2016 Winter;10(1):18–21.
18. Pichi F, Nucci P, Srivastava SK. Choroidal macrovessel. *Ophthalmology*. March 2016;123(3):531.
19. Kovach JL. Unilateral choroidal macrovessel. *JAMA Ophthalmology*. March 2016;134(3):e153678.
20. Mahroo OA, Puertas RA, Saihan Z. Swept source OCT image of a choroidal macrovessel. *Ophthalmology Retina*. 2017 Jul-Aug;1(4):281.
21. Mori H, Yamada H, Sato Y, Takahashi K. Optical coherence tomographic angiography and ultra-widefield indocyanine green angiography of a choroidal macrovessel. *American Journal of Ophthalmology Case Reports*. January 30, 2020;18:100612.
22. Mopuru R, Liu TYA, Arevalo JF. Choroidal macrovessel diagnosed on multimodal imaging, including swept-source optical coherence tomography angiography. *Case Reports in Ophthalmology*. March 24, 2022;13(1):215–219.
23. Tapia Quijada HE, Pareja Ríos A, Serrano García M, Mesa Lugo FI. Choroidal macrovessel and persistent metamorphopsia after epiretinal membrane surgery. *Archivos de la Sociedad Española de Oftalmología*. June 2022;97(6):344–349.

Section VI

Vascular diseases associated with systemic inflammatory diseases

Chapter 26

Systemic lupus erythematosus

Danny A. Mammo and Aleksandra V. Rachitskaya
Cole Eye Institute, Cleveland Clinic, Cleveland, OH, United States

Introduction

Systemic lupus erythematosus (SLE) is a chronic autoimmune disease that affects multiple organ systems in the body. It is characterized by inflammation and tissue damage and can cause a wide range of symptoms, including fatigue, joint pain, skin rashes, and organ dysfunction. SLE is a complex disease with a multifactorial etiology, as a combination of genetic, environment, and lifestyle factors likely contribute to its development. Despite advances in our understanding of SLE, the exact causes of the disease remain unknown. Treatment is focused on suppressing immune activity, managing symptoms, and preventing complications.[1]

SLE can cause a number of ocular complications, such as retinal vasculitis and choroiditis, which can lead to visual impairment and blindness if not treated promptly. This chapter will provide an overview of SLE, a review of the various posterior segment manifestations of SLE, and discuss current systemic and local management strategies.

Systemic diagnosis

The diagnosis of SLE can be challenging due to its wide range of symptoms and the fact that they can be similar to those of other rheumatic conditions. Before 2019, the American College of Rheumatology (ACR) criteria necessitated that a patient must meet at least 4 of 11 criteria for a diagnosis of SLE. The criteria include findings such as malar rash, discoid rash, photosensitivity, oral ulcers, arthritis, serositis, presence of renal disorder, neurological disorder, hematologic disorder, immunologic disorder, and positive antinuclear antibodies (ANA) test.[1] In clinical practice, the diagnosis is frequently made after corroborating clinical and laboratory findings and excluding alternative diagnoses.

In 2019 the European League Against Rheumatism (EULAR) and the ACR published updated classification criteria for SLE to help improve diagnosis. The updated 2019 criteria required a positive ANA test for diagnosis, rather than just as one of many possible criteria that could be used to establish the diagnosis. In addition, various symptoms were weighted from 2 to 10 points and patients with 10 or more points based on their symptoms fulfill diagnostic criteria for SLE. These criteria are useful for both physicians and researchers to improve the diagnosis and study of SLE.[2]

Pathophysiology

The pathophysiology of SLE is complex and not fully understood. Recent research has found that most single nucleotide polymorphisms (SNPs) associated with SLE fall within DNA regions of immune response-related genes, and certain SNPs contribute to abnormal T-cell function. However, currently identified loci only contribute to about 15% of the heritability of SLE, and other factors such as environmental, hormonal, and possible viral influences remain at play.[3] Exposure to ultraviolet light and certain medications have been implicated in the development of SLE.[1] The presence of X-chromosomes has been shown to increase the severity of SLE, partially explaining the increased prevalence of SLE in women.[4] The global prevalence of SLE is estimated to be 43.7 per 100,000 persons. When stratified by sex, prevalence in women and men is 78.73 and 9.26 per 100,000 persons, respectively.[5]

Immune complexes play a key role in tissue injury in SLE, either directly or through secondary mechanisms that increase thrombosis. An abundance of nuclear material exists in blood and tissues, and these nuclear materials, especially in damaged tissue, can be bound by ANA. Due to damaged complement receptors, these immune complexes cannot be

Retinal and Choroidal Vascular Diseases of the Eye. https://doi.org/10.1016/B978-0-443-15583-3.00026-3

cleared sufficiently in SLE patients, leading to kidney injury, cutaneous manifestations, central nervous system involvement, damage to the cardiac conduction system, and ocular manifestations.[1] For example, keratinocytes in the skin exposed to ultraviolet light become apoptic and release nuclear material that becomes bound to ANA. Subsequently, these complexes cannot be cleared, leading to characteristic rashes seen in some SLE patients and further to continued stimulation of the immune system.[6] SLE patients with antibodies to phospholipids and B_2-glycoprotein 1 form immune complexes that interfere with the coagulation system and endothelial cell function, inducing tissue factor production and the downstream formation of thrombi.

Many of the retinal and choroidal manifestations occur due to a combination of these mechanisms. Ocular manifestations present as a small vessel microangiopathy or severe retinal vaso-occlusion. Immune-complex formation and antiphospholipid antibodies can directly lead to small vessel occlusions or retinal vasculitis. Alternatively, ocular findings could be secondary to hypertension due to lupus nephritis or from antiphospholipid syndrome resulting in retinal or choroidal thrombosis.

Most commonly, posterior segment manifestations of SLE are the result of retinal or choroidal vascular damage to small blood vessels through vaso-occlusions.[7] Electron microscopy in SLE patients with retinal involvement has revealed irregular endothelium in capillaries with obliterated lumens, leading to clinical findings of microaneurysms, cotton-wool spots, and retinal hemorrhages.[8] In addition to small-vessel occlusive disease, an inflammatory vascular condition affecting the retinal and choroidal vessels has also been confirmed in histopathological studies. The severity of the vasculitis is thought to be associated with the level of immune complex deposition in vessel walls. Autopsy studies have revealed immune complex deposits in the walls of retinal and choroidal vessels in SLE patients. Fibrin deposition, a common sign in many antibody-mediated vasculitides, has been seen in choroidal vessels and within Bruch's membrane.[9] Interestingly, one patient with SLE and significant choroidal involvement was found to have mast cells in the choroid seen on electron microscopy, leading to the hypothesis that their release increases choroidal vessel permeability, leading to further leakage of nuclear material and immune-complex formation and deposition.[10] Sometimes, occlusions are so severe that patients present with vaso-occlusive retinopathy; these cases are thought to be due to large vessel immune-complex deposition within the vessel walls. These types of occlusions can be seen in the absence of inflammation. These cases have also been associated with antiphospholipid syndrome (APS) and thrombotic events. However, whether the underlying pathophysiology of occlusive vasculitis is secondary to large vessel immune-complex deposition or APS, or a combination of the two, is difficult to ascertain.[7]

Some tissues affected by SLE have stronger associations with specific antibody-immune complexes than with others. For example, antidouble strand (ds) DNA and antinucleosome antibodies contribute to lupus nephritis, anti-NMDAR antibodies contribute to neurocognitive defects, and anti-Ro antibodies contribute to cardiac conduction deficits.[1] However, when it comes to the retina and choroid it is unknown which antibody-based immune complexes are specifically involved, although antinuclear antibodies are thought to be a primary factor. In addition, the antibodies associated with APS play a role in some manifestations.[7]

Antiphospholipid syndrome

APS can occur without SLE and they are two distinct entities. However, there is an established strong association between APS, SLE, and coexisting retinal and choroidal involvement. APS is characterized by the presence of antiphospholipid antibodies (aPLs) in the blood and a history of thrombotic or obstetric complications. These antibodies can target phospholipids, a type of fat that is important for the normal functioning of cell membranes, leading to blood clotting and inflammation.[11]

The exact relationship between SLE and APS is not fully understood, but it is thought that aPLs may play a role in the development of SLE. Studies have shown that aPLs are found in up to 40% of SLE patients, and their presence is associated with an increased risk of thrombotic complications, as well as a poorer prognosis. Additionally, aPLs can also be present in patients with SLE without any clinical symptoms of APS, indicating that aPLs may contribute to the pathogenesis of SLE.[2]

The management of APS in SLE patients is complex, and it requires a multidisciplinary approach. Anticoagulation therapy with warfarin or heparin is the mainstay of treatment for thrombotic complications, while aspirin is often used for obstetric complications. Additionally, immunosuppressive therapy may be used to control the underlying SLE although its use is controversial. It is important to monitor patients closely for any signs of thrombotic or obstetric complications and to adjust treatment as needed.

Overall, the association between SLE and APS highlights the complexity of SLE and the need for a comprehensive approach to its management. The identification of aPLs in SLE patients can help identify those at increased risk for thrombotic complications and guide appropriate management.

Retinal and choroidal manifestations

Posterior segment involvement secondary to SLE is estimated to occur in 7%–29% of SLE patients. As described earlier, it can present as a small vessel microangiopathy or severe retinal vaso-occlusion. Involvement can be asymmetric and also asymptomatic if the posterior pole is spared. Fundus exam findings include retinal vascular sheathing, cotton wool spots, retinal hemorrhages, and vascular occlusions (Fig. 26.1).[12] When vasculitis is present, SLE tends to have an arterial predominance.[13] Sometimes, peripapillary capillaries can also be affected by endothelial cell dysfunction and optic nerve edema can be seen.[14] At times, choroidal involvement due to predominant choroidal infarction can be greater than retinal involvement, and whitening of the choroid can be seen with minimal overlying retinal changes (Fig. 26.2).

Additional retinal or choroidal manifestations can occur due to secondary causes such as malignant hypertension secondary to lupus nephritis. In addition to the retinal changes already discussed, patients can present with dilated and tortuous vessels, Elschnig spots (choriocapillaris infarcts), Siegrist streaks (choriocapillaris necrosis), and Amalric infarcts due to choroidal lobule infarction.[15,16] Further discussion about the retinal and choroidal changes in hypertensive disease can be found in Chapter 33 (Hypertensive retinopathy). Choroidal infarcts can also be a presentation of severe thrombosis secondary to APS with or without SLE association.[17]

Rarely, severe retinal and choroidal thrombotic disease can present with a combination of widespread retinal and choroidal occlusions, cotton wool spots, retinal hemorrhages, and optic nerve edema and be associated with catastrophic APS (CAPS). CAPS is a rare but life-threatening form of APS that results in multiorgan failure so early diagnosis can be critical and vision loss may serve as the presenting symptom (Fig. 26.3).[18]

Rarely, sporadic cases of SLE have been associated with TREX1 mutations, which play a vital role in DNA degradation and DNA repair. These cases can be tough to diagnose and could be suspected when multifocal areas of obliterative vasculitis are noted in the posterior pole of a patient with other systemic SLE symptoms.[19] TREX1 mutations are also implicated in retinal vasculopathy with cerebral leukodystrophy (RVCL) which is a rare genetic disorder with microvascular complications that involve the cerebrum and kidney most commonly in addition to the retina.

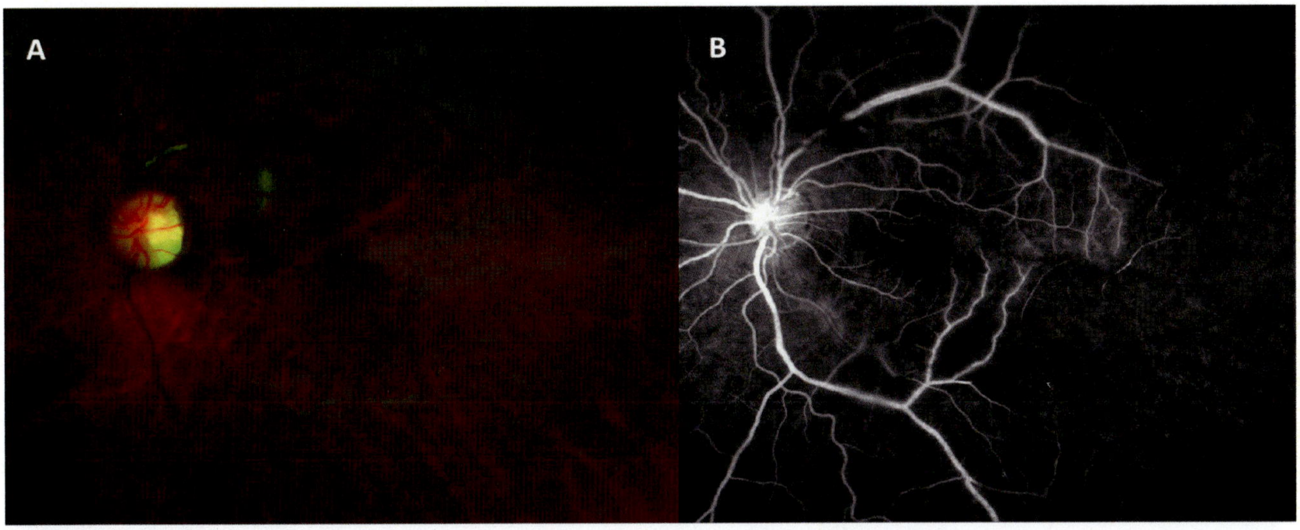

FIGURE 26.1 53-Year-old woman with left eye vision loss with (A) fundus photo of the left posterior pole demonstrating superior intraretinal hemorrhage and severe retinal arterial vascular sheathing superior to the optic nerve and an inferior macula; temporal nonperfusion is seen and (B) fluorescein angiography demonstrating mild nasal disc leakage, superior blockage in the area of the intraretinal hemorrhage, severe temporal nonperfusion, and decreased perfusion of the significantly sheathed arteries.

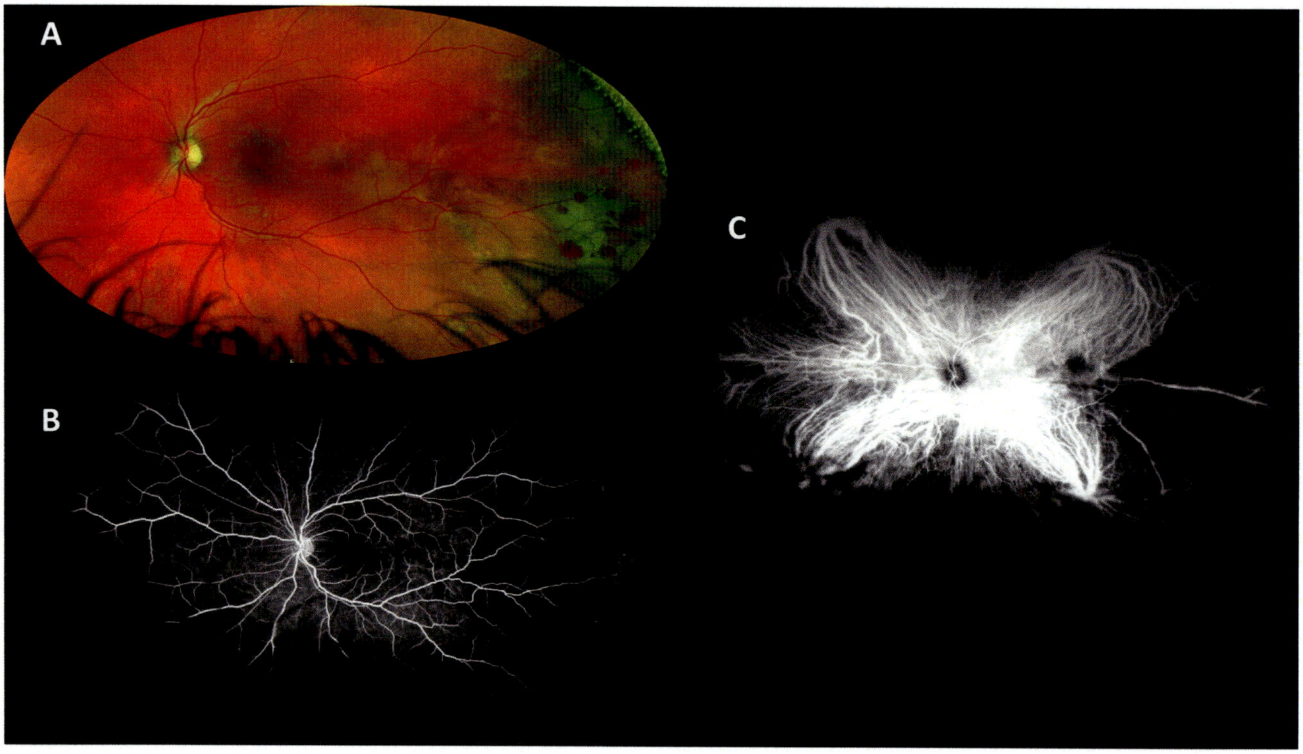

FIGURE 26.2 A 23-year-old woman presented with a nasal scotoma in the left eye as well as photopsias. (A) Widefield color fundus imaging of the left eye revealed choroidal whitening in a triangular distribution starting at the temporal macula out to the periphery, as well as large intraretinal hemorrhages temporally. No sheathing of retinal vessels was seen. (B) Widefield fluorescein angiography midphase demonstrates the absence of the background choroidal flush in a triangular distribution temporally with some overlying retinal capillary remodeling. (C) Widefield indocyanine green angiography demonstrates temporal hypofluorescence in an Amalric distribution, representing infarction of a temporal choroidal lobule.

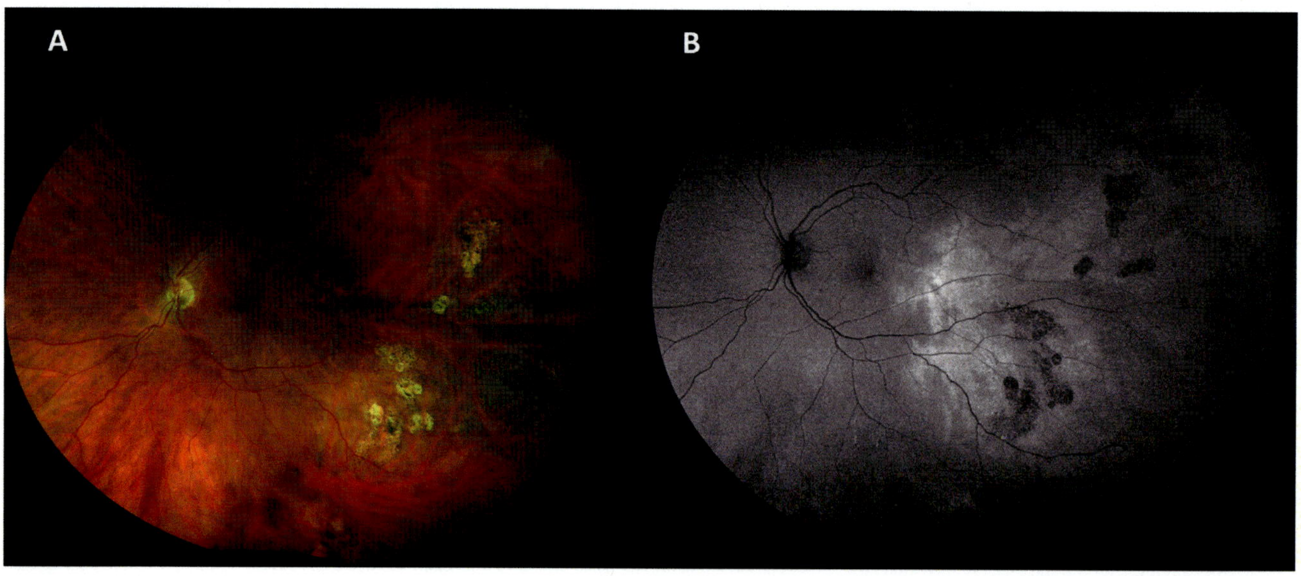

FIGURE 26.3 A 39-year-old female with primary catastrophic antiphospholipid syndrome (CAPS) presented with CAPS interestingly only involving the left side of her body including the left lung, spleen, and kidney. (A) Widefield fundus photo of the left eye demonstrates dramatic temporal pigmentary and atrophic changes in areas of retinal and choroidal infarction and (B) fundus autofluorescence demonstrates mixed hyper and hypoautofluorescence in these areas. *Images courtesy of Aristomenis Thanos MD.*

Imaging

Multimodal imaging plays an important role in diagnosing retinal and choroidal involvement of SLE. Widefield fluorescein angiography (FA) is a useful tool for detecting early signs of SLE-related retinal vasculitis.[20,21]

FA findings include leakage from blood vessels, occlusive disease, arterial versus venous predominance, and neovascularization. The significant choroidal occlusive disease can sometimes also be noted on FA due to a filling defect in the normal background choroidal flush (Fig. 26.2). Indocyanine green angiography (ICG) is also used to evaluate the choroid and can detect choroidal permeability and occlusive disease (Fig. 26.2). Hypofluorescent areas on ICG can be seen even without findings on FA, indicating choroidal infarction, and these findings may be more correlated with patients with associated nephropathy.[22,23] Optical coherence tomography (OCT) is useful to assess the degree of retinal thinning or involvement in occlusive disease and detect the presence of cystoid macular edema. OCT is also used to look for signs of hydroxychloroquine toxicity since many SLE patients concurrently take this medication. These signs include macular volume thinning, outer nuclear layer thinning, and ellipsoid zone changes.[24] OCT angiography (OCT-A) can visualize the microvasculature of the retina and choroid. Occlusive disease can be visible on this modality with appropriate segmentation (Fig. 26.4).[25]

Efforts have been made to try to correlate the degree of OCT-A microvascular changes with systemic disease activity.[26] In the future, algorithms to help assess the degree of perivascular retinal and choroidal thickening on OCT-A may be a useful surrogate to FA in identifying inflammatory foci that would correspond to leakage on FA.[27] Currently, however, widefield FA remains the gold standard in identifying areas of active vasculitis.[28]

Treatment considerations

The treatment of SLE is challenging and there is a lack of randomized trials to guide therapy. However, experiential case series have helped to establish some general treatment principles. Due to the complexity of SLE, a multidisciplinary approach that includes rheumatologists, ophthalmologists, and other specialists is often necessary to achieve the best outcomes. Other specialists may commonly include dermatologists, nephrologists, and cardiologists.

When active ocular manifestations of SLE are noted they may be a surrogate indicator of potentially serious systemic disease activity. Depending on the severity of the retinal and choroidal disease, intravenous methylprednisolone pulse therapy or high-dose prednisone is commonly used. Long-term steroid-sparing immunosuppressive agents commonly used include rituximab, mycophenolate mofetil, and azathioprine. Cyclophosphamide has been used as well but more recent immunosuppressive agents are favored due to significant adverse side effects of cyclophosphamide. It can still be used in severe, treatment-resistant cases of SLE.[12] Antitumor necrosis factor-alpha medications warrant close monitoring due to the possible formation of drug-induced antibodies in these medications, more commonly with infliximab than with adalimumab. Concurrent use of mycophenolate mofetil or methotrexate can limit the production of these antibodies.[29] Many SLE patients are prescribed hydroxychloroquine to reduce systemic flares but hydroxychloroquine is not thought to help with ocular manifestations and if used, concurrent plaquenil retinal toxicity monitoring must be performed.

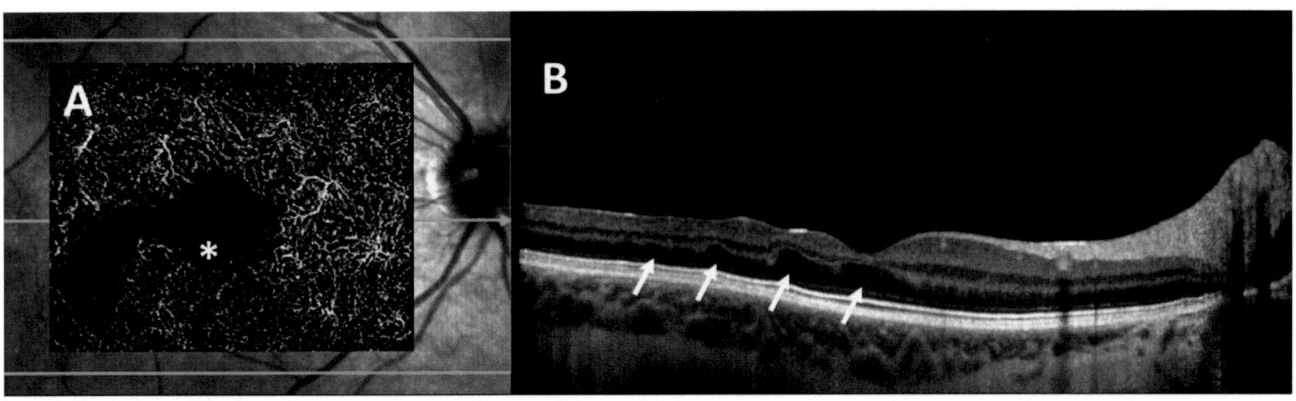

FIGURE 26.4 A 32-year-old male with primary antiphospholipid syndrome who presented with right eye vision loss. (A) OCT-angiography with segmentation of the deep capillary plexus (DCP) demonstrates flow void (*) consistent with DCP infarction and (B) B-scan OCT through the infarct demonstrates sequential retinal infarcts of perivascular lesions (RIPL) sign (*arrows*), highlighted by inner retinal dimpling and outer nuclear layer enlargement indicative of cardiovascular insult. *Images courtesy of Aristomenis Thanos MD.*

For local ocular treatment secondary to neovascular complications from lupus vasculitis, antivascular endothelial growth factor (VEGF) and/or pan-retinal photocoagulation is used. As many SLE patients may be women in child-bearing age, the risk of pregnancy should be assessed before administering intravitreal anti-VEGF medications. In cases where retinal ischemia is seen without concurrent neovascularization, treatment is up to the discretion of the provider and no consensus guidelines exist. Many of these patients may never develop neovascularization with appropriate treatment and rarely, if the occlusive disease is found acutely, reperfusion can occur with aggressive systemic treatment. Therefore, it is the current practice pattern to apply photocoagulation treatment only if neovascularization occurs. At times, local anti-inflammatory treatment with periocular or intraocular steroids can be used to control inflammatory vascular disease. Given that SLE is a systemic disease, these cases are reserved for acute flare-ups (vasculitis, occlusive disease, or cystoid macular edema), when the patient cannot receive further systemic treatment for some reason. Sometimes pregnant patients who cannot tolerate further systemic immunotherapy but have vision-threatening disease might benefit from local treatment. When patients develop occlusive disease thought secondary to APS, as discussed previously, anticoagulation is usually used and the use of additional immunotherapy in these cases is controversial.

Conclusion

In conclusion, SLE is a chronic autoimmune disease that can significantly impact the retina and choroid. The effects of SLE on the eye may lead to various vision-threatening complications, including retinal and choroidal ischemia with or without vasculitis, thrombosis, macular edema, and neovascularization. Early detection and treatment of ocular involvement in SLE is crucial to prevent or minimize vision loss. Interdisciplinary management is crucial as systemic treatment is often needed; however, for vision-threatening complications not controlled with systemic treatment, local ocular treatment may be necessary as well. Patients with SLE should receive prompt medical attention for any changes in their vision. Patients taking hydroxychloroquine should undergo regular eye exams with appropriate screening.

References

1. Tsokos GC. Systemic lupus erythematosus. *New England Journal of Medicine*. 2011;365(22):2110–2121. https://doi.org/10.1056/nejmra1100359.
2. Aringer M, Costenbader K, Daikh D, et al. 2019 European League against rheumatism/American College of Rheumatology classification criteria for systemic lupus erythematosus. *Arthritis & Rheumatology*. 2019;71(9):1400–1412. https://doi.org/10.1002/art.40930.
3. Manolio TA, Collins FS, Cox NJ, et al. Finding the missing heritability of complex diseases. *Nature*. 2009;461(7265):747–753. https://doi.org/10.1038/nature08494.
4. Smith-Bouvier DL, Divekar AA, Sasidhar M, et al. A role for sex chromosome complement in the female bias in autoimmune disease. *Journal of Experimental Medicine*. 2008;205(5):1099–1108. https://doi.org/10.1084/jem.20070850.
5. Tian J, Zhang D, Yao X, Huang Y, Lu Q. Global epidemiology of systemic lupus erythematosus: a comprehensive systematic analysis and modelling study. *Annals of the Rheumatic Diseases*. 2023;82(3):351–356. https://doi.org/10.1136/ard-2022-223035.
6. Manderson AP, Botto M, Walport MJ. The role of complement in the development of systemic lupus erythematosus. *Annual Review of Immunology*. 2004;22:431–456. https://doi.org/10.1146/annurev.immunol.22.012703.104549.
7. Nag TC, Wadhwa S. Vascular changes of the retina and choroid in systemic lupus erythematosus: pathology and pathogenesis. *Current Neurovascular Research*. 2006;3(2):159–168. https://doi.org/10.2174/156720206776875821.
8. Jabs DA, Fine SL, Hochberg MC, Newman SA, Heiner GG, Stevens MB. Severe retinal vaso-occlusive disease in systemic lupus erythematosus. *Archives of Ophthalmology*. 1986;104(4):558–563. https://doi.org/10.1001/archopht.1986.01050160114025.
9. Jabs DA, Hanneken AM, Schachat AP, Fine SL. Choroidopathy in systemic lupus erythematosus. *Archives of Ophthalmology*. 1988;106(2):230–234. https://doi.org/10.1001/archopht.1988.01060130240036.
10. Nag TC, Wadhwa S. Histopathological changes in the eyes in systemic lupus erythematosus: an electron microscope and immunohistochemical study. *Histology and Histopathology*. 2005;20(2):373–382.
11. Meroni PL, Tsokos GC. Editorial: systemic lupus erythematosus and antiphospholipid syndrome. *Frontiers in Immunology*. 2019;10. https://doi.org/10.3389/fimmu.2019.00199.
12. E C-B, Lupus AA, Vasculitis. *Current Rheumatology Reports*. 2020;22.
13. Sanders MD, Graham EM. Retinal vasculitis. *Postgraduate Medical Journal*. 1988;64(753):488–496. https://doi.org/10.1136/pgmj.64.753.488.
14. Santra G, Das I. Systemic lupus erythematosus presenting as neuroretinitis. *Journal of Association of Physicians of India*. 2015;63(2015):77–78.
15. de Azevedo LGB, Biancardi AL, Silva RA, et al. Lupus retinopathy: epidemiology and risk factors. *Arquivos Brasileiros de Oftalmologia*. 2021;84(4):395–401. https://doi.org/10.5935/0004-2749.20210076.
16. Nemiroff J, Phasukkijwatana N, Vaclavik V. The spectrum of amalric triangular choroidal infarction. *Retinal Cases and Brief Reports*. 2017;11:S113–S120.
17. Ang LPK, Yap EY, Fam HB. Bilateral choroidal infarction in a patient with antiphospholipid syndrome: a case report. *Clinical and Experimental Ophthalmology*. 2000;28(4):326–328. https://doi.org/10.1046/j.1442-9071.2000.00322.x.

18. In Yun Y, Kim JH, Lim SH, et al. Case of catastrophic antiphospholipid syndrome presenting as neuroretinitis and vaso-occlusive retinopathy. *BMC Ophthalmology*. 2020;20(1). https://doi.org/10.1186/s12886-020-01755-9.
19. Hosseini, Labilloy A. *Genetics., Mutations*Vol. 2022. StatPearls Publishing; 2023.
20. Tolba DA, El-Fayoumi DMS, Abdelaziz MS, Nabih MH. Fluorescein angiographic findings in patients with active systemic lupus erythematosus. *Ocular Immunology and Inflammation*. 2017;25(6):884–890. https://doi.org/10.1080/09273948.2016.1192653.
21. Fouad SA, Esmat Mahmoud Ali SM, Rezk Alnaggar ARL, Mahfouz S, Essam M, El-Gendy H. Structural retinal assessment using optical coherence tomography and fundus fluorescein angiography in systemic lupus erythematosus patients. *Journal of Clinical Rheumatology*. 2021;27(1):34–39. https://doi.org/10.1097/RHU.0000000000001162.
22. Baglio V, Gharbiya M, Balacco-Gabrieli C, et al. Choroidopathy in patients with systemic lupus erythematosus with or without nephropathy. *Journal of Nephrology*. 2011;24(4):522–529. https://doi.org/10.5301/JN.2011.6244.
23. Dhingra S, Stavrou P. Indocyanine green angiography in systemic lupus erythematosus-associated uveitis. *Ocular Immunology and Inflammation*. 2004;12(1):69–73. https://doi.org/10.1076/ocii.12.1.69.28068.
24. Marmor MF, Kellner U, Lai TYY, Melles RB, Mieler WF, Lum F. Recommendations on screening for chloroquine and hydroxychloroquine retinopathy (2016 revision). *Ophthalmology*. 2016;123(6):1386–1394. https://doi.org/10.1016/j.ophtha.2016.01.058.
25. Pichi F, Woodstock E, Hay S, Neri P. Optical coherence tomography angiography findings in systemic lupus erythematosus patients with no ocular disease. *International Ophthalmology*. 2020;40(8):2111–2118. https://doi.org/10.1007/s10792-020-01388-3.
26. Ermurat S, Koyuncu K. Evaluation of subclinical retinal microvascular changes in systemic lupus erythematosus patients using optical coherence tomography angiography and its relationship with disease activity. *Lupus*. 2022;31(5):541–554. https://doi.org/10.1177/09612033221084222.
27. Abucham-Neto JZ, Torricelli AAM, Lui ACF, Guimarães SN, Nascimento H, Regatieri CV. Comparison between optical coherence tomography angiography and fluorescein angiography findings in retinal vasculitis. *International Journal of Retina and Vitreous*. 2018;4(1). https://doi.org/10.1186/s40942-018-0117-z.
28. Herbort CP, Papasavvas, Tugal-Tutkun I. Benefits and limitations of OCT-A in the diagnosis and follow-up of posterior intraocular inflammation in current clinical practice: a valuable tool or a deceiver? *Diagnostics*. 2022;12.
29. Sazonovs A, Kennedy NA, Moutsianas L, et al. HLA-DQA1*05 carriage associated with development of anti-drug antibodies to infliximab and adalimumab in patients with Crohn's disease. *Gastroenterology*. 2020;158(1):189–199. https://doi.org/10.1053/j.gastro.2019.09.041.

Chapter 27

Retina and choroidal vascular changes in Behçet disease

Shaikha Aljneibi[1] **and Francesco Pichi**[1,2]
[1]*Eye Institute, Cleveland Clinic Abu Dhabi, Abu Dhabi, United Arab Emirates;* [2]*Cleveland Clinic Lerner College of Medicine, Case Western Reserve University, Cleveland, OH, United States*

Introduction

Behçet disease is a chronic systemic vascular disease involving different organ systems with a broad clinical spectrum. This recurrent relapsing inflammatory disorder of unknown etiology was described in the early 20th century by Adamantiades and Behçet as a triad of oral aphthous ulcers, genital ulcers, and ocular manifestations that most often consist of severe bilateral panuveitis.[1,2] Behçet disease is seen worldwide, but it is most common in the Middle and Far East. Age of onset is typically in the second to fourth decade of life, with an equal number of both genders affected but a more severe clinical course in males.[1,2]

Beçhet disease diagnosis

Systemic diagnostic criteria

Without a diagnostic laboratory test, Behçet disease remains a clinical diagnosis based on multiple systemic findings. Different classification criteria have been developed through the years. The International Study Group is one of the most clinically accepted criteria for diagnosing Behçet disease.[3] In 1990, the International Study Group compared the clinical findings of 914 patients with a history of aphthous ulcers with those of controls.[3] According to their recommendation, to confirm the diagnosis of Behçet disease, the oral aphthous ulcer is considered essential. In addition, at least two of the following must be demonstrated: recurrent painful genital ulcers that heal with scarring; ophthalmic lesions including anterior or posterior uveitis, hypopyon, or retinal vasculitis; skin lesions including erythema nodosum or positive results from pathergy skin testing.[3]

Ocular diagnostic criteria

Multiple sets of ocular Behçet diagnostic criteria have been proposed in the past, with the International Study Group for Behçet disease being the most widely used. Tugal-Tutkun et al. proposed an algorithm for diagnosing Behçet disease uveitis by using only characteristics of ocular findings.[4] Following an initial survey among uveitis experts, they collected retrospective data on patients with Behçet disease uveitis and other uveitis, then identified ocular findings with a high diagnostic odd ratio for Behçet uveitis. Subsequently, they collected prospective data on patients with Behçet uveitis and controls and developed a diagnostic tree approach.[4] Among the identified variables that provide the highest accuracy for the diagnosis of Behçet disease uveitis, the most reliable is the presence of superficial retinal infiltrate or its sequelae, retinal nerve fiber layer defect, signs of occlusive retinal vasculitis, and diffuse retinal capillary leakage on fluorescein angiography (FA) as well as the absence of granulomatous anterior uveitis or choroiditis in patients with vitritis.[4] In 2021, the standardization of uveitis nomenclature (SUN) working group applied the International Study Group criteria to 248 cases of Behçet disease, of which 194 achieved supermajority agreement during the selection phase and, therefore, were retained in the final database and used in a machine learning phase.[3] The key features of the new SUN criteria included a compatible uveitic syndrome, either anterior uveitis, anterior and intermediate or posterior uveitis/panuveitis with evidence

Retinal and Choroidal Vascular Diseases of the Eye. https://doi.org/10.1016/B978-0-443-15583-3.00027-5

of retinal vascular involvement or focal infiltrates, and systemic Behçet disease.[5] No case had either focal or multifocal choroiditis, so the identification of choroiditis as a feature is no longer in the definition of eye lesions.[4,5]

This criteria for Behçet disease by the SUN working group seem to perform well for use as classification criteria in clinical research, as it has a low misclassification rate, indicating good discriminatory performance against other panuveitis.[4,5]

Retina vascular changes in adult Behçet patients

Most clinical manifestations of Behçet disease are attributed to vasculitis that involves both the arteries and veins, which can affect blood vessels of any size leading to endothelial damage and abnormal blood flow, resulting in hypercoagulability and an increased risk of thrombosis.

Behçet disease is one of the most severe causes of noninfectious uveitis and occurs in up to 50%–90% of patients. It can be an initial symptom in 10%–20% of cases and lead to blindness in 16%–25% of patients.[6] The primary destructive mechanism of Behçet disease is vasculitis, and areas of retinal nonperfusion may result in irreversible vision loss, especially if it causes macular ischemia.

In clinical practice, FA is a valuable tool for evaluating the posterior segment manifestations of Behçet disease. The leakage on FA may serve as an essential marker for inflammatory activity in Behçet disease uveitis. Although fluorescein angiogram remains an essential diagnostic tool in Behçet disease uveitis, it is invasive with limited depth resolution. Optical coherence tomography angiography (OCTA) provides valuable information and high-resolution images of both retinal and choroidal structures. OCTA is a reliable imaging technique that provides quantitative parameters, including the foveal avascular zone area (FAZ), capillary nonperfusion areas, and vessel densities in both superficial and deep capillary plexuses. OCTA can improve our understanding of the pathophysiology and provide clinically useful information in evaluating and monitoring retinal microvascular networks in patients with Behçet disease. It allows ophthalmologists to measure FAZ and nonperfusion areas bilaterally, also discriminating the superficial and deep vascular layers (Table 27.1).

Retina vascular changes in adult Beçhet patients with ocular involvement

Inactive

Recurrent uveitis attacks may eventually cause retinal vessel occlusion resulting in poor visual acuity because of the qualitative and quantitative changes in the macular area. Eser-Ozzturk et al. presented a cross-sectional study on patients with inactive Behçet uveitis using OCTA images to evaluate the FAZ and capillary vessel density. They found that uveitis attacks affecting the macula are associated with decreased capillary vessel density (Fig. 27.1).

Also, they observed an expansion in the FAZ area in Behçet uveitis, but statistically significant results could not be obtained.[7]

Using OCTA, Cheng et al. investigated and studied the association of retinal vascular changes in the superficial and deep layers around the macula in patients with Behçet disease in remission. The decreased vessel density was more significant in the deep plexus layer and the inferior region[8] which can be a valuable way of assessing involvement in eyes with quiescent uveitis.

Pei et al., while assessing with OCTA the characteristics of retinal microvascular abnormalities in Behçet disease uveitis, noted a vessel density reduction in the quiescent fellow eye during a unilateral uveitis attack. The consistent bilateral vessel density changes during unilateral uveitis relapse indicate macular circulation disturbances in fellow quiescent eyes, which might be a result of neurohumoral regulation and an increase in some systemic inflammatory markers in ocular Behçet disease.[9]

Accorinti et al. studied OCTA of subjects with inactive Behçet uveitis. The inactive ocular disease was defined in those patients who, being on the same therapy or without therapy for at least 6 months before evaluation, displayed no sign of inflammation in the anterior chamber, vitreous or retina, and no vascular leakage or macular edema according to ultrawidefield FA. Inactive patients were divided into subgroups according to the time from the last ocular relapse and the number of acute episodes previously diagnosed in the examined eye.[10] The only statistically significant difference between active and inactive Behçet disease was found in the superficial capillary plexus. The vessel density in the superficial capillary plexus was significantly reduced in subjects with active ocular Behçet disease versus inactive ocular disease.[10] The vessel density reduction in the superficial capillary plexus was related to the active stage of the disease. In patients with inactive disease for more than 5 years, it was possible to observe a significant reduction in vessel density of the deep plexus as well, probably related to the number of relapses experienced before reaching the inactive stage.

Overall, in inactive Behçet uveitis, vessel density seems to be inversely related to the number of ocular relapses and cannot be restored over time.

TABLE 27.1 Summary of the FA and OCTA parameters of Behçet disease reported in the literature.

Study name	No. of patients	BD active uveitis	BD inactive uveitis	No ocular involvement	Mean FAZ area	Mean SCP vessel density	Mean DCP vessel density	Mean choroidal thickness
Optical Coherence Tomography Angiography Findings in Active and Inactive Ocular Behçet Disease	23 patients	8 patients	15 patients	NA	NA**Active:** 50% **Inactive:** 20% "FAZ irregularity is present more frequently in the acute than inactive stage"	**Active:** 3 × 3 mm = 0.4326 × 6 mm = 0.460 **Inactive:** 3 × 3 mm = 0.4766 × 6 mm = 0.467	**Active:** 3 × 3 mm = 0.4366 × 6 mm = 0.458 **Inactive:** 3 × 3 mm = 0.4646 × 6 mm = 0.471	NA
Comparison of Optical Coherence Tomography Angiography Findings in Patients with Behçet's Disease and Healthy Controls	94 eyes of 49 patients	43 eyes		51 eyes	**Ocular:** 0.94 mm^2 **Nonocular:** 0.32 mm2	**Ocular:** 41.9% **Nonocular:** 46.2%	**Ocular:** 44.4% **Nonocular:** 52.4%	NA
The Relationship between Fluorescein Angiography Leakage after Infliximab Therapy and Relapse of Ocular Inflammatory Attacks in Ocular Behçet's Disease Patients	55 eyes of 29 patients	**Presence of ocular attacks:** 39 eyes of 21 patients		**Absence of ocular attacks:** 16 eyes of 18 patients	NA**Total FA Score:** Before infliximab treatment: 10.7After initiation of infliximab treatment: 6.82			
Quantitative measurements with optical coherence tomography angiography in Behçet uveitis	80 eyes of 41 patients	NA	42 eyes of 22 patients	NA	**Inactive:** Superficial: 283 $microm^2$ Deep: 332.9 $microm^2$ **Healthy:** Superficial: 254.2 $microm^2$ Deep: 317.6 $microm^2$	**Inactive:** Parafoveal = 20.89% **Healthy:** Parafoveal = 21.58%	**Inactive:** Parafoveal = 21.17% **Healthy:** Parafoveal = 21.71%	NA
Indocyanine green video angiography and color Doppler imaging in Behçet's disease	31 patients: ICGVideo angiography: 53 eyesDoppler ultrasonography: 61 eyes	31 patients	NA	NA	NA	NA	NA	NADoppler ultrasonography: Decreased flow velocitiesICG video angiography: reduction in choroidal vascular perfusion

Continued

TABLE 27.1 Summary of the FA and OCTA parameters of Behçet disease reported in the literature.—cont'd

Study name	No. of patients	BD active uveitis	BD inactive uveitis	No ocular involvement	Mean FAZ area	Mean SCP vessel density	Mean DCP vessel density	Mean choroidal thickness
Retinal microvasculature features in patients with Behçet's disease: a systematic review and meta-analysis	599 eyes with Behçet disease622 control eyes	599 eyes with Behçet disease	NA	NA	**Active:** Superficial: MD = 0.06 Deep: MD = 0.12	**Active:** Superficial: MD = −1.5 Deep: MD = −4.25	**Active:** Deep: MD = −4.25	NA
Optical coherence tomography angiography of subclinical ocular features in pediatric Behçet disease	46 patients	22 eyes of 22 patients	NA	22 eyes of 22 patients	NA	**Nonocular:** Capillary non-perfusion areas: 6 out of 22 eyes	**Nonocular:** Capillary non-perfusion areas: 10 out of 22 eyes	NA
Analysis of Parafoveal Microvascular Abnormalities in Behcet's Uveitis Using Projection-Resolved Optical Coherence Tomographic Angiography	102 eyes of 60 patients124 eyes of 62 volunteers	102 eyes of 60 patients	NA	NA	**Active:** 0.37 mm^2 **Healthy:** 0.30 mm^2	**Active:** 44.85% **Healthy:** 49.15%	**Active:** 45.63% **Healthy:** 52.83%	NA
Quantitative Analysis of Structural Alterations in the Choroid of Patients With Active Behçet Uveitis	28 patients	56 eyes of 28 patients	NA	NA	NA	NA	NA	**Active:** CS:CVL = 0.413 Choroidal thickness = 393.9 Central foveal thickness = 352.7 **Healthy:** CS:CVL = 0.351 Choroidal thickness = 394.2 Central foveal thickness = 263.5

Optical Coherence Tomography Angiography Findings in Patients With Nonocular Behçet Disease	280 eyes of 140 patients	280 eyes of 140 patients	NA	280 eyes of 140 patients	**Active:** Mean area = 0.32 mm^2	**Active:** 80%	**Active:** 85%	NA
Optical Coherence Tomography Angiography in Patients With Behçet Uveitis	44 eyes of 25 patients	44 eyes of 25 patients	NA	NA	NA	**Active:** Capillary nonperfusion = 63.6% Capillary abnormalities = 54.5%	**Active:** Capillary nonperfusion = 81.8% Capillary abnormalities = 84.1%	NA
Parafoveal Microvascular Alterations in Ocular and Non-Ocular Behçet's Disease Evaluated With Optical Coherence Tomography Angiography	68 patients46 patients22 healthy	46 patients	NA	23 patients	**Ocular:** 0.5 ± 0.3 mm^2 **Nonocular:** 0.4 ± 0.1 mm^2	**Ocular:** 30 +/− 9 **Nonocular:** 36 ± 4	**Ocular:** 25 +/− 7 **Nonocular:** 30 ± 4	NA
Quantitative Analysis of Optical Coherence Tomography Angiography Features in Patients with Nonocular Behcet's Disease	50 eyes of 25 patients	50 eyes of 25 patients	NA	50 eyes of 25 patients	**Active:** Mean: 0.32 ± 0.08 **Healthy:** Mean: 0.26 ± 0.08	**Active:** Mean: 50.83 ± 2.91 **Healthy:** Mean: 51.68 ± 2.18	**Active:** Mean: 53.02 ± 4.95 **Healthy:** Mean: 52.49 ± 5.59	NA
Inner Retinal Microvasculature Damage Correlates With Outer Retinal Disruption During Remission in Behçet's Posterior Uveitis by Optical Coherence Tomography Angiography	19 eyes of 19 patients25 eyes: healthy subjects	NA	19 eyes of 19 patients	NA	**Active:** Mean: superficial: 0.33 ± 0.08 Deep: 0.49 ± 0.16 **Healthy:** Mean: superficial: 0.30 ± 0.08 Deep: 0.44 ± 0.09	**Active:** 47.30 ± 10.9 **Healthy:** 55.30 ± 6.1	**Active:** 54.5 ± 19.4 **Healthy:** 71.0 ± 5.9	NA
Quantitative Evaluation of the Retinal Vascular Parameters With OCTA in Patients With Behçet Disease Without Ocular Involvement	23 patients29 healthy	23 patients	NA	23 patients	**Active:** Mean: superficial: 0.331 ± 0.121 Deep: 0.352 ± 0.126 **Healthy:** Mean: superficial: 0.24 ± 0.07 Deep: 0.257 ± 0.07	**Active:** 54.5 **Healthy:** 55.6	**Active:** 63.9 **Healthy:** 64.1	NA

Continued

TABLE 27.1 Summary of the FA and OCTA parameters of Behçet disease reported in the literature.—cont'd

Study name	No. of patients	BD active uveitis	BD inactive uveitis	No ocular involvement	Mean FAZ area	Mean SCP vessel density	Mean DCP vessel density	Mean choroidal thickness
Quantitative analysis of retinal microcirculation in optical coherence tomography angiography in cases with Behçet's disease without ocular involvement	42 eyes of 21 patients40 eyes of healthy subjects	42 eyes of 21 patients	NA	42 eyes of 21 patients	**Active:** Mean: superficial: 0.31 ± 0.11 Deep: 0.35 ± 0.12 **Healthy:** Mean: superficial: 0.28 ± 0.05 Deep: 0.33 ± 0.04	**Active:** 50.48 ± 3.24 **Healthy:** 55.17 ± 2.57	**Active:** 56.91 ± 4.15 **Healthy:** 61.59 ± 2.54	NA
Changes in subfoveal choroidal thickness associated with uveitis activity in patients with Behçet's	23 eyes of 13 patients	23 eyes of 13 patients	NA	NA	NA	NA	NA	**EDI-OCT-based choroidal thickness** Active stage: 338 ± 94.5 Remission stage: 267 ± 90

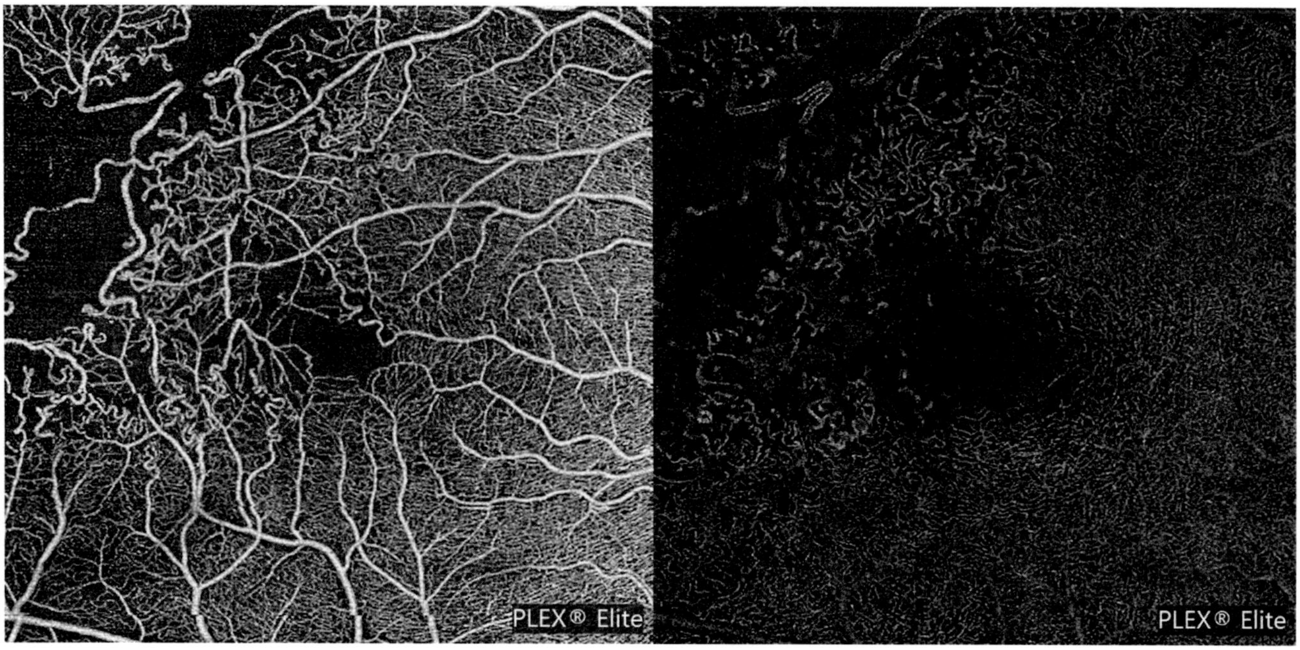

FIGURE 27.1 Macular 6 × 6 OCTA of inactive Behçet uveitis patients with inactive Behçet vasculitis. OCTA shows reduced decorrelation signal in the superficial capillary plexus (A) but mostly in the deep plexus (B).

Active

Fluorescein angiography

Shirahama et al. studied the relationship between leakage on FA before and after the initiation of infliximab therapy and the recurrence of ocular inflammatory attacks in patients with ocular Behçet disease.[11] By comparing the FA scores, between eyes with and without ocular attacks after infliximab therapy. The authors showed that leakage from the optic disc, large retinal vessels, and capillary vessels was significantly reduced after infliximab therapy initiation, and the persistence of leakage after infliximab therapy was strongly related to the presence of ocular inflammatory relapses in patients with ocular Beçhet's disease.[11]

Optical coherence tomography angiography

Multiple studies have demonstrated areas of hypoperfusion using OCTA in patients suffering from Behçet disease with ocular involvement. They indicated that the deep capillary plexus is more involved than the superficial venous plexus in uveitic Behçet patients (Fig. 27.2).

Khairallah et al. were the first to evaluate perifoveal microvascular changes in eyes with Behçet uveitis using OCTA and compared them with fluorescein angiogram findings.[12] They concluded that in eyes with active Behçet disease, OCTA allows better visualization and characterization of perifoveal microvascular changes than FA. Conversely, FA remains indispensable in detecting and evaluating peripheral retinal capillary nonperfusion and neovascularization. Moreover, FA is better at showing retinal vascular and optic disc leakage, clear signs of active Behçet uveitis.[12] The changes detected by OCTA included areas of retinal capillary nonperfusion, rarefied, dilated, or shunting perifoveal capillary vessels, capillary abnormalities, disorganization of the typical architecture of the capillary network, enlargement of the FAZ, and reduction of the capillary vessel density. Their OCTA assessment of superficial and deep capillary plexus separately allowed for the detection of more severe involvement of the deep plexus. The Authors hypothesized that the superficial capillary plexus changes are more strictly related to the active stage of the disease.[12]

Koca et al. detected a significant reduction in OCTA in vascular density in both the superficial and deep capillary plexuses in Behçet disease with ocular involvement compared to nonocular Behçet disease and healthy controls.[13] Additionally, FAZ irregularity was more common in the ocular involvement group. The Authors proposed that in patients with ocular Behçet disease, retinal vascular density and perifoveal microvascular changes (hypoperfusion, nonperfusion, and disorganization) and FAZ irregularity might be used as valuable biomarkers for determining disease activity and prognosis, as well as decision-making for the choice of treatment.[13]

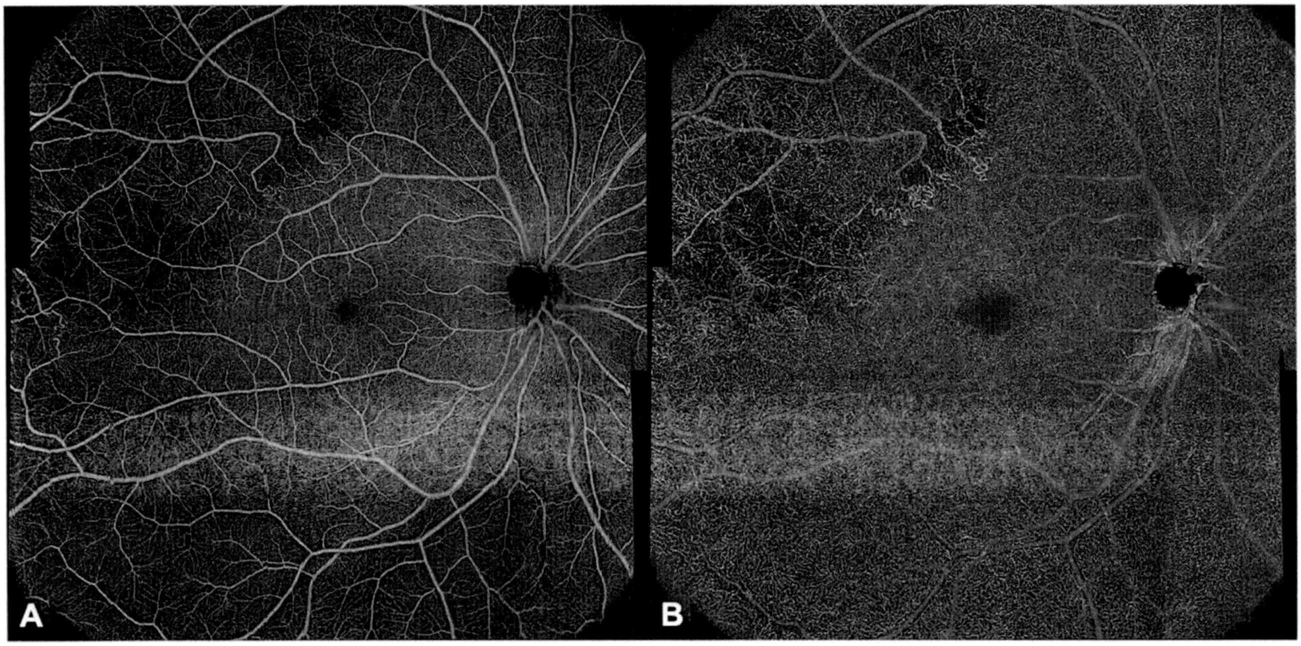

FIGURE 27.2 90-degree OCTA composite of a patient with active Behçet disease. A Behçet disease patient with active superotemporal vasculitis; the affected area shows on widefield OCTA a decrease in the signal (A and B), which is much more pronounced in the deep plexus (B).

Pei et al. showed on OCTA that vessel density of superficial and deep capillary plexuses and FAZ area and perimeter were all significantly different between active Behçet uveitis eyes and controls. In this study, parafoveal capillary disruptions were observed in about one-third of Behçet disease uveitis eyes, and they were more frequently presented in the superficial capillary plexus. The superficial vessel density also had a much stronger correlation with visual acuity than deep vessel density.[9]

Accorinti et al. sought to identify OCTA changes in patients with ocular Behçet disease and compare the different findings according to the clinical status of the patients and healthy controls. They found that patients with the active stage of the disease exhibited reduced vessel density compared to healthy controls in both superficial and deep capillary plexus in all examination frames. Considering all involved eyes, the deep capillary plexus was more frequently affected, suggesting that at this level, the insult of the disease might exert its worse effect.[10]

To sum up these OCTA results, Ji et al. conducted a meta-analysis to determine retinal microvasculature features in active Behçet disease. Thirteen available studies were analyzed, including 599 eyes with Behçet disease and 622 control eyes.[6] The meta-analysis revealed that macular whole en face vessel density, foveal density, and parafoveal density were significantly lower in active eyes. The FAZ was considerably larger in patients with Behçet disease with and without ocular involvement. The authors hypothesized that the impaired vascular endothelial cells might further facilitate immune-mediated vasculitis, leading to vascular obliteration or nonperfusion and subsequently reduced vessel density.[6] The lower foveal vessel density may lead to a larger FAZ, which can be attributed to repeated ocular attacks aggravating retinal ischemia. In addition, the deep foveal vessel density tended to be more severely affected by retinal capillary hypoperfusion or nonperfusion than superficial foveal density.[6]

Retina vascular changes in adult Beçhet patients without ocular involvement

As Behçet disease is a systemic vasculitis affecting multiple organs, Smid et al. hypothesized that retinal microvascular changes might also occur in patients with nonocular Behçet disease.[14] They included 68 subjects from 46 Behçet disease patients and 22 control subjects in their analysis. Of the patients with Behçet disease, 23 had no ocular involvement or any sign of past ocular inflammation and 23 were diagnosed with uveitis.[14] The Authors investigated the vessel density and FAZ area changes on OCTA among the three groups of patients: Behçet disease with and without ocular involvement, and healthy volunteers. The FAZ area was significantly larger in nonuveitic Behçet disease patients compared to healthy volunteers. Also, there was a statistically significant difference in vessel density among the three groups: while vessel density was lowest in the Behçet uveitic group, it was statistically decreased in nonuveitis Beçhet patients, confirming that the parafoveal microvasculature damage resulting from subclinical vasculitis occurs in all Behçet disease patients, even in the absence of clinical signs of uveitis.[14]

Raafat et al. correlated these OCTA microvascular changes that occur in patients diagnosed with Behçet disease without evidence of ocular involvement with macular function determined by central 10-2 perimetry.[5] The study included 20 eyes of 10 patients, of whom all had a normal ocular examination. Their OCTA findings showed microvascular changes (including capillary nonperfusion area, perifoveal arcade disruption, vessel rarefaction, and perifoveal capillary dilatation with and without telangiectasia) in the superficial and deep capillary plexus of nonuveitic Behçet patients, with predominant involvement of the deep capillary plexus. Interestingly, Raafat et al. identified functional changes related to microvascular structural insult on central 10-2 perimetry in 12 eyes that showed relative central scotomata in the pattern deviation map.[5]

In reporting lower capillary vessel densities on OCTA of nonuveitis Behçet eyes, particularly in the foveal region, Goker et al. hypothesized a subclinical form of ocular involvement in such cases.[15] Moreover, Degirmenci et al. found that the FAZ areas in both the superficial and deep capillary plexuses were significantly larger in Behçet patients without ocular involvement compared to healthy controls.[16] To strengthen this "subclinical" hypothesis, Comez et al. showed that retinal microvascular structures could be affected well before the emergence of the eye's apparent clinical signs of Behcet disease. They also determined a reduction in the mean vessel density in the deep capillary networks and the choriocapillaris flow area and an increase in the subfoveal choroidal thickness of the nonocular Behçet disease patients compared to the control group.[17]

Retinal vascular changes in pediatric Behçet patients

Retina vascular changes in pediatric Beçhet patients without ocular involvement

Few previous studies described the microvascular changes of pediatric Behçet disease patients. Although mucocutaneous involvement is a predominant feature in patients with juvenile-onset Behçet disease, ocular involvement may occur in 14.1%–66.2% of cases.[2] Yelmiz et al. were the first to assess macular and peripapillary microvascular features in pediatric Behçet patients with no ocular involvement via OCTA.[2] Their study revealed that vessel density in deep capillary plexus and perifoveal morphological structure was significantly affected even in pediatric nonocular Behçet patients with a short duration of disease.[2] This could be considered a subclinical ocular manifestation in pediatrics uveitis Behçet disease. Conversely, no statistically significant difference was noted between nonocular Behçet pediatric patients and healthy subjects regarding the FAZ area in the superficial capillary plexus.

Choroidal vascular changes in Behçet patients

Although the retinal vascular component is the predominant feature in Behçet disease, some abnormalities in the choroidal circulation can be seen, raising the question that choroidal involvement is of importance in uveitis Behçet disease. Atmaca et al. evaluated the status of the choroid in Behçet's patients and studied the blood flow velocity.[18] They performed ICG video-angiography in 53 eyes with active disease. Hyperfluorescence was observed in the early and late phases of Indocyanine green angiography (ICGA). This hyperfluorescence can be regarded as an essential sign of abnormal choroidal vascular hyperpermeability, secondary to a reduced choroidal vascular perfusion because of obstructive vasculitis.[18] However, Onal et al., who analyzed the morphology of subfoveal choroid during an acute attack of Behçet disease, using enhanced depth imaging optical coherence tomography (EDI-OCT), FA and ICGA, concluded that ICGA is not reliable in assessing the extent of choroidal inflammation.[19] Onal et al. applied quantitative segmentation of the choroid into the choroidal stroma and choroidal vessel lumen through binarization of the EDI-OCT images.[19] They used a software to calculate the choroidal stroma-to-choroidal vessel lumen ratio (CV:CVL). Their study demonstrated increased CV:CVL ratio without significant thickening of the subfoveal choroid in a homogenous group of patients with an acute attack of Behçet uveitis at an early stage of the disease, 4 years or less of disease duration, as measured by EDI-OCT.[19] Overall, they concluded that there was no significant change in the subfoveal choroidal thickness. However, there was a substantial change in the choroidal stromal-to-vessel lumen ratio. These results may suggest that choroidal involvement is not clinically evident in Behçet disease, although previous studies showed choroidal thickening in Behçet uveitis eyes, secondary to diffuse infiltration of choroid with lymphocytes and macrophages.[19]

A third study by Ishikawa et al. looked at the changes in the subfoveal choroidal thickness in patients with Behçet uveitis before and after treatment with infliximab.[20] Their results confirmed that choroidal thickness correlates significantly with anterior inflammation score and total and posterior scores. The choroid of acute Behçet uveitis patients was considerably thicker than in the remission phase, and EDI OCT measured choroidal thickness decreased after initiating infliximab treatment.[20]

Conclusion

Ocular involvement of Behçet disease is characterized by bilateral nongranulomatous panuveitis with retinal vasculitis. As Behçet is systemic vasculitis, it might be expected to lead to primary involvement of the choroid vessel. Recurrent inflammatory attacks of variable severity involving the posterior segment determine the visual prognosis. Therefore, the activity of ocular involvement needs to be identified and closely monitored using different imaging modalities to assess the extent of posterior segment activity and control ocular inflammation. FA remains the gold standard in monitoring inflammation in Behçet disease. Moreover, OCTA is another valuable tool that can help clinicians diagnose and monitor the status of Behçet patients with and without ocular involvement early in the disease course. The major microvascular changes observed in Behçet patients are large FAZ and lower vessel density in foveal and parafoveal areas at the level of deep capillary plexus.

References

1. Chapter 9: Posterior Uveitis and Panuveitis. Aao Website. https://www.aao.org/bcscsnippetdetail.aspx?id=7249c272-3cb7-4d60-92f0-0c0994596043.
2. Yõlmaz Tuğan B, Sönmez HE. Optical coherence tomography angiography of subclinical ocular features in pediatric Behçet disease. *Journal of American Association for Pediatric Ophthalmology and Strabismus*. February 2022;26(1):24.e1–24.e6. https://doi.org/10.1016/j.jaapos.2021.10.004. Epub 2022 Jan 12. PMID: 35032652.
3. Standardization of Uveitis Nomenclature (SUN) Working Group. Classification criteria for Behçet disease uveitis. *American Journal of Ophthalmology*. August 2021;228:80–88. https://doi.org/10.1016/j.ajo.2021.03.058. Epub 2021 May 11. PMID: 33845008; PMCID: PMC8545705.
4. Tugal-Tutkun I, Onal S, Stanford M, et al. An algorithm for the diagnosis of Behçet disease uveitis in adults. *Ocular Immunology and Inflammation*. August 18, 2021;29(6):1154–1163. https://doi.org/10.1080/09273948.2020.1736310. Epub 2020 Apr 14. PMID: 32286112.
5. Raafat KA, Allam RSHM, Medhat BM. Optical coherence tomography angiography findings in patients with nonocular Behçet disease. *Retina*. August 2019;39(8):1607–1612. https://doi.org/10.1097/IAE.0000000000002177. PMID: 29689026.
6. Ji KB, Hu Z, Zhang QL, Mei HF, Xing YQ. Retinal microvasculature features in patients with Behcet's disease: a systematic review and meta-analysis. *Scientific Reports*. January 14, 2022;12(1):752. https://doi.org/10.1038/s41598-021-04730-6. PMID: 35031636; PMCID: PMC8760269.
7. Eser-Ozturk H, Ismayilova L, Yucel OE, Sullu Y. Quantitative measurements with optical coherence tomography angiography in Behçet uveitis. *European Journal of Ophthalmology*. May 2021;31(3):1047–1055. https://doi.org/10.1177/1120672120920214. Epub 2020 Apr 28. PMID: 32345049.
8. Cheng D, Shen M, Zhuang X, et al. Inner retinal microvasculature damage correlates with outer retinal disruption during remission in Behçet's posterior uveitis by optical coherence tomography angiography. *Investigative Ophthalmology & Visual Science*. March 1, 2018;59(3):1295–1304. https://doi.org/10.1167/iovs.17-23113. PMID: 29625452.
9. Pei M, Zhao C, Gao F, et al. Analysis of parafoveal microvascular abnormalities in Behcet's uveitis using projection-resolved optical coherence tomographic angiography. *Ocular Immunology and Inflammation*. April 3, 2021;29(3):524–529. https://doi.org/10.1080/09273948.2019.1685108. Epub 2019 Nov 19. PMID: 31743043.
10. Accorinti M, Gilardi M, De Geronimo D, Iannetti L, Giannini D, Parravano M. Optical coherence tomography angiography findings in active and inactive ocular Behçet disease. *Ocular Immunology and Inflammation*. May 18, 2020;28(4):589–600. https://doi.org/10.1080/09273948.2019.1612452. Epub 2019 Sep 27. PMID: 31560572.
11. Shirahama S, Kaburaki T, Matsuda J, et al. The relationship between fluorescein angiography leakage after infliximab therapy and relapse of ocular inflammatory attacks in ocular Behçet's disease patients. *Ocular Immunology and Inflammation*. November 16, 2020;28(8):1166–1170. https://doi.org/10.1080/09273948.2019.1641611. Epub 2019 Aug 14. PMID: 31411926.
12. Khairallah M, Abroug N, Khochtali S, et al. Optical coherence tomography angiography in patients with Behçet uveitis. *Retina*. September 2017;37(9):1678–1691. https://doi.org/10.1097/IAE.0000000000001418. PMID: 28002270.
13. Koca S, Onan D, Kalaycõ D, Allõ N. Comparison of optical coherence tomography angiography findings in patients with Behçet's disease and healthy controls. *Ocular Immunology and Inflammation*. July 3, 2020;28(5):806–813. https://doi.org/10.1080/09273948.2019.1635167. Epub 2019 Aug 15. PMID: 31414919.
14. Smid LM, Vermeer KA, Missotten TOAR, van Laar JAM, van Velthoven MEJ. Parafoveal microvascular alterations in ocular and non-ocular Behçet's disease evaluated with optical coherence tomography angiography. *Investigative Ophthalmology & Visual Science*. March 1, 2021;62(3):8. https://doi.org/10.1167/iovs.62.3.8. PMID: 33666648; PMCID: PMC7938019.

15. Goker YS, Yõlmaz S, Kõzõltoprak H, Tekin K, Demir G. Quantitative analysis of optical coherence tomography angiography features in patients with nonocular Behcet's disease. *Current Eye Research*. February 2019;44(2):212–218. https://doi.org/10.1080/02713683.2018.1530361. Epub 2018 Oct 9. PMID: 30261150.
16. Değirmenci MFK, Temel E, Yalçõndağ FN. Quantitative evaluation of the retinal vascular parameters with OCTA in patients with Behçet disease without ocular involvement. *Ophthalmic Surgery, Lasers and Imaging Retina*. December 1, 2019;51(1):31–34. https://doi.org/10.3928/23258160-20191211-04. PMID: 31935300.
17. Çömez A, Beyoğlu A, Karaküçük Y. Quantitative analysis of retinal microcirculation in optical coherence tomography angiography in cases with Behçet's disease without ocular involvement. *International Ophthalmology*. October 2019;39(10):2213–2221. https://doi.org/10.1007/s10792-018-1059-z. Epub 2019 Mar 15. PMID: 30875015.
18. Atmaca LS, Batioglu F. Indocyanine green videoangiography and color Doppler imaging in Behçet's disease. *Acta Ophthalmologica Scandinavica*. August 1999;77(4):444–447. https://doi.org/10.1034/j.1600-0420.1999.770418.x. PMID: 10463419.
19. Onal S, Uludag G, Oray M, et al. Quantitative analysis of structural alterations in the choroid of patients with active Behçet uveitis. *Retina*. April 2018;38(4):828–840. https://doi.org/10.1097/IAE.0000000000001587. PMID: 28272284.
20. Ishikawa S, Taguchi M, Muraoka T, Sakurai Y, Kanda T, Takeuchi M. Changes in subfoveal choroidal thickness associated with uveitis activity in patients with Behçet's disease. *British Journal of Ophthalmology*. November 2014;98(11):1508–1513. https://doi.org/10.1136/bjophthalmol-2014-305333. Epub 2014 Jun 17. PMID: 24939422.

Further reading

1. Paovic J, Paovic P, Sredovic V. Behcet's disease: systemic and ocular manifestations. *BioMed Research International*. 2013;2013:247345. https://doi.org/10.1155/2013/247345. Epub 2013 Oct 3. PMID: 24199188; PMCID: PMC3808725.
2. Zając H, Turno-Krecicka A. Ocular manifestations of Behçet's disease: an update on diagnostic challenges and disease management. *Journal of Clinical Medicine*. November 5, 2021;10(21):5174. https://doi.org/10.3390/jcm10215174. PMID: 34768694; PMCID: PMC8584626.

Chapter 28

Multiple sclerosis

Stela Vujosevic[1,2]
[1]Department of Biomedical Surgical and Dental Sciences, University of Milano, Milano, Italy; [2]Eye Clinic, IRCCS MultiMedica, Milano, Italy

Definition

Multiple sclerosis (MS) is the most common inflammatory and demyelinating disease of the central nervous system (CNS).[1,2] It is an acquired autoimmune disorder with variable clinical and pathologic phenotypes, primarily affecting the oligodendroglia, which forms the myelin in CNS.[1,2] Despite demyelination being the hallmark of the disease, it is well established that axons are injured, even from the early stage of MS, representing a major cause of neurological disabilities and symptoms.[2–4]

Epidemiology

Among CNS diseases, MS is the most common nontraumatic cause of permanent disability in young adults.[5] The incidence rate is greater in females than males, with an estimated female-to-male ratio of 2–3/1[6]; the mean age of MS onset is around 30 years.[7] The prevalence of MS is not uniform geographically and increases with latitude from south to north,[8] with higher frequency in northern Europe (100–200 per 100,000) and northern United States (100–150 per 100,000).[9,10] The incidence and prevalence of MS are increasing globally,[11] the underlying reason for which remains unknown.

Etiology

The etiology of MS is complex and multifactorial. According to the most widely held theory, the interaction between environmental and other unidentified factors in individuals with a genetic predisposition may trigger a chronic immuno-mediated inflammation sustained by auto-reactive T cells directed against components of myelin.[1,2] It results in demyelination and variable degrees of axonal damage; later, the disorder is mainly characterized by a neurodegenerative phase that is immune-independent.[1,2]

- *Genetic factors.* MS is an acquired disorder; however, its etiology has a well-recognized genetic component. Large-scale mapping studies have identified over 200 gene loci with polymorphisms predisposing to MS. Each variant is associated with an increased risk of MS but is not necessary or sufficient to cause the disease. Among the highest risk association are the genetic variants located in the major histocompatibility complex (MHC), implicating a pivotal role of immune cells in the pathogenesis of MS.[12]
- *Environmental factors.* Viral infections, low exposure to sunlight, and vitamin D deficiency have been extensively studied to play significant roles in the etiology of MS via its effects on the immune system. Epstein–Barr virus (EBV) infection is the strongest known risk factor among viruses. The hazard of developing MS is approximately 30-fold higher among subjects infected with EBV causing mononucleosis in adolescence. Proof of causality has not yet been established.[13]

Low sunlight exposure could be a relevant aspect of MS epidemiology explaining the well-documented association of MS prevalence with latitude. The literature supports an inverse relationship between sunlight exposure, ultraviolet radiation exposure, or serum vitamin D levels and the risk of MS.[14,15] Other environmental agents unrelated to geography, including tobacco smoking[16] and childhood or adolescent obesity[17] may be involved as risk factors for MS. Despite several

Retinal and Choroidal Vascular Diseases of the Eye. https://doi.org/10.1016/B978-0-443-15583-3.00028-7

hypotheses regarding vaccinations as possible triggers for MS, no association between vaccines and the risk of developing MS and optic neuritis has been found.[18]

Neuropathology

The neuropathologic hallmark of MS is the presence of focal plaques of demyelination lesions within the CNS associated with variable inflammation, gliosis, and axonal loss.[19] These lesions are predominantly located in the optic nerves, brainstem, juxtacortical and periventricular white matter, and spinal cord.[20] Demyelinated plaques are also identified in the cortical gray matter (GM).[21] Although MS has traditionally been considered a disease of multifocal white matter lesions, neuroimaging studies have provided evidence that the spectrum of MS pathology includes a wider range of alterations within the so-called normal-appearing white matter (NAWM) and normal-appearing gray matter (NAGM).[22] These latter two have been associated with progressive loss of brain volume.[22]

Advances in basic immunopathology and neurobiology, as well as developments in magnetic resonance imaging (MRI) and spectroscopic (MRS) techniques,[23] shed light on the understanding of the pathogenesis and gave rise to the concept that neurodegeneration may occur as an independent process concerning inflammation.[22–24]

Disease onset and clinical patterns

The pathological processes of inflammation, demyelination, and axonal damage account for the clinical evolution of the disease. Demyelinated axons lack trophic support and insulation, causing impairment in saltatory conduction of the action potential along the affected nerves. Depolarization might be lost in specific pathways, with consequent neurological manifestations of MS, or might traverse the lesion at reduced velocity, explaining the characteristic delay of evoked potentials.[1] In the early stages of MS, neurologic symptom recovery might suggest a restoration of conduction failure. This is presumably due to the wane of inflammation and recruitment of undifferentiated oligodendrocyte precursors that have the potential to remyelinate naked axons.[25] Cycles of demyelination and remyelination apparently exhaust the capacity for tissue repair.[25]

Progression in MS and irreversible neurological damage are due to a cumulative loss of axons, initiated and maintained by a complex interplay of active inflammation and neurodegeneration, and the increasing vulnerability of intact axons. Clinical phenotypes, course, and prognosis of MS have significant heterogeneity. The current classification of disease course in MS consists of two core phenotypes: relapsing-remitting and progressive disease.[26] Each phenotype is further defined into different clinical subtypes as follows.

- *Clinically isolated syndrome (CIS).* A first, monosymptomatic attack, presumably due to relapsing-remitting MS, is referred to as a CIS and typically presents in adults younger than 40 years as unilateral optic neuritis, a spinal cord syndrome, or a brainstem syndrome, sensory disturbances, motor weakness, and visual complaints (either monocular visual loss or diplopia). As defined by the McDonald criteria, an MS attack is *a monophasic clinical episode with patient-reported symptoms and objective findings typical of MS, reflecting a focal or multifocal inflammatory demyelinating event in the central nervous system, developing acutely or subacutely, with a duration of at least 24 h, with or without recovery, and in the absence of fever or infection.*[27] The onset of the attack is usually acute or subacute over hours to days, and the remission is usually spontaneous over the ensuing weeks to months.[27]
- *Relapsing-remitting MS.* It is the most common clinical phenotype of MS at disease onset, affecting 85%−90% of patients who are typically young (mean age at onset 30 years) and female (male-to-female ratio 1:3).[28] It is characterized by well-defined attacks or relapses followed by complete or incomplete remission.[26,27]
- *Primary progressive MS.* It occurs approximately in 10%−15% of patients and has an insidious presentation, with slow neurologic worsening and accumulation of disability, such as from asymmetric spastic paraparesis that evolves over months or years, progressive hemiparesis, or cerebellar ataxia.[29] Compared to relapsing-remitting MS, it is usually without relapses, presents at an older age (40 years old as mean age of onset), and without sex bias.
- *Secondary progressive MS.* It presents as an initial relapsing-remitting MS disease course followed by the development of a progressive course with time (i.e., secondary progressive MS).[26] There is a gradual worsening and increase in disability with or without relapses. The transition is unpredictable and usually occurs 10−20 years after the disease onset, and the diagnosis is often made retrospectively.[26]

Two important modifiers of these categories are the clinical assessments of disease activity detected by clinical relapses or contrast-enhancing MRI lesions and the progression of disability over a given time period.[26] The clinical disability in MS is commonly measured by the Kurtzke disability status scale or Expanded Disability Status Scale (EDSS), which

TABLE 28.1 Manifestations of multiple sclerosis.

Common clinical features of multiple sclerosis	Atypical clinical features not expected in multiple sclerosis
General	
Relapses and remission, onset between age 15 and 50 years	Steady progression, onset <10 age or > 50 years
Optic nerve	
Unilateral optic neuritis	Progressive optic neuropathy
Brainstem/cerebellum	
Bilateral internuclear ophthalmoplegia, incoordination, vertigo	Complete external ophthalmoplegia, vascular territory syndrome, focal dystonia
Spinal cord	
Partial myelopathy (presenting with sensory and/or motor symptoms, bowel/bladder symptoms, or sexual dysfunction), Lhermitte sign (an electric shock-like transient sensory symptom radiating down the spine or into the limbs most often after movement of the head or neck)	Anterior spinal artery territory lesion, cauda equina symptoms, complete Brown-Sequard syndrome, acute urinary retention
Cerebral hemispheres	
Mild subcortical cognitive impairment	Early dementia, cortical deficits, cortical blindness, encephalopathy, convulsions
Others	
Fatigue (physical exhaustion unrelated to the amount of activity performed), depression, heat sensitivity, or Uhthoff phenomenon (temporary worsening of neurologic signs and symptoms when the body is overheated)	Deficit developing within minutes, fever, weight loss, night sweats, livedo reticularis, early trimester abortions

ranges from 0 (no clinical signs/symptoms) to 10 (death). Scores above 5 refer to the loss of ambulation capabilities.[30] The disease is characterized by a great variety of neurological symptoms that result from the injury of the myelinated nerve fibers involved in sensory, motor, visual, and brainstem pathways. Instead, neuropsychological symptoms associated with cortical dysfunction are less common. There are no clinical manifestations that are pathognomonic to MS, but some are more typical, whereas others are less common and suggestive of an alternative diagnosis (Table 28.1).

Ocular manifestations of MS in adults

Visual symptoms

Disorders of the visual system are very frequent in MS and, in many cases, represent the only symptoms at onset.[31–34] Optic neuritis is the most common ocular finding in patients with MS, occurring in 50% of patients at some point during the disease course, and is the presenting clinical manifestation of MS in 15%–20% of patients. The pathologic basis for optic neuritis is an inflammatory, demyelinating damage to the optic nerve. It usually presents as acute and unilateral visual loss that typically develops over a period of hours to days. It can be preceded and accompanied by unilateral eye pain, often worsened with eye movement. Bilateral symptoms are relatively uncommon, occurring in approximately 10% of cases,[35] and should suggest an alternative etiology, including Leber hereditary optic atrophy, toxic optic neuropathy, or neuromyelitis optica spectrum disease. Bilateral optic neuritis in MS is characterized by an asymmetric onset and different levels of severity in the two eyes.[35] On neurological examination, a relative afferent pupillary defect, that is, Marcus-Gunn pupil, is present if the contralateral eye is uninvolved. This is evaluated by the swinging flashlight test, with the patient fixating on a distant target. The lesion of the optic nerve is retrobulbar in two-thirds of these patients, and consequently, the funduscopic examination is normal in the acute stage; one-third of patients show papillitis with swelling of the optic disk and blurring of disk margins. Visual field test typically shows central or cecocentral scotoma with enlargement of the blind spot. Hemianopsia or quadrantopsia are uncommon but can be seen in MS due to the involvement of optic chiasm or optic radiations. Other signs of optic neuritis can include photopsias,[36] abnormal color vision,[36] and perivenous sheathing or periphlebitis retinae on funduscopic examination.[37] Later, pallor of the optic nerve can result due to axonal loss and gliosis, predominantly in the temporal segment of the disk. Recovered patients often report a variable degree of residual visual loss, deficits in color vision, contrast sensitivity, stereo-acuity, and desaturation of bright colors, particularly red. Prognostic concerns in patients with MS and optic neuritis at onset are recovery of vision and recurrence of optic neuritis. Ninety percent of patients regain 20/40 or better vision at 1 year after an acute episode of optic neuritis, and visual field defects usually resolve.[38]

Eye movement abnormalities

MS can present with eye movement abnormalities that can result in variable degrees of diplopia or oscillopsia. These include alterations of voluntary gaze and nystagmus, which are both very common, alterations of slow phase eye movements, paroxysmal disorders, and, rarely, isolated ocular motor nerve palsies.[34,39] Among the abnormalities of voluntary gaze, internuclear ophthalmoplegia is characteristic of MS. It refers to paresis of ipsilateral eye adduction in horizontal gaze, horizontal nystagmus of the abducting eye, but preserved convergence (differentiating from third cranial nerve palsy). It results from a lesion in the medial longitudinal fasciculus in the brainstem on the side of diminished adduction, which coordinates the abduction of one eye with the adduction of the other. It can be unilateral or bilateral.[34,39] In the course of the disease, 2%–4% of patients with MS can develop acquired pendular nystagmus, which is characterized by rapid, small-amplitude pendular oscillations of the eyes in the primary position. It is accompanied by oscillopsia, often monocular and vertical in direction. It is attributed to a lesion at the level of the cerebellar nuclei or their fiber connections with the brainstem.[40]

Diagnosis

Clinical and radiologic heterogeneity often makes a diagnosis of MS insidious and challenging[41] as many diseases can mimic its appearance. The development of McDonald criteria, as revised in 2017, has enabled earlier and more accurate diagnosis of MS and allowed equally early treatment.[27,42] MS is primarily a clinical diagnosis based on the assessment of neurologic symptoms and signs and confirmed by objective evidence of characteristic CNS involvement by MRI, which is highly sensitive.[27] Advances in MRI and ancillary tests, including serology and genetics, have improved the accuracy of MS diagnosis and the understanding of common causes of MS misdiagnosis.[27] Clinical diagnosis of MS is based on the assessment of the dissemination of CNS lesions in space, that is, in a distinct anatomical location within the CNS, and time, as well as "no better explanation" other than MS.[42] In other words, MS should be suspected when clinical presentation is suggestive of two separate attacks of at least 3 months with neurological signs of two or more CNS lesions.[43]

Diagnostic criteria and their clinical application

The McDonald criteria, as revised in 2017, are intended for the diagnosis of patients in whom MS is clinically suspected and have a typical CIS suggesting the onset of relapsing-remitting MS, which is by far the most common type (Table 28.2).

For the diagnosis of primary progressive MS, these criteria recommend including testing for CSF abnormalities in addition to MRI to increase the diagnostic sensitivity[44] (Table 28.3). The radiologically isolated syndrome (RIS) describes asymptomatic MRI brain lesions compatible with MS in patients with no clinical symptoms.[29,45] These incidental findings, due to the widespread use of MRI should be meticulously differentiated from other conditions that could account for the MRI findings, including aging or vascular-related abnormalities. The Okuda 2009 criteria are generally used as diagnostic criteria for RIS and can be applied in a clinical setting.[46] One-third of asymptomatic subjects meeting 2009 RIS Criteria will develop a seminal acute or progressive clinical event in 5 years of follow-up.[47]

Ancillary tests

Examination of cerebrospinal fluid (CSF), neurophysiological testing of evoked potentials in visual, sensory, or auditory pathways, optical coherence tomography (OCT), and laboratory investigations, including autoantibody tests, are not usually necessary when clinical MRI findings are sufficient to make a diagnosis. However, these exams can be requested as part of the diagnostic work-up to provide supportive evidence of MS and exclude mimics of MS. CSF findings consistent with MS are mildly raised lymphocytic pleocytosis (usually <25 cells/cm^3) and protein (usually <100 mg/dL), a raised IgG index, and IgG oligoclonal bands not present in serum by isoelectric focusing.[48] Oligoclonal bands are found in up to 90% of people with clinically definite MS and strongly predict conversion from CIS to MS.[49,50] However, oligoclonal band positivity is also associated with other neuroinflammatory disorders, so its value needs to be carefully investigated.

TABLE 28.2 The 2017 McDonald criteria for diagnosis of multiple sclerosis in patients with an attack at onset.

Number of clinical attacks	Number of lesions with objective clinical evidence	Additional data needed for a diagnosis of multiple sclerosis
≥2	≥2	None[a]
≥2	1 (as well as clear-cut historical evidence of a previous attack involving a lesion in a distinct anatomical location[b])	None[a]
≥2	1	Dissemination in space demonstrated by an additional clinical attack implicating a different CNS site or by MRI
1	≥2	Dissemination in time demonstrated by an additional clinical attack or by MRI **OR** demonstration of CSF-specific oligoclonal bands[c]
1	1	Dissemination in space demonstrated by an additional clinical attack implicating a different CNS site or by MRI **AND** Dissemination in time demonstrated by an additional clinical attack or by MRI **OR** demonstration of CSF-specific oligoclonal bands[c]

CSF, cerebrospinal fluid; *MRI*, magnetic resonance imaging. If the 2017 McDonald Criteria are fulfilled and there is no better explanation for the clinical presentation, the diagnosis is multiple sclerosis. If multiple sclerosis is suspected by virtue of a clinically isolated syndrome but the 2017 McDonald Criteria are not completely met, the diagnosis is possible multiple sclerosis. If another diagnosis arises during the evaluation that better explains the clinical presentation, the diagnosis is not multiple sclerosis.

[a]*No additional tests are required to demonstrate dissemination in space and time. However, unless MRI is not possible, brain MRI should be obtained in all patients in whom the diagnosis of multiple sclerosis is being considered. In addition, spinal cord MRI or CSF examination should be considered in patients with insufficient clinical and MRI evidence supporting multiple sclerosis, with a presentation other than a typical clinically isolated syndrome, or with atypical features. If imaging or other tests (e.g., CSF) are undertaken and are negative, caution needs to be taken before making a diagnosis of multiple sclerosis, and alternative diagnoses should be considered.*

[b]*Clinical diagnosis based on objective clinical findings for two attacks is most secure. Reasonable historical evidence for one past attack, in the absence of documented objective neurological findings, can include historical events with symptoms and evolution characteristics for a previous inflammatory demyelinating attack; at least one attack, however, must be supported by objective findings. In the absence of residual objective evidence, caution is needed.*

[c]*The presence of CSF-specific oligoclonal bands does not demonstrate dissemination in time per se but can substitute for the requirement for demonstration of this measure.*

Reproduced from Thompson AJ, Banwell BL, Barkhof F, Carroll WM, Coetzee T, Comi G, Correale J, Fazekas F, Filippi M, Freedman MS, Fujihara K, Galetta SL, Hartung HP, Kappos L, Lublin FD, Marrie RA, Miller AE, Miller DH, Montalban X, ... Cohen JA. Diagnosis of multiple sclerosis: 2017 revisions of the McDonald criteria. The Lancet Neurology. 2018;17(2):162–173. https://doi.org/10.1016/S1474-4422(17)30470–2. Table used with the permission of Elsevier Inc.

TABLE 28.3 2017 McDonald criteria for diagnosis of multiple sclerosis in patients with a disease course characterized by progression from onset (primary progressive multiple sclerosis).

Primary progressive multiple sclerosis can be diagnosed in patients with:

- 1 year of disability progression (retrospectively or prospectively determined) independent of clinical relapse

Plus two of the following criteria:

- One or more T2-hyperintense lesions characteristic of multiple sclerosis in one or more of the following brain regions: Periventricular, cortical or juxtacortical, or infratentorial
- Two or more T2-hyperintense lesions in the spinal cord
- Presence of CSF-specific oligoclonal bands

CSF, cerebrospinal fluid.

Reproduced from Thompson AJ, Banwell BL, Barkhof F, Carroll WM, Coetzee T, Comi G, Correale J, Fazekas F, Filippi M, Freedman MS, Fujihara K, Galetta SL, Hartung HP, Kappos L, Lublin FD, Marrie RA, Miller AE, Miller DH, Montalban X, ... Cohen JA. Diagnosis of multiple sclerosis: 2017 revisions of the McDonald criteria. The Lancet Neurology. 2018;17(2):162–173. https://doi.org/10.1016/S1474-4422(17)30470–2. Table used with the permission of Elsevier Inc.

Evoked potentials are electric potentials generated in the CNS by peripheral stimuli of a sense organ, for example, visual evoked potentials triggered by a light flash or changing pattern on a monitor. They provide measures of amplitude and latencies, reflecting the integrity of axons and myelin. The most common finding in MS on evoked potential is a prolonged latency and well-preserved waveform, although it is not specific. Visual-evoked responses are abnormal in 50%−90% of patients with clinically definite MS. They represent the most helpful characteristic to define spatial dissemination of clinically silent lesions in the CNS and thus provide supportive evidence of MS.[51]

Clinical assessment for the anterior visual system in MS

Visual impairment affects at least 50% of patients with MS over the course of the disease.[52] In suspect of optic neuritis or other visual symptoms, a routine clinical assessment includes best corrected visual acuity, low-contrast letter acuity, contrast sensitivity measure, color vision, pupillary reflex, funduscopic examination, visual fields, evoked potentials, MRI techniques and OCT.[52,53] Fluorescein angiography (FA) is not routinely performed to evaluate optic neuritis as it is often within normal limits.[54] However, Lightman et al.[37] showed that in patients presenting with isolated optic neuritis and with retinal abnormalities detected by FA, including perivenous sheathing and dye leakage in different parts of the retinal periphery, conversion to MS was more likely and happened faster. Among these visual acuity measures and clinical exams, OCT is emerging as noninvasive, reliable, and reproducible imaging of the retina for potential use as a biomarker in MS to quantify the neurodegeneration within the retina and the effects of regenerative strategies in clinical trials.[53,55−58] Compared to other imaging or electrodiagnostic techniques, OCT allows distinguishing between demyelinating effects and axonal damage by quantifying NFL that contains axons and glia but not myelin.[53,55−58] Thus, it has the potential to open a diagnostic window for monitoring neurodegeneration, and this is crucial to determine how to prevent neurological disability in MS and monitor early-stage clinical trials of experimental therapeutics.[55] Currently, OCT is used in the clinical management of MS to follow disease progression and as an exploratory outcome in treatment trials and studies to better understand the relationship between objective measurements of retinal changes and clinically relevant outcomes of visual function.[53,55−58] OCT should be seen as a complement to other clinical, radiographical, or neurophysiological assessments and not as a substitute.

OCT and retinal neurodegeneration

Since 1991, OCT has brought clinicians the ability to safely and noninvasively image the retina and choroid at high resolution and quantify the thickness of the retinal nerve fiber layer (NFL).[59] Advances in ocular imaging techniques for the multimodal assessment of optic nerve processes and retinal changes concomitant to CNS disorders have yielded some insights into the pathophysiology of these disorders and clinically relevant information for future therapeutic investigation.[60,61] As an extension of the CNS, the retina is of particular interest, as neurodegenerative and neuroinflammatory processes that have been characterized in CNS disorders have displayed manifestations in the retina. Therefore, the accessibility and organization of the retina demonstrate its potential role as a fundamental tool in the study of neurodegenerative diseases, including MS.[62]

Retinal ganglion cells, which constitute the anterior visual pathway, have their somas in the ganglion cell layer (GCL) and axons in the NFL. These axons are unmyelinated until they leave the eye at the level of the lamina cribrosa, and form the optic nerves, chiasm, and optic tracts. As axonal and neuronal degeneration is a core feature in MS, assessment of the NFL and GCL thicknesses using OCT potentially allows to visualize the processes of neurodegeneration in relation to MS.[58,63,64] In other words, neuronal and axonal components of pathological changes in MS can be objectively monitored by OCT via the inner retina.

The first use of OCT in MS

The earliest application of OCT technology in MS was reported by Parisi et al.[65] in 1999 to assess the correlation between NFL thickness and the retinal or visual pathway function in patients with MS previously affected by optic neuritis. Fourteen patients with a diagnosis of definite MS who had been affected by optic neuritis with complete recovery of visual acuity were examined. The study showed for the first time a significant reduction in the NFL thickness compared to healthy subjects and its correlation with pattern electroretinogram changes in these eyes. These results can be understood as a loss of integrity of retinal ganglion cells and optic tract demyelination and consequent retrograde degeneration of unmyelinated retinal nerve fiber layer axons. Since then, an increased interest in the application of OCT technology for the assessment of retinal changes in patients with MS with and without optic neuritis has led to a growing number of studies.

OCT findings in MS

Next to the well-investigated NFL, the retinal layers that have been systematically analyzed by OCT in MS include GCL, inner plexiform layer (IPL), inner nuclear layer (INL), outer plexiform layer (OPL), and outer nuclear layer (ONL).[66] OCT detects thinning of the NFL in 85% of MS patients with[67] or without optic neuritis.[68] NFL loss is evidenced by OCT approximately 3 months after optic neuritis (Figs. 28.1 and 28.2).[56]

A recent meta-analysis by Petzold et al.[69] showed an estimated average peripapillary NFL (pNFL) thinning after optic neuritis in MS of −20.10 μm (95% CI −22.76 to −17.44; $P < 0{\cdot}0001$) and of −7.41 μm (95% CI −8.98 to −5.83; $P < 0{\cdot}0001$) in MS without optic neuritis. OCT can also detect NFL thinning at the level of the macula (macular NFL). Compared to controls, macular NFL showed a thinning of −6.18 μm (−8.07 to −4.28; $P < 0{\cdot}0001$) in eyes with optic neuritis and −2.15 μm (−3.15 to −1.15; $P < 0{\cdot}0001$) in eyes without optic neuritis. This outcome emphasizes the role of pNFL as a measure for axonal degeneration in MS and its potential use as an outcome measure in clinical trials. This is fundamental because axonal loss, in contrast with demyelination, is not reversible, representing an important cause of sustained disability. Longitudinal studies investigated annual pNFL atrophy rate, but the results are inconsistent partly due to differences in the demographic data.[69] Patients with MS in the early stages of disease, irrespective of previous episodes of optic neuritis, showed the highest annual atrophy rate of the pNFL. Conversely, patients with longer disease duration showed an attenuation of atrophy, emphasizing the importance of early intervention to prevent such injury.[70]

It has been found that temporal peripapillary NFL (T-pNFL) thinning is more sensitive to reflect the neurodegenerative aspect of MS compared to the average pNFL, especially in patients with PPMS. A possible explanation could be that the small parvocellular axons that constitute this region are likely to be more susceptible to damage and preferential loss.[71,72] A limitation of pNFL is that it increases in thickness in up to 82% of cases, possibly due to optic disc swelling at presentation.[73]

Recently, OCT has allowed the segmentation and measurement of GCL thickness, which could potentially reflect ganglion cell degeneration.[63] The macular ganglion cells are the first-order sensory neurons of the visual pathway, and their axons form the pNFL. Most current algorithms measure the thickness of the composite GCL and IPL (GCIPL) as resulted

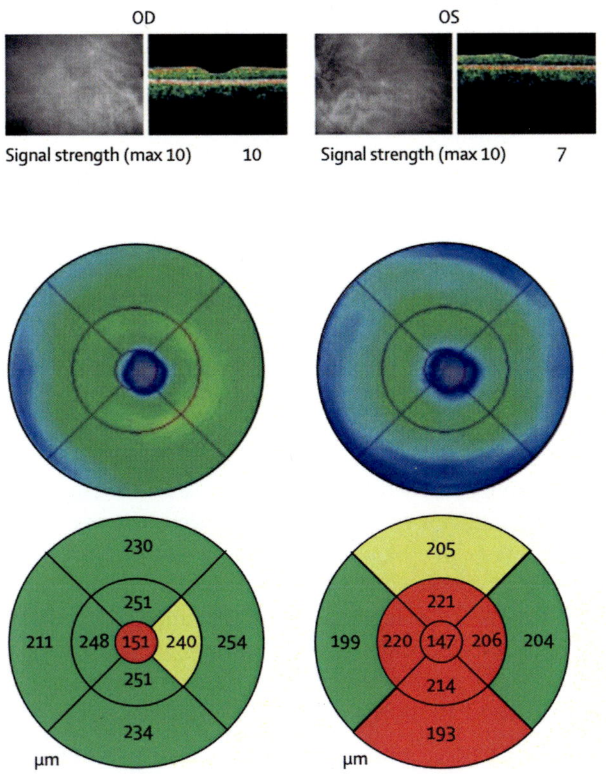

	Parameter	OD	OS	Diff (OD-OS)
Thickness	Foveal minimum	122	123	−1
Mean retinal thickness (μm)	Foveal	151	147	4
	Temporal inner macula	248	206	42
	Superior inner macula	251	221	30
	Nasal inner macula	240	220	20
	Inferior inner macula	251	214	37
	Temporal outer macula	211	204	7
	Superior outer macula	230	205	25
	Nasal outer macula	254	199	55
	Inferior outer macula	234	193	41
	Superior/inferior outer	0·983	1·062	−0·79
	Temporal/nasal inner	1·033	0·936	0·097
	Temporal/nasal outer	0·831	1·025	−0·194
Volume (cm³)	Fovea	0·118	0·115	0·003
	Temporal inner macula	0·39	0·323	0·067
	Superior inner macula	0·395	0·347	0·048
	Nasal inner macula	0·378	0·345	0·033
	Inferior inner macula	0·395	0·337	0·058
	Temporal outer macula	1·122	1·084	0·038
	Superior outer macula	1·219	1·088	0·131
	Nasal outer macula	1·351	1·058	0·293
	Inferior outer macula	1·24	1·023	0·217
	Total macula volume	6·614	5·725	0·889

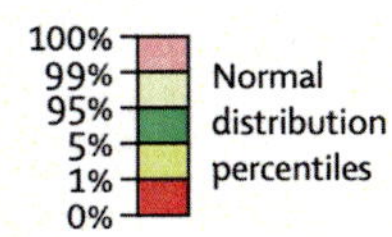

FIGURE 28.1 OCT OCT assessment from a patient with a remote (>2 years) history of acute optic neuritis. Bottom: we present corresponding data on the reduced total macular volume from the same eye, signifying ganglion cell loss. Note that the fovea centralis is reduced in volume on both sides suggesting photoreceptor loss. The right side shows occult involvement. This patient had severe reduction in low-contrast letter acuity (Sloan 1·25% chart) on the affected side.

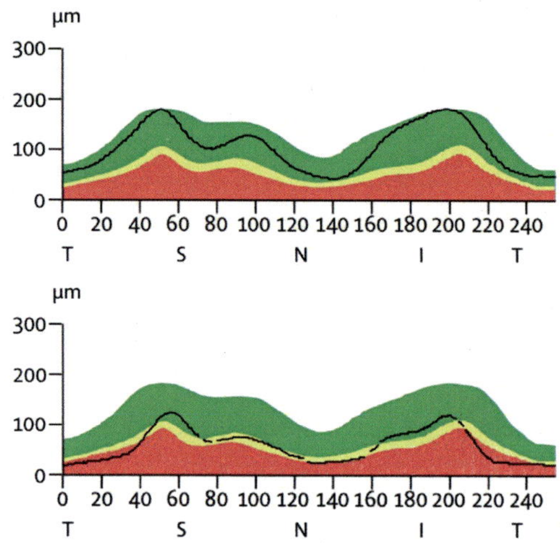

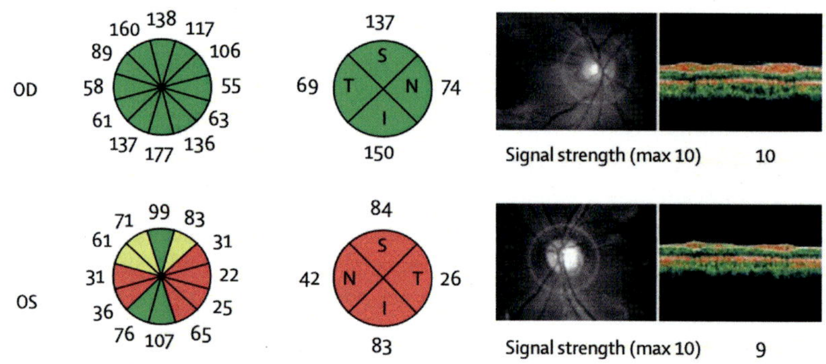

	OD (n=3)	OS (n=3)	OD-OS
Imas/Smax	1·02	0·96	0·06
Smax/Imax	0·98	1·04	−0·06
Smax/Tavg	2·63	4·77	−2·14
Imax/Tavg	2·68	4·58	−1·90
Smax/Navg	2·43	2·93	−0·51
Max-min	138·00	106·00	32·00
Smax	181·00	124·00	57·00
Imax	184·00	119·00	65·00
Savg	137·00	84·00	53·00
Iavg	150·00	83·00	67·00
Mean thickness	107·62	58·83	48·78

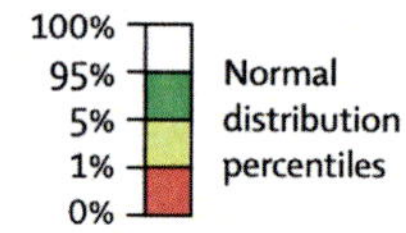

FIGURE 28.2 OCT assessment from a patient with a remote (>2 years) history of acute optic neuritis. Note the reduction in the average RNFL on the affected side compared with the unaffected eye. *I*, inferior; *N*, nasal; *OD*, right eye; *OS*, left eye; *S*, superior; *T*, temporal.

more reliable.[63] Atrophy of the GCIPL was −16.42 µm (−19.23 to −13.60; $P < 0{\cdot}0001$) and −6.31 µm (−7.75 to −4.87; $P < 0{\cdot}0001$) for eyes with and without optic neuritis, respectively, compared with control eyes.[69] Similar to NFL thinning, GCIPL atrophy was found to be more pronounced early in the course of the disease.[70] However, the GCIPL atrophy becomes detectable much earlier than pNFL thinning, already at 1 month after optic neuritis, and does not appear to be affected by edema of the optic disk.[73,74] Therefore, GCIPL could be considered a more sensitive biomarker for early atrophy and neurodegeneration in the visual pathway compared with the pNFL, especially during or following acute stages of optic neuritis.[69,70]

INL showed no atrophy but a small degree of thickening in eyes with optic neuritis compared with healthy eyes (0.77 µm, 0.25−1.28; $P = 0.03$).[75] Increased INL thickness on OCT is associated with the presence of microcystic macular edema (MME) in patients without other reasons for macular edema (that can occur in patients with MS treated with FTY-720 (Fingolimod) or in the context of comorbid clinical uveitis). The presence of MME is mostly transient[75,76] and suggests retinal inflammation due to blood−retinal barrier breakdown and tight junction integrity disruption with fluid accumulation.[77] Another study postulated that MME is caused by retrograde axonal degeneration of the inner retinal layers and consequently impaired fluid absorption in the macula.[78] A third theory suggests that these microcystic-like changes are associated with vitreous traction and do not reflect true edema, excluding the hypothesis of either inflammation or retrograde transsynaptic degeneration.[79] The pathogenic mechanism underlying the MME in MS is still under debate; however, MME is a sign of optic neuropathy, not specific for demyelinating disease,[78] helping to rule out other causes of macular edema associated with diabetes mellitus, neovascular age-related macular degeneration, retinal vein occlusion, and uveitis.[58] Increased INL thickness on OCT is associated with disease activity and worse disability in MS.[75,77] Thus, INL thickness could capture inflammatory disease activity and be a valuable predictor of disease progression in patients with MS and treatment response in clinical trials.[69]

The thickness of the combined ONL and OPL does not show significant alteration in MS, irrespective of the history of optic neuritis,[69] suggesting a potential anatomic and mechanistic limit to neurodegeneration within the retina that deserves further research. In eyes with acute optic neuritis, a small degree of ONL-OPL thickening is observed and is proportional to GCL-IPL thinning. Recognition of deep retinal layer variations is likely to be more relevant for furthering our knowledge of the biological and anatomical sequelae resulting from optic neuritis and its differential diagnosis from other optic neuritis (Figs. 28.1 and 28.2).[80,81]

Retinal hyper-reflecting foci (HRF) have been investigated as novel biomarkers in MS. The most likely hypothesis on their origin, supported by recent studies, is that the HRF might be constituted by clusters of activated microglia, the resident immune cell of the brain and retina, that is, spread throughout the white and gray matter in MS. Indeed, in relapsing−remitting MS, retinal HRF is increased compared to healthy subjects and strongly associated with MRI cortical lesions.[82,83] In addition, HRF count has been linked with soluble markers of microglial origin in CSF and MRI parameters of cortical pathology.[84] These findings support the retinal HRF as promising biomarkers of neuroinflammation worthy of further investigation in MS and in inflammatory and neurodegenerative brain disorders.

Enhanced-depth-imaging (EDI) OCT (EDI-OCT) provides insight into the choroid, which could be altered due to the vascular dysregulation associated with MS.[85–87] Controversial findings have been reported in the literature, with some studies suggesting vascular thinning of the choroid in MS patients,[88–91] while others reporting no difference in the choroidal thickness (CT) between patients with MS and healthy controls, regardless of history of optic neuritis.[91,92] Future studies are needed to clarify the possible role of the choroid in MS, mindful that CT is variably influenced by several factors, including age, intraocular pressure, blood pressure, and axial length.

The thickness of Bruch's membrane opening minimum rim width (BMO-MRW) is a novel measure being examined as additional utility beyond GCIP thickness in MS. It has been considered a structural surrogate for the neuroretinal rim.[93] BMO-MRW and pNFL thicknesses are reduced in MS patients and associated with visual function and disability; however, GCIPL thickness is a more robust marker of visual impairment.[94]

Peripapillary hyper-reflective ovoid mass-like structures (PHOMS) are a novel finding in MS pathology and occur in 10%–18% of eyes affected by MS.[95–97] Current evidence suggests that these PHOMS originate from axoplasmic stasis, congestion in the prelaminar optic nerve head, or impairment of the glymphatic system from the eye through the optic nerve.[95–97] PHOMS have been observed in a broad spectrum of conditions, including papilledema, nonarteritic anterior ischemic optic neuropathy, optic disc drusen, central retinal vein occlusion, optic neuritis, and myopic/tilted optic discs. Thus, they are undoubtedly novel OCT-based biomarkers in many common neuro-ophthalmic disorders, but their study is still in the early stages, and a deeper understanding awaits.

In summary, the segmentation of the retinal layers allowed by OCT shows the progression of neurodegeneration with atrophy affecting axons and neurons. Assessment of pNFL at least 3 months after optic neuritis and macular GCIPL should be included for diagnosis, monitoring, and research on neurodegeneration in MS; assessment of INL could be proven useful in studies focusing on neuroinflammation; macular NFL could be complemental to image the optic disc, although is the least sensitive measure.[69]

OCT-angiography findings in MS

First introduced in 2014, OCT-A is a technique that allows the noninvasive visualization of the retinal and choroidal microvasculature. Therefore, it may supplement OCT for the further understanding of MS (Table 28.4). Current evidence suggests a reduced vessel density (VD), especially in the superficial capillary plexus (SCP) and deep capillary plexus

TABLE 28.4 OCT and OCT-A findings in MS.

OCT	OCT-A
pNFL and macular NFL thinning, greater in MS with ON[74]	Reduced SCP and DCP in both macular and peripapillary areas[104–108]
T-pNFL thinning[76,77]	Reduced ONH perfusion[109]
	Increased volumetric vessel density
GCIPL atrophy, greater in MS with ON[74]	
Annual pNFL and GCIPL atrophy rate greater in the early stages of disease[74]	
INL thickening, associated with MME[81]	
Increased retinal HRF, associated with MRI cortical lesions[88,89]	
BMO-MRW thinning[99,100]	
PHOMS in 10%–18% eyes affected by MS[101–103]	

BMO-MRW, Bruch's membrane opening minimum rim width; *GCIPL*, ganglion cell-inner plexiform layer; *HRF*, hyperreflecting foci; *INL*, inner nuclear layer; *OCT*, optical coherence tomography; *OCT-A*, OCT angiography; *ON*, optic neuropathy; *MME*, microcystic macular edema; *MRI*, magnetic resonance imaging; *MS*, multiple sclerosis; *PHOMS*, Peripapillary hyper-reflective ovoid masslike structures; *p-NFL*, peripapillary nerve fiber layer; *T-pNFL*, temporal peripapillary nerve fiber layer.

(DCP), in the macular and peripapillary areas in eyes affected with MS,[98–102] but further studies are required to support these findings.

Spain et al.[103] studied optic nerve damage in patients with MS with and without a history of optic neuritis and found both retinal structural loss using OCT and decreased optic nerve head perfusion using OCT-A, suggesting OCT-A as a useful tool for additional retinal evaluation.

In addition, OCT-A has been used as a potential marker for disability and for monitoring disease progression.[98–100] Jiang et al.[99] studied the volumetric vessel density (VVD) in the intraretinal layers of eyes in patients with MS and showed that VVD was increased in these patients. Also, they found that VVD in the retinal vascular network and SCP were related to visual function and disability, suggesting a potential use for tracking disease progression. However, studies on the assessment of SCP, DCP, choriocapillaris, and FAZ parameters yielded contradictory findings.[102,104] Further research is required to better define these parameters and their correlation with the visual function and disability status in MS. The applications of OCT-A in MS are rising, and the field of implementation is wide; however, it is unquestionable that the existing data on OCTA-based biomarkers are still limited and targets for future research.

Future directions

In recent years, artificial intelligence (AI) and machine learning (ML) models have been tested to predict, classify, and support the clinical management of different eye diseases.[105] AI has also been explored to predict systemic, nonocular disorders from retinal or ophthalmic images, representing a nascent field called *oculomics*.[106] Indeed, the retinal microvasculature provides an indirect representation of systemic vascular conditions, giving clinically valuable insights to predict neurodegenerative diseases, including multiple sclerosis (MS), Alzheimer's disease and Parkinson's disease, and macrovascular disorders such as stroke and ischemic heart disease.[105,106]

Recent studies have investigated the application of ML models with OCT in MS with promising results[107–110] and we anticipate that the development of automatic AI-based algorithms will become more relevant for the early diagnosis, differential diagnosis, and prediction of disability progression in MS.

Future advances in this field could include the development of digital twins for the implementation of individualized, innovative management of MS that is necessary to cope with the chronic and multidimensional aspects of this disease.[111] AI-based analysis of large amounts of patient data regarding clinical outcomes, biomarkers, treatment plans, multi-omics, and relevant information about health status can create a digital twin paired with an individual patient. This revolutionary tool will contribute to realizing precision medicine and patient-centered care.[111]

Overview of treatment

Treatment of MS includes acute exacerbations treatment, disease-modifying therapies (DMTs), symptom management, comorbidity management, rehabilitation, psychological support, and lifestyle modification.[112]

Treatment of acute exacerbations of multiple sclerosis in adults

Treatment of MS attack is indicated with the primary goal of speeding recovery time from the exacerbation. However, it apparently cannot improve long-term disability or reduce the risk of subsequent relapses.[113]Given that MS attacks lead to a measurable and sustained effect on disability,[114] patients who manifest these clinical episodes should begin DMTs. Typically, initial therapy is with glucocorticoids after having ruled out acute infection. Intravenous methylprednisolone 500–1000 mg or oral prednisone 1000–1250 mg daily for a 3–7-day course is usually first-line treatment. Oral administration of high-dose steroids appears to be not inferior to intravenous ones for improvement of disability and with a similar safety profile.[115] Plasma exchange is considered a second-line treatment, if severe neurologic deficits caused my MS relapses persist despite initial treatment with high-dose glucocorticoids.[116] It is administered daily for five to 7 days or every 2 days for a total of three to seven treatments. Alternative therapies include intramuscular or subcutaneous injection gel of adrenocorticotropic hormone (ACTH)[117] and immunoadsorption,[118] that is, a method of apheresis to bind and remove circulating autoantibodies selectively.

Disease-modifying therapies for multiple sclerosis

There is evidence that DMTs for MS reduce the frequency of MS attacks, decreasing the disability progression over time.[112,119–121]

Given the autoimmune inflammatory dysregulations underlying the complex pathogenesis of the disease, DMTs are immunomodulating and immunosuppressive agents that primarily target neuroinflammation. DMTs might also slow down neurodegeneration in MS, but data are still limited.[112]

The DMTs for relapsing-remitting MS can be categorized by route of administration into infusional, oral, and injectable agents.[121]

- Monoclonal antibodies (mAb)[121] include natalizumab, ocrelizumab (for patients with progressive MS[122]), rituximab (off-label), ofatumumab (subcutaneous injection), and alemtuzumab. mAbs have higher efficacy than oral therapies and injectable drugs with a greater reduction of relapse rate.
- Oral therapies for MS[121] include fumarated (dimethyl fumarate, diroximel fumarate, monomethyl fumarate, teriflunomide), sphingosine 1-phosphate (S1P) receptor modulators (fingolimod, siponimod, ozanimod, ponesimod), and cladribine.[123]
- Older injectable (intramuscular and subcutaneous) therapies[121] are recombinant human interferon β-1b, recombinant human interferon β-1a, and glatiramer acetate. These drugs were the first DMTs approved for relapsing-remitting MS.

A number of issues should be considered for the DMT choice, including drug efficacy, safety, tolerability, and individual patient's preferences. Novel treatment strategies under evaluation include hematopoietic and mesenchymal stem-cell therapies and remyelination therapies for the potential improvement of MS.[112]

Conclusions

MS is the most common inflammatory and demyelinating disease of the CNS.[1,2] Among CNS diseases, MS is the first cause of nontraumatic and permanent disability in young adults.[5] Although MS etiology is not fully understood, it is a complex disease in which complex gene—environment interactions certainly play a significant role. The incidence of MS is increasing worldwide, as the impact on socioeconomic aspects and people's quality of life. Early diagnosis and follow-up are fundamental for optimal disease management, aiming at delaying MS progression and disability.

MS displays ocular signs and symptoms due to direct and/or retrograde degeneration of neurons within the visual pathways.[124] Optic neuritis is one of the most frequent clinical manifestations of MS, often representing an early sign. The retina and optic nerve are properly considered an embryological extension of the brain, with neural and vascular similarities, and therefore it is of particular interest to leverage the rapidly advancing retinal imaging technology to detect and monitor axonal loss or related neurodegeneration processes linked to MS. Imaging devices such as OCT and OCT-A can provide such information rapidly, noninvasively, at a high spatial resolution, at a relatively low cost, and in clinical settings, compared to conventional neuroimaging technologies, including MRI or positron emission tomography.[125]

Therefore, if appropriate imaging biomarkers are identified, retinal imaging might prove useful in at least three different settings.[125] First, it may provide new insights into the underlying MS etiopathogenesis. Second, it may be used as a screening tool, detecting preclinical alterations in the disease process earlier than neuroimaging devices, and it would be essential to obtain effective therapies. Third, it can be used as an endpoint in clinical trials, if the biomarkers are predictive of therapeutic treatments.[125]

Beyond this, AI has the potential to be exploited not only to assist in the diagnosis and monitoring of retinal alterations reported on OCT and other imaging modalities in MS, but also to detect novel retinal and vascular features not discoverable by the human eye.[126]

In conclusion, ophthalmologists, neurologists, and clinician-scientists should be aware of the great importance lying in the extensive work that has been done on retinal and optic nerve imaging as a biomarker of MS. The potential of retinal imaging in MS and other CNS disorders is very promising but has yet to be fully realized.

References

1. Compston A, Coles A. Multiple sclerosis. *The Lancet*. 2008;372(9648):1502—1517. https://doi.org/10.1016/S0140-6736(08)61620-7.
2. Weiner HL. Multiple sclerosis is an inflammatory T-cell—mediated autoimmune disease. *Archives of Neurology*. 2004;61(10):1613. https://doi.org/10.1001/archneur.61.10.1613.
3. Lassmann H. Axonal injury in multiple sclerosis. *Journal of Neurology Neurosurgery and Psychiatry*. 2003;74(6):695—697. https://doi.org/10.1136/jnnp.74.6.695.
4. Bitsch A, Schuchardt J, Bunkowski S, Kuhlmann T, Brück W. Acute axonal injury in multiple sclerosis. Correlation with demyelination and inflammation. *Brain*. 2000;123(6):1174—1183. https://doi.org/10.1093/brain/123.6.1174.

5. Ramagopalan SV, Sadovnick AD. Epidemiology of multiple sclerosis. *Neurologic Clinics*. 2011;29(2):207−217. https://doi.org/10.1016/j.ncl.2010.12.010.
6. Yeh, Marrie RA, Reginald y J, et al. Canadian Pediatric Demyelinating Disease Network. Functional-structural correlations in the afferent visual pathway in pediatric demyelination. *Neurology*. 2014;83(23). https://doi.org/10.1212/WNL.0000000000001046. Epub.
7. Goodin DS. The epidemiology of multiple sclerosis: insights to disease pathogenesis. *Handbook of Clinical Neurology*. 2014;122.
8. Simpson S, Wang W, Otahal P, Blizzard L, Van Der Mei IAF, Taylor BV. Latitude continues to be significantly associated with the prevalence of multiple sclerosis: an updated meta-analysis. *Journal of Neurology, Neurosurgery and Psychiatry*. 2019;90(11):1193−1200. https://doi.org/10.1136/jnnp-2018-320189.
9. Dilokthornsakul P, Valuck RJ, Nair KV, Corboy JR, Allen RR, Campbell JD. Multiple sclerosis prevalence in the United States commercially insured population. *Neurology*. 2016;86(11):1014−1021. https://doi.org/10.1212/WNL.0000000000002469.
10. Global, regional, and national burden of multiple sclerosis 1990-2016: a systematic analysis for the Global Burden of Disease Study. *GBD 2016 Multiple Sclerosis Collaborators*. 2016;18.
11. Browne P, Chandraratna D, Angood C, et al. Atlas of multiple sclerosis 2013: a growing global problem with widespread inequity. *Neurology*. 2014;83(11):1022−1024. https://doi.org/10.1212/WNL.0000000000000768.
12. Goris A, Vandebergh M, McCauley JL, Saarela J, Cotsapas C. Genetics of multiple sclerosis: lessons from polygenicity. *The Lancet Neurology*. 2022;21(9):830−842. https://doi.org/10.1016/S1474-4422(22)00255-1.
13. Ascherio A. Environmental factors in multiple sclerosis. *Expert Review of Neurotherapeutics*. 2013;13(12):3−9. https://doi.org/10.1586/14737175.2013.865866.
14. Ascherio A, Munger KL. Environmental risk factors for multiple sclerosis. Part I: the role of infection. *Annals of Neurology*. 2007;61(4):288−299. https://doi.org/10.1002/ana.21117.
15. Lucas RM, Ponsonby AL, Dear K, et al. Sun exposure and vitamin D are independent risk factors for CNS demyelination. *Neurology*. 2011;76(6):540−548. https://doi.org/10.1212/WNL.0b013e31820af93d.
16. Riise T, Nortvedt MW, Ascherio A. Smoking is a risk factor for multiple sclerosis. *Neurology*. 2003;61(8):1122−1124. https://doi.org/10.1212/01.WNL.0000081305.66687.D2.
17. Balk LJ, Cruz-Herranz A, Albrecht P, et al. Timing of retinal neuronal and axonal loss in MS: a longitudinal OCT study. *Journal of Neurology*. 2016;263(7):1323−1331. https://doi.org/10.1007/s00415-016-8127-y.
18. DeStefano F, Verstraeten T, Jackson LA, et al. Vaccinations and risk of central nervous system demyelinating diseases in adults. *Archives of Neurology*. 2003;60(4):504−509. https://doi.org/10.1001/archneur.60.4.504.
19. Popescu BFG, Pirko I, Lucchinetti CF. Pathology of multiple sclerosis: where do we stand? *CONTINUUM Lifelong Learning in Neurology*. 2013;19(4):901−921. https://doi.org/10.1212/01.CON.0000433291.23091.65.
20. Popescu BFG, Lucchinetti CF. Pathology of demyelinating diseases. *Annual Review of Pathology: Mechanisms of Disease*. 2012;7:185−217. https://doi.org/10.1146/annurev-pathol-011811-132443.
21. Saidha S, Sotirchos ES, Ibrahim MA, et al. Microcystic macular oedema, Thickness of the inner nuclear layer of the retina, and disease characteristics in multiple sclerosis: a retrospective study. *The Lancet Neurology*. 2012;11(11):963−972. https://doi.org/10.1016/S1474-4422(12)70213-2.
22. Kutzelnigg A, Lassmann H. Pathology of multiple sclerosis and related inflammatory demyelinating diseases. *Handbook of Clinical Neurology*. 2014;122:15−58. https://doi.org/10.1016/B978-0-444-52001-2.00002-9.
23. Lassmann H. Multiple sclerosis pathology: evolution of pathogenetic concepts. *Brain Pathology*. 2005;15(3):217−222. https://doi.org/10.1111/j.1750-3639.2005.tb00523.x.
24. Pirko I, Lucchinetti CF, Sriram S, Bakshi R. Gray matter involvement in multiple sclerosis. *Neurology*. 2007;68(9):634−642. https://doi.org/10.1212/01.wnl.0000250267.85698.7a.
25. Chandran S, Hunt D, Joannides A, Zhao C, Compston A, Franklin RJM. Myelin repair: the role of stem and precursor cells in multiple sclerosis. *Philosophical Transactions of the Royal Society B: Biological Sciences*. 2008;363(1489):171−183. https://doi.org/10.1098/rstb.2006.2019.
26. Lublin FD, Reingold SC, Cohen JA, et al. Defining the clinical course of multiple sclerosis: the 2013 revisions. *Neurology*. 2014;83(3):278−286. https://doi.org/10.1212/WNL.0000000000000560.
27. Thompson AJ, Banwell BL, Barkhof F, et al. Diagnosis of multiple sclerosis: 2017 revisions of the McDonald criteria. *The Lancet Neurology*. 2018;17(2):162−173. https://doi.org/10.1016/S1474-4422(17)30470-2.
28. Hirst C, Ingram G, Pickersgill T, Swingler R, Compston DAS, Robertson NP. Increasing prevalence and incidence of multiple sclerosis in south east wales. *Journal of Neurology, Neurosurgery and Psychiatry*. 2009;80(4):386−391. https://doi.org/10.1136/jnnp.2008.144667.
29. Miller DH, Leary SM. Primary-progressive multiple sclerosis. *Lancet Neurology*. 2007;6(10):903−912. https://doi.org/10.1016/S1474-4422(07)70243-0.
30. Kurtzke JF. Rating neurologic impairment in multiple sclerosis: an expanded disability status scale (EDSS). *Neurology*. 1983;33(11):1444−1452. https://doi.org/10.1212/wnl.33.11.1444.
31. Arnold AC. Evolving management of optic neuritis and multiple sclerosis. *American Journal of Ophthalmology*. 2005;139(6):1101−e1. https://doi.org/10.1016/j.ajo.2005.01.031.
32. Balcer LJ. Clinical practice. Optic neuritis. *The New England Journal of Medicine*. 2006;354.
33. Foroozan R, Buono LM, Savino PJ, Sergott RC. Acute demyelinating optic neuritis. *Current Opinion Ophthalmology*. 2002;13(6):375−380. https://doi.org/10.1097/00055735-200212000-00006.

34. Frohman EM, Frohman TC, Zee DS, McColl R, Galetta S. The neuro-ophthalmology of multiple sclerosis. *Lancet Neurology*. 2005;4(2):111–121. https://doi.org/10.1016/S1474-4422(05)00992-0.
35. De La Cruz J, Kupersmith MJ. Clinical profile of simultaneous bilateral optic neuritis in adults. *British Journal of Ophthalmology*. 2006;90(5):551–554. https://doi.org/10.1136/bjo.2005.085399.
36. The clinical profile of optic neuritis. Experience of the optic neuritis treatment trial. Optic neuritis study group. *Archives of Ophthalmology*. 1991;109.
37. Lightman S, Mcdonald WI, Bird AC, et al. Retinal venous sheathing in optic neuritis: its significance for the pathogenesis of multiple sclerosis. *Brain*. 1987;110(2):405–414. https://doi.org/10.1093/brain/110.2.405.
38. Costello F, Coupland S, Hodge W, et al. Quantifying axonal loss after optic neuritis with optical coherence tomography. *Annals of Neurology*. 2006;59(6):963–969. https://doi.org/10.1002/ana.20851.
39. Barnes D, McDonald WI. The ocular manifestations of multiple sclerosis 2 Abnormalities of eye movements. *Journal of Neurology, Neurosurgery and Psychiatry*. 1992;55(10):863–868. https://doi.org/10.1136/jnnp.55.10.863.
40. Barton JJS, Cox TA. Acquired pendular nystagmus in multiple sclerosis: clinical observations and the role of optic neuropathy. *Journal of Neurology Neurosurgery and Psychiatry*. 1993;56(3):262–267. https://doi.org/10.1136/jnnp.56.3.262.
41. Reich DS, Lucchinetti CF, Calabresi PA. Multiple sclerosis. *New England Journal of Medicine*. 2018;378(2):169–180. https://doi.org/10.1056/NEJMra1401483.
42. Schumacher GA, Beebe G, Kibler RF, et al. Problems of experimental trials of therapy in multiple sclerosis: report by the panel on the evaluation of experimental trials of therapy in multiple sclerosis. *Annals of the New York Academy of Sciences*. 1965;122(1):552–568. https://doi.org/10.1111/j.1749-6632.1965.tb20235.x.
43. Poser CM, Paty DW, Scheinberg L, et al. New diagnostic criteria for multiple sclerosis: guidelines for research protocols. *Annals of Neurology*. 1983;13(3):227–231. https://doi.org/10.1002/ana.410130302.
44. Brownlee WJ, Miller DH. Clinically isolated syndromes and the relationship to multiple sclerosis. *Journal of Clinical Neuroscience*. 2014;21(12):2065–2071. https://doi.org/10.1016/j.jocn.2014.02.026.
45. Montalban X, Sastre-Garriga J, Filippi M, et al. Primary progressive multiple sclerosis diagnostic criteria: a reappraisal. *Multiple Sclerosis*. 2009;15(12):1459–1465. https://doi.org/10.1177/1352458509348422.
46. Okuda DT, Mowry EM, Beheshtian A, et al. Incidental MRI anomalies suggestive of multiple sclerosis: the radiologically isolated syndrome. *Neurology*. 2009;72(9):800–805. https://doi.org/10.1212/01.wnl.0000335764.14513.1a.
47. Okuda DT, Siva A, Kantarci O, et al. Radiologically isolated syndrome: 5-year risk for an initial clinical event. *PLoS One*. 2014;9(3):e90509. https://doi.org/10.1371/journal.pone.0090509.
48. Freedman MS, Thompson EJ, Deisenhammer F, et al. Recommended standard of cerebrospinal fluid analysis in the diagnosis of multiple sclerosis: a consensus statement. *Archives of Neurology*. 2005;62(6):865–870. https://doi.org/10.1001/archneur.62.6.865.
49. Brownlee WJ, Hardy TA, Fazekas F, Miller DH. Diagnosis of multiple sclerosis: progress and challenges. *Lancet*. 2017;389(10076):1336–1346. https://doi.org/10.1016/S0140-6736(16)30959-X.
50. Alonso A, Hernán MA. Temporal trends in the incidence of multiple sclerosis: a systematic review. *Neurology*. 2008;71(2):129–135. https://doi.org/10.1212/01.wnl.0000316802.35974.34.
51. Leocani L, Comi G. Clinical neurophysiology of multiple sclerosis. *Handbook of Clinical Neurology*. 2014;122:671–679. https://doi.org/10.1016/B978-0-444-52001-2.00028-5.
52. Trobe JD, Beck RW, Moke PS, Cleary PA. Contrast sensitivity and other vision tests in the optic neuritis treatment trial. *American Journal of Ophthalmology*. 1996;121(5):547–553. https://doi.org/10.1016/S0002-9394(14)75429-7.
53. Frohman E, Costello F, Zivadinov R, et al. Optical coherence tomography in multiple sclerosis. *Lancet Neurology*. 2006;5(10):853–863. https://doi.org/10.1016/S1474-4422(06)70573-7.
54. Osborne B, Balcer LJ. Optic neuritis: pathophysiology, clinical features, and diagnosis.
55. Bermel RA. Unravelling neurodegeneration in multiple sclerosis. *The Lancet Neurology*. 2017;16(10):764–766. https://doi.org/10.1016/S1474-4422(17)30302-2.
56. Galetta KM, Calabresi PA, Frohman EM, Balcer LJ. Optical coherence tomography (OCT): imaging the visual pathway as a model for neurodegeneration. *Neurotherapeutics*. 2011;8(1):117–132. https://doi.org/10.1007/s13311-010-0005-1.
57. Petzold A, de Boer JF, Schippling S, et al. Optical coherence tomography in multiple sclerosis: a systematic review and meta-analysis. *The Lancet Neurology*. 2010;9(9):921–932. https://doi.org/10.1016/S1474-4422(10)70168-X.
58. Vujosevic S, Parra MM, Hartnett ME, et al. Optical coherence tomography as retinal imaging biomarker of neuroinflammation/neurodegeneration in systemic disorders in adults and children. *Eye (Basingstoke)*. 2022. https://doi.org/10.1038/s41433-022-02056-9.
59. Huang D, Swanson EA, Lin CP, et al. Optical coherence tomography. *Science*. 1991;254(5035):1178–1181. https://doi.org/10.1126/science.1957169.
60. Ziemssen T, Ziemssen F. Perspectives of an innovative ophthalmological technology: optical coherence tomography (OCT) - what should be of interest to the neurologist? *Clinical Neurology and Neurosurgery*. 2013;115(1):S55–S59. https://doi.org/10.1016/j.clineuro.2013.09.022.
61. London A, Benhar I, Schwartz M. The retina as a window to the brain - from eye research to CNS disorders. *Nature Reviews Neurology*. 2013;9(1):44–53. https://doi.org/10.1038/nrneurol.2012.227.
62. Satue M, Obis J, Rodrigo MJ, et al. Optical coherence tomography as a biomarker for diagnosis, progression, and prognosis of neurodegenerative diseases. *Journal of Ophthalmology*. 2016;2016. https://doi.org/10.1155/2016/8503859.

63. Britze J, Pihl-Jensen G, Frederiksen JL. Retinal ganglion cell analysis in multiple sclerosis and optic neuritis: a systematic review and meta-analysis. *Journal of Neurology*. 2017;264(9):1837−1853. https://doi.org/10.1007/s00415-017-8531-y.
64. Frohman EM, Fujimoto JG, Frohman TC, Calabresi PA, Cutter G, Balcer LJ. Optical coherence tomography: a window into the mechanisms of multiple sclerosis. *Nature Clinical Practice Neurology*. 2008;4(12):664−675. https://doi.org/10.1038/ncpneuro0950.
65. Parisi V, Manni G, Spadaro M, et al. Correlation between morphological and functional retinal impairment in multiple sclerosis patients. *Investigative Ophthalmology and Visual Science*. 1999;40(11):2520−2527.
66. Cruz-Herranz A, Balk LJ, Oberwahrenbrock T, et al. The APOSTEL recommendations for reporting quantitative optical coherence tomography studies. *Neurology*. 2016;86(24):2303−2309. https://doi.org/10.1212/WNL.0000000000002774.
67. Gelfand JM, Nolan R, Schwartz DM, Graves J, Green AJ. Microcystic macular oedema in multiple sclerosis is associated with disease severity. *Brain*. 2012;135(6):1786−1793. https://doi.org/10.1093/brain/aws098.
68. Jacobs BM, Noyce AJ, Giovannoni G, Dobson R. BMI and low vitamin D are causal factors for multiple sclerosis: a Mendelian Randomization study. *Neurology(R) neuroimmunology & neuroinflammation*. 2020;7(2). https://doi.org/10.1212/NXI.0000000000000662.
69. Petzold A, Balcer L, Calabresi PA, et al. Retinal layer segmentation in multiple sclerosis: a systematic review and meta-analysis. *The Lancet Neurology*. 2017;16(10):797−812. https://doi.org/10.1016/S1474-4422(17)30278-8.
70. Garcia-Martin E, Jarauta L, Pablo LE, et al. Changes in peripapillary choroidal thickness in patients with multiple sclerosis. *Acta Ophthalmologica*. 2019;97(1):e77−e83. https://doi.org/10.1111/aos.13807.
71. Evangelou N, Konz D, Esiri MM, Smith S, Palace J, Matthews PM. Size-selective neuronal changes in the anterior optic pathways suggest a differential susceptibility to injury in multiple sclerosis. *Brain*. 2001;124(9):1813−1820. https://doi.org/10.1093/brain/124.9.1813.
72. Birkeldh U, Manouchehrinia A, Hietala MA, et al. The temporal retinal nerve fiber layer thickness is the most important optical coherence tomography estimate in multiple sclerosis. *Frontiers in Neurology*. 2017;8. https://doi.org/10.3389/fneur.2017.00675.
73. Kupersmith MJ, Garvin MK, Wang JK, Durbin M, Kardon R. Retinal ganglion cell layer thinning within one month of presentation for optic neuritis. *Multiple Sclerosis*. 2016;22(5):641−648. https://doi.org/10.1177/1352458515598020.
74. Gabilondo I, Martínez-Lapiscina EH, Fraga-Pumar E, et al. Dynamics of retinal injury after acute optic neuritis. *Annals of Neurology*. 2015;77(3):517−528. https://doi.org/10.1002/ana.24351.
75. Pengo M, Miante S, Franciotta S, et al. Retinal hyperreflecting foci associate with cortical pathology in multiple sclerosis. *Neurology - Neuroimmunology Neuroinflammation*. 2022;9(4):e1180. https://doi.org/10.1212/nxi.0000000000001180.
76. Burggraaff MC, Trieu J, de Vries-Knoppert WAEJ, Balk L, Petzold A. The clinical spectrum of microcystic macular edema. *Investigative Ophthalmology and Visual Science*. 2014;55(2):952−961. https://doi.org/10.1167/iovs.13-12912.
77. Calabrese M, Filippi M, Gallo P. Cortical lesions in multiple sclerosis. *Nature Reviews Neurology*. 2010;6(8):438−444. https://doi.org/10.1038/nrneurol.2010.93.
78. Abegg M, Dysli M, Wolf S, Kowal J, Dufour P, Zinkernagel M. Microcystic macular edema: retrograde maculopathy caused by optic neuropathy. *Ophthalmology*. 2014;121(1):142−149. https://doi.org/10.1016/j.ophtha.2013.08.045.
79. Barboni P, Carelli V, Savini G, Carbonelli M, La Morgia C, Sadun AA. Microcystic macular degeneration from optic neuropathy: not inflammatory, not trans-synaptic degeneration. *Brain*. 2013;136(7):e239. https://doi.org/10.1093/brain/awt014.
80. Petzold A, Wattjes MP, Costello F, et al. The investigation of acute optic neuritis: a review and proposed protocol. *Nature Reviews Neurology*. 2014;10(8):447−458. https://doi.org/10.1038/nrneurol.2014.108.
81. Martinez-Lapiscina EH, Sepulveda M, Torres-Torres R, et al. Usefulness of optical coherence tomography to distinguish optic neuritis associated with AQP4 or MOG in neuromyelitis optica spectrum disorders. *Therapeutic Advances in Neurological Disorders*. 2016;9(5):436−440. https://doi.org/10.1177/1756285616655264.
82. Pilotto E, Miante S, Torresin T, et al. Hyperreflective foci in the retina of active relapse-onset multiple sclerosis. *Ophthalmology*. 2020;127(12):1774−1776. https://doi.org/10.1016/j.ophtha.2020.03.024.
83. Dobson R, Ramagopalan S, Davis A, Giovannoni G. Cerebrospinal fluid oligoclonal bands in multiple sclerosis and clinically isolated syndromes: a meta-analysis of prevalence, prognosis and effect of latitude. *Journal of Neurology, Neurosurgery and Psychiatry*. 2013;84(8):909−914. https://doi.org/10.1136/jnnp-2012-304695.
84. Puthenparampil M, Torresin T, Franciotta S, et al. Hyper-reflecting foci in multiple sclerosis retina associate with macrophage/microglia-derived cytokines in cerebrospinal fluid. *Frontiers in Immunology*. 2022;13. https://doi.org/10.3389/fimmu.2022.852183.
85. Flammer J, Pache M, Resink T. Vasospasm, its role in the pathogenesis of diseases with particular reference to the eye. *Progress in Retinal and Eye Research*. 2001;20(3):319−349. https://doi.org/10.1016/S1350-9462(00)00028-8.
86. Konieczka K, Erb C. Diseases potentially related to Flammer syndrome. *EPMA Journal*. 2017;8(4):327−332. https://doi.org/10.1007/s13167-017-0116-4.
87. Konieczka K, Koch S, Binggeli T, Schoetzau A, Kesselring J. Multiple sclerosis and primary vascular dysregulation (Flammer syndrome). *EPMA Journal*. 2016;7(1). https://doi.org/10.1186/s13167-016-0062-6.
88. Esen E, Sizmaz S, Demir T, Demirkiran M, Unal I, Demircan N. Evaluation of choroidal vascular changes in patients with multiple sclerosis using enhanced depth imaging optical coherence tomography. *Ophthalmologica*. 2016;235(2):65−71. https://doi.org/10.1159/000441152.
89. Osborne B, Balcer LJ. *Optic Neuritis: Prognosis and Treatment*; 2022. https://www-uptodate-com.pros2.lib.unimi.it/contents/optic-neuritis-prognosis-and-treatment?search=optic%20neuritis&source=search_result&selectedTitle=2~150&usage_type=default&display_rank=2. Accessed December 15, 2022.

90. Garcia-Martin E, Jarauta L, Vilades E, et al. Ability of swept-source optical coherence tomography to detect retinal and choroidal changes in patients with multiple sclerosis. *Journal of Ophthalmology*. 2018;2018. https://doi.org/10.1155/2018/7361212.
91. Yuksel B, Dogan B, Koctekin B, et al. Color vision testing versus pattern visual evoked potentials and optical coherence tomography parameters in subclinical optic nerve involvement in multiple sclerosis. *Journal of Clinical Neuroscience*. 2019;61:48–53. https://doi.org/10.1016/j.jocn.2018.11.011.
92. Masala A, Di Mola I, Cellerino M, et al. Choroidal thickness in multiple sclerosis: an optical coherence tomography study. *Journal of Clinical Neurology*. 2022;18(3):334–342. https://doi.org/10.3988/jcn.2022.18.3.334.
93. Reis ASC, O'Leary N, Yang H, et al. Influence of clinically invisible, but optical coherence tomography detected, optic disc margin anatomy on neuroretinal rim evaluation. *Investigative Ophthalmology & Visual Science*. 2012;53(4):1852–1860. https://doi.org/10.1167/iovs.11-9309.
94. Nguyen J, Rothman A, Gonzalez N, et al. Macular ganglion cell and inner plexiform layer thickness is more strongly associated with visual function in multiple sclerosis than Bruch membrane opening–minimum rim width or peripapillary retinal nerve fiber layer thicknesses. *Journal of Neuro-Ophthalmology*. 2019;39(4):444–450. https://doi.org/10.1097/wno.0000000000000768.
95. Fraser JA, Sibony PA, Petzold A, Thaung C, Hamann S. Peripapillary hyper-reflective ovoid mass-like structure (PHOMS): an optical coherence tomography marker of axoplasmic stasis in the optic nerve head. *Journal of Neuroophthalmology*. 2021. https://doi.org/10.1097/WNO.0000000000001203.
96. Petzold A, Coric D, Balk LJ, et al. Longitudinal development of peripapillary hyper-reflective ovoid masslike structures suggests a novel pathological pathway in multiple sclerosis. *Annals of Neurology*. 2020;88(2):309–319. https://doi.org/10.1002/ana.25782.
97. Wicklein R, Wauschkuhn J, Giglhuber K, et al. Association of peripapillary hyper-reflective ovoid masslike structures and disease duration in primary progressive multiple sclerosis. *European Journal of Neurology*. 2021;28(12). https://doi.org/10.1111/ene.15056.
98. Feucht N, Maier M, Lepennetier G, et al. Optical coherence tomography angiography indicates associations of the retinal vascular network and disease activity in multiple sclerosis. *Multiple Sclerosis Journal*. 2019;25(2):224–234. https://doi.org/10.1177/1352458517750009.
99. Jiang H, Gameiro GR, Liu Y, et al. Visual function and disability are associated with increased retinal volumetric vessel density in patients with multiple sclerosis. *American Journal of Ophthalmology*. 2020;213:34–45. https://doi.org/10.1016/j.ajo.2019.12.021.
100. Murphy OC, Kwakyi O, Iftikhar M, et al. Alterations in the retinal vasculature occur in multiple sclerosis and exhibit novel correlations with disability and visual function measures. *Multiple Sclerosis Journal*. 2020;26(7):815–828. https://doi.org/10.1177/1352458519845116.
101. Ulusoy MO, Horasanlõ B, Isõk-Ulusoy S. Optical coherence tomography angiography findings of multiple sclerosis with or without optic neuritis. *Neurology Research*. 2020:319–326.
102. Yilmaz H, Ersoy A, Icel E. Assessments of vessel density and foveal avascular zone metrics in multiple sclerosis: an optical coherence tomography angiography study. *Eye*. 2020;34(4):771–778. https://doi.org/10.1038/s41433-019-0746-y.
103. Spain RI, Liu L, Zhang X, et al. Optical coherence tomography angiography enhances the detection of optic nerve damage in multiple sclerosis. *British Journal of Ophthalmology*. 2018;102(4):520–524. https://doi.org/10.1136/bjophthalmol-2017-310477.
104. Aly L, Noll C, Wicklein R, et al. Dynamics of retinal vessel loss after acute optic neuritis in patients with relapsing multiple sclerosis. *Neurology - Neuroimmunology Neuroinflammation*. 2022;9(3):e1159. https://doi.org/10.1212/nxi.0000000000001159.
105. Betzler BK, Rim TH, Sabanayagam C, Cheng CY. Artificial intelligence in predicting systemic parameters and diseases from ophthalmic imaging. *Frontiers in Digital Health*. 2022;4. https://doi.org/10.3389/fdgth.2022.889445.
106. Wagner SK, Fu DJ, Faes L, et al. Insights into systemic disease through retinal imaging-based oculomics. *Translational Vision Science and Technology*. 2020;9(2). https://doi.org/10.1167/tvst.9.2.6.
107. López-Dorado A, Ortiz M, Satue M, et al. Early diagnosis of multiple sclerosis using swept-source optical coherence tomography and convolutional neural networks trained with data augmentation. *Sensors*. 2022;22(1):167.
108. Nabizadeh F, Ramezannezhad E, Kargar A, Sharafi AM, Ghaderi A. Diagnostic performance of artificial intelligence in multiple sclerosis: a systematic review and meta-analysis. *Neurological Sciences*. 2023;44(2):499–517. https://doi.org/10.1007/s10072-022-06460-7.
109. Palomar \tPrez del, Cegoñino J, et al. Swept source optical coherence tomography to early detect multiple sclerosis disease. The use of machine learning techniques. *PLoS One*. 2019;14(5).
110. Toosy AT, Eshaghi A. Machine learning utility for optical coherence tomography in multiple sclerosis: is the future now? *Neurology*. 2022;99(11):453–454. https://doi.org/10.1212/WNL.0000000000200862.
111. Voigt I, Inojosa H, Dillenseger A, Haase R, Akgün K, Ziemssen T. Digital twins for multiple sclerosis. *Frontiers in Immunology*. 2021;12. https://doi.org/10.3389/fimmu.2021.669811.
112. McGinley MP, Goldschmidt CH, Rae-Grant AD. Diagnosis and treatment of multiple sclerosis: a review. *JAMA - Journal of the American Medical Association*. 2021;325(8):765–779. https://doi.org/10.1001/jama.2020.26858.
113. Brusaferri F, Candelise L. Steroids for multiple sclerosis and optic neuritis: a meta-analysis of randomized controlled clinical trials. *Journal of Neurology*. 2000;247(6):435–442. https://doi.org/10.1007/s004150070172.
114. Lublin FD, Baier M, Cutter G. Effect of relapses on development of residual deficit in multiple sclerosis. *Neurology*. 2003;61(11):1528–1532. https://doi.org/10.1212/01.WNL.0000096175.39831.21.
115. LePage E, Veillard D, Laplaud DA, Investigators, et al. West Network for Excellence in Neuroscience. Oral versus intravenous high-dose methylprednisolone for treatment of relapses in patients with multiple sclerosis (COPOUSEP):arandomised,controlled, double-blind, non-inferiority trial. *Lancet*. 2015;386(9997):974–981.
116. Arrambide G, Iacobaeus E, Amato MP, ECTRIMS Focused Workshop Group, et al. Aggressive multiple sclerosis (2): treatment. *Multiple Sclerosis*. 2018;26.

117. Thompson AJ, Kennard C, Swash M, et al. Relative efficacy of intravenous methylprednisolone and ACTH in the treatment of acute relapse in MS. *Neurology*. 1989;39(7):969–971. https://doi.org/10.1212/wnl.39.7.969.
118. Dorst J, Fangerau T, Taranu D, et al. Safety and efficacy of immunoadsorption versus plasma exchange in steroid-refractory relapse of multiple sclerosis and clinically isolated syndrome: a randomised, parallel-group, controlled trial. *EClinicalMedicine*. 2019;16:98–106. https://doi.org/10.1016/j.eclinm.2019.10.017.
119. Amato MP, Fonderico M, Portaccio E, et al. Disease-modifying drugs can reduce disability progression in relapsing multiple sclerosis. *Brain*. 2020;143(10):3013–3024. https://doi.org/10.1093/brain/awaa251.
120. Claflin SB, Broadley S, Taylor BV. The effect of disease modifying therapies on disability progression in multiple sclerosis: a systematic overview of meta-analyses. *Frontiers in Neurology*. 2019;10. https://doi.org/10.3389/fneur.2018.01150.
121. Olek M,J, Jonathan DO, Howard MD. Treatment of acute exacerbations of multiple sclerosis in adults. *Uptodate*. 2022. last updated.
122. FDA approves new drug to treat multiple sclerosis. https://www-fda-gov.pros2.lib.unimi.it/NewsEvents/Newsroom/PressAnnouncements/ucm549325.htm. Accessed 20 December 2022.
123. Derfuss T, Mehling M, Papadopoulou A, Bar-Or A, Cohen JA, Kappos L. Advances in oral immunomodulating therapies in relapsing multiple sclerosis. *The Lancet Neurology*. 2020;19(4):336–347. https://doi.org/10.1016/S1474-4422(19)30391-6.
124. Gupta S, Zivadinov R, Ramanathan M, Weinstock-Guttman B. Optical coherence tomography and neurodegeneration: are eyes the windows to the brain? *Expert Review of Neurotherapeutics*. 2016;16(7):765–775. https://doi.org/10.1080/14737175.2016.1180978.
125. Kashani AH, Asanad S, Chan JW, et al. Past, present and future role of retinal imaging in neurodegenerative disease. *Progress in Retinal and Eye Research*. 2021;83. https://doi.org/10.1016/j.preteyeres.2020.100938.
126. Patil AD, Biousse V, Newman NJ. Artificial intelligence in ophthalmology: an insight into neurodegenerative disease. *Current Opinion in Ophthalmology*. 2022;33(5):432–439. https://doi.org/10.1097/ICU.0000000000000877.

Chapter 29

Sarcoidosis

Aniruddha Agarwal[1,2,3] and Jyotirmay Biswas[4]
[1]Department of Ophthalmology, Cleveland Clinic Abu Dhabi, Abu Dhabi, United Arab Emirates; [2]Cleveland Clinic Lerner College of Medicine, Case Western Reserve University, Cleveland, OH, United States; [3]Department of Ophthalmology, Maastricht University Medical Center+, Maastricht, the Netherlands; [4]Department of Uveitis and Ocular Pathology, Sankara Nethralaya, Chennai, Tamil Nadu, India

Introduction

Sarcoidosis is a multisystem chronic inflammatory disease characterized by the presence of noncaseating granulomas (in contrast to tuberculosis). These granulomas affect the respiratory system, lymph nodes, central nervous system, skin, and various ocular tissues. Sarcoidosis was first recognized by Hutchinson in 1869 and the term "sarkoid" was coined by a Norwegian dermatologist, Caeser Boeck in 1899. The term "sarcoid" was initially used to describe the skin lesions of the disease. Ocular sarcoidosis was first described as a condition presenting with features of facial nerve palsy and uveoparotid fever (Heerfordt's syndrome). As sarcoidosis is a systemic disease, it can present with acute severe inflammation such as Löfgren's syndrome (manifesting as fever, polyarthritis, and erythema nodosum).[1]

Ocular sarcoidosis may occur in up to one-third of patients with the disease at any time during the natural history of the condition (sometimes preceding systemic manifestations). Bilateral uveitis is common in this condition. The hallmark of sarcoidosis is the presence of granulomas, which may involve the periorbita, lacrimal gland, extraocular muscles, and adnexa (conjunctiva and lid). The spectrum of disease manifestations is very wide; neuroophthalmologic signs such as optic neuropathy or Horner's syndrome are also well-known in sarcoidosis.[1–3]

The diagnosis of sarcoidosis is often challenging due to its protean manifestations. The diagnostic criteria for ocular sarcoidosis by the first International Workshop on Sarcoidosis (IWOS) is provided in Table 29.1.[4,5]

Epidemiology of sarcoidosis

Sarcoidosis accounts for nearly 5% of all adult uveitis cases and 1% of pediatric cases (sometimes confused with an entity—Blau's syndrome). Ocular involvement is common in sarcoidosis and as many as 40% of cases may have manifestations in the eye. Therefore, routine screening of patients with sarcoidosis could be helpful. African

TABLE 29.1 Diagnostic criteria[a] for ocular sarcoidosis by the first international workshop on ocular sarcoidosis

Definite ocular[b] sarcoidosis	Biopsy-supported diagnosis with a compatible uveitis
Presumed ocular[b] sarcoidosis	Biopsy not done; presence of BHL with a compatible uveitis
Probable ocular[b] sarcoidosis	Biopsy not done; negative BHL; presence of three suggestive intraocular signs and two positive investigational tests
Possible ocular[b] sarcoidosis	Biopsy negative; four suggestive intraocular signs positive; two positive investigational tests

BHL, bilateral hilar lymphadenopathy.
[a]*All other causes of uveitis, especially tuberculosis must be ruled out in all cases.*
[b]*The term "ocular" is used for both intraocular inflammatory lesions in patients with systemic disease and in patients with disease seemingly limited to the eye without any clinically detectable involvement of any systemic organ.*

Retinal and Choroidal Vascular Diseases of the Eye. https://doi.org/10.1016/B978-0-443-15583-3.00029-9

American race has a higher incidence of sarcoidosis compared to Caucasians. The incidence in the African American population is 35.5 per 100,000 individuals compared to 10.9 per 100,000 individuals in Caucasians. In addition, the African American population is more likely to have a chronic, progressive course with a higher number of patients developing ocular manifestations. Permanent visual loss is also more common in this ethnic group. The incidence of sarcoidosis peaks between the age groups of 20 and 39 years. The incidence of the disease is higher among women compared to men.[1,3,6]

Etiology and pathogenesis

The etiology of Sarcoidosis remains unknown. Due to multisystem involvement, environmental causes including occupational influences such as irritants, inorganic particulate matter, and metals in furnaces, have been evaluated as possible causative agents. Cigarette smoking and tobacco exposure have also been proposed as independent risk factors for ocular sarcoidosis. Several ubiquitous environmental influences may play a role in the alteration of immune response in sarcoidosis.[7,8]

A cooperative multicenter study, A Case Control Etiologic Study of Sarcoidosis (ACCESS), enrolled 736 biopsy-confirmed cases and suggested a genetic predisposition for sarcoidosis. In this study, evidence for the allelic variation at the HLA-DRB1 locus as a contributing factor to the disease was presented. The most common HLA associated with sarcoidosis is HLA B8. Genetic factors located in the HLA class II region, that is, DRB1 and DQB1 have been also implicated in the pathogenesis. HLA DQB1*0201 and DRB1*0301 have been associated with a better prognosis. In addition, the familial pattern of disease occurrence has also been described.[9] Sarcoidosis is also noted to occur after organ transplantation.

CD4 T helper cell population that interacts with antigen-presenting B cells may play a central role in the pathology of sarcoidosis. Th1 cells produce IL-2, IFN-γ, and TNF-α resulting in the formation of multisystem granulomas.

Clinical features

Ocular lesions described in sarcoidosis cover a wide spectrum of conditions. The manifestations of sarcoidosis include anterior uveitis, iris nodules, conjunctival nodules, scleral nodules, band keratopathy, or interstitial keratitis. The posterior segment disease is an important manifestation of sarcoidosis and can present with chorioretinitis, periphlebitis, chorioretinal nodules, vitreous inflammation, and retinal neovascularization (Fig. 29.1); and orbital disease, including involvement of the lacrimal gland, nasolacrimal duct, optic nerve, and orbital granulomas. The disease may also cause orbital inflammatory disease (*pseudotumor*), myositis, and cranial nerve palsies.[6,10,11]

Anterior segment

Sarcoidosis most commonly presents with anterior uveitis. Anterior uveitis may be nongranulomatous or chronic granulomatous with large mutton-fat keratic precipitates, iridocyclitis, and attacks of glaucoma. Interstitial keratitis and conjunctival nodules may be noted. Granulomatous inflammation is more common in ocular sarcoidosis. The prognosis is worse for those with chronic disease, who may develop complications such as secondary glaucoma, cataracts, macular edema, and visual loss. Iris nodules are occasionally seen in association with anterior uveitis in patients with sarcoidosis.[12,13]

Posterior segment

Chorioretinal lesions and retinal vasculitis are reported to occur in approximately 14%–28% of patients. Posterior segment disease may present with vitreous accumulations and debris. Vitreous inflammation may present with either "snowballs" or a "string of pearls". There may be multiple, round, punched-out choroidal lesions associated with macular edema. Choroidal granulomas are observed as whitish elevated mass. Usually, the sarcoid granulomas are multiple and distributed throughout the posterior pole and midperiphery. However, rarely, sarcoid granulomas can present as solitary lesions with vascularization and exudation (Fig. 29.2). Optic nerve leakage may be seen on fluorescein angiography. There may be papillitis and macroaneurysms. Exudative retinal detachments are rarely seen in patients with sarcoid uveitis but do occur in those patients with large nodular chorioretinal granulomas.[1,6,10,11,14]

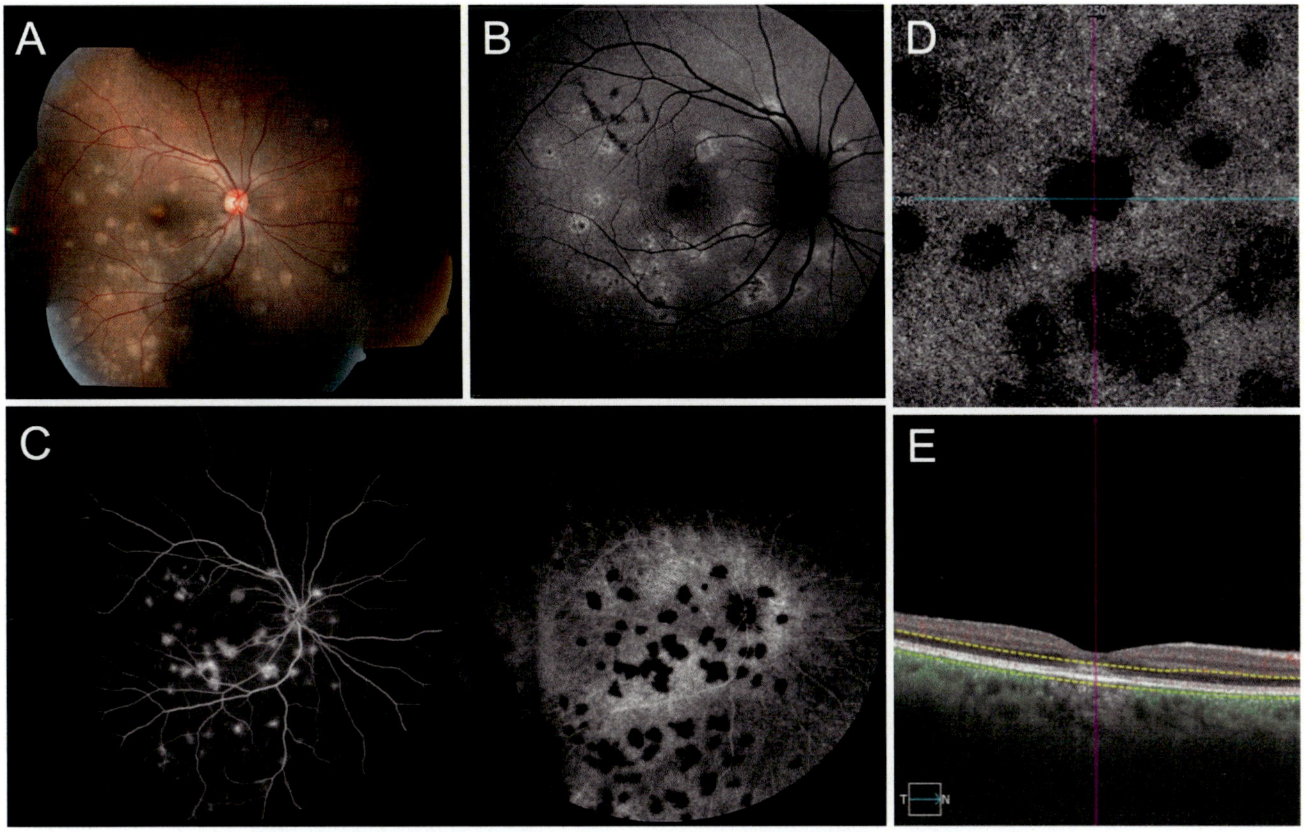

FIGURE 29.1 Multimodal imaging of a patient with sarcoid-related posterior uveitis. The fundus photograph of the right eye shows multifocal choroiditis lesions (A) which appear to be very active. The fundus autofluorescence imaging (B) shows active lesions characterized by hyperautofluorescence. The combined fluorescein angiography (FA) and indocyanine green angiography (ICGA) in the late phase show hyperfluorescence (on FA) due to leakage from the choroidal lesions and hypofluorescence on ICGA indicating the presence of full-thickness choroidal granulomas (C). Corresponding to the FA and ICGA, the optical coherence tomography angiography (OCTA) (D) and OCT (E) show choroidal granulomas manifesting as flow deficit lesions on OCTA and choroidal hyporeflective lesions on OCT.

Retinal vasculitis

Perivascular sheathing is the second most common finding in ocular sarcoidosis and occurs in 10%−17% of patients. The classical perivascular sheathing in sarcoidosis is termed as *candle wax drippings*. It generally presents as a midperipheral periphlebitis without significant vascular occlusion. Few series have documented the occurrence of occlusive retinal vascular disease, particularly branch retinal vein occlusion (Fig. 29.3). Histologic studies have demonstrated vascular compromise by the granulomatous inflammatory material. Arteriolar involvement is very unlikely in patients with sarcoidosis. Cases with "acute retinopathy of sarcoidosis" with extensive perivascular sheathing, vascular occlusion, and intraretinal hemorrhages have been described in the literature. These cases were complicated by subsequent retinal neovascularization.[6,14−16]

Peripheral retinal neovascularization or neovascularization of the optic nerve head can occur in the ocular sarcoid. This condition can be associated with significant visual loss due to vitreous hemorrhage. Peripheral retinal neovascularization is seen in patients with branch retinal vein occlusion or other vascular occlusions. The peripheral neovascular lesions may even simulate a sea fan like that seen in sickle-cell disease. Neovascularization of the disc may also develop after a branch or central vein occlusion.[1,10,17,18]

Neurological manifestations

Neurological manifestations are known to be associated with sarcoidosis and these include encephalopathy, hypothalamic dysfunction, cranial nerve palsies, and pituitary disorders. There could be other discrete manifestations such as chiasmal syndromes and motility disorders. Direct infiltration of the sarcoid mass can affect the optic nerve leading to atrophy. Facial nerve palsy can occur leading to exposure keratopathy.[2,19]

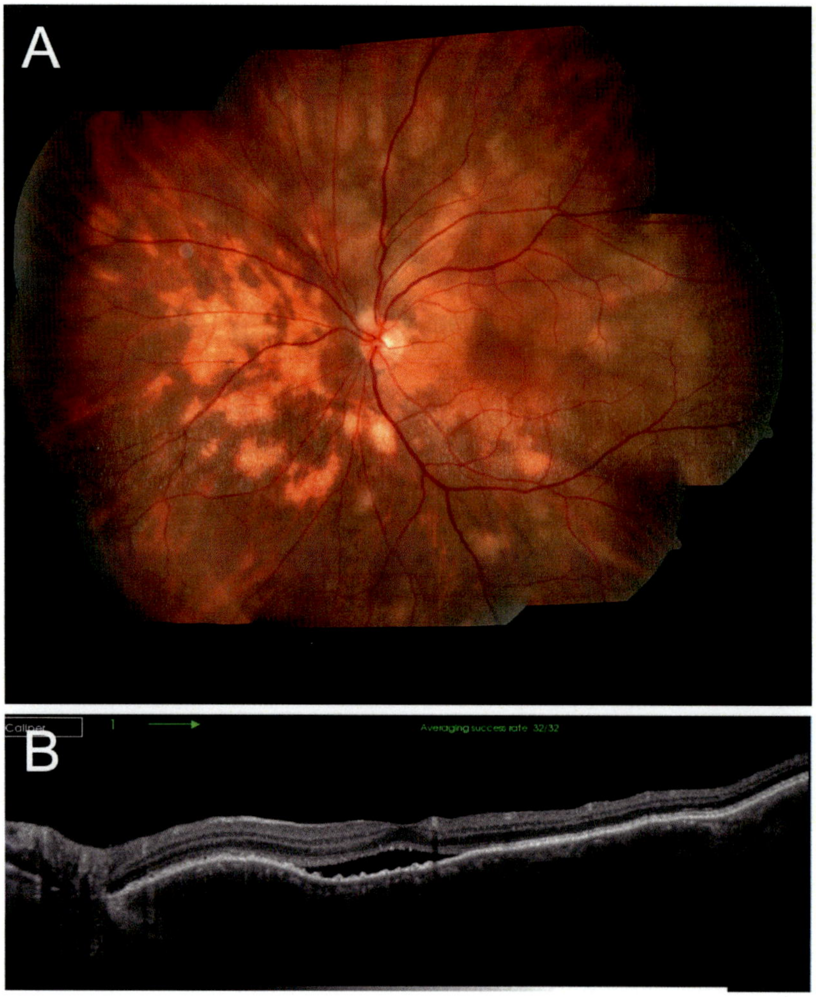

FIGURE 29.2 Fundus imaging and optical coherence tomography (OCT) of a patient with ocular sarcoidosis and extensive choroiditis lesions. The fundus (A) shows extensive deep dull yellow choroiditis lesions. The OCT (B) shows choroidal thickening with focal elevation of the retinal pigment epithelium due to the underlying choroidal granuloma. In addition, there is subretinal fluid accumulation.

Diagnosis of sarcoidosis

Laboratory and imaging evaluation

The single best initial test for the evaluation of patients with suspected sarcoidosis is the chest X-ray as it may be abnormal in 90% of cases. High-resolution computed tomography (CT) can reveal parenchymal findings, large and small airway findings, and mediastinal findings. In the parenchyma, the presence of diffuse nodular opacities (1−5 mm) in the perilymphatic distribution in the upper and middle lung zones is the most common finding. The classic radiologic finding in the mediastinum of a sarcoid patient is symmetric bilateral hilar adenopathy with some form of paratracheal adenopathy. Gallium scanning has been suggested as a useful diagnostic test for sarcoidosis. However, the test is nonspecific, and uptake of gallium is seen in other diseases such as Sjögren syndrome, tuberculosis, and lymphoma. Magnetic resonance imaging (MRI) and positive emission tomography (PET) scan have gained importance in the diagnosis of neurosarcoid and systemic sarcoidosis.[1,4,6]

The serum ACE level is commonly performed in subjects suspected to have sarcoidosis. The serum ACE level is frequently abnormal in patients with active disease and appears to reflect the total-body granuloma content in such patients. However, it is not diagnostic and appears to be of limited utility in the diagnostic dilemma of patients with possible sarcoid uveitis but a normal chest X-ray. Other tests such as serum calcium, urinary calcium, serum lysozyme, and serum immunoglobulins are also nonspecific and nondiagnostic.

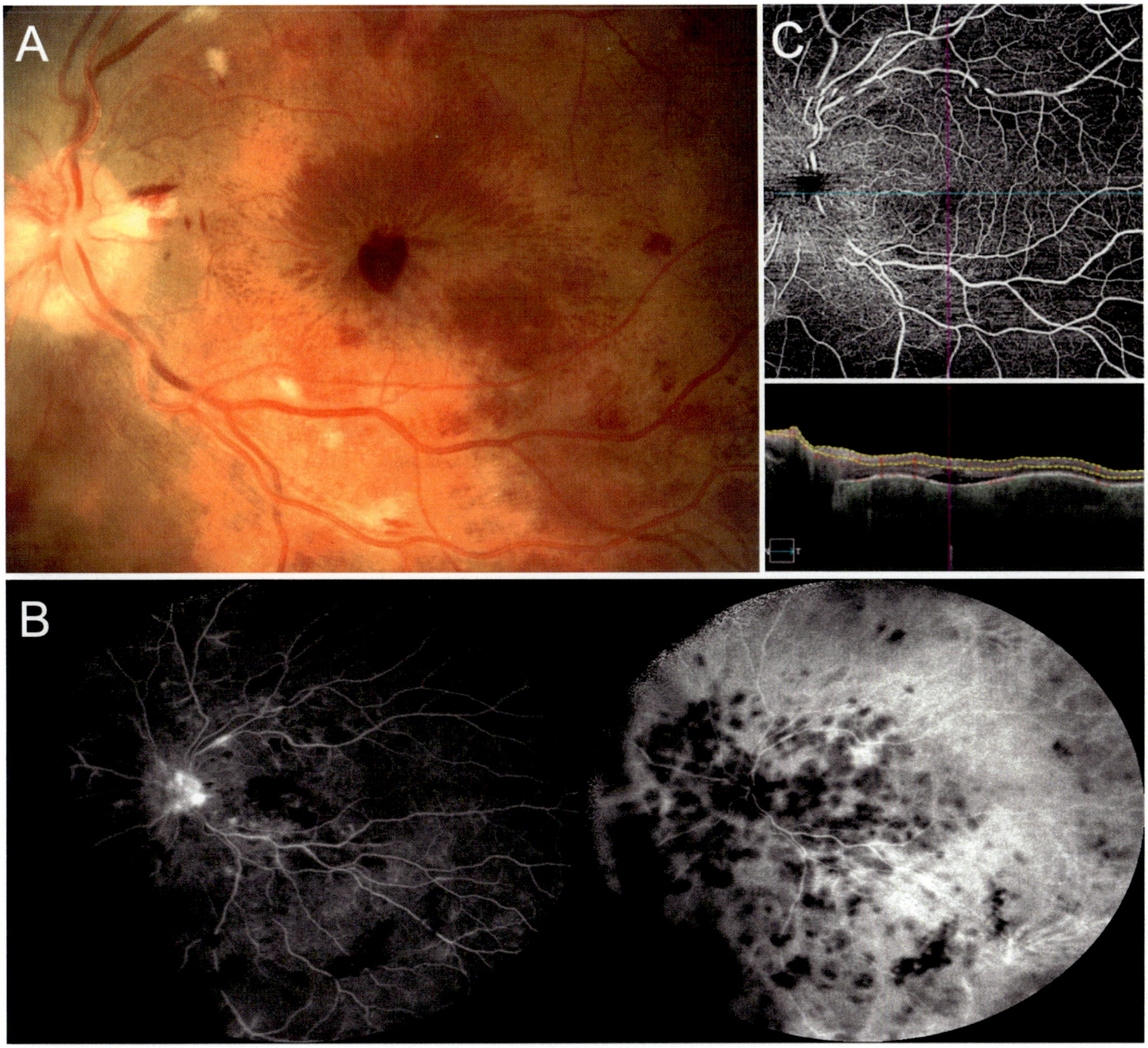

FIGURE 29.3 Patient with retinal vasculitis related to ocular sarcoidosis. The fundus photograph (A) shows extensive retinal vasculitis (occlusive in nature), optic nerve head edema, preretinal (and premacular) hemorrhages, and deep (yellowish) choroidal lesions. The combined fluorescein angiography (FA) and indocyanine green angiography (ICGA) show focal vasculitis (periphlebitis), optic nerve head leakage, blocked fluorescence due to overlying retinal hemorrhages, and hypofluorescence on ICGA due to choroidal involvement. The optical coherence tomography angiography (OCTA) shows tortuosity of the vessels in the superficial retinal vascular plexus and subretinal fluid can be appreciated on the accompanying OCT B-scan.

Histological confirmation is the gold standard in sarcoidosis. Biopsy of clinically evident skin lesions or palpable lymph nodes is frequently performed because of the high yield and low morbidity. Fiberoptic bronchoscopy with transbronchial lung biopsy is positive in 80% of patients with intrapulmonary sarcoidosis. Procedures such as endoscopic ultrasound-guided fine-needle aspiration (EUS-FNA) and endobronchial ultrasound-guided transbronchial needle aspiration (EBUS-TBNA) are relevant tests employed in the diagnosis of sarcoidosis.

Ocular imaging and evaluation

IWOS and other international investigators have worked extensively on the diagnostic criteria for ocular sarcoidosis.[4,5] The investigators of the IWOS evaluated 884 uveitis patients of which 180 patients concluded that features most associated with this condition included bilaterality (86%), snowballs or string of pearls (50%), mutton-fat keratic precipitates, iris nodules, or both (46%), and multiple chorioretinal peripheral lesions (45%). Further, a consensus meeting of the international uveitis specialists was held in 2017 to review the shortcomings of the 2009 criteria (Table 29.2). Subsequently, revised criteria for diagnosing ocular sarcoidosis were published in 2019. A Japanese study group has performed a validation analysis of the 2017 revised IWOS criteria in 323 subjects (including 51 biopsy-proven patients) and concluded that further modifications could be considered (especially in the presumed or probable arm).

Fundus imaging techniques such as fluorescein angiography (FA) and indocyanine green angiography (ICGA) play an important role in the diagnosis and management of ocular sarcoidosis. On FA, retinal vasculitis appears as leakage from retinal arteries, veins, or capillaries (most commonly veins). Segmental phlebitis is characteristic of this condition. Rarely, vasculitis can present as an inflammatory retinal vein occlusion. Other features of FA include cystoid macular edema (CME) and disc hyperfluorescence. FA is also useful in detecting choroiditis lesions that appear hypofluorescent in the early frame and become hyperfluorescent in the late phase. Sarcoid granulomas also show early hypofluorescence followed by late hyperfluorescence (they could be rarely vascular and have intense leakage). ICGA is very useful in demonstrating the deeper choroidal pathology. The findings on ICGA in sarcoidosis could be similar to ocular tuberculosis. The choroidal lesions are typically hypofluorescent in the early phase. Granulomas usually appear iso- or hypofluorescent in the late phase depending upon the thickness of the granuloma (partial thickness vs. full thickness).[1,6,20–23]

Optical coherence tomography (OCT) is useful to determine the presence of CME. Other findings on OCT in sarcoidosis include the detection of choroiditis lesions, and choroidal granulomas that appear well-defined, homogenous, round-shaped, and nonlobulated.

TABLE 29.2 Revised diagnostic criteria proposed by the International Workshop on Ocular Sarcoidosis (2017).

1. Other causes of granulomatous uveitis ruled out
2. Clinical signs suggestive of ocular sarcoidosis
a. Mutton-fat keratic precipitates or iris nodules (Koeppe or Busacca)
b. Trabecular meshwork nodules/tent-shaped peripheral anterior synechiae
c. Snowballs/string-of-pearls vitreous opacities
d. Multiple chorioretinal peripheral lesions (both active/atrophic)
e. Nodular/segmental periphlebitis (candle wax drippings) and/or macroaneurysms
f. Optic disc nodules/granulomas or choroidal nodules
g. Bilateral inflammation
3. Systemic investigations
a. Bilateral hilar lymphadenopathy on chest imaging
b. Negative tuberculin skin test or interferon-gamma release assay
c. Elevated ACE levels
d. Elevated serum lysozyme
e. Abnormal accumulation of gallium-67 scintigraphy or 18F-fluorodeoxyglucose positron emission tomography imaging
f. Elevated CD4/CD8 ratio (>3.5) in bronchoalveolar lavage
g. Lymphopenia
h. Parenchymal lung changes consistent with sarcoidosis
4. Diagnostic criteria of ocular sarcoidosis
Definite: diagnosis supported by biopsy and compatible uveitis
Presumed: diagnosis not supported by biopsy but bilateral hilar lymphadenopathy present with two intraocular signs
Probable: diagnosis not supported by biopsy and bilateral hilar lymphadenopathy absent, but three intraocular signs and two systemic investigations selected from b to h are present

Treatment of ocular sarcoidosis

Treatment of the anterior uveitis associated with sarcoidosis is generally achieved with intensive topical corticosteroids. Patients may require several months of treatment with topical corticosteroids to reach quiescence. In some patients with refractory anterior uveitis, the judicious use of oral corticosteroids may be necessary to suppress the inflammation. Prolonged use of topical corticosteroids carries the risk of development of cataract and glaucoma.[1,6,14]

Posterior segment lesions in sarcoidosis generally require the use of systemic corticosteroids. Dosages are often in the range of oral prednisone 40–80 mg/day initially depending on the body weight. Maintenance therapy with systemic corticosteroids should be less than 5 mg/day of prednisone. Although corticosteroids are very effective in controlling inflammation initially, their long-term use is often associated with significant complications and side effects. The use of steroid-sparing agents is indicated if chronic steroid usage ($\geq$5 mg of prednisone or equivalent daily) is required to control the inflammation. Corticosteroid-sparing agents used in the management of sarcoidosis include hydroxychloroquine, methotrexate, azathioprine, mycophenolate mofetil, or cyclosporine.

Local intravitreal therapies have also been employed in the management of ocular sarcoidosis. Dexamethasone implant (Ozurdex) has been used in the management of patients with intractable vitritis, vasculitis, and cystoid macular edema. The dexamethasone implant is safe and efficacious in reducing anterior chamber inflammation, vitreous haze, and cystoid macular edema in patients with sarcoidosis-related uveitis.[24–26]

Biologics, antagonists of tumor necrosis factor (TNF) such as adalimumab (humanized monoclonal antibody), infliximab (chimeric monoclonal antibody), and etanercept (soluble TNF receptor antagonist) have been reported in small studies to benefit patients with chronic uveitis. However, one must exercise caution when using these agents in patients with sarcoidosis. Use of anti-TNF factor agents may result in a sarcoidosis-like reaction with ocular inflammation. Although the mechanism of this phenomenon is not yet fully understood, several reports suggest that a cautious approach should be taken when using these agents to treat ocular sarcoidosis.[1,6,10,25]

References

1. Liu D, Birnbaum AD. Update on sarcoidosis. *Current Opinion in Ophthalmology*. 2015;26(6):512–516. https://doi.org/10.1097/ICU.0000000000000207.
2. Desai A, Chaon B, Berkenstock M. Neurosarcoidosis and ocular inflammation: a case series and literature review. *Journal of Neuro-Ophthalmology*. 2021;41(2):e259–e266. https://doi.org/10.1097/WNO.0000000000001117.
3. James G. Blau's syndrome and sarcoidosis. *Lancet*. 1999;354(9183):1035. https://doi.org/10.1016/S0140-6736(05)76645-9.
4. Mochizuki M, Smith JR, Takase H, et al. Revised criteria of international workshop on ocular sarcoidosis (IWOS) for the diagnosis of ocular sarcoidosis. *British Journal of Ophthalmology*. 2019;103(10):1418–1422. https://doi.org/10.1136/bjophthalmol-2018-313356.
5. Takase H, Acharya NR, Babu K, et al. Recommendations for the management of ocular sarcoidosis from the international workshop on ocular sarcoidosis. *British Journal of Ophthalmology*. 2021;105(11):1515–1519. https://doi.org/10.1136/bjophthalmol-2020-317354.
6. Sève P, Pacheco Y, Durupt F, et al. Sarcoidosis: a clinical overview from symptoms to diagnosis. *Cells*. 2021;10(4):766. https://doi.org/10.3390/cells10040766.
7. Saidha S, Sotirchos ES, Eckstein C. Etiology of sarcoidosis: does infection play a role? *Yale Journal of Biology and Medicine*. 2012;85(1):133–141.
8. Moller DR, Chen ES. What causes sarcoidosis? *Current Opinion in Pulmonary Medicine*. 2002;8(5):429–434. https://doi.org/10.1097/00063198-200209000-00015.
9. Semenzato G. ACCESS: a case control etiologic study of sarcoidosis. *Sarcoidosis, Vasculitis and Diffuse Lung Diseases*. 2005;22(2):83–86.
10. Takase H. Characteristics and management of ocular sarcoidosis. *Immunological Medicine*. 2022;45(1):12–21. https://doi.org/10.1080/25785826.2021.1940740.
11. Suzuki K, Ishihara M, Namba K, et al. Clinical features of ocular sarcoidosis: severe, refractory, and prolonged inflammation. *Japanese Journal of Ophthalmology*. 2022;66(5):447–454. https://doi.org/10.1007/s10384-022-00927-y.
12. Reddy HS, Khurana RN, Rao NA, Chopra V. Granuloma annulare anterior uveitis. *Ocular Immunology and Inflammation*. 2008;16(1):55–57. https://doi.org/10.1080/09273940801899814.
13. Dammacco R, Biswas J, Kivelä TT, et al. Ocular sarcoidosis: clinical experience and recent pathogenetic and therapeutic advancements. *International Ophthalmology*. 2020;40(12):3453–3467. https://doi.org/10.1007/s10792-020-01531-0.
14. Bonfioli AA, Orefice F. Sarcoidosis. *Seminars in Ophthalmology*. 2005;20(3):177–182. https://doi.org/10.1080/08820530500231938.
15. Androudi S, Dastiridou A, Symeonidis C, et al. Retinal vasculitis in rheumatic diseases: an unseen burden. *Clinical Rheumatology*. 2013;32(1):7–13. https://doi.org/10.1007/s10067-012-2078-1.
16. Patterson M, Bergstrom CS, Clark WL, Grossniklaus HE, Payne JF. Severe panuveitis, retinal vasculitis, and optic disk granuloma secondary to sarcoidosis. *Retinal Cases & Brief Reports*. 2016;10(4):341–344. https://doi.org/10.1097/ICB.0000000000000254.
17. Asdourian GK, Goldberg MF, Busse BJ. Peripheral retinal neovascularization in sarcoidosis. *Archives of Ophthalmology*. 1975;93(9):787–791. https://doi.org/10.1001/archopht.1975.01010020681004.

18. Madigan JC, Gragoudas ES, Schwartz PL, Lapus JV. Peripheral retinal neovascularization in sarcoidosis and sickle cell anemia. *Case Reports in Ophthalmological Medicine*. 1977;83(3):387–391. https://doi.org/10.1016/0002-9394(77)90738-3.
19. Lin MY, Wang Q, Newman NJ, Dattilo M. An unusual presentation of neurosarcoidosis: concurrent optic perineuritis and optic neuritis. *Taiwan Journal of Ophthalmology*. 2021;11(1):104–107. https://doi.org/10.4103/tjo.tjo_84_20.
20. Mahendradas P, Maruyama K, Mizuuchi K, Kawali A, Kitaichi N. Multimodal imaging in ocular sarcoidosis. *Ocular Immunology and Inflammation*. 2020;28(8):1205–1211. https://doi.org/10.1080/09273948.2020.1751210.
21. Calandriello L, Walsh SLF. Imaging for sarcoidosis. *Seminars in Respiratory and Critical Care Medicine*. 2017;38(4):417–436. https://doi.org/10.1055/s-0037-1603765.
22. Régis C, Benali K, Rouzet F. FDG PET/CT imaging of sarcoidosis. *Seminars in Nuclear Medicine*. 2023;53(2):258–272. https://doi.org/10.1053/j.semnuclmed.2022.08.004.
23. Larici AR, Glaudemans AW, Del Ciello A, Slart RH, Calandriello L, Gheysens O. Radiological and nuclear medicine imaging of sarcoidosis. *The Quarterly Journal of Nuclear Medicine and Molecular Imaging*. 2018;62(1):14–33. https://doi.org/10.23736/S1824-4785.17.03046-1.
24. Kim M, Kim SA, Park W, Kim RY, Park YH. Intravitreal dexamethasone implant for treatment of sarcoidosis-related uveitis. *Advances in Therapy*. 2019;36(8):2137–2146. https://doi.org/10.1007/s12325-019-00989-4.
25. Hwang DK, Sheu SJ. An update on the diagnosis and management of ocular sarcoidosis. *Current Opinion in Ophthalmology*. 2020;31(6):521–531. https://doi.org/10.1097/ICU.0000000000000704.
26. Chan WM, Lim E, Liu DTL, Law RWK, Lam DSC. Intravitreal triamcinolone acetonide for choroidal granuloma in sarcoidosis. *American Journal of Ophthalmology*. 2005;139(6):1116–1118. https://doi.org/10.1016/j.ajo.2004.11.055.

Chapter 30

Granulomatosis/polyarteritis nodosa associated disorders

Irini Chatziralli and Konstantinos Pappelis
National and Kapodistrian University of Athens, 2nd Department of Ophthalmology, Athens, Greece

Granulomatosis with polyangiitis

Disease entity

Granulomatosis with polyangiitis (GPA), formerly known as Wegener's granulomatosis, is an autoimmune disorder, histopathologically characterized by necrotizing inflammation of small- or medium-sized vessels and presence of granulomas. Along with microscopic polyangiitis (MPA) and eosinophilic granulomatosis with polyangiitis (eGPA or Churg-Strauss syndrome), GPA is classified as one of the antineutrophil cytoplasmic antibody (ANCA)-associated vasculitides (AAVs), owing to the presence of such circulating antibodies.[1]

The enzymes proteinase 3 (PR3) and myeloperoxidase (MPO), both predominantly expressed in neutrophil granulocytes and participating in neutrophil extracellular traps (NETs), constitute the two main ANCA antigens. Generally, as many as 10%–20% of AAV cases may be found ANCA-negative on testing, depending on disease activity.[2] Nevertheless, it has been long established that cytoplasmic staining ANCAs (c-ANCAs) or anti-PR3 are highly specific (97%) for GPA, while also considerably sensitive (93%) during active disease.[3] In a minority of GPA cases (10%–20%), perinuclear staining (p-ANCAs) or anti-MPO are found.[4]

Etiology and pathogenesis

As in the majority of autoimmune diseases, the exact causal mechanisms behind GPA are poorly understood. Although a strong genetic predisposition has been ruled out, genome-wide association studies (GWASs) have uncovered loci positively or negatively associated with GPA. In particular, major histocompatibility complex class II region HLA-DP is shown to be strongly associated with a c-ANCA pattern (odds ratio [OR] = 7.0 for c-ANCA AAVs; OR = 5.4 for GPA specifically).[5] Familial aggregation has also been observed, further hinting toward a genetic component; however, it is possible that clustering can be attributed, at least partially, to the influence of environmental factors.[6,7] Epigenetically induced expression of PR3 and MPO through DNA methylation has additionally been linked to AAV activity.[8]

With regard to external factors, indirect evidence implicates microbial pathogens, especially toxins secreted by *Staphylococcus aureus*, in GPA relapses.[9,10] Several environmental and chemical factors have been also deemed as risk factors for GPA, the most studied among whom being the presence of airborne silica particles (OR = 3.6, 95% confidence interval [CI]: 1.9–6.8).[11]

ANCAs are considered to be at least partly responsible for the generalized inflammation in GPA through a complex process. In vitro studies have demonstrated that ANCA-mediated induction of degranulating neutrophils can cause significant damage to the vascular endothelium.[12] This observation has been corroborated by the fact that c-ANCA titer correlates with disease activity, although not all studies have been able to replicate this finding.[3,13,14] Endothelial damage is, on the one hand, consequential to cytokine activation leading to the release of lytic enzymes and reactive oxygen species (ROS). On the other hand, AAVs are characterized by decreased rates of NET degradation by DNase I. Therefore, a vicious self-sustaining cycle is established, during which dysregulation of NETs promotes increased ANCA production. This, in turn, results in excessive neutrophil activation causing additional NET formation.[5,15] Recent studies have clarified

Retinal and Choroidal Vascular Diseases of the Eye. https://doi.org/10.1016/B978-0-443-15583-3.00030-5

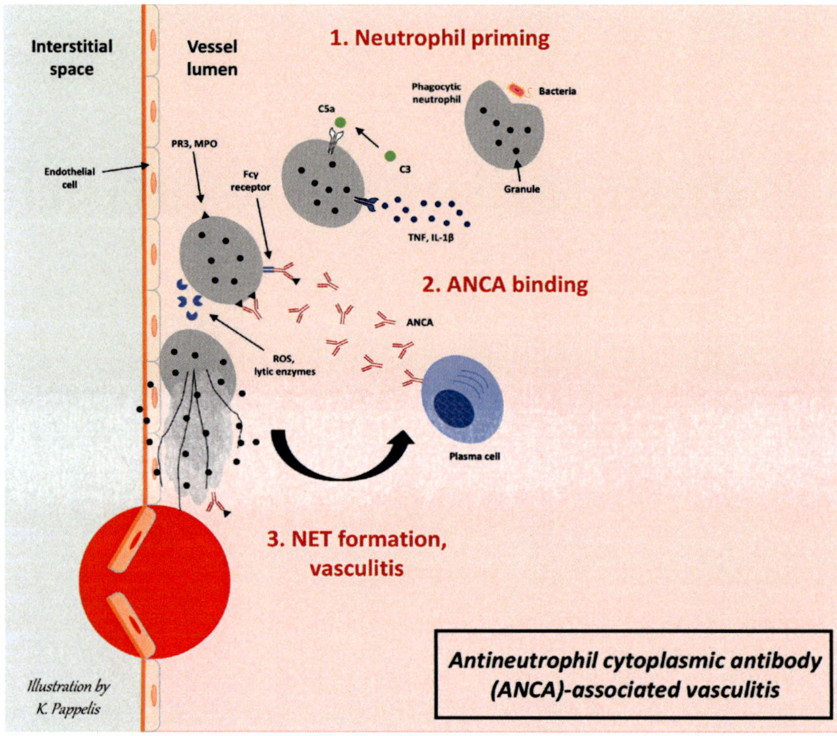

FIGURE 30.1 Antineutrophil cytoplasmic antibody (ANCA)-associated vasculitis pathogenetic mechanisms in antineutrophil cytoplasmic antibody (ANCA)-associated vasculitis. Phagocytic neutrophils are primed by complement activation and proinflammatory cytokines, triggered by an immune response. Primed neutrophils express proteinase 3 (PR3), myeloperoxidase (MPO), and Fcγ receptors on their cytoplasmic membrane, to which c-ANCAs, produced by circulating plasma cells, bind. This leads to increased neutrophil activation, reactive oxygen species (ROS) and lytic enzyme release, and neutrophil extracellular trap (NET) formation. NETs contain DNA-dissociated histones and matrix metalloproteinases (MMPs), which induce endothelial damage and angiopathy. Simultaneously, the low NET degradation rates occurring in AAV result in increased anti-PR3 and anti-MPO production, creating a feedback loop.

that NETs are also able to cause endothelial damage themselves through the production of vasculopathic molecules and complement activation.[16] In GPA, granulomas are formed around the necrotic lesions, due to the cytokine-induced aggregation of neutrophils and other leukocytes.[5] Fig. 30.1 outlines the pathways leading to the development of AAV.

Epidemiology

GPA is a rare disease, with a continuously increasing worldwide incidence ranging from 0.4 to 34 cases per million inhabitants and a meta-analyzed pooled prevalence of 96.8 (95% CI: 92.2−102.0) per million inhabitants.[17−19] Among all AAVs, GPA is predominant in white European populations, while MPA is more prevalent in Asian populations; this is also reflected in the observed ethnic differences in allele frequencies extracted from genomic analyses.[5,19] There appears to be a small sex predilection in AAVs, with the male-to-female ratio ranging between 1.07:1 and 1.48:1, but this seems to be mostly owed to predominance of MPA, rather than GPA, in males.[17,19] Since more than 2 decades, the average age at the time of diagnosis has increased to over 65 years old, compared to the age range of 40−55 years reported in studies of the previous century, possibly owing to the increasing availability of ANCA testing.[4,19]

Clinical presentation and diagnosis

GPA is a systemic disease of diverse and diffuse manifestations, including (but not limited to) upper or lower respiratory tract involvement, sensorineural hearing loss, and glomerulonephritis, the latter being associated with increased disease severity.[20] Although initially present in approximately 20% of patients with GPA, kidney involvement will eventually occur in as many as 80% of cases, if left untreated.[4] Signs, symptoms, and test results, including screening for ANCA, are not highly specific; therefore, establishing a diagnosis for GPA is a multimodal procedure, involving the exclusion of alternate diagnoses mimicking vasculitides and the identification of several clinical, imaging, laboratory, and biopsy findings.[9]

In 2022, the American College of Rheumatology (ACR) and European Alliance of Associations for Rheumatology developed and validated revised criteria for GPA diagnosis in patients already diagnosed with vasculitis (Table 30.1). Although the purpose of this weighted score is the generation of an objective classification system for patient inclusion in randomized control trials (RCTs), it can be helpful with GPA diagnosis, owing to its high sensitivity (92.5%) and

TABLE 30.1 Granulomatosis with polyangiitis classification criteria.

Criteria	Score
Clinical	
Nasal inflammation, bloody or purulent discharge, ulcers, septal involvement	+3
Ear or airway cartilage inflammation, saddle nose deformity, hoarse voice or stridor	+2
Hearing loss (conductive or sensorineural)	+1
Laboratory, imaging, and biopsy	
Positive for antineutrophil cytoplasmic antibodies (c-ANCA) or anti-PR3	+5
Chest radiography with pulmonary nodules, infiltrates, or cavities	+2
Presence of granulomas in biopsy	+2
Imaging-confirmed nasal/paranasal sinus or mastoid involvement	+1
Pauci-immune glomerulonephritis on biopsy	+1
Positive for perinuclear antineutrophil cytoplasmic antibodies (p-ANCA) or anti-MPO	−1
Blood eosinophil count $\geq 1 \times 10^9$/L	−4

Revised by the American College of Rheumatology/European Alliance of Associations for Rheumatology, 2022.

specificity (93.8%) in differentiating GPA from other vasculitides.[1] The score does not include any ophthalmological parameters.

Management and prognosis

GPA is a potentially fatal disease, if left untreated, with a 2-year survival not exceeding 10%, as data from the previous century suggest.[9] With current treatments, 5-year survival exceeds 70%.[21] The revised Five-Factor Score (FFS) can been used to finetune prognosis according to disease severity, the five risk factors being (1) age greater than 65 years, (2) cardiovascular involvement, (3) gastrointestinal involvement, (4) stabilized peak creatinine ≥150 μmol/L, and (5) absence of ear and upper airway tract symptoms.[22,23]

The standard therapeutic strategy in GPA is a combination of high-dose glucocorticoids with high-dose cyclophosphamide (CYC) for 3−6 months, while maintenance of remission is managed with other immuno-suppressants, such as methotrexate and azathioprine for a period of at least 2 years, with concomitant glucocorticoid tapering.[9,24] However, RCTs suggest that treating with rituximab (RTX), a chimeric monoclonal anti-CD20, is at least noninferior to CYC in inducing remission.[25,26] New therapeutic targets for AAV are currently under development, aiming to target cells and molecules related to disease pathogenesis, including (but not limited to) T- or B-cells, plasma cells, cytokines, and the complement system.[5]

Ocular involvement

GPA may affect almost every structure of the eye. The pooled prevalence of ocular complications in GPA is estimated at 26% (95% CI: 19%−34%), but has been reported as high as 58%, in some cohorts.[27] They are three to four times more likely to appear in GPA, compared to MPA or eGPA.[28] The particularly rare chorioretinal manifestations are analyzed in the next subsection while, in this subsection, we outline the involvement of the rest of the ocular structures.

In the population described by Rothschild et al., eye redness was the most common ophthalmic-related symptom of GPA and was mostly linked to episcleritis, conjunctivitis, or combination thereof (39.3% and 52.1%, respectively).[28] Strikingly, in other studies, these percentages were much lower and, instead, orbital involvement was the most common manifestation, appearing in 15%−20% of cases.[4,29,30] Orbital complications present with variable symptom severity, ranging from proptosis, lid involvement, and lacrimal involvement to orbital myositis with diplopia and compressive ischemic optic neuropathy. Interestingly, the coexistence of proptosis with findings from the airway tracts or kidneys almost immediately implies a GPA diagnosis.[4,15,24]

Anterior or posterior scleritis was not as common in the population described by Rothschild et al., but GPA has been associated with the nodular, diffuse, or necrotizing forms, due to inflammation of the deep episcleral vascular plexus.[4,15,24] In a study from Bangladesh, GPA was the second most common systemic association of scleritis, after rheumatoid arthritis (5% in a population of 79 patients with scleritis).[31] Scleral inflammation can additionally cause secondary keratopathy, which usually manifests as peripheral ulcerative keratitis (PUK) and can be accompanied by uveitis.[24] GPA should also be part of the differential diagnosis in very rare cases of uveitis with concomitant meningeal inflammation (uveo-meningeal syndrome).[32]

A good response to RTX has been observed in noninfectious scleritis and noninfectious orbital inflammation caused by GPA (positive therapeutic response of 97.3% in a sample of 75 cases and 84.8% in a sample of 99 patients, respectively).[33,34] Limited evidence suggests that treating GPA-related scleritis and uveitis with RTX is not only more effective, but also requires less interruptions and dose adjustments, compared to CYC.[35,36]

GPA in chorioretinal disease

Attributing retinal vasculitis (RV) to GPA is no simple task, mainly for three reasons. First, the conditions are very rarely associated; indeed, most of the existing literature has the form of case presentations, which likely leads to significant report bias. Second, while it is less likely that an infectious process may mimic a systemic vasculitis with diffuse manifestation, it is of utmost importance that the possibility of ophthalmic infection is definitively ruled out, before establishing a diagnosis of noninfectious RV and initiating any immunosuppressive treatment.[37] Last, RV is an ill-defined entity itself; the retinal vessels are not accessible for biopsy, whereas the diagnosis and classification of systemic vasculitides are, at least partly, dependent on histopathologic features.

Ophthalmologists frequently define RV as the presence of perivascular sheathing and vascular leakage or occlusion in fluorescein angiography (FA).[37–39] This is in accordance with the "provisional" consensus reached by The Standardization of Uveitis Nomenclature Working Group in 2004.[40] Other nonspecific findings, such as cotton-wool spots, retinal and preretinal hemorrhages, cystoid macular edema (CME), retinal pigment epithelium (RPE) disruptions, and presence of vitreous cells may be observed in RV, but they are unlikely (albeit not impossible) to be the sole ophthalmic manifestation of GPA.[30,41–43] RV presents with decreased or blurred vision, metamorphopsia, and visual field deficits, but can also be asymptomatic.[29]

Although reports of GPA and RV are present in the literature, it must be noted that these cases are extremely rare. RV appears in less than 3% of GPA patients and, vice versa, an RV finding seldom is GPA-related.[4,28–30] The latter was clearly shown in one of the largest studies on patients with RV to date (Rosenbaum et al., $n = 207$), among whom only 3 (1.4%) had a concomitant systemic vasculitis diagnosis and only 1 (0.5%) had a GPA diagnosis.[37] In an older study by Graham et al., of the 150 patients with noninfectious vasculitis, 3 (2%) were diagnosed with GPA. These findings suggest that RV alone has a very low positive predictive value for GPA and the latter should not sit at the top of the differential diagnosis list. In addition, seldom is RV the sole piece of evidence of an underlying GPA. That having said, in one single RV case, the presence of GPA was only revealed post hoc via autopsy (negative testing for ANCA, no evidence of systemic vasculitis, and no granulomas on pulmonary biopsy).[44]

Despite still being rare in absolute numbers, occlusive RV is the most sight-threatening and common manifestation of GPA. According to a recent study, out of 42 patients with noninfectious occlusive RV, 4 were diagnosed with GPA (9.5%), which was also the most common systemic association.[38] Therefore, when infectious occlusive RV has been ruled out, GPA should be considered. The ability of leukocytes to infiltrate the choroid has been histopathologically demonstrated (Fig. 30.2).[45–47] Chorioretinal vasculitis following leukocyte infiltration may lead to inflammatory central or branch retinal arteriolar occlusions (CRAO, BRAO), sclerosis of large retinal and choroidal vessels, and infarcts of the choriocapillaris (Fig. 30.3). Midphase FA snapshots may reveal staining of choroidal lesions with a lack of choroidal fluorescence in the submacular region, while early phases of indocyanine green angiography (ICGA) may show filling defects of the choroidal vessels and late phases may show dye leakage.[48–55] Central or branch retinal vein occlusion (CRVO, BRVO) cases with or without neovascular glaucoma have also been reported on rare occasions, even in the absence of other signs of vascular inflammation.[36,56–59] A single case of Purtscher's-like retinopathy secondary to GPA was recently described, suggesting the possibility of embolic occlusion of precapillary arterioles.[60]

At this point, it must be acknowledged that, although FA and ICGA have been widely used in the assessment of GPA-related RV, a few studies also advocate in favor of optical coherence tomography angiography (OCTA).[61–63] Despite its limitations, OCTA B-scans or en face images can be used to better localize, characterize, and objectively quantify lesions

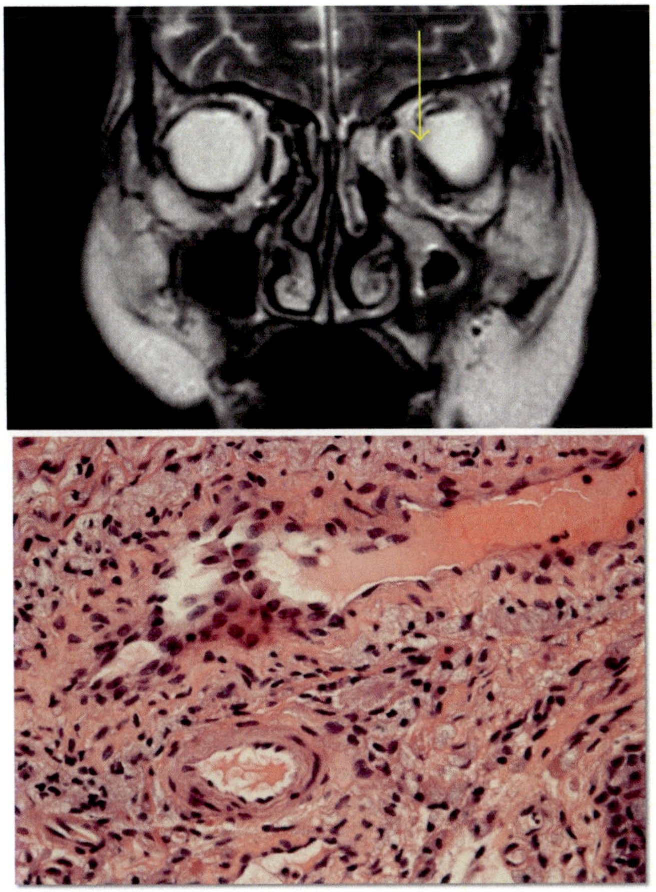

FIGURE 30.2 Choroidal mass in a case of granulomatosis with polyangiitis A case of an extremely rare choroidal mass in an 83-year-old female with granulomatosis with polyangiitis. *Top panel*: Coronal T2-weighted magnetic resonance image displaying the choroidal tumor and lesions in the left paranasal sinus. *Bottom panel*: Hematoxylin and eosin staining of a biopsy specimen of the tumor. Evidence of granulomatous choroiditis, necrotizing vasculitis, and inflammatory cell aggregation. *Recreated from Masuda T, Izumi Y, Takeshita H, Kawahara C, Tsuji Y, Kurohama H, Iwanaga N, Inamoto M, Kase K, Ito M, Kawakami A, Migita K. Granulomatosis with polyangiitis presenting as a choroidal tumor. Case Reports in Rheumatology, 2015:1–5. https://doi.org/10.1155/2015/271823*

in the superficial, deep, and choriocapillaris plexus (Fig. 30.4). In addition, the noninvasive and dyeless nature of OCTA arguably renders it a better imaging option for patients with renal dysfunction.

Other rare manifestations of GPA reported in the literature include arteritic anterior ischemic optic neuropathy (AAION), which can happen simultaneously with arterial occlusion, serous retinal and choroidal detachment, frosted branch angiitis, and acute posterior multifocal placoid pigment epitheliopathy (APMPPE).[64–71] Ophthalmologists should also be alert that opportunistic ocular infections in patients with GPA might be mimicking GPA-related chorioretinal disease. Indeed, viral necrotizing retinitis has been reported in GPA patients treated with immuno-suppressants (usually when CD4 counts are below 50 cells/mm^3).[37,72–76] Last, sometimes caution should also be taken when differentiating tumors from GPA-related granulomas presenting as large masses (Fig. 30.2).[46,77–79].

Polyarteritis nodosa

Disease entity

Polyarteritis nodosa (PAN) is an autoimmune vasculitis, predominantly affecting medium-sized, muscular arteries. It is an obscure entity, characterized by stimulation of the classical immune response from unidentified antigens, likely of viral origin.[80] Contrary to GPA, PAN does not cause granulomatous inflammation, but the inflammatory infiltrates lead to the formation of small aneurysms, resembling "beads of a rosary."[81] According to the 2012 Revised International Chapel Hill

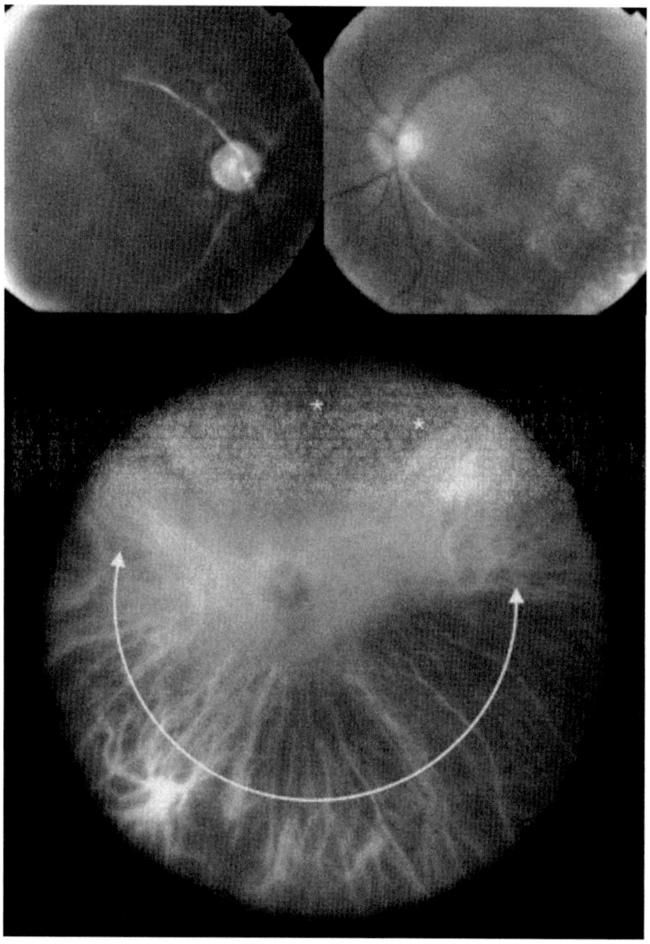

FIGURE 30.3 Bilateral arterial occlusion in granulomatosis with polyangiitis A case of bilateral arterial occlusion in a 64-year-old male with granulomatosis with polyangiitis with bilateral presence of choroidal infarcts (right eye: central retinal artery occlusion, left eye: branch retinal artery occlusion). In the top panel, fundus photographs reveal sclerotic and obstructed arteries, as well as inferior chorioretinal atrophy with pigmentary changes in the left eye. In the bottom panel, an early phase snapshot of wide-angle indocyanine green angiography depicts a large choroidal filling defect in the inferior fundus (*arrow*) and smaller defects in the superior fundus of the left eye (*). *Recreated from Iida T, Spaide RF, Kantor J. Retinal and choroidal arterial occlusion in Wegener's granulomatosis. American Journal of Ophthalmology, 2002;133(1):151–152. https://doi.org/10.1016/S0002-9394(01)01251-X, with permission from Elsevier.*

Consensus Conference Nomenclature for Vasculitides, PAN is defined as a "necrotizing arteritis of medium or small arteries without glomerulonephritis or vasculitis in arterioles, capillaries, or venules, and not associated with ANCAs."8[82]

Etiology and pathogenesis

The exact etiology of PAN is yet to be disentangled. Although the majority of occurrences are idiopathic, hepatitis-B virus (HBV) infection has traditionally been associated with a considerable proportion of PAN cases. Indeed, a decline in HBV-associated PAN from 1 in 3 to approximately 1 in 20 PAN cases was observed after the introduction of HBV vaccination.[83,84] HBV infection can damage the endothelial cells directly, by means of viral replication, or indirectly, due to the formation and deposition of immune complexes.[83] Antibodies against endothelial cells have also been recently implicated in PAN development, due to endothelial activation and perpetuation of inflammatory and chemotactic factor production.[80,85] There exists no pronounced genetic component in PAN pathogenesis; however, it was recently discovered that a subgroup of immunodeficient and cytopenic PAN patients presented with mutations in the ADA2 gene and adenosine deaminase 2 deficiency.[86–88]

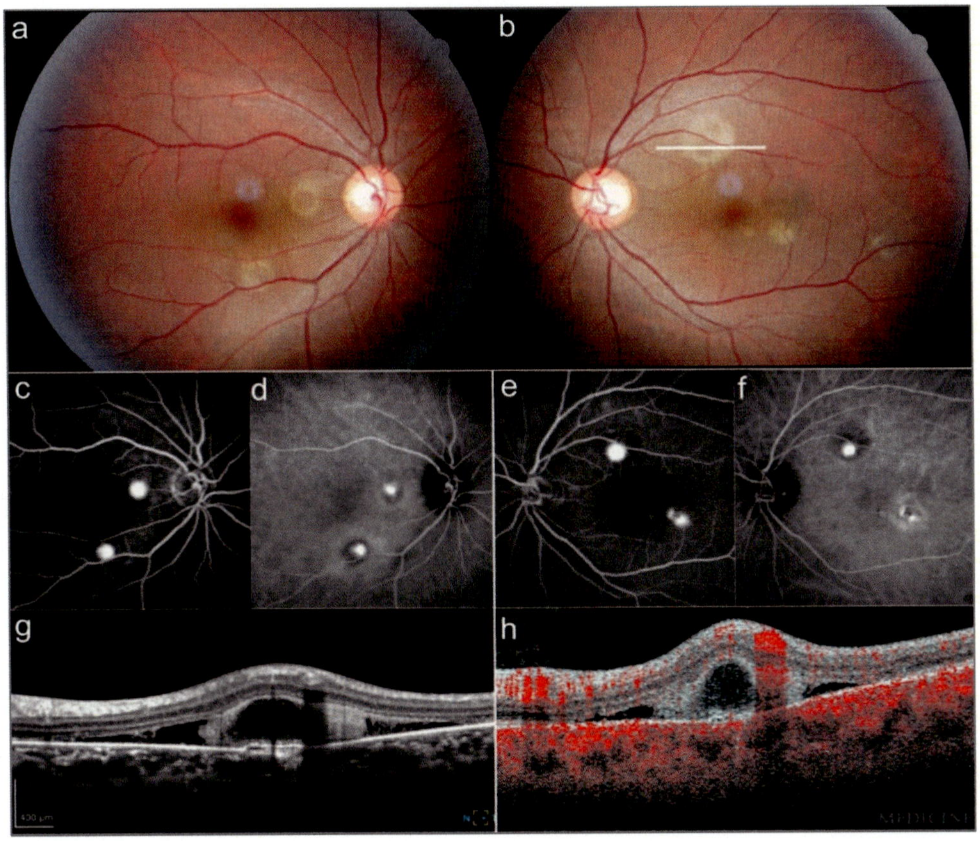

FIGURE 30.4 Central serous chorioretinopathy with fibrin accumulations in granulomatosis with polyangiitis under treatment A 36-year-old male with granulomatosis with polyangiitis under treatment. White-yellow lesions appear in the fundus photographs of both eyes (A, B), which correspond to hyperfluorescent leakages on fluorescein angiography (C, E) and to choroidal vascular hyperpermeability on indocyanine green angiography (D, F). The optical coherence tomography (OCT) B-scan (G) corresponding to the white horizontal line in (B) detected a hyperreflective protrusion with a hypo-reflective bubble-like structure within the serous retinal detachment area. Absence of flow (*red dots*) in the OCT angiography B scan (H) rules out choroidal neovascularization. A diagnosis of central serous chorioretinopathy with fibrin accumulations was set and steroid dosage was reduced. *Reproduced from Takashi N, Nakamura A, Kataoka K, Usui Y, Ito Y, Kaneko H. Optical coherence tomography angiography for the diagnosis of granulomatosis with polyangiitis with serous retinal detachment: A case report. Medicine, 2021;100(7):e24789. https://doi.org/10.1097/MD.0000000000024789*

Epidemiology

Just like GPA, PAN is a rare disease, with a worldwide incidence ranging from 0.9 to 8 cases per million inhabitants and a prevalence of 31 cases per million inhabitants.[19,84,89,90] Due to HBV vaccination, as well as changes in disease nomenclature, there has been a decline in both the incidence and prevalence of PAN.[87] Although it occurs in all ethnic groups, PAN is more common in European populations, while children and adult males are more likely to be affected by the disease (male- to female ratio ranging between 1.6:1 and 1.9:1).[84,91,92] In general, PAN affects younger individuals than GPA, with the average age at the time of diagnosis being between 40 and 50 years.[84,92–95]

Clinical presentation and diagnosis

There currently exists no single reliable indicator for establishing a PAN diagnosis; therefore, a multimodal approach is necessary. The rosary sign is the hallmark of the disease but is neither a necessary nor a sufficient criterion. Based on its diffuse systemic manifestations, the ACR proposed that the existence of 3 out of 10 listed clinical signs is indicative of a PAN diagnosis (Table 30.2).[96] According to the 2012 Chapel Hill criteria, PAN can be diagnosed by the presence of medium-sized artery aneurysms seen on imaging or the recognition of necrotizing arteritis on biopsy, together with negative ANCA testing, no mucocutaneous lymph node syndrome (Kawasaki disease), and absence of glomerulonephritis.[82] A consensus combining ACR and Chapel Hill criteria has been proposed to guide clinical practice and

TABLE 30.2 Polyarteritis nodosa diagnostic criteria.

Presence of 3 out of 10 required:
1. Weight loss ≥4.5 kg
2. Purplish skin discoloration (livedo reticularis)
3. Testicular pain/tenderness
4. Muscle pain/tenderness, weakness
5. Neuropathy
6. Diastolic hypertension (≥90 mm Hg)
7. Blood urea nitrogen ≥40 mg/dL or creatinine ≥1.5 mg/dL
8. Positive testing for hepatitis B
9. Aneurysms (rosary sign) or presence of vasculitis on angiograms
10. Arteritis confirmed by biopsy

Proposed by the American College of Rheumatology, 1990.

epidemiologic surveillance.[97,98] PAN can sometimes mimic other small- or large-vessel vasculitides, but the presence of pulmonary involvement is extremely rare and almost immediately rules out a PAN diagnosis.[87]

Management and prognosis

PAN is also a potentially fatal disease if left untreated; indeed 5-year survival used to be below 15%, according to reports from the previous century.[99] With current treatments, 5-year survival has been reported as high as 88% in adults and even higher in children.[91,92] Relapse rates are estimated at 25% within 5 years but have sometimes been reported as even higher.[92,100] The revised FFS can be used, again, to finetune prognosis according to disease severity, but in the case of PAN the risk factors are four, namely (1) age greater than 65 years, (2) cardiovascular involvement, (3) gastrointestinal involvement, and (4) stabilized peak creatinine ≥150 μmol/L.[83]

Treatment regimens are similar to GPA: glucocorticoid monotherapy or in combination with other immunosuppressants, such as methotrexate and azathioprine, tend to be administered in uncomplicated cases (FFS = 0), while high doses of glucocorticoids accompanied by intensive CYC administration for at least 1 year are considered for severe disease (FFS ≥ 1).[101,102] Maintenance immunosuppression with glucocorticoid tapering follows for a period of 18 months, after achieving remission.[83,87] TNFα inhibitors have also shown promise, especially for DADA2 patients, albeit with conflicting evidence.[83,101] It is evident that in HBV-associated PAN cases, HBV infection management should take place, as well.

Ocular involvement

According to very few studies that have addressed the matter, ocular manifestations in PAN are considerably less frequent than in GPA. In the population described by Rothschild et al., 42 out of 393 (10.7%) PAN cases presented with ophthalmologic manifestations (to be compared with 34.1% in GPA). Together with chorioretinal involvement (described in the next subsection), eye redness was the most common finding in this sample and was attributed, in order of frequency, to conjunctivitis, episcleritis, or keratitis. Uveitis was reported in 12% of cases.[28] In the series of Pagnoux et al., ocular involvement was described in 30 out of 328 (8.6%) PAN patients, with conjunctivitis, keratitis, and uveitis being the most frequent diagnoses, aside from chorioretinal vasculitis.[92]

PAN in chorioretinal disease

Attributing RV to PAN is bound to similar limitations as in GPA, that is, the scarcity of evidence, the vast majority of which exists in anecdotal form, the need for a clear differentiation from infectious causes, and the lack of histopathologic confirmation.[37] Again, the presence of RV will mainly be based on the existence of perivascular sheathing and vascular leakage or occlusion in FA, with or without cotton-wool spots, exudates, retinal hemorrhages, or vitreous cells.[37–39]

Nonspecific symptoms such as decreased or blurred vision, metamorphopsia, and visual field deficits are typical, but not necessary.[29]

PAN patients exhibiting RV are extremely rare. In a large population comprising PAN and AAV patients, RV appeared in 7 out of 393 (1.8%) PAN patients (to be compared with 3 out of 343 [0.9%] GPA patients, see section GPA in chorioretinal disease).[28] However, there have been smaller PAN series with a higher prevalence of RV (8 out of 17 patients reported by Travers et al.).[103] The reason for this discrepancy is not clear but is likely related to comorbidity, such as arterial hypertension, and to selection bias, favoring the inclusion of more severe cases.[104] In the series of Rosenbaum et al., none of the 207 patients with RV were diagnosed with PAN (to be compared with 1 case of GPA).[37] In another series by Graham et al., of the 150 patients with noninfectious RV, 3 (2%) had a PAN diagnosis (to be compared with 3 GPA cases, as well).[105] Therefore, we hereby highlight again the fact that the presence of RV alone has a very low positive predictive value for PAN.

Exudates are the most common finding in PAN and appear more frequently than in GPA (2% of PAN cases vs. 0.6% of GPA cases in the population of Rothschild et al.). In the same study, occlusion appeared in 1% of PAN cases, compared to 0.3% of GPA cases.[28] CRAO, CRVO, or branch occlusions have been reported to coincide with PAN; however, among 42 patients with noninfectious occlusive RV, none was diagnosed with PAN (to be compared with 4 [9.5%] GPA cases).[38,106–111] Other ischemic pathologies, such as choroidal infarction and AION, have also been reported to occur independently, or in combination with retinal or cilioretinal artery occlusion (Fig. 30.5).[112–116].

Involvement of the choroid might be considerably more common than currently assumed and does not solely occur not in the form of the aforementioned infarcts.[117] Increased subfoveal choroidal thickness was reported in a small pediatric PAN sample ($n = 11$), compared to healthy controls, which could indicate mild subclinical inflammation.[118] Choroidal vasculitis with mid- and late-phase segmental choroidal vessel staining and leakage in ICGA has also been noticed in the presence of nodular scleritis.[119] Selective involvement of the choroid was observed in a PAN case presenting with macula-sparing monocular blackouts likely related to reduced choroidal perfusion and was confirmed with blanching of disc capillaries in FA, without concomitant collapse of the retinal vessels.[120]

Similar to GPA, on rare occasions PAN has been associated with retinal detachment, frost branch angiitis, and APMPPE.[65,114,121–123] A single peculiar case of PAN accompanied by homonymous hemianopsia and Roth spots was recently reported for the first time.[124] Table 30.3 summarizes the most important chorioretinal vascular manifestations encountered in both diseases.

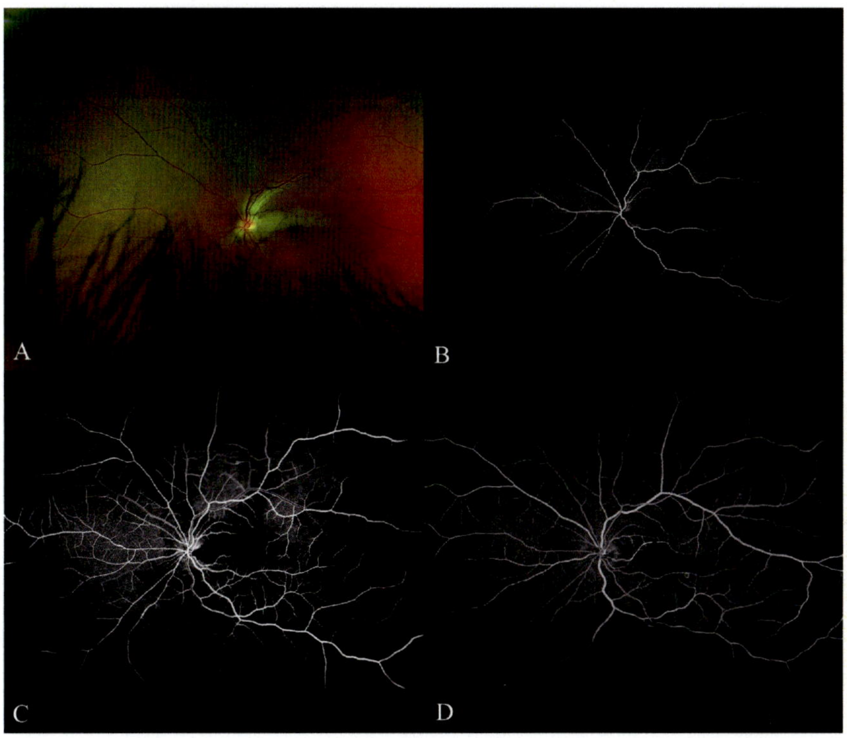

FIGURE 30.5 Cilioretinal artery occlusion and anterior ischemic optic neuropathy in a case of polyarteritis nodosa A case of cilioretinal artery occlusion in a 23-year-old male with polyarteritis nodosa and concomitant anterior ischemic optic neuropathy. Appearance of a pale white area in widefield fundus imaging (A) is accompanied by an overlapping nonfilling of the cilioretinal artery during the arterial phase of fluorescein angiography (B). A wide area of choroidal ischemia can be observed in the arteriovenous phase (C) with subsequent filling of the cilioretinal artery and emergence of the rosary sign, during the late phase (D). *Reproduced from Sener H, Sevim DG, Unlu M. Anterior ischemic optic neuropathy and cilioretinal artery occlusion secondary to polyarteritis nodosa: A case report. Photodiagnosis and Photodynamic Therapy, 2021;34. https://doi.org/10.1016/j.pdpdt.2021.102299, with permission from Elsevier.*

TABLE 30.3 Chorioretinal vascular manifestations associated with granulomatosis with polyangiitis and polyarteritis nodosa.

	Granulomatosis with polyangiitis/ polyarteritis nodosa
Chorioretinal vasculitis	Possible
CRAO/BRAO	
Exudates	
Cotton-wool spots	Infrequent
Hemorrhages	
CRVO/BRVO	
Choroidal infarction	
AION	Very rare
Serous detachment	
Frost branch angiitis	
APMPPE	

AION, anterior ischemic optic neuropathy; *APMPPE*, acute posterior multifocal placoid pigment epitheliopathy; *BRAO*, branch retinal artery occlusion; *BRVO*, branch retinal vein occlusion; *CRAO*, central retinal artery occlusion; *CRVO*, central retinal vein occlusion.

Conclusions

Both GPA and PAN are rare, autoimmune, systemic disorders, with even rarer ophthalmic manifestations. In both diseases, chorioretinal pathology has been described in multiple occasions in the literature, albeit scarcer than the involvement of other ocular structures. In general, RV, defined as perivascular sheathing and leakage or occlusion in FA, occurs in both GPA and PAN, but other nonspecific vascular findings may be present. Clinicians should be alert for the presence of noninfectious occlusive vasculitis of the retinal arteries because it constitutes the most severe manifestation of both diseases. Ruling out infection is crucial for the diagnostic procedure because it can guide treatment decisions, especially with regards to the use of immuno-suppressants. Although vascular imaging by means of FA, ICGA, or OCTA can confirm in most cases the presence of chorioretinal vasculitis, attributing this to GPA or PAN should always depend on a multimodal approach, including clinical, laboratory, imaging, and histological work-up.

References

1. Robson JC, Grayson PC, Ponte C, et al. 2022 American college of rheumatology/European Alliance of associations for rheumatology classification criteria for granulomatosis with polyangiitis. *Annals of the Rheumatic Diseases*. 2022;81(3):315−320. https://doi.org/10.1136/annrheumdis-2021-221795.
2. Bossuyt X, Cohen Tervaert J-W, Arimura Y, et al. Revised 2017 international consensus on testing of ANCAs in granulomatosis with polyangiitis and microscopic polyangiitis. *Nature Reviews Rheumatology*. 2017;13(11):683−692. https://doi.org/10.1038/nrrheum.2017.140.
3. Van Der Woude FJ, Lobatto S, Permin H, et al. Autoantibodies against neutrophils and monocytes: tool for diagnosis and marker of disease activity in WEGENER'S granulomatosis. *Lancet*. 1985;325(8426):425−429. https://doi.org/10.1016/S0140-6736(85)91147-X.
4. Tarabishy AB, Schulte M, Papaliodis GN, Hoffman GS. Wegener's granulomatosis: clinical manifestations, differential diagnosis, and management of ocular and systemic disease. *Survey of Ophthalmology*. 2010;55(5):429−444. https://doi.org/10.1016/j.survophthal.2009.12.003.
5. Nakazawa D, Masuda S, Tomaru U, Ishizu A. Pathogenesis and therapeutic interventions for ANCA-associated vasculitis. *Nature Reviews Rheumatologyl*. 2019;15(2):91−101. https://doi.org/10.1038/s41584-018-0145-y.
6. Abdou NI, Kullman GJ, Hoffman GS, et al. Wegener's granulomatosis: survey of 701 patients in North America. Changes in outcome in the 1990s. *Journal of Rheumatology*. 2002;29(2):309−316.
7. Knight A, Sandin S, Askling J. Risks and relative risks of wegener's granulomatosis among close relatives of patients with the disease. *Arthritis & Rheumatism*. 2008;58(1):302−307. https://doi.org/10.1002/art.23157.
8. Jones BE, Yang J, Muthigi A, et al. Gene-specific DNA methylation changes predict remission in patients with ANCA-associated vasculitis. *Journal of the American Society of Nephrology*. 2017;28(4):1175−1187. https://doi.org/10.1681/ASN.2016050548.

9. Berden A, Goceroglu A, Jayne D, et al. Diagnosis and management of ANCA associated vasculitis. *British Medical Journal*. 2012;344. https://doi.org/10.1136/bmj.e26.
10. Popa ER, Stegeman CA, Abdulahad WH, et al. Staphylococcal toxic-shock-syndrome-toxin-1 as a risk factor for disease relapse in Wegener's granulomatosis. *Rheumatology*. 2007;46(6):1029–1033. https://doi.org/10.1093/rheumatology/kem022.
11. Gómez-Puerta JA, Gedmintas L, Costenbader KH. The association between silica exposure and development of ANCA-associated vasculitis: systematic review and meta-analysis. *Autoimmunity Reviews*. 2013;12(12):1129–1135. https://doi.org/10.1016/j.autrev.2013.06.016.
12. Falk RJ, Terrell RS, Charles LA, Jennette JC. Anti-neutrophil cytoplasmic autoantibodies induce neutrophils to degranulate and produce oxygen radicals in vitro. *Proceedings of the National Academy of Sciences of the United States of America*. 1990;87(11):4115–4119. https://doi.org/10.1073/pnas.87.11.4115.
13. Sanders JSF, Huitma MG, Kallenberg CGM, Stegeman CA. Prediction of relapses in PR3-ANCA-associated vasculitis by assessing responses of ANCA titres to treatment. *Rheumatology*. 2006;45(6):724–729. https://doi.org/10.1093/rheumatology/kei272.
14. Finkielman JD, Merkel PA, Schroeder D, et al. Antiproteinase 3 antineutrophil cytoplasmic antibodies and disease activity in Wegener granulomatosis. *Annals of Internal Medicine*. 2007;147(9):611–619. https://doi.org/10.7326/0003-4819-147-9-200711060-00005.
15. Sivaraman BM, Majumder PD, Mahendradas P, Somanath A, Pyare R, Patil A. Current concepts in granulomatosis poly angiitis and the eye. *Current Opinion in Ophthalmology*. 2022;33(6):564–573. https://doi.org/10.1097/ICU.0000000000000906.
16. Schreiber A, Rousselle A, Becker JU, Von Mässenhausen A, Linkermann A, Kettritz R. Necroptosis controls NET generation and mediates complement activation, endothelial damage, and autoimmune vasculitis. *Proceedings of the National Academy of Sciences of the United States of America*. 2017;114(45):E9618–E9625. https://doi.org/10.1073/pnas.1708247114.
17. Chung SA, Seo P. Microscopic polyangiitis. *Rheumatic Disease Clinics of North America*. 2010;36(3):545–558. https://doi.org/10.1016/j.rdc.2010.04.003.
18. Redondo-Rodriguez R, Mena-Vázquez N, Cabezas-Lucena AM, Manrique-Arija S, Mucientes A, Fernández-Nebro A. Systematic review and metaanalysis of worldwide incidence and prevalence of antineutrophil cytoplasmic antibody (ANCA) associated vasculitis. *Journal of Clinical Medicine*. 2022;11(9):2573. https://doi.org/10.3390/jcm11092573.
19. Watts RA, Hatemi G, Burns JC, Mohammad AJ. Global epidemiology of vasculitis. *Nature Reviews Rheumatology*. 2022;18(1):22–34. https://doi.org/10.1038/s41584-021-00718-8.
20. Comarmond C, Cacoub P. Granulomatosis with polyangiitis (Wegener): clinical aspects and treatment. *Autoimmunity Reviews*. 2014;13(11):1121–1125. https://doi.org/10.1016/j.autrev.2014.08.017.
21. Heijl C, Mohammad AJ, Westman K, Höglund P. Long-term patient survival in a Swedish population-based cohort of patients with ANCA-associated vasculitis. *RMD Open*. 2017;3(1):e000435. https://doi.org/10.1136/rmdopen-2017-000435.
22. Guillevin L, Pagnoux C, Seror R, Mahr A, Mouthon L, Toumelin PL. The five-factor score revisited: assessment of prognoses of systemic necrotizing vasculitides based on the French vasculitis study group (FVSG) cohort. *Medicine*. 2011;90(1):19–27. https://doi.org/10.1097/MD.0b013e318205a4c6.
23. Millet A, Pederzoli-Ribeil M, Guillevin L, Witko-Sarsat V, Mouthon L. Antineutrophil cytoplasmic antibody-associated vasculitides: is it time to split up the group? *Annals of the Rheumatic Diseases*. 2013;72(8):1273–1279. https://doi.org/10.1136/annrheumdis-2013-203255.
24. Sfiniadaki E, Tsiara I, Theodossiadis P, Chatziralli I. Ocular manifestations of granulomatosis with polyangiitis: a review of the literature. *Ophthalmology and Therapy*. 2019;8(2):227–234. https://doi.org/10.1007/s40123-019-0176-8.
25. Jones RB, Tervaert JWC, Hauser T, et al. Rituximab versus cyclophosphamide in ANCA-associated renal vasculitis. *New England Journal of Medicine*. 2010;363(3):211–220. https://doi.org/10.1056/NEJMoa0909169.
26. Stone JH, Merkel PA, Spiera R, et al. Rituximab versus cyclophosphamide for ANCA-associated vasculitis. *New England Journal of Medicine*. 2010;363(3):221–232. https://doi.org/10.1056/NEJMoa0909905.
27. Turk MA, Hayworth JL, Nevskaya T, Pope JE. Ocular manifestations in rheumatoid arthritis, connective tissue disease, and vasculitis: a systematic review and metaanalysis. *Journal of Rheumatology*. 2021;48(1):25–34. https://doi.org/10.3899/JRHEUM.190768.
28. Rothschild PR, Pagnoux C, Seror R, Brézin AP, Delair E, Guillevin L. Ophthalmologic manifestations of systemic necrotizing vasculitides at diagnosis: a retrospective study of 1286 patients and review of the literature. *Seminars in Arthritis and Rheumatism*. 2013;42(5):507–514. https://doi.org/10.1016/j.semarthrit.2012.08.003.
29. Androudi S, Dastiridou A, Symeonidis C, et al. Retinal vasculitis in rheumatic diseases: an unseen burden. *Clinical Rheumatology*. 2013;32(1):7–13. https://doi.org/10.1007/s10067-012-2078-1.
30. Bullen CL, Liesegang TJ, McDonald TJ, DeRemee RA. Ocular complications of wegener's granulomatosis. *Ophthalmology*. 1983;90(3):279–290. https://doi.org/10.1016/S0161-6420(83)34574-7.
31. Rahman Z, Jafarullah O, Huq N, Hasan K, Magesan K, Dutta Majumder P. Clinical Profile of patients with scleritis in Bangladesh. *Ocular Immunology and Inflammation*. 2021;29(6):1102–1105. https://doi.org/10.1080/09273948.2020.1716026.
32. Newman NJ, Slamovits TL, Friedland S, Wilson WB. Neuro-ophthalmic manifestations of meningocerebral inflammation from the limited form of Wegener's granulomatosis. *American Journal of Ophthalmology*. 1995;120(5):613–621. https://doi.org/10.1016/S0002-9394(14)72208-1.
33. Ng CC, Sy A, Cunningham ET. Rituximab for non-infectious uveitis and scleritis. *Journal of Ophthalmic Inflammation and Infection*. 2021;11(1). https://doi.org/10.1186/s12348-021-00252-4.
34. Ng CC, Sy A, Cunningham ET. Rituximab for treatment of non-infectious and non-malignant orbital inflammatory disease. *Journal of Ophthalmic Inflammation and Infection*. 2021;11(1). https://doi.org/10.1186/s12348-021-00253-3.
35. Ahmed A, Foster CS. Cyclophosphamide or rituximab treatment of scleritis and uveitis for patients with granulomatosis with polyangiitis. *Ophthalmic Research*. 2019;61(1):44–50. https://doi.org/10.1159/000486791.

36. Lin M, Anesi SD, Ma L, Ahmed A, Small K, Foster CS. Characteristics and visual outcome of Refractory retinal vasculitis associated with antineutrophil Cytoplasm antibody–associated vasculitides. *American Journal of Ophthalmology*. 2018;187:21–33. https://doi.org/10.1016/j.ajo.2017.12.004.
37. Rosenbaum JT, Ku J, Ali A, Choi D, Suhler EB. Patients with retinal vasculitis rarely suffer from systemic vasculitis. *Seminars in Arthritis and Rheumatism*. 2012;41(6):859–865. https://doi.org/10.1016/j.semarthrit.2011.10.006.
38. Huvard MJ, Pecen PE, Palestine AG. The clinical characteristics of Noninfectious occlusive retinal vasculitis. *Ophthalmology Retina*. 2022;6(1):43–48. https://doi.org/10.1016/j.oret.2021.04.012.
39. Ku JH, Ali A, Suhler EB, Choi D, Rosenbaum JT. Characteristics and visual outcome of patients with retinal vasculitis. *Archives of Ophthalmology*. 2012;130(10):1261–1266. https://doi.org/10.1001/archophthalmol.2012.1596.
40. Nussenblatt RB, Rosenbaum JT, Atmaca LS, et al. Standardization of uveitis nomenclature for reporting clinical data. Results of the first international workshop. *American Journal of Ophthalmology*. 2005;140(3):509–516. https://doi.org/10.1016/j.ajo.2005.03.057.
41. Bataduwaarachchi VR, Galappaththi R, Tissera N. Blindness in a Sri Lankan woman with bilateral breast lumps: a case report. *Journal of Case Reports in Medicine*. 2015;9(1). https://doi.org/10.1186/s13256-015-0792-4.
42. Mangouritsas G, Ulbig M. Cotton-wool spots as the initial ocular manifestation in Wegener's granulomatosis. *German Journal of Ophthalmology*. 1994;3(2):68–70.
43. Matlach J, Freiberg FJ, Gadeholt O, Göbel W. Vasculitis-like hemorrhagic retinal angiopathy in Wegener's granulomatosis. *BMC Research Notes*. 2013;6(1). https://doi.org/10.1186/1756-0500-6-364.
44. Saati S, Sadda SR. Seronegative granulomatous polyangiitis with central retinal artery occlusion as the initial manifestation. *Retinal Cases & Brief Reports*. 2014;8(4):236–239. https://doi.org/10.1097/ICB.0000000000000109.
45. Proia AD. Granulomatous choroiditis in Wegener granulomatosis. *Archives of Ophthalmology*. 2011;129(4):520–521. https://doi.org/10.1001/archophthalmol.2011.62.
46. Masuda T, Izumi Y, Takeshita H, et al. Granulomatosis with polyangiitis presenting as a choroidal tumor. *Case Reports in Rheumatology*. 2015;2015:1–5. https://doi.org/10.1155/2015/271823.
47. Margoni M, Barbareschi M, Rozzanigo U, Chioffi F, Paolazzi G, Marangoni S. Isolated choroid plexus involvement in a case of granulomatosis with polyangiitis negative for antineutrophil cytoplasmic antibodies (ANCA). *Journal of the Neurological Sciences*. 2019;398:128–130. https://doi.org/10.1016/j.jns.2019.01.045.
48. Wong SC, Boyce RL, Dowd TC, Fordham JN. Bilateral central retinal artery occlusion in Wegener's granulomatosis and α1 antitrypsin deficiency [6]. *British Journal of Ophthalmology*. 2002;86(4):476. https://doi.org/10.1136/bjo.86.4.476.
49. Toh T, Cooper RL, Parker A, Vote BJ. Wegener's granulomatosis presenting with painless bulbar-conjunctival ulcer and central retinal artery occlusion. *Clinical & Experimental Ophthalmology*. 2006;34(1):51–53. https://doi.org/10.1111/j.1442-9071.2006.01144.x.
50. Shim KY, Sohn S-I, Kim YC. Central retinal arterial occlusion in granulomatosis with polyangiitis. *Korean Journal of Ophthalmology*. 2018;32(6):519. https://doi.org/10.3341/kjo.2018.0019.
51. Shenoy R, Elagib eNM, al Siyabi H, Amer A. Limited Wegener's granulomatosis presenting as multiple retinal vascular occlusions. *Indian Journal of Ophthalmology*. 2002;50(2):135–137.
52. Peng YJ, Fang PC, Huang WT, Chen YJ. Central retinal artery occlusion in Wegener's granulomatosis: a case report and review of the literature. *Canadian Journal of Ophthalmology*. 2004;39(7):785–789. https://doi.org/10.1016/S0008-4182(04)80076-0.
53. Lamprecht P. Retinal artery occlusion in Wegener's granulomatosis. *Rheumatology*. 2000;39(8):928–929. https://doi.org/10.1093/rheumatology/39.8.928.
54. Kinyoun JL, Kalina RE, Klein ML. Choroidal involvement in systemic necrotizing vasculitis. *Archives of Ophthalmology*. 1987;105(7):939–942. https://doi.org/10.1001/archopht.1987.01060070075032.
55. Iida T, Spaide RF, Kantor J. Retinal and choroidal arterial occlusion in Wegener's granulomatosis. *American Journal of Ophthalmology*. 2002;133(1):151–152. https://doi.org/10.1016/S0002-9394(01)01251-X.
56. Venkatesh P, Chawla R, Tewari HK. Hemiretinal vein occlusion in Wegener's granulomatosis. *European Journal of Ophthalmology*. 2003;13(8):722–725. https://doi.org/10.1177/112067210301300811.
57. Stavrou P, Deutsch J, Rene C, Laws DE, Luqmani RA, Murray PI. Ocular manifestations of classical and limited wegener's granulomatosis. *Quarterly Journal of Medicine*. 1993;86(11):719–725. https://doi.org/10.1093/oxfordjournals.qjmed.a068752.
58. Spalton DJ, Graham EM, Page NGR, Sanders MD. Ocular changes in limited forms of Wegener's granulomatosis. *British Journal of Ophthalmology*. 1981;65(8):553–563. https://doi.org/10.1136/bjo.65.8.553.
59. Wang M, Khurana RN, Sadda SR. Central retinal vein occlusion in Wegener's granulomatosis without retinal vasculitis [6]. *British Journal of Ophthalmology*. 2006;90(11):1435–1436. https://doi.org/10.1136/bjo.2006.095703.
60. Hupy ML, Pecen PE, Kolfenbach JR, Lau M, Palestine AG. Purtscher'slike retinopathy secondary to granulomatosis with polyangiitis following autologous stem cell transplantation. *Retinal Cases & Brief Reports*. 2022;16(4):444–447. https://doi.org/10.1097/ICB.0000000000001002.
61. Takashi N, Nakamura A, Kataoka K, Usui Y, Ito Y, Kaneko H. Optical coherence tomography angiography for the diagnosis of granulomatosis with polyangiitis with serous retinal detachment: a case report. *Medicine*. 2021;100(7):e24789. https://doi.org/10.1097/MD.0000000000024789.
62. Mehta S, Chitnis N, Medhekar A. Utility of optical coherence tomography angiography (OCTA) in granulomatosis with polyangiitis. *Cureus*. 2022. https://doi.org/10.7759/cureus.22612.
63. Güven YZ, Akay F, Akmaz B, Solmaz D, Gercik Ö, Akar S. Subclinical alterations in retinal layers and microvascular structures with OCTA in ANCA-associated vasculitides. *Ocular Immunology and Inflammation*. 2022:1–6. https://doi.org/10.1080/09273948.2022.2038206.

64. Yonekawa Y, Thanos A, Faia LJ. Ultra-widefield imaging in granulomatosis with polyangiitis. *JAMA Ophthalmology*. 2016;134(7):e160034. https://doi.org/10.1001/jamaophthalmol.2016.0034.
65. Hsu CT, Harlan JB, Goldberg MF, Dunn JP. Acute posterior multifocal placoid pigment epitheliopathy associated with a systemic necrotizing vasculitis. *Retina*. 2003;23(1):64–68. https://doi.org/10.1097/00006982-200302000-00010.
66. Quist MS, Bhullar PK, Meekins LC, Gupta D. Granulomatosis with polyangiitis-associated ischemic optic neuropathy in a previously healthy 50-year-old female. *American Journal of Ophthalmology Case Reports*. 2020;20. https://doi.org/10.1016/j.ajoc.2020.100923.
67. Leveille AE, Morse PH. Combined detachments in Wegener's granulomatosis. *British Journal of Ophthalmology*. 1981;65(8):564–567. https://doi.org/10.1136/bjo.65.8.564.
68. Kim DY, Jeong J, Kim JY. Frosted branch angiitis secondary to granulomatosis with polyangiitis. *Korean Journal of Ophthalmology*. 2019;33(5):485. https://doi.org/10.3341/kjo.2019.0033.
69. Khurma V, Appen R, Wolf MD, Hansen KE. Wegener granulomatosis presenting as bilateral loss of vision. *Journal of Clinical Rheumatology*. 2005;11(5):267–269. https://doi.org/10.1097/01.rhu.0000182168.78683.16.
70. Sandhu RK, Adams T, Sibley C, Suhler EB, Kim DH. Granulomatosis with polyangiitis (Gpa) presenting with frosted branch angiitis. *Retinal Cases & Brief Reports*. 2016;10(3):249–251. https://doi.org/10.1097/ICB.0000000000000242.
71. Damato EM, Moriarty JT, Harper SJ, Dick AD, Bailey CC. Plasmapheresis in the management of choroidal vasculitis associated with c-anca positive renal vasculitis. *Retinal Cases & Brief Reports*. 2010;4(4):356–360. https://doi.org/10.1097/ICB.0b013e3181c597a4.
72. Reddy S, Tyagi M, Behera S, Pappuru RR. Cytomegalovirus retinitis in a patient of granulomatosis with polyangiitis on long-term immunosuppressants. *BMJ Case Reports*. 2021;14(2):e236632. https://doi.org/10.1136/bcr-2020-236632.
73. Murray PI. Wegener's granulomatosis, pituitary adenoma and BARN. *International Ophthalmology*. 1994;18(6):361–362. https://doi.org/10.1007/BF00930316.
74. Georgalas I. Cytomegalovirus retinitis presenting as vasculitis in a patient with Wegener's granulomatosis. *Clinical Ophthalmology*. 2008:961. https://doi.org/10.2147/opth.s4022.
75. Bertelmann E, Liekfeld A, Pleyer U, Hartmann C. Cytomegalovirus retinitis in Wegener's granulomatosis: case report and review of the literature. *Acta Ophthalmologica Scandinavica*. 2005;83(2):258–261. https://doi.org/10.1111/j.1600-0420.2005.00410.x.
76. AlQahtani GMS, AlSayed AAD, Gangadharan S, Adhi MI. Fungal endophthalmitis in a case of granulomatosis with polyangitis. *Saudi Journal of Ophthalmology*. 2018;32(3):261–265. https://doi.org/10.1016/j.sjopt.2018.01.001.
77. Yamashita K, Kobayashi S, Kondo M, Ishikura H, Shibuya Y, Hayasaka S. Elevated anti-neutrophil cytoplasmic antibody titer in a patient with Atypical orbital Pseudotumor. *Ophthalmologica*. 1995;209(3):172–175. https://doi.org/10.1159/000310606.
78. Janknecht P, Mittelviefhaus H, Löffler KU. Sclerochoroidal granuloma in wegener's granulomatosis simulating a uveal melanoma. *Retina*. 1995;15(2):150–153. https://doi.org/10.1097/00006982-199515020-00011.
79. Agostini HT, Bräutigam P, Löffler KU. Subretinal tumour in a patient with a limited form of Wegener's granulomatosis. *Acta Ophthalmologica Scandinavica*. 1995;73(5):460–463. https://doi.org/10.1111/j.1600-0420.1995.tb00310.x.
80. Wu W, Chaer RA. Nonarteriosclerotic vascular disease. *Surgical Clinics*. 2013;93(4):833–875. https://doi.org/10.1016/j.suc.2013.04.003.
81. Lassiter FD. The string-of-beads sign. *Radiology*. 1998;206(2):437–438. https://doi.org/10.1148/radiology.206.2.9457197.
82. Jennette JC, Falk RJ, Bacon PA, et al. 2012 Revised International Chapel Hill consensus conference nomenclature of vasculitides. *Arthritis & Rheumatism*. 2013;65(1):1–11. https://doi.org/10.1002/art.37715.
83. De Virgilio A, Greco A, Magliulo G, et al. Polyarteritis nodosa: a contemporary overview. *Autoimmunity Reviews*. 2016;15(6):564–570. https://doi.org/10.1016/j.autrev.2016.02.015.
84. Mahr A, Guillevin L, Poissonnet M, Aymé S. Prevalences of polyarteritis nodosa, microscopic polyangiitis, wegener's granulomatosis, and Churg-Strauss syndrome in a French Urban Multiethnic population in 2000: a Capture-Recapture estimate. *Arthritis Care & Research*. 2004;51(1):92–99. https://doi.org/10.1002/art.20077.
85. Legendre P, Régent A, Thiebault M, Mouthon L. Anti-endothelial cell antibodies in vasculitis: a systematic review. *Autoimmunity Reviews*. 2017;16(2):146–153. https://doi.org/10.1016/j.autrev.2016.12.012.
86. Zhou Q, Yang D, Ombrello AK, et al. Early-onset stroke and vasculopathy associated with mutations in ADA2. *New England Journal of Medicine*. 2014;370(10):911–920. https://doi.org/10.1056/NEJMoa1307361.
87. Springer JM, Byram K. Polyarteritis nodosa: an evolving primary systemic vasculitis. *PGM (Postgraduate Medicine)*. 2022:1–8. https://doi.org/10.1080/00325481.2022.2088940.
88. Elkan PN, Pierce SB, Segel R, et al. Mutant adenosine deaminase 2 in a polyarteritis nodosa vasculopathy. *New England Journal of Medicine*. 2014;370(10):921–931. https://doi.org/10.1056/NEJMoa1307362.
89. Mohammad AJ, Jacobsson LTH, Westman KWA, Sturfelt G, Segelmark M. Incidence and survival rates in Wegener's granulomatosis, microscopic polyangiitis, Churg-Strauss syndrome and polyarteritis nodosa. *Rheumatology*. 2009;48(12):1560–1565. https://doi.org/10.1093/rheumatology/kep304.
90. Watts RA, Lane SE, Bentham G, Scott DGI. Epidemiology of systemic vasculitis: a ten-year study in the United Kingdom. *Arthritis & Rheumatism*. 2000;43(2):414–419. https://doi.org/10.1002/1529-0131(200002)43:2<414::AID-ANR23>3.0.CO;2-0.
91. Ozen S, Anton J, Arisoy N, et al. Juvenile polyarteritis: results of a multicenter survey of 110 children. *The Journal of Pediatrics*. 2004;145(4):517–522. https://doi.org/10.1016/j.jpeds.2004.06.046.

92. Pagnoux C, Seror R, Henegar C, et al. Clinical features and outcomes in 348 patients with polyarteritis nodosa: a systematic retrospective study of patients diagnosed between 1963 and 2005 and entered into the French Vasculitis Study Group database. *Arthritis & Rheumatism*. 2010;62(2):616–626. https://doi.org/10.1002/art.27240.
93. Mohammad AJ, Jacobsson LTH, Mahr AD, Sturfelt G, Segelmark M. Prevalence of Wegener's granulomatosis, microscopic polyangiitis, polyarteritis nodosa and Churg-Strauss syndrome within a defined population in southern Sweden. *Rheumatology*. 2007;46(8):1329–1337. https://doi.org/10.1093/rheumatology/kem107.
94. Kanecki K, Nitsch-Osuch A, Gorynski P, Wierzba W, Tarka P, Tyszko P. Polyarteritis nodosa: decreasing incidence in Poland. *Archives of Medical Science*. 2019;15(5):1308–1312. https://doi.org/10.5114/aoms.2017.68407.
95. Gonzalez-Gay MA, Garcia-Porrua C, Guerrero J, Rodriguez-Ledo P, Llorca J. The epidemiology of the primary systemic vasculitides in northwest Spain: implications of the Chapel Hill Consensus Conference definitions. *Arthritis Care & Research*. 2003;49(3):388–393. https://doi.org/10.1002/art.11115.
96. Lightfoot RW, Michel BA, Bloch DA, et al. The American college of rheumatology 1990 criteria for the classification of polyarteritis nodosa. *Arthritis & Rheumatism*. 1990;33(8):1088–1093. https://doi.org/10.1002/art.1780330805.
97. Hernández-Rodríguez J, Alba MA, Prieto-González S, Cid MC. Diagnosis and classification of polyarteritis nodosa. *Journal of Autoimmunity*. 2014;48–49:84–89. https://doi.org/10.1016/j.jaut.2014.01.029.
98. Watts R, Lane S, Hanslik T, et al. Development and validation of a consensus methodology for the classification of the ANCA-associated vasculitides and polyarteritis nodosa for epidemiological studies. *Annals of the Rheumatic Diseases*. 2007;66(2):222–227. https://doi.org/10.1136/ard.2006.054593.
99. Frohnert PP, Sheps SG. Long-term follow-up study of periarteritis nodosa. *American Medical Journal*. 1967;43(1):8–14. https://doi.org/10.1016/0002-9343(67)90144-1.
100. Samson M, Puéchal X, Devilliers H, et al. Long-term follow-up of a randomized trial on 118 patients with polyarteritis nodosa or microscopic polyangiitis without poor-prognosis factors. *Autoimmunity Reviews*. 2014;13(2):197–205. https://doi.org/10.1016/j.autrev.2013.10.001.
101. Chung SA, Gorelik M, Langford CA, et al. 2021 American college of rheumatology/vasculitis foundation guideline for the management of polyarteritis nodosa. *Arthritis Care & Research*. 2021;73(8):1061–1070. https://doi.org/10.1002/acr.24633.
102. Langford CA. Treatment of polyarteritis nodosa, microscopic polyangiitis, and Churg-Strauss syndrome: where do we stand? *Arthritis & Rheumatism*. 2001;44(3):508–512. https://doi.org/10.1002/1529-0131(200103)44:3<508::AID-ANR96>3.0.CO;2-B.
103. Travers RL, Allison DJ, Brettle RP, Hughes GRV. Polyarteritis nodosa: a clinical and angiographic analysis of 17 cases. *Seminars in Arthritis and Rheumatism*. 1979;8(3):184–199. https://doi.org/10.1016/S0049-0172(79)80007-4.
104. Aristodemou P, Stanford M. Therapy insight: the recognition and treatment of retinal manifestations of systemic vasculitis. *Nature Clinical Practice Rheumatology*. 2006;2(8):443–451. https://doi.org/10.1038/ncprheum0268.
105. Graham EM, Stanford MR, Sanders MD, Kasp E, Dumonde DC. A point prevalence study of 150 patients with idiopathic retinal vasculitis: 1. Diagnostic value of ophthalmological features. *British Journal of Ophthalmology*. 1989;73(9):714–721. https://doi.org/10.1136/bjo.73.9.714.
106. Rodríguez-González F, Reyes-Rodríguez M, Francisco-Hernández F, Martínez-Quintana E. Papillophlebitis as manifestation of polyarteritis nodosa. *Archivos de la Sociedad Espanola de Oftalmologia*. 2019;94(6):300–303. https://doi.org/10.1016/j.oftal.2018.12.007.
107. Leung ACT, McLay A, Boulton-Jones JM. Polyarteritis presenting with thrombocytosis and central retinal vein thrombosis. *Scottish Medical Journal*. 1987;32(1):24–26. https://doi.org/10.1177/003693308703200112.
108. Håskjold E, Frøland S, Egge K. Ocular polyarteritis nodosa: report of a case. *Acta Ophthalmologica*. 1987;65(6):749–751. https://doi.org/10.1111/j.1755-3768.1987.tb07076.x.
109. Thakker AD, Gajre M, Khubchandani R, Mendadkar R, Doshi A, Jadhav A. Bilateral central retinal artery occlusion: an Unusual presentation of polyarteritis nodosa. *Indian Journal of Pediatrics*. 2014;81(12):1401–1402. https://doi.org/10.1007/s12098-014-1438-z.
110. Curi ALL, Freeman G, Pavesio C. Aggressive retinal vasculitis in polyarteritis nodosa [3]. *Eye*. 2001;15(2):229–231. https://doi.org/10.1038/eye.2001.68.
111. Al-Balbeesi AO, Alhallaf RA, Alsubeeh NA, Fathaddin AA, Bedaiwi AA, Omair MA. Coexistence of polyarteritis nodosa of the Vulva and Retina in a Behçet's disease patient: a case report. *Cureus*. 2021. https://doi.org/10.7759/cureus.16096.
112. Vazquez-Romo KA, Rodriguez-Hernandez A, Paczka JA, Nuño-Suarez MA, Rocha-Muñoz AD, Zavala-Cerna MG. Optic neuropathy secondary to polyarteritis nodosa, case report, and diagnostic challenges. *Frontiers in Neurology*. 2017;8. https://doi.org/10.3389/fneur.2017.00490.
113. Sener H, Sevim DG, Unlu M. Anterior ischemic optic neuropathy and cilioretinal artery occlusion secondary to polyarteritis nodosa: a case report. *Photodiagnosis and Photodynamic Therapy*. 2021;34. https://doi.org/10.1016/j.pdpdt.2021.102299.
114. Kostina-O'Neil Y, Jirawuthiworavong GV, Podell DN, Lesser RL. Choroidal and optic nerve infarction in hepatitis C-associated polyarteritis nodosa. *Journal of Neuro-Ophthalmology*. 2007;27(3):184–188. https://doi.org/10.1097/WNO.0b013e31814b1d29.
115. Hsu CT, Kerrison JB, Miller NR, Goldberg MF. Choroidal infarction, anterior ischemic optic neuropathy, and central retinal artery occlusion from polyarteritis nodosa. *Retina*. 2001;21(4):348–351. https://doi.org/10.1097/00006982-200108000-00009.
116. Emad Y, Basaffar S, Ragab Y, Zeinhom F, Gheita T. A case of polyarteritis nodosa complicated by left central retinal artery occlusion, ischemic optic neuropathy, and retinal vasculitis. *Clinical Rheumatology*. 2007;26(5):814–816. https://doi.org/10.1007/s10067-006-0270-x.
117. Morgan CM, Foster CS, D'amico DJ, Gragoudas ES. Retinal vasculitis in polyarteritis nodosa. *Retina*. 1986;6(4):205–209. https://doi.org/10.1097/00006982-198606040-00002.
118. Baytaroğlu A, Kadayifçilar S, Ağin A, et al. Choroidal vascularity index as a biomarker of systemic inflammation in childhood Polyarteritis Nodosa and adenosine deaminase-2 deficiency. *Pediatric Rheumatology*. 2020;18(1). https://doi.org/10.1186/s12969-020-0417-3.

119. Ramtohul P, Comet A, Denis D. Segmental choroidal vasculitis in nodular posterior scleritis associated with polyarteritis nodosa. *Retina*. 2020;40(10):e58–e59. https://doi.org/10.1097/iae.0000000000002907.
120. Newman NM, Hoyt WF, Spencer WH. Macula-sparing monocular blackouts: clinical and pathologic Investigations of Intermittent choroidal vascular Insufficiency in a case of periarteritis nodosa. *Archives of Ophthalmology*. 1974;91(5):367–370. https://doi.org/10.1001/archopht.1974.03900060379006.
121. Kielar RA. Exudative retinal detachment and scleritis in polyarteritis. *American Journal of Ophthalmology*. 1976;82(5):694–698. https://doi.org/10.1016/0002-9394(76)90004-0.
122. Kennedy F, Kapelow R, Kalyon BD, Roth NC, Rishi A, Barilla-LaBarca ML. A rare case of Polyarteritis Nodosa associated with autoimmune hepatitis: a case report. *BMC Rheumatology*. 2021;5(1). https://doi.org/10.1186/s41927-021-00188-1.
123. Stefani FH, Brandt F, Pielsticker K. Periarteritis nodosa and thrombotic thrombocytopenic purpura with serous retinal detachment in siblings. *British Journal of Ophthalmology*. 1978;62(6):402–407. https://doi.org/10.1136/bjo.62.6.402.
124. Jogimahanti AV, Raviskanthan S, Mortensen PW, Kazzaz SA, Lee AG. Roth spot and a homonymous hemianopsia as the presenting manifestation of polyarteritis nodosa. *Journal of Neuro-Ophthalmology: The Official Journal of the North American Neuro-Ophthalmology Society*. 2022;42(2):e482–e484. https://doi.org/10.1097/WNO.0000000000001303.

Chapter 31

Retinal manifestations of Takayasu's arteritis

Sophie Bonnin[1] **and Alain Gaudric**[2]
[1]*Department of Ophthalmology, Rothschild Foundation Hospital, Paris, France;* [2]*Université Paris Cité, Department of Ophthalmology, Hôpital Lariboisière, APHP, Paris, France*

The ocular findings of Takayasu's disease were described by a Japanese ophthalmologist, Mikito Takayasu, in 1908. The term "pulseless disease" was proposed in 1948 by Shimizu and Sano but, ultimately, the term "Takayasu's arteritis" (TAK) has been accepted worldwide (see details in Sugiyama et al.[1]).

Takayasu's arteritis

Takayasu's disease belongs, along with giant cell arteritis, to the group of large vessel vasculitis, as formulated in the 2012 Revised International Chapel Hill Consensus Conference Nomenclature of Vasculitides.[2] TAK is a chronic disease, defined by granulomatous inflammation involving the aorta and its primary branches. Chronic inflammation and endothelial damage usually lead to wall thickening, thrombus formation, and stenotic and occlusive lesions, while acute inflammation results in the destruction of the muscularis and elastic layers leading to dilation and aneurysms.[3] Complications of vascular damage may result in substantial morbidity, including stroke, myocardial infarction, mesenteric ischemia, and limb claudication.[4] Stenosis of the aortic arch and its branches results in a low blood flow in the ophthalmic artery (OA) and retinal vascular anomalies known as Takayasu's retinopathy (TR).

The disease usually appears before the age of 50 years, the male-to-female ratio is 1:9, and the age of onset is about 20 years in most women.[1]

Takayasu's disease affects patients worldwide, but most cases have been reported in Asia and in the Middle and Near East.

TAK pathogenesis

The pathogenesis of TAK is incompletely understood. It has been shown that all layers of the vascular wall were affected in samples collected during surgery or autopsy.

The initial site of inflammation is the medio-adventitial junction, where the vasa vasorum penetrates the artery wall.

The main pathological features are granulomatous inflammatory panarteritis with intimal fibrocellular hyperplasia, disruption of the elastic lamina, and thrombus formation. This inflammation results in a reduction in large vessel lumen, mainly in the subclavian arteries and aortic arch branches, but it may occur in any segment. Ultimately, degenerative changes lead to a muscular layer weakening and aneurysm formation.[5,6]

Acute vascular inflammation observed could be due to the activation of the MICA/NKG2D pathway in response to an unknown stimulus, resulting in the activation of T and NK cells recruited to the vascular wall. The secondary production of proinflammatory cytokines and interferon-γ by Th1 lymphocytes induces the activation of macrophages, the release of VEGF and PDGF, and intimal proliferation.[7,8] There is no specific marker of the disease.

A genetic contribution is supported by the association between TAK and HLA-B52 and B67 in Japanese and other populations.[9]

Retinal and Choroidal Vascular Diseases of the Eye. https://doi.org/10.1016/B978-0-443-15583-3.00031-7

TAK features and classification

The criteria for the diagnosis of TAK have recently been reclassified by the American College of Rheumatology.[4] The absolute requirements for the diagnosis are age below 60 years at the time of diagnosis and signs of vasculitis on imaging. Additional clinical criteria include the female gender, angina or ischemic cardiac pain, arm or leg claudication, reduced pulse in an upper extremity, carotid artery abnormality, and a difference in systemic blood pressure between the arms $\geq$20 mm Hg. Additional imaging criteria include the number of affected territories, the symmetric involvement of paired arteries, and an abdominal aorta involvement with renal or mesenteric involvement. A score is given to each sign.

A common systemic complication of TAK is arterial hypertension either due to the involvement of the renal artery or to coarctation of the aorta.

An angiographic classification into five types has been proposed based on the distribution of the lesions.[10–12] In type I TAK, the inflammatory process is localized to the aortic arch and its branches. In type II TAK, the lesions involve the ascending aorta, aortic arch, and its branches. These two types may lead to TR. The other types involve the other parts of the aorta.

The natural history of the vascular lesions in TAK is variable. Curiously, arteritis targets some portions of the vascular tree while sparing others. This suggests that there is a vessel-specific pattern in the response of the various arterial beds to inflammatory stimuli.[8,13,14]

Takayasu's retinopathy

Prevalence

TR and hypertensive retinopathy are the most common ocular manifestations of TAK.[15] The reported prevalence ranges between 8.1% and 68%[16] and varies over time.[17] In the 1990s, TR was reported in more than 30% of TAK cases.[18,19] Twenty years later, the reported prevalence has decreased to less than 20%,[16,20] and more recently to only 6% of TAK cases.[17] However, in the latter publication, the diagnosis of TR was made based on color photographs of the posterior pole and fluorescein angiography (FA) was performed to confirm the diagnosis in only 4 out of the 65 patients. The prevalence of TR might thus have been underestimated. In a recent study by Poignet et al., microaneurysms were visible in 18.5% of eyes on fundus examination alone. This percentage increased to 24.4% of eyes on ultra-widefield color fundus photography (UWF-CFP) and to 94.4% of eyes on UWF-FA.[21]

Pathophysiology

The pathophysiology of TR remains controversial. The supraaortic stenosis could induce retinal hypoperfusion and subsequent microaneurysm development in response to retinal hypoxia. Poignet et al.[21] have found a significant correlation between the number of microaneurysms and the mean blood flow velocity (BFV) in the central retinal artery (CRA) ($r = -0.61$, $P = .003$), supporting the role of hypoperfusion.

The occurrence of microaneurysms in various retinal vascular diseases, including diabetic retinopathy, ocular ischemic syndrome, and TAK, could be related to a decrease in arterial blood flow and retinal hypoxia. In 1986, Kahn et al.[22] have observed a relatively greater loss of pericytes compared to endothelial cells and a marked loss of pericytes and microaneurysm formation at the equator and in the periphery in eyes with hypotensive retinopathy due to carotid occlusion after retinal trypsin digestion. Interestingly, Poignet et al.[21] have also shown a predominant involvement of the peripheral retina compared to the posterior pole in TAK.

In TAK, the relationship between the decrease in retinal arterial pressure and microaneurysms has been shown in 1983 by Ishikawa et al.[23]

The relationship between hypoxia and the formation of microaneurysms has been shown experimentally by Linsenmeier et al.[24] in a study of retinal hypoxia in long-term diabetic cats. They have shown that inner retinal layer hypoxia may be compatible with any capillary dropout detectable on FA or flat mounts, but that microaneurysms developed in the inner retina where the PO_2 was especially low. Tissue responses to ischemia, such as the production of VEGF, do not appear to require large areas of visible capillary dropout and may result in microaneurysm formation.

In normal eyes of cynomolgus monkeys, intravitreal injections of VEGF have been shown to induce the formation of microaneurysms in the whole peripheral retina,[25] which could explain the pathophysiology of the ocular ischemic syndrome, as seen in TAK. The predominant location of retinal microaneurysms around the venules, where the PO_2 is physiologically lower, also supports the hypothesis of the role of chronic retinal hypoxia.

Clinical features, color fundus photo, and fluorescein angiography findings

A loss of visual acuity (VA) is rarely seen as the first symptom. In a series of 78 patients, Chun et al. have found a VA within the normal range but the most common complaint of patients was a transient subjective deterioration of vision that they have attributed to amaurosis fugax in 25% of their patients.[20]

The ophthalmoscopic signs of TAK have been first summarized by Hirose in 1963. Using the current nomenclature, they consisted in tortuosity and dilation of retinal veins, microaneurysms, arteriovenous anastomosis, capillary dilation, and granular flow in the retinal vessels.[26]

Shimizu[27,28] has been the first to report FA findings in 1965. He considered retinal microaneurysms as the basic pathological condition of the early stage and reported a prolonged duration of the laminar flow, sludging, a slower blood flow, dilated retinal vessels, fluorescein leakage due to increased vascular permeability, and arteriovenous anastomosis.

UWF-FA now easily shows microaneurysms in the retinal periphery (especially in the temporal periphery) associated with slow and poor peripheral perfusion (Figs. 31.1 and 31.2). Leakage and dilation of the retinal veins could be due to a very slow retinal vessel perfusion (Fig. 31.3). Multiple peripheral microaneurysms seem to be the most prevalent and specific sign of TR.

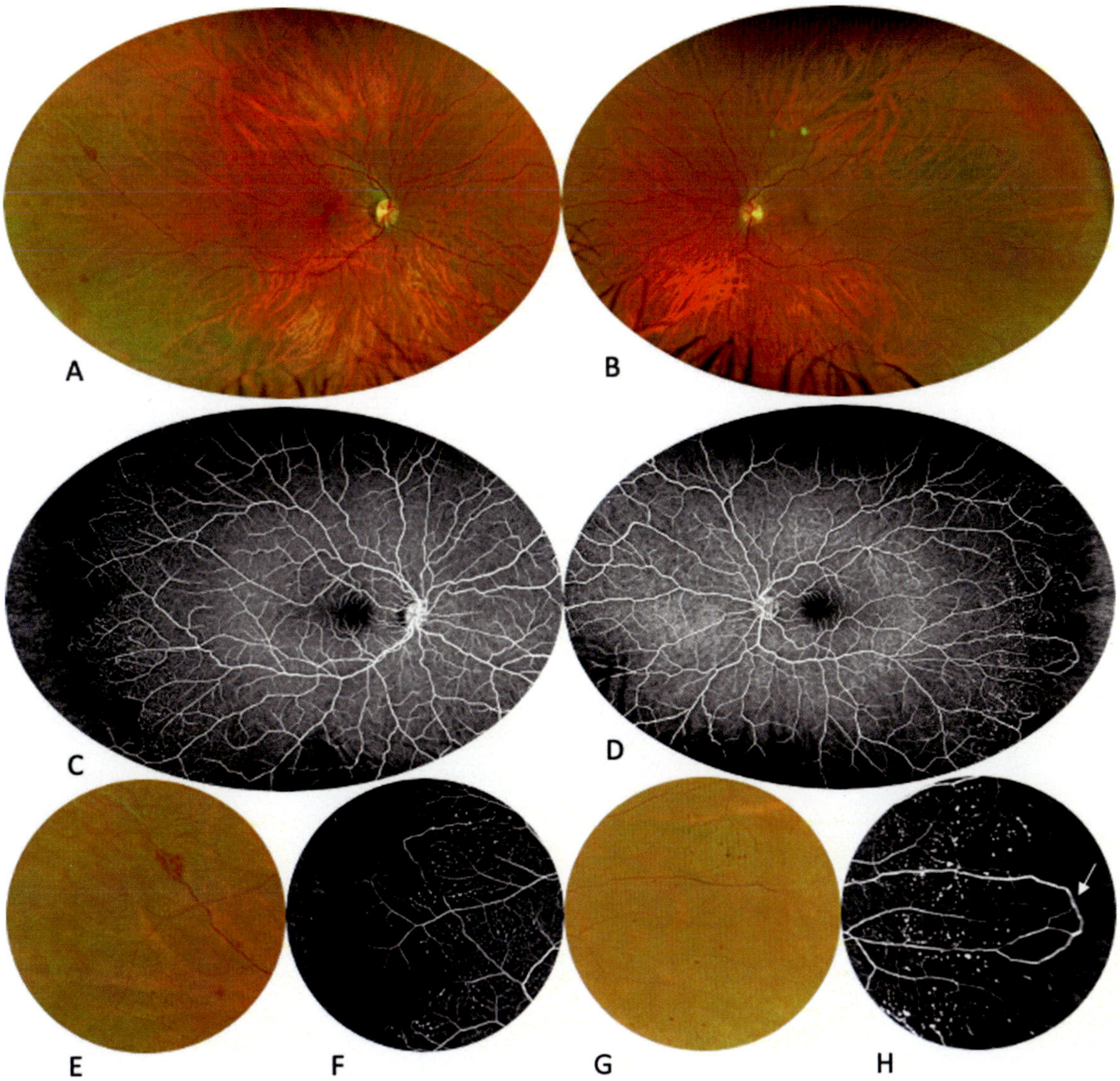

FIGURE 31.1 Takayasu's retinopathy with peripheral microaneurysms and peripheral venovenous shunts. (A and B) Ultra-widefield (UWF) color fundus photography of both eyes showing retinal hemorrhages in the temporal periphery of the right eye (A, detail in E), and microaneurysms in the temporal periphery of the left eye (B, detail in H). (C and D) mid-phase UWF fluorescein angiography showing numerous microaneurysms in the temporal periphery of both eyes behind a large area of peripheral nonperfusion (details in F and I), and a peripheral venovenous anastomosis in the left eye.

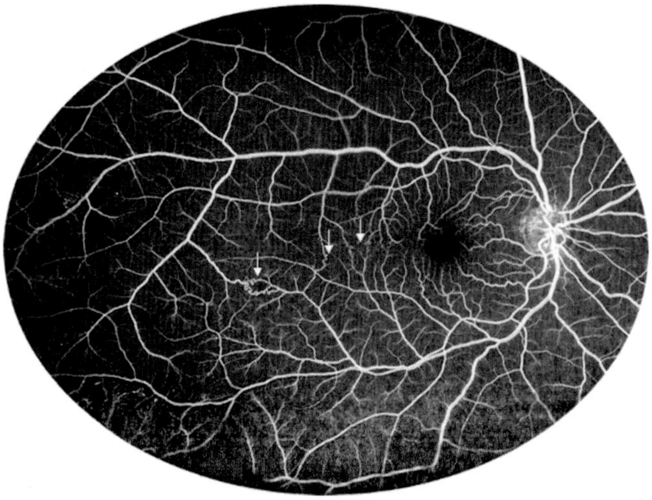

FIGURE 31.2 Takayasu's retinopathy with several venovenous anastomoses (*arrows*) along the temporal raphe despite no sign of branch vein occlusion.

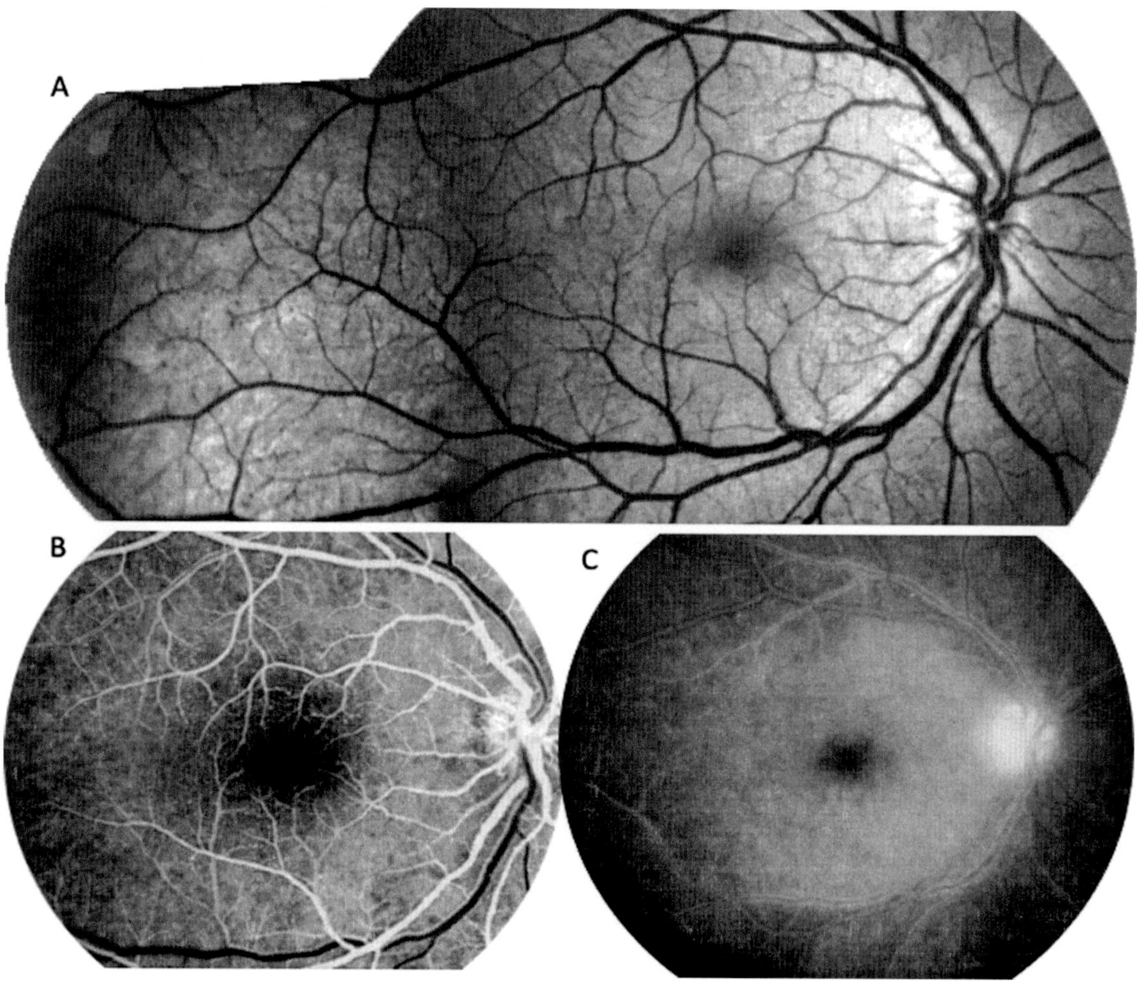

FIGURE 31.3 Takayasu's retinopathy with venous dilation and leakage. The red-free image of the posterior pole and temporal midperiphery showed numerous microaneurysms and a venous dilation (A), confirmed in the early phase of fluorescein angiography (B). Leakage from the veins and optic disc on the late-phase angiogram (C) due to the retinal hypoperfusion and global retinal hypoxia.

Tanaka and Shimizu[29] have reported that shunts may form at arteriovenous crossings or develop through the growth of preferential channels. On UWF-FA, arteriovenous or venovenous shunts have been found in 22% of cases (Figs. 31.1, 31.2, and 31.4).[21] The shunts could form as a result of a functional vascular occlusion due to a local hypoperfusion, or an imbalance of the retinal arterial perfusion in two adjacent areas.

Lastly, some cases of Vogt–Koyanagi–Harada syndrome, and anterior or posterior scleritis associated with TR have also been reported.[17]

Classification and severity of TR

A classification has been proposed by Uyama and Asayama[15] to stage TR, in which Stage 1 is characterized by a distension of the veins, Stage 2 by microaneurysm formation (Figs. 31.1 and 31.2), Stage 3 by the formation of arteriovenous anastomoses (Fig. 31.3), and Stage 4 by the presence of ocular complications such as cataract, proliferative retinopathy, neovascular glaucoma and vitreous hemorrhage.

Due to advances in TAK treatment, the late stages of this classification are now rare, and it has been noted that the retinopathy tends to be less severe.[17] In a series of 27 cases of TAK, the only signs observed were microaneurysms, arteriovenous shunts, and sequelae of branch artery occlusion in a single case. No case with Stage 4 was observed.[21]

However, the use of UWF-FA has allowed identifying a higher proportion of cases with peripheral microaneurysms than the 6%–20% previously reported based on FA.[17,30] In a recent publication including 27 patients with TAK, microaneurysms were found in 18.5% of eyes on fundus examination, and this proportion was increased to 24.4% on UWF-CFP and to 94.4% on UWF-FA.[21]

Blood flow velocity

The dynamic aspects of the retinal circulation have also been investigated by FA. Chun et al.[20] have found a delayed arm-to-retina circulation time in all cases of TR, but a delayed arteriovenous filling time has mainly been found in moderate and severe TR. However, the circulation times are difficult to quantify on FA, and more reliable parameters may be recorded on Doppler ultrasonography of the OA and CRA.

Some publications have found variable results regarding ocular findings and BFV in TAK patients. Matos et al.[31] have shown a statistically significant correlation between the peak systolic velocity in the OA and FA findings, according to the Uyama classification. Esen et al.[17] have also assessed the ocular blood flow in the OA and CRA using Doppler ultrasonography in 65 patients with TAK, but only 6% of them underwent FA to confirm the diagnosis of TR. They have found a reduced ocular blood flow and a significantly higher OA resistance index (RI) in TR patients, but OA RI was not associated with the disease duration. However, no statistical analysis was performed due to the small number of patients in the TR group.

Poignet et al.[21] have found a particularly low BFV in the CRA in patients with the highest number of microaneurysms. The mean BFV in the CRA significantly and negatively correlated with the total number of microaneurysms per eye. However, no significant correlation was found between the mean BFV in the OA and the total number of microaneurysms. Of note, the OA supplies not only the CRA but also all intraorbital organs such as the muscles and lacrimal glands, as well as the skin at the nose root. Therefore, BFV changes in the OA might not specifically reflect changes in the retinal vascular network, while BFV changes in the CRA can.

Hypertensive retinopathy

Hypertensive retinopathy is commonly reported in TAK, arterial hypertension being associated or not with TR.[20] Hypertension may develop in TAK cases either due to the involvement of the renal artery or to coarctation of the aorta. The proportion of TAK patients with hypertensive retinopathy varies according to the studies from 13.5%[20] or 15%[16] to 28.6%[21] and 34%.[17] However, the signs of TR may be difficult to differentiate from those of hypertensive retinopathy.

Other ocular complications

Anterior ischemic optic neuropathy has been reported in TAK.[32,33]

Some case reports have also described an association between TAK and other ocular inflammatory disorders such as anterior uveitis, Vogt-Koyanagi-Harada syndrome, keratoscleritis, and anterior or posterior scleritis (cited by Esen[17]).

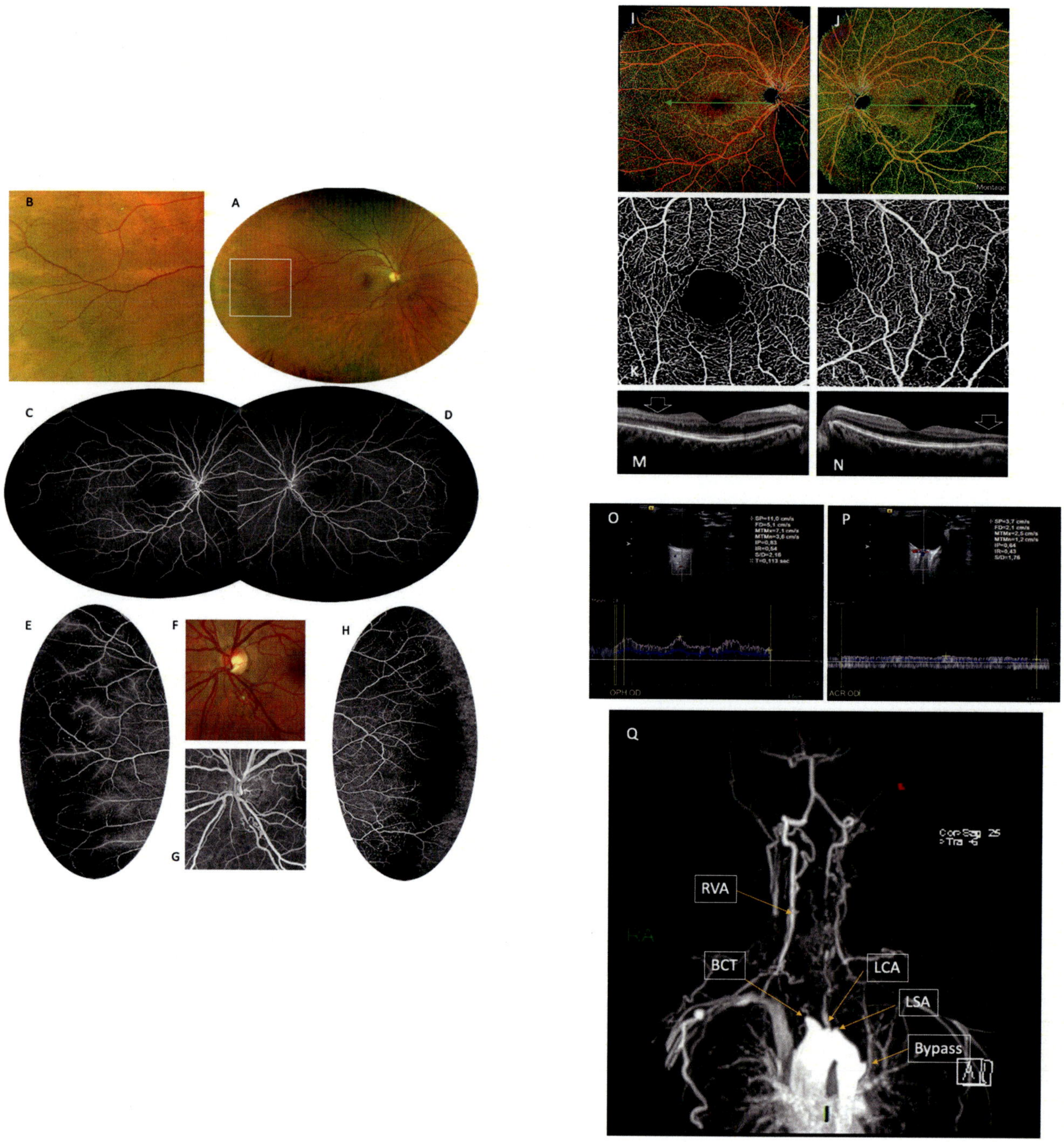

FIGURE 31.4 Case report of a 46 year-old woman with Takayasu's disease and severe stenosis of the two internal carotid arteries. The photos are detailed in the corresponding section of the chapter. (A) Ultra-widefield (UWF) color fundus photography; the white rectangle corresponds to the magnification shown in (B), showing punctuate hemorrhages and microaneurysms. (C and D) UWF fluorescein angiography of both eyes and magnification of the temporal periphery in (E and H), showing multiple microaneurysms and leakage from the most peripheral venules in the right eye (E). (F and G) Detail of the arterioarterial shunt on the inferotemporal artery bypassing a focal obstruction. (I and J) Widefield (12 mm) optical coherence tomography angiography (OCTA) showing wide areas of capillary hypodensity appearing in *dark green*. *Green arrows* correspond to the B-scan shown below. (K and L) Superficial capillary plexus in the macula of both eyes, normal in the right eye (K) and showing an area of capillary drop-out temporal to the macula in the left eye (L). (M and N) OCT B-scan showing a localized thinning of the inner retina more pronounced in the left eye (N) than in the right eye (M). (O) Color Doppler ultrasound (right eye) of the ophthalmic artery with color registration in red (*top*). In the lower part, the velocimetric registry (cm/s) shows an attenuated pulsatile arterial component and a very low peak systolic velocity at 11 cm/s. (O) Color Doppler ultrasound (right eye) with color registration in red and blue in the center of the optic nerve, corresponding to the central retinal artery (CRA) and vein (*top*). In the lower part, the velocimetric registry (cm/s) shows no pulsatile arterial component and a very low peak systolic velocity at 3.5 cm/s. (Q) Magnetic resonance angiography. Only the right vertebral artery (RVA) is patent. *BCT*, brachiocephalic trunk; *LCA*, left carotid artery; *LSA*, left subclavian artery; bypass: occluded from the descending aorta to the left vertebral artery.

Case report (Fig. 31.4)

A 46 year-old woman with TAK presented with an 8 year history of medical and surgical treatment, including four iterative bypass procedures in 5 years, that were all occluded. At the time of examination, the right vertebral artery was the only patent artery (Fig. 31.4Q). The patient was treated with 10 mg of methotrexate/week +30 mg of corticosteroids/day +5 mg of infliximab/kg/day.

Her VA was 2/20 OU. She complained of rare episodes of transient loss of vision, especially during physical exercise. The anterior segments were normal. Fundus examination and UWF-CFP showed several punctiform hemorrhages in the temporal periphery of the right eye (A), that are detailed in (B), but not in the left eye. Mid-phase UWF-FA revealed numerous peripheral microaneurysms in both eyes (C, D), and the details are shown in (E, H). In the peripheral retina of the right eye, a mild leakage was present around the terminal venules (E). In the left eye, an arterioarterial shunt was seen on the inferotemporal artery at the posterior pole, shunting a focal narrowing of the artery (F, G).

Wide-field Optical Coherence Tomography Angiography (WF-OCTA) showed large areas of capillary density loss in the posterior pole of both eyes, that were not suspected on UWF-FA. In the right eye, an area of capillary hypoperfusion was observed in the inferonasal part of the optic disc (I). The capillary density was normal in the macula (K). In the left eye, a large area of capillary hypoperfusion was observed in the inferotemporal part of the posterior pole (J), probably as a sequela of an occlusion of the inferotemporal artery, despite the resumption of the flow by an arterioarterial shunt (F, G). The OCT B-scan (corresponding to the green arrows in I and J) showed an inner retinal thinning in areas of capillary dropout in the left eye (arrows).

Doppler ultrasound of the left OA and CRA showed a strong reduction in BFV that was more pronounced in the CRA than in the OA (O–R).

Magnetic resonance angiography showed occlusion of the brachiocephalic trunk (BCT), left carotid artery (LCA), right vertebral artery (RVA), and left subclavian artery (LSA), as well as of the bypass from the descending aorta to the vertebral artery. As the brain and the eyes are only irrigated by the RVA, this explains the low ocular blood flow.

Treatment

Early diagnosis and treatment of TAK by physicians and surgeons have substantially reduced the number of advanced cases and severe TR cases.[17] TAK treatment is the first step for managing TR.

Treatment of active TAK is based on daily high-dose glucocorticoids and the introduction of immunosuppressive agents. When vascular changes are identified on imaging, escalating immunosuppression may be required and surgery may be considered in case of progressive ischemia.[34]

An effective treatment aiming at reducing the risk of arterial stenosis, and especially supraaortic arterial stenosis, should have a beneficial effect on the progression of TR. A publication has shown that percutaneous endovascular stenting of aortic arch arteries, affected in TA, can lead to reversal of retinopathy changes and restoration of vision.[35] However, Poignet et al. have recently shown that, while there is a significant association between the presence of a supraaortic arterial stenosis and the number of microaneurysms, it is not related to the percentage of stenosis.[21]

In a systematic review on the management of ocular ischemia in TR, Zeng et al. have collected data from 31 patients who underwent open or endovascular surgery and found that although medical therapy alone seemed to be effective on TR, surgery combined with perioperative medications and local therapy should be preferred in case of both acute and chronic vision loss.[36]

In patients with very advanced TR with neovascular complications, intravitreal anti-VEGF injections combined with panretinal laser photocoagulation is the appropriate treatment as for other neovascular retinopathies.

In Conclusion, Takayasu's disease is a rare chronic, granulomatous arteritis involving the aorta and its main branches, complicated by retinopathy when the flow in the carotid arteries is strongly reduced. The earlier sign of TAK is microaneurysm, found in more than 90% of cases on UWF-FA. Medical treatment of TAK combining glucocorticosteroids and immunosuppressive agents and open or endovascular surgery have allowed for reducing the progression to severe stages of TR. If needed, intravitreal anti-VEGF injections and panretinal laser photocoagulation may be used to control intraocular neovascularization.

References

1. Sugiyama K, Ijiri S, Tagawa S, Shimizu K. Takayasu disease on the centenary of its discovery. *Japanese Journal of Ophthalmology*. March 2009;53(2):81–91. https://doi.org/10.1007/s10384-009-0650-2.

2. Jennette JC, Falk RJ, Bacon PA, et al. 2012 revised International Chapel Hill Consensus conference nomenclature of vasculitides. *Arthritis Rheumatism*. January 2013;65(1):1–11. https://doi.org/10.1002/art.37715.
3. Russo RAG, Katsicas MM. Takayasu arteritis. *Frontiers in Pediatrics*. 2018;6:265. https://doi.org/10.3389/fped.2018.00265.
4. Grayson PC, Ponte C, Suppiah R, et al. American college of rheumatology/EULAR classification criteria for Takayasu arteritis. *Annals of the Rheumatic Diseases*. 2022;81(12):1654–1660. https://doi.org/10.1136/ard-2022-223482. PMID - 36351705.
5. Rose AG, Sinclair-Smith CC. Takayasu's arteritis. A study of 16 autopsy cases. *Archives of Pathology & Laboratory Medicine*. May 1980;104(5):231–237.
6. Walker M, Gallagher PJ. The surgical pathology of large vessel disease. *Diagnostic Histopathology*. 2010;16(1):10–16.
7. Saadoun D, Garrido M, Comarmond C, et al. Th1 and Th17 cytokines drive inflammation in Takayasu arteritis. *Arthritis & Rheumatology*. May 2015;67(5):1353–1360. https://doi.org/10.1002/art.39037.
8. Arnaud L, Haroche J, Mathian A, Gorochov G, Amoura Z. Pathogenesis of Takayasu's arteritis: a 2011 update. *Autoimmunity Reviews*. November 2011;11(1):61–67. https://doi.org/10.1016/j.autrev.2011.08.001.
9. Terao C, Yoshifuji H, Ohmura K, et al. Association of Takayasu arteritis with HLA-B 67:01 and two amino acids in HLA-B protein. *Rheumatology*. October 2013;52(10):1769–1774. https://doi.org/10.1093/rheumatology/ket241.
10. Arend WP, Michel BA, Bloch DA, et al. The American college of rheumatology 1990 criteria for the classification of Takayasu arteritis. *Arthritis Rheumatology*. August 1990;33(8):1129–1134. https://doi.org/10.1002/art.1780330811.
11. Hata A, Noda M, Moriwaki R, Numano F. Angiographic findings of Takayasu arteritis: new classification. *International Journal of Cardiology*. August 1996;54 Suppl:S155–S163. https://doi.org/10.1016/s0167-5273(96)02813-6.
12. Moriwaki R, Noda M, Yajima M, Sharma BK, Numano F. Clinical manifestations of Takayasu arteritis in India and Japan— new classification of angiographic findings. *Angiology*. 1997;48(5):369–379. https://doi.org/10.1177/000331979704800501.
13. Chung JW, Kim HC, Choi YH, Kim SJ, Lee W, Park JH. Patterns of aortic involvement in Takayasu arteritis and its clinical implications: evaluation with spiral computed tomography angiography. *Journal of Vascular Surgery*. May 2007;45(5):906–914. https://doi.org/10.1016/j.jvs.2007.01.016.
14. Pryshchep O, Ma-Krupa W, Younge BR, Goronzy JJ, Weyand CM. Vessel-specific toll-like receptor profiles in human medium and large arteries. *Circulation*. September 16, 2008;118(12):1276–1284. https://doi.org/10.1161/CIRCULATIONAHA.108.789172.
15. Uyama M, Asayama K. *Retinal Vascular Changes in Takayasu's Disease (Pulseless Disease), Occurrence and Evolution of the Lesion*. Springer; 1976:549–554.
16. Peter J, David S, Danda D, Peter JV, Horo S, Joseph G. Ocular manifestations of Takayasu arteritis: a cross-sectional study. *Retina*. June 2011;31(6):1170–1178. https://doi.org/10.1097/IAE.0b013e3181fe540b.
17. Esen F, Ergelen R, Alibaz-Oner F, Celik G, Direskeneli H, Kazokoglu H. Ocular findings and blood flow in patients with Takayasu arteritis: a cross-sectional study. *British Journal of Ophthalmology*. July 2019;103(7):928–932. https://doi.org/10.1136/bjophthalmol-2018-312580.
18. Sagar S, Kar S, Gupta A, Sharma BK. Ocular changes in Takayasu's arteritis in India. *Japanese Journal of Ophthalmology*. 1994;38(1):97–102.
19. Kiyosawa M, Baba T. Ophthalmological findings in patients with Takayasu disease. *International Journal of Cardiology*. October 1, 1998;66 Suppl 1:S141–S147. https://doi.org/10.1016/s0167-5273(98)00162-4. discussion S149.
20. Chun YS, Park SJ, Park IK, Chung H, Lee J. The clinical and ocular manifestations of Takayasu arteritis. *Retina*. 2001;21(2):132–140. https://doi.org/10.1097/00006982-200104000-00006.
21. Poignet B, Bonnin P, Gaudric J, et al. Correlation between ultra-wide-field retinal imaging findings and vascular supra-aortic changes in Takayasu arteritis. *Journal of Clinical Medicine*. October 24, 2021;10(21). https://doi.org/10.3390/jcm10214916.
22. Kahn M, Green RW. Ocular features of carotid occlusive disease. *Retina*. 1986;6(4):239–252.
23. Ishikawa K, Uyama M, Asayama K. Occlusive thromboaortopathy (Takayasu's disease): cervical arterial stenoses, retinal arterial pressure, retinal microaneurysms and prognosis. *Stroke*. September–October 1983;14(5):730–735. https://doi.org/10.1161/01.str.14.5.730.
24. Linsenmeier RA, Braun RD, McRipley MA, et al. Retinal hypoxia in long-term diabetic cats. *Investigative Ophthalmology and Visual Science*. 1998;39(9):1647–1657.
25. Tolentino MJ, McLeod DS, Taomoto M, Otsuji T, Adamis AP, Lutty GA. Pathologic features of vascular endothelial growth factor-induced retinopathy in the nonhuman primate. *American Journal of Ophthalmology*. March 2002;133(3):373–385. https://doi.org/10.1016/s0002-9394(01)01381-2.
26. Hirose K. A study of fundus changes in the early stages of Takayasu-Ohnishi (pulseless) disease. *Amerian Journal of Ophthalmology*. February 1963;55:295–301. https://doi.org/10.1016/0002-9394(63)92686-2.
27. Shimizu K. Fluorescein fundus photography in Takayasu's disease. *Rinsho Ganka*. 1965;19:1107–1108.
28. Shikano S, Shimizu K. *Atlas of Fluorescence Fundus Angiography*. WB Saunders; 1968.
29. Tanaka T, Shimizu K. Retinal arteriovenous shunts in Takayasu disease. *Ophthalmology*. November 1987;94(11):1380–1388. https://doi.org/10.1016/s0161-6420(87)33279-8.
30. Chotard G, Diwo E, Coscas F, Butel N, Saadoun D, Bodaghi B. Fluorescein and OCT angiography features of Takayasu disease. *Oculular Immunology and Inflammation*. 2019;27(5):774–780. https://doi.org/10.1080/09273948.2018.1455876.
31. Matos KT, Arantes T, Souza AW, Ramos MH, Allemann N, Muccioli C. Retinal angiography and colour Doppler of retrobulbar vessels in Takayasu arteritis. *Canadian Journal of Ophthalmology*. February 2014;49(1):80–86. https://doi.org/10.1016/j.jcjo.2013.09.014.
32. Tian G, Chen Q, Wang W. Takayasu arteritis presenting as isolated anterior ischemic optic neuropathy. *Internationa Ophthalmology*. April 2018;38(2):823–828. https://doi.org/10.1007/s10792-017-0516-4.

33. Kimura H, Masai H, Kashii S. Anterior ischemic optic neuropathy associated with pulseless disease. *Ophthalmologica*. 1995;209(6):346–348. https://doi.org/10.1159/000310656. PMID - 8751345.
34. Maz M, Chung SA, Abril A, et al. American college of rheumatology/vasculitis foundation guideline for the management of giant cell arteritis and Takayasu arteritis. *Arthritis Rheumatology*. 2021;73(8):1349–1365. https://doi.org/10.1002/art.41774. PMID - 34235884.
35. Dogra M, Kumar A, Bajgai P, et al. Outcome of endovascular procedures on ocular findings in Takayasu arteritis. *Retina*. June 2019;39(6):1142–1148. https://doi.org/10.1097/IAE.0000000000002103.
36. Zeng Y, Duan J, Ge G, Zhang M. Therapeutic management of ocular ischemia in Takayasu's arteritis: a case-based systematic review. *Frontiers in Immunology*. 2022;12:791278. https://doi.org/10.3389/fimmu.2021.791278. PMID - 35095866.

Section VII

Vascular diseases associated with systemic conditions

Chapter 32

Hypertensive retinopathy

Abdulla R. Shaheen and Jayanth Sridhar
Department of Ophthalmology, Bascom Palmer Eye Institute, University of Miami Miller School of Medicine, Miami, FL, United States

Overview bullets

- Hypertensive eye disease includes a spectrum of three primary diseases: hypertensive retinopathy, hypertensive choroidopathy, and hypertensive optic neuropathy.
- Hypertensive retinopathy is the most common form of hypertensive eye disease.
- Hypertensive retinopathy progresses through three phases: vasoconstrictive, sclerotic, and exudative phases.
- Hypertensive retinopathy manifests as two broad changes: arteriolar wall changes, such as narrowing of the blood vessels, and retinal macrovascular lesions, such as hemorrhages and microaneurysms.
- The Wong–Mitchell grading system classifies hypertensive retinopathy as:
  - None: no detectable signs.
  - Mild: arteriolar narrowing, arteriovenous nicking, or arteriolar wall opacification.
  - Moderate: mild signs plus retinal hemorrhage, microaneurysm, cotton-wool spots, hard exudate.
  - Severe: moderate signs plus optic disc swelling.
- The management of hypertensive eye disease involves controlling blood pressure to prevent or slow the progression of the condition.
- The visual outcomes for patients with hypertensive retinopathy depend on the severity of the condition and associated complications.

Case studies

Case 1

A 33-year-old female patient, with no previous history of hypertension, presents at 33 weeks of pregnancy with a headache, edema in her lower extremities, and blurred vision with best-corrected visual acuity (BCVA) 20/200 OD and 20/100 OS. Her blood pressure was measured at 210/140 mmHg, and a urine analysis revealed proteinuria (3 g in 24 h). Laboratory tests, including platelet count, liver enzyme levels, and hemolysis tests, are all within normal limits. Given her symptoms and laboratory results, the patient is diagnosed with preeclampsia and is scheduled for an emergency cesarean section. The baby is born alive and healthy, and the placenta was removed in its entirety and found to be calcified. On postpartum day 3, the patient continues to experience hypertension, edema in her lower extremities, proteinuria, and blurred vision with BCVA 20/400 OD and 20/200 OS. Ophthalmologic examination showed bilateral generalized arteriolar copper wiring and exudative retinal detachment (Fig. 32.1A and B). The patient was started on a combination of antihypertensive medications to control her hypertension and was instructed to remain on bed rest. Her blood pressure was brought under control and her vision gradually improved. Follow-up ophthalmologic evaluations in 1 week reveal the improvement of BCVA to 20/70 OD and 20/80 OS with the improvement of the exudative retinal detachment (Fig. 32.1C and D). As her condition improved, the patient's medications were gradually reduced and discontinued within 6 weeks of delivery, and she was also advised to maintain a healthy lifestyle. Ten months later, the BCVA was 20/25 OD and 20/30 OS, and the retinal detachment resolved (Fig. 32.1E and F). She was instructed to keep regular follow-up with the ophthalmologist and her obstetrician.

Retinal and Choroidal Vascular Diseases of the Eye. https://doi.org/10.1016/B978-0-443-15583-3.00032-9

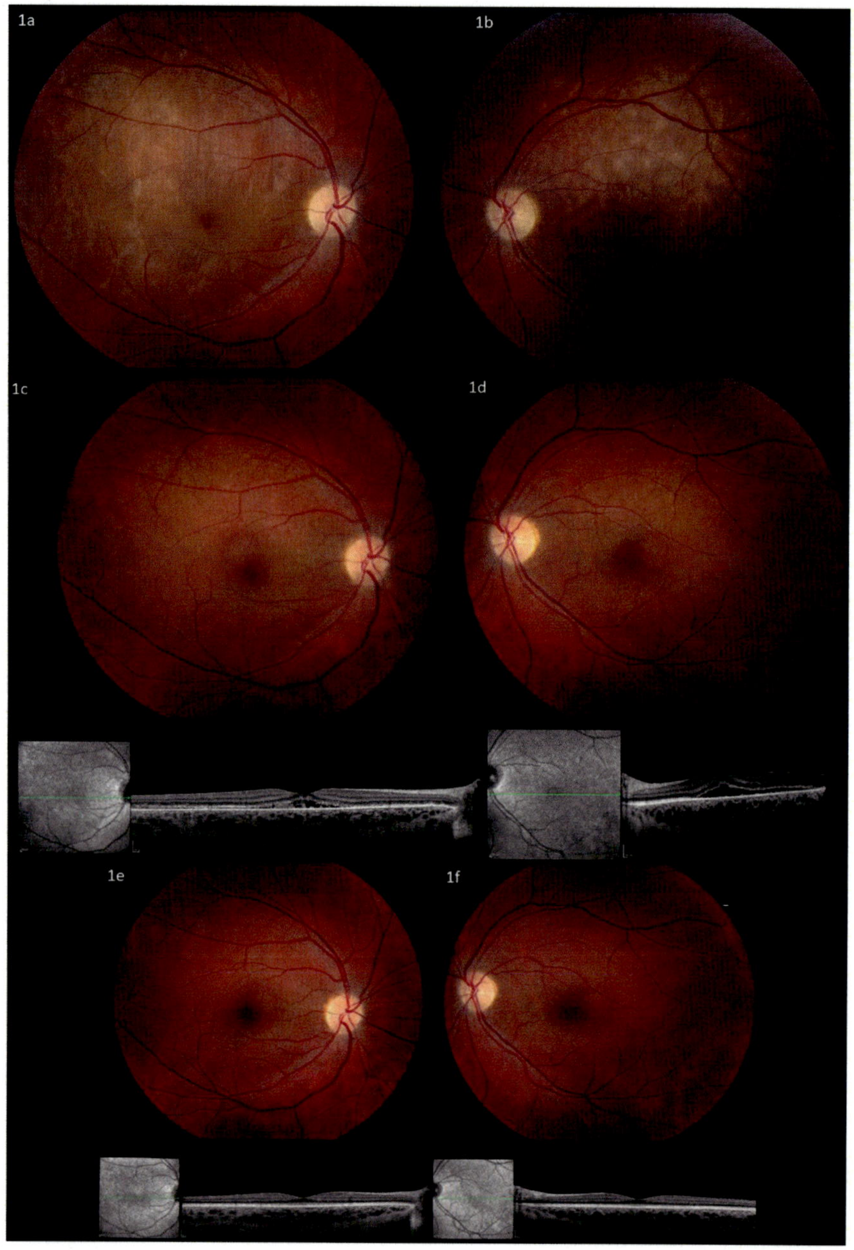

FIGURE 32.1 Illustrating the imaging findings of case 1 preeclamptic patient. Right eye (A) and left eye (B) color fundus photos (CFPs) show bilateral generalized arteriolar narrowing, copper wiring, and exudative retinal detachment at initial presentation. (C and D) CFPs demonstrate improvement in exudative retinal detachment at 1-week follow-up with optical coherence tomography (OCT) showing subretinal fluid at the macula of both eyes. (E and F) show resolution of retinal detachment and subretinal fluid at 10-month follow-up on CFP and OCT, respectively.

Introduction

Chronically elevated blood pressure is associated with structural and functional damage in various organs, including the heart, brain, kidneys, and the eye.[1] The crude mortality rate associated with essential hypertension and hypertensive renal disease has been estimated to be 12.7 per 100,000.[2] Hypertensive eye disease encompasses a spectrum of changes to three primary structures in the eye: hypertensive retinopathy, when the manifestations are in the retina; hypertensive choroidopathy in the choroid and hypertensive optic neuropathy, when the effect is on the optic nerve.

Hypertension also is a risk factor for other ocular diseases, including retinal artery and vein occlusion, nonarteritic anterior ischemic optic neuropathy, and retinal microaneurysm.[3] Hypertensive retinopathy, the focus of this chapter, manifests as two broad changes: arteriolar wall changes, such as narrowing of the blood vessels, and retinal macrovascular lesions, such as hemorrhages and microaneurysms. This chapter will discuss the epidemiology, pathogenesis, clinical features, differential diagnosis, imaging findings and future imaging modalities, recent clinical trials, and management strategies for hypertensive retinopathy.

Epidemiology

Hypertensive retinopathy is a common condition affecting 11% of the nondiabetic population.[4–7] Hypertensive retinopathy seems more common in males and certain racial groups, such as African Americans and Asians, than in White females.[4,6,8–10] These observed demographic differences could be attributed to the different methodologies implied by different population studies, or they could reflect patients' compliance with treatment and access to care, with males less likely to get medical care earlier and non-Whites having limited access to care. There are several risk factors associated with hypertensive retinopathy. These include elevated blood pressure, a history of cardiovascular disease, thickening of the carotid arteries on imaging, obesity, dyslipidemia, diabetes, elevated creatinine levels, and microalbuminuria.[4,10–12]

The incidence of specific signs of hypertensive retinopathy is as follows: arteriovenous nicking occurs in 1.4% of cases (3.3% prevalence), focal arteriolar narrowing in 4.1% of cases (13.5% prevalence), and dot, blot, or flame-shaped hemorrhages and microaneurysms (HMA) in 3.3% of cases (7.8% prevalence).[7,13] There is no difference in the incidence of HMA between males and females, but males have a higher incidence of focal arteriolar narrowing and AVN.[14] Some signs of hypertensive retinopathy, such as focal arteriolar narrowing and AVN, may regress over time 15. For example, a study found that 50.3% of cases of focal arteriolar narrowing, 40.7% of cases of AVN, and 65.9% of cases of HMA regressed over 10 years.[15] Hypertensive retinopathy signs can also indicate the stage of systemic hypertension. For example, retinal arteriolar narrowing, both focal and generalized, can indicate subclinical hypertension[16–21] and masked hypertension, which is characterized by normal blood pressure in the clinic or office (less than 140/90 mmHg), but elevated blood pressure outside of the clinic (ambulatory daytime blood pressure or home blood pressure greater than 135/85 mmHg).[20,22–24]

Retinal venular dilation is a recently recognized manifestation of hypertensive retinopathy that has not traditionally been considered part of the condition (Fig. 32.2). Studies have shown that retinal venular dilation is associated with current hypertension, the future incidence of hypertension, and the future incidence of cardiovascular disease.[16,25]

Hypertensive retinopathy is associated with damage to various target organs. Mild to moderate hypertensive retinopathy has been associated with subclinical cerebral, cardiac, and large vessel changes.[26–35] Evidence suggests that patients with hypertensive retinopathy are at an increased risk of developing chronic kidney disease, although this has not yet been definitively confirmed.[36] Hypertensive retinopathy also increases the risk of stroke,[37] and patients with retinal venular dilations are particularly prone to this complication.[38] Retinal arteriolar narrowing has been associated with lacunar stroke,[39] while retinal HMA has been associated with cerebral hemorrhage.[40,41] Patients with hypertensive retinopathy are three times more likely to develop congestive heart failure than those without the condition.[42] Retinal venular widening has been associated with incident coronary heart disease.[43] Hypertensive retinopathy also increases the risk of all-cause and cardiovascular mortality, with patients with moderate hypertensive retinopathy having a higher cardiovascular mortality[44] and those with retinal HMA being more likely to die from coronary heart disease at risk similar to that of individuals with diabetes mellitus.[15,45]

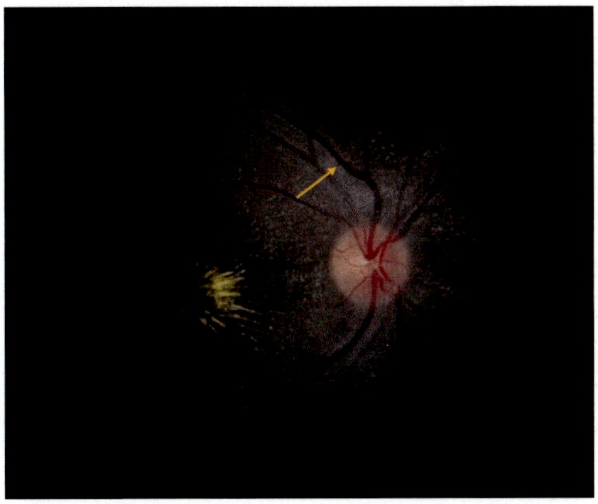

FIGURE 32.2 Severe hypertensive retinopathy with venular dilatation. The image shows the presence of venular engorgement and dilatations (indicated by *arrows*) in the retina, a recently recognized fundus finding in hypertensive retinopathy.

Pathogenesis of hypertensive eye disease

Hypertensive retinopathy progresses through several phases: vasoconstrictive, sclerotic, and exudative.[46] During the vasoconstrictive phase, elevated BP causes a physiological spasm of the arterial wall and narrowing of the arterioles. In the sclerotic phase, persistent elevation of BP over time leads to thickening of the intima, hyperplasia of the media, and hyaline degeneration. In the exudative phase, chronically sustained elevated BP causes tissue ischemia, causing necrosis of the intima and media layer and disrupting the blood—retinal barrier. This disruption leads to the exudation of blood and lipids and retinal nerve fiber layer (RNFL) ischemia. Clinically, the sclerotic phase manifests as generalized and focal areas of retinal arteriolar narrowing (from thickening of the vessel's wall) and silver or copper wiring or altered arteriolar light reflexes (from hyaline degeneration). The exudative phase manifests as retinal hemorrhage and exudation, hard exudates, and cotton wool spots.[47] Several factors trigger elevations in BP, including the sympathetic nervous system, inflammation, the immune system (both the innate and adaptive branches), the renin—angiotensin—aldosterone system (RAAS), the endothelin system, and other mediators. Hypertensive retinopathy does not always progress through the 3 phases in sequence, and elevated BP is not the sole factor for the changes. Other factors possibly involved in the pathogenesis include inflammation, endothelial dysfunction from an impaired generation of nitric oxide, angiogenesis, and oxidative stress.[48,49] Genetic factors may play a role, but current studies provide inconclusive results. Genetic factors may contribute to an individual's risk of developing retinal microcirculatory disorders, but the current evidence is inconclusive. Several single nucleotide polymorphisms may be associated with central retinal artery equivalent (CRAE) narrowing and central venular equivalent (CRVE) dilatation, which correlate with higher mean arterial pressure.[50] Other genetic factors that may play a role include polymorphisms in the ALK-1 and endoglin, which regulate blood pressure and vascular homeostasis.[50] Further research is needed to fully understand the role of genetics in retinal microcirculatory disorders.[50]

The pathogenesis of hypertensive choroidopathy is thought to occur when the elevated BP exceeds the autonomic nervous system regulation capacity, leading to fibrin-platelet deposition and obstruction of the choroid arteries and choriocapillaris, necrosis of the overlying retinal pigment epithelium, and fibrinous exudation.[51] Optical coherence tomography angiography allowed direct assessment of the choroidal blood vessels during different perfusion levels and helped better explain the pathogenesis of hypertensive choroidopathy. It has been found that an increase in pulse pressure leads to increased pulsatility in the small penetrating blood vessels, which can cause microvascular injury and lead to remodeling, endothelial dysfunction, and ischemia. The ischemic damage to the endothelium reduces the density of smaller arterioles and capillaries. Thus, hyperperfusion can manifest as a choroidal thickening, while hypoperfusion and ischemia are as choroidal thinning.[52,53] The pathophysiology of optic disc swelling in hypertensive optic neuropathy is not well understood, but some research suggests that ischemia may be a significant factor.[54]

Clinical features

Hypertensive retinopathy is diagnosed based on the presence of classical fundus signs and a history of hypertension. There are two primary grading systems for hypertensive retinopathy, the Keith—Wagener—Baker system (grades 1—4)[55] (and the subsequently modified system by Scheie et al.)[56] and the Wong—Mitchell system (mild, moderate, severe)[57] (Table 32.1). The latter is more straightforward, reliable, and repeatable and correlates retinal microvascular signs with incident cardiovascular risk supporting its use in clinical practice.[59]

The classification of hypertensive retinopathy proposed by the Keith—Wagener—Baker system is as follows.

- Grade 1: Mild generalized arteriolar narrowing or sclerosis.
- Grade 2: Definite focal narrowing and arteriovenous crossings; moderate to marked sclerosis of the retinal arterioles.
- Grade 3: Signs of grade 2 retinopathy plus retinal hemorrhages, exudates, and cotton wool spots.
- Grade 4: Severe grade 3 retinopathy plus papilledema.

The classification of hypertensive retinopathy proposed by Wong and Mitchell in 2004 is as follows (Fig. 32.3).[57]

- None: no detectable signs
- Mild: one or more of the following: generalized arteriolar narrowing, focal arteriolar narrowing, arteriovenous nicking, opacity ("copper wiring") of the arteriolar wall (Fig. 32.3A).
- Moderate: one or more of the following: retinal hemorrhage (blot, dot, or flame shaped), microaneurysm, cotton-wool spots, hard exudate (Fig. 32.3B).
- Severe: moderate retinopathy plus swelling of the optic disc (Fig. 32.3C).

TABLE 32.1 Classification systems for hypertensive retino.

Classification	Category	Clinical findings
Keith–Wagener–Barker[55]	1	Mild generalized arteriolar narrowing or sclerosis
	2	Definite focal narrowing and arteriovenous crossings; moderate to marked sclerosis of the retinal arterioles
	3	Signs of grade 2 retinopathy plus retinal hemorrhages, exudates, and cotton wool spots
	4	Severe grade 3 retinopathy plus papilledema
Wong and Mitchell[57]	Mild	One or more of the following: Generalized arteriolar narrowing, focal arteriolar narrowing, arteriovenous nicking, opacity ("copper wiring") of the arteriolar wall
	Moderate	One or more of the following: retinal hemorrhage (blot, dot, or flame shaped), microaneurysm, cotton-wool spots, hard exudate
	Malignant	Moderate retinopathy plus swelling of the optic disc
Ahn–Woo–Park[a,58]	Mild to moderate	Keith–Wagener–Barker grade 1, 2, or 3 with or without subretinal fluid
	Malignant without SRF	Severe grade 3 retinopathy plus papilledema without subretinal fluid on OCT
	Malignant with SRF	Severe grade 3 retinopathy plus papilledema with subretinal fluid on OCT

OCT, optical coherence tomography.
[a]*Ahn–Woo–Park classification system applies for patients with severe hypertension (systolic blood pressure ≥180 mm Hg or diastolic blood pressure ≥110 mm Hg).*

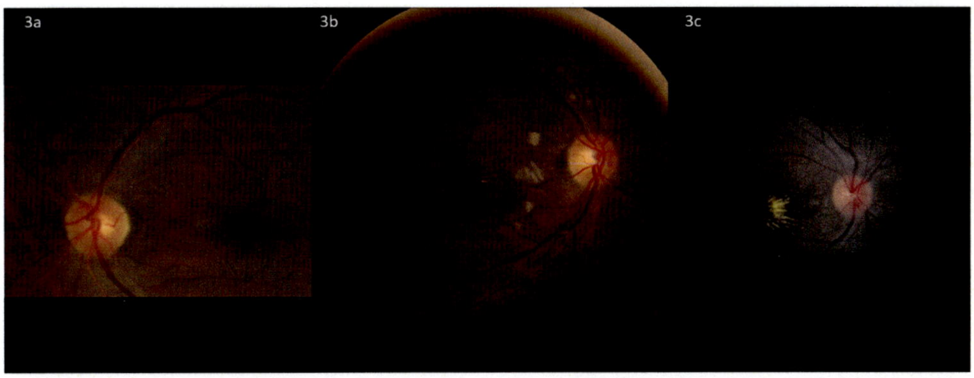

FIGURE 32.3 Showing the color fundus photographs of the different stages of hypertensive retinopathy according to the Wong and Mitchell classification system. (A) Generalized arteriolar narrowing, arteriovenous nicking, and copper wiring of the arteriolar wall characteristics of mild hypertensive retinopathy. (B) Dot, blot, and flame-shaped hemorrhages, cotton-wool spots, and hard exudate characteristics of moderate hypertensive retinopathy. (C) Grade 1 optic disc swelling, in addition to hard exudates, a few retinal hemorrhages, cotton wool spots, and a star-shaped macular lesion caused by leaking retinal vessels.

A recent classification proposed in 2014 incorporates optical coherence tomography (OCT) findings of subretinal fluid for patients with severe hypertension (systolic blood pressure ≥180 mm Hg or diastolic blood pressure ≥110 mm Hg)[58] (Table 32.1). This classification correlates significantly to the final best-corrected visual acuity. It includes the categories of mild-moderate retinopathy, malignant retinopathy without subretinal fluid, and malignant retinopathy with subretinal fluid.[58]

- Mild to moderate: findings in the mild/moderate category in Wong and Michell classification with/without subretinal fluid.
- Malignant: findings in the malignant category in Wong and Michell classification without subretinal fluid.
- Malignant: findings in the malignant category in Wong and Michell classification with subretinal fluid.

It is difficult to determine the full impact of vision loss on the quality of life in individuals with hypertensive eye disease. Patients with mild hypertensive retinopathy often have no symptoms and experience only mild vision loss. In contrast, patients with malignant hypertensive retinopathy can have multiorgan damage.[60] The diagnosis of malignant hypertension is often delayed because symptoms tend to appear only in later stages when organ damage ensues.[60]

The resolution of malignant hypertensive retinopathy over a relatively short period makes retrospective diagnosis difficult. However, the target organ damage in these patients persists, leading to complications and a poorer prognosis than in patients with nonmalignant hypertension.[60]

Hypertensive choroidopathy manifests as Elschnig spots, deep, round, gray-yellow lesions at the retinal pigment epithelium, and Siegrist streaks, linear hyperpigmented streaks along the choroidal arterioles.[3] Fundus autofluorescence imaging better visualizes the choroidopathy lesions than fundus exam or color fundus photo.[3] This condition is more common in young people, as their choroidal vessels are more pliable than sclerotic.[61]

Differential diagnosis

It is crucial to accurately diagnose hypertensive retinopathy to treat and manage the condition properly. Several signs of hypertensive retinopathy, such as retinal arteriolar wall changes and optic disc swelling, may also be present in the general population. This ambiguity of the signs can make it challenging to distinguish hypertensive retinopathy from other conditions, such as diabetic retinopathy. Retinal arteriolar wall changes, arteriovenous nicking (AVN), and optic disc swelling are more specific for hypertensive retinopathy than diabetic retinopathy.[62] On the other hand, retinal hard exudates, macular edema, and neovascularization are more specific for diabetic retinopathy.[62] Other conditions in the differential diagnosis of hypertensive retinopathy include retinal arterial occlusion, retinal vein occlusion, macular aneurysms, retinal artery emboli, ocular ischemic syndrome, and radiation retinopathy. The patient's history of hypertension and the absence of systemic signs of the other conditions distinguish them apart.

Imaging findings and future image studies

Optical coherence tomography scans are valuable for evaluating hypertensive retinopathy and choroidopathy patients. In patients with hypertensive retinopathy, OCT may show reduced retinal and choroidal thickness. In patients with hypertensive choroidopathy, OCT may show macular serous retinal detachment (SRD) or choroidal ischemic damage to the retinal pigment epithelium (RPE).[63,64] Optical coherence tomography angiography shows reduced retinal flow as reduced vessel density or focal areas of capillary sparsity, microangiomas, and focal areas of capillary nonperfusion.[65,66] There is currently no substantial evidence that the changes in blood vessel flow and density detected using OCTA can predict clinical cardiovascular disease (CVD) events in patients with hypertension. More research is needed to determine if OCTA assessment can help manage hypertension and reduce the risk of CVD events.

Artificial intelligence (AI) has contributed to medical image interpretation for different retinal diseases, including detecting diabetic retinopathy using deep learning and convolutional neural networks (CNNs).[67] In the future, AI will likely be used to detect hypertensive retinopathy and predict which individuals are at risk of developing hypertension before even elevated blood pressure is detected. This early detection will allow for implementing lifestyle changes to prevent the development of hypertension.

Newer, noninvasive modalities are being investigated to evaluate the media thickness to internal lumen ratio of small resistance arteries (MLR) of retinal arterioles, including scanning laser Doppler flowmetry (SLDF) and adaptive optics (AO).[68] Both approaches estimate retinal arterioles' wall-to-lumen ratio (WLR), and AO showed a higher correlation with the MLR of small subcutaneous arteries compared to SLDF.[68] Another newer modality, videomicroscopy/capillaroscopy, can noninvasively obtain basal and total capillary density. The ability to noninvasively assess microvascular morphology in a clinical setting may be a significant advancement as wire or pressure micromyography, the current gold standard for measuring the morphologic characteristics of the small resistance arteries are invasive and used for subcutaneous vessels instead of retinal vessels, has significant limitations due to the invasive nature of the procedure.[68] Other newer modalities in retinal vascular imaging include retinal oximetry[69] and dynamic retinal vessel analysis using flicker light-induced vasodilation.[70] Retinal oximetry measures the oxygen saturation of hemoglobin in the retina's blood vessels. Retinal oximetry is highly reproducible with low variability in test–retest studies and healthy populations. Retinal oximetry aims to uncover new pathophysiological principles and novel biomarkers in various retinal diseases and investigates potential applications in the study of systemic and brain disorders. Flicker light-induced retinal vasodilation as an analysis for the dynamics of retinal vessels might be a valuable noninvasive measure of endothelial dysfunction.[70]

Recent clinical trial updates

There are relatively few clinical trials on hypertensive eye disease. One notable ongoing study that began in January 2015 is investigating the noninvasive retinal blood flow measurement using vessel analysis and Fourier Domain OCT.[67] This interventional trial enrolls 60 participants and will use a parallel assignment intervention model with single masking for the outcome's assessor.[67] The study's primary purpose is to advance basic scientific knowledge in this area and is expected to be completed in December 2023.[67]

Management strategies

The 2020 International Society of Hypertension Global Hypertension Practice Guidelines recommends the following: "Fundoscopy is particularly important in cases of hypertensive emergencies, as it can help to detect signs of severe hypertension such as retinal hemorrhage, microaneurysms (abnormal blood vessels), and papilledema (swelling of the optic disc). It is recommended that fundoscopy be performed in patients with grade 2 hypertension (slightly elevated blood pressure) by experienced examiners or, if available, using alternative techniques such as digital fundus cameras".[71]

Patients with hypertension should undergo retinal evaluation by a trained healthcare provider as part of their routine care. If no signs of hypertensive retinopathy are detected, routine care is sufficient. Patients with mild hypertensive retinopathy should have close monitoring of their cardiovascular disease (CVD) risk by their provider, be screened for diabetes, sent to an ophthalmologist, and receive treatment to control the high blood pressure to prevent or halt damage to the retinal and choroidal blood vessels. Patients with malignant hypertensive retinopathy and malignant choroidopathy should be treated for hypertension immediately and referred to an ophthalmologist for visual acuity tests and retinal imaging.

Evidence suggests that blood pressure control can lead to regression of signs of hypertensive retinopathy, both in patients with hypertension and in patients with hypertension and diabetes who have hypertensive retinopathy.[72–74] Blood pressure control also decreases the risk of developing hypertensive retinopathy.[13] The initial goal for blood pressure control in patients with hypertension is to achieve systolic blood pressure (SBP) of less than 140 mmHg and diastolic blood pressure (DBP) of less than 90 mmHg.[75] If the patient tolerates the treatment well, it is recommended to aim for a blood pressure of 130/80 mmHg, with efforts to achieve an even lower blood pressure in individuals between the ages of 18 and 65.[75] Lowering BP can be achieved with the primary care provider through lifestyle changes, such as exercising regularly, eating a healthy diet, managing stress, and taking medications. There is no robust evidence that treatments such as anti-VEGF or steroids help manage hypertensive retinopathy. It is also unclear whether certain antihypertensive medications are superior to others in managing this condition. One study suggests that ACE inhibitors and ARBs may be superior because they prevent vessel remodeling and endothelial dysfunction, which is common in hypertensive retinopathy.[75] However, more research is needed to confirm these findings. It is also unknown whether specific medications can reverse established hypertensive eye disease and if regression in the severity of ocular changes indicates better systemic and visual outcomes.

The visual outcomes in individuals with hypertensive retinopathy depend on the condition's severity and other complications. Mild or moderate hypertensive retinopathy may not cause symptoms and does not increase the risk of visual impairment. However, patients with malignant hypertensive retinopathy may experience profound vision loss, particularly in hypertensive optic neuropathy.[76,77] Earlier control of BP may improve vision outcomes after the regression of retinopathy and choroidopathy.[76] However, it is unclear which patients will recover their vision or sustain a visual loss or other complications, including macular edema and retinal ischemia, which impact visual outcomes.

Acknowledgment

The authors would like to acknowledge Jesse Sengillo MD for providing us with Fig. 32.3A.

References

1. Schmieder RE. End organ damage in hypertension. *Deutsches Ärzteblatt International.* 2010;107(49):866–873.
2. Centers for Disease Control and Prevention, National Center for Health Statistics. *National Vital Statistics System, Mortality 1999–2020 on CDC WONDER Online Database, Released in 2021. Data are from the Multiple Cause of Death Files, 1999–2020, as Compiled from Data Provided by the 57 Vital Statistics Jurisdictions Through the Vital Statistics Cooperative Program*; 2022. Published 2021 http://wonder.cdc.gov/ucd-icd10.html. Accessed December 24, 2022.
3. Fraser-Bell S, Symes R, Vaze A. Hypertensive eye disease: a review. *Clinical & Experimental Ophthalmology.* 2017;45(1):45–53.

4. Ojaimi E, Nguyen TT, Klein R, et al. Retinopathy signs in people without diabetes: the multi-ethnic study of atherosclerosis. *Ophthalmology*. 2011;118(4):656–662.
5. Kawasaki R, Wang JJ, Rochtchina E, et al. Cardiovascular risk factors and retinal microvascular signs in an adult Japanese population: the Funagata study. *Ophthalmology*. 2006;113(8):1378–1384.
6. Wong TY, Klein R, Duncan BB, et al. Racial differences in the prevalence of hypertensive retinopathy. *Hypertension*. 2003;41(5):1086–1091.
7. Klein R, Klein BEK, Moss SE, Wang Q. Hypertension and retinopathy, arteriolar narrowing, and arteriovenous nicking in a population. *Archives of Ophthalmology*. 1994;112(1):92–98.
8. Chao JR, Lai MY, Azen SP, et al. Retinopathy in persons without diabetes: the Los Angeles Latino eye study. *Investigative Ophthalmology & Visual Science*. 2007;48(9):4019–4025.
9. Peng XY, Wang FH, Liang YB, et al. Retinopathy in persons without diabetes. The Handan eye study. *Ophthalmology*. 2010;117(3):531–537.e532.
10. Zhu Z, Wang W, Scheetz J, Zhang J, He M. Prevalence and risk profile of retinopathy in non-diabetic subjects: national health and nutrition examination survey 2005 to 2008. *Clinical & Experimental Ophthalmology*. 2019;47(9):1173–1181.
11. Wong TY, Barr ELM, Tapp RJ, et al. Retinopathy in persons with impaired glucose metabolism: the Australian diabetes obesity and lifestyle (AusDiab) sudy. *American Journal of Ophthalmology*. 2005;140(6):1157–1159.
12. Wong TY, Klein R, Sharrett AR, et al. The prevalence and risk factors of retinal microvascular abnormalities in older persons: the cardiovascular health study. *Ophthalmology*. 2003;110(4):658–666.
13. Wang S, Xu L, Jonas JB, et al. Five-year incidence of retinal microvascular abnormalities and associations with arterial hypertension: the Beijing eye study 2001/2006. *Ophthalmology*. 2012;119(12):2592–2599.
14. Klein R, Klein BEK, Moss SE, Tasman W. The relation of systemic hypertension to changes in the retinal vasculature: the Beaver Dam eye study. *Transactions of the American Ophthalmological Society*. 1997;95:329–350.
15. Liew G, Campbell S, Klein R, et al. Ten-year longitudinal changes in retinal microvascular lesions: the atherosclerosis risk in communities study. *Ophthalmology*. 2011;118(8):1612–1618.
16. Ikram MK, Witteman JCM, Vingerling JR, Breteler MMB, Hofman A, De Jong PTVM. Retinal vessel diameters and risk of hypertension: the Rotterdam study. *Hypertension*. 2006;47(2):189–194.
17. Kawasaki R, Cheung N, Wang JJ, et al. Retinal vessel diameters and risk of hypertension: the multiethnic study of atherosclerosis. *Journal of Hypertension*. 2009;27(12):2386–2393.
18. Klein R, Klein BEK, Moss SE, Wong TY. The relationship of retinopathy in persons without diabetes to the 15-year incidence of diabetes and hypertension: Beaver Dam eye study. *Transactions of the American Ophthalmological Society*. 2006;104:98–106.
19. Wang JJ, Rochtchina E, Liew G, et al. The long-term relation among retinal arteriolar narrowing, blood pressure, and incident severe hypertension. *American Journal of Epidemiology*. 2008;168(1):80–88.
20. Wong TY, Klein R, Sharrett AR, et al. Retinal arteriolar diameter and risk for hypertension. *Annals of Internal Medicine*. 2004;140(4):248–255+I238.
21. Wong TY, Shankar A, Klein R, Klein BEK, Hubbard LD. Prospective cohort study of retinal vessel diameters and risk of hypertension. *British Medical Association*. 2004;329(7457):79–82.
22. Pickering TG, Eguchi K, Kario K. Masked hypertension: a review. *Hypertension Research*. 2007;30(6):479–488.
23. Leung H, Jie JW, Rochtchina E, Wong TY, Klein R, Mitchell P. Impact of current and past blood pressure on retinal arteriolar diameter in an older population. *Journal of Hypertension*. 2004;22(8):1543–1549.
24. Wong TY, Hubbard LD, Klein R, et al. Retinal microvascular abnormalities and blood pressure in older people: the cardiovascular health study. *British Journal of Ophthalmology*. 2002;86(9):1007–1013.
25. Ding J, Wai KL, McGeechan K, et al. Review: Retinal vascular caliber and the development of hypertension: a meta-analysis of individual participant data. *Journal of Hypertension*. 2014;32(2):207–215.
26. Cheung N, Mosley T, Islam A, et al. Retinal microvascular abnormalities and subclinical magnetic resonance imaging brain infarct: a prospective study. *Brain*. 2010;133(7):1987–1993.
27. Cooper LS, Wong TY, Klein R, et al. Retinal microvascular abnormalities and MRI-defined subclinical cerebral infarction: the atherosclerosis risk in communities study. *Stroke*. 2006;37(1):82–86.
28. Kawasaki R, Cheung N, Mosley T, et al. Retinal microvascular signs and 10-year risk of cerebral atrophy: the atherosclerosis risk in communities (ARIC) study. *Stroke*. 2010;41(8):1826–1828.
29. Wong TY, Klein R, Richey Sharrett A, et al. Cerebral white matter lesions, retinopathy, and incident clinical stroke. *Journal of the American Medical Association*. 2002;288(1):67–74.
30. Cheung N, Bluemke DA, Klein R, et al. Retinal arteriolar narrowing and left ventricular remodeling. The multi-ethnic study of atherosclerosis. *Journal of the American College of Cardiology*. 2007;50(1):48–55.
31. Cuspidi C, Meani S, Valerio C, et al. Prevalence and correlates of advanced retinopathy in a large selected hypertensive population. The Evaluation of Target Organ Damage in Hypertension (ETODH) study. *Blood Pressure*. 2005;14(1):25–31.
32. Kim GH, Youn HJ, Kang S, Choi YS, Moon JI. Relation between grade II hypertensive retinopathy and coronary artery disease in treated essential hypertensives. *Clinical and Experimental Hypertension*. 2010;32(7):469–473.
33. Tapp RJ, Owen CG, Barman SA, et al. Associations of retinal microvascular diameters and tortuosity with blood pressure and arterial stiffness United Kingdom biobank. *Hypertension*. 2019;74(6):1383–1390.

34. Tikellis G, Arnett DK, Skelton TN, et al. Retinal arteriolar narrowing and left ventricular hypertrophy in African Americans. The Atherosclerosis Risk in Communities (ARIC) study. *American Journal of Hypertension*. 2008;21(3):352–359.
35. Triantafyllou A, Anyfanti P, Gavriilaki E, et al. Association between retinal vessel caliber and arterial stiffness in a population comprised of normotensive to early-stage hypertensive individuals. *American Journal of Hypertension*. 2014;27(12):1472–1478.
36. Yip W, Sabanayagam C, Ong PG, et al. Joint effect of early microvascular damage in the eye & kidney on risk of cardiovascular events. *Scientific Reports*. 2016;6(1):27442.
37. Ong YT, Wong TY, Klein R, et al. Hypertensive retinopathy and risk of stroke. *Hypertension*. 2013;62(4):706–711.
38. Wieberdink RG, Ikram MK, Koudstaal PJ, Hofman A, Vingerling JR, Breteler MMB. Retinal vascular calibers and the risk of intracerebral hemorrhage and cerebral infarction. *Stroke*. 2010;41(12):2757–2761.
39. Yatsuya H, Folsom AR, Wong TY, Klein R, Klein BE, Sharrett AR. Retinal microvascular abnormalities and risk of lacunar stroke: atherosclerosis risk in communities study. *Stroke*. 2010;41(7):1349–1355.
40. Baker ML, Hand PJ, Liew G, et al. Retinal microvascular signs may provide clues to the underlying vasculopathy in patients with deep intracerebral hemorrhage. *Stroke*. 2010;41(4):618–623.
41. Baker ML, Hand PJ, Wong TY, et al. Retinopathy and lobar intracerebral hemorrhage: insights into pathogenesis. *Archives of Neurology*. 2010;67(10):1224–1230.
42. Wong TY, Rosamond W, Chang PP, et al. Retinopathy and risk of congestive heart failure. *Journal of the American Medical Association*. 2005;293(1):63–69.
43. Chandra A, Seidelmann SB, Claggett BL, et al. The association of retinal vessel calibres with heart failure and long-term alterations in cardiac structure and function: the Atherosclerosis Risk in Communities (ARIC) Study. *European Journal of Heart Failure*. 2019;21(10):1207–1215.
44. Wong TY, Klein R, Nieto FJ, et al. Retinal microvascular abnormalities and 10-year cardiovascular mortality: a population-based case-control study. *Ophthalmology*. 2003;110(5):933–940.
45. Sairenchi T, Iso H, Yamagishi K, et al. Mild retinopathy is a risk factor for cardiovascular mortality in Japanese with and without hypertension: the Ibaraki Prefectural Health Study. *Circulation*. 2011;124(23):2502–2511.
46. Tso MO, Jampol LM. Pathophysiology of hypertensive retinopathy. *Ophthalmology*. 1982;89(10):1132–1145.
47. Cuspidi C, Sala C, Grassi G. Updated classification of hypertensive retinopathy: which role for cardiovascular risk stratification? *Journal of Hypertension*. 2015;33(11):2204–2206.
48. Delles C, Michelson G, Harazny J, Oehmer S, Hilgers KF, Schmieder RE. Impaired endothelial function of the retinal vasculature in hypertensive patients. *Stroke*. 2004;35(6):1289–1293.
49. Tsai WC, Li YH, Huang YY, Lin CC, Chao TH, Chen JH. Plasma vascular endothelial growth factor as a marker for early vascular damage in hypertension. *Clinical Science*. 2005;109(1):39–43.
50. Dziedziak J, Zaleska-Żmijewska A, Szaflik JP, Cudnoch-Jedrzejewska A. Impact of arterial hypertension on the eye: a review of the pathogenesis, diagnostic methods, and treatment of hypertensive retinopathy. *Medical Science Monitor: International Medical Journal of Experimental and Clinical Research*. 2022;28:e935135.
51. Bourke K, Patel MR, Prisant LM, Marcus DM. Hypertensive choroidopathy. *Journal of Clinical Hypertension*. 2004;6(8):471–472.
52. Saito M, Noda K, Saito W, Hirooka K, Hashimoto Y, Ishida S. Increased choroidal blood flow and choroidal thickness in patients with hypertensive chorioretinopathy. *Graefe's Archive for Clinical and Experimental Ophthalmology*. 2020;258(2):233–240.
53. Mulè G, Vadalà M, Sinatra N, et al. Relationship of choroidal thickness with pulsatile hemodynamics in essential hypertensive patients. *Journal of Clinical Hypertension*. 2021;23(5):1030–1038.
54. Hayreh SS, Servais GE, Virdi PS. Fundus lesions in malignant hypertension: V. Hypertensive optic neuropathy. *Ophthalmology*. 1986;93(1):74–87.
55. Keith NM, Wagener HP, Barker NW. Somme different types of essential hypertension: their course and prognosis. *American Journal of the Medical Sciences*. 1939;197:332–343.
56. Scheie HG. Evaluation of ophthalmoscopic changes of hypertension and arteriolar sclerosis. *A M A Archives of Ophthalmology*. 1953;49(2):117–138.
57. Wong TY, Mitchell P. Hypertensive retinopathy. *N Engl J Med*. 2004;351(22):2310–2317.
58. Ahn SJ, Woo SJ, Park KH. Retinal and choroidal changes with severe hypertension and their association with visual outcome. *Investigative Ophthalmology & Visual Science*. 2014;55(12):7775–7785.
59. Downie LE, Hodgson LA, Dsylva C, et al. Hypertensive retinopathy: comparing the Keith-Wagener-Barker to a simplified classification. *Journal of Hypertension*. 2013;31(5):960–965.
60. Shantsila A, Lip GYH. Malignant hypertension revisited-does this still exist? *American Journal of Hypertension*. 2017;30(6):543–549.
61. Luo BP, Brown GC. Update on the ocular manifestations of systemic arterial hypertension. *Current Opinion in Ophthalmology*. 2004;15(3):203–210.
62. Grosso A, Cheung N, Veglio F, Wong TY. Similarities and differences in early retinal phenotypes in hypertension and diabetes. *Journal of Hypertension*. 2011;29(9):1667–1675.
63. Rotsos T, Andreanos K, Blounas S, Brouzas D, Ladas DS, Ladas ID. Multimodal imaging of hypertensive chorioretinopathy by swept-source optical coherence tomography and optical coherence tomography angiography: case report. *Medicine (Baltim)*. 2017;96(39):e8110.
64. Hirano Y, Yasukawa T, Ogura Y. Bilateral serous retinal detachments associated with accelerated hypertensive choroidopathy. *International Journal of Hypertension*. 2010;2010:964513.

65. Hua D, Xu Y, Zeng X, et al. Use of optical coherence tomography angiography for assessment of microvascular changes in the macula and optic nerve head in hypertensive patients without hypertensive retinopathy. *Microvascular Research.* 2020;129:103969.
66. Lee WH, Park JH, Won Y, et al. Retinal microvascular change in hypertension as measured by optical coherence tomography angiography. *Scientific Reports.* 2019;9(1):156.
67. Huang X, Wang H, She C, et al. Artificial intelligence promotes the diagnosis and screening of diabetic retinopathy. *Frontiers in Endocrinology.* 2022;13:946915.
68. Rizzoni D, Agabiti Rosei C, De Ciuceis C, Semeraro F, Rizzoni M, Docchio F. New methods to study the microcirculation. *American Journal of Hypertension.* 2018;31(3):265–273.
69. Stefánsson E, Olafsdottir OB, Eliasdottir TS, et al. Retinal oximetry: metabolic imaging for diseases of the retina and brain. *Progress in Retinal and Eye Research.* 2019;70:1–22.
70. Lim M, Sasongko MB, Ikram MK, et al. Systemic associations of dynamic retinal vessel analysis: a review of current literature. *Microcirculation.* 2013;20(3):257–268.
71. Unger T, Borghi C, Charchar F, et al. International society of hypertension global hypertension practice guidelines. *Hypertension.* 2020;75(6):1334–1357, 2020.
72. Hughes AD, Stanton AV, Jabbar AS, Chapman N, Martinez-Perez ME, Thom SAMcG. Effect of antihypertensive treatment on retinal microvascular changes in hypertension. *Journal of Hypertension.* 2008;26(8):1703–1707.
73. Thom S, Stettler C, Stanton A, et al. Differential effects of antihypertensive treatment on the retinal microcirculation. *Hypertension.* 2009;54(2):405–408.
74. McGill JB. Improving microvascular outcomes in patients with diabetes through management of hypertension. *Postgraduate Medicine.* 2009;121(2):89–101.
75. Whelton PK, Carey RM, Mancia G, Kreutz R, Bundy JD, Williams B. Harmonization of the American College of Cardiology/American Heart Association and European Society of Cardiology/European Society of Hypertension Blood Pressure/Hypertension Guidelines: comparisons, reflections, and recommendations. *Circulation.* 2022;146(11):868–877.
76. Wong W, Gopal L, Yip CC. Hypertensive retinopathy and choroidopathy. *CMAJ (Canadian Medical Association Journal).* 2020;192(14):E371.
77. Cortina G, Hofer J, Giner T, Jungraithmayr T. Severe visual loss caused by unrecognized malignant hypertension in a 15-year-old girl. *Pediatrics International.* 2015;57(2):e42–e44.

Chapter 33

Pregnancy-induced hypertension

Anthony A. Jones, John K. Lippincott, Monica P. Gonzalez and Judy E. Kim
Department of Ophthalmology, Medical College of Wisconsin, Milwaukee, WI, United States

Introduction

Approximately 8% of pregnancies are complicated by pregnancy-induced hypertension (PIH). It comprises one of four conditions: chronic hypertension, gestational hypertension, preeclampsia–eclampsia, and chronic hypertension with superimposed preeclampsia. Any of the preceding classifications of PIH can cause changes in various organ systems of the mother and can contribute to maternal and fetal morbidity and mortality. This multisystem disorder often involves the kidneys, liver, hematopoietic system, placenta, and eyes—specifically, the retinal and choroidal vasculature.[1] The retinal and choroidal vasculature is unique in that its vessels can be directly and noninvasively observed. Therefore, the ophthalmologic evaluation is an invaluable tool that should be used as a diagnostic aid in the assessment of preeclampsia.

This chapter reviews PIH, its associated changes to the retinal and choroidal vasculature, and pursuant sequelae.

Terminology and classification

Hypertensive disorder of pregnancy terminology

Pregnancy-induced hypertension is defined as systolic blood pressure (SBP) >140 mmHg and diastolic blood pressure (DBP) >90 mmHg. It is classified as mild (SBP 140–149 and DBP 90–99 mmHg), moderate (SBP 150–159 and DBP 100–109 mmHg), and severe (>SBP 160 and >DBP 110 mmHg).[2,3]

Hypertensive disorder of pregnancy classification

- *Chronic hypertension:* the presence of hypertension before conception, the development of elevated blood pressure before 20 weeks' gestational age, or the persistence of hypertension beyond 12 weeks after delivery.
- *Gestational hypertension:* hypertension arising de novo after 20 weeks' gestation in the absence of proteinuria and without biochemical or hematological abnormalities. It is a provisional diagnosis that can change to chronic hypertension if the blood pressure remains elevated beyond 12 weeks postpartum. Approximately one in four women with gestational hypertension will progress to preeclampsia. Risk of progression to preeclampsia is inversely related to the gestational age at which gestational hypertension is diagnosed.
- *Preeclampsia–eclampsia*: new-onset hypertension that develops after 20 weeks of gestation and is accompanied by proteinuria or evidence of end-organ dysfunction such as thrombocytopenia, liver or renal impairment, pulmonary edema, or neurologic or visual dysfunction.

HELLP syndrome (hemolysis, elevated liver enzymes, low platelets) is a severe manifestation of preeclampsia–eclampsia and is considered a variant, not a separate disorder.

Eclampsia is defined as seizures in a pregnant woman with preeclampsia with no other identifiable cause.

- *Chronic hypertension with superimposed preeclampsia:* preeclampsia in a pregnant patient with known chronic hypertension after 20 weeks' gestation. This category of PIH carries the highest rate of maternal and fetal complications, higher than with either condition alone.[2–4]

Retinal and Choroidal Vascular Diseases of the Eye. https://doi.org/10.1016/B978-0-443-15583-3.00033-0

As we focus on ocular changes that occur in the setting of preeclampsia, it is important to note that the etiology of preeclampsia is often multifactorial, but its pathogenesis is generally accepted to originate from reduced placental perfusion. This leads to abnormal vascular remodeling and maternal vascular endothelial dysfunction systemically.[4]

The retinal and choroidal pathology related to preeclampsia is diverse and includes changes such as hypertensive retinopathy and choroidopathy, exacerbation of diabetic retinopathy, exudative retinal detachments, retinal vascular occlusions, retinal and vitreous hemorrhages, and many other findings. Depending on which retinal and choroidal changes occur, as well as to what severity, will determine the secondary visual symptoms experienced by the patient. Commonly associated visual symptoms include decreased vision, photopsia, diplopia, and visual field defects. Often, the onset of visual changes in a pregnant patient may herald looming seizure onset. Therefore, these symptoms should be investigated immediately by an obstetrician to rule out life-threatening disease. Ophthalmologic evaluation can proceed once the patient is deemed stable. As profound as these changes may be, the visual prognosis is usually good, and permanent, severe vision loss or blindness is uncommon.

Overview of choroidal and retinal vasculature

Choroidal vasculature

The choroid is a layered vascular network that receives its blood supply from the short and long posterior ciliary arteries, which are branches of the ophthalmic artery. The cardinal duty of the choroid is to supply oxygen and nutrients to the RPE and outer one-third of the retina. Its terminal layer—the choriocapillaris network—has the highest capillary density and the highest blood flow per unit weight compared to any other tissue in the human body. The capillaries of the choriocapillaris have a larger lumen than those in other areas of the body, and they are fenestrated to allow for material exchange. This exchange is mediated through the systemic hemodynamic system, as well as through metabolic and regulatory signals from the surrounding environment.[5–9]

One important consideration is that of the blood–retina barrier (BRB), which is composed of both an inner (iBRB) and outer barrier (oBRB). The oBRB is formed by tight junctions between neighboring RPE cells and functions in the regulation of the movement of solutes and nutrients from the choroid to the subretinal space.[6–9] Dysregulation of the choriocapillaris can lead to RPE damage and loss of integrity of the oBRB. Subsequently, protein and fluid leakage into the retinal neuroepithelial layer can disrupt retinal architecture and directly injure photoreceptors and support cells.

Retinal vasculature

The blood supply to the inner two-thirds of the retina is supplied by the central retinal artery, which is also a branch of the ophthalmic artery. Retinal blood vessels are unique in two ways. First is the iBRB. While the oBRB is formed by the RPE, the iBRB is contained within the inner retinal microvasculature. Much like the blood-brain barrier, the iBRB is composed of tight junctions between the nonfenestrated inner endothelial cell lining of the vessels.[6–9] The second unique characteristic is the retinal vasculatures' lack of sympathetic innervation and its reliance on autoregulation to modulate blood flow.[6–9] The endothelial lining of these vessels rests on a basal lamina composed of pericytes and the processes of Muller cells and astrocytes. This composition of the basal lamina influences the activity of the iBRB through regulatory signals used to indicate changes in its microenvironment.[6–9] An example of autoregulation can be appreciated in states of hypertension, such as preeclampsia, when systemic and local (intraocular) levels of the renin-aldosterone system (RAAS) increase. The main effector of RAAS, angiotensin II (Ang II), has been identified within retinal vessels, and when stimulated, can cause profound cell proliferation and fibrosis. This leads to a cascade of disruption which ultimately leads to the findings of hypertensive retinopathy, including those discussed within this chapter. On the other hand, Mas-receptors (Mas-R), which are also located in the eye, and stimulated by the product of the breakdown of Ang II, cause antiinflammatory effects, inhibition of cell proliferation, antifibrosis, and vasodilatation.[10]

Preeclampsia/eclampsia-associated hypertensive choroidopathy and retinopathy

End organ damage caused by preeclampsia is nearly entirely ascribed to disease of the microvasculature. Therefore, it is logical for the choroid, retina, and supporting tissues to be at increased risk during episodes of acute hypertension. The classic pathologic changes can be attributed to hypertensive choroidopathy or retinopathy and may be manifestations of new disease or worsening underlying chronic disease.[11]

Hypertensive choroidopathy

Pathophysiology

Choroidal blood flow is highly dependent on systemic perfusion and is extremely sensitive to abnormalities in systemic hemodynamics. During episodes of acute and severely elevated blood pressure, such as those in preeclampsia, there are alterations to vascular endothelial responses, lumen diameter, and ultimately impaired blood flow dynamics.[6,7] The main mechanism by which these changes occur is the same mechanism that ensues during any episode of acute elevation in blood pressure and is not unique to preeclampsia. Such episodes cause acutely increased systemic vascular resistance (which typically is decreased in pregnancy due to normal hormonal changes allowing for higher blood flow to the choroid[12]) and a reflexive vasospasm that causes choroidal arterioles, or their supplying ciliary arteries, to undergo acute constriction. This leads to decreased perfusion, or in severe cases, nonperfusion of the choriocapillaris.[8,13–15] Overall pathogenesis involves an elevation in blood pressure that exceeds the autonomic regulatory capacity of the choroidal blood vessels. This leads to endothelial damage which signals a cascade of inflammatory mediators that can cause vasospasm, leakage, and fibrin platelet adhesions that obstruct choroidal arteries and the choriocapillaris.[14] Furthermore, given its nutritive and metabolic supply to the RPE and outer retina, any abnormal alteration in blood flow to the choroid will consequently affect these tissue layers as well. Clinically, this can be observed as an array of pigmentary changes including mottling, migration, hyperpigmentation, and hypopigmentation (Fig. 33.1).

Clinical features

Elschnig spots: compromise of choroidal vasculature due to acute hypertension leads to a cascade of dysfunction involving the RPE and photoreceptors. Without the choroid to supply oxygen and nutrients, the RPE becomes ischemic, and necrosis of the tissue rapidly sets in. This derangement can cause a slew of pathologic changes, each of which can present with a different appearance on examination. One common manifestation of hypertensive choroidopathy is an Elschnig spot. These are pale yellow spots with well-defined margins that appear in areas of damaged RPE secondary to underlying choriocapillaris nonperfusion. Once the acute insult has subsided, these spots develop a central pigment in the surrounding area of injury.[14,16,17] Histopathologic descriptions of Elschnig spots in an experimental model show occlusion of terminal choroidal arterioles and choriocapillaris by fibrinous exudate and desquamated endothelial cells with surrounding edematous tissue.[11,18]

Siegrist streak: a Siegrist streak is a linear hyperpigmented streak overlying choroidal arteries. These typically appear after acute insult to the RPE, leading to atrophy in some areas as well as reactive RPE hyperplasia along the margins of the area of insult.[19]

Serous retinal detachment: central serous chorioretinopathy (CSC) is a well-documented, albeit rare, phenomenon that is known to occur in uncomplicated pregnancies. During normal pregnancy, levels of endogenous cortisol rise by 2–4 times,[20] and this elevation has been implicated as a major contributor to the development of CSC.[21] It is known that cortisol is a vasoconstrictor and thus may cause decreased perfusion to the choroid, subsequent RPE damage with oBRB disruption, and the development of subretinal exudation.[22–24] As previously discussed, through similar perfusion defects, hypertensive chorioretinopathy causes oBRB breakdown which also allows for subretinal exudation. When these two mechanisms are combined, a severe form of CSC with serous retinal detachment can occur. These are typically bilateral (though unilateral cases have been reported) and can occur at any stage of pregnancy (even postpartum). The onset of symptoms is abrupt and can include blurry vision with hyperopic shifts, metamorphopsias, and visual field defects. It is important to note that photopsias and floaters, as seen in rhegmatogenous retinal detachments, are typically absent.[22,25–34]

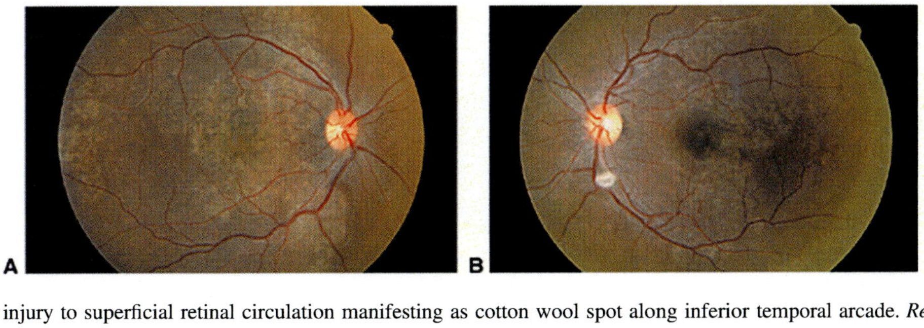

FIGURE 33.1 Preeclampsia-induced fundus changes fundus photos of a 32-year-old female presenting with decreased vision in both eyes after undergoing emergency cesarean section for preeclampsia. (A) Right eye, and (B) left eye, fundus photos demonstrate patchy pigmentary changes secondary to choroidal perfusion abnormalities with subsequent injury to the overlying retinal pigment epithelium and retina. Left eye fundus photo also shows injury to superficial retinal circulation manifesting as cotton wool spot along inferior temporal arcade. *Reproduced, with permission, from Newman SL, 2001–02 Basic and Clinical Science Course, Section 5, Neuro-Ophthalmology. American Academy of Ophthalmology, 2001.*

Imaging findings

Angiography: early studies used intravenous fluorescein angiography (FA) to investigate the pathogenesis of preeclampsia-induced choroidopathy, but it was not until the early 1980s that two investigators were able to accurately describe the delayed filling of the choriocapillaris as the root cause of the choroidal disease.[35] Shortly after this publication, several other studies supporting these findings emerged.[17,31,35–37] As such, classic findings on FA that would indicate hypertensive chorioretinopathy include delayed perfusion of the choroidal capillaries, nonperfusion areas, and late slow fluorescein leakage between the subretinal and pigmented epithelial spaces (Fig. 33.2). Some consider FA to be the gold standard for the diagnosis of preeclamptic hypertensive choroidopathy. However, given its invasiveness, it is not a widely utilized imaging modality by most ophthalmologists in pregnant patients.

Another less commonly used imaging modality used to evaluate PIH is indocyanine green angiography (ICGA).[38] It has been shown that ICG does not cross the placenta, and its safety in pregnancy has been well established.[25,38–41] Furthermore, ICG is better at studying the choroidal vasculature than FA, which is better for evaluating the retinal vasculature. During episodes of preeclampsia, the findings seen on ICGA mirror those on FA and include areas of delayed choroidal perfusion or nonperfusion, staining of choroidal vasculature, and subretinal leakage.[38,42]

Optical coherence tomography (OCT)/optical coherence tomography angiography (OCTA): choroidal thickness has been studied using OCT with enhanced depth imaging (EDI) in both noncomplicated pregnancy, as well as in patients with preeclampsia. Although it varies in each stage of pregnancy, it has been noted that the choroidal thickness is significantly increased in uncomplicated pregnancy as compared to nonpregnant individuals. Like other changes in the choroid during pregnancy, this increase in thickness is thought to be multifactorial and likely due to hormonal changes and increased volume throughout the circulatory system.[43,44] In preeclampsia, there have been conflicting studies as to whether the choroidal thickness is increased or decreased when compared to normal pregnancy. However, it generally remains thicker than in nonpregnant women.[45–47] Similarly, OCTA has been used to study the choriocapillaris blood flow in this population, and it showed that the flow area in question was significantly higher in both normal pregnancies and in preeclampsia when compared to nonpregnant women.[48,49]

Aside from the general preeclampsia-related choroidal changes on OCT and OCTA, there are also distinctive OCT features representative of Elschnig spots and serous retinal detachments. Elschnig spots on OCT commonly show V-shaped adhesions with focal elevation of a thick reflective band of RPE and choriocapillaris. At the center of these lesions, there is disruption of the external limiting membrane and the ellipsoid zone, or a hyporeflective shadow behind the elevation.[16,50] These findings parallel the histologic features of Elschnig spots, with the greatest RPE damage occurring at the center of the lesion where the RPE is necrotic on histology.[11,16,18,50]

Serous retinal detachments in preeclampsia can range from small focal neuroepithelial detachments that are difficult to detect on fundus examination to large detachments involving the majority of the macula. In either case, OCT reveals an elevation of the neurosensory retina off the RPE with a clear and optically empty space between, indicating serous fluid. The detached retina may appear normal, show elongated photoreceptors, or show some areas of subretinal hyperreflective material suggestive of fibrin deposition.[51,52] Given its noninvasive nature, OCT is generally recommended for patients with

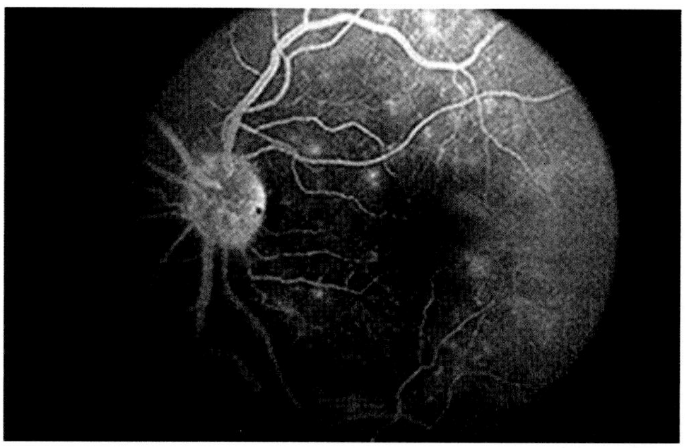

FIGURE 33.2 Fluorescein angiography in preeclampsia fluorescein angiogram of the left eye from the same patient in Fig. 33.1 shows choroidal perfusion abnormalities in a similar distribution to pigmentary changes on fundus photo (Fig. 33.1B). Areas of focal late choroidal staining can also be appreciated throughout the posterior pole. *Reproduced, with permission, from Newman SL, 2001–02 Basic and Clinical Science Course, Section 5, Neuro-Ophthalmology. American Academy of Ophthalmology, 2001.*

preeclampsia with visual symptoms. Most studies using OCT show resolution of the serous detachments within 4–6 weeks.[52]

Hypertensive retinopathy

Pathophysiology

Much like the choroid, the retina undergoes physiologic changes during normal pregnancy which then become deranged in preeclampsia. It has been shown that in uncomplicated pregnancy the retinal and macular thickness increases during the second and third trimesters.[8,53,54] It is believed that the mechanism behind these changes is through the same hemodynamic process previously discussed for the choroid. Most studies do not show a significant difference between retinal nerve fiber layer (RNFL) thickness in normal pregnancy and preeclampsia.[55,56] However, it has been shown that in the 2-month postnatal period, the difference in RNFL thickness was significantly thinner in mothers who had experienced severe preeclampsia.[55] This thinning of the RNFL is thought to result from permanent hypoxic injury to the retinal layers due to acute microcirculatory vasoconstriction.[57]

Furthermore, during preeclampsia, there is a disruption in the balance of circulating angiogenic factors, which play a key role in the pathogenesis of PIH-associated hypertensive retinopathy. In brief, there is increased expression of inflammatory cytokines such as tumor necrosis factor (TNF-α), soluble fms-like tyrosine kinase-1 (sFlt1), and vascular endothelial growth factor (VEGF).[58,59] These cytokines induce extracellular matrix remodeling, destruction of the endothelial blood–retinal barrier, and promotion of differentiation, proliferation, and migration of endothelial cells.[60,61] Ultimately, vascular leakage and subsequent significant retinal impairment manifest clinically in the form of hypertensive retinopathy.[61]

Clinical features

In preeclampsia-associated hypertensive retinopathy, the first clinically relevant change is a spastic phase in which arteriolar spasm creates elongated, spindle-shaped arterioles with lumen constriction. This vasospasm typically begins focally in the peripheral superonasal arterioles and slowly spreads toward the disc. If left uncontrolled, the vasospasms may generalize to the entire retinal vascular network.[62–66] At times, this constriction may be so severe that there is a complete collapse of the lumen leading to temporary infarcts in the retina. This has been documented in several cases and manifests symptomatically as amaurosis fugax.[24,67–69] The condition's spastic nature leads to variation in the degree and location of appreciable pathology on each subsequent exam. Normalization of blood pressure typically leads to restoration of a normal arterial lumen. If blood pressure is not controlled at this stage, then further manifestations of hypertensive retinopathy will ensue. Classic findings include microaneurysms, retinal hemorrhages, cotton wool spots, retinal edema, and lipid exudates.[64,70] More severe cases may also cause hypertensive optic neuropathy which is seen on examination as optic disc swelling.

Aneurysms: aneurysms of the retinal arterioles may develop in preeclampsia and, like those seen in malignant hypertension, are usually of the miliary form. They commonly occur in small vessels at sites of arterio-venous crossing and near cotton-wool spots.[65,71,72]

Hemorrhages: both superficial flame-shaped hemorrhages and deeper dot blot hemorrhages (DBH) can occur in preeclampsia. Flame hemorrhages characteristically arise from retinal capillaries near the optic disc and fan out along the striate course of the RNFL. They typically self-resolve in 6–8 weeks. Dot blot hemorrhages occur within deeper layers of the retina (deeper nuclear or plexiform layers) and usually are indicative of a more severe stage of hypertensive retinopathy. They tend to take longer to resolve than flame-shaped hemorrhages.[63,65,66,72]

Cotton wool spots: cotton wool spots are whitish-gray patches seen in the superficial retina on the nerve fiber layer (Fig. 33.1B). These can signify severe hypertension, as there is an interruption of the axoplasmic flow of the retinal ganglion cells, leading to axoplasmic debris accumulation and fibrinoid necrosis of the terminal and precapillary arterioles.[65,66,72,73]

Edema: retinal edema secondary to hypertensive retinopathy typically begins near the optic disc and courses along the retinal vessels toward the macula.[62,63,65,66,72] On fundoscopy, it will appear as a milky surface with faint striations running in the direction of nerve fiber layers. In cases of severe edema, it will appear as frank retinal elevation with abnormal retinal contour. Studies suggest fluid accumulation within Muller cells to be the earliest histopathologic manifestation of macular edema. If the accumulation becomes severe, the Muller cells become damaged and die, resulting in the formation of large cystoid cavities, or cystoid macular edema.[74]

Hard exudates: hard exudates are yellow flecks located in the outer plexiform layer or Henle's layer. They are composed of macrophages filled with lipid residues and indicate excessive retinal vascular leakage from damaged capillaries. They are usually located within the macula and may form a ring-like figure known as a "macular star."[66,72,75,76]

Hypertensive optic neuropathy: optic disc swelling in the setting of preeclampsia is known as hypertensive optic neuropathy. This finding is limited to the most severe cases of preeclampsia.[66,72]

Imaging findings

OCT/OCTA: as previously discussed, preeclampsia causes a thinning of the RNFL. This change is noted on OCT and becomes apparent during the first several postnatal months. Otherwise, OCT findings in hypertensive retinopathy associated with preeclampsia are the same as would be expected in any case of hypertensive retinopathy. Such findings include microaneurysms, which appear as small capsular structures with a high signal intensity completely circumscribed in round or oval hyporeflective spaces. Flame-shaped hemorrhages on OCT appear as highly reflective spots at the RNFL layer causing shadowing. Dot blot hemorrhages on the other hand are observed in several patterns, most commonly as oval-shaped hyperreflective echoes in the inner retinal layers causing shadowing of deeper layers. Cotton wool spots appear as hyperreflective, nodular, or elongated lesions in the RNFL and sometimes the ganglion cell layer (GCL), also with shadowing of deeper layers. Retinal edema produces classic cystic spaces that appear as optically empty spaces involving the layers from the outer nuclear layer to GCL. Lastly, hard exudates appear as highly reflective echoes with shadowing on the posterior retinal layers.[74,77,78]

Studies using OCTA have analyzed the effects of preeclampsia on retinal circulation. It has been shown that, in preeclampsia, there is a significant decrease in the vascular density within the superficial and deep foveal layers. Specifically, there is a decrease in the parafoveal, deep temporal, and deep superior vascular densities. Overall, the deep capillary plexus was more affected in preeclampsia. This is believed to be due to high metabolic activity and complex vascular structure of the outer retina causing susceptibility to poor perfusion from generalized vasospasm. Some of these preeclampsia-related vascular changes were noted in the setting of no retinopathy on the fundus exam.[79] This demonstrates the value of OCTA in understanding the pathogenesis of ocular disease in preeclampsia.

Fluoresceine angiography: similar to OCT findings, the retinal changes from preeclampsia are not unique to this disease and are the same as those found in other cases of hypertensive retinopathy. Interestingly, there are often cases where severe choroidal and retinal pathology can be seen, but the retinal vasculature remains seemingly normal on FA.[35] When retinal vascular disease is present, it reveals microaneurysms, which appear as hyperfluorescent dots with some accompanied by focal leakage of dye.[35,80] Venous congestion is common and there is typically associated delay in peripapillary choroidal perfusion, as well as petaloid leakage in the maculae.[35,81] Peripheral retinal vascular leakage in late-phase frames has also been demonstrated by ultra-widefield fluorescein angiography.[81] Staining of the subretinal spaces is common and was discussed in the section regarding serous retinal detachments secondary to choroidal pathology. Occasionally, where the RPE is disrupted or damaged by hypertensive retinopathy, there are early hyper-fluorescent changes that represent window defects.[35,80]

Management of hypertensive choroidopathy and hypertensive retinopathy

The management of hypertensive choroidopathy/retinopathy in the setting of fulminant preeclampsia is directed immediately toward life-saving measures for the mother and fetus. In brief, treatment protocols for preeclampsia include stroke prevention through antihypertensive therapy, seizure prophylaxis through intravenous magnesium sulfate or, if indicated, benzodiazepines. The definitive treatment of preeclampsia is the delivery of the fetus.[1]

In most cases, the retinal and choroidal changes associated with acute hypertension reverse after delivery and resolution of preeclampsia, although focal areas of retinal thinning and/or choroidal atrophy may persist. Interestingly, the use of magnesium for preeclampsia has been shown to dilate retinal arterioles and consequently prevent visual loss from vasospasm.[82]

Exacerbation of preexisting diabetic retinopathy during PIH

Pregnancy is considered an independent risk factor for the development and progression of diabetic retinopathy (DR) in both type 1 and type 2 diabetes; therefore, increased ophthalmologic surveillance is warranted.[83–89] This risk is further exacerbated in the setting of PIH and especially in preeclampsia.[86,89–92] According to The American Academy of Ophthalmology Preferred Practice Patterns,[83] pregnant patients with diabetes should undergo dilated fundus examination in the first trimester, with subsequent follow-up determined by the severity of retinopathy. Gestational diabetes does not require eye examinations during pregnancy. However, moderate nonproliferative diabetic retinopathy (NPDR) requires reexamination every 3–6 months, whereas severe NPDR or worse should be reexamined every 1–3 months. Heightened surveillance should continue during the first year postpartum. These recommendations do not include specifics with regards to PIH or preeclampsia but given the recognized worsening of DR in these hypertensive states, the ophthalmologist should carry a heightened concern for this unique patient population.[86,93]

Management

Management of DR in the patient with preeclampsia involves normalization of blood pressure as well as treatments geared directly toward the DR. With regards to laser surgery for DR in the pregnant patient, several studies have shown decreased progression or regression of proliferative diabetic retinopathy (PDR).[84,94–97] Although spontaneous regression of DR and PDR during the postpartum period has been shown,[86] some studies advocate for early use of laser for PDR in pregnant patients given worse outcomes in observation groups.[88,94,95] Furthermore, for diabetic macular edema (DME) requiring treatment, certain groups advocate avoidance of anti-VEGF therapy in favor of focal laser photocoagulation or intravitreal corticosteroid injections due to safety concerns regarding anti-VEGF therapy during pregnancy.[91]

The safety of Intravitreal anti-VEGF use in pregnancy has not been substantiated through large prospective studies and available information is limited. Most anti-VEGF medications are listed as category C or B and their use is only recommended if clearly needed, and with exercised precautions due to concerns regarding teratogenicity.[16] With that said, several small studies have reported on the use of such medications during pregnancy, most of which reported no adverse outcomes.[98–102] In rare cases where intravitreal anti-VEGF therapy was administered, there were reported adverse outcomes of miscarriage, preeclampsia, and infant bradycardia and respiratory distress.[103,104] Whether these poor outcomes were related to the use of intravitreal anti-VEGF or a coincidence remains unknown. Similarly, there is little literature geared toward the use and safety profile of intravitreal steroids for DME in pregnancy. Several small studies have reported their use, none of which showed any deleterious effects on the pregnancy or toward the fetus.[105,106]

Lastly, DR-related indications for vitreoretinal surgery are the same for both pregnant and nonpregnant patients and include nonclearing vitreous hemorrhage, sight-threatening tractional retinal detachments, and neovascular glaucoma.[14] When surgery is indicated, it should be performed under local anesthesia with careful consideration of the anesthetic agent. These drugs should be limited to FDA category A and B medications such as lidocaine, etidocaine, and prilocaine hydrochloride. On the other hand, bupivacaine and mepivacaine, class C category drugs, should be avoided as they can cause fetal bradycardia. Furthermore, subtenon injection is generally preferred over retrobulbar block, as the latter poses the risk of accidental injection of medication into the systemic circulation which can cause severe harm to the mother and fetus. In general, it is advised to wait until the postpartum period for any such intervention, but if this is not possible, a surgical plan should be devised with the help of the obstetrician.[86,91,107]

Retinal vascular occlusion in the setting of PIH

Vasospasm: as previously mentioned, preeclampsia has the potential to cause retinal arterial spasms that can lead to the collapse of the arterial lumen—a form of transient retinal artery occlusion.[24,59,68,69] Similarly, one study used a Doppler assessment of orbital vasculature during episodes of visual disturbances in the setting of preeclampsia to further elucidate vascular flow dynamics. Doppler assessment demonstrated decreased blood flow velocity and increased resistance to blood flow within the central retinal artery, and it was concluded that the visual disturbances were directly related to retinal artery vasospasm and ischemia.[108] On the other hand, two other studies employed Doppler flow in preeclampsia and showed a decreased impedance to flow with orbital hyperperfusion.[109,110] It should be noted, that in these latter studies, such contradictory findings were only statistically significant in cases comparing "severe preeclampsia" to preeclampsia and normal pregnancy.

Hypercoagulability: during the third trimester and postpartum period of normal pregnancy, there is an induced state of hypercoagulability and subsequent increased risk of 0.05%–1.8% for retinal venous thromboembolism.[86,111] One study showed that in comparison with normal pregnancy, preeclampsia increased the likelihood of hypercoagulability by 4.89%.[112] Inference suggests that preeclampsia increases the risk of retinal venous thromboembolism above these recognized values. However, studies are limited and no direct analysis has been performed.

Several case reports have documented retinal vascular occlusions in preeclamptic patients.[113–116] A purported unifying theory posits preeclampsia as a condition that exacerbates all three components of Virchow's triad: vessel wall abnormality, abnormal blood flow, and enhanced coagulability of blood.[116] Although all reported cases occurred in the setting of preeclampsia, the numbers are small, and all but one noted more likely explanatory pathology for the vascular occlusions than preeclampsia. One occurred in severe preeclampsia with HELLP syndrome, which is a known thrombotic microangiopathic vasculopathy. The other occurred in a patient who was found to have familial thrombophilia. Given the paucity of literature, the connection between vascular occlusion and preeclampsia is not fully understood.

Purtscher-like retinopathy: Purtscher retinopathy occurs in the setting of trauma (usually compressive) and is characterized by a constellation of retinal findings including cotton-wool spots, retinal hemorrhages, optic disc edema, and Purtscher flecken (areas of inner retinal whitening).[117] There are several systemic conditions, including preeclampsia, that have been reported to cause similar findings, and in these cases when trauma is absent, they are referred to as Purtscher-like retinopathy. Although exceedingly rare, preeclampsia-related Purtscher-like retinopathy has been documented in several cases.[16,118–124] The pathogenesis of this disorder in preeclampsia is unknown but it has been postulated to involve multilayered precapillary retinal arterial obstruction by complement-induced leukoemboli.[16,118] Unlike other preeclampsia-induced ophthalmic diseases that have near complete visual recovery, Purtscher-like retinopathy often causes permanent visual loss.[108]

Conclusion

In summary, the retinal and choroidal vascular changes observed in the setting of preeclampsia represent critical manifestations of this multisystem disorder. This delicate network of vasculature undergoes notable modifications in response to preeclampsia, including arteriolar narrowing, vessel leakage, hemorrhaging, vasospasm, and endothelial dysfunction. These changes often reflect the severity of preeclampsia, and ophthalmologists, in collaboration with obstetricians, play a crucial role in the interdisciplinary care of preeclamptic patients. With early detection of retinal and choroidal changes, valuable insights into the overall systemic vascular status can be gained which help optimize maternal and fetal outcomes. Further research is needed to elucidate the underlying mechanism behind this complex disease and to help advance therapeutic interventions specifically targeting the retinal and choroidal vascular alterations that may occur. By deepening our understanding of these ocular changes, we can strive to improve the overall care and outcomes for women affected by this complex and potentially devastating condition.

References

1. WHO Guidelines Approved by the Guidelines Review Committee, in WHO Recommendations for Prevention and Treatment of Pre-eclampsia and Eclampsia; 2011.
2. Brown MA, Magee LA, Kenny LC, et al. Hypertensive disorders of pregnancy: ISSHP classification, diagnosis, and management recommendations for international practice. *Hypertension.* 2018;72(1):24–43. https://doi.org/10.1161/HYPERTENSIONAHA.117.10803.
3. Kintiraki E, Papakatsika S, Kotronis G, Goulis DG, Kotsis V. Pregnancy-Induced hypertension. *Hormones.* 2015;14(2):211–223. https://doi.org/10.14310/horm.2002.1582.
4. Gathiram P, Moodley J. Pre-eclampsia: its pathogenesis and pathophysiolgy. *Cardiovascular Journal of Africa.* 2016;27(2):71–78. https://doi.org/10.5830/CVJA-2016-009.
5. Lejoyeux R, Benillouche J, Ong J, et al. Choriocapillaris: fundamentals and advancements. *Progress in Retinal and Eye Research.* 2022;87. https://doi.org/10.1016/j.preteyeres.2021.100997.
6. Campbell M, Humphries P. The blood-retina barrier tight junctions and barrier modulation. *Advances in Experimental Medicine and Biology.* 2013;763:70–84. https://doi.org/10.1007/978-1-4614-4711-5_3.
7. Cunha-Vaz J, Bernardes R, Lobo C. Blood-retinal barrier. *European Journal of Ophthalmology.* 2011;21(6):3–9. https://doi.org/10.5301/EJO.2010.6049.
8. He X, Ji Y, Yu M, Tong Y, Oishi A. Chorioretinal alterations induced by preeclampsia. *Journal of Ophthalmology.* 2021;2021:1–9. https://doi.org/10.1155/2021/8847001.
9. Kiel JW. Integrated systems physiology: from molecule to function to disease. In: *The Ocular Circulation.* Morgan & Claypool Life Sciences; 2010.
10. Dziedziak J, Zaleska-Żmijewska A, Szaflik JP, Cudnoch-Jedrzejewska A. Impact of arterial hypertension on the eye: a review of the pathogenesis, diagnostic methods, and treatment of hypertensive retinopathy. *Medical Science Monitor.* 2022;28. https://doi.org/10.12659/msm.935135.
11. Hayreh SS, Servais GE, Virdi PS, Marcus ML, Rojas P, Woolson RF. Fundus lesions in malignant hypertension: III. Arterial blood pressure, biochemical, and fundus changes. *Ophthalmology.* 1986;93(1):45–59. https://doi.org/10.1016/S0161-6420(86)33775-8.
12. Azuma K, Okubo A, Suzuki T, et al. Assessment of the choroidal structure in pregnant women in the first trimester. *Scientific Reports.* 2021;11(1). https://doi.org/10.1038/s41598-021-84204-x.
13. Tsukikawa M, Stacey AW. A review of hypertensive retinopathy and chorioretinopathy. *Clinical Optometry.* 2020;12. ISSN: 11792752:67–73. https://doi.org/10.2147/OPTO.S183492.
14. Bourke K, Patel MR, Prisant LM, Marcus DM. Hypertensive choroidopathy. *Journal of Clinical Hypertension.* 2004;6(8):471–472. https://doi.org/10.1111/j.1524-6175.2004.3749.x.

15. Tso MOM, Jampol LM. Pathophysiology of hypertensive retinopathy. *Ophthalmology*. 1982;89(10):1132–1145. https://doi.org/10.1016/S0161-6420(82)34663-1.
16. AlTalbishi Ala'a, Khateb S, Amer R. Elschnig's spots in the acute and remission stages in preeclampsia: spectral-domain optical coherence tomographic features. *European Journal of Ophthalmology*. 2015;25(5):e84–e87. https://doi.org/10.5301/ejo.5000586.
17. Fastenberg DM, Fetkenhour CL, Choromokos E, Shoch DE. Choroidal vascular changes in toxemia of pregnancy. *American Journal of Ophthalmology*. 1980;89(3):362–368. https://doi.org/10.1016/0002-9394(80)90005-7.
18. Kishi S, Tso MOM, Hayreh SS. Fundus lesions in malignant hypertension: I. A pathologic study of experimental hypertensive choroidopathy. *Archives of Ophthalmology*. 1985;103(8):1189–1197. https://doi.org/10.1001/archopht.1985.01050080101029.
19. Pohl ML. Siegrist's streaks in hypertensive choroidopathy. *Journal of the American Optometric Association*. 1988;59(5):372–376.
20. Mastorakos G, Ilias I. Maternal and fetal hypothalamic-pituitary-adrenal axes during pregnancy and postpartum. *Annals of the New York Academy of Sciences*. 2003;997(1):136–149. https://doi.org/10.1196/annals.1290.016.
21. Liang ZQ, Huang LZ, Qu JF, Zhao MW. Association between endogenous cortisol level and the risk of central serous chorioretinopathy: a meta-analysis. *International Journal of Ophthalmology*. 2018;11(2):296–300. https://doi.org/10.18240/ijo.2018.02.19.
22. Day H, Burns J, Bosio P. A case of bilateral serous retinal detachments in severe pre-eclampsia. *Journal of Obstetrics and Gynaecology*. 2008;28(5):534–535. https://doi.org/10.1080/01443610802234467.
23. Dinn RB, Harris A, Marcus PS. Ocular changes in pregnancy. *Obstetrical and Gynecological Survey*. 2003;58(2):137–144. https://doi.org/10.1097/00006254-200302000-00024.
24. Roos NM, Wiegman MJ, Jansonius NM, Zeeman GG. Visual disturbances in (pre)eclampsia. *Obstetrical and Gynecological Survey*. 2012;67(4):242–250. https://doi.org/10.1097/OGX.0b013e318250a457.
25. Auger N, Fraser WD, Paradis G, Healy-Profitós J, Hsieh A, Rhéaume M-A. Preeclampsia and long-term risk of maternal retinal disorders. *Obstetrics & Gynecology*. 2017;129(1):42–49. https://doi.org/10.1097/aog.0000000000001758.
26. Dornan KJ, Mallek DR, Wittmann BK. The sequelae of serous retinal detachment in preeclampsia. *Obstetrics and Gynecology*. 1982;60(5):657–663.
27. Hutchings K, Sangalli M, Halliwell T, Tuohy J. Bilateral retinal detachment in pregnancy. *The Australian and New Zealand Journal of Obstetrics and Gynaecology*. 2002;42(4):411–413. https://doi.org/10.1111/j.0004-8666.2002.409_3.x.
28. Maslovitz S, Lessing JB, Kupferminc MJ. Bilateral retinal detachment in preeclamptic women with thrombophilia. *International Journal of Gynecology & Obstetrics*. 2005;91(1):65–66. https://doi.org/10.1016/j.ijgo.2005.06.016.
29. Omoti AE, Waziri-Erameh JM, Okeigbemen VW. A review of the changes in the ophthalmic and visual system in pregnancy. *African Journal of Reproductive Health*. 2008;12(3):185–196.
30. Prado RSdo, Figueiredo EL, Magalhães TVB. Retinal detachment in preeclampsia. *Arquivos Brasileiros de Oftalmologia*. 2002;79(2):185–186. https://doi.org/10.1590/S0066-782X2002001100011.
31. Raposo JTBV, Melo BCDS, Maciel NFBB, Leite SD, Rebelo ÓRC, Lima AMF. Serous retinal detachment in pre-eclampsia: case report and literature review. *Revista Brasileira de Ginecologia e Obstetrícia / RBGO Gynecology and Obstetrics*. 2020;42(11):772–773. https://doi.org/10.1055/s-0040-1718448.
32. Sherman D, Lurie S, Turetz J, Bukovsky I. Central serous retinopathy and severe preeclampsia. *Journal of Obstetrics and Gynaecology*. 2000;20(3):311–312. https://doi.org/10.1080/01443610050009728.
33. Tuckey J. Serous retinal detachment — an unusual complication of preeclampsia. *International Journal of Obstetric Anesthesia*. 1995;4(4):248–250. https://doi.org/10.1016/0959-289x(95)82919-2.
34. Watanabe H, Hamada H, Fujiki Y, Takeuchi S, Urushigawa K, Yoshikawa H. Third nerve palsy and serous retinal detachment with preeclampsia. *Hypertension in Pregnancy*. 2006;25(1):33–35. https://doi.org/10.1080/10641950500543889.
35. Mabie WC, Ober RR. Fluorescein angiography in toxaemia of pregnancy. *British Journal of Ophthalmology*. 1980;64(9):666–671. https://doi.org/10.1136/bjo.64.9.666.
36. Hage R. Retinal detachment in preeclampsia: a series of three cases. *J Fr Ophtalmol*. 2012;35(10).
37. Sathish S, Arnold JJ. Bilateral choroidal ischaemia and serous retinal detachment in pre-eclampsia. *Clinical and Experimental Ophthalmology*. 2000;28(5):387–390. https://doi.org/10.1046/j.1442-9071.2000.00342.x.
38. Valluri S, Adelberg DA, Curtis RS, Olk RJ. Diagnostic indocyanine green angiography in preeclampsia. *American Journal of Ophthalmology*. 1996;122(5):672–677. https://doi.org/10.1016/s0002-9394(14)70485-4.
39. Fineman MS, Maguire JI, Fineman SW, Benson WE. Safety of indocyanine green angiography during pregnancy: a survey of the retina, macula, and vitreous societies. *Archives of Ophthalmology*. 2001;119(3):353–355. https://doi.org/10.1001/archopht.119.3.353.
40. Obana A, Miki T, Hayashi K, et al. Survey of complications of indocyanine green angiography in Japan. *American Journal of Ophthalmology*. 1994;118(6):749–753. https://doi.org/10.1016/s0002-9394(14)72554-1.
41. Probst P, Paumgartner G, Caucig H, Fröhlich H, Grabner G. Studies on clearance and placental transfer of indocyanine green during labor. *Clinica Chimica Acta*. 1970;29(1):157–160. https://doi.org/10.1016/0009-8981(70)90237-8.
42. Iida T, Hagimura N, Otani T, Ikeda F, Muraoka K. Choroidal vascular lesions in serous retinal detachment viewed with indocyanine green angiography. *Journal of Japanese Ophthalmological Society*. 1996;100(10):817–824.
43. Dadaci Z, Alptekin H, Oncel Acir N, Borazan M. Changes in choroidal thickness during pregnancy detected by enhanced depth imaging optical coherence tomography. *British Journal of Ophthalmology*. 2015;99(9):1255–1259. https://doi.org/10.1136/bjophthalmol-2014-306343.

44. Goktas S, Basaran A, Sakarya Y, et al. Measurement of choroid thickness in pregnant women using enhanced depth imaging optical coherence tomography. *Arquivos Brasileiros de Oftalmologia*. 2014;77(3):148–151. https://doi.org/10.5935/0004-2749.20140038.
45. Ataş M, Açmaz G, Aksoy H, et al. Evaluation of the macula, retinal nerve fiber layer and choroid in preeclampsia, healthy pregnant and healthy non-pregnant women using spectral-domain optical coherence tomography. *Hypertension in Pregnancy*. 2014;33(3):299–310. https://doi.org/10.3109/10641955.2013.877924.
46. Benfica CZ, Zanella T, Farias LB, Oppermann MLR, Canani LHS, Lavinsky D. Choroidal thickness in preeclampsia measured by spectral-domain optical coherence tomography. *International Ophthalmology*. 2019;39(9):2069–2076. https://doi.org/10.1007/s10792-018-1043-7.
47. Jiang MS, Xu XL, Yang T, Li F, Zhang XD. Comparison of choroidal thickness in preeclamptic, healthy pregnant, and nonpregnant women: a systematic review and meta-analysis. *Ophthalmic Research*. 2019;62(1):1–10. https://doi.org/10.1159/000499916.
48. Chanwimol K, Balasubramanian S, Nassisi M, et al. Retinal vascular changes during pregnancy detected with optical coherence tomography angiography. *Investigative Ophthalmology and Visual Science*. 2019;60(7):2726–2732. https://doi.org/10.1167/iovs.19-26956.
49. Urfalõoglu S, Bakacak M, Özdemir G, Güler M, Beyoglu A, Arslan G. Posterior ocular blood flow in preeclamptic patients evaluated with optical coherence tomography angiography. *Pregnancy Hypertension*. 2019;17:203–208. https://doi.org/10.1016/j.preghy.2019.07.001.
50. Theodossiadis PG, Kollia AK, Gogas P, Panagiotidis D, Moschos M, Theodossiadis GP. Retinal disorders in preeclampsia studied with optical coherence tomography. *American Journal of Ophthalmology*. 2002;133(5):707–709. https://doi.org/10.1016/S0002-9394(01)01430-1.
51. Neudorfer M, Spierer O, Goder M, et al. The prevalence of retinal and optical coherence tomography findings in preeclamptic women. *Retina*. 2014;34(7):1376–1383. https://doi.org/10.1097/iae.0000000000000085.
52. Somfai GM, Miháltz K, Tulassay E, Rigó J. Diagnosis of serous neuroretinal detachments of the macula in severe preeclamptic patients with optical coherence tomography. *Hypertension in Pregnancy*. 2006;25(1):11–20. https://doi.org/10.1080/10641950500543848.
53. Cankaya C, Bozkurt M, Ulutas O. Total macular volume and foveal retinal thickness alterations in healthy pregnant women. *Seminars in Ophthalmology*. 2013;28(2):103–111. https://doi.org/10.3109/08820538.2012.760628.
54. Demir M, Oba E, Can E, et al. Foveal and parafoveal retinal thickness in healthy pregnant women in their last trimester. *Clinical Ophthalmology*. 2011;5(1):1397–1400. https://doi.org/10.2147/opth.s23944.
55. Arab M, Entezari M, Ghamary H, et al. Peripapillary retinal nerve fiber layer thickness in preeclampsia and eclampsia. *International Ophthalmology*. 2018;38(6):2289–2294. https://doi.org/10.1007/s10792-017-0718-9.
56. Tok A, Beyoğlu A. Antenatal and postpartum comparison of HD-OCT findings of macula, retinal nerve fiber layer, ganglion cell density between severe preeclampsia patients and healthy pregnant woman. *Hypertension in Pregnancy*. 2020;39(3):252–259. https://doi.org/10.1080/10641955.2020.1758938.
57. Alizadeh Ghavidel L, Mousavi F, Bagheri M, Asghari S. Preeclampsia induced ocular change. *International Journal of Women's Health and Reproduction Sciences*. 2018;6(2):123–126. https://doi.org/10.15296/ijwhr.2018.20.
58. Irani RA, Zhang Y, Zhou CC, et al. Autoantibody-mediated angiotensin receptor activation contributes to preeclampsia through tumor necrosis factor-α signaling. *Hypertension*. 2010;55(5):1246–1253. https://doi.org/10.1161/HYPERTENSIONAHA.110.150540.
59. Maynard SE, Karumanchi SA. Angiogenic factors and preeclampsia. *Seminars in Nephrology*. 2011;31(1):33–46. https://doi.org/10.1016/j.semnephrol.2010.10.004.
60. Garg A, Wapner RJ, Ananth CV, et al. Choroidal and retinal thickening in severe preeclampsia. *Investigative Opthalmology & Visual Science*. 2014;55(9). https://doi.org/10.1167/iovs.14-14143.
61. Penn JS, Madan A, Caldwell RB, Bartoli M, Caldwell RW, Hartnett ME. Vascular endothelial growth factor in eye disease. *Progress in Retinal and Eye Research*. 2008;27(4):331–371. https://doi.org/10.1016/j.preteyeres.2008.05.001.
62. Chandran JR, Narayanan IB, Rajan J. Ocular manifestations: are they significant in hypertensive disorders of pregnancy? *Journal of Obstetrics and Gynecology of India*. 2021;71(2):118–123. https://doi.org/10.1007/s13224-020-01385-7.
63. Hallum AV. Eye changes in hypertensive toxemia of pregnancy: a study of three hundred cases. *Journal of the American Medical Association*. 1936;106(19):1649–1651. https://doi.org/10.1001/jama.1936.02770190025009.
64. Reddy SC, Nalliah S, Rani S, George K, Who TS. Fundus changes in pregnancy induced hypertension. *International Journal of Ophthalmology*. 2012;5(6):694–697. https://doi.org/10.3980/j.issn.2222-3959.2012.06.08.
65. Wagener HP. Arterioles of the retina in toxemia of pregnancy. *Journal of the American Medical Association*. 1933;101(18):1380–1384. https://doi.org/10.1001/jama.1933.02740430026009.
66. Warad C, Midha B, Pandey U, et al. Ocular manifestations in pregnancy-induced hypertension at a tertiary level hospital in Karnataka, India. *Cureus*. 2023. https://doi.org/10.7759/cureus.34887.
67. Mathew M, Rajani CK, Gulati N. Amaurosis in toxaemia of pregnancy. *Indian Journal of Ophthalmology*. 1975;23(1):25–26.
68. Mourelo M, Galeiras R, Freire D, Álvarez M, Díaz JL, García T. Postpartum amaurosis in a woman with severe preeclampsia. *Indian Journal of Critical Care Medicine*. 2011;15(4):227–229. https://doi.org/10.4103/0972-5229.92077.
69. Swende TZ, Abwa T. Reversible blindness in fulminating preeclampsia. *Annals of African Medicine*. 2009;8(3):189–191. https://doi.org/10.4103/1596-3519.57247.
70. Chaine G, Kohner EM. LA retinopathie hypertensive. *Journal Francais d'Ophtalmologie*. 1983;6(12):995–1005.
71. Ballantyne AJ, Loewenstein A. Retinal micro-aneurysms and punctate haemorrhages. *British Journal of Ophthalmology*. 1944;28(12):593–598. https://doi.org/10.1136/bjo.28.12.593.
72. Uma MS, Bhuvana S, Annamalai R, Muthayya M. Visual morbidity and spectrum of ophthalmic changes in pregnancy induced hypertension. *Journal of Family Medicine and Primary Care*. 2022;11(6). https://doi.org/10.4103/jfmpc.jfmpc_1716_21.

73. Brown GC, Brown MM, Hiller T, Fischer D, Benson WE, Magargal LE. Cotton-wool spots. *Retina*. 1985;5(4):206–214. https://doi.org/10.1097/00006982-198500540-00003.
74. Kim BY, Smith SD, Kaiser PK. Optical coherence tomographic patterns of diabetic macular edema. *American Journal of Ophthalmology*. 2006;142(3):405-e1. https://doi.org/10.1016/j.ajo.2006.04.023.
75. Cheung CY. Hypertensive eye disease. *Nature Reviews Disease Primers*. 2022;8(1):14.
76. Cusick M, Chew EY, Chan C-C, Kruth HS, Murphy RP, Ferris FL. Histopathology and regression of retinal hard exudates in diabetic retinopathy after reduction of elevated serum lipid levels. *Ophthalmology*. 2003;110(11):2126–2133. https://doi.org/10.1016/j.ophtha.2003.01.001.
77. Gella L, Raman R, Rani PK, Sharma T. Spectral domain optical coherence tomography characteristics in diabetic retinopathy. *Oman Journal of Ophthalmology*. 2014;7(3). https://doi.org/10.4103/0974-620X.142594.
78. Horii T, Murakami T, Nishijima K, Sakamoto A, Ota M, Yoshimura N. Optical coherence tomographic characteristics of microaneurysms in diabetic retinopathy. *American Journal of Ophthalmology*. 2010;150(6):840–848.e1. https://doi.org/10.1016/j.ajo.2010.06.015.
79. Ciloglu E, Okcu NT, Dogan NÇ. Optical coherence tomography angiography findings in preeclampsia. *Eye (Basingstoke)*. 2019;33(12):1946–1951. https://doi.org/10.1038/s41433-019-0531-y.
80. Kenny GS, Cerasoli JR. Color fluorescein angiography in toxemia of pregnancy. *Archives of Ophthalmology*. 1972;87(4):383–388. https://doi.org/10.1001/archopht.1972.01000020385003.
81. Lin P, Hahn P, Fekrat S. Peripheral retinal vascular leakage demonstrated by ultra-widefield fluorescein angiography in preeclampsia with HELLP syndrome. *Retina*. 2012;32(8):1689–1690. https://doi.org/10.1097/IAE.0b013e318261a757.
82. Belfort MA, Saade GR, Moise KJ. The effect of magnesium sulfate on maternal retinal blood flow in preeclampsia: a randomized placebo-controlled study. *American Journal of Obstetrics and Gynecology*. 1992;167(6):1548–1553. https://doi.org/10.1016/0002-9378(92)91737-U.
83. Diabetes Control and Complications Trial Research Group. Effect of pregnancy on microvascular complications in the diabetes control and complications trial. The Diabetes Control and Complications Trial Research Group. *Diabetes Care*. 2000;23(8):1084–1091. https://doi.org/10.2337/diacare.23.8.1084.
84. Chandrasekaran PR, Madanagopalan VG, Narayanan R. Diabetic retinopathy in pregnancy - a review. *Indian journal of ophthalmology*. 2021;69(11):3015–3025. https://doi.org/10.4103/ijo.IJO_1377_21.
85. Chew EY, Mills JL, Metzger BE, et al. Metabolic control and progression of retinopathy: the diabetes in early pregnancy study. *Diabetes Care*. 1995;18(5):631–637. https://doi.org/10.2337/diacare.18.5.631.
86. Errera MH, Kohly RP, da Cruz L. Pregnancy-associated retinal diseases and their management. *Survey of Ophthalmology*. 2013;58(2):127–142. https://doi.org/10.1016/j.survophthal.2012.08.001.
87. Klein BEK, Moss SE, Klein R. Effect of pregnancy on progression of diabetic retinopathy. *Diabetes Care*. 1990;13(1):34–40. https://doi.org/10.2337/diacare.13.1.34.
88. Mallika PS, Tan AK, Aziz S, Asok T, Syed Alwi SAR, Intan G. Diabetic retinopathy and the effect of pregnancy. *Malaysian Family Physician*. 2010;5(1):2–5.
89. Schultz KL, Birnbaum AD, Goldstein DA. Ocular disease in pregnancy. *Current Opinion in Ophthalmology*. 2005;16(5):308–314. https://doi.org/10.1097/01.icu.0000179803.42218.cc.
90. Rosenn B, Miodovnik M, Kranias G, et al. Progression of diabetic retinopathy in pregnancy: association with hypertension in pregnancy. *American Journal of Obstetrics and Gynecology*. 1992;166(4):1214–1218. https://doi.org/10.1016/s0002-9378(11)90608-5.
91. Rosenthal JM, Johnson MW. Management of retinal diseases in pregnant patients. *Journal of Ophthalmic and Vision Research*. 2018;13(1):62–65. https://doi.org/10.4103/jovr.jovr_195_17.
92. Wykoff CC, Brown DM. The effect of pregnancy on diabetic retinopathy. *Retina Today*. 2012:77–79.
93. Flaxel CJ, Adelman RA, Bailey ST, et al. Diabetic retinopathy preferred practice Pattern®. *Ophthalmology*. 2020;127(1):P66–P145. https://doi.org/10.1016/j.ophtha.2019.09.025.
94. Agardh E. A case of progression of diabetic retinopathy during pregnancy. *Acta Ophthalmologica Scandinavica*. 2002;80(5):524–530. https://doi.org/10.1034/j.1600-0420.2002.800512.x.
95. Chan WC, Lim LT, Quinn MJ, Knox FA, McCance D, Best RM. Management and outcome of sight-threatening diabetic retinopathy in pregnancy. *Eye*. 2004;18(8):826–832. https://doi.org/10.1038/sj.eye.6701340.
96. Hercules BL, Wozencroft M, Gayed II, Jeacock J. Peripheral retinal ablation in the treatment of proliferative diabetic retinopathy during pregnancy. *British Journal of Ophthalmology*. 1980;64(2):87–93. https://doi.org/10.1136/bjo.64.2.87.
97. Sunness JS. The pregnant woman's eye. *Survey of Ophthalmology*. 1988;32(4):219–238. https://doi.org/10.1016/0039-6257(88)90172-5.
98. Jouve L, Akesbi J, Nordmann JP. Safety and efficacy of ranibizumab for pregnant women in idiopathic choroidal neovascularization. *Acta Ophthalmologica*. 2015;93(7):e597–e598. https://doi.org/10.1111/aos.12611.
99. Rosen E, Rubowitz A, Ferencz JR. Exposure to verteporfin and bevacizumab therapy for choroidal neovascularization secondary to punctate inner choroidopathy during pregnancy. *Eye*. 2009;23(6). https://doi.org/10.1038/eye.2008.218.
100. Sarhianaki A, Katsimpris A, Petropoulos IK, Livieratou A, Theoulakis PE, Katsimpris JM. Intravitreal administration of ranibizumab for idiopathic choroidal neovascularization in a pregnant woman. *Klinische Monatsblatter fur Augenheilkunde*. 2012;229(4):451–453. https://doi.org/10.1055/s-0031-1299207.
101. Sarmad A, Lip PL. Intravitreal anti-vascular endothelial growth factor in early pregnancy and the complex management of advance diabetic retinopathy and maculopathy during pregnancy. *Acta Ophthalmologica*. 2016;94(8):e812–e813. https://doi.org/10.1111/aos.13083.

102. Tarantola RM, Folk JC, Boldt HC, Mahajan VB. Intravitreal bevacizumab during pregnancy. *Retina*. 2010;30(9):1405–1411. https://doi.org/10.1097/IAE.0b013e3181f57d58.
103. Petrou P, Georgalas I, Giavaras G, Anastasiou E, Ntana Z, Petrou C. Early loss of pregnancy after intravitreal bevacizumab injection. *Acta Ophthalmologica*. 2010;88(4). https://doi.org/10.1111/j.1755-3768.2009.01572.x.
104. Sullivan L, Kelly SP, Glenn A, Williams CPR, McKibbin M. Intravitreal bevacizumab injection in unrecognised early pregnancy. *Eye*. 2014;28(4):492–494. https://doi.org/10.1038/eye.2013.311.
105. Concillado M, Lund-Andersen H, Mathiesen ER, Larsen M. Dexamethasone intravitreal implant for diabetic macular edema during pregnancy. *American Journal of Ophthalmology*. 2016;165:7–15. https://doi.org/10.1016/j.ajo.2016.02.004.
106. Fazelat A, Lashkari K. Off-label use of intravitreal triamcinolone acetonide for diabetic macular edema in a pregnant patient. *Clinical Ophthalmology*. 2011;5(1):439–441. https://doi.org/10.2147/OPTH.S14584.
107. Raman P, Livingstone BI. Advanced diabetic eye disease in pregnancy. *Journal of Clinical Gynecology and Obstetrics*. 2018;7(3–4):72–75. https://doi.org/10.14740/jcgo487w.
108. Belfort MA, Saade GR. Retinal vasospasm associated with visual disturbance in preeclampsia: color flow doppler findings. *American Journal of Obstetrics and Gynecology*. 1993;169(3):523–525. https://doi.org/10.1016/0002-9378(93)90612-M.
109. Diniz ALD, Moron AF, dos Santos MC, Sass N, Pires CR, Debs CL. Ophthalmic artery Doppler as a measure of severe pre-eclampsia. *International Journal of Gynecology & Obstetrics*. 2008;100(3):216–220. https://doi.org/10.1016/j.ijgo.2007.07.013.
110. Silverman RH. Plane-wave ultrasound doppler of the eye in preeclampsia. *Translation Vision Science & Technology*. 2020;9(10).
111. Panagiotis T. *Thrombophilia and Pregnancy: Diagnosis and Management*. Rijeka: IntechOpen; 2019. Ch. 6.
112. Essome H, Idjem MSN, Njamen TN, et al. Hypercoagulability in the context of pre-eclampsia: case-control study at the laquintinie douala hospital (Cameroon). *Open Journal of Obstetrics and Gynecology*. 2020;10(12):1708–1727. https://doi.org/10.4236/ojog.2020.10120155.
113. Gonzalvo FJ, Abecia E, Pinilla I, Izaguirre LB, Olivan JM, Honrubia FM. Central retinal vein occlusion and HELLP syndrome. *Acta Ophthalmologica Scandinavica*. 2000;78(5):596–598. https://doi.org/10.1034/j.1600-0420.2000.078005596.x.
114. Kurtz WS, Glueck CJ, Hutchins RK, Sisk RA, Wang P. Retinal artery and vein thrombotic occlusion during pregnancy: markers for familial thrombophilia and adverse pregnancy outcomes. *Clinical Ophthalmology*. 2016;10:935–938. https://doi.org/10.2147/OPTH.S106164.
115. Lara-Torre E, Lee MS, Wolf MA, Shah DM. Bilateral retinal occlusion progressing to long-lasting blindness in severe preeclampsia. *Obstetrics & Gynecology*. 2002;100(5, Part 1):940–942. https://doi.org/10.1097/00006250-200211000-00019.
116. Rahman I, Saleemi G, Semple D, Stanga P. Pre-eclampsia resulting in central retinal vein occlusion. *Eye*. 2006;20(8):955–957. https://doi.org/10.1038/sj.eye.6702065.
117. Agrawal A, McKibbin M. Purtscher's retinopathy: epidemiology, clinical features and outcome. *British Journal of Ophthalmology*. 2007;91(11):1456–1459. https://doi.org/10.1136/bjo.2007.117408.
118. Blodi BA, Johnson MW, Gass JDM, Fine SL, Joffe LM. Purtscher's-like retinopathy after childbirth. *Ophthalmology*. 1990;97(12):1654–1659. https://doi.org/10.1016/S0161-6420(90)32365-5.
119. Cernea D, Dragoescu A, Novac M. HELLP syndrome complicated with postpartum subcapsular ruptured liver hematoma and purtscher-like retinopathy. *Case Reports in Obstetrics and Gynecology*. 2012;2012:1–4. https://doi.org/10.1155/2012/856135.
120. Fairbanks JL, Lloyd RE, Dempsey SP. Case report: postpartum purtscher-like retinopathy. *Optometry and Vision Science : Official Publication of the American Academy of Optometry*. 2021;98(11):1304–1308. https://doi.org/10.1097/OPX.0000000000001797.
121. Shaikh S, Ruby AJ, Piotrowski M. Preeclampsia-related chorioretinopathy with Purtscher's-like findings and macular ischemia. *Retina*. 2003;23(2):247–250. https://doi.org/10.1097/00006982-200304000-00021.
122. Shukla D, Maheshwari R, Ramchandani B, Kanungo S. PURTSCHER-LIKE retinopathy with serous retinal detachment in preeclampsia of pregnancy: complications and management. *RETINAL Cases & Brief Reports*. 2010;4(4):332–335. https://doi.org/10.1097/icb.0b013e3181c33389.
123. Singh K, Jain D, Wallang B. Purtscher's retinopathy in pre-eclampsia: a blinding combination. *International Ophthalmology*. 2014;34(1):103–106. https://doi.org/10.1007/s10792-013-9739-1.
124. Stewart MW, Brazis PW, Guier CP, Thota SH, Wilson SD. Purtscher-like retinopathy in a patient with HELLP syndrome. *American Journal of Ophthalmology*. 2007;143(5):886–887. https://doi.org/10.1016/j.ajo.2006.12.005.

Chapter 34

Sickle cell retinopathy

Abhinaya Bharath[1], Adrienne W. Scott[2] and Sally S. Ong[3]
[1]Drexel University College of Medicine, Philadelphia, PA, United States; [2]Retina Division, Wilmer Eye Institute, Johns Hopkins University School of Medicine, Baltimore, MD, United States; [3]Department of Ophthalmology, Wake Forest School of Medicine, Winston–Salem, NC, United States

Pathogenesis

This section will cover the pathophysiology of sickle cell disease (SCD) and risk factors for sickle cell retinopathy (SCR).

SCD is an autosomal recessive disorder caused by a point mutation in the hemoglobin beta gene (HBB) found on chromosome 11. HBA and HBB are the wild-type alleles, each transcribed into α-globin and β-globin respectively. Normal adult hemoglobin usually consists of two α-globin chains and two β-globin chains attached by a central iron-containing heme molecule.[1] When there is a single nucleotide mutation (GAG to GTG), glutamic acid is substituted to valine in the β-globin chain, resulting in sickle hemoglobin (HbS).[1,2] People with two copies of this mutation have the homozygous form of the disease known as HbSS.[1] Compound heterozygous disease is manifested when an individual has one copy of HbS and another hemoglobin variant (e.g., HbC or β-thalassemia).[1] The HbC variant is caused by the GAG to AAG mutation leading to a substitution of glutamic acid to lysine in the B-globin chain.[1] In β-thalassemia, patients either have reduced amounts of B-globin (B+ thalassemia) or no B-globin chains present (B0 thalassemia).[1] When present with the HbS variant, this is known as hemoglobin sickle-beta thalassemia (HbSthal). When there is only one copy of HbS and one copy of normal hemoglobin A, individuals are considered to have the sickle cell trait phenotype.[3] Heterozygosity of the sickle cell trait is present in about one in eight African Americans. The abnormal sickle gene has been selectively inherited in the sub-Saharan African region because the sickle cell trait is thought to protect against malarial infection during childhood.[3–5]

During deoxygenation, the defective HbS polymerizes into long chains, causing red blood cells to become distorted into the eponymous sickle shape. Sickled red blood cells express adhesion molecules $\alpha_4\beta_1$, CD36, band 3 protein, sulfated glycolipid, lutheran protein, phosphatidylserine, and integrin-associated protein.[1,6–12] These adhesion molecules facilitate binding of the sickled cells to the endothelial walls of the vasculature,[1,13] causing aggregation of sickled cells in the microcirculation, reduced deformability of the red blood cells, and stagnant blood flow leading to thrombosis and ischemia. The abnormal sickled cells also hemolyze rapidly and can cause local endothelial dysfunction.[14] Vaso-occlusive events and hemolysis-related endothelial dysfunction can affect any organ system. Acute pain, acute chest syndrome, avascular necrosis, stroke, pulmonary hypertension, priapism, and leg ulcers can lead to significant morbidity and early mortality with average life expectancy reduced by 20 years compared to the general population in the United States.[14] All ocular and orbital structures can be affected by ischemic events in SCD. However, the most common cause of vision loss in SCD is SCR, which is the focus of this chapter.[15]

Epidemiology

Individuals of Black and/or African descent especially are at high risk for SCD in addition to certain populations of Caribbean, Mediterranean, East Indian, and South and Central American descent.[1] Globally, approximately 300,000 infants are born annually with HbSS disease, with more than half of these born in Nigeria, India, and the Democratic Republic of Congo.[16,17] Due to the trans-Atlantic slave trade as well as past and ongoing waves of immigration, individuals with SCD are now present in most countries. However, even for patients with the same genotype, there is significant

Retinal and Choroidal Vascular Diseases of the Eye. https://doi.org/10.1016/B978-0-443-15583-3.00034-2

phenotypic variability in the manifestations of SCD caused by factors such as oxidative stress, environmental factors, genetics, inflammation, the proportion of sickled cells, ability to generate fetal hemoglobin, and sensitivity of hypoxia-induced angiogenesis.[1,18]

Risk factors

Risk factors for developing vision-threatening complications from SCD, most commonly from proliferative sickle retinopathy (PSR), include genotype and older age. While patients with HbSS are more likely to experience increased vaso-occlusive sickle cell pain crises and have higher systemic morbidity and mortality rates, individuals with HbSC and HbSthal are more likely to develop PSR and vision loss. For example, during a routine screening program in Mali of sickle cell patients with an average age of 24.2 (range 10–64 years), Dembele and coauthors found that the prevalence of sickle cell retinopathy was 12.4% (91/731) in HbSC, 5.2% (38/734) in HbSS, 9.4% (5/53) in $HbSthal^0$ and 9.3% (8/86) in $HbSthal^+$ patients. Furthermore, among those with retinopathy, PSR was more common in HbSC (64.1%) than in HbSS/HbSthal patients (30.2%).[19] In a prospective longitudinal study of patients followed from birth, Downes et al. found that by age 24–26, 43% of individuals with HbSC have developed PSR compared to only 14% of patients with HbSS.[20]

The reasons for this difference are incompletely understood. It is hypothesized that in HbSS, retinal vessels undergo complete occlusion resulting in complete vascular infarction and anoxia, with decreased production of angiogenic factors, like vascular endothelial growth factor and hypoxia-inducible factor-1a. In HbSC or sickle thalassemia, partial occlusion of retinal vasculature leads to tissue hypoxia, which stimulates the chronic indolent production of proangiogenic factors by neighboring cells, leading to pathologic neovascularization.[21] Additionally, patients with the HbSC genotype have higher blood viscosity and hematocrit compared to those with HbSS, which promotes stagnant blood flow and thrombosis in the retinal precapillary arterioles.[1,21] In a Jamaican cohort study, risk factors for developing PSR were found to include elevated hemoglobin levels in males and decreased levels of fetal hemoglobin (HbF) in both males and females in HbSS, as well as increased mean cell volume and low HbF in both sexes, and high total hemoglobin and high mean corpuscular hemoglobin concentration in men in HbSC.[22] Other studies have shown that a 15% increase in HbF levels can reduce the odds of developing PSR by up to 50%.[23]

Individuals with the sickle cell trait (SCT) typically remain visually asymptomatic and do not usually develop SCR. However, those with SCT can develop sickling complications at high altitudes, in colder climates, and with dehydration.[1,24,25] A case of an individual with sickle cell trait who developed PSR after mild ocular blunt trauma has also been described.[26] Vision loss in SCT is very rare, although PSR with resultant retinal detachment has been rarely observed (Fig. 34.1).

The prevalence of PSR increases with older age. In HbSS, prevalence peaks between age 25 and 39 years in both men and women, whereas in HbSC, it peaks earlier from 15 to 24 years in men and from 20 to 39 years in women.[1,27,28] Many studies from different countries have corroborated the association between age and progression to PSR.[1]

Protective factors

Novel drug therapies that target the pathophysiology of SCD may prevent ocular complications. Hydroxyurea is a ribonucleotide reductase inhibitor that increases the production of HbF, which normally decreases shortly after birth.[23,29] HbF prevents the polymerization of defective HbS, thus decreasing red cell rigidity and hemolysis, and improving anemia. Hydroxyurea also increases the production of the potent vasodilator, nitric oxide, reduces red cell adhesion, and suppresses leukocytes, thus decreasing vaso-occlusive events.[14] In a study of 123 children with HbSS, Estepp and coauthors reported that children with <15% HbF had 7.1-fold increased odds of developing retinopathy. In children who were receiving hydroxyurea therapy, a lower HbF level was found in those who developed retinopathy (9%) compared to those who did not (16%).[30] A retrospective cross-sectional study of 300 adults with HbSS treated with hydroxyurea similarly showed that HbF levels >15% decreased the odds of developing retinopathy by 50%.[23] The role of hydroxyurea in preventing retinopathy in HbSC patients is less known. Hydroxyurea treatment has also been shown to reduce the rates of retinal thinning on optical coherence tomography (OCT) in patients with SCR.[31]

Hydroxyurea is now a recommended first-line treatment for all patients with HbSS and $HbSthal^0$ beginning at 9 months old based on evidence from randomized clinical trials showing decreased incidence of acute pain episodes and acute chest syndrome with hydroxyurea use.[32–34] However, only 28% of eligible children were prescribed hydroxyurea according to a recent report,[35] and only 23% of adults with more than three pain crises a year filled a prescription for hydroxyurea.[36] Lack of clinician expertise in SCD and mistrust from a historically underserved and marginalized patient population are some of

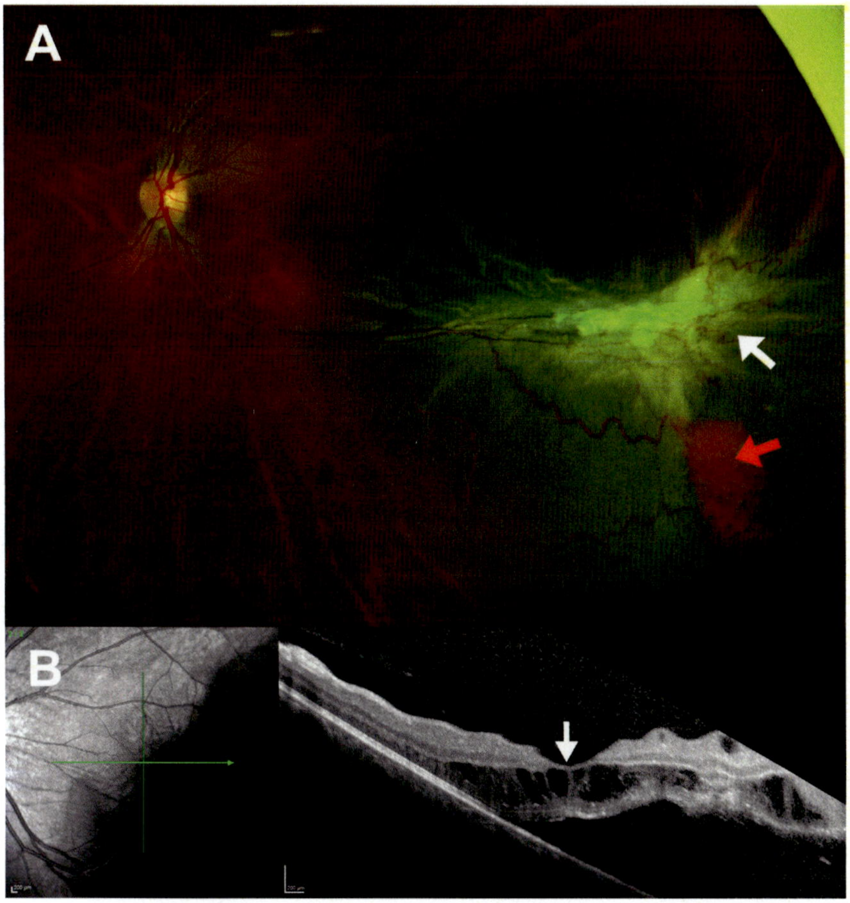

FIGURE 34.1 Retinal detachment in a patient with sickle cell trait. A 38-year-old man with a history of 1 month of painless vision loss presented for urgent evaluation. A macula involving retinal detachment was noted with fibrous proliferation in the temporal periphery (*white arrow*) on fundus photography (A). A salmon patch hemorrhage was also seen (*red arrow*). Optical coherence tomography (B) confirmed that the retinal detachment involved the fovea (*white arrow*). A hemoglobin panel confirmed sickle cell trait.

the factors that have been thought to contribute to low hydroxyurea use.[37–39] Exchange transfusion is another systemic therapy that is thought to decrease the risk of progression to PSR.[27,40]

Clinical features

Sickle cell retinopathy is categorized into nonproliferative sickle retinopathy (NPSR) and proliferative sickle retinopathy (PSR).

Nonproliferative sickle cell retinopathy

In NPSR, clinical findings include salmon patches, iridescent spots, black sunbursts, and angioid streaks.[41] Salmon patches are reddish-orange preretinal hemorrhages between the retina and the internal limiting membrane that are caused by bleeding from superficial blood vessels near areas of nonperfusion (Fig. 34.2). They occur when sickled cells abruptly block superficial arterioles. Originally red, the patches turn salmon-colored due to the hemolysis of red blood cells.[41,42] Over time, the hemorrhage is resorbed and leaves behind iridescent spots, which are deposits of hemosiderin-laden macrophages beneath the internal limiting membrane.[41] A black sunburst describes a dark chorioretinal scar, which is often a result of hypertrophy and hyperplasia of retinal pigment epithelium cells that migrate and proliferate as a response to hemorrhage. These can be found near areas of nonperfusion.[41,43] Linear breaks in the elastic lamina of Bruch's membrane can result in the characteristic pattern of pigmented fissures from the disc seen in angioid streaks, which have been reported to be more commonly seen in HbSS compared to HbSC (Fig. 34.3).[44,45] The exact etiology of angioid streaks in SCD remains unknown. Progression of angioid streaks in SCD to the macula with neovascularization and disciform changes is thought to be relatively uncommon compared to other conditions associated with angioid streaks like pseudoxanthoma elastica.[1,46]

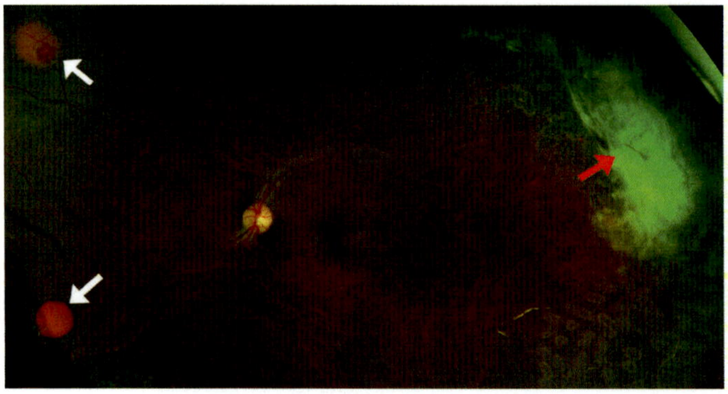

FIGURE 34.2 Salmon patch hemorrhages. Salmon patch hemorrhages were observed in the nasal periphery (*white arrows*) in a 21-year-old man with HbSC and a history of scatter laser for extensive sea fan neovascularization with partial autoinfarction (*red arrow*).

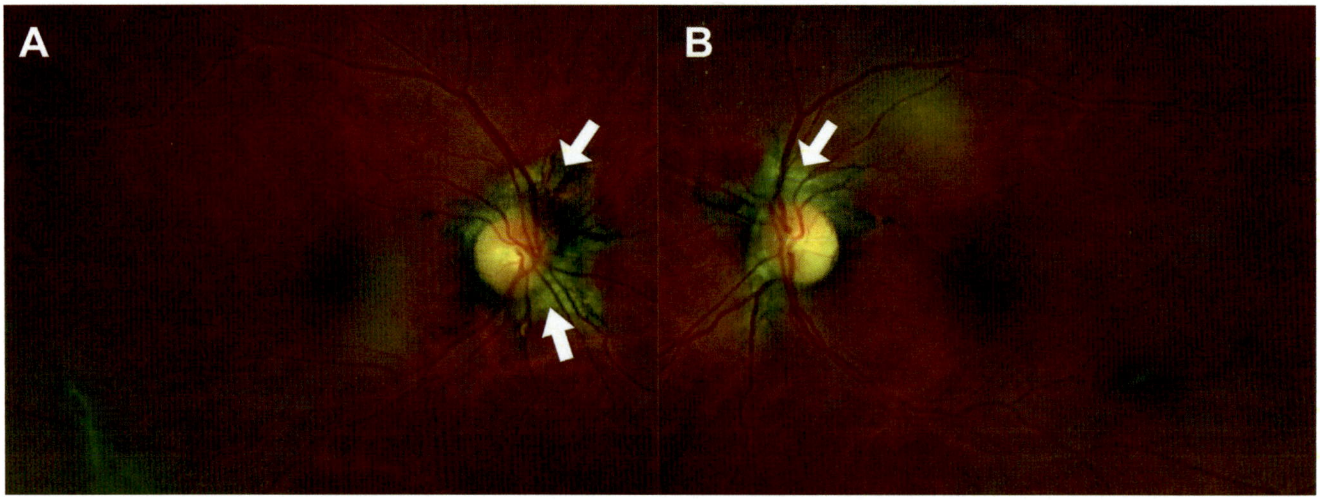

FIGURE 34.3 Angioid streaks. Angioid streaks (*white arrows*) in the right (A) and left (B) eyes of a 41-year-old woman with HbSS who remained visually asymptomatic.

Proliferative sickle cell retinopathy

Sickled and defective red blood cells in the peripheral retinal vasculature can lead to a cascade of events that result in the eponymous "sea fan" neovascular fronds seen in proliferative sickle cell retinopathy (PSR). Vaso-occlusions of the peripheral retinal vasculature result in vascular remodeling, which leads to arteriovenous anastomoses. Angiographic studies have shown that sea fan neovascularization arises from these arteriovenous anastomoses bordering avascular retina. Using high-resolution histopathologic analysis, McLeod et al. showed that preretinal neovascularization in PSR can also arise from hairpin loops, preretinal vascular loops, and at arteriovenous crossings.[47] The pattern of neovascularization in PSR is termed "sea fan" due to its similar appearance to the oceanic species *Gorgonia flabellum*.[47] As the sea fans progress, they can hemorrhage, be pulled off the retinal surface by adherent vitreous, or auto-infarct. The first two events can lead to vitreous hemorrhage and retinal detachment, which are the most common causes of vision loss in PSR.[48] In a longitudinal study of 313 sickle cell patients followed up to 8 years in Jamaica, spontaneous regression or auto-infarction of sea fans was observed in up to 33/68 (49%) of patients with PSR on initial examination; however, blinding complications from PSR still occurred in 14/119 (12%) of eyes.[49] PSR remains the most important cause of vision loss in SCD.

Goldberg staging

In 1971, Goldberg[50] proposed a classification system for PSR after he evaluated the natural course of untreated PSR under indirect ophthalmoscopy and fluorescein angiography (Table 34.1).

TABLE 34.1 PSR staging.

Stage 1	Peripheral retinal arteriolar occlusions
Stage 2	Peripheral arteriovenous anastomoses (hair-pin loop)
Stage 3	Neovascular and fibrous proliferation (sea fan neovascularization)
Stage 4	Vitreous hemorrhage
Stage 5	Retinal detachment

PSR is a progressive disease with patients between 20 and 39 at the greatest risk of PSR progression.[51] In Stage 1, the peripheral arterioles occlude with nonperfusion of the retina anterior to these arterioles. Over time, these occluded arterioles turn from dark red to have a silver-wiring appearance. In Stage 2, the occluded arterioles form connections with terminal venules (Fig. 34.4).

Goldberg Stage 3 marks the onset of proliferative disease with new vessel formation at the junction of the vascular and avascular retina. Sea fan fronds can most commonly be found in the superotemporal and inferotemporal quadrants.[48] These sea fan neovascular fronds can hemorrhage, cause traction on the retina, or auto-infarct.[48,52] Up to 30%–60% of sea fan neovascularization can auto-infarct, which assumes a more whitish appearance due to fibrosis as opposed to red when they first develop.[20,47,49,50,53–55] Spontaneous regression is more commonly seen in HbSS (40%) than HbSC (20%) disease.[51]

Most patients remain visually asymptomatic until PSR progresses to Stage 4, in which vitreous contraction on the fragile sea-fan neovascular fronds results in vitreous hemorrhage. If the vitreous hemorrhage is severe, individuals can experience rapid-onset vision loss. This is more commonly seen in HbSC than in HbSS patients.[24,44,45] Individuals with over 60 degrees of neovascularization are at increased risk of recurring vitreous hemorrhage.[56,57] Chronic vitreous contraction on the sea-fan neovascular complexes can cause persistent traction on the peripheral retina resulting in tractional retinal detachment or cause a full-thickness retinal break resulting in combined rhegmatogenous/tractional retinal detachment. The peripheral avascular retina in PSR is atrophic and prone to retinal breaks. In Stage 5, tractional or combined rhegmatogenous/tractional retinal detachment (TRD) occurs.

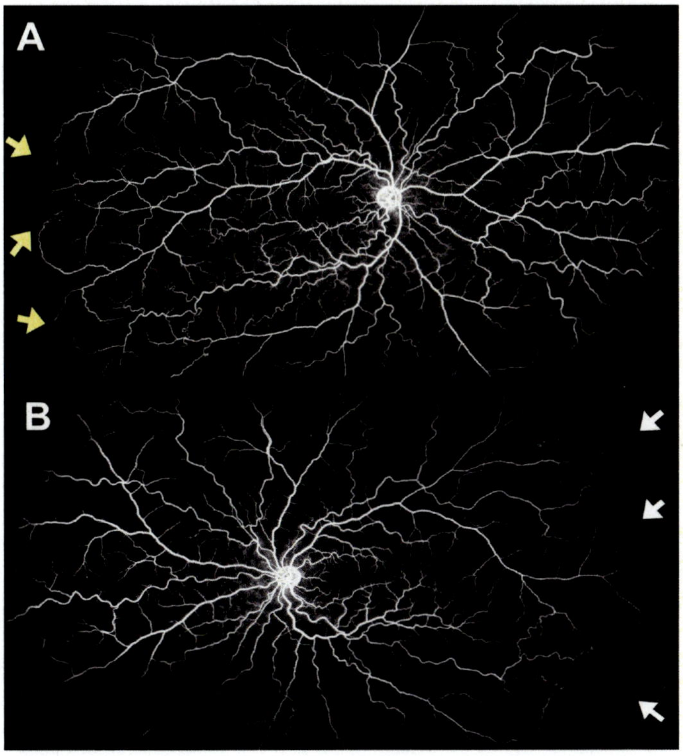

FIGURE 34.4 Peripheral arteriolar occlusions and arteriovenous anastomoses in sickle cell retinopathy. Fluorescein angiography of a 24 year-old-man with HbSS showed peripheral arteriovenous anastomoses (*yellow arrows*) in the right eye (A) and arteriolar occlusions (*white arrows*) in the left eye (B).

In 1994, Penman and colleagues proposed a separate classification system based on peripheral vasculature changes after studying a cohort of Jamaican patients with SCD.[58] Type I peripheral vascular changes were described as smooth, qualitatively normal borders while type II changes included abrupt "moth-eaten" terminations of small and medium vessels. If capillary stumps were observed to extend into the nonperfused retina, type II changes were subcategorized as type IIa; if not, then as type IIb. Type III referred to indeterminate changes whereby a type II pattern had reverted to normal after reperfusion of the vascular bed but can be distinguished from type I given the abrupt cessation of fluorescein dye in an arteriole with a hypofluorescent column beyond it.

Other posterior segment ocular manifestations

Vascular occlusions in the posterior pole are less common causes of acute vision loss in SCD. These include branch retinal artery occlusion, central retinal artery occlusion, and submacular choroidal infarctions. Central retinal artery occlusions have been reported to occur spontaneously in both HbSC and HbSS disease,[59−61] and have also been shown to occur after retrobulbar anesthesia.[62,63] In central retinal artery occlusion, patients present with sudden-onset, painless vision loss. The degree of permanent vision loss often depends on duration, site of the ischemia and whether reperfusion occurs; however, visual prognosis is generally poor. These situations require immediate assessment and treatment. Acute ocular decompression can be considered to increase blood flow. These patients should be referred for urgent hematology consultation with hydration, oxygenation, and consideration of possible red cell exchange transfusion.

Many patients with SCD have also been shown to have chronic macular flow voids on OCT angiography (OCT-A). These macular flow voids are most commonly observed temporally, which is known as the vascular watershed zone, and have been noted to affect the deep capillary plexus to a greater degree than the superficial capillary plexus. These macular flow voids may also lead to remodeling and enlargement of the foveal avascular zone.[64,65] Areas of macular flow voids are also thought to precede structural OCT thinning since decreased vascular density but not decreased retinal thickness was observed in children with SCD compared to race and age-matched controls,[66] while decreased vascular density was reported to correspond to retinal thinning in adults.[67,68] Patients with structural retinal thinning often do not report vision abnormalities but microperimetry studies have detected abnormal changes that correspond to these areas of thinning.[69]

Rarely, individuals with SCD may develop acute vision changes with central or paracentral scotomas due to paracentral acute middle maculopathy (PAMM) or acute macular neuroretinopathy (AMN). PAMM is thought to occur due to flow loss in the deep and/or intermediate retinal capillary plexus, and has been postulated to precede macular thinning in sickle cell maculopathy.[70,71] The deep capillary plexus may be particularly susceptible to ischemic insults due to its high oxygen requirements, and due to the vascular supply originating from a watershed zone of the retinal circulation. AMN is hypothesized to occur due to vascular impairments in the deep capillary plexus or choroid, and is observed as a hyperreflective lesion affecting the outer retinal layers on OCT.[72] Even though PAMM and AMN lesions typically improve with time, patients often complain of persistent central or paracentral scotomas.

Currently, there is no available treatment for SCD maculopathy.[65,73,74]

Imaging studies

Goldman and Penman's classification systems were completed using 30-degree fundus cameras as well as fluorescein angiography with peripheral sweeps. A major limitation of these imaging modalities is the difficulty in capturing very peripheral lesions, which predominate in SCR. For example, 13 out of 24 eyes that developed PSR in the Penman study had a vascular border that was too peripheral to be imaged on 30-degree fundus photography before becoming abnormal.[58]

Since then, the development of ultra-widefield (UWF) imaging has substantially improved our ability to image peripheral lesions in SCR. UWF imaging allows for a 200-degree field of view of the retina in a single image. Compared to standard field imaging, UWF imaging presents several benefits: (1) it is less reliant on photographer skill and patient cooperation with extreme gaze direction to capture the peripheral retina and thus reduces imaging variations due to technical skill; (2) no montaging is required to form a composite image; (3) it allows for a wider retinal area to be imaged at the same time, especially critical when examining multiple lesions of interest during timing dependent testing like fluorescein angiography, and when examining children or patients with developmental delay with a short attention span.

A small study of 12 eyes showed that in most eyes ($n = 11$), UWF imaging detected peripheral vascular changes not visualized in standard seven-field photographs.[75] A larger study of 70 eyes showed that UWF fluorescein angiography led to more eyes being categorized as Goldberg Stage 2 than Stage 1 when compared to UWF fundus photos and clinical examination, which points to the difficulty of picking up subtle peripheral anastomoses on examination.[76] Importantly, the study also found that an experienced retina specialist was able to clinically detect all areas of neovascularization and that

UWF imaging did not change Goldberg staging from 2 to 3 compared to clinical examination. Since clinicians will often consider treatment only when a patient has Goldberg stage 3 or worse, the authors concluded that UWF imaging therefore did not impact the clinical management of PSR.[76]

Application of OCT

More recently, spectral domain (SD) OCT has revolutionized the management of posterior segment ophthalmic conditions. The device is especially useful in evaluating the macula and has been of tremendous value in studying maculopathy in SCD. Multiple studies have shown that, compared to age, race, and gender-matched controls, patients with SCD display subclinical macular thinning and foveal splaying.[73,77–79] "Foveal splaying" refers to a flattening of the foveal shape, and may be due to occlusions of the parafoveal capillaries.[73] Macular thinning also appears to be worse in the temporal region, known as a watershed zone of retinal vasculature, and to progress over time (Fig. 34.5).[78]

While some studies have shown that retinal thinning is worse in the inner retina,[77,78] others have shown that the outer retina is also affected.[73] Macular thinning was also found to be worse with increasing age and stage of sickle cell retinopathy and was most severe in patients with the HbSS genotype.[79]

The advent of OCT angiography (OCT-A) has added to the armamentarium of noninvasive testing that can now be used to visualize the vasculature of the posterior pole. These images can capture information on the choriocapillaris, radial peripapillary capillaries, and capillary network of the retina.[80] In comparison to fluorescein angiography, OCT-A is more sensitive in detecting microangiopathy in SCD, especially in the macula.[67] A small study of 18 eyes showed that most microvascular abnormalities were located in the temporal region, and affected both the superficial and deep capillary plexi.[67] These microvascular changes may then cause the corresponding areas of macular thinning seen on OCT.[67] Another study of 36 SCR and 26 normal OCT-A images found that quantitative parameters on OCT-A like FAZ size and FAZ contour irregularity and avascular density in the temporal region were more sensitive than OCT thickness changes in detecting SCR.[81]

Multimodal imaging, incorporating UWF imaging, OCT, and/or OCT-A, is a suggested approach for monitoring retinal circulation in SCD patients over time (Fig. 34.5).

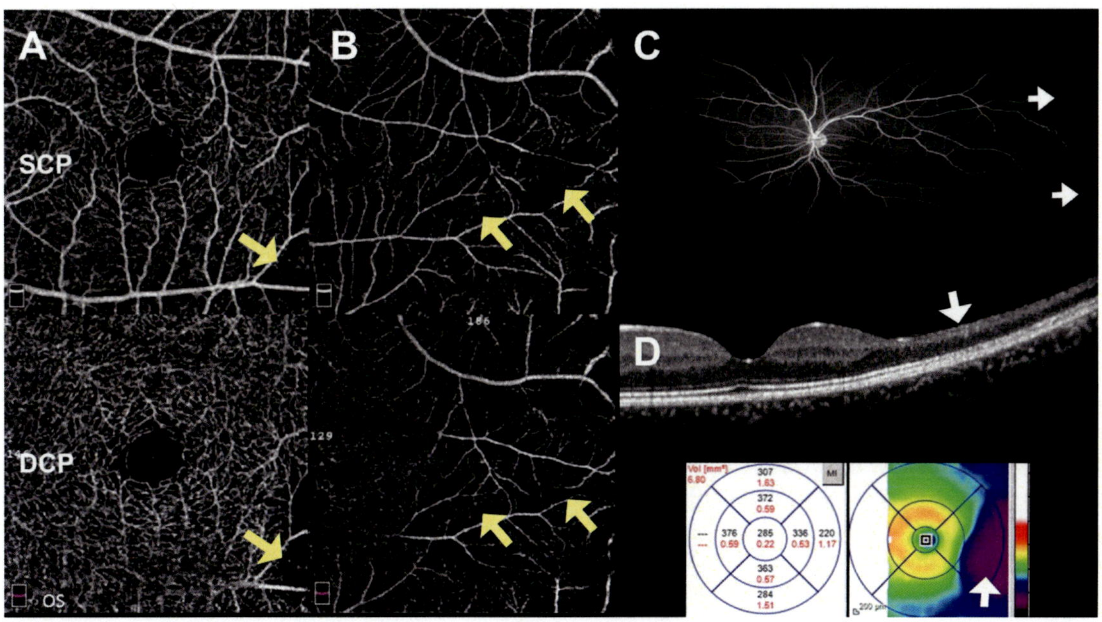

FIGURE 34.5 Temporal flow voids and thinning in the macula in a patient with sickle cell retinopathy. A 15-year-old with HbSS presented for routine sickle cell retinopathy surveillance examination. On the 3 × 3 mm optical coherence tomography angiography centered on the fovea (A), flow voids (*yellow arrows*) were noted temporally in the superficial capillary plexus (SCP) and deep capillary plexus (DCP). These flow voids were observed to be more severe in the temporal macula on the 6 × 6 mm optical coherence tomography angiography (B). Fluorescein angiography (C) showed mild nonperfusion (*white arrows*) in the retinal periphery while optical coherence tomography (D) and the corresponding thickness map (inset) illustrated temporal macular thinning (*white arrows*).

Management strategies

The goals of treatment in PSR are to prevent visual threatening complications from vitreous hemorrhage and retinal detachment. Patients with Goldberg Stages 1 and 2 usually do not have any visual sequelae and can be safely observed. Treatment is usually considered when patients have Stage 3 or worse PSR.

Observation

Current evidence does not support the treatment of small sea-fan neovascularization since up to 32% of eyes demonstrated auto-infarction of neovascular complexes, according to a longitudinal study of patients with PSR followed from birth to age 20 years.[20] In addition, permanent visual loss was uncommon in patients observed up to 26 years of age. One of the two randomized clinical trials performed in SCR to date by Farber and colleagues further reported that laser photocoagulation did not decrease the incidence of new sea-fan neovascularization and did not impact the rate of spontaneous regression of sea fans.[53] Therefore, the general consensus is to observe small sea-fan complexes that are asymptomatic.

Laser photocoagulation

Before the 1990s, the goal of therapy was to occlude the feeder vessel or obliterate the neovascular complex by diathermy, cryotherapy, or laser photocoagulation but these techniques were limited by complications. Diathermy was studied by Hannon in 1956 with initial favorable results in two patients but the long-term outcomes were unknown.[82] Condon and Serjeant used diathermy in 13 patients with promising initial results but one patient developed anterior segment ischemia, which is a devastating complication.[83] Goldbaum and coauthors found that triple freeze–thaw cryotherapy successfully closed 28 sea-fan retinal neovascularizations in nine eyes but two eyes developed retinal detachments presumably due to increased vitreoretinal traction.[84] Jampol et al. published the results of one of the only two clinical trials in SCR to date using Xenon arc or argon laser photocoagulation directly on feeder vessels and showed that feeder vessel photocoagulation successfully reduced the incidence of vitreous hemorrhage and vision loss.[85,86] However, the same authors and others have also reported that feeder vessel photocoagulation is technically challenging to perform and has been associated with retinal breaks, retinal detachment,[85] vitreous hemorrhage,[84] choroidal ischemia,[87] and choroidal-vitreal neovascularization.[84,88] Given the complications associated with feeder vessel ablation, these techniques have been abandoned.

Scatter laser photocoagulation is now the mainstay of treatment in PSR.[53] The purpose of scatter laser is to ablate the ischemic retina and suppress the proangiogenic milieu that drives new vessel formation. Additionally, the destruction of ischemic areas of the retina can help shunt blood flow and nutrients to healthier tissue, helping maintain it.[89] There are two types of scatter laser that can be performed—sectoral or circumferential. Sectoral laser refers to a laser that is applied in the avascular retina surrounding sea-fan complexes while circumferential laser refers to the broad application of peripheral laser in nonperfused retina including those areas distant to the neovascular complexes. In the clinical trial by Farber and coauthors, sectoral scatter laser was performed in 99 eyes versus 75 control eyes.[53] The incidences of vitreous hemorrhage and prolonged vision loss were both statistically significantly reduced in the treated eyes and no complications were associated with the argon scatter laser used.[53] A Cochrane review based on the two clinical trials by Farber et al.[53] and Jampol et al.,[85] concluded that laser photocoagulation should be considered to prevent vision loss and vitreous hemorrhage even though the regression and development of new PSR were not affected by laser treatment.[86] However, circumferential scatter laser has not been studied in a clinical trial and has not been compared to sectoral scatter laser in terms of efficacy in PSR; therefore, there lacks expert agreement on which technique is superior.

A recent study on autopsy eyes showed that vascular endothelial cells lining the sea fans as well as the ischemic retina surrounding the sea fans and distant to the sea fans all secrete proangiogenic factors like vascular endothelial growth factor (VEGF) and hypoxia-inducible factor (HIF-1 α).[90] The authors thus suggested that broad sectoral or circumferential scatter photocoagulation may be necessary to quench the proangiogenic growth factors that drive PSR. In the same study, Rodrigues and colleagues showed that VEGF and HIF-1 α expression was also upregulated in the marginal and transitional zones.[90] The marginal zone is the region between perfused and nonperfused retina extending 1 mm anterior and 1 mm posteriorly to the border between perfused and nonperfused retina. The transitional zone is the region posterior to the marginal zone with mostly perfused retina and focal areas of capillary dropout and can extend up to 2 mm posterior to the marginal zone. The authors, therefore, postulated that broad sectoral or circumferential laser to all three zones (nonperfused, marginal, and transitional) may be necessary for PSR.[90] We recommend this approach in scatter laser treatment for PSR (Fig. 34.6).

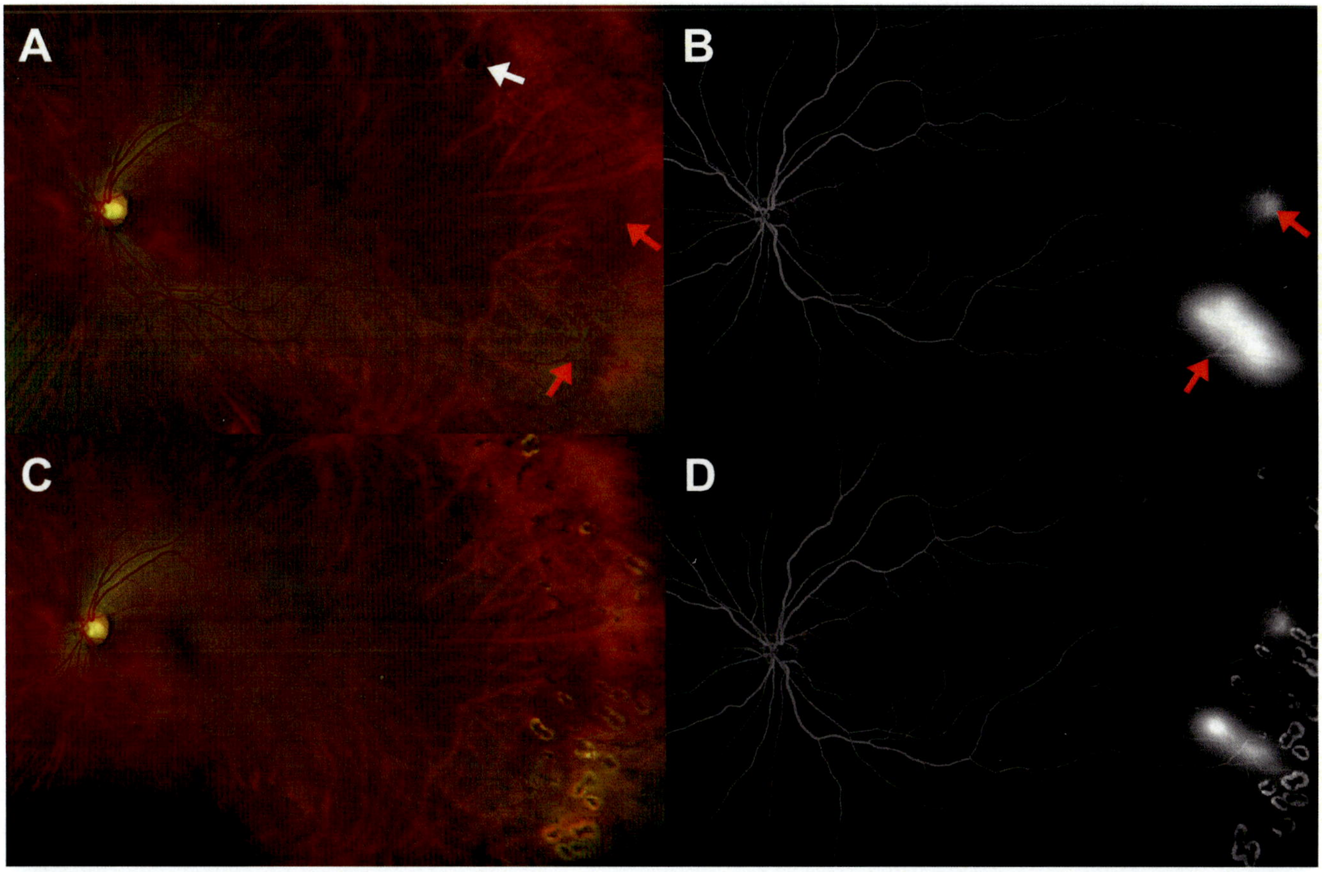

FIGURE 34.6 Scatter laser in the treatment of proliferative sickle cell retinopathy. A 57-year-old woman with HbSC noted a mild increase in floaters. At presentation, she was found to have new sea fan neovascular fronds (*red arrows*) as shown on fundus photography (A) and fluorescein angiography (B). A black sunburst was also noted on exam (*white arrow*). Scatter laser was performed. At the 6-month follow-up, the neovascular fronds had regressed on fundus photography (C), and fluorescein angiography (D) showed decreased leakage from the sea fan complexes.

Antivascular endothelial growth factor injections

Given the association between vascular endothelial growth factor (VEGF) and sea fan formation in PSR,[91,92] intravitreal anti-VEGF injections may play a role in the treatment of PSR. The use of anti-VEGF injections in Stage 4 PSR has been studied in several case reports and case series.[93–95] A cumulative six patients who presented with nonclearing vitreous hemorrhage or recurrent vitreous hemorrhage despite prior scatter laser were treated with anti-VEGF injections, which allowed the involution of sea fan neovascularization and resolution of hemorrhage. After the clearance of vitreous hemorrhage, some of these patients were treated with scatter laser photocoagulation while others were observed. The authors of these studies then concluded that anti-VEGF therapy may allow patients with stage 4 PSR to avoid vitrectomy surgery.

Another case report examined the use of anti-VEGF as a preoperative adjunct to tractional-rhegmatogenous retinal detachment repair in Stage 5 PSR.[96] Intravitreal bevacizumab was given 3 days before the scheduled surgery. Intraoperatively, the sea-fan complexes appeared less vascular and more fibrotic compared to preoperatively. The authors believed that the use of preoperative bevacizumab facilitated the segmentation and dissection of neovascular membranes and reduced both intraoperative and postsurgical bleeding.[96] This experience reflects the findings of other surgeons who utilized preoperative anti-VEGF in the repair of tractional retinal detachments in proliferative diabetic retinopathy.[97–99] Despite these promising results, one case of hyphema that developed several days after intravitreal anti-VEGF injection has been reported, therefore emphasizing the importance of patient education of concerning signs to look out for post-procedure.[100] Future studies are necessary to explore the use of anti-VEGF in PSR as either primary or adjunctive therapy.

Vitrectomy

Indications for vitrectomy include nonclearing vitreous hemorrhage, macula involving tractional retinal detachment, rhegmatogenous retinal detachment, and symptomatic epiretinal membrane or macular hole (Fig. 34.7).

With the introduction of anti-VEGF injections, surgery for nonclearing vitreous hemorrhage is now usually reserved for monocular patients or those with bilateral vitreous hemorrhage.

Historically, anterior segment ischemia was a feared complication after surgical repair of retinal detachment in patients with HbSC disease.[101] In particular, the risk factors included placement of a high encircling scleral buckle, a significant rise of intraocular pressure, and disruption of ciliary blood flow (e.g., by disinserting the extraocular muscles). Anterior segment ischemia was also reported in a patient who underwent vitrectomy without scleral buckling for repair of traction rhegmatogenous retinal detachment.[102] The authors hypothesized that extensive laser photocoagulation, systemic hypotension, and ocular hypertension during scleral depression may have combined with intravascular thrombosis from sickling to severely disrupt choroidal blood flow and induce anterior segment ischemia.

To minimize the risk of anterior segment ischemia, Ryan and Goldberg recommended avoiding high and broad encircling scleral buckles, not disinserting extraocular muscles, avoiding retrobulbar blocks and the use of analgesics with sympathomimetics, not using carbonic anhydrase inhibitors, avoiding excessive cryopexy, maximizing hydration, oxygenation, analgesia, and strict intraocular pressure control.[101] Exchange transfusion was also advocated by the authors although now not routinely done due to associated risks, albeit uncommon, including allergic reactions, blood-borne infections, and immune hemolytic reactions. In the modern era, the advent of small gauge instruments and valved cannulas have allowed the tight control of intraocular pressure, and with the insight gained from previously published studies, anterior segment ischemia has become extremely rare.[103]

Even though anterior segment ischemia has become less of a concern these days, surgery to repair retinal detachments in PSR remains fraught with challenges. The neovascular complexes in PSR tend to be located in the peripheral retina

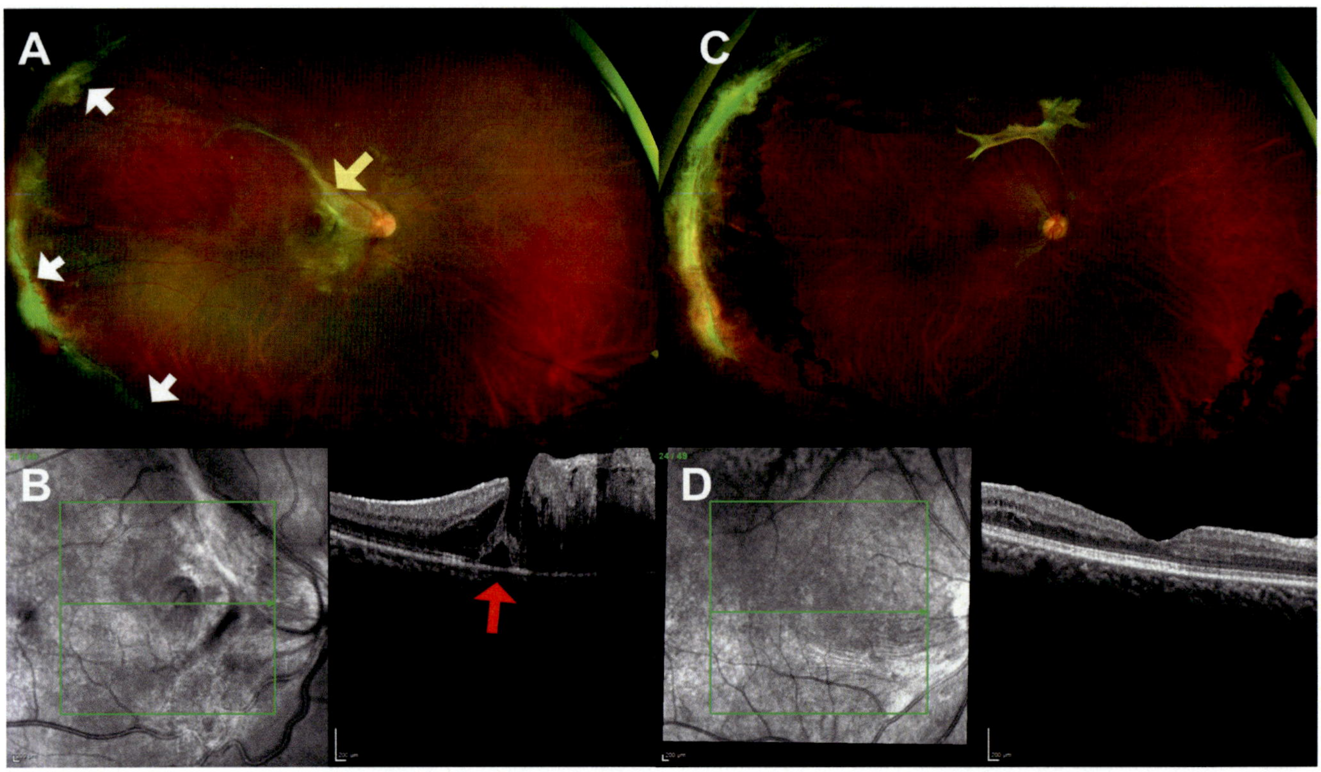

FIGURE 34.7 Epiretinal membrane, macular edema, and a partial thickness macular hole in a patient with proliferative sickle cell retinopathy A 23-year-old woman with HbSC presented with decreased vision in the right eye to 20/400. Fundus photography (A) showed the extensive fibrovascular proliferation in the macula (*yellow arrow*) with auto-infarcted sea fan neovascularization (*white arrows*) seen on clinical examination. Optical coherence tomography (B) demonstrated the presence of epiretinal membrane, macular edema, with a partial thickness macular hole (*red arrow*) in the right eye. Vitrectomy with membrane peeling and endolaser was performed. Two years postoperatively, fundus photography (C) and optical coherence tomography (D) showed normalization of the macular contour, with vision improvement to 20/40.

which is very thin, atrophic, and prone to retinal breaks. Iatrogenic retinal breaks are common during surgery, especially during delamination, which has spurred some authors to abandon delamination and instead use segmentation techniques to remove traction around neovascular fronds.[104] One of 15 patients in Jampol's series (1982) developed iatrogenic retinal breaks during tractional-rhegmatogenous retinal detachment repair[105] while seven of 18 eyes (39%) in Williamson's study (2009) developed iatrogenic breaks[104] during surgery for PSR complications. In Chen's study (2013), single-surgery anatomic success was achieved in only 50% (4/8) of eyes who underwent retinal detachment repair.[103] Of note, all four eyes that redetached had presented with combined tractional-rhegmatogenous retinal detachments.[103]

Screening

Current screening guidelines are derived from expert consensus recommendations.[106] In 2014, an expert panel through the National Heart, Lung, and Blood Institute issued guidelines on the management of acute and chronic complications of SCD.[3] These guidelines specified that all SCD patients should be referred to an ophthalmologist for an annual dilated fundoscopic exam starting at age 10. If the dilated exam is unremarkable, then rescreening should be performed every 1–2 years. Patients with suspected retinopathy should be referred to a retinal specialist for evaluation and management.[107] While these guidelines are adequate based on the epidemiology of SCR, there remains a need for updated and contemporary screening guidelines, with consideration for the incorporation of sickle cell maculopathy, and/or other systemic risk factors that inform an individual patient's risk for vision-threatening PSR.

In primary care and hematology visits, SCD patients should be queried about visual changes such as flashing lights or new or significant floaters.[27] Although there lacks expert consensus on the ideal screening methodology to detect SCR, at a minimum, slit lamp biomicroscopy, and indirect ophthalmoscopy with a high-power condensing fundus lens through dilated pupils are necessary to allow for thorough examination of the peripheral retina, where NPSR and PSR lesions typically present.[106] Undilated exams are insufficient for the monitoring of PSR. Other tests performed during this exam may include ultrawide-field imaging such as fundus photography and fluorescein angiography, especially if PSR is suspected. These imaging tools have the added advantage of allowing permanent documentation of the clinical findings.

Since HbSC and sickle thalassemia patients between the ages of 20–39 are at an increased risk of developing PSR compared to those with HbSS, additional effort should be made to ensure that they receive dilated eye exams. As patients with potentially sight-threatening PSR may be visually asymptomatic, it is critical to educate SCD patients and care providers about the importance of annual dilated retinal exams to monitor their eyes to identify and prevent vision-compromising complications. While newborns since 2006 have been screened for SCD, it is important to also consider hemoglobin electrophoresis testing for adults who present with ophthalmic findings that may be consistent with SCR.[108]

References

1. Elagouz M, Jyothi S, Gupta B, Sivaprasad S. Sickle cell disease and the eye: old and new concepts. *Survey of Ophthalmology*. 2010;55(4):359–377. https://doi.org/10.1016/j.survophthal.2009.11.004.
2. Ashley-Koch A, Yang Q, Olney RS. Sickle hemoglobin (Hb S) allele and sickle cell disease: a HuGE review. *American Journal of Epidemiology*. 2000;151(9):839–845. https://doi.org/10.1093/oxfordjournals.aje.a010288.
3. Yawn BP, Buchanan GR, Afenyi-Annan AN, et al. Management of sickle cell disease: summary of the 2014 evidence-based report by expert panel members. *JAMA - Journal of the American Medical Association*. 2014;312(10):1033–1048. https://doi.org/10.1001/jama.2014.10517.
4. Dewundara S, Nassiri N, Kim JM, et al. Retinal and choroidal vascular occlusion following aqueous misdirection syndrome in a patient with sickle cell trait. *Retin Cases & Brief Reports*. 2019;13(4):343–347. https://doi.org/10.1097/ICB.0000000000000590.
5. Elguero E, Délicat-Loembet LM, Rougeron V, et al. Malaria continues to select for sickle cell trait in central Africa. *Proceedings of the National Academy of Sciences*. 2015;112(22):7051–7054. https://doi.org/10.1073/pnas.1505665112.
6. Sugihara K, Sugihara T, Mohandas N, Hebbel RP. Thrombospondin mediates adherence of CD36+ sickle reticulocytes to endothelial cells. *Blood*. 1992;80(10):2634–2642. https://doi.org/10.1182/blood.v80.10.2634.2634.
7. Swerlick RA, Eckman JR, Kumar A, Jeitler M, Wick TM. Alpha 4 beta 1-integrin expression on sickle reticulocytes: vascular cell adhesion molecule-1-dependent binding to endothelium. *Blood*. 1993;82(6):1891–1899. https://doi.org/10.1182/blood.v82.6.1891.1891.
8. Wun T, Paglieroni T, Field CL, et al. Platelet-erythrocyte adhesion in sickle cell disease. *Journal of Investigative Medicine*. 1999;47(3):121–127.
9. Hebbel RP. Adhesive interactions of sickle erythrocytes with endothelium. *Journal of Clinical Investigation*. 1997;99(11):2561–2564. https://doi.org/10.1172/jci119442.
10. Hebbel RP. Blockade of adhesion of sickle cells to endothelium by monoclonal antibodies. *New England Journal of Medicine*. 2000;342(25):1910–1912. https://doi.org/10.1056/NEJM200006223422512.
11. Joneckis CC, Shock DD, Cunningham ML, Orringer EP, Parise LV. Glycoprotein IV-independent adhesion of sickle red blood cells to immobilized thrombospondin under flow conditions. *Blood*. 1996;87(11):4862–4870. https://doi.org/10.1182/blood.v87.11.4862.bloodjournal87114862.

12. Lasky LA. Selecting: interpreters of cell-specific carbohydrate information during inflammation. *Science*. 1992;258(5084):964–969. https://doi.org/10.1126/science.1439808.
13. Moriarty BJ, Acheson RW, Condon PI, Serjeant GR. Patterns of visual loss in untreated sickle cell retinopathy. *Eye*. 1988;2(3):330–335. https://doi.org/10.1038/eye.1988.62.
14. Kavanagh PL, Fasipe TA, Wun T. Sickle cell disease: a review. *JAMA*. 2022;328(1):57–68. https://doi.org/10.1001/jama.2022.10233.
15. Abdalla Elsayed MEA, Mura M, Al Dhibi H, et al. Sickle cell retinopathy. A focused review. *Graefe's Archive for Clinical and Experimental Ophthalmology*. 2019;257(7):1353–1364. https://doi.org/10.1007/s00417-019-04294-2.
16. Piel FB, Hay SI, Gupta S, Weatherall DJ, Williams TN, Osrin D. Global burden of sickle cell anaemia in children under five, 2010–2050: modelling based on demographics, excess mortality, and interventions. *PLoS Medicine*. 2013;10(7). https://doi.org/10.1371/journal.pmed.1001484.
17. Kato GJ, Piel FB, Reid CD, et al. Sickle cell disease. *Nature Reviews Disease Primers*. 2018;4. https://doi.org/10.1038/nrdp.2018.10.
18. Kato GJ, Hebbel RP, Steinberg MH, Gladwin MT. Vasculopathy in sickle cell disease: biology, pathophysiology, genetics, translational medicine, and new research directions. *American Journal of Hematology*. 2009;84(9):618–625. https://doi.org/10.1002/ajh.21475.
19. Dembélé AK, Toure BA, Sarro YS, et al. Prévalence et facteurs de risque de la rétinopathie drépanocytaire dans un centre de suivi drépanocytaire d'Afrique subsaharienne. *Revue de Medecine Interne*. 2017;38(9):572–577. https://doi.org/10.1016/j.revmed.2017.01.010.
20. Downes SM, Hambleton IR, Chuang EL, Lois N, Serjeant GR, Bird AC. Incidence and natural history of proliferative sickle cell retinopathy: observations from a cohort study. *Ophthalmology*. 2005;112(11):1869–1875. https://doi.org/10.1016/j.ophtha.2005.05.026.
21. Lutty GA, Phelan A, McLeod DS, Fabry ME, Nagel RL. A rat model for sickle cell-mediated vaso-occlusion in retina. *Microvascular Research*. 1996;52(3):270–280. https://doi.org/10.1006/mvre.1996.0064.
22. Fox PD, Dunn DT, Morris JS, Serjeant GR. Risk factors for proliferative sickle retinopathy. *British Journal of Ophthalmology*. 1990;74(3):172–176. https://doi.org/10.1136/bjo.74.3.172.
23. Mian UK, Tang J, Allende APM, et al. Elevated fetal haemoglobin levels are associated with decreased incidence of retinopathy in adults with sickle cell disease. *British Journal of Haematology*. 2018;183(5):807–811. https://doi.org/10.1111/bjh.15617.
24. Welch RB, Goldberg MF. Sickle-cell hemoglobin and its relation to fundus abnormality. *Archives of Ophthalmology*. 1966;75(3):353–362. https://doi.org/10.1001/archopht.1966.00970050355008.
25. Nagpal KC, Asdourian GK, Patrianakos D, et al. Proliferative retinopathy in sickle cell trait: report of seven cases. *Archives of Internal Medicine*. 1977;137(3):325–328. https://doi.org/10.1001/archinte.1977.03630150035011.
26. Mehta JS, Whittaker KW, Tsaloumas MD. Latent proliferative sickle cell retinopathy in sickle cell trait. *Acta Ophthalmologica Scandinavica*. 2001;79(1):81–82. https://doi.org/10.1034/j.1600-0420.2001.079001081.x.
27. Duan XJ, Lanzkron S, Linz MO, Ewing C, Wang J, Scott AW. Clinical and ophthalmic factors associated with the severity of sickle cell retinopathy. *American Journal of Ophthalmology*. 2019;197:105–113. https://doi.org/10.1016/j.ajo.2018.09.025.
28. Saidkasimova S, Shalchi Z, Mahroo OA, et al. Risk factors for visual impairment in patients with sickle cell disease in London. *European Journal of Ophthalmology*. 2016;26(5):431–435. https://doi.org/10.5301/ejo.5000767.
29. Ware RE. How I use hydroxyurea to treat young patients with sickle cell anemia. *Blood*. 2010;115(26):5300–5311. https://doi.org/10.1182/blood-2009-04-146852.
30. Estepp JH, Smeltzer MP, Wang WC, Hoehn ME, Hankins JS, Aygun B. Protection from sickle cell retinopathy is associated with elevated HbF levels and hydroxycarbamide use in children. *British Journal of Haematology*. 2013;161(3):402–405. https://doi.org/10.1111/bjh.12238.
31. Lim JI, Niec M, Sun J, Cao D. Longitudinal assessment of retinal thinning in adults with and without sickle cell retinopathy using spectral-domain optical coherence tomography. *JAMA Ophthalmology*. 2021;139(3):330–337. https://doi.org/10.1001/jamaophthalmol.2020.6525.
32. Charache S, Terrin ML, Moore RD, et al. Effect of hydroxyurea on the frequency of painful crises in sickle cell anemia. *New England Journal of Medicine*. 1995;332(20):1317–1322. https://doi.org/10.1056/NEJM199505183322001.
33. Thornburg CD, Files BA, Luo Z, et al. Impact of hydroxyurea on clinical events in the BABY HUG trial. *Blood*. 2012;120(22):4304–4310. https://doi.org/10.1182/blood-2012-03-419879.
34. Wang WC, Ware RE, Miller ST, et al. Hydroxycarbamide in very young children with sickle-cell anaemia: a multicentre, randomised, controlled trial (BABY HUG). *Lancet*. 2011;377(9778):1663–1672. https://doi.org/10.1016/S0140-6736(11)60355-3.
35. Brousseau DC, Richardson T, Hall M, et al. Hydroxyurea use for sickle cell disease among medicaid-enrolled children. *Pediatrics*. 2019;144(1). https://doi.org/10.1542/peds.2018-3285.
36. Stettler N, McKiernan CM, Melin CQ, Adejoro OO, Walczak NB. Proportion of adults with sickle cell anemia and pain crises receiving hydroxyurea. *JAMA - Journal of the American Medical Association*. 2015;313(16):1671–1672. https://doi.org/10.1001/jama.2015.3075.
37. Walsh KE, Cutrona SL, Kavanagh PL, et al. Medication adherence among pediatric patients with sickle cell disease: a systematic review. *Pediatrics*. 2014;134(6):1175–1183. https://doi.org/10.1542/peds.2014-0177.
38. Loiselle K, Lee JL, Szulczewski L, Drake S, Crosby LE, Pai ALH. Systematic and meta-analytic review: medication adherence among pediatric patients with sickle cell disease. *Journal of Pediatric Psychology*. 2016;41(4):406–418. https://doi.org/10.1093/jpepsy/jsv084.
39. Badawy SM, Thompson AA, Holl JL, Penedo FJ, Liem RI. Healthcare utilization and hydroxyurea adherence in youth with sickle cell disease. *Pediatric Hematology and Oncology*. 2018;35(5–6):297–308. https://doi.org/10.1080/08880018.2018.1505988.
40. Khwarg SG, Feldman S, Ligh J, Straatsma BR. Exchange transfusion in sickling maculopathy. *Retina*. 1985;5(4):227–229. https://doi.org/10.1097/00006982-198500540-00007.
41. Romayananda N, Goldberg MF, Green WR. Histopathology of sickle cell retinopathy. *Transactions of the American Academy of Ophthalmology and Otolaryngology*. 1973;77(5):652–676.

42. Gagliano DA, Goldberg MF. The evolution of salmon-patch hemorrhages in sickle cell retinopathy. *Archives of Ophthalmology*. 1989;107(12):1814–1815. https://doi.org/10.1001/archopht.1989.01070020896034.
43. Emerson GG, Lutty GA. Effects of sickle cell disease on the eye: clinical features and treatment. *Hematology/Oncology Clinics of North America*. 2005;19(5):957–973. https://doi.org/10.1016/j.hoc.2005.07.005.
44. Condon PI, Serjeant GR. Ocular findings in homozygous sickle cell anemia in Jamaica. *American Journal of Ophthalmology*. 1972;73(4):533–543. https://doi.org/10.1016/0002-9394(72)90005-0.
45. Clarkson JG. The ocular manifestations of sickle-cell disease: a prevalence and natural history study. *Transactions of the American Ophthalmological Society*. 1992;90:481–504.
46. Nagpal KC, Asdourian G, Goldbaum M, Apple D, Goldberg MF. Angioid streaks and sickle haemoglobinopathies. *British Journal of Ophthalmology*. 1976;60(1):31–34. https://doi.org/10.1136/bjo.60.1.31.
47. McLeod DS, Merges C, Fukushima A, Goldberg MF, Lutty GA. Histopathologic features of neovascularization in sickle cell retinopathy. *American Journal of Ophthalmology*. 1997;124(4):455–472. https://doi.org/10.1016/S0002-9394(14)70862-1.
48. Goldberg MF. Retinal neovascularization in sickle cell retinopathy. *Transactions - American Academy of Ophthalmology and Otolaryngology*. 1977;83(3):OP409–431.
49. Condon PI, Serjeant GR. Behaviour of untreated proliferative sickle retinopathy. *British Journal of Ophthalmology*. 1980;64(6):404–411. https://doi.org/10.1136/bjo.64.6.404.
50. Goldberg MF. Natural history of untreated proliferative sickle retinopathy. *Archives of Ophthalmology*. 1971;85(4):428–437. https://doi.org/10.1001/archopht.1971.00990050430006.
51. Fox PD, Vessey SJ, Forshaw ML, Serjeant GR. Influence of genotype on the natural history of untreated proliferative sickle retinopathy—an angiographic study. *British Journal of Ophthalmology*. 1991;75(4):229–231. https://doi.org/10.1136/bjo.75.4.229.
52. Nagpal KC, Goldberg MF, Rabb MF. Ocular manifestations of sickle hemoglobinopathies. *Survey of Ophthalmology*. 1977;21(5):391–411. https://doi.org/10.1016/0039-6257(77)90042-X.
53. Farber MD, Jampol LM, Fox P, et al. A randomized clinical trial of scatter photocoagulation of proliferative sickle cell retinopathy. *Archives of Ophthalmology*. 1991;109(3):363–367. https://doi.org/10.1001/archopht.1991.01080030065040.
54. Goldberg MF, Jampol LM. Treatment of neovascularization, vitreous hemorrhage, and retinal detachment in sickle cell retinopathy. *Transactions of the New Orleans Academy of Ophthalmology*. 1983;31:53–81.
55. Nagpal KC, Patrianakos D, Asdourian GK, Goldberg MF, Rabb M, Jampol L. Spontaneous regression (autoinfarction) of proliferative sickle retinopathy. *American Journal of Ophthalmology*. 1975;80(5):885–892. https://doi.org/10.1016/0002-9394(75)90285-8.
56. Condon PI, Whitelocke RAF, Bird AC, Talbot JF, Serjeant GR. Recurrent visual loss in homozygous sickle cell disease. *British Journal of Ophthalmology*. 1985;69(9):700–706. https://doi.org/10.1136/bjo.69.9.700.
57. Mcleod DS, Goldberg MF, Lutty GA. Dual-perspective analysis of vascular formations in sickle cell retinopathy. *Archives of Ophthalmology*. 1993;111(9):1234–1245. https://doi.org/10.1001/archopht.1993.01090090086026.
58. Penman AD, Talbot JF, Chuang EL, Thomas P, Serjeant GR, Bird AC. New classification of peripheral retinal vascular changes in sickle cell disease. *British Journal of Ophthalmology*. 1994;78(9):681–689. https://doi.org/10.1136/bjo.78.9.681.
59. Fine LC, Petrović V, Irvine AR, Bhisitkul RB. Correction-spontaneous central retinal artery occlusion in hemoglobin SC disease. *American Journal of Ophthalmology*. 2000;130(6):906–907. https://doi.org/10.1016/s0002-9394(00)00798-4.
60. Jampol LM, Condon P, Dizon-Moore R, Serjeant G, Schulman JA. Salmon-patch hemorrhages after central retinal artery occlusion in sickle cell disease. *Archives of Ophthalmology*. 1981;99(2):237–240. https://doi.org/10.1001/archopht.1981.03930010239002.
61. Marcus RE, Bolger JP, Roderick PJ, Rowson NJ. Central retinal artery occlusion in homozygous sickle cell (SS) disease. *Clinical & Laboratory Haematology*. 1988;10(4):467–470. https://doi.org/10.1111/j.1365-2257.1988.tb01196.x.
62. Klein ML, Jampol LM, Condon PI, Rice TA, Serjeant GR. Central retinal artery occlusion without retrobulbar hemorrhage after retrobulbar anesthesia. *American Journal of Ophthalmology*. 1982;93(5):573–577. https://doi.org/10.1016/S0002-9394(14)77371-4.
63. Roth SE, Magargal LE, Kimmel AS, Augsburger JJ, Morrison DL. Central retinal-artery occlusion in proliferative sickle-cell retinopathy after retrobulbar injection. *Annals of Ophthalmology*. 1988;20(6):221–224.
64. Asdourian GK, Nagpal KC, Busse B, et al. Macular and perimacular vascular remodelling sickling haemoglobinopathies. *British Journal of Ophthalmology*. 1976;60(6):431–453. https://doi.org/10.1136/bjo.60.6.431.
65. Stevens TS, Busse B, Lee CB, Woolf MB, Galinos SO, Goldberg MF. Sickling hemoglobinopathies: macular and perimacular vascular abnormalities. *Archives of Ophthalmology*. 1974;92(6):455–463. https://doi.org/10.1001/archopht.1974.01010010469002.
66. Ong SS, Linz MO, Li X, Liu TYA, Han IC, Scott AW. Retinal thickness and microvascular changes in children with sickle cell disease evaluated by optical coherence tomography (OCT) and OCT angiography. *American Journal of Ophthalmology*. 2020;209:88–98. https://doi.org/10.1016/j.ajo.2019.08.019.
67. Minvielle W, Caillaux V, Cohen SY, et al. Macular microangiopathy in sickle cell disease using optical coherence tomography angiography. *American Journal of Ophthalmology*. 2016;164:137–144. https://doi.org/10.1016/j.ajo.2015.12.023.
68. Han IC, Tadarati M, Scott AW. Macular vascular abnormalities identified by optical coherence tomographic angiography in patients with sickle cell disease. *JAMA Ophthalmology*. 2015;133(11):1337–1340. https://doi.org/10.1001/jamaophthalmol.2015.2824.
69. Chow CC, Genead MA, Anastasakis A, Chau FY, Fishman GA, Lim JI. Structural and functional correlation in sickle cell retinopathy using spectral-domain optical coherence tomography and scanning laser ophthalmoscope microperimetry. *American Journal of Ophthalmology*. 2011;152(4):704–711. https://doi.org/10.1016/j.ajo.2011.03.035.

70. Hussnain SA, Coady PA, Stoessel KM. Paracentral acute middle maculopathy: precursor to macular thinning in sickle cell retinopathy. *BMJ Case Reports*. 2017;2017. https://doi.org/10.1136/bcr-2016-216124.
71. Jung JJ, Chen MH, Frambach CR, Rofagha S, Lee SS. Spectral domain versus swept source optical coherence tomography angiography of the retinal capillary plexuses in sickle cell maculopathy. *Retinal Cases and Brief Reports*. 2018;12(2):87−92. https://doi.org/10.1097/ICB.0000000000000448.
72. Ong SS, Ahmed I, Scott AW. Association of acute macular neuroretinopathy or paracentral acute middle maculopathy with sickle cell disease. *Ophthalmology Retina*. 2021;5(11):1146−1155. https://doi.org/10.1016/j.oret.2021.01.003.
73. Hoang QV, Chau FY, Shahidi M, Lim JI. Central macular splaying and outer retinal thinning in asymptomatic sickle cell patients by spectral-domain optical coherence tomography. *American Journal of Ophthalmology*. 2011;151(6):990−994.e1. https://doi.org/10.1016/j.ajo.2010.12.010.
74. Mathew R, Bafiq R, Ramu J, et al. Spectral domain optical coherence tomography in patients with sickle cell disease. *British Journal of Ophthalmology*. 2015;99(7):967−972. https://doi.org/10.1136/bjophthalmol-2014-305532.
75. Cho M, Kiss S. Detection and monitoring of sickle cell retinopathy using ultra wide-field color photography and fluorescein angiography. *Retina*. 2011;31(4):738−747. https://doi.org/10.1097/IAE.0b013e3181f049ec.
76. Han IC, Zhang AY, Liu TYA, Linz MO, Scott AW. Utility of ultra-widefield retinal imaging for the staging and management of sickle cell retinopathy. *Retina*. 2019;39(5):836−843. https://doi.org/10.1097/IAE.0000000000002057.
77. Witkin AJ, Rogers AH, Ko TH, Fujimoto JG, Schuman JS, Duker JS. Optical coherence tomography demonstration of macular infarction in sickle cell retinopathy. *Archives of Ophthalmology*. 2006;124(5):746−747. https://doi.org/10.1001/archopht.124.5.746.
78. Cai CX, Han IC, Tian J, Linz MO, Scott AW. Progressive retinal thinning in sickle cell retinopathy. *Ophthalmology Retina*. 2018;2(12):1241−1248.e2. https://doi.org/10.1016/j.oret.2018.07.006.
79. Lim JI, Cao D. Analysis of retinal thinning using spectral-domain optical coherence tomography imaging of sickle cell retinopathy eyes compared to age- and race-matched control eyes. *American Journal of Ophthalmology*. 2018;192:229−238. https://doi.org/10.1016/j.ajo.2018.03.013.
80. Spaide RF, Klancnik JM, Cooney MJ. Retinal vascular layers imaged by fluorescein angiography and optical coherence tomography angiography. *JAMA Ophthalmology*. 2015;133(1):45−50. https://doi.org/10.1001/jamaophthalmol.2014.3616.
81. Alam M, Thapa D, Lim JI, Cao D, Yao X. Quantitative characteristics of sickle cell retinopathy in optical coherence tomography angiography. *Biomedical Optics Express*. 2017;8(3):1741−1753. https://doi.org/10.1364/BOE.8.001741.
82. Hannon JF. Vitreous hemorrhages: associated with sickle cell-hemoglobin C disease. *American Journal of Ophthalmology*. 1956;42(5):707−712. https://doi.org/10.1016/0002-9394(56)90511-6.
83. Condon PI, Serjeant GR. Photocoagulation and diathermy in the treatment of proliferative sickle retinopathy. *British Journal of Ophthalmology*. 1974;58(7):650−662. https://doi.org/10.1136/bjo.58.7.650.
84. Goldbaum MH, Goldberg MF, Nagpal K, Asdourian GK, Galinos SO. Proliferative sickle retinopathy. In: *Current Diagnosis and Management of Chorioretinal Diseases*. 1977, 1977.
85. Jampol LM, Condon P, Farber M, Rabb M, Ford S, Serjeant G. A randomized clinical trial of feeder vessel photocoagulation of proliferative sickle cell retinopathy: I. Preliminary results. *Ophthalmology*. 1983;90(5):540−545. https://doi.org/10.1016/S0161-6420(83)34537-1.
86. Myint KT, Sahoo S, Thein AW, Moe S, Ni H. Laser therapy for retinopathy in sickle cell disease. *Cochrane Database of Systematic Reviews*. 2015;2015(10). https://doi.org/10.1002/14651858.CD010790.pub2.
87. Goldbaum MH, Galinos SO, Apple D, et al. Acute choroidal ischemia as a complication of photocoagulation. *Archives of Ophthalmology*. 1976;94(6):1025−1035. https://doi.org/10.1001/archopht.1976.03910030521016.
88. Galinos SO, Asdourian GK, Woolf MB, Goldberg MF, Busse BJ. Choroido-vitreal neovascularization after argon laser photocoagulation. *Archives of Ophthalmology*. 1975;93(7):524−530. https://doi.org/10.1001/archopht.1975.01010020540012.
89. Ryan SJ. *Retina*. 2013, 2013. Elsevier.
90. Rodrigues M, Kashiwabuchi F, Deshpande M, et al. Expression pattern of HIF-1α and VEGF supports circumferential application of scatter laser for proliferative sickle retinopathy. *Investigative Ophthalmology and Visual Science*. 2016;57(15):6739−6746. https://doi.org/10.1167/iovs.16-19513.
91. Cao J, Mathews MK, McLeod DS, Merges C, Hjelmeland LM, Lutty GA. Angiogenic factors in human proliferative sickle cell retinopathy. *British Journal of Ophthalmology*. 1999;83(7):838−846. https://doi.org/10.1136/bjo.83.7.838.
92. Kim SY, Mocanu C, Mcleod DS, et al. Expression of pigment epithelium-derived factor (PEDF) and vascular endothelial growth factor (VEGF) in sickle cell retina and choroid. *Experimental Eye Research*. 2003;77(4):433−445. https://doi.org/10.1016/S0014-4835(03)00174-X.
93. Shaikh S. Intravitreal bevacizumab (Avastin) for the treatment of proliferative sickle retinopathy. *Indian Journal of Ophthalmology*. 2008;56(3). https://doi.org/10.4103/0301-4738.40380.
94. Mitropoulos PG, Chatziralli IP, Parikakis EA, Peponis VG, Amariotakis GA, Moschos MM. Intravitreal Ranibizumab for stage IV proliferative sickle cell retinopathy: a first case report. *Case Reports in Ophthalmological Medicine*. 2014;2014:1−6. https://doi.org/10.1155/2014/682583.
95. Cai CX, Linz MO, Scott AW. Intravitreal bevacizumab for proliferative sickle retinopathy: a case series. *Journal of VitreoRetinal Diseases*. 2018;2(1):32−38. https://doi.org/10.1177/2474126417738627.
96. Moshiri A, Ha NK, Ko FS, Scott AW. Bevacizumab presurgical treatment for proliferative sickle-cell retinopathy-related retinal detachment. *Retinal Cases and Brief Reports*. 2013;7(3):204−205. https://doi.org/10.1097/ICB.0b013e3182845d31.
97. Guthrie G, Hall AB, Dhalla K, Davis RM, Steel DH. Bevacizumab as an adjunct to vitreoretinal surgery for diabetic retinopathy in East Africa. *Eye*. 2013;27(11):1263−1268. https://doi.org/10.1038/eye.2013.182.
98. Pokroy R, Desai UR, Du E, Li Y, Edwards P. Bevacizumab prior to vitrectomy for diabetic traction retinal detachment. *Eye*. 2011;25(8):989−997. https://doi.org/10.1038/eye.2011.149.

99. Rizzo S, Genovesi-Ebert F, Di Bartolo E, Vento A, Miniaci S, Williams G. Injection of intravitreal bevacizumab (Avastin) as a preoperative adjunct before vitrectomy surgery in the treatment of severe proliferative diabetic retinopathy (PDR). *Graefe's Archive for Clinical and Experimental Ophthalmology*. 2008;246(6):837–842. https://doi.org/10.1007/s00417-008-0774-y.
100. Babalola OE. Intravitreal bevacizumab (Avastin) associated with secondary hyphaema in a case of proliferative sickle cell retinopathy. *Case Reports*. 2010;2010. https://doi.org/10.1136/bcr.11.2009.2441.
101. Ryan SJ, Goldberg MF. Anterior segment ischemia following scleral buckling in sickle cell hemoglobinopathy. *American Journal of Ophthalmology*. 1971;72(1):35–50. https://doi.org/10.1016/0002-9394(71)91588-1.
102. Leen JS, Ratnakaram R, Del Priore LV, Bhagat N, Zarbin MA. Anterior segment ischemia after vitrectomy in sickle cell disease. *Retina*. 2002;22(2):216–219. https://doi.org/10.1097/00006982-200204000-00014.
103. Chen RWS, Flynn HW, Lee W-H, et al. Vitreoretinal management and surgical outcomes in proliferative sickle retinopathy: a case series. *American Journal of Ophthalmology*. 2014;157(4):870–875.e1. https://doi.org/10.1016/j.ajo.2013.12.019.
104. Williamson TH, Rajput R, Laidlaw DAH, Mokete B. Vitreoretinal management of the complications of sickle cell retinopathy by observation or pars plana vitrectomy. *Eye*. 2009;23(6):1314–1320. https://doi.org/10.1038/eye.2008.296.
105. Jampol LM, Green JL, Goldberg MF, Peyman GA. An update on vitrectomy surgery and retinal detachment repair in sickle cell disease. *Archives of Ophthalmology*. 1982;100(4):591–593. https://doi.org/10.1001/archopht.1982.01030030593008.
106. Amissah-Arthur KN, Mensah E. The past, present and future management of sickle cell retinopathy within an African context. *Eye (Basingstoke)*. 2018;32(8):1304–1314. https://doi.org/10.1038/s41433-018-0162-8.
107. Rosenberg JB, Hutcheson KA. Pediatric sickle cell retinopathy: correlation with clinical factors. *Journal of AAPOS*. 2011;15(1):49–53. https://doi.org/10.1016/j.jaapos.2010.11.014.
108. Benson JM, Therrell BL. History and current status of newborn screening for hemoglobinopathies. *Seminars in Perinatology*. 2010;34(2):134–144. https://doi.org/10.1053/j.semperi.2009.12.006.

Chapter 35

Leukemic retinopathy

Zeinab El Moussawi[1], Hana A. Mansour[1] and Ahmad M. Mansour[2,a]
[1]Department of Ophthalmology, American University of Beirut, Beirut, Lebanon; [2]Department of Ophthalmology, Retina Service, American University of Beirut, Beirut, Lebanon

Introduction

Leukemia is a rare disease group of neoplastic diseases of high mortality. They are caused by the noncontrolled and irregular development of the hematopoietic stem cells of the bone marrow.[1] Based on the origin of those hematopoietic stem cells, leukemia can be further divided into myeloid or lymphoid.[1] Leukemia can be acute with rapid development or chronic with indolent progression. The former is fast and uncontrolled, forming immature and nonfunctional leukocytes. These leukocytes fail their immune function, leading to severe illness.[1] On the other hand, chronic leukemia has a gradual progression with more mature cells in the bloodstream, yet these cells may transform into a more aggressive form of the disease.[1] Accordingly, Leukemic cells can migrate and invade much more freely than tumor cells from solid tumors. In solid tumors, the tumor cells require mutations to disseminate and metastasize, while Leukemic cells are highly spreadable, which makes treatment more challenging.[2] Leukemia starts in the bone marrow where the blood cell precursors reside, but often moves quickly into the blood.[1] Signs of the disease are also seen in other tissues like lymph nodes, spleen, liver, and central nervous system.[1]

Four major groups of leukemia are identified: acute myeloid leukemia (AML), chronic myeloid leukemia (CML), acute lymphoblastic leukemia (ALL), and chronic lymphocytic leukemia (CLL). ALL is considered the most common type of leukemia in young children while CML is the most common type in middle-aged adults.[1] General signs and symptoms of the four types of leukemia include persistent fatigue and weakness, frequent infections, weight loss, easy bleeding, and bruising along with other signs.[1]

Ocular involvement of leukemia was first reported by Liebreich in 1861.[3] Ocular manifestations are frequent and can be noted in up to 50% of patients at diagnosis.[4] Most ocular involvement can be asymptomatic. However, once symptomatic, it can be the primary sign of the disease. Those symptoms include blurry vision (unilateral or bilateral), floaters, and some defects in the visual field. Thus, taking ocular complaints seriously and assessing them is essential as they may be the primary marker for leukemia.[4] Choroidal and retinal tissues are the most affected sites in Leukemia.[4] Patients with leukemic retinopathy seem to have a reduced survival rate, as hematological parameters correlated with retinal changes.[4] In this chapter, we will explore the various manifestations of leukemic retinopathies and propose known pathogenesis and most recent management strategies.

Pathogenesis

Ocular involvement in leukemia can be divided into primary and secondary involvement.[4] Primary involvement refers to leukemic infiltration while secondary involvement relates to the systemic abnormalities caused by leukemia, including hyperviscosity, anemia, thrombocytopenia, opportunistic infections, and treatment-related complications.[4]

The Leukemic cells are known to invade ocular tissues, such as the retina, the choroid, and the optic nerve, and the vitreous. Histologically, the choroid is the most frequently involved organ, and as the choroid is highly vascularized, it

[a] The authors have no financial interest to disclose.

Retinal and Choroidal Vascular Diseases of the Eye. https://doi.org/10.1016/B978-0-443-15583-3.00035-4

offers a unique place for leukemic involvement.[3] Clinically, the retina is the most common site of involvement, and often leukemic retinopathy can be the primary presenting sign of the disease or of a relapse after a period of remission.[3]

Leukemic retinopathy

Retinal leukemic infiltrates can be seen in around 3% of leukemia patients.[3] Neoplastic cells can infiltrate the retina at any level, and more commonly the inner layers (Figs. 35.1 and 35.2). When direct infiltration of the leukemic cells occurs, structural destruction is expected. Leukemic retinopathy can be present in multiple forms. The infiltrates appear as gray to white nodules of varying sizes.[4] Small infiltrates are usually seen in the perivascular areas while larger massive diffuse infiltrates are accompanied by retinal detachment. Perivascular sheathing or grayish-white lines around retinal vessels represent leukemic infiltration. Another manifestation is the white center hemorrhage or Roth's spots whose center harbors platelet-fibrin aggregates and often a collection of leukemic cells.

The severity of the systemic disease has been correlated with the direct leukemic infiltrates.[5] Also, complete or partial tissue destruction of the involved retinal tissues has been seen on histological examination. Those infiltrates are reversible with systemic treatment and functional recovery can be seen.[5]

Below we present a case of direct leukemic infiltration.[6] It is the case of a 15-year-old male who was newly diagnosed with acute lymphoblastic leukemia (ALL) and had just started treatment. He complained of blurry vision a few days after the diagnosis was made. A dilated fundus exam revealed retinal hemorrhages due to severe thrombocytopenia (platelet count 51,000 per microliter) (Fig. 35.1). Despite systemic chemotherapy, vision deteriorated with the retinal hemorrhages becoming diffuse and covering the whole posterior pole.[6] This was accompanied by optic disc elevation and serous macular detachment, raising the suspicion of CMV retinitis.[6] On further follow-up, retinal hemorrhages decreased and leukemic subretinal infiltrates were clearly noted on fundus examination (Fig. 35.2). Those findings denoted failure of the first-line chemotherapy. Following a change in the protocol and subtenon corticosteroid injection, the retinal infiltrates decreased.[6] Because of good clinical response, a second subtenon injection was given. The patient passed away a few weeks later, in line with the general rule: the presence of a strong correlation between the severity of the systemic disease with the presence of leukemic infiltrates.

Leukemic choroidopathy

Among ocular tissues, the choroid is the most frequently involved because the choroid has a high rate of blood flow and oxygen supply, making accessibility and survival of the leukemic cells easy.[7] When the choroid is involved, blood flow through the choriocapillaris is disrupted. Consequently, the retinal pigment epithelium continuity is disrupted and subretinal fluid with subsequent retinal detachment is seen. This invasion of the choroid by the leukemic cells can be subtle or severe to the extent of an evident choroidal mass.[4]

Choroidal infiltrates generally present as serous retinal detachment. Detachment is mostly seen in the posterior pole and manifests as a shallow retinal detachment. This common involvement of the choroid is seen most consistently on histopathological examinations. Those choroidal changes are commonly seen in the perivascular area in a patchy or in a more

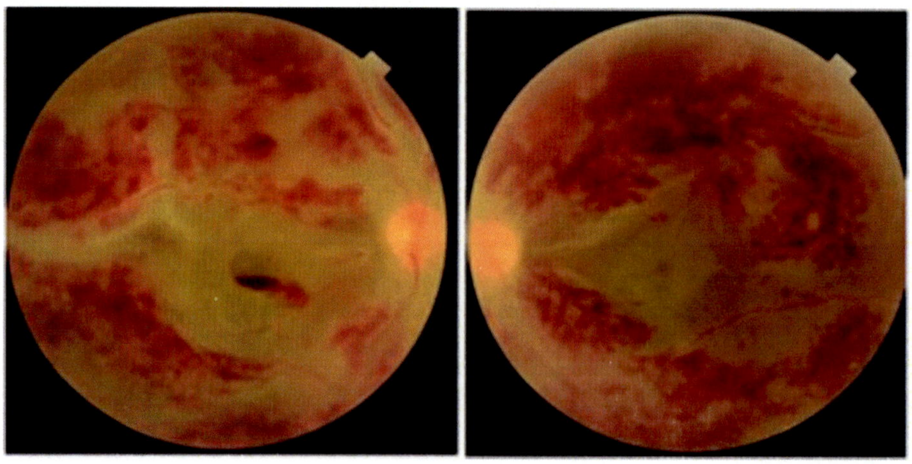

FIGURE 35.1 Initial and 2 weeks after ganciclovir. This 15-year-old male was newly diagnosed with acute lymphoblastic leukemia (ALL) and was just started treatment. He complained of blurry vision a few days after the diagnosis was made. A dilated fundus exam revealed retinal hemorrhages due to severe thrombocytopenia (platelet count 51,000 per microliter) (Fig. 35.1). Despite systemic chemotherapy, vision deteriorated with the retinal hemorrhages covering most of the arcades. Fig. 35.1 shows diffuse superficial perivascular retinal hemorrhages with underlying creamy looking etina suggesting CMM retinitis.

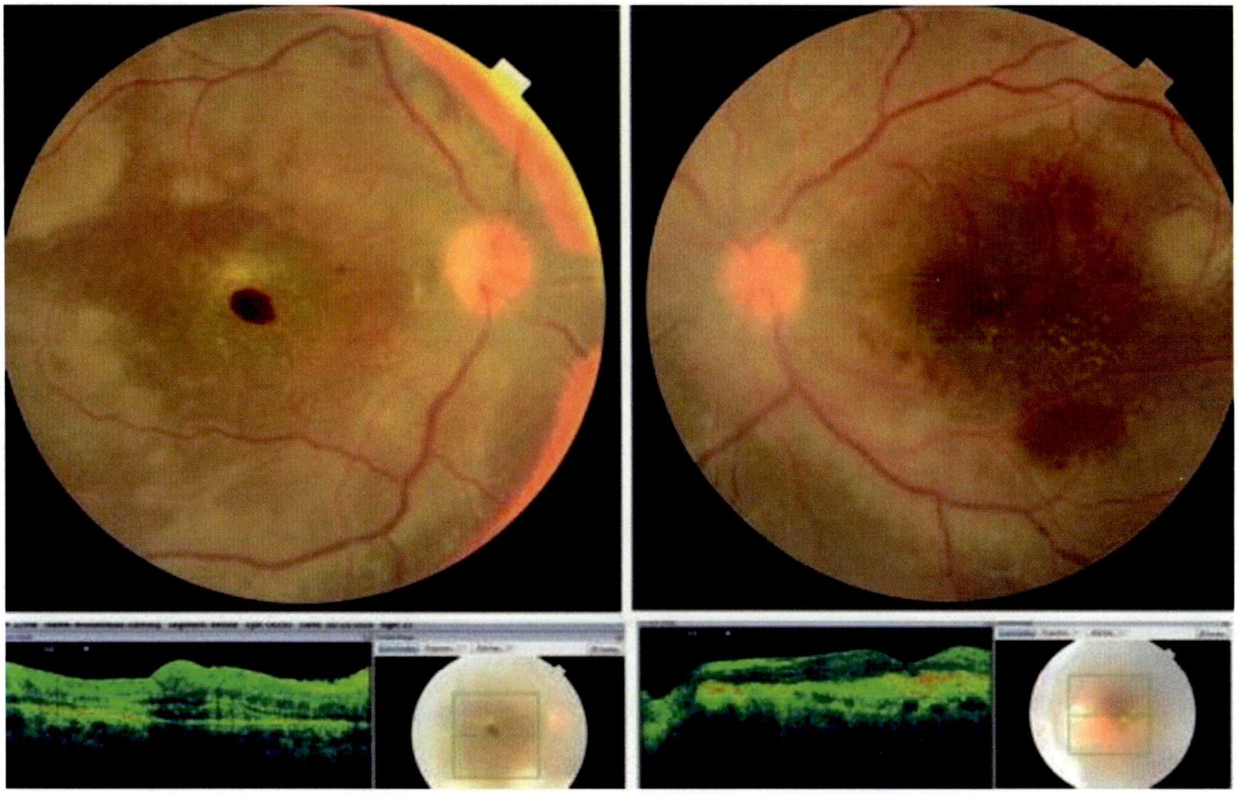

FIGURE 35.2 After 2 weeks of ganciclovir, the hemorrhages resolved with evidence of diffuse retinal infiltrates throughout the posterior pole with localization of the creamy infiltrates to the subretinal space by OCT.

diffuse fashion.[4,8] In addition, the choroid is commonly thickened as seen on the OCT.[4] As for the retinal pigment epithelium (RPE), secondary changes may be seen. Those changes range between atrophy and hypertrophy.[4,8] The photoreceptors might be affected with some loss.[4,8]

Leukemic optic neuropathy

Optic disc swelling is seen in 5%−13% of leukemia, predominantly ALL,[9] and results most often from direct infiltration and less commonly from either retrolaminar infiltration or elevated intracranial pressure.[4] Pathologically, the nerve head shows massive leukemic infiltrates and hemorrhagic necrosis (Fig. 35.3).[10] Elevated intracranial pressure can result from cerebral venous sinus thrombosis (hyperviscosity, brain infiltration, CNS infection secondary to immunosuppression) and from the use of corticosteroids and/or vitamin A in the various treatment protocols. Optic nerve infiltration is an ominous sign of poor survival and can be the site of recurrence of the disease.[8,9]

Vitreous infiltration

As for the vitreous, it is postulated that the internal limiting membrane forms a barrier in front of the leukemic cell infiltrates. Thus, the vitreous is not commonly affected. However, vitreous seeds have been seen in some leukemic cases and it has been theorized that the leukemic cell can invade the vitreous from the optic nerve head.[8]

Secondary retinal manifestations

Posterior pole involvement not related to leukemic invasion, can be noted in around 39% of leukemia patients.[4] Retinal hemorrhages are the most common retinal finding in leukemia (Fig. 35.4). They occur secondary to the combination of thrombocytopenia and anemia, especially hemoglobin levels <5 g/dL and platelet count <50,000 cells/mm^3, like findings in aplastic anemia.[11] The hemorrhages can take several forms such as preretinal, superficial, dot and blot, Roth spots (white

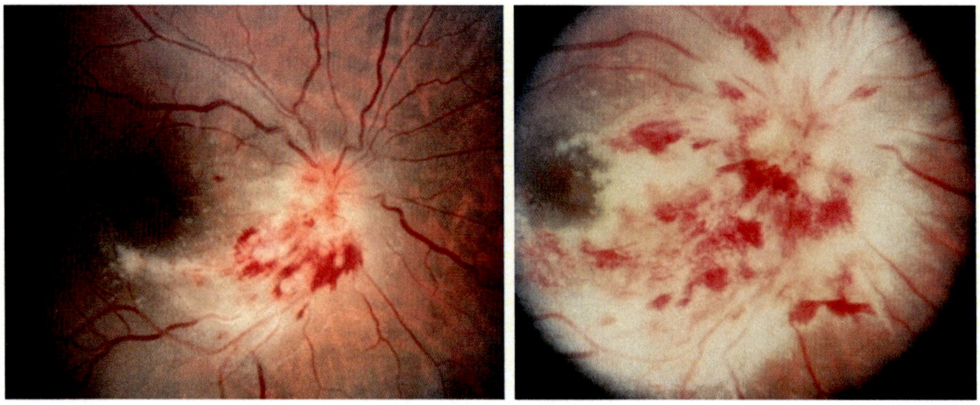

FIGURE 35.3 This 22-year-old man with AML had leukemic optic nerve infiltration that worsened to involve the whole posterior pole in a fulminant fashion. *Courtesy of late Maurice F Rabb, MD, Chicago, USA from the Retina Image Bank with permission.*

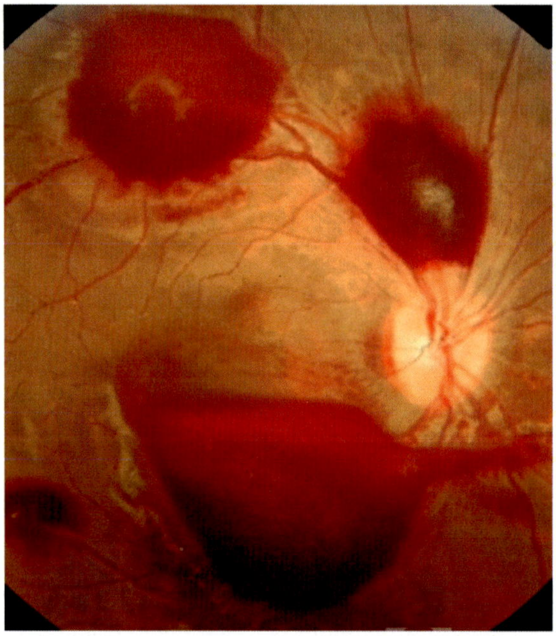

FIGURE 35.4 Multifocal Subhyaloid hemorrhages not blocking central vision in a 21-year-old patient with AML on chemotherapy with low platelet count of 17,000. *Courtesy of Dr Mallika Goyal, Hyderabad, India and Retina Image Bank, American Society of Retina Specialists, with permission.*

centered hemorrhages), subinternal limiting membrane, and subhyaloid. There is the topographic predilection of the hemorrhages to the posterior pole and inner retinal layers. Often, the hemorrhages can also take a perivascular location associated with leukemic cells.[4] The prevalence of retinal hemorrhage in leukemic patients was significantly higher in adults as compared to pediatrics[12] and was associated with worse prognosis of the cases.[5]

A rare case of peripapillary hemorrhages followed subarachnoid hemorrhage in a case of AML a few days after initiation of chemotherapy. This case report by Ghanbarnia et al.[13] had blurry vision and a change in the level of consciousness from left frontal lobe subarachnoid hemorrhage.[13] Another similar interesting finding in cases of acute promyelocytic leukemia) is Terson syndrome consisting of intracranial hemorrhages, and retinal hemorrhages either due to the leukemia or due to the induction chemotherapy.

In chronic leukemia, prolonged hyperviscosity develops with findings of venous stasis retinopathy (venous dilation and tortuosity).[4] Microaneurysms, especially at the retinal periphery, and peripheral proliferative retinopathy will eventually develop.[4] Cotton-wool spots are also commonly seen in the posterior pole.[4]

Opportunistic retinal infections

Ocular infections in immunocompromised patients such as leukemia patients are not uncommon. This is mainly due to reduced immunity either from the primary disease and thus malfunctioning leukocytes or from the secondary side effects of chemotherapy. Viral retinitis as well as ocular toxoplasmosis are commonly seen. Unfortunately, endogenous fungal endophthalmitis has also been reported.

Viral retinitis can be caused by one of the following viruses: cytomegalovirus, herpes simplex virus, varicella-zoster virus, and measles morbillivirus. Cytomegalovirus retinitis is encountered commonly in ALL and presents in three patterns: granular pattern, frosted branch angiitis, and fulminant hemorrhagic pattern. The last two patterns can mimic leukemic retinopathy (Figs. 35.1 and 35.2).

Ocular toxoplasmosis was also reported in CML presenting as multifocal retinochoroiditis and retinal vasculitis.[14]

Chemotherapy-related retinal changes

Ocular tissues are highly sensitive to the systemic anticancer therapies administered in all oncology cases, including leukemia.[15] Thus, it is essential that the primary treating team are familiar with the signs and symptoms of chemotherapy ocular toxicity for early intervention and proper consultation and management. Unfortunately, those toxicities and side effects are inevitable and nonpreventable, and they depend on the class of the agent used.

Chemotherapy-induced ocular side effects can involve any tissue or site in the eye. They can involve the adnexa (eyelids, eyebrows, periorbital tissues), anterior segment (cornea and conjunctiva), or posterior segment (retina and optic nerve head).[15]

The most used lines of chemotherapy include vincristine, doxorubicin, daunorubicin, cytarabine, methotrexate, 6-mercaptopurine, L-asparaginase, and cyclophosphamide.[15,16] Identifying a causative relationship between a specific chemotherapeutic agent and retinal toxicity is not easy as in most cases, multiple therapeutic agents are used.

Vincristine is a natural alkaloid that inhibits microtubule assembly during mitosis halting cell division.[16] This agent has a long half-life and with long treatment, it accumulates in the vascular system leading to vascular attenuation.[16,17] Ischemia of the optic nerve head or retina was reported from vincristine use resulting in visual decline or nyctalopia.

Imaging findings

Optical coherence tomography (OCT) is instrumental in identifying leukemic infiltrates (outer hyper-reflective retinal lesions) and assessing the amount of increase in the thickness of the choroidal layer), retinal hemorrhages (inner hyper-reflective lesions),[4] serous retinal detachment, and other vascular changes.

On OCT, leukemic infiltrates will be seen as outer hyper-reflective lesions with an increase in the thickness of the choroidal layer. When those infiltrates resolve, disruption of the continuity of the retinal layer involved, especially the photoreceptor layer, can be identified.

Decrease in vascular flow, decreased vascular density have been reported using OCT angiography by Cicinelli et al.[4] OCT angiography of leukemic eyes detected universal vascular changes of the posterior pole, that is, lower vessel density of the superficial capillary plexus irrespective of the presence of leukemic retinopathy.[18] Those changes were associated with lower hemoglobin level and lower platelet count. These OCT angiographic findings were reversible following successful chemotherapy.[18]

Fluorescein angiography enables us to identify anomalies of the vascular flow, areas of nonperfusion, foci of neovascularization, and the presence of aneurysms. A specific angiographic pattern seen in ALL is known as the "leopard spot retinopathy"[19]: round lesions with peripheral dark dots. These lesions are at the level of the retinal pigment epithelium (RPE) and are caused by atrophy of the RPE at the center of the spot with fibrosis and hyperplastic changes at the periphery.[20] Although it is not exclusive to leukemic retinopathies, it is common as we mentioned in ALL cases. Most of the reported leopard spot retinopathy cases were young ALL patients in remission but presented with blurry vision.

Treatment

Medical intervention depends on the retinopathy severity. In addition, it requires a multidisciplinary approach with collaboration between Ophthalmology and Hematology/Oncology teams. Thus, the primary target of treatment and management must be to control the primary disease, and to decrease the risk of secondary infections, to reverse the hematological abnormalities and eventually reverse the retinopathy and choroidopathy. Changing the line of chemotherapy and induction of radiotherapy and possible immunotherapy might be needed in those cases, or even hematopoietic stem cell transplant in cases that meet certain criteria.[4]

As early detection of any type of ocular involvement, especially retinopathy, is vital, it is recommended that frequent ophthalmic assessments are done regularly in leukemia patients regardless of the symptoms and more frequently when the patient reports any ocular complaints. A reasonable regimen of ophthalmic exam would be that all patients with leukemia in all age groups should be screened once at presentation and then undergo routine examination every 6 months (or earlier if there are ocular complaints).[21] Thus, the primary team must be aware that visual changes reported by patients shall be taken seriously and the Ophthalmology team shall be informed for earlier and more frequent assessment.[4]

The role of the ophthalmology team does not stop here at frequent follow-ups. Intervention shall be done in some cases that we will be discussing, depending on the ophthalmic anomaly that we are dealing with. When we are referring to mild cases (i.e., retinal hemorrhages; cotton-wool spots; Roth spots; dilated tortuous vessels; retinal vessels sheathing and optic disc hyperemia and edema), spontaneous resolution is expected and no further intervention is warranted or indicated. On the other hand, moderate cases of retinopathy (i.e., proliferative retinopathy; retinal vein occlusion; vitreous hemorrhage) and severe or advanced cases (i.e., optic disc infiltration; exudative retinal detachment; choroidal infiltration and hemorrhage) indicate that both ophthalmological and systemic interventions and treatment are needed to avoid irreversible vision loss.

Proliferative retinopathy is seen in leukemic cases, intervention is no different from proliferation seen in diabetic retinopathies and retinal vein occlusion for instance. The treatment will include panretinal laser photocoagulation with possible intravitreal anti-VEGF agent injection.[4] Similarly, macular edema is treated with the conventional approach to any case of macular edema. Surgical interventions are indicated in cases of vitreous hemorrhage due to proliferative retinopathy, subretinal hemorrhages, and retinal detachment (especially when traction is present).

As for optic neuropathy secondary to leukemic infiltration, the treatment approach of choice is a relevant concern. This is because chemotherapy is not an appropriate choice due to the blood—ocular barrier.[22] Consequently, alternative radiotherapy shall be used to treat the intraorbital portion of the optic nerve that is involved with secondary infiltrates.

Chemotherapy-induced toxic retinopathy and visual changes are inevitable. Discontinuing the insulting agent was reported in some cases to reverse visual loss. However, this is not the case always, and irreversible damage to the retinal layers is sometimes seen and expected regardless of discontinuing the chemotherapeutic agent.[16]

As we have mentioned so far, the current indicated treatment of patients with ocular leukemic involvement includes systemic chemotherapy and biological treatments. Due to inadequate penetration of many drugs to the leukemic foci,[21] radiotherapy might be an option, with the risk of radiation retinopathy or optic neuropathy. Thus, intravitreal injections of chemotherapeutic agents are being investigated. An example of a chemotherapeutic agent that can be injected intravitreally is methotrexate (MTX).[21] The idea behind using MTX for leukemic cases is based on the positive response to this agent in ocular lymphoma.[17,21] This agent is specifically used for leukemic retinopathy. The injections can be repeated, with a standard concentration of 400 mg/0.1 mL and a standard regimen: biweekly for the first month, weekly for the second and third months, and finally monthly for the fourth month.[21] Such regimen of intravitreal MTX is promising with reversal of visual loss and resolution of the ocular infiltrates (retinal, choroidal, optic nerve head and anterior chamber inflammation).[21]

References

1. Owaidhah NA, Khawaji ZY, Alahmadi LA, Badawi AS, Mogharbel GH, Makhdoom ON. Epidemiological trends and clinical characteristics of childhood leukemia in Saudi Arabia: a review. *Cureus*. 2022;14(8):e28178. https://doi.org/10.7759/cureus.28178.
2. Whiteley AE, Price TT, Cantelli G, Sipkins DA. Leukaemia: a model metastatic disease. *Nature Reviews Cancer*. 2021;21(7):461—475. https://doi.org/10.1038/s41568-021-00355-z.
3. Hui VWK, Szeto SKH. Clinical and imaging features of leukemic retinopathy [Working Title] [Preprint]. In: *Leukemia - from Biology to Diagnosis and Treatment*. 2022. https://doi.org/10.5772/intechopen.107649.
4. Adaniya A, Luciano AD, Alvarado-Villacorta R, et al. Serous retinal detachment as a sign of leukemic choroidopathy: a systematic review. *Survey of Ophthalmology*. 2022;67(1):149—167. https://doi.org/10.1016/j.survophthal.2021.04.007.
5. Benvenuto F, Sgroi M, Guillen SS, Ancona D, Fandiño A. Ocular findings in children with acute leukemia at a tertiary care center in South America. *Oman Journal of Ophthalmology*. 2022;15(2):159—162. https://doi.org/10.4103/ojo.ojo_347_20.
6. Mansour AM, Sbeity N. Leukemic retinopathy mimicking Cytomegalic Virus Retinitis ending in total blindness. *The Journal of Middle East and North Africa Sciences*. 2017;3(6):13—16. https://doi.org/10.12816/0037730.
7. Grant MB, Lutty GA. Retinal and choroidal vasculature. *Retina*. 2013:433—446. https://doi.org/10.1016/b978-1-4557-0737-9.00018-7.
8. Sharma T, Grewal J, Gupta S, Murray PI. Ophthalmic manifestations of acute leukaemias: the ophthalmologist's role. *Eye*. 2004;18(7):663—672. https://doi.org/10.1038/sj.eye.6701308.

9. Alrobaian MA, Henderson AD. Neuro-ophthalmic manifestations of acute leukemia. *Journal of Neuro-Ophthalmology*. 2020;41(4). https://doi.org/10.1097/wno.0000000000001071.
10. Brown GC, Shields JA, Augsburger JJ, Serota FT, Koch P. Leukemic optic neuropathy. *International Ophthalmology*. 1981;3(2):111–116. https://doi.org/10.1007/bf00133423.
11. Mansour AM, Lee JW, Yahng SA, et al. Ocular manifestations of idiopathic aplastic anemia: retrospective study and literature review. *Clinical ophthalmology [Auckland, N.Z.]*. 2014;8:777–787. https://doi.org/10.2147/OPTH.S62163.
12. Sayadi J, Gouider D, Allouche R, et al. Ophthalmic manifestations of newly diagnosed acute leukemia patients in a Tunisian cohort. *Clinical Ophthalmology*. 2022;16:3425–3435. https://doi.org/10.2147/opth.s365648.
13. Ghanbarnia MJ, Sedaghat S, Rasoulinejad SA. Leukemic retinopathy presenting as concurrent bilateral subhyaloid hemorrhage and subarachnoid hemorrhage in a patient with acute monocytic leukemia: a case report. *Journal of Medical Case Reports*. 2022;16(1). https://doi.org/10.1186/s13256-022-03700-4.
14. Hazan A, Patel RM, Levinson D, et al. A typical bilateral *Toxoplasma* retinochoroiditis in a bone marrow transplant patient with negative serum titers. *Journal of Ophthalmic Inflammation and Infection*. 2013;3:23. https://doi.org/10.1186/1869-5760-3-23.
15. Omoti AE, Omoti CE. Ocular toxicity of systemic anticancer chemotherapy. *Pharmacy Practice*. 2006;4(2):55–59.
16. Adhikari S, Dongol RM, Hewett Y, Shah BK. Vincristine-induced blindness: a case report and review of literature. *Anticancer Research*. 2014;34(11):6731–6733.
17. Smith J. Role of intravitreal methotrexate in the management of primary central nervous system lymphoma with ocular involvement historical image. *Ophthalmology*. 2002;109(9):1709–1716. https://doi.org/10.1016/s0161-6420[02]01125-9.
18. Cicinelli MV, Mastaglio S, Menean M, et al. Retinal microvascular changes in patients with acute leukemia. *Retina*. 2022;42(9):1762–1771. https://doi.org/10.1097/iae.0000000000003504.
19. Robert MPA, Faure C, Reman O, et al. Leopard spot retinopathy: an early clinical marker of leukaemia recurrence? *Annals of Hematology*. 2008;87:927–929. https://doi.org/10.1007/s00277-008-0510-z.
20. Cohen SY, Massin P, Quentel G. Unilateral, idiopathic leopard-spot lesion of the retinal pigment epithelium. *Archives Ophthalmology*. 2002;120(4):512–516.
21. Vishnevskia-Dai V, Sella King S, Lekach R, Fabian RD, Zloto O. Ocular manifestations of leukemia and results of treatment with Intravitreal Methotrexate. *Scientific Reports*. 2020;10(1). https://doi.org/10.1038/s41598-020-58654-8.
22. Barcelos R, Roisman L, Tochetto LB, Tochetto BB. Retinal changes in patients with acute lymphocytic leukemia [ALL]: discussion of topic from a case report. *Revista Brasileira de Oftalmologia*. 2021;80(1). https://doi.org/10.5935/0034-7280.20210007.

Further reading

1. Leonardy NJ, et al. Analysis of 135 autopsy eyes for ocular involvement in leukemia. *American Journal of Ophthalmology*. 1990;109(4):436–444. https://doi.org/10.1016/s0002-9394[14]74610-0.
2. Balubaid MM, Alqahtani AS. A case of acute promyelocytic leukemia with retinal hemorrhages beneath internal limiting membrane during clinical remission. *Cureus*. 2021;13(2):e13387. https://doi.org/10.7759/cureus.13387.
3. Rodrigues GR, Mendonca TM. Retinal hemorrhages in leukemia. *New England Journal of Medicine*. 2022;386(19). https://doi.org/10.1056/nejmicm2116866.
4. Mejia-Vergara AJ, et al. Papilledema and intracranial hypertension in leukemia: case series and review. *Canadian Journal of Ophthalmology*. 2022;57(2). https://doi.org/10.1016/j.jcjo.2021.06.022.
5. Steinhorst UH, et al. Ocular toxicity of daunomycin: effects of subdivided doses on the rabbit retina after vitreous gas compression. *Graefe's Archive for Clinical and Experimental Ophthalmology*. 1993;231(10):591–594. https://doi.org/10.1007/bf00936524.
6. Voest EE, et al. Phase I study using deferoxamine and iron sorbitol citrate in an attempt to modulate the iron status of tumor cells to enhance doxorubicin activity. *Cancer Chemotherapy and Pharmacology*. 1993;31(5):357–362. https://doi.org/10.1007/bf00686148.
7. Szeto SKH, et al. Massive retinal infiltrates as the presenting sign of chronic myeloid leukemia: clinical and imaging features of Leukemic Retinopathy. *European Journal of Ophthalmology*. 2021;33(1). https://doi.org/10.1177/11206721211048792.

Chapter 36

Purstcher retinopathy

Jonathan T. Caranfa and Caroline R. Baumal
Tufts Medical Center/New England Eye Center, Boston, MA, United States

Overview:

- Purtscher retinopathy is a rare, retinal microvasculopathy typically affecting both eyes with posterior pole retinal findings.
- Initially described in the setting of trauma (cranial, crush injuries, chest compression, long bone fractures) but has been associated with other nontraumatic systemic disorders (acute pancreatitis, renal failure, connective tissue disease, malignancy, coagulopathies, pregnancy).
- Fundoscopic findings include cotton wool spots, superficial retinal hemorrhages, and Purtscher flecken (polygonal areas of retinal whitening with clearing surrounding adjacent retinal vessels).
- Visual impairment is variable with some degree of spontaneous improvement in many cases.
- High-dose corticosteroids may hasten visual recovery and should be considered on a case-by-case basis: however, substantial evidence is lacking to support widespread indiscriminate use.

Introduction

Purtscher retinopathy, also known as traumatic retinal angiopathy, lymphorrhagia retinae, or retinal teletraumatism is an uncommon, severe occlusive retinal microvasculopathy that occurs after an inciting nonocular traumatic event. It was first described in 1910 by Otmar Purtscher after a fundoscopic examination of a middle-aged man who suffered cranial trauma and vision loss revealed multiple areas of retinal whitening and hemorrhage around the optic nerve and in posterior poles bilaterally.[1] Purtscher retinopathy has since been described after numerous causes of indirect, nonocular trauma, often manifesting as bilateral painless vision loss. This condition is classically associated with compressive trauma such as crush injuries and thoracic compression: however, cases of Purtscher retinopathy following head injury, long bone fracture(s), and shaken baby syndrome have been reported. The term Purtscher-like retinopathy is used when the pathognomonic fundoscopic findings of Purtscher retinopathy are associated with a nontraumatic etiology. Purtscher-like retinopathy has been reported from systemic disorders such as acute pancreatitis, connective tissue disease (systemic lupus erythematosus (SLE), scleroderma, dermatomyositis), coagulopathies (thrombotic thrombocytopenic purpura (TTP), hemolytic uremic syndrome (HUS)), chronic renal failure, cryoglobulinemia associated with hepatitis C, embolic syndromes (fat, air, amniotic fluid), pregnancy (sepsis postcesarean, gestosis, preeclampsia, hemolysis, elevated liver enzymes and low platelets (HELLP syndrome), malignancy (pancreatic and lung adenocarcinoma, multiple myeloma) and following retrobulbar anesthesia (Table 36.1).[2] As these entities share a common pathogenesis and clinical features, the remainder of this chapter will collectively use Purtscher retinopathy to describe both traumatic and nontraumatic causes. Despite the many potential underlying etiologies of the disorder, Purtscher retinopathy remains a rare entity with a reported 0.24 cases per million per year. The exact incidence may be higher, as patients with trauma or severe illnesses may not be able to report or lack symptoms altogether.[2,3] The two most common causes of Purtscher and Purtscher-like retinopathy are chest compression and long bone fracture(s) followed by acute pancreatitis.

Clinical features

Classically, patients with Purtscher retinopathy present with painless vision loss in one or both eyes. When occurring concomitantly with a traumatic event, visual decline can be at the time of presentation or delayed by 24—48 h. Reduction in

Retinal and Choroidal Vascular Diseases of the Eye. https://doi.org/10.1016/B978-0-443-15583-3.00036-6

TABLE 36.1 Causes of Purtscher and Purtscher-like retinopathy.

Traumatic (Purtscher retinopathy)	Nontraumatic (Purtscher-like retinopathy)
Head trauma	Acute pancreatitis
Chest/thoracic compression	Chronic renal failure
Long bone fracture(s)	Malignancy • Pancreas • Lung • Multiple myeloma
Crush injury	Connective tissue disorders • Systemic lupus erythematosus (SLE) • Scleroderma • Dermatomyositis
Shaken baby syndrome	Cryoglobulinemia in hepatitis C
Valsalva	Coagulopathies • Thrombotic thrombocytopenic purpura (TTP) • Hemolytic uremic syndrome (HUS)
	Embolism • Fat • Air • Amniotic fluid
	Pregnancy-related • Sepsis postcesarean • Gestosis • Preeclampsia • Hemolytic anemia, elevated liver enzymes, low platelets (HELLP syndrome)
	Retrobulbar anesthesia

visual acuity is variable from minimal or no impairment to counting fingers which is often accompanied by central visual field loss in the form of central, paracentral, or arcuate scotoma with the peripheral visual field being preserved. Acute fundoscopic findings include cotton wool spots (93% of cases), pre and intraretinal hemorrhages (65%), and Purtscher flecken (63%) (Fig. 36.1). Optic nerve edema, macular edema, and retinal arterial occlusion have also been observed. In patients with a history of Purtscher retinopathy, chronic posterior pole changes such as retinal pigment epithelial mottling, optic disc pallor, and attenuation/sheathing of the retinal vessels can be seen.[4]

The pathognomonic finding of Purtscher flecken consists of discrete areas of polygonal superficial retinal whitening with clear demarcation between the affected retina and adjacent retinal vessels. The flecken are variable in size ranging from one quarter to several disc diameters. If the retinal whitening surrounds the fovea, a pseudocherry red spot may be noted. Cotton wool spots are often confluent, large, and numerous, while pre- and intraretinal hemorrhages are more limited in number ($n = 1-10$).[4] These funduscopic findings are localized in the posterior pole, namely surrounding the optic disc and macula. Bilateral disease is estimated to occur in up to 60% of all cases and in the majority of cases, close fundoscopic examination will yield subtle findings in the fellow eye.[2,4–8]

Imaging findings

Fluorescein angiography

Fluorescein angiography (FA) is a useful tool for the diagnosis of Purtscher retinopathy, demonstrating blockage of choroidal fluorescence by the retinal ischemic whitening or hemorrhage in the early/acute phase of the disorder. In mild cases, late leakage from the retinal arterioles, capillaries, and venules may be present. In severe forms, occluded retinal arterioles, macular ischemia, capillary nonperfusion, and optic nerve leakage can be observed.[9,10]

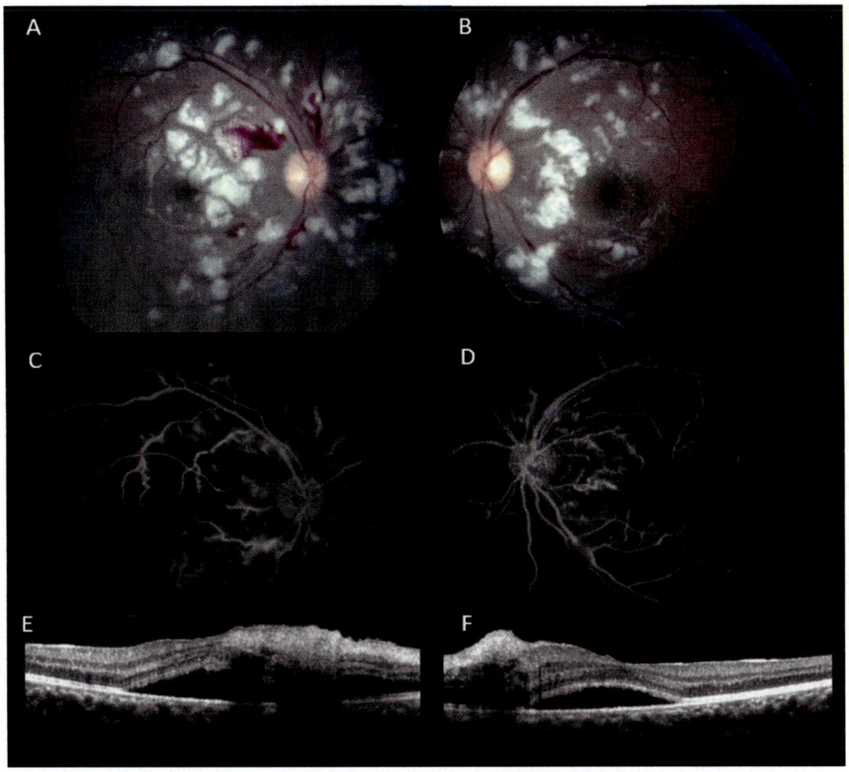

FIGURE 36.1 16-year-old female with Purtscher-like retinopathy secondary to systemic lupus erythematosus presented with acute loss of vision. Visual acuity measured counting fingers right eye and 20/200 left eye. (A, B) color fundus photograph ((A) right eye, (B) left eye, Topcon) demonstrates numerous peripapillary and perifoveal cotton-wool spots and superficial retinal hemorrhages in the nerve fiber layer. (C, D) Fluorescein angiography ((A) right eye 4.3 min, (B) left eye 2.6 min, Topcon) reveals capillary nonperfusion in the distribution of the cotton-wool spots, blocked fluorescence from retinal hemorrhages, and staining of the perifoveal veins. (E, F) optical coherence tomography ((E) right eye, (F) left eye OCT zeiss) reveals subfoveal fluid, hyperreflectivity of the inner retina, and retinal thickening nasal to the fovea.

Indocyanine green angiography

Indocyanine green angiography (ICG) findings that have been reported include choroidal hypocyanescence in the paramacular area, which can persist up to 5 months after onset. This implies that the choroidal vasculature may play a role in the underlying pathophysiology and contribute to poor visual outcomes in severe cases of Purtscher retinopathy.[9]

Electroretinography

Electroretinography (ERG) data in Purtscher retinopathy is limited. Haq and colleagues describe a case of a 20-year-old male with Purtscher retinopathy secondary to acute pancreatitis. Multifocal ERG (mfERG) analysis demonstrated decreased response amplitudes in both A and B waves in the areas of the affected retina on initial evaluation. Despite the resolution of clinical retinal findings and a significant improvement in visual acuity at the 6 month follow-up, the mfERG lacked gross improvement. However, when comparing the ratio of responses for the central 2-degree area to an area extending from 2 to 6 degrees (used to reduce test—retest variability when comparing a series of consecutive mfERG responses), a progressive increase in the ratio of amplitudes was seen. Initial involvement and subsequent slight recovery of both the A and B wave amplitudes implies that patients with Purtscher retinopathy sustain a functional loss in both the inner and outer retinal layers consistent with the findings of other imagining and histopathological evaluations.[11]

Optical coherence tomography and angiography

In the acute phase, optical coherence tomography (OCT) imagining shows inner retinal hyperreflectivity, specifically in the inner plexiform (IPL), inner nuclear (INL), and outer plexiform layers (OPL) with thickening of the RNFL and ganglion cells layers. In rare cases, his hyperreflectivity has been observed to extend to the outer nuclear layer (ONL).[12] OCT findings of paracentral acute middle maculopathy (PAMM), characterized by hyperreflectivity of the INL, connote involvement of the intermediate and deep retinal capillary plexuses.[13,14] Macular edema with the presence of intra and subretinal fluid and disruption of the subfoveal outer retinal segments may be seen during the initial phases of the disorder. In late/chronic cases of Purtscher retinopathy, the retinal hyperreflectivity and edema on OCT as well as the cotton-wool spots and retinal hemorrhages seen in the acute phase will have resolved. Inner and outer retinal atrophy along with photoreceptor loss become the chronic OCT predominant features.[12,15]

Studies reporting optical coherence tomography and angiography (OCTA) findings confirm capillary nonperfusion at the level of superficial and deep capillary plexuses and these changes may persist even after the resolution of other retinal findings.[15] These areas of ischemia correlate to the location of and extension beyond the clinically visible cotton wool spots.[16] While outer retinal and choriocapillaris perfusion alterations are not typically not observed on OCTA, these data may be limited given the severity of systemic disease at presentation making ocular imaging difficult to attain.[17] Additionally, given a patient's history and classic clinical picture, OCT(A) imagining is not needed to make the diagnosis, which further limits the availability of OCTA data.

Visual field

Visual field (VF) testing demonstrates central, paracentral, or arcuate scotomas with preservation of the peripheral visual field.[4] In a recent analysis, Miguel et al. found that 93% of Purtscher retinopathy patients suffered from a centrally occurring visual obscuration that remained stable or slightly improved over time.[13]

Additional diagnostic evaluation

The diagnosis of Purtscher retinopathy is a clinical one with a history of present illness (traumatic event or known systemic disease) in conjunction with pathognomonic fundoscopic findings being sufficient to make the diagnosis. However, in cases where there is evidence of retinal pathology without a known inciting event, additional investigation may be warranted. Laboratory analysis including serum amylase or lipase with appropriate use of abdominal computer tomography (CT) imaging can help diagnose pancreatitis. Additionally, serum markers for connective tissue disease such as complete blood count (CBC), antinuclear antibodies (ANA), complement C5 protein, antidouble-stranded DNA, antiphospholipid antibody, rheumatoid factor, creatine phosphokinase, and myoglobin may help aid in unmasking an underlying systemic disease causing Purtscher retinopathy. Finally, a detailed review of systems in addition to CT or positron emission tomography (PET) imaging may prove useful if underlying malignancy is suspected.

Pathogenesis

The characteristic finding of Purtscher flecken in Purtscher retinopathy is described as superficial retinal whitening with a clear demarcation line between the affected retina and adjacent retinal arteries and veins. The clearing of approximately 50 μm on either side of the retinal vessels corresponds to the capillary free zone suggesting the primary pathological process responsible for Purtscher retinopathy involves the precapillary arterioles.[18] In contrast, proximal occlusion by larger embolic particles would produce confluent retinal whitening as often seen following branch or central retinal artery occlusions. Distal capillary occlusions from smaller emboli would be expected to produce cotton wool spots which has been shown experimentally in animal models.[19] In 1992, Behrens-Baumann et al. induced retinal changes characteristic of Purtscher retinopathy after injection of fibrin clots ranging from 0.15 to 1.0 mm into the ophthalmic artery of a pig demonstrating that intermediate-sized emboli are required to produce lesions resembling Purtscher flecken.[20]

The characteristic appearance of Purtscher retinopathy may be related to the unique anatomy of the retina and more specifically the vascular anatomy of the peripapillary region and macula. Throughout most of the retina, several capillary layers exist (superficial, intermediate, and deep) providing adequate perfusion with anastomotic potential in the setting of ischemia. However, the peripapillary capillaries (which extend 2-disc diameters nasal to the optic disc and 4-disc diameters temporal to the disc in the same distribution of Purtscher flecken) have fewer arteriolar feeds and anastomoses compared with the capillary beds of other parts of the retina, which may make them more susceptible to embolic occlusion.[18,21,22]

The underlying pathogenesis of Purtscher retinopathy has been widely debated and there is likely some variability depending on cause. In the original case report, Purtscher theorized that his observed retinal findings were a result of extravasation of lymphatic matter from the retinal vessels as a consequence of the sudden increase in intracranial pressure following a traumatic head injury.[1] In contrast, capillary endothelial damage by oxygen free radicals as well as inner retinal layer rheological events leading to vascular endothelial dysregulation have been described in recent case reports.[23,24] Similar to Purtscher, several authors believe a sudden increase in intracranial pressure is the inciting event; however, it is not the lymph extravasation that causes the pathology but instead precapillary occlusion of the retinal arteries leading to the observed findings in Purtscher retinopathy.

In patients with thoracic cavity compression (or Valsalva), the resultant reduced venous return is theorized to cause an acute expansion of the retinal veins and therefore fundoscopic findings such as cotton wool spots, retinal hemorrhages, and Purtscher flecken.[25] In acute pancreatitis, lipase and/or free fatty acid–induced vasculitis leading to thrombosis and vascular occlusion has been postulated.[26,27] Moreover, thrombotic microangiopathy, a clinical feature seen in both TTP and HUS, caused by a reduction in the activity of von Willebrand factor cleaving protease ADAMTS-13 leading to a hypercoagulable state may play a key role in the pathogenesis of Purtscher retinopathy.[23,28]

Currently, the most well-accepted theory of the underlying pathogenesis of Purtscher retinopathy is developed microembolization resulting in precapillary occlusion and microvascular infarcts of the inner retina. Potential sources of microemboli include air (after chest compression), fat (pancreatic enzymatic degradation of omental fat or release following long bone injury), amniotic fluid (during or after childbirth), leukocyte aggregates, platelets, or fibrin. Moreover, medication particles can be the embolic source as seen in Purtscher retinopathy following retrobulbar anesthesia.[7]

Given the presence of Purtscher retinopathy in many proinflammatory conditions, complement activation–induced leukocyte aggregation is believed to play an essential role in the development of this disorder. The leukocyte aggregates have been measured up to 50 μm in size, which is sufficient to induce retinal arteriolar occlusion.[29] Lai et al. attempted to mimic Purtscher retinopathy in cats via complement activation–induced small retinal arteriolar occlusion via leukocyte aggregates. While their experimental model demonstrated the ability to form leukocyte aggregates large enough to induce transient retinal vascular obstructions, they failed to show Purtscher-like ischemic retinal infarcts indicating that other factors in addition to complement activation are essential in the pathophysiology of Purtscher retinopathy.[30]

Fat emboli have been reported in cases of long bone fracture, orthopedic surgery, crush injuries, and in the setting of pancreatitis. Several authors have proposed these emboli make their way into the venous circulation via arteriovenous pulmonary shunts, a patent foramen ovale, or persistent ductus arteriosus ultimately causing retinal occlusions. However, fat emboli are generally only 5–10 μm in size, which would be too small to cause occlusion of the 45 μm diameter precapillary arterioles but instead induce infarction in the smaller 5 μm retinal capillaries leading to cotton wool spots and retinal hemorrhage, but not Purtscher flecken.[18,26] This is in contradistinction to fat embolic syndrome where the release of fat particles into the systemic circulation triggers other systemic effects including saponification of lipid stones via an elevation in plasma lipase, vasculitis, hypovolemia, complement activation, and a hypercoagulable state.[31] It is this constellation of physiological effects that may therefore contribute to the development of Purtscher retinopathy as opposed to the emboli themselves.

While the exact mechanism is not clear and may be cause dependent, the presence of multifocal lesions in the posterior pole, bilateral involvement, and retinal vascular occlusion with corroboratory fluorescein and OCTA imaging support that microvascular occlusion is the most likely cause of Purtscher retinopathy. There may not be a single unifying inciting factor that explains all of the causes of Purtscher retinopathy; however small and intermediate-sized emboli, from multiple origins appear capable of producing the pathognomonic retinal findings.

Histopathology

Histopathological evidence of Purtscher retinopathy is limited. Kincaid et al. noted focal areas of retinal edema, cystoid degeneration, and disruption of the inner retinal layers with abrupt transition to normal retina on histopathology from a retina specimen of a patient with pancreatitis.[32] Retinal arterioles and choroidal vessels demonstrated occlusion with fibrin-like material and electron microscopy showed small arterioles with narrowed lumens containing proteinaceous material consistent with recanalized thrombi. Some areas of photoreceptors were lost; however, the underlying retinal pigment epithelium and choroid were unaffected. In 1993 Hollo and Bobek performed histology on an eye with Purtscher retinopathy secondary to acute necrotizing pancreatitis and noted an absence of retinal vascular occlusions suggesting that vascular embolization is not solely responsible for the retinal findings in Purtscher retinopathy.[33]

Management

Due to the rarity of this disorder and the heterogeneity of underlying inciting events, there are no randomized controlled trials to guide treatment, with retrospective evaluation and published case studies driving therapeutic decisions. Prompt diagnosis of Purtscher or Purtscher-like retinopathy may help to narrow the differential diagnosis of the underlying etiology, with subsequent treatment of the inciting condition as the first major consideration.

Observation is the most often utilized strategy, as many patients will demonstrate spontaneous improvement in visual acuity.[4,13]

There is limited evidence supporting the role of corticosteroids in the treatment of Purtscher retinopathy with two case reports demonstrating favorable visual improvement and outcomes following intravenous (IV) methylprednisolone. Wang et al. described a 17-year-old female with Purtscher retinopathy 2 days following a traffic accident. Visual acuity improvement from baseline counting fingers to 20/40 was noted 2 weeks later following corticosteroid therapy (methylprednisolone 1 gram (g) daily for 3 days followed by an oral steroid tapered over 3 weeks).[10] Vision subsequently returned to baseline at her 10 month follow-up. Atabay and colleagues reported a 25-year-old male with Purtscher retinopathy following a motor vehicle accident. The patient was started on oral prednisone 60 mg/day by an outside provider and upon presentation to the authors 3 weeks later, visual acuity was measured at 20/800. The patient was promptly started on IV methylprednisolone 1 g daily for three doses and was noted to have rapid vision improvement to 20/70 after 3 days.[34] These authors hypothesized the beneficial effects of corticosteroids may include the stabilization of damaged neuronal membranes and microvascular channels, which allows for partial recovery of retinal nerve fiber layers. Additionally, corticosteroids inhibit tissue damaging complement activation and granulocyte aggregation while improving retinal anatomy.

While incidental case reports hypothesize a visual benefit from corticosteroids for eyes with Purtscher retinopathy, the only cumulative analysis to date did not support an overall benefit for this treatment. In a review from Miguel et al., 68 cases of Purtscher retinopathy were identified between 1980 and 2010, from which 20 had received corticosteroid treatment.[13] The treatment regimens varied greatly both in type (prednisone, methylprednisolone, dexamethasone), dose, and administration (oral, intravenous). Despite the heterogeneity of the therapeutic approach, the authors concluded that there was no statistically significant difference between visual acuity improvement for patients treated with corticosteroids at 1, 2, or 6 months compared with observation. To better understand the role of corticosteroids in Purtscher retinopathy, an analysis with homogeneous treatment regimens would be beneficial.

Other treatment modalities have been reported including the use of nonsteroidal antiinflammatory drugs (NSAIDs), specifically indomethacin, bevacizumab in the setting of macular edema and eculizumab (monoclonal antibody directed against complement C5) in a patient with HUS, thrombotic microangiopathy, and Purtscher retinopathy.[35–37] While these cases reported favorable outcomes with therapy directed at the underlying cause, evidence for these therapeutic approaches is limited and should only be considered on a case-by-case basis.

Prognosis and outcomes

Given the rarity of this disorder, heterogeneity in the underlying cause of Purtscher retinopathy, and variable initial presentation, the prognosis is variable but many patients show some degree of spontaneous visual improvement with subtle long-term effects.

In 2007, Agrawal and McKibbin published a prospective observational study identifying 15 cases (24 eyes) from April 2004 and March 2005 of Purtscher retinopathy from various underlying causes. Of the reported cases, 26% had resolution of retinal findings (cotton wool spots, Purtscher flecken) within 1 month of presentation and within 6 months all eyes showed resolution of clinical fundoscopic pathology. Without treatment, 50% of eyes demonstrated visual improvement of two Snellen lines and 23% showed improvement by four Snellen lines at final follow-up. There was only 1 out of 24 eyes that demonstrated a final visual acuity worse than on presentation. The authors noted a nonsignificant trend where early resolution of acute retinal changes was associated with better initial and final visual recovery.[2]

In a larger retrospective analysis, Miguel et al. identified 68 cases of Purtscher retinopathy with a mean presenting visual acuity <20/200 (range: 20/20 to light perception). The authors found a mean final visual acuity improvement of 0.91 logMAR (SD = 0.86) with the greatest improvement in visual outcomes noted in patients diagnosed with acute pancreatitis and trauma.[13]

Overall, prognosis and visual recovery are closely related to the extent and location of the retinal ischemia, amount of fovea and optic nerve involvement, and underlying cause. Additional prognostic factors reported elsewhere include optic disc swelling or leakage, choroidal hypoperfusion, and involvement of the outer retina.[9,10]

Summary

Purtscher and Purtscher-like retinopathy share a common underlying clinical appearance and potential pathology but differ by etiology with the former due to a nonocular traumatic event. Diagnosis is based on history and clinical features, where imaging may be helpful to confirm the diagnosis and understand the pathogenesis. Visual disability ranges from minimal to severe impairment with the majority of patients experiencing some level of visual recovery without intervention. High-dose corticosteroids have demonstrated efficacy in select cases of the disorder; however, lack of substantial evidence limits widespread indiscriminate use. In cases where a specific systemic etiology is known, treatment should always be directed at the underlying.

References

1. Purtscher O. Noch unbekannte befunde nach schadeltrauma. *Bericht Deutsche Ophthalmologische Gesellschaft.* 1910;36:294−301.
2. Miguel AI, Henriques F, Azevedo LF, Loureiro AJ, Maberley DA. Systematic review of Purtscher's and Purtscher-like retinopathies. *Eye.* 2013;27(1):1−13.
3. Agrawal A, McKibbin M. Purtscher's retinopathy: epidemiology, clinical features and outcome. *The British Journal of Ophthalmology.* 2007;91(11):1456−1459.
4. Nor-Masniwati S, Azhany Y, Zunaina E. Purtscher-like retinopathy following valsalva maneuver effect: case report. *Journal of Medical Case Reports.* 2011;5.
5. Agrawal A, McKibbin MA. Purtscher's and Purtscher-like retinopathies: a review. *Survey of Ophthalmology.* 2006;51(2):129−136.
6. Burton TC. Unilateral Purtscher's retinopathy. *Ophthalmology.* 1980;87(11):1096−1105.
7. Kocak N, Kaynak S, Kaynak T, Oner HF, Cingil G. Unilateral Purtscher-like retinopathy after weight-lifting. *European Journal of Ophthalmology.* 2003;13(4):395−397.
8. Lemagne JM, Michiels X, van Causenbroeck S, Snyers B. Purtscher-like retinopathy after retrobulbar anesthesia. *Ophthalmology.* 1990;97(7):859−861.
9. Wilkinson WS, Morgan CM, Baruh E, Gitter KA. Retinal and choroidal vascular occlusion secondary to corticosteroid embolisation. *The British Journal of Ophthalmology.* 1989;73(1):32−34.
10. Holló G. Frequency of Purtscher's retinopathy. *The British Journal of Ophthalmology.* 2008;92(8):1156.
11. Wang AG, Yen MY, Liu JH. Pathogenesis and neuroprotective treatment in Purtscher's retinopathy. *Japanese Journal of Ophthalmology.* 1988;42(4):318−322.
12. Harrison TJ, Abbasi CO, Khraishi TA. Purtscher retinopathy: an alternative etiology supported by computer fluid dynamic simulations. *Investigative Ophthalmology and Visual Science.* 2011;52(11):8102−8107.
13. Yaylali SA, Bromand N, Kilic G. A new OCT finding in Purtscher-like retinopathy. *Ophthalmic Surgery Lasers and Imaging Retina.* 2018;49(1):64−69.
14. Haq F, Vajaranant TS, Szlyk JP, Pulido JS. Sequential multifocal electroretinogram findings in a case of Purtscher-like retinopathy. *American Journal of Ophthalmology.* 2002;134(1):125−128.
15. Kumar V, Tomar A. Optical coherence tomography angiography imaging of Purtscher retinopathy. *Canadian Journal of Ophthalmology.* 2017;52(6):e201−e203.
16. Gomez-Ulla F, Fente B, Torreiro MG, Salorio MS, Gonzalez F. Choroidal vascular abnormality in Purtscher's retinopathy shown by indocyanine green angiography. *American Journal of Ophthalmology.* 1996;122(2):261−263.
17. Vezzola D, Allegrini D, Romano MR, et al. Optical coherence tomography angiography in Purtscher-like retinopathy associated with dermatomyositis: a case report. *Journal of Medical Case Reports.* 2019;13(1):1−6.
18. Michaelson IC, Campbell AC. The anatomy of the finer retinal vessels. *Transactions of the Ophthalmological Societies of the United Kingdom.* 1940;60:71−111.
19. Ashton N, Henkind P. Experimental occlusion of retinal arterioles (using graded glass Ballotini). *The British Journal of Ophthalmology.* 1965;49(5):225−234.
20. Behrens-Baumann W, Scheurer G, Schroer H. Pathogenesis of Purtscher's retinopathy - an experimental study. *Graefe's Archive for Clinical and Experimental Ophthalmology.* 1992;230(3):286−291.
21. Henkind P. Radial peripapillary capillaries of the retina. I. Anatomy: human and comparative. *The British Journal of Ophthalmology.* 1967;51(2):115−123.
22. Dollery CT, Henkind P, Paterson JW, Ramalho PS, Hill DW. I. ophthalmoscopic and circulatory changes in focal retinal Ischaemia. *The British Journal of Ophthalmology.* 1966;50(6):285−324.
23. Chapman K, Seldon M, Richards R. Thrombotic microangiopathies, thrombotic thrombocytopenic purpura, and ADAMTS-13. *Seminars in Thrombosis and Hemostasis.* 2012;38(1):47−54.
24. Stewart MW, Brazis PW, Guier CP, Thota SH, Wilson SD. Purtscher-like retinopathy in a patient with HELLP syndrome. *American Journal of Ophthalmology.* 2007;143(5):886−887.
25. Gil P, Raimundo M, Marques JP, Póvoa J, Silva R. Optical coherence tomography angiography characterization of acute and late stage Purtscher retinopathy. *European Journal of Ophthalmology.* 2018;28(4):NP1−NP6.
26. Chuang EL, Miller FS, Kalina RE. Retinal lesions following long bone fractures. *Ophthalmology.* 1985;92(3):370−374.
27. Roden D, Fitzpatrick G, O'Donoghue H, Phelan D. Purtscher's retinopathy and fat embolism. *The British Journal of Ophthalmology.* 1989;73(8):677−679.
28. Shapiro I, Jacob HS. Leukoembolization in ocular vascular occlusion. *Annals of Ophthalmology.* 1982;14(1):60−62.
29. Kelly, Hc. Traumatic chorioretinopathies. *Retina.* 2001;2:1815.

30. Chen X, Rahimy E, Sergott RC, et al. Spectrum of retinal vascular diseases associated with paracentral acute middle maculopathy. *American Journal of Ophthalmology*. 2015;160(1):26–34.
31. Mellor A, Soni N. Fat embolism. *Anaesthesia*. 2001;56(2):145–154.
32. Lai JC, Johnson MW, Martonyi CL, Till GO. Complement-induced retinal arteriolar occlusions in the cat. *Retina*. 1997;17(3):239–246.
33. Holló G, Bobek I. Clinicopathology of a case with retinopathy of pancreatitis. *Acta Ophthalmologica*. 1993;71(3):422–425.
34. Atabay C, Kansu T, Nurlu G. Late visual recovery after intravenous methylprednisolone treatment of Purtscher's retinopathy. *Annals of Ophthalmology*. 1993;25(9):330–333.
35. Álvarez S, Camps S, Alvarez A, García O. Optic coherence tomography angiography follow-up in a case of Purtscher-like retinopathy due to atypical hemolytic uremic syndrome. *European Journal of Ophthalmology*. 2020;30(3):NP14–NP17.
36. Kincaid MC, Green WR, Knox DL, Mohler C. A clinicopathological case report of retinopathy of pancreatitis. *The British Journal of Ophthalmology*. 1982;66(4):219–226.
37. Nesmith BL, Bitar MS, Schaal S. The anatomical and functional benefit of bevacizumab in the treatment of macular edema associated with Purtscher-like retinopathy. *Eye*. 2014;28(8):1038–1040.

Chapter 37

Radiation retinopathy

Thibaud Mathis[1,2], Laurent Kodjikian[1,2] and Juliette Thariat[3]
[1]Service d'Ophtalmologie, Hôpital Universitaire de la Croix-Rousse, Hospices Civils de Lyon, Lyon, France; [2]Laboratoire UMR-CNRS 5510 MATEIS, INSA, Université Lyon 1, Villeurbanne, France; [3]Laboratoire de Physique Corpusculaire/IN2P3-CNRS UMR 6534 — ARCHADE, Unicaen — Université de Normandie, Caen, France

Overview bullets

- Radiation retinopathy is one of the main causes of irreversible decreases in visual acuity following ocular and periocular irradiation.
- Noninvasive imaging techniques such as OCT-angiography are increasingly frequently used to help detect radiation retinopathy.
- Patient management is based on laser photocoagulation of the areas of peripheral retinal ischemia and anti-VEGF intravitreal injections or dexamethasone implants for macular edema.
- The main means of prevention are using targeted radiotherapy techniques to avoid any unnecessary irradiation of the retina and administering prophylactic anti-VEGF injections immediately following radiotherapy.

Introduction

Radiation retinopathy describes all retinal lesions occurring when the ocular globe is exposed to external radiation. It usually develops following irradiation of the eye or structures in the periocular region such as the orbit, the ear nose throat (ENT) region, or the brain. Despite the precautions taken to limit the dose and the surface area of the retina affected by the radiation, in some instances its irradiation is unavoidable. This is the case when treating uveal melanoma, for example, which by definition is located near the retina.[1] Despite the use of targeted radiotherapy techniques such as proton beam therapy, brachytherapy, or stereotactic radiotherapy, radiation retinopathy is still one of the most common complications resulting from ocular irradiation and the main risk factor for decreased visual acuity secondary to radiotherapy.[2] Moreover, abnormal neovascularization can occur following retinal ischemia secondary to radiation retinopathy[3] and lead to neovascular glaucoma which ends in secondary enucleation in a nonnegligeable number of cases.[4] Radiation maculopathy is the macular equivalent of radiation retinopathy. It can be expressed in an ischemic form, similar to radiation retinopathy as described above, and/or an edematous form, leading to an increase in macular retinal thickness explained by the edema. Generally speaking, the term radiation maculopathy refers to the latter form.

Although radiation retinopathy has been described for over a century, our understanding of the disease is still incomplete, unlike diabetic retinopathy with which it shares numerous pathophysiological and clinical characteristics.[5] Consequently, most of the treatments offered to patients with radiation retinopathy are based on the findings of larger-scale studies investigating diabetic retinopathy. However, there are a number of differences and nuances between the two diseases, and these should be taken into account in radiation retinopathy management.

Basics and pathogenesis

The initial postradiation damage to the retina occurs in the endothelial cells.[5] These radiation-induced vascular alterations have been most extensively studied in digestive pathologies. It has been shown that exposure to irradiation leads to the rapid apoptosis of endothelial cells.[6] The surviving cells present modifications to the surface membrane (endothelial cell dysfunction)[7] specifically an increase in the expression of molecules involved in recruiting leukocytes and thrombocytes,

Retinal and Choroidal Vascular Diseases of the Eye. https://doi.org/10.1016/B978-0-443-15583-3.00037-8

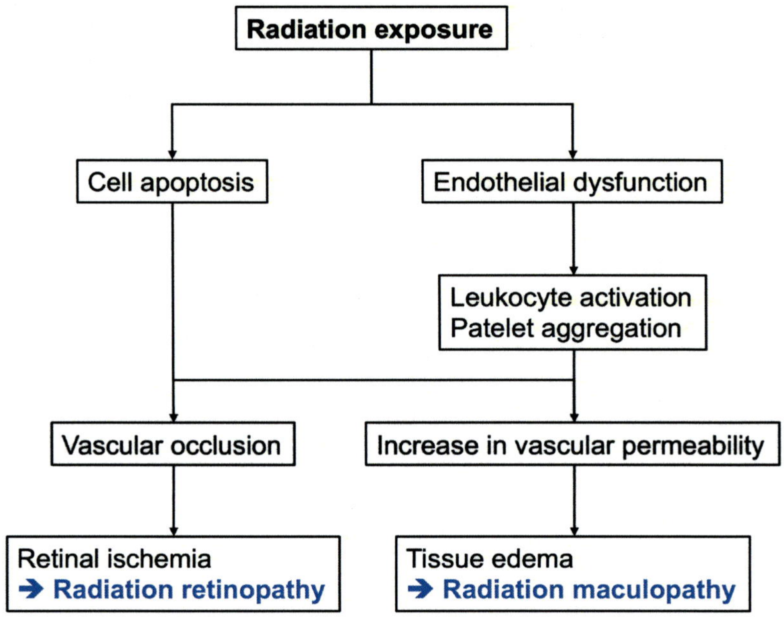

FIGURE 37.1 Pathophysiology of radiation-induced retinal damage.

activating both an inflammatory and a coagulation response. This radiation-induced cell death and inflammatory/thrombotic changes modify the architectural structure of blood vessel walls.[8] The loss of endothelial cells causes vascular dilatation, which in turn causes microaneurysms to form, as well as localized occlusion of the retinal capillaries. On the one hand, this mechanism will progressively cause ischemia in the retina downstream of the vascular occlusion. The hypoxic retina will then lose its visual function and secrete proangiogenic factors such as vascular endothelial growth factor (VEGF). This abnormal secretion of VEGF and proinflammatory signals can lead to retinal neovascularization, which characterizes the proliferative stage of retinopathy. In the final stage, the proliferation of vessels extends to the iris and the iridocorneal angle, covering the trabecular meshwork and thus preventing the flow of aqueous humor. This leads to a significant increase in intraocular pressure, causing severe orbital pain and, in some cases, leading to devastating neovascular glaucoma.[9] On the other hand, capillary loss leads to an increase in vascular permeability resulting in the extravasation of the vascular content. In clinical terms, this manifests as retinal edema and exudate, mainly located in the highly vascularized structures of the eye, including the macula (Fig. 37.1).

Although the initial cause is different, the ischemic and edematous processes in radiation retinopathy are almost identical to those observed in diabetic retinopathy.[10] Having established the parallel between these two pathologies, some authors have turned their attention to the inflammatory component of this postradiation pathology, also evidenced in diabetic retinopathy.[11] Immune cells recruited from the bloodstream have been shown to be present in the retina of fully irradiated animals.[12] This inflammatory recruitment probably occurs following the activation of the resident microglia in healthy retinas, which are activated when a nonphysiological condition is detected, such as oxidative stress following irradiation. The activation of these microglial cells leads to the secretion of factors contributing to the recruitment of circulating immune cells, thus generating adverse effects in the retina.[13] Interestingly, hyperreflective foci have also been found on optical coherence tomography (OCT) in the retina of patients treated with brachytherapy for choroidal melanoma. These hyperreflective foci were more frequently detected in patients presenting with macular edema as opposed to those with no macular edema. In light of the supposed significance of this finding in diabetic macular edema and its probable correlation with more severe inflammation, the same conclusions can be drawn for radiation maculopathy.[14] It would therefore appear that ocular immune privilege is broken after ocular irradiation, which partially explains the clinical symptoms associated with radiation retinopathy.

Risk factors

Numerous factors have been described as being associated with an increased incidence of radiation retinopathy. Most of these factors concern the radiation technique used, the type of tumor irradiated, and the patient's comorbidities and radiosensitivity. These factors should be taken into account when deciding what to propose in terms of treatment, follow-up, and prophylactic treatment to prevent the onset of this type of complication.

Radiation technique

One of the main factors affecting the onset of radiation retinopathy is the total dose delivered to the retina. The higher the dose the greater the risk of a retinal complication. The risk of developing radiation retinopathy has been shown to start when doses of 15 Gy or more are delivered to the retina and becomes frequent at doses of 45 Gy and above for normofractionated radiotherapy.[15] Some studies found no cases of radiation retinopathy at doses below 40 Gy.[15,16]

Dose fractionation is also a predictive factor for the onset of radiation retinopathy. Some authors found much higher rates of radiation retinopathy when ocular irradiation was delivered in fractions of over 1.9 Gy, with a total dose of 45–55 Gy.[16] Others showed that when treating nasal tumors, hyperfractionation significantly reduced the incidence of radiation retinopathy, especially for patients receiving large doses (>50 Gy).[17]

The surface area of the retina irradiated has also been shown to be associated with the incidence of radiation retinopathy. For an equivalent dose, the larger the area of the retina irradiated, the higher the risk of developing radiation retinopathy.[15]

These different irradiation parameters invite oncologists to define the appropriate treatment technique in consultation with the radiation oncologist. For example, when treating radioresistant tumors (such as uveal melanoma) located near radiosensitive structures, the radiotherapy should be capable of delivering high doses with a steep dose gradient passing from a tumoricidal dose to a dose tolerated by the optic nerve and the macula, in the space of around a millimeter. In this respect, conventional external radiotherapy (photons delivered via tridimensional conformal radiotherapy [3D-CRT] or intensity-modulated radiotherapy [IMRT]) is not appropriate, but plaque brachytherapy[18] and proton beam therapy are validated techniques,[19] and helium or carbon ions can also be used.[20] Unlike electrons, which lose energy consistently and expand their dose laterally, heavy ions (hadrons) lose energy exponentially, the beam depositing more energy the deeper it penetrates with a small lateral dose spread of up to 15 cm in tissue. This culminates in the delivery of the maximum dose at the maximum depth of penetration, a point known as the Bragg peak. The depth of the Bragg peak increases in proportion to the energy in the beam, which means the kinetic energy can be carefully selected to deliver the highest possible dose into the volume of the tumor at a given depth, with much lower entrance doses and negligible exit doses. Finally, stereotactic radiotherapy (Gamma Knife, CyberKnife, forms of photon beam radiotherapy) can only ever be an option at best, as the intermediate dose spillage is high. The choice of radiotherapy technique is therefore guided by the required tumoricidal dose, the site of the tumor, its volume, and finally the availability of the technique at a facility close to the patient's home (Fig. 37.2).

Tumor

The site of the tumor is one of the most important parameters in assessing the risk of developing radiation retinopathy. Generally speaking, the closer the irradiated lesion is to the retina, the greater the risk of developing radiation retinopathy.[21] This is the case for orbital tumors, which are more likely to lead to this type of complication than tumors located lower down the ENT region. It is also the case for ocular tumors: tumors in the anterior chamber of the eye, such as conjunctival tumors or tumors of the iris present a lower risk than choroidal tumors located close to the retina.[22,23] At the latter location, the position of the lesion on the retina can also affect the risk of radiation retinopathy and of radiation maculopathy in particular. Paramacular tumors located at least 3 mm from the macula are associated with a higher risk of decreased visual acuity due to radiation maculopathy.[17]

The radiosensitivity of the tumor is intrinsically linked to the dose required to treat it. Uveal melanoma, considered to be radioresistant, requires doses of 60 Gy (RBE) or more to achieve tumor control.[1] This dose is theoretically above the tolerance threshold for retinal tissue.[24] In contrast, choroidal hemangiomas that are benign vascular tumors can be controlled with a dose of around 20 Gy (RBE). The retina tolerates these lower doses very well, which explains the minimal postradiation complications and good long-term visual acuity outcomes for these patients.[25]

Tumor size (thickness and diameter) can also independently play a role in the onset of radiation retinopathy and of radiation maculopathy in particular.[26] Administering a high dose at the base of the tumor has also been found to be associated with the onset of radiation maculopathy.

Patient

In addition to the radiation technique and site of the lesion, other patient-related factors can increase the risk of developing radiation retinopathy. It has been shown that some patients are more sensitive to radiation than others; for example, some genotypes have been found to be associated with an increased risk of postradiation complications.[27] Furthermore, diabetes and, to an even greater extent, diabetic retinopathy predating the radiotherapy, have been shown to significantly increase both the risk of developing radiation retinopathy and the severity of the disease.[16,28–30] The alteration to the endothelial

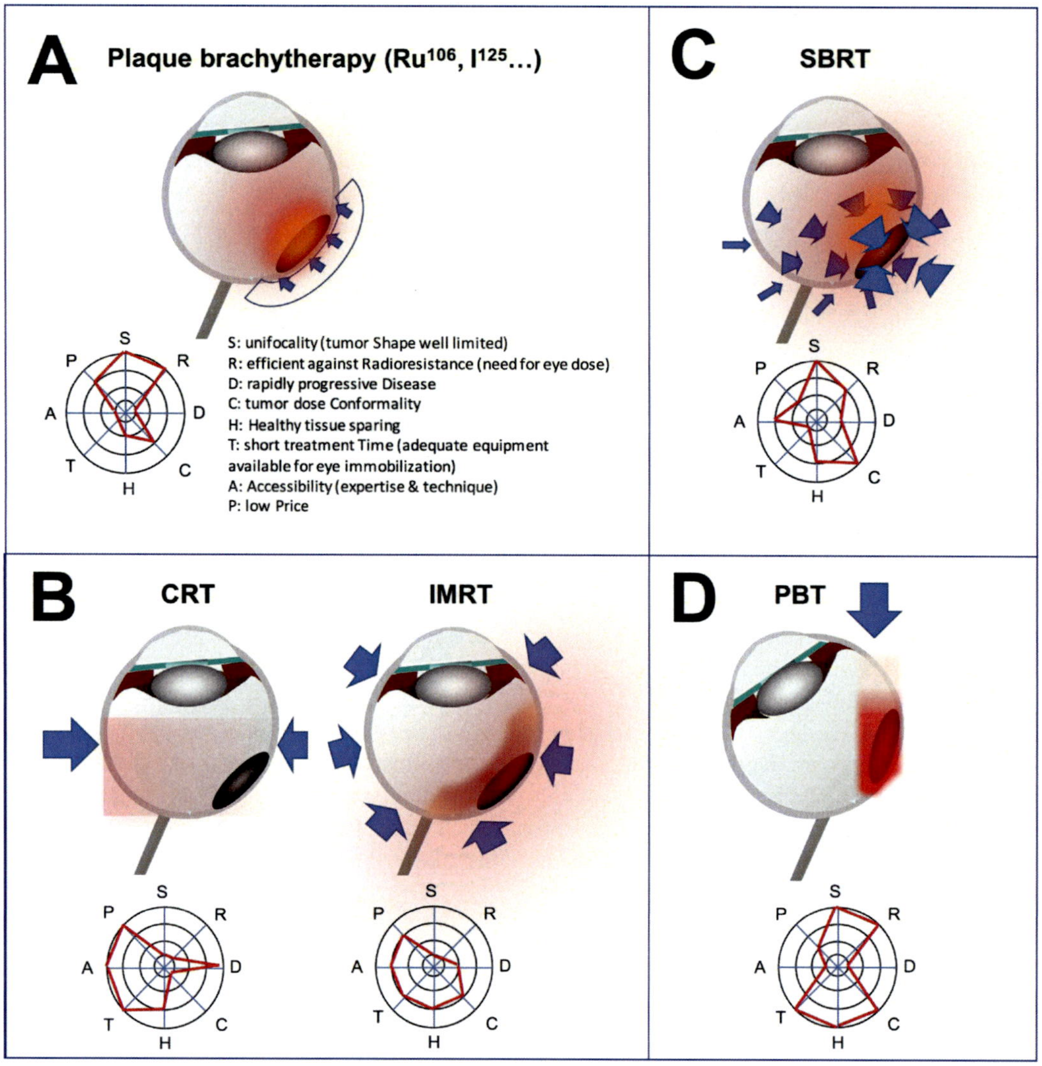

FIGURE 37.2 Radiotherapy techniques and corresponding radar charts. (A) Brachytherapy involves applying a radioactive isotope to the sclera, which delivers radiation to the tissue over a short distance. The plaque delivers a heterogeneous dose from the sclera to the apex of the tumor. (B) 3D conformal radiation therapy (CRT) delivers a uniform dose to the eye using 1–3 fields. Intensity-modulated radiotherapy (IMRT) uses 5–9 fields and a multileaf collimator, which allows for the complex concave distribution of the dose of radiation. (C) Stereotactic body radiotherapy (SBRT) delivers radiation over several sessions via noncoplanar beams converging on the tumor. (D) Proton beam therapy (PBT) deposits energy in high doses on a very precise target thanks to the physical properties of protons. The energy is deposited at the Bragg peak thus protecting the surrounding tissue. *Adapted from Mathis T, Jardel P, Loria O, et al. New concepts in the diagnosis and management of choroidal metastases. Progress in Retinal and Eye Research. 2019;68:144–176. https://doi.org/10.1016/j.preteyeres.2018.09.003.*

cells (in radiation retinopathy) and to the pericytes (in diabetic retinopathy) have a combined effect that destroys the retinal capillaries thus leading to their closure.[10]

Regarding age, the findings are inconsistent: some studies conclude that young age is a risk factor for developing radiation retinopathy,[29–31] while others find the exact opposite.[16,28] Regarding cancer care management, treatment with combined radiotherapy and chemotherapy has been described as a risk factor for the onset of radiation retinopathy. Indeed, chemotherapy increases the number of oxygen-derived free radicals, which are also pathogenic for endothelial cells (see above).[16] However, the effects of new targeted therapies have not yet been specifically assessed.[32]

Clinical features

The onset of radiation retinopathy is always delayed in relation to the time of irradiation, with the literature reporting a mean delay ranging from 18 to 24 months,[26,33] although some studies found even longer delays. The diagnosis is very rarely made beyond 5 years postradiotherapy for patient cohorts with long-term follow-up.[15,16] The prevalence of radiation retinopathy also varies significantly between studies depending on the radiation technique used and the site of the tumor (see above). Finally, these variations might be explained by other factors relating to the detection methods and diagnostic criteria used. It is important to differentiate between the symptoms and clinical signs associated with radiation retinopathy and those associated with radiation maculopathy.

Symptoms

In cases of damage to the periphery, there may be no sign of the ocular symptoms associated with radiation retinopathy. The disease is usually diagnosed fortuitously at a routine check-up. The patient may notice a slight reduction in their field of vision or hemeralopia if the ischemia is located close to the posterior pole and the macula. However, damage to the macula causes more marked symptoms including a decrease in visual acuity (often progressing very rapidly) or central or paracentral scotoma due to photoreceptor degeneration. In cases of vitreous hemorrhage associated with retinal neovascularization (characterizing proliferative radiation retinopathy), the patient may also experience myodesopsia. In a neovascular glaucoma complication, the symptoms resulting from the significant increase in intraocular pressure are extremely painful red eyes, associated with a significant decrease in visual acuity due to the corneal edema. In cases of radiation maculopathy, depending on the extent of the macular edema, the patient may experience metamorphopsia and decreased visual acuity.

Fundus examination

The first clinical signs of ischemic injury resulting from radiation retinopathy can be seen on the fundus examination as microaneurysms, retinal hemorrhage, cotton-wool spots, telangiectasia, and venous beading (Fig. 37.3).[2,33] These signs may affect the macular region (ischemic maculopathy) or the retinal periphery (ischemic peripheral retinopathy). In proliferative cases, neovascularization can be observed in the retina and/or iris (Fig. 37.4) introducing its own set of complications (vitreous hemorrhage, tractional retinal detachment, neovascular glaucoma). In parallel to the ischemic injury, the possible onset of radiation maculopathy should be noted, which sometimes manifests in the form of hard exudate associated with cystoid macular edema (Fig. 37.5). This radiation maculopathy often precedes the clinical signs of ischemia visible on the fundus.[2,34]

As well as screening for radiation retinopathy and radiation maculopathy, the clinical examination may also reveal other radiation-induced ocular injuries such as cataract or optic neuropathy.[35,36] The latter is another delayed-onset sight-threatening complication also characterized by vascular occlusion localized at the head of the optic nerve (Fig. 37.6).[36]

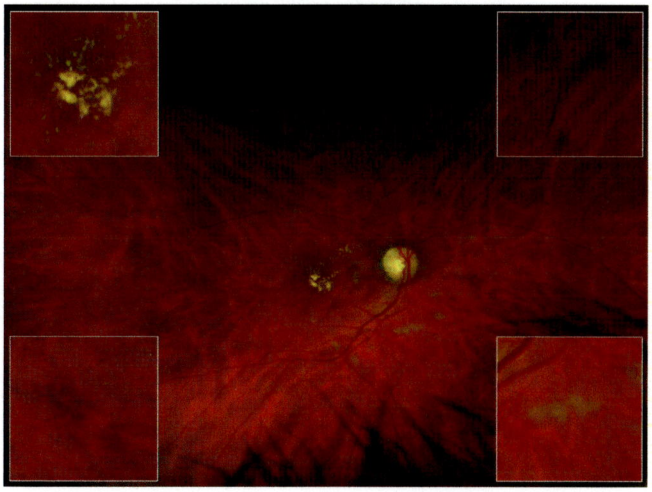

FIGURE 37.3 Cardinal features of radiation retinopathy. Hard exudate on macular area (*upper left inset*), venous beading and intraretinal telangiectasia (*upper right inset*), retinal hemorrhage and microaneurysms (*lower left inset*), cotton-wool spots (*lower right inset*).

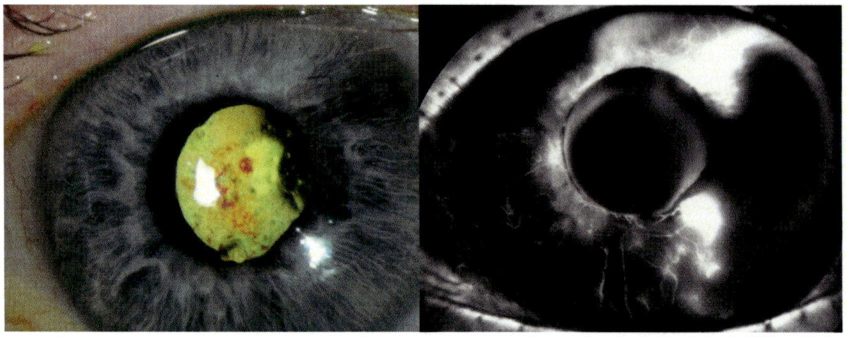

FIGURE 37.4 Neovascular glaucoma occurring 3 years after treatment of a large choroidal melanoma with proton beam therapy. Iris angiography showing fluorescein diffusion from immature neovessels.

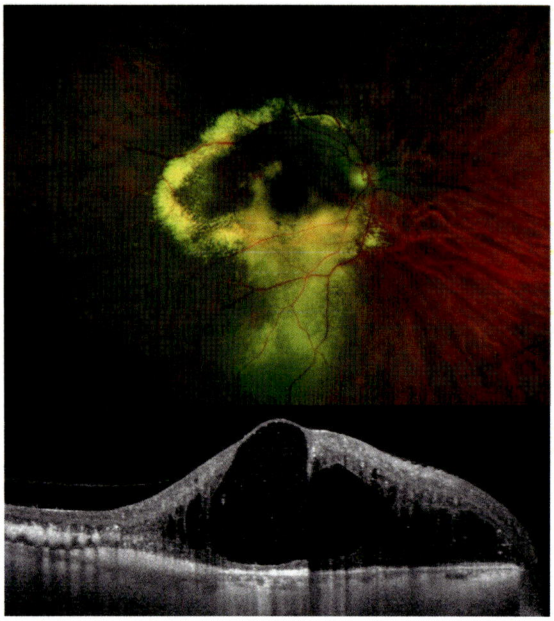

FIGURE 37.5 Radiation maculopathy occurring 18 months after Carbon-12 hadron therapy to treat nasopharyngeal cancer. Note the presence of a large area of hard exudate, resulting in cystoid macular edema.

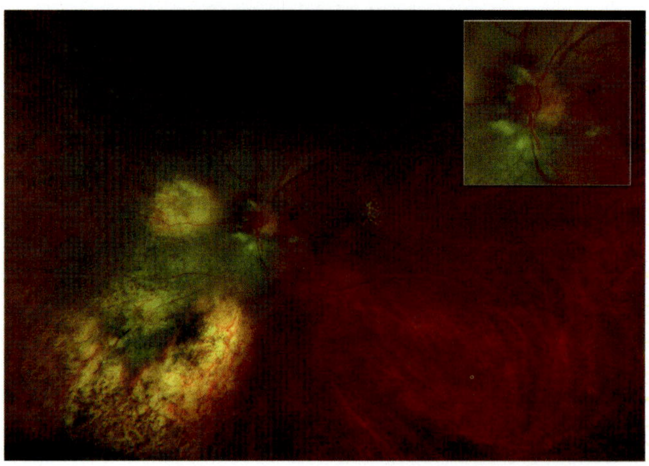

FIGURE 37.6 Radiation-induced optic neuropathy occurring 2 years after proton beam therapy for choroidal melanoma. There is an optic nerve edema accompanied by cotton-wool spot and juxta-papillary hemorrhage (inset).

Imaging findings and future imaging

For many years, fluorescein angiography was the only technique available to visualize the retinal and choroidal network; it is therefore the examination of choice when investigating vascular retinal pathologies and remains the gold standard for detecting radiation-induced retinal damage.[37] In radiation retinopathy, fluorescein angiography highlights the nonperfused areas of the retina, thus revealing retinal ischemia. In addition, vascular malformations such as microaneurysms or telangiectasia are also highlighted by fluorescence and are therefore very clearly visible. Fluorescein angiography can also detect subclinical edema as dye leakage reveals blood-retinal barrier breakdown associated with fluid extravasation. However, this examination requires the injection of fluorescein which can cause an allergic reaction, although anaphylactic shock is rare.

In recent years, OCT has been introduced, revolutionizing retinal pathology management. This imaging technique produces a quasi-histological image of the retina, enabling a very detailed analysis of its architecture. This examination can also detect the macular edema that defines radiation maculopathy at an earlier stage. However, it has limitations when screening for retinal ischemia, especially if located in the periphery, an area that is difficult to access for this examination of the macula. OCT-angiography (OCTA) is an even more recent technology that can be used to visualize vascular flow without injecting a contrast medium. OCTA can thereby detect ischemic areas, defined by the absence of blood flow. As this imaging technique has become part of routine clinical practice, numerous studies have investigated the analysis of radiation retinopathy using OCTA.[38,39,43] This rapid examination is particularly useful for diagnosing retinal damage as it detects vascular anomalies at a very early stage,[38,39] but has limitations when investigating the peripheral retina (Fig. 37.7).[40]

Several studies have analyzed retinal damage on OCTA following ocular irradiation using brachytherapy and proton beam therapy. The types of damage observed included decreased perfusion of the superficial and deep retinal capillary plexus, atrophy of the choriocapillaris layer at the tumor boundaries, and expansion of the foveal avascular zone in the macula of irradiated eyes compared to nonirradiated eyes.[38,39] It has recently been shown that ischemic alterations to the choriocapillaris precede the clinical stages of radiation retinopathy, thus defining an early stage of radiation choroidopathy. It is well known that the capillaries are the first network to be affected by vascular pathologies such as diabetes or radiation retinopathy, thus explaining the early alteration of the choriocapillaris which is composed of a fine, but very dense, network of capillaries. The alteration to the choriocapillaris after ocular irradiation is, therefore, a new OCTA finding, reflecting the overall ocular vascularization dysfunction, visible before the onset of the disease.[39] Highlighting the markers of alterations to the retina and choroid can help to identify the patients at risk of developing the disease and offer them preventive treatment and closer follow-up. However, it is important to note that analyzing the OCTA findings for treated eyes can be complicated by the presence of numerous artifacts. Patients with radiation maculopathy often have very low visual acuity which makes fixation difficult and signal loss more common.[41] These difficulties may be exacerbated by secondary dryness inducing increased blinking, and/or disorders affecting the ocular media (cataracts, intravitreal hemorrhage, etc.). Furthermore, the thickness of the macular edema can make it impossible to analyze the retinal and choroidal vascularization.

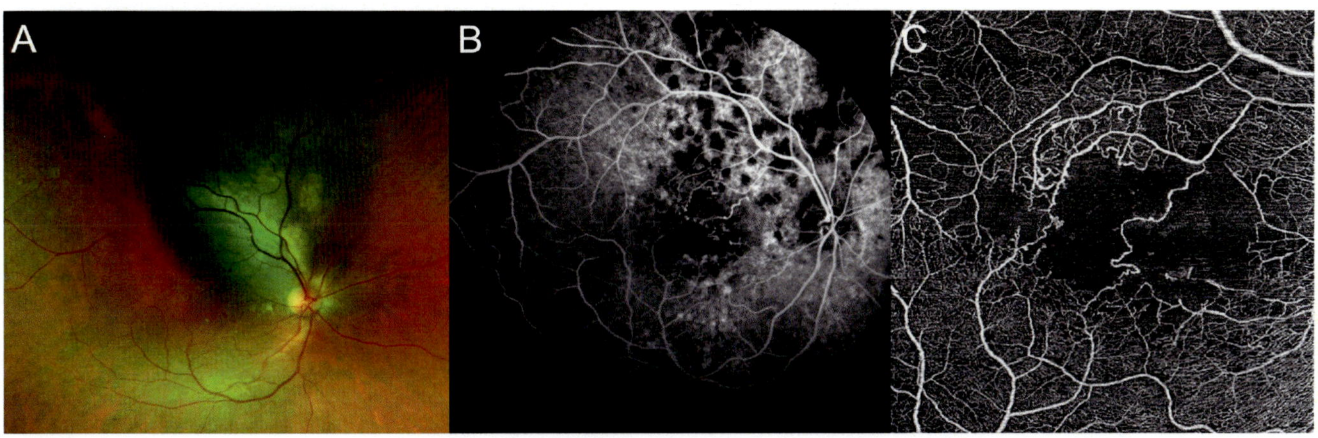

FIGURE 37.7 Macular ischemia occurring 4 years after proton beam therapy for choroidal melanoma. (A) Retinography showing small macular hemorrhage and aneurysms below the tumor. (B) Fluorescein angiography showing areas of nonperfusion in the macula, as well as telangiectasia and microaneurysms. (C) Optical coherence tomography more clearly showing areas of nonperfusion and presence of telangiectasia.

OCTA is now used in routine practice to assess macular ischemia, but fluorescein angiography is still required to examine the entire surface of the retina.

Classification of radiation retinopathy

Improved imaging techniques and new treatment options have improved the diagnosis and management of patients with radiation retinopathy. These advances have encouraged ocular oncologists to grade radiation retinopathy as accurately as possible to optimize its management at each stage. Unfortunately, these classifications are incomplete and rarely used outside of clinical trials. The Common Terminology Criteria for Adverse Events (CTCAE) v5 classification describes retinopathy as an ocular complication of radiotherapy. However, the staging is not specific and only takes into account the visual acuity score which does not reflect the surface area and size of the peripheral ischemia. However, in cases of radiation maculopathy, the decrease in visual acuity may be associated with an increase in macular edema, but the relationship is not linear. More accurate scales are required for ophthalmological purposes to propose treatment and make prognoses according to the severity of the disease. The first specific classification of radiation retinopathy described by Finger in 2005 was only based on the clinical findings on the fundus and their angiographic manifestations.[42] Although this classification is progressive in terms of the severity of the clinical symptoms, it does not distinguish between the ischemic and edematous features of radiation retinopathy and radiation maculopathy, respectively, which have different prognoses and treatments, nor does it recommend any specific care management. In 2010, Horgan et al. proposed a new classification incorporating OCT, for radiation maculopathy only.[34] This classification is tripartite and distinguishes between: (1) clinically significant macular edema on fundus examination; (2) angiographic alteration (separating macular ischemia from diffuse or focal macular edema); (3) OCT-evident macular edema using a prior classification defining the severity and extent of the macular edema in five grades.[26] Finally, the most recent classification proposed by Veverka et al. includes OCTA findings; the first stage of radiation retinopathy being defined by anomalies only detected on OCTA with no other clinical or paraclinical signs.[43] This classification then combines the previously described classifications (Table 37.1).[34,42] The particularity of this classification is that it treats the OCTA findings as a noninvasive, early diagnosis tool for the detection of the disease.

Treatment

Treating ischemia in radiation retinopathy

The treatment of radiation retinopathy is based on laser photocoagulation of the retinal ischemia, which prevents the proliferation of retinal neovascularization and neovascular glaucoma.[44] Destroying the damaged retina reduces the secretion of VEGF and increases oxygenation of the nonphotocoagulated areas of the retina.[45] This treatment is indicated in cases where there is a large surface area of ischemia (>25%–30% of the retina) with a corresponding risk of proliferative complications (Fig. 37.8). Anti-VEGF treatment can be proposed for proliferative forms of radiation retinopathy with a view to facilitating photocoagulation. However, the findings of studies investigating diabetic retinopathy suggest that in the longer term replacing photocoagulation with treatment using antiangiogenic molecules is an unwise choice due to the risk of patients not responding to anti-VEFG treatment or being lost to follow-up.[46]

TABLE 37.1 Radiation retinopathy classification.

Grade	Characteristics
0	No signs of radiation retinopathy
1	OCTA: presence of microaneurysms
2	OCT: increased retinal thickness, no edema
3	OCT: cystoid macula edema
4	Clinical signs of radiation retinopathy on fundus
5	OCTA impossible due to macular edema

OCT, optical coherence tomography; *OCTA*, OCT-angiography.
Adapted and simplified from Veverka KK, Abouchehade JE, Iezzi R, Pulido JS. Noninvasive grading of radiation retinopathy: the use of optical coherence tomography angiography. Retina. 2015;35(11):2400e2410. https://doi.org/10.1097/IAE.0000000000000844.

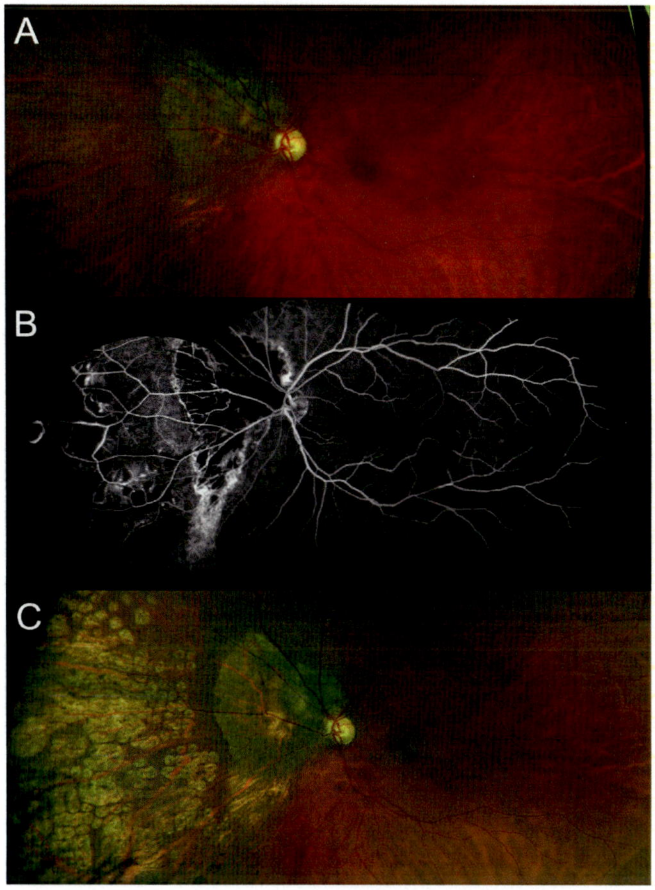

FIGURE 37.8 Management of ischemic peripheral retinopathy. (A) Retinography 2 years after proton beam therapy for juxta-papillary melanoma showing nasal peripheral vascular occlusion and retinal hemorrhages. (B) Corresponding fluorescein angiography confirming nasal peripheral areas of nonperfusion. (C) Retinography 2 years after pan-retinal photocoagulation showing no other signs of radiation retinopathy.

Treating edema in radiation maculopathy

For many years, there was no effective treatment for macular edema. Since 2006, when marketing authorization was granted to use anti-VEGF to treat age-related macular degeneration, numerous studies have investigated their efficacy for treating edema associated with radiation maculopathy. However, these treatments require injections into the vitreous cavity, a minimally invasive procedure, which nonetheless is associated with a rare risk of endophthalmitis or ocular trauma (retinal detachment, vitreous hemorrhage, etc.) in eyes already weakened by irradiation. Furthermore, for intraocular tumors, a potential risk of the cancer spreading was initially raised, but recent studies have proven that this is not the case.[47,48] Soon after the introduction of anti-VEGF, dexamethasone implants were also granted marketing authorization to treat certain retinal pathologies and were also used to treat edema in radiation maculopathy, as well as exudative complications caused by tumors.[49]

Intravitreal anti-VEGF injections

VEGF is a molecule that plays a key role in promoting angiogenesis. Blocking this pathway with intravitreal anti-VEGF injections is a common approach to treating diabetic maculopathy and edema in cases of retinal vein occlusion. Several drugs (bevacizumab, ranibizumab, and aflibercept) have been found to be superior to placebo or laser photocoagulation in treating macular complications of radiotherapy. Numerous studies investigating radiation maculopathy have also demonstrated the benefits of these treatments,[22,47,50,51] in particular for reducing macular edema and improving visual acuity.[52]

Although no direct comparisons have been made, lower reductions in visual acuity over time have been found for patients treated with these anti-VEGF molecules as opposed to observation only. Indeed, several studies investigated the long-term visual acuity of patients undergoing radiotherapy treatment for choroidal melanoma, before the introduction of intravitreal therapies. The results of these studies varied widely between cohorts, with 40%–80% of irradiated eyes found to have visual acuity of more than 20/200 after 5 years.[53,54] Recently, visual acuity was found to be equal to or higher than 2/200 for almost all eyes at the end of the follow-up period in a large cohort of patients with radiation maculopathy treated over a 10-year period with regular, continuous anti-VEGF injections (every 4–12 weeks).[47] In addition to its findings on visual acuity, the results of this particularly long study also showed a reduction in the clinical signs of radiation retinopathy (hemorrhage, exudate) and a clinical reduction in the height of the macular edema. Relatively few adverse events were observed in this study, making it possible to treat and follow almost all of the cohort over the full duration.

Although anti-VEGF treatment is promising in terms of protecting the vision of patients with radiation maculopathy, it only delays the inevitable and the treatment usually requires numerous injections (often monthly) to control the edema.[55] Unlike for AMD and diabetic macular edema, there is no consensus on anti-VEGF treatment protocols in terms of the number and frequency of injections. Nonetheless, new anti-VEGF therapies with molecules that deliver higher doses, are more targeted, or have an increased half-life will probably lead to a reduced treatment burden for these patients while maintaining high safety levels, as already seen for diabetic macular edema.[56]

Intravitreal injections of dexamethasone implant

Corticosteroids' mechanism of action consists of reducing the secretion of proinflammatory cytokines, including VEGF, by the irradiated retina, thus reducing the impact of inflammation on vascular permeability.[49,57] The main advantage of the dexamethasone implant is that it releases an effective steroid dose over a 4-month period which is sufficient to control the edema in most cases. However, it can also cause local adverse effects such as the development of steroid-induced cataracts or ocular hypertension. To date, no randomized studies have compared the efficacy in terms of visual acuity of anti-VEGF and dexamethasone implants in treating radiation maculopathy. However, real-life studies have shown that the two treatments are equivalent in functional terms, although dexamethasone implants require less injections, thus reducing the therapeutic burden and improving compliance.[58,59] Recently, a small nonrandomized case series showed that fluocinolone acetonide implant treatment following treatment with dexamethasone implant is effective in reducing the number of injections required.[60]

Other available therapies

As for other edematous macular etiologies, focal laser treatment of certain vascular anomalies (such as microaneurysms or some telangiectasia) can lead to the resorption of exudate or focal edema and ultimately lead to improved vision.[61,62] This treatment can be performed alone or as a complement to intravitreal drug treatments.

Radiation retinopathy prevention

The first line of radiation retinopathy prevention is to limit the dose of radiation delivered to the retina, as well as the surface area irradiated. Targeted radiotherapy techniques such as proton beam therapy, brachytherapy, and, to a lesser extent, radiosurgery or stereotactic radiotherapy concentrate the radiation beams on the lesion, thus limiting the volume of healthy tissue exposed to high doses. If maintaining a fixed ocular position during the planning scan and treatment sessions is not sufficient to protect the retina, the macular area can be protected by means of immobilization, fixation, or optimized dosimetry. The latter requires locating the fovea (and the surrounding macula), which on 7T magnetic resonance imaging (MRI) with correction of kinetic ocular artifacts, can be seen as a notch on the retina. However, the 1.5 or 3T high-resolution ocular MRI scans currently available for use in routine practice cannot be used to visualize the macula. Nevertheless, if the topology of the posterior pole is known, the position of the fovea can be "reconstituted" by measuring 4.5 mm temporal from the center of the optic disc. The macula can then be delineated as the zone 5.5 mm around the fovea (Fig. 37.9) on the uvea, as the retina cannot be distinguished from the other two layers on a planning scanner or 1.5 or 3T MRI. A dose constraint can then also be applied to the macula structure if this is reasonable from an oncological perspective, that is, it does not reduce the chances of controlling the tumor.

Even with these precautions, complications may occur and ocular oncologists and radiotherapists have therefore looked for different ways of limiting the dose to the macula. The irradiation margins can be reduced, although this can be associated with an increased risk of local recurrence.[63] In brachytherapy, some authors have studied the effectiveness of eccentric positioning of the plaque to avoid irradiating the macula, but this does not in fine reduce the irradiated surface.[64]

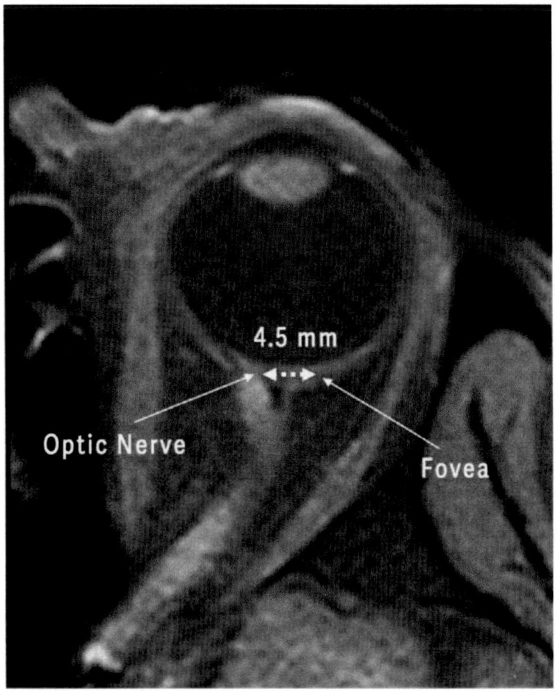

FIGURE 37.9 Locating the macula on a 1.5T MRI of the orbit performed before irradiation. The presumed location of the fovea is measured 4.5 mm temporal from the center of the optic disc.

Such adaptations are controversial given the risk of local recurrence and the fact that they do not completely eliminate the onset of radiation radiotherapy.

A more recent approach to preventing and reducing the onset of radiation retinopathy and radiation maculopathy involves systematically administering anti-VEGF injections every two to four months. This strategy has been shown to obtain a reduction in signs of radiation retinopathy and radiation maculopathy.[65,66] This preventive regimen could be beneficial for tumors with a high risk of postradiation complications (thick tumors, close to the macula) but there is no consensus on how it should be delivered, especially as it requires long-term treatment.

Conclusion and future directions

Radiation retinopathy is a disease with very severe consequences that can lead to blindness and even eye loss. In recent years, it has become easier to identify at-risk patients thanks to developments in screening methods and improved retinal and choroidal imaging. However, unlike for diabetic retinopathy, there is no clearly established classification for radiation retinopathy that makes follow-up for these patients less consensual. An improved classification of patients would make it possible to implement more established treatment regimens evaluated in randomized studies. The identification of early markers on OCTA is also currently being studied as a way of classifying at-risk patients requiring closer follow-up. In the future, finding these markers could potentially trigger the administration of preventive treatment. Once again, there is currently no consensus on the treatment regimens or the duration of treatment for these patients. However, numerous specialists propose proactive treatment regimens based on the results found for similar pathologies such as diabetic retinopathy and retinal vein occlusion to limit macular edema fluctuations which are known to decrease patients' vision in the long term. From another perspective, advances in research into therapeutics could help reduce the prevalence of radiation retinopathy. The potential benefits of the recently discovered FLASH radiotherapy in terms of sparing healthy tissue is one notable example.

In conclusion, all specialists providing care for patients irradiated in the head and neck region (ophthalmologists, oncologists, radiotherapists) should be familiar with radiation retinopathy as it can lead to serious ocular complications. As a consequence of exposure to radiation, its onset is always delayed in relation to the exposure. It is only on diagnosis that the most appropriate treatment can be determined, making early diagnosis of the utmost importance.

References

1. Mathis T, Cassoux N, Tardy M, et al. Management of uveal melanomas, guidelines for oncologists. *Bull Cancer*. 2018;105(10):967–980. https://doi.org/10.1016/j.bulcan.2018.07.011.
2. Guyer DR, Mukai S, Egan KM, Seddon JM, Walsh SM, Gragoudas ES. Radiation maculopathy after proton beam irradiation for choroidal melanoma. *Ophthalmology*. 1992;99(8):1278–1285. https://doi.org/10.1016/S0161-6420(92)31832-9.
3. Mathis T, Levron A, Pommier P, Denis P, Thariat J, Kodjikian L. Intra- and subretinal neovascularization following radiation therapy: contribution of OCT-angiography. *Journal Français d'Ophtalmologie*. 2018;41(10):e481–e483. https://doi.org/10.1016/j.jfo.2018.04.007.
4. Egger E, Zografos L, Schalenbourg A, et al. Eye retention after proton beam radiotherapy for uveal melanoma. *International Journal of Radiation Oncology Biology Physics*. 2003;55(4):867–880. https://doi.org/10.1016/S0360-3016(02)04200-1.
5. Archer DB, Amoaku WMK, Gardiner TA. Radiation retinopathy—clinical, histopathological, ultrastructural and experimental correlations. *Eye*. 1991;5(2):239–251. https://doi.org/10.1038/eye.1991.39.
6. Rotolo JA, Maj JG, Feldman R, et al. Bax and Bak do not exhibit functional redundancy in mediating radiation-induced endothelial apoptosis in the intestinal mucosa. *International Journal of Radiation Oncology Biology Physics*. 2008;70(3):804–815. https://doi.org/10.1016/j.ijrobp.2007.11.043.
7. Mollà M, Panés J. Radiation-induced intestinal inflammation. *World Journal of Gastroenterology*. 2007;13(22). https://doi.org/10.3748/wjg.v13.i22.3043.
8. Spaide RF, Fujimoto JG, Waheed NK, Sadda SR, Staurenghi G. Optical coherence tomography angiography. *Progress in Retinal and Eye Research*. 2018;64:1–55. https://doi.org/10.1016/j.preteyeres.2017.11.003.
9. Thariat J, Martel A, Matet A. Non-cancer effects following ionizing irradiation involving the eye and orbit. *Cancers*. 2022;14.
10. Mathis T, Jardel P, Loria O, et al. New concepts in the diagnosis and management of choroidal metastases. *Progress in Retinal and Eye Research*. 2019;68:144–176. https://doi.org/10.1016/j.preteyeres.2018.09.003.
11. Dong N, Xu B, Wang B, Chu L. Study of 27 aqueous humor cytokines in patients with type 2 diabetes with or without retinopathy. *Molecular Vision*. 2013;19:1734–1746.
12. Kaneko H, Nishiguchi KM, Nakamura M, Kachi S, Terasaki H. Characteristics of bone marrow-derived microglia in the normal and injured retina. *Investigative Ophthalmology and Visual Science*. 2008;49(9):4162–4168. https://doi.org/10.1167/iovs.08-1738.
13. Chen M, Zhao J, Luo C, et al. Para-inflammation-mediated retinal recruitment of bone marrow-derived myeloid cells following whole-body irradiation is CCL2 dependent. *Glia*. 2012;60(5):833–842. https://doi.org/10.1002/glia.22315.
14. Parrozzani R, Midena G, Frizziero L, Marchione G, Midena E. Radiation maculopathy is anticipated by OCT hyperreflective retinal FOCI: a large, prospective, confirmation study. *Retina*. 2022;42(4):752–759. https://doi.org/10.1097/IAE.0000000000003359.
15. Takeda A, Shigematsu N, Suzuki S, et al. Late retinal complications of radiation therapy for nasal and paranasal malignancies: relationship between irradiated-dose area and severity. *International Journal of Radiation Oncology Biology Physics*. 1999;44(3):599–605. https://doi.org/10.1016/s0360-3016(99)00057-7.
16. Parsons JT, Bova FJ, Fitzgerald CR, Mendenhall WM, Million RR. Radiation retinopathy after external-beam irradiation: analysis of time-dose factors. *International Journal of Radiation Oncology Biology Physics*. 1994;30(4):765–773. https://doi.org/10.1016/0360-3016(94)90347-6.
17. Monroe AT, Bhandare N, Morris CG, Mendenhall WM. Preventing radiation retinopathy with hyperfractionation. *International Journal of Radiation Oncology Biology Physics*. 2005;61(3):856–864. https://doi.org/10.1016/j.ijrobp.2004.07.664.
18. Nag S, Quivey JM, Earle JD, Followill D, Fontanesi J, Finger PT. The American Brachytherapy Society recommendations for brachytherapy of uveal melanomas. *International Journal of Radiation Oncology Biology Physics*. 2003;56(2):544–555. https://doi.org/10.1016/S0360-3016(03)00006-3.
19. Caujolle JP, Mammar H, Chamorey E, Pinon F, Herault J, Gastaud P. Proton beam radiotherapy for uveal melanomas at nice teaching hospital: 16 years' experience. *International Journal of Radiation Oncology Biology Physics*. 2010;78(1):98–103. https://doi.org/10.1016/j.ijrobp.2009.07.1688.
20. Mishra KK, Quivey JM, Daftari IK, et al. Long-term results of the UCSF-LBNL randomized trial: charged particle with helium ion versus iodine-125 plaque therapy for choroidal and ciliary body melanoma. *International Journal of Radiation Oncology Biology Physics*. 2015;92(2):376–383. https://doi.org/10.1016/j.ijrobp.2015.01.029.
21. Mashayekhi A, Schönbach E, Shields CL, Shields JA. Early subclinical macular edema in eyes with uveal melanoma: association with future cystoid macular edema. *Ophthalmology*. 2015;122(5):1023–1029. https://doi.org/10.1016/j.ophtha.2014.12.034.
22. Murray TG, Latiff A, Villegas VM, Gold AS. Aflibercept for radiation maculopathy study. *Ophthalmology Retina*. 2019;3(7):561–566. https://doi.org/10.1016/j.oret.2019.02.009.
23. Thariat J, Rahmi A, Salleron J, et al. Proton beam therapy for Iris melanomas in 107 patients. *Ophthalmology*. 2018;125(4):606–614. https://doi.org/10.1016/j.ophtha.2017.10.009.
24. Thariat J, Racadot S, Pointreau Y, et al. Radiothérapie conformationnelle avec modulation d'intensité des cancers des voies aérodigestives supérieures: dose de tolérance de l'œil et des voies optiques. *Cancer/Radiothérapie*. 2016;20(6–7):467–474. https://doi.org/10.1016/j.canrad.2016.07.079.
25. Mathis T, Maschi C, Mosci C, et al. Comparative effectiveness of proton beam versus photodynamic therapy to spare the vision in circumscribed choroidal hemangioma. *Retina*. 2021;41(2):277–286. https://doi.org/10.1097/iae.0000000000002843.
26. Horgan N, Shields CL, Mashayekhi A, Teixeira LF, Materin MA, Shields JA. Early macular morphological changes following plaque radiotherapy for uveal melanoma. *Retina*. 2008;28(2):263–273. https://doi.org/10.1097/IAE.0b013e31814b1b75.
27. Foray N, Colin C, Bourguignon M. Radiosensibilité L'évidence d'un facteur individuel. *Medecine/Sciences*. 2013;29(4):397–403. https://doi.org/10.1051/medsci/2013294013.

28. Kaushik M, Pulido JS, Schild SE, Stafford S. Risk of radiation retinopathy in patients with orbital and ocular lymphoma. *International Journal of Radiation Oncology Biology Physics*. 2012;84(5):1145–1150. https://doi.org/10.1016/j.ijrobp.2011.12.097.
29. Bianciotto C, Shields CL, Pirondini C, Mashayekhi A, Furuta M, Shields JA. Proliferative radiation retinopathy after plaque radiotherapy for uveal melanoma. *Ophthalmology*. 2010;117(5):1005–1012. https://doi.org/10.1016/j.ophtha.2009.10.015.
30. Gragoudas ES, Li W, Lane AM, Munzenrider J, Egan KM. Risk factors for radiation maculopathy and papillopathy after intraocular irradiation. *Ophthalmology*. 1999;106(8):1571–1578. https://doi.org/10.1016/S0161-6420(99)90455-4.
31. Cicinelli MV, Nicola DM, Gigliotti RC. Predictive factors of radio-induced complications in 194 eyes undergoing gamma knife radiosurgery for uveal melanoma. *Acta Ophthalmology*. 2021;2021.
32. Landers MB, Stefansson E, Wolbarsht ML. Panretinal photocoagulation and retinal oxygenation. *Retina*. 1982;2(3):167–175. https://doi.org/10.1097/00006982-198200230-00007.
33. Brown GC, Shields JA, Sanborn G, Augsburger JJ, Savino PJ, Schatz NJ. Radiation retinopathy. *Ophthalmology*. 1982;89(12):1494–1501. https://doi.org/10.1016/S0161-6420(82)34611-4.
34. Horgan N, Shields CL, Mashayekhi A, Shields JA. Classification and treatment of radiation maculopathy. *Current Opinion in Ophthalmology*. 2010;21(3):233–238. https://doi.org/10.1097/ICU.0b013e3283386687.
35. Mathis T, Rosier L, Meniai F, et al. The lens opacities classification system III grading in irradiated uveal melanomas to characterize proton therapy-induced cataracts. *American Journal of Ophthalmology*. 2019;201:63–71. https://doi.org/10.1016/j.ajo.2019.01.025.
36. Mohamed Ali A, Mathis T, Bensadoun RJ, Thariat J. Radiation induced optic neuropathy: does treatment modality influence the risk? *Bulletin du Cancer*. 2019;106(12):1160–1176. https://doi.org/10.1016/j.bulcan.2019.09.008.
37. Hayreh SS. Post-radiation retinopathy. A fluorescence fundus angiographic study. *British Journal of Ophthalmology*. 1970;54(11):705–714. https://doi.org/10.1136/bjo.54.11.705.
38. Shields CL, Say EAT, Samara WA, Khoo CTL, Mashayekhi A, Shields JA. Optical coherence tomography angiography of the macula after plaque radiotherapy of choroidal melanoma: comparison of irradiated versus nonirradiated eyes in 65 patients. *Retina*. 2016;36(8):1493–1505. https://doi.org/10.1097/IAE.0000000000001021.
39. Veverka KK, Abouchehade JE, Iezzi R, Pulido JS. Noninvasive grading of radiation retinopathy: the use of optical coherence tomography angiography. *Retina*. 2015;35(11):2400–2410. https://doi.org/10.1097/IAE.0000000000000844.
40. Gilli C, Kodjikian L, Loria O, et al. Quantitative analysis of choriocapillaris alterations in swept-source optical coherence tomography-angiography during radiation retinopathy. *Retina*. 2022;42(10):1995–2003. https://doi.org/10.1097/iae.0000000000003559.
41. Viebahn M, Barricks ME, Osterloh MD. Synergism between diabetic and radiation retinopathy: case report and review. *British Journal of Ophthalmology*. 1991;75(10):629–632. https://doi.org/10.1136/bjo.75.10.629.
42. Say EAT, Ferenczy S, Magrath GN, Samara WA, Khoo CTL, Shields CL. Image quality and artifacts on optical coherence tomography angiography. *Retina*. 2017;37(9):1660–1673. https://doi.org/10.1097/IAE.0000000000001414.
43. Finger PT. Laser photocoagulation for radiation retinopathy after ophthalmic plaque radiation therapy. *British Journal of Ophthalmology*. 2005;89(6):730–738. https://doi.org/10.1136/bjo.2004.052159.
44. Kinyoun JL, Lawrence B, Barlow WE. Proliferative radiation retinopathy. *Archives of Ophthalmology*. 1996;114(9):1097–1100. https://doi.org/10.1001/archopht.1996.01100140299007.
45. Amoaku WMK, Frew L, Mahon GJ, Gardiner TA, Archer DB. Early ultrastructural changes after low-dose X-irradiation in the retina of the rat. *Eye*. 1989;3(5):638–646. https://doi.org/10.1038/eye.1989.98.
46. Gonzalez-Cortes JH, Martinez-Pacheco VA, Gonzalez-Cantu JE, et al. Current treatments and innovations in diabetic retinopathy and diabetic macular edema. *Pharmaceutics*. 2023;15(1). https://doi.org/10.3390/pharmaceutics15010122.
47. Finger PT, Chin KJ, Semenova EA. Intravitreal anti-VEGF therapy for macular radiation retinopathy: a 10-year study. *European Journal of Ophthalmology*. 2015;26(1):60–66. https://doi.org/10.5301/ejo.5000670.
48. el Filali M, Ly LV, Luyten GPM, et al. Bevacizumab and intraocular tumors: an intriguing paradox. *Molecular Vision*. 2012;18:2454–2467.
49. Malclès A, Nguyen A-M, Mathis T, Grange J-D, Kodjikian L. Intravitreal dexamethasone implant (Ozurdex) for exudative retinal detachment after proton beam therapy for choroidal melanoma. *European Journal of Ophthalmology*. 2017;27(5):596–600. https://doi.org/10.5301/ejo.5000940.
50. Finger PT. Intravitreous ranibizumab (Lucentis) for radiation maculopathy. *Archives of Ophthalmology*. 2010;128(2). https://doi.org/10.1001/archophthalmol.2009.376.
51. Fallico M, Reibaldi M, Avitabile T, et al. Intravitreal aflibercept for the treatment of radiation-induced macular edema after ruthenium 106 plaque radiotherapy for choroidal melanoma. *Graefe's Archive for Clinical and Experimental Ophthalmology*. 2019;257(7):1547–1554. https://doi.org/10.1007/s00417-019-04347-6.
52. Shah SU, Shields CL, Bianciotto CG, et al. Intravitreal bevacizumab at 4-month intervals for prevention of macular edema after plaque radiotherapy of uveal melanoma. *Ophthalmology*. 2014;121(1):269–275. https://doi.org/10.1016/j.ophtha.2013.08.039.
53. Fontanesi J, Meyer D, Xu S, Tai D. Treatment of choroidal melanoma with I-125 plaque. *International Journal of Radiation Oncology Biology Physics*. 1993;26(4):619–623. https://doi.org/10.1016/0360-3016(93)90278-4.
54. Finger PT, Chin KJ, Duvall G. Palladium-103 for choroidal melanoma study group. palladium-103 ophthalmic plaque radiation therapy for choroidal melanoma: 400 treated patients. *Ophthalmology*. 2009;116:790–796.
55. Shields CL, Demirci H, Dai V, et al. Intravitreal triamcinolone acetonide for radiation maculopathy after plaque radiotherapy for choroidal melanoma. *Retina*. 2005;25(7):868–874. https://doi.org/10.1097/00006982-200510000-00009.

56. Mathis T, Kodjikian L. Do new drugs for diabetic macular edema offer a safer option? *Expert Opinion on Drug Safety*. 2022;21(12):1453–1457. https://doi.org/10.1080/14740338.2022.2160442.
57. Jermak CM, Dellacroce JT, Heffez J, Peyman GA. Triamcinolone acetonide in ocular therapeutics. *Survey of Ophthalmology*. 2007;52(5):503–522. https://doi.org/10.1016/j.survophthal.2007.06.004.
58. Seibel I, Hager A, Riechardt AI, Davids AM, Böker A, Joussen AM. Antiangiogenic or corticosteroid treatment in patients with radiation maculopathy after proton beam therapy for uveal melanoma. *American Journal of Ophthalmology*. 2016;168:31–39. https://doi.org/10.1016/j.ajo.2016.04.024.
59. Russo A, Reibaldi M, Avitabile T, et al. Dexamethasone intravitreal implant vs ranibizumab in the treatment of macular edema secondary to brachytherapy for choroidal melanoma. *Retina*. 2018;38(4):788–794. https://doi.org/10.1097/iae.0000000000001585.
60. Zimmermann L, Kneifel C, Grajewski L, Ciernik IF, Krause L. Treatment of radiation-induced maculopathy with fluocinolone acetonide. *Graefe's Archive for Clinical and Experimental Ophthalmology*. 2020;258(11):2535–2539. https://doi.org/10.1007/s00417-020-04804-7.
61. Amoaku WMK, Archer DB. Fluorescein angiographic features, natural course and treatment of radiation retinopathy. *Eye*. 1990;4(5):657–667. https://doi.org/10.1038/eye.1990.93.
62. Chaperon M, Kodjikian L, Agard E, et al. Screening of telangiectatic capillaries in chronic macular edema based on multimodal imaging: a study of 101 eyes. LyoMAC1 study. *Graefe's Archive for Clinical and Experimental Ophthalmology*. 2022;260(8):2501–2508. https://doi.org/10.1007/s00417-022-05592-y.
63. Caujolle JP, Paoli V, Chamorey E, et al. Local recurrence after uveal melanoma proton beam therapy: recurrence types and prognostic consequences. *International Journal of Radiation Oncology Biology Physics*. 2013;85(5):1218–1224. https://doi.org/10.1016/j.ijrobp.2012.10.005.
64. Russo A, Laguardia M, Damato B. Eccentric ruthenium plaque radiotherapy of posterior choroidal melanoma. *Graefe's Archive for Clinical and Experimental Ophthalmology*. 2012;250(10):1533–1540. https://doi.org/10.1007/s00417-012-1962-3.
65. Eandi CM, Polito MS, Schalenbourg A, Zografos L. Eighteen-month results of intravitreal anti–vascular endothelial growth factor on vision and microcirculation in radiation maculopathy. *Retina*. 2021;41(9):1883–1891. https://doi.org/10.1097/iae.0000000000003105.
66. Shields CL, Dalvin LA, Chang M, et al. Visual outcome at 4 years following plaque radiotherapy and prophylactic intravitreal bevacizumab (every 4 months for 2 years) for uveal melanoma: comparison with nonrandomized historical control individuals. *JAMA Ophthalmology*. 2020;138(2):136–146. https://doi.org/10.1001/jamaophthalmol.2019.5132.

Chapter 38

Giant cell arteritis—associated ocular vascular diseases

Marie-Helene Errera[1], Thomas M. Pearce[2] and Gabrielle Bonhomme[1]

[1]*Division of Ophthalmology, University of Pittsburgh Medical Center, Pittsburgh, PA, United States;* [2]*Division of Neuropathology, University of Pittsburgh Medical Center, Pittsburgh, PA, United States*

Overview bullets

- GCA is most common in people of Northern Europe and of Scandinavian descent patients aged >50.
- GCA is considered a medical emergency due to the potential risk of permanent visual loss.
- GCA is a disease of misdirected adaptive immunity. Innate immunity in GCA: monocytes and macrophages act as disease drivers in GCA resulting in macrophages (histocytes) and T cells forming granulomatous infiltrates in the adventitia and media. T cells encounter dendritic cells that lack the immunoinhibitory ligand PD-L1 and enter unopposed and persistent activation. The chronically stimulated T cells and macrophages result in the continuous production of chemokines and cytokines, tissue-damaging mediators (ROS, MMP-9), and growth factors (VEGF, PDGF), eliciting a maladaptive response-to-injury presenting as vessel wall remodeling, with wall vascularization and intimal hyperplasia.
- The American College of Rheumatology (ACR) guidelines for GCA recommend temporal artery biopsy as the first diagnostic test. In histology, GCA is characterized by pan arteritis of medium- to large-sized arteries.
- Recently, the European Alliance for Associations of Rheumatology (EULAR) guidelines have recommended color doppler ultrasound (CDU) of temporal and axillary arteries or magnetic resonance imaging (MRI) as the first diagnostic test to confirm GCA at centers with appropriate equipment and sufficient radiologic expertise.
- The British Society for Rheumatology (BSR) has recommended the use of a confirmatory test for GCA, which can either be "a temporal artery biopsy at least 1 cm in length or an ultrasound of the temporal and axillary arteries, or both."
- The halo sign detected by the color doppler ultrasound (CDU) of the temporal arteries has a good diagnosis performance for GCA.
- Treatments: Vascular complications of GCA can be abrupt and irreversible, for example, visual loss and cerebrovascular accidents, but these are not inevitable if GCA is diagnosed early and treated appropriately.
- High-dose glucocorticoids are the cornerstone of treatment in GCA.
- From clinical trials, anti-interleukin (IL)-6 receptor antibody, tocilizumab, significantly enhances the chances of achieving remission, preventing flares, and reducing glucocorticosteroid requirements in newly diagnosed and relapsing patients with GCA (Low evidence 1b).

Introduction

Giant cell arteritis (GCA), also known as "temporal arteritis," is an autoimmune disease that exclusively affects individuals over 50 years of age. The disease preferentially involves the thoracic aorta and its major branch vessels, including the temporal artery and vessels supplying the optic nerve and the retina. The clinical manifestations of GCA include aortic dissection, aortic aneurysm, and blindness due to arteritic ischemic optic neuropathy (ION) or retinal artery occlusion. More rarely, cerebral infarction may occur due to direct extension of vasculitis involving carotid or vertebrobasilar circulation.[1]

High disease risk in Northern European populations has supported the concept that both genetic and environmental factors shape disease susceptibility. Polymorphisms in the major histocompatibility complex (MHC), specifically the Human leukocyte antigen (HLA)-DR region confers the highest risk.[2] Among non-HLA regions, plasminogen (PLG) and

Retinal and Choroidal Vascular Diseases of the Eye. https://doi.org/10.1016/B978-0-443-15583-3.00038-X

prolyl 4-hydroxylase subunit alpha-2 (P4HA2), involved in vascular remodeling and neoangiogenesis, appear to play a role as risk determinants suggesting the relevance of these processes in GCA pathogenesis.[3]

The inflammatory activity of vascular lesions is driven by adaptive immune responses. The initiation of inflammation in GCA is considerably complex, with a critical role played by vessel-residing dendritic cells (DCs) localized at the adventitia-media border of normal medium-sized arteries acting in response to pathogen-derived macromolecules and exhibiting a gatekeeper function in regulating T-cell recruitment and retention to the arterial adventitia.[4] MMP-9 produced by monocytes and macrophages of GCA patients plays a critical role in the granulomatous inflammation in established and in late disease by digesting the basement membrane to breach the barrier between vasa vasorum capillaries and extracellular tissue.[5]

The recently published 2022 European League Against Rheumatism (EULAR)/American College of Rheumatology (ACR) classification criteria for GCA have a sensitivity of 87.0% (95% CI 82.0%−91.0%) and a specificity of 94.8% (95% CI 91.0%−97.4%) as compared to a sensitivity of only 37.1% for the 1990 ACR criteria in the diagnosis of GCA with large vessel involvement.[6]

Rapid and precise diagnosis of GCA is crucial for preventing irreversible damage such as ischemic visual loss. Since patients need long-term glucocorticoid therapy there is a significant burden of therapy-related complications, making accurate diagnosis important to avoid adverse events.

Pathogenesis

The three-layered walls of the muscular and elastic arteries are, in a healty state, free of inflammatory cells and protected by immune privilege. In the case of GCA, the immune privilege is lost, and both innate as well as adaptive immune cells can enter the privileged site.[7] In GCA, monocytes produce high amounts of the metalloproteinase MMP-9 and digest the basement membrane to overcome the barrier between vasa vasorum capillaries and extracellular tissue, and eventually, tissue-damaging multinucleated giant cells emerge.[5] Under physiologic conditions, PD-1 on T cells recognizes PD-L1 on antigen-presenting cells (e.g., macrophages), resulting in dampening of T-cell activation. GCA dendritic cells (DC) drive vasculitis by expressing CD86, amplifying disease-relevant enhancement of T-cell immunity.[8] Furthermore, in GCA, T cells encounter DC that lacks the immunoinhibitory ligand PD-L1 disrupting this negative signal and boosting T-cell effector functions. Therefore, T cells enter unopposed and persistent activation, especially since DC provides chemokines and cytokines to enhance immune cell recruitment. Chronically, stimulated CD4 T cells differentiate into multifunctional effector cells, undergo clonal expansion in the vessel wall, and release cytokines. In GCA patients, CD4+ T cells aberrantly express NOTCH1. These CD4+NOTCH1+ T cells from GCA patients recognize aberrantly expressed JAGGED1 on the surface of vasa vasorum endothelial cells, facilitating their invasion into the vessel wall.[9] Macrophages in the vessel wall produce chemokines and cytokines, tissue-damaging mediators (ROS, MMP-9), and growth factors (VEGF, PDGF). Finally, stimulated T cells and macrophages elicit a maladaptive response to injury presenting as vessel wall remodeling, with wall vascularization and intimal hyperplasia.[7] Tissue-resident memory CD4+ T cells render the lesion autonomous and ensure the chronicity of the disease.

The master regulators of the faulty immune response are CD4+ T cells that enter a protected tissue niche, take tissue residence, gain autonomy, and differentiate into multiple classes of differentiated effector T cells. A multitude of T-cell effector cytokines have been mapped to the lesions, including IL-2, IFN-γ, IL-17, IL-21, IL-9, IL-22, and GM-CSF. IFN-γ-producing CD4+ T cells represent the dominant T-cell subset in inflamed temporal arteries.[10] IFN-γ activates macrophages, DC, and endothelial cells and inhibits the proliferation of mesenchymal cells.

In GCA, immune responses driving the disease process create an inflammatory "response to injury" of the arterial wall in which excessive proliferation of myofibroblasts leads to thickening of the intima. Eventually, the arterial vessel lumen becomes occluded, blood flow is disrupted, and dependent organ structures suffer from ischemia resulting in massive ischemic damage to organs like the eye, the brain, and the heart.

Description

Giant cell arteritis may present with nonspecific symptoms such as malaise, unintentioned weight loss, depression, and fatigue or with the classic symptoms of new or temporal headache, scalp tenderness, jaw claudication, visual changes, or polymyalgia rheumatica. Irreversible blindness, the most feared complication, results from necrosis of the posterior ciliary branch of the ophthalmic artery.

Polymyalgia rheumatica, which can occur with or without giant cell arteritis, is characterized by painful stiffness of shoulder and pelvic girdle muscles.

Given its relative rarity, often nonspecific initial symptoms of pain, malaise, and headache, diverse symptomatology, and predilection for rapid progression to bilateral blindness, the timely diagnosis of GCA is often challenging.

Demographics

GCA is a disease of older people. The average age at onset is 72, and almost all people with the disease are over the age of 50. Women are afflicted with the disease 2–3 times more commonly than men. While GCA may occur in every racial group, it is most common in people of Northern Europe and of Scandinavian descent (14.6–43.6 per 100,000 aged >50 years).[11]

Classic symptoms of giant cell arteritis

The most common symptoms of GCA are headache, pain in the shoulders and hips (called polymyalgia rheumatica), pain, heaviness, or fatigue in the jaw after chewing (called jaw claudication), fever, and visual loss. Polymyalgia rheumatica can occur with or without giant cell arteritis. Other symptoms can include temporal tenderness, cough, throat pain, tongue pain, weight loss, depression, stroke, claudication in the masseter muscles, and, less often, diminished pulse or unequal blood pressure measurements in different extremities. Irreversible blindness, the most commonly feared complication, results from necrosis of the posterior ciliary branch of the ophthalmic artery. This is the most feared complication—inexorably develops in both eyes if the disease is not treated in a timely fashion.

Rates of vision loss as high as 35%–50% have been reported.[12] When a visual loss occurs due to GCA, real visual improvement occurs only rarely, in an estimated 5% of patients.[13]

Because the disease is relatively uncommon and because the disease can cause so many different symptoms, the diagnosis of GCA can be difficult to make. With appropriate therapy, GCA is an eminently treatable, controllable, and often curable disease.

Clinical ophthalmic findings

The rate of ophthalmic involvement is reported to be as high as 72%.[14]

Ocular involvement in GCA includes arteritic anterior ischemic optic neuropathy (A-AION) (Fig. 38.1) and retinal artery occlusion (Fig. 38.2), and more uncommonly: extraocular muscle palsies, episcleritis, scleritis, anterior segment ischemia, tonic pupil, ischemic iridoplegia, uveitis, retinal nerve fiber layer infarcts, and choroidal ischemia.[15]

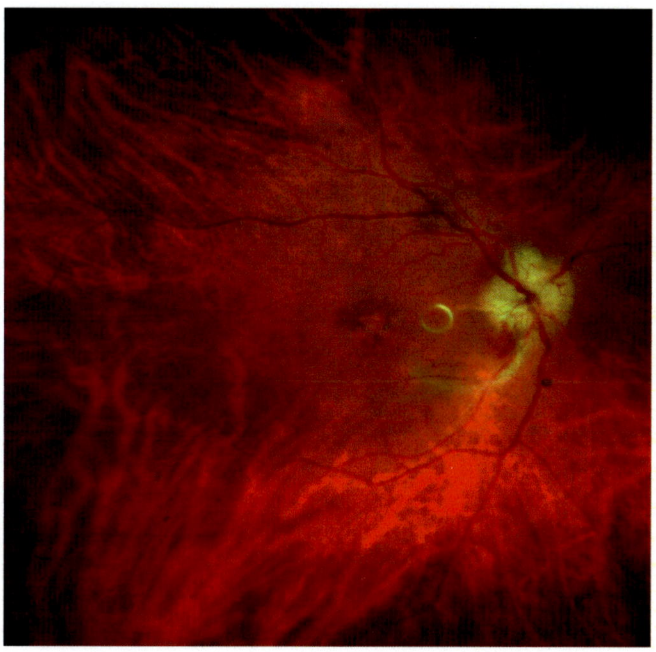

FIGURE 38.1 Case 1. Fundus color picture of a 91-year-old male with giant cell arteritis (GCA) proven by a temporal artery biopsy showing pallid disc edema and peripapillary hemorrhage.

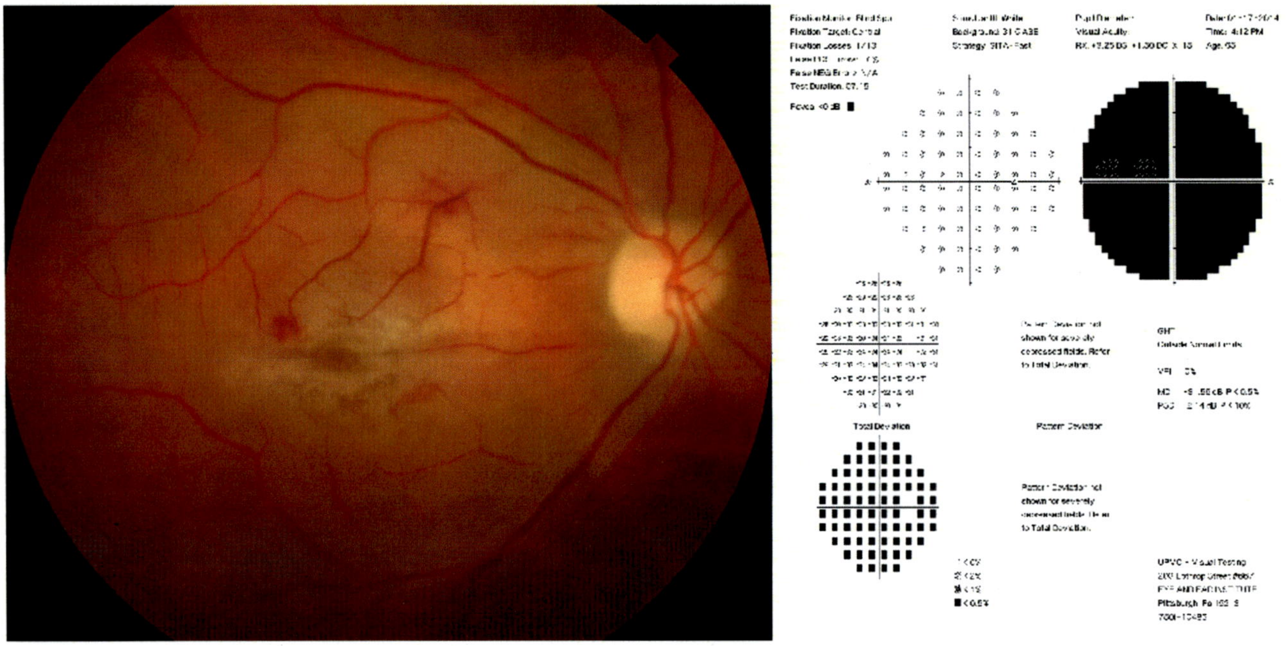

FIGURE 38.2 Case 2. Fundus color picture of a 65-year-old female with giant cell arteritis (GCA) proven by a temporal artery biopsy showing sectoral pale disc edema and retinal ischemia (whitening due to hypoperfusion vasculitis—associated arterial occlusion). There was no residual visual field.

A-AION is the product of inflammation, thrombosis, and the consequent ischemia of the short posterior ciliary arteries (PCAs) originating from the ophthalmic artery. Most people have two PCAs, whereas some have only one and others as many as five.[16] Fluorescein fundus angiography studies have shown that the medial PCA is more frequently occluded, which results in infarction of the optic nerve head.[17] Ocular manifestations include a rapidly developing, mostly unilateral loss of vision associated with severely reduced visual acuity and/or altitudinal visual field defects. Ophthalmoscopy reveals a pale, edematous, chalky-white optic disc that less commonly may show hyperemic swelling, an ophthalmoscopic picture quite similar to that of patients with non-A-AION. In less ominous situations, only a segment of the optic disc is affected, but this occurs in <25%—30% of patients.[18] The outcome of A-AION is the onset of optic atrophy.

Arteritic posterior ischemic optic neuropathy (A-PION)

This insidious presentation of GCA accounts for ≤10% of all IONs. As a consequence of the ischemia affecting the retrobulbar portion of the optic nerve, the patients complain of acute and painless reduction in visual acuity and color vision as well as visual field defects associated with a relative afferent pupillary defect. No abnormalities of the anterior segment and intraocular pressure are detected, and the optic nerve head is funduscopically normal in the acute phase, optic nerve atrophy with optic rim pallor becomes clearly evident later on.[17] Although A-PION more frequently involves the PCAs, ischemia of other orbital arteries may also be responsible for its occurrence.[16]

Central retinal artery occlusion (CRAO)

In the majority of patients, the central retinal artery and one or more of the PCAs derive from a common trunk of the ophthalmic artery. The fundus examination of an eye with CRAO shows diffuse intraretinal hemorrhages, a pallid retina, and optic disc edema of the affected eyes. A cherry red spot against a pale retina is also observed. The fluorescein column of the central retinal artery is almost completely absent by fluorescein angiography.

A previous study including 245 new cases of GCA identified described that the annual incidence of arteritic ION from GCA among individuals aged 50 or older was 1.3 per 100,000 population. Ischemic optic neuropathy was reported in 6.9% of GCA patients, central retinal artery occlusion in 1.6%, and cilioretinal artery occlusion in 0.4%. Permanent visual loss developed in 8.2% of GCA patients.[19]

Antonio-Santos et al. have retrospectively analyzed 37 patients with A-AION associated with GCA and reported that 9 had preserved visual acuity of 20/40 or better in the affected eye and color vision, suggesting that investigations for GCA should be performed not only in ION patients with profound visual loss but also in those presenting with good visual acuity, especially if they have giant cell arteritis systemic symptoms. The authors highlighted the indices that should be taken into consideration (i.e., age >50 years old, jaw or tongue claudication, new headache, abnormal visual fields, elevated erythrocyte sedimentation rate, and C-reactive protein (CRP). Notably, one of their patients developed involvement of the fellow eye despite parenteral high-dose corticosteroid treatment, suggesting latency of steroid response and continued risk of visual deterioration due to optic nerve head hypoperfusion.[20] High-dose corticosteroid therapy takes approximately 1 week to abort the arteritic process in the posterior ciliary artery walls; studies have shown that any visual deterioration often occurred within the first week of corticosteroid initiation.[21] Patients presenting with unilateral ION may also have a subclinical involvement of the fellow eye, evidenced by optic nerve sheath enhancement on MRI. All patients with preserved visual acuity had initial visual field defects that spared the central field.[20]

Among other studies, Dammacco et al. have reported an ocular involvement in 19 patients in the form of arteritic anterior and posterior ischemic optic neuropathy in 11 patients (57.9%) and one patient (5.3%), respectively, cotton wool spots in three patients (15.8%), central retinal artery occlusion in two patients (10.5%), and anterior segment ischemia and multifocal choroidal ischemia in one patient each (5.3%).[17]

Cotton Wool Spots have been reported to occur in almost one-third of GCA patients suffering visual loss in the early phase of the disease.[22] The spots are due to localized accumulations of axoplasmic debris as a result of focal inner retinal ischemia. Platelet microembolization and/or hypoperfusion of the terminal portions of the retinal vasculature are thought to play a role in their pathogenesis. Cotton wool spots may be an isolated finding (Fig. 38.3) or associated with other ocular features, such as optic disc edema.[23]

Anterior segment ischemia is a rare occurrence in cohorts of patients with GCA. It is more commonly diagnosed in patients with general ocular ischemic syndrome with involvement of the anterior ciliary arteries and sometimes of the long branches of the PCAs.[24] Corneal edema, anterior uveitis, and Descemet's folds with later rubeosis iridis can be detected.[17]

Visible Choroidal Ischemia is rarely seen with GCA, as often the retina may appear normal on ophthalmoscopic examination. Fundus fluorescein angiography reveals these perfusion abnormalities with delayed choroidal filling.

Although posterior[25] vasculitis in the form of retinal vasculitis associated with low-grade uveitis is uncommon as a presenting feature of GCA, it has been reported.[26] Panuveitis has been rarely described. One case of anterior uveitis was reported occurring 3 weeks before the diagnosis of GCA was made by TAB with an A-AION in the contralateral eye.[27]

This study emphasizes that treatment is not aimed at improving vision, but rather at preventing further visual deterioration in both eyes and in controlling the systemic vasculitic process.

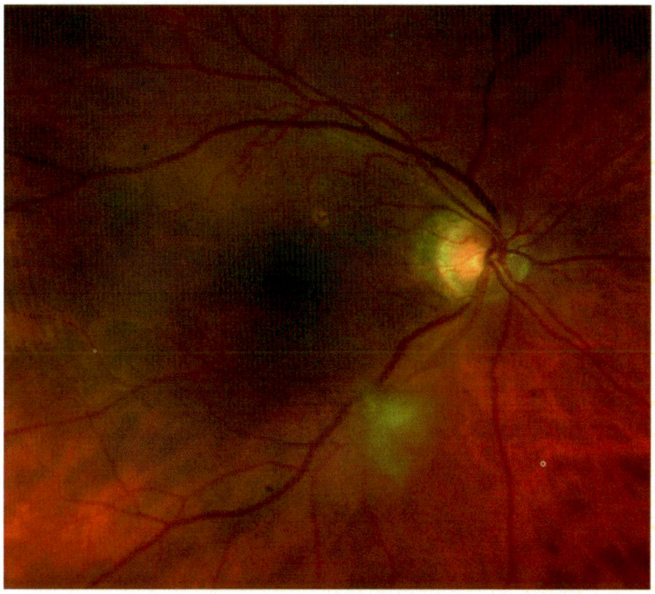

FIGURE 38.3 Case 2. Fundus color picture of a 75-year-old female with GCA proven by a bilateral temporal artery biopsy and bilateral cotton-wool spots, without optic disc edema, are seen in the right eye.

Other complications of GCA

The rate of occurrence of vascular complications in GCA is high within the first year from diagnosis (5 events per 100 person-years). The risk of development of aortic aneurysms/dissection increases over time, particularly after 5 years from diagnosis, and has been reported to occur in up to 20%–30% of patients.[28]

Aortic involvement is associated with a fivefold increased risk of death. Severe infections and GC dose above 10 mg/day of prednisone after 12 months of treatment represent a significant mortality risk (HR 3.19, 95% CI 1.76–5.53; $P = .025$ and HR 1.93, 95% CI 1.08–3.47; $P = .0001$, respectively). An increased risk of infection-related mortality was confirmed in diabetic patients, further enhancing the need for careful management of comorbid conditions in patients with GCA.[29]

How is giant cell arteritis diagnosed?

There is no independent gold standard diagnosis for GCA. The reference diagnosis used is based on classification criteria for GCA that include clinical features at presentation and biopsy results.

- **Biopsy of temporal artery (TAB)**
  The American College of Rheumatology (ACR) guidelines for GCA recommend TAB as the first diagnostic test.[30] The term "giant cell arteritis" is often used because when one looks at biopsies of inflamed temporal arteries under a microscope, one often sees large multinucleated "giant" cells.
  In histology, GCA is characterized by pan arteritis of medium- to large-sized arteries, often with fragmented internal elastic lamina, especially in the extracranial branches of the carotid artery. The vasculitic lesions of GCA are composed of tissue-infiltrating and tissue-resident innate and adaptive immune cells; mostly, CD4+ T cells, dendritic cells, macrophages, histiocytes, and multinucleated giant cells (Fig. 38.4).[31] Giant cells and histiocytes are found in 75% of specimens. Other less frequent inflammatory patterns in GCA include vasa vasorum vasculitis, in which inflammation is limited to the adventitial vasa vasorum without extension to the media.[32]
  Almost all patients have a markedly elevated sedimentation rate (ESR), averaging about 100 mm/h. However, fluctuations in the measurement of ESR and CRP can occur despite the persistence of vessel wall inflammation.[33] The lack of reliable disease biomarkers capturing vessel wall inflammation is problematic in managing GCA.
  Most patients have mild normochromic normocytic anemia, and 20%–30% have mildly elevated serum alkaline phosphatase. The leukocyte count at presentation is usually normal, a point favoring giant cell arteritis over infection or malignancy.
  In patients with giant cell arteritis, arterial involvement is patchy: therefore, studies recommend maximizing the chance of diagnosis by obtaining an optimal length cut-off of 0.7–1.5 cm segment and examining multiple sections.[34] Positive biopsy specimens show infiltration of the vessel wall with mononuclear inflammatory cells and giant cells, intimal proliferation, and thrombosis.
- **Color Doppler ultrasound (CDU)**
  Associations of Rheumatology (EULAR) guidelines have recommended CDU of temporal and axillary arteries or MRI as the first diagnostic test to confirm GCA at centers with appropriate equipment and sufficient radiologic expertise Doppler ultrasound (US) of the temporal artery or magnetic resonance imaging (MRA).[35] In one of the guidelines, the assessment of the axillary artery is recommended to be performed only in case of a normal CDU of temporal arteries.[36] The British Society for Rheumatology (BSR) proposed the use of a confirmatory test for GCA which can either be, "a temporal artery biopsy at least 1 cm in length, or an ultrasound of the temporal and axillary arteries, or both.[37] However, CDU is not considered a standard of care in the United States.
  CDU of the temporal arteries is a rapid, noninvasive method to diagnose GCA. The detection of the halo sign, that is, a dark area surrounding the vessel lumen by CDU of temporal arteries has good accuracy for the diagnosis of cranial GCA.[38] The compression sign in temporal arteries and the addition of axillary arteries assessment improves the diagnostic performance of CDU for GCA. The cutoff of intima-media thickness of the artery wall representing active inflammation in the temporal and axillary artery is $\geq$0.4 mm for temporal, facial, and occipital arteries.[39]
  In patients suspected of GCA, CDUS of the temporal artery tree on both sides should be performed as soon as possible by an experienced examiner using a machine equipped with a high-frequency linear or hockey stick probe (15 MHz or more).[40] Scanning axillary and subclavian arteries adds about 10% to the diagnostic yield of ultrasound because some patients with GCA have isolated extracranial vasculitis.[41]

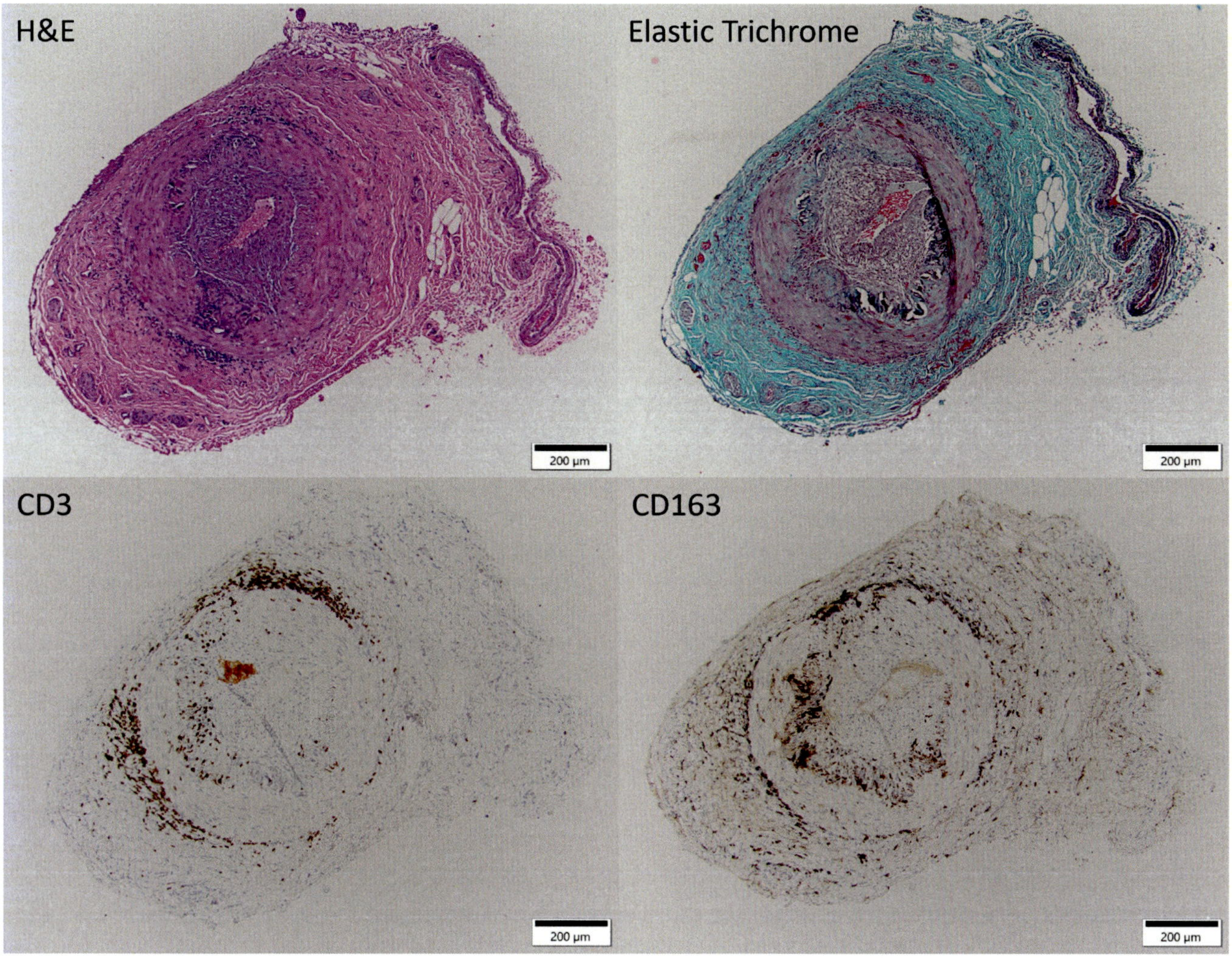

FIGURE 38.4 Histopathologic findings from case 2. Microscopic examination revealed a muscular artery with inflammation involving the media and adventitia. Hematoxylin and eosin (H&E) demonstrates the celluar infitrate. Elastic trichome demonstrates fragmentation of the internal elastica lamina. CD3 immunohistochemical staining highlights T cells. CD163 immunohistochemical staining highlights histiocytes. Scale bars: 200 μm.

A survey study evaluated 22 studies including 2893 participants with suspected GCA who underwent temporal artery CDU. The primary analysis results showed a sensitivity of 0.76 (95% confidence interval [95 CI] 0.69–0.81) and specificity of 0.93 (95 CI 0.89–0.95) when the halo sign was compared to clinical diagnosis. A sensitivity value of 0.84 (95 CI 0.72–0.92) and specificity of 0.95 (95 CI 0.88–0.98) were found in five studies involving 1037 participants that analyzed the halo sign and temporal artery compression sign. A sensitivity of 0.86 (95 CI 0.78–0.91) and specificity of 0.95 (95 CI 0.89–0.98) were found in four studies with 603 participants where the halo sign was evaluated by CDU on temporal and axillary arteries.[42]

Comparing TAB and CDU

Luqmani et al. analyzed 381 patients who underwent both ultrasound and TAB within 10 days of starting treatment for suspected GCA. The sensitivity of biopsy was 39% [95% confidence interval (CI) 33%–46%], which was significantly lower than previously reported and inferior to CDU (54%, 95% CI 48%–60%); the specificity of TAB (100%, 95% CI 97%–100%) was superior to CDU (81%, 95% CI 73%–88%). When performing biopsies only on negative cases, sensitivity increased to 65% and specificity was maintained at 81%, reducing the need for biopsies by 43%. Strategies combining clinical judgment (clinician's assessment at 2 weeks) with the tests showed sensitivity and specificity of 91% and 81%, respectively, for TAB and 93% and 77%, respectively, for ultrasound; cost-effectiveness (incremental net monetary benefit) was £485 per patient in favor of ultrasound with both cost savings and a small health gain. Interrater analysis revealed moderate agreement among sonographers (intraclass correlation coefficient 0.61, 95% CI 0.48−0.75), similar to pathologists (0.62, 95% CI 0.49−0.76).[43]

Other modalities for GCA diagnosis

Other modalities for the diagnosis of large-vessel vasculitis (LVV), like computerized tomography angiography (CTA), magnetic resonance angiography (MRA), and fluorodeoxyglucose positron emission tomography (FDG-PET) for the diagnosis of LVV were acknowledged in the 2022 EULAR/ACR GCA classification criteria.[6]

MRA of temporal arteries versus CDU have comparable sensitivity and specificity (69%−73% and 88%−91%, respectively).[44]

Treatments

What makes the treatment complex in GCA relates to the functional heterogeneity of vasculitic effector cells, which lends stability to the inflammatory lesions and renders them resistant to targeted immunosuppression.[7]

High-dose glucocorticoids (GCs) are the mainstay of treatment in GCA along with the recently approved anti-interleukin (IL)-6 receptor antibody, tocilizumab.[45]

GCs are highly effective in suppressing extravascular GCA, flattening the acute phase response, clinical symptoms, and abnormal laboratory parameters.[46] GCs should be initiated once GCA is suspected and before a definite diagnosis.

When started at a high dose of 1 mg/kg, GC induces rapid improvement in both clinical symptoms and inflammatory markers, minimizing the odds of GCA complications. Patients are frequently treated with a high dose of 1000 mg i.v. methylprednisolone for three consecutive when ocular symptoms or visual loss threats. High-dose GC should be maintained for 4 weeks. Glucocorticoids should be slowly tapered after that, by 10 mg every 2 weeks until 20 mg/day is reached, then reduced by 2.5 mg every 2−4 weeks to 10 mg and afterward by 1 mg every 1−2 months according to clinical response, with the ultimate goal of reaching ≤5 mg prednisolone daily after 1 year.[46] EULAR recommends that high-dose glucocorticoid therapy (40−60 mg/day prednisone-equivalent) should be initiated immediately for induction of remission in active GCA.[47]

The 2020 British Society of Rheumatology (BSR) guidelines also recommend an initial dose of 40−60 mg of prednisolone daily, tapering to complete GC cessation over a period of 12−18 months.[48]

Relapses of GCA are common during tapering and after the discontinuation of therapy, with a recent meta-analysis demonstrating an overall prevalence of relapses of around 47% in those treated with GC monotherapy.[49]

The toxicity of long-term GC treatment in GCA leads to find steroid-sparing agents.

Tocilizumab became the first biologic to receive US Food and Drug Administration approval for GCA. Subsequently, ACR recommends TCZ with GC as first-line treatment for GCA, while EULAR recommends it as second-line therapy.[50] EULAR recommends: "adjunctive therapy in selected patients with GCA (refractory or relapsing disease, presence of an increased risk for glucocorticoid-related adverse events or complications) using TCZ at a recommended dose of 162 mg. Methotrexate at a minimum dose of weekly 15 mg may be used as an alternative."[47]

The efficacy and safety of TCZ for the treatment of both newly diagnosed and refractory/relapsing GCA have been demonstrated in two randomized, placebo-controlled trials.[51] One RCT ($n = 251$ patients) (Trial of Tocilizumab in Giant Cell Arteritis (GiACTA)) tested two schemes of TCZ s.c. (162 mg every week or every other week) with 2 GC tapering protocols (26 weeks or 52 weeks). The other study ($n = 30$ patients) assessed the use of intravenous TCZ (8 mg/kg/monthly). The primary outcome was met in both studies. In the GiACTA trial, sustained GC-free remission at week 52 was 56% (TCZ/weekly) and 53% (TCZ/every other week) versus 14% (26-week GC taper) or 18% (52-week GC taper); $P < .001$. The HR for flares was 0.23 (99% CI 0.11−0.46) versus 0.28 (0.12−0.66); $P < .001$ compared with patients receiving a 26-week taper (of GC alone). The weekly TCZ dose was more efficacious in preventing flares in relapsing disease compared with the every-other-week regimen. The cumulative median GC dose over 52 weeks was significantly lower for the group receiving TCZ (1862 mg) compared with both GC schemes (3818 mg for the 52-week taper and 3296 for the 26-week taper); $P < .001$ for both comparisons. In the trial from Villiger et al., complete remission was achieved by week 12 in 12/20 (85%) patients treated with TCZ (at a GC dose 0.1 mg/kg/day) versus 4/10 (40%) in the comparator group; risk difference was 45% (95% CI 11%−79%; $P = .03$). The cumulative prednisolone dose after 52 weeks was 43 mg/kg in the TCZ group versus 110 mg/kg in the placebo group ($P = .005$). No safety issues arose from the two RCTs.[52]

However, apart from the IL-6 pathway, the more promising targets in GCA treatment are the IL12/IL23 axis antagonists, IL17 inhibitors, modulators of T lymphocytes, and inhibitors of either the JAK/STAT pathway, the granulocyte-macrophage colony-stimulating factor, or the endothelin.

IL-12/IL-23 Pathway Inhibition: Ustekinumab (UST) is a therapeutic human immunoglobulin (Ig) G1 kappa clonal antibody that binds to the p40 subunit common to both unbound IL-12 and IL-23. So far, separate groups have conducted two open-label single-center prospective trials, showing conflicting results. Currently, there is a new phase 2, randomized, parallel assignment, open-label study (NCT03711448) with s.c. UST 90 mg, at weeks 0, 4, 126, and 28, plus GC taper versus GC taper in relapsing GCA, with the objective of evaluating patients in remission at week 52. This study is still in the recruitment period with an estimated completion date of January 2025. Moreover, there is currently a phase 2 clinical trial examining guselkumab for the treatment of GCA (ClinicalTrials.gov Identifier: NCT04633447). This trial examines whether 12 months of guselkumab, in combination with 6 months of prednisone, is effective in keeping GCA in remission.

T-Cell Modulation. Abatacept (ABA), a CTLA4-Ig small molecule fusion protein, binds to CD80/CD86, thus preventing the engagement of CD28 with its ligand, which ultimately inhibits T-cell activation. One randomized control trial (NCT04474847), a phase 3, randomized, parallel assignment, compared s.c. ABA 125 mg every week for 12 months, versus s.c. placebo every week for 12 months in 41 patients with newly diagnosed or relapsing GCA (20 randomized in the ABA arm). In this trial, ABA was superior in increasing relapse-free survival at 12 months versus placebo (48% vs. 31%; $P = .049$). The median duration of remission was also higher in the ABA trial compared to placebo (9.9 vs. 3.9 months; $P = .023$).[53] A recent real-world study comparing ABA versus TCZ conducted in 33 patients with biopsy-proven GCA showed a complete or partial clinical response in 62% of patients treated with ABA, versus 88% of those treated with TCZ. However, the percentage of patients receiving prednisone at 12 months, not exceeding 7.5 mg/day as maintenance in any of the groups, was 43% in the ABA group compared to 100% in the group treated with TCZ.[54] These results favor the possible use of ABA as a rescue therapy in the management of GCA. Contradictory results are reported by a recent meta-analysis of six randomized clinical trials, including 260 GCA patients and 193 controls, showing no significant differences between anti-TNF, ABA, and placebo.[55]

IL-1 Inhibition. In a retrospective study, which analyzed GCA patients treated with anakinra (ANK), all six patients exhibited complete clinical and biological remission after a median duration of ANK therapy of 19 (18–32) months.[56] One randomized, controlled, double-blind study of ANK against placebo in addition to steroids in CA (GiAnT; NCT02902731) with ANK 100 mg/day by s.c. injection from day 1 until the end of week 16 is still in the recruitment period.

References

1. Larivière D, Sacre K, Klein I, et al. Extra- and intracranial cerebral vasculitis in giant cell arteritis: an observational study. *Medicine (Baltimore)*. 2014;93(28):e265. https://doi.org/10.1097/MD.0000000000000265.
2. Weyand CM, Hicok KC, Hunder GG, Goronzy JJ. The HLA-DRB1 locus as a genetic component in giant cell arteritis. Mapping of a disease-linked sequence motif to the antigen binding site of the HLA-DR molecule. *Journal of Clinical Investigation*. 1992;90(6):2355–2361.
3. Carmona FD, Vaglio A, Mackie SL, et al. A genome-wide association study identifies risk alleles in plasminogen and P4HA2 associated with giant cell arteritis. *The American Journal of Human Genetics*. 2017;100(1):64–74.
4. Coit P, et al. An update on the role of epigenetics in systemic vasculitis. *Current Opinion in Rheumatology*. January 2018;30(1):4–15.
5. Watanabe R, Maeda T, Zhang H, et al. MMP (matrix metalloprotease)-9-producing monocytes enable T cells to invade the vessel wall and cause vasculitis. *Circulation Research*. 2018;123(6):700–715.
6. Ponte C, Grayson PC, Robson JC, et al. 2022 American College of Rheumatology/EULAR classification criteria for giant cell arteritis. *Annals of the Rheumatic Diseases*. 2022;81:1647–1653. https://doi.org/10.1136/ard-2022-223480. Hunder GG, Bloch DA, Michel BA, et al. The American College of Rheumatology 1990 criteria for the classification of giant cell arteritis. *Arthritis & Rheumatology*. 1990;33:1122–1653. https://doi.org/10.1002/art.1780330810. Balbir-Gurman A, Awisat A, Keret S, Silawy A, Kaly L, Rosner I, Rozenbaum M, Boulman N, Shouval A, Rimar D, Slobodin G. Giant cell arteritis: state of the art in diagnosis, monitoring, and treatment. *Rambam Maimonides Medical Journal*. April, 2023;14(2).
7. Akiyama M, Ohtsuki S, Berry GJ, Liang DH, Goronzy JJ, Weyand CM. Innate and adaptive immunity in giant cell arteritis. *Frontiers in Immunology*. February 25, 2021;11:621098.
8. Krupa WM, Dewan M, Jeon MS, et al. Trapping of misdirected dendritic cells in the granulomatous lesions of giant cell arteritis. *American Journal of Pathology*. 2002;161(5):1815–1823.
9. Wen Z, Shen Y, Berry G, et al. The microvascular niche instructs T cells in large vessel vasculitis via the VEGF-Jagged1-Notch pathway. *Science Translational Medicine*. 2017;9(399):eaal3322.
10. Weyand CM, Watanabe R, Zhang H, Akiyama M, Berry GJ, Goronzy JJ. Cytokines, growth factors and proteases in medium and large vessel vasculitis. *Clinical Immunology*. 2019;206:33–41.
11. Sharma A, Mohammad AJ, Turesson C. Incidence and prevalence of giant cell arteritis and polymyalgia rheumatica: a systematic literature review. *Seminars in Arthritis and Rheumatism*. 2020;50:1040–1048.

12. Birkhead NC, Wagener HP, Shick RM. Treatment of temporal arteritis with adrenal corticosteroids. *JAMA*. 1975;163:821–827. Graham E, Holland A, Avery A, et al. Prognosis in giant-cell arteritis. *British Medical Journal*. 1981;282:269–271.
13. Hayreh SS, Zimmerman B, Kardon RH. Visual improvement with corticosteroid therapy in giant cell arteritis. Report of a large study and review of literature. *Acta Ophthalmologica Scandinavica*. 2002;80:355–367.
14. Figus M, Talarico R, Posarelli C, et al. Ocular involvement in giant cell arteritis. *Clinical & Experimental Rheumatology*. 2013;31(1 Suppl 75):S96.
15. Thurtell MJ, Longmuir RA. Third nerve palsy as the initial manifestation of giant cell arteritis. *Journal of Neuro-Ophthalmology*. 2014;34:243–245. Fytili C, Bournia VK, Korkou C, et al. Multiple cranial nerve palsies in giant cell arteritis and response to cyclophosphamide: a case report and review of the literature. *Rheumatology International*. 2015;35:773–776. Shakir R, Steuer A. An unusual cause of visual loss in a patient with giant cell arteritis. *International Journal of Rheumatic Diseases*. 2013;16:359–360. Foroozan R, Buono LM, Savino PJ, Sergott RC. Tonic pupils from giant cell arteritis. *British Journal of Ophthalmology*. 2003;87:510–512. Slemp SN, Martin SE, Burgett RA, Hattab EM. Giant cell arteritis presenting with uveitis. *Ocular Immunology and Inflammation*. 2014;22:391–393. Sane M, Selvadurai A, Reidy J, et al. Transient visual loss due to reversible 'pending' central retinal artery occlusion in occult giant cell arteritis. *Eye (Lond)*. 2014;28:1387–1390. Thanos A, Esmaili DD, Eliott D. Cotton-wool spots associated with mild nonprogressive visual loss. *JAMA Ophthalmology*. 2015;133:1471–1472. Hafidi Z, Handor H, Elmoussaif H, et al. Unilateral multifocal choroidal ischemia revealing a giant cell arteritis: a case report. *Pan African Medical Journal*. 2014;19:228. Olali C, Aggarwal S, Ahmed S, Gupta M. Giant cell arteritis presenting as macular choroidal ischaemia. *Eye (Lond)*. 2011;25:121–123.
16. Hayreh SS. Posterior ischaemic optic neuropathy: clinical features, pathogenesis, and management. *Eye (Lond)*. 2004;18(11):1188–1206. https://doi.org/10.1038/sj.eye.6701562.
17. Dammacco R, Alessio G, Giancipoli E, et al. Giant cell arteritis: the experience of two collaborative referral centers and an overview of disease pathogenesis and therapeutic advancements. *Clinical Ophthalmology*. March 11, 2020;14:775–793.
18. De Smit E, O'Sullivan E, Mackey DA, Hewitt AW. Giant cell arteritis: ophthalmic manifestations of a systemic disease. *Graefes Archive for Clinical and Experimental Ophthalmology*. 2016;254(12):2291–2306. Dammacco R, Alessio G, Giancipoli E, Leone P, Cirulli A, Resta L, Vacca A, Dammacco F. Giant cell arteritis: the experience of two collaborative referral centers and an overview of disease pathogenesis and therapeutic advancements. *Clinical Ophthalmology*. March 11, 2020;14:775–793.
19. Chen JJ, Leavitt JA, Fang C, et al. Evaluating the incidence of arteritic ischemic optic neuropathy and other causes of vision loss from giant cell arteritis. *Ophthalmology*. 2016;123:1999–2003.
20. Antonio-Santos AA, Murad-Kejbou SJ, Foroozan R, Yedavally S, Kaufman DI, Eggenberger ER. Preserved visual acuity in anterior ischemic optic neuropathy secondary to giant cell (temporal) arteritis. *Journal of Vascular and Interventional Neurology*. January 2016;8(5):17–21.
21. Hayreh SS, Zimmerman B. Visual deterioration in giant cell arteritis patients while on high doses of corticosteroid therapy. *Ophthalmology*. 2003;110:120–215. Cornblath WT, Eggenberger ER. Progressive visual loss from giant cell arteritis despite high-dose intravenous methylprednisolone. *Ophthalmology*. 1997;104:854–858.
22. Hayreh SS, Podhajsky PA, Zimmerman B. Occult giant cell arteritis: ocular manifestations. *American Journal of Ophthalmology*. 1998;125(4):521–526. https://doi.org/10.1016/S0002-9394(99)80193-7.
23. De Smit E, O'sullivan E. Cotton-wool spots in giant cell arteritis. *CMAJ (Canadian Medical Association Journal)*. 2013;185(9):796. https://doi.org/10.1503/cmaj.120540.
24. De Smit E, O'Sullivan E, Mackey DA, Hewitt AW. Giant cell arteritis: ophthalmic manifestations of a systemic disease. *Graefes Archive for Clinical and Experimental Ophthalmology*. 2016;254(12):2291–2306.
25. Rajesh CV, Cole M. Panuveitis as a presenting feature of giant cell arteritis. *British Journal of Ophthalmology*. 2000;84:340.
26. Moschos MM, Guex-Crosier Y. Giant cell arteritis: a rare cause of posterior vasculitis. *Clinical Ophthalmology*. 2009;3:111–115. Epub 2009 Jun 2. Nguyen NV, Karkhur S, Yuksel M, Hassan M, Halim MS, Nguyen QD, Hasanreisoglu M. Posterior uveitis associated with large vessel giant cell arteritis. *Ocular Immunology and Inflammation*. October–November 2022;30(7–8):2019–2115.
27. Bandini F, Benedetti L, Ceppa P, Corallo G. Uveitis as a presenting sign of giant cell arteritis. *Journal of Neuro-Ophthalmology*. 2005;25(3):247–248.
28. García-Martínez A, Arguis P, Prieto-González S, et al. Prospective long term follow-up of a cohort of patients with giant cell arteritis screened for aortic structural damage (aneurysm or dilatation). *Annals of the Rheumatic Diseases*. 2014;73:1826–1832.
29. Kermani TA, Warrington KJ, Crowson CS, et al. Large-Vessel involvement in giant cell arteritis: a population-based cohort study of the incidence-trends and prognosis. *Annals of the Rheumatic Diseases*. 2013;72:1989–1994. Schmidt J, Smail A, Roche B, et al. Incidence of severe infections and infection-related mortality during the course of giant cell arteritis: a multicenter, prospective, Double-Cohort study. *Arthritis & Rheumatology*. 2016;68:1477–1482. Monti S, Águeda AF, Luqmani RA, Buttgereit F, Cid M, Dejaco C, Mahr A, Ponte C, Salvarani C, Schmidt W, Hellmich B. Systematic literature review informing the 2018 update of the EULAR recommendation for the management of large vessel vasculitis: focus on giant cell arteritis. *RMD Open*. September 16, 2019;5(2):e001003.
30. Maz M, Chung S, Abril A, et al. American College of Rheumatology/Vasculitis Foundation Guidelines for the management of giant cell arteritis and Takayasu arteritis. *Arthritis & Rheumatology*. 2021;73(2021):1349–1365.
31. Weyand CM, Goronzy JJ. Immune mechanisms in medium and large-vessel vasculitis. *Nature Reviews Rheumatology*. 2013;9(12):731–740.
32. Cavazza A, Muratore F, Boiardi L, et al. Inflamed temporal artery: histologic findings in 354 biopsies, with clinical correlations. *The American Journal of Surgical Pathology*. 2014;38:1360–1370.
33. Maleszewski JJ, Younge BR, Fritzlen JT, et al. Clinical and pathological evolution of giant cell arteritis: a prospective study of follow-up temporal artery biopsies in 40 treated patients. *Modern Pathology*. 2017;30(6):788–796.

34. Ypsilantis E, Courtney ED, Chopra N, et al. Importance of specimen length during temporal artery biopsy. *British Journal of Surgery*. 2011;98:1556–1560. Oh LJ, Wong E, Gill AJ, McCluskey P, Smith JEH. Value of temporal artery biopsy length in diagnosing giant cell arteritis. *ANZ Journal of Surgery*. 2018;88:191–1560.
35. Dejaco C, Ramiro S, Duftner C, et al. EULAR recommendations for the use of imaging in large vessel vasculitis in clinical practice. *Annals of the Rheumatic Diseases*. 2018;77:636–643. https://doi.org/10.1136/annrheumdis-2017-212649. MacKie SL, Dejaco C, Appenzeller S, Camellino D, Duftner C, GonzalezChiappe S, Mahr A, Mukhtyar C, Reynolds G, de Souza AWS, Brouwer E, Bukhari M, Buttgereit F, Byrne D, Cid MC, Cimmino M, Direskeneli H, Gilbert K, Kermani TA, Khan A, Lanyon P, Luqmani R, Mallen C, Mason JC, Matteson EL, Merkel PA, Mollan S, Neill L, Sullivan EO, Sandovici M, Schmidt WA, Watts R, Whitlock M, Yacyshyn E, Ytterberg S, Dasgupta B. British Society for Rheumatology guideline on diagnosis and treatment of giant cell arteritis. *Rheumatology (United Kingdom)*. 2020;59:E1–643.
36. Dejaco C, Ramiro S, Duftner C, et al. EULAR recommendations for the use of imaging in large vessel vasculitis in clinical practice. *Annals of the Rheumatic Diseases*. 2018;77:636–643. https://doi.org/10.1136/annrheumdis-2017-212649.
37. British Society of Rheumatology. *BSR Guideline on Diagnosis and Treatment of Giant Cell Arteritis: Executive Summary, Draft Revision*. London: British Society of Rheumatology; 2019. https://www.rheumatology.org.uk/Portals/0/Documents/Guidelines/GCA/Open_Consultation_GCA_GL_Executive_Summary.pdf?ver=2019-07-02-152636-127.
38. Schmidt WA, Blockmans D. Investigations in systemic vasculitis—the role of imaging. *Best Practice & Research Clinical Rheumatology*. 2018;32:63–82.
39. Ješe R, Rotar Ž, Tomšič M, Hočevar A. The cut-off values for the intima-media complex thickness assessed by colour Doppler sonography in seven cranial and aortic arch arteries. *Rheumatology*. 2021;60:1346–1352.
40. Balbir-Gurman A, Awisat A, Keret S, et al. Giant cell arteritis: state of the art in diagnosis, monitoring, and treatment. *Rambam Maimonides Medical Journal*. April 30, 2023;14(2).
41. Schmidt WA, Seifert A, Gromnica-Ihle E, Krause A, Natusch A. Ultrasound of proximal upper extremity arteries to increase the diagnostic yield in large-vessel giant cell arteritis. *Rheumatology*. 2008;47:96–101.
42. Nakajima E, Moon FH, Junior NC, Macedo CR, de Souza AWS, Iared W. Accuracy of Doppler ultrasound in the diagnosis of giant cell arteritis: a systematic review and meta-analysis. *Advances in Rheumatology*. February 8, 2023;63(1):5.
43. Luqmani R, Lee E, Singh S, et al. The role of ultrasound compared to biopsy of temporal arteries in the diagnosis and treatment of giant cell arteritis (TABUL): a diagnostic accuracy and cost-effectiveness study. *Health Technology Assessment*. November 2016;20(90):1–238.
44. Bley TA, Reinhard M, Hauenstein C, et al. Comparison of duplex sonography and high-resolution magnetic resonance imaging in the diagnosis of giant cell (temporal) arteritis. *Arthritis & Rheumatism*. 2008;58:2574–2578. Duftner C, Dejaco C, Sepriano A, Falzon L, Schmidt WA, Ramiro S. Imaging in diagnosis, outcome prediction and monitoring of large vessel vasculitis: a systematic literature review and meta-analysis informing the EULAR recommendations. *RMD Open*. 2018;4:e000612.
45. Akiyama M, Kaneko Y, Takeuchi T. Tocilizumab in isolated polymyalgia rheumatica: a systematic literature review. *Seminars in Arthritis and Rheumatism*. 2020;50(3):521–525.
46. Monti S, Agueda A, Luqmani R, et al. Diagnosis, general management, drug and surgical treatment for large vessel vasculitis (with a focus on giant cell arteritis): a systematic literature review informing the 2018 update of the EULAR recommendation for the management of large vessel vasculitis. *RMD Open*. 2019;5(2):e001003.
47. Hellmich B, Agueda A, Monti S, et al. 2018 Update of the EULAR recommendations for the management of large vessel vasculitis. *Annals of the Rheumatic Diseases*. January 2020;79(1):19–30.
48. Mackie SL, Dejaco C, Appenzeller S, et al. British Society for Rheumatology guideline on diagnosis and treatment of giant cell arteritis. *Rheumatology*. 2020;59:e1–e23.
49. Mainbourg S, Addario A, Samson M, et al. Prevalence of giant cell arteritis relapse in patients treated with glucocorticoids: a meta-analysis. *Arthritis Care & Research*. 2020;72:838–849.
50. Maz M, Chung SA, Abril A, et al. 2021 American College of rheumatology/vasculitis foundation guideline for the management of giant cell arteritis and takayasu arteritis. *Arthritis Care & Research*. 2021;73:1071–1087. https://doi.org/10.1002/acr.24632. Hellmich B, Agueda A, Monti S, et al. 2018 Update of the EULAR recommendations for the management of large vessel vasculitis. *Annals of the Rheumatic Diseases*. 2020;79:19–1087. https://doi.org/10.1136/annrheumdis-2019-215672.
51. Villiger PM, Adler S, Kuchen S, et al. Tocilizumab for induction and maintenance of remission in giant cell arteritis: a phase 2, randomised, double-blind, placebo-controlled trial. *Lancet*. 2016;387:1921–1927. Stone JH, Tuckwell K, Dimonaco S, Klearman M, Aringer M, Blockmans D, Brouwer E, Cid MC, Dasgupta B, Rech J, et al. Trial of tocilizumab in giant-cell arteritis. *New England Journal of Medicine*. 2017;377:317–328. https://doi.org/10.1056/NEJMoa1613849.
52. Villiger PM, Adler S, Kuchen S, et al. Tocilizumab for induction and maintenance of remission in giant cell arteritis: a phase 2, randomised, double-blind, placebo-controlled trial. *Lancet*. 2016;387:1921–1927. Stone JH, Tuckwell K, Dimonaco S, Klearman M, Aringer M, Blockmans D, Brouwer E, Cid MC, Dasgupta B, Rech J, et al. Trial of tocilizumab in giant-cell arteritis. *The New England Journal of Medicine*. 2017;377:317–328. 10.1056/NEJMoa1613849. Monti S, Águeda AF, Luqmani RA, Buttgereit F, Cid M, Dejaco C, Mahr A, Ponte C, Salvarani C, Schmidt W, Hellmich B. Systematic literature review informing the 2018 update of the EULAR recommendation for the management of large vessel vasculitis: focus on giant cell arteritis. *RMD Open*. September 16, 2019;5(2):e001003.
53. Langford CA, Cuthbertson D, Ytterberg SR, et al. Vasculitis clinical research consortium. A randomized, double-blind trial of abatacept (CTLA-4Ig) for the treatment of giant cell arteritis. *Arthritis & Rheumatology*. 2017;69:837–845.

54. Castañeda S, Prieto-Peña D, Vicente-Rabaneda EF, et al. Advances in the treatment of giant cell arteritis. *Journal of Clinical Medicine*. March 13, 2022;11(6):158. Rossi D, Cecchi I, Sciascia S, Naretto C, Alpa M, Roccatello D. An agent-to-agent real life comparison study of tocilizumab versus abatacept in giant cell arteritis. *Clinical and Experimental Rheumatology*. 2021;39(Suppl. S129):125–128.
55. Castañeda S, Prieto-Peña D, Vicente-Rabaneda EF, et al. Advances in the treatment of giant cell arteritis. *Journal of Clinical Medicine*. March 13, 2022;11(6):1588. Song GG, Lee YH. Efficacy and safety of biological agents in patients with giant cell arteritis: a meta-analysis of randomized trials. *International Journal of Clinical Pharmacology and Therapeutics*. 2020;58:504–510.
56. Deshayes S, Ly KH, Rieu V, et al. French Study Group for Large Vessel Vasculitis (GEFA). Steroid-sparing effect of anakinra in giant-cell arteritis: a case series with clinical, biological and iconographic long-term assessments. *Rheumatology*. 2021;61:400–406.

Section VIII

Vascular tumors

Chapter 39

Retinal capillary hemangioblastoma and cavernous hemangioma

Lourdes Vidal-Oliver[1], Santiago Montolío-Marzo[2], Francisco Espejo-Arjona[3], Jorge Ruiz-Medrano[4], Roberto Gallego-Pinazo[2] and J. Fernando Arévalo[5]

[1]*Department of Retina, Fisabio Clinic, Valencia, Spain;* [2]*Department of Macula and Clinical Trials, Oftalvist Clinic, Valencia, Spain;* [3]*Department of Ophthalmology, University Hospital Virgen Macarena, Sevilla, Spain;* [4]*Department of Ophthalmology, University Hospital Puerta de Hierro, Madrid, Spain;* [5]*Wilmer Eye Institute, Johns Hopkins University School of Medicine, Baltimore, MD, United States*

Overview bullets

- Retinal capillary hemangioblastoma is a red/pink rounded lesion with tortuous feeding vessels. Retinal cavernous hemangioma has a "bunch-of-grapes" appearance.
- As opposed to retinal cavernous hemangioma, retinal capillary hemangioblastoma can cause exudation.
- von Hippel–Lindau disease must be ruled out in cases of retinal capillary hemangioblastoma.
- Neuroimaging should be performed in all cases of retinal cavernous hemangioma.

Retinal capillary hemangioblastoma

Introduction

Retinal capillary hemangioblastoma (RCH) is a vascular hamartoma that can be found isolated or as part of von Hippel–Lindau disease.[1] They are most commonly identified in individuals between 25 and 30 years old.[2]

von Hippel–Lindau disease (VHL) is an autosomal dominant hereditary disease caused by a partial defect in the short arm of chromosome 3. Mutations in the VHL gene cause von Hippel–Lindau syndrome. The VHL gene is a tumor suppressor gene, which means it keeps cells from growing and dividing too rapidly or in an uncontrolled way. Mutations in this gene prevent the production of the VHL protein or lead to the production of an abnormal version of the protein. RCH is the most common lesion in VHL and an early finding in this syndrome. Furthermore, 84% of RCH lesions are caused by VHL.[3] Isolated RCHs are diagnosed with VHL in 25%.[4] When RCH is present, it is essential to rule out VHL syndrome and related life-threatening lesions such as subtentorial hemangioblastoma, renal cell carcinoma, and pheochromocytoma.

Symptoms

RCH might be completely asymptomatic in the early stages. Most lesions progress slowly over time and a small percentage may even regress spontaneously.[2,5] As they grow, they may become more symptomatic and even vision threatening. Furthermore, tumor size influences treatment outcomes as small tumors (<1.5 mm) have been reported an eradication rate of up to 99.5%, whereas larger tumors (>1.5 mm) show a lower success, 68% eradication rate. RCH location also influences its behavior with juxtapapillary lesions remaining stable for longer than peripheral lesions.[2]

Decreased visual acuity is mainly caused by subretinal or intraretinal fluid due to exudation from the proliferating tumor or epiretinal membrane, tractional retinal detachment secondary to glial proliferation, and neovascular glaucoma. Vitreous hemorrhage might cause sudden vision loss.

Retinal and Choroidal Vascular Diseases of the Eye. https://doi.org/10.1016/B978-0-443-15583-3.00039-1

Diagnosis

RCH is seen as a red/pinkish bulging round mass in retinography (Fig. 39.1). It is usually placed in the periphery of the retina and in fewer cases juxtapapillary (17%, most commonly temporal to the optic disc).[2,6] A dilated tortuous artery feeds the tumor and an engorged vein drains it (feeder vessels). Yellowish exudates are usually found around the tumor. Due to its exudative appearance, it may resemble the exudation seen in Coat's disease.

RCH is usually a unilateral solitary tumor even in VHL. Up to one-third of patients with VHL may have multiple retinal hemangioblastomas, and half of the VHL patients may present with bilateral involvement.[7,8] RCH shows three different growth patterns: exophytic, endophytic (most frequent in VHL), and sessile.[6,9] Exophytic growth will not show the classic bulging round mass and multimodal imaging will help us guide the diagnosis.[10]

Fluorescein angiography is the most commonly used diagnostic method due to the high vascularization of the tumor. It shows rapid filling, early hyperfluorescence, and variable late staining and leakage. It will also help differentiate feeding and draining vessels (Figs. 39.1 and 39.2).[8]

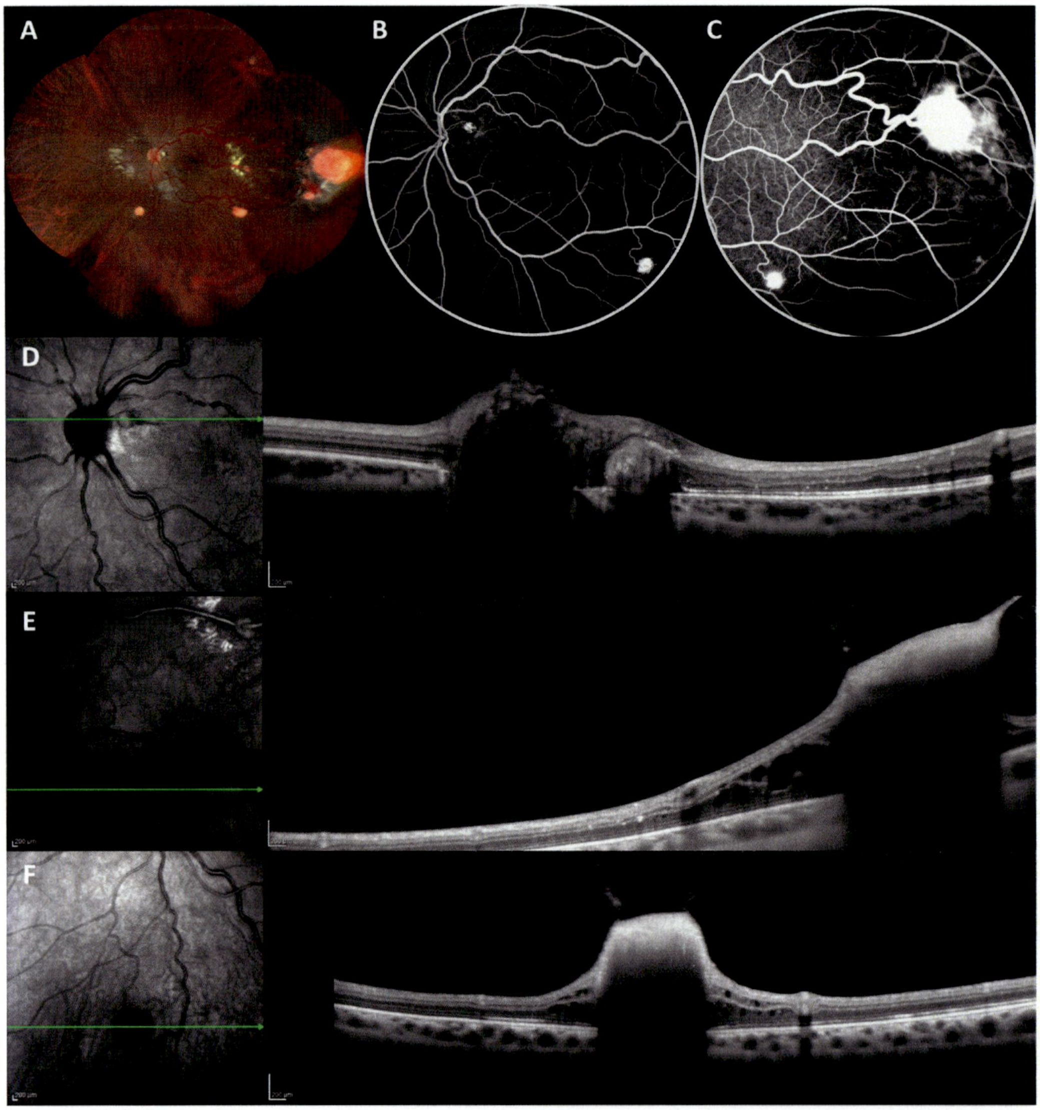

FIGURE 39.1 Multimodal imaging analysis of a case of retinal hemangioblastoma in a patient with von Hippel–Lindau syndrome. Color photograph (A) reveals the presence of lipid exudates in the peripapillary region and temporal to the fovea, with three small vascular lesions in the inferior region and a larger one in the temporal retina. Fluorescein angiography (B and C) shows the significant hyperfluorescence of the vascular lesions, with tortuosity of the feeding vessels. The structural optical coherence tomography scans over some of these lesions (D–F) enhance the typical architectural changes of retinal hemangioblastomas.

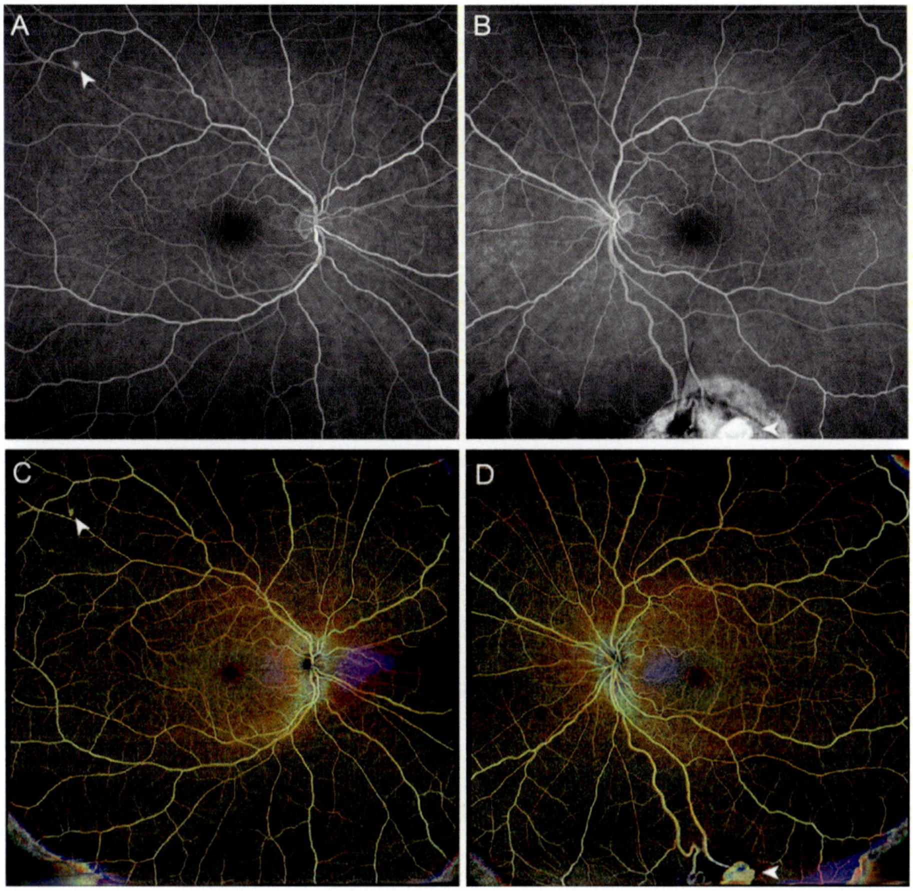

FIGURE 39.2 OCT-A imaging of retinal hemangioblastomas in von Hippel–Lindau syndrome. A 34-year-old woman with a history of pheochromocytoma, pancreatic tumors, and multiple CNS (cerebellar and spinal cord) and retinal hemangioblastomas secondary to von Hippel–Lindau syndrome presented for follow-up and underwent OCT-angiography (OCT-A) imaging. The patient was first diagnosed at an outside hospital at age 24 based on ocular manifestations and the diagnosis of VHL was confirmed with genetic testing despite both parents testing negative. Visual acuity was 20/20 bilaterally at the time of presentation and has remained stable. (A) Fluorescein angiography (FA) image of right eye demonstrating small untreated capillary hemangioblastoma (see *arrowhead*) superior temporal to the macula. (B) FA image of left eye with large inferior hemangioblastoma (previously treated with cryotherapy 9 years before image) with persistent flow. (C and D) 12 × 12 OCT-A image montages reveal corresponding lesions consistent with the hemangioblastomas identified in FA images in (A and B). *Courtesy of Amir H. Kashani, MD, PhD.*

Autofluorescence shows a uniform hypoautofluorescent lesion.[10]

Indocyanine green may help us distinguish small RCH from redundant leakage points in fluorescein angiography.[11]

Ultrasonography is essential in the presence of opaque media. It can help us measure the tumor size, its internal reflectivity, the presence of subretinal fluid, and complications such as retinal detachment.[8]

Optical coherence tomography (OCT) will show a hyperreflective lesion with shadowing and intraretinal fluid, whereas subretinal fluid is far less common (Fig. 39.1).[10] OCT is especially useful to asses macular edema, presence of fluid, and thickness changes to measure treatment response.[8]

OCT angiography has been used to study RCH in the macular region showing dense capillary net matching the lesion and lesion flow reduction after treatment (Fig. 39.2).[12] However, peripheral and larger lesions cannot be measured with angio-OCT machines.[8]

We must rule out VHL and associated systemic lesions in every patient showing RCH. Genetic testing may give some information about the course of the disease. Missense mutations are the most common while protein-truncating mutations have a higher probability for a severe course.

Several entities might be mistaken by RCH. RCH as an orange or red deep retinal lesion can be mistaken for peripapillary choroidal neovascularization or a small subretinal hemorrhage. Juxtapapillary RCH may obscure the optic disc margin and resemble optic disc edema or optic neuritis. We may include a differential diagnosis of juxta papillary lesions combined hamartoma of the retina and retinal pigment epithelium, astrocytic hamartoma, optic disc granuloma, and optic disc metastasis.[10]

Treatment

The treatment of RCH is based on tumor size, location, and secondary complications. The patient can be observed without treatment. Visual complaints, macular threatening, or lesion progression are the main reasons to treat these lesions. After treatment, a temporal increase in exudation or a secondary epiretinal membrane might be found.[6] Peripheral lesions can be treated with lower chances of visual loss.[13]

Observation: it is an option for asymptomatic, nonprogressive, nonvision threatening small tumors (diameter lower than 0.5 mm), especially juxtapapillary lesions. Nonetheless, a close follow-up is required to treat if needed.

Laser therapy: it is the most widely used treatment modality for small tumors not located in the far periphery showing no complications or subretinal fluid before treatment.[14] The objective is to reduce the caliber of the feeding and draining vessels. Complete disappearance is possible if the tumor is smaller than one disc diameter and improvement is possible up to 2.5 disc diameters.[15,16] Laser therapy might be repeated, specially guided by increased leakage of the lesion in fluorescein angiography. A tumor eradication rate of 89.9% has been reported with a good safety profile for peripheral tumors.[14] Hemorrhages and fluid accumulation are eventual complications after treatment.[14]

Photodynamic therapy: it is a good alternative to laser therapy when treating lesions with subretinal fluid. It is preferentially used for juxta papillary lesions compared to laser as it is less harmful to the perilesional tissue.[6] Its efficacy is limited in larger tumors as verteporfin may only be activated in the most superficial part of the lesion. A transient exudative response appears after treatment in 24% of the patients.[2,13] Its efficacy decays in peripheral tumors to 64.3%.[14] It is also more expensive than other alternatives.[1] Some authors consider that there is no enough evidence to recommend photodynamic therapy alone but some series encourage to keep using it either alone or combined with other therapies as it has shown promising results.[6,13]

Intravitreal or subtenon steroid injection: steroid treatment is helpful to control inflammation and exudation. They are also helpful to control inflammation secondary to other therapies.[2,17]

Cryotherapy: it may be useful especially for larger lesions, RCH in the far periphery or accompanied by a serous retinal detachment (although serous fluid will decrease success rate). It should be applied until the ice completely encloses the RCH.[18] It can cause a transient increase in exudate or proliferative vitreoretinopathy. A resolution rate of up to 70.2% has been reported for peripheral tumors, lower than other treatment strategies.[14]

Anti-VEGF therapy: they reduce vascular permeability and exudation, may stop RCH growth, and even prevent the appearance of new lesions.[1,8,17] They show the best safety profile among the available treatments, especially for juxtapapillary lesions.[6] They do not increase the incidence of cataracts in young patients compared to intravitreal steroids but show a lower success if inflammation or macular edema are severe, compared to steroids.[2] Their effect may be limited in large exophytic lesions as the drug is delivered intravitreal and may not reach the whole tumor.[6] Systemic administration of anti-VEGF drugs has shown no benefits.[19] Furthermore, full tumor control cannot be achieved with this therapy.[14]

Transpupillary thermotherapy: it is more suitable for posterior lesions which might be more accessible. Its efficacy has been questioned but a tumor eradication rate of up to 80% has been reported for peripheral tumors using high fluences.[14]

Plaque radiotherapy: it has shown to be useful, particularly for large tumors. It is technically more difficult to treat juxta papillary lesions and the macula might be easily damaged by the treatment in the most posterior cases.[2] A success rate of up to 96.3% has been reported for peripheral tumors.[14,20] It should be avoided in the presence of exudative retinal detachment.[14] External beam radiotherapy has shown no success in tumor eradication.[14]

Pars plana vitrectomy: it is considered for tumor excision in refractory lesions or to solve complications such as the presence of epiretinal membrane, macular hole, tractional retinal detachment, serous retinal detachment, or vitreous hemorrhage.[6,21] Feeder vessel ligation previous to tumor resection has been described to reduce the incidence of vitreous hemorrhage during surgery.[22] Tumor eradication rate has been reported up to 100% in peripheral tumors.[14]

Decisions must be taken according to the patient-specific characteristics, based on multimodal imaging. Secondary effects from the therapy will greatly influence our treatment approach. Juxtapapillary lesions can be observed if small, asymptomatic, and nonprogressive. They may be treated with photodynamic therapy and anti-VEGF as a first line to reduce secondary effects near the macula. Vitreoretinal surgery will be preferred if epiretinal membrane, vitreous hemorrhage, or retinal detachment are present or in refractory lesions.[6] Small peripheral RCH might be treated with laser therapy if accessible. Larger peripheral or inaccessible RCH can be treated with radiotherapy, cryotherapy, and transpupillary thermotherapy. Vitreoretinal surgery will be used in juxtapapillary lesions or in larger RCH over 4.5 mm.[14] Furthermore, the combination of different treatment strategies increases the chances for success and reduces the risk of iatrogenic damage.[11]

Oral HIF-2α inhibitor belzutifan *for ocular von Hippel—Lindau (VHL) disease*: In an open-label, single-arm, phase 2 study (NCT03401788), patients with VHL disease-associated clear cell renal cell carcinoma (VHL-RCC) received belzutifan, 120 mg orally daily.[23] Each eye was evaluated for RCH number/size/location; degree of feeding/draining vessel dilation; and presence of intraretinal/preretinal/vitreous hemorrhage, exudation, and/or fibrosis. Eyes were assigned a summary grade of improved, stable, or progressed at the last follow-up. A total of 61 patients were enrolled between 2018 and 2019. As of June 2020, the median follow-up was 69 weeks (range, 18—105). Confirmed ORR for VHL-RCC was

36% (95% CI, 24−49), with a disease control rate of 98% (95% CI, 91−100). The most common adverse events were anemia (90%) and fatigue (61%) for the entire study population. Color fundus images were obtained for 29 eyes (16 patients) having ≥1 RCH. 16/29 eyes (55%) were graded as improved, 12/29 eyes (41%) were stable, and 1/29 eyes (3%) were not evaluable (poor image quality). No eye was graded as progressed and no new RCH occurred. 1/29 eyes (3%) had a decrease in visual acuity >15 letters equivalent, secondary to retinal detachment in the setting of RH, epiretinal membranes, and myopia. The study concluded that Belzutifan demonstrated promising activity, including reduction in size, vascularity, and/or exudation in a subset of RCH, in addition to beneficial effects on RCC in patients with VHL disease. There are also some recently published case reports addressing its benefits.[24,25]

Retinal cavernous hemangioma

Introduction

Cavernous hemangioma of the retina (CHR) is a congenital vascular hamartoma arising from the inner retina, usually located in the periphery, mainly unilateral and unifocal. The diagnosis has been described from birth to 74 years, with no sex preponderance. Most cases are sporadic, although an autosomal dominant hereditary pattern in 20% of the reported cases has been described.

CHR can present either as an isolated lesion or as a part of a neuro-oculo-cutaneous syndrome. Up to 14% of patients with a diagnosis of CHR also have cerebral vascular hamartomas, so due to their morbidity risk, neuroimaging must be warranted in all patients to rule out cerebral lesions.[26]

The diagnosis of CHR is mainly clinical, with the hallmark being a cluster of "grape-like" aneurysms with perilesional grayish glial tissue. They are usually stable and asymptomatic if they do not involve the optic nerve or the macula, so observation is the preferred management.[27] Exudation is very rare and suggests an alternative diagnosis.

Clinical presentation

Up to 50% of reported cases are asymptomatic, being detected usually by chance.[26] This is mainly due to the typical nonexudative and nonprogressive nature of CHR and its peripheral location in 70% of cases. However, vision loss secondary to vitreous hemorrhage can be the presenting sign (25%),[28] caused by mechanical traction of the overlying glial tissue or epiretinal membrane to the walls of the aneurysms in the absence of a posterior vitreous detachment.[29]

In other cases, the presence of symptoms depends on tumor location.

Optic nerve tumors cause an enlarged blind spot depending on the size of the lesion.[30] They also tend to have higher incidences of vitreous hemorrhage, which can be either spontaneous or triggered by coagulation disorders or pregnancy.[30,31]

Macular CHR usually has a worse visual prognosis, causing a variable vision loss that can lead to strabismus in children.[32]

Another infrequent presentation is the anterior retina, where the tumor can overhang the ciliary body, involve the lens, and cause heterochromia.[33] Shields et al. reported a case that ended up in enucleation due to terminal glaucoma secondary to recurrent hyphemas and vitreous hemorrhage 52 years after the diagnosis.

CHR, although very rare, can also affect the entire retina, causing profound and irreversible visual loss.[34,35]

Although CHR is typically nonprogressive, some changes in the follow-up can be seen. The most frequent finding is the contraction of the overlying fibroglial tissue causing vascular tortuosity or even obliteration of the aneurysms. Also, although rare, there are reported cases of significant lesion growth.[36]

Diagnosis

The diagnosis of CHR is clinical. Typical retinography findings include aneurysms forming clusters with a "grape-like" appearance (Fig. 39.3). The periphery of the retina is the most common location (70%), followed by the optic nerve and macula.[26] Color range can vary depending on the degree of blood oxygenation inside the aneurysms, the extent of RPE hyperplasia, and the amount of fibroglial tissue on the surface of the lesion. Deoxygenated blood usually has a red-blueish color, RPE hyperplasia sometimes gives a darker appearance and fibroglial tissue is usually white or gray. CHR size can vary from just a few aneurysms to a pan-retinal lesion, but it is usually 1−2 disk diameters in width and nonprogressive. Due to the slow vascular flow inside the lesion, a separation of erythrocytes and plasma occurs, forming a layering or pseudohypopion inside the aneurysms. As the lesion is independent of the retinal circulation, the rest of the vasculature of

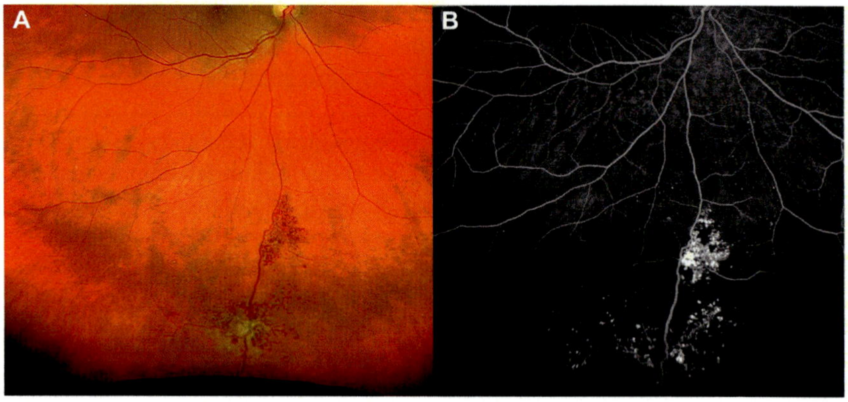

FIGURE 39.3 Typical appearance in the *color* photograph (A) and the fluorescein angiography (B) of retina cavernous hemangioma. Aneurysms forming clusters with a "*grape*-like" appearance can be observed.

the retina remains unaffected.[27] There are no feeder arteries nor engorgement of the surrounding vessels although there are four patients reported in the literature with a coexistence of CHR and retinal macrovessel.[37–39].

Histopathology. Biopsy of the lesion is only useful in cases of atypical tumor appearance. CHR is composed of interconnected blood-filled vascular spaces separated by thin fibrous septa with glial cells. Walls are made of non-fenestrated endothelial cells and pericytes, so there is no leakage of the intravascular content. Inside these spaces, a separation between red blood cells and plasma can be seen.[40]

These vascular hamartomas originate from the inner retina but can involve the full retina thickness. However, the choroid remains intact.[26]

Fluorescein angiography is the most valuable ancillary test to diagnose a CHR. It is hypofluorescent in the early phase, with a slow and incomplete late filling of the lesion without vascular leakage.[41] A characteristic "Fluorescein cap" can be seen inside of the aneurysms in the late phases, with hyperfluorescence of the upper half containing the plasma and hypofluorescence of the lower part due to blockage caused by the red blood cells (Fig. 39.3).[42]

Fundus autofluorescence shows an intense hypoautofluorescence of the lesion with some overlying areas of hyper-autofluorescence corresponding to the fibroglial tissue. Sometimes, the plasma inside of the cystic spaces can also form an "hyperautofluorescent cap."[41,43]

Indocyanine green is not useful in these cases because there is no choroidal involvement.

Ultrasound is preferred in cases of vitreous hemorrhage or other media opacities. CHR can be seen by ultrasound as a homogeneous mass with high internal reflectivity on the A-scan. The absence of focal choroidal excavation and acoustic shadow allows the differential diagnosis of malignant melanoma.[41] Yu et al. described the presence of intralesional phleboliths in a child presenting with echodense signals.[44]

OCT shows the depth of the lesion. It arises from the inner retina although it usually also affects the outer retina. The structure of the tumor can be studied with micrometer range resolution: thin-walled saccular structures with or without hyperreflective material inside. The vesicle boundaries are clear and intact.[41] An epiretinal membrane on the surface of the CHR is frequently seen, with or without traction.[45] Dense, hyperreflective, fibrous membranes surround the aneurysms.[43] En-face image shows the typical "bunch of grapes" appearance.

OCT angiography (OCTA) has many limitations in these cases. The blood flow inside the lesion is usually too slow to reach the threshold of the OCTA machine and can be depicted as vascularization.[46] In the cases where the flow is fast enough, an irregular saccular dilatation inside of the tumor has been described.[47] Segmentation artifacts are also usually present because of the retinal disorganization caused by the lesion. However, if the signal is strong enough, a normal choriocapillaris appearance can be confirmed.[48]

MultiColor. Mazzini et al. described the superficial fibroglial component seen with a greenish color on MultiColor imaging.[41]

Diagnosis is usually easy with multimodal imaging, although there are few entities that should be included in the differential diagnosis. Coats disease or macular telangiectasia type 1 is formed by aneurysms with exudation, with an abnormal surrounding vascular tree and it is usually progressive. CHR is a hamartoma, separated from a normal vasculature of the retina, without exudation and nonprogressive.[27]

There are also reports of CHR simulating malignant tumors. Secondary to the slow blood flow within the CHR, phleboliths can be formed, which appears in ultrasound and CT scan as calcified lesions, a characteristic sign of retinoblastoma.[44] In other cases with vitreous hemorrhage, malignant melanoma should be ruled out by ultrasonography.

When a diagnosis of CHR is made, further studies must be run to rule out systemic involvement. As 14% of cases have a concomitant intracranial lesion, neuroimaging is mandatory even without symptoms. Others can also have cutaneous lesions. The majority of cases are sporadic (80%), so testing family members is usually not necessary if there is only one eye affected. However, if CHR is found bilaterally, family screening is advisable, as 70% of autosomal dominant CHR are bilateral.[26]

Management

CHR is a nonprogressive, nonexudative lesion, so the preferred management in the majority of cases is observation. Further treatment strategies are reserved for complications such as vitreous hemorrhage, traction, or diagnostic doubts.

Pars plana vitrectomy is useful in cases of persistent or recurrent vitreous hemorrhage, release of traction, and cauterization of bleeding spots. Subtotal tumor excision has also been achieved with good outcome.[49]

Enucleation or evisceration is reserved for blind painful eyes and suspected malignancy of the lesion.

Other less frequent treatment options reported in single case reports include photodynamic therapy, causing tumor regression in a CHR of the disc,[31] and proton beam therapy, which was used to treat a peripheral CHR with collapse of the aneurysms. The utility of infliximab is controversial. There are two reports of its utilization in CHR, only one with improvement of the lesion.[50,51]

Conclusions

RCH and CHR are two vascular entities affecting the retina that can be easily differentiated by multimodal imaging. Whereas RCH is a homogeneous mass with feeder vessels and likely exudation, CHR has a "bunch-of-grapes" appearance, with normal surrounding retinal vascularization and without exudation.

Isolated RCHs are related to VHL in 25% of the patients. We must rule out VHL diagnosis and eventual life-threatening lesions related to this syndrome. RCHs are classically red/pinkish bulging round masses with rapid filling, early hyperfluorescence, and variable late staining and leakage in fluorescein angiography.

Observation is an option for asymptomatic nonprogressive, nonvision-threatening small RCH. Juxtapapillary RCH might be preferentially treated with photodynamic therapy and anti-VEGF to preserve the macula. Peripheral lesions might be treated with laser therapy, cryotherapy, transpupillary thermotherapy, or radiotherapy. Vitreoretinal surgery shows an eradication rate of 100% but it is used for refractory RCH, large RCH (>4.5 mm), or if other complications are present (vitreous hemorrhage, epiretinal membrane, or concomitant retinal detachment).

CHRs have coexistent intracranial lesions in 14% of patients, so neuroimaging should be performed in all cases. Family screening is often necessary if both eyes are involved. Retinography shows a cluster of aneurysms with fibroglial tissue overlying and surrounding the lesion. Fluorescein angiography is the most useful ancillary test to diagnose CHR, showing slow filling of the tumor and no fluorescein leakage. In the late phase, a "fluorescein cap" can be seen, representing the separation between erythrocytes and plasma inside the aneurysms. Observation is the preferred management modality because it is usually nonprogressive. Overall, 25% of patients suffer from vitreous hemorrhage that can require vitrectomy if persistent or recurrent.

References

1. Won YK, Lee MW, Shin YI, Kim JY, Jo YJ. Clinical results of various treatments for retinal capillary hemangioma. *Korean Journal of Ophthalmology*. April 2020;34(2):133–142.
2. Chelala E, Dirani A, Fadlallah A. Intravitreal anti-VEGF injection for the treatment of progressive juxtapapillary retinal capillary hemangioma: a case report and mini review of the literature. *Clinical Ophthalmology*. 2013;7:2143–2146.
3. Lee SH, Park KH, Woo SJ, Park SJ, Joo K. Clinical and genetic characteristics of retinal capillary hemangioblastoma in Korean patients. *Korean Journal of Ophthalmology*. October 2022;36(6):543–549.
4. Singh A, Shields J, Shields C. Solitary retinal capillary hemangioma: hereditary (von Hippel-Lindau disease) or nonhereditary? *Archives of Ophthalmology*. February 2001;119(2):232–234.
5. Milewski SA. Spontaneous regression of a capillary hemangioma of the optic disc. *Archives of Ophthalmology*. August 2002;120(8):1100–1101.
6. Saitta A, Nicolai M, Giovannini A, Mariotti C. Juxtapapillary retinal capillary hemangioma: new therapeutic strategies. *Medical Hypothesis, Discovery, and Innovations Ophthalmology*. 2014;3(3):71–75.
7. Singh AD, Shields CL, Shields JA. von Hippel-Lindau disease. *Survey of Ophthalmology*. 2001;46(2):117–142.
8. Otero-Marquez O, Chui TY, Pinhas A, et al. 3-D OCT angiographic evidence of anti-VEGF therapeutic effects on retinal capillary hemangioma. *American Journal of Ophthalmology Case Reports*. 2022;25:101394.

9. Al-Essa RS, Helmi HA, Alkatan HM, Maktabi AMY. Juxtapapillary retinal capillary hemangioma: a clinical and histopathological case report. *International Journal of Surgery Case Reports*. 2021;79:227–230.
10. Russell JF, Villegas VM, Schwartz SG, et al. Multimodal imaging in the diagnosis of exophytic juxtapapillary retinal capillary hemangioblastoma. *American Journal of Ophthalmology*. May 2021;225:128–136.
11. Guo J, Du L, Zhou P, Guo X, Dai F, Jin X. Combined therapy guided by multimodal imaging of fifteen retinal capillary hemangioblastomas in a monocular Von Hippel- Lindau syndrome case report. *BMC Ophthalmology*. May 2022;22(1):205.
12. Ouederni M, Maamouri R, Sassi H, Nouri S, Cheour M. [Multimodal imaging in the diagnosis of retinal capillary hemangioblastoma]. *Journal Français d'Ophtalmologie*. June 2021;44(6):912–914.
13. Di Nicola M, Williams BKJ, Hua J, et al. Photodynamic therapy for retinal hemangioblastoma: treatment outcomes of 17 consecutive patients. *Ophthalmology Retina*. January 2022;6(1):80–88.
14. Hajjaj A, van Overdam KA, Gishti O, Ramdas WD, Kiliç E. Efficacy and safety of current treatment options for peripheral retinal haemangioblastomas: a systematic review. *Acta Ophthalmologica*. February 2022;100(1):e38–e46.
15. Lane CM, Turner G, Gregor ZJ, Bird AC. Laser treatment of retinal angiomatosis. *Eye (Lond)*. 1989;3(Pt 1):33–38.
16. Shields JA. The expanding role of laser photocoagulation for intraocular tumors. The 1993 H. Christian Zweng Memorial Lecture. *Retina*. 1994;14(4):310–322.
17. Ach T, Thiemeyer D, Hoeh AE, Schaal KB, Dithmar S. Intravitreal bevacizumab for retinal capillary haemangioma: longterm results. *Acta Ophthalmologica*. 2010;88:e137–e138.
18. Singh AD, Nouri M, Shields CL, Shields JA, Perez N. Treatment of retinal capillary hemangioma. *Ophthalmology*. October 2002;109(10):1799–1806.
19. von Buelow M, Pape S, Hoerauf H. Systemic bevacizumab treatment of a juxtapapillary retinal haemangioma. *Acta Ophthalmologica Scandinavica*. February 2007;85(1):114–116.
20. Dalbah S, Bechrakis NE, Thomasen H, et al. Brachytherapy for peripheral retinal capillary haemangioblastoma in von Hippel-Lindau disease. *Klin Monbl Augenheilkd*. July 2021;238(7):781–787.
21. Parra MM, Arias JD, Viteri EJ, Bolivar I, Vega LY, Hartnett ME. Vitrectomy-assisted excision in the treatment of juxtapapillary retinal capillary hemangioma. *Ophthalmic Surgery, Lasers and Imaging Retina*. October 2022;53(10):570–573.
22. Weng CY. Transvitreal feeder vessel ligation and En Bloc resection of a retinal capillary hemangioblastoma. *American Journal of Ophthalmology*. May 2022;237:e3–e5.
23. Wiley H, Coleman HR, Jonasch E, et al. Oral HIF-2α inhibitor belzutifan for ocular von Hippel-Lindau (VHL) disease. *Investigative Ophthalmology and Visual Science*. 2021;62(8):32.
24. Jones AA, Schloemer NJ, Wirostko WJ. Successful treatment of von Hippel-Lindau (VHL) disease-associated retinal capillary hemangioblastoma (RCH) with belzutifan in a pediatric patient. *Retinal Cases and Brief Reports*. July 13, 2023. https://doi.org/10.1097/ICB.0000000000001452.
25. Grimes JM, Gershkovich A, Bogomolny D, Marr BP. Two cases of von Hippel- Lindau syndrome-associated retinal hemangioblastoma treated with belzutifan. *Retinal Cases & Brief Reports*. November 16, 2022. https://doi.org/10.1097/ICB.0000000000001376.
26. Wang W, Chen L. Cavernous hemangioma of the retina: a comprehensive review of the literature (1934-2015). *Retina*. April 2017;37(4):611–621.
27. Gass JD. Cavernous hemangioma of the retina. A neuro-oculo-cutaneous syndrome. *American Journal of Ophthalmology*. April 1971;71(4):799–814.
28. Fowler BJ, Simon L, Scott NL, Negron CI, Berrocal AM. Case report: vitreous hemorrhage as the presenting sign of retinal cavernous hemangioma in a newborn. *American Journal of Ophthalmology Case Reports*. 2021;23:101174.
29. Pringle E, Chen S, Rubinstein A, Patel CK, Downes S. Optical coherence tomography in retinal cavernous haemangioma may explain the mechanism of vitreous haemorrhage. *Eye (London, England)*. 2009;23:1242–1243.
30. Vela JI, Garcia-Vilaró M, Buil JA. Haemorrhage of a cavernous haemangioma of the optic disc in pregnancy. *BMJ Case Reports*. December 2010;2010.
31. Shanmugam MP, Ramanjulu R, Dwivedi S, Barigali A, Havanje A. Therapeutic surprise! photodynamic therapy for cavernous haemangioma of the disc. *Indian Journal of Ophthalmology*. August 2017;65(8):754–757.
32. Knani L, Mahjoub A, Nouri H, et al. [Childhood strabismus secondary to posterior pole cavernous hemangioma of the retina]. *Journal Français d'Ophtalmologie*. 2021;44:e555–e557.
33. Schiller ER, Kalafatis NE, Shields CL. Retinal cavernous hemangioma in a newborn: an unusual presentation of a rare tumor. *Journal of American Association for Pediatric Ophthalmology and Strabismus*. 2022;26:330–333.
34. Yannuzzi NA, Fallas B, Berrocal AM. Diffuse retinal cavernous hemangioma in a 2-year-old girl. *JAMA Ophthalmology*. May 2020;138(5):e191696.
35. Hewick S, Lois N, Olson JA. Circumferential peripheral retinal cavernous hemangioma. *Archives of Ophthalmology*. October 2004;122(10):1557–1560.
36. Kushner MS, Jampol LM, Haller JA. Cavernous hemangioma of the optic nerve. *Retina*. 1994;14(4):359–361.
37. Bruè C, Vance SK, Yannuzzi LA, Freund KB. Cavernous hemangioma associated with retinal macrovessels. *Retina Cases Brief Reports*. 2011;5(4):323–325.
38. Thanos A, Randhawa S, Drenser KA. Macular retinal cavernous hemangioma associated with congenital retinal macrovessel. *JAMA Ophthalmology*. September 2016;134(9):e161683.
39. Zbiba W, Elleuch IE, Sayadi S, Laadheri M. Multimodal imaging in a case of a congenital retinal macrovessel associated with a retinal cavernous hemangioma: a case report. *BMC Ophthalmology*. February 2020;20(1):46.

40. Messmer E, Font RL, Laqua H, Höpping W, Naumann GO. Cavernous hemangioma of the retina. Immunohistochemical and ultrastructural observations. *Archives of Ophthalmology*. March 1984;102(3):413–418.
41. Mazzini C, Vicini G, Nicolosi C, et al. Multimodal imaging of a retinal cavernous hemangioma. *European Journal of Ophthalmology*. November 2020:1120672120971549.
42. Kumar M, Reddy N, Konana VK, Kanakamedla A, Ruia S, Gudimetla J. Fluorescein cap: fluorescein angiographic feature of retinal cavernous hemangioma. *Indian Journal of Ophthalmology*. October 2018;66(10):1473–1474.
43. Li J, Li Y, Li H. New interpretation of multimodality fundus imaging for retinal cavernous hemangioma. *Current Eye Research*. April 2019;44(4):423–427.
44. Yu MD, Dalvin LA, Shields CL. Retinal cavernous hemangioma with intralesional phleboliths. *Retina Cases Brief Reports*. 2020;14(4):301–304.
45. Singh E, Sen A. A case of retinal cavernous hemangioma analyzed with optical coherence tomography angiography. *Indian Journal of Ophthalmology*. 2019;67:561–563.
46. Spaide RF, Fujimoto JG, Waheed NK, Sadda SR, Staurenghi G. Optical coherence tomography angiography. *Progress in Retinal and Eye Research*. May 2018;64:1–55.
47. Kalevar A, Patel KH, McDonald HR. Optical coherence tomography angiography of retinal cavernous hemangioma. *Retina*. May 2017;37(5):e50–e51.
48. Cennamo G, Amoroso F, Solari D, Alfieri M, de Crecchio G. Optical coherence tomography angiography in retinal cavernous hemangioma. *International Journal of Ophthalmology*. 2017;10(12):1945–1946.
49. Haller JA, Knox DL. Vitrectomy for persistent vitreous hemorrhage from a cavernous hemangioma of the optic disk. *American Journal of Ophthalmology*. 1993;116:106–107.
50. Japiassú RM, Moura Brasil OF, de Souza EC. Regression of macular cavernous hemangioma with systemic infliximab. *Ophthalmic Surgery, Lasers Imaging*. March 2010:1–3.
51. Alsulaiman SM, Abouammoh MA, Al-Dahmash SA, Abu El-Asrar AM. Is systemic infliximab therapy effective for retinal cavernous hemangioma? *Saudi Medical Journal*. September 2014;35(9):1127–1130.

Chapter 40

Choroidal hemangioma

Alexandre Matet[1,2], Denis Malaise[2], Laurence Rosier[3] and Nathalie Cassoux[1,2]

[1]Université Paris Cité, Paris, France; [2]Institut Curie, Department of Ocular Oncology, Paris, France; [3]Centre Rétine Gallien, Bordeaux, France

Overview bullets

- Choroidal hemangioma are benign vascular lesions of the ocular fundus
- Circumscribed hemangioma is sporadic and presents as a red-orange lesion with subretinal fluid.
- The pathognomonic sign of circumscribed choroidal hemangioma is "wash-out" on late ICG. Its management relies on PDT or targeted irradiation for thicker lesions
- Diffuse hemangioma is often associated with Sturge–Weber syndrome
- It is a diffuse choroidal thickening that causes subretinal exudation
- It is best treated by external radiotherapy of the whole choroid

Introduction

Choroidal hemangiomas are benign vascular tumors that are divided into two types. The circumscribed choroidal hemangioma is a localized enlargement of the choroid that has no extraocular associations, while the diffuse type can be associated with facial naevus flammeus or variations of Sturge–Weber syndrome. Both manifest with exudation and subretinal fluid accumulation.

Circumscribed choroidal hemangioma

Pathogenesis

Circumscribed choroidal hemangioma is a benign vascular tumor of the choroid. It is a rare vascular malformation with an unknown incidence because small asymptomatic lesions are probably underdiagnosed. Circumscribed choroidal hemangioma is usually diagnosed when it manifests with a decrease in visual acuity due to macular subretinal fluid.

The lesions consist of dilation of choroidal vasculature with intermixed proliferation of small and large blood vessels. The histomorphology is characterized by thin-walled vascular channels with endothelial cells and scant stroma cells.[1] Recent genetic data on rare tissue material identified somatic mutations in the *GNAQ* gene, especially on the Q209 codon.[1,2] Interestingly, this gene is also involved in the tumorigenesis of diffuse choroidal hemangioma and melanocytic lesions, such as choroidal naevi and melanoma.

Clinical features

Circumscribed choroidal hemangioma is often asymptomatic and discovered during a routine fundus examination. The diagnosis is most frequent in middle-aged adults, between 40 and 60 years old. Symptoms include decreased visual acuity, metamorphopsia, phosphenes, and visual field changes. Due to retinal bulging, it can cause hyperopia that exacerbates the perceived visual blur.

Clinically, circumscribed choroidal hemangioma presents as a round or slightly oval choroidal bulge, typically orange-red in color (Fig. 40.1). It may be difficult to distinguish on a fundus color photograph and the elevation may be more visible on fundus examination. It appears translucent on retro-illumination. The base of the hemangioma is sometimes

Retinal and Choroidal Vascular Diseases of the Eye. https://doi.org/10.1016/B978-0-443-15583-3.00040-8

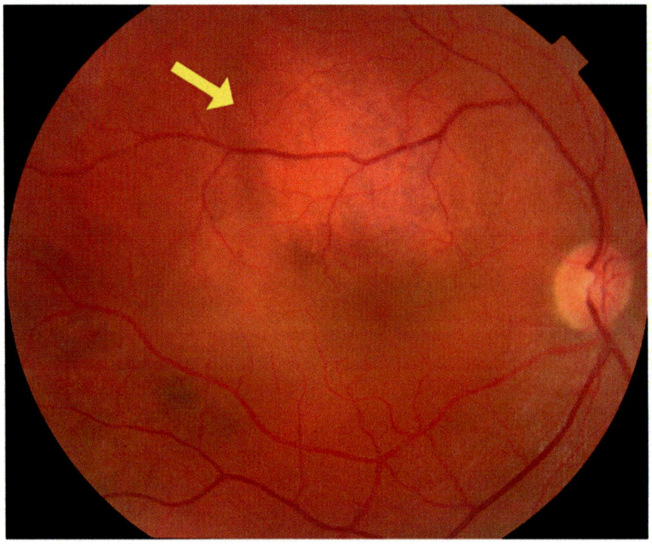

FIGURE 40.1 Red-orange choroidal lesion superior to the macula corresponding to a circumscribed choroidal hemangioma (*yellow arrow*), with exudative serous macular detachment.

surrounded by a dark, fuzzy choroidal ring corresponding to a condensation of uveal melanocytes compressed by the tumor, more visible in dark-skinned individuals. The localization of circumscribed choroidal hemangioma is unilateral, usually at the posterior pole, and more specifically in the macular or perimacular area.[3]

Significant alterations of the pigment epithelium (atrophy, depigmentation, or hyperplasia) are sometimes present, especially in long-standing lesions after a long period of silent, mild exudation (Fig. 40.2). Exudative retinal serous detachment is very frequent in symptomatic lesions, most often perilesional or localized at its inferior border. In neglected lesions, this serous retinal detachment may become bullous (Fig. 40.3) and may exceptionally lead to neovascular glaucoma. Subretinal fibrosis and orange pigment on the tumor surface may also be observed in chronic lesions.[4]

Imaging findings

Multimodal imaging findings are necessary to confirm the diagnosis and assess the visual potential to guide treatment decisions.

Wide-field color fundus photography

We emphasize the importance of using a fundus camera that renders true color. Otherwise, the risk of misdiagnosis, especially between hemangioma and choroidal melanoma, is significant because amelanotic lesions are darker with certain imaging systems.

Ocular ultrasonography

In A mode, the internal reflectivity is very high, homogeneous, and regular without attenuation.

In B mode, the choroidal hemangioma appears as an iso- or hyper-echogenic biconvex choroidal mass with soft and indistinct contours, homogeneous with regular internal structure. There is no associated choroidal excavation (Fig. 40.4). It may be associated with a thin hyperechogenic layer at its internal border, corresponding to an exudative retinal detachment (Fig. 40.5).

In addition, B-mode ultrasonography allows the assessment of tumor dimensions (greatest height and diameter).

Autofluorescence

Choroidal hemangioma is inherently iso-autofluorescent. Extrinsic hyperautofluorescent areas correspond to surface orange pigment or recent subretinal fluid, whereas hypoautofluorescent patches correspond to areas of pigment epithelial hyperplasia, fibrous metaplasia, or atrophy (Fig. 40.6).[5]

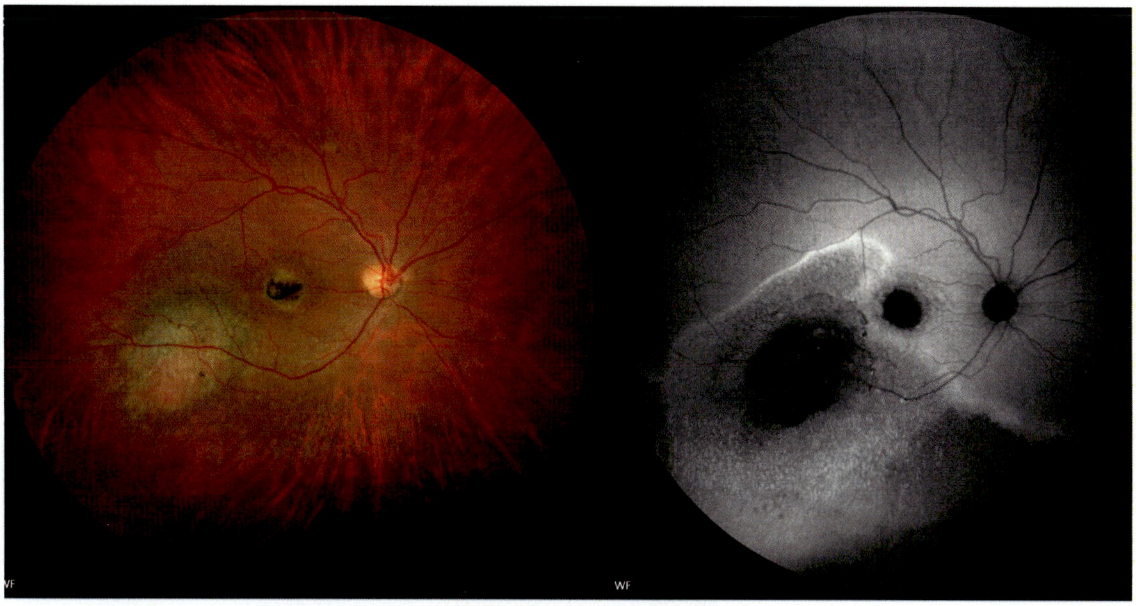

FIGURE 40.2 Long-standing circumscribed choroidal hemangioma associated with perilesional atrophy (*left*) and pigment epithelial alterations best visible on autofluorescence (*right*).

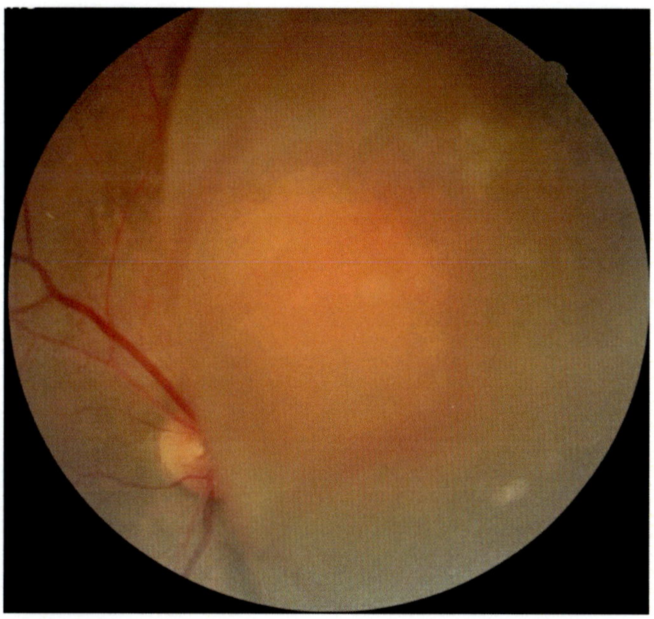

FIGURE 40.3 Bullous retinal detachment overlying an advanced circumscribed choroidal hemangioma.

Inferior flowing subretinal fluid may result in gravitational epitheliopathy that is clearly visible in autofluorescence (Fig. 40.2).

Fluorescein angiography

Fluorescein angiography demonstrates a hyperfluorescent lesion with the internal vascular network visible in the early stages, sometimes before the arterial phase. Thereafter, the fluorescence increases, with diffusion and filling phenomena within the tumor that becomes hyperfluorescent (Fig. 40.7). When present, the pigmented peritumoral ring appears hypofluorescent.

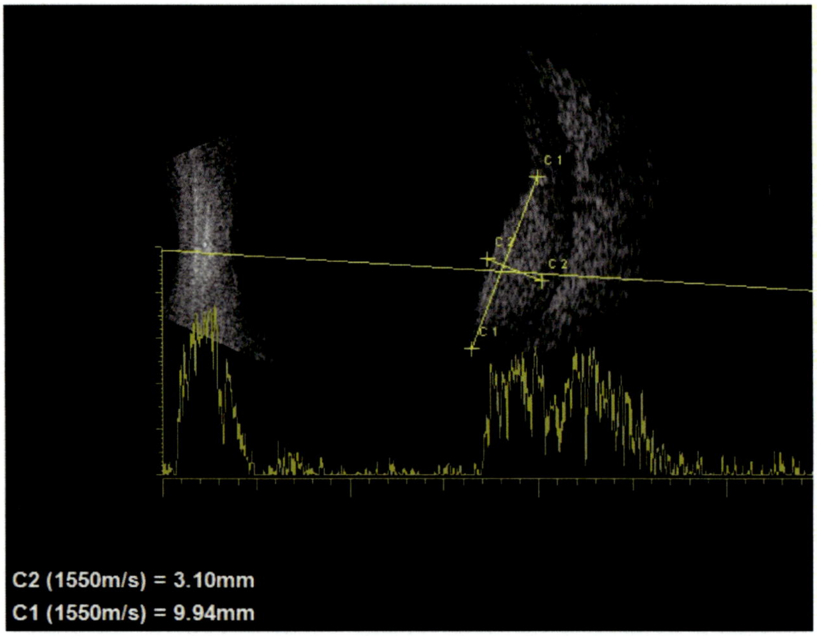

FIGURE 40.4 Ultrasonography of circumscribed choroidal hemangioma showing a dome-shaped, iso-echogenic lesion.

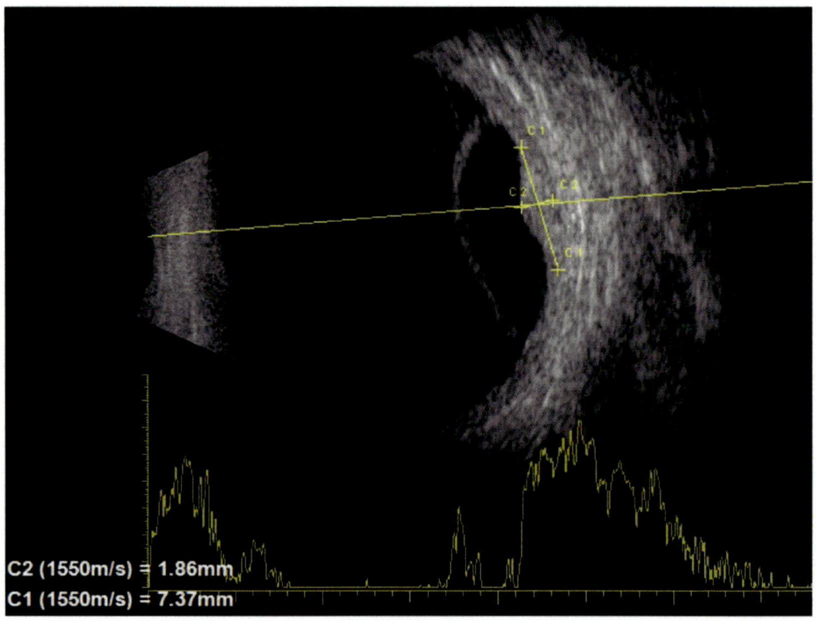

FIGURE 40.5 Ultrasonography of circumscribed choroidal hemangioma showing a slightly elevated iso-echogenic lesion associated with a bullous retinal detachment.

Indocyanine green angiography

Indocyanine green angiography demonstrates rapid and early hyperfluorescence with good visibility of the tumor feeder vessels. The fluorescence of the lesion increases to a maximum reached between 3 and 5 min after intravenous dye injection, and hypofluorescence appears at the late phase 20–30 min after dye injection. This hypofluorescence probably results from the rapid clearance of the contrast media and is referred to as the pathognomonic "wash-out" phenomenon (Fig. 40.8).[6] An hyperfluorescent contour around the lesion is often associated with the late angiographic phase.

High-resolution indocyanine green angiography (using a high-resolution digital photography system) is critical to diagnosing circumscribed choroidal hemangioma. The fluorescence pattern differs between amelanotic choroidal lesions, especially choroidal metastases that remain hypofluorescent along all phases, amelanotic choroidal melanoma that displays

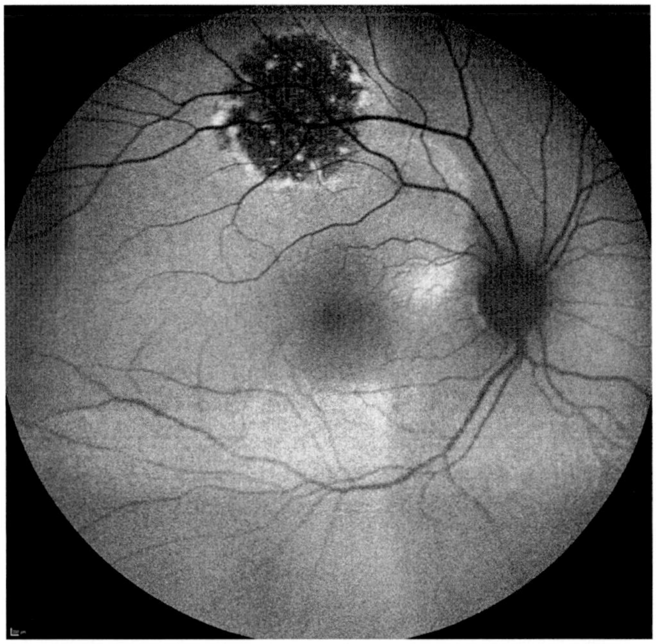

FIGURE 40.6 Autofluorescence showing hyperautofluorescent spots on the lesion and diffuse hyperautofluorescence corresponding to the serous retinal detachment.

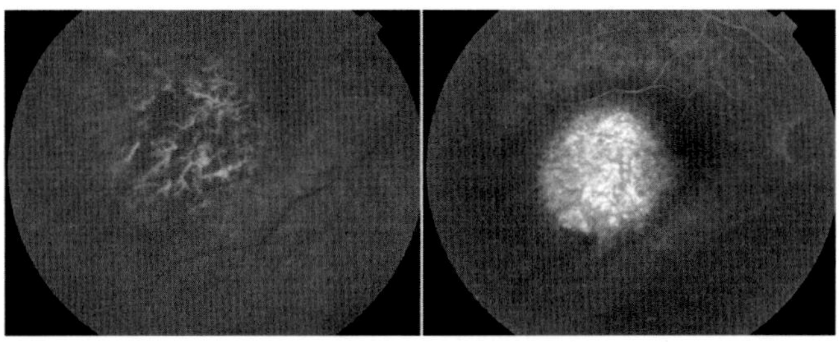

FIGURE 40.7 Fluorescein angiography showing internal vascularization at early phase (*left*) and late diffuse hyperfluorescence (*right*).

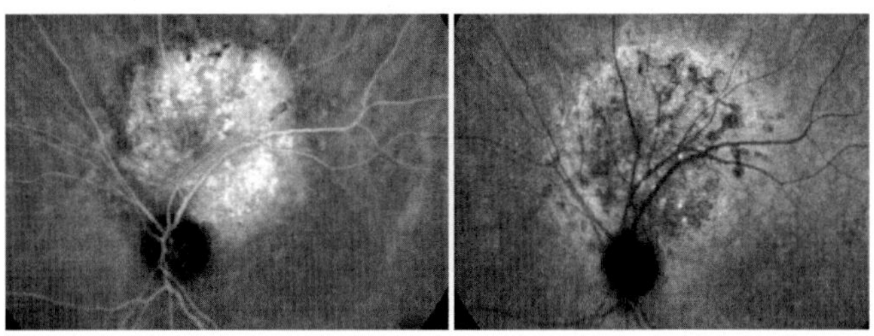

FIGURE 40.8 ICG angiography showing early hyperfluorescence (*left*) and late hypofluorescence, described as the pathognomonic "wash out" phenomenon (*right*).

the "double vascularization" sign (visualization of intratumoral and retinal vessels), and choroidal hemangioma that shows early hyperfluorescence and late "wash out."

Optical coherence tomography

Optical coherence tomography (OCT) allows the visualization of the lesions itself and explores potentially associated subretinal or intraretinal exudation (Figs. 40.9 and 40.10).

OCT coupled with an enhanced-depth imaging of swept-source mode allows the appreciation of the posterior border in shallow hemangiomas (thickness less than 1 mm).[7] The tumor choroidal tissue is dense and distinct from the normal

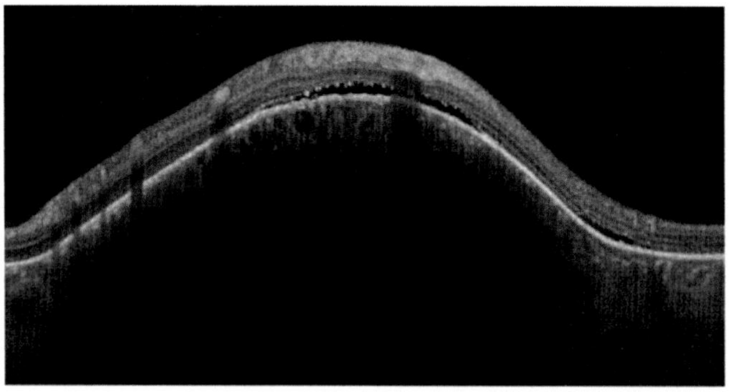

FIGURE 40.9 OCT showing a dome-shaped choroidal lesion with overlying and juxta-lesional serous retinal detachment. The signal attenuation makes it difficult to evaluate the lesion thickness, which is best assessed by ultrasonography.

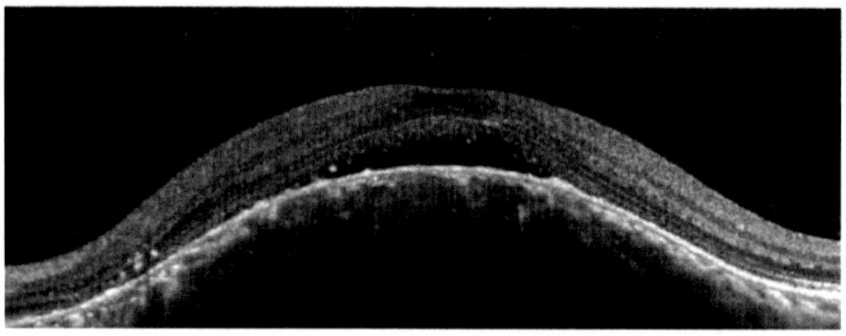

FIGURE 40.10 OCT showing a dome-shaped choroidal lesion with overlying serous retinal detachment. Pigment retinal alterations are visible at the fovea.

choroid. The overlying choriocapillaris appears compressed and altered. Sometimes, dilatation of the choroidal tumor vessels can be seen, which may appear more visible after treatment. These tumor vessels appear to extend from the choriocapillaris through the Sattler–Haller layer.

Subretinal fluid may be present on the surface of the hemangioma, around the lesion, or in a gravitational distribution. Intraretinal edema with a cystoid appearance or, in long-standing cases, intraretinal cystic degeneration may also be seen (Fig. 40.11).

OCT is critical to monitor the regression of macular edema and subretinal fluid after treatment. It also allows to distinguish recent from chronic forms with retinal atrophy that have a worse visual prognosis. In particular, the analysis of the outer retina with the photoreceptor signal (inner/outer segment junction) and the integrity of the pigment epithelium allows us to assess the potential for functional recovery after treatment.[8]

OCT-angiography

Recent advances have improved the OCT-angiography (OCT-A) technology. It demonstrates an abnormal vascular structure with large, interconnected, dilated vessels within the hemangioma, corresponding to the large, anarchic, interspaced intralesional capillaries (Fig. 40.12).[9]

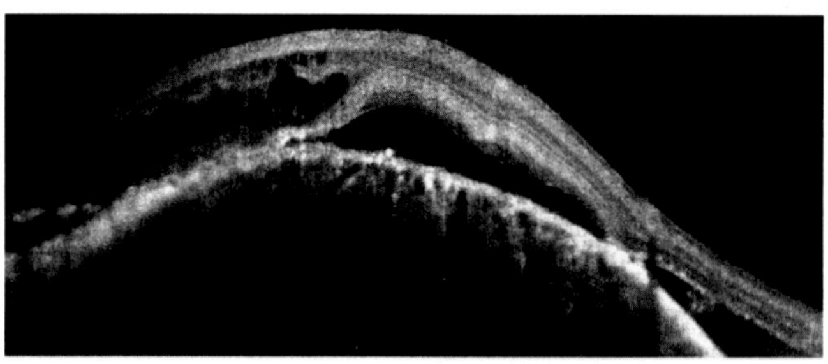

FIGURE 40.11 OCT in a long-standing case showing intraretinal cystoid degeneration overlying the lesion.

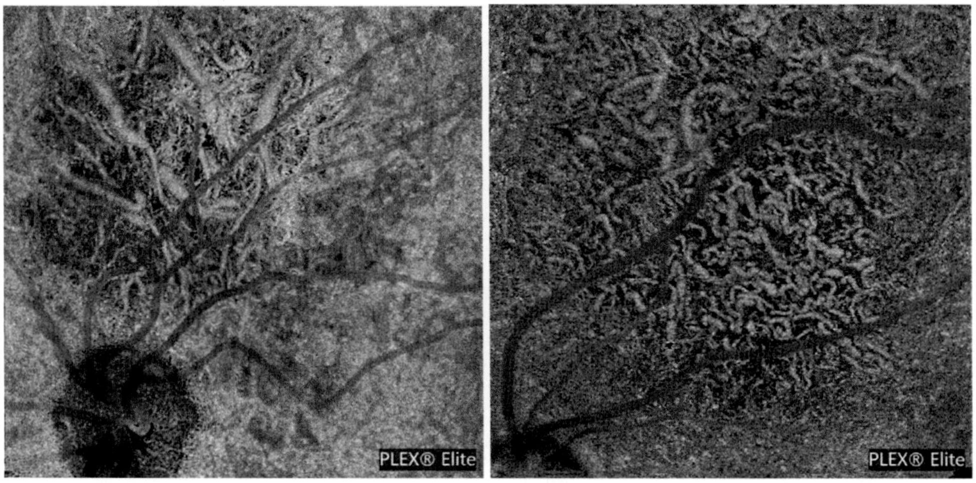

FIGURE 40.12 OCT angiography showing intralesional vascular flow with interconnected abnormal choroidal vessels.

Management strategies

Choroidal hemangioma is a benign tumor, treated only if it becomes symptomatic due to exudation with visual loss or metamorphopsias. The preservation of visual function is the aim of the treatment.

Treatment remains indicated in long-standing cases with very low vision, to prevent massive exudative retinal detachment and neovascular glaucoma.

Currently, two therapeutic strategies are available: photodynamic therapy (PDT) with verteporfin and low-dose radiotherapy, preferably with proton therapy.

Low-dose irradiation: Example of fractionated proton therapy

Irradiation is the first-line treatment of choice for large hemangiomas (>3.5 mm thickness) or those with massive exudation, and the second-line treatment in case of recurrence after two PDTs. Here, we will discuss proton beam therapy but many centers use low-dose brachytherapy with favorable results. Proton therapy delivers a precise dose to a target tissue. For choroidal hemangioma, the selected dose is 20 cobalt Gray equivalent, delivered in eight fractions to reduce complications. The delivered dose is homogeneous and strictly limited to the tumor volume, which is critical, since most hemangiomas are close to the papilla and/or the macula.

As with choroidal melanoma, a surgical procedure is required before irradiation, to place tantalum clips onto the sclera, indicating the tumor location. The distances between the clips, the tumor, and the limbus are measured. Given that circumscribed choroidal hemangioma are often posterior, the lesion is visualized via indirect ophthalmoscopy and not scleral transillumination.

Irradiation is delivered in a specialized proton beam therapy facility with a 2.5-mm safety margin around the lesion required to compensate for mispositioning or ocular movements.

Anatomical recovery is obtained in a few months (Fig. 40.13), and visual improvement is observed depending on the initial visual acuity and the delay between the first symptoms and the treatment.[10] In a retrospective series of 71 patients, improvement in visual acuity, resolution of subretinal fluid, and lesion thinning were observed in 52%, 100%, and 91.5% of patients, respectively. In addition, 75% of patients treated before 6-month after symptom onset experienced visual acuity improvement of two ETDRS lines or more at 52-month follow-up.[11]

The main adverse effects are radiation complications (radiation maculopathy and neuropathy, cataract, and dry eye syndrome),[12] which are, however, far less frequent than after treatment of choroidal melanoma because of the lower radiation doses used. In particular, the modification of the delivery protocol from four to eight fractions allowed to further reduce the rate of these complications.[13]

The disadvantages of proton therapy are primarily the need for surgery, the need to travel to a biomedical ocular proton beam facility, and the financial cost.

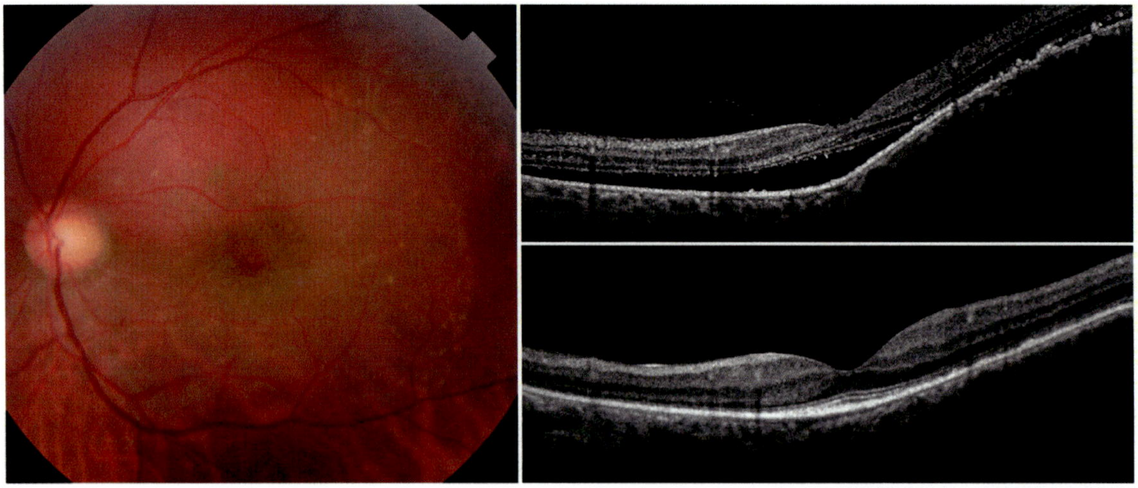

FIGURE 40.13 Response to proton beam therapy. Circumscribed choroidal hemangioma before treatment (fundus, *left*) and OCT showing macular subretinal fluid (*upper right*). OCT 6 months after treatment showing regression of macular subretinal fluid (*lower right*).

Photodynamic therapy

Since the publication of the first two cases of choroidal hemangiomas treated with off-label verteporfin PDT, it has become a promising option, especially for choroidal hemangioma less than 3.5-mm thick (Fig. 40.14).[14–17]

PDT is minimally invasive and easily organized. The current protocol is adapted from the TAP study for macular degeneration. The patient first receives verteporfin as a slow intravenous infusion (6 mg/m^2) over a period of 10 min, then after an interval of 5 min, exposure to the diode laser (692 nm) is performed, at a dose of 50 J/cm^2 with an intensity of 600 mW/cm^2 and an application time of 83 s.

The photosensitizer is activated by 692-nm light radiation and will induce photochemical reactions mediated by tissue oxygen, responsible for endothelial cell damage, thus generating selective vascular occlusion of abnormal vessels. The primary goal of PDT is the resorption of subretinal and intraretinal fluid. In some studies, patients have received multiple courses of treatment until complete tumor resolution, with the occurrence of choroidal atrophy and retinal neurosensory degeneration in many cases.

In the series of Blasi et al. including 25 patients, a rapid resorption of subretinal fluid and a decrease in tumor volume were obtained. Visual acuity increased by two lines in 76% of patients and by more than three lines in 40% of patients at 5

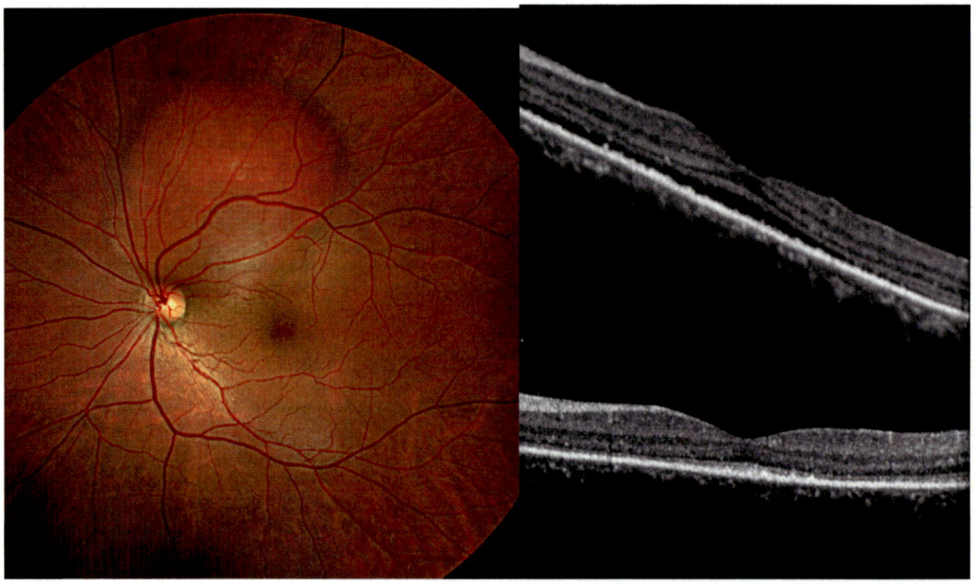

FIGURE 40.14 Response to PDT. Circumscribed choroidal hemangioma before treatment (fundus, *left*) and OCT showing macular subretinal fluid (*upper right*). OCT 2 months after treatment showing regression of macular subretinal fluid (*lower right*).

years with one treatment session in most patients. The visual improvement was obtained during the first 3 months in most patients and was maintained thereafter. No complications have been reported.[15]

The duration of symptoms and the initial visual acuity level have been identified as predictive factors for the functional benefit of PDT for circumscribed choroidal hemangioma. In addition, a higher number of PDT sessions correlates with poorer visual outcomes.[15,16]

In a recent literature review, 90% of patients presented tumor volume reduction and visual acuity improvement or stabilization, but almost a quarter of patients required multiple treatment sessions.[18]

Few complications are reported, consisting most often in pigment epithelial proliferation, choroidal atrophy, transient choroidal effusion, and moderate subretinal fibrosis. Choroidal ischemia and atrophy have been reported after three or more PDT sessions. In all cases, it is preferable to wait 4–6 months after PDT to obtain resolution of subretinal fluid before repeating treatment.

Many protocols are used depending on the investigators, with variables affecting the dose of verteporfin, the time of passage of verteporfin (flash administration vs. slow 10-min infusion), the laser power (50 or 100 J/cm^2), the number of laser spots (1 or more), and the number of treatment sessions (1–5).

It is also possible to use semifluence for the macular spot, but in the series by Blasi et al.,[15] the best results were obtained with the diode laser at a power of 100 J/cm^2 without recurrence or complications at 5 years.

Conclusion

Circumscribed choroidal hemangioma is a benign vascular tumor, revealed by macular sub- and intraretinal exudation. The diagnosis is essentially clinical, confirmed by ultrasound, fluorescein/ICG angiography, and OCT. The two therapeutic strategies are irradiation (by either proton or brachytherapy) for larger lesions, and PDT for smaller ones.

Diffuse choroidal hemangioma

Pathogenesis

Diffuse choroidal hemangioma (DCH) is a benign vascular tumor of the choroid, distinguished from the circumscribed form by its larger size, often poorly delineated borders, diagnosis in children, and frequent association with Sturge–Weber syndrome. Its early diagnosis is important to rapidly manage exudative complications and ocular hypertension in the case of Sturge–Weber syndrome.

Pathology

Diffuse choroidal hemangioma is characterized by an intermixed proliferation of small and large blood vessels and is usually classified as a mixed hemangioma. Fibrous transformation of the proliferated RPE is observed in over 50% of diffuse choroidal hemangiomas in Sturge–Weber syndrome. Histologically, diffuse choroidal hemangioma has the same characteristics as circumscribed choroidal hemangioma, except for its larger size and more blurred boundaries. In the choroid, there is vascular engorgement as well as proliferation of vessels of different calibers.[19]

Sturge–Weber syndrome

Sturge–Weber syndrome is a rare nonhereditary congenital anomaly (estimated incidence between 1:20,000 and 50,000 births), variably and nonsystematically associating meningeal, cutaneous, and ocular angiomatous vascular malformations. Some authors termed this syndrome "encephalotrigeminal angiomatosis."2[20] An anomaly in the embryonic development of neural crest precursor cells is thought to be responsible. A mutation of the GNAQ gene is frequently found. Interestingly, a mutation of the GNAQ gene can also be found in pigmentovascular phacomatoses, wine stains, circumscribed choroidal hemangiomas, and ocular melanocytosis, and it predisposes to the development of uveal melanoma.[21,22]

In the skin, a flat hemangioma is present in the territory of the trigeminal nerve, also called nevus flammeus or port wine stain (Fig. 40.15). When associated with Sturge–Weber syndrome, this cutaneous hemangioma is bilateral in 20%–30% of cases.[23] However, only a minority (10%) of facial cutaneous hemangiomas are associated with Sturge–Weber syndrome. The meningeal involvement is characterized by a leptomeningeal hemangioma ipsilateral to the skin involvement. The venous abnormalities cause ischemic cortical lesions, cortical atrophy, and calcifications, responsible in some cases for convulsions in children and significant motor or cognitive developmental retardation. Cerebral damage can also affect the

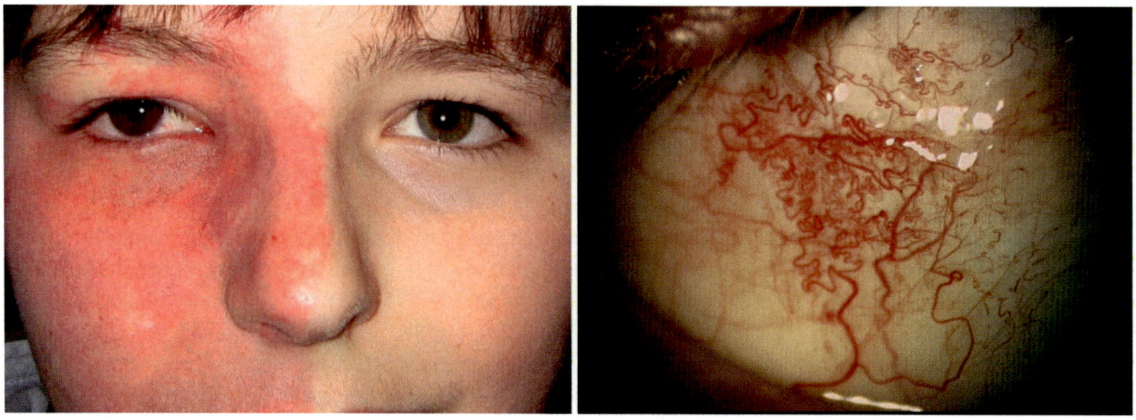

FIGURE 40.15 Cutaneous naevus flammeus (port wine stain) (*left*) and conjunctival telangiectasia (*right*) in patients with Sturge–Weber syndrome.

visual field. At the ophthalmological level, in addition to diffuse choroidal hemangioma, congenital or juvenile glaucoma, conjunctival or episcleral vessel abnormalities, heterochromia, and retinal vessel dilatation are mainly observed.[24]

Glaucoma is the most frequent ocular manifestation of Sturge–Weber syndrome: it is present in 50%–70% of cases, ipsilateral to the skin involvement. Several causes have been described.[25] A congenital malformation of the iridocorneal angle causes an increase in resistance to the evacuation of aqueous humor and is responsible for congenital forms of glaucoma, with or without associated megalocornea or buphthalmia. The occurrence of glaucoma sometimes later (in children and young adults) is probably related to the progressive increase in episcleral venous pressure within episcleral hemangiomas. The treatment is medical or surgical but remains difficult, and associated with a poorer visual prognosis than other causes of secondary glaucoma.

Clinical features

Diffuse choroidal hemangioma is present in 30%–50% of patients with Sturge–Weber syndrome.[26–28] It may be asymptomatic or cause decreased vision (hyperopia, amblyopia, foveolar deformity, glaucoma, or exudative complications), strabismus due to loss of fixation, or red pupil reflex. In addition to the glaucoma described above, conjunctival or episcleral telangiectasias (Fig. 40.15) may be associated. On the fundus, there is a large, diffuse, red-orange choroidal mass that may sometimes involve the entire choroid, ipsilateral to the skin lesion (Fig. 40.16). The red appearance of the fundus

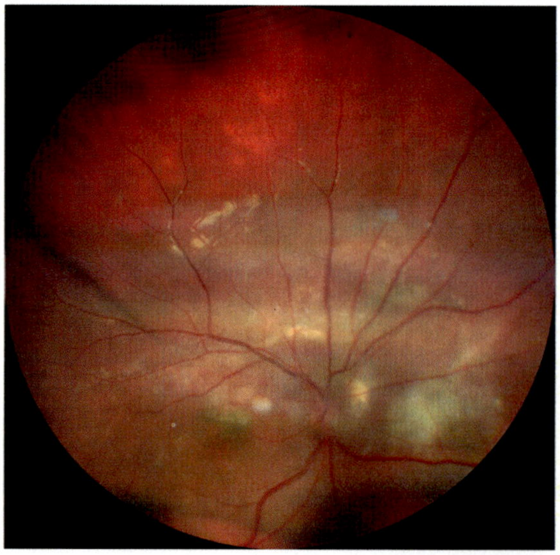

FIGURE 40.16 Wide-field color fundus photograph showing subtotal exudative retinal detachment and subretinal fibrosis due to diffuse choroidal hemangioma in an 8-year-old child with Sturge–Weber syndrome.

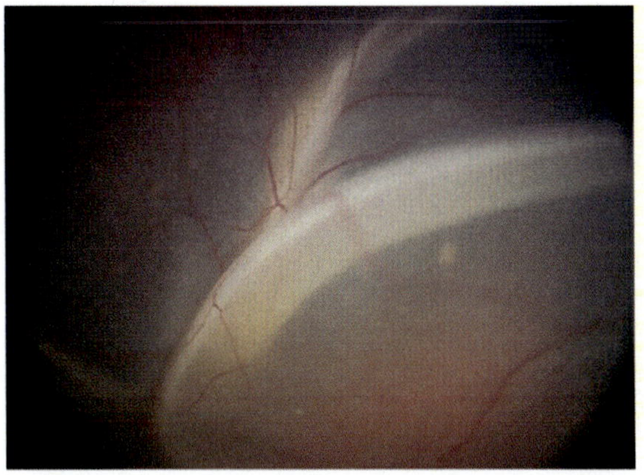

FIGURE 40.17 Bullous total exudative retinal detachment complicating a diffuse choroidal hemangioma.

is best seen in comparison with the contralateral eye. The retinal vessels may be tortuous and dilated. Hemangioma may be complicated by retinal atrophy, pigment epithelial metaplasia, fibrous remodeling, cystic retinal degeneration, localized retinal serous detachment, and in most advanced cases, total exudative retinal detachment (Fig. 40.17) and neovascular glaucoma. Although diffuse choroidal hemangioma is congenital, the onset of exudative complications is variable, from childhood to teenage years and young adulthood.

Imaging findings

Fluorescein and ICG angiography

On fluorescein angiography, the diffuse hyperfluorescence of the lesion is observed from the early phase (Fig. 40.18) and increases until the late phase. On ICG angiography, tumor vascularization is visible at the early phases, with "washout" in the late phase, which is not systematically observed.

Ultrasonography

B-mode ultrasound shows a diffuse choroidal thickening, with hyperechogenicity, sometimes associated with retinal detachment (Fig. 40.19).

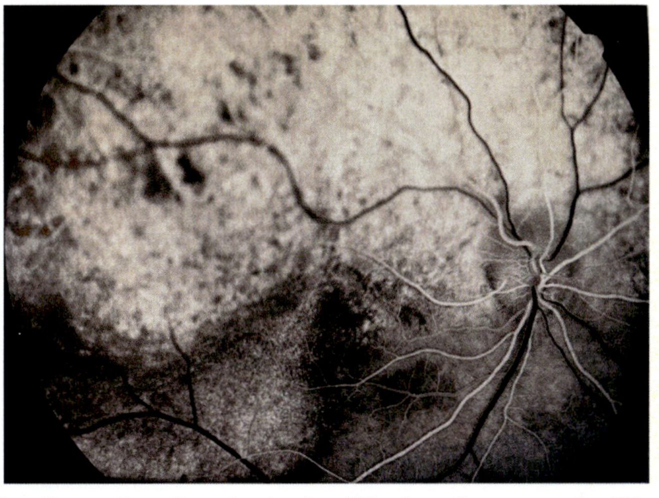

FIGURE 40.18 Early-phase fluorescein angiography showing diffuse hyperfluorescence of a diffuse choroidal hemangioma.

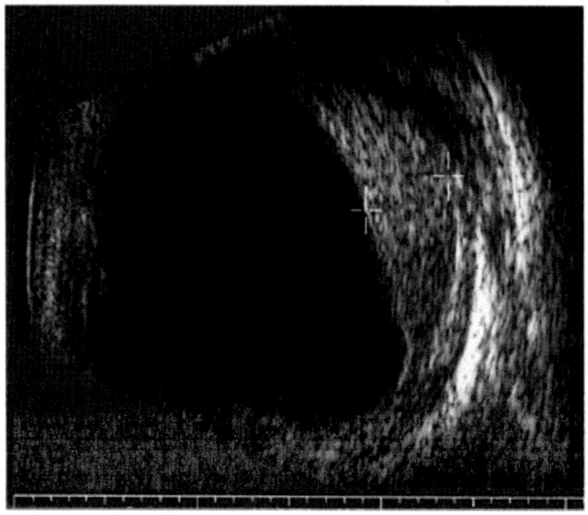

FIGURE 40.19 Ocular B-mode ultrasonography showing a diffuse iso-echogenic choroidal thickening corresponding to a diffuse choroidal hemangioma.

OCT

On OCT, diffuse choroidal hemangioma appears as an area of diffuse choroidal thickening, where the choroidal vessels appear dilated and irregular. Cystoid retinal degeneration or intraretinal schisis is frequently found overlying the lesion. OCT is critical to identify subretinal exudation and guide treatment indications (Fig. 40.20).

MRI

On MRI, diffuse choroidal hemangioma is hyperintense relative to the vitreous on T1 sequences, enhanced after contrast injection (unlike the adjacent retinal detachment) and is isointense on T2 sequences.

Management strategies

The goal of treatment of diffuse choroidal hemangioma is to induce regression of the hemangioma and of the resulting exudation, to preserve visual acuity, and to limit severe long-term complications such as total retinal detachment or neovascular glaucoma, which may warrant enucleation of the affected globe. However, the treatment of diffuse choroidal hemangioma is not systematic. Nonexudative diffuse choroidal hemangioma can sometimes be observed, with optical correction of hyperopia or amblyopia if necessary.

The conventional treatment is irradiation of the lesion by external photon radiotherapy. Certain centers have also reported the use of brachytherapy for partial diffuse choroidal hemangioma. Low-dose external choroidal radiotherapy (20 Gy in 10 fractions with lens sparing) has been shown to be effective in reducing the thickness of the hemangioma and inducing retinal reattachment in 96% and 92% of cases, respectively (26 eyes, mean follow-up 47 months) (Fig. 40.21).[29] Other treatments have also been described, depending on the teams and the characteristics of the diffuse hemangioma: PDT, proton therapy, intravitreal anti-VEGF injections[30] or, more rarely, systemic treatment with propranolol, which remains controversial in this indication.[31,32] Visual recovery remains uncertain and is limited by the pretreatment visual level and the characteristics of the hemangioma (macular localization, duration of exudation before treatment), but also by the cooccurrence of other ocular manifestations of Sturge-Weber syndrome (glaucoma, cerebral visual field damage).[33]

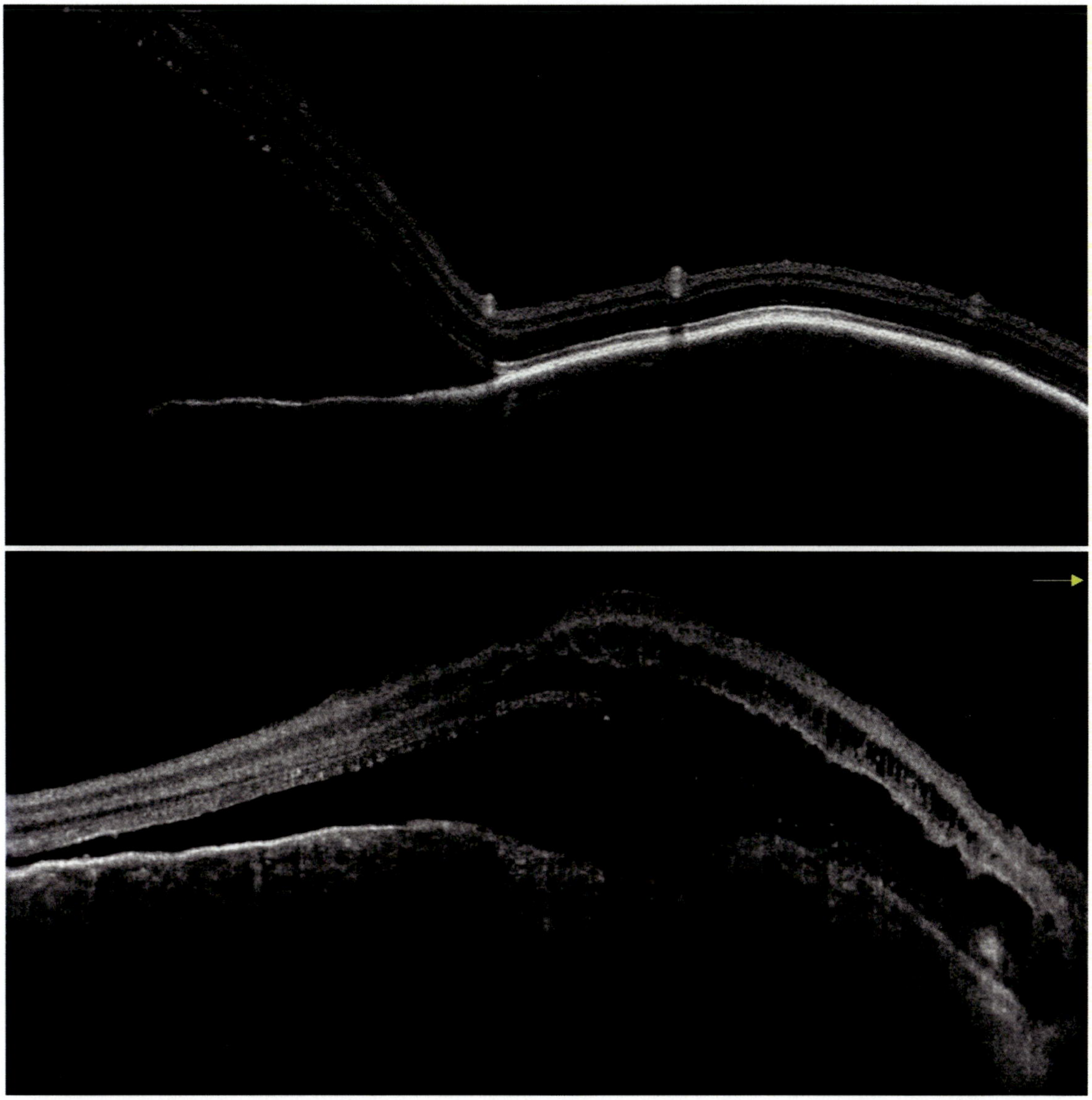

FIGURE 40.20 OCT of diffuse choroidal hemangioma in two different patients, showing irregular choroidal thickening and overlying subretinal fluid (*top*), and combined sub- and intraretinal exudation (*bottom*).

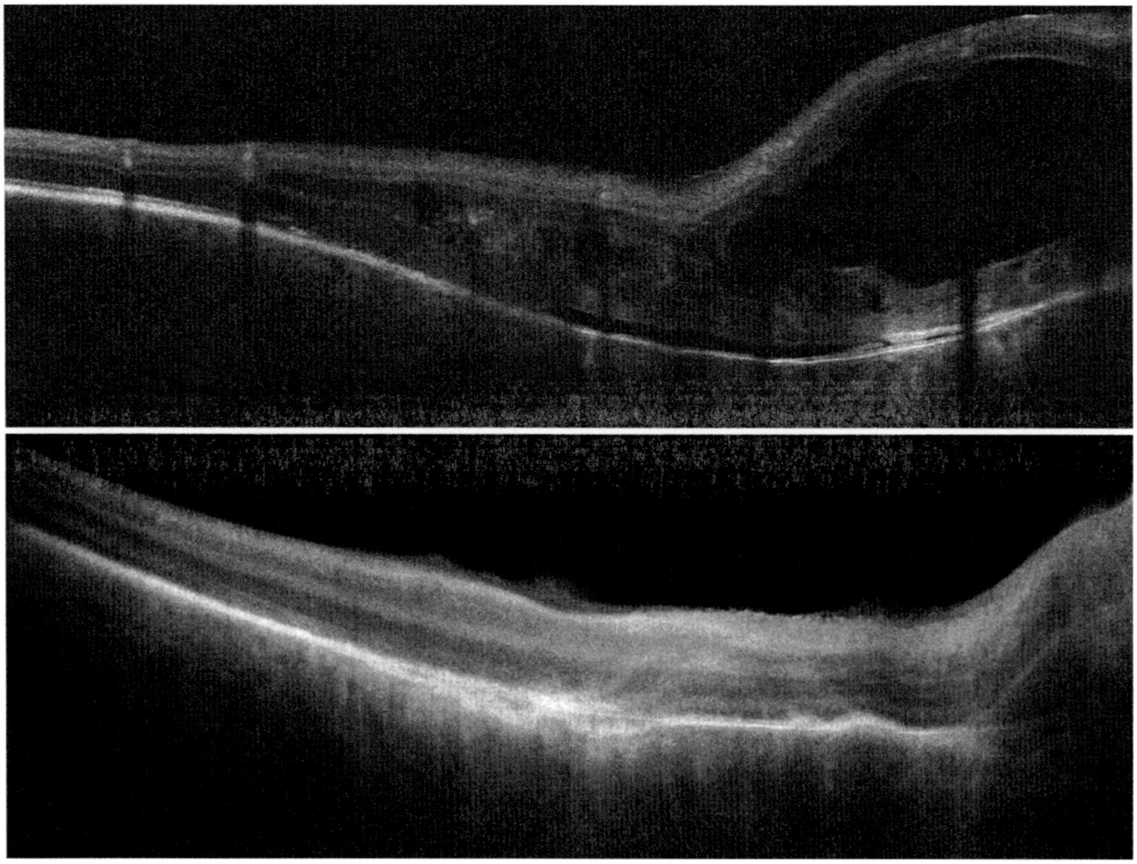

FIGURE 40.21 OCT of diffuse choroidal hemangioma in the 8-year-old child presented in Fig. 40.16, showing subretinal fluid and intraretinal edema at diagnosis (*top*) and resolution 6 months after external-beam photon radiotherapy (*bottom*).

Conclusion

Diffuse choroidal hemangioma is most often associated with Sturge–Weber syndrome. Its main complication is subretinal exudation. The preferred treatment is low-dose, fractionated external beam radiotherapy but visual recovery is uncertain and depends on multiple factors.

References

1. Guin L, Metz K, Horst Kreis S, et al. GNAQ Q209R mutations are highly specific for circumscribed choroidal hemangioma. *Cancers (Basel)*. 2019;11(7):1031. https://doi.org/10.3390/cancers11071031.
2. Francis JH, Milman T, Grossniklaus H, et al. GNAQ mutations in diffuse and solitary choroidal hemangiomas. *Ophthalmology*. 2019;126(5):759–763. https://doi.org/10.1016/j.ophtha.2018.12.011.
3. Mashayekhi A, Shields CL. Circumscribed choroidal hemangioma. *Current Opinion in Ophthalmology*. 2003;14(3):142–149. https://doi.org/10.1097/00055735-200306000-00006.
4. Karimi S, Nourinia R, Mashayekhi A. Circumscribed choroidal hemangioma. *Journal of Ophthalmic & Vision Research*. 2015;10(3):320–328.
5. Ramasubramanian A, Shields C, Harmon S, Shields J. Autofluorescence of choroidal hemangioma in 34 consecutive eyes. *Retina*. 2010;30(1):16–22.
6. Schalenbourg A, Piguet B, Zografos L. Indocyanine green angiographic findings in choroidal hemangiomas: a study of 75 cases. *Ophthalmologica*. 2000;214(4):246–252.
7. Torres VLL, Brugnoni N, Kaiser PK, Singh AD. Optical coherence tomography enhanced depth imaging of choroidal tumors. *American Journal of Ophthalmology*. 2011;151(4):586–593.
8. Zhang Y, Liu W, Fang Y, et al. Photodynamic therapy for symptomatic circumscribed macular choroidal hemangioma in Chinese patients. *American Journal of Ophthalmology*. 2010;150(5):710–715. https://doi.org/10.1016/j.ajo.2010.05.016.
9. Takkar B, Azad S, Shakrawal J. Blood flow pattern in a choroidal hemangioma imaged on swept-source-optical coherence tomography angiography. *Indian Journal of Ophthalmology*. 2017;65(11).

10. Mathis T, Maschi C, Mosci C, Espensen CA, Rosier L. Comparative effectiveness of proton beam versus photodynamic therapy to spare the vision in circumscribed choroidal hemangioma. *Retina*. 2021;41(2):1–10.
11. Levy-Gabriel C, Lumbroso-Le Rouic L, Plancher C. Long-term results of low-dose proton beam therapy for circumscribed choroidal hemangiomas. *Retina*. 2009;29(2):170–175.
12. Zeisberg A, Seibel I, Cordini D. Long-term (4 years) results of choroidal hemangioma treated with proton beam irradiation. *Graefes Archive for Clinical and Experimental Ophthalmology*. 2014;252(7):1165–1170.
13. Mahdjoubi A, Dendale R, Desjardins L. Treatment of exudative circumscribed choroidal hemangioma : efficacy of fractionated proton therapy (20 Gray relative biological effectiveness in 8 fractions. *Retina*. 2019;39(4):692–699.
14. Boixadera A, Arumí JG, Martínez-Castillo V. Prospective clinical trial evaluating the efficacy of photodynamic therapy for symptomatic circumscribed choroidal hemangioma. *Ophthalmology*. 2009;116(1):100–105 1.
15. Blasi MA, Tiberti AC, Scupola A. Photodynamic therapy with verteporfin for symptomatic circumscribed choroidal hemangioma : five-year outcomes. *Ophthalmology*. 2010;117(8):1630–1637.
16. Jurklies B, Anastassiou G, Ortmans S. Photodynamic therapy using verteporfin in circumscribed choroidal haemangioma. *British Journal of Ophthalmology*. 2003;87(1):84–89.
17. Singh AD. Photodynamic therapy of circumscribed choroidal haemangioma. *British Journal of Ophthalmology*. 2004;88(11):1414–1418.
18. Cerman E, Cekiç O. Clinical use of photodynamic therapy in oculars tumors. *Survey of Ophthalmology*. 2015;60:557–574.
19. Shanmugam P, Ramanjulu R. Vascular tumors of the choroid and retina. *Indian Journal of Ophthalmology*. 2015;63(2):133. https://doi.org/10.4103/0301-4738.154387.
20. Pascual-Castroviejo I, Pascual-Pascual SI, Velazquez-Fragua R, Viaño J. Sturge-Weber syndrome: study of 55 patients. *The Canadian Journal of Neurological Sciences*. 2008;35(3):301–307.
21. Tan W, Nadora DM, Gao L. The somatic GNAQ mutation (R183Q) is primarily located within the blood vessels of port wine stains. *Journal of the American Academy of Dermatology*. 2016;74(2):380–383.
22. Shirley MD, Tang H, Gallione CJ. Sturge-Weber syndrome and port-wine stains caused by somatic mutation in GNAQ. *New England Journal of Medicine*. 2013;368(21):1971–1979.
23. Formisano M, Abdolrahimzadeh B, Mollo R. Bilateral diffuse choroidal hemangioma in Sturge Weber syndrome : a case report highlighting the role of multimodal imaging and a brief review of the literature. *Journal of Current Ophthalmology*. 2018;31(2):242–249.
24. Silverstein M, Salvin J. Ocular manifestations of Sturge-Weber syndrome. *Current Opinion in Ophthalmology*. 2019;30(5):301–305.
25. Javaid U, Ali MH, Jamal S, Butt NH. Pathophysiology, diagnosis, and management of glaucoma associated with Sturge-Weber syndrome. *International Ophthalmology*. 2018;38(1):409–416.
26. Sullivan TJ, Clarke MP, Morin JD. The ocular manifestations of the Sturge-Weber syndrome. *Journal of Pediatric Ophthalmology & Strabismus*. 1992;29(6):349–356.
27. Celebi S, Alagoz G, Aykan U. Ocular findings in Sturge-Weber syndrome. *European Journal of Ophthalmology*. 2000;10(3):239–243.
28. Singh AD, Kaiser PK, Sears JE. Choroidal hemangioma. *Ophthalmology Clinics of North America*. 2005;18(1):151–161.
29. Randon M, Lévy-Gabriel C, Abbas R. Results of external beam radiotherapy for diffuse choroidal hemangiomas in Sturge-Weber syndrome. *Eye*. 2018;32(6):1067–1073.
30. Paulus YM, Jain A, Moshfeghi DM. Resolution of persistent exudative retinal detachment in a case of Sturge-Weber syndrome with anti-VEGF administration. *Ocular Immunology and Inflammation*. 2009;17(4):292–294.
31. Krema H, Yousef YA, Durairaj P, Santiago R. Failure of systemic propranolol therapy for choroidal hemangioma of Sturge-Weber syndrome: a report of 2 cases. *JAMA Ophthalmology*. 2013;131(5):681–683.
32. Thapa R, Shields CL. Oral propranolol therapy for management of exudative retinal detachment from diffuse choroidal hemangioma in Sturge-Weber syndrome. *European Journal of Ophthalmology*. 2013;23(6):922–924.
33. Koenraads Y, Egmond-Ebbeling MB, Boer JH. Visual outcome in Sturge-Weber syndrome: a systematic review and Dutch multicentre cohort. *Acta Ophthalmologica*. 2016;94(7):638–645.

Chapter 41

Choroidal melanoma

Chiara M. Eandi[1,2] and Andrea Montesel[1]

[1]*University of Lauanne, Jules Gonin Eye Hospital, Fondation Asile des Aveugles, Lausanne, Switzerland;* [2]*University of Torino, Department of Surgical Sciences, Torino, Italy*

Overview bullets

- 85%–90% of intraocular melanomas arise from the posterior uvea
- Most frequent primary malignant intraocular tumor in adults (six cases/million/year)
- Pigmented mass, well-circumscribed, located under the RPE
- Pigmentation varies from creamy-white to dark chocolate-brown
- Multiple shapes: dome or mushroom shaped, flat, multinodular, or small
- Size is based on the tumor thickness, including small (<3.0 mm), medium (3.1–8.0 mm), and large (8.1 mm or greater)
- Asymptomatic until it interferes with the visual structures (retina, optic nerve)
- Diagnosis: clinical fundus examination, fundus photography, and ultrasonography
- Treatment: conservative radiation therapy (brachytherapy or proton beam)
- Survival depends on the development of liver metastasis
- Around 20%–30% of UM patients die within 5 years and 45% die within 15 years
- Up to 50% of patients will suffer from metastases (mostly to liver)
- Cytogenetic mutations (monosomy 3) are risk factors for distant metastases

Pathogenesis

Choroidal melanoma (CM) is a rare malignancy that arises from the malignant transformation of melanocytes within the choroid. It represents 85%–90% of all uveal melanomas and is the most frequent primary intraocular malignancy in adults, occurring in approximately six individuals per million population annually.[1] The peak age for diagnosis is between 70 and 79 years, with an approximate median age of diagnosis at 63 years.[2]

It is considered to be a sporadic event, with a slight male preponderance, and is more common in the white population.[3] However, certain risk factors have been identified, including light-pigmented eyes, fair skin, inability to tan, ocular melanocytosis, dysplastic nevus syndrome, as well as BAP1 and MBD4 gene mutations.[4–6]

Clinical features and symptoms

Choroidal melanoma typically appears as a gray-brown choroidal lesion, with a range from creamy-white to dark chocolate-brown, even within one tumor, but it can present as partially pigmented (30% of the cases) and amelanotic (15% of the cases).[7] The funduscopic evaluation typically reveals a pigmented mass, well-circumscribed, located under the retinal pigment epithelium. The two most frequent types appearing on B-scan ultrasonography are dome-shaped (Fig. 41.1) if the axial tumor growth is contained by the Bruch's membrane and the sclera and are mushroom-shaped (Fig. 41.2) if the melanoma creates a Bruch's membrane rupture allowing the tumor cells to develop unrestrained under the neuro-retina.

Other possible growth patterns are diffuse melanoma, which has a mainly horizontal growth and is an infiltrative flat or slightly raised melanoma, and the multinodular melanoma (Fig. 41.3), which results from intratumoral metadifferentiation, leading to different cell clones with varying growth rates.[8] Choroidal melanoma is clinically grouped into three sizes based on the tumor thickness, including small (<3.0 mm), medium (3.1–8.0 mm), and large (8.1 mm or greater).[9] In a review of

Retinal and Choroidal Vascular Diseases of the Eye. https://doi.org/10.1016/B978-0-443-15583-3.00041-X

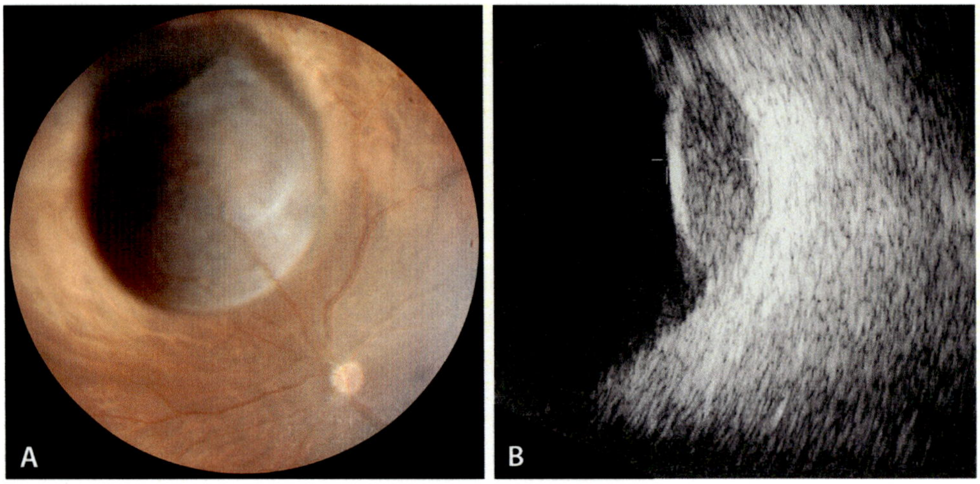

FIGURE 41.1 Dome-shaped choroidal melanoma. (A) Color fundus photography (Panoret-1000, CMT Medical Technologies Inc, Valley Stream, NY) of a pigmented mass within the nasal superior choroid in a 73-year-old female. A rupture of the RPE at the peripheral surface of the lesion can be noticed. VA = 1.0. (B) On 10-MHz B-scan ultrasonography, the tumor is dome-shaped and associated with subretinal fluid, explaining the flashes of light for which the patient initially consulted. $H = 4.9$ mm. H = height. *Courtesy of Dr A. Schalenbourg.*

7256 cases of choroidal melanoma, the mean basal diameter at presentation was 11.3 mm and the mean tumor thickness was 5.5 mm.[10]

Clinical manifestations of choroidal melanoma depend on the location and size of the tumor. Symptoms are nonspecific, often multiple, and tend to appear sequentially. It typically presents with painless loss of vision or metamorphopsia in most patients. However, larger tumors may also cause peripheral visual field loss and serous retinal detachment, leading to the perception of flashing or flickering lights (photopsia). Other manifestations include secondary glaucoma, cystoid macular edema, and moderate inflammatory reaction producing a Tyndall effect in the vitreous or anterior chamber.[11] Extraocular extension through the vortex veins and the posterior ciliary arteries, or optic nerve infiltration via the nerve portion close to the posterior eye wall are described as rare manifestations.[12,13] In a minority of cases with more peripheral tumors located outside the visual axis, CM may be entirely asymptomatic and the tumor is only detected during routine ophthalmic screening.[14]

Diagnosis

The diagnosis of choroidal melanoma is based on clinical examination of the patient with dilated fundus biomicroscopy and B-scan ultrasonography (10, 20, 35, and/or 50 MHz), usually without the need for invasive procedures or tissue sampling.

Multimodal imaging is helpful to confirm the diagnosis and enable clinicians to plan appropriate treatment decisions in specific conditions that present with similar clinical characteristics, such as small choroidal malignant melanoma and benign choroidal nevus.[15]

Different systems have been proposed to help clinicians in detecting CM. Shields et al. developed a well-known mnemonic "To Find Small Ocular Melanoma" acronym, encapsulating the following risk factors: thickness >2 mm, presence of subretinal fluid, symptoms of vision loss, orange pigment and tumor margin within 3 mm of the disc.[16] As recent years have seen advances in multimodal imaging techniques, this mnemonic has been revised to "To Find Small Ocular Melanoma Doing IMaging," with the "M" representing "Melanoma hollow" on ultrasound and "DIM" representing DIaMeter >5 mm' on fundus photography.[17] Using the TFSOM-DIM score system, the mean 5 year estimate of nevus growth into melanoma was reported as 1.1% for those with zero risk factors, 11% with one factor, 22% with two factors, 34% with three factors, 51% with four factors, and 55% with five risk factors.[18]

Damato et al. developed the MOLES acronym and scoring system, to help nonocular oncology experts to estimate the likelihood of malignancy in melanocytic choroidal tumors and to manage patients accordingly.[19] The acronym stands for Mushroom shape, Orange pigment, Large size, Enlargement, and Subretinal fluid, all of which are well-known to be indicators of malignancy. Each of these features is given a score of 0, 1, or 2 according to whether it is absent, borderline, or present. Roelofs et al. validated the MOLES system by reviewing the imaging of 451 treated choroidal melanomas and retrospectively gave these MOLES scores, found a sensitivity of 99.8% for indicating malignancy in melanocytic choroidal tumors (MOLES ≥3).[20]

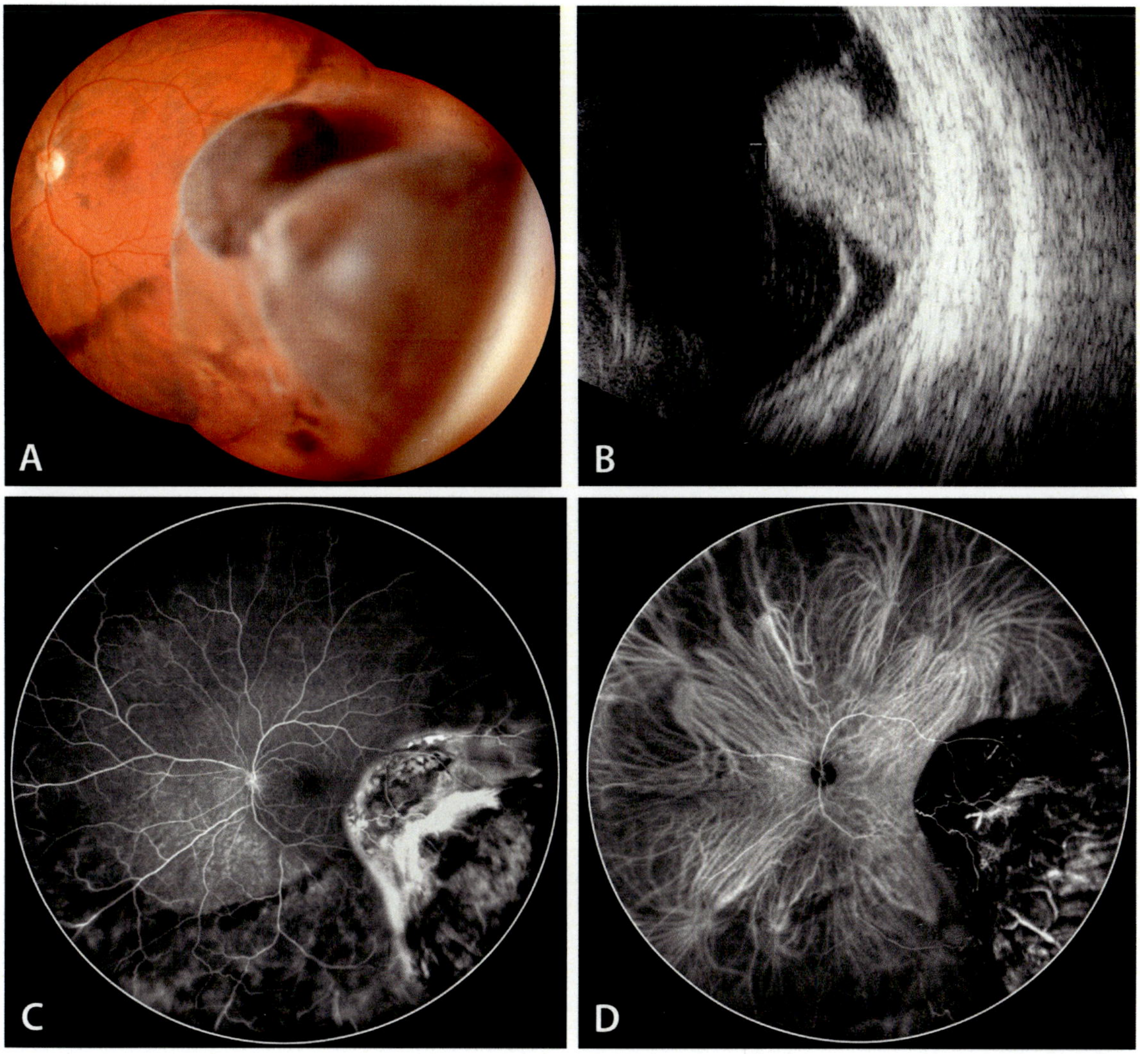

FIGURE 41.2 Large mushroom-shaped choroidal melanoma. (A) Composed color fundus photography (Panoret-1000, CMT Medical Technologies Inc., Valley Stream, NY) of a pigmented mass rising from the peripheral temporal inferior choroid in a 54-year-old male, complaining of a progressive nasal superior scotoma. VA = 1.0. An associated semicircular watershed line of subretinal pigment dispersion can be observed. (B) On 10 MHz B-scan ultrasonography, the tumor is mushroom shaped and complicated by an extensive retinal detachment. $H = 8.7$ mm. (C) On panoramic FA (Heidelberg Engineering spectralis widefield imaging using a staurenghi 150-degree contact lens), the mass presents an increasing hyperfluorescent diffusion from its surface. Signs of ischemia can be seen within the secondary inferior retinal detachment, even before radiation therapy. In consequence, intravitreal anti-VEGFs are also initiated, to avoid neovascular complications.[24] (D) On panoramic ICG-A (Heidelberg Engineering spectralis widefield imaging using a staurenghi 150-degree contact lens), the hypocyanescent mass presents a dilatation of its intratumoral vessels, and the associated inferior retinal detachment provokes a masking defect of the underlying choroidal vascularization, caused by an increased density of the subretinal fluid. *FA*, fluorescein angiography; *H*, height; *ICG-A*, indocyanine green angiography; *VA*, visual acuity; *VEGF*, vascular endothelial growth factor. *Courtesy of Dr A Schalenbourg.*

Differential diagnosis

According to the Collaborative Ocular Melanoma Study (COMS), the clinical misdiagnosis rate is 0.48%, supporting the fact that most intraocular tumors may be diagnosed clinically.[21] Shields and colleagues[22] reported that among 400 referrals for evaluation for posterior uveal melanoma, 8% were diagnosed with hemangiomas, 9.5% were diagnosed with retinal pigment-epithelium hypertrophy, 23.5% were diagnosed with disciform degeneration, and 26.5% were diagnosed with choroidal nevi.

Other differential diagnosis includes optic disk melanocytoma, a heavily pigmented dark tumor, typically located at the optic disk; congenital hypertrophy of the RPE (CHRPE), a benign pigmented fundus lesion with a similar diameter but flat, darker, and with a sharply delineated, often double contour; peripheral exudative hemorrhagic chorioretinopathy (PEHCR),

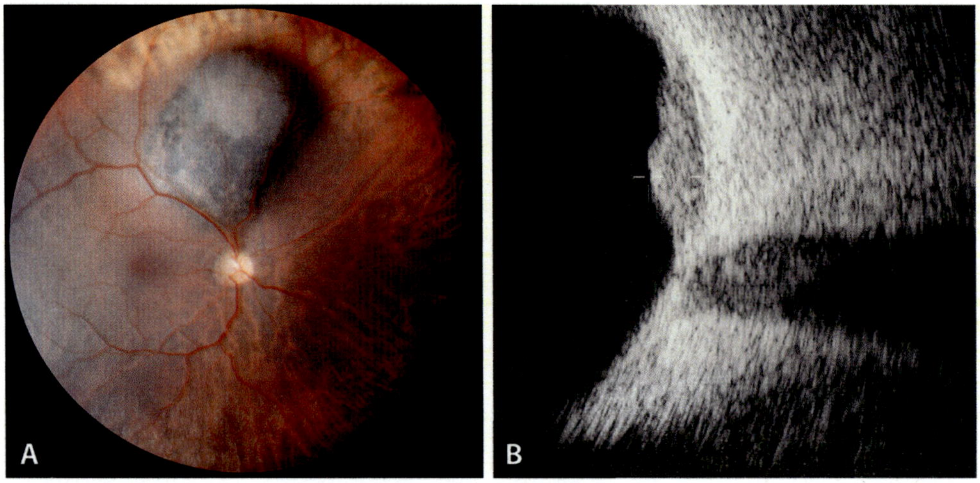

FIGURE 41.3 Multinodular choroidal melanoma. (A) Color fundus photography (Panoret-1000, CMT Medical Technologies Inc., Valley Stream, NY) of a binodular partially pigmented mass within the suprapapillary choroid in a 58-year-old female. Orange pigment is present on the surface of the inferior, more pigmented part. The superior lobe has a creamier color. VA = 0.7. Patient had been investigated over the past year and treated with oral steroids by neurologists because of "unexplained" photopsia. (B) On 10 MHz B-scan ultrasonography, the binodular shape can be observed, with a maximal *H* of 3.2 mm, as well as the infralesional subretinal fluid. *H* (height). *Courtesy of Dr A Schalenbourg.*

a peripheral retinal degeneration of the elderly manifesting with exudative and hemorrhagic changes, and subretinal or suprachoroidal hemorrhages.[9] Furthermore, amelanotic melanoma needs to be differentiated from solitary metastasis. However, probably the most challenging differential diagnosis is the small posterior melanomas (Fig. 41.4), which need to be differentiated from choroidal nevi. Small posterior melanomas usually present associated symptoms and risk factors such as subretinal fluid, contact of the tumor with the optic disc, absence of drusen, presence of hyperautofluorescent orange pigment, presence of pinpoints on fluorescein angiography and a thickness of more than 2 mm.[16,23,24] Documented tumor growth following periodic observation confirms the diagnosis.

Current research is focused on developing artificial intelligence (AI) models to identify risk factors for the malignant transformation of choroidal nevus based on ocular multimodal imaging features.[25,26] The concept of AI-based screening of ophthalmic imaging for both the presence of choroidal nevus and the factors that suggest potential transformation into melanoma has the potential to greatly enhance the detection and early treatment of melanoma within the eye.[27]

Imaging findings

Color fundus photography

Color fundus photography (FP) represents the gold standard imaging modality in the management of choroidal tumors. FP is very useful at baseline to define clinical characteristics of the tumor and during follow-up for monitoring local tumor control after treatment. In the last years, several new fundus cameras have been developed with different advantageous technical characteristics such as digital imaging. Moreover, the introduction of wide-field cameras allows to document not only posterior tumors but also peripheral lesions and their relationships with adjacent ocular structures.[28] However, it is fundamental to be aware of the limitations of these different cameras which might mislead the diagnosis and management of retinal diseases and ocular tumors in particular. The ultra-widefield (200 degrees) Optos camera (Optos PLC, Dunfermline, UK), a confocal scanning laser imaging technology, has the advantage of being a noncontact and nonmydriatic camera, but it is restricted in its ultra-wide angle, mainly to the horizontal axis, by the eyelashes sometimes obscuring fundus details. Moreover, color images are recomposed with only two reflected monochromatic laser images (red: 633 nm and green: 532 nm), resulting in color distortion due to the green shade over the image[29] (Fig. 41.4B). The last released Clarus 500 (Carl Zeiss Meditech Inc., Dublin, CA) has a potential 133 degrees the widest field of view with less color distortion, adding a grayish shade over the image. However, images present peripheral artifacts and aberrations due to the reconstruction process. The spectralis multicolor modality (Heidelberg Engineering Inc., Heidelberg, Germany) is also based on confocal scanning laser technology with a viewing angle limited to 30 degrees. The color images are the result of three simultaneously acquired color laser images reflected in the blue, green, and infrared spectrum, resulting in a false color. These "reconstitute" fundus color images, that do not reflect reality can seriously lead to misdiagnosis and eventually

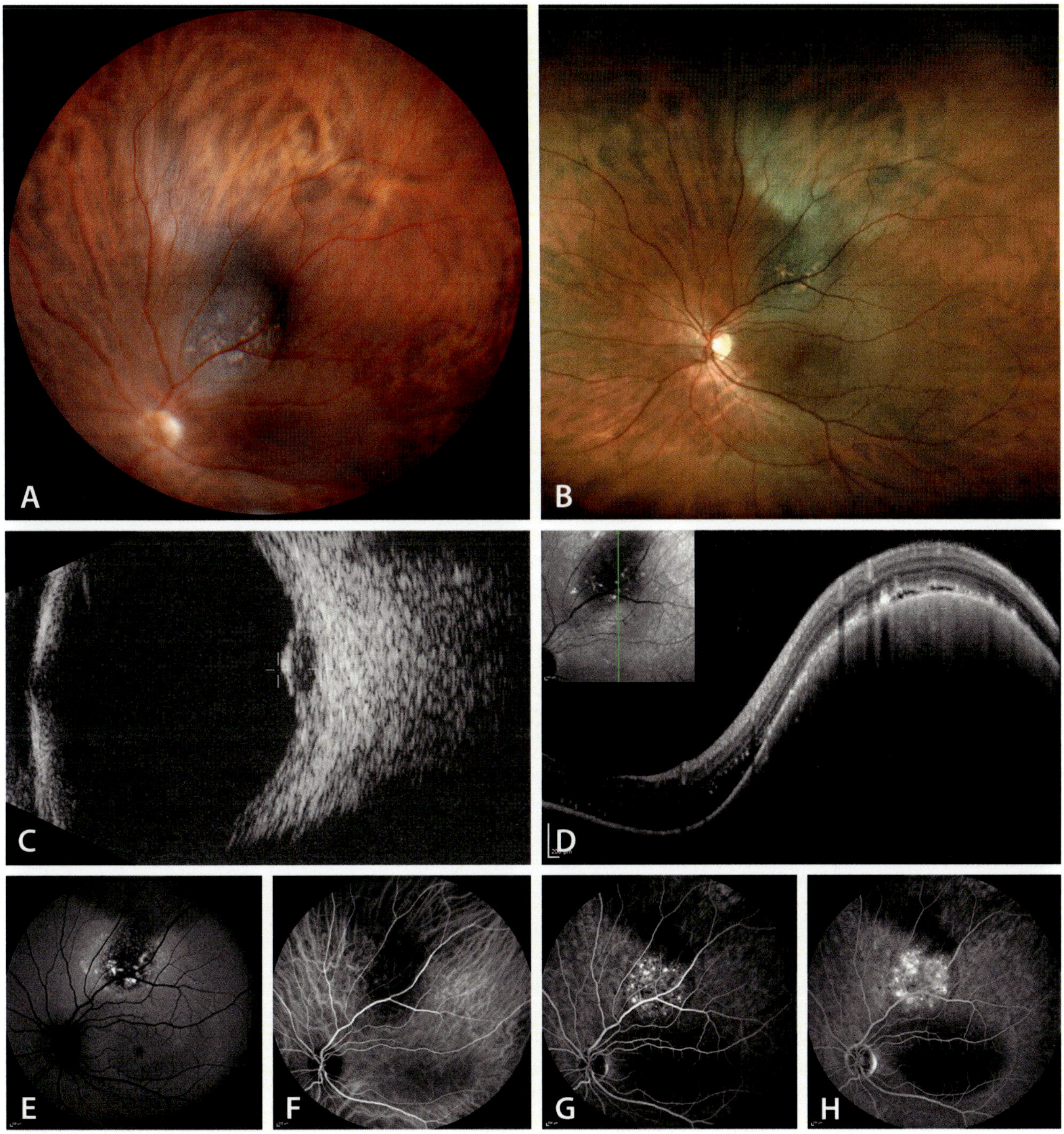

FIGURE 41.4 Small choroidal melanoma. (A) Color fundus photography (Panoret-1000, CMT Medical Technologies Inc., Valley Stream, NY) of a small pigmented mass within the supramacular choroid in a 62-year-old male. VA = 0.6. Patient had observed a relative grayish inferior scotoma over the past weeks. (B) Color fundus photography (Optos, PLC, Dunfermline, UK) of the same lesion, at the same day, but with a different camera. A green shade interferes with the actual tumor colors, and its margins are less in contrast with the surrounding choroid. There is also a lesser impression of a three-dimensional lesion (C) on 10 MHz B-scan ultrasonography, the tumor is dome-shaped. $H = 3.0$ mm (D) on B-scan OCT (Heidelberg Engineering spectralis), "shaggy" photoreceptors at the tumor's surface can be observed. Subretinal fluid surrounds the mass. (E) Orange pigment at the surface appears auto-fluorescent (Heidelberg Engineering spectralis). (F) On ICG-A (Heidelberg Engineering spectralis), the tumor provokes a masking defect. (G) On early FA, the orange pigment appears hypofluorescent. (H) On late FA (Heidelberg Engineering spectralis), the pinpoints, already visible at the early stage, present a progressive diffusion at the surface, causing a macular detachment. *FA*, fluorescein angiography; *H*, height; *ICG-A*, indocyanine green angiography; *OCT*, optical coherence tomography; *VA*, visual acuity; *VEGF*, vascular endothelial growth factor. *Courtesy of Dr A Schalenbourg.*

to the wrong treatment. In fact, color is one of the most important diagnostic features of choroidal tumors. Melanomas tend to be brown, gray black, or white. Hemangiomas are usually orange-red. Most metastases are white, with the exception of renal and carcinoid secondaries, which can be orange.[30]

The Panoret-1000 (CMT Medical Technologies Inc, Valley Stream, NY) is a contact lens-based imaging technology using transscleral illumination, as opposed to the transpupillary illumination of the other cameras, with a 100-degree field of view. The comprehensive color images are digitally recomposed from the three primary colors (red, green, and blue) reflected by the fundus. Fundus imaging is possible despite media opacities and a pupil diameter as small as 2.5 mm. Transscleral illumination allows a more accurate definition of the tumor extent because it provides better contrast between tumor margins and surrounding tissues (Fig. 41.4A).[29]

Optical coherence tomography

Optical coherence tomography (OCT) has a limited application in visualizing choroidal melanoma, but it can be helpful in detecting concomitant lesions in the retina such as subretinal fluid and/or exudative retinal detachment (Fig. 41.4D). A more modern technique, enhanced depth imaging (EDI) OCT, can visualize melanoma as choroidal shadowing, the degree of which depends on the pigmentation of the malignancy, due to deeper penetration of the light beam.[31] EDI-OCT can provide high-resolution images of the choroid and allow for accurate measurement of tumor thickness, which can be more accurate than ultrasonography in small-sized lesions.[32,33] This is secondary to difficulty on ultrasound in detecting the precise sclerochoroidal junction, poorer resolution of the overlying retina leading to an inadvertent inclusion with thickness measurement, and gross estimation with ultrasonographic calipers.[34] EDI-OCT features of choroidal melanomas include homogeneous optical reflectivity along the anterior surface, with shadowing throughout deeper layers of the tumors, loss of overlying photoreceptors, and compression and thinning of the choriocapillaris. A "shaggy" appearance of the photoreceptors overlying small choroidal melanoma has also been described, and it could represent edematous photoreceptors or macrophages with lipofuscin on the posterior surface of the detached retina.[33]

OCT angiography

OCT angiography (OCTA) has limitations in detecting the whole tumor vascular content details, probably due to the high pigment content of these lesions visible with the laser wavelength used and the common presence of image artifacts.[35] Nonetheless, recent advancements in OCTA technology, such as swept-source (SS) OCTA, have expanded its range of applications. SS-OCTA uses an increased laser wavelength and scan speeds allowing for increased penetration through the RPE and larger scan areas.[36] SS-OCTA has been proposed for the differential diagnosis of choroidal melanomas and choroidal nevi. Significant differences between these conditions were found for the pattern of reflectivity and presence/absence of avascular zones and vascular anomalies between choroidal melanomas, elevated choroidal nevi, and flat choroidal nevi.[37]

Recently, Pellegrini et al. assessed the role of OCTA in the investigation of the vascular network in different choroidal melanomas.[38] In their series of 22 eyes, using an SS-OCTA, they could image and obtain flow information in small to large melanomas regardless of their location, postulating that SS-OCTA could represent a useful tool for noninvasive assessment of CM intrinsic characteristics at baseline and allowing for an enhanced visualization of the changes occurring in response to treatment. Moreover, OCTA-derived quantitative metrics may offer promising tools in identifying structural and microvascular factors contributing to the visual outcome[39] and to the monitoring of intravitreal anti-VEGF treatment of eyes with radiation maculopathy after proton beam therapy.

Fundus fluorescein angiography and indocyanine green angiography

Fundus fluorescein angiography (FA) and indocyanine green angiography (ICGA) show no pathognomonic pattern and neither FA nor ICGA are considered essential in concluding the diagnostic process, but they may help distinguish choroidal melanoma from underlying pathologies. The most common findings on FA are an intrinsic tumor circulation ("dual" or "double" circulation) because of the superimposition of the intravascular fluorescence of the intact retinal vasculature over the fluorescence of large caliber vessels within choroidal tumor, mottled fluorescence during the arteriovenous phase, late progressive leakage, lesion staining, and multiple pinpoint leaks (hot spots) at the level of the retinal pigment epithelium (Fig. 41.4G,H).[40,41] Evidence of retinal vascular abnormalities such as microaneurysms, tortuous intraretinal capillaries, and zones of nonperfusion has also been reported on FFA.[24,42] CMs are usually hypofluorescent on ICGA (Fig. 41.4F), but some variations might depend on tumor thickness and pigmentation.[43,44] ICGA is mainly used to study intratumoral microcirculation and changes in the surrounding choroidal vasculature, and it might potentially differentiate nonpigmented choroidal melanoma from choroidal hemangioma.[27]

FAF

Choroidal melanoma is associated with several degenerative changes in the overlying tissues, which has been named as "Tumor-Associated Retinal Pigment Epitheliopathy."[45] This secondary epitheliopathy is characterized by the shedding of photoreceptor outer segments and a localized detachment of the RPE, which together with an intense phagocytic activity is responsible for changes in fundus coloration and lipofuscin accumulation. Blue-light fundus autofluorescence (FAF) detects fluorophores and mostly lipofuscin, accumulating at the level of RPE cells and photoreceptor outer segments that appear as orange pigment at fundus examination.[46] Therefore, FAF could play a role in detecting early and small clumps of orange pigment, which is a well-known risk factor for small melanomas, and in differentiating between CM and naevus.[47] In choroidal nevi, FAF shows a normal pattern of background fundus fluorescence with no corresponding areas of hyper or hypofluorescence over the nevi, while CM usually appears as a confluent plaque-like hyperautofluorescent area with central hypoautofluorescence (Fig. 41.4E).[48] There is consensus that choroidal nevi rarely show brightly increased FAF due to orange pigment, whereas remarkably increased FAF (ranging from mildly to intensely) overlying choroidal melanomas derives from lipofuscin accumulation in RPE cells and in macrophages of malignant choroidal tumors usually seen as orange pigment over the lesion[49] with an excellent correlation between increased FAF and orange pigment.[50] However, increased FAF on the tumor surface does not necessarily indicate orange pigment but it may also result from pigmentation, fibrous metaplasia, or sub-RPE drusen.[51]

Ultrasonography

Ultrasonography (US) is the most valued diagnostic test. The combination of A-mode (8 MHz) and B-mode (10 MHz) ultrasonography is essential for measuring tumor dimensions, evaluating the extent, shape, and internal structure of the tumor, planning treatment, and monitoring the response to treatment.[7] CM typically demonstrates low to medium reflectivity due to their cellular homogeneous architecture. On A-scan ultrasonography, choroidal melanoma shows a high initial spike with medium to low internal echoes with smooth attenuation (angle kappa sign).[52,53] Vascular pulsations within the tumor can also be seen with this modality. On B-scan, classic features of choroidal melanoma are an acoustically silent zone within the melanoma, choroidal excavation (a depression free of echos lying beneath the tumor and behind the retinal surface overlying the tumor), and shadowing of orbital fat behind the lesion.[53,54] Lesions can appear as dome-shaped (Fig. 41.1B), mushroom-shaped (i.e., collar button) (Fig. 41.2B), or irregular in shape depending on the degree of Bruch's membrane rupture and axial tumor growth. The estimation of melanoma thickness by combined A-mode and B-mode ultrasonography provides reproducible and concordant values of tumor thickness, which is the most reliable and important measurement. Measurement of tumor diameter and tumor height can vary among trained operators, especially for small lesions. Nevertheless, various reports have suggested that up to 96% of choroidal melanomas can be correctly diagnosed with a combination of A and B-scan ultrasonography.[55] With US is also clearly visible the retinal detachment (Fig. 41.2B), a concomitant condition with uveal melanoma, seen as an accumulation of subretinal fluid that splits the layers (choroid and retina). Moreover, B-mode ultrasonography can also detect and highlight the presence of extrascleral extension of melanomas, and it is considered superior to MRI in the detection of extrascleral extension.[56] The protruding portion of the tumor has lower reflectivity than the orbital fat, making it identifiable.[57] Thus, the US is suitable for assessing tumor extrascleral extension and visualizing lesions even as small as 2 mm.[8] Color Doppler (CDI) is a noninvasive method that shows the blood flow in vessels with the flow direction.[55] CDI examination of CM will show evidence of perfusion in most instances, while choroidal nevi are minimally vascular.[58]

Clinical trials

The COMS is the largest study ever performed in ocular oncology, with 43 participating centers, more than 2000 recruited patients, and 28 publications.[59] CMOS aimed to determine the best management for uveal melanomas, depending on the size and comparing the treatment of radiation and enucleation.[60]

It was designed as a three-arm trial including two multicenter randomized clinical trials comparing the effectiveness of iodine 125 (I-125) brachytherapy to enucleation for the treatment of medium-size choroidal melanomas, and the effectiveness of enucleation with and without preoperative external-beam radiotherapy for large choroidal melanomas. The third arm was an observational study of the natural history of small choroidal melanomas.

The primary outcome measure was time to death from all-cause mortality. Secondary outcome measures included metastasis-free survival, cancer-free survival, and years of useful vision. The trial of enucleation alone versus preenucleation radiotherapy included 1003 patients with large choroidal melanoma (>10 mm in apical height and >16 mm in basal diameter) randomized to either enucleation or enucleation preceded by external-beam radiation. No difference was found between the two treatment arms (the 5 year all-cause mortality was 43% and 38%; 5 year tumor-related mortality was 28% and 26%; and 10 year tumor-related mortality was 40% and 45% in enucleation and enucleation with preoperative radiation, respectively), thus reporting no survival advantage attributable to preenucleation radiation and reassuring that primary enucleation does not accelerate death from metastatic melanoma.[61,62]

In 1317 patients with medium-sized choroidal melanoma (2.5−10 mm in apical height and ≤16 mm in basal diameter), 660 were randomized to enucleation and 657 to iodine-125 brachytherapy, reporting that mortality with histopathologically confirmed melanoma metastasis after brachytherapy with iodine-125 was no worse than following enucleation.[63] By 12 years, cumulative all-cause mortality was 43% in the brachytherapy arm and 41% in the enucleation arm. The 5-, 10-, and 12-year mortality with histopathologically confirmed melanoma metastasis was 10%, 18%, and 21%, respectively, in the brachytherapy arm, and 11%, 17%, and 17%, respectively, in the enucleation arm.[62,63] These results confirmed no survival differences between patients whose tumors were treated with I-125 brachytherapy and those treated with enucleation.

Overall, 204 patients were enrolled in the COMS natural history study of small melanomas (1.5−2.4 mm in apical height and 5−16 mm in basal diameter); 2- and 5-year Kaplan−Meier estimates of tumor growth was 21% and 31%, respectively. The 5 year all-cause mortality was 6% and tumor-related mortality was 1% in patients with small choroidal melanomas under observation.[64]

COMS results have had a positive impact on the standardization and accuracy of diagnosis and the clinical care of uveal melanoma. However, the original categorization of COMS patients was based solely on tumor size, although today uveal melanoma prognosis is known to be multifactorial and there is increased evidence that genetic profile governs prognosis and not just the modality of treatment. Given the currently available evidence, it would be indicated to continue to locally treat choroidal melanoma optimally to conserve the eye and vision when possible. Patients should be categorized based on their established clinical profile, histopathological characteristics, and gene expression profiles (GEP), and those at high risk for systemic metastasis should be considered for adjuvant therapy once its benefits have been well-established.[65]

Managements strategies

After a diagnosis of choroidal melanoma, treatment is focused on cancer control, followed by eye and vision preservation.[66] Before initiating ocular therapy for choroidal melanoma, it is essential to perform a systemic workup to ensure that there is no evidence of metastasis. Choroidal melanoma has a well-documented capacity to spread hematogenously, with the liver being the first and only site of metastasis in most patients (90%), followed by lungs (24%), bone (16%), skin, and central nervous system.[67,68] Once it is confirmed that the disease is confined to the eye, local ophthalmic therapy can be directed toward the primary tumor. It is uncommon for extraocular metastasis to occur at the time of initial ocular presentation, affecting less than 5% of patients.[69] However, if hematogenous spread is detected, then local therapy for the eye may be deferred in favor of systemic treatment, depending on the patient's symptomatology and overall clinical status. It is crucial to note that the management of uveal melanoma is highly individualized, and treatment decisions should be made based on a single patient's unique circumstances and overall health. Therefore, a comprehensive evaluation and discussion of treatment options with a multidisciplinary team are recommended to ensure the best possible outcome for the patient.[69]

Treatment of localized UM can currently be classified into eye salvage treatment or enucleation. As discussed in the previous paragraph, according to the COMS, there is no difference in survival whether choroidal melanoma is treated with enucleation or radiation therapy.[59] As a consequence, enucleation is only indicated when conservative treatment is either unsafe or unreasonable. The patient's preference to undergo or not undergo enucleation should also be taken into consideration.[8]

Radiation therapy

Local radiotherapy for choroidal melanoma is performed using two different sources, either a radioactive plaque or an external beam. The first technique, called brachytherapy, involves affixing a dish-shaped source on the sclera over the base of the intraocular tumor defined by transillumination.[70] Currently, the most frequently used plaques are ruthenium-106 ($^{106}Ru/^{106}Rh$) and iodine-125 (^{125}I), emitting beta and gamma irradiation, respectively.[8] Brachytherapy may be used for tumors less than 18 mm in basal diameter and up to 12 mm thickness in which there is a reasonable chance of salvaging vision. Plaques are removed during a second intervention, once the required dose is delivered at the apex of the intraocular tumor (85–100 Gy for uveal melanoma). Brachytherapy indication is limited when the height and/or diameter of the tumor exceeds the capacity of the plaque or when the tumor location is unsuitable, for example, around the optic disk or on the iris.

External beam radiotherapy uses charged heavy particles such as protons, as a source of external beam irradiation. Protons produce low irradiation when penetrating a mass until they stop, at which point they deliver the maximum of their energy, called the "Bragg peak". By modulating the proton beam, the Bragg peaks can be accumulated, allowing delivery of a uniform radiation dose (60 Gy in four daily fractions for uveal melanoma) to a given target volume. Preceding the irradiation, tantalum clips are surgically sowed upon the sclera as points of reference with regard to the tumor base.[8,71] This treatment method enables the delivery of a concentrated dose of radiation to the tumor while minimizing exposure to the surrounding tissues. It is typically used for larger or posteriorly located tumors. Literature data suggest that brachytherapy and external beam radiotherapy are comparable for tumor control, visual outcome, and systemic prognosis, even if no modern randomized clinical trial exists comparing these modalities.[66,71,72] A study of all stages of UM revealed that, with a medium follow-up of 5 years, proton-beam-targeted therapy has a 96.4% control rate locally and preserves the eyes in 95% of cases.[73] The limitation of this technique is that it requires a cyclotron to accelerate the protons, adapted, with a stereotactic chair, for ocular treatment.[8]

Radiation therapy is associated with severe sight-limiting side effects. The severity, location, and incidence of radiation-induced complications are related to the type of radiation used, its method of delivery, the size of the tumor treated, and the amount of radiation delivered to normal ocular structures. Radiation effects are often delayed and therefore complications have been noted to increase over time. Complications include radiation-induced cataracts, retinopathy, maculopathy, optic neuropathy, toxic tumor syndrome, and secondary glaucoma.[74,75]

Enucleation

Choroidal melanomas are currently treated by eye-preserving radiotherapies. In the past, enucleation was the primary treatment for patients with CM before radiation therapy became available. However, in the present day, enucleation is indicated in the cases of large tumor diameter or a tumor target volume requiring more than half of the eye to be irradiated (defined using the COMS criteria) because of the high risk of secondary enucleation after conservative treatment in these cases, circumferential tumor invasion, when the limits of the tumor cannot be determined with confidence, in the presence of (sub)total retinal detachment, extrascleral spread, optic nerve invasion, patients suffering from complete vision loss, and pain due to secondary glaucoma.[76] Additional relative indications for enucleation in selected cases include the loss of visual function following radiation.

Prognosis and follow-up

Patients with UM are at lifetime risk of having metastasis even with treatment of primary malignancy. The survival of uveal melanoma patients depends on whether hematogenous spread has occurred before the treatment of the primary tumor. Nearly 50% of uveal melanoma patients will suffer from metastases, with the liver being the most common initially affected organ. Around 20%–30% of UM patients die within 5 years and 45% die within 15 years,[61,71] but the overall risk varies according to the clinical, histopathologic, and cytogenetic risk factors, and increases with advanced patient age and local tumor recurrence.[77]

Adverse cytogenetic mutations include, most importantly, a loss of the entire chromosome 3 (monosomy 3) and, secondly, an amplification of the long arm of chromosome 8 (8q gain).[78] The principal tumor suppressor gene lost in the former has been identified as the gene encoding BRCA1-associated protein 1 (BAP1) on chromosome 3p21.[79]

Although there are no consensus guidelines concerning optimal surveillance tests after initial management, MRI has demonstrated the highest sensitivity in identifying small liver lesions. Based on gene-expression profiling or cytogenetics, patients with low-risk disease should be considered for routine imaging, including chest CT, and abdomen and pelvis MRI every 6–12 months. In contrast, patients with a higher metastatic recurrence risk warrant a tighter follow-up, with imaging taken every 3–6 months.[80]

The risk of local tumor recurrence due to incomplete irradiation has been reported at 1.1%–4% at 5 years after proton therapy[81] at 4.2%–15% after iodine-125[82,83] and at up to 19% after ruthenium-106 plaque therapy.[84] The risk of neovascular glaucoma, complicating ischemic serous retinal detachment and a major cause of secondary enucleation after conservative radiotherapy, has considerably decreased with the prophylactic use of intravitreal antivascular endothelial growth factor, followed by laser treatment once the retina is reattached.[24] Visual prognosis depends on the anatomical location of the tumor, its size, and the extent of the retinal detachment at diagnosis.[71] Other tumor and radiation-induced side effects responsible for vision loss include radiation-induced optic neuropathy and ischemic retinopathy and maculopathy.

Acknowledgment

The Authors would like to thank Dr. Ann Schalenbourg for providing the figures and Mr. Yann Leuba for his precious assistance in preparing the figures.

References

1. Krantz BA, Dave N, Komatsubara KM, Marr BP, Carvajal RD. Uveal melanoma: epidemiology, etiology, and treatment of primary disease. *Clinical Ophthalmology*. 2017;11:279–289.
2. Damato EM, Damato BE. Detection and time to treatment of uveal melanoma in the United Kingdom: an evaluation of 2384 patients. *Ophthalmology*. 2012;119:1582–1589.
3. Vasalaki M, Fabian ID, Reddy MA, Cohen VML, Sagoo MS. Ocular oncology: advances in retinoblastoma, uveal melanoma and conjunctival melanoma. *British Medical Bulletin*. 2017;121:107–119.
4. Shields CL, Kaliki S, Livesey MG, et al. Association of ocular and oculodermal melanocytosis with the rate of uveal melanoma metastasis: analysis of 7872 consecutive eyes. *JAMA Ophthalmology*. 2013;131:993–1003.
5. Weis E, Shah CP, Lajous M, Shields JA, Shields CL. The association between host susceptibility factors and uveal melanoma: a meta-analysis. *Archives of Ophthalmology*. 2006;124:54–60.
6. Derrien A-C, Rodrigues M, Eeckhoutte A, et al. Germline MBD4 mutations and Predisposition to uveal melanoma. *JNCI: Journal of the National Cancer Institute*. 2021;113:80–87.
7. Branisteanu DC, Bogdanici C, Branisteanu D, et al. Uveal melanoma diagnosis and current treatment options (review). *Experimental and Therapeutic Medicine*. 2021;22:1428.
8. Schalenbourg A, Zografos L. Uveal melanocytic tumors. In: Singh AD, Seregard S, eds. *Ocular Tumors. ESASO Course Series*. vol 7. Basel: Karger; 2016:5–19.
9. Shields CL, Manalac J, Das C, Ferguson K, Shields JA. Choroidal melanoma: clinical features, classification, and top 10 pseudomelanomas. *Current Opinion in Ophthalmology*. 2014;25:177–185.
10. Shields CL, Furuta M, Thangappan A, et al. Metastasis of uveal melanoma millimeter-by-millimeter in 8033 consecutive eyes. *Archives of Ophthalmology*. 2009;127:989–998.
11. Shields CL, Materin M, Shields JS, Gershenbaum E, Singh AD, Smith A. Factors associated with elevated intraocular pressure in eyes with iris melanoma. *British Journal of Ophthalmology*. 2001;85:666 LP–669.
12. Papadopoulou D, Moulin AP, Zografos L, Schalenbourg A. How would you manage this small melanocytic choroidal tumour? small choroidal melanoma hiding a large extrascleral extension. *British Journal of Ophthalmology*. 2012;96(12):1539–1540.
13. Szalai E, Wells JR, Grossniklaus HE. Mechanisms of optic nerve invasion in primary choroidal melanoma. *Ocular Oncology and Pathology*. 2017;3:267–275.
14. Eskelin S, Kivelä T. Mode of presentation and time to treatment of uveal melanoma in finland. *British Journal of Ophthalmology*. 2002;86:333–338.
15. Shields CL, Demirci H, Materin MA, et al. Clinical factors in the identification of small choroidal melanoma. *Canadian Journal of Ophthalmology*. 2004;39:351–357.
16. Shields CL, Shields JA, Kiratli H, De Potter P, Cater JR. Risk factors for growth and metastasis of small choroidal melanocytic lesions. *Ophthalmology*. 1995;102:1351–1361.
17. Shields CL, Dalvin LA, Michael DY, et al. Choroidal nevus transformation into melanoma per millimeter increment in thickness using multimodal imaging in 2355 cases: the 2019 Wendell L. Hughes lecture. *Retina*. 2019;39.
18. Shields CL, Lim L-AS, Dalvin LA, Shields JA. Small choroidal melanoma: detection with multimodal imaging and management with plaque radiotherapy or AU-011 nanoparticle therapy. *Current Opinion in Ophthalmology*. 2019;30:206–214.
19. Damato BE. Can the MOLES acronym and scoring system improve the management of patients with melanocytic choroidal tumours? *Eye (Lond)*. 2023;37:830–836.
20. Roelofs KA, O'Day R, Al Harby L, et al. The moles system for planning management of melanocytic choroidal tumors: is it safe? *Cancers*. 2020;12.
21. Accuracy of diagnosis of choroidal melanomas in the collaborative ocular melanoma study: COMS report no. 1. *Archives of Ophthalmology*. 1990;108:1268–1273.
22. Shields JA, Augsburger JJ, Brown GC, Stephens RF. The differential diagnosis of posterior uveal melanoma. *Ophthalmology*. 1980;87:518–522.
23. Singh AD, Kalyani P, Topham A. Estimating the risk of malignant transformation of a choroidal nevus. *Ophthalmology*. 2005;112:1784–1789.

24. Mantel I, Schalenbourg A, Bergin C, et al. Prophylactic use of bevacizumab to avoid anterior segment neovascularization following proton therapy for uveal melanoma. *American Journal of Ophthalmology*. 2014;158:693–701.e2.
25. Zabor EC, Raval V, Luo S, Pelayes DE, Singh AD. A prediction model to discriminate small choroidal melanoma from choroidal nevus. *Ocular Oncology and Pathology*. 2022;8:71–78.
26. Oganov AC, Seddon I, Jabbehdari S, et al. Artificial intelligence in retinal image analysis: development, advances, and challenges. *Survive of Ophthalmology*. 2023;68(5):905–919. https://doi.org/10.1016/j.survophthal.2023.04.001.
27. Shields CL, Lally SE, Dalvin LA, et al. White paper on ophthalmic imaging for choroidal nevus identification and transformation into melanoma. *Translational Vision Science and Technology*. 2021;10:24.
28. Callaway NF, Mruthyunjaya P. Widefield imaging of retinal and choroidal tumors. *International Journal of Retina and Vitreous*. 2019;5:49.
29. Schalenbourg A, Zografos L. Pitfalls in colour photography of choroidal tumours. *Eye*. 2013;27:224–229.
30. Elfalah M, AlRyalat SA, Abu-Yaghi NE, et al. The impact of color vision deficiency on the capability of ophthalmologists to diagnose benign and malignant choroidal tumors. *Journal of Clinical Medicine*. 2023;12.
31. Obuchowska I, Konopińska J. Importance of optical coherence tomography and optical coherence tomography angiography in the imaging and differentiation of choroidal melanoma: a review. *Cancers*. 2022;14.
32. Shah SU, Kaliki S, Shields CL, et al. Enhanced depth imaging optical coherence tomography of choroidal nevus in 104 cases. *Ophthalmology*. 2012;119:1066–1072.
33. Shields CL, Kaliki S, Rojanaporn D, Ferenczy SR, Shields JA. Enhanced depth imaging optical coherence tomography of small choroidal melanoma: comparison with choroidal nevus. *Archives of Ophthalmology*. 2012;130:850–856.
34. Shields CL, Manalac J, Das C, Saktanasate J, Shields JA. Review of spectral domain enhanced depth imaging optical coherence tomography of tumors of the choroid. *Indian Journal of Ophthalmology*. 2015;63:117–121.
35. Say EAT, Ferenczy S, Magrath GN, Samara WA, Khoo CTL, Shields CL. Image quality and artifacts on optical coherence tomography angiography: comparison of pathologic and paired fellow eyes in 65 patients with unilateral choroidal melanoma treated with plaque radiotherapy. *Retina*. 2017;37.
36. Naseripour M, Ghasemi Falavarjani K, Mirshahi R, Sedaghat A. Optical coherence tomography angiography (OCTA) applications in ocular oncology. *Eye (Lond)*. 2020;34:1535–1545.
37. Greig EC, Laver N, Mendonca LSM, et al. Swept-source optical coherence tomography angiography in small choroidal melanomas and choroidal nevi. *Retina*. 2021;41:1182–1192.
38. Pellegrini M, Corvi F, Invernizzi A, Ravera V, Cereda A, Staurenghi G. Swept-source optical coherence tomography angiography in choroidal melanoma: an analysis of 22 consecutive cases. *Retina*. 2019;39:1510–1519.
39. Matet A, Daruich A, Zografos L. Radiation maculopathy after proton beam therapy for uveal melanoma: optical coherence tomography angiography alterations influencing visual acuity. *Investigative Ophthalmology and Visual Science*. 2017;58:3851–3861.
40. Pettit TH, Barton A, Foos RY, Christensen RE. Fluorescein angiography of choroidal melanomas. *Archives of Ophthalmology*. 1970;83:27–38.
41. Augsburger JJ, Golden MI, Shields JA. Fluorescein angiography of choroidal malignant melanomas with retinal invasion. *Retina*. 1984;4:232–241.
42. Cantrill HL, Douglas Cameron J, Ramsay RC, Knobloch WH. Retinal vascular changes in malignant melanoma of the choroid. *American Journal of Ophthalmology*. 1984;97:411–418.
43. Bakhoum MF, Knight DK, Goldbaum MH. Defining extent of diffuse choroidal melanoma with indocyanine green angiography. *Ophthalmology*. 2021;128:834.
44. Sallet G, Amoaku WMK, Lafaut BA, Brabant P, De Lacy JJ. Indocyanine green angiography of choroidal tumors. *Graefe's Archive for Clinical and Experimental Ophthalmology*. 1995;233:677–689.
45. Damato BE, Foulds WS. Tumour-associated retinal pigment epitheliopathy. *Eye*. 1990;4:382–387.
46. Pole C, Ameri H. Fundus autofluorescence and clinical applications. *Journal of Ophthalmic and Vision Research*. 2021;16:432–461.
47. Solnik M, Paduszyńska N, Czarnecka AM, et al. Imaging of uveal melanoma-current standard and methods in development. *Cancers*. 2022;14.
48. Lavinsky D, Belfort Jr RN, Navajas E, Torres V, Martins MC, Belfort R. Fundus autofluorescence of choroidal nevus and melanoma. *British Journal of Ophthalmology*. 2007;91:1299–1302.
49. Bindewald-Wittich A, Holz FG, Ach T, Fiorentzis M, Bechrakis NE, Willerding GD. Fundus autofluorescence imaging in patients with choroidal melanoma. *Cancers*. 2022;14.
50. Amselem L, Pulido JS, Gunduz K, et al. Changes in fundus autofluorescence of choroidal melanomas following treatment. *Eye*. 2009;23:428–434.
51. Gündüz K, Pulido JS, Ezzat K, Salomao D, Hann C. Review of fundus autofluorescence in choroidal melanocytic lesions. *Eye*. 2009;23:497–503.
52. Wang T-J, Yang CH, Liao SL, et al. Characteristic ultrasonographic findings of choroidal tumors. *Journal of Medical Ultrasound*. 2003;11:55–59.
53. Sobottka B, Kreissig I. Ultrasonography of metastases and melanomas of the choroid. *Current Opinion in Ophthalmology*. 1999;10:164–167.
54. Singh P, Singh A. Choroidal melanoma. *Oman Journal of Ophthalmology*. 2012;5:3–9.
55. Char DH, Stone RD, Irvine AR, et al. Diagnostic modalities in choroidal melanoma. *American Journal of Ophthalmology*. 1980;89:223–230.
56. Jacobsen BH, Ricks C, Harrie RP. Ocular ultrasound versus MRI in the detection of extrascleral extension in a patient with choroidal melanoma. *BMC Ophthalmology*. 2018;18:320.
57. Zografos L. *Tumeurs Intraoculaires*. Masson; 2002:159–166.
58. Hilborn MD, Munk PL, Lin DT, Vellet AD, Poon PY. Sonography of ocular choroidal melanomas. *AJR American Journal of Roentgenology*. 1993;161:1253–1257.
59. Salowe R, O'Brien JM. The collaborative ocular melanoma study randomized trial of iodine 125 brachytherapy for choroidal melanoma. *Ophthalmology*. 2020;127:S146–S147. https://doi.org/10.1016/j.ophtha.2019.09.047. Preprint at.

60. Margo CE. The collaborative ocular melanoma study: an overview. *Cancer Control.* 2004;11:304–309.
61. The collaborative ocular melanoma study (COMS) randomized trial of pre-enucleation radiation of large choroidal melanoma II: initial mortality findings. COMS report no. 10. *American Journal of Ophthalmology.* 1998;125:779–796.
62. Hawkins BS. The collaborative ocular melanoma study (COMS) randomized trial of pre-enucleation radiation of large choroidal melanoma: IV. Ten-year mortality findings and prognostic factors. COMS report number 24. *American Journal of Ophthalmology.* 2004;138:936–951.
63. Diener-West M, The Collaborative Ocular Melanoma Study Group. The COMS randomized trial of iodine 125 brachytherapy for choroidal melanoma, III: initial mortality findings. COMS report No. 18. *Archives of Ophthalmology.* 2001;119:969–982.
64. Mortality in patients with small choroidal melanoma. COMS report no. 4. The collaborative ocular melanoma study group. *Archives of Ophthalmology.* 1997;115:886–893.
65. Honavar SG. Is Collaborative ocular melanoma study (COMS) still relevant? *Indian Journal of Ophthalmology.* 2018;66:1385–1387. https://doi.org/10.4103/ijo.IJO_1588_18. Preprint at.
66. Abrams MJ, Gagne NL, Melhus CS, Mignano JE. Brachytherapy vs. external beam radiotherapy for choroidal melanoma: survival and patterns-of-care analyses. *Brachytherapy.* 2016;15:216–223.
67. Bedikian AY, Kantarjian H, Young SE, Bodey GP. Prognosis in metastatic choroidal melanoma. *Southern Medical Journal.* 1981;74:574–577.
68. Diener-West M, The Collaborative Ocular Melanoma Study Group. Development of metastatic disease after enrollment in the COMS trials for treatment of choroidal melanoma: collaborative ocular melanoma study group report No. 26. *Archives of Ophthalmology.* 2005;123:1639–1643.
69. Chattopadhyay C, Kim DW, Gombos DS, et al. Uveal melanoma: from diagnosis to treatment and the science in between. *Cancer.* 2016;122:2299–2312.
70. Finger PT. Radiation therapy for choroidal melanoma. *Survey of Ophthalmology.* 1997;42:215–232.
71. Gragoudas ES. Proton beam irradiation of uveal melanomas: the first 30 years the Weisenfeld lecture. *Investigative Ophthalmology and Visual Science.* 2006;47:4666–4673.
72. Kaliki S, Shields CL. Uveal melanoma: relatively rare but deadly cancer. *Eye (Lond).* 2017;31:241–257.
73. Seibel I, Cordini D, Rehak M, et al. Local recurrence after primary proton beam therapy in uveal melanoma: risk factors, retreatment approaches, and outcome. *American Journal of Ophthalmology.* 2015;160:628–636.
74. Wen JC, Oliver SC, McCannel TA. Ocular complications following I-125 brachytherapy for choroidal melanoma. *Eye.* 2009;23:1254–1268.
75. McCannel TA. Post-brachytherapy tumor endoresection for treatment of toxic maculopathy in choroidal melanoma. *Eye (Lond).* 2013;27:984–988.
76. Gambrelle J, Grange JD, Devouassoux M, et al. Survival after primary enucleation for choroidal melanoma: changes induced by the introduction of conservative therapies. *Graefes Archive for Clinical and Experimental Ophthalmology.* 2007;245:657–663.
77. Soliman N, Mamdouh D, Elkordi A. Choroidal melanoma: a mini review. *Medicines.* 2023;10. https://doi.org/10.3390/medicines10010011. Preprint at.
78. van den Bosch T, van Beek JGM, Vaarwater J, et al. Higher percentage of FISH-determined monosomy 3 and 8q amplification in uveal melanoma cells relate to poor patient prognosis. *Investigative Ophthalmology and Visual Science.* 2012;53:2668–2674.
79. Harbour JW, Onken MD, Roberson ED, et al. Frequent mutation of BAP1 in metastasizing uveal melanomas. *Science.* 2010;330:1410–1413.
80. Yang J, Manson DK, Marr BP, Carvajal RD. Treatment of uveal melanoma: where are we now? *Therapeutic Advances in Medical Oncology.* 2018;10:1758834018757175.
81. Egger E, Schalenbourg A, Zografos L, et al. Maximizing local tumor control and survival after proton beam radiotherapy of uveal melanoma. *International Journal of Radiation Oncology, Biology, Physics.* 2001;51:138–147.
82. Harbour JW, Char DH, Kroll S, Quivey JM, Castro J. Metastatic risk for distinct patterns of postirradiation local recurrence of posterior uveal melanoma. *Ophthalmology.* 1997;104:1783–1785.
83. Wilson MW, Hungerford JL. Comparison of episcleral plaque and proton beam radiation therapy for the treatment of choroidal melanoma. *Ophthalmology.* 1999;106:1579–1587.
84. Summanen P, Immonen I, Heikkonen J, Tommila P, Laatikainen L, Tarkkanen A. Survival of patients and metastatic and local recurrent tumor growth in malignant melanoma of the uvea after ruthenium plaque radiotherapy. *Ophthalmic Surgery.* 1993;24:82–90.

Further reading

1. Eandi CM, Polito MS, Schalenbourg A, Zografos L. Eighteen-month results of intravitreal anti-vascular endothelial growth factor on vision and microcirculation in radiation maculopathy. *Retina.* 2021;41:1883–1891.
2. Frizziero L, et al. Uveal melanoma biopsy: a review. *Cancers.* 2019;11. https://doi.org/10.3390/cancers11081075. Preprint at.
3. Eide N, Walaas L. Fine-needle aspiration biopsy and other biopsies in suspected intraocular malignant disease: a review. *Acta Ophthalmology.* 2009;87:588–601.
4. Maheshwari A, Finger PT. Laser treatment for choroidal melanoma: current concepts. *Survey of Ophthalmology.* 2023;68:211–224.
5. Bai H, Bosch JJ, Heindl LM. Current management of uveal melanoma: a review. *Clinical and Experimental Ophthalmology.* 2023;51(5):484–494. https://doi.org/10.1111/ceo.14214.
6. Harbour JW, Meredith TA, Thompson PA, Gordon ME. Transpupillary thermotherapy versus plaque radiotherapy for suspected choroidal melanomas. *Ophthalmology.* 2003;110:2207–2214. discussion 2215.
7. Desjardins L, et al. Combined proton beam radiotherapy and transpupillary thermotherapy for large uveal melanomas: a randomized study of 151 patients. *Ophthalmic Research.* 2006;38:255–260.

8. Mashayekhi A, et al. Primary transpupillary thermotherapy for choroidal melanoma in 391 cases: importance of risk factors in tumor control. *Ophthalmology*. 2015;122:600−609.
9. Roelofs KA, Fabian ID, Arora AK, Cohen VML, Sagoo MS. Long-term outcomes of small pigmented choroidal melanoma treated with primary photodynamic therapy. *Ophthalmology Retina*. 2021;5:468−478.
10. Yordi S, Soto H, Bowen RC, Singh AD. Photodynamic therapy for choroidal melanoma: what is the response rate? *Survey of Ophthalmology*. 2021;66:552−559.

Chapter 42

Retinal vasoproliferative tumors

Martin A. Mainster[1], Alan Chew Bonilla[2] and Gerardo Ledesma-Gil[2]

[1]*Department of Ophthalmology, University of Kansas School of Medicine, Kansas City, KS, United States;* [2]*Retina Department, Institute of Ophhtalmology Fundacion Conde de Valenciana, Mexico City, Mexico*

Overview bullets

- Retinal vasoproliferative tumors (VPTs) are yellowish-pink, peripheral gliovascular masses.
- VPTs are classified as secondary if there is a known associated systemic or ophthalmic problem or primary if they are idiopathic.
- There is a general consensus that VPTs are reactive tumors so the use of the term primary for idiopathic lesions is potentially etiologically misleading.
- Differentiating between primary and secondary lesions is useful clinically because secondary lesions occur at a younger age and are more likely to be multiple, aggressive, and visually debilitating than "primary" VPTs.
- Newly diagnosed VPTs require careful systemic and ophthalmic evaluations to rule out associated disorders, especially in younger patients.
- VPTs cause serious vitreoretinal problems including intraretinal and subretinal exudation, exudative retinal detachment, and remote epiretinal membranes (ERMs) and cystoid macular edema.
- VPTs can be difficult to identify because of their peripheral location, indistinct borders, and color similarity to adjacent peripheral retina. We hypothesize that unrecognized or subclinical tumors may be linked pathogenically to otherwise idiopathic epiretinal membranes, CME, and vitreoretinal traction.
- The differential diagnosis for VPTs includes retinal capillary hemangioblastoma, Coats disease, peripheral exudative hemorrhagic chorioretinopathy, and choroidal melanoma.
- Ultra-widefield photographic and OCT imaging systems, ultrasonography, and tissue biopsy are useful for identifying, analyzing, and monitoring VPTs.
- VPTs are monitored clinically by sequential ophthalmoscopy and retinal imaging.
- There are no current biomarkers for reliably predicting VPT expansion.
- Small, asymptomatic VPTs can be followed by observation; symptomatic tumors are usually treated with cryopexy or laser photocoagulation; episcleral plaque therapy is often used for large VPTs.
- Pars plana vitrectomy may be needed to remove ERMs or vitreous hemorrhage.
- Anti-VEGF, anti-TNF, or steroid medications may be combined with other treatment modalities or used as monotherapy.
- There have been no randomized, prospective controlled trials to date for treating VPTs because they are uncommon and have variable natural histories.

Introduction

Retinal vasoproliferative tumors (VPTs) are yellowish-pink retinal masses that often occur in the inferior-temporal peripheral retina. They usually have a dome or nodule shape but may present as irregular, diffuse, or mushroom-shaped lesions.[1–5] VPTs cause intraretinal and subretinal exudation, vitreous and retinal hemorrhages, cystoid macular edema (CME), macular fibrosis, and central epiretinal membranes (ERMs).[6] Exudation can be localized or extend progressively centrally to involve the macula. VPTs are also known as reactive retinal astrocytic tumors (RRATs) because histopathological studies demonstrate that they are composed predominantly of astrocytic tissue with less prominent vascular and retinal pigment epithelial elements, even early in their development.[7–15]

Retinal and Choroidal Vascular Diseases of the Eye. https://doi.org/10.1016/B978-0-443-15583-3.00042-1

Disruptions of chorioretinal homeostasis due to trauma, inflammation, or other stresses can cause reactive proliferative retinochoroidopathies in conditions including VPT, proliferative diabetic retinopathy,[16] reactive retinal pigment epithelial hyperplasia,[17] proliferative vitreoretinopathy,[9,18] peripheral exudative hemorrhagic chorioretinopathy,[19,20] and benign reactive lymphoid hyperplasia of the choroid.[21,22] The morphology and growth of these reactive lesions are governed by the combinations of chorioretinal cell types that undergo hyperplasia and metaplasia. Confusion arises when the predominant cell type inferred from clinical studies conflicts with histopathologic and related immunohistochemical and molecular diagnostic findings.[10,11,23,24] This conundrum has led to the current, concurrent use of *vasoproliferative tumor* and *reactive retinal astrocytic tumor* to identify the same clinical disorder.[14,15,25] Histopathology and its confirmatory studies often supersede clinical findings in the choice of disease identification terminology,[11] but we use the older clinical term VPT in this chapter because of its clinical utility, with the caveat that *reactive retinal astrocytic tumor* more accurately describes current histopathological understanding of VPT microscopic anatomy and pathogenesis.[10–12] Additional names used for VPTs in the past have included *peripheral retinal angioma with exudative retinopathy*,[26] *peripheral nodular retinal telangiectasis*,[27] *presumed acquired retinal hemangioma*,[28] and *reactionary retinal glioangiosis*.[8]

VPTs are reactive cellular responses to degenerative, inflammatory, ischemic, or other vitreoretinal stresses, with genetic predispositions yet to be elucidated.[4,5,13,29] They are classified as secondary when there is a suspected inciting stimulus and primary when they are idiopathic.[4,5] Small, peripheral asymptomatic VPTs can be managed by periodic observation but more advanced symptomatic lesions require treatment to prevent serious vision loss and complications leading to neovascular glaucoma, phthisis, and enucleation.[6,30] VPTs are uncommon disorders with highly variable natural courses so prospective trials to establish evidence-based management guidelines for recalcitrant cases have not been feasible.[25,31–33]

Pathogenesis

Current histopathologic evidence supports a glial origin for VPTs, with inciting stimuli inducing local, benign, astrocytic hyperplastic proliferation with secondary vascular and retinal pigment epithelial responses.[7,8,10–15,34] Retinal astrocytic proliferation can be divided into diagnostic groups based on lesion size, with VPT (aka *focal nodular gliosis*), submassive gliosis, and massive gliosis occupying <25%, 25%–50% or >50% of the vitreous cavity, respectively.[11,35] This tumor spectrum shares common histopathologic features that include: (1) a predominant component of pilocytic fibrous spindle-shaped astrocytes with low Ki-67 proliferative indices, (2) abnormal telangiectatic vasculature including large sclerotic vessels with hyalinized walls, and (3) retinal pigment epithelial fibrous and osseous metaplasia ultimately forming an outer calcification rim.[10,12,35,36] Additionally, genes significantly upregulated in an excised VPT specimen include astrocytes but not vascular or angiogenesis-related genes.[36] An additional case report documented the progression of VPT from early focal nodular gliosis to massive retinal gliosis unresponsive to therapy. Histopathology confirmed that the lesion was reactive retinal gliosis.[11,12]

Clinical data supporting an alternative retinal neovascular etiology for VPTs includes the presence of a feeding retinal artery and draining vein early in tumor development and visibly or angiographically prominent blood vessels in more developed lesions.[5,23,31,36,37] VEGF immunoreactivity is identifiable in hyalinized blood vessels and adjacent glial cells of excised VPT specimens,[14,38] consistent with retinal neovascularization potentially initiating the growth of the cellular mass.[23] Inflammatory conditions including intermediate uveitis could cause retinal vascular incompetence facilitating neovascularization and resultant glial proliferation.[39] Moreover, disorders associated with and possibly a cause of secondary VPTs include nongliotic retinal vasculopathies such as Coats disease, familial exudative vitreoretinopathy, tuberculosis, and retinal vasculitis.[5,14,15,40] Resolving the issue of whether VPT formation is driven primarily by vascular or glial proliferation would help guide future management of these tumors.

Bilateral primary VPT has been described in monozygotic twins, indicating a potential genetic or epigenetic predisposition for this disorder.[29] Additionally, a novel mutation of the *NR2E3* (aka *PNR*) gene was identified in a family with VPT and Goldmann-Favre phenotypes.[41] Retinal vasculature plays a primary role in VPT pathogenesis, but the choroid[3,4,25,26,39,42] and vitreous humor[25,43–46] are also involved, as in the other reactive proliferative retinochoroidopathies discussed above. Additionally, the vitreous humor plays a significant role in the distal (noncontiguous) effects of smaller VPT lesions including CME and ERM because of its roles in vitreoretinal traction and as a repository for mediators of inflammation and fibrosis.[47,48]

Clinical features

VPTs typically occur in a patient's third or fourth decade of life, causing vision loss, photopsia, and floaters. Decreased vision results from CME, ERM, intraretinal and subretinal exudation, exudative or tractional retinal detachment, and macular exudation.[4,5,43,49] Tumors usually have an irregular, circumscribed dome or globular shape, as shown in Figs. 42.1 and 42.2. Their ophthalmoscopic visualization may be difficult because of their peripheral location, indistinct border, and similarity in color to adjacent peripheral retina.[4,5,31] Fewer than 10% of tumors have an alternative diffuse, multinodular morphology with a lateral extent greater than 10 mm.[4,5] Most tumors are solitary (87%) but multiple lesions can occur, usually in bilateral cases.[4,5] A telangiectatic vascular network with aneurysms of variable size may be apparent on the inner surface of the tumor, with retinal vessels of largely normal diameter entering the lesion at its posterior border.[4,5,31] Visualization of the tumor may be obscured by overlying vitreous hemorrhage. Subretinal and/or intraretinal exudation is associated with most cases (80%), with subretinal fluid in 50% and adjacent RPE hyperplasia in 10% of VPTs.[5,50]

VPTs were classified originally as idiopathic (74%) or secondary to associated ocular disease (26%).[5] The percentage of secondary tumors is higher (50%–68%) in more recent studies.[25,42] The term primary[4] rather than idiopathic[5] is used currently but it is potentially misleading etiologically because there is a general consensus that VPTs are reactive rather than primary lesions.[10,13,23,24] Conversely, as shown in Fig. 42.1, the "primary" (idiopathic) versus secondary classification is useful clinically because secondary tumors generally occur at an earlier age, have poorer vision, and are more likely to be multiple, bilateral, and larger than "primary" VPTs without a known associated disorder.

Secondary VPTs are associated with a broad spectrum of ocular and systemic disorders including retinitis pigmentosa, intermediate uveitis, Coats disease (cf., Fig. 42.1), prior ocular surgery, retinal vasculitis, familial exudative vitreoretinopathy, toxoplasmosis, toxocariasis, trauma, retinopathy of prematurity, tuberculosis, toxoplasmosis, sickle cell S–C disease, myelinated nerve fibers and the Waardenburg, and Goldmann-Favre syndromes.[2–5,15,40,41,50–54] As with other uncommon diseases, it is difficult to prove that these associations are causal rather than coincidental. Nonetheless, if a VPT is identified, thorough systemic and ophthalmic evaluations are indicated to rule out associated disorders, especially in younger patients (Fig. 42.2).

The differential diagnosis of VPT includes retinal capillary hemangioblastoma, Coats disease, peripheral exudative hemorrhagic chorioretinopathy (PEHCR), and choroidal melanoma. VPTs lack the prominent, tortuous draining and feeding vessels of hemangioblastomas and do not cause the macular star exudates that may accompany hemangioblastomas.[2] Coats disease usually occurs in younger people than VPTs. It produces prominent "light bulb telangiectasias" and capillary nonperfusion, and also macular circinate and stellate exudates that are not characteristic of VPTs.[55] Additionally, Coats disease even in very advanced cases does not cause the nodular retinal masses found in VPT.[2,5,56] Peripheral exudative hemorrhagic chorioretinopathy occurs in older people than VPTs and they often have age-related macular degeneration and angiographically demonstrable eccentric choroidal neovascularization.[20] Choroidal melanomas seldom have the surrounding exudates common in VPTs. Additionally, ultrasonography documents VPTs to be solid

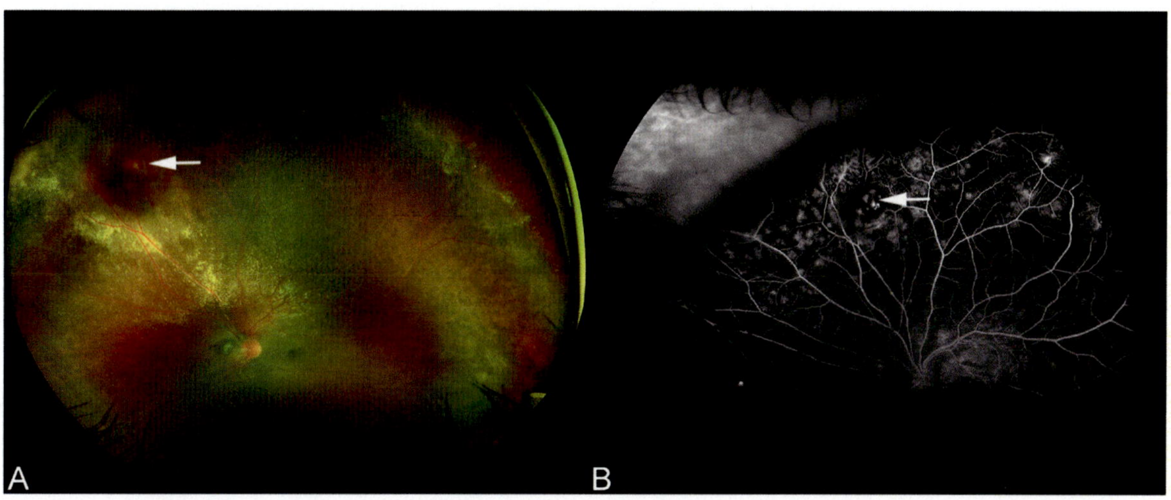

FIGURE 42.1 A 20-year-old woman with adult-onset Coats disease and a secondary vasoproliferative tumor. (A) Ultra-widefield pseudocolor imaging shows a yellowish-pink, dome-shaped mass (*arrow*) encircled by gravitating lipid exudation. (B) Fluorescein angiography shows the tumor's vascularity (*arrow*) and also widespread telangiectasia with areas of peripheral nonperfusion.

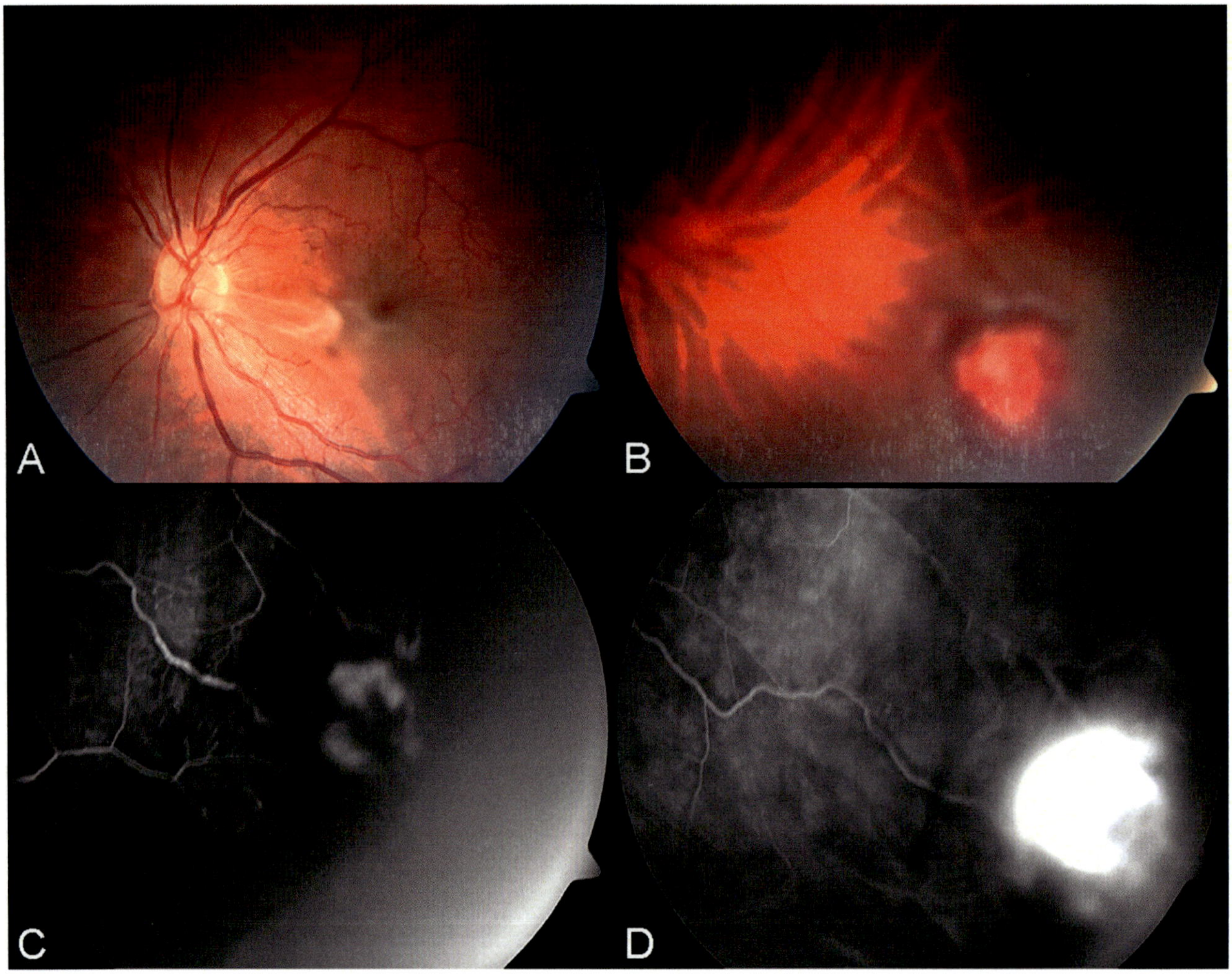

FIGURE 42.2 A 38-year-old woman with an idiopathic vasoproliferative tumor. Fundus color imaging shows (A) an epiretinal membrane with macular traction and (B) a yellowish-pink dome-shaped, inferotemporal, midperipheral retinal mass. Fluorescein angiography demonstrates (C) rapid early filling of tumor vessels and (D) diffuse late leakage of the gliovascular mass.

tumors with high internal reflectivity and no choroidal excavation (similar to PEHCR[20]), whereas melanomas have low internal reflectivity and acoustic hollowing.[30]

Diagnostic evaluation

The peripheral location of most VPTs makes it difficult to perform fluorescein angiography or/and optical coherence tomography (OCT) with high-resolution, standard-field imaging devices. Nonetheless, useful angiographic information can be obtained with ultra-widefield imaging systems when lesions are not obscured by vitreous hemorrhage or other media opacities, as shown in Fig. 42.1.[6]

Fluorescein angiography. Fluorescein angiography usually reveals early-phase rapid filling of tumor vessels, hyperfluorescent capillary abnormalities and diffuse late-phase tumor leakage and staining (Figs. 42.1 and 42.2).[4–6] Fluorescein leakage from tumor vessels persists even after tumor activity decreases so it is not useful for monitoring lesion regression.[45] Fluorescein angiography and structural OCT document tumor-related CME and ERMs in clear ocular media.[4–6,57].

Short-wavelength autofluorescence imaging predicably shows hypoautofluorescence in areas of RPE atrophy without lipofuscin production and hyperautofluorescence of regions of active lipofuscin generation due to inflammation. Thus hyperautofluorescence occurs in areas of new tumor growth and hypoautofluorescence is a valuable measure of postoperative tumor inactivity and regression.[43]

Indocyanine green (ICG) angiography. ICG binds more strongly to plasma proteins than fluorescein, reducing extravascular dye leakage and staining and thereby improving visualization of vascular structures including VPT telangiectasis[45] and choroidal circulation and neovascularization.[58,59] ICG angiography provides early-phase detailed views of tumor vessels and late-phase tissue staining.[37] It also offers useful documentation of ophthalmoscopic findings although these findings have not been correlated with tumor regression.[45]

OCT angiography (OCTA). OCTA documents blood flow in tumor and adjacent vessels without the advantages or interference of exogenous dye leakage and staining. Recent studies have demonstrated that OCTA flow signals disappear after effective cryopexy for VPT. Thus, OCTA is a potentially useful noninvasive method for analyzing postoperative results of VPT treatment.[37]

Ultrasonography. Ultrasonography is a valuable diagnostic tool for studying chorioretinal tumors, especially when the diagnosis is in doubt, location is very peripheral or lesions are obscured by vitreous hemorrhage or cataract.[60,61] As noted above, VPTs are high internal reflectivity, solid tumors lacking choroidal excavation.[30] They generally have a well-defined dome or nodule shape but less frequently are irregular, diffuse, or mushroom-shaped masses.[1–5] Ultrasonography is potentially useful for documenting a tumor's postoperative increase in internal reflectivity suggestive of fibrosis or reduction in size, both indicators of tumor regression.[62]

Biopsy. Fine needle aspiration or intraocular biopsy can provide valuable data when tumor diagnosis remains in doubt despite clinical findings, especially when vision is threatened by widespread exudation. As noted above, reactive astrocytic proliferation is the predominant finding in advanced lesions from eyes that have required enucleation. The primary finding of spindle-shaped astrocytes is accompanied by abnormal telangiectatic or sclerotic vessels and fibrous or osseous retinal pigment epithelial metaplasia.[10,12,35,36] Biopsies of less advanced VPTs also show abundant spindle-shaped glial cells with eosinophilic cytoplasm with small uniform nuclei and no evidence of pleomorphism.[2,8,9,14,36] Additional findings include capillary proliferation, dilated hyalinized blood vessels, RPE hyperplasia and metaplasia at the tumor base, and LCA-positive leukocytes and CD3-positive T cells.[2,8,9,14,36]

Treatment and prognosis

There are no reliable current diagnostic methodologies for predicting tumor growth. In the absence of an evidence-based treatment protocol for VPT,[25,31–33] management is usually based on visual symptom severity as well as tumor size, location, progression, and ocular complications. In general, macular involvement and treatment complications are associated with a poor visual prognosis.[2,63] The primary goals of treatment are tumor involution and vision preservation. Tumors of different types have been managed by observation, cryotherapy, photocoagulation, photodynamic therapy, plaque brachytherapy, and intraocular surgery, often with adjunct intravitreal anti-VEGF or steroid therapy.[25,30,49,64]

Observation. Small peripheral tumors not causing macular or visual problems can be managed conservatively using periodic examinations. VPTs can cause severe vision loss, so early laser photocoagulation or cryopexy is often used to prevent tumor growth and its potentially serious ocular complications.[6,30,31]

Cryotherapy. Transconjunctival double or triple freeze—thaw cryotherapy is used widely to treat small- to medium-sized VPTs.[30,32,42,49,64] Single session treatment may be effective but multiple sessions are often required to achieve full tumor regression.[30,42,49,64] Cryopexy is seldom used for large tumors because the heavier cryotherapy needed would increase the risk of complications including increasing the severity and extent of existing exudative retinal detachment.[30,64] Rhegmatogenous retinal detachment is an occasional complication of cryotherapy for VPT.[42] A potential benefit of cryotherapy in addition to tumor control is ERM release, reported in 63% of cases in one study, probably because treatment induces a posterior vitreous detachment separating ERM from the internal retinal membrane.[32,49]

Photocoagulation. Slitlamp laser photocoagulation targeting retinal telangiectasis has been shown to induce regression of VPT-associated exudation and CME.[65] Moreover, 532-nm laser radiation was used in one study because of the excellent absorption of green light in hemoglobin.[65,66] Long-duration, high-irradiance laser exposures were delivered in successive clinical sessions that were continued until telangiectatic vessels were judged closed. The treatment was relatively painless, probably because of low laser light penetration to the choroid due to VPT thickness and the reduced tissue penetration depth of short wavelength laser radiation.[65,66] Similar results were achieved with a single session of endophotocoagulation during pars plana vitrectomy for VPT-related ERM.[65] Moreover, 532-nm green laser light was also used for endophotocoagulation, with scleral depression of the VPT and higher power, longer duration laser exposures than those used in multiple slit-lamp photocoagulation sessions.[65] Subretinal and intraretinal exudate may take over a year to resolve with laser photocoagulation.[65]

Photodynamic therapy. Standard dose with standard or increased fluence 689 nm (red light) photodynamic therapy (PDT) can also induce VPT regression and decrease retinal and subretinal exudation.[67–70] Red light has higher tissue penetration than shorter wavelengths used for photocoagulation, potentially reaching deeper VPT vessels containing verteporfin.[66] Exposures can be applied at one or more tumor sites depending on VPT size.[70] PDT has been combined successfully with an intravitreal dexamethasone implant to reduce PDT-related inflammation in a patient with perifoveal exudate.[71]

Episcleral plaque radiation therapy (brachytherapy). Iodine-125 (low energy gamma radiation) and ruthenium-106 (short range β radiation) plaque brachytherapy are both effective methods for treating VPTs. In general, plaque therapy is used to treat (1) large tumors with bases wider than 6 mm and thicknesses greater than 2 mm or (2) tumors that are unresponsive to cryotherapy or photocoagulation.[33,63,72–74] ^{106}Ru β radiation has a steeper dose gradient in tissue than ^{125}I gamma radiation, with roughly a 5-mm hypothetical limit of tissue penetration.[75,76] Despite that estimate, ruthenium-106 brachytherapy has been used successfully for uveal melanomas up to 8 mm in thickness.[77] Episcleral brachytherapy is an effective primary or rescue treatment for symptomatic VPTs.[33,63,72,74] Adverse effects are uncommon but they include radiation retinopathy, ERM formation, cataract, and neovascular glaucoma.[63,72,74]

Other VPT therapies. Pars plana vitrectomy is frequently used to remove VPT-associated ERMs and manage a vitreous hemorrhage, sometimes combined with other procedures including surgical tumor resection, episcleral cryotherapy, and endophotocoagulation.[44,46,65,72,73] Small studies have also shown the potential value of monotherapies with intravitreal injections of anti-VEGF drugs[78–80] or systemic administration of tocilizumab[81] or infliximab.[82] Robotic-assisted radiosurgery (CyberKnife) in a small trial induced tumor fibrosis and VPT regression.[62]

Concluding summary

Retinal vasoproliferative tumors are uncommon, benign, vascularized, yellowish-pink retinal masses that occur in the third or fourth decade of life. They are classified as secondary when an ocular or systemic problem is associated with them or primary when they are idiopathic. This primary versus secondary classification is heuristically problematic because VPTs are generally considered to be reactive lesions. Nonetheless, the primary-secondary distinction is useful clinically because secondary lesions are usually more widely distributed, aggressive, and visually debilitating than primary VPTs. As with other uncommon diseases, it is difficult to prove secondary VPT associations are causal rather than coincidental. Nonetheless, if a VPT is identified, thorough systemic and ophthalmic evaluations are indicated to rule out concomitant disorders, especially in younger patients.

There is still disagreement regarding whether VPTs initially are (1) vascular lesions that induce astrocytic proliferation, as supported by clinical data or (2) astrocytic proliferations that induce vascularization and collateral retinal pigment epithelial responses, as indicated by histopathological evidence. Regardless of their origin, VPTs usually cause serious vitreoretinal problems including vitreous hemorrhage, extensive intraretinal and subretinal exudation, and remote ERMs and CME.

VPTs cause visual problems because they produce intraretinal and subretinal exudation, exudative retinal detachment, and remote epiretinal membranes (ERMs) and cystoid macular edema. We hypothesize that unrecognized or subclinical tumors may be involved in the pathogenesis of otherwise idiopathic epiretinal membranes, CME, and vitreoretinal traction or related interface problems. Ultra-widefield photographic and OCT imaging systems are useful for analyzing VPTs and following their progression and treatment response. Complementary ultrasonography and supplemental biopsies are especially valuable for very peripheral lesions and/or those that are diagnostically ambiguous. There are no reliable biomarkers at present for predicting lesion expansion, regression, or exudation, so repeated ophthalmoscopy and imaging is an essential part of VPT clinical management.

Small, asymptomatic VPTs can be followed by observation. Symptomatic lesions are usually treated with cryopexy, laser photocoagulation, or photodynamic therapy. Larger lesions are often treated with episcleral iodine-125 or ruthenium-106 episcleral plaque therapy. Pars plana vitrectomy is used when indicated for ERM and vitreous hemorrhage removal. Alternative therapies used independently or in conjunction with the preceding ones include anti-VEGF, anti-TNF, and steroid medications.

Three decades of research have produced a plethora of sometimes conflicting scientific reports on VPTs without the prospective, randomized controlled trials needed to optimize their clinical management. This broad range of investigations has provided the foundation for multiinstitutional VPT research programs but organizing and implementing them will be problematic because of the variable clinical course of these uncommon, difficult-to-treat tumors and the current diversity of preferences for managing them.

References

1. Ayres B, Bretana ME, Dubovy S, Schefler AC. Mushroom-shaped secondary vasoproliferative tumor. *Ophthalmic Surg Lasers Imaging*. October 4, 2012;43:e104–e106. https://doi.org/10.3928/15428877-20120927-03.
2. Heimann H, Bornfeld N, Vij O, et al. Vasoproliferative tumours of the retina. *British Journal of Ophthalmology*. October 2000;84(10):1162–1169. https://doi.org/10.1136/bjo.84.10.1162.
3. Pineiro-Ces A, Blanco-Teijeiro MJ, Mera-Yanez MP, Capeans-Tome C. Ultrasound diagnosis in vasoproliferative tumours of the ocular fundus. *Archivos de la Sociedad Espanola de Oftalmologi*. August 2011;86(8):247–253. https://doi.org/10.1016/j.oftal.2011.04.008. Diagnostico ecografico de los tumores vasoproliferativos del fondo de ojo.
4. Shields CL, Kaliki S, Al-Dahmash S, et al. Retinal vasoproliferative tumors: comparative clinical features of primary vs secondary tumors in 334 cases. *JAMA Ophthalmology*. March 2013;131(3):328–334. https://doi.org/10.1001/2013.jamaophthalmol.524.
5. Shields CL, Shields JA, Barrett J, De Potter P. Vasoproliferative tumors of the ocular fundus. Classification and clinical manifestations in 103 patients. *Archives of Ophthalmology*. May 1995;113(5):615–623. https://doi.org/10.1001/archopht.1995.01100050083035.
6. Shields CL, Shields JA. Retinal vascular tumors. In: Ehlers JP, ed. *The Retina Illustrated*. Thieme; 2020:381–387. Chap. 96.
7. Khawly JA, Matthews JD, Machemer R. Appearance and rapid growth of retinal tumor (reactive astrocytic hyperplasia?). *Graefes Archive for Clinical and Experimental Ophthalmology*. January 1999;237(1):78–81. https://doi.org/10.1007/s004170050198.
8. Irvine F, O'Donnell N, Kemp E, Lee WR. Retinal vasoproliferative tumors: surgical management and histological findings. *Archives of Ophthalmology*. April 2000;118(4):563–569. https://doi.org/10.1001/archopht.118.4.563.
9. Hiscott P, Mudhar H. Is vasoproliferative tumour (reactive retinal glioangiosis) part of the spectrum of proliferative vitreoretinopathy? *Eye (Lond)*. September 2009;23(9):1851–1858. https://doi.org/10.1038/eye.2008.351.
10. Poole Perry LJ, Jakobiec FA, Zakka FR, et al. Reactive retinal astrocytic tumors (so-called vasoproliferative tumors): histopathologic, immunohistochemical, and genetic studies of four cases. *American Journal of Ophthalmology*. March 2013;155(3):593–608 e1. https://doi.org/10.1016/j.ajo.2012.09.002.
11. Grossniklaus HE, Lenis TL, Jakobiec FA. Retinal reactive astrocytic tumor (focal nodular gliosis): the entity also known as vasoproliferative tumor. *Ocular Oncology and Pathology*. September 2017;3(3):161–163. https://doi.org/10.1159/000455149.
12. Hudson LE, Mendoza PR, Yan J, Grossniklaus HE. Reactive retinal astrocytic tumor (focal nodular gliosis): a case report. *Ocular Oncology and Pathology*. January 2017;3(1):1–7. https://doi.org/10.1159/000448042.
13. Singh AD, Soto H, Bellerive C, Biscotti CV. Reactive retinal astrocytic tumor (focal nodular gliosis): report of the clinical spectrum of 3 cases. *Ocular Oncology and Pathology*. September 2017;3(3):235–239. https://doi.org/10.1159/000455151.
14. Hashimoto I, Takase H, Kase S, Iwasaki Y, Kobayashi D, Ohno-Matsui K. Clinicopathological analysis of secondary retinal vasoproliferative tumor/reactive retinal astrocytic tumor successfully treated by endoresection. *Retin Cases Brief Reports*. September 5, 2022. https://doi.org/10.1097/ICB.0000000000001310.
15. Kandasamy S, Agarwal D, Chawla R. Multimodal imaging of reactive retinal astrocytic vasoproliferative tumour in a case of systemic tuberculosis. *BMJ Case Reports*. January 10, 2023;16(1). https://doi.org/10.1136/bcr-2021-244642.
16. Korhonen A, Gucciardo E, Lehti K, Loukovaara S. Proliferative diabetic retinopathy transcriptomes reveal angiogenesis, anti-angiogenic therapy escape mechanisms, fibrosis and lymphatic involvement. *Scientific Reports*. September 22, 2021;11(1):18810. https://doi.org/10.1038/s41598-021-97970-5.
17. Azarcon CP, Craven CM, Wells JR. Reactive retinal pigment epithelium hyperplasia in presumed ocular histoplasmosis syndrome. *Ophthalmology Retina*. January 2022;6(1):36. https://doi.org/10.1016/j.oret.2021.10.002.
18. Idrees S, Sridhar J, Kuriyan AE. Proliferative vitreoretinopathy: a review. *International Ophthalmology Clinics. Winter*. 2019;59(1):221–240. https://doi.org/10.1097/IIO.0000000000000258.
19. Mantel I, Uffer S, Zografos L. Peripheral exudative hemorrhagic chorioretinopathy: a clinical, angiographic, and histologic study. *American Journal of Ophthalmology*. December 2009;148(6):932–938 e1. https://doi.org/10.1016/j.ajo.2009.06.032.
20. Safir M, Zloto O, Fabian ID, Moroz I, Gaton DD, Vishnevskia-Dai V. Peripheral exudative hemorrhagic chorioretinopathy with and without treatment-clinical and multimodal imaging characteristics and prognosis. *PLoS One*. 2022;17(9):e0275163. https://doi.org/10.1371/journal.pone.0275163.
21. Chang TS, Byrne SF, Gass JD, Hughes JR, Johnson RN, Murray TG. Echographic findings in benign reactive lymphoid hyperplasia of the choroid. *Archives of Ophthalmology*. June 1996;114(6):669–675. https://doi.org/10.1001/archopht.1996.01100130661005.
22. Cockerham GC, Hidayat AA, Bijwaard KE, Sheng ZM. Re-evaluation of "reactive lymphoid hyperplasia of the uvea": an immunohistochemical and molecular analysis of 10 cases. *Ophthalmology*. January 2000;107(1):151–158. https://doi.org/10.1016/s0161-6420(99)00025-1.
23. Shields JA, Shields CL. Reactive retinal astrocytic tumors (so-called vasoproliferative tumors): histopathologic, immunohistochemical, and genetic studies of four cases. *American Journal of Ophthalmology*. July 2013;156(1):202–203. https://doi.org/10.1016/j.ajo.2013.03.016.
24. Jakobiec FA, Grossniklaus HE. Reply: to PMID 23219067. *American Journal of Ophthalmology*. July 2013;156(1):203. https://doi.org/10.1016/j.ajo.2013.03.015.
25. Kiri H, Raval V, Ali H, Das AV, Kaliki S. Vasoproliferative retinal tumors: clinical presentation and treatment outcome. *European Journal of Ophthalmology*. December 27, 2022:11206721221149075. https://doi.org/10.1177/11206721221149075.
26. Henkind P, Morgan G. Peripheral retinal angioma with exudative retinopathy in adults (coats's lesion). *British Journal of Ophthalmology*. January 1966;50(1):2–11. https://doi.org/10.1136/bjo.50.1.2.

27. Baines PS, Hiscott PS, McLeod D. Posterior non-vascularized proliferative extraretinopathy and peripheral nodular retinal telangiectasis. *Transactions of the Ophthalmological Societies of the United Kingdom*. 1982;102(Pt 4):487–491.
28. Shields JA, Decker WL, Sanborn GE, Augsburger JJ, Goldberg RE. Presumed acquired retinal hemangiomas. *Ophthalmology*. November 1983;90(11):1292–1300. https://doi.org/10.1016/s0161-6420(83)34389-x.
29. Wachtlin J, Heimann H, Jandeck C, et al. Bilateral vasoproliferative retinal tumors with identical localization in a pair of monozygotic twins. *Archives of Ophthalmology*. June 2002;120(6):860–862.
30. Salvi SM, Rundle PA, Rennie I, Singh AD. Retinal vascular tumors. In: Singh AD, Damato BE, eds. *Clinical Ophthalmic Oncology*. Springer Nature; 2019:19–38 Chapter 3.
31. Rennie IG. Retinal vasoproliferative tumours. *Eye (Lond)*. March 2010;24(3):468–471. https://doi.org/10.1038/eye.2009.305.
32. Diafas A, Toumanidou V, Kassos I, et al. Macular sequelae in vasoproliferative tumors: results of surgical approach. *International Ophthalmology*. October 2021;41(10):3515–3522. https://doi.org/10.1007/s10792-021-01918-7.
33. Case MJ, Lentz C, Duffy EW, Magrath GN, Cooper SL. Large vasoproliferative retinal tumor refractory to cryotherapy treated with salvage I-125 plaque radiation therapy. *Advances in Radiation Oncology*. September-October 2022;7(5):100972. https://doi.org/10.1016/j.adro.2022.100972.
34. Yanoff M, Fine BS. *Ocular Pathology, a Text and Atlas*. 2nd ed. Harper & Row, Publishers; 1982.
35. Jakobiec FA, Thanos A, Stagner AM, Grossniklaus HE, Proia AD. So-called massive retinal gliosis: a critical review and reappraisal. *Survey Ophthalmology*. May-Jun 2016;61(3):339–356. https://doi.org/10.1016/j.survophthal.2015.12.002.
36. Shehri M, Bouhenni R, Ghazi NG, Abu Safieh L, Edward DP. Retinal reactive astrocytic tumor: gene expression profiling. *JAMA Ophthalmology*. June 2014;132(6):773–775. https://doi.org/10.1001/jamaophthalmol.2014.299.
37. Tanimukai T, Noda K, Hirooka K, Kase S, Ishida S. Noninvasive imaging of a vasoproliferative retinal tumor treated with cryopexy. *Case Reports Ophthalmology*. May-Aug 2022;13(2):611–616. https://doi.org/10.1159/000525939.
38. Saito W, Kase S, Fujiya A, Dong Z, Noda K, Ishida S. Expression of vascular endothelial growth factor and intravitreal anti-VEGF therapy with bevacizumab in vasoproliferative retinal tumors. *Retina*. October 2013;33(9):1959–1967. https://doi.org/10.1097/IAE.0b013e3182923490.
39. Pichi F, Neri P, Agarwal A, et al. Vasoproliferative tumors in intermediate uveitis. *Retina*. September 2020;40(9):1765–1773. https://doi.org/10.1097/IAE.0000000000002656.
40. Shields JA, Reichstein D, Mashayekhi A, Shields CL. Retinal vasoproliferative tumors in ocular conditions of childhood. *Journal of American Association for Pediatric Ophthalmology and Strabismus*. February 2012;16(1):6–9. https://doi.org/10.1016/j.jaapos.2011.10.010.
41. Manayath GJ, Namburi P, Periasamy S, Kale JA, Narendran V, Ganesh A. A novel mutation in the NR2E3 gene associated with Goldmann-Favre syndrome and vasoproliferative tumor of the retina. *Molecular Vision*. 2014;20:724–731.
42. Walinjkar JA, Sharma US, Rishi P, Rishi E, Gopal L, Sharma T. Clinical features and treatment outcomes of vasoproliferative tumors in Indian participants. *Indian Journal Ophthalmology*. February 2018;66(2):246–251. https://doi.org/10.4103/ijo.IJO_210_17.
43. Heimann H, Jmor F, Damato B. Imaging of retinal and choroidal vascular tumours. *Eye (Lond)*. February 2013;27(2):208–216. https://doi.org/10.1038/eye.2012.251.
44. Zheng B, Chen Y, Chen L, et al. Comparative study on the efficacy and safety of tumor resection in vitrectomy for retinal vasoproliferative tumors. *Journal of Ophthalmology*. 2019;2019:7464123. https://doi.org/10.1155/2019/7464123.
45. Matsuura S, Ishida T, Tobe M, Uchida M, Souma R, Ohno-Matsui K. Indocyanine green angiographic findings in seven eyes with vasoproliferative retinal tumor. *American Journal Ophthalmology Case Reports*. September 2020;19:100831. https://doi.org/10.1016/j.ajoc.2020.100831.
46. Zhang W, Qiang Z, Song H, et al. Management of vasoproliferative tumors of the retina with macular complications by pars plana vitrectomy combined with episcleral cryotherapy. *Journal of Ophthalmology*. 2021;2021:6667755. https://doi.org/10.1155/2021/6667755.
47. Ledesma-Gil G, Fine HF, Fernandez-Avellaneda P, Kaden TR, Engelbert M. The vitreous: making the invisible visible. *Ophthalmic Surg Lasers Imaging Retina*. June 1, 2020;51(6):312–319. https://doi.org/10.3928/23258160-20200603-01.
48. Dos Santos FM, Ciordia S, Mesquita J, et al. Vitreous humor proteome: unraveling the molecular mechanisms underlying proliferative and neovascular vitreoretinal diseases. *Cellular and Molecular Life Sciences*. December 31, 2022;80(1):22. https://doi.org/10.1007/s00018-022-04670-y.
49. Manjandavida FP, Shields CL, Kaliki S, Shields JA. Cryotherapy-induced release of epiretinal membrane associated with retinal vasoproliferative tumor: analysis of 16 cases. *Retina*. August 2014;34(8):1644–1650. https://doi.org/10.1097/IAE.0000000000000137.
50. Ledesma-Gil G, Moreno Andrade AB, Shields CL. Secondary vasoproliferative tumor in adult-onset Coats disease. *Canadian Journal Ophthalmology*. February 2022;57(1):69–70. https://doi.org/10.1016/j.jcjo.2021.03.002.
51. Galinos SO, Smith TR, Brockhurst RJ. Angioma-like lesion in hemoglobin sickle cell disease. *Annals of Ophthalmology*. October 1979;11(10):1549–1552.
52. Muakkassa NW, Schwaber EJ, Laver N, Duker JS. Bilateral myelinated retinal nerve fibers associated with bilateral vitreous cysts and solitary vasoproliferative tumor: a new syndrome. *Ophthalmic Surg Lasers Imaging Retina*. May 1, 2018;49(5):356–359. https://doi.org/10.3928/23258160-20180501-09.
53. Honavar SG. Retinal vasoproliferative tumor - a proposal for classification. *Indian Journal Ophthalmology*. February 2018;66(2):185–186. https://doi.org/10.4103/ijo.IJO_128_18.
54. Mori K, Ohta K, Murata T. Vasoproliferative tumors of the retina secondary to ocular toxocariasis. *Canadian Journal Ophthalmology*. October 2007;42(5):758–759. https://doi.org/10.3129/i07-137.
55. Sen M, Shields CL, Honavar SG, Shields JA. Coats disease: an overview of classification, management and outcomes. *Indian Journal Ophthalmology*. June 2019;67(6):763–771. https://doi.org/10.4103/ijo.IJO_841_19.
56. Haik BG. Advanced coats' disease. *Transactions of the American Ophthalmological Society*. 1991;89:371–476.

57. Freund KB, Sarraf D, Mieler WF, Yannuzzi LA, Shields CL. *The Retina Atlas*. 2nd ed. Elsevier; 2017.
58. Freeman WR, Bartsch DU, Mueller AJ, Banker AS, Weinreb RN. Simultaneous indocyanine green and fluorescein angiography using a confocal scanning laser ophthalmoscope. *Archives of Ophthalmology*. April 1998;116(4):455–463. https://doi.org/10.1001/archopht.116.4.455.
59. Spaide RF, Gemmy Cheung CM, Matsumoto H, et al. Venous overload choroidopathy: a hypothetical framework for central serous chorioretinopathy and allied disorders. *Progress in Retinal and Eye Research*. January 2022;86:100973. https://doi.org/10.1016/j.preteyeres.2021.100973.
60. Singh AD, Hayden BC. *Ophthalmic Ultrasonography*. 2012 207.
61. Bhende M, Kamat H, Krishna T, et al. *The Sankara Nethralaya Atlas of Ophthalmic Ultrasound and Ultrasound Biomicroscopy*. 2nd ed. ed. Jaypee Brothers Medical Publishers; 2013:414.
62. Schmelter V, Fuerweger C, Muacevic A, Priglinger SG, Foerster P, Liegl R. Robotic-assisted radiosurgery for the treatment of vasoproliferative tumours. *Acta Ophthalmologica*. June 2021;99(4):456–460. https://doi.org/10.1111/aos.14609.
63. Anastassiou G, Bornfeld N, Schueler AO, et al. Ruthenium-106 plaque brachytherapy for symptomatic vasoproliferative tumours of the retina. *British Journal of Ophthalmology*. April 2006;90(4):447–450. https://doi.org/10.1136/bjo.2005.081422.
64. Marback EF, Guerra RL, Maia Junior Ode O, Marback RL. Retinal vasoproliferative tumor. *Arquivos Brasileiros de Oftalmologia*. May-June 2013;76(3):200–203. https://doi.org/10.1590/s0004-27492013000300016.
65. Krivosic V, Massin P, Desjardins L, Le Hoang P, Tadayoni R, Gaudric A. Management of idiopathic retinal vasoproliferative tumors by slit-lamp laser or endolaser photocoagulation. *American Journal Ophthalmology*. July 2014;158(1):154–161 e1. https://doi.org/10.1016/j.ajo.2014.03.005.
66. Mainster MA. Wavelength selection in macular photocoagulation. Tissue optics, thermal effects, and laser systems. *Ophthalmology*. July 1986;93(7):952–958. https://doi.org/10.1016/s0161-6420(86)33637-6.
67. Blasi MA, Scupola A, Tiberti AC, Sasso P, Balestrazzi E. Photodynamic therapy for vasoproliferative retinal tumors. *Retina*. April 2006;26(4):404–409. https://doi.org/10.1097/01.iae.0000238554.61165.fd.
68. Chan RP, Lai TY. Photodynamic therapy with verteporfin for vasoproliferative tumour of the retina. *Acta Ophthalmology*. September 2010;88(6):711–712. https://doi.org/10.1111/j.1755-3768.2008.01447.x.
69. Hussain RN, Jmor F, Damato B, Heimann H. Verteporfin photodynamic therapy for the treatment of retinal vasoproliferative tumors. *Ophthalmology*. November 2015;122(11):2361–2363. https://doi.org/10.1016/j.ophtha.2015.05.026.
70. Mazloumi M, Dalvin LA, Abtahi SH, et al. Photodynamic therapy in ocular oncology. *Journal of Ophthalmic and Vision Research*. October-December 2020;15(4):547–558. https://doi.org/10.18502/jovr.v15i4.7793.
71. Cebeci Z, Oray M, Tuncer S, Tugal Tutkun I, Kir N. Intravitreal dexamethasone implant (Ozurdex) and photodynamic therapy for vasoproliferative retinal tumours. *Canadian Journal Ophthalmology*. August 2014;49(4):e83–e84. https://doi.org/10.1016/j.jcjo.2014.04.006.
72. Cohen VM, Shields CL, Demirci H, Shields JA. Iodine I 125 plaque radiotherapy for vasoproliferative tumors of the retina in 30 eyes. *Archives of Ophthalmology*. September 2008;126(9):1245–1251. https://doi.org/10.1001/archopht.126.9.1245.
73. Garcia-Arumi J, Distefano LN, Fonollosa A, Quijano C, Corcostegui B. Management of vision-threatening complications of vasoproliferative tumors of the retina. *Ophthalmic Research*. 2015;54(1):34–40. https://doi.org/10.1159/000430955.
74. Abolfathzadeh N, Naseripour M, Jaberi R, et al. Ru-106 plaque radiotherapy for vasoproliferative tumors of retina: a 15-year single-center experience. *International Ophthalmology*. August 2020;40(8):2095–2102. https://doi.org/10.1007/s10792-020-01386-5.
75. Ghassemi F, Sheibani S, Arjmand M, et al. Comparison of iodide-125 and ruthenium-106 brachytherapy in the treatment of choroidal melanomas. *Clinical Ophthalmology*. 2020;14:339–346. https://doi.org/10.2147/OPTH.S235265.
76. Fili M, Trocme E, Bergman L, et al. Ruthenium-106 versus iodine-125 plaque brachytherapy of 571 choroidal melanomas with a thickness of >/=5.5 mm. *British Journal of Ophthalmology*. January 2020;104(1):26–32. https://doi.org/10.1136/bjophthalmol-2018-313419.
77. Jiang P, Purtskhvanidze K, Kandzia G, et al. 106)Ruthenium eye plaque brachytherapy in the management of medium sized uveal melanoma. *Radiation Oncology*. July 29, 2020;15(1):183. https://doi.org/10.1186/s13014-020-01621-4.
78. Kenawy N, Groenwald C, Damato B. Treatment of a vasoproliferative tumour with intravitreal bevacizumab (Avastin). *Eye (Lond)*. June 2007;21(6):893–894. https://doi.org/10.1038/sj.eye.6702782.
79. Chen KJ, Wang NK, Chao AN. Epimacular membrane detachment following intravitreal bevacizumab for retinal vasoproliferative tumor. *Ophthalmology*. May 2017;124(5):742. https://doi.org/10.1016/j.ophtha.2016.10.019.
80. Chu T-W, Tsai SH, Chen L-J. Retinal vasoproliferative tumor regression after intravitreal aflibercept. *Taiwan Journal Ophthalmology*. 2023;13(2):249–252.
81. Adan A, Mesquida M, Llorenc V, Modesto C. Tocilizumab for retinal vasoproliferative tumor secondary to juvenile idiopathic arthritis-associated uveitis: a case report. *Graefes Archive for Clinical and Experimental Ophthalmology*. January 2014;252(1):163–164. https://doi.org/10.1007/s00417-013-2466-5.
82. Japiassu RM, Brasil OF, Cunha AL, de Souza EC. Regression of vasoproliferative tumor with systemic infliximab. *Ophthalmic Surg Lasers Imaging*. July-August 2008;39(4):348–349. https://doi.org/10.3928/15428877-20080701-09.

Section IX

Miscellaneous

Chapter 43

Susac syndrome

Lihteh Wu[1,2], Miguel Cruz-Pimentel[3], Carol E. Schwartz[4], Andres Wu[1] and Max Wu[1]
[1]Asociados de Macula, Vitreo y Retina de Costa Rica, Primer Piso Torre Mercedes Paseo Colón, San José, Costa Rica; [2]Illinois Eye and Ear Infirmary, Department of Ophthalmology and Visual Sciences, University of Illinois Chicago College of Medicine, Chicago, IL, United States; [3]Department of Ophthalmology and Vision Sciences, University of Toronto, Toronto, ON, Canada; [4]Sunnybrook Hospital, University of Toronto, Toronto, ON, Canada

Introduction

Occlusive retinal arterial disease secondary to retinal emboli, vascular inflammation, and vasospasm often heralds the onset of an underlying systemic condition. Typical conditions include atrial fibrillation, carotid stenosis, interatrial or interventricular communications, and, in the appropriate setting, temporal arteritis. Susac's syndrome (SuS), a rare condition that is often overlooked and misdiagnosed, causes multiple recurrent retinal arterial occlusions. A high index of suspicion is needed to make the appropriate diagnosis. Prevention of deafness, dementia, and blindness progression relies on an early diagnosis and treatment.

Historical

Over 40 years ago, Susac and colleagues[1] described two female patients with personality changes and paranoid psychosis who subsequently developed visual loss, neurologic symptoms, and loss of hearing. The loss of vision was due to multiple branch retinal arterial occlusions (BRAO). Due to the lack of embolism or inflammation in both cases, the authors concluded that microangiopathy of the brain and retina was responsible for the clinical findings.[1] Other descriptive terms such as retinocochleocerebral vasculopathy,[2] RED-M (retinopathy, encephalopathy, deafness-associated microangiopathy),[2] and SICRET (small infarcts of cochlear, retinal, and encephalic tissue)[3] were used to describe similar cases. William F Hoyt coined the term Susac's syndrome in 1986 to describe the clinical triad of retinal arterial occlusions, hearing loss, and encephalopathy.[4]

Epidemiology

Since its description in 1979, there have been over 500 cases reported in the literature.[5] Most of these involved Caucasian individuals from North America and Europe. The overall incidence is unknown. In the Central European population, the prevalence has been estimated to be 0.14 cases per 100,000 people.[6] A recent Israeli study suggests that the incidence of SuS has increased 5.4 times.[7] SuS tends to affect female patients with a sex ratio of 3:1 and does not show racial preference. Typically male patients develop monocyclic disease, whereas female patients have an equal distribution of monocyclic and polycyclic disease.[4,5] Although the age range in SuS has been reported from 2 to 72 years old, the majority are between the third and fifth decades of life upon diagnosis.[5] We believe that SuS is currently underdiagnosed and misdiagnosed.

Clinical findings

The clinical triad of BRAO, encephalopathy, and neurosensorial hearing loss defines SuS. However, the complete clinical triad is observed in less than 15% of patients at presentation delaying the diagnosis of SuS.[5,8] In the largest series to date, encephalopathy was present in 67%, ophthalmic involvement in 40%, and vestibulocochlear involvement in 37% of cases

Retinal and Choroidal Vascular Diseases of the Eye. https://doi.org/10.1016/B978-0-443-15583-3.00043-3

at presentation. Eventually, 85% of all patients completed the triad.[5] Although the average time from symptom onset to triad completion was 21 weeks, some cases took up to 10 years to complete the triad.[9]

Neurological

The most common symptom of SuS is headache, which is present in almost 80% of cases.[5,8] The European Susac Consortium recognizes headaches as a sign of cerebral involvement in SuS.[10] The headache that characterizes patients with SuS has a sudden onset and does not precede other symptoms by more than 6 months. These headaches may chronify or exacerbate before the manifestation of SuS. Headaches are typified as the migrainous or oppressive type, not the trigemino-autonomous type.[10] The characteristic headache has a varied presentation, which can be intermittent, with worsening or sudden onset with a severity that has been described from mild to severe. In most cases, it is described as frontal or holocephalic, pulsating, unilateral, aggravated by physical activity, accompanied by nausea, absence of ipsilateral cranial symptoms, with or without photophobia or phonophobia, with or without vomiting, which may be accompanied by visual disturbances, which it is not clear if it is a scintillating scotoma or a consequence of a BRAO.[3,11–13] The headaches may increase in frequency after the presentation of SuS.[14] In patients where the headache occurs with the onset of the symptoms of SuS presentation, it is characterized by being more severe and accompanied by an aura, in contrast to cases where the headache does not present as an initial symptom but rather as a symptom of recurrence or exacerbation in most chronic cases.[11,12] Unlike migraine headaches, which are usually unilateral and accompanied by aura, the headaches experienced by patients with SuS can present with bilateral involvement. However, both types of headaches can occur in patients under 40 years of age.[15]

Headaches are almost always associated with encephalopathy.[8] Headaches are probably due to inflamed leptomeningeal vessels and damage from precapillary arterioles infarcts.[16,17] Occasionally, headaches in patients with SuS coupled with abnormalities in the leptomeninges and the brain seen on MRI images may suggest metastasis.[8] Headaches usually respond to analgesics.[14] Headache exacerbations may herald increased disease activity.[18] Immunomodulatory therapy improves headaches, and headaches in SuS can be used as a parameter to understand disease recurrences or exacerbations.[12,13,19]

Encephalopathy is by far the most common initial manifestation of SuS, present in 75% of cases. However, it must be recognized that encephalopathy may prevent patients from complaining of visual and hearing loss.[5,20] Symptoms may develop acutely in a stroke-like fashion or subacutely.[5,21] Encephalopathy is characterized by the abrupt onset of neuropsychiatric disturbances or multifocal neurologic deficits. Encephalopathy may present with progressive cognitive decline characterized by the loss of executive functions, short-term memory, and concentration.[5] Other manifestations include confusion often involving disorientation, emotional disturbance, behavioral changes, psychosis with paranoid ideations, fluctuating apathy, and reduction of vigilance including coma.[5,22] Other neurological symptoms such as ataxia, vertigo, gait abnormality, and sensory disturbance were reported in 25% of cases.[5] Paresis and upper motor signs such as spasticity, hyperreflexia, and extensor plantar response were seen in a fifth of cases.[5] Other less commonly reported symptoms include urinary dysfunction, dysarthria, nausea and vomiting, seizures, oculomotor dysfunction, and diplopia.[1,5,20,22,23]

Vestibulocochlear

Initially, bilateral or unilateral low-frequency sensorineural hearing loss is typical, and as the disease progresses, high-frequency loss ensues.[24] Sensorineural hearing loss has been reported in 86% of reported cases. Nystagmus, a roaring tinnitus, and peripheral vertigo are often associated. Tinnitus was reported in almost 30% of cases.[5] SuS patients exhibit normal brain stem auditory evoked responses that are not consistent with a brain stem or vestibulocochlear nerve origin for the hearing loss.[20,25,26] Instead, arteriolar microinfarctions of the cochlear apex and vestibular labyrinth are most likely responsible for these findings.[20] The hearing loss may become permanent and severe.[27] Cochlear implantation may be necessary.[11,28] Infarction of the membranous labyrinth can lead to jerk nystagmus.[4,15]

Ophthalmological

The most common ophthalmic complaints of patients with SuS include decreased visual acuity, scintillating scotomas, photopsias, and visual field defects.[12,13,16] Patients may be asymptomatic if the BRAO is in the far retinal periphery.[4,17]

The hallmark ophthalmic feature in SuS is the presence of recurrent multiple BRAO in the absence of intraocular inflammation (Fig. 43.1). Ocasionally ocular manifestations may be the sole presenting sign of SuS. Other patients have

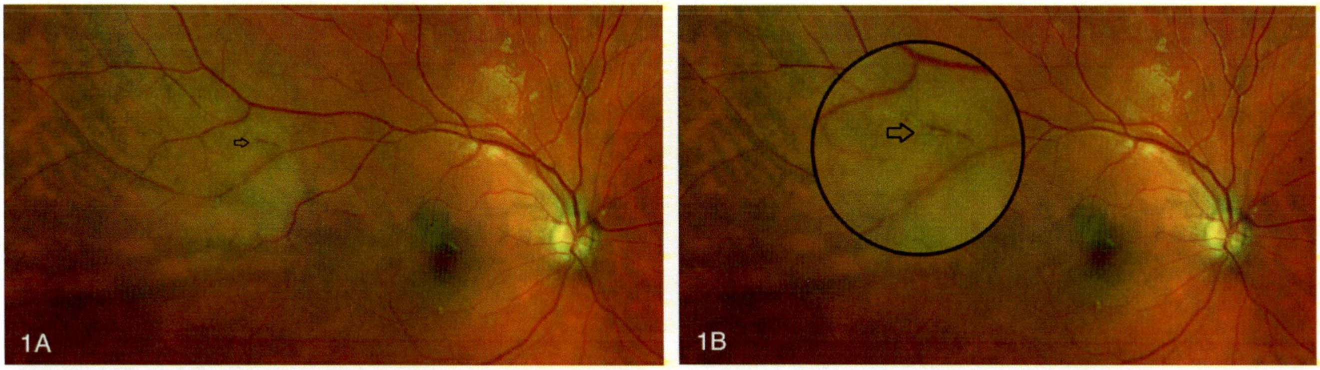

FIGURE 43.1 Ultra-widefield pseudocolor photo of a patient with SuS, showing (A) retinal whitening in areas of BRAO and Gass plaques (*arrow*) along the lumen of the superior temporal arterial vessels; (B) high magnification of Gass plaques.

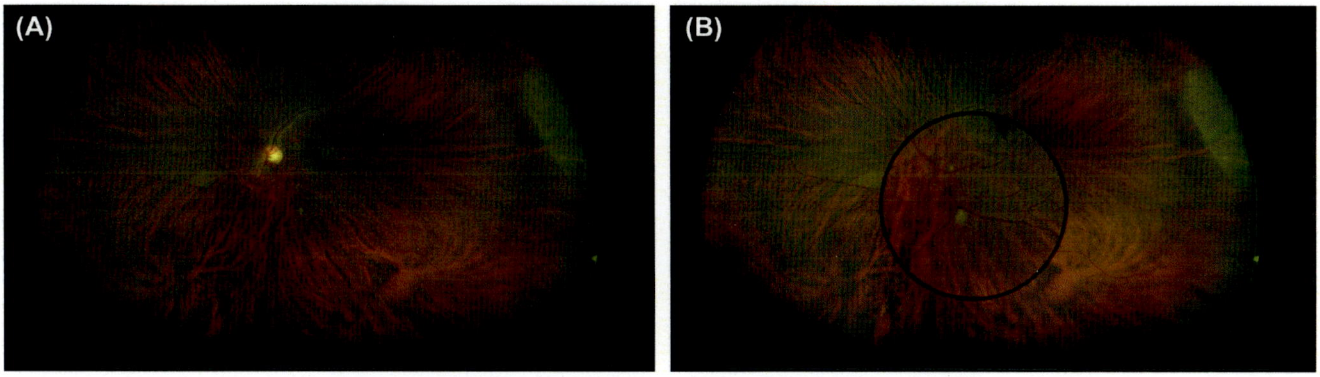

FIGURE 43.2 Ultra-widefield pseudocolor fundus photo of patient with SuS, showing (A) retinal cotton wool spots at the bifurcation of two vessels and (B) high magnification of cotton wool spots.

recurrent BRAO that are accompanied by vestibuloauditory and sensorimotor symptoms.[13,16,20] Ophthalmic findings in SuS include focal retinal whitening (Fig. 43.1), box-car segmentation, arteriolar narrowing, cotton wool spots (Fig. 43.2), optic disc hyperemia, optic disc pallor, perivascular sheathing, peripheral capillary closure, and arterial silver wiring.[16,19,29,30] Rarely, amputation of the peripheral retinal vasculature, retinal neovascularization, and microaneurysms are the sole ophthalmic findings (Fig. 43.3).[31] SuS patients typically manifest yellow to yellow-white nonrefractile or refractile retinal arterial wall plaques (Gass plaques), which are found at midarteriolar segments away from the arterial bifurcations (Fig 43.1).[13,16,19,32] These should not be confused with Hollenhorst, platelet-fibrin or calcium emboli and attributed to embolic disease. Gass plaques are not pathognomonic for SuS. They may be also found in toxoplasmosis, acute retinal necrosis, primary vitreoretinal lymphoma, and arterial macroaneurysms.[33] Gass[13,16] speculated that these plaques consist of blood lipid atheromata that have extravasated into the damaged arterial wall. These retinal arterial wall plaques may resolve over time.[19] Central retinal artery occlusion is a rare manifestation of SuS. In these cases, the typical cherry-red spot is seen acutely.[34–39] The presence of retinal arterioarterial collaterals away from the optic nerve in a young patient with a history of BRAO is suggestive of SuS.[40] Acute macular neuroretinopathy may also develop.[41] Neovascular complications may occur in up to 25% of eyes.[16]

Other systemic findings

Skin involvement has been reported in less than 5% of patients with SuS.[5] Some patients have presented with livedo racemosa and others with livedo reticularis.[1,2,23,26,35,36,42–47] Livedo refers to a whitish central core of skin that is surrounded by a purplish netlike skin mottling.[48] Livedo reticularis is thought to be benign and often reversible, whereas livedo racemosa is pathologic. In livedo reticularis, the discoloration is characterized by symmetrical, regular, and unbroken rings. In contrast, in livedo racemosa, the discoloration consists of irregularly broken rings.[48] Sneddon's syndrome is typically associated with livedo racemosa and must be differentiated from SuS.[42,49]

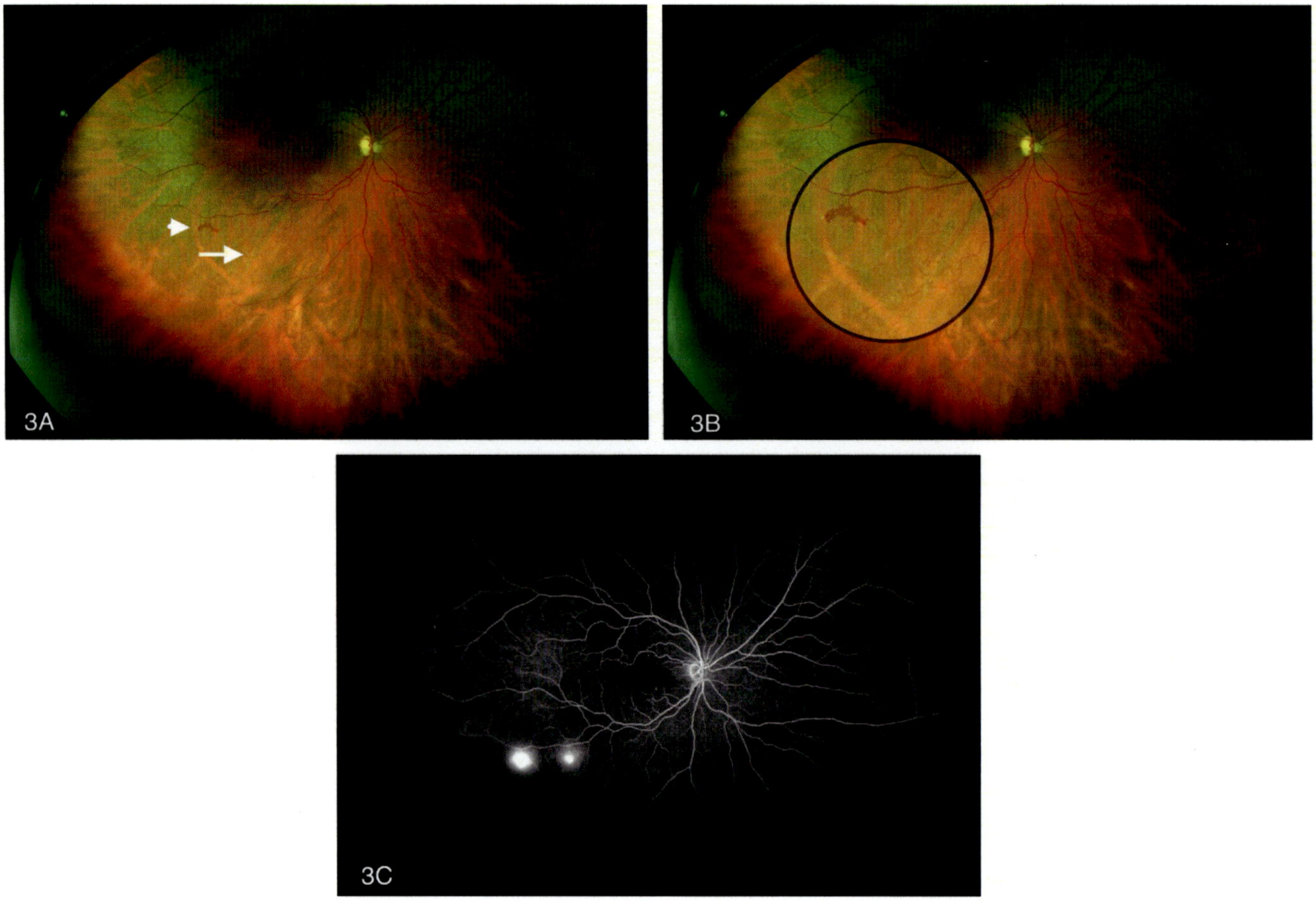

FIGURE 43.3 Ultra-widefield pseudocolor fundus image showing (A) retinal neovascularization (*arrowhead*) and ghost vessels (*arrow*) in a patient with SuS; (B) a high magnification image highlights the presence of neovascularization and ghost vessels; (C) UWF FA shows the leakage from the retinal neovascularization and significant capillary dropout in the fundus periphery.

Some patients with SuS suffer from myalgias and arthralgias. Muscle biopsy demonstrates microangiopathic changes even in asymptomatic patients, underscoring the subclinical microangiopathy of this condition in other organs.[5,23]

Clinical course

The presence or absence of encephalopathy during the first 2 years will determine the clinical course.[50] Previously, it was considered that if encephalopathy presented itself during the first 2 years, the disease was self-limited (monocyclic).[5] However, the occurrence of late recurrences up to 30 years later suggests that many patients with the monocyclic course were not followed long enough to diagnose their recurrences.[16,50–54] The monocyclic course was considered to be the most common and comprises about 54% of the cases in the literature.[5] On the other hand, if patients with BRAO and hearing loss do not develop encephalopathy during the initial 2 years, encephalopathy will most likely not develop. However, these patients will be prone to recurrent BRAO and hearing loss (polycyclic).[5,50] Around 42% of reported cases followed the polycyclic course of the disease.[5] Although encephalopathy is not manifested, some of these patients may have MRI with callosal changes.[50] Rarely (4% of reported cases),[5] some patients embark on a chronic progressive and potentially fatal course.[5,55–57] Chronic marijuana abuse may hasten the clinical course leading to death in some cases.[56]

In about 50% of patients, recovery from the neurological symptoms occur and patients are able to resume their normal life. The other 50% suffer from mild to moderate dementia.[4] Some patients are incapacitated by global cerebral atrophy.

The visual outcomes have been described in a series of 16 patients with polycyclic SuS that were followed for an average of 9 years.[16] Snellen visual acuity was maintained in most eyes. Only 9% of eyes lost $\geq$2 lines of Snellen visual acuity and only one ended with a visual acuity of $\leq$20/60. However, almost 30% of eyes developed extensive peripheral field loss and central field defects. Neovascularization developed in 50% of eyes. Six eyes developed retinal neovascularization, four eyes developed neovascularization of the disc and two eyes developed anterior segment neovascularization.[16] One of the eyes that developed anterior segment neovascularization experienced neovascular glaucoma.[16,58] In a recent study, the retinal lesion evolution was studied with serial coregistered spectral domain OCT and FA.[59]

Permanent visual, neurologic, and hearing sequelae are common if treatment is delayed or inadequate.[20,26,60]

Diagnostic criteria

Diagnostic criteria for SuS were developed by the European Susac Consortium hoping to prevent misdiagnosis and delay of treatment, which is a major cause of irreversible damage to the affected organs.[10] Documentation of brain, retinal, and vestibulocochlear involvement confirms a definite diagnosis of SuS.

Brain involvement was defined as behavioral changes, and/or new cognitive impairment, and/or new headache and/or new focal neurological symptoms. Headaches are of the migrainous or oppressive type, not trigemino-autonomous type. In addition, headaches cannot precede other symptoms by more than 6 months. Brain involvement is confirmed by typical T2 or FLAIR weighted MRI sequences where typical corpus callosum findings such as hyperintense, multifocal, and round small lesions are found.[10] Documentation of BRAO by fundoscopic clinical examination, fluorescein angiography, or SD-OCT confirms retinal involvement. Finally, the onset of hearing loss and/or peripheral vertigo and/or tinnitus suggests vestibulocochlear involvement. Vestibular vertigo needs to be confirmed by vestibular-evoked myogenic responses and caloric testing. The hearing loss needs to be confirmed by an audiogram.[10]

When only two of the three systems are involved, a probable SuS can be diagnosed. Finally, possible SuS refers to those patients with some findings of the above triad but not fulfilling the rest of the criteria.[10]

Differential diagnosis

Other publications have reviewed the differential diagnosis of SuS at length.[5,21,25] The neurological findings in SuS are quite heterogeneous, variable, and nonspecific. Due to the similarities in MRI images, SuS is most often confused with multiple sclerosis (MS) and acute disseminated encephalomyelitis (ADEM).[15,36,61] Patients may also be misdiagnosed with infectious encephalitis,[15] central nervous system vasculitis,[62] psychosis, migraine with aura, malignant tumors,[15] stroke,[15] cerebral autosomal dominant arteriopathy with subcortical infarcts and leukoencephalopathy (CADASIL), Creutfeld-Jakob disease,[15] and myopathy encephalopathy lactate acidosis and stroke-like episodes (MELAS).[62] The presence of BRAO should point to SuS since BRAO is rarely seen in these other conditions.[50]

Bilateral hearing loss and tinnitus are not typically part of the clinical picture of MS. Fluctuating hearing loss in some SuS patients may mimic Meniere's disease.[50] MS and SuS need to be differentiated from each other since treatment with interferon beta can exacerbate SuS.[63]

The retinal findings from SuS have been mistaken for tuberculosis, syphilis, systemic lupus erythematosus (SLE), systemic vasculitis, toxoplasmosis, sarcoidosis, primary vitreoretinal lymphoma, pars planitis, or Behcet's disease.[2,17,22,44,64] Two patients with multiple nonembolic retinal arteriole occlusions accompanied by central nervous system involvement without any evidence of ocular inflammation or embolic disease were diagnosed with micro-angiopathy of the retinal arterioles secondary to SLE.[65] However, Susac et al.[1] called into question the diagnosis by pointing out that SLE was diagnosed despite the negative LE preparations and the lack of ANA determinations.

Ancillary testing

Since the vast majority of patients with SuS do not present with the classical triad coupled with the fact that a substantial number of patients with SuS suffer from subclinical pathology, underscores the need to completely assess the possibility of SuS with a brain MRI, ultra-widefield fluorescein angiography (UWF-FA) and a pure tone audiogram.[5]

Lumbar puncture

Cerebrospinal fluid (CSF) findings may help differentiate between infectious conditions or MS or SuS. Oligoclonal bands are rarely seen in SuS. Typical CSF findings in SuS include a mild lymphocytic pleocytosis (up to 20 cells/μL) and an increased protein of up to 2 g/L.[15,23,44,66]

Neuroimaging

MRI is an integral part of the workup of SuS, particularly if encephalopathy is present. MRI demonstrates a triad of widespread abnormalities in the gray matter, white matter, and leptomeninges.[67] The white matter lesions are typically supratentorial and almost always involve the corpus callosum (Fig. 43.4).[15,68] According to a review of more than 300 patients, SuS involved the supratentorial white matter and the corpus callosum in 98% and 78% of cases respectively. Focal gadolinium enhancement was seen in 51% of cases compared to 23% of leptomeningeal enhancement.[5] Leptomeningeal enhancement does not occur with ADEM or MS.[67] The severity of encephalopathy is directly proportional to the presence of the MRI triad. The more severe the encephalopathy, the greater the likelihood of having the full triad. Patients with mild cases will tend to have only white matter involvement.

MRI findings of central callosal lesions are pathognomonic for SuS (Fig. 43.4). Acutely, these are best seen as hyperintensities on sagittal T2 fluid-attenuated inversion recovery (FLAIR).[69] These are characterized by the involvement of the central fibers with the sparing of the peripheral fibers.[67] Acutely they are referred to as "snowballs," which eventually evolve into central callosal holes. The central callosal holes are best seen as hypointense punched-out lesions in sagittal T1 images.[70] Central callosal involvement is not seen in MS and ADEM. The roof of the corpus callosum is also frequently involved in SuS. When the roof is affected, the imaged MRI lesions are described as having "icicle and spoke" configurations. In patients with long tract signs, the internal capsule is frequently affected by microinfarcts that are lined in a "string of pearls" configuration.[69]

Sophisticated MRI techniques such as diffusion tensor imaging (DTI) or ultrahigh field MRI may help differentiate MS from SuS in difficult cases.[70] DTI evaluates the micro- and macrostructural fiber integrity. DTI revealed that there is widespread fiber damage in the prefrontal white matter and in the genu of the corpus callosum. These findings are not found in MS.[71,72] Ultrahigh field MRI using 7 T field strength reveals that the white matter lesions of MS consist of a hypointense perilesional ring with a centrally located vein. These are not typically found in SuS.[70]

The white matter in the cerebellum, brainstem, and middle cerebellar peduncles may be involved in up to 43% of patients with SuS as well.[5] Parenchymal brain involvement is another common finding, also seen in about 42% of patients.[5] Even though clinical findings suggest that in some patients the spinal cord may be affected in SuS, there has been no MRI documentation of spinal cord involvement.[5,21]

Arteriography is usually normal because the involved precapillary arterioles measure only 100 μm in diameter.[4,66] In selected cases, arteriography may be useful in differentiating SuS from other vasculopathies. CT is usually unremarkable and of little value in the diagnosis of SuS.[4,23] Electroencephalograms usually show prominent generalized slowing.[20,23]

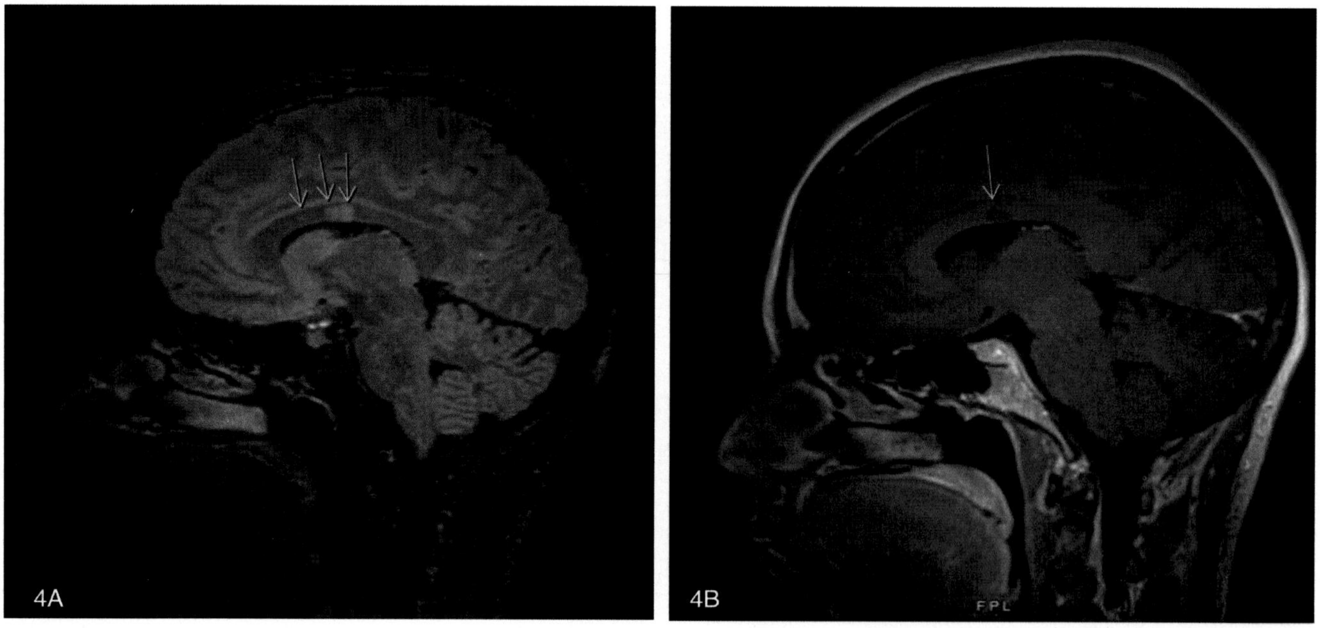

FIGURE 43.4 (A) Sagittal MRI T2 projection (FLAIR) sequence of a 43-year-old female with SuS, showing the characteristic hyperintense lesions called snowballs (*arrows*) at the level of the corpus callosum; (B) T1 projection of the same patient. Notice the different intensity appearance (*arrow*) of these lesions.

Ophthalmic imaging

Multimodal ophthalmic imaging, particularly UWF-FA, is critical in the diagnosis and follow-up of patients with SuS. Some BRAO may be asymptomatic therefore all suspected patients of having SuS should undergo fluorescein angiography (FA). UWF-FA has definite advantages over conventional FA due to its ability to capture more peripheral retinal areas. Its use should be mandatory in patients suspected of having SuS.[73] Typically fluorescein hyperfluorescence stains the arterial walls of involved segments acutely (Fig. 43.5).[12,19,74] The hyperfluorescence stays within the arterial walls and fluorescein does not leak into the adjacent retinal tissue. There is no widespread breakdown of the blood—retinal barrier making overt vasculitis an unlikely mechanism of disease.[12,59,75] This arterial wall hyperfluorescence may occur at a distant site from the occluded arteriole (Fig. 43.5).[12,16,75—77] Segmental arteriolar wall staining that migrates to the retinal periphery has been observed in one case.[77] Some of this arterial wall hyperfluorescence may evolve into a BRAO but many do not.[59] FA demonstrates areas of capillary nonperfusion, collateral vessels, and retinal neovascularization (Fig. 43.6). FA has shown that retinal veins may be involved in SuS as well.[32,78]

FA also plays an important role in assessing the response to treatment. Reperfusion of the retinal vessels, resolution of arteriolar fluorescein hyperfluorescence, or even relapses may occur during treatment. Therefore, changes in UWF-FA may assist the clinician in decision-making to reduce or step up the dose of immunomodulatory treatment.

Optical coherence tomography (OCT), particularly spectral domain OCT complements FA. As the main retinal finding in SuS is that of multiple BRAO, the OCT findings will depend on the location and degree of occlusion of the BRAO. Acutely BRAO causes retinal edema and manifests itself on OCT as a thickened hyperreflective band in the inner retina along the distribution of the occluded vascular segment. With time, the BRAO resolves causing the retina to thin (Fig. 43.7). At the same time, the retina may become reperfused. An FA at this time may not detect abnormalities associated with prior BRAO despite inner retinal thinning. In contrast, the OCT will clearly detect inner retinal thinning (Figs. 43.7—43.9).[73,79—81] If the ischemia only affects the deep capillary plexus, hyperreflective bands in the inner nuclear

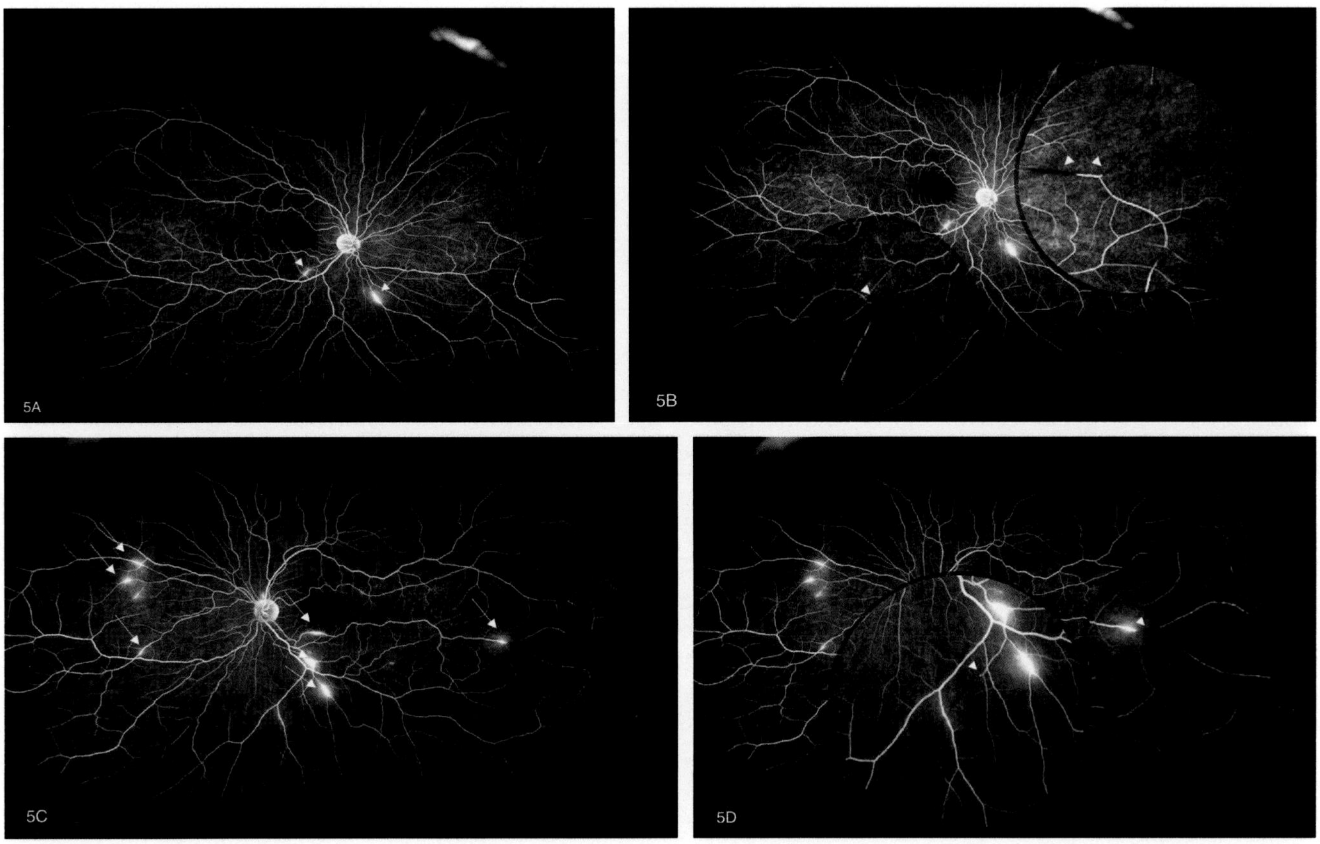

FIGURE 43.5 UWF fluorescein angiography performed during the acute phase of a patient with SuS syndrome showing retinal wall hyper-autofluorescence in both eyes (A and C) (*arrows*). High magnification shows multiple BRAOs in the nasal and inferior temporal periphery of the right eye (B) and along the inferior temporal and temporal periphery of the left eye (D); in both eyes (*arrowheads*), BRAOs were located distal to the areas of arteriole wall hyperfluorescence.

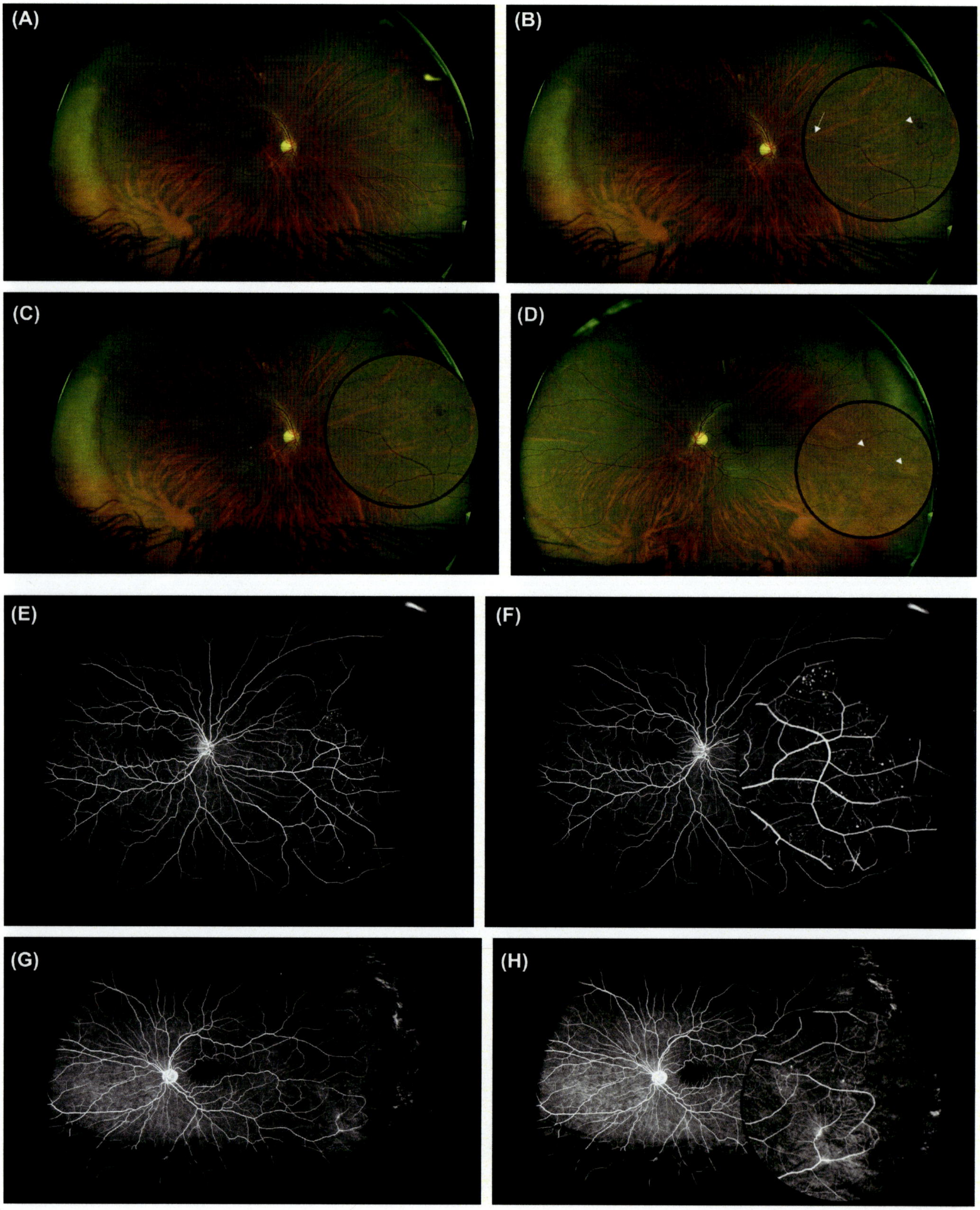

FIGURE 43.6 Three-year follow-up of the patient shown in Fig. 43.5.Ultra-widefield pseudocolor fundus photo of a patient with SuS showing (A) intraluminal refractile deposits in the segment of a superior nasal arteriole (*arrow*) and (B) microaneurysms in the nasal periphery (*arrowhead*) of the right eye and microaneurism in the temporal periphery of the left eye (*arrowhead*). UWF FA shows microaneurisms and collateral vessels along areas of nonperfusion in both eyes (E, F, G, H). These findings were seen 3 years after the initial onset of SuS of the same patient shown in Fig. 43.5.

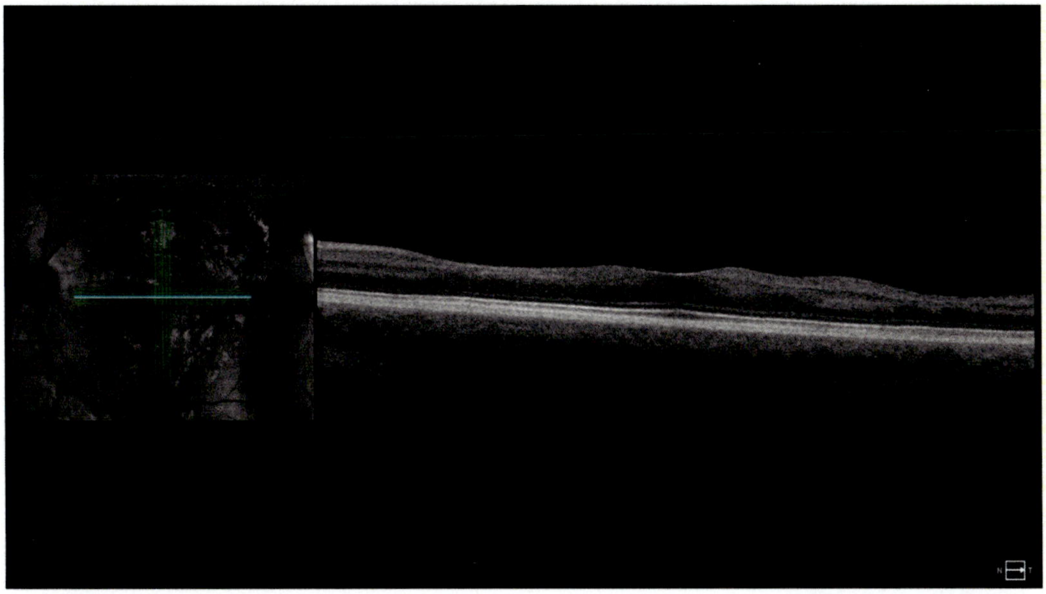

FIGURE 43.7 OCT showing thinning of the retinal nerve fiber layer, ganglion cell layer, and a loss of the foveal contour in a patient with SuS, 1 year after the onset of the disease.

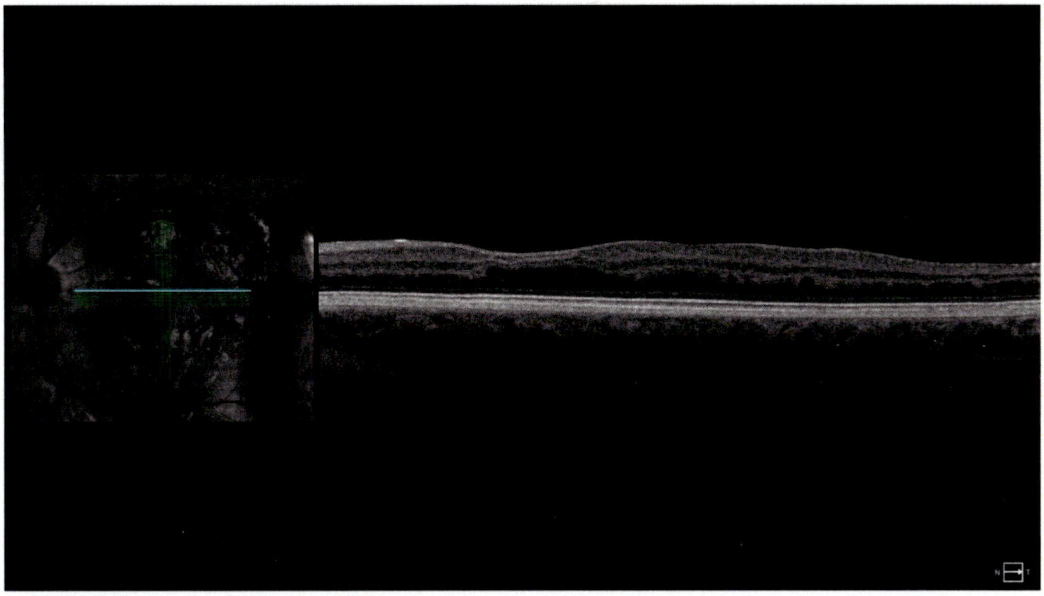

FIGURE 43.8 OCT of a patient with SuS showing undulations and a nodular pattern at the level of the external plexiform layer (*white box*), which is hypothesized as ischemic edema of bipolar cells.

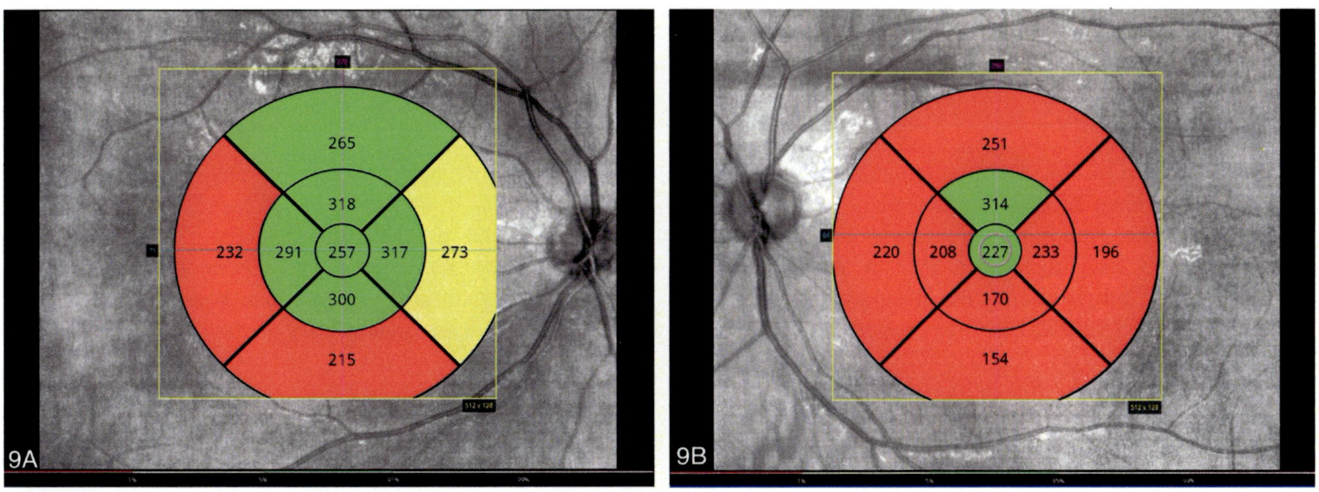

FIGURE 43.9 OCT macular thickness map of a patient with SuS showing (A) sectoral macular thinning of the right eye and (B) generalized reduction in the central macular thickness of the left eye (B) 1 year after the onset of the disease.

layer extending into the inner plexiform layer consistent with paracentral acute middle maculopathy may appear in the OCT.[82] Outer plexiform layer undulations and nodularity have been interpreted as ischemic swelling of bipolar cells (Fig. 43.8).[81] OCT has recently been shown to help differentiate between multiple sclerosis and SuS.[59,70] In SuS, the resolved BRAO will show focal areas of inner retinal thinning on OCT, whereas in MS the retinal thinning is diffuse in nature and mostly affects the temporal retinal nerve fiber layer.[83] In addition, patients with SuS may experience loss of the foveal contour whereas patients with MS rarely ever have this finding on OCT.[84]

Optical coherence tomography angiography (OCT-A) complements the information gained from FA. Different degrees of capillary ischemia have been reported in both the deep and superficial capillary plexus.[73,85–87] Not surprisingly the areas of capillary ischemia seen on OCTA-A map out to the same areas where microperimetry and visual field testing denote low sensitivities.[86]

Until recently involvement of the choroidal circulation was in doubt.[75] However, more recently, indocyanine green angiography (ICG-A) has documented small circumscribed areas of choriocapillaris nonperfusion and segmental staining of choroidal vessels in eyes with SuS.[75,88]

Vestibulocochlear testing

SuS patients typically manifest an asymmetric bilateral neurosensorial hearing loss characterized by low to moderate and poor speech discrimination in pure tone audiograms. A deficit of caloric response in the affected ear may be revealed on videonystagmography.[22,28] Auditory brain responses range from normal cochlear and brainstem function to findings consistent with abnormal cochlear nerve function.[17,20,23–26,43,60,89] Similarly, electrocochleography findings have been variable. In one case, the findings suggested endolymphatic hydrops, whereas in another case the results were normal.[17,24]

However, for patients who cannot undergo audiometry, an objective measure of the auditory function is brainstem auditory-evoked potentials (BAEPs), which allow for auditory pathway function from the auditory nerve to the midbrain. In SuS, BAEP may show a diminished response in wave I. If the patient is fit for audiometry, it will be found that low tones are initially affected, reflecting a vulnerability of the cochlear apex to microinfarcts.[8]

Histopathology

Skin, brain, and muscle biopsies demonstrate precapillary arteriolar deposition of complement components, particularly C4d, accompanied by an occasional perivascular infiltration of lymphocytes.[1,4,20,23,25,43,89–92] Antibody-mediated microvascular injury is characterized by C4d deposition, which leads to obliterative microangiopathy. Histopathological findings include endothelial cell denudement and necrosis but an absence of fibrinoid necrosis, necrotizing vasculitis, or demyelination. There is fibrin deposition and reduplication of the vascular basement membrane zone.[25] Juvenile dermatomyositis (JDM), characterized by skin and muscle microinfarcts, shares similar findings.[15,90,93]

The sole ocular histopathological report of SuS in the literature confirms that SuS is an endotheliopathy and that both arteries and veins are involved in the disease process.[29] In this case, the lumen of involved vessels was compressed by eosinophilic serous material that extended from the internal limiting membrane to the vessel's outer wall. In areas without eosinophilic spaces, the vessel walls were thickened causing vessel lumen compression as well. Unlike diabetes mellitus, the vessel wall thickening was not uniform. No thrombi were seen in any of the occluded vessels. It is presumed that endothelial cell dysfunction and death cause focal breakdown of the blood–retinal barrier, which allows the extravasation of serous fluid into the spaces mentioned above. These eosinophilic material-filled spaces are thought to be responsible for the arterial wall hyperfluorescence seen in FA.[29]

Pathophysiology

The pathogenesis of this condition remains elusive. Coagulopathy,[26,94,95] vasospasm,[96] viral infections,[17,44,97] and immune mechanisms[98] have all been implicated.

As a large proportion of patients with SuS are women of childbearing age, pregnancy is of interest.[26,95] About 5% of female patients have developed SuS in association with their pregnancy.[5] Some cases have been diagnosed during pregnancy and others during the postpartum period.[23,26,44,95,99,100] Exacerbation of disease has been reported during pregnancy and in subsequent pregnancies.[101] Some investigators believe that pregnancy may induce a hypercoagulable state. Based on a case report of a pregnant young woman with a patent foramen ovale that improved with anticoagulation, Gordon et al.[95] proposed that SuS may be secondary to microembolism associated with a transient hypercoagulable state and the presence of right to left cardiac communication. Others have proposed that the procoagulant Factor VIII and von Willebrand factor antigen are abnormally elevated in the serum and may play a role in the pathogenesis of SuS. Both of these have been proposed to serve as markers of endothelial damage.[15,23] Despite these observations, no procoagulant state has been consistently documented in SuS.[23,26] Furthermore only 9 out of 304 patients had elevated Factor VIII in their serum.[5] Other patients have had recurrences while anticoagulated rendering this hypothesis less likely.[74]

Subclinical viral infections have been speculated as possible triggers of immune-mediated damage to the endothelium which in turn may cause thrombosis.[2,17,25,44,97] However, only 19 patients out of more than 300 cases suffered an infection before the first onset of disease suggesting that the above hypothesis is also unlikely.[5]

The leading hypothesis implicates an autoimmune inflammatory endotheliopathy that causes a microangiopathy that leads to vessel occlusion and ischemia affecting the precapillary arterioles of the retina, the inner ear, and the brain.[98] Leptomeningeal vessel damage is probably the cause of the migrainous headaches.[102] Antiendothelial cell antibodies (AECA) have been isolated from the serum of patients with SS.[103,104] However, AECAs are not specific to SuS and may be found in other conditions such as vasculitides, connective tissue disorders, and other inflammatory diseases.[105] These antibodies bind to different endothelial cell antigens.[106] It is currently unclear if these autoantibodies are etiologically related to the disease process or represent an epiphenomenon. In a series of 11 patients with SuS, the sera of nine patients tested positive for AECA by indirect immunofluorescence and Western blot assays. A distinctive band of 50 kDa protein on Western blot was identified in eight patients. This 50 kDA protein was not seen in other autoimmune conditions suggesting that perhaps this particular AECA may be somewhat related to the pathogenesis of SuS. The exact identity of this protein remains to be elucidated.[90] AECA are not associated with disease severity.[103]

Recent immunohistochemical studies from transgenic mouse models and tissue from SuS patients reveal that an unknown endothelial cell antigen triggers cytolytic $CD8^+$ cells to undergo clonal expansion and differentiation into $CD8_{Temra}$ cells.[107] $CD8_{Temra}$ cells express granzyme B and perforin, which when recognized by HLA complexes polarize their cytolytic vacuoles toward the basal membrane of endothelial cells. Endothelial cell apoptosis mediated by perforin and granzyme B ensues damaging the blood–brain barrier or the blood–retinal barrier. As a result, microhemorrhages and small vessel occlusion leads to small ischemic lesions with loss of astrocytes, oligodendrocytes, neurons, and axons in the corpus callosum, the cerebellum, the retina, and the inner ear. Eventually, these ischemic lesions become gliotic through infiltration by surrounding astrocytes.[107]

The HLA analysis in patients with SuS found a high incidence of HLA-C*06 and HLA-C*07. It is believed that a genetic affinity is behind the immunological mechanisms that activate CD8 cells to generate an autoimmune response. The specific tissue antigenic stimulus that produces this clonal activation of CD8 lymphocytes in endothelial cells remains to be determined in the retina, brain, and cochlea.[107]

Treatment

To date, the optimal treatment of SuS is unknown since there have been no clinical trials. Furthermore, the variable natural history makes treatment assessment difficult to judge. A big confounder has been that a few patients have been documented to have spontaneous remission.[4,89] Despite these shortcomings, it is strongly recommended that the disease be rapidly and completely suppressed since the target organs may be irreversibly damaged leading to blindness, dementia, and deafness.[108]

Immunosuppresion is the cornerstone of treatment but the best regimen still needs to be defined. Due to the variable course of the disease, treatment duration remains unclear. As the histopathological findings of SuS are very similar to those found in JDM, treatment recommendations are modeled after the treatment of JDM.[108]

Treatment is initiated with a pulse of methylprednisolone (1 g/day for 5 days) followed by a daily dose of 1 mg/kg for 2–4 weeks depending on the severity of the disease. Tapering of the steroids is done slowly, 10%–15% of the dose every 2 weeks until reaching 10–15 mg per day at around 6 months. In the next 6 months, there is an even slower tapering of 2.5 mg every month.[62] In certain cases, the disease process is resistant to high-dose steroids.[16,23,25,26,74,101,109] Intravenous immunoglobulin for these corticosteroid-resistant cases should be considered.[91,109–112] It must be borne in mind that immunoglobulin-related thrombotic complications in patients with autoimmune diseases receiving immunoglobulin have been reported in up to 13% of patients.[113] Alternatively plasma exchange may be considered instead.[23,89,91,92,109,110]

Removal of humoral factors by plasma exchange, including the putative pathogenic antibodies, from the patient's blood may ameliorate the clinical picture.[109] In a retrospective case series of 29 patients with SuS,[109] 11 were treated with intravenous immunoglobulin. Despite treatment, two patients had a worsening of their symptoms. In the same series, nine patients were treated with plasma exchange and none of them worsened.[109] In more severe cases, further immunosuppression with cyclophosphamide, rituximab, mycophenolate mofetil, or anti-TNF α agents such as infliximab should be considered.[34,62,101]

Treatment is guided according to the severity of the CNS involvement which is classified into extremely severe, severe, moderate, and mild manifestations.[108] Extremely severe CNS involvement requires intravenous pulsed methylprednisolone followed by high-dose oral prednisone combined with intravenous immunoglobulin, cyclophosphamide, mycophenolate mofetil, tacrolimus, and rituximab. For severe CNS disease, treatment guidelines are similar to the extremely severe category with the exception of the need for cyclophosphamide which is optional. Moderate CNS disease is treated with intravenous pulsed methylprednisolone followed by high-dose oral prednisone combined with intravenous immunoglobulin, mycophenolate mofetil, and rituximab. In mild CNS cases, intravenous pulsed steroids are optional. High-dose oral steroids are combined with intravenous immunoglobulin, mycophenolate mofetil, and rituximab. Treatment usually lasts more than 2 years for all patients regardless of severity category.[108]

Natalizumab, an anti-α4 integrin, partially attenuated the inflammatory activity in four SuS patients.[107]

If BRAO is the predominant clinical picture, methylprednisolone daily pulses of 1000 mg for three consecutive days are recommended. Thereafter, oral prednisone in a dose of 40−60 mg for 1−2 weeks is given. Afterward, the prednisone is tapered 5−10 mg every 1−2 weeks. Intravenous immunoglobulin at an initial dose of 2 g/kg followed by 1 g/kg every 2 weeks is recommended as well. Mycophenolate mofetil is also recommended. Serial FA findings and whether or not CNS involvement develops will determine further treatment decisions. If stability is achieved, oral steroids can be tapered off; intravenous immunoglobulin can be also tapered after 6 months and mycophenolate mofetil after 6−12 months. Some patients may not respond to this treatment regimen and may need cyclophosphamide and or rituximab.[108]

Platelet antiaggregants such as aspirin or clopidogrel have been recommended as adjuvants to prevent procoagulant states.[23,51,75,114,115] Prophylactic treatment with anticoagulants or platelet antiaggregants do not appear to be effective and probably should not be prescribed. Multiple recurrent BRAOs were observed in patients who were taking these drugs. Furthermore, serious hemorrhagic complications in patients taking warfarin were recorded.[16]

Nimodipine in combination with aspirin has been used by some to relieve the possible arteriolar vasospasm.[4,114]

Some authors have recommended hyperbaric oxygen therapy to treat the retinal occlusions.[28,31,116] Yepez et al.[117] have suggested using intravitreal triamcinolone as an adjunct therapy. In their case report, the ocular manifestations resolved following intravitreal triamcinolone. These therapies are difficult to assess given the small numbers and the limited follow-up. Ocular neovascularization should be treated accordingly with anti-VEGF, laser photocoagulation, vitreous surgery, or a combination of these.[16,31,58]

Clinical case

A healthy 19-year-old female patient with a prior history of herpes simplex 1 presented to the emergency room (ER) complaining of headaches, nausea, and lightheadedness. Initial work-up ruled out COVID-19. She was treated with an intravenous dose of ketorolac and discharged home with a dose of 10 mg po every 6 h. She developed dizziness during the next 3 days despite the ketorolac treatment. She returned to the ER where she was assessed neurologically. Both Kernig and Brudzinki signs were absent. A lumbar puncture showed a mild white blood count, normal glucose, and a mildly elevated protein level (1.24 mg/dL). An encephalitis and meningitis DNA and RNA panel was run on the cerebrospinal fluid and turned out to be negative for a number of bacteria and viruses. A head CT and CT angiogram were read as within the normal limits. She was sent home. Over the next few days, she felt lethargic and tired. One month after the onset of symptoms, the patient reported decreased hearing on the right side. She also developed confusion, poor concentration, and cognitive problems. Concurrently, her mother observed changes in her behavior and inappropriate speech.

The patient returned to the ER where a brain MRI revealed white matter lesions. The patient was diagnosed with multiple sclerosis and started with immunomodulatory therapy with glatiramer acetate 20 mg subcutaneously per day. Despite this treatment, over the next few days, the patient developed an abnormal gait balance. A month after being

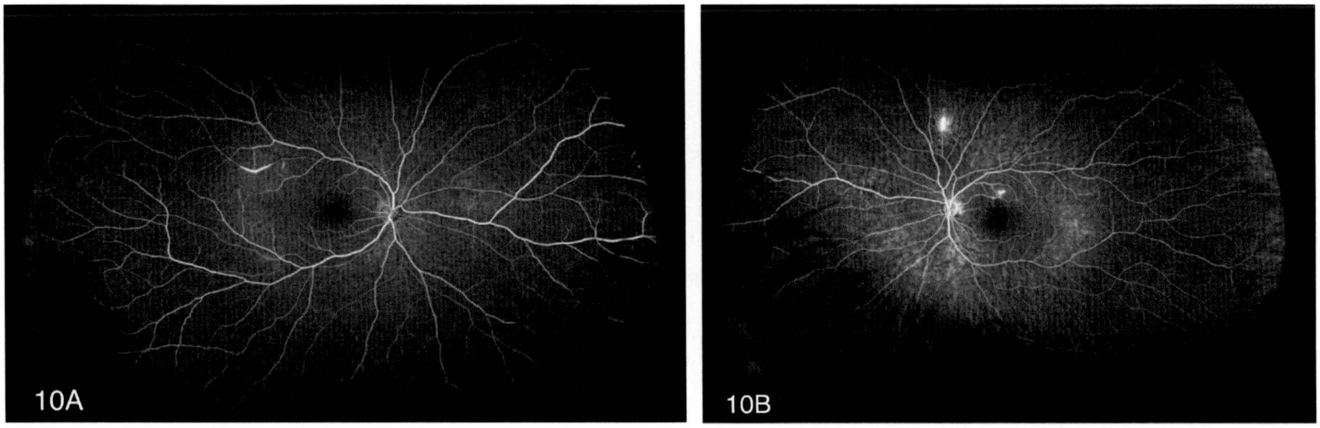

FIGURE 43.10 UWF-FA showing (A) superior temporal BRAO in the right eye and (B) vessel wall hyperautofluorescence in the left eye.

initiated on treatment with glatiramer acetate, the patient awoke with painless visual loss in the superior visual field of her right eye. At the same time, the patient also noticed a blocking sensation in her left ear. Ophthalmic examination showed a normal optic nerve and multiple arteriolar occlusions. The presence of areas of hyperfluorescence in the arteriolar vessel wall using fluorescein angiography (Fig. 43.10).

A repeat brain MRI with a FLAIR sequence revealed hyperintensities in the brain parenchyma involving the corpus callosum, the supratentorial white matter including the periventricular white matter, the lentiform nucleus and the bilateral thalamus, brainstem, middle peduncles, and cerebellum (Fig. 43.11).

With these findings, the diagnosis of SuS was finally made. The glatiramer acetate was discontinued. The patient was started on treatment with intravenous methylprednisolone 1 g per day for 5 days and then was switched to oral prednisolone 55 mg for several months. The patient was treated with intravenous immunoglobulin for 10 months. Mycophenolate mofetil was also started and escalated up to a dose of 750 mg per day. The patient continued with the monthly intravenous immunoglobulin treatment for several months and reported an improvement in symptoms with no recurrences. Her defect in the visual field has been maintained over time (Fig. 43.12).

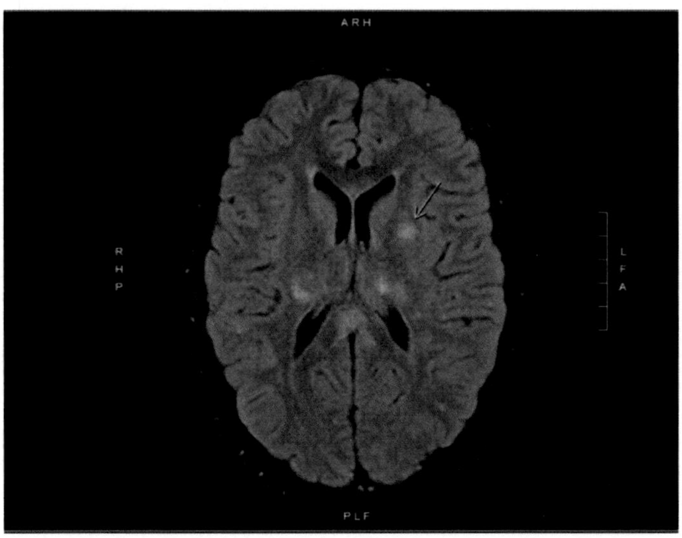

FIGURE 43.11 MRI axial projection (FLAIR) showing white matter lesions (*arrow*).

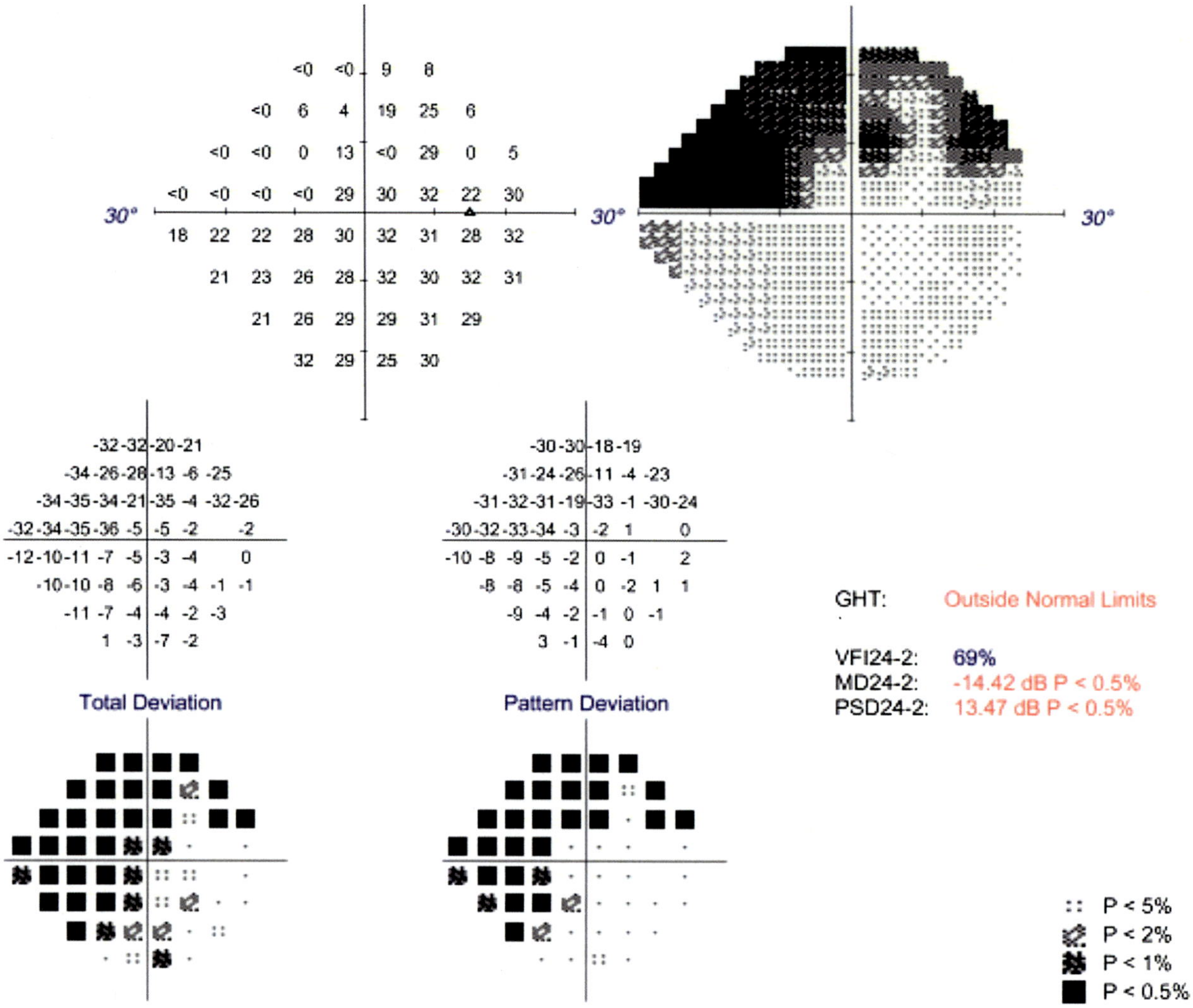

12A

FIGURE 43.12 Visual field testing demonstrating persistent defects in the right eye (A) and the left eye (B) after a year of disease onset.

Summary

SuS is a relatively rare condition that is often overlooked and misdiagnosed. A high index of suspicion is needed to correctly diagnose the patient. Most patients will not present with the clinical triad of encephalopathy, BRAO, and sensorineural hearing loss. MRI of the brain, UWF-FA, and a pure tone audiogram are essential tests in the workup of a patient suspected to have SuS. SuS is triggered by an unknown endothelial antigen that produces a clonal activation of CD8 lymphocytes that targets the endothelium. Prompt diagnosis is critical to diminish the possible sequelae of blindness, deafness, or dementia. Treatment is based on immunosuppressive therapy.

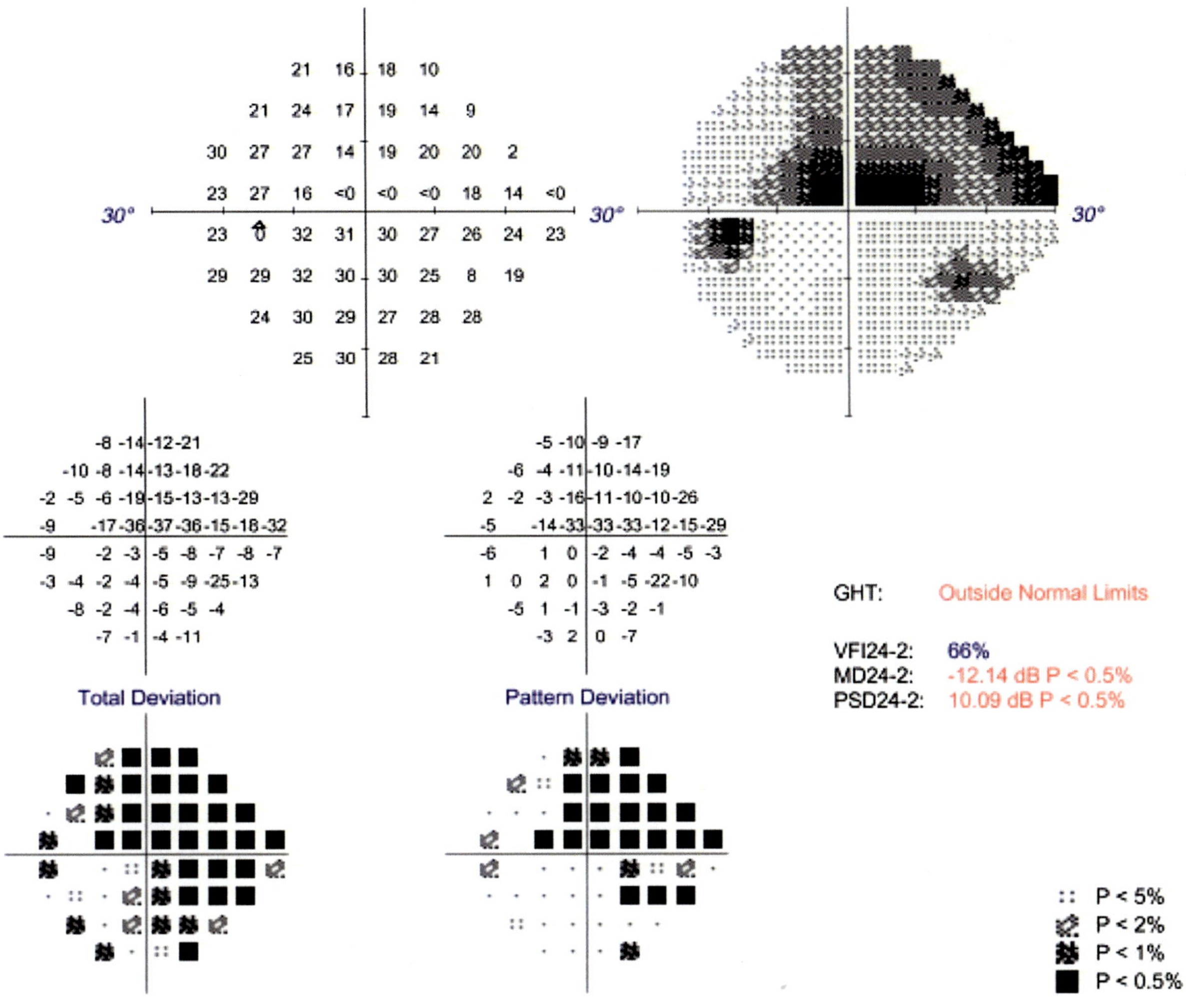

12B

FIGURE 43.12 cont'd

References

1. Susac JO, Hardman JM, Selhorst JB. Microangiopathy of the brain and retina. *Neurology*. 1979;29(3):313–316.
2. Nicolle MW, McLachlan RS. Microangiopathy with retinopathy, encephalopathy, and deafness (RED-M) and systemic features. *Seminars in Arthritis and Rheumatism*. 1991;21(3):123–128.
3. Mala L, Bazard MC, Berrod JP, et al. [Small retinal, cochlear, and cerebral infarctions in the young patient, "SICRET" syndrome of Susac syndrome]. *Journal Français d'Ophtalmologie*. 1998;21(5):375–380.
4. Susac JO. Susac's syndrome: the triad of microangiopathy of the brain and retina with hearing loss in young women. *Neurology*. 1994;44(4):591–593.
5. Dorr J, Krautwald S, Wildemann B, et al. Characteristics of Susac syndrome: a review of all reported cases. *Nature Reviews Neurology*. 2013;9(6):307–316.
6. Seifert-Held T, Langner-Wegscheider BJ, Komposch M, et al. Susac's syndrome: clinical course and epidemiology in a Central European population. *International Journal of Neuroscience*. 2017;127(9):776–780.
7. Wilf-Yarkoni A, Elkayam O, Aizenstein O, et al. Increased incidence of Susac syndrome: a case series study. *BMC Neurology*. 2020;20(1):332.
8. Vodopivec I, Venna N, Rizzo 3rd JF, Prasad S. Clinical features, diagnostic findings, and treatment of Susac syndrome: a case series. *Journal of the Neurological Sciences*. 2015;357(1–2):50–57.
9. Ben David C, Sharif K, Watad A, et al. Susac syndrome: a rare cause of a confusional state. *The Israel Medical Association Journal (IMAJ)*. 2017;19(10):651–653.
10. Kleffner I, Dorr J, Ringelstein M, et al. Diagnostic criteria for Susac syndrome. *Journal of Neurology, Neurosurgery & Psychiatry*. 2016;87(12):1287–1295.

11. Garcia-Carrasco M, Mendoza-Pinto C, Cervera R. Diagnosis and classification of Susac syndrome. *Autoimmunity Reviews*. 2014;13(4–5):347–350.
12. Egan RA, Hills WL, Susac JO. Gass plaques and fluorescein leakage in Susac Syndrome. *Journal of the Neurological Sciences*. 2010;299(1–2):97–100.
13. Gass JD, Tiedeman J, Thomas MA. Idiopathic recurrent branch retinal arterial occlusion. *Ophthalmology*. 1986;93(9):1148–1157.
14. Obelieniene D, Macaityte R, Balnyte R, et al. Characteristics of headache in relation to the manifestation of Susac syndrome. *Medicina (Kaunas)*. 2017;53(6):420–425.
15. Susac JO, Egan RA, Rennebohm RM, Lubow M. Susac's syndrome: 1975-2005 microangiopathy/autoimmune endotheliopathy. *Journal of the Neurological Sciences*. 2007;257(1–2):270–272.
16. Johnson MW, Thomley ML, Huang SS, Gass JD. Idiopathic recurrent branch retinal arterial occlusion. Natural history and laboratory evaluation. *Ophthalmology*. 1994;101(3):480–489.
17. Saw VP, Canty PA, Green CM, et al. Susac syndrome: microangiopathy of the retina, cochlea and brain. *Clinical and Experimental Ophthalmology*. 2000;28(5):373–381.
18. Dawe JA, Green AL. Headache in Susac's syndrome. *Current Pain and Headache Reports*. 2021;25(4):25.
19. Egan RA, Ha Nguyen T, Gass JD, et al. Retinal arterial wall plaques in Susac syndrome. *American Journal of Ophthalmology*. 2003;135(4):483–486.
20. O'Halloran HS, Pearson PA, Lee WB, et al. Microangiopathy of the brain, retina, and cochlea (Susac syndrome). A report of five cases and a review of the literature. *Ophthalmology*. 1998;105(6):1038–1044.
21. Dorr J, Radbruch H, Bock M, et al. Encephalopathy, visual disturbance and hearing loss-recognizing the symptoms of Susac syndrome. *Nature Reviews Neurology*. 2009;5(12):683–688.
22. Papo T, Biousse V, Lehoang P, et al. Susac syndrome. *Medicine (Baltim)*. 1998;77(1):3–11.
23. Petty GW, Engel AG, Younge BR, et al. Retinocochleocerebral vasculopathy. *Medicine (Baltim)*. 1998;77(1):12–40.
24. Bateman ND, Johnson IJ, Gibbin KP. Susac's syndrome: a rare cause of fluctuating sensorineural hearing loss. *Journal of Laryngology & Otology*. 1997;111(11):1072–1074.
25. Monteiro ML, Swanson RA, Coppeto JR, et al. A microangiopathic syndrome of encephalopathy, hearing loss, and retinal arteriolar occlusions. *Neurology*. 1985;35(8):1113–1121.
26. MacFadyen DJ, Schneider RJ, Chisholm IA. A syndrome of brain, inner ear and retinal microangiopathy. *The Canadian Journal of Neurological Sciences*. 1987;14(3):315–318.
27. Aubart-Cohen F, Klein I, Alexandra JF, et al. Long-term outcome in Susac syndrome. *Medicine (Baltim)*. 2007;86(2):93–102.
28. Cubillana Herrero JD, Soler Valcarcel A, Albaladejo Devis I, et al. [Susac syndrome as a cause of sensorineural hearing loss]. *Acta Otorrinolaringológica Española*. 2002;53(5):379–383.
29. McLeod DS, Ying HS, McLeod CA, et al. Retinal and optic nerve head pathology in Susac's syndrome. *Ophthalmology*. 2011;118(3):548–552.
30. Sanchez-Vicente JL, Molina-Socola FE, Medina-Tapia A, et al. Retinal and optic disc vasculitis in Susac's syndrome. *Archivos de la Sociedad Espanola de Oftalmologia*. 2018;93(5):246–250.
31. Avino JA, Espana E, Peris-Martinez C, Menezo JL. [Atypical ophthalmic findings in Susac syndrome]. *Archivos de la Sociedad Espanola de Oftalmologia*. 2007;82(3):179–181.
32. Johnson MW, Flynn Jr HW, Gass JD. Idiopathic recurrent branch retinal arterial occlusion. *Archives of Ophthalmology*. 1989;107(5):757.
33. Gass JD, Trattler HL. Retinal artery obstruction and atheromas associated with non-Hodgkin's large cell lymphoma (reticulum cell sarcoma). *Archives of Ophthalmology*. 1991;109(8):1134–1139.
34. Buelens T, Herode L, Nubourgh I, et al. Central retinal artery occlusion and Susac syndrome: a case report. *Retin Cases Brief Reports*. 2014;8(3):187–192.
35. Adatia FA, Sheidow TG. Central retinal artery occlusion as the initial ophthalmic presentation of Susac's syndrome. *Canadian Journal of Ophthalmology*. 2004;39(3):288–291.
36. Murata Y, Inada K, Negi A. Susac syndrome. *American Journal of Ophthalmology*. 2000;129(5):682–684.
37. Egan RA. Unilateral central retinal artery occlusion as the sole presenting sign of Susac syndrome in a young man: case report. *Arquivos Brasileiros de Oftalmologia*. 2013;76(5):328.
38. Apostolos-Pereira SL, Kara-Jose LB, Marchiori PE, Monteiro ML. Unilateral central retinal artery occlusion as the sole presenting sign of Susac syndrome in a young man: case report. *Arquivos Brasileiros de Oftalmologia*. 2013;76(3):192–194.
39. Monteiro ML. Reply: fibrin-like material in the retinal arterioles and fluorescein angiography findings in Susac syndrome. *Arquivos Brasileiros de Oftalmologia*. 2013;76(5):328–329.
40. Egan RA, Jirawuthiworavong G, Lincoff NS, et al. Retinal arterio-arterial collaterals in susac syndrome. *Journal of Neuro-Ophthalmology*. 2018;38(4):459–461.
41. Yang YS, Zhang L, Asdaghi N, et al. Acute macular neuroretinopathy in susac syndrome: a new association. *Retin Cases Brief Reports*. 2018;14(4):310–314.
42. Turc G, Monnet D, Dupin N, et al. Skin involvement in Susac's syndrome. *Journal of the Neurological Sciences*. 2011;305(1–2):152–155.
43. Heiskala H, Somer H, Kovanen J, et al. Microangiopathy with encephalopathy, hearing loss and retinal arteriolar occlusions: two new cases. *Journal of the Neurological Sciences*. 1988;86(2–3):239–250.

44. Coppeto JR, Currie JN, Monteiro ML, Lessell S. A syndrome of arterial-occlusive retinopathy and encephalopathy. *American Journal of Ophthalmology*. 1984;98(2):189–202.
45. Matamoros N, BenEzra D. Bilateral retinopathy and encephalopathy. *Graefes Archive for Clinical and Experimental Ophthalmology*. 1989;227(1):39–41.
46. Bucciarelli S, Cervera R, Martinez M, et al. Susac's syndrome or catastrophic antiphospholipid syndrome? *Lupus*. 2004;13(8):607–608.
47. Grover N, Whiteside OJ, Ramsden JD. Susac syndrome: outcome of unilateral cochlear implantation. *Journal of Laryngology & Otology*. 2011;125(8):856–858.
48. Dean SM. Livedo reticularis and related disorders. *Current Treatment Options in Cardiovascular Medicine*. 2011;13(2):179–191.
49. Egan RA. Sneddon's syndrome versus Susac syndrome. *Korean Journal of Ophthalmology*. 2014;28(1):113.
50. Rennebohm RM, Egan RA, Susac JO. Treatment of Susac's syndrome. *Current Treatment Options in Neurology*. 2008;10(1):67–74.
51. Petty GW, Matteson EL, Younge BR, et al. Recurrence of Susac syndrome (retinocochleocerebral vasculopathy) after remission of 18 years. *Mayo Clinic Proceedings*. 2001;76(9):958–960.
52. Robles-Cedeno R, Ramio-Torrenta L, Laguillo G, et al. Long-term clinical and radiological evolution in one case of Susac's syndrome. *Neurological Sciences*. 2012;33(6):1407–1410.
53. Feresiadou A, Eriksson U, Larsen HC, et al. Recurrence of susac syndrome following 23 Years of remission. *Case Reports in Neurology*. 2014;6(2):171–175.
54. Garcia-Basterra I, Aranda IB, Garcia-Ben A, Garcia-Campos JM. Susac's syndrome: isolated retinal artery occlusion after 10 years of remission. *Arquivos Brasileiros de Oftalmologia*. 2018;81(2):144–147.
55. Saux A, Niango G, Charif M, et al. Susac's syndrome, a rare, potentially severe or lethal neurological disease. *Journal of the Neurological Sciences*. 2010;297(1–2):71–73.
56. Shields RA, Kleinman RA, Smith SJ, et al. A fatal case of Susac syndrome: the importance of ophthalmic examination in confirming the diagnosis. *American Journal of Ophthalmology Case Reports*. 2018;12:18–20.
57. Hantson P, Di Fazio V, Del Mar Ramirez Fernandez M, et al. Susac-like syndrome in a chronic cocaine abuser: could levamisole play a role? *Internet Journal of Medical Toxicology*. 2015;11(1):124–128.
58. Capone Jr A, Meredith TA. Profound central visual loss and ocular neovascularization in idiopathic recurrent branch retinal arterial occlusion. *Retina*. 1990;10(4):265–268.
59. Brandt AU, Oberwahrenbrock T, Costello F, et al. Retinal lesion evolution in Susac syndrome. *Retina*. 2016;36(2):366–374.
60. Schwitter J, Agosti R, Ott P, et al. Small infarctions of cochlear, retinal, and encephalic tissue in young women. *Stroke*. 1992;23(6):903–907.
61. Susac JO. Susac's syndrome. *AJNR American Journal Neuroradiology*. 2004;25(3):351–352.
62. Kleffner I, Duning T, Lohmann H, et al. A brief review of Susac syndrome. *Journal of the Neurological Sciences*. 2012;322(1–2):35–40.
63. Algahtani H, Shirah B, Amin M, et al. Susac syndrome misdiagnosed as multiple sclerosis with exacerbation by interferon beta therapy. *The Neuroradiology Journal*. 2018;31(2):207–212.
64. Papadaki M, Lefebvre P, Janssens S, et al. Bilateral retinal ischemic vasculopathy in a pregnant patient. *Retin Cases Brief Reports*. 2015;9(2):185–189.
65. Pfaffenbach DD, Hollenhorst RW. Microangiopathy of the retinal arterioles. *JAMA*. 1973;225(5):480–483.
66. Bitrian E, Sanchez-Dalmau B, Gilbert ME, et al. Retinal infarcts in a patient with an acute confusional syndrome. *Survey of Ophthalmology*. 2009;54(4):503–506.
67. Susac JO, Murtagh FR, Egan RA, et al. MRI findings in Susac's syndrome. *Neurology*. 2003;61(12):1783–1787.
68. Uyhazi KE, Tamhankar MA, Liu GT, et al. Diagnostic and therapeutic challenges. *Retina*. 2017;37(6):1209–1214.
69. Rennebohm R, Susac JO, Egan RA, Daroff RB. Susac's syndrome–update. *Journal of the Neurological Sciences*. 2010;299(1–2):86–91.
70. Dorr J, Ringelstein M, Duning T, Kleffner I. Update on Susac syndrome: new insights in brain and retinal imaging and treatment options. *Journal of Alzheimer's Disease*. 2014;42 Suppl 3:S99–S108.
71. Kleffner I, Deppe M, Mohammadi S, et al. Diffusion tensor imaging demonstrates fiber impairment in Susac syndrome. *Neurology*. 2008;70(19 Pt 2):1867–1869.
72. Kleffner I, Deppe M, Mohammadi S, et al. Neuroimaging in Susac's syndrome: focus on DTI. *Journal of the Neurological Sciences*. 2010;299(1–2):92–96.
73. Ammar MJ, Kolomeyer AM, Bhatt N, et al. Recurrent branch retinal artery occlusion from susac syndrome: case report and review of literature. *Retin Cases Brief Reports*. 2018;14(4). https://doi.org/10.1097/ICB.0000000000000751.
74. Notis CM, Kitei RA, Cafferty MS, et al. Microangiopathy of brain, retina, and inner ear. *Journal of Neuro-Ophthalmology*. 1995;15(1):1–8.
75. Martinet N, Fardeau C, Adam R, et al. Fluorescein and indocyanine green angiographies in Susac syndrome. *Retina*. 2007;27(9):1238–1242.
76. Milbratz GH, Marquardt FA, Guimaraes Neto HP, et al. Retinal vasculitis in Susac syndrome: case report. *Arquivos Brasileiros de Oftalmologia*. 2009;72(3):397–399.
77. Kawano K, Ito Y, Haga F, et al. Migration of segmental arteriolar fluorescein staining peripherally in eye with idiopathic recurrent branch retinal arterial occlusion. *American Journal of Ophthalmology Case Reports*. 2017;7:80–82.
78. Klufas MA, Dinkin MJ, Bhaleeya SD, et al. Ultra-widefield fluorescein angiography reveals retinal phlebitis in Susac's syndrome. *Ophthalmic Surg Lasers Imaging Retina*. 2014;45(4):335–337.
79. Leung CK, Tham CC, Mohammed S, et al. In vivo measurements of macular and nerve fibre layer thickness in retinal arterial occlusion. *Eye (Lond)*. 2007;21(12):1464–1468.

80. Takahashi H, Iijima H. Sectoral thinning of the retina after branch retinal artery occlusion. *Japanese Journal of Ophthalmology*. 2009;53(5):494–500.
81. Agarwal A, Soliman M, Sarwar S, et al. Spectral-domain optical coherence tomography evaluation of retinal structure in patients with Susacs syndrome. *Retin Cases Brief Reports*. 2017;11(2):123–125.
82. Haider AS, Viswanathan D, Williams D, Davies P. Paracentral acute middle maculopathy in Susac syndrome. *Retin Cases Brief Reports*. 2017;14(2). https://doi.org/10.1097/ICB.0000000000000633.
83. Brandt AU, Zimmermann H, Kaufhold F, et al. Patterns of retinal damage facilitate differential diagnosis between Susac syndrome and MS. *PLoS One*. 2012;7(6):e38741.
84. Bernard JT, Romero R, Agrawal K, et al. Optical coherence tomography in Susac's syndrome. *Multiple Sclerosis and Related Disorders*. 2014;3(1):110–116.
85. Spiess K, Martinez JRG. OCT angiography: assessment of retinal ischemia in Susac's syndrome. *Ophthalmic Surg Lasers Imaging Retina*. 2017;48(6):505–508.
86. Azevedo AGB, Lima LH, Muller L, et al. Anatomical and functional correlation in Susac syndrome: multimodal imaging assessment. *International Journal of Retina Vitreous*. 2017;3:39.
87. Garcia-Serrano JL, Munoz de Escalona-Rojas JE, Callejas-Rubio JL, Barrero-Hernandez FJ. Optical coherence tomography angiography in the early diagnosis of Susac syndrome. *Neurologia*. 2017;35(1):62–63. https://doi.org/10.1016/j.nrl.2017.07.014.
88. Balaskas K, Guex-Crosier Y, Borruat FX. Indocyanine-green angiography findings in Susac's syndrome. *Klin Monbl Augenheilkd*. 2012;229(4):426–427.
89. Bogousslavsky J, Gaio JM, Caplan LR, et al. Encephalopathy, deafness and blindness in young women: a distinct retinocochleocerebral arteriolopathy? *Journal of Neurology, Neurosurgery & Psychiatry*. 1989;52(1):43–46.
90. Crowson AN, Magro CM. The role of microvascular injury in the pathogenesis of cutaneous lesions of dermatomyositis. *Human Pathology*. 1996;27(1):15–19.
91. Do TH, Fisch C, Evoy F. Susac syndrome: report of four cases and review of the literature. *AJNR (American Journal of Neuroradiology)*. 2004;25(3):382–388.
92. Vila N, Graus F, Blesa R, et al. Microangiopathy of the brain and retina (Susac's syndrome): two patients with atypical features. *Neurology*. 1995;45(6):1225–1226.
93. Rennebohm RM, Lubow M, Rusin J, et al. Aggressive immunosuppressive treatment of Susac's syndrome in an adolescent: using treatment of dermatomyositis as a model. *Pediatric Rheumatology Online Journal*. 2008;6:3.
94. Barker RA, Anderson JR, Meyer P, et al. Microangiopathy of the brain and retina with hearing loss in a 50 year old woman: extending the spectrum of Susac's syndrome. *Journal of Neurology, Neurosurgery & Psychiatry*. 1999;66(5):641–643.
95. Gordon DL, Hayreh SS, Adams Jr HP. Microangiopathy of the brain, retina, and ear: improvement without immunosuppressive therapy. *Stroke*. 1991;22(7):933–937.
96. Flammer J, Kaiser H, Haufschild T. Susac syndrome: a vasospastic disorder? *European Journal of Ophthalmology*. 2001;11(2):175–179.
97. Digre KB, Blodi CF, Bale JF. Cytomegalovirus infection in a healthy adult associated with recurrent branch retinal artery occlusion. *Retina*. 1987;7(4):230–232.
98. Magro CM, Poe JC, Lubow M, Susac JO. Susac syndrome: an organ-specific autoimmune endotheliopathy syndrome associated with anti-endothelial cell antibodies. *American Journal of Clinical Pathology*. 2011;136(6):903–912.
99. Antulov R, Holjar Erlic I, Perkovic O, et al. Susac's syndrome during pregnancy - the first Croatian case. *Journal of the Neurological Sciences*. 2014;341(1–2):162–164.
100. Pawate S, Agarwal A, Moses H, Sriram S. The spectrum of Susac's syndrome. *Neurological Sciences*. 2009;30(1):59–64.
101. Hardy TA, Garsia RJ, Halmagyi GM, et al. Tumour necrosis factor (TNF) inhibitor therapy in Susac's syndrome. *Journal of the Neurological Sciences*. 2011;302(1–2):126–128.
102. Greco A, De Virgilio A, Gallo A, et al. Susac's syndrome–pathogenesis, clinical variants and treatment approaches. *Autoimmunity Reviews*. 2014;13(8):814–821.
103. Jarius S, Neumayer B, Wandinger KP, et al. Anti-endothelial serum antibodies in a patient with Susac's syndrome. *Journal of the Neurological Sciences*. 2009;285(1–2):259–261.
104. Bitra RK, Eggenberger E. Review of Susac syndrome. *Current Opinion in Ophthalmology*. 2011;22(6):472–476.
105. Jarius S, Kleffner I, Dorr JM, et al. Clinical, paraclinical and serological findings in Susac syndrome: an international multicenter study. *Journal of Neuroinflammation*. 2014;11:46.
106. Belizna C, Duijvestijn A, Hamidou M, Tervaert JW. Antiendothelial cell antibodies in vasculitis and connective tissue disease. *Annals of the Rheumatic Diseases*. 2006;65(12):1545–1550.
107. Gross CC, Meyer C, Bhatia U, et al. CD8(+) T cell-mediated endotheliopathy is a targetable mechanism of neuro-inflammation in Susac syndrome. *Nature Communications*. 2019;10(1):5779.
108. Rennebohm RM, Asdaghi N, Srivastava S, Gertner E. Guidelines for treatment of Susac syndrome - an update. *International Journal of Stroke*. 2018;15(5):1747493017751737.
109. Mateen FJ, Zubkov AY, Muralidharan R, et al. Susac syndrome: clinical characteristics and treatment in 29 new cases. *European Neurological Journal*. 2012;19(6):800–811.

110. Hahn JS, Lannin WC, Sarwal MM. Microangiopathy of brain, retina, and inner ear (Susac's syndrome) in an adolescent female presenting as acute disseminated encephalomyelitis. *Pediatrics*. 2004;114(1):276–281.
111. Fox RJ, Costello F, Judkins AR, et al. Treatment of Susac syndrome with gamma globulin and corticosteroids. *Journal of the Neurological Sciences*. 2006;251(1–2):17–22.
112. Rennebohm RM, Susac JO. Treatment of Susac's syndrome. *Journal of the Neurological Sciences*. 2007;257(1–2):215–220.
113. Marie I, Maurey G, Herve F, et al. Intravenous immunoglobulin-associated arterial and venous thrombosis; report of a series and review of the literature. *British Journal of Dermatology*. 2006;155(4):714–721.
114. Wildemann B, Schulin C, Storch-Hagenlocher B, et al. Susac's syndrome: improvement with combined antiplatelet and calcium antagonist therapy. *Stroke*. 1996;27(1):149–151.
115. Lubow M. Re: Susac syndrome in four male patients. *Retina*. 2007;27(8):1174 (author reply).
116. Li HK, Dejean BJ, Tang RA. Reversal of visual loss with hyperbaric oxygen treatment in a patient with Susac syndrome. *Ophthalmology*. 1996;103(12):2091–2098.
117. Yepez JB, Murati FA, Pettito M, Arevalo JF. Intravitreal triamcinolone in susac syndrome. *Retin Cases Brief Reports*. 2017;11(3):236–239.

Chapter 44

Retinal vascular regeneration

Varun Pathak[1], Pietro M. Bertelli[1], Jasenka Guduric-Fuchs[1], Jessica Eyre[1], Emma Reid[2], Noemi Lois[1], Alan W. Stitt[1] and Reinhold J. Medina[1,3]

[1]*Wellcome-Wolfson Institute for Experimental Medicine, Faculty of Medicine, Health, and Life Sciences, Queen's University Belfast, Belfast, United Kingdom;* [2]*School of Cardiovascular & Metabolic Health, University of Glasgow, Glasgow, United Kingdom;* [3]*Department of Eye and Vision Science, Institute of Life Course and Medical Sciences, University of Liverpool, Liverpool, United Kingdom*

Overview bullets

- Regeneration of the retinal vasculature is a homeostatic process to maintain vascular function that supports retinal metabolic needs.
- Spontaneous vascular reperfusion of avascular areas in ischemic eye disease has been reported.
- New therapeutic approaches to promote retinal vascular regeneration are: exogenous cell therapies and modulation of molecular pathways that enhance endogenous vascular repair.
- Senolytic drugs remove senescent cells in retinopathies and promote retinal vascular regeneration.

Introduction

Retinal function is reliant on adequate blood perfusion, which reflects its unique oxygen-dependent metabolism. In particular, the photoreceptors have exceptionally high bioenergetic demands,[1] which are met by the vasculature in the eye, delivering oxygen and nutrients. This has been demonstrated by elegant studies showing that light-induced changes in oxygen and glucose consumption are evident in the outer but not in the inner retina.[2] This underscores the potential to harness vascular regeneration in the retina to supply metabolic needs in ischemic retinopathies. The retina is perfused by two distinct vascular beds; the central retinal artery perfuses the inner retina while the choroidal vasculature perfuses the outer retina.[3] The inner retinal vasculature is composed of three interconnected vascular networks. The superficial vascular plexus enters the retina alongside the optic nerve, distributes radially to the peripheral retina, as well as diving into the inner retina to form the medial and deeper vascular plexus. These retinal vascular networks are shown in the mouse retina (Fig. 44.1).

Supplying the outer retina, the choroidal vasculature is derived from the posterior ciliary arteries and is divided into large nonfenestrated vessels in the outer choroid known as the Haller's layer, the Sattler's layer in the inner choroid, and the fenestrated choriocapillaris separated from the retinal pigmented epithelium by the Bruch's membrane. A tight control of retinal capillary perfusion is essential to accommodate for the dynamic temporal and spatial metabolic demands in the functioning retina.[4] Similarly, impaired perfusion under pathological conditions is locally associated with decreased light sensitivity,[5] and sites of retinal vein occlusion lesions showed decreased retinal sensitivity.[6]

Vascular regeneration or reparative angiogenesis is a highly debated topic as some consider the vascular system to possess low regenerative capacity, akin to neurons and cardiomyocytes.[7] It is important to note that angiogenesis in the context of retinal disease is classically linked with pathology and commonly described as pathological retinal neovascularization. While this concept is firmly established and backed up by the therapeutic efficacy of anti-VEGFs (vascular endothelial growth factors), there is a positive side for retinal angiogenesis when it relates to vascular repair and reperfusion of ischemic retinas, usually in the early stages of the disease. Effective repair of the retinal vasculature would prevent disease progression into late stages associated with pathological neovascularization. In addition, enhancing vascular regeneration in retinal ischemic disease can protect retinal neurons from cell death, and potentially preserve/restore visual function. There is a therapeutic window of opportunity when vasoregenerative approaches are most

Retinal and Choroidal Vascular Diseases of the Eye. https://doi.org/10.1016/B978-0-443-15583-3.00044-5

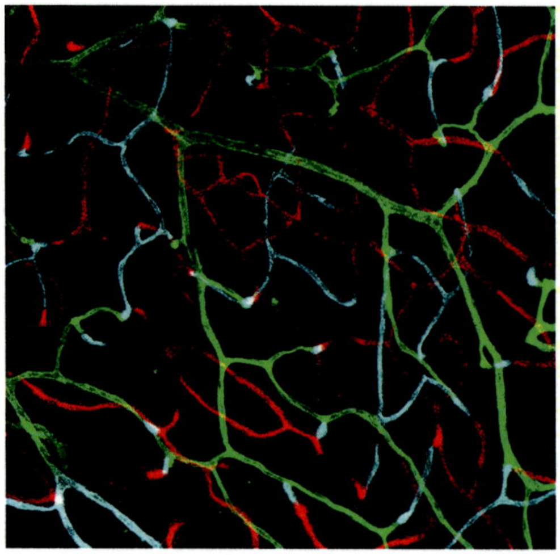

FIGURE 44.1 Three interconnected retinal vascular network layers. Mouse retinal vasculature stained with Isolectin B4 to identify blood vessels and imaged using a confocal microscope. Image was pseudocolored in relation to retinal depth to visualize the superficial, middle, and deep retinal vascular plexuses, in green, cyan, and, red respectively.

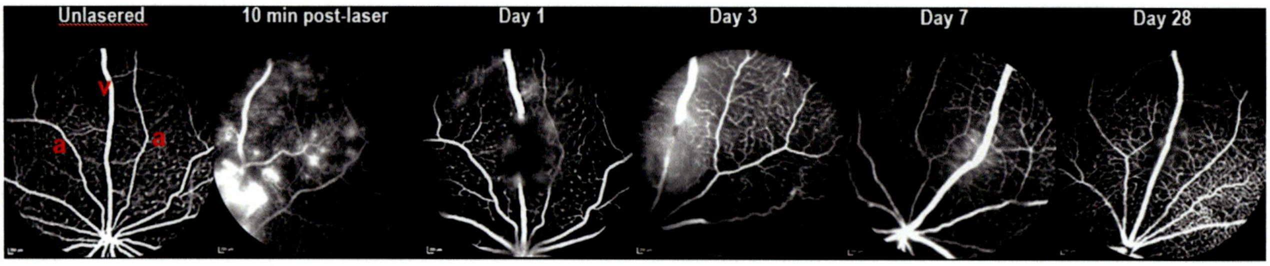

FIGURE 44.2 Vascular reperfusion in the mouse retina. Effective retinal reperfusion in the mouse retina after laser-induced retinal vein occlusion. Representative images of different time points to show vascular remodeling and repair after laser injury.

beneficial in the ischemic retina. If applied too late when retinal neurons are already lost, no vision improvements can be expected, and there may be a risk of contributing to pathological neovascularization. There is also controversy surrounding the existence of endogenous vascular stem cells or endothelial progenitor cells and their potential role in ischemic retinopathies.[8] In the mouse retina, however, endogenous vascular repair is unequivocally demonstrated. Evidence for this comes from the oxygen-induced retinopathy model (OIR), which mimics retinopathy of prematurity. The mouse OIR model exhibits major retinal central vaso-obliteration at postnatal day 8 (P8), pathological preretinal neovascularization at P16, and importantly, a full recovery to normal vasculature at P25.[9] Moreover, in a mouse model of retinal vein occlusion, we have consistently found that after 3 days of confirming laser-induced vein occlusion, there is vascular remodeling and perfusion recovery of ischemic areas (Fig. 44.2).

Unfortunately, this exceptional capacity to repair and regenerate the retinal microvasculature, observed in the mouse, is not found in humans. The cause and severity of the retinal vaso-obliteration will likely determine if spontaneous repair through endogenous mechanisms is sufficient. Nevertheless, the human retinal microvasculature does effectively respond to proangiogenic signals as shown in diabetic retinopathy (DR). The diabetic retina exhibits hypoxia, leading to VEGF upregulation that typically promotes the formation of new blood vessels, but also vascular leakage.[10] These new retinal vessels migrate toward the vitreous and are prone to leakage and hemorrhage; this stage of DR is known as proliferative DR. Understanding the dysregulated angiogenic response in the diabetic retina and elucidating means to direct vascular regeneration to occur in the right place and at the right time could lead to innovative therapies for DR. This chapter will review the scientific evidence for retinal vascular regeneration in relation to clinical conditions, cell therapies, molecular pathways, and senolytic drugs.

Clinical evidence for reperfusion of the ischemic retina

There have been reports of spontaneous reperfusion of previously nonperfused retina.[11] A case study in a 33-year-old patient with a 24-year history of diabetes and proliferative DR, with repeated photocoagulation, showed effective revascularization in a previously avascular area.[12] Moreover, a retrospective study of 94 eyes from 74 patients with diabetic retinopathy, followed up for 2 years, highlighted reperfusion in 69% of the eyes based on fluorescein angiography.[13] Ultra-widefield fundus fluorescein angiography (UWF-FFA) enables the visualization of areas of retinal nonperfusion and new vessels in nearly the entire retina. Using UWF-FFA, retinal reperfusion of previously nonperfused areas after anti-VEGF injections has been shown in some patients with DR[14]; however, other studies have failed to confirm this finding.[15] Comparison of retinal images from optical coherence tomography angiography (OCT-A) with en face OCT in three patients with branch retinal vein occlusion suggested that vascular structures found in en face OCT, but not in OCT-A represent "ghost" vessels caused by remaining basement membrane.[16] This was supported by the immunohistochemistry for Collagen IV findings in a postmortem retinal flat mount from one of the patients. Interestingly, such nonperfused retinal areas with vascular basement membranes still preserved vascular lumens, but without endothelial and mural cells. On this basis, reendothelialization of these vascular basement membranes may be possible through endogenous vasoreparative endothelial cells. Pericytes will also be needed to support endothelial cell viability and function. Many markers for endothelial stem cells have been proposed.[17] Among them, $CD157^+$ cells have been successfully isolated from mouse retina and characterized as exhibiting major angiogenic potency and, therefore, named as the putative vascular endothelial stem cells.[18]

Cell therapies to regenerate the retinal vasculature

While endogenous vascular repair of the ischemic human retina is a possibility, it has been proposed that cell therapies can significantly enhance and accelerate the regeneration of the retinal vasculature.[19] This is particularly critical under conditions such as diabetes, where there is an intrinsic impairment of endogenous vasoreparative mechanisms that would benefit from exogenous cell therapy. Therefore, there are active preclinical research and clinical trials investigating the therapeutic potential of various stem cells for retinopathies,[20] and here we will discuss the most promising cell therapy candidates (Fig. 44.3).

Endothelial colony-forming cells

A cell replacement strategy to regenerate the retinal vasculature requires potent vasoreparative endothelial cells. Cumulative preclinical evidence in ischemic retinopathy models has shown that endothelial colony-forming cells (ECFCs) effectively revascularize the ischemic retina in the OIR[21] and the streptozotocin (STZ)-induced[22] DR models. ECFCs have

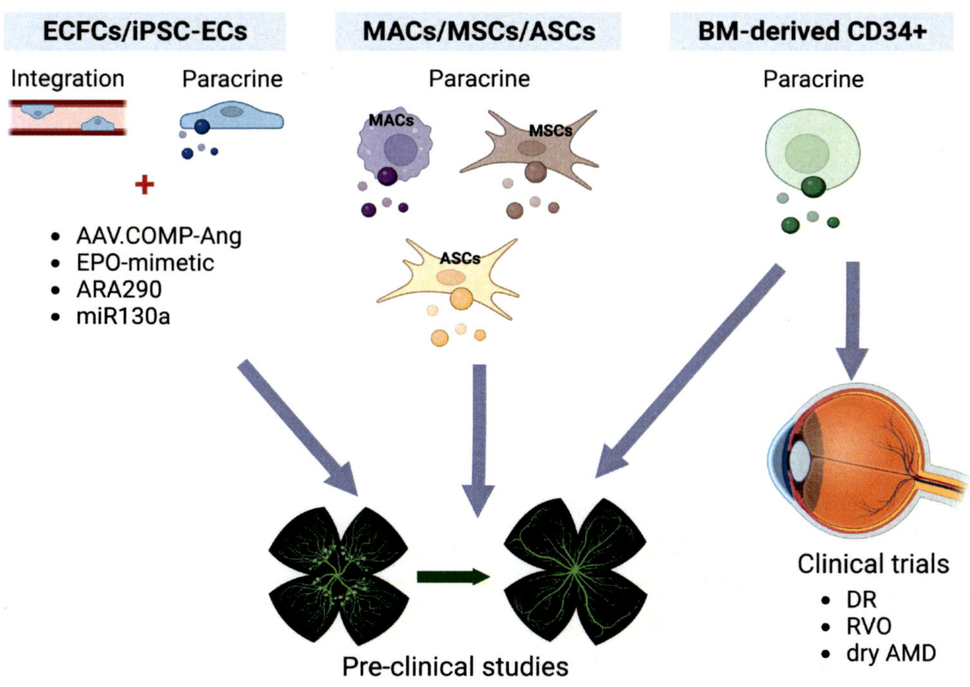

FIGURE 44.3 Cytotherapies for promoting vascular regeneration in the retina. Cell therapies that have been investigated in eye disease to promote vascular regeneration. *Created with BioRender.com.*

a well-defined molecular identity as endothelial progenitor cells characterized by a cell surface profile depicted as $CD31^+$ $CD105^+$ $CD45^-$ $CD14^-$ $CD90^-$. ECFCs must display key endothelial progenitor functionality features, including high proliferative and angiogenic activity as demonstrated by clonogenics, 3D-tubulogenesis, and sprouting assays. The high purity and potency render them an ideal cell population for cell therapy.[23] ECFCs are isolated from human cord blood (CB) using standardized protocols for a potential allogeneic cell therapy approach. Dose studies in a mouse OIR model have demonstrated therapeutic efficacy with as low as 1000 cells delivered intravitreally without inducing an inflammatory response.[24] The vasoreparative potential of ECFCs can be further augmented by combination with AAV.COMP-Ang1,[22] EPO mimetic ARA290,[25] and miR-130a activation.[26] The mechanisms of action to explain ECFCs' vasoreparative properties in the retina are twofold. The primary mechanism of action is direct integration and engraftment into the host retinal vasculature,[21] where ECFCs act as a cell replacement for damaged and lost endothelium. A secondary mechanism of action is the release of paracrine factors such as insulin-like growth factor binding proteins (IGFBPs), which provide trophic support for the retinal neurovascular unit.[27]

Pluripotent stem cells

Pluripotent stem cells such as embryonic stem cells (ESCs) and induced pluripotent stem cells (iPSCs) have also been extensively studied as a source of endothelial cells to regenerate the retinal vasculature. A protocol was developed to manufacture iPSC-derived ECFCs, and these iPSC-ECFCs were as effective as CB-derived ECFCs in reducing avascular areas in the OIR model.[28] Vascular progenitors generated from CB-derived iPSCs, defined as $CD31^+$ $CD146^+$ cells, have enhanced engraftment into retinal vasculature of the ischemia-reperfusion model in immunodeficient NOD-SCID mice that lack T and B lymphocytes, when compared to fibroblast-derived iPSCs.[29] These $CD31^+$ $CD146^+$ iPSC vascular progenitors migrated through to the deep retinal layers and incorporated readily into damaged blood vessels. Interestingly, results suggested that these vascular progenitors could give rise to both endothelial cells and pericytes. Moreover, considering that iPSCs generated from diabetic patients cells have reduced efficacy compared to those from nondiabetic individuals, possibly because of epigenetic changes relating to bouts of hyperglycemia, new protocols to restore functionality in diabetic iPSCs have been investigated. Reprogramming to a naïve human pluripotent stem cell state, using tankyrase inhibitors, before endothelial differentiation was shown to improve survival and engraftment of diabetic iPSC-derived endothelial cells in the mouse retinal ischemia-reperfusion injury model.[30] There are expectations that these CB-ECFCs or iPSC-endothelial cells will translate into a first-in-man clinical trial within the next 5 years; however, the translational pathway for cell therapies to regenerate the eye vasculature is currently led by $CD34^+$ progenitor cells.

$CD34^+$ progenitor cells

These $CD34^+$ cells are isolated from bone marrow and are considered enriched for endothelial progenitors. Preclinical evidence showed the capacity of GMP-grade bone marrow $CD34^+$ cells to repair the mouse retinal vasculature without tumorigenicity after 8 months.[31] A phase I clinical trial for autologous $CD34^+$ cells in six patients with irreversible vision loss secondary to a variety of causes and, with a 6-month follow-up, found that cell therapy was well tolerated and without ocular complications.[32] Following on from this, the clinical trial NCT01736059 is currently evaluating the therapeutic efficacy of autologous bone marrow $CD34^+$ cells delivered intravitreally into patients with DR, retinal vein occlusion, and nonexudative age-related macular degeneration. Interestingly, combinatorial progenitor cell therapy using $CD34^+$ cells and ECFCs was tested in the OIR mouse model. Results confirmed that ECFCs but not $CD34^+$ cells engrafted into the retinal vasculature; however, the combined therapy was the most effective in rescuing ischemia and preventing retinal pathological neovascularization.[33]

Mesenchymal stem cells, adipose stem cells, and myeloid angiogenic cells

An alternative way to promote retinal vascular regeneration is through the manipulation of cell-to-cell communication via paracrine factors, extracellular vesicles, and immunomodulation. Among the cell types with these mechanisms of action are mesenchymal stem cells (MSCs), adipose stem cells (ASCs), and myeloid angiogenic cells (MACs). ASCs or their conditioned media were protective of retinal function in the Ins2Akita mouse by decreasing inflammation and vascular leakage.[34] ASCs were also shown to become pericytes when integrated into the ischemic retinal vasculature,[35] promoting tissue repair in the OIR model. MACs delivered intravitreally in the OIR model promoted vascular regeneration by IL8 secretion,[36] and these proreparative effects were impaired when MACs were isolated from people with diabetes.[37]

Preclinical evidence highlights the significant potential of cell therapies to regenerate the retinal vasculature and provide new therapeutic options for ischemic retinopathies in the future. While the translational development of cell therapies for eye disease should be supported, the primary concern when developing this approach must be first and foremost patient safety. This requires strict adherence to requirements from regulatory agencies such as the US Food and Drug Administration (FDA), the European Medicines Agency (EMA), and the UK Medicines and Healthcare Products Regulatory Agency (MHRA), to obtain ethical and marketing approvals. Lack of regulatory oversight in some stem cell clinics has led to alarming outcomes such as vision loss following intravitreal injection of cells.[38]

Molecular pathways that modulate retinal vascular repair

Multiple molecular pathways have been linked to vascular regeneration in the retina. Here, we will discuss the most frequently reported (Fig. 44.4). There is extensive evidence that links transcriptional activation of **hypoxia-inducible factor (HIF) 1α and 2α** in the oxygen-deprived retina as a key initiator of proangiogenic pathways that drive pathological neovascularization. By contrast, less is known about the role of HIFs in retinal vascular regeneration. It has been shown that while HIFs derived from retinal myeloid cells are not essential for pathological neovascularization, they are

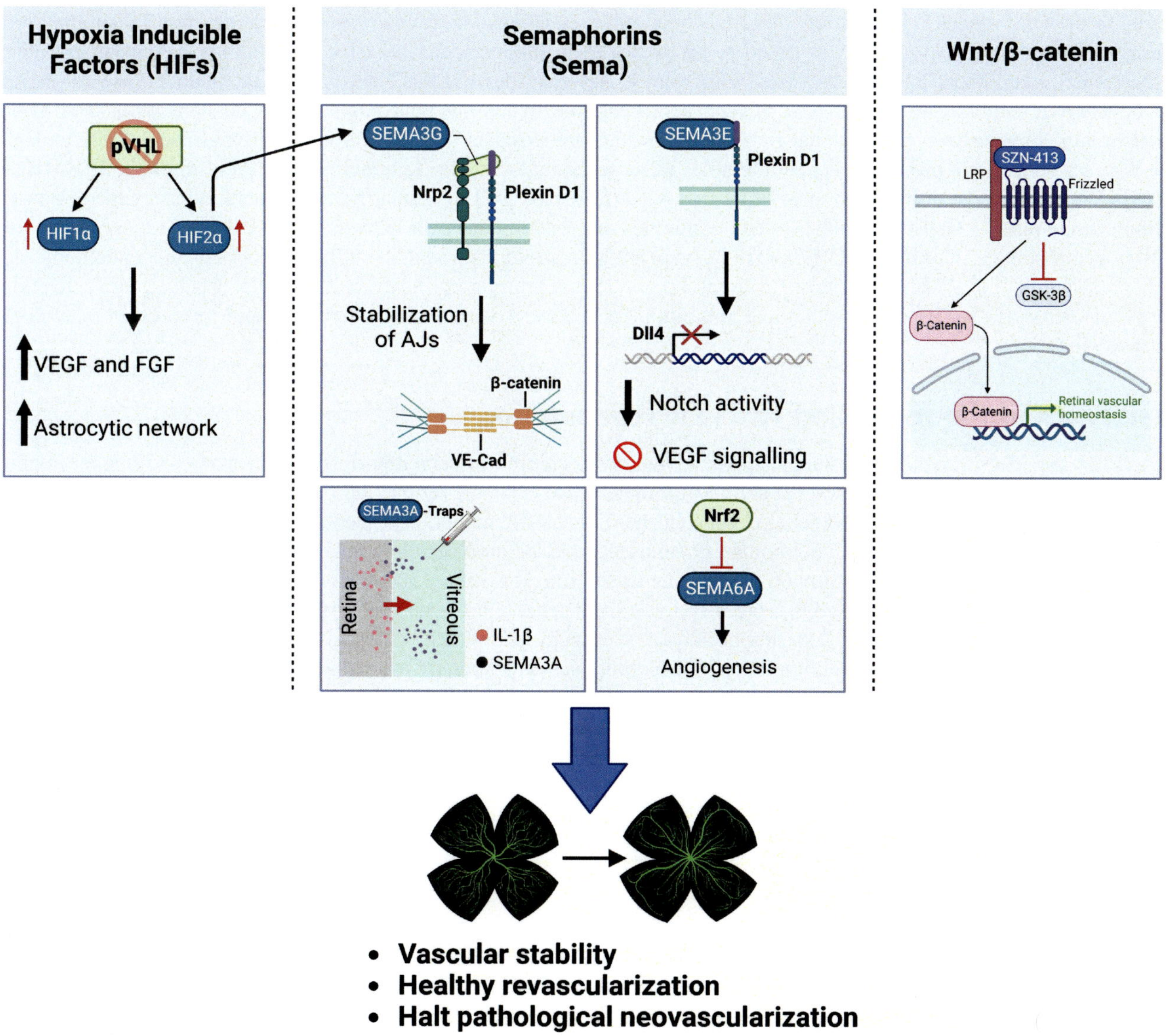

FIGURE 44.4 Molecular players in retinal vasoreparative processes. Molecular pathways that modulate retinal vascular regeneration. *Created with BioRender.com.*

important for vascular regeneration.[39] In the OIR model, activation of HIF1α and HIF2α by deletion of Von Hippel-Lindau protein (pVHL) in retinal myeloid cells led to enhanced revascularization of ischemic areas through a localized increase in VEGF and fibroblast growth factor (FGF) associated with an improved astrocytic network. In addition, there is cross-talk between hypoxia and **Semaphorins (Sema)**. Sema are a family of secreted proteins involved in neuronal guidance and vascular formation that have been associated with retinal vascular regeneration. Increased **Sema3G** was found in the vitreous of patients with proliferative DR and in OIR mouse retinas. Sema3G expression was modulated by HIF2α. Endothelium-derived Sema3G was needed to modulate retinal vascular remodeling by driving regression of pathological neovascularization while enhancing reparative vascular regeneration.[40] Sema3G stabilized the interaction between β-catenin and VE-cadherin in retinal vasculature via the Neuropilin-2(Nrp2)-PlexinD1 signaling pathway. **Sema3A** expression was found to be increased in mouse OIR avascular areas associated with the upregulation of inflammatory cytokine IL1β. Intravitreal injection of Sema3A-traps ameliorated vascular regeneration and suppressed pathological neovascularization in the OIR, STZ diabetes, and choroidal neovascularization mouse models.[41] This further suggested Sema3A as a potential target to enhance vascular regeneration in the ischemic retina. **Sema3E**, predominantly secreted by neuronal cells, binds PlexinD1 expressed in sprouting vasculature, and controls VEGF signaling by inhibiting delta-like canonical Notch ligand 4 (Dll4) expression and Notch activity. In the OIR model, Sema3E intravitreal injection blocked VEGF-induced vascular outgrowth, enabling effective regeneration of the retinal vasculature.[42] Similarly, retinal ganglion cell-derived Sema3E is a critical mediator for retinal vascular regeneration induced by MSCs or their conditioned media in the OIR model.[43] In addition, **Nrf2** has been reported to mediate retinal vascular regeneration by suppression of **Sema6A**.[44] The Nrf2 expression was found to be increased in the avascular areas of OIR retinas, mainly in endothelial cells, retinal ganglion cells, and astrocytes. Importantly, neuronal-derived Nrf2 was shown to mediate vascular regeneration, whereas endothelial-derived Nrf2 played a critical role in halting pathological neovascularization. The **Wnt/β-catenin** signaling pathway is important for retinal vascular homeostasis. The development of pharmacological molecules for Wnt activation was proposed as new therapies to target retinal vascular remodeling. The Wnt mimetic SZN-413 was tested in OIR. Intravitreal injection of SZN-413 showed decreased OIR avascular areas by facilitating regeneration of the retinal vasculature.[45] Cell cross-talk between astrocytes and endothelial cells plays a role in the pathogenesis of proliferative retinopathies. **Connexin 43 (Cx43)** is a gap junction protein present in retinal astrocytes and endothelial cells. Phosphorylation of Cx43 leads to astrocyte apoptosis and retinal injury amplification. In the OIR model, genetic deletion of Cx43 in astrocytes or its pharmacological inhibition, accelerated vascular regeneration, and suppressed pathological neovascularization.[46]

Senolytics in the retina and vascular regeneration

Senolytics are drugs that eliminate senescent cells from tissues and have been reported to promote healthy aging in animal models. Evidence suggests that senescent cells accumulate in the ischemic retina[47] and that diabetes induced an accelerated senescence program in retinal endothelial cells.[48] In the mouse OIR model, senescence was primarily observed in retinal ganglion cells in avascular areas.[47] Inflammatory senescence-associated secretory factors from senescent retinal ganglion cells promoted abnormal vascular growth and induced paracrine senescence in retinal microglia and endothelial cells via Sema3A. A follow-up study revealed that the secretome from senescent retinal vasculature engaged neutrophils to produce neutrophil extracellular traps (NETs) that removed senescent endothelial cells to remodel unhealthy vasculature.[49] Therefore, clearing up senescent endothelial cells was proposed as a strategy to enhance reparative vascular regeneration. Pathological new retinal vessels in the mouse OIR model are enriched with senescent endothelial cells. Senolysis using the p16-INK-ATTAC mice or BCL-xL inhibitors demonstrated decreased pathological angiogenesis coupled with enhanced vascular repair.[50] These findings have led to clinical trials to evaluate the utility of senolytics in diabetic macular edema (DME) and age-related macular degeneration (AMD). After showing safety in a Phase 1 study, the clinical trials NCT04857996 BEHOLD and NCT05275205 ENVISION are evaluating the therapeutic efficacy of the senolytic drug UBX-1325 for DME and neovascular AMD, respectively.

Concluding remarks

Vascular repair is a homeostatic mechanism that can be harnessed to address retinal vascular disease. Understanding the molecular pathways that control vascular regeneration will facilitate the development of new therapies to halt the progression of ischemic eye disease, and potentially reverse the natural history of disease. In addition, understanding the location (retinal-macular vs. retinal-peripheral or choroidal) and mechanisms by which vascular damage occurs in the eye in different diseases is essential. For example, mechanisms of vascular damage may not be the same in acute pathology, such as retinal vein occlusion, when compared with those in chronic diseases such as diabetes. In cases where endogenous vasoreparative mechanisms are severely impaired, there will be a need for cell therapies to repair and regenerate the ischemic retinal or choroidal vasculature.

Acknowledgments

The research work in the laboratory of RJM is supported by Diabetes UK (20/0006162), the Dunhill Medical Trust (RPGF1910\199), the Macular Society, the Leverhulme Trust (RPG-2023-149), and the Department for the Economy Northern Ireland under the US-Ireland R&D Partnership Program with Ref. USI 158.

References

1. Hurley JB, Lindsay KJ, Du J. Glucose, lactate, and shuttling of metabolites in vertebrate retinas. *Journal of Neuroscience Research*. 2015;93(7):1079−1092. https://doi.org/10.1002/jnr.23583.
2. Linsenmeier RA, Zhang HF. Retinal oxygen: from animals to humans. *Progress in Retinal and Eye Research*. 2017;58:115−151. https://doi.org/10.1016/j.preteyeres.2017.01.003.
3. Sun Y, Smith LEH. Retinal vasculature in development and diseases. *Annual Review of Vision Science*. 2018;4:101−122. https://doi.org/10.1146/annurev-vision-091517-034018.
4. Yu DY, Cringle SJ, Yu PK, et al. Retinal capillary perfusion: spatial and temporal heterogeneity. *Progress in Retinal and Eye Research*. 2019;70:23−54. https://doi.org/10.1016/j.preteyeres.2019.01.001.
5. Iijima H. Reduced light sensitivity due to impaired retinal perfusion in branch retinal vein occlusion. *Japanese Journal of Ophthalmology*. 2018;62(2):151−157. https://doi.org/10.1007/s10384-017-0546-5.
6. Khayat M, Perais J, Wright DM, Williams M, Lois N. Anatomic−functional correlates in lesions of retinal vein occlusion. *Investigative Ophthalmology & Visual Science*. 2021;62, 10-10.
7. Brunet A, Goodell MA, Rando TA. Ageing and rejuvenation of tissue stem cells and their niches. *Nature Reviews Molecular Cell Biology*. 2023;24(1):45−62. https://doi.org/10.1038/s41580-022-00510-w.
8. Stitt AW, O'Neill CL, O'Doherty MT, Archer DB, Gardiner TA, Medina RJ. Vascular stem cells and ischaemic retinopathies. *Progress in Retinal and Eye Research*. 2011;30(3):149−166. https://doi.org/10.1016/j.preteyeres.2011.02.001.
9. Connor KM, Krah NM, Dennison RJ, et al. Quantification of oxygen-induced retinopathy in the mouse: a model of vessel loss, vessel regrowth and pathological angiogenesis. *Nature Protocols*. 2009;4(11):1565−1573. https://doi.org/10.1038/nprot.2009.187.
10. Stitt AW, Curtis TM, Chen M, et al. The progress in understanding and treatment of diabetic retinopathy. *Progress in Retinal and Eye Research*. 2016;51:156−186. https://doi.org/10.1016/j.preteyeres.2015.08.001.
11. Mohite AA, Perais JA, McCullough P, Lois N. Retinal ischaemia in diabetic retinopathy: understanding and overcoming a therapeutic challenge. *Journal of Clinical Medicine*. 2023;12(6). https://doi.org/10.3390/jcm12062406.
12. Mohan R, Kohner EM. Retinal revascularisation in diabetic retinopathy. *British Journal of Ophthalmology*. 1986;70(2):114−117. https://doi.org/10.1136/bjo.70.2.114.
13. Takahashi K, Kishi S, Muraoka K, Shimizu K. Reperfusion of occluded capillary beds in diabetic retinopathy. *American Journal of Ophthalmology*. 1998;126(6):791−797. https://doi.org/10.1016/S0002-9394(98)00242-6.
14. Levin AM, Rusu I, Orlin A, et al. Retinal reperfusion in diabetic retinopathy following treatment with anti-VEGF intravitreal injections. *Clinical Ophthalmology*. 2017;11:193−200. https://doi.org/10.2147/OPTH.S118807.
15. Bonnin S, Dupas B, Lavia C, et al. Anti-vascular endothelial growth factor therapy can improve diabetic retinopathy score without change in retinal perfusion. *Retina*. 2019;39(3):426−434. https://doi.org/10.1097/IAE.0000000000002422.
16. Powner MB, Sim DA, Zhu M, et al. Evaluation of nonperfused retinal vessels in ischemic retinopathy. *Investigative Ophthalmology and Visual Science*. 2016;57(11):5031−5037. https://doi.org/10.1167/iovs.16-20007.
17. Chambers SEJ, Pathak V, Pedrini E, et al. Current concepts on endothelial stem cells definition, location, and markers. *Stem Cells Translational Medicine*. 2021;10(2):S54−S61. https://doi.org/10.1002/sctm.21-0022.
18. Wakabayashi T, Naito H, Suehiro Ji, et al. CD157 marks tissue-resident endothelial stem cells with homeostatic and regenerative properties. *Cell Stem Cell*. 2018;22(3):384−397. https://doi.org/10.1016/j.stem.2018.01.010.
19. Bertelli PM, Pedrini E, Guduric-Fuchs J, et al. Vascular regeneration for ischemic retinopathies: hope from cell therapies. *Current Eye Research*. 2020;45(3):372−384. https://doi.org/10.1080/02713683.2019.1681004.
20. Park SS. Cell therapy applications for retinal vascular diseases: diabetic retinopathy and retinal vein occlusion. *Investigative Opthalmology & Visual Science*. 2016;57(5):ORSFj1. https://doi.org/10.1167/iovs.15-17594.
21. Medina RJ, O'Neill CL, Humphreys MW, Gardiner TA, Stitt AW. Outgrowth endothelial cells: characterization and their potential for reversing ischemic retinopathy. *Investigative Ophthalmology and Visual Science*. 2010;51(11):5906−5913. https://doi.org/10.1167/iovs.09-4951.
22. Cahoon JM, Rai RR, Carroll LS, et al. Intravitreal AAV2.COMP-Ang1 prevents neurovascular degeneration in a murine model of diabetic retinopathy. *Diabetes*. 2015;64(12):4247−4259. https://doi.org/10.2337/db14-1030.
23. Banno K, Yoder MC. Tissue regeneration using endothelial colony-forming cells: promising cells for vascular repair. *Pediatric Research*. 2018;83(1−2):283−290. https://doi.org/10.1038/pr.2017.231.

24. Reid E, Guduric-Fuchs J, O'Neill CL, et al. Preclinical evaluation and optimization of a cell therapy using human cord blood-derived endothelial colony-forming cells for ischemic retinopathies. *Stem Cells Translational Medicine*. 2018;7(1):59–67. https://doi.org/10.1002/sctm.17-0187.
25. O'Leary OE, Canning P, Reid E, et al. The vasoreparative potential of endothelial colony-forming cells in the ischemic retina is enhanced by cibinetide, a non-hematopoietic erythropoietin mimetic. *Experimental Eye Research*. 2019;182:144–155. https://doi.org/10.1016/j.exer.2019.03.001.
26. Guduric-Fuchs J, Pedrini E, Lechner J, et al. miR-130a activates the VEGFR2/STAT3/HIF1α axis to potentiate the vasoregenerative capacity of endothelial colony-forming cells in hypoxia. *Molecular Therapy - Nucleic Acids*. 2021;23:968–981. https://doi.org/10.1016/j.omtn.2021.01.015.
27. Sakimoto S, Marchetti V, Aguilar E, et al. CD44 expression in endothelial colony-forming cells regulates neurovascular trophic effect. *JCI Insight*. 2017;2(2). https://doi.org/10.1172/jci.insight.89906.
28. Prasain N, Lee MR, Vemula S, et al. Differentiation of human pluripotent stem cells to cells similar to cord-blood endothelial colony-forming cells. *Nature Biotechnology*. 2014;32(11):1151–1157. https://doi.org/10.1038/nbt.3048.
29. Park TS, Bhutto I, Zimmerlin L, et al. Vascular progenitors from cord blood-derived induced pluripotent stem cells possess augmented capacity for regenerating ischemic retinal vasculature. *Circulation*. 2014;129(3):359–372. https://doi.org/10.1161/CIRCULATIONAHA.113.003000.
30. Park TS, Zimmerlin L, Evans-Moses R, et al. Vascular progenitors generated from tankyrase inhibitor-regulated naïve diabetic human iPSC potentiate efficient revascularization of ischemic retina. *Nature Communications*. 2020;11(1). https://doi.org/10.1038/s41467-020-14764-5.
31. Park SS, Caballero S, Bauer G, et al. Long-term effects of intravitreal injection of GMP-grade bone-marrow-derived CD34 + cells in NOD-SCID mice with acute ischemia-reperfusion injury. *Investigative Ophthalmology and Visual Science*. 2012;53(2):986–994. https://doi.org/10.1167/iovs.11-8833.
32. Park SS, Bauer G, Abedi M, et al. Intravitreal autologous bone marrow cd34+ cell therapy for ischemic and degenerative retinal disorders: preliminary phase 1 clinical trial findings. *Investigative Ophthalmology and Visual Science*. 2015;56(1):81–89. https://doi.org/10.1167/iovs.14-15415.
33. Calzi SL, Shaw LC, Moldovan L, et al. Progenitor cell combination normalizes retinal vascular development in the oxygen-induced retinopathy (OIR) model. *JCI Insight*. 2019;4(21). https://doi.org/10.1172/jci.insight.129224.
34. Elshaer SL, Evans W, Pentecost M, et al. Adipose stem cells and their paracrine factors are therapeutic for early retinal complications of diabetes in the Ins2Akita mouse. *Stem Cell Research and Therapy*. 2018;9(1). https://doi.org/10.1186/s13287-018-1059-y.
35. Mendel TA, Clabough EBD, Kao DS, et al. Pericytes derived from adipose-derived stem cells protect against retinal vasculopathy. *PLoS One*. 2013;8(5). https://doi.org/10.1371/journal.pone.0065691.
36. Medina RJ, O'Neill CL, O'Doherty TM, et al. Myeloid angiogenic cells act as alternative M2 macrophages and modulate angiogenesis through interleukin-8. *Molecular Medicine*. 2011;17(9–10):1045–1055. https://doi.org/10.2119/molmed.2011.00129.
37. Chambers SEJ, O'Neill CL, Guduric-Fuchs J, et al. The vasoreparative function of myeloid angiogenic cells is impaired in diabetes through the induction of IL1β. *Stem Cell*. 2018;36(6):834–843. https://doi.org/10.1002/stem.2810.
38. Kuriyan AE, Albini TA, Townsend JH, et al. Vision loss after intravitreal injection of autologous \stem Cells\ for AMD. *New England Journal of Medicine*. 2017;376(11):1047–1053. https://doi.org/10.1056/NEJMoa1609583.
39. Villacampa P, Liyanage SE, Klaska IP, et al. Stabilization of myeloid-derived HIFs promotes vascular regeneration in retinal ischemia. *Angiogenesis*. 2020;23(2):83–90. https://doi.org/10.1007/s10456-019-09681-1.
40. Chen DY, Sun NH, Chen X, et al. Endothelium-derived semaphorin 3G attenuates ischemic retinopathy by coordinating β-catenin-dependent vascular remodeling. *Journal of Clinical Investigation*. 2021;131(4). https://doi.org/10.1172/JCI135296.
41. Binet F, Beaulieu N, Beauchemin K. Semaphorin3A-Trap accelerates vascular regeneration in ischemic retinopathy and reduces retinal edema. *Investigative Ophthalmology & Visual Science*. 2019;60.
42. Fukushima Y, Okada M, Kataoka H, et al. Sema3E-PlexinD1 signaling selectively suppresses disoriented angiogenesis in ischemic retinopathy in mice. *Journal of Clinical Investigation*. 2011;121(5):1974–1985. https://doi.org/10.1172/JCI44900.
43. Noueihed B, Rivera JC, Dabouz R, et al. Mesenchymal stromal cells promote retinal vascular repair by modulating Sema3E and IL-17A in a model of ischemic retinopathy. *Frontiers in Cell and Developmental Biology*. 2021;9. https://doi.org/10.3389/fcell.2021.630645.
44. Wei Y, Gong J, Xu Z, et al. Nrf2 in ischemic neurons promotes retinal vascular regeneration through regulation of semaphorin 6A. *Proceedings of the National Academy of Sciences of the United States of America*. 2015;112(50):E6927–E6936. https://doi.org/10.1073/pnas.1512683112.
45. Nguyen H, Chen H, Vuppalapaty M, et al. SZN-413, a FZD4 agonist, as a potential novel therapeutic for the treatment of diabetic retinopathy. *Translational Vision Science & Technology*. 2022;11(9):19. https://doi.org/10.1167/tvst.11.9.19.
46. Slavi N, Toychiev AH, Kosmidis S, et al. Suppression of connexin 43 phosphorylation promotes astrocyte survival and vascular regeneration in proliferative retinopathy. *Proceedings of the National Academy of Sciences of the United States of America*. 2018;115(26):E5934–E5943. https://doi.org/10.1073/pnas.1803907115.
47. Oubaha M, Miloudi K, Dejda A, et al. Senescence-associated secretory phenotype contributes to pathological angiogenesis in retinopathy. *Science Translational Medicine*. 2016;8(362). https://doi.org/10.1126/scitranslmed.aaf9440.
48. Bertelli PM, Pedrini E, Hughes D, et al. Long term high glucose exposure induces premature senescence in retinal endothelial cells. *Frontiers in Physiology*. 2022;13. https://doi.org/10.3389/fphys.2022.929118.
49. Binet F, Cagnone G, Crespo-Garcia S, et al. Neutrophil extracellular traps target senescent vasculature for tissue remodeling in retinopathy. *Science*. 2020;369(6506). https://doi.org/10.1126/science.aay5356.
50. Crespo-Garcia S, Tsuruda PR, Dejda A, et al. Pathological angiogenesis in retinopathy engages cellular senescence and is amenable to therapeutic elimination via BCL-xL inhibition. *Cell Metabolism*. 2021;33(4):818–832. https://doi.org/10.1016/j.cmet.2021.01.011.

Chapter 45

Vascular changes after vitreoretinal surgery

Isabela Martins Melo[1] and Rajeev H. Muni[1,2]
[1]*Department of Ophthalmology, St. Michael's Hospital Unity Health Toronto, Toronto, ON, Canada;* [2]*Department of Ophthalmology and Vision Sciences, University of Toronto, Toronto, ON, Canada*

Vascular changes following vitreoretinal surgery can be divided into two major categories: change in vessel morphology, which includes vessel conformation, tortuosity, caliber, and location or perfusion changes, which include blood flow, vessel and perfusion density, vascularity index, and vessel luminal area. These changes can present over a spectrum from subclinical, when they are only detected as mild changes in the analysis of optical coherence tomography angiography (OCTA), to severe, when patients present with clear clinical signs of vascular occlusions.

Morphologic vascular changes

Retinal displacement

In recent years, the concept of retinal structural "integrity" was introduced to define how the retina was reattached following rhegmatogenous retinal detachment (RRD) repair. As surgical techniques evolved and reattachment rates improved, there has been a paradigm shift to consider how well the retina has reattached and the corresponding anatomic and functional outcomes.[1] One important parameter of retinal integrity is retinal displacement, first described in 2010 by Shiragami et al. as hyperautofluorescent lines that had similar caliber and contour to corresponding adjacent retinal blood vessels on fundus autofluorescence (FAF) imaging. These abnormal hyperautofluorescent lines were hypothesized to represent the original location of the shifted retinal vessels before the retinal detachment,[2] being later defined as "retinal vessel printings" (RVPs) by dell'Omo et al. in 2012.[3] This hypothesis has recently been proven by Brosh et al., who had a unique case of macula-off RRD, which was treated with pars plana vitrectomy (PPV) and had pre-RRD ultra-widefield FAF images available. The overlay of the pre- and post-RRD repair images revealed severe temporal displacement with perfect alignment of the RVPs with the pre-RRD retinal vessels, confirming it was their previous location (Fig. 45.1). Other landmarks, such as choroidal vessels, pigmentary changes, and nasal retinal vessels in the area unaffected by the detachment, were also in perfect alignment.[4] It is thought that this phenomenon of RVPs occurs because the retinal pigment epithelium (RPE) cells that were previously shielded from light have altered metabolic activity or a different composition of fluorophores, which results in the observed hyperautofluorescence.[2,3,5]

Shigarami et al. were also the first to propose that the direction of retinal displacement could be related to the position taken immediately after the operation, hypothesizing that the gravitational shift of residual subretinal fluid would be the cause of retinal displacement.[6] Later, in 2017, dell'Omo et al. proposed that the pathophysiology of displacement could be, in fact, related to the different physical properties of the tamponades, as they observed differences in the incidence of displacement under sulfur hexafluoride (SF6) and silicone oil (SO).[7] Although variations in surgical techniques were proposed to try to reduce the risk of retinal displacement, these modifications neither eliminated nor substantially reduced the overall risk.[7,8] Using the pre- and post-RRD ultra-widefield FAF images, Brosh et al. also noticed that the retina becomes increasingly displaced with increasing distance from the optic nerve.[4] Thus, together with Lee et al. in 2013, they hypothesized that retinal displacement in some cases could occur as a rotational stretch around absolute and relative points of fixation, such as the optic nerve and attached areas of the retina.[4,5]

Retinal and Choroidal Vascular Diseases of the Eye. https://doi.org/10.1016/B978-0-443-15583-3.00045-7

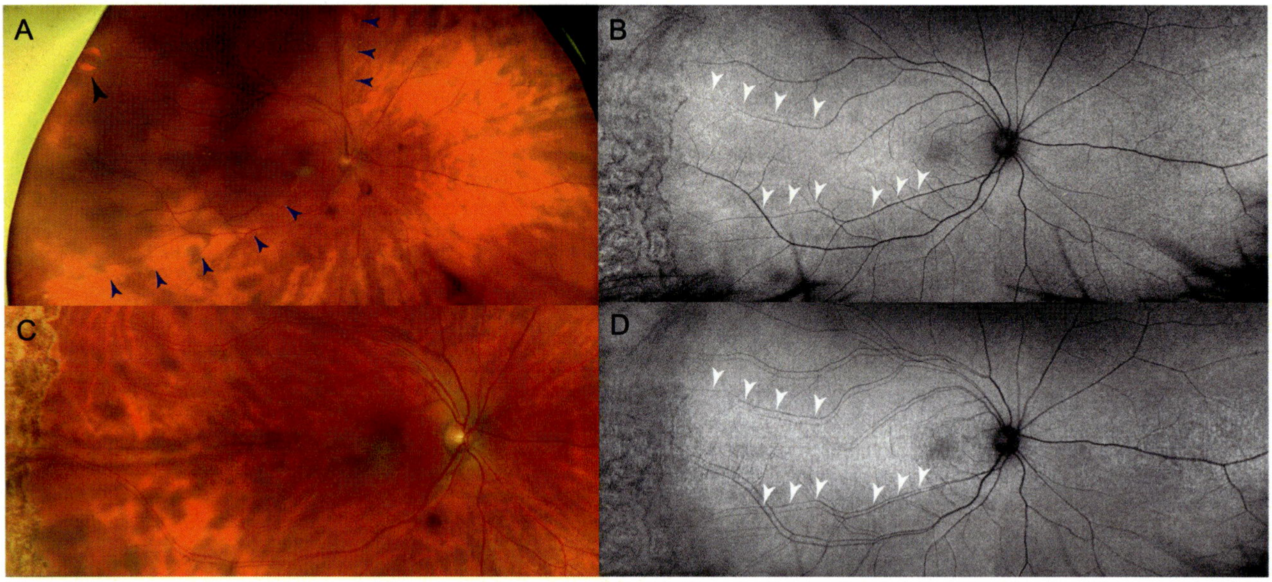

FIGURE 45.1 A) Ultra-widefield color image showing macula-off retinal detachment (small *arrowheads* highlighting the edges of the detachment); (B) post retinal detachment (RD) repair fundus autofluorescence (FAF) demonstrating retinal vessel printings (RVP: *white arrow*); (C and D) overlay of post-RD with 50% transparent pre-RD color (C) and FAF (D) image, proving that the RVP represents the prior location of the retinal vessel and that the true extent of retinal displacement is far greater than what is seen on FAF. *Reproduced from Brosh K, Roditi E, Muni RH. Retinal vessel printings on fundus autofluorescence imaging represent retinal displacement: proof of prior hypothesis.* American Journal of Ophthalmolohy. *June 2022;238:e3–e4. https://doi.org/10.1016/j.ajo.2022.01.024. Epub 2022 Feb 14. PMID: 35176253.*

In a multicenter retrospective study comparing the incidence of displacement following pneumatic retinopexy (PnR) versus pars plana vitrectomy (PPV), Brosh et al. 2020 were the first to show that the mechanism of retinal reattachment and, consequentially, the surgical technique influenced the risk of retinal displacement. In this study, patients who underwent PnR had a 7% risk (8/114) of retinal displacement compared to a 44.4% risk (55/124) in those undergoing PPV ($P < .001$). Additionally, if patients had retinal displacement observed on FAF, they would refer to these cases as having a Low-Integrity Retinal Attachment (LIRA) (Fig. 45.2).[9] The study that followed, the Retinal Displacement after Pneumatic Retinopexy versus Vitrectomy for Rhegmatogenous Retinal Detachment (ALIGN) study, subsequentially confirmed that patients who underwent PPV had a greater risk of retinal displacement when compared to PnR. In this multicenter nonrandomized trial, patients who underwent PPV for RRD repair had a 50.7% risk of retinal displacement, while the PnR group had a 14.7% risk ($P < .001$). Another interesting finding of the ALIGN trial was that aniseikonia scores were worse in eyes following PPV versus PnR ($P = .011$), and in patients with retinal displacement versus those without ($P = .017$), suggesting that retinal displacement may result in altered spacing of the photoreceptor mosaic, affecting image size perception.[10]

Recognizing for the first time that the risk of retinal displacement varied between surgical techniques led to further research to determine its cause. Studies demonstrated that the cause of retinal displacement was related to the type and size of the associated tamponade and patient positioning.[11,12] Several authors have demonstrated that the retina is most frequently displaced inferiorly.[2,3,8,9] Casswell et al. (2019) corroborated this hypothesis after demonstrating that immediate face-down positioning caused less displacement when compared to positioning to the break.[13] Regarding the type of tamponade, in 2017 dell'Omo et al. reported a higher rate of retinal displacement following PPV with SF6 compared to SO (41.2% vs. 14.3%),[7] a similar result to what Codenotti et al. observed in 2013 when 71.4% of patients undergoing PPV with gas had postoperative RVPs, while it was only seen in 22.2% of those with SO.[14]

Several clinical observations also led to a greater understanding of the mechanism of action in retinal displacement. In one particular case, perfluorocarbon liquid (PFCL) was injected into the vitreous cavity at the end of an RRD repair and left in place instead of a gas tamponade. While the PFCL was injected, subretinal fluid (SRF) was displaced inferiorly and temporally, which corresponded to the exact direction of the severe displacement noticed postoperatively (Fig. 45.3).[15] Mathematical models and computational simulations demonstrated that the flow of subretinal fluid under the thin elastic retina occurs from the posterior to the anterior aspect of the posterior segment in a direction related to gravity as the patient elevates their head after surgery (in most cases). This movement of fluid, induced by the buoyant force of the gas

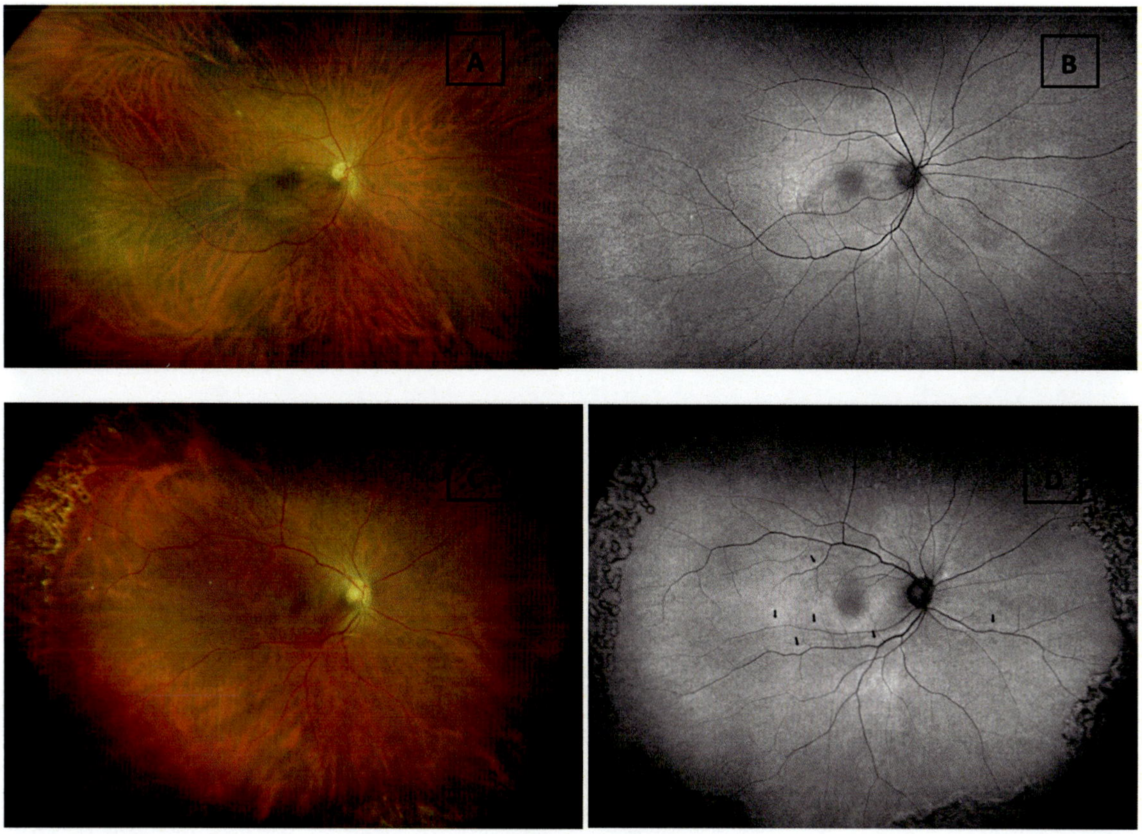

FIGURE 45.2 Ultra-widefield color photo and fundus autofluorescence (FAF) imaging after retinal detachment repair. (A and B) Demonstrating anatomic reattachment and a high-integrity retinal attachment (HIRA) with no retinal displacement. (C and D) Imaging of another patient demonstrating anatomic reattachment with a low-integrity retinal attachment (LIRA) with retinal displacement on FAF; retinal vessel printings are seen (*arrowheads*), indicating that the retina has been displaced compared to its original location. *Reproduced from Brosh K, Francisconi CLM, Qian J, Sabatino F, Juncal VR, Hillier RJ, Chaudhary V, Berger AR, Giavedoni LR, Wong DT, Altomare F, Kadhim MR, Newsom RB, Marafon SB, Muni RH. Retinal displacement following pneumatic retinopexy versus pars plana vitrectomy for rhegmatogenous retinal detachment.* JAMA Ophthalmology. *June 1, 2020; 138(6):652–659. https://doi.org/10.1001/jamaophthalmol.2020.1046. PMID: 32324204; PMCID: PMC7290408.*

tamponade, leads to a stretch of the retina, which then reattaches to the RPE in the displaced/stretched position.[11,12] In the case mentioned above, the PFCL likely maintained the retina in the displaced stretched state while it was reattached.[15]

Therefore, to minimize retinal displacement, the iatrogenic flow of SRF induced by the tamponade must be reduced. One way of doing so can be achieved by using a very small volume of gas at the end of the surgery with no air–fluid exchange. Muni et al. demonstrated a case of minimal gas vitrectomy, in which an intravitreal injection of 0.6 mL pure SF6 was performed after closing the sclerotomies and removing 0.3 mL of aqueous from the anterior chamber. The FAF performed 1 month following surgery demonstrated no evidence of RVPs that would be suggestive of retinal displacement.[16] Understanding how the buoyant force of the gas bubble can physically displace SRF is a new concept. In 2021 Lee et al. performed a clinical experiment where patients with superior bullous retinal detachment were imaged preoperatively and immediately following PnR. They demonstrated how a small 0.6 mL bubble of SF6 could cause an immediate displacement of a large quantity of SRF (Fig. 45.4).[17] After this study, they calculated the total force that the gas tamponade exerts on the retina in the setting of a PPV and found that it is greater than in PnR by a factor of 10.

Finally, another important concept is that internal drainage has a relatively minor impact on retinal displacement. This is supported by the fact that RVPs following PPV are almost always noticed just above the current location of the vessel, suggesting inferior displacement and an effect related to gravity[2,3,8,9] even when draining from a superior break, and secondly, that it can sometimes occur following PnR where there is no drainage.[10] Even if there is a transient displacement during the drainage procedure, the later effect that occurs when the patient changes head position immediately after surgery, in the presence of a large tamponade, likely nullifies any prior displacement.

Although more commonly observed after RRD repair, retinal displacement may also occur following PPV for surgical repair of full-thickness macular hole (FTMH) and epiretinal membrane (ERM). In such cases, the shift of the vessels

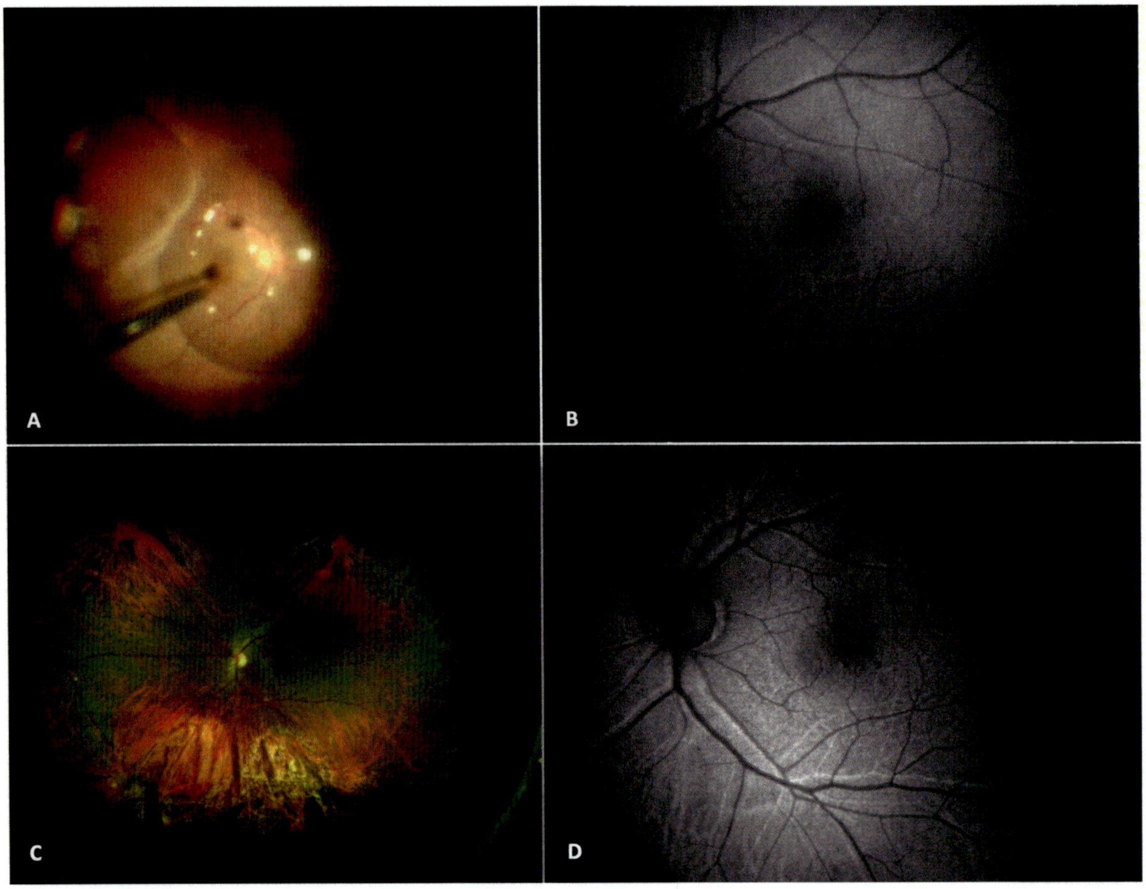

FIGURE 45.3 A 61-year-old male had retinal displacement with perfluorocarbon liquid (PFCL) without air–fluid exchange. (A and B) One week postoperatively, while PFCL was in the eye as a short-term tamponade, the retina was attached but with significant retinal displacement on fundus autofluorescence (FAF) imaging in the direction of subretinal fluid flow. (C) PFCL removal was performed 1 week later. (D) The retinal displacement on FAF did not resolve. *Reproduced from Marafon SB, Juncal VR, Muni RH. Retinal shift with perfluorocarbon liquid without air-fluid exchange.* Ophthalmology. *2020;127(5):598. https://doi.org/10.1016/j.ophtha.2020.01.053.*

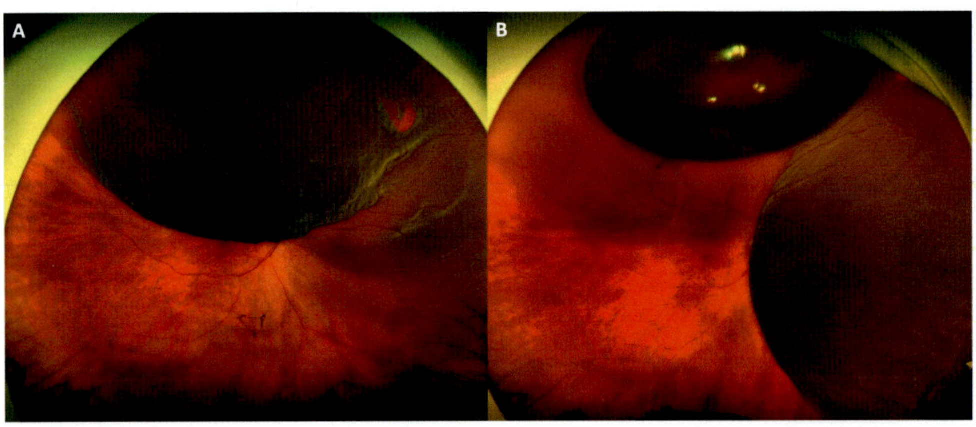

FIGURE 45.4 Ultra-widefield fundus photo of a 59-year-old female before and after pneumatic retinopexy (PnR) demonstrating immediate displacement of subretinal fluid. (A) Pre-PnR fundus photo of the right eye shows right superior bullous rhegmatogenous retinal detachment extending from 10 to 4 o'clock with three retinal breaks superiorly. (B) Fundus photo of the right eye performed immediately, 1–2 min after the bubble injection, shows massive displacement of subretinal fluid inferiorly. *Reproduced from Lee WW, Ramachandran A, Hamli H, Escaf LC, Bansal A, Muni RH. Immediate subretinal fluid displacement from the buoyant force of a small gas bubble in pneumatic retinopexy: insights into the potential mechanism of retinal displacement following retinal detachment repair.* Retinal Cases & Brief Reports. *2021. https://doi.org/10.1097/ICB.0000000000001187.*

generally occurs in opposite directions. While the retina relaxes after ERM removal,[18] it has been shown to contract after FTMH closure. Rodrigues et al. were the first to compare retinal displacement in both etiologies prospectively and were able to show that the retina consistently relaxed in 100% of eyes with ERM and contracted in 75% of eyes with FTMH.[19] Interestingly, other groups also noticed that internal limiting membrane peel for FTMH led to a movement of the fovea toward the optic nerve.[20–22] As RVPs are already present preoperatively,[23,24] these groups used pre- and postoperative comparative measurements between vascular landmarks to assess retinal displacement.[19–22]

Vessel conformation and position are not the only morphological parameters that have been studied. The area and the circularity of the foveal avascular zone (FAZ) using optical coherence tomography angiography (OCTA) has also been a subject of study. However, studies that analyzed only the FAZ area presented conflicting results. Woo et al. and Nam et al. demonstrated FAZ enlargement after successful PPV for RRD repair,[25,26] while Petrou et al. and Lyssek-Boron et al., on the contrary, demonstrated a reduction postoperatively.[27,28] These inconsistent findings are likely due to high intersubject variability, as the FAZ area varies according to central retinal thickness, choroidal thickness, axial length, age, and gender.[29,30] On the contrary, Shiihara et al. demonstrated that FAZ shape-related parameters such as circularity, axial ratio, and roundness have better reproducibility and repeatability as they have smaller intersubject coefficients of variation (0.08, 0.09, and 0.08, respectively) than those related to its area/size (area = 0.35, perimeter = 0.18 and diameter = 0.17), as geometric indices are less affected by individual variations.[29] Therefore, these metrics may be particularly useful following RRD repair, as structural changes to the retina may be more likely to change the shape of the FAZ rather than its size.

In 2022, in a post hoc analysis of the ALIGN trial, Melo et al. found that FAZ circularity, delineated using the superficial capillary plexus segmentation of the OCTA, was reduced in patients who underwent PPV versus PnR for macula off RRDs and in patients that had retinal displacement in the macular region versus those without (Fig. 45.5), while there was no significant difference in the FAZ area between groups.[31] These findings suggest that different imaging modalities of the retinal and choroidal vasculature could also provide an assessment of retinal displacement. Roditi et al. have shown how FAF imaging severely underestimates the presence and extent of retinal displacement. They were able to perfectly overlay OCT infrared images performed before the occurrence of RRD and after the RRD repair with vitrectomy using fixed landmarks in the choroidal vessels, with the assistance of homography. While RVPs were noticed only in 43.7% of the postoperative FAFs, all patients had retinal displacement when assessed with homography (Fig. 45.6).[32] This highlighted that our imaging modalities currently limit us and that it is the extent of retinal displacement rather than the presence/absence of retinal displacement that is most important.

Another morphologic parameter assessed in retinal vessels pre- and postoperatively in patients treated for epiretinal membrane is vessel tortuosity. Miyazawa et al. calculated the distortion of the perifoveal vessels using the actual vessel length and the direct vessel branching point distance in the nasal, temporal, superior, and inferior quadrants of the macula. They hypothesized that the strain on retinal vessels would be decreased after PPV by the release of tractional forces, resulting in an increased vessel length.[33] However, there is a potential bias in the selection of vessels, and we know that ERMs cause an asymmetrical shift of the retina,[34] which can make the results unpredictable and not reproducible. In 2012, Kofod et al. demonstrated that multiple centers of traction existed in eyes with ERM and that the retinal movement did not have a consistent relationship to the fovea,[35] a concept also shown earlier by Gupta et al., implying that variable retinal vessel movement could occur in all macular subfields.[36]

Interestingly, Kofod et al. used choroidal landmarks to validate the OCT overlay technique used to analyze the movement of the retina at different time points in patients with ERM.[35] Although the choroid seems to be relatively stable preoperatively, after PPV, there have been some reports of choroidal changes. Although mean choroidal thickness does not seem to differ between eyes with ERM and controls, it was shown to decrease in the postoperative period of patients who underwent PPV with an internal limiting membrane (ILM) peeling,[37] which was also observed by Rizo et al. in 2018, when they assessed choroidal vascularity index (CVI) pre- and postoperatively (Fig. 45.7). The CVI is the proportion of the luminal area in relation to the cross-sectional choroid area. It was found to decrease following ERM removal when compared to the fellow eyes, suggesting structural vascular changes would occur after surgery.[38] It was hypothesized that a severe ERM could cause retinal traction with consequential distortion of the choroidal vessels, which would then tend to normalize after surgery.[39]

Perfusion vascular changes

Methods for the evaluation of vascular perfusion and blood flow

Several techniques have been used for measuring retinal blood flow. In this section, we will focus our discussion on vascular evaluations performed by fluorescein angiography, laser Doppler ultrasonography, laser speckle flowgraphy, and high-resolution OCTA. We will start with a brief overview of these techniques.

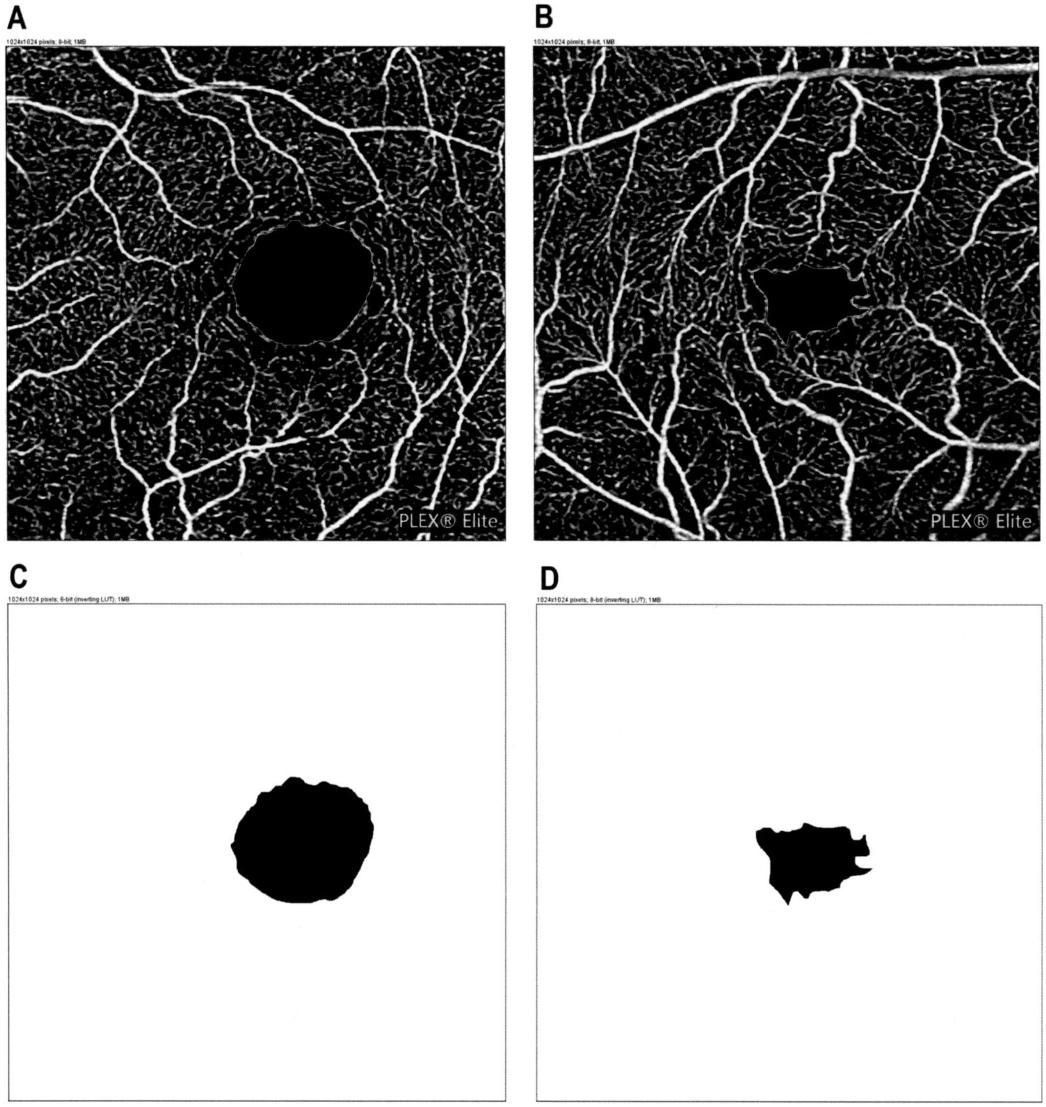

FIGURE 45.5 Illustrative example of the foveal avascular zone (FAZ) delineation on a 3 × 3-mm optical coherence tomography angiography scan, using ImageJ software in two different subjects: one who underwent pneumatic retinopexy (images A and C) and the other who underwent pars plana vitrectomy (images B and D). FAZ area (mm^2), axial ratio, circularity, and roundness were calculated based on the delineation of the inner border of the superficial capillary plexus FAZ. *Reproduced from Melo IM, Francisconi CLM, Marafon SB, Figueiredo NA, Juncal VR, Bhambra N, Ta Kim DT, Brosh K, Berger AR, Giavedoni LR, Wong DT, Altomare F, Chow DR, Hillier RJ, Muni RH. Optical coherence tomography angiography morphology following rhegmatogenous retinal detachment repair: ALIGN post-hoc analysis.* Retina. *December 21, 2022. https://doi.org/10.1097/IAE.0000000000003713. Epub ahead of print. PMID: 36729013.*

Fluorescein angiography is a testing modality in which an orange water-soluble dye is injected into the peripheral circulation reaching the retinal and choroidal vasculature, while sequential fundus photographs are taken with a camera equipped with an excitation and barrier filter. Densitometric analysis of these frames or digitized video can be used to measure the flow of the dye through the vascular bed (hyperfluorescence) and assess zones of capillary nonperfusion (hypofluorescence), and vascular filling defects can be timed and quantified.[40,41] Although it is generally considered a safe diagnostic test, fluorescein angiography is still invasive and might lead to severe and lethal reactions in extremely rare cases (1:222,000).[42]

Laser Doppler imaging is a noninvasive technique in which backscattered light is generated by the disturbances that moving blood cells cause on the laser beam. This reflected light is then converted into an electrical signal that is processed proportionally to the tissue perfusion in the scanned area. Once completed, it generates an image that illustrates vascular perfusion.[43] Similarly, laser speckle flowgraphy is a technique that measures the relative blood flow on the optic nerve head (ONH), through the detection of a pattern produced by the interference of the laser light that is scattered by the movement of red blood cells. The relative blood flow of the ONH is then calculated and expressed as the mean blur rate.[44]

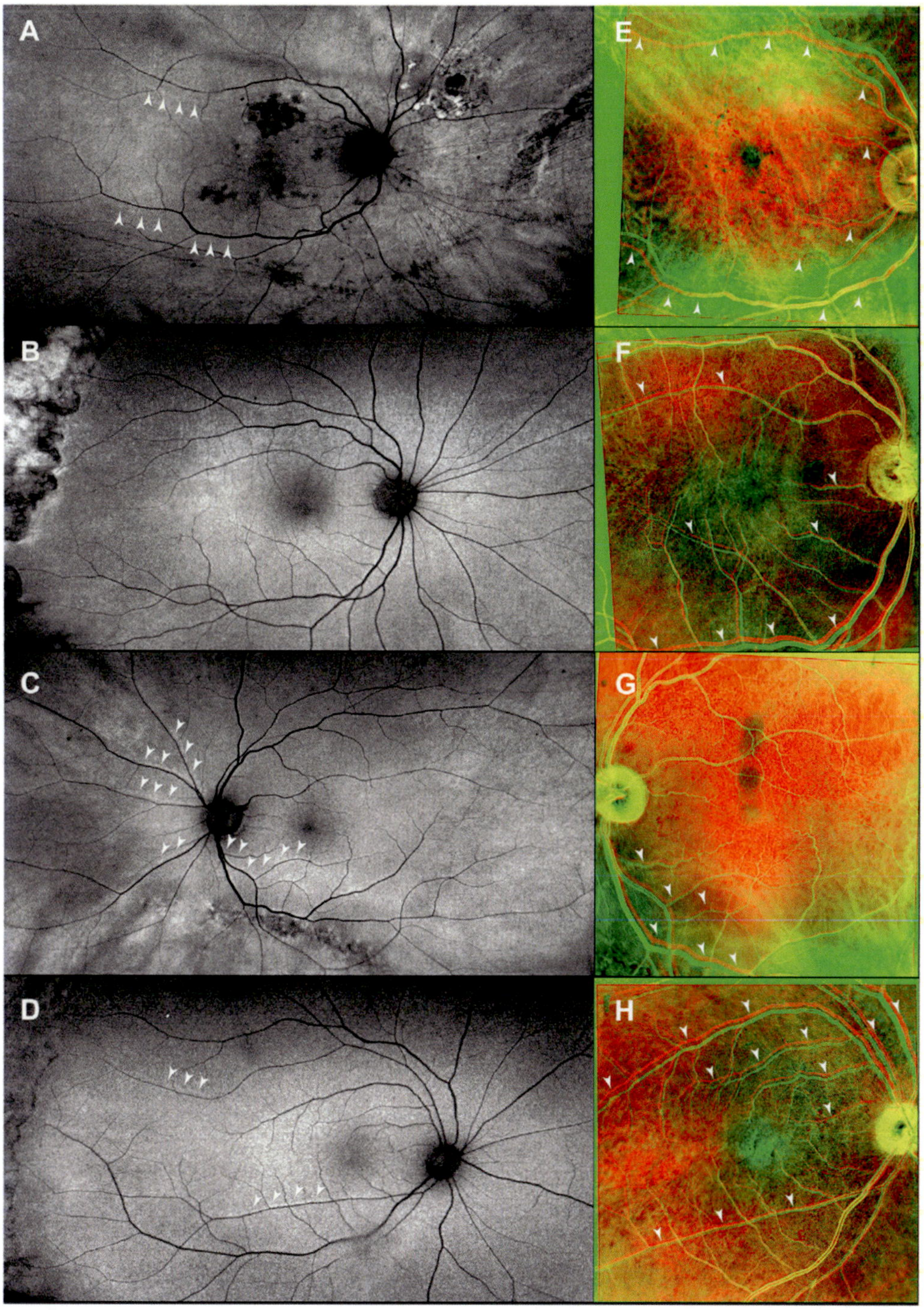

FIGURE 45.6 Ultrawide-field fundus autofluorescence (*left*) compared with infrared overlay images after homography (*right*). (A, C, D) Subtle intermittent retinal vessel printings (*white arrows*) are appreciated corresponding to the location of retinal displacement seen with infrared overlay (E, G, H), but limited only to large vessels. (B) No retinal vessel printings are appreciated, while the infrared overlay (F) shows clear displacement of both temporal arcades (*white arrows*). *Reproduced from Roditi E, Bansal A, Potter MJ, Totah H, Strassman I, Magal I, Hanhart J, Muni RH, Brosh K. Fundus autofluorescence underestimates retinal displacement.* Ophthalmology Retina. *December 26, 2022:S2468-6530(22)00636-4. https://doi.org/10.1016/j.oret.2022.12.010. Epub ahead of print. PMID: 36581080.*

OCTA also shares some similarities, as it is an image based on retinal blood flow using the movement of erythrocytes above a minimum threshold, determined by consecutive B-scans. The OCTA allows a noninvasive assessment of the capillary networks[45] and has been widely used in different retinal diseases. To evaluate retinal perfusion, several

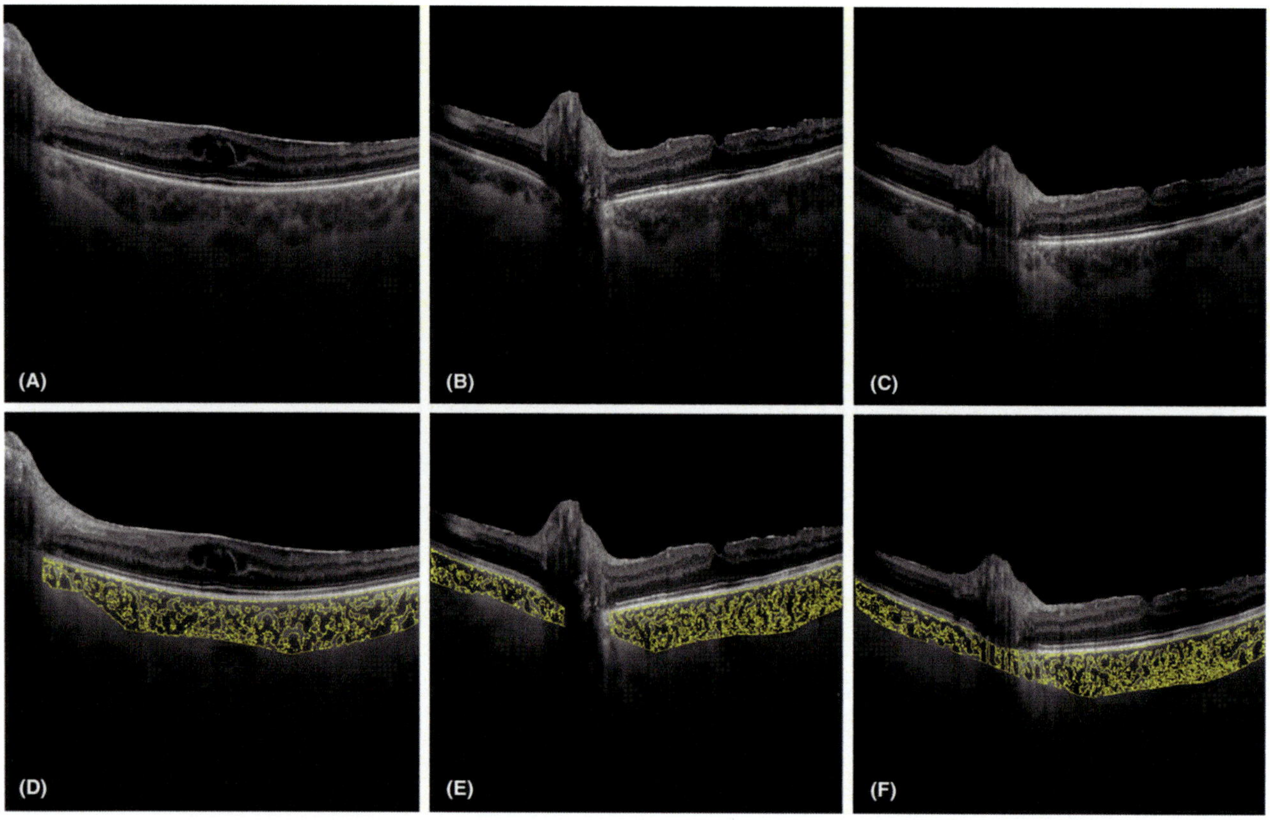

FIGURE 45.7 Swept-source optical coherence tomography scans of patients with epiretinal membrane, preoperatively and postoperatively. (A and D) Preoperative scans with choroidal vasculature (A) being nonbinarized and (D) postbinarization. (B and E) Pre- and postbinarization, respectively, at 1 month after surgery; (C and F) pre- and postbinarized scans at 3 months after surgery. *Reproduced from Rizzo S, Savastano A, Finocchio L, Savastano MC, Khandelwal N, Agrawal R. Choroidal vascularity index changes after vitreomacular surgery.* Acta Ophthalmologica. *December 2018; 96(8):e950–e955. https://doi.org/10.1111/aos.13776. Epub 2018 May 31. PMID: 29855162.*

biomarkers have been proposed,[46] of which vessel length density (VLD) and perfusion density (PD) will be the ones explored in this section. The VLD and PD of the superficial capillary plexus are OCTA-derived quantitative metrics that can be obtained with high levels of repeatability and reproducibility.[47]

Pars plana vitrectomy

Epiretinal membrane and macular hole

Only a few studies have investigated retinal perfusion in eyes following ERM surgery, and although ERM is not primarily a vascular disease, severe inner retinal distortion could be expected to influence retinal perfusion to some extent. However, the reported retinal perfusion changes in eyes with ERM are quite variable. Iwase et al. demonstrated that there were no significant differences in the retinal blood flow at the ONH, assessed by laser speckle flowgraphy, in eyes that underwent PPV for an ERM combined with cataract surgery versus those who had only cataract surgery, which indicates that PPV alone is unlikely to influence postoperative blood flow at the ONH.[44] However, when assessing retinal microvasculature with fluorescein angiography at an imaging angle of 20 degrees, early images of the macula demonstrated a significant improvement in blood flow velocity postoperatively.[48,49] It has also been reported that tortuous perifoveal capillary vessels are recanalized after surgery, and microaneurysms decrease in size.[49] Similarly, when Mastropasqua et al. studied the early retinal vascular changes using swept-source (SS) OCTA after PPV for ERM, parafoveal VLD and PD increased at 6 months postoperatively, likely due to a reopening of suboccluded vessels secondary to traction release. However, the overall PD did not show any significant difference compared to preoperative values, which was hypothesized to be due to a concurrent decrease in central vessel crowding.[50] In 2021, D'Aloisio et al. compared early vascular changes after PPV for ERM and macular hole (MH) and found that VLD increased significantly only in the MH group, suggesting that the duration of follow-up (3 months) was likely insufficient for postoperative vessel structure restoration in this case.[51] Told et al. also observed an increase in VLD postoperatively in eyes treated for ERM, which was particularly important in patients with a less severe baseline ERM stage. However, they advised careful interpretations of VLD measurements since image processing varies greatly, which in some cases may not be fully understood by the user.[52]

Although inner retinal layer alterations have gained more attention in eyes with ERM, choroidal involvement has also been reported. Choriocapillaris imaging with SS-OCT has been investigated in eyes with FTMH, and the results suggest that a variation in its flow may be involved in vitreomacular interface diseases.[53] Bacherini et al. found a significant negative correlation between the presenting ERM stage,[54] the choroidal perfusion and vessel density, and the choriocapillaris perfusion density. However, the same group found longitudinal changes in the choroidal plexus postoperatively. Specifically, they noted a significant decrease in the VLD and PD of the choriocapillaris from baseline to the first month postoperatively, followed by an increase in PD at the 6-month follow-up.[55] Similarly, Li et al. observed a reversible reduction of choriocapillaris circulation within the macular area in the subjects who underwent PPV with ILM peeling. The proposed explanation was that after PPV, retinal arteriolar saturation increases due to vitreous removal, which would then reduce the retinal oxygen consumption with a temporary decrease in choriocapillaris VLD and PD.[56] In another study, the choriocapillaris perfusion was noted to increase 3 months after surgery, while that of the Sattler layer significantly decreased, suggesting there could be a shift of blood flow from the deeper to more superficial choroidal vessels.[57]

Retinal detachment

Pars plana vitrectomy

Previous studies have demonstrated that vascular disturbances are present at baseline in RRD eyes when measured at the major retinal arteries, even if the macula is not involved. Eshita et al. detected a reduction in the blood flow using the laser Doppler technique and found a significant correlation with the extent of the detachment. In addition, when comparing the blood-flow rates in arteries supplying the detached versus the attached retina, in the same patient, the difference was approximately 30% lower in the detached area.[58] In a previous study by Satoh, the retinal mean circulation time in the detached area, measured by video fluorescence angiography, was longer when compared to the attached retina, and both were longer than those of normal subjects, suggesting that circulatory disturbances might occur in the entire retina.[59] While some studies assessing the central retinal artery (CRA) and the ophthalmic artery, found normal ocular blood flow in RRD patients,[60,61] Roldán-Pallarés et al. detected a lower CRA peak systolic velocity and end-diastolic velocity in retinal detachment eyes compared to normal controls.[62]

Concerning postoperative vascular changes, there have been only a few studies examining the effects of PPV on the ocular blood flow in eyes with RRD. Over a long period of evaluation, Sato et al. assessed tissue blood flow at the macula and the neuroretinal rim region more than 6 months after the RRD repair (mean of 18.9 ± 16.9 months) using a scanning laser Doppler flowmetry. They reported that the ocular microcirculation at the macula and optic nerve head was not affected 6 months after PPV for RRD. However, the use of gas tamponade was correlated with a reduced blood flow rate at the neuroretinal disk rim, which could be a subclinical adverse effect of intraocular gas.[63] On the contrary, Iwase et al. found an increase in the blood flow velocity measured at the ONH, with increasing time following PPV. They hypothesized that the preoperative measurement of blood flow would be disturbed by the tortuous vessels running vertically in a bullous detachment. Once the retina was reattached after the PPV, the blood flow rate would then recover to the normal state.[44] When evaluated with OCTA, a significant increase in the retinal superficial and deep capillary plexus flow density as well as in the choriocapillaris was observed following successful PPV for RRD repair during an evaluation period of 3 months. In this study, macular perfusion parameters were assessed in the parafoveal region and automatically evaluated by the AngioVue OCTA software.[64]

OCTA vessel density is another parameter that might reliably relate to perfusion changes after retinal detachment repair. In patients with recent onset macula-on RRD undergoing vitreoretinal surgery, no changes in macular vessel density were observed with OCTA.[25,51,65] On the contrary, Gironi et al. assessed long-term vascular changes in the macula after RRD repair and found a reduction in vessel density in operated eyes (Fig. 45.8), irrespective of surgical technique or preoperative extent of the detachment.[66] Bonfiglio et al. found reduced parafoveal superficial vascular density as well as reduced foveal and parafoveal deep vascular density values in patients who underwent PPV for macula off RRD.[67] In a retrospective study measuring VD after 2–10 years of the RRD repair, a significant reduction of the macular VD could still be detected by OCTA even several years after the surgery. The deep vasculature seemed to be more affected in the early postoperative period, and the superficial vessels and the extent of the nonflow area are more involved after a longer postoperative time.[68] When evaluating different tamponades, Roohipoor et al. demonstrated decreased retinal VD in macula off RRDs with SO when compared to healthy controls.[69] Similarly, Lee et al. found a reduced VD, especially in the deep capillary plexus, in patients who had more than 3 months of follow-up after SO removal.[70]

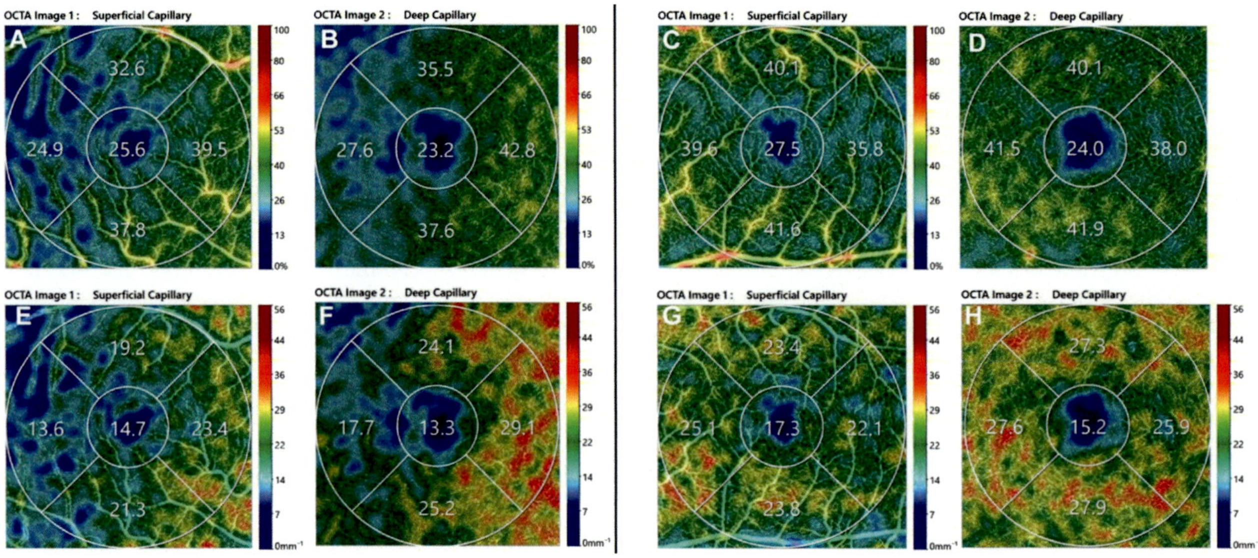

FIGURE 45.8 Optical coherence tomography angiography 3 × 3 images of postoperative perfusion parameters in macula-on (*left*) and macula-off (*right*) rhegmatogenous retinal detachment eye. (*Above*) SCP VD of macula-on (A) and macula-off (C) and DCP VLD of macula-on (B) and macula-off (D); (*below*) DCP VD of macula-on (E) and macula-off (G) and DCP VLD of macula-on (F) and macula-off (H). *DCP*, deep capillary plexus; *SCP*, superficial capillary plexus; *VD*, vessel density; *VLD*, vessel length density. *Reproduced from Gironi M, D'Aloisio R, Verdina T, Vivarelli C, Leonelli R, Kaleci S, Toto L, Mastropasqua R. Long-term macular vascular changes after primary rhegmatogenous retinal detachment surgery resolved with different tamponade or different surgical techniques.* Life (Basel). *September 30, 2022; 12(10):1525. https://doi.org/10.3390/life12101525. PMID: 36294960; PMCID: PMC9605421.*

Scleral buckle

Ocular circulatory abnormalities have been long associated with buckling and encircling procedures. Ischemic syndromes of the anterior segment following scleral buckle are rare but severe vascular complications can occur.[71] In the sixties, Boniuk and Zimmerman noted necrosis of the iris and ciliary body due to iatrogenic damage to the blood supply in 21 out of 150 eyes enucleated following retinal detachment operations.[72] It was hypothesized that these vascular changes could be attributed to (1) the disinsertion of the rectus muscles, with consequential damage of the anterior ciliary vessels, (2) the partial obstruction of long posterior ciliary arteries due to excessive diathermy, and (3) the constricting effect of an encircling band, with or without scleral dissection.[72,73] However, Robertson showed years later that anterior segment ischemia could also develop when using only a segmental buckle.[74] Besides a possible partial occlusion of the long ciliary artery or vortex system by the radial buckle, he proposed other contributing factors, such as long-standing cardiovascular disease, a recognized cause of spontaneous anterior segment ischemia.[74–76] Sickle cell hemoglobinopathy has also been historically associated with high rates of anterior segment ischemic events following buckling procedures.[73] However, more modern series did not demonstrate an increased risk,[77] which may likely be explained by changes in surgical technique and preoperative preventive measures.[78]

However, not all vascular changes following the scleral buckle are dramatic. Even when not clinically evident, it has been shown that retinal and choroidal blood flow is frequently compromised in buckling procedures and that there is a direct relationship between the extent of the band and the postoperative circulatory changes.[79] In 1992 Ogasawara et al., using the laser Doppler technique, found that the blood flow rates in retinal arteries of eyes that had undergone uncomplicated scleral buckling and encircling surgery were on average 50% lower than the controls.[80] Additionally, the intraocular pressure was proven to provide indirect evidence that there is an underlying deficiency in the anterior uveal perfusion in such cases. The explanation is that altered perfusion of the anterior choroid, ciliary body, and ciliary processes decreases aqueous humor formation, resulting in chronically reduced intraocular pressure.[81] These findings suggest that subclinical ischemia may be more frequent than what has been reported.

When specifically assessing choroidal vascular changes after buckling procedures, the literature shows somewhat contradictory results. When evaluating choroidal thickness (CT), Iwase et al. did not find a difference when measuring subfoveal CT after postoperative week 4, although it temporarily increased in the first week.[82] This was similar to Kimura et al., who reported a temporary thickening of the CT which returned to normal values after 3 months. However, both groups studied changes after segmental buckling.[83] On the contrary, Odrobina et al. in 2013 found that subfoveal CT remained thickened for much longer when using an encircling band,[84] which suggests that factors such as the extent, size, and width of the buckle and postoperative inflammation likely influence long-term CT changes,[83] a finding that other groups corroborated.[85,86] In 2021, Bernabei assessed other OCT choroidal biomarkers, such as vessel luminal area (LA) and stromal area (SA), and the CVI, which corresponds to the proportion of the LA in relation to the cross-sectional choroidal area. They found increased LA in operated eyes,[86] which was once hypothesized to represent blood flow stasis secondary to impaired circulation.[87] However, CVI was not different between groups, suggesting that either LA and SA increased in the same proportion or there was an adaptation process of the stromal tissue after the placement of the encircling band. When directly assessing choroidal ocular blood with either bidirectional laser Doppler, laser speckle flowmetry, or color Doppler imaging, impaired chorioretinal circulation was consistently found in most studies,[58,79,82,88,89] especially when evaluating the buckled side,[90] although it seems to return to baseline rates after three to 6 months.[91,92]

Although ocular blood flow can be decreased after scleral buckling procedures, patients usually do not develop frank vascular occlusions following surgery. Nevertheless, events such as retinal vein and artery occlusion and anterior ischemic optic neuropathy have been reported.[71] Several triggering mechanisms could be involved in such cases. Firstly, anesthesia is a known risk factor. This could be related to direct causes such as an iatrogenic lesion after retrobulbar injection, due to retrobulbar hemorrhage, optic nerve sheath hematoma or retrograde embolic propagation,[93–95] or indirectly when the anesthetic agent induces vascular spasm.[93,96] Another contributing factor is intraocular pressure elevation that may happen due to ciliary body edema and choroidal swelling caused by extensive cryotherapy in scleral buckling.[71] And finally, as already mentioned, systemic diseases such as hypertension, arrhythmia, arteriosclerosis, diabetes, and other blood diseases play an important role in the occurrence of occlusive vascular events,[97] as a severely damaged vascular system has impaired autoregulation of the retinal blood flow.

Conclusions

Although most vascular changes following vitreoretinal surgery may be mild, as assessed with multimodal imaging, they can have a substantial impact on functional outcomes such as reduced visual acuity, metamorphopsia, and aniseikonia. Retinal displacement has been extensively studied over the past 15 years, and now it is considered an important anatomic biomarker following RRD repair. It is currently detected through the observation of vascular abnormalities (mainly retinal vessel printings) of first- or second-order retinal vessels on fundus autofluorescence. However, newer techniques such as OCT homography and analysis of OCT-angiography have substantial promise to assist with the detection of retinal displacement. Regarding vascular perfusion, fortunately, our current surgical techniques seem to induce mostly minimal or transient changes. Nevertheless, severe vascular events such as anterior segment ischemia and retinal vascular occlusions can still occur, especially if patients have predisposing clinical comorbidities.

References

1. Muni RH, Lee WW, Bansal A, Ramachandran A, Hillier RJ. A paradigm shift in retinal detachment repair: the concept of integrity. *Progress in Retinal and Eye Research*. November 2022;91:101079. https://doi.org/10.1016/j.preteyeres.2022.101079. Epub 2022 Oct 14. PMID: 36253216.
2. Shiragami C, Shiraga F, Yamaji H, et al. Unintentional displacement of the retina after standard vitrectomy for rhegmatogenous retinal detachment. *Ophthalmology*. January 2010;117(1):86–92.e1. https://doi.org/10.1016/j.ophtha.2009.06.025. Epub 2009 Nov 6. PMID: 19896189.
3. dell'Omo R, Mura M, Lesniak Oberstein SY, Bijl HTH. Early simultaneous fundus autofluorescence and optical coherence tomography features after pars plana vitrectomy for primary hematogenous retinal detachment. *Retina*. 2012;32(4):719e28. https://doi.org/10.1097/IAE.0b013e31822c293e.
4. Brosh K, Roditi E, Muni RH. Retinal vessel printings on fundus autofluorescence imaging represent retinal displacement: proof of prior hypothesis. *American Journal of Ophthalmology*. June 2022;238:e3–e4. https://doi.org/10.1016/j.ajo.2022.01.024. Epub 2022 Feb 14. PMID: 35176253.
5. Lee E, Williamson TH, Hysi P, et al. Macular displacement following rhegmatogenous retinal detachment repair. *British Journal of Ophthalmology*. October 2013;97(10):1297–1302. https://doi.org/10.1136/bjophthalmol-2013-303637. Epub 2013 Jul 18. PMID: 23868959.
6. Shiragami C, Fukuda K, Yamaji H, Morita M, Shiraga F. A method to decrease the frequency of unintentional slippage after vitrectomy for rhegmatogenous retinal detachment. *Retina*. 2015;35(4):758–763. https://doi.org/10.1097/IAE.0000000000000383.
7. dell'Omo R, Scupola A, Viggiano D, et al. Incidence and factors influencing retinal displacement in eyes treated for rhegmatogenous retinal detachment with vitrectomy and gas or silicone oil. *Investigative for Ophthalmology & Visual Science*. 2017;58(6):BIO191–BIO199. https://doi.org/10.1167/iovs.17-21466.
8. Guber J, Schawkat M, Lang C, Scholl HPN, Valmaggia C. How to prevent retinal shift after rhegmatogenous retinal detachment repair: a prospective, randomized study. *Ophthalmology Retina*. 2019;3(5):417–421. https://doi.org/10.1016/j.oret.2019.01.010.
9. Brosh K, Francisconi CLM, Qian J, et al. Retinal displacement following pneumatic retinopexy vs pars plana vitrectomy for rhegmatogenous retinal detachment. *JAMA Ophthalmology*. June 1, 2020;138(6):652–659. https://doi.org/10.1001/jamaophthalmol.2020.1046. PMID: 32324204; PMCID: PMC7290408.

10. Francisconi CLM, Marafon SB, Figueiredo NA, et al. Retinal displacement after pneumatic retinopexy versus vitrectomy for rhegmatogenous retinal detachment (ALIGN). *Ophthalmology*. April 2022;129(4):458–461. https://doi.org/10.1016/j.ophtha.2021.12.007. Epub 2021 Dec 17. PMID: 34929256.
11. Farahvash A, Marafon SB, Juncal VR, Figueiredo N, Ramachandran A, Muni RH. Understanding the mechanism of retinal displacement following rhegmatogenous retinal detachment repair: a computer simulation model. *Acta Ophthalmologica*. 2022;100(5):e1163–e1171. https://doi.org/10.1111/aos.15050. PMID: 34694071.
12. Farahvash A, Marafon SB, Juncal VR, Figueiredo N, Ramachandran A, Muni RH. Impact of tamponade agent on retinal displacement following pars plana vitrectomy for rhegmatogenous retinal detachment repair: a computer simulation model. *Acta Ophthalmologica*. 2022;100(7):e1470–e1478. https://doi.org/10.1111/aos.15118. PMID: 35257484.
13. Casswell EJ, Heeren TFC, Lee E, Khabra K, Yorston D, Charteris DG. Postretinal detachment retinal displacement: how best to detect it? *Ophthalmologica*. 2019;243(4):280–287. https://doi.org/10.1159/000504713. Published online 2019.
14. Codenotti M, Fogliato G, Iuliano L, et al. Influence of intraocular tamponade on unintentional retinal displacement after vitrectomy for rhegmatogenous retinal detachment. *Retina*. 2013;33(2):349–355. https://doi.org/10.1097/IAE.0b013e318263d180.
15. Marafon SB, Juncal VR, Muni RH. Retinal shift with perfluorocarbon liquid without air-fluid exchange. *Ophthalmology*. 2020;127(5):598. https://doi.org/10.1016/j.ophtha.2020.01.053.
16. Muni RH, Felfeli T, Figueiredo N, Marafon SB, Escaf LS, Juncal VR. Minimal gas vitrectomy technique for reducing risk of retinal displacement following rhegmatogenous retinal detachment repair. *Retin Cases & Brief Reports*. 2020b;16.
17. Lee WW, Ramachandran A, Hamli H, Escaf LC, Bansal A, Muni RH. Immediate subretinal fluid displacement from the buoyant force of a small gas bubble in pneumatic retinopexy: insights into the potential mechanism of retinal displacement after retinal detachment repair. *Retin Cases & Brief Reports*. 2023;17(3):251–255. https://doi.org/10.1097/ICB.0000000000001187. PMID: 34468443.
18. Yoon YS, Woo JM, Woo JE, Min JK. Superficial foveal avascular zone area changes before and after idiopathic epiretinal membrane surgery. *International Journal of Ophthalmology*. October 18, 2018;11(10):1711–1715. https://doi.org/10.18240/ijo.2018.10.21. PMID: 30364246; PMCID: PMC6192960.
19. Rodrigues IA, Lee EJ, Williamson TH. Measurement of retinal displacement and metamorphopsia after epiretinal membrane or macular hole surgery. *Retina*. April 2016;36(4):695–702. https://doi.org/10.1097/IAE.0000000000000768. PMID: 26398687.
20. Liu J, Hu ZZ, Zheng XH, et al. Displacement of the retina after idiopathic macular hole surgery with different internal limiting membrane peeling patterns. *International Journal of Ophthalmology*. September 18, 2021;14(9):1408–1412. https://doi.org/10.18240/ijo.2021.09.17. PMID: 34540618; PMCID: PMC8403843.
21. Ishida M, Ichikawa Y, Higashida R, Tsutsumi Y, Ishikawa A, Imamura Y. Retinal displacement toward optic disc after internal limiting membrane peeling for idiopathic macular hole. *American Journal of Ophthalmology*. 2014;157(5):971–977.
22. Kawano K, Ito Y, Kondo M, et al. Displacement of foveal area toward optic disc after macular hole surgery with internal limiting membrane peeling. *Eye*. 2013;27(7):871–877.
23. Nitta E, Shiraga F, Shiragami C. Displacement of the retina and its recovery after vitrectomy in idiopathic epiretinal membrane. *American Journal of Ophthalmology*. 2013;155:1014–1020.
24. dell'Omo R, Cifariello F, dell'Omo E, et al. Influence of retinal vessel printings on metamorphopsia and retinal Architectural abnormalities in eyes with idiopathic macular epiretinal membrane. *Investigative Ophthalmological & Visual Science*. 2013;54(12):7803–7811. https://doi.org/10.1167/iovs.13-12817.
25. Woo JM, Yoon YS, Woo JE, Min JK. Foveal avascular zone area changes analyzed using OCT angiography after successful rhegmatogenous retinal detachment repair. *Current Eye Research*. May 2018;43(5):674–678. https://doi.org/10.1080/02713683.2018.1437922. Epub 2018 Feb 16. PMID: 29451996.
26. Nam SH, Kim K, Kim ES, Kim DG, Yu SY. Longitudinal microvascular changes on optical coherence tomographic angiography after macula-off rhegmatogenous retinal detachment repair surgery. *Ophthalmologica*. 2021;244(1):34–41. https://doi.org/10.1159/000508464. Epub 2020 May 7. PMID: 32380497.
27. Petrou Sr P, Angelidis CD, Andreanos K, et al. Reduction of foveal avascular zone after vitrectomy demonstrated by optical coherence tomography angiography. *Cureus*. March 7, 2021;13(3):e13757. https://doi.org/10.7759/cureus.13757. PMID: 33717769; PMCID: PMC7943050.
28. Lyssek-Boroń A, Wylegała A, Krysik K, et al. Assessment of vascular changes in patients after pars plana vitrectomy surgery due to macula-off rhegmatogenous retinal detachment. *Journal of Clinical Medicine*. October 28, 2021;10(21):5054. https://doi.org/10.3390/jcm10215054.
29. Shiihara H, Terasaki H, Sonoda S, Kakiuchi N, Sakamoto T. Evaluation of shape of foveal avascular zone by optical coherence tomography angiography in eyes with branch retinal vein occlusion. *Journal of VitreoRetinal Diseases*. 2018;2(3):138–145.
30. Iafe NA, Phasukkijwatana N, Chen X, Sarraf D. Retinal capillary density and foveal avascular zone area are age-dependent: quantitative analysis using optical coherence tomography angiography. *Investigative Ophthalmology & Visual Science*. 2016;57:5780–5787.
31. Melo IM, Francisconi CLM, Marafon SB, et al. Optical coherence tomography angiography morphology following rhegmatogenous retinal detachment repair: ALIGN post-hoc analysis. *Retina*. December 21, 2022. https://doi.org/10.1097/IAE.0000000000003713. Epub ahead of print. PMID: 36729013.
32. Roditi E, Bansal A, Potter MJ, et al. Fundus autofluorescence underestimates retinal displacement. *Ophthalmology Retina*. December 26, 2022;7. https://doi.org/10.1016/j.oret.2022.12.010. Epub ahead of print. PMID: 36581080.
33. Miyazawa K, Sakimoto S, Kanai M, et al. Vascular tortuosity analysis in eyes with epiretinal membrane imaged by optical coherence tomography angiography. *BMC Ophthalmology*. 2022;22:198. https://doi.org/10.1186/s12886-022-02420-z.

34. Momota A, Iwase T, Akahori T, et al. Association between displacement and thickness of macula after vitrectomy in eyes with epiretinal membrane. *Science Reports*. 2020;10:13227.
35. Kofod M, la Cour M. Quantification of retinal tangential movement in epiretinal membranes. *Ophthalmology*. 2012;119(9):1886–1891.
36. Gupta P, Sadun AA, Sebag J. Multifocal retinal contraction in macular pucker analyzed by combined optical coherence tomography/scanning laser ophthalmoscopy. *Retina*. March 2008;28(3):447–452. https://doi.org/10.1097/IAE.0b013e318160a754. PMID: 18327137.
37. Ruan K, Zhang Y, Cheng D, et al. Short-term postoperative changes in the choroidal vascularity index in patients with a unilateral epiretinal membrane. *Research Square*. 2022;23(1). https://doi.org/10.21203/rs.3.rs-2195529/v1.
38. Rizzo S, Savastano A, Finocchio L, Savastano MC, Khandelwal N, Agrawal R. Choroidal vascularity index changes after vitreomacular surgery. *Acta Ophthalmologica*. December 2018;96(8):e950–e955. https://doi.org/10.1111/aos.13776. Epub 2018 May 31. PMID: 29855162.
39. Michalewska Z, Michalewski J, Adelman RA, Zawiślak E, Nawrocki J. Choroidal thickness measured with swept source optical coherence tomography before and after vitrectomy with internal limiting membrane peeling for idiopathic epiretinal membranes. *Retina*. March 2015;35(3):487–491. https://doi.org/10.1097/IAE.0000000000000350. PMID: 25322467.
40. Ruia S, Tripathy K. Fluorescein Angiography. Updated 2022 Nov 26. In: *StatPearls*. Treasure Island (FL): StatPearls Publishing; 2022 Jan-. Available from: https://www.ncbi.nlm.nih.gov/books/NBK576378/.
41. Tomic L, Mäepea O, Sperber GO, Alm A. Comparison of retinal transit times and retinal blood flow: a study in monkeys. *Investigative Ophthalmology & Visual Science*. 2001;42(3):752–755.
42. Yannuzzi LA, Rohrer KT, Tindel LJ, et al. Fluorescein angiography complication survey. *Ophthalmology*. 1986;93(5):611–617. ISSN 0161-6420.
43. Wardell K, Jakobsson A, Nilsson GE. Laser Doppler perfusion imaging by dynamic light scattering. *IEEE (Inst Electr Electron Eng) Transaction on Biomedical Engineering*. 1993;40(4):309–316. PMID8375866. S2CID23260143 10.1109/10.222322.
44. Iwase T, Kobayashi M, Yamamoto K, Yanagida K, Ra E, Terasaki H. Changes in blood flow on optic nerve head after vitrectomy for rhegmatogenous retinal detachment. *Investigative Ophthalmology & Visual Sciene*. 2016;57(14):6223–6233. https://doi.org/10.1167/iovs.16-20577.
45. de Carlo TE, Romano A, Waheed NK, et al. A review of optical coherence tomography angiography (OCTA). *International Journal of Retina and Vitreous*. 2015;1:5. https://doi.org/10.1186/s40942-015-0005-8.
46. Kalra G, Zarranz-Ventura J, Chahal R, Bernal-Morales C, Lupidi M, Chhablani J. Optical coherence tomography (OCT) angiolytics: a review of OCT angiography quantitative biomarkers. *Survey of Ophthalmology*. 2022 July-August;67(4):1118–1134. https://doi.org/10.1016/j.survophthal.2021.11.002. Epub 2021 Nov 5. PMID: 34748794.
47. Lei J, Durbin MK, Shi Y, et al. Repeatability and reproducibility of superficial macular retinal vessel density measurements using optical coherence tomography angiography en face images. *JAMA Ophthalmology*. October 1, 2017;135(10):1092–1098. https://doi.org/10.1001/jamaophthalmol.2017.3431. PMID: 28910435; PMCID: PMC5710485.
48. Kadonosono K, Itoh ND, Nomura EMD, Ohno S. Capillary blood flow velocity in patients with idiopathic epiretinal membranes. *Retina*. 1999;19(6):536–539.
49. Yagi T, Sakata K, Funatsu H, Hori S. Evaluation of perifoveal capillary blood flow velocity before and after vitreous surgery for epiretinal membrane. *Graefes Archives for Clinical and Experimental Ophthalmology*. 2012;250:459–460. https://doi.org/10.1007/s00417-011-1618-8LETTERTOTHEEDITOR.
50. Mastropasqua R, D'Aloisio R, Viggiano P, et al. Early retinal flow changes after vitreoretinal surgery in idiopathic epiretinal membrane using swept source optical coherence tomography angiography. *Journal of Clinical Medicine*. 2019;8(12):2067. https://doi.org/10.3390/jcm8122067.
51. D'Aloisio R, Carpineto P, Aharrh-Gnama A, et al. Early vascular and functional changes after vitreoretinal surgery: a comparison between the macular hole and epiretinal membrane. *Diagnostics*. June 3, 2021;11(6):1031. https://doi.org/10.3390/diagnostics11061031. PMID: 34205224; PMCID: PMC8226563.
52. Told R, Georgopoulos M, Reiter GS, Wassermann L, Aliyeva L, Baumann L, Abela-Formanek C, Pollreisz A, Schmidt-Erfurth U, Sacu S. Intraretinal microvascular changes after ERM and ILM peeling using SSOCTA. *PLoS One*. 2020;15(12). https://doi.org/10.1371/journal.pone.0242667. Published: December 1, 2020.
53. Ahn J, Yoo G, Kim JT, Kim SW, Oh J. Choriocapillaris layer imaging with swept-source optical coherence tomography angiography in lamellar and full-thickness macular hole. *Graefes Archives for Clinial and Experimental Ophthalmology*. January 2018;256(1):11–21. https://doi.org/10.1007/s00417-017-3814-7. Epub 2017 Oct 14. PMID: 29032413.
54. Govetto A, Lalane 3rd RA, Sarraf D, Figueroa MS, Hubschman JP. Insights into epiretinal membranes: presence of ectopic inner foveal layers and a new optical coherence tomography staging scheme. *American Journal of Ophthalmology*. March 2017;175:99–113. https://doi.org/10.1016/j.ajo.2016.12.006. Epub 2016 Dec 18. PMID: 27993592.
55. Bacherini D, Dragotto F, Caporossi T, et al. The role of OCT angiography in the assessment of epiretinal macular membrane. *Journal of Ophthalmology*. 2021;2021:1–9. https://doi.org/10.1155/2021/8866407.
56. Li Z, Zhang J, Lin T, Peng W, Lu L, Hu J. Macular vascular circulation and retinal oxygen saturation changes for idiopathic macular epiretinal membrane after vitrectomy. *Acta Ophthalmologica*. 2019;97:296–302. https://doi.org/10.1111/aos.14066.
57. Rommel F, Brinkmann MP, Sochurek JAM, Prasuhn M, Grisanti S, Ranjbar M. Ocular blood flow changes impact visual acuity gain after surgical treatment for idiopathic epiretinal membrane. *Journal of Clinical Medicine*. June 7, 2020;9(6):1768. https://doi.org/10.3390/jcm9061768. PMID: 32517301; PMCID: PMC7356826.
58. Eshita T, Shinoda K, Kimura I, et al. Retinal blood flow in the macular area before and after scleral buckling procedures for rhegmatogenous retinal detachment without macular involvement. *Japanese Journal of Ophthalmology*. July-August, 2004;48(4):358–363. https://doi.org/10.1007/s10384-004-0096-5. PMID: 15295662.

59. Satoh Y. Retinal circulation in rhegmatogenous retinal detachment demonstrated by videofluorescence angiography and image analysis. I.The condition of retinal circulation before retinal detachment surgery. *Nippon Ganka Gakkai Zasshi*. 1989;93:1002–1008.
60. Hanioglu-Kargi S, Yazar Z, Ziraman I, Gursel E. Effects of scleral buckling on the retrobulbar haemodynamic changes. *Eye*. 1999;14:165–171.
61. Regillo CD, Sergott RC, Brown GC. Successful scleral buckling procedures decrease central retinal artery blood flow velocity. *Ophthalmology*. 1993;100:1044–1049.
62. Roldán-Pallarés M, Musa AS, Hernández-Montero J, Bravo-Llatas Sc C, Fernández-Durango R. Retinal detachment and proliferative vitreoretinopathy: central retinal artery blood velocities, intraocular pressure, and endothelin 1. *Retina*. September 2013;33(8):1528–1539. https://doi.org/10.1097/IAE.0b013e318285cbef. PMID: 23974951.
63. Sato EA, Shinoda K, Kimura I, Ohtake Y, Inoue M. Microcirculation in eyes after rhegmatogenous retinal detachment surgery. *Current Eye Research*. 2007;32(9):773–779. https://doi.org/10.1080/02713680701531108.
64. Wang H, Xu X, Sun X, Ma Y, Sun T. Macular perfusion changes assessed with optical coherence tomography angiography after vitrectomy for rhegmatogenous retinal detachment. *Graefes Archives for Clinical and Experimental Ophthalmology*. April 2019;257(4):733–740. https://doi.org/10.1007/s00417-019-04273-7. Epub 2019 Feb 22. PMID: 30796563.
65. Yoshikawa Y, Shoji T, Kanno J, et al. Evaluation of microvascular changes in the macular area of eyes with rhegmatogenous retinal detachment without macular involvement using swept-source optical coherence tomography angiography. *Clinical Ophthalmology*. 2018;12:2059–2067.
66. Gironi M, D'Aloisio R, Verdina T, et al. Long-term macular vascular changes after primary rhegmatogenous retinal detachment surgery resolved with different tamponade or different surgical techniques. *Life*. September 30, 2022;12(10):1525. https://doi.org/10.3390/life12101525. PMID: 36294960; PMCID: PMC9605421.
67. Bonfiglio V, Ortisi E, Scollo D, et al. Vascular changes after vitrectomy for rhegmatogenous retinal detachment: optical coherence tomography angiography study. *Acta Ophthalmologica*. August 2020;98(5):e563–e569. https://doi.org/10.1111/aos.14315. Epub 2019 Nov 26. PMID: 31773840.
68. Resch MD, Balogh A, Lászik G, Nagy ZZ, Papp A. Association between retinal vessel density and postoperative time after primary repair of rhegmatogenous retinal detachment. *PLoS One*. October 1, 2021;16(10):e0258126. https://doi.org/10.1371/journal.pone.0258126. PMID: 34597349; PMCID: PMC8486093.
69. Roohipoor R, Tayebi F, Riazi-Esfahani H, et al. Optical coherence tomography angiography changes in macula-off rhegmatogenous retinal detachments repaired with silicone oil. *International Ophthalmology*. December 2020;40(12):3295–3302. https://doi.org/10.1007/s10792-020-01516-z. Epub 2020 Jul 27. PMID: 32720170.
70. Lee JH, Park YG. Microvascular changes on optical coherence tomography angiography after rhegmatogenous retinal detachment vitrectomy with silicone tamponade. *PLoS One*. March 12, 2021;16(3):e0248433. https://doi.org/10.1371/journal.pone.0248433. PMID: 33711059; PMCID: PMC7954302.
71. Fischer C, Bruggemann A, Hager A, Callizo Planas J, Roider J, Hoerauf H. Vascular occlusions following ocular surgical procedures: a clinical observation of vascular complications after ocular surgery. *Journal of Ophthalmology*. 2017;2017:9120892. https://doi.org/10.1155/2017/9120892. Epub 2017 Jul 11. PMID: 28781891; PMCID: PMC5525065.
72. Boniuk M, Zimmerman LE. Necrosis of uvea, sclera, and retina following operations for retinal detachment. *Archives of Ophthalmology 1%1*. 1961;66:318–326.
73. Ryan SJ, Goldberg MF. Anterior segment ischemia following scleral buckling in sickle cell hemoglobinopathy. *American Journal of Ophthalmology*. 1971;72(1):35–50. https://doi.org/10.1016/0002-9394(71)91588-1.
74. Robertson DM. Anterior segment ischemia after segmental episcleral buckling and cryopexy. *American Journal of Ophthalmology*. 1975;79(5):871–874. https://doi.org/10.1016/0002-9394(75)90748-5.
75. Pineles SL, Chang MY, Oltra EL, et al. Anterior segment ischemia: etiology, assessment, and management. *Eye*. February 2018;32(2):173–178. https://doi.org/10.1038/eye.2017.248.Epub2017Nov17. PMID: 29148529; PMCID: PMC5811735.
76. Knox DL. Ischemic ocular inflammation*. *American Journal of Ophthalmology*. 1965;60(6):995–1002. https://doi.org/10.1016/0002-9394(65)92806-0.
77. Chen RW, Flynn Jr HW, Lee WH, et al. Vitreoretinal management and surgical outcomes in proliferative sickle retinopathy: a case series. *American Journal of Ophthalmology*. April 2014;157(4):870–875.e1. https://doi.org/10.1016/j.ajo.2013.12.019. Epub 2013 Dec 31. PMID: 24384527; PMCID: PMC3959584.
78. Nithianandan H, Sridhar J. Surgical and medical perioperative management of sickle cell retinopathy: a literature review. *International Ophthalmology Clinics*. 2020 Fall;60(4):77–87. https://doi.org/10.1097/IIO.0000000000000323. PMID: 33093318; PMCID: PMC7586501.
79. Yoshida A, Hirokawa H, Ishiko S, Ogasawara H. Ocular circulatory changes following scleral buckling procedures. *British Journal of Ophthalmology*. September 1992;76(9):529–531. https://doi.org/10.1136/bjo.76.9.529. PMID: 1420055; PMCID: PMC504341.
80. Ogasawara H, Feke GT, Yoshida A, Milbocker MT, Weiter JJ, McMeel JW. Retinal blood flow alterations associated with scleral buckling and encircling procedures. *British Journal of Ophthalmology*. May 1992;76(5):275–279. https://doi.org/10.1136/bjo.76.5.275. PMID: 1390509; PMCID: PMC504255.
81. Dobbie JG. Circulatory changes in the eye associated with retinal detachment and its repair. *Transaction of the American Ophthalmology Society*. 1980;78:503–566. PMID: 7257069; PMCID: PMC1312153.
82. Iwase T, Kobayashi M, Yamamoto K, Yanagida K, Ra E, Terasaki H. Change in choroidal blood flow and choroidal morphology due to segmental scleral buckling in eyes with rhegmatogenous retinal detachment. *Scientific Reports*. July 20, 2017;7(1):5997. https://doi.org/10.1038/s41598-017-05126-1. PMID: 28729551; PMCID: PMC5519709.

83. Kimura M, Nishimura A, Yokogawa H, et al. Subfoveal choroidal thickness change following segmental scleral buckling for rhegmatogenous retinal detachment. *American Journal of Ophthalmology*. 2012;154:893–900. https://doi.org/10.1016/j.ajo.2012.05.010.
84. Odrobina D, Laudańska-Olszewska I, Gozdek P, Maroszyński M, Amon M. Influence of scleral buckling surgery with encircling band on subfoveal choroidal thickness in long-term observations. *BioMed Research International*. 2013. https://doi.org/10.1155/2013/586894.
85. Tsen CL, Sheu SJ, Chen SC, Wu TT. Imaging analysis with optical coherence tomography angiography after primary repair of macula-off rhegmatogenous retinal detachment. *Graefe's Archieve for Clinical and Experimental Ophthalmology*. 2019;257:1847–1855. https://doi.org/10.1007/s00417-019-04381-4.
86. Bernabei F, Pellegrini M, Taroni L, et al. Choroidal vascular changes after encircling scleral buckling for rhegmatogenous retinal detachment. *Eye*. September 2021;35(9):2619–2623. https://doi.org/10.1038/s41433-020-01307-x. Epub 2020 Nov 23. PMID: 33223533; PMCID: PMC8376964.
87. Pellegrini M, Veronese C, Bernabei F, et al. Choroidal vascular changes in multiple evanescent white dot syndrome. *Ocular Immunology and Inflammation*. February 17, 2021;29(2):340–345. https://doi.org/10.1080/09273948.2019.1678650. Epub 2019 Nov 7. PMID: 31697565.
88. Yoshida A, Feke GT, Green GJ, et al. Retinal circulatory changes after scleral buckling procedures. *American Journal of Ophthalmology*. February 1983;95(2):182–188. https://doi.org/10.1016/0002-9394(83)90012-0. PMID: 6824050.
89. Ito Y, Sasoh M, Ido M, Osawa S, Wakitani Y, Uji Y. Effects of scleral buckling without encircling procedures on retrobulbar hemodynamics as measured by color Doppler imaging. *Archives of Ophthalmology*. July 2005;123(7):950–953. https://doi.org/10.1001/archopht.123.7.950. PMID: 16009837.
90. Nagahara M, Tamaki Y, Araie M, Eguchi S. Effects of scleral buckling and encircling procedures on human optic nerve head and retinochoroidal circulation. *British Journal of Ophthalmology*. January 2000;84(1):31–36. https://doi.org/10.1136/bjo.84.1.31. PMID: 10611096; PMCID: PMC1723226.
91. Sugawara R, Nagaoka T, Kitaya N, et al. Choroidal blood flow in the foveal region in eyes with rhegmatogenous retinal detachment and scleral buckling procedures. *British Journal of Ophthalmology*. November 2006;90(11):1363–1365. https://doi.org/10.1136/bjo.2006.097485. Epub 2006 Aug 16. PMID: 16914476; PMCID: PMC1857488.
92. Yokota H, Mori F, Nagaoka T, Sugawara R, Yoshida A. Pulsatile ocular blood flow: changes associated with scleral buckling procedures. *Japanese Journal of Ophthalmology*. March-April, 2005;49(2):162–165. https://doi.org/10.1007/s10384-004-0161-0. PMID: 15838736.
93. Hørven I. Ophthalmic artery pressure during retrobulbar anaesthesia. *Acta Ophthalmology*. 1978;56(4):574–586. https://doi.org/10.1111/j.1755-3768.1978.tb01370.x. PMID: 735768.
94. Sullivan KL, Brown GC, Forman AR, Sergott RC, Flanagan JC. Retrobulbar anesthesia and retinal vascular obstruction. *Ophthalmology*. 1983;90(4):373–377. ISSN 0161-6420.
95. Jung EH, Park KH, Woo SJ. Iatrogenic central retinal artery occlusion following retrobulbar anesthesia for intraocular surgery. *Korean Journal of Ophthalmology*. August 2015;29(4):233–240. https://doi.org/10.3341/kjo.2015.29.4.233. Epub 2015 Jul 21. PMID: 26240507; PMCID: PMC4520866.
96. Vinerovsky A, Rath EZ, Rehany U, Rumelt S. Central retinal artery occlusion after peribulbar anesthesia. *Journal of Cataract & Refract Surgery*. 2004;30:913–915.
97. Hayreh SS, Zimmerman B, McCarthy MJ, Podhajsky P. Systemic diseases associated with various types of retinal vein occlusion. *American Journal of Ophthalmology*. January 2001;131(1):61–77. https://doi.org/10.1016/s0002-9394(00)00709-1. PMID: 11162981.

Chapter 46

Retinal and choroidal vascular diseases associated with COVID-19

Paul S. Micevych and Jay M. Stewart
University of California San Francisco, Department of Ophthalmology, San Francisco, CA, United States

Overview bullets

- COVID-19, caused by the SARS-CoV-2 virus, may impact the retina and choroid in addition to more common systemic and ocular surface effects
- Pathogenesis of SARS-CoV-2 remains an area of ongoing study, but involves binding to the ACE-2 receptor expressed in retinal vascular endothelium and may involve secondary proinflammatory, immunogenic, and prothrombotic effects
- The most common posterior segment findings in COVID-19 are cotton-wool spots, intraretinal hemorrhages, and vascular tortuosity
- The most reported secondary retinal diseases in the setting of COVID-19 disease include retinal vein occlusion, retinal arterial occlusion, acute macular neuroretinopathy, and paracentral acute middle maculopathy
- The literature surrounding COVID-19 posterior segment disease is rapidly expanding and encompasses a wide spectrum of retinal, choroidal, and uveitic findings

Overview and pathophysiology

The novel coronavirus disease in 2019 (COVID-19), caused by the severe acute respiratory syndrome coronavirus 2 (SARS-CoV-2), carries a wide range of systemic and ocular manifestations, including retinal and choroidal vascular diseases. Current literature suggests ocular involvement in COVID-19 stands at approximately 10%.[1] The most reported ocular finding in COVID-19 is conjunctivitis, often a presenting sign of systemic COVID-19 infection.[2–5] By contrast, posterior segment involvement is much less common, and existing literature is composed largely of case reports and series. Posterior segment involvement in COVID-19 represents a highly diverse set of retinal and choroidal diseases. In general, it is challenging to delineate true and coincidental associations between COVID-19 and posterior segment sequelae, so future studies of a prospective nature will be increasingly important. In this chapter, we will discuss theories of pathogenesis, summarize the existing literature, and describe several representative examples.

The pathophysiology of COVID-19 involves infection of human cells by the SARS-CoV-2 virus. Penetration into human cells involves binding of viral spike protein to the angiotensin-converting enzyme-2 (ACE-2) receptor in the presence of transmembrane serine protease 2.[6] ACE-2 is highly expressed in vascular endothelium in pulmonary tissue,[6] and has been previously identified on the ocular surface,[7] aqueous humor,[8] retina,[9] and brain.[10] In the rodent retina, the ACE-2 receptor has been identified in both Muller cells and vascular endothelial cells.[11] Furthermore, RNA from SARS-Co-V-2 has been identified in the posterior segment. Postmortem human biopsies in patients diagnosed with COVID-19 have revealed SARS-CoV-2 viral RNA in the retina, vitreous humor, and optic nerve.[12–15] The location of the virus within the retina and its potential for replication remains poorly understood.[14] Proposed theories of viral tropism to the retina include direct spread from the ocular surface and anterior segment, hematogenous spread from the breakdown of the inner and outer blood–retina barriers, or neuronal retrograde transport along the optic nerve.[16] Fig. 46.1 illustrates various proposed routes of viral tropism to the retina and represents possible immunogenic, prothrombotic, and proinflammatory effects of systemic COVID-19 infection.

Retinal and Choroidal Vascular Diseases of the Eye. https://doi.org/10.1016/B978-0-443-15583-3.00046-9

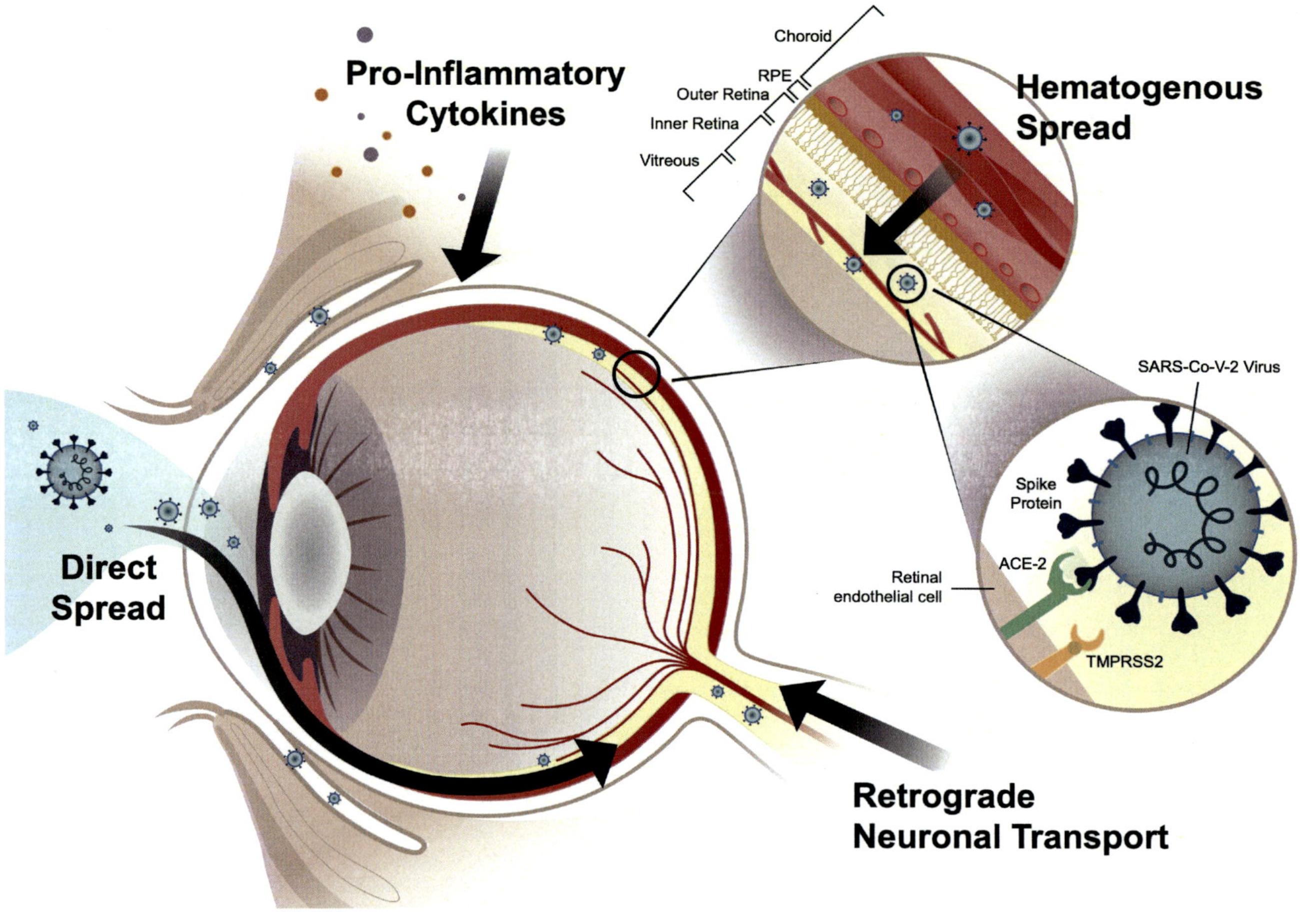

FIGURE 46.1 Pathophysiology of SARS-CoV-2 ocular involvement. Graphic representation of the anatomic localization of virus within various segments of the eye, cellular binding and penetration mechanisms, proinflammatory effects of systemic COVID-19 infection, and possible proposed routes of viral tropism (direct spread through the anterior segment, hematogenous spread across the inner and outer blood–retina–barriers, and retrograde neuronal transport along the optic nerve). *ACE-2*, angiotensin-converting enzyme-2 receptor; *TMPRSS2*, transmembrane serine protease 2; *SARS-Co-V-2*, severe acute respiratory syndrome coronavirus 2; *COVID-19*, novel coronavirus disease in 2019.

Primary COVID-19-related disease

The most common posterior segment involvement in COVID-19 is a nonspecific retinal microangiopathy, detailed in Table 46.1. COVID-19-associated microangiopathy typically presents with intraretinal hemorrhages, cotton-wool spots, and vascular tortuosity, as isolated findings or in combination.[17–23] The first reported case of COVID-19 microangiopathy is shown in Fig. 46.2.[24] In most cases, microangiopathy has been characterized by a subclinical or mildly symptomatic

TABLE 46.1 COVID-19-associated retinal microangiopathy: clinical and imaging features.

Clinical Exam	Cotton-wool spots, intraretinal hemorrhages, tortuous vessels
OCT	Ellipsoid zone and interdigitation zone disruption, retinal pigment epithelium irregularity, vessel diameter alterations, focal choroidal thickening, choroidal caverns
OCTA	Reduced vascular density in superficial/deep capillary plexuses and choriocapillaris, enlarged foveolar avascular zone
Fundus autofluorescence	Hypo-, hyper-, and mixed autofluorescence patterns
ICGA	Hypofluorescent spots, intervortex shunts, "hemangioma-like" lesions

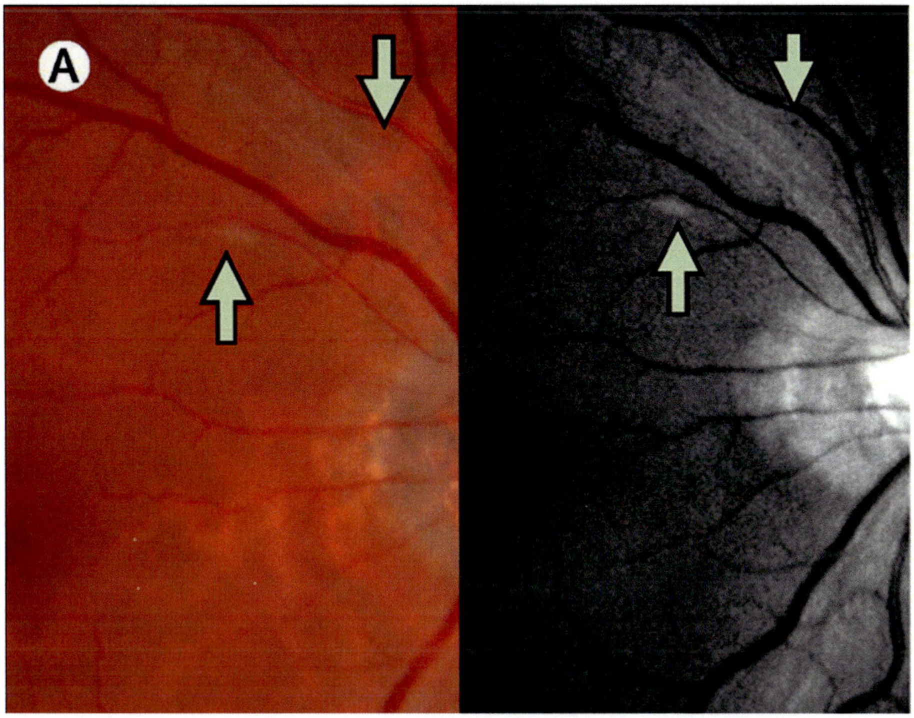

FIGURE 46.2 First reported case of COVID-19 retinal involvement. Fundus photography (*left*) and red-free imaging (*right*) show a cotton-wool spot along the superior arcade and a subtle intraretinal microhemorrhage (*arrows*). This case and figure are presented with the permission of the original authors and publisher. The article was published in the *Lancet*, Marinho PM, Marcos AAA, Romano AC et al., Retinal findings in patients with COVID-19, Copyright Elsevier (2020). *From Marinho PM, Marcos AAA, Romano AC, Nascimento H, Belfort R. Retinal findings in patients with COVID-19. The Lancet. 2020;395(10237):1610. https://doi.org/10.1016/S0140-6736(20)31014-X.*

presentation with a self-limited course and minimal impact on visual acuity. As these clinical findings are highly nonspecific to COVID-19, further work has been done to characterize the true association between microangiopathy and COVID-19. Recent work has demonstrated a nearly 9 times greater likelihood of retinal microangiopathy in patients with COVID-19 compared with healthy controls.[25] Others have demonstrated that COVID-19 microangiopathy may be specific to the convalescent phase of the disease.[17] To date, management of COVID-19 microangiopathy and vascular alterations has been largely supportive with serial follow-up indicated to evaluate for improvement or resolution.

Multimodal imaging characteristics

Noninvasive retinal imaging can demonstrate a variety of retinal vasculature changes in patients with clinical and subclinical disease. Optical coherence tomography (OCT) has demonstrated that retinal vessel diameter may be altered in patients with COVID-19.[26] Changes in retinal layer thickness in patients with COVID-19 include increased central macular thickness and thinning of ganglion cells and inner nuclear layers.[27] Qualitative changes on OCT include ellipsoid zone and indigitation zone disruption, retinal pigment epithelium irregularity, possible hyperreflective intraretinal lesions, focal choroidal thickening, and choroidal caverns.[24,28,29] OCT angiography (OCTA) has also demonstrated vascular alterations in patients with COVID-19, characterized by decreased vascular density in the superficial and deep capillary plexuses and enlargement of the foveolar avascular zone.[25,30,31] There is also evidence to suggest that these vascular changes from COVID-19 may persist longer-term. Flow deficits within the superficial and deep capillary plexuses and choriocapillaris have been shown to persist upon 6-month follow-up.[32] The clinical impact of these persistent vascular abnormalities after COVID-19 infection is currently unknown. Fundus autofluorescence (AF) patterns in patients with COVID-19 are highly varied, and patients may present with hyper-AF, hypo-AF, or mixed hyper/hypo-AF, most often located adjacent to vascular structures.[29] Indocyanine green angiography (ICGA) findings in patients with COVID-19 include hypocyanescent spots, intervortex shunts, and "hemangioma-like" lesions.[28]

TABLE 46.2 Reported COVID-19 disease-related posterior segment disorders.

Retina	*Choroid and Uveitis*
Retinal vein occlusion	Panuveitis ± optic neuritis and vasculitis
Retinal artery occlusion	Posterior uveitis ± vasculitis
Acute macular neuroretinopathy	Intermediate uveitis ± vasculitis
Paracentral acute middle maculopathy	Inflammatory chorioretinopathies
Purtscher-like retinopathy	Serpiginous choroiditis
Fovealitis	
Acute viral retinitis	
Pachychoroid-spectrum disease	

Secondary COVID-19-related disease

Retinal involvement

Secondary retinal and choroidal diseases have also been reported in association with COVID-19, as listed in Table 46.2. The prothrombotic and proinflammatory state induced by COVID-19 has been postulated to promote retinal thromboembolic events, including retinal vein occlusion (RVO) and retinal artery occlusion (RAO). COVID-19 has been associated with impending-, branch-, hemi-, and central RVOs,[33–38] as well as branch- and central RAOs,[39–46] each in the absence of traditional explanatory risk factors. A recent large prospective cohort study determined that the incidence of RVO increased in the 6 months after COVID-19, with peak incidence 6–8 weeks after diagnosis.[47]

The proinflammatory and prothrombotic environment in COVID-19 has also been postulated to promote microthrombosis and microvascular ischemia heralding the development of acute macular neuroretinopathy (AMN) and paracentral acute middle maculopathy (PAMM).[48–50] AMN and PAMM are poorly understood, acquired retinal vascular disorders with pathophysiology believed to involve retinal or choroidal microvascular ischemia.[51,52] AMN is characterized by dark red petaloid lesions around the fovea, confirmed by hyporeflective parafoveal wedge-shaped lesions on infrared scanning laser ophthalmoscopy (SLO) with hyperreflective lesions at the level of the outer plexiform layer or outer nuclear layer on OCT, often with underlying ellipsoid zone disruption. Lesions in PAMM may be distinguished from those in AMN with cross-sectional OCT, by their more superficial location at the level of the inner nuclear layer.[53] Since the pandemic in 2020, numerous cases of AMN[49,54–63] and PAMM[45,50,57,62,64] have been reported in association with COVID-19. In one report, AMN was the presenting manifestation of COVID-19.[61] The incidence of AMN increased in 2020 compared with prepandemic levels in a recent small study, supporting the possible emerging association.[48] Fig. 46.3 demonstrates a clinical case of COVID-19-associated AMN with characteristic findings on multimodal retinal imaging.

Rare reports of association between COVID-19 and other forms of retinal involvement also warrant brief mention. There have been individual reports of COVID-19 in association with Purtscher-like retinopathy,[65] bilateral fovealitis characterized by refractile foveal deposits,[66] and acute viral retinitis.[67] There have also been reports of pachychoroid-spectrum disease including central serous chorioretinopathy; however, concomitant steroid use may explain some of these cases.[54,64,68]

Choroidal involvement and uveitis

Choroidal involvement and uveitis have also been reported in association with COVID-19. Reports of anterior, intermediate, and posterior uveitis,[69] panuveitis,[70–72] retinal vasculitis,[35] inflammatory chorioretinopathy,[73] and serpiginous choroiditis[74] have been attributed to COVID-19 infection. COVID-19-associated panuveitis has been reported to present with both granulomatous[71,75] and nongranulomatous[70,72] patterns. In one study, panuveitis associated with optic neuritis was the presenting sign of COVID-19 disease and ultimately resulted in permanent vision loss.[70] A small series of COVID-19-associated anterior, intermediate/posterior, and panuveitis cases demonstrated responsiveness to topical and oral corticosteroid treatment. It is important to note that corticosteroids must be used with caution in COVID-19 and taken in the context of ongoing systemic disease with a multidisciplinary approach. These reports of COVID-19-associated uveitis

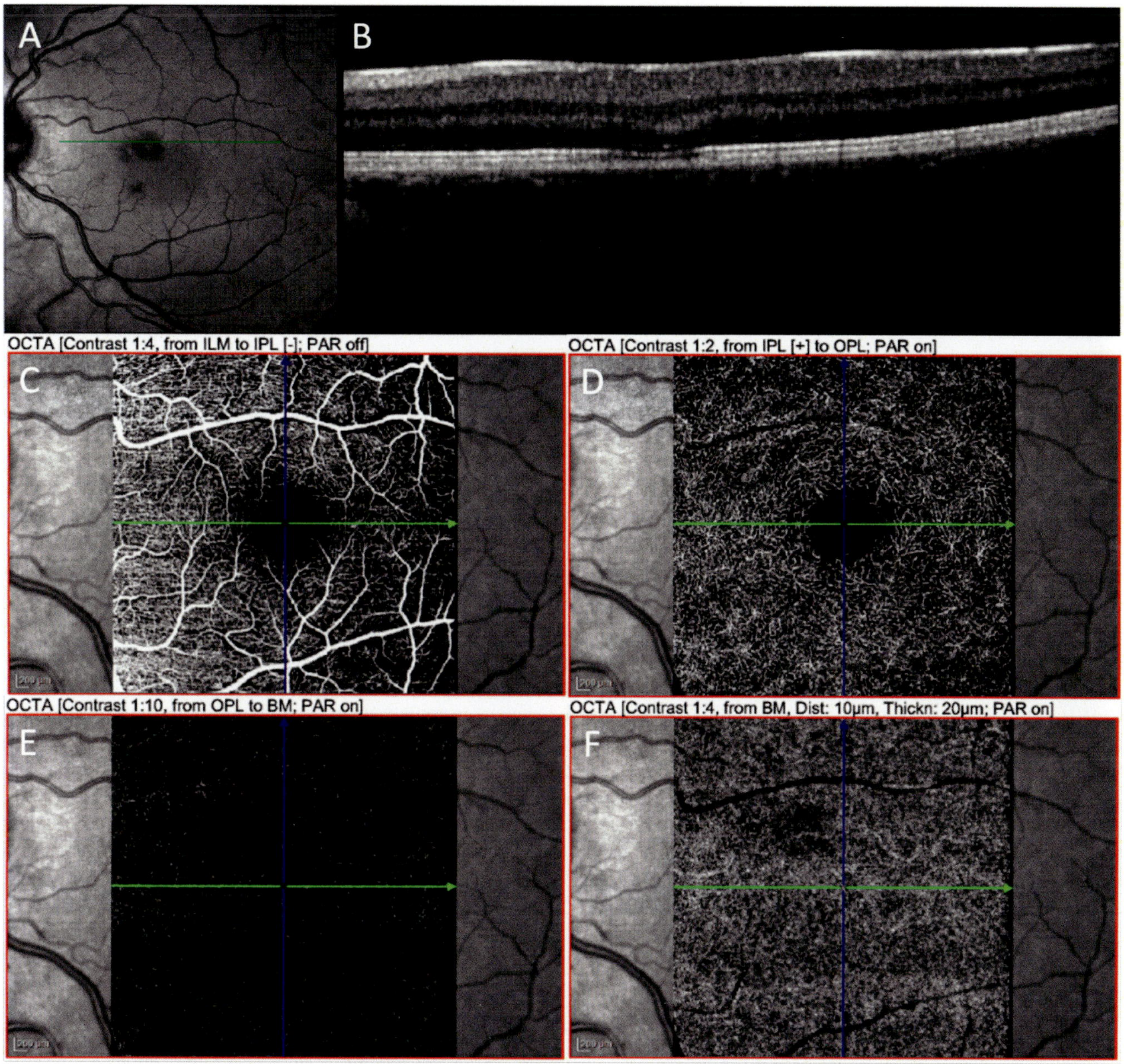

FIGURE 46.3 Case study of COVID-19-associated AMN A 13-year-old patient presented with headaches and bilateral visual scotomas the same day as the onset of COVID-19 symptoms. Visual acuity was 20/20 in both eyes. OCT-revealed hyporeflective parafoveal lesions on the en face infrared images (A) corresponding to the location of the outer retinal layer disruptions in cross-section (B). OCTA of superficial and deep capillary plexuses (C, D) demonstrated no apparent flow voids, whereas OCTA of the outer retina (E) and choriocapillaris (F) revealed flow voids corresponding to the location of the hyporeflective infrared lesions. Within 1 month, the patient's scotomas spontaneously resolved. This case and figure are presented with the permission of the original authors and publisher. The article was published in *Ophthalmology Retina*, Dinh RH, Tsui E, Wieder MS et al., Acute Macular Neuroretinopathy and Coronavirus Disease 2019, Copyright Elsevier (2022). *From Dinh RH, Tsui E, Wieder MS, Barash A, Park MM, Rahimy E, Mruthyunjaya P, Lu LJ, Michalak SM, Shah RJ, Sierpina D, Winter TW, Shields RA, Uchiyama E, Lee GD, Komati R, Lee E, Kasi SK, Do BK. Acute Macular Neuroretinopathy and Coronavirus Disease 2019. Ophthalmology Retina. 2023;7(2):198–200. https://doi.org/10.1016/j.oret.2022.09.005.*

TABLE 46.3 Reported COVID-19 vaccine-related posterior segment disorders.

Retina	*Choroid and Uveitis*
Retinal vein occlusion	Panuveitis ± vasculitis
Acute zonal occult outer retinopathy	Posterior uveitis ± vasculitis
Acute macular neuroretinopathy	Intermediate uveitis ± vasculitis
Paracentral acute middle maculopathy	Multifocal choroiditis
Central serous retinopathy	Multiple evanescent white dot syndrome
Acute retinal necrosis	Vogt–Koyanagi–Harada disease

suggest a possible immunological trigger induced by COVID-19; however, further work and larger studies will be important in the future to delineate the degree of association between each of these disorders and COVID-19.

COVID-19 vaccine-related disease

The COVID-19 mRNA vaccines have also been implicated in rare reports of posterior segment disease. Reports of retinal findings include RVO,[76] acute zonal occult outer retinopathy,[77] AMN and PAMM,[78,79] central serous chorioretinopathy,[80] and acute retinal necrosis secondary to varicella zoster virus reactivation.[81] Reports of choroidal or uveitic findings include multifocal choroiditis,[82,83] multiple evanescent white dot syndrome,[84–86] panuveitis,[87,88] intermediate or posterior uveitis,[89] and Vogt–Koyanagi–Harada disease.[90,91] Greater description of these reports and their level of association is beyond the scope of this chapter, but they nonetheless deserve mention in the context of COVID-19-related disease and therefore are listed together in Table 46.3. It remains important to note that mRNA vaccination has been highly effective in COVID-19 disease prevention and mortality reduction in the general population. Important questions remain regarding causality and coincidence for these retinal diseases in the setting of COVID-19.[92]

Further prospective and randomized work is needed to better understand the pathophysiology, longer-term impact, and incidence of COVID-19 retinal involvement. Currently, no randomized controlled trials exist to understand the natural history of disease or guide management. However, several clinical trials are currently ongoing. One trial aims to evaluate the rate and extent of ocular involvement in COVID-19 (NCT04665752) and a second strives to build a noninvasive retinal imaging database to better understand retinal damage after COVID-19 (NCT04881253). It also remains unknown whether different variants of COVID-19 may present with different risk profiles or degrees of severity for retinal disease, but this is a valuable area for future study.

Summary

While less common than ocular surface or systemic disease, COVID-19 posterior segment involvement has been increasingly reported in the literature. Retinal and choroidal involvement may be a presenting sign of systemic disease or develop at a later stage. The highly varied spectrum of COVID-19 retinal and choroidal involvement includes subclinical disease, mild nonspecific microangiopathy, and a host of secondary posterior segment disorders. Long-term implications of COVID-19 retinal and choroidal diseases remain unknown.

References

1. Nasiri N, Sharifi H, Bazrafshan A, Noori A, Karamouzian M, Sharifi A. Ocular manifestations of COVID-19: a systematic review and meta-analysis. *Journal of Ophthalmic and Vision Research*. 2021;16(1):103–112. https://doi.org/10.18502/jovr.v16i1.8256.
2. Al-Sharif E, Strianese D, AlMadhi NH, et al. Ocular tropism of coronavirus (CoVs): a comparison of the interaction between the animal-to-human transmitted coronaviruses (SARS-CoV-1, SARS-CoV-2, MERS-CoV, CoV-229E, NL63, OC43, HKU1) and the eye. *International Ophthalmology*. 2021;41(1):349–362. https://doi.org/10.1007/s10792-020-01575-2.
3. Chen L, Deng C, Chen X, et al. Ocular manifestations and clinical characteristics of 535 cases of COVID-19 in Wuhan, China: a cross-sectional study. *Acta Ophthalmologica*. 2020;98(8):e951–e959. https://doi.org/10.1111/aos.14472.
4. Guan W, Ni Z, Hu Y, et al. Clinical characteristics of coronavirus disease 2019 in China. *New England Journal of Medicine*. 2020;382(18):1708–1720. https://doi.org/10.1056/NEJMoa2002032.

5. Wu P, Duan F, Luo C, et al. Characteristics of ocular findings of patients with coronavirus disease 2019 (COVID-19) in hubei province, China. *JAMA Ophthalmology*. 2020;138(5):575–578. https://doi.org/10.1001/jamaophthalmol.2020.1291.
6. Castro P, Palomo M, Moreno-Castaño AB, et al. Is the endothelium the missing link in the pathophysiology and treatment of COVID-19 complications? *Cardiovascular Drugs and Therapy*. 2022;36(3):547–560. https://doi.org/10.1007/s10557-021-07207-w.
7. Sungnak W, Huang N, Bécavin C, et al. SARS-CoV-2 entry factors are highly expressed in nasal epithelial cells together with innate immune genes. *Nature Medicine*. 2020;26(5):681–687. https://doi.org/10.1038/s41591-020-0868-6.
8. Holappa M, Valjakka J, Vaajanen A. Angiotensin(1-7) and ACE2, "the hot spots" of renin-angiotensin system, detected in the human aqueous humor. *Open Ophthalmology Journal*. 2015;9(1):28–32. https://doi.org/10.2174/1874364101509010028.
9. Senanayake PDS, Drazba J, Shadrach K, et al. Angiotensin II and its receptor subtypes in the human retina. *Investigative Ophthalmology and Visual Science*. 2007;48(7):3301–3311. https://doi.org/10.1167/iovs.06-1024.
10. Natoli S, Oliveira V, Calabresi P, Maia LF, Pisani A. Does SARS-Cov-2 invade the brain? Translational lessons from animal models. *European Journal of Neurology*. 2020;27(9):1764–1773. https://doi.org/10.1111/ene.14277.
11. Tikellis C, Johnston CI, Forbes JM, et al. Identification of angiotensin converting enzyme 2 in the rodent retina. *Current Eye Research*. 2004;29(6):419–427. https://doi.org/10.1080/02713680490517944.
12. Araujo-Silva CA, Marcos AAA, Marinho PM, et al. Presumed SARS-CoV-2 viral particles in the human retina of patients with COVID-19. *JAMA Ophthalmology*. 2021;139(9):1015–1021. https://doi.org/10.1001/jamaophthalmol.2021.2795.
13. Casagrande M, Fitzek A, Püschel K, et al. Detection of SARS-CoV-2 in human retinal biopsies of deceased COVID-19 patients. *Ocular Immunology and Inflammation*. 2020;28(5):721–725. https://doi.org/10.1080/09273948.2020.1770301.
14. Casagrande M, Fitzek A, Spitzer M, et al. Detection of SARS-CoV-2 genomic and subgenomic RNA in retina and optic nerve of patients with COVID-19. *British Journal of Ophthalmology*. 2022;106(9):1313–1317. https://doi.org/10.1136/bjophthalmol-2020-318618.
15. Sawant OB, Singh S, Wright RE, et al. Prevalence of SARS-CoV-2 in human post-mortem ocular tissues. *Ocular Surface*. 2021;19:322–329. https://doi.org/10.1016/j.jtos.2020.11.002.
16. de Figueiredo CS, Raony Í, Giestal-de-Araujo E. SARS-CoV-2 targeting the retina: host–virus interaction and possible mechanisms of viral tropism. *Ocular Immunology and Inflammation*. 2020;28(8):1301–1304. https://doi.org/10.1080/09273948.2020.1799037.
17. Bansal R, Markan A, Gautam N, et al. Retinal involvement in COVID-19: results from a prospective retina screening program in the acute and convalescent phase. *Frontiers in Medicine*. 2021;8. https://doi.org/10.3389/fmed.2021.681942.
18. Caporossi T, Bacherini D, Tartaro R, Virgili G, Peris A, Giansanti F. Retinal findings in patients affected by COVID 19 intubated in an intensive care unit. *Acta Ophthalmologica*. 2021;99(7):e1244–e1245. https://doi.org/10.1111/aos.14734.
19. Invernizzi A, Torre A, Parrulli S, et al. Retinal findings in patients with COVID-19: results from the SERPICO-19 study. *EClinicalMedicine*. 2020;27:100550. https://doi.org/10.1016/j.eclinm.2020.100550.
20. Landecho MF, Yuste JR, Gándara E, et al. COVID-19 retinal microangiopathy as an in vivo biomarker of systemic vascular disease? *Journal of Internal Medicine*. 2021;289(1):116–120. https://doi.org/10.1111/joim.13156.
21. Lani-Louzada R, Ramos Cdo VF, Cordeiro RM, Sadun AA, Lewin AS. Retinal changes in COVID-19 hospitalized cases. *PLoS One*. 2020;15(12):e0243346. https://doi.org/10.1371/journal.pone.0243346.
22. Pereira LA, Soares LCM, Nascimento PA, et al. Retinal findings in hospitalised patients with severe COVID-19. *British Journal of Ophthalmology*. 2022;106(1):102–105. https://doi.org/10.1136/bjophthalmol-2020-317576.
23. Savastano MC, Gambini G, Cozzupoli GM, et al. Retinal capillary involvement in early post-COVID-19 patients: a healthy controlled study. *Graefe's Archive for Clinical and Experimental Ophthalmology*. 2021;259(8):2157–2165. https://doi.org/10.1007/s00417-020-05070-3.
24. Marinho PM, Marcos AAA, Romano AC, Nascimento H, Belfort R. Retinal findings in patients with COVID-19. *The Lancet*. 2020;395(10237):1610. https://doi.org/10.1016/S0140-6736(20)31014-X.
25. Teo KY, Invernizzi A, Staurenghi G, Cheung CMG. COVID-19-Related retinal micro-vasculopathy – a review of current evidence. *American Journal of Ophthalmology*. 2022;235:98–110. https://doi.org/10.1016/j.ajo.2021.09.019.
26. Aşõkgarip N, Temel E, Hõzmalõ L, Örnek K, Sezgin FM. Retinal vessel diameter changes in COVID-19 infected patients. *Ocular Immunology and Inflammation*. 2021;29(4):645–651. https://doi.org/10.1080/09273948.2020.1853783.
27. Oren B, Aksoy Aydemõr G, Aydemõr E, et al. Quantitative assessment of retinal changes in COVID-19 patients. *Clinical and Experimental Optometry*. 2021;104(6):717–722. https://doi.org/10.1080/08164622.2021.1916389.
28. Abdelmassih Y, Azar G, Bonnin S, et al. Covid-19 associated choroidopathy. *Journal of Clinical Medicine*. 2021;10(20). https://doi.org/10.3390/jcm10204686.
29. Marinho PM, Marcos AAA, Branco AMC, et al. Retinal autofluorescence findings after COVID-19. *International Journal of Retina and Vitreous*. 2021;7(1). https://doi.org/10.1186/s40942-021-00341-5.
30. Abrishami M, Emamverdian Z, Shoeibi N, et al. Optical coherence tomography angiography analysis of the retina in patients recovered from COVID-19: a case-control study. *Canadian Journal of Ophthalmology*. 2021;56(1):24–30. https://doi.org/10.1016/j.jcjo.2020.11.006.
31. Cetinkaya T, Kurt MM, Cetinkaya H, Akpolat C. Analysis of microvasculature in children recovered from COVID-19 using swept-source OCT/OCTA technology. *Ocular Immunology and Inflammation*. 2022:1–7. https://doi.org/10.1080/09273948.2022.2054431.
32. Turker IC, Dogan CU, Dirim AB, Guven D, Kutucu OK. Evaluation of early and late COVID-19-induced vascular changes with OCTA. *Canadian Journal of Ophthalmology*. 2022;57(4):236–241. https://doi.org/10.1016/j.jcjo.2021.05.001.
33. Finn AP, Khurana RN, Chang LK. Hemi-retinal vein occlusion in a young patient with COVID-19. *American Journal of Ophthalmology Case Reports*. 2021;22. https://doi.org/10.1016/j.ajoc.2021.101046.

34. Invernizzi A, Pellegrini M, Messenio D, et al. Impending central retinal vein occlusion in a patient with coronavirus disease 2019 (COVID-19). *Ocular Immunology and Inflammation*. 2020;28(8):1290–1292. https://doi.org/10.1080/09273948.2020.1807023.
35. Sheth JU, Narayanan R, Goyal J, Goyal V. Retinal vein occlusion in COVID-19: a novel entity. *Indian Journal of Ophthalmology*. 2020;68(10):2291–2293. https://doi.org/10.4103/ijo.IJO_2380_20.
36. Shiroma HF, Lima LH, Shiroma YB, et al. Retinal vascular occlusion in patients with the Covid-19 virus. *International Journal of Retina and Vitreous*. 2022;8(1). https://doi.org/10.1186/s40942-022-00371-7.
37. Shroff D, Kumar S, Naidu A, Gupta C, Shroff CM. Retinal vasoocclusive spectrum following COVID-19. *Indian Journal of Ophthalmology*. 2022;70(4):1412–1415. https://doi.org/10.4103/ijo.IJO_2837_21.
38. Walinjkar JA, Makhija SC, Sharma HR, Morekar S, Natarajan S. Central retinal vein occlusion with COVID-19 infection as the presumptive etiology. *Indian Journal of Ophthalmology*. 2020;68(11):2572–2574. https://doi.org/10.4103/ijo.IJO_2575_20.
39. Acharya S, Diamond M, Anwar S, Glaser A, Tyagi P. Unique case of central retinal artery occlusion secondary to COVID-19 disease. *IDCases*. 2020;21:e00867. https://doi.org/10.1016/j.idcr.2020.e00867.
40. Ateş O, Yõldõrõm M, Yõldõrõm K. Branch retinal artery occlusion in patient with COVID-19: case report. *Korean Journal of Ophthalmology*. 2021;35(6):484–485. https://doi.org/10.3341/kjo.2021.0008.
41. Bapaye MM, Nair AG, Bapaye CM, Bapaye MM, Shukla JJ. Simultaneous bilateral central retinal artery occlusion following COVID-19 infection. *Ocular Immunology and Inflammation*. 2021;29(4):671–674. https://doi.org/10.1080/09273948.2021.1891262.
42. Montesel A, Bucolo C, Mouvet V, Moret E, Eandi CM. Case report: central retinal artery occlusion in a COVID-19 patient. *Frontiers in Pharmacology*. 2020;11. https://doi.org/10.3389/fphar.2020.588384.
43. Murchison AP, Sweid A, Dharia R, et al. Monocular visual loss as the presenting symptom of COVID-19 infection. *Clinical Neurology and Neurosurgery*. 2021;201:106440. https://doi.org/10.1016/j.clineuro.2020.106440.
44. Sweid A, Hammoud B, Weinberg JH, et al. Letter: thrombotic neurovascular disease in covid-19 patients. *Neurosurgery*. 2020;87(3):E400–E406. https://doi.org/10.1093/neuros/nyaa254.
45. Turedi N, Onal Gunay B. Paracentral acute middle maculopathy in the setting of central retinal artery occlusion following COVID-19 diagnosis. *European Journal of Ophthalmology*. 2022;32(3):NP62–NP66. https://doi.org/10.1177/1120672121995347.
46. Uzun A, Keles Sahin A, Bektas O. A unique case of branch retinal artery occlusion associated with a relatively mild coronavirus disease 2019. *Ocular Immunology and Inflammation*. 2021;29(4):715–718. https://doi.org/10.1080/09273948.2021.1933071.
47. Modjtahedi BS, Do D, Luong TQ, Shaw J. Changes in the incidence of retinal vascular occlusions after COVID-19 diagnosis. *JAMA Ophthalmology*. 2022;140(5):523–527. https://doi.org/10.1001/jamaophthalmol.2022.0632.
48. Azar G, Bonnin S, Vasseur V, et al. Did the covid-19 pandemic increase the incidence of acute macular neuroretinopathy? *Journal of Clinical Medicine*. 2021;10(21). https://doi.org/10.3390/jcm10215038.
49. Dinh RH, Tsui E, Wieder MS, et al. Acute macular neuroretinopathy and coronavirus disease 2019. *Ophthalmology Retina*. 2023;7(2):198–200. https://doi.org/10.1016/j.oret.2022.09.005.
50. Padhy SK, Dcruz RP, Kelgaonkar A. Paracentral acute middle maculopathy following SARS-CoV-2 infection: the D-dimer hypothesis. *BMJ Case Reports*. 2021;14(3):e242043. https://doi.org/10.1136/bcr-2021-242043.
51. Iovino C, Au A, Ramtohul P, et al. Coincident PAMM and AMN and insights into a common pathophysiology. *American Journal of Ophthalmology*. 2022;236:136–146. https://doi.org/10.1016/j.ajo.2021.07.004.
52. Scharf J, Freund KB, Sadda SV, Sarraf D. Paracentral acute middle maculopathy and the organization of the retinal capillary plexuses. *Progress in Retinal and Eye Research*. 2021;81:100884. https://doi.org/10.1016/j.preteyeres.2020.100884.
53. Moura-Coelho N, Gaspar T, Ferreira JT, Dutra-Medeiros M, Cunha JP. Paracentral acute middle maculopathy—review of the literature. *Graefe's Archive for Clinical and Experimental Ophthalmology*. 2020;258(12):2583–2596. https://doi.org/10.1007/s00417-020-04826-1.
54. Mahjoub A, Dlensi A, Romdhane A, et al. Bilateral central serous chorioretinopathy post-COVID-19. *Journal Francais d'Ophtalmologie*. 2021;44(10):1484–1490. https://doi.org/10.1016/j.jfo.2021.10.001.
55. Aidar MN, Gomes TM, de Almeida MZH, de Andrade EP, Serracarbassa PD. Low visual acuity due to acute macular neuroretinopathy associated with covid-19: a case report. *American Journal of Case Reports*. 2021;22(1). https://doi.org/10.12659/AJCR.931169.
56. David JA, Fivgas GD. Acute macular neuroretinopathy associated with COVID-19 infection. *American Journal of Ophthalmology Case Reports*. 2021;24. https://doi.org/10.1016/j.ajoc.2021.101232.
57. Gascon P, Briantais A, Bertrand E, et al. Covid-19-Associated retinopathy: a case report. *Ocular Immunology and Inflammation*. 2020;28(8):1293–1297. https://doi.org/10.1080/09273948.2020.1825751.
58. Giacuzzo C, Eandi CM, Kawasaki A. Bilateral acute macular neuroretinopathy following COVID-19 infection. *Acta Ophthalmologica*. 2022;100(2):e611–e612. https://doi.org/10.1111/aos.14913.
59. Jalink MB, Bronkhorst IHG. A sudden rise of patients with acute macular neuroretinopathy during the COVID-19 pandemic. *Case Reports in Ophthalmology*. 2022:96–103. https://doi.org/10.1159/000522080.
60. Masjedi M, Pourazizi M, Hosseini NS. Acute macular neuroretinopathy as a manifestation of coronavirus disease 2019: a case report. *Clinical Case Reports*. 2021;9(10). https://doi.org/10.1002/ccr3.4976.
61. Preti RC, Zacharias LC, Cunha LP, Monteiro MLR. Acute macular neuroretinopathy as the presenting manifestation of COVID-19 infection. *Retinal Cases and Brief Reports*. 2022;16(1):12–15. https://doi.org/10.1097/ICB.0000000000001050.
62. Virgo J, Mohamed M. Paracentral acute middle maculopathy and acute macular neuroretinopathy following SARS-CoV-2 infection. *Eye*. 2020;34(12):2352–2353. https://doi.org/10.1038/s41433-020-1069-8.

63. Zamani G, Ataei Azimi S, Aminizadeh A, et al. Acute macular neuroretinopathy in a patient with acute myeloid leukemia and deceased by COVID-19: a case report. *Journal of Ophthalmic Inflammation and Infection*. 2020;10(1). https://doi.org/10.1186/s12348-020-00231-1.
64. Goyal M, Murthy SI, Annum S. Retinal manifestations in patients following COVID-19 infection: a consecutive case series. *Indian Journal of Ophthalmology*. 2021;69(5):1275–1282. https://doi.org/10.4103/ijo.IJO_403_21.
65. Bottini AR, Steinmetz S, Blinder KJ, Shah GK, Ferreras A. Purtscher-like retinopathy in a patient with COVID-19. *Case Reports in Ophthalmological Medicine*. 2021;2021:1–5. https://doi.org/10.1155/2021/6661541.
66. Kumar A, Kumar P, Kaushik J, Dubbaka S, Manumala LK. COVID-19 induced maculopathy. *Clinical and Experimental Optometry*. 2021;104(6):734–735. https://doi.org/10.1080/08164622.2021.1896947.
67. Liu L, Cai D, Huang X, Shen Y. COVID-2019 associated with acquired monocular blindness. *Current Eye Research*. 2021;46(8):1247–1250. https://doi.org/10.1080/02713683.2021.1874027.
68. Abrishami M, Daneshvar R, Shoeibi N, et al. Pachychoroid spectrum disorder findings in patients with coronavirus disease 2019. *Case Reports in Ophthalmological Medicine*. 2021;2021:1–9. https://doi.org/10.1155/2021/4688764.
69. Iriqat S, Yousef Q, Ereqat S. Clinical profile of COVID-19 patients presentinwith uveitis – a short case series. *International Medical Case Reports Journal*. 2021;14:421–427. https://doi.org/10.2147/IMCRJ.S312461.
70. Benito-Pascual B, Gegúndez JA, Díaz-Valle D, et al. Panuveitis and optic neuritis as a possible initial presentation of the novel coronavirus disease 2019 (COVID-19). *Ocular Immunology and Inflammation*. 2020;28(6):922–925. https://doi.org/10.1080/09273948.2020.1792512.
71. Eser-Ozturk H, Izci Duran T, Aydog O, Sullu Y. Sarcoid-like uveitis with or without tubulointerstitial nephritis during COVID-19. *Ocular Immunology and Inflammation*. 2022:1–8. https://doi.org/10.1080/09273948.2022.2032195.
72. François J, Collery AS, Hayek G, et al. Coronavirus disease 2019-associated ocular neuropathy with panuveitis: a case report. *JAMA Ophthalmology*. 2021;139(2):247–249. https://doi.org/10.1001/jamaophthalmol.2020.5695.
73. Ortiz-Seller A, Martínez Costa L, Hernández-Pons A, Valls Pascual E, Solves Alemany A, Albert-Fort M. Ophthalmic and neuro-ophthalmic manifestations of coronavirus disease 2019 (COVID-19). *Ocular Immunology and Inflammation*. 2020;28(8):1285–1289. https://doi.org/10.1080/09273948.2020.1817497.
74. Providência J, Fonseca C, Henriques F, Proença R. Serpiginous choroiditis presenting after SARS-CoV-2 infection: a new immunological trigger? *European Journal of Ophthalmology*. 2022;32(1):NP97–NP101. https://doi.org/10.1177/1120672120977817.
75. Ozdemir Yalcinsoy K, Ozdamar Erol Y, Ozdal PC. Bilateral granulomatous panuveitis in A patient with COVID-19. *Ocular Immunology and Inflammation*. 2022:1–4. https://doi.org/10.1080/09273948.2022.2061523.
76. Romano D, Morescalchi F, Romano V, Semeraro F. COVID-19 AdenoviralVector vaccine and central retinal vein occlusion. *Ocular Immunology and Inflammation*. 2022;30(5):1286–1288. https://doi.org/10.1080/09273948.2022.2079534.
77. Maleki A, Look-Why S, Manhapra A, Foster CS. COVID-19 recombinant mRNA vaccines and serious ocular inflammatory side effects: real or coincidence? *Journal of Ophthalmic and Vision Research*. 2021;16(3):490–501. https://doi.org/10.18502/jovr.v16i3.9443.
78. Mambretti M, Huemer J, Torregrossa G, Ullrich M, Findl O, Casalino G. Acute macular neuroretinopathy following coronavirus disease 2019 vaccination. *Ocular Immunology and Inflammation*. 2021;29(4):730–733. https://doi.org/10.1080/09273948.2021.1946567.
79. Bøhler AD, Strøm ME, Sandvig KU, Moe MC, Jørstad ØK. Acute macular neuroretinopathy following COVID-19 vaccination. *Eye*. 2022;36(3):644–645. https://doi.org/10.1038/s41433-021-01610-1.
80. Fowler N, Mendez Martinez NR, Pallares BV, Maldonado RS. Acute-onset central serous retinopathy after immunization with COVID-19 mRNA vaccine. *American Journal of Ophthalmology Case Reports*. 2021;23. https://doi.org/10.1016/j.ajoc.2021.101136.
81. Mishra SB, Mahendradas P, Kawali A, Sanjay S, Shetty R. Reactivation of varicella zoster infection presenting as acute retinal necrosis post COVID 19 vaccination in an Asian Indian male. *European Journal of Ophthalmology*. 2023;33(1):NP32–NP36. https://doi.org/10.1177/11206721211046485.
82. Kakarla PD, Venugopal RYC, Manechala UB, Rijey J, Anthwal D. Bilateral multifocal choroiditis with disc edema in a 15-year-old girl following COVID-19 vaccination. *Indian Journal of Ophthalmology*. 2022;70(9):3420–3422. https://doi.org/10.4103/ijo.IJO_861_22.
83. Goyal M, Murthy SI, Annum S. Bilateral multifocal choroiditis following COVID-19 vaccination. *Ocular Immunology and Inflammation*. 2021;29(4):753–757. https://doi.org/10.1080/09273948.2021.1957123.
84. Seong HJ, Lee CS. Multiple evanescent white dot syndrome with submacular fluid in dome-shaped macula following COVID-19 vaccination: a case report. *Korean Journal of Ophthalmology*. 2022;36(5):471–473. https://doi.org/10.3341/kjo.2022.0077.
85. Ninet L, Comet A, Denis D, David T. Multiple evanescent white dot syndrome following BioNTech SARS-CoV2 mRNA vaccination. *Journal Francais d'Ophtalmologie*. 2022;45(7):e306–e308. https://doi.org/10.1016/j.jfo.2022.03.002.
86. Lin KS, Hsieh MH. Multiple evanescent white dot syndrome following medigen vaccine biologics corporation COVID-19 vaccination. *Ocular Immunology and Inflammation*. 2022;30(5):1278–1281. https://doi.org/10.1080/09273948.2022.2062388.
87. Mudie LI, Zick JD, Dacey MS, Palestine AG. Panuveitis following vaccination for COVID-19. *Ocular Immunology and Inflammation*. 2021;29(4):741–742. https://doi.org/10.1080/09273948.2021.1949478.
88. Kung TPH, Zhang C, Sieminski SF. Acute panuveitis after COVID-19 mRNA booster vaccination following cataract surgery. *American Journal of Ophthalmology Case Reports*. 2022;28. https://doi.org/10.1016/j.ajoc.2022.101726.

89. Testi I, Brandão-de-Resende C, Agrawal R, et al. Ocular inflammatory events following COVID-19 vaccination: a multinational case series. *Journal of Ophthalmic Inflammation and Infection*. 2022;12(1). https://doi.org/10.1186/s12348-021-00275-x.

90. Saraceno JJF, Souza GM, dos Santos Finamor LP, Nascimento HM, Belfort R. Vogt-Koyanagi-Harada Syndrome following COVID-19 and ChAdOx1 nCoV-19 (AZD1222) vaccine. *International Journal of Retina and Vitreous*. 2021;7(1). https://doi.org/10.1186/s40942-021-00319-3.

91. Papasavvas I, Herbort CP. Reactivation of Vogt-Koyanagi-Harada disease under control for more than 6 years, following anti-SARS-CoV-2 vaccination. *Journal of Ophthalmic Inflammation and Infection*. 2021;11(1). https://doi.org/10.1186/s12348-021-00251-5.

92. Jampol LM, Tauscher R, Schwarz HP. COVID-19, COVID-19 vaccinations, and subsequent abnormalities in the retina: causation or coincidence? *JAMA Ophthalmology*. 2021;139(10):1135−1136. https://doi.org/10.1001/jamaophthalmol.2021.3483.

Index

'*Note:* Page numbers followed by "f" indicate figures and "t" indicate tables.'

D

I

K

L

N

O

P

S

T